Mosquitoes

A LABORATORY MANUAL

EDITED BY

Laura B. Duvall
Columbia University

Benjamin J. Matthews
The University of British Columbia

COLD SPRING HARBOR LABORATORY PRESS
Cold Spring Harbor, New York • www.cshlpress.org

MOSQUITOES
A LABORATORY MANUAL

Printed in the United States of America

Publisher	John Inglis
Executive Editor	Alejandro Montenegro-Montero
Acquisition Editor	Maria Smit
Managing Editors	Maria Smit and Christin Munkittrick
Developmental Editors	Rebecca Barr, Miriam Bergeret, Shannon Coleman, and Brittany Dodson
Senior Project Manager	Inez Sialiano and Barbara Acosta
Editorial Assistant	Danett Gil
Permissions Coordinator	Carol Brown
Production Editor	Kathleen Bubbeo
Production Manager and Cover Designer	Denise Weiss
Director of Product Development & Marketing	Wayne Manos

Cover art: Female *Aedes aegypti* mosquito blood feeding on a human host. Image courtesy of Alex Wild.

Library of Congress Cataloging-in-Publication Data

Names: Duvall, Laura B., 1985- editor. | Matthews, Benjamin J., 1983- editor.
Title: Mosquitoes: a laboratory manual / edited by Laura B. Duvall, Columbia University, and Benjamin J. Matthews, the University of British Columbia.
Description: Cold Spring Harbor, New York : Cold Spring Harbor Laboratory Press, [2024] | Includes bibliographical references and index. |
Summary: "Mosquitoes serve as important models for studies of insect biology and behavior; they are also common human pests that spread infectious diseases (e.g., malaria). This comprehensive laboratory manual provides guidance for establishing and maintaining healthy mosquito colonies, step-by-step protocols for investigating their genetics, development, physiology, and behavior, and tools for developing more effective mosquito control modalities (e.g., traps and repellents)"-- Provided by publisher.
Identifiers: LCCN 2023050080 (print) | LCCN 2023050081 (ebook) |
ISBN 9781621824411 (paperback) | ISBN 9781621824428 (epub)
Subjects: LCSH: Mosquitoes as laboratory animals. |
Mosquitoes--Research--Methodology--Laboratory manuals. |
Mosquitoes--Behavior. | Animal models in research.
Classification: LCC SF407.M67 M67 2024 (print) | LCC SF407.M67 (ebook) |
DDC 628.9/657--dc23/eng/20231201
LC record available at https://lccn.loc.gov/2023050080
LC ebook record available at https://lccn.loc.gov/2023050081

10 9 8 7 6 5 4 3 2 1

Mosquitoes

A LABORATORY MANUAL

ALSO FROM COLD SPRING HARBOR LABORATORY PRESS

RELATED TITLES

Emerging Model Organisms

Malaria: Biology in the Era of Eradication

OTHER LABORATORY MANUALS

Antibodies: A Laboratory Manual, Second Edition

*African Turquoise Killifish (*Nothobranchius furzeri*): A Laboratory Manual*

Budding Yeast: A Laboratory Manual

Cell Death Techniques: A Laboratory Manual

CRISPR–Cas: A Laboratory Manual

Experiments in Bacterial Genetics: A Laboratory Manual

Fission Yeast: A Laboratory Manual

Molecular Cloning: A Laboratory Manual, Fourth Edition

Mouse Phenotypes: Generation and Analysis of Mutants: A Laboratory Manual, Second Edition

Xenopus: A Laboratory Manual

WEBSITE

www.cshprotocols.org

Contents

SECTION 9. IMAGING AND ANALYSIS OF GENE EXPRESSION

SECTION 10. INSECTICIDE RESISTANCE

Preface

Mosquitoes have had, and continue to have, an immense impact on human health, ecology, and the course of history. Accordingly, the study of mosquitoes and the diseases they transmit has been a long-standing pursuit of scientists and researchers around the world. In the past decade, there has been increased awareness of new and reemerging mosquito-borne diseases, with an urgent need for innovative control strategies, magnified by looming challenges like climate change and increasing instances of insecticide resistance. At the same time, the development of advanced gene editing technologies has drawn diverse research groups to focus their efforts on understanding mosquito biology, including novel genetic strategies to control mosquito populations or render them unable to transmit pathogens. In addition to vector control, these tools will continue to facilitate basic discovery science about mosquito physiology, behavior, and interactions with pathogens. Concomitantly, there has been an increase in researchers trained in diverse areas working on mosquitoes, greatly expanding research perspectives and available techniques brought to bear on these important animals.

This collection is designed to serve as a valuable resource for researchers, students, and professionals interested in mosquito biology, physiology, and behavior. It brings together knowledge and expertise from a broad group of scientists and is intended for use both by research groups that already have established mosquito colonies in their laboratories and groups who would like to start their own colony. The collection covers a wide array of topics, including the basics of mosquito rearing and colony maintenance, and protocols to evaluate mosquito behavior, visualize transcript and protein expression, quantify physiological responses to sensory stimuli, and score insecticide resistance. As mentioned above, one of the most exciting and rapidly evolving areas of mosquito research is the use of genome-editing technologies, such as CRISPR–Cas9, which have revolutionized researchers' ability to precisely manipulate the mosquito genome. As such, this collection also features protocols for gene editing, validating genetic lesions, and genotyping genetically edited animals.

Many of the protocols included here have been developed in *Aedes aegypti* or *Anopheles gambiae*; however, we anticipate that they will be useful to researchers studying many different species, with minor modifications. Indeed, it is our hope that this collection serves as a resource to apply these techniques in a broad range of mosquito species.

As editors of this manual, we are immensely grateful to the contributing authors and researchers who have generously shared their expertise and protocols. Their collective efforts have resulted in a comprehensive and up-to-date compilation of methodologies. We would also like to thank Maria Smit, Alejandro Montenegro-Montero, Christin Munkittrick, Inez Sialiano, Barbara Acosta, Danett Gil, Kathleen Bubbeo, and Richard Sever at Cold Spring Harbor Laboratory Press for their invaluable help with and enthusiasm about this collection.

Finally, we acknowledge that this manual is not exhaustive, as the field of mosquito research is continuously evolving. However, it is our hope that it will serve as a foundation for future investigations and inspire new generations of scientists.

Laura B. Duvall
Benjamin J. Matthews

GENERAL SAFETY AND HAZARDOUS MATERIAL INFORMATION

Cold Spring Harbor Laboratory Manuals should be used by laboratory personnel with experience in laboratory and chemical safety or students under the supervision of such trained personnel. The procedures, chemicals, and equipment referenced in these manuals are hazardous and can cause serious injury unless performed, handled, and used with care and in a manner consistent with safe laboratory practices. Students and researchers using the procedures in this manual do so at their own risk. It is essential for your safety that you consult the appropriate Material Safety Data Sheets, the manufacturers' manuals provided with the relevant equipment, and your institution's Environmental Health and Safety Office (hereafter referred to as safety office), as well as the General Safety and Disposal Cautions in the Appendix for proper handling of hazardous materials. Cold Spring Harbor Laboratory makes no representations or warranties with respect to the material set forth in its manuals and has no liability in connection with the use of these materials.

All registered trademarks, trade names, and brand names mentioned in this book are the property of the respective owners. Readers should please consult individual manufacturers and other resources for current and specific product information.

Appropriate sources for obtaining safety information and general guidelines for laboratory safety are provided in the General Safety and Hazardous Material Information Appendix.

CHAPTER 1

Overview of *Aedes aegypti* and Use in Laboratory Studies

Margot P. Wohl[1] and Conor J. McMeniman[1,2,3]

[1]*W. Harry Feinstone Department of Molecular Microbiology and Immunology, Johns Hopkins Malaria Research Institute, Johns Hopkins Bloomberg School of Public Health, Johns Hopkins University, Baltimore, Maryland 21205, USA;* [2]*The Solomon H. Snyder Department of Neuroscience, Johns Hopkins University School of Medicine, Baltimore, Maryland 21205, USA*

The yellow fever mosquito *Aedes aegypti* is a prolific disease vector. This mosquito has been the subject of scientific investigation for more than a century. Continued research into *Aedes aegypti* biology is crucial for understanding how to halt the suite of major arthropod-borne viral diseases this mosquito transmits. Here, we provide an introductory overview of *Aedes aegypti* life cycle; evolutionary history, biology, and ecology; genetics and sex differences; vector competence; and laboratory colonization and considerations for rearing this robust mosquito species for use in laboratory research.

LIFE CYCLE AND REQUIREMENTS FOR VITALITY

Aedes aegypti is a holometabolous, dipteran insect that undergoes complete metamorphosis with egg, larval, pupal, and adult stages (Fig. 1A–C). Egg, larval, and pupal stages are aquatic, whereas adult stages are terrestrial. Adult male and female *Ae. aegypti* can both feed on plant nectar to obtain energy, whereas only females consume blood that contains nutrients needed to trigger vitellogenesis and support egg production (Scott and Takken 2012). H_2O from nectar, blood, or other sources is also required for hydration.

After their initial emergence from the aquatic pupal stage, adult female and male *Ae. aegypti* require 2–4 d to fully mature to enable mating and host seeking (Hernández-Martínez et al. 2015; de Carvalho et al. 2021; Hill et al. 2021). Progressively over this period, female *Ae. aegypti* become capable of seeking out vertebrate hosts including humans on which to blood feed and the male genitalia rotates to enable copulation with females (Christophers 1960; Munstermann 1997).

Mating typically occurs before the first blood meal, and a single mating event results in the transfer of sperm from a male into the female spermathecae for fertilization of all eggs during an adult female's lifetime (Craig 1967). However, polyandry does occur occasionally (Richardson et al. 2015). *Ae. aegypti* females are anautogenous, which means a blood meal is required to initiate each gonotrophic cycle of egg production (Attardo et al. 2005). *Ae. aegypti* is a diurnal mosquito species, with peak periods of biting activity typically occurring in the hours after sunrise and before sunset during dusk (Trpis et al. 1973).

About 3–5 d after a blood meal, females lay their eggs within human-made or natural containers that are partially filled with standing H_2O (Nelson 1986). After being laid, eggs will develop for ~4 d, at which point they are fully embryonated and ready to hatch. *Ae. aegypti* eggs are covered by a melanized chorion that is desiccation-resistant (Fig. 1A). When eggs are dried, they can enter a quiescent, dormant state for months at a time (Christophers 1960). After full embryonation, eggs may hatch after being submerged in H_2O (e.g., after a rain shower).

---

[3]Correspondence: cmcmeni1@jhu.edu

Cite this introduction as *Cold Spring Harb Protoc*; doi:10.1101/pdb.top107651

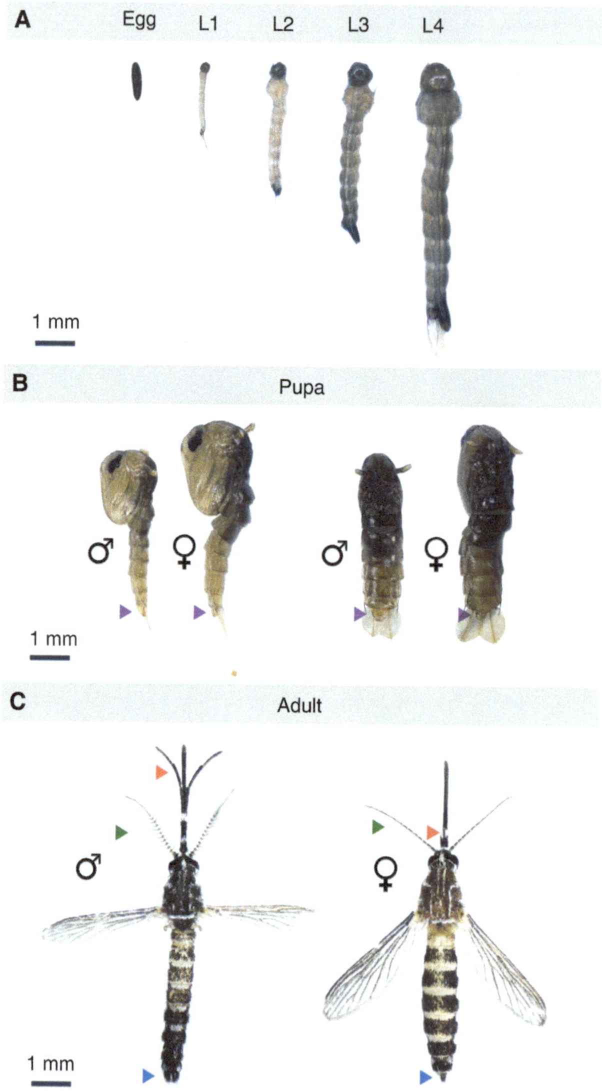

FIGURE 1. Life cycle stages of the yellow fever mosquito *Aedes aegypti*. (*A*) Egg and four larval instar (L1–L4) stages. (*B*) Pupal stage. Side view (*left*) and dorsal view (*right*) of male and female pupal stage. The most visible sexually dimorphic feature other than body size, the genital lobe, is denoted with purple arrows. (*C*) Adult stage. Dorsal view of male (*left*) and female (*right*) adults. The most visible sexually dimorphic features other than size are denoted with arrows. Orange, maxillary palps; green, antennae; blue, genitals/claspers. Legs are not shown.

After hatching, *Ae. aegypti* larvae filter feed on organic and microbial matter within the H_2O source. Larval development as well as adult survival, vector competence, fecundity, fertility, and behavior can all be affected by microbiota and nutrition during the larval stage (Coon et al. 2014; Dickson et al. 2017; Souza et al. 2019). *Ae. aegypti* has four larval instars (L1–L4) that undergo successive molts as they grow from ~1 mm to 8 mm long (Fig. 1A). Under optimal conditions, *Ae. aegypti* spends ~1 d each as L1, L2, and L3 instars and 3 d as an L4 instar before pupating.

Pupation typically occurs ~1 wk after hatching. Pupae look like a comma, with their abdomen usually curled below their cephalothorax. Notably, pupae do not require food. Within 2–3 d post-pupation, the pupae emerge as adults. The typical time it takes to develop from egg to adult under optimal conditions in the laboratory is 9–12 d, although this timing can vary particularly in response to changes in larval nutrition and rearing densities (Couret et al. 2014).

 Cite this introduction as *Cold Spring Harb Protoc*; doi:10.1101/pdb.top107651

Throughout the life cycle of *Ae. aegypti*, temperature is very important both for the optimal development of larva and pupa (Couret and Benedict 2014; Couret et al. 2014) and for adult behaviors such as wingbeat frequency, host seeking, and blood feeding (Reinhold et al. 2018). *Ae. aegypti* typically performs best at temperatures closest to their native habitats of ~25°C–35°C and with high absolute humidity (Dickens et al. 2018).

EVOLUTIONARY HISTORY, BIOLOGY, AND ECOLOGY

Ae. aegypti (Linnaeus) is part of the largest tribe of Culicidae known as Aedini. *Ae. aegypti sensu stricto* —namely, the species type form (commonly denoted as *Ae. aegypti*)—is a highly anthropophilic mosquito species that commonly lives around humans. *Ae. aegypti* is believed to have diverged from its subspecies *Ae. aegypti formosus* (Walker) during the domestication of humans (Brown et al. 2011; Rose et al. 2020).

Ae. aegypti females lay their eggs in standing sources of H_2O accumulated in containers found within and near housing. In this setting, females feed frequently and preferentially on humans (Scott et al. 1993; Scott et al. 2000; Ponlawat and Harrington 2005; Harrington et al. 2014). In contrast, *Ae. aegypti formosus* lives predominantly in the forests of sub-Saharan Africa, lays its eggs in tree holes or ground pools, and prefers to feed on nonhuman animals (Trpis and Hausermann 1978; Lounibos 1981; Gloria-Soria et al. 2016).

These two subspecies can be distinguished by their color, with *Ae. aegypti* appearing paler and with more white scales on the first abdominal tergite relative to *Ae. aegypti formosus* (McClelland 1974). *Ae. aegypti* likely found their way from Africa to other continents on slave-trading vessels (Christophers 1960; Crawford et al. 2017; Soghigian et al. 2020). At present, *Ae. aegypti* is found globally within tropical and subtropical regions mostly between 35° north and south latitudes (Kraemer et al. 2015) where its presence poses a continual threat to public health. *Ae. aegypti formosus* is still mostly confined to forested areas in Africa (Gloria-Soria et al. 2016; Rose et al. 2020).

The related species *Aedes albopictus* is often studied alongside *Ae. aegypti* because they can reside together in some geographical locations (Kraemer et al. 2019) and can act as vectors for the same viruses (Kamgang et al. 2020). The two species can be distinguished by markings on the scutum, the dorsal part of the thorax. *Ae. albopictus* has a white streak through the middle of the scutum parallel to the anterior–posterior axis, whereas *Ae. aegypti* has markings reminiscent of a lyre, the ancient, U-shaped harp (Nelson 1986).

GENETICS AND SEX DIFFERENCES

Ae. aegypti is diploid and harbors three autosomal chromosomes, with a genome size of ~1.25 Gb (Nene et al. 2007; Matthews et al. 2018). The first autosome contains a dominant nonrecombining portion containing sex determination alleles, called the M locus (McClelland 1966). Within this male-determining locus, the dominant gene *Nix* is responsible for the downstream regulation of sex determination transcription factors *fruitless* and *doublesex* (Hall et al. 2015), and its expression is sufficient to convert otherwise female mosquitoes into fertile, nonbiting males (Aryan et al. 2020).

Sex differentiation can be seen in both behavior and body plan. Arguably, the biggest difference is behavioral, with strong preferential host seeking and subsequent blood feeding by female *Ae. aegypti* on humans (McMeniman et al. 2014). Male *Ae. aegypti* can also be attracted to humans (Amos et al. 2021); however, they do not blood feed and likely only congregate near humans as a strategy to find females for mating.

Differences in body plans between male and female mosquitoes are most easily visible at the pupal stage (Fig. 1B). Female pupa can be ~20%–40% larger than male pupa, although the size of each sex is normally distributed and can overlap, so caution is advised when using size to sort the sexes (Wormington and Juliano 2014; Zacarés et al. 2018). Another dimorphism is the morphology of pupal genital lobes. Male genital lobes are more rounded and wider than female lobes.

As adults, females remain larger than males (Fig. 1C). The easiest way to distinguish the sexes is to inspect the morphology of the maxillary palps and antennae on the head. Males have plumose antennae, and their palps are much longer, extending nearly the length of the proboscis.

VECTOR COMPETENCE

Ae. aegypti is a globally important disease vector for many arthropod-borne viruses (arboviruses) of major public health significance, especially in urban environments because of its propensity to live near humans (Scott and Takken 2012). The genetic ability of a mosquito to acquire and transmit a pathogen is known as vector competence. As its colloquial name suggests, *Ae. aegypti* was first implicated as the primary vector for yellow fever virus (YFV; family Flaviviridae, genus Flavivirus) (Reed and Carroll 1901). *Ae. aegypti* is also the primary urban vector for dengue virus (DENV; family Flaviviridae, genus *Flavivirus*) (Bancroft 1906), Zika virus (ZIKV; family Flaviviridae, genus *Flavivirus*) (Dick et al. 1952), and chikungunya virus (CHIKV; family Togaviridae, genus *Alphavirus*) (Ross 1956). Different geographic strains of *Ae. aegypti* have also been used extensively to study the genetics of mosquito vector competence for filarial nematodes afflicting humans, including *Brugia malayi* and *Wuchereria bancrofti* (Macdonald 1962a,b; Macdonald and Ramachandran 1965; Severson et al. 1994; Juneja et al. 2015; Edgerton et al. 2020).

Ae. aegypti–borne pathogens are transmitted to humans by female mosquitoes. When host seeking, females take a blood meal from an infected individual, and these pathogens typically enter the midgut and then disseminate throughout the mosquito body, undergoing a period of development from ingestion to infectivity known as the extrinsic incubation period. Subsequently, these pathogens are deposited via the act of feeding into the skin and bloodstream of the next human host that is bitten. Therefore, *Ae. aegypti* has historically been used as a powerful laboratory model to understand the etiology of mosquito-borne diseases (Reed and Carroll 1901; Siler et al. 1925; Siler 1926) and the genetic underpinnings of mosquito susceptibility to arboviral and filarial infections (Macdonald 1967; Black et al. 2002; Souza-Neto et al. 2019).

Given its importance in disease transmission, *Ae. aegypti* vector competence is a key target for vector control strategies that aim to quell the viruses this mosquito transmits. For example, some strategies have recently used mosquito infection with various strains of the obligate intracellular bacterium *Wolbachia pipientis* to block arboviral transmission. *Ae. aegypti* appears to be naturally uninfected by *Wolbachia* in the field, but this mosquito species has successfully been transinfected with various *Wolbachia* strains derived from related dipterans, including *Ae. albopictus* and *Drosophila melanogaster*, using embryonic microinjection (Xi et al. 2005; McMeniman et al. 2009; Walker et al. 2011). The ability of numerous *Wolbachia* strains to potently block replication and dissemination of flaviruses (DENV, YFV, and ZIKV) and alphaviruses (CHIKV) in *Ae. aegypti* and induce the crossing sterility cytoplasmic incompatibility, in which the sperm of infected males and eggs of uninfected females do not make viable zygotes, is currently being used as an applied tool for mosquito population replacement (Ryan et al. 2019; Utarini et al. 2021) and suppression (Zheng et al. 2019; Crawford et al. 2020) to decrease the capacity of wild *Ae. aegypti* populations to transmit arboviruses.

LABORATORY STRAINS

Aedes aegypti has been a major source of scientific research dating back to when it was first discovered that this mosquito species was the primary vector for YFV (Reed and Carroll 1901). Part of the appeal of *Ae. aegypti* as a study organism relates to its ease of laboratory rearing, general hardiness, and desiccation-resistant eggs that facilitate straightforward strain establishment, colony maintenance, and storage. As such, numerous strains of *Ae. aegypti* have been established and maintained by different laboratories over time.

A few common strains of *Ae. aegypti* are used in the laboratory, each with their own origin story. The Rockefeller (ROCK) strain was first isolated over a century ago in Havana, Cuba, and the Orlando strain was established from mosquitoes in Florida around 1939 (Kuno 2010). The origin of the Liverpool strain and derivative stocks is quite the saga—beginning with mosquitoes from West Africa of mixed filaria susceptibility that were, through a series of pair crossings, bred to all be filaria-susceptible (Macdonald 1962a,b; Kuno 2010).

Notably, since their establishment, many of these *Ae. aegypti* strains have diverged genetically across laboratories (Gloria-Soria et al. 2019). Therefore, the genetic background of mosquitoes used for study should be evaluated based on what traits are important for the objective of investigation, and careful consideration should be given to whether the laboratory strain at hand achieves that goal. Toward that end, establishment of more recently collected field strains is still executed on a routine basis for *Ae. aegypti* and *Ae. aegypti formosus* (McBride et al. 2014; Ross et al. 2019).

The generation of many new strains of *Ae. aegypti* has recently been facilitated by the successful deployment of CRISPR–Cas9 tools in this species including knockouts, knock-ins, and driver/responder lines that use binary expression systems (Kistler et al. 2015; Li et al. 2017; Zhao et al. 2021). The easiest method to identify transgenic mosquitoes is through the integration of flourescent markers driven by a constitutive body-wide (i.e., polyubiquitin, OpIE2) or eye-specific (3XP3) promotor. As genetic tools for *Ae. aegypti* continue to be optimized and the number of laboratory strains rapidly expands, establishment of a central rearing facility such as the Bloomington Stock Center for *Drosophila melanogaster* could serve to distribute genetic reagents widely and foster scientific growth.

CONCLUSION

Laboratory studies of *Ae. aegypti* have been used to investigate countless facets of mosquito biology and control over past decades, providing valuable insights into diverse fields, such as mosquito genetics, insecticide resistance, vector competence, host–microbe interactions, physiology, and the development of next-generation control strategies. Optimized protocols for laboratory rearing of *Ae. aegypti*, a high-quality genome sequence (Matthews et al. 2018), a suite of genome engineering methods (Coates et al. 1998; Kistler et al. 2015; Li et al. 2017), and high-resolution transcriptomic approaches (Cui and Franz 2020) for this mosquito are valuable tools that can be used in an integrative fashion to rapidly accelerate our understanding of this major disease vector in the future.

By mimicking natural conditions and supplying necessary nutrients for development and reproduction, one can maintain *Ae. aegypti* with ease in laboratory settings. Protocols for rearing strains of *Ae. aegypti* in the laboratory are available for both routine colony maintenance (see Protocol 1: Batch Rearing *Aedes aegypti* [Wohl and McMeniman 2022a]) and group crosses (see Protocol 2: Single-Pair and Small-Group Crosses of *Aedes aegypti* [Wohl and McMeniman 2022b]).

REFERENCES

Amos BA, Hoffmann AA, Staunton KM, Lau MJ, Burkot TR, Ross PA. 2021. Long-range but not short-range attraction of male *Aedes aegypti* (Diptera: Culicidae) mosquitoes to humans. *J Med Entomol* **59:** 83–88. doi:10.1093/jme/tjab164

Aryan A, Anderson MAE, Biedler JK, Qi Y, Overcash JM, Naumenko AN, Sharakhova MV, Mao C, Adelman ZN, Tu Z. 2020. Nix alone is sufficient to convert female *Aedes aegypti* into fertile males and myo-sex is needed for male flight. *Proc Natl Acad Sci* **117:** 17702–17709. doi:10.1073/pnas.2001132117

Attardo GM, Hansen IA, Raikhel AS. 2005. Nutritional regulation of vitellogenesis in mosquitoes: implications for anautogeny. *Insect Biochem Mol Biol* **35:** 661–675. doi:10.1016/j.ibmb.2005.02.013

Bancroft TL. 1906. On the aetiology of dengue fever. *Austral Med Gaz* **25:** 17–18.

Black WC, Bennett KE, Gorrochótegui-Escalante N, Barillas-Mury CV, Fernández-Salas I, de Lourdes Muñoz M, Farfán-Alé JA, Olson KE, Beaty BJ. 2002. Flavivirus susceptibility in *Aedes aegypti*. *Arch Med Res* **33:** 379–388. doi:10.1016/S0188-4409(02)00373-9

Brown JE, McBride CS, Johnson P, Ritchie S, Paupy C, Bossin H, Lutomiah J, Fernandez-Salas I, Ponlawat A, Cornel AJ, et al. 2011. Worldwide patterns of genetic differentiation imply multiple 'domestications' of *Aedes aegypti*, a major vector of human diseases. *Proc Biol Sci* **278:** 2446–2454.

Christophers SR. 1960. *Aedes aegypti (L.) the yellow fever mosquito; its life history, bionomics, and structure*. University Press, Cambridge, England.

Coates CJ, Jasinskiene N, Miyashiro L, James AA. 1998. Mariner transposition and transformation of the yellow fever mosquito, *Aedes aegypti*. *Proc Natl Acad Sci* **95:** 3748–3751. doi:10.1073/pnas.95.7.3748

Coon KL, Vogel KJ, Brown MR, Strand MR. 2014. Mosquitoes rely on their gut microbiota for development. *Mol Ecol* **23:** 2727–2739. doi:10.1111/mec.12771

Couret J, Benedict MQ. 2014. A meta-analysis of the factors influencing development rate variation in *Aedes aegypti* (Diptera: Culicidae). *BMC Ecol* **14:** 3. doi:10.1186/1472-6785-14-3

Couret J, Dotson E, Benedict MQ. 2014. Temperature, larval diet, and density effects on development rate and survival of *Aedes aegypti* (Diptera: Culicidae). *PLoS ONE* **9:** e87468. doi:10.1371/journal.pone.0087468

Craig GB Jr. 1967. Mosquitoes: female monogamy induced by male accessory gland substance. *Science* **156:** 1499–1501. doi:10.1126/science.156.3781.1499

Crawford JE, Alves JM, Palmer WJ, Day JP, Sylla M, Ramasamy R, Surendran SN, Black W, Pain A, Jiggins FM. 2017. Population genomics reveals that an anthropophilic population of *Aedes aegypti* mosquitoes in West Africa recently gave rise to American and Asian populations of this major disease vector. *BMC Biol* **15:** 16. doi:10.1186/s12915-017-0351-0

Crawford JE, Clarke DW, Criswell V, Desnoyer M, Cornel D, Deegan B, Gong K, Hopkins KC, Howell P, Hyde JS, et al. 2020. Efficient production of male *Wolbachia*-infected *Aedes aegypti* mosquitoes enables large-scale suppression of wild populations. *Nat Biotechnol* **38:** 482–492. doi:10.1038/s41587-020-0471-x

Cui Y, Franz AWE. 2020. Heterogeneity of midgut cells and their differential responses to blood meal ingestion by the mosquito, *Aedes aegypti*. *Insect Biochem Mol Biol* **127:** 103496. doi:10.1016/j.ibmb.2020.103496

de Carvalho SS, Rodovalho CM, Gaviraghi A, Mota MBS, Jablonka W, Rocha-Santos C, Nunes RD, Sá-Guimarães TDE, Oliveira DS, Melo ACA, et al. 2021. *Aedes aegypti* post-emergence transcriptome: unveiling the molecular basis for the hematophagic and gonotrophic capacitation. *PLoS Negl Trop Dis* **15:** e0008915. doi:10.1371/journal.pntd.0008915

Dick GWA, Kitchen SF, Haddow AJ. 1952. Zika Virus (I). Isolations and serological specificity. *Trans R Soc Trop Med Hyg* **46:** 509–520. doi:10.1016/0035-9203(52)90042-4

Dickens BL, Sun H, Jit M, Cook AR, Carrasco LR. 2018. Determining environmental and anthropogenic factors which explain the global distribution of *Aedes aegypti* and *Ae. albopictus*. *BMJ Glob Health* **3:** e000801. doi:10.1136/bmjgh-2018-000801

Dickson LB, Jiolle D, Minard G, Moltini-Conclois I, Volant S, Ghozlane A, Bouchier C, Ayala D, Paupy C, Moro CV, et al. 2017. Carryover effects of larval exposure to different environmental bacteria drive adult trait variation in a mosquito vector. *Sci Adv* **3:** e1700585. doi:10.1126/sciadv.1700585

Edgerton EB, McCrea AR, Berry CT, Kwok JY, Thompson LK, Watson B, Fuller EM, Nolan TJ, Lok JB, Povelones M. 2020. Activation of mosquito immunity blocks the development of transmission-stage filarial nematodes. *Proc Natl Acad Sci* **117:** 3711–3717. doi:10.1073/pnas.1909369117

Gloria-Soria A, Ayala D, Bheecarry A, Calderon-Arguedas O, Chadee DD, Chiappero M, Coetzee M, Elahee KB, Fernandez-Salas I, Kamal HA, et al. 2016. Global genetic diversity of *Aedes aegypti*. *Mol Ecol* **25:** 5377–5395. doi:10.1111/mec.13866

Gloria-Soria A, Soghigian J, Kellner D, Powell JR. 2019. Genetic diversity of laboratory strains and implications for research: the case of *Aedes aegypti*. *PLoS Negl Trop Dis* **13:** e0007930. doi:10.1371/journal.pntd.0007930

Hall AB, Basu S, Jiang X, Qi Y, Timoshevskiy VA, Biedler JK, Sharakhova MV, Elahi R, Anderson MA, Chen XG, et al. 2015. A male-determining factor in the mosquito *Aedes aegypti*. *Science* **348:** 1268–1270. doi:10.1126/science.aaa2850

Harrington LC, Fleisher A, Ruiz-Moreno D, Vermeylen F, Wa CV, Poulson RL, Edman JD, Clark JM, Jones JW, Kitthawee S, et al. 2014. Heterogeneous feeding patterns of the dengue vector, *Aedes aegypti*, on individual human hosts in rural Thailand. *PLoS Negl Trop Dis* **8:** e3048. doi:10.1371/journal.pntd.0003048

Hernández-Martínez S, Rivera-Perez C, Nouzova M, Noriega FG. 2015. Coordinated changes in JH biosynthesis and JH hemolymph titers in *Aedes aegypti* mosquitoes. *J Insect Physiol* **72:** 22–27. doi:10.1016/j.jinsphys.2014.11.003

Hill SR, Taparia T, Ignell R. 2021. Regulation of the antennal transcriptome of the dengue vector, *Aedes aegypti*, during the first gonotrophic cycle. *BMC Genomics* **22:** 71. doi:10.1186/s12864-020-07336-w

Juneja P, Ariani CV, Ho YS, Akorli J, Palmer WJ, Pain A, Jiggins FM. 2015. Exome and transcriptome sequencing of *Aedes aegypti* identifies a locus that confers resistance to *Brugia malayi* and alters the immune response. *PLoS Pathog* **11:** e1004765. doi:10.1371/journal.ppat.1004765

Kamgang B, Vazeille M, Tedjou A, Yougang AP, Wilson-Bahun TA, Mousson L, Wondji CS, Failloux AB. 2020. Different populations of *Aedes aegypti* and *Aedes albopictus* (Diptera: Culicidae) from Central Africa are susceptible to Zika virus infection. *PLoS Negl Trop Dis* **14:** e0008163. doi:10.1371/journal.pntd.0008163

Kistler KE, Vosshall LB, Matthews BJ. 2015. Genome engineering with CRISPR-Cas9 in the mosquito *Aedes aegypti*. *Cell Rep* **11:** 51–60. doi:10.1016/j.celrep.2015.03.009

Kraemer MU, Sinka ME, Duda KA, Mylne AQ, Shearer FM, Barker CM, Moore CG, Carvalho RG, Coelho GE, Van Bortel W, et al. 2015. The global distribution of the arbovirus vectors *Aedes aegypti* and *Ae. albopictus*. *Elife* **4:** e08347. doi:10.7554/eLife.08347

Kraemer MUG, Reiner RC, Brady OJ, Messina JP, Gilbert M, Pigott DM, Yi D, Johnson K, Earl L, Marczak LB, et al. 2019. Past and future spread of the arbovirus vectors *Aedes aegypti* and *Aedes albopictus*. *Nat Microbiol* **4:** 854–863. doi:10.1038/s41564-019-0376-y

Kuno G. 2010. Early history of laboratory breeding of *Aedes aegypti* (Diptera: Culicidae) focusing on the origins and use of selected strains. *J Med Entomol* **47:** 957–971. doi:10.1603/ME10152

Li M, Bui M, Yang T, Bowman CS, White BJ, Akbari OS. 2017. Germline Cas9 expression yields highly efficient genome engineering in a major worldwide disease vector, *Aedes aegypti*. *Proc Natl Acad Sci* **114:** E10540–E10549. doi:10.1073/pnas.1711538114

Lounibos LP. 1981. Habitat segregation among African treehole mosquitoes. *Ecol Entomol* **6:** 129–154. doi:10.1111/j.1365-2311.1981.tb00601.x

Macdonald WW. 1962a. The genetic basis of susceptibility to infection with semi-periodic *Brugia malayi* in *Aëdes aegypti*. *Ann Trop Med Parasitol* **56:** 373–382. doi:10.1080/00034983.1962.11686135

Macdonald WW. 1962b. The selection of a strain of *Aëdes aegypti* susceptible to infection with semi-periodic *Brugia malayi*. *Ann Trop Med Parasitol* **56:** 368–372. doi:10.1080/00034983.1962.11686134

Macdonald WW. 1967. Genetics of vectorial capacity. *Bull World Health Organ* **36:** 566–568.

Macdonald WW, Ramachandran CP. 1965. The influence of the gene *fm* (filarial susceptibility, *Brugia malayi*) on the susceptibility of *Aedes aegypti* to seven strains of *Brugia*, *Wuchereria* and *Dirofilaria*. *Ann Trop Med Parasitol* **59:** 64–73. doi:10.1080/00034983.1965.11686284

Matthews BJ, Dudchenko O, Kingan SB, Koren S, Antoshechkin I, Crawford JE, Glassford WJ, Herre M, Redmond SN, Rose NH, et al. 2018. Improved reference genome of *Aedes aegypti* informs arbovirus vector control. *Nature* **563:** 501–507. doi:10.1038/s41586-018-0692-z

McBride CS, Baier F, Omondi AB, Spitzer SA, Lutomiah J, Sang R, Ignell R, Vosshall LB. 2014. Evolution of mosquito preference for humans linked to an odorant receptor. *Nature* **515:** 222–227. doi:10.1038/nature13964

McClelland GA. 1966. Sex-linkage at two loci affecting eye pigment in the mosquito *Aedes aegypti* (Diptera: culicidae). *Can J Genet Cytol* **8:** 192–198. doi:10.1139/g66-024

McClelland GAH. 1974. A worldwide survey of variation in scale pattern of the abdominal tergum of *Aedes aegypti* (L.) (Diptera: Culicidae). *Trans R Entomol Soc Lond* **126:** 239–259. doi:10.1111/j.1365-2311.1974.tb00853.x

McMeniman CJ, Lane RV, Cass BN, Fong AW, Sidhu M, Wang YF, O'Neill SL. 2009. Stable introduction of a life-shortening *Wolbachia* infection into the mosquito *Aedes aegypti*. *Science* **323:** 141–144. doi:10.1126/science.1165326

McMeniman CJ, Corfas RA, Matthews BJ, Ritchie SA, Vosshall LB. 2014. Multimodal integration of carbon dioxide and other sensory cues drives mosquito attraction to humans. *Cell* **156:** 1060–1071. doi:10.1016/j.cell.2013.12.044

Munstermann LE. 1997. Care and maintenance of Aedes mosquito colonies. In *The molecular biology of insect disease vectors: a methods manual* (ed. Crampton JM, Beard CB, Louis C), pp. 13–20. Springer Netherlands, Dordrecht.

Nelson MJ. 1986. *Aedes aegypti*: biology and ecology. *Revista Panamericana De Salud Publica-pan American Journal of Public Health.*

Nene V, Wortman JR, Lawson D, Haas B, Kodira C, Tu ZJ, Loftus B, Xi Z, Megy K, Grabherr M, et al. 2007. Genome sequence of *Aedes aegypti*, a major arbovirus vector. *Science* **316:** 1718–1723. doi:10.1126/science.1138878

Ponlawat A, Harrington LC. 2005. Blood feeding patterns of *Aedes aegypti* and *Aedes albopictus* in Thailand. *J Med Entomol* **42:** 844–849. doi:10.1093/jmedent/42.5.844

Reed W, Carroll J. 1901. The prevention of yellow fever. *Public Health Pap Rep* **27:** 113–129.

Reinhold JM, Lazzari CR, Lahondère C. 2018. Effects of the environmental temperature on *Aedes aegypti* and *Aedes albopictus* mosquitoes: a review. *Insects* **9:** 158. doi:10.3390/insects9040158

Richardson JB, Jameson SB, Gloria-Soria A, Wesson DM, Powell J. 2015. Evidence of limited polyandry in a natural population of *Aedes aegypti*. *Am J Trop Med Hyg* **93:** 189–193. doi:10.4269/ajtmh.14-0718

Rose NH, Sylla M, Badolo A, Lutomiah J, Ayala D, Aribodor OB, Ibe N, Akorli J, Otoo S, Mutebi JP, et al. 2020. Climate and urbanization drive mosquito preference for humans. *Curr Biol* **30:** 3570–3579.e3576. doi:10.1016/j.cub.2020.06.092

Ross RW. 1956. The Newala epidemic. III. The virus: isolation, pathogenic properties and relationship to the epidemic. *J Hyg (Lond)* **54:** 177–191. doi:10.1017/S0022172400044442

Ross PA, Endersby-Harshman NM, Hoffmann AA. 2019. A comprehensive assessment of inbreeding and laboratory adaptation in *Aedes aegypti* mosquitoes. *Evol Appl* **12:** 572–586. doi:10.1111/eva.12740

Ryan PA, Turley AP, Wilson G, Hurst TP, Retzki K, Brown-Kenyon J, Hodgson L, Kenny N, Cook H, Montgomery BL, et al. 2019. Establishment of wMel *Wolbachia* in *Aedes aegypti* mosquitoes and reduction of local dengue transmission in Cairns and surrounding locations in northern Queensland, Australia. *Gates Open Res* **3:** 1547. doi:10.12688/gatesopenres.13061.1

Scott TW, Takken W. 2012. Feeding strategies of anthropophilic mosquitoes result in increased risk of pathogen transmission. *Trends Parasitol* **28:** 114–121. doi:10.1016/j.pt.2012.01.001

Scott TW, Chow E, Strickman D, Kittayapong P, Wirtz RA, Lorenz LH, Edman JD. 1993. Blood-feeding patterns of *Aedes aegypti* (Diptera: Culicidae) collected in a rural Thai village. *J Med Entomol* **30:** 922–927. doi:10.1093/jmedent/30.5.922

Scott TW, Amerasinghe PH, Morrison AC, Lorenz LH, Clark GG, Strickman D, Kittayapong P, Edman JD. 2000. Longitudinal studies of *Aedes aegypti* (Diptera: Culicidae) in Thailand and Puerto Rico: blood feeding frequency. *J Med Entomol* **37:** 89–101. doi:10.1603/0022-2585-37.1.89

Severson DW, Mori A, Zhang Y, Christensen BM. 1994. Chromosomal mapping of two loci affecting filarial worm susceptibility in *Aedes aegypti*. *Insect Mol Biol* **3:** 67–72. doi:10.1111/j.1365-2583.1994.tb00153.x

Siler JF. 1926. *Dengue: its history, epidemiology, mechanism of transmission, etiology, clinical manifestations, immunity, and prevention*. Bureau of Printing, Manila.

Siler JF, Hall MMW, Hitchens MAP. 1925. Transmission of dengue fever by mosquitoes. *Proc Soc Exp Biol Med* **23:** 197–201. doi:10.3181/00379727-23-2890

Soghigian J, Gloria-Soria A, Robert V, Le Goff G, Failloux AB, Powell JR. 2020. Genetic evidence for the origin of *Aedes aegypti*, the yellow fever mosquito, in the southwestern Indian Ocean. *Mol Ecol* **29:** 3593–3606. doi:10.1111/mec.15590

Souza RS, Virginio F, Riback TIS, Suesdek L, Barufi JB, Genta FA. 2019. Microorganism-based larval diets affect mosquito development, size and nutritional reserves in the yellow fever mosquito *Aedes aegypti* (Diptera: Culicidae). *Front Physiol* **10:** 152. doi:10.3389/fphys.2019.00152

Souza-Neto JA, Powell JR, Bonizzoni M. 2019. *Aedes aegypti* vector competence studies: a review. *Infect Genet Evol* **67:** 191–209. doi:10.1016/j.meegid.2018.11.009

Trpis M, Hausermann W. 1978. Genetics of house-entering behaviour in East African populations of *Aedes aegypti* (L.) (Diptera: Culicidae) and its relevance to speciation. *Bull Entomol Res* **68:** 521–532. doi:10.1017/S0007485300009494

Trpis M, McClelland GA, Gillett JD, Teesdale C, Rao TR. 1973. Diel periodicity in the landing of *Aedes aegypti* on man. *Bull World Health Organ* **48:** 623–629.

Utarini A, Indriani C, Ahmad RA, Tantowijoyo W, Arguni E, Ansari MR, Supriyati E, Wardana DS, Meitika Y, Ernesia I, et al. 2021. Efficacy of Wolbachia-infected mosquito deployments for the control of dengue. *N Engl J Med* **384:** 2177–2186. doi:10.1056/NEJMoa2030243

Walker T, Johnson PH, Moreira LA, Iturbe-Ormaetxe I, Frentiu FD, McMeniman CJ, Leong YS, Dong Y, Axford J, Kriesner P, et al. 2011. The wMel *Wolbachia* strain blocks dengue and invades caged *Aedes aegypti* populations. *Nature* **476:** 450–453. doi:10.1038/nature10355

Wohl MP, McMeniman CJ. 2022a. Batch rearing *Aedes aegypti*. *Cold Spring Harb Protoc* doi:108017

Wohl MP, McMeniman CJ. 2022b. Single-pair and small-group crosses of *Aedes aegypti*. *Cold Spring Harb Protoc* doi:108018

Wormington JD, Juliano SA. 2014. Sexually dimorphic body size and development time plasticity in *Aedes* mosquitoes (Diptera: Culicidae). *Evol Ecol Res* **16:** 223–234.

Xi Z, Khoo CC, Dobson SL. 2005. *Wolbachia* establishment and invasion in an *Aedes aegypti* laboratory population. *Science* **310:** 326–328. doi:10.1126/science.1117607

Zacarés M, Salvador-Herranz G, Almenar D, Tur C, Argilés R, Bourtzis K, Bossin H, Pla I. 2018. Exploring the potential of computer vision analysis of pupae size dimorphism for adaptive sex sorting systems of various vector mosquito species. *Parasit Vectors* **11:** 656. doi:10.1186/s13071-018-3221-x

Zhao Z, Tian D, McBride CS. 2021. Development of a pan-neuronal genetic driver in *Aedes aegypti* mosquitoes. *Cell Rep Methods* **1:** 100042. doi:10.1016/j.crmeth.2021.100042

Zheng X, Zhang D, Li Y, Yang C, Wu Y, Liang X, Liang Y, Pan X, Hu L, Sun Q, et al. 2019. Incompatible and sterile insect techniques combined eliminate mosquitoes. *Nature* **572:** 56–61. doi:10.1038/s41586-019-1407-9

Protocol 1

Batch Rearing *Aedes aegypti*

Margot P. Wohl[1] and Conor J. McMeniman[1,2,3]

[1]*W. Harry Feinstone Department of Molecular Microbiology and Immunology, Johns Hopkins Malaria Research Institute, Johns Hopkins Bloomberg School of Public Health, Johns Hopkins University, Baltimore, Maryland 21205, USA;* [2]*The Solomon H. Snyder Department of Neuroscience, Johns Hopkins University School of Medicine, Baltimore, Maryland 21205, USA*

Standardized rearing methods for the yellow fever mosquito *Aedes aegypti* are critical to facilitate controlled laboratory studies. This protocol describes a batch rearing protocol for *Aedes aegypti* stocks that yields healthy eggs, larvae, pupae, and adults in the laboratory for long-term colony maintenance and experimental manipulation. Foundational principles for the rearing and containment of these life cycle stages, as well as steps for mating and blood feeding *Aedes aegypti* to yield viable eggs for continuous culture or storage, are detailed.

MATERIALS

It is essential that you consult the appropriate Material Safety Data Sheets and your institution's Environmental Health and Safety Office for proper handling of equipment and hazardous material used in this protocol.

RECIPES: Please see the end of this protocol for recipes indicated by <R>. Additional recipes can be found online at http://cshprotocols.cshlp.org/site/recipes.

Reagents

Aedes aegypti mosquitoes

Bleach (10%)

Broth solution for hatching *Ae. aegypti* eggs <R>

Eggs will also hatch in preboiled and subsequently cooled H_2O or vacuum-treated H_2O. It is most important that the H_2O is deoxygenated (Judson 1960).

Ethanol (70% and 100%)

H_2O (distilled)

Larval diet (e.g., TetraMin Tropical Tablets; Tetra 16110)

Other larval diets such as liver powder mixed with brewer's yeast or alternate commercial brands of fish, rabbit, cat, and dog food may be used as a substitute according to user preferences and local availability. Feeding regimens should be optimized accordingly to maximize growth and viability (Benedict et al. 2020). Store larval diets in a cool, dry place for maximum shelf life.

Mice (e.g., Swiss Webster)

Blood from laboratory-bred mice supports excellent Ae. aegypti *egg yields, promoting the long-term viability of colonies. Alternative hosts commonly used in the laboratory for blood feeding* Ae. aegypti *include rats, rabbits,*

[3]Correspondence: cmcmeni1@jhu.edu

Cite this protocol as *Cold Spring Harb Protoc;* doi:10.1101/pdb.prot108017

chickens, and guinea pigs. All animal use should follow institutional animal care and use committee guidelines. Another option is feeding on humans (Harrington et al. 2020) with appropriate institutional review board approval. Artificial feeding methods including providing animal blood or artificial diets (Kogan 1990; Costa-da-Silva et al. 2013; Baughman et al. 2017; Gonzales et al. 2018; Sri-In et al. 2020) supplemented with ATP through artificial or natural membranes are also alternatives. Investigators should carefully consider the suitability and long-term feasibility of each of these blood feeding methods to support sustained colony maintenance. This can be achieved by monitoring the reproductive fitness of mosquitoes maintained transgenerationally on each blood source and recording metrics such as blood feeding rates, fecundity, and egg viability.

Mouse sedation mixture <R>
Sucrose solution (10%) for adult mosquito maintenance <R>

Equipment

Aspirator with high-efficiency particulate absorbing filter (optional, see Troubleshooting; e.g., John W. Hock Company 612)
Bottle for sucrose solution in cage (2-oz, round-glass; e.g., Fisher FB02911944)
Bouffant caps, disposable (nonpleated, white polypropylene, 21″; e.g., VersaPro 210)
Cage (aluminum, 8″ × 8″ × 8″; e.g., BioQuip 1450A)

Other types of cages, such as plastic (e.g., BugDorm-1 insect rearing cage) or netted versions with frames of different dimensions are also suitable.

Climate-controlled room (e.g., Environmental Growth Chambers EGC TC2)

Alternatively, reduced quantities of immature and adult stages may be reared in standup incubators (e.g., Percival I41VL) or on shelving or the bench in the laboratory at room temperature with appropriate humidity supplementation as described below.

Cold plate (optional, see Step 12; e.g., Torrey Pines Scientific Echoterm IC2)

Alternatively, one can use a Petri dish placed on ice.

Cold room (optional, see Step 12; 4°C)
Cotton balls, medium (optional, see Step 8; e.g., Fisher Scientific 22-456-880)
Cups (high-density polyethylene, 250-mL, 8.5-oz; e.g., VWR 89009-664)
Filter paper (55-mm; Whatman 1001-090)
Forceps
Mosquito suction trap (recommended, see Step 12; e.g., Biogents BG Sentinel 2 trap)
Paintbrush (fine, soft-bristle; e.g., AIT Products B07J6F3MQH)
Paper towels (e.g., Scott Multi-Fold Towels)
Polycarbonate lid (e.g., ½ Size Clear Polycarbonate Flat Food Pan Lid; Vigor 247FP12LFCL)

Polycarbonate lids (not airtight) minimize surface biofilm formation in pan H_2O; biofilms may suffocate larvae, particularly if they are overfed. This is especially important if rearing at <23°C because surface biofilms may form if cool air constantly blows over the H_2O surface. Lids further act as a barrier to prevent cross-contamination between strains and escape/entry of mosquitoes.

Polycarbonate pan (e.g., ½ Size Clear Polycarbonate Food Pan, 2 ½″-deep; Vigor 247FP122CL)

We use polycarbonate pans for rearing immature aquatic stages because they are stackable and space efficient, freezable to kill desiccated eggs, and can be heat treated or bleached (10% solution) to avoid cross-contamination. However, other pans or vessel types with different dimensions (e.g., mason jars or buckets) can be used according to user preferences and availability.

Scissors
Tape (see Step 8)
Tea strainer (with openings of 0.5–1 mm)
Transfer pipette (polyethylene; e.g., Fisher 13-711-7M)
Tupperware (6.5″ × 11.5″ × 3.5″)
Waste receptacle for killing immature aquatic *Ae. aegypti* life cycle stages (e.g., 5-gallon bucket with lid)

To kill aquatic immature Ae. aegypti *life cycle stages (i.e., eggs, larvae and pupae) after rearing or experimental use, many insectary facilities may alternatively use dedicated "kill tanks" or similar insect waste disposal mech-*

anisms. They serve as common waste receptacles that allow users to easily pump temperature or bleach-inactivated wastewater directly into sewage after treatment in accordance with institutional environmental health and safety guidelines.

Ziploc bags (gallon size)

METHOD

Maintain Ae. aegypti *strains in an insectary facility at 27°C and 80% relative humidity with a 12:12 h light:dark photoperiod using a walk-in climate-controlled room.*

If a climate-controlled room is not available, reduced quantities of immature stages and adults can be maintained in a standup incubator with temperature and humidity control or alternatively on shelving or the laboratory bench at room temperature (~25°C ± 1°C). To maintain a high relative humidity for adults in the absence of a humidified chamber, mosquito cages can be placed in large, transparent or slightly translucent plastic bags alongside an open container containing cotton balls saturated with distilled H_2O. Alternatively, wet towels may be draped over the mosquito cages and heat lamps used to provide warmth.

Fastidious laboratory practices should be used to avoid the cross-contamination of mosquito strains and rearing equipment. Handling one strain at a time, wearing a separate pair of disposable laboratory gloves per genotype, wiping the work bench down with 70% ethanol between genotypes, implementing standard operating procedures for cleaning used rearing equipment, and mitigating accidental mosquito releases into the insectary facility are recommended preventative measures. We recommend routinely genotyping all strains using phenotypic and molecular diagnostics when possible (i.e., every other generation).

Personal protective equipment (long pants, enclosed shoes, gloves, and a laboratory coat) should always be used to prevent accidental exposure to mosquito bites or other laboratory hazards in the insectary facility. Specific care should be taken when handling or entering cages to prevent mosquito bites or escapees. Personnel should monitor their health daily if performing research locally at an insectary facility in a disease endemic region or returning from travel from an area with active or past transmission of Aedes*-transmitted viruses. Consult with your institutional occupational health and safety office for recommended best practices regarding health surveillance in your insectary facility.*

There are four larval instars (L1–L4) that can be identified by watching for molting skin if a particular stage is needed for experiments. On average, stages L1–L3 last 1 d each, whereas L4 and the pupal stage last 2.5–3 d. When rearing Ae. aegypti *using this protocol, the time from hatching stored egg stocks to obtaining fully embryonated eggs to establish the next generation is ~3 wk (Fig. 1). Mosquito development times can vary due to many factors including strain, larval density, food supply, hydration, temperature, and humidity. Timing of developmental stages in this protocol is approximate. The life span of mosquitoes varies, and females often outlast males. In the laboratory, females can live for over 3 mo, but about half will die by the 7-wk mark (McMeniman et al. 2009).*

It may be possible to rear other Aedes *species using this protocol or modified versions thereof (Munstermann and Wasmuth 1985); however, we have not tried this.*

Hatching Eggs

1. On day 1, rinse polycarbonate pans and lids with distilled H_2O to remove any cleaning solution residue.
2. Add 500 mL of a broth solution for rearing *Ae. aegypti* to a clean polycarbonate pan. Submerge egg papers (Fig. 2A) in the broth, with the eggs facing up.

   For the routine colony maintenance of a strain, we initially hatch 2 desiccated 55-mm-diameter egg papers that were sourced from an 8.5-oz oviposition cup containing six of these papers total from a stock cage with approximately 200 gravid females. In our experience working with many Ae. aegypti *genotypes, hatching this number of eggs provides sufficient larvae for downstream steps to avoid population bottlenecks, even after eggs from a strain are desiccated for 4–6 mo.*

3. On day 2, add 500 mL of distilled H_2O to the pan to bring the total volume to 1 L. Then, add 1 g of finely milled TetraMin.

   Hatched first-instar larvae (L1) should be apparent. If desiccated eggs have been stored for prolonged periods (e.g., >4 mo), they may make take an additional day or two to hatch. In this case, adjust the time line of the following rearing steps accordingly.

Cite this protocol as *Cold Spring Harb Protoc*; doi:10.1101/pdb.prot108017

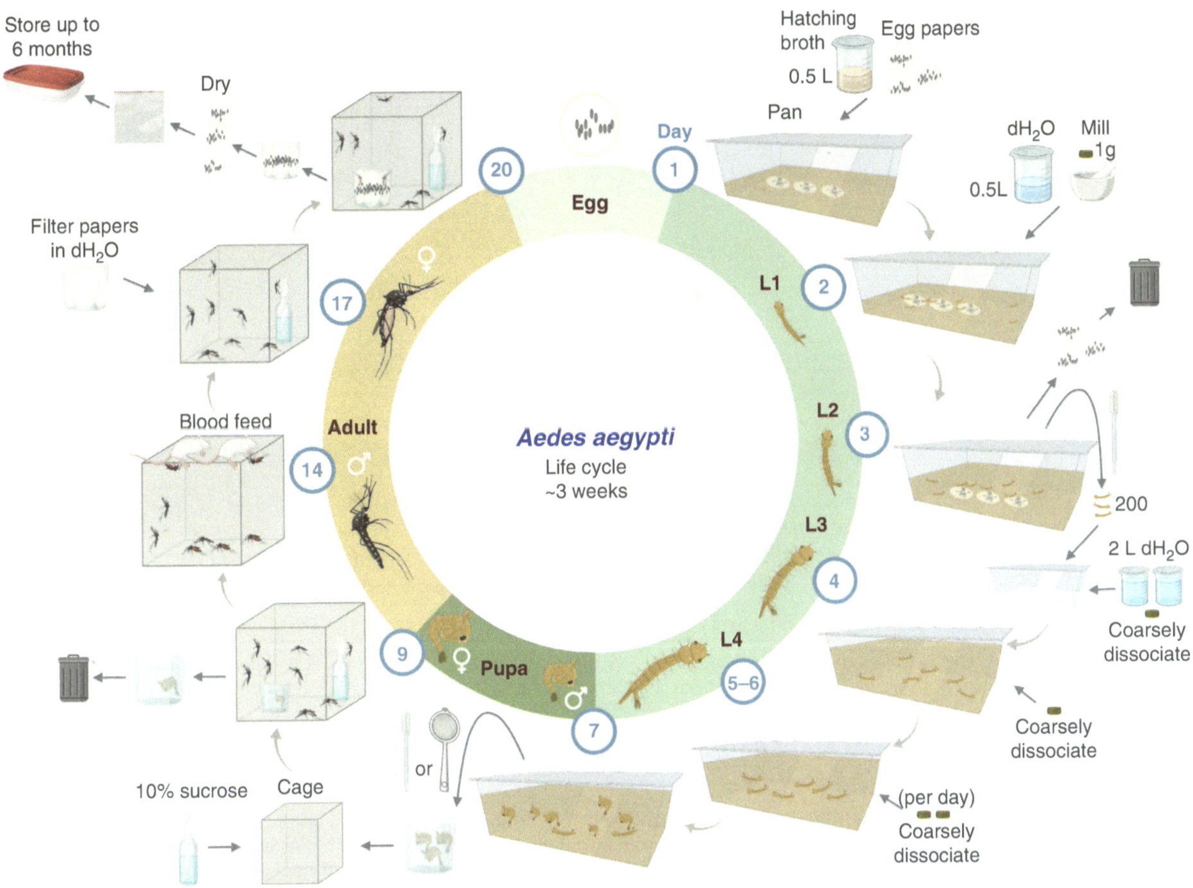

FIGURE 1. *Aedes aegypti* batch rearing protocol. This graphical summary illustrates our laboratory protocol for maintaining *Ae. aegypti* colonies. Created with BioRender.com.

Thinning and Rearing Larvae

4. On day 3, thin hatched larvae to a set density as follows.

   Second-instar larvae (L2) should be apparent.

   i. Fill a new pan with 2 L of fresh distilled H_2O. Coarsely dissociate by hand into quarters a 1-g TetraMin tablet and distribute it throughout the pan.

      Multiple pans may be needed according to the final target number of mosquitoes required for investigation. For routine stock maintenance, we maintain two pans per genotype.

   ii. Remove oviposition papers with forceps.

   iii. With scissors, cut ~1.5 cm off the end of the transfer pipette tip to ensure the bore is wide enough to facilitate the uptake of larvae without physically damaging them (Fig. 2B). Use one disposable pipette per genotype to avoid cross-contamination.

   iv. Count and remove larvae with the cut disposable plastic transfer pipette. Place 200 larvae into the new pan.

      Higher larval densities can lead to an excess of waste and nutrient competition, of which both can negatively affect larval development. One can rear Ae. aegypti *in smaller containers at higher densities, but in that case, daily exchanges of H_2O should be performed to remove waste products in the H_2O. Feeding should be optimized according to larval density in these cases. Overfeeding promotes the overproliferation of bacteria and biofilm formation and therefore should be avoided.*

5. On day 4, add one coarsely dissociated 1-g TetraMin tablet to each pan with larvae.

   Third-instar larvae (L3) should be apparent.

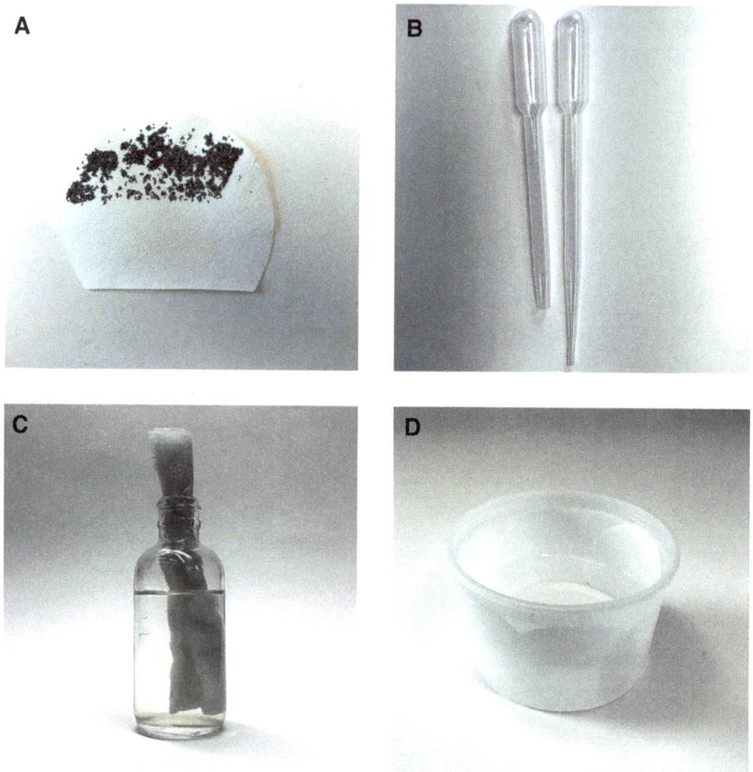

FIGURE 2. Common items used for *Aedes aegypti* batch rearing in the laboratory. (*A*) Oviposition paper with a suitable number of conditioned eggs for colony maintenance. (*B*) Disposable plastic pipette cut for transferring L2–L4 larvae and pupa and uncut for handling L1 larvae. (*C*) Glass bottle with sucrose solution and rolled paper towel wick. (*D*) Oviposition cup for egg deposition during batch rearing.

6. On day 5 to ~7, add two coarsely dissociated 1-g TetraMin tablets per day to each pan until pupation, which typically occurs by day 7.

   Fourth-instar larvae (L4) should be apparent by day 5, and larvae should pupate fully by ~day 7. Males develop quicker and will often pupate before females, so make sure to wait until enough females pupate before collecting. Variation in the timing of pupation may arise due to temperature, larval density, and food availability (Couret et al. 2014). If timing of larval development and/or pupation is much slower, see Troubleshooting.

   Pupae can be differentiated from larvae by their comma-like shape. After this stage, they do not require food until they emerge as adults ~2–3 d after pupation. As soon as pupation is observed, securely affix a disposable 21″ bouffant cap to the top of the open polycarbonate pan. Positioned beneath the pan lid around the pan rim, this breathable white polypropylene barrier serves as an additional barrier for escape of adult mosquitoes after pupal eclosion.

Collecting Pupae

7. At day 1 after pupation (~day 7 after hatching), collect pupae.

   At this time point, ~75% of the larvae will have pupated.

   If pupation appears asynchronous (i.e., <75% of the larvae have pupated) or adult emergence is seen in pans before pupal collection, see Troubleshooting.

   i. Position a tea strainer over a waste receptacle.

   ii. Pour pan contents through the tea strainer to collect the pupae.

   iii. Rinse pupae and any residual L4 larvae collected in the strainer with distilled H_2O three times.

Cite this protocol as *Cold Spring Harb Protoc*; doi:10.1101/pdb.prot108017

This rinse step is critical to ensure a decreased bacterial density in the eclosion cup that may otherwise lead to anoxic H_2O conditions and pupal lethality. You may combine pupae from two pans into a single cup to yield approximately 400 adults per cage. Residual L4 larvae will pupate within the next day or two and can remain in the cup.

iv. Transfer immature mosquitoes into an 8.5-oz cup filled halfway with fresh distilled H_2O.

As an alternative to straining, pupae can also be collected one at a time by using a truncated plastic disposable transfer pipette. This manual method is much slower but facilitates pupal sexing using size-based dimorphism, which may be needed for experiments or crossing schemes.

v. Chemically treat strained pan H_2O and any immature stages that are not required by adding 10% bleach to the waste receptacle. Allow the solution to sit for 48 h and then discard.

Alternatively, wastewater can be heat treated or frozen fully and then discarded.

8. Prepare a cage as follows.

i. Visually check the cage for holes and securely plug them with tape.

This step mitigates the potential for escapees.

ii. Provide a source of sucrose solution for nutrition for the emerging adults (Fig. 2C).

We prefer to use an autoclaved 2-oz glass bottle filled with 10% sucrose and a paper towel folded lengthwise, rolled up, and used as a wick inside the vial. Alternatively, cotton balls soaked in a 10% sucrose solution can be used, but they need to typically be changed daily to prevent them from drying out and to mitigate bacterial and mold growth. Under experimental conditions that require the removal of sucrose, a source of H_2O should always be maintained with the cage to prevent dehydration of adult stages.

9. Place the cup with pupae in the cage prepared in Step 8.

Rearing Adults

10. At day 3 postpupation (~day 9 posthatching), remove the cup originally filled with pupae from the cage when most have emerged as adults and dispose of contents as outlined in Step 7v.

Adult males tend to drown in open H_2O if these cups are kept for long periods in the cage.

Take precautionary measures when moving pupal cups into and out of cages to prevent adult escape. Cap used pupal cups securely with a lid before removal from the cage to enclose stray adults. Also, during cup removal, blow down the cage sleeve to physically move mosquitoes back into the cage.

If an accidental escape occurs, kill any loose mosquitoes immediately. Escapees may be knocked down by spraying with 100% ethanol in a spray bottle; however, it does not kill them. Squashing, freezing, or electrocution are validated methods to kill adult Ae. aegypti.

Adult females and males should be of uniform size. If noticeably growth-stunted adult mosquitoes or large variation in adult sizing are noted, see Troubleshooting.

11. Allow mosquitoes to have constant access to a sucrose solution except when blood feeding.

Replace sucrose bottles on a weekly basis to mitigate the potential for the growth of bacteria or fungi in the sucrose solution or wicks drying out. If mold or mites in the insectary facility are apparent, see Troubleshooting.

Take precautionary measures when moving sucrose bottles into and out of cages to mitigate accidental adult releases. Brush off or squash adult mosquitoes adhered to the sucrose bottle and wicks when moving them. Additionally, blowing down the cage sleeve helps to physically move mosquitoes back into the cage and prevent escape.

After adult emergence, it takes ~24 h for male genitalia to rotate 180° for copulation and for females to become sexually receptive. As such, 2 d are typically needed postemergence to ensure successful copulation. Females become capable of host feeding at ~2 d postemergence and will strongly host seek at 5 d postemergence in the laboratory for blood feeding.

12. (Recommended) Use precautionary measures to catch any loose adult mosquitoes and prevent their escape from the laboratory.

- Install mosquito suction traps in the laboratory.

- Sort mosquitoes for behavioral analysis or sexing using cold anesthetization in a walk-in cold room or using a cold plate.

  Use a fine soft-bristle paintbrush to manipulate anesthetized adult Ae. aegypti *without damaging them.*

Blood Feeding Female Mosquitoes

13. On day 5–10 postemergence (~day 14 posthatching), remove the sucrose solution at least 4 h before blood feeding on a mouse.

    This step encourages blood feeding.

    If starving mosquitoes of sucrose for longer periods (e.g., overnight), be sure to maintain a H_2O source in the cage to prevent colony dehydration. H_2O can be placed in similar vials as described for sucrose with a paper towel wick.

14. Sedate mice with a 7.5 µL/g intraperitoneal injection of the mouse sedation mixture (e.g., 0.15 mL per 20g mouse). Wait until the mice are no longer moving as assessed through toe pinching.
15. Place the mouse on top of the cage (~1 mouse for every 100 mosquitoes in a cage) and allow mosquitoes to feed until they are fully engorged (~15–30 min). Return the sucrose solution to the cage.

    Blowing into the cage can help activate mosquitoes if feeding is slow, as mosquitoes are attracted to CO_2.

16. After mosquitoes have been fed, place mice back into their cages and monitor to make sure they return to normal activity.

Providing Ovipositing Materials

17. On day 3 post-blood feeding (~day 17 posthatching), fill an 8.5-oz cup with ~100 mL of distilled H_2O and place six filter papers around the H_2O line of the cup overlapping each other (Fig. 2D).

    The bottom half of the papers should be submerged in H_2O and the top, exposed to air, should be moist.

18. Place the cup in the cage with blood-fed adults. Females will lay most of their eggs directly above the H_2O line.

    From 3 d post–blood feeding onward, gravid mosquitoes may also lay eggs on the sucrose wick or any wet cotton found in or on the cage. If egg deposition on these other substrates is a concern, remove them to restrict egg laying to the oviposition cup.

Conditioning and Storing Eggs

19. On day 5 post–blood feeding (~day 19 posthatching), remove the egg-laying cup from the cage. Pour the remaining H_2O into the waste bucket.
20. Gently rinse the egg papers in the oviposition cup by gently filling the center of the cup with distilled H_2O and dispose of this H_2O as outlined in Step 7v. Repeat three times to wash away any mosquito carcasses and other detritus while being careful not to physically detach eggs from the papers.

    This step minimizes the buildup of bacterial and fungal contamination during the egg-conditioning procedure. However, for stocks that are being hatched within the short term (e.g., 1 wk to 1 mo) or where egg loss during washing should be minimized, this step is not necessary. Forceps may also be used to carefully remove mosquito carcasses from the wet oviposition papers.

21. Remove each paper from the cup with clean forceps and place it onto a double layer of unfolded paper towel to dry for 5 min.
22. Place up to three egg papers on a new, unfolded, paper towel. Then, fold the papers up like a "parcel." Place up to two parcels inside a 1-gallon Ziploc bag along with two unfolded paper towels, which further distributes the remaining moisture in the bag. Ensure the bag is sealed.
23. Store sealed Ziploc bags inside Tupperware at room temperature.

Cite this protocol as *Cold Spring Harb Protoc*; doi:10.1101/pdb.prot108017

Eggs take 96–120 h postoviposition to fully embryonate, after which time they are capable of hatching. Synchronous hatching may be achieved any time after ~day 21 by following Steps 1–3. Desiccated eggs maintain viability for up to 6 mo at room temperature; however, hatch rates may decrease over time. For routine colony maintenance, we recommend hatching and rearing each stock (flipping stocks) every 4–5 mo. Some laboratories store desiccated eggs at lower temperatures (e.g., 15°C–18°C) and report a longer shelf life (>6 mo).

24. (Optional) After collecting eggs from adults, blood feed females again for more rounds of oviposition if needed.

    One can commonly get females to lay clutches of eggs for as many as five gonotrophic cycles in the laboratory; however, reproductive output decreases with mosquito age (Joy et al. 2010).

Cleaning Equipment and Discarding Mosquitoes

25. To kill adults in capped pupal eclosion cups or cages, freeze them overnight at −20°C, and then discard the carcasses.
26. Freeze all used polycarbonate pans, lids, and plasticware overnight at −20°C before cleaning for reuse. Disposable transfer pipettes as well as bouffant caps used to cover the polycarbonate pans during pupation are single use and can be frozen overnight and then discarded.
27. Wash cages, pans, lids, 8.5-oz cups, and strainers used for rearing and handling immature stages in 10% bleach; rinse thoroughly with distilled H_2O; and then air-dry.

DISCUSSION

This optimized batch-rearing method can be used to generate healthy *Ae. aegypti* life cycle stages for both colony maintenance and experimental manipulation including transgenesis (Coates et al. 1998; Kistler et al. 2015).

Major points to consider during the care and maintenance of *Ae. aegypti* in the laboratory include the use of fastidious practices when handling desiccated eggs, larvae, pupae, and adults to prevent strain cross-contamination; the application of standardized rearing densities and feeding regimens for larval stages to maximize growth and achieve synchronized pupation on the chosen diet (Couret et al. 2014); and the use of containment procedures to mitigate the release of adult *Ae. aegypti* into the laboratory. Routine cleaning of the insectary facilities including walk-in environmental rooms, incubators, and shelving with a 10% bleach solution and rinsing of eggs with H_2O before storage are also critical to minimize the build-up of fungi or bacteria that thrive in the humid environments required to rear *Ae. aegypti.*

The general hardiness of *Ae. aegypti* and the desiccation-resistant nature of its eggs offer advantages for the laboratory culture of rapidly expanding collections of wild-type and transgenic strains of this globally important disease vector. When using this protocol for long-term colony maintenance, we recommend that an inventory of stored egg stocks be maintained to ensure that *Ae. aegypti* strains under care are flipped periodically to maximize their transgenerational viability.

TROUBLESHOOTING

Problem (Step 6): Larval or pupal development is stunted.
Solution: Rear larval stages at the recommended density per pan to avoid overcrowding and follow the prescribed feeding regimen. If this suggestion does not help, you can also try optimizing the amount of larval diet provided each day or change the diet source to develop a custom feeding regimen. To increase developmental speed, try increasing the ambient temperature of the laboratory rearing environment.

Problem (Step 7): Pupation is asynchronous.
Solution: As per troubleshooting recommendations for Step 6.

Problem (Step 7): Adults have emerged in larval pans before pupal collection.
Solution: Carefully open the lid or pan cover and insert your aspirator carefully beneath the disposable hair net covering the pan. Aspirate mosquitoes into a holding cage. Alternatively, if these mosquitoes are not needed, spray the adults with 100% ethanol to knock them down on the surface of the H_2O. Collect them for disposal by skimming the surface of the H_2O with a paper towel. Freeze the paper towel with the collected mosquitoes for 24 h to kill them.

Problem (Step 10): Adult mosquitoes are very small or a large variation in adult sizing is noted.
Solution: As per Troubleshooting recommendations for Steps 6 and 7.

Problem (Step 11): Mold or mites are apparent in the insectary facility.
Solution: Every 4 mo, clean all wipeable surfaces associated with walk-in environmental rooms, incubators, and shelving with a 10% bleach solution and then H_2O. Clean sucrose spills up immediately to minimize supplementary sugar sources that may further promote mite survival.

RECIPES

Broth Solution for Hatching Aedes aegypti *Eggs*

Finely mill a larval diet source (e.g., TetraMin Tropical Tablets, Tetra 16110) using a mortar and pestle.
Add 1% w/v of milled larval diet to distilled H_2O in an autoclavable bottle.
Autoclave the solution for 30 min at 121°C.

Store at room temperature in a sealed bottle until use.

Mouse Sedation Mixture

Reagent	Stock concentration	Amount
Ketamine	100 mg/mL	2 mL
Acepromazine	10 mg/mL	1 mL
NaCl for injection (sterile)	0.9%	7 mL

Combine the reagents and mix well. Makes a total of 10 mL.
Store at room temperature in the dark. Mouse sedation mixture may be stored for the period specified by the manufacturer expiry dates of the reagents used.

Sucrose Solution (10%) for Adult Mosquito Maintenance

Create a solution of 10% w/v of sucrose dissolved in distilled H_2O.

Store at 4°C in a sterile sealed container.

ACKNOWLEDGMENTS

This protocol was adapted from many others (Gerberg et al. 1994; Munstermann 1997; Zheng et al. 2015) and generational knowledge handed down from pioneering *Ae. aegypti* researchers that long

 Cite this protocol as *Cold Spring Harb Protoc*; doi:10.1101/pdb.prot108017

preceded us. We thank the National Institutes of Health (R21 AI146450-01), Innovative Vector Control Consortium, Johns Hopkins Malaria Research Institute, and Bloomberg Philanthropies for generous funding to C.J.M. and a postdoctoral fellowship to M.P.W. from The Molecular and Cellular Basis of Infectious Diseases (MCBID) Program (T32A1007417) funded by the National Institutes of Health.

REFERENCES

Baughman T, Peterson C, Ortega C, Preston SR, Paton C, Williams J, Guy A, Omodei G, Johnson B, Williams H, et al. 2017. A highly stable blood meal alternative for rearing *Aedes* and *Anopheles* mosquitoes. *PLoS Negl Trop Dis* **11:** e0006142. doi:10.1371/journal.pntd.0006142

Benedict MQ, Hunt CM, Vella MG, Gonzalez KM, Dotson EM, Collins CM. 2020. Pragmatic selection of larval mosquito diets for insectary rearing of *Anopheles gambiae* and *Aedes aegypti*. *PLoS ONE* **15:** e0221838. doi:10.1371/journal.pone.0221838

Coates CJ, Jasinskiene N, Miyashiro L, James AA. 1998. Mariner transposition and transformation of the yellow fever mosquito, *Aedes aegypti*. *Proc Natl Acad Sci* **95:** 3748–3751. doi:10.1073/pnas.95.7.3748

Costa-da-Silva AL, Navarrete FR, Salvador FS, Karina-Costa M, Ioshino RS, Azevedo DS, Rocha DR, Romano CM, Capurro ML. 2013. Glytube: a conical tube and Parafilm M–based method as a simplified device to artificially blood-feed the dengue vector mosquito, *Aedes aegypti*. *PLoS ONE* **8:** e53816. doi:10.1371/journal.pone.0053816

Couret J, Dotson E, Benedict MQ. 2014. Temperature, larval diet, and density effects on development rate and survival of *Aedes aegypti* (Diptera: Culicidae). *PLoS ONE* **9:** e87468. doi:10.1371/journal.pone.0087468

Gerberg EJ, Barnard DR, Ward RA. 1994. *Manual for mosquito rearing and experimental techniques*. American Mosquito Control Association, Lake Charles, Louisiana.

Gonzales KK, Rodriguez SD, Chung H-N, Kowalski M, Vulcan J, Moore EL, Li Y, Willette SM, Kandel Y, Van Voorhies WA, et al. 2018. The effect of SkitoSnack, an artificial blood meal replacement, on *Aedes aegypti* life history traits and gut microbiota. *Sci Rep* **8:** 11023. doi:10.1038/s41598-018-29415-5

Harrington LC, Foy BD, Bangs MJ. 2020. Considerations for human blood-feeding and arthropod exposure in vector biology research: an essential tool for investigations and disease control. *Vector Borne Zoonotic Dis* **20:** 807–816. doi:10.1089/vbz.2020.2620

Joy TK, Arik AJ, Corby-Harris V, Johnson AA, Riehle MA. 2010. The impact of larval and adult dietary restriction on lifespan, reproduction and growth in the mosquito *Aedes aegypti*. *Exp Gerontol* **45:** 685–690. doi:10.1016/j.exger.2010.04.009

Judson CL. 1960. The physiology of hatching of aedine mosquito eggs: hatching stimulus. *Ann Entomol Soc Am* **53:** 688–691. doi:10.1093/aesa/53.5.688

Kistler KE, Vosshall LB, Matthews BJ. 2015. Genome engineering with CRISPR–Cas9 in the mosquito *Aedes aegypti*. *Cell Rep* **11:** 51–60. doi:10.1016/j.celrep.2015.03.009

Kogan PH. 1990. Substitute blood meal for investigating and maintaining *Aedes aegypti* (Diptera: Culicidae). *J Med Entomol* **27:** 709–712. doi:10.1093/jmedent/27.4.709

McMeniman CJ, Lane RV, Cass BN, Fong AW, Sidhu M, Wang YF, O'Neill SL. 2009. Stable introduction of a life-shortening *Wolbachia* infection into the mosquito *Aedes aegypti*. *Science* **323:** 141–144. doi:10.1126/science.1165326

Munstermann LE. 1997. Care and maintenance of *Aedes mosquito* colonies. In *The molecular biology of insect disease vectors: a methods manual* (ed. Crampton JM, Beard CB, Louis C), pp. 13–20. Springer Netherlands, Dordrecht.

Munstermann LE, Wasmuth LM. 1985. *Aedes aegypti*. In *Handbook of insect rearing* (ed. Singh P, Moore RF), pp. 7–14. Elsevier, Amsterdam.

Sri-In C, Weng SC, Shiao SH, Tu WC. 2020. A simplified method for blood feeding, oral infection, and saliva collection of the dengue vector mosquitoes. *PLoS ONE* **15:** e0233618. doi:10.1371/journal.pone.0233618

Zheng ML, Zhang DJ, Damiens DD, Lees RS, Gilles JR. 2015. Standard operating procedures for standardized mass rearing of the dengue and chikungunya vectors *Aedes aegypti* and *Aedes albopictus* (Diptera: Culicidae)—II—Egg storage and hatching. *Parasit Vectors* **8:** 348. doi:10.1186/s13071-015-0951-x

Protocol 2

Single-Pair and Small-Group Crosses of *Aedes aegypti*

Margot P. Wohl[1] and Conor J. McMeniman[1,2,3]

[1]*W. Harry Feinstone Department of Molecular Microbiology and Immunology, Johns Hopkins Malaria Research Institute, Johns Hopkins Bloomberg School of Public Health, Johns Hopkins University, Baltimore, Maryland 21205, USA;* [2]*The Solomon H. Snyder Department of Neuroscience, Johns Hopkins University School of Medicine, Baltimore, Maryland 21205, USA*

Aedes aegypti is an emerging model insect species used in genetics studies because of its ease of laboratory rearing, desiccation-resistant eggs, expanding genetic toolkit, and high-quality reference genome. Here, we describe procedures to isolate and sex virgin female and male mosquitoes and establish successful mating crosses. We also detail how to blood feed mosquitoes from these crosses, isolate individual or small groups of females for egg laying, condition these eggs for storage and hatching, and verify female mating status.

MATERIALS

It is essential that you consult the appropriate Material Safety Data Sheets and your institution's Environmental Health and Safety Office for proper handling of equipment and hazardous material used in this protocol.

RECIPES: Please see the end of this protocol for recipes indicated by <R>. Additional recipes can be found online at http://cshprotocols.cshlp.org/site/recipes.

Reagents

Aedes aegypti pupae
Bleach (10%)
Broth solution for hatching *Ae. aegypti* eggs <R>
H_2O (distilled)
Ice
Mice housed in cages in a vivarium (e.g., Swiss Webster)
Mouse sedation mixture <R>
Phosphate-buffered saline (PBS; optional, see Step 20; Fisher Scientific BP399500)
Sucrose solution (10%) for adult mosquito maintenance <R>

Equipment

Aluminum cage (8″ × 8″ × 8″; BioQuip 1450A)
Aspirator (e.g., John W. Hock Company 612)
Climate-controlled room (e.g., Environmental Growth Chambers EGC TC2)

[3]Correspondence: cmcmeni1@jhu.edu

Cite this protocol as *Cold Spring Harb Protoc*; doi:10.1101/pdb.prot108018

Cold plate (optional, see Step 22; e.g., Torrey Pines Scientific Echoterm IC2)
Alternatively, one can use a Petri dish placed on ice.

Cotton balls, medium (e.g., Fisher Scientific 22-456-880)
Cups (8.5-oz, VWR 89009-664; and 12-oz, Solo 412SIN-0041)
Dissecting microscope
Drosophila vials (Genesee 32–114) and plugs (Genesee 49–101) (optional; see Steps 4 and 15)
Filter paper (55-mm; Whatman 1001-090)
Forceps
Ice bucket
Kimwipes (Kimberly-Clark 34155)
Mating verification (optional, see Steps 21–27)
  Compound microscope (with 40×–60× objective)
  Coverslips (Fisher Scientific 12-545-AP)
  Dissection dish (glass)
  Microscope slides (e.g., Fisher Scientific 12-544-2)
Netting with a small mesh size (<1-mm; preferably polyester, e.g., white tulle)
Paintbrush (fine, soft-bristle; e.g., AIT Products B07J6F3MQH)
Paper towels
Petri dish with lid
Pipette (polyethylene; e.g., Fisher 13-711-7M)
Polycarbonate pan (4-L; e.g., Cambro 24CW135)
Rubber bands
Scissors
Tea strainer (with openings of 0.5–1 mm)
Tupperware (6 1/2″ × 11 1/2″ × 3 1/2″)
Waste receptacle and lid (5-gallon size)
Ziploc bags (1-gallon size)

METHOD

To obtain virgin female and male adults for single-pair and small-group crosses of Ae. aegypti, *it is most convenient to sex individuals at the pupal stage. Female and male pupae with pronounced size-based dimorphism to facilitate sexing can be obtained by following the standardized batch-rearing procedure outlined in Protocol 1: Batch Rearing* Aedes aegypti *(Wohl and McMeniman 2022a).*

Maintain Aedes aegypti strains in an insectary facility at 27°C and 80% relative humidity with a 12:12 h light:dark photoperiod using a walk-in climate-controlled room.

Pupal Sexing

1. Place a tea strainer over a waste receptacle and pour pupae from a rearing pan into the strainer.
2. Place the pupae in a fresh volume of distilled H_2O in an 8.5-oz cup or polycarbonate pan.
   Placing pupae in clean H_2O allows for a clearer visualization of size differences between pupal sexes.
   Chemically treat the discarded rearing pan H_2O in the waste receptacle by adding 10% bleach and allowing the solution to sit for 48 h. Discard the waste. Alternatively, wastewater can be heat treated or frozen fully and then discarded.
3. Add 100 mL of fresh distilled H_2O to two new 12-oz pupal eclosion cups. One is for virgin males and one for virgin females.
4. Sort male and female pupae into their designated cups by using a polyethylene pipette with the end cut off with scissors to make it large enough for pupae to flow without obstruction.

- Differentiate the sexes visually by size, with the females being larger and broader than the males.
- For more precision, pipette the pupa on a Petri dish laying on top of ice and use a dissecting microscope to inspect the shape of the genital lobe (see Fig. 1B in Introduction: **Overview of *Aedes aegypti* and Use in Laboratory Studies** [Wohl and McMeniman 2022b]). Use a paintbrush to gently manipulate the pupa while inspecting the genital lobe.

  Pupae can be sorted into small groups (~25) into 12-oz eclosion cups or alternatively placed individually into single vials to eclose. Generally, do not allow groups of pupae for sexing to eclose in large cages or at too high a density, as it is more difficult to classify males and females upon emergence if crowded and to find stray individuals of the unwanted sex.

5. Place mesh over the 12-oz pupal eclosion cup and add two rubber bands to prevent emerging adults from escaping.

   Use an appropriately sized piece of mosquito netting to cover the cup. This is necessary to prevent the escape of adult Ae. aegypti *that may aggressively seek to exit holding containers if gaps in coverage are present around the cup lip or if a large mesh size is used. Additionally, pulling the mesh taut around the margin of the rubber bands prevents mosquitoes from entering side pockets where physical damage or escape may occur during attempts to transfer them via aspiration. Finally, the use of two rubber bands on cups is a good precautionary practice in case one breaks.*

Maintaining the Virginity of Adults

6. To ensure mosquito virginity, inspect pupal eclosion cups every morning and evening for emerging adults.
7. Use an aspirator to remove emerging adults and identify them based on sexually dimorphic features such as the morphology of the antennae, maxillary palps, and genitalia (see Fig. 1C in Introduction: **Overview of *Aedes aegypti* and Use in Laboratory Studies** [Wohl and McMeniman 2022b]).

   The tip of the aspirator can be plugged transiently with your finger or a cotton ball to allow close visual inspection. Sexually dimorphic features of adults are visible by eye without magnification. Very rarely, gynandromorphs with hybrid female and male features may be found and can be discarded.

   Avoid aspirating mosquitoes that have not fully eclosed from their pupal casing or newly emerged mosquitoes floating on the surface of the H_2O*, as these mosquitoes are often fragile. Adults that have fully emerged will be resting on the surface of the cup walls or mesh and are suitable for aspiration.*

8. Immediately remove any stray adults from the nonrequired sex and discard.

   Once fertilized, females can use sperm from a single male for her lifetime; however, males may continue to copulate with multiple females. Therefore, one should maintain the virginity of females.

9. Transfer virgin adults of each sex that have eclosed to new 12-oz holding cups to allow them to sexually mature for 2 d. Place a sucrose-soaked cotton ball on top of the mesh covering of each cup.

   To facilitate aspirator entry into the pupal eclosion or adult holding cups without mosquito escape, gently pull a small margin of the affixed mesh out from underneath the rubber bands to create a finger sized opening where the aspirator tube can be inserted. If mosquitoes try to escape during the aspiration process, gently pull the mesh taught again and repeat the process.

Establishing Mating Crosses

10. Establish mating crosses between virgin male and female mosquitoes in 12-oz cups (single pair) or 12-oz cups or cages (small group) according to preferences and experimental conditions (Fig. 1).

    The smallest cup size we have determined that reliably facilitates enough room for successful copulation between a male and female Ae. aegypti *is a 12-oz Solo cup.* Ae. aegypti *will not mate in* Drosophila *vials but will oviposit in them.*

    Typically, single-pair matings between one virgin male and one virgin female lead to successful insemination and production of offspring. However, if matings between single pairs are not absolutely required for your

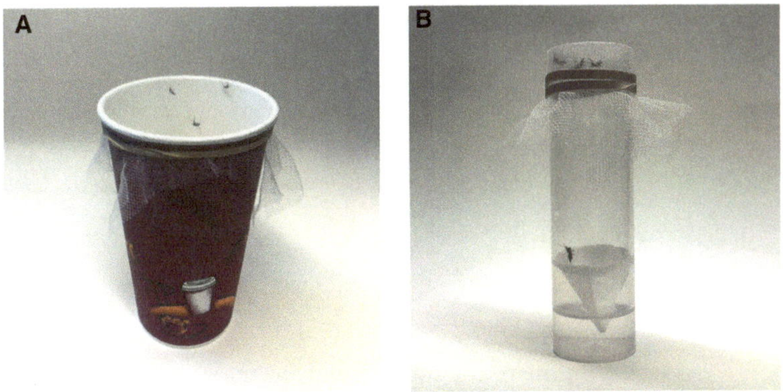

FIGURE 1. Equipment for single-pair and small-group mating crosses with *Aedes aegypti*. (*A*) Mosquito holding cup for mating consisting of a 12-oz solo cup covered with fine white mesh netting suitable to contain multiple *Ae. aegypti*. (*B*) Oviposition vial capped with mesh containing a moist filter paper cone partially submerged in distilled H_2O onto which female *Ae. aegypti* lay eggs on its interior surface.

experimental application, add a higher ratio of males to females (e.g., 3:1) to maximize the probability of insemination.

11. Leave mosquitoes to mate for 24 h.

    For routine experimentation, a mating period of 24 h is more than sufficient to ensure successful insemination.

    In instances when you would like to trial a shorter mating period or mating at set times of day, one can visually confirm mating attempts and verify insemination by terminally dissecting the spermathecae of females as described in Steps 21–27.

Blood Feeding

12. At least 4 h before blood feeding of females from crosses is scheduled to occur, remove the sucrose source from the 12-oz cups or cages to ensure more vigorous feeding.
13. Anesthetize mice with an intraperitoneal injection of the mouse sedation mixture and apply the sedated mice to the cup or cage harboring females.
14. Blood feed females for 30 min or until females are fully engorged with blood. See Protocol 1: Batch Rearing *Aedes aegypti* [Wohl and McMeniman 2022a] for more details and for alternatives to using mice as a blood source.

    Mating a group of virgin females with a group of virgin males and subsequently isolating individual females for oviposition are common practices during the establishment of transgenic or Wolbachia*-transinfected lines. In this case, females can be blood-fed as a group and then moved individually to oviposition vials.*

    Likewise, multiple crosses between individual males and groups of virgin females are routine practice. To save time and avoid having to blood feed many individual cups or cages of this cross type, virgin females may be blood fed as a group 24 h before mating crosses. Alternatively, females may be blood-fed postmating.

15. Return mice to their cages and verify recovery from sedation before transporting them back to the vivarium.
16. Return sucrose sources to the mosquito cups or cages.

    If egg deposition on sucrose-soaked wicks or cotton balls supplied to cages or cups is a concern, remove them 48 h post–blood feeding. As a substitute, use a gloved finger to manually streak the sucrose solution periodically every 6–12 h onto the cage mesh or cup mesh. This will provide enough moisture and sucrose to sustain the mosquitoes until oviposition substrates are introduced for egg laying.

Oviposition

17. Wait 72 h after blood feeding and then introduce oviposition substrates.

For Small-Group Mating in a Cage

i. Fill an 8.5-oz cup with ~100 mL of distilled H_2O.

ii. Line the interior of the cup with six overlapping filter papers and place in the cage.

For Small-Group Mating or Single-Pair Crossings in 12-oz Cups

Other containers can be used as well.

iii. Add 1.5 cm of distilled H_2O to a *Drosophila* vial.

iv. Take one filter paper and fold it in half two times so that you are left with a 90° arc.

v. Open one of the outside folds to make a cone.

vi. Place the apex of the cone in the bottom of the vial ~0.5 cm below the H_2O line so that all the surfaces of the cone are moist. Make sure that there are no gaps between the cone and the inner wall of the vial.

vii. Gently aspirate females into a *Drosophila* vial or knockout females with cold and transfer with a paintbrush being careful not to submerge the females in the H_2O.

viii. Put a plug or mesh secured with a rubber band on the vial to prevent mosquito escape.

For Embryonic Microinjection Experiments

ix. Aspirate groups of five to 10 gravid females into a *Drosophila* oviposition vial with a moist filter paper cone (Fig. 1B).

x. Place it in a dark place, such as a cupboard or drawer for 60 min, for on-demand egg laying.

These eggs can be used subsequently for injection at the preblastoderm stage. One can use the melanization of eggs to standardize the developmental stage at which to inject.

18. Wait 2 d to allow females to lay eggs on oviposition papers. During this period, the eggs can stay wet on the oviposition papers.

After eggs are laid, females may be removed and killed for genotyping if desired or blood feeding repeated as per Steps 12–16 to yield additional rounds of oviposition.

If using a molecular diagnostic tool for genotyping, investigators may instead choose to genotype newly eclosed virgin adults by single-leg polymerase chain reaction (PCR) before establishment of the mating crosses in Steps 10 and 11. This can accelerate the establishment of homozygous strains between females and males carrying target alleles.

Egg Conditioning, Storing, and Hatching

19. Condition the eggs for storage.

For Small-Group Mating in a Cage

i. Remove adult female(s) via aspiration.

ii. Remove papers from the oviposition cup with forceps and let them dry on two folded paper towels for 5 min.

iii. Place papers inside one unfolded paper towel and fold inside like a "parcel."

iv. Place into a 1-gallon-sized Ziploc bag padded with two unfolded paper towels to distribute remaining moisture. Seal the Ziploc bag.

v. Store the bag in a Tupperware for up to 6 mo at room temperature.

Cite this protocol as *Cold Spring Harb Protoc*; doi:10.1101/pdb.prot108018

For Small-Group Mating or Single-Pair Crossings in 12-oz Cups

vi. Remove adult female(s) from the *Drosophila* vial via aspiration.

vii. Remove the filter paper using forceps and dry on two folded paper towels for 5 min.

viii. Pour out remaining H_2O from vial.

ix. Dry the inside of the vial quickly with a Kimwipe to soak up any residual moisture.

x. Put the egg paper back into vial and put a plug back on to securely cap the vial.

xi. Multiple labeled vials can be stored at room temperature in sealed Ziploc bags. Eggs may remain viable for up to 6 mo.

After conditioning, eggs will be fully embryonated for hatching by 96–120 h postoviposition.

For maximal yield of larvae, conditioned eggs should be hatched within 120–240 h postoviposition. Stored eggs can maintain viability for up to 6 mo; however, during storage, hatch rates may decrease over time.

20. To hatch conditioned eggs, submerge oviposition papers in broth solution for hatching *Ae. aegypti* eggs. Monitor hatching after 48 h.

Oviposition papers with eggs derived from single-pair crossings and small-group matings may be hatched in 8.5-oz cups, polycarbonate pans, or other vessel types (e.g., jars or other cup sizes) according to user preferences and experimental design. Eggs will also hatch in preboiled and subsequently cooled distilled H_2O or vacuum-treated distilled H_2O. For hatching large numbers of oviposition papers derived from single-pair crossings (e.g., >30), we find submersion of each paper containing eggs in ~150 mL broth solution in individual 8.5-oz cups to be the convenient. Partially attaching a lid to the 8.5-oz cups such that it is securely fastened, but not airtight, also allows cups to be stacked on shelving and is therefore space-efficient.

See Protocol 1: Batch Rearing Aedes aegypti *(Wohl and McMeniman 2022a) for more details on hatching eggs.*

(Optional) Mating Verification via the Spermathecae

In some cases, it may be desirable to verify that the female Ae. aegypti *used in crosses successfully mated with the supplied male(s). Blood-fed females may lay a few unfertilized eggs when unmated. Incompatible cross types with* Wolbachia*-infected* Ae. aegypti *males may also yield eggs that do not hatch because of cytoplasmic incompatibility.*

Please note that this process is terminal.

21. Knock out a female mosquito using cold anesthetization by placing a holding cup containing the mosquito into an ice bucket containing ice.
22. After the female is immobilized, place it on a cold plate or on a Petri dish resting upon ice.
23. Place the female mosquito under examination into a glass dissection dish well.
24. Remove its abdomen under a dissecting microscope with sharp forceps.
25. Under a drop of PBS, next remove the last two segments of the abdomen using sharp forceps. Open these segments to find the three brown spheres of the spermathecae, the sperm storage organ.
26. Using a pipette, transfer the spermathecae to a microscope slide with a fresh small drop of PBS and apply a coverslip.
27. View the spermathecae under a compound microscope at 40–60×.

If inseminated, the female spermathecae should contain sperm, which look like a mass of swirling tiny threads (Rosay 1969). If not inseminated, the three spheres of the spermathecae will appear as empty translucent brown capsules.

DISCUSSION

During genetic studies with *Ae. aegypti*, routine yet critical procedures are setting up single-pair and small-group crosses between the sexes to generate heterozygous and homozygous mosquitoes of a desired genotype (McBride et al. 2014; McMeniman et al. 2014). After interspecific transfer, single-pair and small-group crosses are also integral for establishing different strains of the obligate intracellular bacterium *Wolbachia pipientis* in *Ae. aegypti* for stock establishment and subsequent phenotypic characterization (Xi et al. 2005; McMeniman et al. 2009; Walker et al. 2011).

Key considerations when establishing both single-pair and small-group crosses include the use of rigorous practices to ensure the virginity of newly emerged adults and that mating is established between males and females that are sexually mature. The provision of holding cups or cages with sufficient space for *Ae. aegypti* to mate is also critical to guarantee successful insemination. As it is often desirable to collect all the deposited eggs from an individual female or small group of females from a cross when screening for transgenics or *Wolbachia* infection, removal of cotton balls or wicks soaked with sucrose solution from the holding cup or cage is also recommended at 48 h post–blood feeding to prevent egg laying by *Ae. aegypti* on these substrates. This practice will direct egg deposition toward oviposition papers for proper collection and storage.

Egg deposition by single or small groups of females in standard *Drosophila* vials gives investigators considerable flexibility to collect and store eggs from defined cohorts of females in a high-throughput fashion. During this process, a lack of successful oviposition and/or egg hatch rates from a cross may be indicative of unsuccessful blood feeding to initiate egg development or alternatively represent problems with mating and insemination. Within the context of *Wolbachia* infection in *Ae. aegypti*, a lack of hatching in crosses between infected males and uninfected females may also reflect the induction of the crossing sterility cytoplasmic incompatibility (Xi et al. 2005; McMeniman et al. 2009; Walker et al. 2011). In these scenarios, if mating pairs or small groups used in the cross are still alive, these procedures can be repeated for subsequent gonotrophic cycles and insemination queried by dissecting the spermathecae according to investigator preferences. Although dissecting spermathecae is a terminal method, one group has reported the ability to view the spermathecae in live females, allowing the confirmation of mating before oviposition (Carrasquilla and Lounibos 2015).

RECIPES

Broth Solution for Hatching Aedes aegypti *Eggs*

Finely mill a larval diet source (e.g., TetraMin Tropical Tablets, Tetra 16110) using a mortar and pestle.
Add 1% w/v of milled larval diet to distilled H_2O in an autoclavable bottle.
Autoclave the solution for 30 min at 121°C.

Store at room temperature in a sealed bottle until use.

Mouse Sedation Mixture

Reagent	Stock concentration	Amount
Ketamine	100 mg/mL	2 mL
Acepromazine	10 mg/mL	1 mL
NaCl for injection (sterile)	0.9%	7 mL

Combine the reagents and mix well. Makes a total of 10 mL.
Store at room temperature in the dark. Mouse sedation mixture may be stored for the period specified by the manufacturer expiry dates of the reagents used.

Cite this protocol as *Cold Spring Harb Protoc*; doi:10.1101/pdb.prot108018

Sucrose Solution (10%) for Adult Mosquito Maintenance

Create a solution of 10% w/v of sucrose dissolved in distilled H_2O.

Store at 4°C in a sterile sealed container.

ACKNOWLEDGMENTS

We acknowledge generous funding to C.J.M. from the National Institutes of Health (R21 AI146450-01), Innovative Vector Control Consortium, Johns Hopkins Malaria Research Institute, and Bloomberg Philanthropies. M.P.W. is supported by a postdoctoral fellowship under The Molecular and Cellular Basis of Infectious Diseases (MCBID) Program (T32A1007417) from the National Institutes of Health.

REFERENCES

Carrasquilla MC, Lounibos LP. 2015. Detection of insemination status in live *Aedes aegypti* females. *J Insect Physiol* **75:** 1–4. doi:10.1016/j.jinsphys.2015.01.015

McBride CS, Baier F, Omondi AB, Spitzer SA, Lutomiah J, Sang R, Ignell R, Vosshall LB. 2014. Evolution of mosquito preference for humans linked to an odorant receptor. *Nature* **515:** 222–227. doi:10.1038/nature13964

McMeniman CJ, Lane RV, Cass BN, Fong AW, Sidhu M, Wang YF, O'Neill SL. 2009. Stable introduction of a life-shortening *Wolbachia* infection into the mosquito *Aedes aegypti*. *Science* **323:** 141–144. doi:10.1126/science.1165326

McMeniman CJ, Corfas RA, Matthews BJ, Ritchie SA, Vosshall LB. 2014. Multimodal integration of carbon dioxide and other sensory cues drives mosquito attraction to humans. *Cell* **156:** 1060–1071. doi:10.1016/j.cell.2013.12.044

Rosay B. 1969. Anatomical indicators for assessing age of mosquitoes: dissection techniques and field application of methods. *Mosquito News* **29:** 419–423.

Walker T, Johnson PH, Moreira LA, Iturbe-Ormaetxe I, Frentiu FD, McMeniman CJ, Leong YS, Dong Y, Axford J, Kriesner P, et al. 2011. The wMel *Wolbachia* strain blocks dengue and invades caged *Aedes aegypti* populations. *Nature* **476:** 450–453. doi:10.1038/nature10355

Wohl MP, McMeniman CJ. 2022a. Batch rearing *Aedes aegypti*. *Cold Spring Harb Protoc* doi:10.1101/pdb.prot108017

Wohl MP, McMeniman CJ. 2022b. Overview of *Aedes aegypti* and use in laboratory studies. *Cold Spring Harb Protoc* doi:10.1101/pdb.prot107651

Xi Z, Khoo CC, Dobson SL. 2005. *Wolbachia* establishment and invasion in an *Aedes aegypti* laboratory population. *Science* **310:** 326–328. doi:10.1126/science.1117607

CHAPTER 2

Considerations for Rearing and Maintaining *Anopheles* in the Laboratory

Laura N. Leite,[1,2] Priscila Bascuñán,[1,2,3] Ellen M. Dotson,[1] and Mark Q. Benedict[1,2,4]

[1]*Entomology Branch, Division of Parasitic Diseases and Malaria, Centers for Disease Control and Prevention (CDC), Atlanta, Georgia 30333, USA;* [2]*CDC Foundation, Atlanta, Georgia 30308, USA*

Anopheles mosquitoes can transmit several human pathogens, including viruses such as o'nyong-nyong and parasites including *Plasmodium* spp. and *Wuchereria* spp., which cause malaria and filariasis, respectively. Rearing *Anopheles* species of medical importance under laboratory conditions allows researchers to carry out experiments to better understand their genetics, physiology, and behavior. However, *Anopheles* species vary in how easily they can be reared in the laboratory, and some species have been difficult to colonize. Once established, members of the important African *Anopheles gambiae* complex thrive following a standard protocol and are predictable in growth and development rates. Here, we provide useful basic information and guidance to successfully maintain colonies of *An. gambiae* and other species of *Anopheles* in a laboratory setting. We also provide an example of a 3-wk rearing schedule that produces sufficient numbers of mosquitoes while minimizing the work required during weekends. In the accompanying protocols, we detail efficient methods and techniques suitable for several species of this genus at the egg, larva, pupae, and adult stages; however, it will be necessary for researchers to adjust methods as needed based on site-specific rearing observations of their particular strains.

BACKGROUND

Because mosquitoes are important vectors of animal and human disease and are nuisance feeders, hundreds of laboratories have conducted research to understand them and the diseases they transmit, usually with the purpose of developing control methods. Because these methods may be chemical, genetic, antiparasitic, and mechanical, diverse scientific disciplines are applied to study them. Among mosquitoes, only *Anopheles* spp. transmit human malaria, the deadliest vector-borne disease to humans, so they are naturally of great research interest (WHO 2022). Unfortunately, in comparison to two other important genera, *Culex* and *Aedes*, they are generally not the most vigorous nor easiest to rear.

All mosquito species have four developmental stages and undergo complete metamorphosis between each stage (Clements 1992). They are egg, larva, pupa, and adult. The egg and pupa stages are nonfeeding and subsist on reserves. Larvae feed on microorganisms and detritus that can easily be substituted for in the laboratory. The larval form is aquatic and consists of four instars during which the body size increases gradually but the general form does not change until turning into a pupa, which

[3]These authors contributed equally to this work.

[4]Correspondence: mbenedict@cdc.gov

Cite this introduction as *Cold Spring Harb Protoc*; doi:10.1101/pdb.top107802

is also an aquatic stage. In the adult stage, females of most species require a blood meal to develop mature eggs, although there are exceptions in which females can develop some eggs using reserves carried into adulthood. After the aquatic stages, adult *Anopheles* mosquitoes become nocturnal, generally resting during the daytime. Males feed only on various sugar sources that can easily be replaced with sucrose, fructose, and glucose in the laboratory, with sucrose being the most readily available and inexpensive. Females usually mate only once but in some circumstances can mate several times. The sperm are stored in a special female organ (the spermatheca) that often contains enough sperm from one mating for a lifetime of egg fertilization. Only in rare circumstances do females or males exhaust their supply of sperm. Females must be blood-fed to develop eggs and mated to fertilize them during oviposition (egg laying).

Mosquito's natural environmental conditions, resting places, blood sources, and larval habitats all differ from what can be achieved in the laboratory. Some suitable conditions such as larval habitat and diet are easily reproduced in the laboratory, but some steps are particularly critical and can be difficult to achieve, notably providing suitable blood meals and stimulating mating. Nonetheless, the methods that are successful for rearing numerous species overlap considerably.

CONSIDERATIONS FOR ALL STAGES

Mosquitoes are usually maintained in an environmental chamber at ~27°C and 80% relative humidity, but simple improvised enclosures or modified rooms can be constructed that are also suitable. Numerous studies have shown that altering the rearing temperature can affect the larval development by speeding it up (higher temperatures up to ~30°C) or slowing it down (lower temperatures). However, temperatures that are too high (~38°C–40°C) or too low (~10°C–12°C) can also reduce larval survival (Bayoh and Lindsay 2004; Christiansen-Jucht et al. 2014). One study showed that *Anopheles gambiae* larvae survive the longest (>20 d) at temperatures ranging 14°C–20°C (Bayoh and Lindsay 2004); however, this increase in larval development time is not practical for routine laboratory rearing. A 12:12 h dark–light cycle is also necessary, with or without a sunrise and sunset period. Numerous and varied light sources and intensities have been used and do not appear to be critical; however, pupation, oviposition, and mating are entrained during the larval stage, so the same light schedule should be maintained for all stages. Note that significant heating of the water can occur in containers with transparent covers due to the lighting, so it is common that the one on the top of a stack will be warmer than the others. Other factors that can influence the water temperature are the location of the larval pans in the rearing room (shelf height) and whether pans are covered or not (evaporative cooling of H_2O is decreased if pans are covered). The rearing room air temperature may need to be adjusted slightly higher to reach the desired H_2O temperature.

After oviposition, it is not necessary to keep adults at 27°C. Adult longevity is increased at somewhat lower temperatures (e.g., 22°C) and can provide adequate backup stock in the event of a rearing failure. However, many laboratories do not have the capacity for different temperature chambers and find it simplest to keep them all at the same temperature being used for larval rearing.

Water Types

A variety of different water types can be used for egg hatching and larval and pupal rearing. Deionized, untreated deep well, distilled, and chlorinated/fluoridated municipal supply water are frequently used to successfully rear larvae. Notably, the water should be free of any toxic chemicals. Test available water types on a small sample of larvae with the desired water type. Determine larval survival before using the selected type for all larval stocks. Insectary staff should also be aware that variations in chlorination in some locations can differ seasonally. If high chlorination is suspected to be an issue, leaving containers of water open overnight allows the chlorine to dissipate to a harmless level.

Cite this introduction as *Cold Spring Harb Protoc*; doi:10.1101/pdb.top107802

Mosquito species that develop in brackish water may require saltwater for optimal development. *Anopheles merus*, for example, thrives in 1.5% (w/v) salt water (e.g., aquarium sea or pond salt). For strains that rear best in salt water, use it for all aquatic stages including oviposition.

EGGS

Mated female mosquitoes are typically ready to lay eggs 48–72 h after taking a blood meal, with some variability among species. Gravid (with fully developed eggs), mated *Anopheles* females lay individual eggs directly on the H_2O or damp paper in the laboratory, mostly at night (dark cycle).

Egg collection must be done with extreme care to avoid egg damage and potential contamination between strains. Disinfecting eggs before egg hatching is a key step to maintaining healthy anopheline colonies for many generations, as it can inhibit the growth of harmful microbes (especially microsporidia) and prevent their transfer between generations. Some methods include soaking eggs in a bleach solution or a solution with bleach and Roccal-D (see Protocol 1: *Anopheles* Egg Collection, Disinfection, and Hatching [Leite et al. 2023a] for information on egg collection and disinfection).

Larval eclosion, or egg hatching, takes place after embryogenesis is complete, which normally occurs 2–3 d postoviposition (Clements 1992; Yaro et al. 2006). The duration of embryonic development can vary by species and other external factors, such as the presence of certain chemical components in the water and temperature, particularly during early embryonic development (Cardozo Gonçalvez de Carvalho et al. 2002; Yaro et al. 2006; Impoinvil et al. 2007). Agitation-induced egg hatching has also been investigated in *An. gambiae*, and it has been shown that early hatchers produce smaller larvae than those that hatch later (Yaro et al. 2006).

En masse oviposition and hatching are the most common methods to maintain a mosquito colony in the insectary; however, individual hatching is an alternative for when the progeny of a single parent need to be monitored (see Protocol 1: *Anopheles* Egg Collection, Disinfection, and Hatching [Leite et al. 2023a] for information on egg collection and hatching).

LARVAE

Container Type

Anopheles larvae can be successfully reared in many types of containers, but containers should be suitable for maximizing the water surface area because larvae rest and feed at the surface most of the time. Generally, this means that relatively shallow, rectangular pans are best. The ideal materials and sizes for rearing pans primarily depend on the scale of the project and the materials available. Many laboratories use locally available shallow plastic pans that are easily procured in shops (restaurant supply shops are good sources) or from plasticware suppliers. Rearing pans must be durable (resistant to near-boiling water without cracking, deforming, molding, or rusting) after numerous washings and from being kept in high-humidity (≥70% relative humidity) rearing rooms or chambers. Stainless steel and hard plastics, such as acrylonitrile butadiene styrene, high-impact polystyrene, and polypropylene, are ideal because they meet all of these criteria.

Container Size Relationship to Larval Density

Several considerations should be made when selecting the size of larval rearing pans. *Anopheles* larvae should be reared in water that is ~1.0–1.5 cm in depth, and the ratio of third-instar (L3) larvae to the amount of water should be one larva per 1–2 mL of water. For example, a pan of surface dimensions 20 cm × 35 cm (700 cm^2) containing 500 mL of water will allow healthy rearing of 250 to 300 larvae. A lid is also recommended to prevent water evaporation, evaporative cooling, and free-flying females from landing and laying eggs in the pan and contaminating the colony.

It is essential that larvae are reared at an appropriate density. Excessively dense numbers of larvae can result in decreased emergence rate, body size, and thus female fecundity (Gilles et al. 2011; Benedict et al. 2020). Estimating larval density can be accomplished by several methods as described in Protocol 2: *Anopheles* Larval Rearing (Leite et al. 2023b).

Food Types for Larvae

Many types of larval food are used by insectaries around the world. They can consist of simple mixtures made by laboratory staff or commercially available premixed food in the form of powders, pellets, or flakes. Locally formulated foods have the advantage that researchers can freely adjust the proportion of each component, whereas commercially produced proportions are fixed (but can be supplemented). Nonetheless, consistency in food quality and quantity is necessary, and the food should be stored in a refrigerator or freezer in a sealed bag or other container.

Factors to consider when choosing a type of larval food are nutritional value, availability, cost, and performance. Several larval diet mixtures have been evaluated under laboratory conditions (Zanin et al. 2019; Benedict et al. 2020); however, various kinds of fish food have been shown to contain the most suitable ingredients to maximize mosquito growth rate, size, and survival. Recently, researchers recommended using a low-cost commercially available fish food for mosquito rearing. The authors showed that mosquito developmental rates and adult longevity were equal to those of other diets, with high survivability to the adult stage (Benedict et al. 2020). Some of the larval foods that are widely used by laboratories rearing *Anopheles* are as follows:

- Damiens diet: consisting of tuna meal and bovine liver powder, developed by Damiens et al. (2012).
- Koi food: a commercial fish food, such as Doctors Foster and Smith Staple Diet Quality Koi and Goldfish Food, which contains high levels of proteins and comes in a pellet form. It is recommended to grind it and either measure and pour the powder directly onto the surface of the water or prepare a slurry and add equal volumes to each pan. If the latter is chosen, make sure to shake the container vigorously while pipetting as the food sediments quickly, resulting in inconsistent dispensing. Koi pellets are ideal for weekend feedings when larvae will not be attended to for several days. The pellets will sink to the bottom of the pans and will slowly break apart, providing a large amount of food to the larvae without causing the water to foul.
- TetraMin Plus Tropical Flakes: a commercial fish food containing shrimp meal.

Additionally, dietary supplements of vitamins and minerals are suitable options to improve larval growth, produce larger males, and reduce development time in females (Damiens et al. 2012; Nascimento et al. 2021). Brewer's or baker's yeast can also be used to feed first-instar larvae after hatching. This is useful to minimize large particles in the water, for instance, for when a larval sorter or dispenser will be used. It is also easier to see the larvae when estimating densities. First-instar larvae require food with a small particle size and typically do fine when hatched with 25 mg of ground koi food that has been sieved for a particle size of <600 µm.

Food Amounts for Larvae

Food quantity affects larval development time, survival, and adult traits (Araújo et al. 2012; Benedict et al. 2020). The most critical relationship is the amount of food/larvae/day over a wide range of densities of larvae/mL of rearing water. As will be noted elsewhere, ranges of densities are suitable. On the low end, it is difficult for larvae to forage for sufficient food and this may affect larval survival and produce small larvae and adults; on the high end, the accumulation of waste due to deteriorating unconsumed food inhibits larval growth. When excessive food accumulates, the water will turn cloudy and smelly, resulting from an abundance of microbial growth, and could lead to larval death. Thus, regulating the amount of food administered per pan as well as the larvae per milliliter of water are key factors for efficiently rearing healthy larvae. Before the food type and feeding method are chosen for routine rearing, a period of testing

Cite this introduction as *Cold Spring Harb Protoc*; doi:10.1101/pdb.top107802

is required to determine the amount and frequency of food that maximize the performance of a given mosquito colony. For this testing, different types and amounts of food should be compared by adjusting the food/larva/day. For *Anopheles* species, 0.1–0.8 mg food/larva/day is a reasonable range to begin the evaluation (Benedict et al. 2020). As a starting point for determining the larval density and food amount, a systematic procedure is recommended (Valero et al. 2016).

PUPAE

The pupal stage of *An. gambiae* lasts ~36 h, and daily adult eclosion is usually somewhat synchronous, occurring when the light level is falling in the evening if pupae are collected from rearing pans before midmorning. For many species, most adults will eclose before the next morning. Because the pupa is nonfeeding, the main requirement beyond a suitable temperature is clean water. Pupae are typically placed in small cups containing water at a depth of 1–3 cm. Pupal density should be kept low enough that all pupae can easily reach the surface to breathe after diving, otherwise death due to asphyxiation results. See Protocol 4: *Anopheles* Pupa Collection and Sex Identification (Leite et al. 2023c) for detailed methods to collect pupae and sort them by sex.

ADULTS

Cage Types and Sizes

Adult caging is one of the least critical choices, and many options are suitable. Nevertheless, several features are desirable. Most cages have at least a couple of sides that are made of screen mesh. The mesh provides ventilation and a surface that the adults can grip for resting. Also, at least one mesh side must be available for access to a blood source. Because anopheline adults are largely nocturnal and rest in dimly lit places, somewhat dark cages are suitable, although it is common to see brightly lit adult caging. Consideration should be given to how the cage will be cleaned. Some are difficult to disassemble and clean thoroughly, so in some cases, researchers have used paper cartons that can be discarded after use.

Commercially produced cages (Fig. 1A) are a great option if funds are available. Constructing cages out of other available materials is another viable option. Large, hard plastic containers such as buckets with lids can be modified to include a mesh top and cloth sleeve (Fig. 1B). Paper cups (e.g., soup or ice cream cups) can similarly be modified (Fig. 1C). These paper cups cannot be washed and thus are not a viable option for continuous reuse, but they can work well for experiments and short-term projects. Another option is to have a metal frame constructed and then stretch a mesh sleeve over the frame (Fig. 1D; Savage and Lowe 1971). The metal frames can be washed and autoclaved, and the mesh can be discarded or rewashed after each use. The frame is 25 × 15 × 20 cm, and the cage can hold about 500 mosquitoes. The size and material of the cage should be determined by material availability, the density of mosquitoes to be contained, and whether the cage is for continuous reuse. For reference, a 30-cm cube Bugdorm (Fig. 1A, left) can house 750 mosquitoes and a smaller 20-cm cube cage (Fig. 1A, right) is more suitable for 300 mosquitoes.

Blood Sources

Feeding *Anopheles* on live animals is becoming less feasible because of ethical and animal facility constraints. As an alternative, fresh blood can be purchased from commercial vendors for use in a membrane feeder (e.g., Hemotek; see Protocol 3: *Anopheles* Adult Anesthesia, Feeding, and Sex Separation (Leite et al. 2023d) for information on feed options). The most commonly used blood types are human, cow, rabbit, chicken, and horse blood. Blood must be mechanically defibrinated or contain an anticoagulant (e.g., sodium citrate, heparin, or EDTA). By preventing the blood from clotting, it can be stored in a refrigerator for several weeks, although a decline in mosquito feeding avidity will occur as the blood ages. Some colonies will feed for ~1 mo on the same lot of blood, whereas other colonies will feed

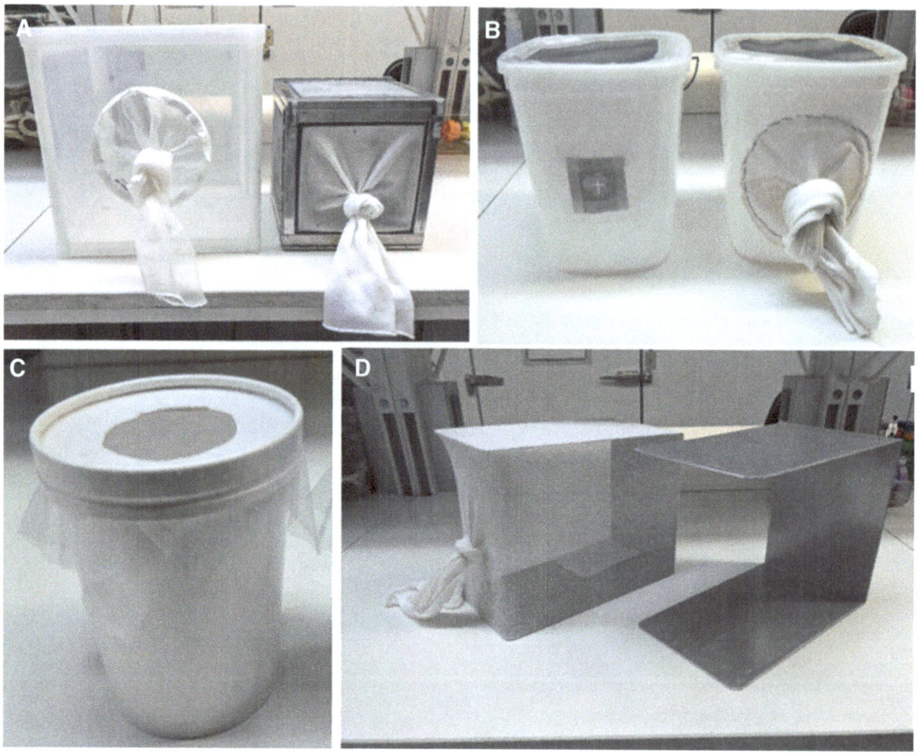

FIGURE 1. Housing for adult mosquitoes. (*A*) Commercially available mosquito cages. (*B*) Hard plastic bucket cages constructed for long-term use. (*C*) Paper cup holding container for short-term use. (*D*) Metal frame cage with mesh sleeve.

only on freshly collected blood. Most colonies fall somewhere between these two extremes. Importantly, verify that the animal from which blood is collected was not treated with ivermectin or any other insecticidal drugs before blood collection because they can cause mosquito mortality.

Mosquitoes can also be membrane fed an artificial blood meal if it satisfies the minimal nutritional requirements for egg production. Some laboratories have developed specific formulations and shown their efficacy (Marques et al. 2019; Da Silva Costa et al. 2020).

Mosquito species and age, blood quality and freshness, and time of day are some of the factors that may affect blood feeding rates. Preliminary tests to learn more about the feeding preferences of the mosquito strain of interest are recommended.

REARING SCHEDULE FOR *ANOPHELES* SPECIES

Although strain variation exists, the schedule we present here in Table 1 should work well for all *An. gambiae* complex species (including *An. gambiae*, *Anopheles coluzzii*, *Anopheles arabiensis*, *Anopheles quadriannulatus*, *An. merus*, and *Anopheles melas*) as well as for *Anopheles stephensi*, *Anopheles albimanus*, and *Anopheles dirus*.

This schedule was created to minimize weekend work and is based on blood-feeding mosquitoes on a Thursday and/or Friday, placing oviposition cups in cages on the following Sunday, bleaching eggs on a Monday, and hatching eggs on Tuesday. The schedule can be shifted as needed. Although the development time is often shorter than the 3-wk schedule presented here, this schedule allows certain activities to be performed predictably on specific days of the week and avoids the most labor-intensive activities from falling on the weekend.

 Cite this introduction as *Cold Spring Harb Protoc*; doi:10.1101/pdb.top107802

TABLE 1. *Anopheles* rearing schedule

Days posthatching	Day of the week	Rearing step
0	Tuesday	Hatch eggs by placing clean, bleached eggs (this is the day following day 20) with 1 mL 2% brewer's or baker's yeast or 25 mg of ground koi food in 300–500 mL of H_2O in a rearing pan. Note: When replacing H_2O or diluting larvae density in pans, always return the H_2O volume to ~1.0–1.5 cm in depth (~300–500 mL in a 22-cm × 35-cm pan).
1	Wednesday	Following the methods described in Protocol 2: *Anopheles* Larval Rearing [Leite et al. 2023b], separate first-instar larvae into pans to the desired density, typically 250–500/pan. Density varies by strain and method followed. Feed ~50 mg of ground koi food. Refer to the Food Type section for more information.
2	Thursday	Observe only.
3	Friday	Dilute larvae to ~250/pan and add 5 pellets or 140–200 mg of ground koi food.
4	Saturday	No action needed.
5	Sunday	Add 2–4 more koi pellets if most food has been consumed.
6	Monday	Clean rearing H_2O (refer to the section Rearing Schedule for *Anopheles* species) and keep larval density at ~250/pan. Feed larvae 150–200 mg of ground koi food. Pick any pupae (refer to Protocol 2: *Anopheles* Larval Rearing [Leite et al. 2023b] for details) and place in a cage containing a sugar H_2O source (refer to the section Cage Types and Sizes).
7	Tuesday	Clean rearing H_2O if needed, collect pupae, and start (or add to) a cage. Ensure the sugar H_2O source is adequate in the cage with emerging pupae. Feed larvae 150–200 mg of ground koi food.
8	Wednesday	
9	Thursday	
10	Friday	Continue to collect pupae and add them to the cage. Discard any remaining larvae if enough pupae have been collected. If too many larvae remain or they have not started pupating, choose one of the following options according to your preference: Option 1: Staff comes in over the weekend. Clean H_2O and feed 150–200 mg of ground koi food on Friday. Continue to follow the daily schedule. Option 2: No one comes in over the weekend. Clean H_2O and feed 6–8 pellets on Friday and cover with lids to prevent any adults from escaping. Resume schedule on day 13 (Monday). Ensure the sugar H_2O source is adequate.
11	Saturday	Clean H_2O if needed, pick pupae, and start (or add to) a cage. Feed larvae 150–200 mg of ground koi food. Ensure the sugar H_2O source is adequate.
12	Sunday	Clean H_2O if needed, pick pupae, and start (or add to) a cage. Feed larvae 150–200 mg of ground koi food. For most strains, pupation will be completed by day 13. If not, repeat this step until all larvae have pupated or enough for a cage has been collected (see Cage Type and Sizes section for more details).
13	Monday	Ensure the sugar H_2O source is adequate.
14	Tuesday	Ensure the sugar H_2O source is adequate.
15	Wednesday	Ensure the sugar H_2O source is adequate.
16	Thursday	Blood feed cages. Refer to the sections on blood feeding in Protocol 3: *Anopheles* Adult Anesthesia, Feeding, and Sex Separation (Leite et al. 2023c).
17	Friday	Ensure the sugar H_2O source is adequate.
18	Saturday	No action needed.
19	Sunday	Place oviposition cup in a cage. Refer to Protocol 1: *Anopheles* Egg Collection, Disinfection, and Hatching [Leite et al. 2023a] for details on oviposition and egg collection methods. Ensure the sugar H_2O source is adequate.
20	Monday	Remove oviposition cups from cages and bleach eggs. Refer to Protocol 1: *Anopheles* Egg Collection, Disinfection, and Hatching [Leite et al. 2023a] for egg bleaching details. Take bleached eggs and place them in an environmental rearing chamber. Repeat all steps with the new generation of eggs.

This rearing guide is an example based on a 22-cm × 35-cm larval pan. Larval density and food amounts should be adjusted proportionally to the number of larvae and pan size. This general *Anopheles* rearing schedule can be adapted as needed.

In Table 1, we refer to "cleaning" water. This consists of filtering larvae out of all or a portion of the water and replacing or diluting it with fresh rearing water at a similar temperature. Cleaning water is a choice made based on the turbidity of the water, odor, and surface film. A foul odor, scum on the surface, and turbid water all indicate the food amount is excessive or the larval density too low. Under ideal circumstances, no water cleaning is necessary. This must be judged by staff.

Members of the *An. gambiae* complex should complete pupation by day 10. If most mosquitoes have not pupated by that time, reevaluate larval densities and food type and/or amounts, as the larval density may be too high or there may be insufficient food.

Schedule Deviations for Other *Anopheles* Species

Some *Anopheles* species (e.g., *Anopheles farauti, Anopheles sinensis, Anopheles funestus, Anopheles minimus, Anopheles atroparvus, Anopheles freeborni, Anopheles quadrimaculatus,* and *An. albimanus*) thrive when reared at higher densities as early instars. The same schedule can be used, but larval densities should be kept at 400–500 per tray until the larvae are L3s. Most larvae will not start pupation until day 10. Species such as *An. funestus* and *An. minimus* most likely will not start pupating until closer to day 13, and pupation will not be complete until later that week. It is also best not to disturb (or as little as possible) these strains and other finicky colonies. Feed larval pans closer to the minimum food amount the rearing schedule suggests and change the water only when necessary. Pupae of these strains can also be sensitive to handling, so it is recommended that if pupal death is observed to avoid the use of a strainer, as this can result in mortality. Some trial and error will be necessary to determine a colony's ideal larval density and feeding regimen that result in the most robust adults in the shortest time frame. A general method for determining this has been reported (Valero et al. 2016).

CONCLUSIONS

With experience, systematic method development, and our guidance here, it will become routine to rear *Anopheles* mosquitoes with high survival rates, good size, and high adult longevity. Keep in mind that the health and size of immature stages determine the longevity and fecundity of adults, so one must work diligently to rear larvae under conditions that are as near optimal as is practical.

ACKNOWLEDGMENTS

We thank all members of the MR4 Vector Activity (Malaria Research and Reference Reagent Resource Center, BEI Resources) and Target Malaria Teams who contributed to this introduction by providing technical information on rearing methods, some of the images, and general insectary support. In particular, we thank Catherine Steele, Ilsse Ortega, and Katelyn Cavender. We also thank Michael Aidoo and Audrey Lenhart for proofreading this manual and providing valuable feedback. The findings and conclusions in this report are those of the authors and do not necessarily represent the official position of the Centers for Disease Control and Prevention.

REFERENCES

Araújo MDS, Gil LHS, E-Silva ADA. 2012. Larval food quantity affects development time, survival and adult biological traits that influence the vectorial capacity of *Anopheles darlingi* under laboratory conditions. *Malar J* **11:** 261. doi:10.1186/1475-2875-11-261

Bayoh MN, Lindsay SW. 2004. Temperature-related duration of aquatic stages of the Afrotropical malaria vector mosquito *Anopheles gambiae* in the laboratory. *Med Vet Entomol* **18:** 174–179. doi:10.1111/j.0269-283X.2004.00495.x

Benedict MQ, Hunt CM, Vella MG, Gonzalez KM, Dotson EM, Matilda Collins C. 2020. Pragmatic selection of larval mosquito diets for insectary rearing of *Anopheles gambiae* and *Aedes aegypti*. *PLoS ONE* **15:** e0221838. doi:10.1371/journal.pone.0221838

Cardozo Gonçalvez de Carvalho S, de Jesus Martins Junior A, Bento Pereira Lima J, Valle D. 2002. Temperature influence on embryonic development of *Anopheles albitarsis* and *Anopheles aquasalis*. *Mem Inst Oswaldo Cruz* **97:** 1117–1120. doi:10.1590/S0074-02762002000800009

Christiansen-Jucht C, Parham PE, Saddler A, Koella JC, Basáñez M-G. 2014. Temperature during larval development and adult maintenance influences the survival of *Anopheles gambiae s.s. Parasit Vectors* **7:** 489. doi:10.1186/s13071-014-0489-3

Clements AN. 1992. *The biology of mosquitoes: development, nutrition, and reproduction*. Chapman & Hall.

Damiens D, Benedict MQ, Wille M, Gilles JRL. 2012. An inexpensive and effective larval diet for *Anopheles arabiensis* (Diptera: Culicidae): eat like a horse, a bird, or a fish? *J Med Entomol* **49:** 1001–1011. doi:10.1603/ME11289

Da Silva Costa G, Rodrigues MMS, Silva ADAE. 2020. Toward a blood-free diet for *Anopheles darlingi* (Diptera: Culicidae). *J Med Entomol* **57:** 947–951. doi:10.1093/jme/tjz217

Gilles JRL, Lees RS, Soliban SM, Benedict MQ. 2011. Density-dependent effects in experimental larval populations of *Anopheles arabiensis* (Diptera: Culicidae) can be negative, neutral, or overcompensatory depending on density and diet levels. *J Med Entomol* **48:** 296–304. doi:10.1603/ME09209

Impoinvil DE, Cardenas GA, Gihture JI, Mbogo CM, Beier JC. 2007. Constant temperature and time period effects on *Anopheles gambiae* egg hatching. *J Am Mosq Control Assoc* **23:** 124–130. doi:10.2987/8756-971X(2007)23[124:CTATPE]2.0.CO;2

Leite LN, Bascuñán P, Dotson EM, Benedict MQ. 2023a. *Anopheles* egg collection, disinfection, and hatching. *Cold Spring Harb Protoc* doi:10.1101/pdb.prot108186

Leite LN, Bascuñán P, Dotson EM, Benedict MQ. 2023b. *Anopheles* larval rearing. *Cold Spring Harb Protoc* doi:10.1101/pdb.prot108187

Leite LN, Bascuñán P, Dotson EM, Benedict MQ. 2023c. *Anopheles* pupa collection and sex identification. *Cold Spring Harb Protoc* doi:10.1101/pdb.prot108189

Leite LN, Bascuñán P, Dotson EM, Benedict MQ. 2023d. *Anopheles* adult anesthesia, feeding, and sex separation. *Cold Spring Harb Protoc* doi:10.1101/pdb.prot108188

Marques J, Cardoso JCR, Félix RC, Power DM, Silveira H. 2019. A blood-free diet to rear anopheline mosquitoes. *J Vis Exp* **2020:** 60144. doi:10.3791/60144

Nascimento DAS, Trindade FTT, Silva ADAE. 2021. Dietary supplementation with vitamins and minerals improves larvae and adult rearing conditions of *Anopheles darlingi* (Diptera: Culicidae). *J Med Entomol* **58:** 71–78. doi:10.1093/jme/tjaa173

Savage KE, Lowe RE. 1971. A one-piece aluminum cage designed for adult mosquitoes. *Mosquito News* **31:** 111–112. Available from https://www.cabdirect.org/cabdirect/abstract/19712902108

Valero L, Collins CM, Lees RS, Benedict MQ. 2016. Benchmarking vector arthropod culture: an example using the African malaria mosquito, *Anopheles gambiae* (Diptera: Culicidae). *Malar J* **15:** 262. doi:10.1186/s12936-016-1288-4

WHO. 2022. https://www.who.int/news-room/fact-sheets/detail/malaria [Accessed November 17, 2022].

Yaro AS, Dao A, Adamou A, Crawford JE, Ribeiro JM, Gwadz R, Traoré SF, Lehmann T. 2006. The distribution of hatching time in *Anopheles gambiae*. *Malar J* **5:** 19. doi:10.1186/1475-2875-5-19

Zanin CRF, Trindade FTT, Silva AAE. 2019. Effect of different food and sugar sources on the larval biology and adult longevity of *Anopheles darlingi* (Diptera: Culicidae). *Trop Biomed* **36:** 569–577.

Protocol 1

Anopheles Egg Collection, Disinfection, and Hatching

Laura N. Leite,[1,3] Priscila Bascuñán,[2,3] Ellen M. Dotson,[1] and Mark Q. Benedict[2,4]

[1]*Entomology Branch, Division of Parasitic Diseases and Malaria, Centers for Disease Control and Prevention (CDC), Atlanta, Georgia 30333, USA;* [2]*CDC Foundation, Atlanta, Georgia 30308, USA*

Gravid (i.e., with fully developed eggs), mated *Anopheles* females typically lay their eggs directly on water ~48–72 h after a blood meal. Unlike some other mosquito species, *Anopheles* eggs cannot be desiccated and stored for long durations, and, hence, colonies must be reared continuously. In this protocol, we discuss methods for egg collection, including individual and *en masse* oviposition; egg disinfection to avoid the transmission of infectious agents to the next generation; and egg hatching for colony maintenance or experimentation. We also include optional methods for estimating life history traits such as fecundity, fertility, and larval mortality rates from egg counts.

MATERIALS

It is essential that you consult the appropriate Material Safety Data Sheets and your institution's Environmental Health and Safety Office for proper handling of equipment and hazardous material used in this protocol.

Reagents

Anopheles mosquitoes in a cage (gravid, mated females that have been housed with males for several days)

H_2O (refer to the Water Types section in the Introduction: Considerations for Rearing and Maintaining *Anopheles* in the Laboratory [Leite et al. 2023a])

Hot H_2O (>70°C)

Ideally, an insectary would have an InSinkErator or comparable instant hot H_2O tap installed. An electric tea pot is a convenient alternative.

Household bleach solution (1%–2%, v/v, in H_2O)

Larval food (2%, w/v, brewers or baker's yeast or 25 mg of ground koi food)

Roccal-D and bleach disinfectant solution (1%–2%, v/v, of household bleach solution plus 0.7% Roccal-D) (optional; see Step 10)

Sugar H_2O (10%, w/v)

Equipment

Camera (see Step 18)

Container with a secure lid (any container that can hold ~50 mL of H_2O)

The lid should have an opening covered with a fine mesh. Alternatively, the container can be covered with fine mesh and an elastic band.

---

[3]These authors contributed equally to this work.

[4]Correspondence: mbenedict@cdc.gov

Cite this protocol as *Cold Spring Harb Protoc*; doi:10.1101/pdb.prot108186

Cotton balls
Dissecting microscope
Dissecting needle (e.g., Fisherbrand 08-960A)
Elastic bands
Environmental chamber or room with temperature, humidity, and light controls

Maintain mosquitoes in an environmental chamber at ~27°C, 80% relative humidity, and a 12 h light:12 h dark cycle, with or without a sunrise and sunset period.

Filter paper (e.g., Whatman Grade 3 Qualitative, 90-mm-diameter circle, Sigma-Aldrich WHA1003090)
Fine mesh (any commercially available mosquito netting)
Forceps
Image software program (e.g., egg counter [Mollahosseini et al. 2012] or ImageJ).

ImageJ is free software that allows the researcher to count eggs one by one using the multipoint tool that adds a number every time a mouse click is made on the image.

Mechanical counter (two-place denominator; optional, see Step 18)
Pencil (optional; see Step 18)
Petri dish (90-mm-diameter)
Rearing pan with lid (lid is optional though strongly recommended; e.g., US Plastics 52051)

*See Introduction: **Considerations for Rearing and Maintaining** Anopheles **in the Laboratory** (Leite et al. 2023a) for more information*

Scissors
Sealed container (see Step 14)
Sponge cloth (optional; see Step 17)
Squat, wide-opening plastic cups (~250-mL)
Stainless-steel forceps
Suction device to collect adult mosquitoes (e.g., Fulton RHM 200 aspirator)
Vacuum filtration system

A Nalgene 500-mL vacuum filtration system can be adapted to this method by breaking off the upper chamber and removing the membrane. Connect it to a vacuum system with a trap, which is composed of an Erlenmeyer filtration flask capped with a two-holed stopper. Insert the tube from the Nalgene filter into one hole and connect the tube to vacuum source to the side of the flask. Use a blue grinding pestle or similar object to plug the second hole to allow for easy release of the vacuum pressure.

Wash bottle (500-mL; e.g., Sigma-Aldrich Z423106)

METHOD

Egg Collection

Mated female mosquitoes are typically ready to lay eggs 48–72 h after taking a blood meal, with some variability among species. Gravid (with fully developed eggs), mated Anopheles *females lay individual eggs directly on the water or damp filter paper in the laboratory, mostly at night (dark cycle). En masse oviposition and hatching are the most used methods to maintain a mosquito colony in the insectary, whereas individual ovipositions are more common when few females are available or for experimental activities such as genetic crosses or when life history traits are being determined.*

1. Provide an oviposition substrate to gravid female *Anopheles* mosquitoes.

Single-Mosquito Oviposition

i. Line the wall of a 50-mL container with a strip of filter paper and fill with H_2O up to 0.5–1 cm above the bottom of the filter paper (Fig. 1A).

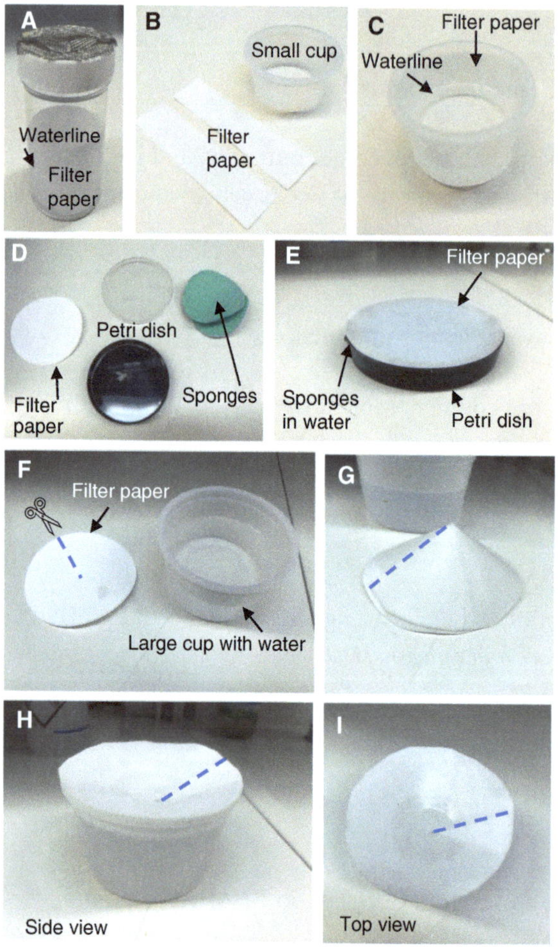

FIGURE 1. Types of containers for egg collection. (*A*) Plastic cup lined with filter paper used for individual females. The cup is filled with enough H_2O to ensure the filter paper remains wet. Mesh at the top of the container allows the female to have access to cotton wetted with sugar H_2O if necessary. (*B,C*) To collect eggs *en masse*, a cup with H_2O and lined with filter paper can be placed in a cage with gravid and mated female mosquitoes or (*D,E*) a filter paper can be placed on top of sponges soaked in H_2O contained in a Petri dish. (*F–I*) Alternatively, a 9-cm-diameter round filter paper can be cut from the edge to the center to create a funnel. The funnel is then placed on top of a cup filled with H_2O touching the paper on which females will land and oviposit.

ii. Cover the opening of the container with fine mesh secured with an elastic band, cotton ball, or other similar item.

iii. With scissors, create a small hole on the mesh (one that can easily be covered with a cotton ball or a piece of tape once the mosquito is in the cup).

iv. Using an aspirator, remove a female from the cage, carefully transfer it into the container through the small hole in the mesh, and plug the container with a cotton ball.

v. Let the females oviposit overnight (dark cycle, ~12 h)

This step does not have to occur during the night. A laboratory may have the dark cycle occurring during the actual day, which will also stimulate females to oviposit.

For one night only, providing sugar H_2O is usually not necessary.

vi. On the following day, examine the container(s) for the presence of eggs. With an aspirator, remove the female from tubes containing eggs. Allow the remaining females an opportunity oviposit another night and provide them sugar H_2O.

If checking fecundity or fertility parameters, allow eggs to hatch (generally ~2–3 d after oviposition occurs) and continue to Step 17; otherwise, proceed to Protocol 2: Anopheles *Larval Rearing (Leite et al. 2023b) for information on how to rear larvae.*

En Masse *Oviposition*

vii. Line the sides of a labeled plastic cup (~250-mL) with filter paper (Fig. 1B,C) and fill to a depth of ~1.0 cm of H_2O such that the paper is halfway covered with H_2O.

Do not leave gaps in the paper because eggs will stick to many plastics and dry out.

 Cite this protocol as *Cold Spring Harb Protoc*; doi:10.1101/pdb.prot108186

viii. Place the cup into a cage with gravid females.

ix. Allow the females to oviposit "overnight" (dark cycle, ~12 h).

This step does not have to occur during the night. A laboratory may have the dark cycle occurring during the day, which will also make females oviposit.

x. Remove the egg cup from the cage. Take care that no adult mosquitoes escape during this step, by puffing briskly down the cage sleeve as the egg dish is being pulled out.

xi. Cover the egg cup with a Petri dish and proceed to Step 2.

Other en masse *oviposition methods include moistening a sponge and placing a filter paper on top (Fig. 1D,E) in a Petri dish or making a filter paper funnel and submerging it partially in H_2O (Fig. 1F–I).*

Egg Disinfection

Although it is possible to disinfect eggs collected from individual females, this method is typically used only for en masse *collections. Disinfecting eggs with bleach is a key step to keeping healthy* Anopheline *colonies for many generations, as it can inhibit the growth of harmful microbes (especially microsporidia) and prevent their transfer between generations. Use the Roccal-D solution along with bleach solution when a colony infection is suspected, as bleach solution alone does not prevent infection.*

2. Ensure that all the necessary materials (Fig. 2A) are clean and available before starting.
3. Remove any dead adult mosquitoes from the oviposition cup(s) or container(s) with clean forceps and wipe the adults from the forceps with a clean cotton ball. Use a fresh cotton ball for each oviposition cup.
4. Kill any eggs remaining on the forceps by submerging the forceps in hot H_2O (>70°C) before and after removing the adults from oviposition cups to prevent colony contamination. If possible, use another clean pair of forceps for each cup.
5. Turn on the vacuum filtration system at low suction.
6. Center a fresh filter paper on the platform with the vacuum applied.
7. Wet the filter paper with purified H_2O using a wash bottle. Increase the vacuum suction until the filter paper disk forms a depression.

   It may be necessary to press the paper down gently to close gaps and achieve suction.

8. Slowly pour H_2O and eggs onto the center of the disk, taking care that the eggs do not spill outside the depression. Rinse the sides of the oviposition cup with purified H_2O via the wash bottle until the desired number of eggs are on the filter paper (Fig. 2B).
9. Release the vacuum.
10. Add the bleach solution (or, if necessary, Roccal-D and bleach disinfectant solution) to the depression (Fig. 2C). Allow the eggs to sit in the bleach for 30 sec.
11. Apply the vacuum to drain the liquid completely.
12. Release the vacuum and rinse the eggs by adding purified H_2O to the entire depression. Allow eggs to sit for 3–5 sec.
13. Apply suction and drain the H_2O completely.
14. Rinse the eggs a third time with H_2O if disinfecting with the Roccal-D and bleach disinfectant solution.
15. Remove the damp egg paper and place it in a labeled sealed container large enough to fit the filter paper (Fig. 2D). Secure the lid. Place in an environmental chamber at the standard rearing conditions.
16. Store the eggs for up to 24 h.

    The 24 h holding time will allow embryos to develop completely before hatching and promotes simultaneous hatching when the eggs are transferred to H_2O. Longer storage periods can reduce embryo viability; however, An. gambiae *eggs have been stored at 4°C and 48% ± 2% relative humidity for up to 5 d without*

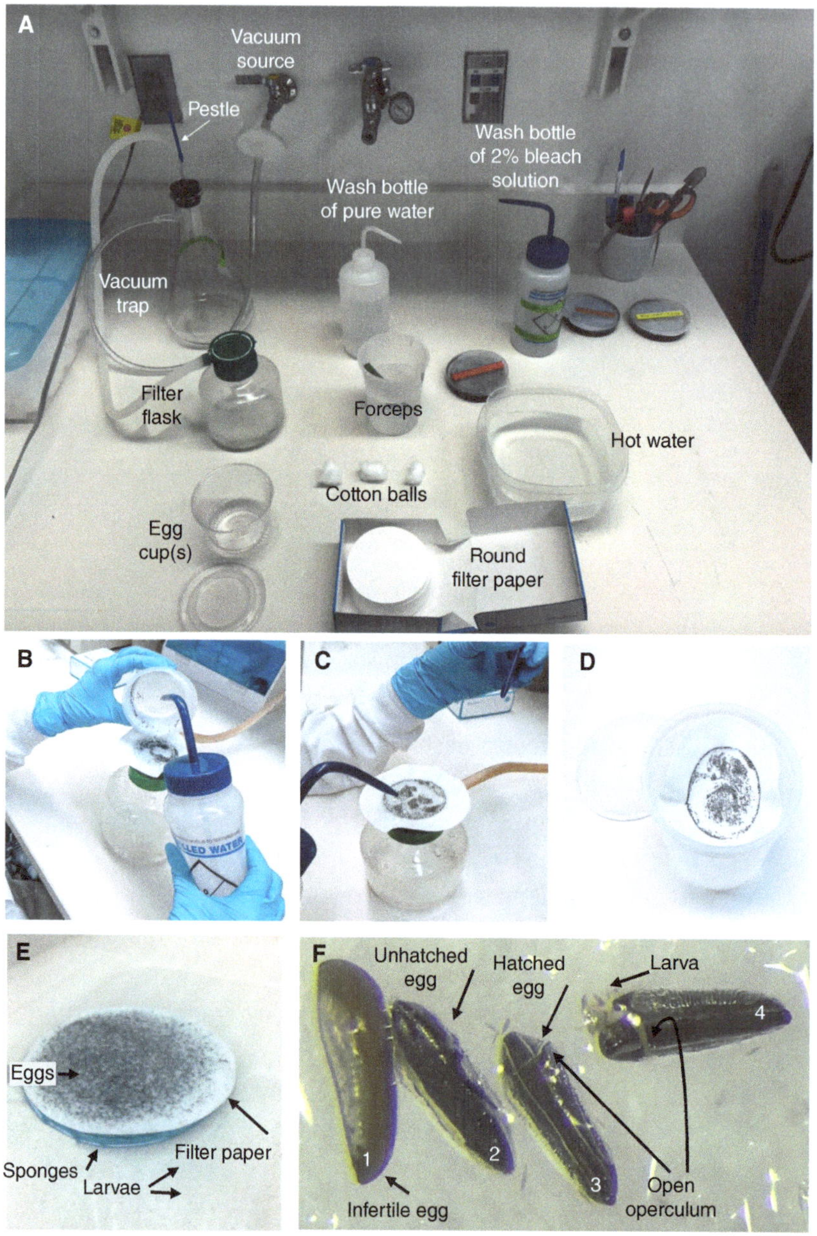

FIGURE 2. Egg disinfection and hatching. (*A*) Components of an egg bleaching station. (*B*) Rinsing eggs from the oviposition cup onto the filter paper using the vacuum filtration system. (*C*) Floating eggs in a bleach solution for disinfection. (*D*) Clean eggs on filter paper placed in a holding container for storage before hatching. (*E*) Setup for hatching on sponge cloth. A filter paper containing eggs is placed on top of a raised bed of sponge cloth, allowing the eggs to stay moist but contained and the larvae to wriggle off. (*F*) Example of an infertile egg (egg 1) and fertile eggs that are either unhatched (egg 2), hatched (egg 3 and 4), or in the process of hatching (larva eclosing, egg 4). The open operculum is observed in eggs 3 and 4.

affecting egg hatching rates, larval development and survival, and adult sex ratio (Mazigo et al. 2019). Trials with other species and strains are recommended before using these procedures.

Egg Hatching for Colony Maintenance or Experimentation

Larval eclosion, or egg hatching, takes place after embryogenesis is complete, which normally occurs 2–3 d post-oviposition (Clements 1992; Yaro et al. 2006). The duration of embryonic development can vary by species and other external factors, such as the presence of certain chemical components in the H_2O and temperature, in particular, during early embryonic development (Cardozo Gonçalvez de Carvalho et al. 2002; Yaro et al. 2006; Impoinvil et al.

Cite this protocol as *Cold Spring Harb Protoc*; doi:10.1101/pdb.prot108186

2007). Agitation-induced egg hatching has also been investigated in An. gambiae, *and it has been shown that early hatchers produce smaller larvae than those that hatch later (Yaro et al. 2006).*

17. Provide a rearing pan containing H_2O for the eggs to hatch and a food source for the newly emerged larvae (2% brewers or baker's yeast or 25 mg ground koi food).

General En Masse *Hatching for Colony Maintenance*

i. Place a larval rearing pan filled with 500 mL (~1–1.5 cm in depth) of rearing H_2O and food on the shelf where the larvae will be reared. If yeast is used, add 1 mL of 2% brewer's or baker's yeast/500 mL of H_2O.

Place the pan on the shelf before adding eggs. If a pan with H_2O and eggs is moved, many eggs will be stranded on the sides of the pan where they will dry out and die unless they are washed back onto the water using a wash bottle.

ii. Wash the eggs off the filter paper with H_2O into the larval pan (see Protocol 2: *Anopheles* Larval Rearing [Leite et al. 2023b]).

iii. Cover the pan with a lid.

Continue to Step 18 if counting eggs; otherwise, proceed to Protocol 2: Anopheles *Larval Rearing (Leite et al. 2023b) for information on how to rear larvae.*

Hatching on a Sponge Platform

This method is suitable for separating eggshells from larvae and is especially useful when the number of eggs, hatching rate, or timing needs to be analyzed.

iv. Place a larval rearing pan filled with 500 mL (~1–1.5 cm in depth) of rearing H_2O on the shelf where the larvae will be reared.

v. Carefully place a filter paper containing eggs (bleached or unbleached) on a raised platform consisting of a stack of sponge cloth disks slightly higher than the H_2O level (Fig. 2E).

This will provide enough moisture for larvae to hatch while preventing the eggs from spreading across the H_2O. The larvae will wriggle off the paper. It is important to use a smooth sponge cloth that prevents larvae from crawling underneath the filter paper rather than swimming away.

vi. (Optional) After hatching is complete, collect the paper with the eggshells and unhatched eggs without disturbing the larvae for the tracking of parameters to estimate several life-history traits of the colony. To assess these parameters, proceed with Step 18; otherwise, proceed to Protocol 2: *Anopheles* Larval Rearing (Leite et al. 2023b) for information on how to rear larvae.

Counting Eggs

18. If needed for experimental data, count the total number of eggs and determine hatching rate to assess fecundity and/or fertility (see Steps 19 and 20). Perform egg counts with one of the following options.

Manual Counting

i. Lay the egg paper(s) on a smooth nonporous surface such as a Petri dish under a dissecting microscope.

ii. (Optional) To make counting easier, draw sections on the filter paper with a pencil in a pattern that makes it possible to ensure all eggs are counted once.

iii. Move the filter paper to each cluster of eggs, pricking them with a dissecting needle if necessary to determine whether the egg has hatched (see Fig. 2F for examples of unfertilized, fertilized/unhatched, and fertilized/hatched eggs).

iv. Record the status of the egg hatch of the desired number of eggs on the denominator.

A mechanical counter can be used when the number of eggs is high.

Automated Counting

This method of counting is not recommended if hatch rate needs to be assessed because hatching rate cannot be reliably determined from a photograph.

v. Lay the filter paper on a flat, evenly illuminated nonporous surface.

This step will help to better identify eggs from other debris when counting.

vi. Take a photo with a camera, filling the photograph with the eggs but not missing any and ensuring good focus over the entire field.

Although a microscope camera can be used, a cell phone camera is adequate. Whichever option is chosen, perform preliminary tests to ensure all eggs are in focus.

vii. Use an image software program such as ImageJ to determine the total number of eggs.

Estimating Life-History Traits

19. Estimate fecundity, which is the number of hatched eggs per female, by dividing the total number of hatched eggs laid on a filter paper by the number of gravid females in the cage.

To determine fertile, infertile, hatched, or unhatched, see Steps 20 and 21 for more details. This estimate should be considered a population estimate because it does not consider whether all females may not have mated or blood fed.

20. Determine fertility rate by dividing the number of fertile (hatched or unhatched) eggs on the filter paper by the total number of eggs laid.

Because some fertile eggs may not have hatched by the time the eggs are analyzed, distinguishing between infertile and fertile/unhatched eggs is essential for a good estimation of this parameter. Use the following visual cues to determine egg fertility:

- *Color: Fertile eggs are dark brown (Fig. 2F, eggs 2–4), whereas infertile eggs are lighter in color (Fig. 2F, egg 1).*
- *Bursting: If the egg was fertile and the larva hatched, a visibly loosened cap (the operculum) on the anterior pole of the egg is usually observed (Fig. 2F, eggs 3 and 4). If uncertain, prod the egg with a dissecting needle.*
- *Contents: If the cap is not loose and the color of the egg is not pale or is uncertain, apply slight pressure on the egg with a dissecting needle to observe the contents. If the egg is fertile, a larva will be present (Fig. 2F, egg 4), whereas a watery yolk will come out of infertile eggs.*

Some mosquito strains may have a delayed hatching. Thus, if the fertility rate at a particular time point is needed, unhatched but fertile eggs should be considered separately from fertile and hatched eggs for the calculation.

21. To estimate larval mortality at a time point, count the total number of larvae alive at a particular larval instar and divide by the total number of eggs that hatched.

This calculation will provide the number of larvae that eclosed but died afterward. Thus, larval mortality at a particular time point can be determined. Early-instar larval mortality may be difficult to assess because dead larvae may be eaten by other larvae and decay rapidly; therefore, determining later-instar larval mortality with this method is recommended.

ACKNOWLEDGMENTS

We thank all members of the MR4 Vector Activity (Malaria Research and Reference Reagent Resource Center, BEI Resources) and Target Malaria Teams who contributed to this protocol by providing technical information on rearing methods, some of the images, and general insectary support. In particular, we thank Catherine Steele, Ilsse Ortega, and Katelyn Cavender. We also thank Michael Aidoo and Audrey Lenhart for proofreading this manual and providing valuable feedback.

Cite this protocol as *Cold Spring Harb Protoc*; doi:10.1101/pdb.prot108186

REFERENCES

Cardozo Gonçalvez de Carvalho S, de Jesus Martins Junior A, Bento Pereira Lima J, Valle D. 2002. Temperature Influence on embryonic development of *Anopheles albitarsis* and *Anopheles aquasalis*. *Mem Inst Oswaldo Cruz* **97:** 1117–1120. doi:10.1590/S0074-02762002000800009

Clements AN. 1992. *The biology of mosquitoes: development, nutrition, and reproduction*. Chapman & Hall, London.

Impoinvil DE, Cardenas GA, Gihture JI, Mbogo CM, Beier JC. 2007. Constant temperature and time period effects on *Anopheles gambiae* egg hatching. *J Am Mosq Control Assoc* **23:** 124–130. doi:10.2987/8756-971X(2007)23[124:CTATPE]2.0.CO;2

Leite LN, Bascuñán P, Dotson EM, Benedict MQ. 2023a. Considerations for rearing and maintaining *Anopheles* in the laboratory. *Cold Spring Harb Protoc* doi:10.1101/pdb.top107802

Leite LN, Bascuñán P, Dotson EM, Benedict MQ. 2023b. *Anopheles* larval rearing. *Cold Spring Harb Protoc* doi:10.1101/pdb.prot108187

Mazigo E, Kidima W, Myamba J, Kweka EJ. 2019. The impact of *Anopheles gambiae* egg storage for mass rearing and production success. *Malar J* **18:** 52. doi:10.1186/s12936-019-2691-4

Mollahosseini A, Rossignol M, Pennetier C, Cohuet A, Dos AA, Chandre F, Shahbazkia HR. 2012. A user-friendly software to easily count *Anopheles* egg batches. *Parasit Vectors* **5:** 122. doi:10.1186/1756-3305-5-122

Yaro AS, Dao A, Adamou A, Crawford JE, Ribeiro JM, Gwadz R, Traoré SF, Lehmann T. 2006. The distribution of hatching time in *Anopheles gambiae*. *Malar J* **5:** 19. doi:10.1186/1475-2875-5-19

Protocol 2

Anopheles Larval Rearing

Laura N. Leite,[1,3] Priscila Bascuñán,[2,3] Ellen M. Dotson,[1] and Mark Q. Benedict[2,4]

[1]*Entomology Branch, Division of Parasitic Diseases and Malaria, Centers for Disease Control and Prevention (CDC), Atlanta, Georgia 30333, USA;* [2]*CDC Foundation, Atlanta, Georgia 30308, USA*

Mosquito larvae are aquatic and go through four development stages (larval instars L1–L4) before pupation. Species vary in the duration of larval development, and a variety of external factors affect the development rate (e.g., water temperature, food type, and larval density), which are discussed more thoroughly elsewhere. Here, we detail how to rear *Anopheles* larvae. This protocol describes appropriate distribution of larvae into rearing pans, feeding of larvae, cleaning of pans, and care until pupation.

MATERIALS

It is essential that you consult the appropriate Material Safety Data Sheets and your institution's Environmental Health and Safety Office for proper handling of equipment and hazardous material used in this protocol.

Reagents

Anopheles larvae (recently hatched, 1-d-old)

Bleach solution in H_2O (<5%, v/v; optional, see Step 6)

Hand dish-washing detergent (optional, see Step 6)

Hot H_2O (>70°C)

Ideally, an insectary would have an InSinkErator or comparable instant hot H_2O tap installed. An electric tea pot is a convenient alternative.

Koi food (ground and whole pellets; refer to Introduction: Considerations for Rearing and Maintaining *Anopheles* in the Laboratory [Leite et al. 2023a] for food types and preparation and storage information).

Rearing H_2O (see Introduction: Considerations for Rearing and Maintaining *Anopheles* in the Laboratory [Leite et al. 2023a] for acceptable H_2O types).

Equipment

Automated larval counter (optional, see Step 1; handheld; e.g., Uline H-7350)

Environmental chamber or room that will achieve ~27°C

Maintain larval mosquitoes in an environmental chamber at ~27°C and a 12 h light:12 h dark light cycle, with or without a sunrise and sunset period.

---

[3]These authors contributed equally to this work.

[4]Correspondence: mbenedict@cdc.gov

Cite this protocol as *Cold Spring Harb Protoc*; doi:10.1101/pdb.prot108187

Fine-mesh bucket filter (100 µm; e.g., US Plastics 10879)
Fine-mesh nylon or polyester filters with opening sizes that prevent the various stages of larvae from passing through.
Heat-resistant beaker (500 mL^{-1}-L polypropylene beaker, see Step 6; e.g., Uline S-21658)
Paper towels
Rearing pan with lid (lid is optional though strongly recommended; e.g., US Plastics 52051)

See Introduction: Considerations for Rearing and Maintaining Anopheles *in the Laboratory (Leite et al. 2023a) for more information*

Sponge (optional, see Step 6)
Strainer (any stainless-steel sieve with a handle; see Step 6)

Strainers with different pore sizes can be used, but it should be small enough to trap the larvae or pupae while allowing the H_2O and excess food to pass through. A variety of strainer sizes should be kept in the laboratory to facilitate straining larvae of varying stages and sizes.

Transfer pipette (cut ~3 mm of the tip off with scissors (polyethylene; e.g., Fisherbrand 137117M)
Wash bottle (e.g., Sigma-Aldrich Z423106)

METHOD

Refer to Introduction: Considerations for Rearing and Maintaining Anopheles *in the Laboratory (Leite et al. 2023a) for a rearing guide and schedule for* A. gambiae *complex strains and information about species deviations. This rearing guide uses a 22-cm×35-cm larval rearing pan and koi food. Larval densities and food amounts will need to be adjusted if a different sized pan or food type is used.*

1. Distribute 1-d-old larvae into rearing pans by following one the methods described below.

   Refer to the Introduction: Considerations for Rearing and Maintaining Anopheles *in the Laboratory (Leite et al. 2023a) for information on optimal rearing densities and significance.*

Visual Estimation

i. Visually estimate and transfer 500 L1 larvae into rearing pans with a cut transfer pipette. Pour excess H_2O from the larval pans into the fine-mesh bucket filter.

   Take care not to pour out and destroy larvae unless an excess of larvae is present. Larvae typically are less buoyant and remain on the bottom of the pan, allowing for the H_2O on top to be removed.

   Use a photograph to refer to when adjusting the density of larvae (Rutledge et al. 1976). See Figure 1A for an example of a pan with 500 L1s. An insectary can make a similar guide using their larval pans with the desired densities. Over time, an experienced technician will be able to visually observe and estimate the density in a healthy range without a photographic reference.

Manual Counting

ii. Manually count and transfer 500 L1 larvae into rearing pans using a cut transfer pipette (Fig. 1A).

   This method can provide a more accurate estimate but is time-consuming and depends on the skills of the technician. The process can be sped up by counting larvae into drops (Fig. 1B) and using a handheld tally counter.

Automated Counting

iii. Use an automated larval counter and transfer 500 L2 larvae per pan (Fig. 1C).

   Benedict et al. (2020) detail the development and use of an automated particle counter (APC) for counting early-instar larvae using an LED light sensor and a capillary. This method is ideal for large-scale projects for which the cost of a larval counter is warranted and necessary to save time for the insectary staff. As recommended by the authors, having a dedicated reservoir, feed tube, and capillary for each strain will avoid contamination among colonies.

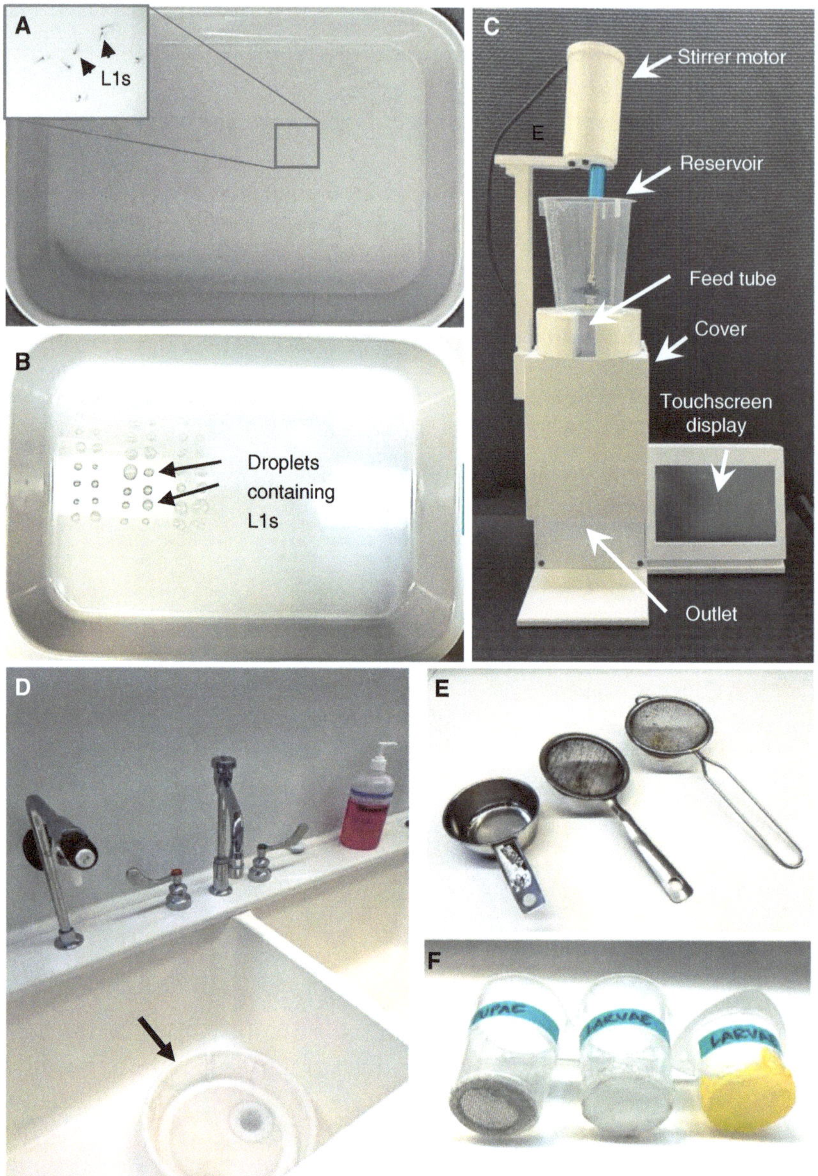

FIGURE 1. Estimating larval density and types of strainers. (*A*) Example of a pan with 500 larvae to use as a reference when visually estimating larval pans. Arrows in the *inset* indicate the presence of first-instar larvae (L1s). (*B*) Example of a manual counting method using droplets containing L1s. (*C*) The automated particle counter (APC) device consists of a reservoir where the initial larvae are contained and a stirrer that keeps the larvae evenly distributed in the container. The APC also contains an LED light source, a sensor, and a glass capillary tube that are protected behind the cover (not shown), and a touchscreen display to adjust settings. (*D*) A fine-mesh filter (black arrow) is located in the sink to prevent larvae and pupae from going down the drain. (*E*,*F*) Smaller strainers of different opening sizes can be used to filter out larvae of different stages.

2. Return the rearing H_2O level to 500 mL/pan.
3. Provide each pan of larvae with ~50 mg of ground koi food.
4. On day 3 post hatching, dilute larvae to ~250/pan and add five pellets or 140–200 mg of ground koi food. Add water to the desired level.
5. On day 5 posthatching, add two to four koi pellets if necessary.

   Feed here if most food has been consumed.

 Cite this protocol as *Cold Spring Harb Protoc*; doi:10.1101/pdb.prot108187

6. On day 6 posthatching, clean H_2O in the larval rearing pans following the method described below.
   i. Place a fine-mesh sink filter on the bottom of the sink (Fig. 1D) to serve as a biosafety measure preventing larvae from going down the drain.
   ii. Heat-decontaminate all rearing equipment (e.g., strainers, transfer pipettes). Submerge the strainer completely in hot H_2O (>70°C) collected in a heat-resistant beaker and then thoroughly rinse with cool H_2O. For transfer pipettes, aspirate enough hot H_2O to coat the inside of the pipettes, invert if necessary, and then expel the H_2O.
   iii. Place a strainer (Fig. 1E,F) over the fine-mesh sink filter and pour the H_2O containing the larvae into it for collection.
   iv. (Optional) Wipe larval pans with a paper towel and rinse with clean H_2O used for rearing if excess scum is observed in the pan. Alternatively, use fresh pans.

      If necessary, hand dishwashing detergent can be used to clean pans. Soak materials in hot H_2O, scrub clean with a sponge, and rinse thoroughly. If a colony infection is suspected, soak pans in a bleach solution overnight (if the pan allows) and then scrub, rinse, and soak the pans in H_2O overnight.
   v. Use a wash bottle containing clean H_2O to rinse the larvae from the strainer back into the larval pan. Return the H_2O level to the desired volume.
   vi. Feed larvae 150–200 mg ground koi food.
7. Continue until all or a sufficient number of larvae have pupated.
8. Proceed to Protocol 4: *Anopheles* Pupa Collection and Sex Identification (Leite et al. 2023b) and Protocol 3: *Anopheles* Adult Anesthesia, Feeding, and Sex Separation (Leite et al. 2023c) for instructions on how to collect pupae and maintain adults, respectively.

ACKNOWLEDGMENTS

We thank all members of the MR4 Vector Activity (Malaria Research and Reference Reagent Resource Center, BEI Resources) and Target Malaria Teams who contributed to this protocol by providing technical information on rearing methods, some of the images, and general insectary support. In particular, we thank Catherine Steele, Ilsse Ortega, and Katelyn Cavender. We also thank Michael Aidoo and Audrey Lenhart for proofreading this manual and providing valuable feedback. The findings and conclusions in this report are those of the authors and do not necessarily represent the official position of the Centers for Disease Control and Prevention.

REFERENCES

Benedict MQ, Bascuñán P, Hunt CM, Aviles EI, Rotenberry RD, Dotson EM. 2020. Trials of the Automated Particle Counter for laboratory rearing of mosquito larvae. *PLoS ONE* **15:** e0241492.

Leite LN, Bascuñán P, Dotson EM, Benedict MQ. 2023a. Considerations for rearing and maintaining *Anopheles* in the laboratory. *Cold Spring Harb Protoc* doi:10.1101/pdb.top107802

Leite LN, Bascuñán P, Dotson EM, Benedict MQ. 2023b. *Anopheles* pupa collection and sex identification. *Cold Spring Harb Protoc* doi:10.1101/pdb.prot108189

Leite LN, Bascuñán P, Dotson EM, Benedict MQ. 2023c. *Anopheles* adult anesthesia, feeding, and sex separation. *Cold Spring Harb Protoc* doi:10.1101/pdb.prot108188

Rutledge LC, Sofield RK, Piper G. 1976. Rapid counting methods for mosquito larvae. *Mosq News* **36:** 537–540.

Protocol 3

Anopheles Adult Anesthesia, Feeding, and Sex Separation

Laura N. Leite,[1,3] Priscila Bascuñán,[2,3] Ellen M. Dotson,[1] and Mark Q. Benedict[2,4]

[1]*Entomology Branch, Division of Parasitic Diseases and Malaria, Centers for Disease Control and Prevention (CDC), Atlanta, Georgia 30333, USA;* [2]*CDC Foundation, Atlanta, Georgia 30308, USA*

The adult stage is the only nonaquatic stage of the *Anopheles* mosquito. Both male and female *Anopheles* mosquitoes require access to a source of sugar to survive. In the insectary, a temperature of ∼27°C and 80% relative humidity and a cycle of 12 h light:12 h dark light, ideally with a sunrise and sunset period, are necessary minimum conditions to mimic their natural environment. Laboratory-reared *Anopheles* can survive for over a month; however, decreased activity and increased mortality may be observed ∼2 wk postemergence depending on the species and health of the colony. Details on how to maintain adults *Anopheles* are discussed here. Information and considerations on blood and sugar feeding are described. This protocol also provides instructions on how to differentiate male and female adult mosquitoes.

MATERIALS

It is essential that you consult the appropriate Material Safety Data Sheets and your institution's Environmental Health and Safety Office for proper handling of equipment and hazardous material used in this protocol.

RECIPES: Please see the end of this protocol for recipes indicated by <R>. Additional recipes can be found online at http://cshprotocols.cshlp.org/site/recipes.

Reagents

Anesthetizing chemical (e.g., chloroform, diethyl ether, or FlyNap)
Anopheles pupae (newly eclosed and placed in a cup of H_2O, in a cage)
Blood (optional, see Step 9 and Introduction: Considerations for Rearing and Maintaining *Anopheles* in the Laboratory [Leite et al. 2023a])
Hand dishwashing detergent
Household bleach (∼5%, v/v)
Ice
Sugar solution (10%, w/v) <R>

Sucrose solution corn syrup, or diluted honey can be used as sugar sources.

[3]These authors contributed equally to this work.
[4]Correspondence: mbenedict@cdc.gov

Cite this protocol as *Cold Spring Harb Protoc;* doi:10.1101/pdb.prot108188

Equipment

Anesthetizing chamber (optional, see Step 4; glass container with tight fitting lid large enough to hold anesthetizing container)

Anesthetizing container (optional, see Step 4; Falcon 50-mL polypropylene conical centrifuge tubes, Fisher Scientific 14-959-49A, or equivalent with the tip removed and both ends covered in mesh held in place with rubber bands)

Do not use polystyrene, as it will dissolve when any of the above chemicals are used.

CO_2 anesthetizing platform with CO_2 source (optional, e.g., Genesee Scientific Flystuff Ultimate Flypad 59-172)

This item is also known as a CO_2 block.

Cotton balls

Cotton wick (optional, see Step 1; e.g., Carolina 51101)

Cup (optional, see Step 1; small plastic cup suitable for holding cotton balls moistened with the sugar solution)

Environmental chamber

Maintain mosquitoes in an environmental chamber at ~27°C with 80% relative humidity and a 12 h light:12 h dark cycle, with or without a sunrise and sunset period.

Erlenmeyer flask (optional, see Step 1; 50-mL)

Filter paper

Freezer (optional, see Step 4)

Fume hood (optional, see Step 4)

Glass bell membrane system (optional, see Step 8)

H_2O bath (optional, see Step 8)

Hemotek blood-feeding system (optional, see Step 9; e.g., Hemotek SP6W1-3)

Paper towels (optional, see Step 1)

Parafilm (optional, see Steps 1 and 4)

Petri dish (90-mm-diameter) (optional, see Step 1 and 4; e.g., Fisherbrand 09-720-500)

Sponge (optional, see Step 4)

Stereomicroscope with a light source (optional, see Step 5)

Suction device to collect adult mosquitoes (e.g., Fulton RHM 200 aspirator)

Vial for sugar solution (optional, see Step 1)

This vial can be constructed out of a plastic tube (50-mL; Falcon) with a cap that can be easily modified by gluing a fine mesh or porous plastic to the cap of the vial (Fig. 1). A stainless-steel wire hook can be added to the vial to secure it to any mesh cage. This can be achieved by punching a hole into the rim of the tube, inserting a wire, and then bending the wire with the use of pliers to secure it in place.

METHOD

Sugar Feeding

Like many other mosquito species, Anopheles *males and females acquire energy by feeding on nectar, fruit juices, or other plant juices that contain sugars. Without access to a sugar source, most adults will die after 2 d. Therefore, it is imperative to provide a reliable sugar source.*

1. Provide a sugar solution for mosquitoes when the first pupae are placed in the cage. Use one of the options detailed below:
   - Moisten cotton balls with the sugar solution and place them on top of the cage, allowing the mosquitoes to feed through the mesh, or place the moistened cotton balls in a small plastic cup such as a medicine cup inside the cage. Remoisten the cotton balls with H_2O daily.

     The cotton balls should always be moist but not oversaturated such that they drip into the cage and trap adult mosquitoes in the sticky solution.

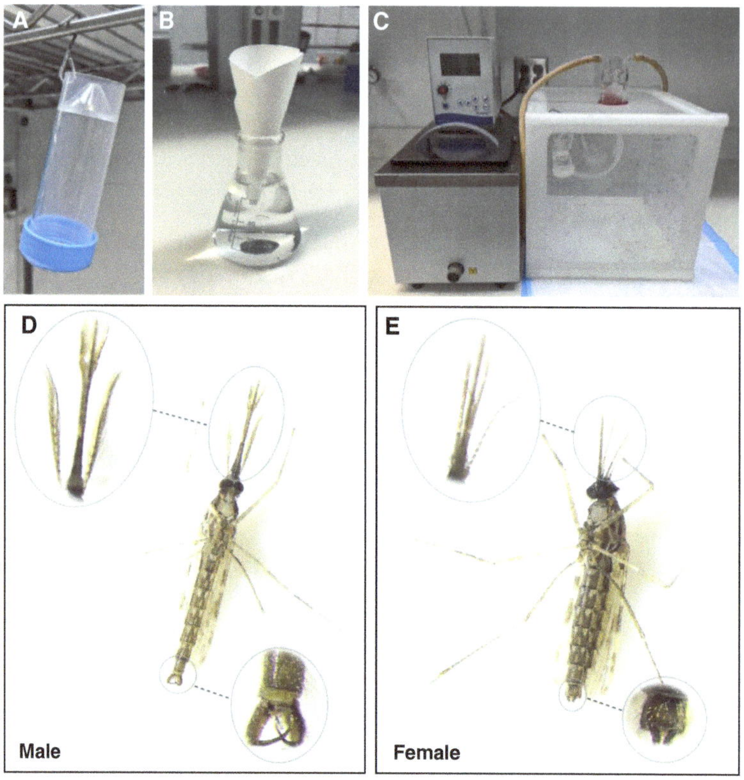

FIGURE 1. Feeding and sorting adult mosquitoes. (*A*) Hanging Falcon tube with a fine mesh glued on the lid and (*B*) 50-mL Erlenmeyer flask with a funnel of filter paper are two examples of containers that can be used to provide adult mosquitoes with a sugar solution. (*C*) An artificial membrane blood-feeding system uses a glass bell with a heated H_2O circulation system. (*D*) Male and (*E*) female *An. gambiae* adults. *Top insets* show feather-like (plumose) antennae with long bristles in males, whereas more filiform-like antennae are present in females. *Bottom insets* show the male genitalia composed of two claspers in males and the absence of them in females.

- Pour the sugar solution into a vial and then hang it inverted inside the cage (Fig. 1A). Place an absorbent material (e.g., paper towel) under the vial in case of a leak.

  If 0.1% (w/v) methylparaben is included, these sugar vials can last up to a month.

- Pour the sugar solution into an Erlenmeyer flask containing a cone or funnel made with filter paper (Fig. 1B) or containing a thick cotton wick inserted into the flask until it reaches the solution.

  The solution will saturate filter paper or wick and should last for about a month, and the flask can be easily refilled if necessary.

  Mosquitoes will feed directly from the filter paper, but some mosquitoes may try to enter the funnel and die. Remove the flask during oviposition because some blood-fed females will use the filter paper or the wick as an oviposition substrate. If the flask is removed, cotton soaked in sugar solution can be temporarily placed on top of the cage and covered with a Petri dish lid or a piece of Parafilm to reduce evaporation. Between uses, soak sugar vials in H_2O with hand dishwashing detergent, rinse, and soak for 24 h in H_2O. The caps can be washed in the same manner but soaked in bleach to prevent microbial growth.

2. Change the sugar source every 3–5 d and replace if discoloration is observed.

   Mold and fungus will grow in a sugar solution very quickly because of the high temperature and high-humidity rearing conditions. The addition of 0.1% (w/v) methylparaben to the sugar solution will prevent mold and fungal growth and keep the sugar from spoilage.

3. Proceed to Step 4 for information on adult observation, manipulation, and sex separation or Step 7 to blood feed mosquitoes.

Cite this protocol as *Cold Spring Harb Protoc*; doi:10.1101/pdb.prot108188

Sedating Adults for Observation, Manipulation, and Sex Separation

Separating adult mosquitoes by sex may be necessary for research projects or to prevent mating from occurring. Although, it is ideal to separate the sexes in the pupa stage instead (refer to Protocol 4: Anopheles *Pupa Collection and Sex Identification (Leite et al. 2023b) to guarantee mating will not occur. It also may be necessary to manipulate or observe adults while they are temporarily anesthetized. This may be for identification purposes. A single leg could also be removed for genetic testing, leaving the adult mosquito still alive.*

4. Sedate mosquitoes by using one of the following options:

Cold Anesthetization

i. Place mosquitoes in a freezer until they are anesthetized (a few minutes) or until death has occurred, according to the need (>20 min depending on species, temperature, and whether living adults are desired).

ii. After the adults are anesthetized and inactive, shake them onto a filter paper in a Petri dish and place on crushed ice.

Chemical Anesthetization

iii. Cover the bottom of an anesthetizing chamber with a sponge.

iv. Working in a fume hood, saturate a sponge or cotton ball with an anesthetizing chemical.

v. Cover the anesthetizing chamber with the lid until ready to use.

vi. Gently aspirate the desired number of mosquitoes into an anesthetizing container. Replace the mesh and rubber band.

vii. Place the anesthetizing container of females into the anesthetizing chamber and close the lid. Leave it for ~10 sec when using diethyl ether and 5 sec when using chloroform.

The amount of time females should be left in the chamber is dependent on the strength of the anesthetizing chemical and the species being evaluated. A good indication that they have been left for a sufficient amount of time is when they have all ceased flying and have been lying at the bottom of the anesthetizing chamber for 2–3 sec.

viii. Promptly remove the container from the anesthetizing chamber, allow it to vent in the fume hood for 10 sec, and then remove it from the fume hood.

The mosquitoes will stay sedated for at least several minutes.

CO_2 Anesthetization

ix. Aspirate mosquitoes from their cage and then place them on a CO_2 anesthetizing platform.

Mosquitoes will need to be sorted on the platform and will quickly wake up once removed. Some species will suffer mortality if CO_2 exposure is prolonged.

5. To separate the sexes, by eye or under a stereoscope, visually inspect adult *Anopheles* mosquitoes for one or more of the following characteristics.
   - Body shape: Males are thinner and tubular, whereas females are wider and rounded (Fig. 1D,E).
   - Antenna: Males have plumose (feather-like) antennae with long bristles, whereas female antennae are filiform and less plumose (pilose) (Fig. 1D,E, top insets).
   - Proboscis: The female proboscis is needle-like and serves to pierce the skin during the blood-feeding process. The male proboscis is wider and branched and is exclusively used to suck nectar or other sugar sources (Fig. 1D,E, top insets).
   - Terminalia: The male terminalia are composed of two claspers that help in grabbing the female during mating. Female terminalia lack this structure (Fig 1D,E, bottom insets).
6. Conduct any other observations or evaluations if necessary.

Blood Feeding

Most mosquito species are anautogenous, meaning they require a blood meal from a vertebrate host to produce eggs. Eggs may be needed to further propagate a colony or for experimental purposes.

Mosquitoes normally require cues such as heat, moisture, CO_2, and host volatiles to locate a blood meal, and therefore, they are not always immediately attracted to artificial blood sources. We provide some tips that can help stimulate mosquitoes to feed. (1) Feed mosquitoes on freshly collected blood. Blood can be stored at 4°C for several weeks postcollection; however, mosquito feeding rates, as well as female fecundity and fertility, may decline over time. (2) Breathe gently into a cage to stimulate host-seeking behavior and feeding by providing heat and CO_2. (3) Additionally, feed mosquitoes at their preferred time. Many species have a strong preference for the time of day that they feed. Monitor new colonies to assess their feeding preference and plan the blood feedings accordingly. If feeding at their preferred time is not feasible, place a dark cloth over the cage before and during feeding. This cloth simulates the night and may help "trick" them into feeding during the day. Alternatively, the day/night schedule in an environmental chamber can be adjusted to make "nighttime" occur when staff are present.

7. (Optional) Remove all sugar H_2O from cages ~6–24 h before blood feeding.

   Temporarily starving females can boost feeding rates of hesitant strains.

8. Provide adult female *Anopheles* with a blood meal using one of the described methods below.

   Females can be blood fed from day 2 after emergence; however, 7 d after emergence will minimize the workload during weekends (see Introduction: Considerations for Rearing and Maintaining Anopheles *in the Laboratory [Leite et al. 2023a] for a detailed rearing schedule and information on blood sources and types).*

Live Animal

A live animal source is the most effective feeding method for most laboratory-reared mosquitoes. However, there are logistical, cost, and ethical concerns with using and housing live animals, and this is not feasible for every insectary. Commonly used laboratory animals are mice, guinea pigs, and rabbits. Researchers wishing to consider animals for blood feeding should consult their institutional animal use and care committee to determine the requirements and conditions under which this can be done.

Hemotek Blood-Feeding System

This system heats the blood via an electrical heating element and allows blood-sucking insects access to blood through a pierceable membrane such as collagen or Parafilm. To include odor components that may entice mosquitoes to blood feed, rub the membrane (e.g., Parafilm) on your forearm before use.

i. Use the Hemotek by following the manufacturer's instructions.

Glass Bell Blood-Feeding System

A standard circulating laboratory H_2O bath heats the bell and keeps the blood warm. It allows blood-sucking insects access to blood through a pierceable membrane (Fig 1C).

ii. Follow a previously published procedure to use this system (BEI_Resources 2015).

9. Feed mosquitoes until engorgement.

   Most Anopheles *feed within 20 min, although some finicky strains may take hours or even may need to feed overnight.*

10. Replace the sugar source if previously removed.

11. Proceed with Protocol 1: *Anopheles* Egg Collection, Disinfection, and Hatching [Leite et al. 2023c] to rear the next generation.

Cite this protocol as *Cold Spring Harb Protoc*; doi:10.1101/pdb.prot108188

RECIPE

Sugar Solution (10%, w/v)

Prepare 10% (w/v) sucrose [human food grade] in H_2O.

Optional: Include 0.1% (w/v) methylparaben (e.g., Spectrum Chemical 99-76-399-76-3).

If the solution includes methylparaben, store at room temperature to prevent methylparaben precipitation; the solution is stable for months.

If the solution does not include methylparaben, store at 4°C; once used at room temperature replace within a week.

ACKNOWLEDGMENTS

We thank all members of the MR4 Vector Activity (Malaria Research and Reference Reagent Resource Center, BEI Resources) and Target Malaria Teams who contributed to this protocol by providing technical information on rearing methods, some of the images, and general insectary support. In particular, we thank Catherine Steele, Ilsse Ortega, and Katelyn Cavender. We also thank Michael Aidoo and Audrey Lenhart for proofreading this manual and providing valuable feedback. The findings and conclusions in this report are those of the authors and do not necessarily represent the official position of the Centers for Disease Control and Prevention.

REFERENCES

BEI_Resources. 2015. Methods in *Anopheles* research. In https://www.beiresources.org/AnophelesProgram/TrainingMethods.aspx, MR4/BEI Resources.

Leite LN, Bascuñán P, Dotson EM, Benedict MQ. 2023a. Considerations for rearing and maintaining *Anopheles* in the laboratory. *Cold Spring Harb Protoc* doi:10.1101/pdb.prot107802

Leite LN, Bascuñán P, Dotson EM, Benedict MQ. 2023b. *Anopheles* pupa collection and sex identification. *Cold Spring Harb Protoc* doi:10.1101/pdb.prot108189

Leite LN, Bascuñán P, Dotson EM, Benedict MQ. 2023c. *Anopheles* egg collection, disinfection, and hatching. *Cold Spring Harb Protoc* doi:10.1101/pdb.prot108186

Protocol 4

Anopheles Pupa Collection and Sex Identification

Laura N. Leite,[1,3] Priscila Bascuñán,[1,2,3] Ellen M. Dotson,[1] and Mark Q. Benedict[1,2,4]

[1]*Entomology Branch, Division of Parasitic Diseases and Malaria, Centers for Disease Control and Prevention (CDC), Atlanta, Georgia 30333, USA;* [2]*CDC Foundation, Atlanta, Georgia 30308, USA*

For most *Anopheles* species, larval–pupal metamorphosis commences ~1 wk after egg hatching. However, depending on the amount of food provided, H_2O temperature, and larval density, the pupation process can be accelerated or delayed. Synchronous pupation is difficult to accomplish consistently, and, thus, pupae need to be separated from larvae daily. Adult emergence will take place 24–48 h after pupation. Most adults will eclose before the next morning (light cycle) in many species. Here, we provide information on some methods available to collect pupae and to sort pupae by sex. Notably, pupa collection and sorting are some of the most time-consuming procedures of the overall mosquito rearing process. Some methods mentioned here attempt to help reduce work effort and time required.

MATERIALS

It is essential that you consult the appropriate Material Safety Data Sheets and your institution's Environmental Health and Safety Office for proper handling of equipment and hazardous material used in this protocol.

Reagents

Anopheles larvae and pupae

H_2O (refer to Water Types in Introduction: Considerations for Rearing and Maintaining *Anopheles* in the Laboratory [Leite et al. 2023a])

Equipment

Cage (e.g., 30-cm cube, BugDorm-1 insect rearing cage)

Cups (e.g., 9-oz disposable plastic, squat drink cup)

Environmental chamber

Maintain pupae in an environmental chamber at ~27°C with 80% relative humidity and a 12 h light:12 h dark light cycle, with or without a sunrise and sunset period.

Fay–Morlan glass plate separator (optional, see Step 1; e.g., John W. Hock Company 5412)

Fine-tipped paintbrush

Rearing pan (e.g., US Plastics 52051)

See Introduction: Considerations for Rearing and Maintaining Anopheles *in the Laboratory (Leite et al. 2023a) for more information.*

---

[3]These authors contributed equally to this work.

[4]Correspondence: mbenedict@cdc.gov

Cite this protocol as *Cold Spring Harb Protoc;* doi:10.1101/pdb.prot108189

Strainer with opening size that allows food and detritus to pass through but retains larvae and pupae
Stereomicroscope and light source
Transfer pipette with the tip cut off with scissors (polyethylene; e.g., Fisherbrand 137117M)

If the tip is too small, it will damage the pupae. Other objects such as a screen scoop can be used as well.

Wash bottle (e.g., Sigma-Aldrich Z423106)

METHOD

Pupa Collection

1. Separate pupae from larvae following one of the methods described below.

    Take care to not overcrowd pupae in the cup after collection. If they are so dense that they have difficulty reaching the surface of the H_2O, they will die because they need to breathe air. In this case, put them in a larger cup or divide them among additional cups.

Manual Separation with Pipettes

This method is the most time-consuming but is convenient to use when there are few pupae in the container or if a sample needs to be removed. For larger numbers of pupae, other methods are recommended.

i. Use a cut transfer pipette to collect each pupa individually and transfer them to a cup with ~100 to 200 mL of H_2O (~1–2 cm in depth).

ii. Place the cup in an adult cage for eclosion.

Separation by Vortexing

Similar methods with small variations such as swirling with chilled H_2O and flask techniques can be found in the Anopheles Research Manual (BEI Resources 2015).

iii. Use a strainer to concentrate the larvae and pupae from several pans in a small volume of H_2O (about half full; Fig. 1A).

iv. Use a transfer pipette to create a circular vortex in the cup by swirling the H_2O rapidly (Fig. 1B).

    The fourth instar larvae and pupae will separate, with the larvae concentrated in the center bottom of the cup and pupae on the surface.

v. When separation has peaked, pipette the larvae from the center (Fig. 1C) of the bottom of the cup and place them into their original container at the desired density.

vi. (Optional) Vortex a second or third time until no more larvae are present (Fig. 1D).

vii. Add more H_2O to the pupal cup so that it contains 100 to 200 mL (~1–2 cm in depth) and place the cup in an adult cage for eclosion.

Separation using a Fay–Morlan Glass Plate Separator

The Fay–Morlan glass plate separator (Fay and Morlan 1959) is commonly used to separate larvae from pupae and to separate the sexes in some mosquito genera, such as Aedes that have dimorphic sex size. The simplicity of this method makes it a good option when working with many pupae.

viii. Place a larval rearing pan at the bottom of the glass plates, underneath the acrylic drip edge to catch H_2O and mosquito immatures that fall through.

ix. Carefully pour the larvae and pupae to be separated into the top, tapered section of the glass plates.

    Approximately 300–500 larvae can be poured through at a time without clogging the plates.

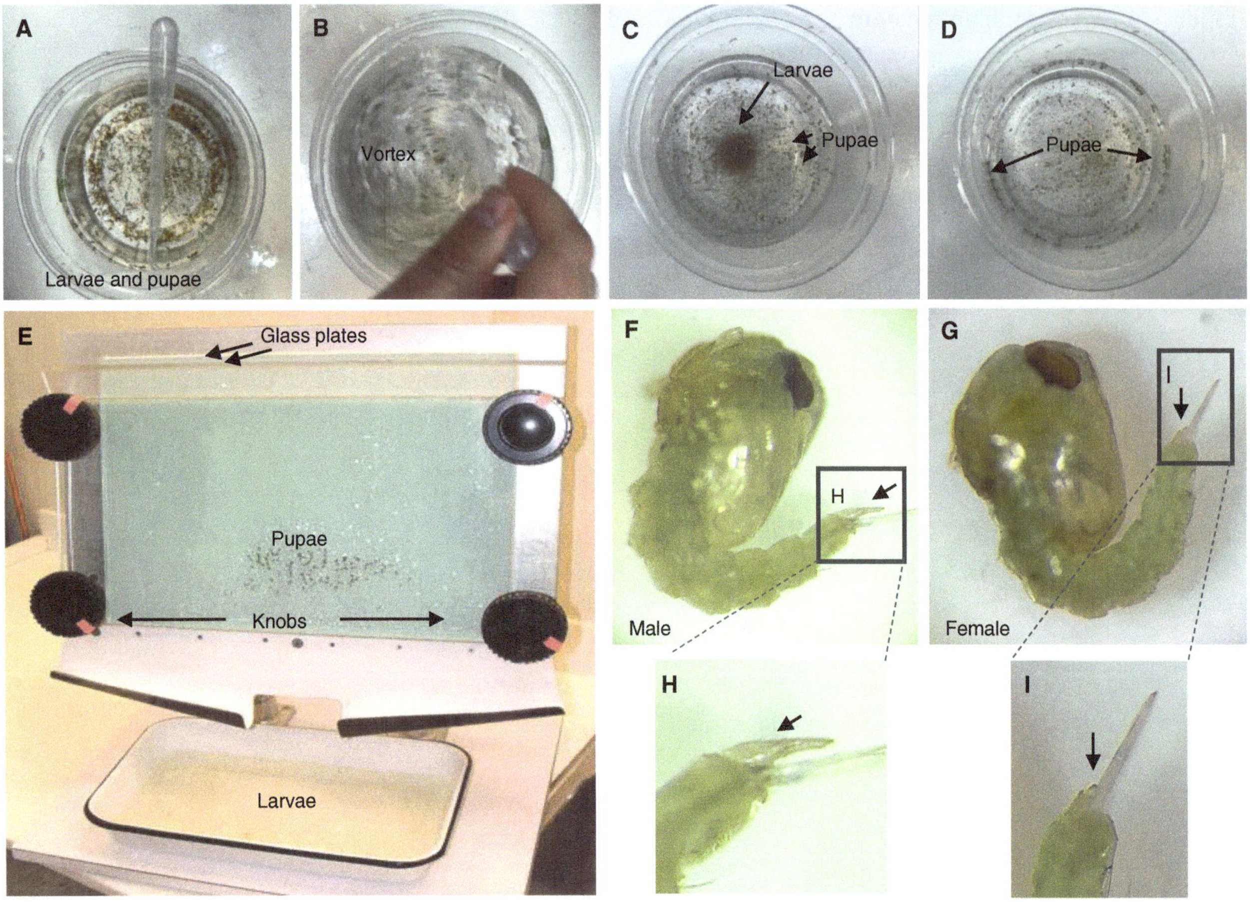

FIGURE 1. Separating and sorting pupae. (*A*) The vortex method for sorting larvae from pupae consists of combining specimens from several rearing pans into a round container, (*B*) creating a vortex by carefully swirling the H_2O with a pipette, (*C*) removing the larvae remaining in the center of the cup, (*D*) and collecting the pupae floating on the edges of the cup, (*E*) The Fay–Morlan separator consists of two glass plates that can separate larvae and pupae according to their size. By carefully adjusting the knobs, larvae and pupae can be dispensed separately into pans. (*F*) Male and (*G*) female pupae can be sorted under a stereoscope by observing the terminalia (*H* and *I*, respectively; black arrows).

Pupae should be concentrated in an upper band, whereas larvae should be concentrated in a lower band between the plates (Fig. 1E).

x. Evenly and slowly rotate the bottom knobs counterclockwise to separate the two plates and release the larvae into a pan, while adding more H_2O from the top. Continue this step until all larvae have been collected.

The extra H_2O helps to flush the larvae out from between the plates without harming them.

Depending on how many larvae and pupae have been poured into the separator, more than one larval rearing pan may be needed.

xi. Place a fresh rearing pan under the drip edge. Turn the knobs counterclockwise to collect the pupae while pouring H_2O from the top (Fig. 1E).

Some larvae may be big enough or curled and be retained with the pupae but most will be flushed through if the separator is not congested with pupae.

xii. Rinse the plate separator thoroughly by allowing the passage of abundant H_2O from the top and placing a pan at the bottom to collect any remaining larvae or pupae.

This step will help prevent cross contamination between colonies. Inspect the device carefully before proceeding to separate mosquitoes from a different stock or species.

Cite this protocol as *Cold Spring Harb Protoc*; doi:10.1101/pdb.prot108189

(Optional) Pupal Sex Identification and Sorting

Sorting pupae by sex may be required under certain circumstances. For example, an experiment may require that females mate with a particular type of male. In this case, females must be sorted at the pupal stage to maintain their unmated status and ensure that the desired mating occurs. Unlike the sexes of some other mosquito genera, such as most Aedes*, male and female pupae of most* Anopheles *species do not differ much in size.*

2. Set up a stereomicroscope and illuminator.
3. Using a cut transfer pipette, place a small pool of pupae on a Petri dish or slide. Remove H_2O using the transfer pipette until the pupae remain on their side and movement is limited.

   Pupae must remain wet to survive.

4. Under the microscope, observe the terminalia.

   Male pupae have more prominent genitalia (Fig. 1F,H) than female pupae (Fig. 1G,I). If the genitalia cannot be easily observed, rotate the pupae with a fine-tipped paintbrush to expose the terminalia.

5. Slide one sex aside with a paintbrush into a cluster.
6. Using a wash bottle, rinse the male and the female pupae into separate cups and place in cages for adult eclosion.

   Adults will have emerged on the following day (after ~24–48 h).

7. Proceed with Protocol 3: *Anopheles* Adult Anesthesia, Feeding, and Sex Separation (Leite et al. 2023b) for information about maintaining adults.

ACKNOWLEDGMENTS

We thank all members of the MR4 Vector Activity (Malaria Research and Reference Reagent Resource Center, BEI Resources) and Target Malaria Teams who contributed to this protocol by providing technical information on rearing methods, some of the images, and general insectary support. In particular, we thank Catherine Steele, Ilsse Ortega, and Katelyn Cavender. We also thank Michael Aidoo and Audrey Lenhart for proofreading this manual and providing valuable feedback. The findings and conclusions in this protocol are those of the authors and do not necessarily represent the official position of the Centers for Disease Control and Prevention.

REFERENCES

BEI_Resources. 2015. Methods in *Anopheles* research. In https://www.beiresources.org/AnophelesProgram/TrainingMethods.aspx, MR4/BEI Resources.

Fay RW, Morlan HB. 1959. A mechanical device for separating the developmental stages, sexes and species of mosquitoes. *Mosq News* **19:** 144–147.

Leite LN, Bascuñán P, Dotson EM, Benedict MQ. 2023a. Considerations for rearing and maintaining *Anopheles* in the laboratory. *Cold Spring Harb Protoc* doi:10.1101/pdb.top107802

Leite LN, Bascuñán P, Dotson EM, Benedict MQ. 2023b. *Anopheles* adult anesthesia, feeding, and sex separation. *Cold Spring Harb Protoc* doi:10.1101/pdb.prot108188

CHAPTER 3

Points to Consider When Establishing and Rearing *Culex* Mosquitoes in the Laboratory

Megan E. Meuti,[1] Alden Siperstein, and Matthew Wolkoff

Entomology Department, The Ohio State University, Columbus, Ohio 43210, USA

Culex mosquitoes transmit several pathogens to humans and animals, including viruses that cause West Nile fever and St. Louis encephalitis and filarial nematodes that cause canine heartworm and elephantiasis. Additionally, these mosquitoes have a cosmopolitan distribution and provide interesting models for understanding population genetics, overwintering dormancy, disease transmission, and other important and ecological questions. However, unlike *Aedes* mosquitoes that produce eggs that can be stored for weeks at a time, no obvious "stopping" point exists in the development of *Culex* mosquitoes. Therefore, these mosquitoes require nearly continuous care and attention. Here, we describe some general considerations when rearing laboratory colonies of *Culex* mosquitoes. We highlight different methods so that readers may choose what works best for their experimental needs and laboratory infrastructure. We hope that this information will enable additional scientists to conduct laboratory research on these important disease vectors.

INTRODUCTION

Culex mosquitoes transmit viruses that cause West Nile fever (Goldberg et al. 2009; Andreadis 2012), St. Louis encephalitis (Bailey et al. 1978), and Japanese encephalitis (Wispelaere et al. 2017), as well as filarial nematodes that cause canine heartworm (Vezzani et al. 2011) and elephantiasis (Harb et al. 1993). Moreover, *Culex* mosquitoes are an important vector of Ross River virus in inland Australia (Russel 2002) and Sindbis virus in Egypt (Turell et al. 2002), and *Culex tarsalis* can transmit Rift Valley fever to humans and cattle in West Africa (Bergren et al. 2021). Admittedly, the global burden on human health by *Culex*-vectored diseases is far less than those transmitted by *Aedes* mosquitoes (e.g., dengue, Zika, yellow fever, and chikungunya) and *Anopheles* mosquitoes (e.g., human, avian, and murine malaria). Therefore, most mosquito research has focused on *Aedes* and *Anopheles*. However, members of the *Culex pipiens* complex that are highly competent vectors of West Nile virus extend further into temperate environments in North America, South America, Africa, Europe, Asia, and Australia (Ciota and Kramer 2013) than *Ae. aegypti* and *Ae. albopictus* (Kraemer et al. 2015). Therefore, a high proportion of the human population is at risk of contracting West Nile virus, as it is endemic or potentially endemic in at least 114 countries (Chancey et al. 2015).

THE CHALLENGES OF WORKING WITH *CULEX*

Apart from the differences in disease burden, another reason why researchers have gravitated to working with *Aedes* mosquitoes is likely because of the ease with which they are reared and maintained

[1]Correspondence: meuti.1@osu.edu

Cite this introduction as *Cold Spring Harb Protoc*; doi:10.1101/pdb.top107823

in the laboratory. The eggs of nearly all *Aedes* species can be stored under humid conditions for weeks, if not months, allowing researchers to hatch an appropriate number of larvae whenever they are ready to begin an experiment (Beckel 1958; Kalpage and Brust 1974; Shroyer and Craig 1983; Parker 1985). Unfortunately, *Culex* eggs are not as hardy and cannot be stored for any length of time. This is likely due to accelerated rates of embryogenesis relative to *Aedes* and *Anopheles* species, as well far lower levels of embryonic desiccation resistance (Vargas et al. 2014). Notably, *Culex* larvae will hatch in 24–72 h depending on temperature (Kiarie-Makara et al. 2015). Therefore, researchers who work with *Culex* mosquitoes must continuously maintain the mosquitoes in the laboratory. Additionally, because most existing laboratory strains of *Culex* have been established relatively recently, they are not as well adapted to artificial laboratory conditions. Finally, most *Culex* species are zoophilic, preferring to feed on live, nonhuman hosts (Schultz 1992; Tuno et al. 2017), and are not nearly as aggressive biters as *Aedes* or *Anopheles* mosquitoes (Hawley 1988; Wanji et al. 2003). These characteristics make rearing *Culex* in the laboratory more challenging, as it is often difficult to achieve high levels of blood feeding and egg production in laboratory colonies.

CURRENT *CULEX*-REARING PROTOCOLS

Despite the challenges, we were able to successfully establish a colony of Northern house mosquitoes, *Cx. pipiens*, from central Ohio in 2013 (Buckeye strain) (see Protocol 1: Establishing a *Culex* Colony from Field-Collected Eggs [Meuti et al. 2023a]) that we have continuously maintained (see Protocol 2: Rearing and Maintaining a *Culex* Colony in the Laboratory [Meuti et al. 2023b]). Several other researchers have also established colonies of *Cx. pipiens*, including the autogenous form *Cx. pipiens molestus* that can generate their first batch of eggs without a blood meal (Roubaud 1930; Dobrotworsky 1955; Spielman 1957; Mutebi and Savage 2009; Gao et al. 2019), the anautogenous form, *Cx. pipiens pipiens* (Paul et al. 2005; Yurchenko et al. 2020), and the Southern house mosquito, *Culex quinquefasciatus* (Sundararaman 1949; de Melo Chalegre et al. 2015). Additionally, Kauffman et al. (2017) have provided an excellent and detailed protocol for rearing *Aedes* and *Culex* mosquitoes and have used their protocol to successfully rear *Cx. pipiens*, *Cx. quinquefasciatus*, and *Cx. tarsalis*. Although this is more difficult, other researchers have established colonies of *Culex salinarius* (Wallis and Spielman 1953; Wallis and Whitman 1968; Eldridge et al. 1972), *Culex territans* (Desser et al. 1973), and *Culex erraticus* (Coon 2006), whereas limited numbers of *Culex restuans* have been reared in a laboratory but only after forcing males and females to copulate (Eldridge et al. 1972). The research that has been done on these species and strains has greatly enhanced our understanding of the physiology, ecology, and behavior of *Culex* mosquitoes. However, several important research questions remain, particularly about physiological and behavioral differences among different populations of *Culex* mosquitoes. Therefore, we need other researchers to rise to the challenge of working with *Culex* mosquitoes.

CONSIDERATIONS WHEN COLLECTING THE DESIRED *CULEX* SPECIES FROM THE FIELD

Different species of *Culex* mosquitoes vary in their geographic and temporal distributions, as well as their microhabitat preferences and which vertebrate hosts they prefer to bite. These factors should be kept in mind when trying to collect and/or establish a desired *Culex* species from the field. Here, we provide recommendations on how to collect the most common and medically important species of *Culex* in North America.

First, to collect *Cx. quinquefasciatus*, *Cx. pipiens*, *Cx. restuans*, and *Culex erraticus*, we would recommend using the gravid H_2O recipe included in our protocol (Protocol 1: Establishing a *Culex* Colony from Field-Collected Eggs [Meuti et al. 2023a]) without modification. *Cx. quinquefasciatus* occupies a southern range, whereas *Cx. pipiens* occupies a more northern range. However, these two

Cite this introduction as *Cold Spring Harb Protoc*; doi:10.1101/pdb.top107823

species have been known to hybridize between 34°N and 46°N (Yurchenko et al. 2020). Therefore, researchers living within the hybrid zone should be aware that they may not collect pure strains of either species and should check for hybridization by using previously described protocols (Smith and Fonseca 2004). Additionally, the time of year can greatly influence which species is collected. For example, east of the Mississippi River, *Cx. restuans* is far more abundant early in the year than *Cx. pipiens*, making mid-May through mid-July ideal for collecting *Cx. restuans* and mid-July through August ideal for collecting mostly *Cx. pipiens* (Jackson and Paulson 2006; Helbing et al. 2015; Meuti laboratory, unpubl. data). Similarly, *Cx. erraticus* is more abundant later in the year (August until mid-October; Meuti laboratory, unpubl. data). Therefore, if researchers wish to establish populations of *Cx. erraticus*, they should time their collections appropriately.

Within the southeastern United States, *Culex nigripalpus* is commonly collected from gravid traps across the urban-to-rural gradient (Hancock and Camp 2022). Notably, the range of this species has been expanding northward into the Midwest (Akaratovic et al. 2021). Therefore, researchers living within the current range can likely collect these mosquitoes using our accompanying protocol (Protocol 1: Establishing a *Culex* Colony from Field-Collected Eggs [Meuti et al. 2023a]) without modification.

To collect *Cx. tarsalis*, which is distributed throughout the western half of North America, we recommend creating more dilute gravid H_2O, using one-half to one-quarter of the ingredients included in Protocol 1: Establishing a *Culex* Colony from Field-Collected Eggs (Meuti et al. 2023a). As *Cx. tarsalis* is more halotolerant than other *Culex* species (Bohart and Washino 1978; Du and Millar 1999), researchers can add up to 0.5% saline (NaCl) to the gravid H_2O to increase the proportion of *Cx. tarsalis* collected relative to other species.

Cx. salinarius can be collected from almost any H_2O source, although they are most commonly found in high concentrations in salt marshes (Murphey 1961). Therefore, researchers who want to establish laboratory colonies of *Cx. salinarius* can easily collect them by placing gravid H_2O traps at these locales. If researchers are trying to collect *Cx. salinarius* outside salt marshes, it is important to note that females that have been raised in fresh H_2O prefer to oviposit in fresh H_2O. However, females of *Cx. salinarius* that have been reared in freshwater will still lay their eggs in H_2O with high salinity (Murphey 1961). Therefore, researchers can collect *Cx. salinarius* in areas where they are less abundant by filling tubs with gravid H_2O containing a ~3% saline solution (NaCl). Although this procedure may reduce the total number of *Culex* egg rafts that are collected, it will increase the proportion, and likely the total number, of *Cx. salinarius* egg rafts.

Cx. territans prefers to feed on amphibians (Burkett-Cadena et al. 2008). Adults of *Cx. territans* are therefore more abundant near creeks that have high frog populations and are easiest to find in late April through mid-May in Columbus, Ohio. During this time, gravid adult females of *Cx. territans* can be collected by aspirating them from culverts (A. Siperstein, pers. observations). Adult females can be placed directly into a cage and transported back to the laboratory where they will likely lay eggs in any nutrient-rich H_2O source that is provided to them.

CONSIDERATIONS WHEN ESTABLISHING REARING PROTOCOLS

Larval Growth and Development

Although *Culex* larvae are container breeders that are adapted to developing in nutrient-rich H_2O, they are highly susceptible to bacterial overgrowth, and entire pans of larvae can die within 24–48 h. To prevent this mortality, it is imperative that the larvae are not "overfed" (too much food is given to the larvae at any point in time). Underfeeding, although not ideal, will cause the mosquitoes to develop more slowly. Therefore, it is almost always better to underfeed (add slightly less food to larval containers) than to overfeed. The exception to this idea is when generating autogenously reproducing mosquitoes, which have higher nutritional requirements during their larval phase and where underfeeding reduces body size and overall fecundity and can increase mortality (Kassim et al.

2012). Please also note that larval growth and development are directly dependent on temperature. In the accompanying protocols (Protocol 1: Establishing a *Culex* Colony from Field-Collected Eggs [Meuti et al. 2023a]; and Protocol 2: Rearing and Maintaining a *Culex* Colony in the Laboratory [Meuti et al. 2023b]), we rear *Culex* mosquitoes under high temperatures (24°C–27°C). We also routinely rear *Cx. pipiens* at lower temperatures (18°C) that, in combination with short photoperiods (light:dark, 8 h:16 h), induce overwintering dormancy or diapause in the mosquitoes (Eldridge 1968; Sanburg and Larsen 1973; Spielman and Wong 1973). *Culex* mosquitoes reared at lower temperatures should be fed less frequently, ideally once every 2 to 4 d.

When feeding the larvae, one should check the color and scent of the mosquito-rearing H_2O. If it is dark brown/cloudy or highly pungent and/or if there are any dead larvae (bloated and white in appearance) floating near the bottom, it is best to immediately filter the mosquito larvae out of their old and dirty container and then place them into a new container with clean H_2O and fresh food. Additionally, if you see excessive biofilm accumulation on the sides of the pan or top of the H_2O, it is also a good idea to filter the H_2O. Please see Protocol 2: Rearing and Maintaining a *Culex* Colony in the Laboratory [Meuti et al. 2023b] for detailed instructions on how to filter mosquito larval water.

Pupal Development

Like most mosquitoes, *Culex* species generally only spend ~36 h at 25°C in the pupal phase (Birley 1979). Therefore, when *Culex* mosquitoes are reared at high temperatures, it is necessary to remove pupae from the larval contains and place them into an adult cage at least once a day, which will ensure that no adult mosquitoes escape. In our experience, the pupal phase is the most sensitive period in mosquito development. Therefore, it is also important to ensure that cups containing pupae are not overcrowded; we recommend placing no more than 150 pupae in a plastic cup with a diameter of 11.4 cm and height of 4.4 cm. Placing small pieces of Styrofoam (~3 cm in diameter) in the pupal cup may also facilitate adult emergence by providing a platform for newly emerged adult mosquitoes, and placing a paper cone on the top of the deli cup may prevent adult mosquitoes from falling in and drowning in the pupal H_2O. Last, the pupal cup should be filled halfway using H_2O from the larval pan, as the use of fresh H_2O can result in high mortality, possibly because of osmotic shock.

Caring for Adult Mosquitoes

Adult mosquitoes require constant access to moisture, and they perform best in humid environments. If you are rearing *Culex* mosquitoes in an environmental chamber that does not have humidity control, the relative humidity within the cage can be increased by placing a wet sponge on top of the cage and covering the cage with a plastic bag. Moreover, both male and female mosquitoes need to have access to both H_2O and sugar sources for reproduction and fertility (Kassim et al. 2012). We have reared *Culex* mosquitoes on 10% sucrose solutions only, and they do quite well (Meuti et al. 2015a,b). However, to promote reproduction, we have found that it is better to offer them additional and more natural sugar sources, such as organic (pesticide-free) raisins and honey. The sucrose solution and H_2O source should be covered to prevent mosquitoes from drowning, and the liquid should be made available to the mosquitoes via a cotton dental wick, piece of cotton roll, or filter paper. This setup will permit adult mosquitoes to feed and hydrate freely, without any risk of drowning. Although adult mosquitoes require substantially less careful monitoring than the larval and pupal phases of mosquitoes, it is necessary to frequently check cages that contain adult mosquitoes to ensure that they have access to food and H_2O. Replace food and H_2O at least once per week to prevent mold growth or fermentation of the sugar solution. After blood feeding, it is best to check the cages every day, as females must have access to a sugar and H_2O source to produce a high number of egg rafts that will hopefully yield abundant viable larvae.

 Cite this introduction as *Cold Spring Harb Protoc*; doi:10.1101/pdb.top107823

Blood Feeding Newly Established and Laboratory Strains of *Culex*

Blood Source

The best way to encourage mosquitoes to blood feed is to offer a live vertebrate host, as it will provide the mosquitoes with the necessary visual, olfactory, and temperature cues. Moreover, *Culex* species are well adapted to specific hosts and generally produce greater numbers of eggs when they are allowed to feed on avian hosts (Shroyer and Siverly 1972; Downe and Archer 1975; Nayar and Sauerman 1975; Lyimo and Ferguson 2009; Richards et al. 2012; Alto et al. 2014). Therefore, if animal use protocols are in place, it would be best to offer an appropriately restrained avian host, as any host movements will further deter biting (Edman et al. 1974; Darbro and Harrington 2007).

Another option is to use live humans. If the need to establish a colony outweighs the risks and annoyances of using human volunteers, we recommend that the volunteer offer an unwashed foot or arm to the mosquitoes. Additionally, we recommend that the volunteer avoid consuming foods that contain garlic or citrus, as they are repellents to mosquitoes (Bhuyan et al. 1974; Giatropoulos et al. 2012) and can be emitted in the sweat/pores of the host (Sato et al. 2020). However, field-collected females may be infected with viral pathogens that were vertically transmitted from their mothers. Generally, the rate of vertical transmission is quite low (0.3%–0.6% for West Nile virus; Goddard et al. 2003), but the potential for infection should be reflected in any biosafety protocols that are in place to protect human volunteers and should be clearly communicated to the person who is planning to put their foot, hand, or arm into the mosquito cage.

If *Culex* species can be induced to take a blood meal from an artificial source, we strongly recommend this option. Avian blood that is heated in an artificial membrane feeding system (e.g., Hemotek) can be offered. Artificial feeding systems are convenient and allow researchers to maintain mosquito colonies without creating extensive biosafety or animal use protocols. Sodium citrate–treated blood from a variety of sources, including chickens, can be easily purchased (e.g., Lampire Biological Laboratories and Pel-Freeze Biologicals). Additionally, Griffith and Turner (1996) reported that it is possible to induce females of *Cx. quinquefasciatus* to consume a blood substitute and continue to lay viable eggs. Note that other laboratories have had success using circulating H_2O baths to warm blood (Deng et al. 2012; Witmer et al. 2018), whereas Kauffman et al. (2017) described how to use sausage casings to warm and offer blood. Although circulating H_2O baths and sausage casings warmed in H_2O are less expensive, they are also slightly more challenging to use. These artificial feeding systems lack many of the necessary cues that mosquitoes need to orient to the blood source. Moreover, the artificial membranes (e.g., Parafilm, cellulose paper, and sausage casings) are quite different from the skin of vertebrate hosts. Therefore, it is not surprising that many if not most field-collected *Culex* females may fail to feed on blood offered through an artificial membrane. To help encourage newly collected *Culex* females to feed on artificial blood feeders, we recommend that the membrane for the system be rubbed against the skin of a bird or human (i.e., the foot and ankle) to transfer some of the host volatiles. We also recommend placing a source of carbon dioxide (e.g., dry ice) near the cage. Finally, we recommend that the blood be supplemented with adenosine triphosphate (ATP; Sigma-Aldrich; ~200 µL of 0.1 M ATP per 15 mL of blood), as it encourages female mosquitoes that initially penetrate the membrane to feed to repletion (Galun 1967; Galun et al. 1985; Griffith and Turner 1996).

Timing of the Blood Meal

Mosquito species vary in the age at which they become competent to take a blood meal. Females of *Aedes aegypti* are ready to bite and lay eggs as early as 24 h after adult emergence (Howard 1923; Armstrong and West 1966) and most will bite 24–72 h after emergence (Christophers 1960). The age at which *Culex* mosquitoes first bite/blood feed has not extensively been investigated, although high rates of biting are achieved 5 d after adult emergence, especially when the mosquitoes have been reared under long photoperiods and high temperatures (Eldridge 1968). In our experience, females of *Cx. pipiens* blood feed well when they are 5–10 d old, and therefore we recommend this time frame when offering the first blood meal to adult mosquitoes.

Most *Culex* mosquitoes are crepuscular, and therefore offering a blood meal, whether avian, human, or artificial, at dawn or dusk will likely stimulate blood feeding. As field-collected *Culex* females will likely need several opportunities to blood feed before they bite, we recommend offering the first blood meal 3 d after adult emergence and every day or two thereafter. We also recommend leaving the blood source in the cage for as long as possible. If artificial blood feeders are being used, we recommend that they remain in the cage for at least 1–2 h to stimulate blood feeding; we have had the most success feeding mosquitoes with an artificial blood source overnight.

Offering Extra Blood Meals

Female *Culex* mosquitoes can take multiple blood meals and produce several batches of eggs (up to four batches according to Awahmukalah and Brooks 1985). Therefore, if researchers need to rapidly build up their populations of *Culex*, they can offer the same cage of *Culex* females multiple blood meals. We recommend waiting at least 1 wk between blood feedings for mosquitoes reared at 24°C–27°C to ensure that every female that initially took a blood meal has enough time to oviposit and reinitiate a subsequent gonotrophic cycle. Moreover, as the sex ratio becomes more male-biased in older females (Qutubuddin 1953; pers. observations) and because most female mosquitoes will only survive ~6 wk after adult emergence at 22°C–26°C (Oda et al. 1999; Alto et al. 2012), we recommend blood feeding each cage a maximum of four times. Additionally, whenever possible, we suggest rearing the offspring of 5- to 10-d-old females that have been blood fed for the first time, as it will yield a high number of larvae with a sex ratio of nearly 1:1 males to females (Gómez et al. 1977).

Working with Multiple Strains, Species, or Populations of *Culex* Mosquitoes

If you are rearing multiple strains or species of *Culex* mosquitoes, it is also very important to ensure that everything is properly labeled so that you do not unintentionally cross laboratory strains. Several members of the *Cx. pipiens* complex (e.g., *Cx. pipiens molestus*, *Cx. pipiens pipiens*, and *Cx. quinquefasciatus*) can produce fertile hybrids (Smith and Fonseca 2004; Bahnck and Fonseca 2006). To avoid such a problem, we recommend using a set of separate and clearly labeled containers to collect egg rafts, rear larvae, and allow pupal emergence within adult cages. We also suggest that separate labeled disposable pipettes are used for different strains and species. Whenever possible, it is best to work with different mosquito strains in separate rearing spaces and/or work with one species/strain at a time.

ACKNOWLEDGMENTS

We thank Robert Aldridge, Josh Benoit, David Denlinger, Leeanne Garrett, Woodbridge Foster, Joe Rinehart, Rebecca Robich, David Shetlar, and Cheolho Sim for helping us to identify locations and trapping techniques to collect mosquitoes and for assistance in developing these protocols. We also thank members of the Meuti Laboratory, past and present, for their diligence in rearing mosquitoes. M.E.M. and M.W. were supported by a National Science Foundation grant (IOS 1944324), and M.E.M. and A.S. were supported by a National Institutes of Health grant (R21-AI144266).

REFERENCES

Akaratovic KI, Kiser JP, Whitt PB, Harrison RL, Harrison BA. 2021. *Culex nigripalpus* distribution expansion: first record in Virginia, new county records in North Carolina, and revised United States map. *J Am Mosq Control Assoc* **37:** 188–197. doi:10.2987/21-7025

Alto BW, Muturi EJ, Lampman RL. 2012. Effects of nutrition and density in *Culex pipiens*. *Med Vet Entomol* **26:** 396–406. doi:10.1111/j.1365-2915.2012.01010.x

Alto BW, Connelly CR, O'Meara GF, Hickman D, Karr N. 2014. Reproductive biology and susceptibility of Florida *Culex coronator* to infection with West Nile virus. *Vector-Borne Zoonotic Dis* **14:** 606–614. doi:10.1089/vbz.2013.1501

Andreadis TG. 2012. The contribution of *Culex pipiens* complex mosquitoes to transmission and persistence of West Nile virus in North America. *J Am Mosq Control Assoc* **28:** 137–151. doi:10.2987/8756-971X-28.4s.137

Armstrong JA, West AS. 1966. Blood feeding activity by newly emerged female mosquitoes. *Mosq News* **25:** 263–268.

Awahmukalah DST, Brooks MA. 1985. Viability of *Culex pipiens pipiens* eggs affected by nutrition and aposymbiosis. *J Invertebr Pathol* **45:** 225–230. doi:10.1016/0022-2011(85)90012-6

Bahnck CM, Fonseca DM. 2006. Rapid assay to identify the two genetic forms of *Culex (Culex) pipiens* L. (Diptera: Culicidae) and hybrid pop-

Cite this introduction as *Cold Spring Harb Protoc*; doi:10.1101/pdb.top107823

ulations. *Am J Trop Med Hyg* **75:** 251–255. doi:10.4269/ajtmh.2006.75.2.0750251

Bailey CL, Eldridge BF, Hayes DE, Watts DM, Tammariello RF, Dalrymple JM. 1978. Isolation of St. Louis encephalitis virus from overwintering *Culex pipiens* mosquitoes. *Science* **199:** 1346–1349. doi:10.1126/science.628843

Beckel WE. 1958. Investigations of permeability, diapause, and hatching in the eggs of the mosquito *Aedes hexodontus* dyar. *Can J Zool* **36:** 541–554. doi:10.1139/z58-050

Bergren NA, Borland EM, Hartman DA, Kading RC. 2021. Laboratory demonstration of the vertical transmission of Rift Valley fever virus by *Culex tarsalis* mosquitoes. *PLoS Negl Trop Dis* **15:** e0009273. doi:10.1371/journal.pntd.0009273

Bhuyan M, Saxena BN, Rao KM. 1974. Repellent property of oil fraction of garlic, *Allium sativum* Linn. *Indian J Exp Biol* **12:** 575–576.

Birley MH. 1979. The estimation and simulation of variable developmental period, with application to the mosquito *Aedes aegypti* (L.). *Popul Ecol* **21:** 68–80. doi:10.1007/BF02512639

Bohart RM, Washino RK. 1978. *Mosquitoes of California (Vol. 4084)*. University of California, Agriculture and Natural Resources.

Burkett-Cadena ND, Graham SP, Hassan HK, Guyer C, Eubanks MD, Katholi CR, Unnasch TR. 2008. Blood feeding patterns of potential arbovirus vectors of the genus *Culex* targeting ectothermic hosts. *Am J Trop Med Hyg* **79:** 809–815. doi:10.4269/ajtmh.2008.79.809

Chancey C, Grinev A, Volkova E, Rios M. 2015. The global ecology and epidemiology of West Nile virus. *BioMed Res Int* **2015:** 376230. doi:10.1155/2015/376230

Christophers SR. 1960. *Aedes aegypti* (L.): *the yellow fever mosquito: its life history, bionomics and structure*. Cambridge University Press, London.

Ciota AT, Kramer LD. 2013. Vector-virus interactions and transmission dynamics of West Nile virus. *Viruses* **5:** 3021–3047. doi:10.3390/v5123021

Coon BR. 2006. "*Field and laboratory studies of* Culex erraticus *(Diptera: Culicidae) ability to detect hosts, habitat identification and attempts at colonization*." Ph.D. dissertation, University of Florida.

Darbro JM, Harrington LC. 2007. Avian defensive behavior and blood-feeding success of the West Nile vector mosquito, *Culex pipiens*. *Behav Ecol* **18:** 750–757. doi:10.1093/beheco/arm043

de Melo Chalegre KD, Tavares DA, Romão TP, de Menezes HS, Nascimento NA, de Oliveira CM, de-Melo-Neto OP, Silva-Filha MH. 2015. Co-selection and replacement of resistance alleles to *Lysinibacillus sphaericus* in a *Culex quinquefasciatus* colony. *FEBS J* **282:** 3592–3602. doi:10.1111/febs.13364

Deng L, Koou S-Y, Ng LC, Lam-Phua SG, Png AB. 2012. A novel mosquito feeding system for routine blood-feeding of *Aedes aegypti* and *Aedes albopictus*. *Trop Biomed* **29:** 169–174.

Desser SS, McIver SB, Ryckman A. 1973. *Culex territans* as a potential vector of *Trypanosoma rotatorium*. I. Development of the flagellate in the mosquito. *J Parasitol* **1:** 353–358. doi:10.2307/3278833

Dobrotworsky NV. 1955. The *Culex pipiens* group in south-eastern Australia. IV Crossbreeding experiments within the *Culex pipiens* group. *Proc Linnean Soc New South Wales* **80:** 33–43.

Downe AER, Archer JA. 1975. The effects of different blood-meal sources on digestion and egg production in *Culex tarsalis* Coq. (Diptera: Culicidae). *J Med Entomol* **12:** 431–437. doi:10.1093/jmedent/12.4.431

Du YJ, Millar JG. 1999. Electroantennogram and oviposition bioassay responses of *Culex quinquefasciatus* and *Culex tarsalis* (Diptera: Culicidae) to chemicals in odors from Bermuda grass infusions. *J Med Entomol* **36:** 158–166. doi:10.1093/jmedent/36.2.158

Edman JD, Webber LA, Schmid AA. 1974. Effect of host defenses on the feeding pattern of *Culex nigripalpus* when offered a choice of blood sources. *J Parasitol* **60:** 874. doi:10.2307/3278923

Eldridge BF. 1968. The effect of temperature and photoperiod on blood-feeding and ovarian development in mosquitoes of the *Culex pipiens* complex. *Am J Trop Med Hyg* **17:** 133–140. doi:10.4269/ajtmh.1968.17.133

Eldridge BF, Bailey CL, Johnson MD. 1972. A preliminary study of the seasonal geographic distribution and overwintering of *Culex restuans* Theobald and *Culex salinarius* Coquillet (Diptera: Culicidae). *J Med Entomol* **9:** 233–238. doi:10.1093/jmedent/9.3.233

Galun R. 1967. Feeding stimuli and artificial feeding. *Bull World Health Organ* **36:** 590–593.

Galun R, Koontz LC, Gwadz RW, Ribeiro JMC. 1985. Effect of ATP analogues on the gorging response of *Aedes aegypti*. *Physiol Entomol* **10:** 275–281. doi:10.1111/j.1365-3032.1985.tb00048.x

Gao Q, Su F, Zhou Y-B, Chu W, Cao H, Song L-L, Zhou J-J, Leng P-E. 2019. Autogeny, fecundity, and other life history traits of *Culex pipiens molestus* (Diptera: Culicidae) in Shanghai, China. *J Med Entomol* **56:** 656–664. doi:10.1093/jme/tjy228

Giatropoulos A, Papachristos DP, Kimbaris A, Koliopoulos G, Polissiou MG, Emmanouel N, Michaelakis A. 2012. Evaluation of bioefficacy of three Citrus essential oils against the dengue vector *Aedes albopictus* (Diptera: Culicidae) in correlation to their components enantiomeric distribution. *Parasitol Res* **111:** 2253–2263. doi:10.1007/s00436-012-3074-8

Goddard LB, Roth AE, Reisen WK, Scott TW. 2003. Vertical transmission of West Nile virus by three California *Culex* (Diptera: Culicidae) species. *J Med Entomol* **40:** 743–746. doi:10.1603/0022-2585-40.6.743

Goldberg TL, Ruiz MO, Hamer GL, Brawn JD, Kitron UD, Hayes DB, Loss SR, Walker ED. 2009. Host selection by *Culex pipiens* mosquitoes and West Nile virus amplification. *Am J Trop Med Hyg* **80:** 268–278. doi:10.4269/ajtmh.2009.80.268

Gómez G, Rabinovich JE, Machado-Allison GE. 1977. Population analysis of *Culex pipiens fatigans* Wied.(Diptera: Culicidae) under laboratory conditions. *J Med Entomol* **13:** 453–463. doi:10.1093/jmedent/13.4-5.453

Griffith JSR, Turner GD. 1996. Culturing *Culex quinquefasciatus* mosquitoes with a blood substitute diet for the females. *Med Vet Entomol* **10:** 265–268. doi:10.1111/j.1365-2915.1996.tb00741.x

Hancock C, Camp JV. 2022. Habitat-specific host selection patterns of *Culex quinquefasciatus* and *Culex nigripalpus* in Florida. *J Am Mosq Control Assoc* **38:** 83–91.

Harb M, Faris R, Gad AM, Hafez ON, Ramzy R, Buck AA. 1993. The resurgence of lymphatic filariasis in the Nile delta. *Bull World Health Organ* **71:** 49–54.

Hawley WA. 1988. The biology of *Aedes albopictus*. *J Am Mosq Control Assoc Suppl* **1:** 1–39.

Helbing CM, Moorhead DL, Mitchell L. 2015. Population dynamics of *Culex restuans* and *Culex pipiens* (Diptera: Culicidae) related to climatic factors in northwest Ohio. *Environ Entomol* **44:** 1022–1028. doi:10.1093/ee/nvv094

Howard LO. 1923. *The yellow-fever mosquito*. US Government Printing Office, Washington, DC.

Jackson BT, Paulson SL. 2006. Seasonal abundance of *Culex restuans* and *Culex pipiens* in southwestern Virginia through ovitrapping. *J Am Mosq Control Assoc* **22:** 206–212. doi:10.2987/8756-971X(2006)22[206:SAOCRA]2.0.CO;2

Kalpage KSP, Brust RA. 1974. Studies on diapause and female fecundity in *Aedes atropalpus*. *Environ Entomol* **3:** 139–145. doi:10.1093/ee/3.1.139

Kassim NF, Webb CE, Russell RC. 2012. The importance of males: larval diet and adult sugar feeding influences reproduction in *Culex molestus*. *J Am Mosq Control Assoc* **28:** 312–316. doi:10.2987/12-6274R.1

Kauffman E, Payne A, Franke MA, Schmid MA, Harris E, Kramer LD. 2017. Rearing of *Culex* spp. and *Aedes* spp. mosquitoes. *Bio Protoc* **7:** e2542. doi:10.21769/BioProtoc.2542

Kiarie-Makara M, Ngumbi PM, Lee D-K. 2015. Effects of temperature on the growth and development of *Culex pipiens* complex mosquitoes (Diptera: Culicidae). *IOSR J Pharm Bio Sci* **10:** 1–10.

Kraemer MU, Sinka ME, Duda KA, Mylne AQ, Shearer FM, Barker CM, Moore CG, Carvalho RG, Coelho GE, Van Bortel W, et al. 2015. The global distribution of the arbovirus vectors *Aedes aegypti* and *Ae. albopictus*. *eLife* **4:** e08347. doi:10.7554/eLife.08347

Lyimo IN, Ferguson HM. 2009. Ecological and evolutionary determinants of host species choice in mosquito vectors. *Trends Parasitol* **25:** 89–196. doi:10.1016/j.pt.2009.01.005

Meuti ME, Stone M, Ikeno T, Denlinger DL. 2015a. Functional circadian clock genes are essential for the overwintering diapause of the Northern house mosquito, *Culex pipiens*. *J Exp Biol* **218:** 412–422. doi:10.1242/jeb.113233

Meuti ME, Short CA, Denlinger DL. 2015b. Mom matters: Diapause characteristics of *Culex pipiens–Culex quinquefasciatus* (Diptera: Culicidae) hybrid mosquitoes. *J Med Entomol* **52:** 131–137. doi:10.1093/jme/tju016

Meuti ME, Siperstein A, Wolkoff M. 2023a. Establishing a *Culex* colony from field-collected eggs. *Cold Spring Harb Protoc* doi:10.1101/pdb.prot108079

Meuti ME, Siperstein A, Wolkoff M. 2023b. Rearing and maintaining a *Culex* colony in the laboratory. *Cold Spring Harb Protoc* doi:10.1101/pdb.108080

Murphey FJ. 1961. "The bionomics of *Culex salinarius* Coquillett." PhD dissertation, University of Delaware, Newark.

Mutebi JP, Savage HM. 2009. Discovery of *Culex pipiens pipiens* form *molestus* in Chicago. *J Am Mosq Control Assoc* **25:** 500–503. doi:10.2987/09-5910.1

Nayar JK, Sauerman DM. 1975. The effects of nutrition on survival and fecundity in Florida mosquitoes. part 3. Utilization of blood and sugar for fecundity. *J Med Entomol* **12:** 220–225. doi:10.1093/jmedent/12.2.220

Oda T, Uchida K, Mori A, Mine M, Eshita Y, Kurokawa K, Kato K, Tahara H. 1999. Effects of high temperature on the emergence and survival of adult *Culex pipiens molestus* and *Culex quinquefasciatus* in Japan. *J Am Mosq Control Assoc* **15:** 153–156.

Parker BM. 1985. Effects of photoperiod on the induction of embryonic diapause in *Aedes taeniorhynchus* (Diptera: Culicidae). *J Med Entomol* **22:** 392–397. doi:10.1093/jmedent/22.4.392

Paul A, Harrington LC, Zhang L, Scott JG. 2005. Insecticide resistance in *Culex pipiens* from New York. *J Am Mosq Control Assoc* **21:** 305–309. doi:10.2987/8756-971X(2005)21[305:IRICPF]2.0.CO;2

Qutubuddin M. 1953. The emergence and sex ratio of *Culex fatigans* Wied. (Diptera, Culicidae) in laboratory experiments. *Bull Entomol Res* **43:** 549–565. doi:10.1017/S000748530002664X

Richards SL, Anderson SL, Yost SA. 2012. Effects of blood meal source on the reproduction of *Culex pipiens quinquefasciatus* (Diptera: Culicidae). *J Vector Ecol* **37:** 1–7. doi:10.1111/j.1948-7134.2012.00194.x

Roubaud E. 1930. Sur l'existence de race biologiques genetiquement distinctes chez moustiques commun *Culex pipiens*. *C R Acad Sci* **191:** 1386–1388.

Russell RC. 2002. Ross River virus: ecology and distribution. *Annu Rev Entomol* **47:** 1–31. doi:10.1146/annurev.ento.47.091201.145100

Sanburg LL, Larsen JR. 1973. Effect of photoperiod and temperature on ovarian development in *Culex pipiens pipiens*. *J Insect Physiol* **19:** 1173–1190. doi:10.1016/0022-1910(73)90202-3

Sato S, Sekine Y, Kakumu Y, Hiramoto T. 2020. Measurement of diallyl disulfide and allyl methyl sulfide emanating from human skin surface and influence of ingestion of grilled garlic. *Sci Rep* **10:** 465. doi:10.1038/s41598-019-57258-1

Schultz GW. 1992. Biting activity of mosquitos (Diptera: Culicidae) at a malarious site in Palawan, Republic of The Philippines. *Southeast Asian J Trop Med Public Health* **23:** 464–469.

Shroyer DA, Craig GB. 1983. Egg diapause in *Aedes triseriatus* (diptera: culicidae): geographic variation in photoperiodic response and factors influencing diapause termination. *J Med Entomol* **20:** 601–607. doi:10.1093/jmedent/20.6.601

Shroyer DC, Siverly RE. 1972. A comparison of egg production of *Culex pipiens pipiens* L. fed on avian and mammalian hosts. *Mosquito News* **32:** 636–637.

Smith JL, Fonseca DM. 2004. Rapid assays for identification of members of the *Culex* (*Culex*) *pipiens* complex, their hybrids, and other sibling species (Diptera: Culicidae). *Am J Trop Med Hyg* **70:** 339–345. doi:10.4269/ajtmh.2004.70.339

Spielman A. 1957. The inheritance of autogeny in the *Culex pipiens* complex of mosquitoes. *Am J Epidemiol* **65:** 404–425. doi:10.1093/oxfordjournals.aje.a119878

Spielman A, Wong J. 1973. Studies on autogeny in natural populations of *Culex pipiens*. 3. Midsummer preparation for hibernation in anautogenous populations. *J Med Entomol* **10:** 319–324. doi:10.1093/jmedent/10.4.319

Sundararaman S. 1949. Biometrical studies on intergradation in the genitalia of certain populations of *Culex pipiens* and *Culex quinquefasciatus* in the United States. *Am J Hyg* **50:** 307–314.

Tuno N, Tsuda Y, Takagi M. 2017. How zoophilic Japanese encephalitis vector mosquitoes feed on humans. *J Med Entomol* **54:** 8–13. doi:10.1093/jme/tjw165

Turell MJ, Morrill JC, Rossi CA, Gad AM, Cope SE, Clements TL, Arthur RR, Wasieloski LP, Dohm DJ, Nash D, et al. 2002. Isolation of West Nile and Sindbis viruses from mosquitoes collected in the Nile Valley of Egypt during an outbreak of Rift Valley fever. *J Med Entomol* **39:** 248–250. doi:10.1603/0022-2585-39.1.248

Vargas HC, Farnesi LC, Martins AJ, Valle D, Rezende GL. 2014. Serosal cuticle formation and distinct degrees of desiccation resistance in embryos of the mosquito vectors *Aedes aegypti*, *Anopheles aquasalis* and *Culex quinquefasciatus*. *J Insect Physiol* **1:** 54–60. doi:10.1016/j.jinsphys.2014.02.001

Vezzani D, Mesplet M, Eiras DF, Fontanarrosa MF, Schnittger L. 2011. PCR detection of *Dirofilaria immitis* in *Aedes aegypti* and *Culex pipiens* from urban temperate Argentina. *Parasitol Res* **108:** 985–989. doi:10.1007/s00436-010-2142-1

Wallis RC, Spielman A. 1953. Laboratory rearing of *Culex salinarius* (Diptera, Culicidae). *Proc Entomol Soc Washington* **55:** 140–142.

Wallis RC, Whitman LO. 1968. Colonization of *Culex salinarius* in the laboratory. *Mosquito News* **28:** 366–368.

Wanji S, Tanke T, Atanga SN, Ajonina C, Nicholas T, Fontenille D. 2003. *Anopheles* species of the mount Cameroon region: biting habits, feeding behaviour and entomological inoculation rates. *Trop Med Int Health* **8:** 643–649. doi:10.1046/j.1365-3156.2003.01070.x

Wispelaere M, Desprès P, Choumet V. 2017. European *Aedes albopictus* and *Culex pipiens* are competent vectors for Japanese encephalitis virus. *PLoS Negl Trop Dis* **11:** e0005294. doi:10.1371/journal.pntd.0005294

Witmer K, Sherrard-Smith E, Straschil U, Tunnicliff M, Baum J, Delves M. 2018. An inexpensive open source 3D-printed membrane feeder for human malaria transmission studies. *Malaria J* **17:** 1–7. doi:10.1186/s12936-018-2436-9

Yurchenko AA, Masri RA, Khrabrova NV, Sibataev AK, Fritz ML, Sharakhova MV. 2020. Genomic differentiation and intercontinental population structure of mosquito vectors *Culex pipiens pipiens* and *Culex pipiens molestus*. *Sci Rep* **10:** 7504. doi:10.1038/s41598-020-63305-z

Cite this introduction as *Cold Spring Harb Protoc*; doi:10.1101/pdb.top107823

Protocol 1

Establishing a *Culex* Colony from Field-Collected Eggs

Megan E. Meuti,[1] Alden Siperstein, and Matthew Wolkoff

Entomology Department, The Ohio State University, Columbus, Ohio 43210, USA

Culex larvae are well adapted to growing and developing in containers, and therefore collecting and rearing field-collected *Culex* to adulthood in the laboratory is relatively straightforward. What is substantially more challenging is simulating natural conditions that encourage *Culex* adults to mate, blood feed, and reproduce in laboratory settings. In our experience, this is the most difficult hurdle to overcome when establishing new laboratory colonies. Here, we detail how to collect *Culex* eggs from the field and establish a colony in the laboratory. Successfully establishing a new colony of *Culex* mosquitoes in the laboratory will allow researchers to evaluate physiologically, behaviorally, and ecologically relevant aspects of their biology and better understand and manage these important disease vectors.

MATERIALS

It is essential that you consult the appropriate Material Safety Data Sheets and your institution's Environmental Health and Safety Office for proper handling of equipment and hazardous material used in this protocol.

RECIPES: Please see the end of this protocol for recipes indicated by <R>. Additional recipes can be found online at http://cshprotocols.cshlp.org/site/recipes.

Reagents

Bleach solution (3%–5%)

Blood source

Large amounts (~500 mL) of sodium-citrate treated chicken's blood can be purchased from biological suppliers and divided into 6-, 9-, or 15-mL aliquots. These aliquots can be frozen for up to 16 mo at −70°C and thawed when needed.

Distilled H_2O

Reverse osmosis (RO) H_2O can also be used.

DNA extraction kit

In our experience, the DNeasy (QIAGEN) and Phire tissue-direct PCR kits (Thermo Fisher Scientific) work well.

Dry fish food (e.g., Tetramin tropical fish flakes pulverized using a coffee grinder)

Grass clippings (freshly cut; one handful of Kentucky bluegrass, crabgrass, or other locally available grasses; see Step 26)

[1]Correspondence: meuti.1@osu.edu

Cite this protocol as *Cold Spring Harb Protoc*; doi:10.1101/pdb.prot108079

Gravid H_2O for Mosquito Larvae Collection <R>

Most protocols to generate gravid H_2O entail placing organic material (usually grass clippings or straw) in H_2O and allowing it to ferment at high temperatures such as outside during summer (Zulueta 1950; Gjullin et al. 1965; Gubler 1971; Kramer and Mulla 1979). We have also had success in allowing the H_2O to ferment in environmental chambers set to 27°C. Yeast increases carbon dioxide production and facilitates the breakdown of organic material, and milk protein increases the amino acid content of the H_2O and supports bacterial growth (Burkett-Cadena and Mullen 2008).

Honey (optional, see Step 20)
PCR master mix (e.g., DreamTaq, Fisher Scientific)
Primers

See Crabtree et al. (1995), Smith and Fonseca (2004), and Bahnck and Fonseca (2006).

Raisins (optional, see Step 20; organic)
Soap (optional, see Step 33; e.g., Dawn or Palmolive dish soap)
Sucrose solution (10%) made from table sugar or laboratory-grade sucrose (e.g., Domino granulated sugar; Fisher Scientific)

Equipment

Black paper or tape (see Step 19)
Black tubs (plastic, 20.5-in × 15-in × 5-in)

Research shows that large containers (≥35-L) are more effective at collecting mosquitoes than small containers (≤6-L; Allan and Kline 2004; Popko and Walton 2016).

Compound light microscope (optional, see Step 21)
Cotton
Dissecting microscope (100×–200× magnification with backlighting)
Environmental chamber

A room where lights are on a photoperiod timer and temperature and/or humidity are controlled can also be used. To stimulate mosquito growth and reproduction, we recommend setting the chamber/room to long-day conditions (>14 h light/d) and temperatures of 24°C–27°C. Culex *mosquitoes will develop at temperatures as low as 15°C, but development will be slower. If you want to induce diapause/overwintering dormancy,* Culex *mosquitoes should be exposed to photoperiods <12 h light/d and temperatures <21°C. Both reproductively active and diapausing* Culex *mosquitoes prefer humid conditions (relative humidity >70%). This can be achieved by using a humidifier and/or placing moist sponges on top of the mosquito cages.*

Forceps (optional, see Step 21)
Large syringe/turkey baster (e.g., Oxo good grips turkey baster)
Mesh cages (30 cm^3 or larger recommended; e.g., BugDorm-1 Insect Rearing Cage)
Paper towels or filter paper
Plastic deli cups with lids (e.g., diameter of 11.4 cm; height of 4.4 cm; holding volume of 250–500 mL; Fabri-Kal Corp)
Plastic shoe-box-sized containers (e.g., Sterlite, 28-cm × 16.7-cm × 6.9-cm)
Plastic transfer pipette (e.g., ~3-mL; Fischer Scientific)
Plastic utensils (e.g., spoon or spatula)
Sponges
Thermocycler

METHOD

Field Collections

Many species of Culex *mosquitoes are container breeders and prefer to lay their egg rafts in nutrient-rich H_2O. Therefore, in the summer, it is relatively easy to collect* Culex *eggs in black tubs containing gravid H_2O.*

Cite this protocol as *Cold Spring Harb Protoc*; doi:10.1101/pdb.prot108079

FIGURE 1. (*A*) Placement of a container of gravid H_2O used to collect *Culex* egg rafts. (*B*) Close-up of *Culex* egg rafts in a pan (egg rafts circled in red).

1. Place a black tub containing 4–11 L (1–3 gallons) of gravid H_2O along the edge of forested or semiforested areas. For best results, place one to three tubs at least 12 m apart within a site and select three to 10 sites (>1.5-km-apart) within a study locale.

   In our experience, this setup allows researchers to collect the highest number of Culex *egg rafts, which will allow for a great degree of genetic diversity and improve your chances of establishing a viable colony. Placing the tubs near vegetation allows researchers to easily access to the traps while ensuring that the traps are close to mosquitoes' resting habitat (Fig. 1A).*

2. Monitor the tubs daily for the presence of eggs, as egg rafts hatch between 24 and 72 h after being laid, depending on temperature (Miller et al. 2015).

3. Remove egg rafts (Fig. 1B) from the gravid traps using plastic utensils.

   White spoons are especially useful, as they provide a strong contrast with the black egg rafts.

   Although mosquito larvae can also be collected from gravid traps using large syringes or turkey basters, researchers will likely find it helpful to collect individual egg rafts as each raft is laid by a single female, meaning that only one to three of the larvae emerging from that egg raft need to be morphologically and/or genetically identified. Moreover, collecting unique egg rafts that belong to a single mosquito species will allow researchers to easily ascertain the effective population size and genetic diversity of their initial colony.

4. Place eggs into plastic deli-cup-sized containers lined with dry paper towels or filter paper.

   As a small amount of the gravid H_2O will also be transferred with the egg rafts, using dry paper towels ensures that the eggs will not be too moist and will adequately stick to the paper material until they are transported to the laboratory.

5. If mosquito larvae are present in the traps, remove them by using a plastic transfer pipette or large syringe/turkey baster and place them into plastic deli cups (250- to 500-mL) and seal with leak-proof lids.

6. After an adequate number of mosquito eggs/larvae have been collected (~500–1000 egg rafts or 2500–150,000 *Culex* larvae), dump the gravid H_2O from the tubs. Bring the mosquitoes and tubs back to the laboratory.

   The last step ensures that researchers are not creating additional breeding habitats for container-breeding mosquitoes that transmit disease.

Species Separation

After Culex *egg rafts or putative* Culex *larvae are collected, the next step to establishing a robust colony is to identify and separate your desired species from other mosquito species.*

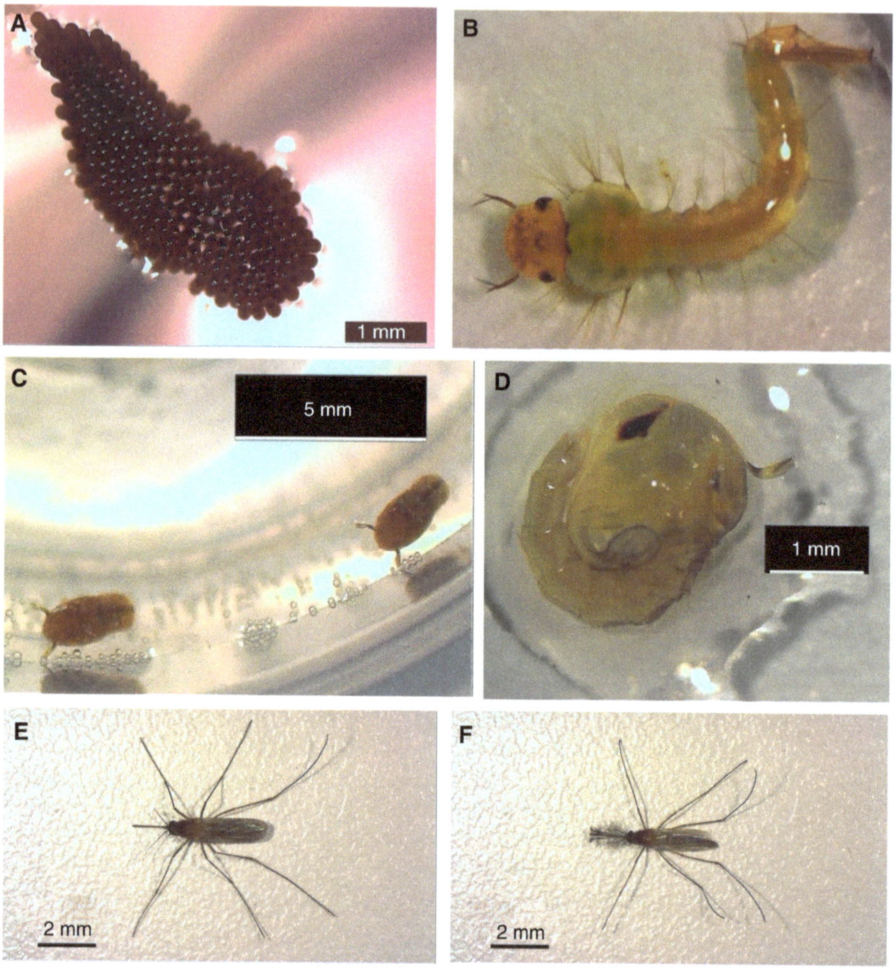

FIGURE 2. Life stages of laboratory-reared *Culex pipiens*. (*A*) A representative egg raft under a microscope. (*B*) A single larva of *Culex pipiens*. (*C*) Dorsal view of two pupae. (*D*) Lateral view of one pupa. (*E*) Nondiapausing female. (*F*) Adult male.

7. If egg rafts were collected, place a single raft into an individual plastic deli cup with ~100 mL distilled H_2O (Fig. 2A).
8. Place the containers with egg rafts in an environmental chamber set to long-day conditions (>13 h light/d) and a high temperature (>24°C).

   These conditions will prevent the mosquitoes from entering their overwintering dormancy upon reaching adulthood (Eldridge 1968; Spielman and Wong 1973).
9. Monitor the containers daily, and upon seeing first-instar larvae, add a small amount of dry fish food (~25 mg) to the container.

   Waiting until the larvae emerge is helpful, as some egg rafts may never hatch, and it is easier to see the larvae in clean H_2O. Additionally, waiting to add the food until larvae are present limits bacterial growth and promotes greater larval survival.
10. Provide the larvae with more food according to the feeding schedule in Protocol 2: Rearing and Maintaining a *Culex* Colony in the Laboratory (Meuti et al. 2023a).

Identifying *Culex* Larvae

Morphological Identification

Although Culex *larvae can be identified as early as the first instar (Dodge 1966; Haeger and O'Meara 1983; Jackson and Paulson 2006), in our experience, larger third- and fourth-instar larvae allow researchers to see diagnostic differences between species more clearly.*

Cite this protocol as *Cold Spring Harb Protoc*; doi:10.1101/pdb.prot108079

11. If *Culex* larvae from individual egg rafts were maintained separately, morphologically identify one to three larvae (Fig. 2B) by using a dissecting microscope and a published key.

    Darsie and Ward (2005) as well as Gaffigan (2021) provide excellent keys for identifying Culex *larvae.*

12. However, if egg rafts were maintained together and/or if putative *Culex* larvae were collected, morphologically identify each larva with a key.

PCR-Based Identification

Although determining the species of Culex *larvae is much easier than identifying pupae and adults (Michener 1947; Harrington and Poulson 2008), researchers will likely benefit from confirming morphological identifications with PCR-based assays to ensure that they have the desired species, especially if researchers are relatively inexperienced in identifying mosquitoes visually. Moreover, some PCR-based assays allow researchers to distinguish between hybrids and/or different subspecies, particularly within the* Cx. pipiens *complex. For example, Crabtree et al. (1995) provide a protocol and primers to distinguish* Cx. pipiens, Cx. restuans, *and* Cx. salinarius; *Smith and Fonseca (2004) describe how to distinguish* Cx. pipiens *from* Cx. quinquefasciatus *and their hybrids; and Bahnck and Fonseca (2006) provide instructions to estimate the proportion of introgression between autogenous* Cx. pipiens molestus *and anautogenous* Cx. pipiens pipiens.

13. Isolate genomic DNA (gDNA) from one to five larvae by using a DNA extraction kit.

    One single larva contains enough gDNA for subsequent PCR assays. However, researchers may elect to isolate RNA from a group of larvae, either originating from the same egg raft or from multiple egg rafts, to confirm morphological identifications.

14. Mix the extracted DNA with PCR master mix and species-specific primers to amplify the genomic fragments of interest in a thermocycler, per manufacturer instructions.
15. After PCR amplification, separate the DNA fragments using gel electrophoresis (see protocols in Smith and Fonseca 2004 and Bahnck and Fonseca 2006 for details).
16. Determine the species identity based on the presence/absence and sizes of various PCR products.

Rearing the Identified Larvae to Adulthood

17. Pool larvae belonging to the desired species by placing them in plastic shoe-box-sized containers.

    This strategy is used because makes it more likely that the larvae will survive to adulthood. Moreover, pooling larvae from multiple egg rafts into larger containers helps to retain the genetic diversity of the population that could otherwise be lost if bacterial growth in one or several small containers of larvae derived from a unique egg raft were to excessively grow, causing all or most of the larvae to die.

18. Estimate the number of mosquito larvae in each container and add an appropriate amount of larval food to the containers, by using the guidelines in Protocol 2: Rearing and Maintaining a *Culex* Colony in the Laboratory (Meuti et al. 2023a).
19. Upon pupation (Fig. 2C,D), with a plastic transfer pipette, carefully move pupae from their larval containers into an open plastic deli cup half-filled with H_2O derived from the larval pan, which is then placed into a mesh cage where the adults can emerge (Fig. 2E,F).

    We recommend placing pupae derived from field-collected eggs or larvae into large cages (>30 cm^3; cages 62 cm^3 are ideal) and placing black paper or tape along the bottom third of the exterior of the cage. This serves as a marker that allows males of Culex *to form swarms within the cage and may increase mating success.*

20. Provide emerging adults in cages with sugar sources that simulate natural sources, such as honey-soaked sponges and/or raisins, in addition to a 10% sucrose solution.

    Offering a variety of sucrose sources should facilitate the health of both male and female mosquitoes and ensure that they successfully mate in the confined conditions of a cage.

21. (Optional) To ensure that mating takes place, use forceps to dissect the spermathecae from a subset of females under a compound light microscope and examine them for the presence of sperm according to Bellamy and Reeves (1963).

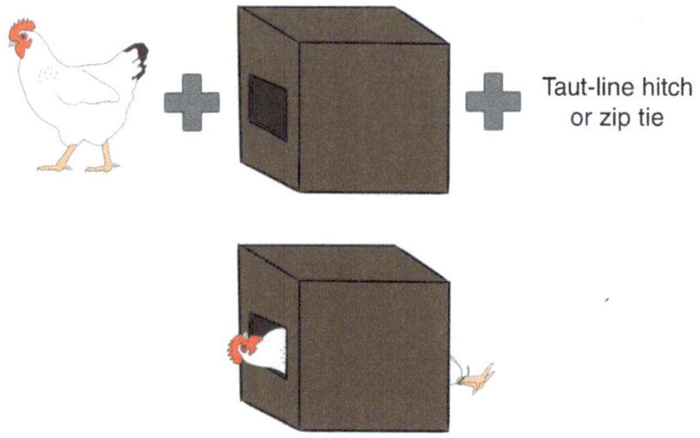

FIGURE 3. Cartoon diagram showing how to restrain a chicken for blood feeding. Supplies that are needed include a small cardboard box (L × W × H of ~25.4-cm × 20-cm × 20-cm) with holes cut on either side, a live rooster or hen, and either plastic ties or a rope tied in a taut line hitch. First place the tie/hitch around the chicken's legs. Then carefully and gently place the head of the chicken through one end of the box and the feet through the other. The chicken should be oriented on its back or side. Once the chicken is safely restrained, its feet can be placed inside the cloth sleeve of a mosquito cage. The sleeve should be secured with an additional zip tie to prevent mosquitoes from escaping.

Blood Feeding Field-Collected Mosquitoes

22. Allow females to age for 2–5 d. Then, to encourage the females to take a blood meal, remove all sucrose solutions for 24–48 h before blood feeding (3–7 d after adult emergence).
23. During this period, provide a H_2O source (e.g., via wet sponge or cotton) to ensure that males and females do not become dehydrated and die.

    However, as dehydrated female mosquitoes are more likely to bite (Hagan et al. 2018), it might be helpful to remove the H_2O source 2–4 h before blood feeding.
24. Blood feed female mosquitoes (see Introduction: Points to Consider When Establishing and Rearing *Culex* Mosquitoes in the Laboratory [Meuti et al. 2023b] for options or Protocol 2: Rearing and Maintaining a *Culex* Colony in the Laboratory [Meuti et al. 2023a] for specifics).

    As most Culex *mosquitoes are crepuscular, offering a blood meal at dawn or dusk will likely stimulate blood feeding. Additionally, we recommend offering the first blood meal 3 d after emergence. Whenever possible, we also recommend offering a live vertebrate host (Galun 1967). Blood meal analyses are common for a wide range of* Culex *in the United States and Canada, and you should investigate the host preferences of your target species in advance. To establish laboratory colonies of* Cx. pipiens, *we have worked with the Institutional Animal Care and Use Committee and Poultry Sciences at The Ohio State University to develop a protocol to restrain a live hen or rooster in a small cardboard box (Fig. 3).*
25. (Optional) If field-collected *Culex* females do not feed well, offer a blood meal every day or two until mosquitoes blood feed.

Collecting Eggs from Blood-Fed Females

26. Two days after blood feeding, create an infusion (oviposition H_2O) of freshly cut grass clippings and 500 mL of distilled H_2O. Age the infusion by setting it for ~48 h at room temperature or in an incubator/growth chamber set to ~25°C.
27. Four days after blood feeding, place ~200 mL of oviposition H_2O in a plastic deli cup inside the mosquito cage.
28. Monitor oviposition H_2O daily for the presence of egg rafts. Transfer egg rafts to clean plastic deli cups with distilled H_2O.

    Clean H_2O makes it easier to count the larvae.
29. Record the number of egg rafts that were collected as an accurate measure of the genetic diversity of the newly established laboratory strain.

Establish the Colony

30. After the larvae have hatched from the egg rafts, count approximately 200 of the first-generation larvae (F_1) and place them in shoe-box-sized plastic containers with ~450 mL distilled H_2O.

Cite this protocol as *Cold Spring Harb Protoc*; doi:10.1101/pdb.prot108079

31. Follow the procedures in Protocol 2: Rearing and Maintaining a *Culex* Colony in the Laboratory (Meuti et al. 2023a) to rear the larvae to adulthood. Place F_1 pupae inside new cages, and blood feed females as previously described in Step 24.
32. Continue rearing and blood feeding to establish a colony, allowing each new generation of mosquitoes to emerge within a new cage.

Wash Mosquito Containers and Cages

33. After any containers are no longer needed for rearing, wash them with a 3%–5% bleach solution, scrubbing the mosquito pan thoroughly and vigorously.

    Alternatively, soap can be used to clean mesh cages where adult mosquitoes are housed. Soap should be used sparingly, and the containers should be rinsed several times to remove all residue.
34. Rinse the containers at least three times with H_2O, as bleach residue will kill mosquito larvae and pupae.

DISCUSSION

As previously mentioned, establishing a laboratory colony of *Culex* mosquitoes from field-collected individuals is quite difficult. Some species of *Culex* are better suited to laboratory rearing than others, likely because of species-specific differences in stenogamy (the ability to mate in confined spaces) and the likelihood that mosquitoes will blood feed. While *Cx. pipiens* and *Cx. quinquefasciatus* are stenogamous and will mate within laboratory cages (Laven 1951; Sebastian and de Meillon 1967; Kim et al. 2018), *Cx. restuans* will not (Eldridge et al. 1972). To facilitate mating, it may be helpful to allow field-collected mosquitoes to emerge in large cages (>60-cm^3). Furthermore, wrapping the bottom third of the exterior of the cage with black paper or tape and/or lining the half of the bottom of the cage with black paper and the other half with white paper may also facilitate mating by providing markers allowing *Culex* males to know where to swarm within the cage. Sebastian and de Meillon (1967) report that for *Cx. pipiens fatigans* ratios of two males to one female facilitate the greatest mating success in the laboratory and that females typically mate 3–4 d after adult emergence within the laboratory. If mating is a concern, researchers can dissect the spermathecae of female *Culex* to examine them for the presence of sperm (Bellamy and Reeves 1963; Kim et al. 2018).

The other major hurdle to establishing a laboratory population of *Culex* is inducing females to blood feed. We discuss this in Introduction: Points to Consider When Establishing and Rearing *Culex* Mosquitoes in the Laboratory (Meuti et al. 2023b), as well as in Protocol 2: Rearing and Maintaining a *Culex* Colony in the Laboratory (Meuti et al. 2023a). Notably, even if researchers are unable to establish a breeding population of *Culex* mosquitoes, following the protocol that we have detailed here will allow scientists to collect *Culex* from the field and rear at least one generation in the laboratory. In doing so, researchers can gather important biological information about different populations and species of *Culex* mosquitoes. This information includes the developmental rates and responsiveness of *Culex* mosquitoes to various temperatures, photoperiodic conditions, and food sources, as well as their mating behavior under controlled conditions.

RECIPE

Gravid H_2O for Mosquito Larvae Collection

Add 17.3 g of milk protein powder, 3.35 g of baker's yeast, and a large handful of moldy or freshly cut grass clippings to 11.35 L (3 gallons) of tap H_2O. Allow the mixture to ferment for 3 d outside in the summertime or at least 4 d at 27°C.

ACKNOWLEDGMENTS

We thank Robert Aldridge, Josh Benoit, David Denlinger, Leeanne Garrett, Woodbridge Foster, Joe Rinehart, Rebecca Robich, David Shetlar, and Cheolho Sim for helping us to identify locations and trapping techniques to collect mosquitoes and for assistance in developing these protocols. We also thank members of the Meuti Laboratory, past and present, for their diligence in rearing mosquitoes. M.E.M. and M.W. were supported by a National Science Foundation grant (IOS 1944324), and M.E.M. and A.S. were supported by a National Institutes of Health grant (R21-AI144266).

REFERENCES

Allan SA, Kline D. 2004. Evaluation of various attributes of gravid female traps for collection of *Culex* in Florida. *J Vector Ecol* **29:** 285–294.

Bahnck CM, Fonseca DM. 2006. Rapid assay to identify the two genetic forms of *Culex (Culex) pipiens L.* (Diptera: Culicidae) and hybrid populations. *Am J Trop Med Hyg* **75:** 251–255. doi:10.4269/ajtmh.2006.75.2.0750251

Bellamy RE, Reeves WC. 1963. The winter biology of *Culex tarsalis* (Diptera: Culicidae) in Kern County, California. *Ann Entomol Soc Am* **56:** 314–323. doi:10.1093/aesa/56.3.314

Burkett-Cadena ND, Mullen GR. 2008. Comparison of infusions of commercially available garden products for collection of container-breeding mosquitoes. *J Am Mosq Control Assoc* **24:** 236–243. doi:10.2987/5597.1

Crabtree MB, Savage HM, Miller BR. 1995. Development of a species-diagnostic polymerase chain reaction assay for the identification of *Culex* vectors of St. Louis encephalitis virus based on interspecies sequence variation in ribosomal DNA spacers. *Am J Trop Med Hyg* **53:** 105–109. doi:10.4269/ajtmh.1995.53.105

Darsie RF, Ward RA Jr. 2005. *Identification and geographical distribution of the mosquitoes of North America, north of Mexico.* American Mosquito Control Association, Mount Laurel, NJ.

Dodge HR. 1966. Studies on mosquito larvae: II. The first-stage larvae of North American Culicidae and of world Anophelinae. *Can Entomol* **98:** 337–393. doi:10.4039/Ent98337-4

Eldridge BF. 1968. The effect of temperature and photoperiod on blood-feeding and ovarian development in mosquitoes of the *Culex pipiens* complex. *Am J Trop Med Hyg* **17:** 133–140. doi:10.4269/ajtmh.1968.17.133

Eldridge BF, Bailey CL, Johnson MD. 1972. A preliminary study of the seasonal geographic distribution and overwintering of *Culex restuans* Theobald and *Culex salinarius* Coquillet (Diptera: Culicidae). *J Med Entomol* **9:** 233–238. doi:10.1093/jmedent/9.3.233

Gaffigan TV. 2021. Identification key to the genera of mosquito larvae for the world. The Walter Reed Biosystematics Unit, retrieved from: http://www.wrbu.org/keys/WW_GN_L/Mosq_Gen_Worldwide_L.html.

Galun R. 1967. Feeding stimuli and artificial feeding. *Bull World Health Organ* **36:** 590–593.

Gjullin CM, Johnson JO, Jr FWP. 1965. The effect of odors released by various waters on the oviposition sites selected by two species of *Culex. Mosq News* **25:** 268–271.

Gubler DJ. 1971. Studies on the comparative oviposition behavior of *Aedes (Stegomyia) albopictus* and *Aedes (stegomyia) polynesiensis* Marks. *J Med Entomol* **8:** 675–682. doi:10.1093/jmedent/8.6.675

Haeger JS, O'Meara GF. 1983. Separation of first-instar larvae of four Florida *Culex (Culex)*. *Mosq News* **43:** 76–77.

Hagan RW, Didion EM, Rosselot AE, Holmes CJ, Siler SC, Rosendale AJ, Hendershot JM, Elliot KSB, Jennings EC, Nine GA, et al. 2018. Dehydration prompts increased activity and blood feeding by mosquitoes. *Sci Rep* **8:** 6804. doi:10.1038/s41598-018-24893-z

Harrington LC, Poulson RL. 2008. Considerations for accurate identification of adult *Culex restuans* (Diptera: Culicidae) in field studies. *J Med Entomol* **45:** 1–8. doi:10.1093/jmedent/45.1.1

Jackson BT, Paulson SL. 2006. Seasonal abundance of *Culex restuans* and *Culex pipiens* in southwestern Virginia through ovitrapping. *J Am Mosq Control Assoc* **22:** 206–212.

Kim S, Trocke S, Sim C. 2018. Comparative studies of stenogamous behaviour in the mosquito *Culex pipiens* complex. *Med Vet Entomol* **32:** 427–435. doi:10.1111/mve.12309

Kramer WL, Mulla MS. 1979. Oviposition attractants and repellents of mosquitoes: oviposition responses of *Culex* mosquitoes to organic infusions. *Environ Entomol* **8:** 1111–1117. doi:10.1093/ee/8.6.1111

Laven H. 1951. Crossing experiments with *Culex* strains. *Evolution* **5:** 370–375. doi:10.1111/j.1558-5646.1951.tb02795.x

Michener CD. 1947. Mosquitoes of a limited area in Southern Mississippi. *Am Midl Nat* **37:** 325. doi:10.2307/2421659

Miller N, Mettelman RC, Baker SC, Harbison JE. 2015. Use of Centers for Disease Control and Prevention gravid trap in catch basins: proof-of-concept trials. *J Am Mosq Control Assoc* **31:** 196–199. doi:10.2987/14-6470R

Meuti ME, Siperstein A, Wolkoff M. 2023a. Rearing and maintaining a *Culex* colony in the laboratory. *Cold Spring Harb Protoc* doi:10.1101/pdb.prot108080

Meuti ME, Siperstein A, Wolkoff M. 2023b. Points to consider when establishing and rearing *Culex* mosquitoes in the laboratory. *Cold Spring Harb Protoc* doi:10.1101/pdb.top107823

Popko DA, Walton WE. 2016. Large-volume gravid traps enhance collection of *Culex* vectors. *J Am Mosq Control Assoc* **32:** 91–102. doi:10.2987/moco-32-02-91-102.1

Sebastian A, de Meillon B. 1967. Experiments on the mating of *Culex pipiens fatigans* in the laboratory. *Bull World Health Organ* **36:** 47.

Smith JL, Fonseca DM. 2004. Rapid assays for identification of members of the *Culex (Culex) pipiens* complex, their hybrids, and other sibling species (Diptera: Culicidae). *Am J Trop Med Hyg* **70:** 339–345. doi:10.4269/ajtmh.2004.70.339

Spielman A, Wong J. 1973. Studies on autogeny in natural populations of *Culex pipiens.* 3. Midsummer preparation for hibernation in anautogenous populations. *J Med Entomol* **10:** 319–324. doi:10.1093/jmedent/10.4.319

Zulueta JD. 1950. Comparative oviposition experiments with caged mosquitoes. *Am J Epidemiol* **52:** 133–142. doi:10.1093/oxfordjournals.aje.a119413

Protocol 2

Rearing and Maintaining a *Culex* Colony in the Laboratory

Megan E. Meuti,[1] Alden Siperstein, and Matthew Wolkoff

Entomology Department, The Ohio State University, Columbus, Ohio 43210, USA

After overcoming the significant obstacle of getting adult *Culex* mosquitoes to reproduce and blood feed in the laboratory, maintaining a laboratory colony is much more achievable. However, great care and attention to detail are still required to ensure that the larvae have adequate food without being overwhelmed by bacterial growth. Additionally, achieving the appropriate densities of larvae and pupae is essential, as overcrowding delays development, prevents pupae from successfully emerging as adults, and/or reduces adult fecundity and alters sex ratios. Finally, adult mosquitoes should have constant access to H_2O and nearly constant access to sugar sources to ensure that both males and females have adequate nutrition and can produce the maximum number of offspring. Here, we describe our methods for maintaining the Buckeye strain of *Culex pipiens* and how other researchers might modify them to suit their specific needs.

MATERIALS

It is essential that you consult the appropriate Material Safety Data Sheets and your institution's Environmental Health and Safety Office for proper handling of equipment and hazardous material used in this protocol.

RECIPES: Please see the end of this protocol for recipes indicated by <R>. Additional recipes can be found online at http://cshprotocols.cshlp.org/site/recipes.

Reagents

Adenosine Triphosphate (ATP) Solution for Mosquito Feeding (0.1 M) <R>

The ATP helps make the blood seem "fresh" to the mosquitoes and encourages them to feed to repletion (Galun 1967).

Bleach

Blood source

The preferred source is commercially purchased chicken blood treated with an anticoagulant (e.g., sodium citrate–treated chicken blood from Pel Freeze or Lampire Biologicals). Other options are human volunteers with appropriate institutional review board and biosafety protocols in place and live chicken or other avian host if proper Institutional Animal Care and Use Committee and biosafety protocols are in place.

Dish soap (optional, see Step 8; e.g., Dawn or Palmolive dish soap)

Grass (e.g., Kentucky bluegrass, crabgrass, or other local grass)

H_2O (distilled and hot tap)

Reverse osmosis H_2O can also be used.

[1]Correspondence: meuti.1@osu.edu

Cite this protocol as *Cold Spring Harb Protoc*; doi:10.1101/pdb.prot108080

Larval food (Tetramin tropical fish flakes that are ground with a coffee grinder)

Rodent or puppy chow can also be used. Additionally, if you need to promote autogenous egg development (e.g., laying the first batch without a blood meal), such as for Cx. pipiens molestus, *protein-rich food sources like a slurry of beef liver powder (2.53% wt/vol solution) or ground brine shrimp should be used to feed mosquito larvae (Gao et al. 2019).*

Mosquitoes (plastic deli-cup container where egg rafts have been placed and allowed to hatch into first-instar larvae)

Sugar source (e.g., raisins, honey, and/or 10% sucrose)

Equipment

Cotton, braided dental rolls (e.g., Pearson, ~16.2-cm) or filter paper (e.g., Whatman Filter Paper, 42.5-mm circle); see step 19

Covered plastic or glass container (e.g., GladWare home mini round food storage containers, 4-oz or 118-mL; see Step 19)

A hole, ~1-cm-diameter, can be drilled through the lid to allow filter paper or a cotton dental wick to fit through.

Environmental chamber

A room where lights are on a photoperiod timer and that is temperature- and/or humidity-controlled can also be used. To stimulate mosquito growth and reproduction, we recommend setting the chamber/room to long-day conditions (>14 h light/d) and temperatures of 24°C–27°C. Culex *mosquitoes will develop at temperatures as low as 15°C, but development will be slower. If you want to induce diapause/overwintering dormancy,* Culex *mosquitoes should be exposed to photoperiods of <12 h light/d and temperatures of <21°C. Both reproductively active and diapausing* Culex *mosquitoes prefer humid conditions (relative humidity of >70%). This can be achieved by using a humidifier and/or placing moist sponges on top of the mosquito cages.*

Graduated plastic cylinder (1-L)

Hemotek artificial blood feeding system

See Introduction: Points to Consider When Establishing and Rearing Culex *Mosquitoes in the Laboratory (Meuti et al. 2023) for options if the strain you are working with will not feed on a Hemotek system.*

Kitchen scrubbing brush (e.g., Vikan Dish Brush)
Labeling tape
Large black trash bag
Manual counter
Marker
Mesh cages (31-cm^3 or larger recommended)
Metal spatula or white plastic spoon (see Step 30)
Microplate sieve with openings <300 µm (e.g., Endecotts)
Paper towels
Parafilm
Petri dishes
Plastic bag (optional, see Step 18)
Plastic deli cups (e.g., diameter of 11.4 cm; height of 4.4 cm; holding volume of 250–500 mL)
Plastic, disposable transfer pipette (e.g., 3-mL; Fisher Scientific; labeled with strain name)
Plastic shoe-box-sized containers (e.g., Sterlite, 11 in. × 6.6 in. × 2.7 in)
Scissors
Sponges
Styrofoam (optional, see Step 17)
White plastic container or a clear plastic container with a white paper towel or piece of paper underneath it

Cite this protocol as *Cold Spring Harb Protoc*; doi:10.1101/pdb.prot108080

METHOD

Rearing *Culex* Larvae to Adulthood

1. Use an appropriately labeled plastic transfer pipette to draw up newly hatched, first-instar larvae from the containers (day 1 of larval life; see Table 1).
2. Pipette the larvae out in small droplets onto a white plastic container, wrapping the droplets around the pan. Ideally, place fewer than 10 larvae per droplet to facilitate easy counting.
3. Use the manual counter to count every mosquito larva in each droplet and continue to count until approximately 220 larvae are in the white plastic container.
4. Using the plastic 1-L graduated cylinder, measure out 450 mL of distilled H_2O.
5. Label a plastic, shoe-box sized container (hereafter "larval rearing pan") with label tape.

   We recommend that labels include the initials of the researcher, the species, strain, cohort/lab generation, rearing conditions, purpose of the mosquitoes (e.g., general colony maintenance; title of the relevant experiment), and the date.
6. Tilt the white plastic container with the mosquito larvae over a larval rearing pan and use the distilled H_2O to transfer all the larvae into the rearing pan.
7. Place ~50 mg of larval food into the rearing pan with the larvae.
8. Repeat until the desired number of pans of larvae have been made. If any larvae remain, kill the larvae by adding bleach (~10% of the total volume) or dish soap.

   Typically, we recommend creating 12 pans of Culex *mosquitoes for each strain and rearing condition (approximately 2640 larvae in total).*
9. Place the containers of mosquito larvae into an environmental chamber that is set to the appropriate conditions (see Introduction: Points to Consider When Establishing and Rearing *Culex* Mosquitoes in the Laboratory [Meuti et al. 2023]).
10. On day 2, add 100 mg of larval food to each of the larval pans.
11. On day 3, monitor the larvae for growth and development. Remove any dead larvae from the pan with a disposable transfer pipette and discard.
12. On day 4, add 250 mg of larval food to each of the larval pans.
13. On day 5, monitor the pans for bacterial growth and filter if necessary.

    Pans should be filtered if there is excessive bacterial growth (e.g., the mats or clouds of bacteria are evident and the H_2O becomes opaque), if the H_2O is highly pungent, and/or if more than 20 dead, white larvae are present on the bottom of the pan.

    i. To filter the larvae, first use the plastic transfer pipette to remove and discard any dead larvae.

    ii. Then, prepare a clean larval pan, measure 400 mL of distilled H_2O, and place the microplate sieve in the sink.

    iii. Carefully pour the pan containing the larvae through the sieve.

    iv. Quickly and gently rinse the larvae with distilled tap H_2O. Then, transfer the larvae into the clean container using the 400 mL of clean distilled H_2O.

    v. If the pan was filtered, add 300 mg of larval food after transferring larvae to their new pan, move the label to the new plan, and add a note of when the pan was filtered.
14. On day 6, monitor the pans for bacterial growth and filter if necessary, using the instructions detailed in Step 13.
15. On day 7, add 350 mg of larval food to each of the larval pans.
16. On day 8, monitor the pans for bacterial growth and filter if necessary, using the instructions mentioned in Step 13.

TABLE 1. Feeding schedule for rearing *Culex* larvae at 24°C–27°C

Pan (50 mg food)					100 mg food		No feed	250 mg food		No feed	No feed	350 mg food		No feed	Pick 1		Pick 2		Cage Info
Date	Number of pans	Initial	Purpose (main colony/apt ID)	Cohort	Date	Initial	Date	Date	Initial	Date	Date	Date	Initial	Date	Date	Initial	Date	Initial	Emergence date/notes
7.25.22	12	AS	Main colony	F42	7.26	MW	7.27	7.28	AS	7.29	7.30	7.31	MEM	8.01	8.02	MEM	8.03	MW	Emerged 08.02
8.04.22	6	MW	FlyBox	F43	8.05	MW	8.06	8.07	MEM	8.08	8.09	8.10	MW	8.11	8.12	AS	8.13	MEM	Placed into small cages

The top row indicates how much food should be given or when mosquito pupae should be picked (Pick 1 and 2). Labels for each column represent other important information that should be recorded starting when the pans of larvae are established and each subsequent day thereafter. Researchers use their initials to indicate who fed the mosquito larvae and when, allowing multiple people to easily and accurately care for laboratory colonies.

Cite this protocol as *Cold Spring Harb Protoc*; doi:10.1101/pdb.prot108080

17. On day 9, monitor the pans to see if any pupae have developed. If so, collect the pupae.
    i. With the label tape and marker, label plastic deli cups with the species, strain, your initials, and date. Fill with ~1″ of the larval rearing H_2O to avoid osmotic shock.
    ii. Use an appropriately labeled plastic transfer pipette to suck up the pupae and place them into the prepared container.
    iii. Once the cup has 150 pupae, start placing them into a new labeled plastic deli-cup container to avoid overcrowding. If desired, add a small piece of Styrofoam (~2.5-cm-diameter) to the cup to provide newly emerged adults a place to climb off the H_2O surface.
18. Place the container into an appropriately labeled cage (e.g., your initials, species, strain, cohort, and the next day's date through the end of adult emergence).

    If the cage is in a non-humidity-controlled environment, place a wet sponge on top of the cage and cover the cage with a plastic bag. This retains high humidity in the cage, which enhances adult mosquito survival.
19. Provision the cage with a H_2O source and food (e.g., 10% sucrose solution, raisins, and/or honey-soaked sponges).

    H_2O and sucrose solutions can be provided to mosquitoes by placing them in covered plastic or glass containers and allowing either filter paper or cotton to wick the solution.
20. On day 10, transfer any pupae from the larval rearing containers into a new labeled, plastic deli cup. Place the deli cup with the pupae into the adult mosquito cage.
21. On day 11, transfer any remaining pupae into a new open container. Place the pupal cup into a previously established adult mosquito cage.

    At this point, fewer than 10% of the pupae should remain. We recommend killing any remaining larvae to synchronize mosquito development, which is useful for subsequent experiments. To kill remaining mosquito larvae, first consolidate them by filtering all the H_2O in the larval pans through the microplate sieve. Then use a small amount of hot tap H_2O to transfer the larvae into a deli cup container and then add an equal amount of bleach to create a 50% final solution.

Blood Feeding Adult *Culex* Females

Anautogenous female Culex *mosquitoes need to be blood-fed so that they can produce eggs and allow the laboratory strain to persist. Only long-day-reared, nondiapausing females will blood feed. For established colonies, we recommend that cages containing adult females that were reared at 24°C–27°C be fed 5–10 d after adult emergence, whereas long-day mosquitoes that are reared at 18°C–21°C should blood feed 14–20 d after adult emergence.*

22. One to two days before blood feeding, remove the sugar sources from the cages. Ensure that the mosquitoes have access to regular H_2O and wet sponges if they are not in a humidity-controlled environment. Remove all plastic deli cups from the cages to prevent mosquitoes from drowning and/or laying eggs prematurely.

    Removing the sugar sources makes it more likely that female mosquitoes will be hungry and ready to take a blood meal.
23. Prepare the Hemotek artificial membrane system to blood feed females.
    i. If blood was previously frozen, thaw it overnight at 4°C or for ~4 h at room temperature.
    ii. With scissors, cut squares of Parafilm for every heating unit of the Hemotek that you plan to use. Rub the squares of Parafilm against human skin (stinky feet work best) so that they pick up human odors, allowing female mosquitoes to find the blood source.
    iii. Gently stretch the Parafilm so that it forms an ~6″ thin square without tearing it.

        If any holes or tears are present, start with a new piece of Parafilm.
    iv. Place one piece of stretched Parafilm over a metal blood feeding cylinder such that the side of the Parafilm that has been rubbed against human skin is facing upward. Secure the Parafilm in place by wrapping it tightly around the metal cylinder and then sliding a black O-ring into the groove. Repeat these steps for each feeding cylinders. Place the cylinders in a Petri dish

such that the Parafilm side is down and roll up any excess Parafilm that has contacted the bottom of the metal blood feeding cylinder.

v. Add 200 µL of a 0.1 M ATP solution to 15 mL of chicken blood using an adjustable pipette or plastic transfer pipette, and gently mix the blood to distribute the ATP.

vi. Using a disposable plastic pipette, transfer the chicken blood to the Hemotek blood-feeding cylinder. Slowly pipette the blood in, being careful to avoid bubbles. Lift the cylinder up to make sure that a layer of blood completely or almost completely covers the bottom of the Parafilm.

vii. Without setting the cylinder on the bench (still hold it with one hand in the air), place plastic plugs in the holes, and screw the blood feeding cylinder into the heating units. Place the heating unit on a Petri dish.

viii. Repeat this process with all blood-feeding cylinders until they are filled with blood and turn on the blood feeding unit.

ix. Place the heaters with the blood on top of each cage containing *Culex* females that are ready to blood feed. Cover the cage and heaters with a large black trash bag to simulate evening. Then exhale into the large mosquito cage to raise the CO_2 levels and stimulate blood feeding.

As Culex *mosquitoes are crepuscular, it may be best to feed at dawn or dusk. If possible, it is likely best to feed the cages within the incubator in which they are reared. If mosquitoes do not feed well, it might be necessary to move the heating units inside the cages, ensuring that the cloth sleeves of the mosquito cage are secured around the electrical cord of the Hemotek so that mosquitoes do not escape. It might also be necessary to try moving the cages to warmer or cooler rooms, and/or removing H_2O sources. If available, dry ice can also be placed near the cages to increase CO_2 levels, and/or dirty and stinky socks can be placed near the feeding units to further stimulate feeding.*

24. Leave the Hemotek heating units on each cage for at least 2 h and up to 14 h.

The heating units can be transferred to other cages as well.

25. Disassemble the Hemotek blood feeder.

i. After the cages have had access to the blood for an appropriate amount of time, turn off the artificial blood feeder. Transfer the metal blood feeding cylinders to Petri dishes.

ii. Disassemble the blood feeding units. Use the gray-handled screw tool (provided with the Hemotek artificial blood feeding system) to remove the plastic plugs. Pipette the old blood back into the chicken blood container or use paper towels to manually remove as much congealed blood from the feeding cylinder as possible and dispose of the old blood and used Parafilm according to approved biosafety protocols.

iii. Place the plugs, cylinders, and the gray screw tool in a container with a 10% bleach solution for 5–10 min. Scrub them with a brush until they are clean, rinse them with hot tap H_2O, and allow them to air dry.

Leaving the blood feeding cylinders or other materials in bleach for long periods of time can damage them. Therefore, they should be soaked in bleach for no more than 15 min and rinsed thoroughly.

iv. Wipe down the bench, Petri dishes, and heating units of the blood feeder with paper towels and a 10% bleach solution.

26. Provision each cage with a sugar solution.

At this stage, we provision our cages with 10% sucrose, raisins, and honey-soaked sponges to help promote egg laying.

27. Monitor the cages daily to ensure that mosquitoes have access to both sugar and H_2O and that mosquitoes are in a humid environment.

If females do not have these resources, they will fail to produce egg rafts or only produce very small ones (e.g., containing only approximately 50 eggs instead of approximately 200 eggs).

28. Two days after blood feeding, make oviposition H_2O.

Cite this protocol as *Cold Spring Harb Protoc*; doi:10.1101/pdb.prot108080

i. Use scissors to cut off a handful of grass tips.

ii. Place the freshly cut grass into a 1-L plastic graduated cylinder and fill the container with 500 mL of distilled H_2O.

Freshly cut grass is best. Avoid placing soil in the oviposition H_2O, as it can contaminate the H_2O with microbes. Also, avoid using grass that may have been treated with fertilizers or pesticides.

iii. Allow the grass and H_2O mixture to rest for ~48 h at room temperature.

29. Four days after blood feeding, place the oviposition H_2O into the cages that have been blood-fed.

i. Label a deli container with the strain, species, date, and cohort.

ii. Pour the oviposition H_2O into the labeled container such that the oviposition H_2O is 5- to 8-cm-deep (~200-mL) and some grass clippings are present.

The grass clippings may give female mosquitoes a surface to land and stand on while ovipositing.

iii. Place the container inside the appropriate cage.

30. Five days after blood feeding, collect *Culex* egg rafts.

i. Label a new clean deli cup and fill it with 2–4 cm of distilled H_2O.

ii. Using a metal spatula or white plastic spoon, carefully scoop out egg rafts and transfer them to the prepared container.

iii. Move the oviposition H_2O back into each cage as it is likely that more females will lay egg rafts the following day.

31. Once the larvae hatch, begin at Step 1 to establish new pans of mosquito larvae.

DISCUSSION

Using this protocol, researchers can continuously maintain *Culex* species in the laboratory. Depending on the needs and interests of the researcher as well as the source population of *Culex* mosquitoes, the photoperiodic and temperature conditions can be adjusted. For example, populations collected from higher latitudes will likely fare better in the laboratory if they are exposed to longer photoperiods (>14 h/light day) and lower temperatures (24°C) compared with mosquitoes that are collected from tropical and subtropical locations.

Researchers can induce diapause in temperate *Culex* mosquitoes by exposing mosquitoes to short photoperiods and lower temperatures (Sanburg and Larsen 1973; Spielman and Wong 1973). If researchers are interested in the differences between diapausing and nondiapausing mosquitoes, we strongly recommend rearing mosquitoes under the same low temperature (18°C or 20°C) and using different photoperiods (Meuti et al. 2015a,b; Fyie et al. 2021). This is because diapause induction is primarily driven by differences in photoperiod, and therefore a high proportion of nondiapausing and reproductively active mosquitoes can be obtained even when mosquitoes are reared at low temperatures (Meuti et al. 2015a,b). Moreover, as mosquitoes are ectothermic organisms, holding temperature constant allows researchers to accurately characterize physiological differences between diapausing and nondiapausing mosquitoes.

Rearing mosquitoes at lower temperatures and/or allowing mosquitoes to enter diapause may be favorable, as this allows researchers to maintain mosquitoes in the laboratory for 2–5 mo with little active care; after adult mosquitoes enter diapause, their food sources can be removed, and they will survive for several months if they are held under humid conditions and have access to H_2O (Meuti et al. 2015a). Diapause can be terminated in the laboratory by exposing mosquitoes to higher temperatures (24°C–27°C) and longer photoperiods (>13 h light/d) for 1–2 wk. Postdiapause females can then take a blood meal and produce the next generation of mosquito offspring. If researchers plan to study diapause long term, regularly allowing mosquitoes to complete diapause and reproduce after diapause termination will ensure that the diapause incidence of the laboratory colony remains high.

RECIPE

Adenosine Triphosphate (ATP) Solution for Mosquito Feeding (0.1 M)

Dissolve 551 mg of adenosine 5′-triphosphate disodium salt hydrate (Sigma-Aldrich) in 8 mL of distilled H_2O. Adjust the pH to 7 using 1 M potassium hydroxide (VWR) and raise the final volume to 10 mL. Store 200 µL aliquots of the solution for up to 1 yr at −20°C.

ACKNOWLEDGMENTS

We thank Robert Aldridge, Josh Benoit, David Denlinger, Megan Fritz, Leeanne Garrett, Woodbridge Foster, Joe Rinehart, Rebecca Robich, David Shetlar, and Cheolho Sim for helping us to identify locations and trapping techniques to collect mosquitoes and for assistance in developing these protocols. We also thank members of the Meuti Laboratory, past and present, for their diligence in rearing mosquitoes. M.E.M. and M.W. were supported by a National Science Foundation grant (IOS 1944324), and M.E.M. and A.S. were supported by a National Institutes of Health grant (R21-AI144266).

REFERENCES

Fyie LR, Gardiner MM, Meuti ME. 2021. Artificial light at night alters the seasonal responses of biting mosquitoes. *J Insect Physiol* **129:** 104194. doi:10.1016/j.jinsphys.2021.104194

Galun R. 1967. Feeding stimuli and artificial feeding. *Bull World Health Organ* **36:** 590–593.

Gao Q, Su F, Zhou Y-B, Chu W, Cao H, Song L-L, Zhou J-J, Leng P-E. 2019. Autogeny, fecundity, and other life history traits of *Culex pipiens molestus* (Diptera: Culicidae) in Shanghai, China. *J Med Entomol* **56:** 656–664. doi:10.1093/jme/tjy228

Meuti ME, Stone M, Ikeno T, Denlinger DL. 2015a. Functional circadian clock genes are essential for the overwintering diapause of the Northern house mosquito, *Culex pipiens*. *J Exp Biol* **218:** 412–422. doi:10.1242/jeb.113233

Meuti ME, Short CA, Denlinger DL. 2015b. Mom matters: diapause characteristics of *Culex pipiens–Culex quinquefasciatus* (Diptera: Culicidae) hybrid mosquitoes. *J Med Entomol* **52:** 131–137. doi:10.1093/jme/tju016

Meuti ME, Siperstein A, Wolkoff M. 2023. Points to consider when establishing and rearing *Culex* mosquitoes in the laboratory. *Cold Spring Harb Protoc* doi:10.1101/pdb.top107823

Sanburg LL, Larsen JR. 1973. Effect of photoperiod and temperature on ovarian development in *Culex pipiens pipiens*. *J Insect Physiol* **19:** 1173–1190. doi:10.1016/0022-1910(73)90202-3

Spielman A, Wong J. 1973. Studies on autogeny in natural populations of *Culex pipiens*. 3. Midsummer preparation for hibernation in anautogenous populations. *J Med Entomol* **10:** 319–324. doi:10.1093/jmedent/10.4.319

Cite this protocol as *Cold Spring Harb Protoc*; doi:10.1101/pdb.prot108080

CHAPTER 4

Establishing Colonies from Field-Collected Mosquitoes: Special Accommodations for Wild Strains

Noah H. Rose,[1,2,6] John J. Shepard,[3,6] and Diego Ayala[4,5,6]

[1]*Department of Ecology and Evolutionary Biology,* [2]*Princeton Neuroscience Institute, Princeton University, Princeton, New Jersey 08544, USA;* [3]*Department of Entomology and Center for Vector Biology and Zoonotic Diseases, The Connecticut Agricultural Experiment Station, New Haven, Connecticut 06511, USA;* [4]*MIVEGEC, Univ. Montpellier, CNRS, IRD, Montpellier BP 64501, 34394, France;* [5]*Medical Entomology Unit, Institut Pasteur de Madagascar, Antananarivo BP1274, 101, Madagascar*

A researcher may have many reasons for wanting to establish new laboratory colonies from field-collected mosquitoes. In particular, the ability to study the diversity found within and among natural populations in a controlled laboratory environment opens up a wide range of possibilities for understanding how and why burdens of vector-borne disease vary over space and time. However, field-collected mosquitoes are often more difficult to work with than established laboratory strains, and considerable logistical challenges are involved in safely transporting field-collected mosquitoes into the laboratory. Here, we provide advice for researchers working with *Aedes aegypti*, *Anopheles gambiae*, and *Culex pipiens*, as well as notes on other closely related species. We provide guidance on each stage of the life cycle and highlight the life stages for which it is easiest to initiate new laboratory colonies for each species. In accompanying protocols, we provide methods detailing *Ae. aegypti* egg collection and hatching as well as how to transport larvae and pupae from the field.

BACKGROUND

Mosquitoes show high diversity within species or species complexes, and populations often differ in key vectorial traits, such as host preference, immunity, local adaptation, and physiological changes (Coluzzi 1984; Fonseca et al. 2004; Kilpatrick et al. 2007; Riehle et al. 2011; Ayala et al. 2019; Aubry et al. 2020; Rose et al. 2020). To better understand why and how a mosquito becomes a vector of human diseases, it is necessary to dissect these key vectorial traits in a controlled environment. To this end, laboratory mosquito colonies established from field collections are a powerful tool. However, these efforts are slightly paradoxical because as soon as mosquitoes arrive in an insectary, they inevitably begin to adapt to the laboratory environment, potentially affecting the traits that should be studied. This "domestication" process must be considered to ensure that experimental protocols and results are indicative of field conditions (Ross et al. 2019). For this reason, early-generation laboratory colonies may be more representative of populations in the field than older, more strongly laboratory-adapted colonies. Furthermore, older laboratory strains frequently show strong signatures of accidental cross-breeding with other laboratory strains and may be entirely unrepresentative of the populations from which they were supposedly derived (Gloria-Soria et al. 2019). However, the lower

[6]Correspondence: noahr@princeton.edu; John.shepard@ct.gov; diego.ayala@ird.fr

Cite this introduction as *Cold Spring Harb Protoc*; doi:10.1101/pdb.top107654

genetic polymorphism of these strains can be an advantage, as they can be used to establish a reference genome, for example (Holt et al. 2002; Matthews et al. 2018). Altogether, these and other aspects need to be taken into account when deciding when to work with laboratory-based mosquito colonies.

Despite these challenges, with patience and careful experimental design, laboratory experiments using diverse field-collected mosquito colonies have great potential to illuminate the drivers of geographical variation in the spread of mosquito-borne disease (Fritz et al. 2008; Ritchie et al. 2014; Ferguson 2018; Ayala et al. 2019; Brady and Hay 2020; Rose et al. 2020). Such experiments can clarify the molecular mechanisms of pathogen transmission (Aubry et al. 2020), the origin and evolution of insecticide resistance in natural populations (Hemingway et al. 2016), or the underlying genetic signatures of local adaptation (Cheng et al. 2018; Ayala et al. 2019). These studies, in turn, can inform a wide variety of strategies to reduce these burdens (Coluzzi 1984; Farajollahi et al. 2011; Bowman et al. 2016; Williams et al. 2018).

Because individual aspects of laboratory maintenance and the study of mosquitoes are covered elsewhere, we focus here on aspects that are specific to collecting fresh materials and establishing new laboratory colonies. We provide guidance for *Aedes aegypti*, *Anopheles gambiae*, and *Culex pipiens*. For each species, we discuss each stage of the life cycle, noting different possible starting points for collecting individuals from the field. In addition, we provide notes about working with close relatives of these species when possible.

AEDES AEGYPTI

Ae. aegypti comprises both an ancestral generalist form (often termed *Ae. aegypti formosus*) and a recently evolved human specialist (often termed *Ae. aegypti aegypti*) form that gave rise to all invasive populations outside its native African range (Mattingly 1957; Powell et al. 2018; Soghigian et al. 2020). However, in the *Ae. aegypti* native range in sub-Saharan Africa, many populations exist on an admixed spectrum between these two extremes (Rose et al. 2020). Care must be taken not to bias collections toward human specialists because their preference for human hosts and habitats makes them comparatively easier to collect and maintain.

Eggs

Field-collected eggs are by far the easiest and most reliable starting point for laboratory colonies of *Ae. aegypti*. The field collections themselves are relatively straightforward and take only a few days to yield a robust sample of local diversity (see Protocol 1: Collecting, Storing, and Hatching *Aedes aegypti* Eggs [Rose et al. 2023]), making it possible to sample animals from many different locations in a reasonable amount of time. An alternative approach is to directly collect dry natural breeding sites and transport them back to the laboratory (Sylla et al. 2013). The ovicups (small plastic cups lined with seed-germination paper and filled with water) described in Protocol 1: Collecting, Storing, and Hatching *Aedes aegypti* Eggs (Rose et al. 2023) serve as an attractive oviposition substrate to both generalist and human-specialist populations of *Ae. aegypti* and can even be used to recover eggs of closely related forest-dwelling species (e.g., *Aedes africanus*) if used in a forest environment. Once collected and dried, *Ae. aegypti* eggs are quite hardy and can be easily transported in airtight plastic bags by mail or in hand-carried luggage with proper permits.

When collecting *Ae. aegypti* eggs for laboratory colonies, it is important to consider the oviposition and flight behavior of this species. *Ae. aegypti* is not typically inclined to fly great distances (rarely more than few hundred meters) and typically lays several eggs in a suitable oviposition site, although they will sometimes allocate eggs across few different sites (Colton et al. 2003). Accordingly, eggs from a single ovicup typically contain many siblings. Conversely, eggs from different ovicups, spaced >100 m apart, rarely contain siblings (Rose et al. 2020). When establishing a new laboratory colony, it is ideal to include equal numbers of individuals (e.g., two to five females) from 10 or more ovicups spaced

Cite this introduction as *Cold Spring Harb Protoc*; doi:10.1101/pdb.top107654

>100 m apart. Conversely, attempting to establish a laboratory colony from many individuals from a single ovicup may result in extensive inbreeding and an unhealthy colony.

Hatching

Although *Ae. aegypti* eggs are stable for long periods of time (up to 6 mo) when dry, they readily hatch when exposed to a deoxygenated nutrient-rich hatch broth (see Protocol 1: Collecting, Storing, and Hatching *Aedes aegypti* Eggs [Rose et al. 2023]). However, field-collected eggs are often more resistant to hatching than those collected from laboratory strains, and hatching may be asynchronous over a few days. Therefore, it is important to keep egg papers submerged and to watch carefully for late-hatching larvae to avoid biasing laboratory colonies against them.

Larvae

Field-collected larvae can serve as a viable starting point for laboratory colonies. One advantage of starting with field-collected larvae is that they come from natural breeding sites and might be less biased against populations or individuals that do not favor ovicups. However, in our experience, ovicups and larval collections yield colonies with similar characteristics. In addition, rearing larvae in the field is time-consuming and transporting healthy larvae back to a laboratory environment can be difficult. Nonetheless, larval collections can play an important role in understanding the ecology of local mosquito populations (i.e., physicochemical parameters associated with breeding sites; see Xia et al. 2021) and are also important for species other than *Ae. aegypti* with less easy-to-collect eggs (see the sections *Anopheles gambiae* Complex and *Culex pipiens*). Rearing field-collected larvae is very similar to rearing larvae in the laboratory, except that in the field one must consider predators, including predatory *Culex* species and *Toxorhynchites* that are particularly common in forest environments. Although larvae can be difficult to identify down to the species level, by learning to identify *Aedes* larvae at least by their swimming behavior (wiggling, forming S-like shapes) and shorter siphons, one can exclude non-*Aedes* potential predators and avoid waking up 1 d to find that the research material has been eaten.

Laboratory rearing larvae from field-collected eggs are not very different from rearing larvae from laboratory colonies. The main difficulty is that hatching is often asynchronous and densities are often lower than in laboratory colonies. Therefore, it is crucial to avoid overfeeding larvae (which can lead to bacterial growth). In addition, larvae may pupate and eclose over several days, requiring more attention than laboratory colonies. It is also important to avoid larvae overpopulation in order to obtain healthy adults.

Adults

Because eggs and larvae may be readily obtained, researchers may neglect the collection of adults as the starting point for a new colony. However, a few advantages to collecting adults can be noted. First, by using adults, we can easily control the number of distinct individuals that started the colony. Second, field-collected adult females have generally already mated, which therefore bypasses the sensitive step of mating under artificial conditions. However, ensuring oviposition is a critical point for obtaining the next generation. Classical capturing methods, such as human-landing catch or animal traps, can introduce bias in the behavioral diversity of the new colony. Other methods such as indoor-resting collections can be challenging and fruitless despite an exhaustive effort. New methods based on traps that maintain live adult male and female mosquitoes have been recently developed and provide a promising alternative (L'Ambert et al. 2021). Such methods, aimed at increasing the longevity of collected adults for virus detection, can be easily adapted to recover live mosquitoes for starting a colony.

Blood Feeding

Many possible approaches are available for blood feeding field-collected *Ae. aegypti* females, and the best approach is likely to be determined by logistical constraints and methods approved for use in a particular research environment. Human volunteers, membrane feeding, or animal feeding can all work. Whatever approach is used, the most important thing is to avoid imposing strong selection on the subset of field-collected individuals that are most willing to accept the method used because it could result in colonies that are not representative of the study location where they were collected. The only remedy for this potential issue is patience. Females that are given access only to water (and not sucrose) for 24 h will generally eventually blood feed even if they are initially reticent. However, no method is perfect, and the selection imposed by laboratory feeding is a strong reason to use early-generation colonies in research projects that investigate diversity among mosquito populations.

Oviposition

Collecting eggs from field-derived females is another step in which it is important, but difficult, to avoid imposing strong selection. Patience is again important. Typically, *Ae. aegypti* females will readily lay on filter-paper cones resting in 5 mL of soil-infused water at the bottom of a *Drosophila* vial ("ovivials") at 5–7 d after blood feeding, but field-derived females may not immediately lay eggs. When establishing a new laboratory colony, it is useful to put each female founder of a new colony in its own vial, covering each rack of vials with a black cloth. For each female, eggs should be dried in a pan lined with paper towels 48 h after oviposition. Eggs dried prematurely may die, and eggs left wet too long may hatch prematurely. Handling each female separately allows one to collect eggs despite variation in oviposition date.

Notes on Other *Aedes* Species

The same general approach can often be used for other *Aedes* species; at a minimum, researchers have reported success with *Aedes albopictus*, *Aedes bromeliae*, *Aedes vittatus*, and *Aedes polynesiensis* (Vega-Rúa et al. 2014; Richard et al. 2016; Mulwa et al. 2018).

ANOPHELES GAMBIAE COMPLEX

An. gambiae belongs to a complex of nine morphologically identical species (Coetzee et al. 2013; Barrón et al. 2019). The complex has received special attention because of its prominent role in malaria transmission across Africa. This complex includes *An. gambiae*, *Anopheles coluzzii*, and *Anopheles arabiensis* (i.e., three major malaria vectors) and other members that show a local or no role in malaria transmission. Moreover, the members of this complex are ecologically divided into fresh- and salt-water-tolerant species (White et al. 2011). Their geographical distribution is determined by this and other ecological features. *An. gambiae*, *An. coluzzii*, and *An. arabiensis* have colonized large parts of sub-Saharan Africa, whereas other members are limited to coastal zones (*Anopheles melas* and *Anopheles merus*) or to some specific forested areas (e.g., *Anopheles bwambae* and *Anopheles fontenillei*). Because of the epidemiological interest of these species, a large effort has been devoted to establishing laboratory colonies. Several species of this complex such as *An. gambiae*, *An. coluzzii*, *An. arabiensis*, *An. merus*, and *Anopheles quadriannulatus* breed in fresh water, lay eggs in filter water paper, feed from artificial feeders, and can be kept under insectarium conditions. They can be freely ordered at BEI Resources (https://www.beiresources.org/).

In most places, several members of the complex are in sympatry. Therefore, before starting a field-based colony, each specimen must be carefully identified. Within the complex, reproductive isolation is not complete. *An. gambiae* and *An. coluzzii* can produce full-fertile offspring, whereas among the other members of the complex, there are partial postzygotic isolation barriers (White et al. 2011). These isolation barriers can lead to hybrid colonies (e.g., in the case of mixing *An. gambiae* and

 Cite this introduction as *Cold Spring Harb Protoc*; doi:10.1101/pdb.top107654

An. coluzzii) or prevent colony establishment. For instance, if the starting point is the eggs of females collected in the field, one will need to perform a molecular identification of the mother (i.e., while the eggs hatch) before pooling the offspring of different females (Fanello et al. 2002; Santolamazza et al. 2008). On the other hand, if the starting point is the larvae, one will need to have females lay eggs individually and carry out molecular identification before pooling the offspring. The advice hereafter accounts for the possibility of sympatry at collection sites.

Eggs

Unlike *Aedes* mosquitoes, it is currently almost impossible to establish a colony from field-collected eggs of *Anopheles* mosquitoes. First, *Anopheles* eggs do not resist desiccation and therefore are difficult to transport from the field. Shililu et al. (2004) showed that egg survival is drastically reduced (by 78%–83%) after 24 h on a dry surface. Second, *An. gambiae* and the other members of the complex lay eggs in open water bodies of variable size (White et al. 2011). This behavior complicates the task of finding eggs and the possibility of collecting them at natural breeding sites. Third, to our knowledge, no oviposition cup trap is known to be reliably effective for collecting *Anopheles*.

Hatching

Another biological feature that complicates the establishment of *Anopheles* colonies from the field is that eggs hatch at least 48 h after oviposition, so they cannot be stored for long periods of time before processing. However, females can be forced to lay eggs on a humid surface (i.e., filter paper, see the section Oviposition) before putting them in water (Nepomichene et al. 2017). This possibility can be exploited to recover eggs from gravid females in the field and to move them to the insectarium within few days (up to 5 d).

Larvae/Pupae

Unlike eggs, *Anopheles* larvae collected in the field can be used to start a colony. *An. gambiae*, *An. coluzzii*, and *An. arabiensis* larvae and pupae can develop in water bodies (i.e., puddles) at high densities. This facilitates the fast collection of hundreds of specimens. A critical point is the transfer from the field to the insectarium. First, it is important to collect the most developed stages of larvae, such as larval instar 3 (L3) or L4 because their chance of survival is higher. Second, larvae need to be carefully collected, for instance by using the dipping technique (Silver 2007). Third, to reduce mortality, larvae must be transported in a recipient full of water and without any air, to avoid larvae movement inside the water, which can result in larval mortality due to shaking (see Protocol 2: Mosquito Larvae and Pupae Transport from the Field [Ayala et al. 2023]). Finally, it is important to carefully check for the presence of aquatic insect predators, including larvae of other mosquito species (Roux and Robert 2019).

Usually, pupae are present at lower densities than larvae; however, they can be an important source of biological material for starting a colony because they have a high emergence rate, thus producing many adults. They are more delicate than larvae, and therefore need to be collected carefully to avoid any damage. Moreover, they need to be almost in constant contact with air. For this reason, their transport varies drastically from that of larvae. Pupae should be introduced carefully in 1.5-mL tubes with a droplet of water to allow them to be in contact with the air and to avoid desiccation and at the same time to reduce shaking during transport (see Protocol 2: Mosquito Larvae and Pupae Transport from the Field [Ayala et al. 2023]).

Laboratory techniques for rearing larvae can be used for field-collected specimens (BEI Resources). Whatever the technique used, it is important to limit osmotic stress (Kengne et al. 2019). Therefore, we suggest that researchers mix distilled water (currently used in laboratories to avoid biotic pathogens) with water from the breeding sites. After a few days, water can be replaced and specimens can be offered some larval food. Another important aspect is to limit the number of larvae per pan. We recommend not exceeding 200–300 larvae per pan (9-cm × 12-cm; BEI Resources), which equates to 1 L4 larva per 1–2 mL of water. Adult mosquitoes resulting from these pans will have higher fitness (BEI Resources). Pupae can be directly placed in cups with distilled water for adult emergence. It is important to immediately provide sugar resources (i.e., cotton with a 10% sugar solution) to increase survival.

Adults

This is the most common starting point from many field-based *Anopheles* colonies. The main advantage of collecting adult females is that almost every female has already mated, thus avoiding the need of mating in insectarium cages that can be challenging (see the section Mating). The anthropophilic (human-biting) and endophilic (house-entering) behavior of *An. gambiae*, *An. coluzzii*, and *An. arabiensis* can be exploited to collect adult females. Two methods that are particularly efficient are human-landing catches and indoor collections with an aspirator (manual or electric) (Silver 2007). For the other species of the complex, animal traps or resting traps, such as the Muirhead–Thompson trap, can be more useful. In those methods, damage to the collected specimens is limited, thus improving their survival. It is important to promptly provide a blood source for feeding completion (see the section Blood Feeding) because it can increase the number of blood-fed females that lay eggs. In the field, adult mosquitoes can be kept in cages for a few days. It is important to provide ideal conditions of humidity, temperature, and protection against wind, sunlight, and predators (i.e., ants). Two to six days after blood feeding, females are at the gravid stage, and they can be forced to lay eggs in 1.5-mL tubes with a paper filter and humid cotton (see Nepomichene et al. 2017 for details). These females can be easily transported to the insectary facilities because they are in tubes.

Once mosquitoes have emerged, it is important not to overcrowd cages to avoid high mortality due to the continuous mosquito movements. A maximum number is 500 specimens (in a standard cage of 30.5-cm × 30.5-cm × 30.5-cm), assuming a 50:50 sex ratio.

Blood Feeding

An. gambiae can be subjected to a strong selection in host choice (i.e., human) during blood feeding, leading to underrepresentation of the genetic polymorphisms present in natural populations. However, anthropophilic behavior is predominant in the main vectors within the *An. gambiae* complex (Takken and Verhulst 2013). Therefore, feeding mosquitoes on human volunteers should be the preferred method, at least during the first generations. Then, for logistic and ethical reasons, it is preferable to shift to artificial feeding systems. We strongly recommend blood feeding mosquitoes in the evening, or at least under dark conditions, for 20 min at most. To increase the number of feeding females, the sugar solution can be removed during the previous 12 h. In general, species within the *An. gambiae* complex show a good feeding rate.

Mating

Female and male mating under laboratory conditions can be challenging, and it is a common source of failure when trying to establish a colony. Members of the *An. gambiae* complex are eurygamous (i.e., they need space for mating) and require large cages (typically >1 m^3) that allow their natural swarming behavior. Large cages are especially important when mating freshly collected females and first-generation mosquitoes. However, after the first generations, strong selection for laboratory viability leads to the emergence of a stenogamous colony (i.e., they can mate in standard 30.5-cm × 30.5-cm × 30.5-cm cages). Moreover, it is important to control the number of males per female. In recently established colonies, the number of males should be slightly higher than the number of females. However, very often the first generations after field collection do not mate. They can be forced to mate by using a very skilled and time-consuming technique that is regularly performed in insectaries (https://www.beiresources.org/Publications/MethodsinAnophelesResearch.aspx). Therefore, it is better to promote natural mating in the first days and consider forced mating as a complementary option if females are not gravid after 1 wk.

 Cite this introduction as *Cold Spring Harb Protoc*; doi:10.1101/pdb.top107654

Oviposition

Well-adapted colonies of the *An. gambiae* complex can oviposit on humid filter paper in Petri dishes (10-cm-diameter). Then, eggs can be transferred to pans with water for hatching. The oviposition rate is generally low in freshly adapted colonies. As only females prone to lay eggs on this artificial substrate will be retained (*Anopheles* normally lay eggs directly in water and not on humid substrates, like *Aedes*), commonly, many colonies start with a huge number of mosquitoes. Another solution, at least temporarily, is to force egg laying (Nepomichene et al. 2017) to allow the survival of the new colony. However, this is very time-consuming and can only be performed with a limited number of females. This method is typically used to establish iso-female lines (progeny of single females). Conversely, to avoid bias and to try to recover the natural polymorphism, pool the hatched larvae of all females in the same pan.

Notes on Related Species

Most of the information presented here can be applied to the other *Anopheles* species. Some differences can be noted in the life cycle length. Some species tend to bite more during the night. However, it is still difficult to establish *Anopheles* colonies in the laboratory, with some exceptions such as *An. darlingi* (Puchot et al. 2022). For instance, no *Anopheles nili* or *Anopheles moucheti* colonies are available, despite the multiple efforts to generate them. Moreover, some *An. gambiae*, *An. coluzzii*, and *An. arabiensis* populations require continuous efforts, sometimes with little success. It is important to keep in mind that not all *Anopheles* species can be established or kept in the laboratory. Also, the necessity to establish a colony from the field should be carefully considered because representative colonies of several *Anopheles* species, including species within the *An. gambiae* complex such as *An. gambiae*, *An. coluzzii*, *An. arabiensis*, *An. quadriannulatus*, and/or *An. merus* are already freely available (see https://www.beiresources.org/Home.aspx). Sometimes we can carry out planned tests (e.g., experimental infections, larval tolerance to different compounds, and host preference) with only the F_1 (larvae or adults). These tests can directly be conducted under field conditions or close laboratory facilities with the necessary logistic efforts.

CULEX PIPIENS

Mosquitoes within the *Cx. pipiens* complex have a global distribution, composed of the following six taxa: *Culex pipiens* f. *pipiens*, *Culex pipiens* f. *molestus*, *Culex quinquefasciatus*, *Culex pipiens pallens*, *Culex austrailicus*, and *Culex globocoxitus* (Aardema et al. 2020). Depending on the geographic region, hybrids may be present, and considerations should be made before starting a colony from a wild-type population. For example, *Cx. pipiens* f. *pipiens* and *Cx. quinquefasciatus* frequently hybridize between 36°N and 39°N, and there are regions where there is interbreeding between *Cx. pipiens* f. *pipiens* (aboveground) and *Cx. pipiens* f. *molestus* (belowground) in North America (Kent et al. 2007; Huang et al. 2008). Furthermore, changes or shifts in mosquito–vertebrate host associations have been reported from hybrids within the *Cx. pipiens* complex (Farajollahi et al. 2011). Under circumstances in which hybrids may be present, microsatellite analysis can be used to differentiate species, or the relative amount of hybridization, and determine the contributions of members of the species complex to the overall population structure in the region where the colony material is collected (Fonseca et al. 2004; Kent et al. 2007; Huang et al. 2008; Aardema et al. 2020).

Eggs

The collection of egg rafts of *Cx. pipiens* may be the preferred method for establishing a colony from field-collected specimens because of their relative seasonal abundance and ease of collection. Because *Cx. pipiens* egg rafts contain 30–200 or more eggs per raft (Means 1987), progeny from multiple females can be collected with relative ease. Containers with large openings can be set out and baited with a gravid trap infusion (e.g., hay-yeast-lactalbumin or rabbit chow) to promote oviposition. Catch basin sumps can also be sampled if openings are large enough or if grates are easily removed (McMillan et al. 2021). These containers can be checked daily or periodically for the presence of egg rafts on the water surface. Egg rafts can be collected readily during the season when adult mosquitoes are active. Large numbers of egg rafts can be transferred to a smaller container (500- to 1000-mL) containing clean water (e.g., reverse osmosis, distilled, deionized, or dechlorinated tap water) for transportation back to the laboratory. Individual egg rafts should be selected using a scoopula or a small piece of aluminum window screen material and placed in individual containers (30- to 100-mL) containing clean water plus a small amount of larval diet for hatching. Care should be taken when transferring individual rafts so that the convex surface of the egg raft is oriented to be in contact with the water surface.

Hatching

Eggs will typically hatch within 24–48 h of collection when maintained at temperatures >22°C in ambient day light in >12 h. Preferably, containers with individual egg rafts should be maintained in an incubator or insectary at 2°C–27°C with a light cycle of 16 h light:8 h dark. Various light:dark regimens can be used, but light should be >13 h to ensure that diapause is not initiated in developing larvae or pupae. In areas where other oviposition by other *Culex* species (e.g., *Culex restuans*, *Culex salinarius*, or *Culex tarsalis*) may be possible in the same container, larval characters can be used to identify the source of the egg raft. In the case of *Cx. pipiens* and *Cx. restuans*, characters of L1 can be used to distinguish between these species (Means 1987), although it may be necessary to identify larvae at the L3 or L4 stage (Darsie and Ward 1981) if *Cx. salinarius* or *Cx. tarsalis* may be present in the habitat.

Rearing Larvae and Pupae

Newly hatched larvae should be transferred to shallow rearing pans (enamel-coated, metal, or plastic) with clean water and larval diet. A variety of sources of larval diet can be used including (but not limited to) a 3:2 mixture of bovine liver powder:brewer's yeast or finely ground cat, dog, rodent, or tropical fish food. If *Cx. pipiens* has been identified, larvae from two to four egg rafts may be combined in a single pan. A maximum density of 0.5–1 larvae/mL of water will allow for complete development; however, a lower density of 0.1–0.2 larvae/mL may be preferred, especially if larger adult specimens or faster development times are desired. Larval diet should be added as a slurry approximately every other day. We suggest adding a sufficient amount of larval diet to promote development through pupation and adjusting the level of feeding to prevent the water from becoming cloudy or foul-smelling. Higher levels of nutrition, for example >3 mg of diet per larva (Kassim et al. 2012) are preferred for *Culex* larval rearing compared to lower nutritional regimens (Dodson et al. 2011; Alto et al. 2012). If the organic levels are excessive, whole or partial water changes may be necessary, with fresh food added to the larvae. Insufficient larval diet will result in slow or halted development of larval instars, indicated by the water in the pan having a very clear appearance. If this is observed, the quantity or frequency of larval diet provided should be increased. Larvae maintained at 25.5°C with a light cycle of 16 h light:8 h dark will typically pupate in 6–10 d, depending on density and the level of larval diet provided. Temperature and light regimens can be altered, but it should be noted that diapause can be induced if larvae and pupae develop under reduced day length (Spielman and Wong 1973; Eldridge and Bailey 1979) alone or with lower temperature conditions (Field et al. 2022).

Pupae should be removed from the larval pans by pipetting or other techniques and transferred to a shallow container (300- to 500-mL, glass or polypropylene), containing fresh water, with a maximum density of 1 pupa/mL. Containers should be placed in a standard cage (30.5-cm × 30.5-cm × 30.5-cm). Eclosion will occur in 24–48 h.

Starting a colony using field-collected larvae can be successful (Robich and Denlinger 2005), but care should be taken to ensure that *Cx. pipiens* larvae can be identified and separated from other mosquito species (primarily *Aedes* or *Culex*) that may be developing in the same container. Larvae can

 Cite this introduction as *Cold Spring Harb Protoc*; doi:10.1101/pdb.top107654

be collected by using a standard dipper. Field-collected water may be augmented with clean water, and larval diet should be used to ensure the larvae develop optimally.

Adults

Adult cages can be maintained at 25.5°C with a light cycle of 16 h light:8 h dark, and a minimum of 70% relative humidity. Temperature and light regimens can be altered if experiments relating to diapause conditions are required. *Cx. pipiens* adapts well to free mating in standard cages but should be given at least 1 wk to ensure mating has occurred before blood feeding is attempted.

Adults should be offered a carbohydrate source. Typically, 10% sucrose is provided using saturated cotton balls or wicks. Natural sources of sugar can be used, such as raisins, apple slices, or honey. If studies related to pesticides are to be initiated, the use of American Chemical Society-grade or certified organic carbohydrate sources may be considered, for colony maintenance.

Diapaused *Cx. pipiens* females can be collected from hibernacula by using a mechanical aspirator and transported back to the laboratory to initiate colonies. These mated females should be allowed to acclimate under long photoperiod conditions for several days before they are offered a blood meal (Eldridge and Bailey 1979).

Autogenous (having the ability to lay eggs without blood-feeding), nondiapausing *Cx. pipiens* f. *molestus* adults can be collected from locations where known populations exist (Kent et al. 2007). Mechanical aspirators are used to collect adults. Under laboratory conditions, caged adults should be offered 10% sucrose and provided a small container containing water and a small amount of larval diet to encourage oviposition. Provided that adequate protein is given as part of the larval diet, *Cx. pipiens* f. *molestus* mosquitoes will produce very small egg rafts compared with *Cx. pipiens* f. *pipiens*, but if larvae are maintained constantly, a *Cx. molestus* colony can be maintained without the need of a blood meal.

Blood Feeding

Approval by the appropriate Institutional Animal Use and Care Committee is required before feeding mosquitoes on live animals.

The use of live domesticated birds may be preferred for *Cx. pipiens* colony establishment and maintenance if proper space is available to maintain these subjects. Small subjects, such as newly hatched domestic chickens (*Gallus gallus*) or adult button quail (*Coturnix chinensis*), are suggested. Birds should be placed in a nylon mesh bag or sleeve that restricts the movement of the bird without causing injury to the animal. The secured mesh bag should be suspended from the top of the cage, which ensures that the bird will remain in an upright position during the feeding. Blood feeding should be encouraged under nighttime conditions and conducted at a minimum length of time to limit stress on the bird. It is often necessary to blood feed two or three times per week until the colony is established. In this instance, multiple birds should be used to limit the stress on individual birds. The use of a live subject may encourage a higher number of females to feed and oviposit, which may increase the genetic diversity of the colony for future generations.

Defibrinated chicken, goose, or mammalian blood can be sourced commercially and used with an artificial membrane feeding system (Hemotek or water-jacketed glass feeder). Initially, it is advised to use poultry skin, hog intestine, or parafilm as the membrane, with the blood warmed to, and maintained at, 39°C while available to the mosquitoes. Additionally, it may be necessary to deprive the colony of sucrose for 24 h before offering an artificial blood meal, with cotton balls saturated with reverse-osmosis (or distilled) water provided to the colony. The membrane feeder can be placed inside or on top of the cage, depending on the type of feeder used. Feeding should be conducted in the dark, with black plastic sheeting used to cover the cage. The use of artificial feeders also results in short exposure times, usually under an hour, because the natural membranes frequently dry out, reducing the ability for the female to pierce the membrane to obtain blood. Therefore, until the colony adapts to the system, artificial sources of blood should be offered more frequently than a live subject.

The use of artificial blood meals will likely reduce the initial genetic diversity in the colony because few wild-type females will feed under such circumstances (Ciota et al. 2014). However, in subsequent generations, the number of females that will readily obtain blood from artificial means will increase and a robust colony will form that does not require live animals for maintenance.

Oviposition

Females will typically oviposit in 3–4 d after obtaining a blood meal. Ideally, 300- to 500-mL glass or polypropylene containers with 100–300 mL of clean water and a small amount of larval diet should be provided for oviposition. Males and females will often die on the surface of the water in the container. To limit this mortality and to synchronize egg collection, consider adding the oviposition containers on the third day following feeding. The largest egg rafts are obtained from the first gonotrophic (egg development and laying) cycle of females that were fully engorged with a blood meal. Smaller egg rafts will indicate that the females in the colony have previously produced a brood of eggs or obtained a partial blood meal. Egg rafts should be transferred to larval rearing pans within 24 h of oviposition.

Notes on Related Species

The collection of *Cx. salinarius* larvae and host-seeking females has been used to establish colonies of this mosquito. Because *Cx. salinarius* requires a large space for free mating, a larger cage should be used for adults (50-cm × 50-cm × 50-cm or 61-cm × 61-cm × 61-cm). Vertebrate hosts (e.g., hatchling chicken or rodents) have been used as the source of blood for these colonies (Wallis and Whitman 1968; Tveten and Meola 1988; Anderson et al. 2012). Autogeny has been documented and evaluated in *Cx. salinarius*, with larval diet and density identified as factors regulating its expression in newly established colonies (Tveten and Meola 1988).

The establishment and maintenance of *Cx. tarsalis* colonies have been successful (Anderson et al. 2012; Reisen et al. 2014). Maintenance of *Cx. tarsalis* in a standard cage (30.5-cm × 30.5-cm × 30.5-cm) is possible (Dodson et al. 2014) but larger cages (45-cm × 45-cm × 45-cm) have also been used (Eberle and Reisen 1986). As with other *Culex* colonies, the use of artificial blood feeding systems is possible with established colonies of *Cx. tarsalis* for use in West Nile virus transmission studies (Danforth et al. 2015). Anautogenous and autogenous strains within *Cx. tarsalis* populations and the methods used to establish colonies of each type have been documented (Eberle and Riesen 1986).

REFERENCES

Aardema ML, vonHoldt BM, Fritz ML, Davis SR. 2020. Global evaluation of taxonomic relationships and admixture within the *Culex pipiens* complex of mosquitoes. *Parasit Vectors* **13:** 8. doi:10.1186/s13071-020-3879-8

Alto BW, Muturi EJ, Lampman RL. 2012. Effects of nutrition and density in *Culex pipiens. Med Vet Entomol* **26:** 396–406. doi:10.1111/j.1365-2915.2012.01010.x

Anderson JF, Main AJ, Cheng G, Ferrandino FJ, Fikrig E. 2012. Horizontal and vertical transmission of West Nile virus genotype NY99 by *Culex salinarius* and genotypes NY99 and WN02 by *Culex tarsalis. Am J Trop Med Hyg* **86:** 134. doi:10.4269/ajtmh.2012.11-0473

Aubry F, Dabo S, Manet C, Filipović I, Rose NH, Miot EF, Martynow D, Baidaliuk A, Merkling SH, Dickson LB, et al. 2020. Enhanced Zika virus susceptibility of globally invasive *Aedes aegypti* populations. *Science* **370:** 991–996. doi:10.1126/science.abd3663

Ayala D, Zhang S, Chateau M, Fouet C, Morlais I, Costantini C, Hahn MW, Besansky NJ. 2019. Association mapping desiccation resistance within chromosomal inversions in the African malaria vector *Anopheles gambiae. Mol Ecol* **28:** 1333–1342. doi:10.1111/mec.14880

Ayala D, Shepard JJ, Rose NH. 2023. Mosquito larvae and pupae transport from the field. *Cold Spring Harb Protoc* doi:10.1101/pdb.prot108184

Barrón MG, Paupy C, Rahola N, Akone-Ella O, Ngangue MF, Wilson-Bahun TA, Pombi M, Kengne P, Costantini C, Simard F. 2019. A new species in the major malaria vector complex sheds light on reticulated species evolution. *Sci Rep* **9:** 14753. doi:10.1038/s41598-019-49065-5

Bowman LR, Donegan S, McCall PJ. 2016. Is dengue vector control deficient in effectiveness or evidence?: systematic review and meta-analysis. *PLoS Negl Trop Dis* **10:** e0004551. doi:10.1371/journal.pntd.0004551

Brady OJ, Hay SI. 2020. The global expansion of dengue: how *Aedes aegypti* mosquitoes enabled the first pandemic arbovirus. *Annu Rev Entomol* **65:** 191–208. doi:10.1146/annurev-ento-011019-024918

Cheng C, Tan JC, Hahn MW, Besansky NJ. 2018. Systems genetic analysis of inversion polymorphisms in the malaria mosquito *Anopheles gambiae. Proc Natl Acad Sci* **115:** E7005–E7014. doi:10.1073/pnas.1806760115

Ciota AT, Matacchiero AC, Kilpatrick AM, Kramer LD. 2014. The effect of temperature on life history traits of *Culex* mosquitoes. *J Med Entomol* **51:** 55–62. doi:10.1603/ME13003

Coetzee M, Hunt RH, Wilkerson R, Della Torre A, Coulibaly MB, Besansky NJ. 2013. *Anopheles coluzzii* and *Anopheles amharicus*, new members of the *Anopheles gambiae* complex. *Zootaxa* **3619:** 246–274. doi:10.11646/zootaxa.3619.3.2

Colton YM, Chadee DD, Severson DW. 2003. Natural skip oviposition of the mosquito *Aedes aegypti* indicated by codominant genetic markers. *Med Vet Entomol* **17:** 195–204. doi:10.1046/j.1365-2915.2003.00424.x

Coluzzi M. 1984. Heterogeneities of the malaria vectorial system in tropical Africa and their significance in malaria epidemiology and control. *Bull World Health Organ* **62:** 107–113.

Danforth ME, Reisen WK, Barker CM. 2015. Extrinsic incubation rate is not accelerated in recent California strains of West Nile virus in *Culex tarsalis* (Diptera: Culicidae). *J Med Entomol* **52:** 1083–1089.

Darsie RF Jr, Ward RA. 1981. *Identification and geographical distribution of the mosquitoes of North America, north of Mexico.* Walter Reed Army Institute of Research, Washington, DC.

Dodson BL, Kramer LD, Rasgon JL. 2011. Larval nutritional stress does not affect vector competence for West Nile virus (WNV) in *Culex tarsalis. Vector Borne Zoonotic Dis* **11:** 1493–1497. doi:10.1089/vbz.2011.0662

Dodson BL, Hughes GL, Paul O, Matacchiero AC, Kramer LD, Rasgon JL. 2014. Wolbachia enhances West Nile virus (WNV) infection in the mosquito *Culex tarsalis. PLoS Negl Trop Dis* **8:** e2965. doi:10.1371/journal.pntd.0002965

Eberle MW, Reisen WK. 1986. Studies on autogeny in *Culex tarsalis*: 1. Selection and genetic experiments. *J Am Mosq Control Assoc* **2:** 38–43.

Eldridge BF, Bailey CL. 1979. Experimental hibernation studies in *Culex pipiens* (Diptera: Culicidae): reactivation of ovarian development and blood-feeding in prehibernating females. *J Med Entomol* **15:** 462–467. doi:10.1093/jmedent/15.5-6.462

Fanello C, Santolamazza FD, Della Torre A. 2002. Simultaneous identification of species and molecular forms of the *Anopheles gambiae* complex by PCR-RFLP. *Med Vet Entomol* **16:** 461–464. doi:10.1046/j.1365-2915.2002.00393.x

Farajollahi A, Fonseca DM, Kramer LD, Marm Kilpatrick A. 2011. "Bird biting" mosquitoes and human disease: a review of the role of *Culex pipiens* complex mosquitoes in epidemiology. *Infect Genet Evol* **11:** 1577–1585. doi:10.1016/j.meegid.2011.08.013

Ferguson NM. 2018. Challenges and opportunities in controlling mosquito-borne infections. *Nature* **559:** 490–497. doi:10.1038/s41586-018-0318-5

Field EN, Shepard JJ, Clifton ME, Price KJ, Witmier BJ, Johnson K, Boze B, Abadam C, Ebel GD, Armstrong PM, et al. 2022. Semi-field and surveillance data define the natural diapause timeline for *Culex pipiens* across the United States. *Commun Biol* **5:** 1300. doi:10.1038/s42003-022-04276-x

Fonseca DM, Keyghobadi N, Malcolm CA, Mehmet C, Schaffner F, Mogi M, Fleischer RC, Wilkerson RC. 2004. Emerging vectors in the *Culex pipiens* complex. *Science* **303:** 1535–1538. doi:10.1126/science.1094247

Fritz ML, Huang J, Walker ED, Bayoh MN, Vulule J, Miller JR. 2008. Ovipositional periodicity of caged *Anopheles gambiae* individuals. *J Circad Rhythms* **6:** 2. doi:10.1186/1740-3391-6-2

Gloria-Soria A, Soghigian J, Kellner D, Powell JR. 2019. Genetic diversity of laboratory strains and implications for research: the case of *Aedes aegypti. PLoS Negl Trop Dis* **13:** e0007930. doi:10.1371/journal.pntd.0007930

Hemingway J, Ranson H, Magill A, Kolaczinski J, Fornadel C, Gimnig J, Coetzee M, Simard F, Roch DK, Hinzoumbe CK, et al. 2016. Averting a malaria disaster: will insecticide resistance derail malaria control? *Lancet* **387:** 1785–1788. doi:10.1016/S0140-6736(15)00417-1

Holt RA, Subramanian GM, Halpern A, Sutton GG, Charlab R, Nusskern DR, Wincker P, Clark AG, Ribeiro JC, Wides R. 2002. The genome sequence of the malaria mosquito *Anopheles gambiae. Science* **298:** 129–149. doi:10.1126/science.1076181

Huang S, Molaei G, Andreadis TG. 2008. Genetic insights into the population structure of *Culex pipiens* (Diptera: Culicidae) in the northeastern United States by using microsatellite analysis. *Am J Trop Med Hyg* **79:** 518–527. doi:10.4269/ajtmh.2008.79.518

Kassim NFA, Webb CE, Russell RC. 2012. Is the expression of autogeny by *Culex molestus Forskal* (Diptera: Culicidae) influenced by larval nutrition or by adult mating, sugar feeding, or blood feeding? *J Vector Ecol* **37:** 162–171. doi:10.1111/j.1948-7134.2012.00213.x

Kengne P, Charmantier G, Blondeau-Bidet E, Costantini C, Ayala D. 2019. Tolerance of disease-vector mosquitoes to brackish water and their osmoregulatory ability. *Ecosphere* **10:** e02783. doi:10.1002/ecs2.2783

Kent RJ, Harrington LC, Norris DE. 2007. Genetic differences between *Culex pipiens* f. *molestus* and *Culex pipiens pipiens* (Diptera: Culicidae) in New York. *J Med Entomol* **44:** 50–59. doi:10.1093/jmedent/41.5.50

Kilpatrick AM, Kramer LD, Jones MJ, Marra PP, Daszak P, Fonseca DM. 2007. Genetic influences on mosquito feeding behavior and the emergence of zoonotic pathogens. *Am J Trop Med Hyg* **77:** 667–671. doi:10.4269/ajtmh.2007.77.667

L'Ambert G, Gendrot M, Briolant S, Nguyen A, Pages S, Bosio L, Palomo V, Gomez N, Benoit N, Savini H. 2021. Mosquito excreta reveals circulation of West Nile virus and its underlying ecosystem. bioRxiv doi:10.1101/2021.12.05.471258

Matthews BJ, Dudchenko O, Kingan SB, Koren S, Antoshechkin I, Crawford JE, Glassford WJ, Herre M, Redmond SN, Rose NH, et al. 2018. Improved reference genome of *Aedes aegypti* informs arbovirus vector control. *Nature* **563:** 501–507. doi:10.1038/s41586-018-0692-z

Mattingly PF. 1957. Genetical aspects of the *Aedes aegypti* problem: I. Taxonomy and bionomics. *Ann Trop Med Parasitol* **51:** 392–408. doi:10.1080/00034983.1957.11685829

McMillan JR, Harden CA, Burtis JC, Breban MI, Shepard JJ, Petruff TA, Misencik MJ, Bransfield AB, Poggi JD, Harrington LC, et al. 2021. The community-wide effectiveness of municipal larval control programs for West Nile virus risk reduction in Connecticut, USA. *Pest Manag Sci* **77:** 5186–5201. doi:10.1002/ps.6559

Means RG. 1987. *Mosquitoes of New York: Part II. Genera of Culicidae other than Aedes. Bulletin Number 430b.* The University of the State of New York, Albany.

Mulwa F, Lutomiah J, Chepkorir E, Okello S, Eyase F, Tigoi C, Kahato M, Sang R. 2018. Vector competence of *Aedes bromeliae* and *Aedes vitattus* mosquito populations from Kenya for chikungunya virus. *PLoS Negl Trop Dis* **12:** e0006746. doi:10.1371/journal.pntd.0006746

Nepomichene TN, Andrianaivolambo L, Boyer S, Bourgouin C. 2017. Efficient method for establishing F_1 progeny from wild populations of *Anopheles* mosquitoes. *Malar J* **16:** 21. doi:10.1186/s12936-017-1681-7

Powell JR, Gloria-Soria A, Kotsakiozi P. 2018. Recent history of *Aedes aegypti*: vector genomics and epidemiology records. *BioScience* **68:** 854–860. doi:10.1093/biosci/biy119

Puchot N, Lecoq MT, Carinci R, Duchemin JB, Gendrin M, Bourgouin C. 2022. Establishment of a colony of *Anopheles darlingi* from French Guiana for vector competence studies on malaria transmission. *Front Trop Dis* 3. doi:10.3389/fitd.2022.949300

Reisen WK, Fang Y, Martinez VM. 2014. Effects of temperature on the transmission of West Nile virus by *Culex tarsalis* (Diptera: Culicidae). *J Med Entomol* **43:** 309–317. doi:10.1093/jmedent/43.2.309

Richard V, Paoaafaite T, Cao-Lormeau V-M. 2016. Vector competence of French Polynesian *Aedes aegypti* and *Aedes polynesiensis* for Zika virus. *PLoS Negl Trop Dis* **10:** e0005024. doi:10.1371/journal.pntd.0005024

Riehle MM, Guelbeogo WM, Gneme A, Eiglmeier K, Holm I, Bischoff E, Garnier T, Snyder GM, Li X, Markianos K, et al. 2011. A cryptic subgroup of *Anopheles gambiae* is highly susceptible to human malaria parasites. *Science* **331:** 596–598. doi:10.1126/science.1196759

Ritchie SA, Gubler D, Ooi EE, Vasudevan S, Farrar J. 2014. *Dengue vector bionomics: why* Aedes aegypti *is such a good vector.* Ch. 24 in *Dengue and dengue hemorrhagic fever.* CAB International, Oxfordshire, United Kingdom.

Robich RM, Denlinger DL. 2005. Diapause in the mosquito *Culex pipiens* evokes a metabolic switch from blood feeding to sugar gluttony. *Proc Natl Acad Sci* **102:** 15912–15917. doi:10.1073/pnas.0507958102

Rose NH, Sylla M, Badolo A, Lutomiah J, Ayala D, Aribodor OB, Ibe N, Akorli J, Otoo S, Mutebi J-P, et al. 2020. Climate and urbanization drive mosquito preference for humans. *Curr Biol* **30:** 3570–3579.e6. doi:10.1016/j.cub.2020.06.092

Rose NH, Shepard JJ, Ayala D. 2023. Collecting, storing, and hatching *Aedes aegypti* eggs. *Cold Spring Harb Protoc* doi:10.1101/pdb.prot108183

Ross PA, Endersby-Harshman NM, Hoffmann AA. 2019. A comprehensive assessment of inbreeding and laboratory adaptation in *Aedes aegypti* mosquitoes. *Evol Appl* **12:** 572–586. doi:10.1111/eva.12740

Roux O, Robert V. 2019. Larval predation in malaria vectors and its potential implication in malaria transmission: an overlooked ecosystem service? *Parasit Vectors* **12:** 217. doi:10.1186/s13071-019-3479-7

Santolamazza F, Mancini E, Simard F, Qi Y, Tu Z, della Torre A. 2008. Insertion polymorphisms of SINE200 retrotransposons within speciation islands of *Anopheles gambiae* molecular forms. *Malar J* **7:** 163. doi:10.1186/1475-2875-7-163

Shililu JI, Grueber WB, Mbogo CM, Githure JI, Riddiford LM, Beier JC. 2004. Development and survival of *Anopheles gambiae* eggs in drying soil: influence of the rate of drying, egg age, and soil type. *J Am Mosquito Control Assoc* **20:** 243–247.

Silver JB. 2007. *Mosquito ecology: field sampling methods.* Springer Science & Business Media, Berlin, Germany.

Soghigian J, Gloria-Soria A, Robert V, Le Goff G, Failloux A-B, Powell JR. 2020. Genetic evidence for the origin of *Aedes aegypti*, the yellow fever mosquito, in the southwestern Indian Ocean. *Mol Ecol* **29:** 3593–3606. doi:10.1111/mec.15590

Spielman A, Wong J. 1973. Environmental control of ovarian diapause in *Culex pipiens. Ann Entomol Soc Am* **66:** 905–907. doi:10.1093/aesa/66.4.905

Sylla M, Ndiaye M, Black WC. 2013. *Aedes* species in treeholes and fruit husks between dry and wet seasons in southeastern Senegal. *J Vector Ecol* **38:** 237–244. doi:10.1111/j.1948-7134.2013.12036.x

Takken W, Verhulst NO. 2013. Host preferences of blood-feeding mosquitoes. *Annu Rev Entomol* **58:** 433–453. doi:10.1146/annurev-ento-120811-153618

Tveten MS, Meola RW. 1988. Autogeny in *Culex salinarius* from Texas, Florida and New Jersey. *J Am Mosquito Control Assoc* **4:** 436–441.

Vega-Rúa A, Zouache K, Girod R, Failloux A-B, Lourenço-de-Oliveira R. 2014. High level of vector competence of *Aedes aegypti* and *Aedes albopictus* from ten American countries as a crucial factor in the spread of Chikungunya virus. *J Virol* **88:** 6294–6306. doi:10.1128/JVI.00370-14

Wallis RC, Whitman L. 1968. Oviposition of *Culiseta morsitans* (Theobald) and comments on the life cycle of the American form. *Mosquito News* **28:** 198–200.

White BJ, Collins FH, Besansky NJ. 2011. Evolution of *Anopheles gambiae* in relation to humans and malaria. *Annu Rev Ecol Evol Syst* **42:** 111–132. doi:10.1146/annurev-ecolsys-102710-145028

Williams YA, Tusting LS, Hocini S, Graves PM, Killeen GF, Kleinschmidt I, Okumu FO, Feachem RGA, Tatarsky A, Gosling RD. 2018. Expanding the vector control toolbox for malaria elimination: a systematic review of the evidence. *Adv Parasitol* **99:** 345–379. doi:10.1016/bs.apar.2018.01.003

Xia S, Dweck HK, Lutomiah J, Sang R, McBride CS, Rose NH, Ayala D, Powell JR. 2021. Larval sites of the mosquito *Aedes aegypti formosus* in forest and domestic habitats in Africa and the potential association with oviposition evolution. *Ecol Evol* **11:** 16327–16343. doi:10.1002/ece3.8332

Cite this introduction as *Cold Spring Harb Protoc*; doi:10.1101/pdb.top107654

Protocol 1

Collecting, Storing, and Hatching *Aedes aegypti* Eggs

Noah H. Rose,[1,2,6] John J. Shepard,[3,6] and Diego Ayala[4,5,6]

[1]*Department of Ecology and Evolutionary Biology,* [2]*Princeton Neuroscience Institute, Princeton University, Princeton, New Jersey 08544, USA;* [3]*Department of Entomology and Center for Vector Biology and Zoonotic Diseases, The Connecticut Agricultural Experiment Station, New Haven, Connecticut 06511, USA;* [4]*MIVEGEC, Univ. Montpellier, CNRS, IRD, Montpellier BP 64501, 34394, France;* [5]*Medical Entomology Unit, Institut Pasteur de Madagascar, Antananarivo BP1274, 101, Madagascar*

Laboratory study of natural populations of mosquitoes can play a key role in determining the underlying causes of variation in burdens of mosquito-borne disease. *Aedes aegypti* is the main vector of the viruses that cause dengue, chikungunya, Zika, and yellow fever, making it a high priority for laboratory study. *Ae. aegypti* eggs provide an ideal starting point for new laboratory colonies. Eggs can be collected using ovicups, which are small plastic cups lined with seed-germination paper and partially filled with leaf-infused H_2O. Once collected, dry eggs will remain viable for months and can be safely transported long distances back to the laboratory as long as they are properly stored. This protocol provides step-by-step instructions for preparing for collecting, storing, and hatching *Ae. aegypti* eggs and has successfully yielded laboratory colonies from locations across both the native and invasive range of this species.

MATERIALS

It is essential that you consult the appropriate Material Safety Data Sheets and your institution's Environmental Health and Safety Office for proper handling of equipment and hazardous material used in this protocol.

Reagents

Mango leaf infusion, freshly prepared

Place 40 dry mango leaves in 10 L of H_2O in a large 20-L bucket. If mango leaves are not available, pesticide/herbicide-free grass clippings or hay are acceptable. Let this infusion sit overnight (8–10 h) or up to 3 d before use.

Reverse-osmosis H_2O

TetraMin fish food (tropical tablets)

Equipment

Airtight 1-gallon plastic bags

Airtight carboy (10-L)

Camera or phone (see Step 6)

Electric kettles (at least three; 1.7-L)

Fine-tipped waterproof marker

[6]Correspondence: noahr@princeton.edu; John.shepard@ct.gov; diego.ayala@ird.fr

Cite this protocol as *Cold Spring Harb Protoc*; doi:10.1101/pdb.prot108183

Global positioning system (GPS) receiver (e.g., Garmin eTrex 10 or cell phone GPS)
Mortar and pestle
Nail or awl (see Step 2)
Nails, wire, or other implements used to hang the cup (optional, see Step 4)
Paper towels
Plastic cups (black, 32-oz; e.g., Executive Advertising item 342057)
Plastic pans (>1-L capacity)
Seed germination paper (15″ × 6″, 76-lb heavy weight)

Other unbleached paper (e.g., unbleached coffee filters) can work, if necessary.

Strap wrench
Whirl-Pak bags (24-oz)

METHOD

Setting Up Ovicups

Figure 1 shows an ovicup in the field.

1. Line the inside of a 32-oz plastic cup with a piece of seed germination paper, pulling the two ends to a slight "peak" so that the paper fits snugly against the sides of the cup. Use a fine-tipped waterproof marker to write a unique identifier on one of the top corners of the paper.
2. Using a nail or an awl, poke holes in the side of each cup ~5 cm from the bottom.

   This step prevents rain from overfilling the cups.
3. Fill the cup with 200 mL of mango leaf infusion.

   Hereafter, the plastic cup with seed paper and H_2O is referred to as an ovicup.
4. Place the ovicup in a location in the field where it is unlikely to be disturbed.

   In some cases, we will nail it or use a loop of wire to attach it to the side of a tree or a post, especially if researchers are concerned about children disturbing the ovicups.
5. Record the date and GPS coordinates (with the GPS receiver) of the ovicup location for later reference.

FIGURE 1. An ovicup in the field (photo by Noah Rose).

Cite this protocol as *Cold Spring Harb Protoc*; doi:10.1101/pdb.prot108183

6. If possible, take a photograph with a phone or camera of the ovicup to show where it is placed.

   In general, we try to place 10–20 different ovicups at least 100 m apart for a given site. This step helps to ensure that the eggs laid in different cups are not related to each other (Rose et al. 2020).

Collecting Papers from Ovicups

7. Leave the ovicups out for 3 d (i.e., place on day 1 and then pick it up on day 4).

   This timing gives mosquitoes plenty of time to lay eggs, while avoiding hatching among the earliest eggs collected. Shorter collections can be effective but reduce yield and risk egg desiccation if eggs are dried too soon. Longer collections risk the earliest-collected eggs hatching.

8. From each ovicup, remove the wet paper by hand (taking care not to disturb any eggs near the water line) and fold it into thirds or quarters so that the eggs are on the inside.

   This step will allow the paper to better fit inside the Whirl-Pak bags after drying (see Step 12).

9. Wrap the wet, folded piece of seed paper with a paper towel and place it inside a plastic bag.
10. Pour out the infusion from each ovicup onto the ground, taking care not to leave standing water. Save the ovicups for future use only if first frozen or autoclaved to prevent cross-contamination.
11. Bring the paper towel-wrapped seed papers back to the laboratory or hotel. Spread them on a table in a single layer to allow each paper to slowly dry for 24 h.

    If the eggs dry too fast, they can die, but this 24-h method reliably yields healthy eggs. Make sure that eggs are not dried in an air-conditioned environment—this is too dry and might kill the embryos.

12. Once the seed papers are dry, place each one in a dry, airtight, plastic Whirl-Pak bag.

    Ensure that the seed papers are entirely dry and that they are packed under ambient, humid conditions. If papers are still wet, condensation can form during transport, leading to premature hatching.

13. Place the Whirl-Pak bags in an airtight 1-gallon plastic bag for transportation.

    These two airtight layers should keep eggs in a humid environment even when transported in dry, air-conditioned environments.

14. Once back in the laboratory, hatch eggs as soon as possible (recommended, especially if they have been exposed to varying temperature and humidity during transportation; proceed with Step 15) or store them under standard conditions (e.g., 15°C, 80% relative humidity) for up to 6 mo (viability will depend on the collection location, storage conditions, and transportation history).

Preparing Hatch Broth

This broth works well with eggs from generalist and human-specialist Ae. aegypti, *for both field-collected materials and from laboratory strains. Note that one can also use a vacuum rather than boiling to deoxygenate broth, but that approach is not discussed here.*

15. Use three 1.7-L electric kettles to boil fresh reverse-osmosis H_2O.

    Use two to five kettles, depending on how many eggs need to be hatched; the number of TetraMin tablets (added in Step 16) should be adjusted accordingly. The number of kettles may slightly affect the deoxygenation levels, but we have not seen any noticeable differences in how well it works for hatching. Three kettles were used when the protocol was developed, so it should be quite reliable.

16. While the H_2O is boiling, finely grind three tablets of TetraMin fish food into powder with a mortar and pestle.

    You can use 0.5 tablets per kettle instead of 1 tablet per kettle if you notice problems with bacterial growth due to late hatching; just make sure that you are ready to add more food promptly after hatching.

17. Once H_2O has boiled, pour it into a carboy.
18. Add the powdered fish food.

    It is better to add the H_2O before the food so that the hot H_2O does not cook the fish food powder onto the bottom of the carboy.

19. Seal the carboy very tightly (we use a strap wrench).

    The cooling H_2O will form a vacuum seal. We find that this step effectively deoxygenates the broth.

20. Shake the carboy gently to mix the broth.
21. Leave the broth overnight (or if you are in a rush, for 5 h in a 4°C cold room)
22. When you are ready to use the broth, open the carboy.

    You should hear a hissing sound as the vacuum is released; if you do not hear it, the seal may have been bad, and broth may not be properly deoxygenated. In this case, prepare a new batch. If the problem recurs, replace the carboy gasket.

    Unopened broth can be used for at least 2 d after preparation.

Hatching Eggs

23. Fully submerge each egg paper in hatch broth in its own clean plastic pan and place the pans in standard insectary conditions (e.g., 26°C, 75% humidity).

    In this hatch broth, generally eggs hatch in 24–48 h, although sometimes there are stragglers.

24. After larvae are hatched, add more reverse-osmosis H_2O and rear and feed as usual. See Introduction: Establishing Colonies from Field-Collected Mosquitoes: Special Accommodations for Wild Strains (Rose et al. 2023).

REFERENCES

Rose NH, Sylla M, Badolo A, Lutomiah J, Ayala D, Aribodor OB, Ibe N, Akorli J, Otoo S, Mutebi J-P, et al. 2020. Climate and urbanization drive mosquito preference for humans. *Curr Biol* **30:** 3570–3579.e6. doi:10.1016/j.cub.2020.06.092

Rose NH, Shepard JJ, Ayala D. 2023. Establishing colonies from field-collected mosquitoes: special accommodations for wild strains. *Cold Spring Harb Protoc* doi:10.1101/pdb.top107654

Protocol 2

Mosquito Larvae and Pupae Transport from the Field

Diego Ayala,[1,2,6] John J. Shepard,[3,6] and Noah H. Rose[4,5,6]

[1]*MIVEGEC, Univ. Montpellier, CNRS, IRD, Montpellier BP 64501, 34394, France;* [2]*Medical Entomology Unit, Institut Pasteur de Madagascar, Antananarivo BP1274, 101, Madagascar;* [3]*Department of Entomology and Center for Vector Biology and Zoonotic Diseases, The Connecticut Agricultural Experiment Station, New Haven, Connecticut 06511, USA;* [4]*Department of Ecology and Evolutionary Biology,* [5]*Princeton Neuroscience Institute, Princeton University, Princeton, New Jersey 08544, USA*

Laboratory study of field-collected mosquitoes can allow researchers to better understand the ways variation within and among mosquito populations shapes burdens of mosquito-borne disease. The *Anopheles gambiae* complex comprises the most important vectors of malaria, but it can be challenging to keep in the laboratory. For some species of mosquitoes, especially *An. gambiae*, it is very difficult to bring viable eggs into the laboratory. Instead, it is preferable to collect larvae or pupae and then transport them as carefully as possible back to the laboratory. This simple protocol allows a researcher to start new laboratory colonies from larvae or pupae collected from natural breeding sites or proceed directly to their planned experiments. The use of natural breeding sites provides additional reassurance that the resulting colonies are representative of natural populations.

MATERIALS

It is essential that you consult the appropriate Material Safety Data Sheets and your institution's Environmental Health and Safety Office for proper handling of equipment and hazardous material used in this protocol.

Equipment

Disposable 3-mL plastic pipettes with the tip cut off to enlarge the hole

The larger hole is necessary to prevent damaging larvae and pupae.

Freezer storage boxes for microcentrifuge tubes (81-well)
Funnel adapted to the water bottle
Larval pan (18-in × 12 ½-in × 2 1/8-in)
Microcentrifuge tubes (1.5-mL)
Plastic container to hold 1.5-L water bottles
Water bottle (1.5-L)
Water dipper, ladle, or fine-mesh sieve (15- to 20-cm-diameter)

[6]Correspondence: noahr@princeton.edu; John.shepard@ct.gov; diego.ayala@ird.fr

Cite this protocol as *Cold Spring Harb Protoc;* doi:10.1101/pdb.prot108184

METHOD

1. Collect larvae or pupae according to the dipping or netting method with a water dipper, ladle, or fine-mesh net. While a dipper or ladle is preferable for small water bodies such as puddles, a fine-mesh net is particularly useful for large breeding sites such as rice fields (Silver 2007).
2. Place larvae or pupae into the larval pan containing breeding site H_2O.
3. Fill half of the water bottle with H_2O from the larval site; avoid introducing vegetation fragments or mud. You can use the mesh sieve to strain the H_2O.
4. Transfer the larvae with the cut plastic pipettes to the 1.5-L water bottle.

   If there are many, first pass the water containing the larvae or pupae through the sieve. Place the sieve in the larval pan, reversing it, and rinse the larvae or pupae through with H_2O. Directly transfer the animals to the bottle using the funnel.
5. Fill the bottle with larval site H_2O until no more air is present and close the bottle. Do this step at the very last moment before leaving the collection site. Place the bottles into a plastic container.
6. Transfer the pupae from the larval pan to 1.5-mL microcentrifuge tubes with a cut plastic pipette.

   This will allow them to safely eclose without drowning.
7. Using a pipette, remove H_2O from the microcentrifuge tube until only a drop of H_2O is left. The H_2O can be from the breeding site or just mineral H_2O.

   This ensures that pupae will not dry out before eclosion.
8. Place the tubes in a freezer storage box (without ice) to avoid large temperature variation.
9. Transport the larvae and pupae to the laboratory. See Introduction: Establishing Colonies from Field-Collected Mosquitoes: Special Accommodations for Wild Strains (Rose et al. 2023) for information on larval care.

DISCUSSION

Here we discuss tips for increasing the success of *Anopheles gambiae* larvae and pupae collection. Optimal breeding sites for *An. gambiae* are exposed to sunlight, temperate, with no vegetation, and not exposed to passing vehicles or frequent running water. Larvae should ideally be collected in the morning. During the afternoon the temperature can be too high, so prevent the larvae from staying on the surface. In the late afternoon it can be difficult to observe larvae because of low sunlight.

Some equipment can be improvised or bought at the collection site. For instance, a funnel can be made by cutting another 1.5-mL bottle and keeping the narrow part as sieve. Moreover, white plastic plates can also be used as a larval pan.

If the transportation time from the site is longer than 10 h, you can remove some H_2O and open the bottle lid to allow the larvae to breathe. Refill the bottle before continuing the trip to the laboratory facilities. For pupae, just ensure that the droplet does not evaporate. Refill if necessary. Mineral water can be used to refill for larvae or pupae transportation. During the trip, avoid the direct light into the bottles and microcentrifuge tubes. This will increase the temperature.

REFERENCES

Rose NH, Shepard JJ, Ayala D. 2023. Establishing colonies from field-collected mosquitoes: special accommodations for wild strains. *Cold Spring Harb Protoc* doi:10.1101/pdb.top107654

Silver JB. 2007. *Mosquito ecology: field sampling methods*. Springer Science & Business Media, Berlin, Germany.

Cite this protocol as *Cold Spring Harb Protoc*; doi:10.1101/pdb.prot108184

CHAPTER 5

Membrane Feeding Devices to Blood Feed Mosquitoes in the Laboratory

Andre Luis Costa-da-Silva[1]

Department of Biological Sciences and Biomolecular Sciences Institute, Florida International University, Miami, Florida 33199, USA

Husbandry protocols for medically important mosquito species in the laboratory have the blood feeding procedure as an inherent step in the maintenance routine for egg production. Artificial feeders have become the main solution for delivering blood meals to these hematophagous insects, circumventing the need for vertebrate hosts. Mimicking the skin, these devices both heat the blood and hold a membrane through which the blood can be ingested by mosquitoes. Several commercial and noncommercial artificial mosquito feeders are available, but obstacles like high cost and complex assembly can restrict their routine use. To address this issue, a number of feeders have been designed that can be built with common materials found in the laboratory. They are cheaper and more accessible, but the absence of built-in temperature control and heating units can limit their efficacy. One of the most accessible mosquito feeders is the Glytube. It consists of a 50-mL conical centrifuge tube with heated glycerol that delivers blood through a membrane. The Glytube has been used successfully to model mosquito blood feeding. Recently, 3D printing has emerged as an alternative method to create customized artificial feeders. This technology is expected to facilitate access to and reduce the cost of these devices with open-source 3D model files that can be shared among laboratories maintaining mosquitoes worldwide.

THE ELEMENTS OF ARTIFICIAL MOSQUITO FEEDERS

The study of anautogenous mosquito vectors in the laboratory requires a blood meal delivery method to feed females and sustain egg production (Kauffman et al. 2017). Because the use of human or other vertebrates as sources of blood must be limited to specific scientific purposes upon bioethical/biosafety approval by an Institutional Review Board (IRB), artificial feeders have become the method of choice for mosquito maintenance and experimentation in laboratory settings. The artificial feeders are in compliance with the replacement, reduction, and refinement guiding principles (3Rs) for the improvement of animal use and welfare (Costa-da-Silva et al. 2014). Using artificial feeders eliminates the high costs of feeding mosquitoes with mammals, which require animal facilities to house them and trained personnel to manipulate, anesthetize, and sacrifice them (Dias et al. 2018). The use of artificial feeders also allows researchers to manipulate the composition of the blood meal, such as by adding pathogens, drugs, peptides, and other molecules (Costa-da-Silva et al. 2017; Duvall et al. 2019). The artificial feeders can even be used to feed non-blood-based artificial meals to mosquitoes (Kandel et al. 2020).

[1]Correspondence: adacosta@fiu.edu

Cite this introduction as *Cold Spring Harb Protoc*; doi:10.1101/pdb.top107655

In principle, an artificial mosquito feeder must have two elements: the feeding element, a piece with the meal reservoir where the blood is offered to the females through a skin-mimicking membrane they can penetrate with their mouthpieces, and the heating element, a part that keeps the blood meal warm within temperature ranges that simulate warm-blooded vertebrate hosts to induce feeding. Both the feeding and heating elements must be in close contact for heat transference to occur.

With these basic requirements in mind, a variety of devices have been developed for blood feeding. Romano and colleagues recently reviewed the literature on artificial membrane feeders to offer blood to the main mosquito vectors and found a diverse range of designs, sizes, feeding capacities, and applications (Romano et al. 2018). However, attempts to assemble and use feeders composed of several customized parts, electronics, and heating elements can often lead to frustration.

As an alternative, it is well known that standard artificial feeding devices, like water-jacketed membrane glass feeders or powered membrane feeding systems, are readily available to order online from a few specific suppliers and are easy to set up. The literature supports their use for blood feeding mosquitoes. However, the costs of these commercially available feeders may be prohibitive for their application (Romano et al. 2018).

BUILDING A MOSQUITO FEEDER USING A CONICAL CENTRIFUGE TUBE

Aiming to develop an artificial mosquito feeder that could be built with common materials found in mosquito research laboratories, Costa-da-Silva et al. (2013) created the Glytube device. Glytube is a simple, affordable, and efficient feeder built with a 50-mL conical centrifuge tube. It offers a provisory or permanent solution for mosquito researchers who need to blood feed female mosquitoes in the laboratory.

The conical centrifuge tube is used as a heating unit that is filled with heated glycerol to keep the blood warm. The cap piece of the tube is modified to hold a Parafilm-M membrane and serve as the blood reservoir. The heating unit containing prewarmed glycerol and the feeding unit filled with prewarmed blood are coupled before feeding (see Protocol 1: Artificial Membrane Feeding Mosquitoes in the Laboratory with Glytube [Costa-da-Silva 2022]).

Glytube works with a passive prewarmed heating element. This element keeps the blood meal within a temperature range that mosquitoes respond to for the time that the females have access to the membrane. For example, *Aedes aegypti* has been shown to have satisfactory feeding response to artificial feeders adjusted from 28°C to 40°C (Cosgrove and Wood 1995). It is recommended that feeding with Glytube should be performed in the insectary or rooms where the temperature is at the optimal response range for the mosquito species being fed (Costa-da-Silva et al. 2013). The lack of constant heat source and adjustable temperature are limitations of this device. Although additional elements could be added to the Glytube's heating unit for temperature control, this adds to the complexity of the device.

Glytube eliminates limitations regarding the number of feeding units that can be used simultaneously. If more units are necessary, Glytube is easily scalable by adding more units onto a cage/container with a large number of mosquitoes, or to offer blood to multiple small containers at the same time as sometimes required by batch feeding in experimental conditions. In several studies, Glytube has been adapted from its original purpose of simply delivering blood to female mosquitoes to delivering artificial blood meals, virus-infected blood meals, or neuropeptides, as well as to studying repellency (Costa-da-Silva et al. 2017; Duvall et al. 2017, 2019; Dennis et al. 2019; Jové et al. 2020). Furthermore, recent improvements created by other authors solved some technical issues that could appear during its assembly and use. For example, instead of simply wrapping a stretched strip of Parafilm around the opening of the tube to hold the sealing film in place, also adding a rubber band can better prevent leaks of glycerol from the 50-mL tube (Jové et al. 2020).

Cite this introduction as *Cold Spring Harb Protoc*; doi:10.1101/pdb.top107655

3D PRINTING TECHNOLOGY AND MOSQUITO FEEDERS

Looking to the next generation of feeders, 3D printing technology is expected to become a game changer in the development and production of 3D-printed artificial mosquito feeders, especially because the designs are easily customizable. Recently, two studies showed that the feeding efficiencies and infection rates of 3D-printed feeding devices were comparable to standard feeders when *Anopheles stephensi* were assayed for *Plasmodium* infection (Witmer et al. 2018; Graumans et al. 2020). Although 3D printers are currently still considered costly for many laboratories in developing countries, the advent of low-cost 3D printers coupled with open-source model files of 3D-printed feeders will facilitate access to this technology, especially for developing countries. The customization of 3D-printed feeders is an advantage that can help reduce costs and modify designs for specific feeding procedures (Ishengoma and Mtaho 2014; Witmer et al. 2018; Graumans et al. 2020). For example, the size of 3D-printed feeders can easily be altered, which facilitates the delivery of small volumes of blood meals to female mosquitoes (Graumans et al. 2020).

Even though 3D-printed feeders can be designed to host a heat source with electronic temperature control, they will likely not provide a solution for a printed built-in heating source (Witmer et al. 2018). Nevertheless, common sources of heat, such as a simple or circulating water bath, can easily continue to function as the heating element for passively heated feeders like Glytube (Costa-da-Silva et al. 2013).

SUMMARY

Research performed with hematophagous mosquitoes is only possible if a blood or other artificial meal can be delivered to mosquitoes in the laboratory. Artificial membrane feeders are devices that allow female mosquitoes to ingest a blood meal without the use of live mammals. By coupling heating and feeding elements, artificial membrane feeders stimulate females to seek a host and offer a pierceable membrane surface through which they can ingest the blood. There are a variety of feeding devices developed for feeding mosquitoes; however, complexity, cost, and efficiency levels compared with live mammals can limit their use. Feeders designed with few components and using accessible materials are desirable, even though they present other limitations like passive heating units. Glytube is an example of a simple and accessible feeder made from a conical centrifuge tube that has been successfully used in several studies to feed mosquitoes with blood or other meals. Moreover, the challenging days of developing simple artificial feeders are numbered since researchers have started designing and customizing low-cost 3D-printed feeders. Shareable open-source models of 3D feeders and the increasing use of 3D printers at universities, research institutes, and laboratories conducting mosquito research will likely lead to the rapid emergence of the next generation of artificial mosquito feeders in the coming decades.

ACKNOWLEDGMENTS

I thank Dr. Matthew DeGennaro for his support, incentive, and review of this protocol. This topic introduction was supported in part by the United States Centers for Disease Control (CDC) Grant 1U01CK000510: Southeastern Regional Center of Excellence in Vector-Borne Diseases: The Gateway Program. The CDC did not have a role in the design of the study, the collection, analysis, or interpretation of data, nor in writing the manuscript.

REFERENCES

Cosgrove JB, Wood RJ. 1995. Probing and gorging responses of three mosquito species to a membrane feeding system at a range of temperatures. *J Am Mosq Control Assoc* **11:** 339–342.

Costa-da-Silva AL. 2022. Artificial membrane feeding mosquitoes in the laboratory with Glytube. *Cold Spring Harbor Protoc* doi:108013

Costa-da-Silva AL, Navarrete FR, Salvador FS, Karina-Costa M, Ioshino RS, Azevedo DS, Rocha DR, Romano CM, Capurro ML. 2013. Glytube: a conical tube and Parafilm M-based method as a simplified device to artificially blood-feed the dengue vector mosquito, *Aedes aegypti*. *PLoS ONE* **8:** e53816. doi:10.1371/journal.pone.0053816

Costa-da-Silva AL, Carvalho DO, Kojin BB, Capurro ML. 2014. Implementation of the artificial feeders in hematophagous arthropod research cooperates to the vertebrate animal use replacement, reduction and refinement (3Rs) principle. *J Clin Res Bioeth* **5:** 167.

Costa-da-Silva AL, Ioshino RS, Araújo HR, Kojin BB, Zanotto PM, Oliveira DB, Melo SR, Durigon EL, Capurro ML. 2017. Laboratory strains of *Aedes aegypti* are competent to Brazilian Zika virus. *PLoS One* **12:** e0171951. doi:10.1371/journal.pone.0171951

Dennis EJ, Goldman OV, Vosshall LB. 2019. *Aedes aegypti* mosquitoes use their legs to sense DEET on contact. *Curr Biol* **29:** 1551–1556.e5. doi:10.1016/j.cub.2019.04.004

Dias L, Bauzer L, Lima J. 2018. Artificial blood feeding for Culicidae colony maintenance in laboratories: does the blood source condition matter? *Rev Inst Med Trop Sao Paulo* **60:** e45. doi:10.1590/s1678-9946201860045

Duvall LB, Basrur NS, Molina H, McMeniman CJ, Vosshall LB. 2017. A peptide signaling system that rapidly enforces paternity in the *Aedes aegypti* mosquito. *Curr Biol* **27:** 3734–3742.e5. doi:10.1016/j.cub.2017.10.074

Duvall LB, Ramos-Espiritu L, Barsoum KE, Glickman JF, Vosshall LB. 2019. Small-molecule agonists of *Ae. aegypti* neuropeptide Y receptor block mosquito biting. *Cell* **176:** 687–701.e5. doi:10.1016/j.cell.2018.12.004

Graumans W, Heutink R, van Gemert GJ, van de Vegte-Bolmer M, Bousema T, Collins KA. 2020. A mosquito feeding assay to examine *Plasmodium* transmission to mosquitoes using small blood volumes in 3D printed nano-feeders. *Parasit Vectors* **13:** 401. doi:10.1186/s13071-020-04269-x

Ishengoma FR, Mtaho AB. 2014. 3D printing: developing countries perspectives. *Int J Comput Appl* **104:** 30–34.

Jové V, Venkataraman K, Gabel TM, Duvall LB. 2020. Feeding and quantifying animal-derived blood and artificial meals in *Aedes aegypti* mosquitoes. *J Vis Exp* **164:** 10.3791/61835.

Kandel Y, Mitra S, Jimenez X, Rodriguez SD, Romero A, Blakely BN, Cho SY, Pelzman C, Hansen IA. 2020. Long-term mosquito culture with SkitoSnack, an artificial blood meal replacement. *PLoS Negl Trop Dis* **14:** e0008591. doi:10.1371/journal.pntd.0008591

Kauffman E, Payne A, Franke MA, Schmid MA, Harris E, Kramer LD. 2017. Rearing of *Culex* spp. and *Aedes* spp. mosquitoes. *Bio Protoc* **7:** e2542. doi:10.21769/BioProtoc.2542

Romano D, Stefanini C, Canale A, Benelli G. 2018. Artificial blood feeders for mosquito and ticks—where from, where to? *Acta Trop* **183:** 43–56. doi:10.1016/j.actatropica.2018.04.009

Witmer K, Sherrard-Smith E, Straschil U, Tunnicliff M, Baum J, Delves M. 2018. An inexpensive open source 3D-printed membrane feeder for human malaria transmission studies. *Malar J* **17:** 282. doi:10.1186/s12936-018-2436-9

Protocol 1

Artificial Membrane Feeding Mosquitoes in the Laboratory with Glytube

Andre Luis Costa-da-Silva[1]

Department of Biological Sciences and Biomolecular Sciences Institute, Florida International University, Miami, Florida 33199, USA

Artificial membrane feeders are devices that replace vertebrate hosts when blood feeding mosquitoes. Created using readily available materials found in mosquito laboratories, Glytube is one of the simplest artificial membrane feeders that scientists can use to provide a blood meal to female mosquitoes in the laboratory. Glytube was optimized for blood feeding the main epidemiologically important species, such as *Aedes aegypti*, *Aedes albopictus*, *Culex quinquefasciatus*, and *Anopheles aquasalis*. Although the Glytube can be used without modifications to blood feed mosquito females, some researchers have modified the device to improve its assembly and to adapt it to different feeding assays that do not involve a basic blood meal, broadening its utility. Here, I describe how to transform a 50-mL conical centrifuge tube into a convenient, versatile, and cost-effective artificial membrane blood feeder.

MATERIALS

It is essential that you consult the appropriate Material Safety Data Sheets and your Institution's Environmental Health and Safety Office for proper handling of equipment and hazardous materials used in this protocol.

Reagents

Commercial defibrinated vertebrate whole blood (Thermo Scientific R54020) or red blood cells concentrate

Do not use expired blood. After the blood is well homogenized, it can be stored in microcentrifuge tubes in aliquots of 1.5–2 mL and kept frozen at −80°C.

Glycerol (100%)

Equipment

Biohazard waste disposal for blood

Parafilm-M

Plastic sealing film (heat-resistant)

Polypropylene conical bottom centrifuge tube with threaded cap (50-mL)

Preferentially, the tubes should be high quality and high strength manufactured tubes to prevent cracks when preparing the tube's cap.

[1]Correspondence: adacosta@fiu.edu

Cite this protocol as *Cold Spring Harb Protoc*; doi:10.1101/pdb.prot108013

Premated adult female mosquitoes that are at least 5-d-old (50 to 100 individuals) placed in an appropriate container/cage (see Step 14)

Aedes aegypti species (Culex quinquefasciatus and Anopheles aquasalis also feed on Glytube; see Costa-da-Silva et al. 2013).

Deprive female mosquitoes of 10% sugar solution for 24 h before feeding (see Step 16).

Scissors

Sharp razor blade

Water baths preset to 37°C and 50°C (see Steps 10 and 11)

If only one water bath is available, preset the temperature to 42°C–45°C (as modified by Jové et al. 2020) (see Step 11).

METHOD

Assembling the Feeding Element of the Glytube

An assembled Glytube and the feeding procedure are shown in Figure 1.

1. Unscrew and remove the cap from a 50-mL conical tube, and then set the tube aside.
2. With the threaded side of the cap facing down on the bench surface, use a sharp razor blade to carefully cut a 2.5-cm diameter hole in the center of the cap.

   Cutting the cap demands some skill and force. As another option, Jové and colleagues suggested using a lathe for better consistency (Jové et al. 2020).
3. Using scissors, cut a square 5.0-cm × 5.0-cm piece of Parafilm. Carefully stretch the Parafilm evenly until the layer becomes translucent so that it is thin enough for mosquitoes to pierce. Stretch the Parafilm over the outer surface of the cap to seal the hole, ensuring the edges of Parafilm adhere well around the lateral friction grooves of the cap. Ensure the layer of Parafilm is tense and visually inspect it for holes to prevent leaks.

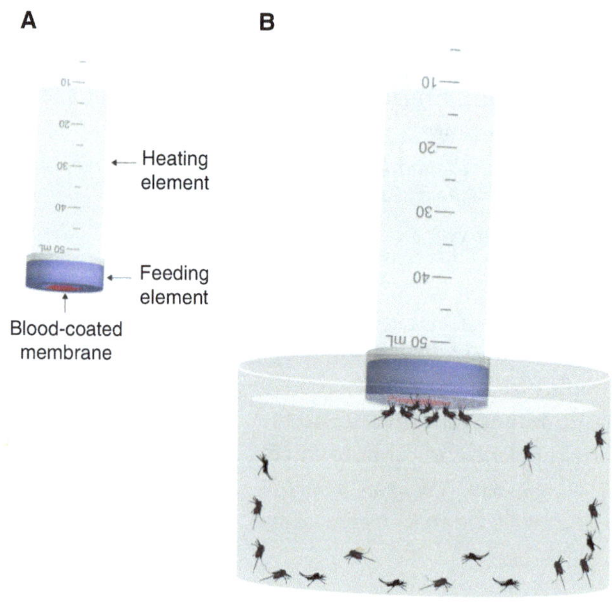

FIGURE 1. Illustration of a Glytube feeder. (*A*) The assembled feeder with blood in the inner cap reservoir; (*B*) mosquito females taking a blood meal through the mesh of a container where they can pierce the membrane on the cap of the Glytube feeder and ingest the blood. Glytube illustrations were angled for a better visualization of the feeding surface.

Cite this protocol as *Cold Spring Harb Protoc*; doi:10.1101/pdb.prot108013

To enhance mosquitoes' attraction to the Glytube feeder, the Parafilm can be odorized by rubbing it on the surface of human skin free of other odors, repellents, or cosmetic products for 30 sec before stretching it; however, odorization is not a required step.

4. With the scissors, cut a 5.0-cm × 2.5-cm strip of Parafilm. While stretching it, wrap it around the grooves of the cap over the edges of the Parafilm membrane created in Step 3. This will help hold the Parafilm membrane firmly in place on the cap.
5. Cut off any excess Parafilm from the cap. Set it aside and be careful not to damage the membrane.

   The feeding element of the Glytube is now complete. The outer part of the cap with the membrane mimics the skin while the inner threaded side forms the meal reservoir that can be filled with 1.0–2.5 mL of blood or other meal.

Assembling the Heating Element of the Glytube

6. Place the 50-mL conical tube upright into a plastic tube rack and fill it with 40 mL of 100% glycerol.
7. Cut a 10-cm × 13-cm piece of heat-resistant plastic sealing film and use it to cover the open top of the 50-mL tube. Fold down the edges of the film against the threads of the tube to create a seal.
8. Using scissors, cut another 5.0-cm × 2.5-cm strip of Parafilm. Hold the folded edges of the sealing film against the lateral threaded area of the tube with the Parafilm strip. Similar to Step 4, roll the Parafilm strip tightly over the threaded area of the tube while stretching it out.

   Alternatively, Jové and colleagues use two 5-cm × 5-cm overlapped pieces of Parafilm to seal the 50-mL conical tube instead of wrapping a second piece of Parafilm around the threaded area of the tube (Jové et al. 2020). When using two 5-cm × 5-cm pieces of Parafilm, they also suggest wrapping a rubber band around the threaded area of the tube to keep the two pieces of Parafilm in place to prevent glycerol leaks.

9. Invert the 50-mL tube to test the heat-resistance sealing film for leaks.

   The heating element of the Glytube feeder is now complete.

Feeding the Mosquitoes

10. Transfer the plastic rack with the sealed tube to the water bath for 15 min at 50°C to prewarm the glycerol.
11. Prewarm an aliquot (1.0–2.5 mL) of commercial defibrinated vertebrate whole blood in the water bath for 15 min at 37°C.

    If only one water bath is available, incubate the sealed glycerol-filled 50-mL conical tube and the whole blood aliquot for 15 min at 42°C–45°C (as modified in Jové et al. 2020).

    Be careful when manipulating the feeding and heating elements when completing Steps 12–17. For a better feeding response, execute Steps 12–16 in no more than 5 min to ensure the whole blood and the heating element do not cool down to room temperature before feeding.

12. Pipette 1.0–2.5 mL of warmed whole blood into the inner part of the 50-mL tube's cap (reservoir) onto the Parafilm membrane surface.

    For more than 50 female mosquitoes, 2.0–2.5 mL of blood is desirable to produce a larger surface of blood coating the membrane through which the females bite.

13. Invert the 50-mL conical tube filled with prewarmed glycerol upside down. With the sealed area of the tube facing the inner part of the cap containing the whole blood, quickly and gently place the tube in the cap and screw the cap about halfway on to hold it in place. To ensure the cap is securely attached to the 50-mL tube, slowly lift the tube off the bench surface.
14. Check the meshed surface side of the cage/container where the females are ready to be fed.

    Ensure that the meshed surface of the cage/container can be pierced by the females' proboscises and that there is sufficient surface area for the females to access the membrane. Verify that the meshed surface is taut and firm before placing the Glytube feeder on it.

15. Place the mounted Glytube onto the mesh such that the Glytube remains upright by itself.

    See Troubleshooting.

16. Ensure the Parafilm membrane of the cap of the Glytube is in contact with the meshed surface of the cage/container and that the female mosquitoes can pierce the Parafilm membrane of the Glytube. Do this by observing whether the females begin feeding and whether they are quickly becoming engorged with blood.

    To improve feeding responses, instead of feeding mosquitoes on 10% sucrose solution, starve the female mosquitoes by only giving them H_2O for 24 h before the blood meal.

17. Allow the female mosquitoes to feed for 30–45 min at 27°C ± 1°C.

    Most females should be fully engorged after this period. Longer feeding times do not increase feeding efficiency.

Removing the Glytube after Feeding

18. After feeding, prepare a sheet of paper towel. Gently remove the Glytube from the meshed surface of the cage/container to avoid leaking blood into the cage/container with the mosquitoes. As soon as you lift the Glytube off the meshed surface, place the paper towel underneath the cap to prevent blood leaks.
19. Discard the Glytube cap (reservoir) with leftover blood and the sealing film from the 50-mL tube in the biohazardous waste bin for blood. The cap can also be reused after being soaked in a bleach solution (1:10 dilution of 5.25% bleach) for 24 h and rinsed in H_2O. The 50-mL tube with glycerol can be recapped with a new screw cap without a hole and kept at −20°C for additional uses.

TROUBLESHOOTING

Problem (Step 15): The meshed surface of the cage/container housing the female mosquitoes before feeding is not firm enough to hold the Glytube. The Glytube tends to tilt over when placed on a loose meshed surface.

Solution: Use adhesive tape to hold the Glytube upright on the cage/container's meshed surface. Stick one end of the tape on the cage/container, pass it over the Glytube, and then stick it to the opposite side of the cage/container. If the mesh is held by a snap top lid, secure the mesh and the lid against the container using adhesive tape. The adhesive tape prevents the mesh from sliding into the container due to the weight of the Glytube feeder.

ACKNOWLEDGMENTS

I thank Dr. Laura Duvall and past and current research groups for the improvements made to the Glytube assembly, as well as Dr. Margareth Capurro and the coauthors of the Glytube manuscript for their support while creating the Glytube feeder. This protocol was supported in part by the United States Centers for Disease Control (CDC) Grant 1U01CK000510: Southeastern Regional Center of Excellence in Vector-Borne Diseases: The Gateway Program. The CDC did not have a role in the design of the study, the collection, analysis, or interpretation of data, nor in writing the manuscript.

REFERENCES

Costa-da-Silva AL, Navarrete FR, Salvador FS, Karina-Costa M, Ioshino RS, Azevedo DS, Rocha DR, Romano CM, Capurro ML. 2013. Glytube: a conical tube and Parafilm M–based method as a simplified device to artificially blood-feed the dengue vector mosquito, *Aedes aegypti*. *PloS ONE* **8:** e53816. doi:10.1371/journal.pone.0053816

Jové V, Venkataraman K, Gabel TM, Duvall LB. 2020. Feeding and quantifying animal-derived blood and artificial meals in *Aedes aegypti* mosquitoes. *J Vis Exp* **164:** 10.3791/61835. doi:10.3791/61835

Cite this protocol as *Cold Spring Harb Protoc*; doi:10.1101/pdb.prot108013

CHAPTER 6

Trypsin, the Major Proteolytic Enzyme for Blood Digestion in the Mosquito Midgut

Andrew B. Nuss[1,2] and Monika Gulia-Nuss[1,3]

[1]*Department of Biochemistry and Molecular Biology,* [2]*Department of Agriculture, Veterinary, and Rangeland Sciences, University of Nevada, Reno, Nevada 89557, USA*

When a female mosquito takes a blood meal, proteolytic activity surges in the midgut. Trypsin-like serine proteases are the major endoproteolytic enzyme induced by feeding in mosquitoes. The mosquito midgut lacks trypsin activity before the blood meal, but in most anautogenous mosquitoes, trypsin activity increases continuously up to 30 h after feeding and subsequently returns to baseline levels by 60 h. Trypsin activity in mosquitoes is restricted entirely to the posterior midgut lumen, where blood is stored and digested. Trypsin enzyme activity can be quantitatively measured using the artificial *N*α-benzoyl-*DL*-arginine 4-nitroanilide hydrochloride substrate, a method described in our associated protocol.

VERTEBRATE BLOOD AS A FOOD SOURCE

Female mosquitoes of several genera such as *Aedes*, *Culex*, and *Anopheles* are known for their ability to transmit many pathogens of human and animal health significance. This pathogen transmission is due to their dependence on a protein-rich vertebrate blood meal that allows oogenesis to proceed. The blood meal is composed of mostly protein (~500 µg in 2 µL of the dry weight of blood), of which 80% consists of hemoglobin (~300 µg), serum albumin (~50 µg), and immunoglobulin (~15 µg) (Isoe et al. 2009a). Blood feeding induces the synthesis and release of proteolytic enzymes in the mosquito midgut, which leads to the degradation of blood meal proteins into peptides and amino acids. Upon feeding, blood is stored in the midgut, the site of both digestion and absorption of nutrients (Fisk 1950). Digestive enzymes are secreted into the midgut lumen within a few hours after feeding. The peak activity of proteolytic enzymes in the yellow fever mosquito, *Aedes aegypti*, is reached after 20–30 h (Gooding 1966a,b; Briegel and Lea 1975). In other species, the precise timing of blood digestion may vary, but their dynamics of protease synthesis are similar to that of *Ae. aegypti.*

PEPTIDASES AND BLOOD DIGESTION

Peptidases are proteolytic enzymes that can be subdivided into endopeptidases represented by serine proteases either trypsin-like (Felix et al. 1991; Barillas-Mury and Wells 1993; Kalhok et al. 1993) or chymotrypsin-like (Jiang et al. 1997; Bian et al. 2008) and exopeptidases that function as aminopep-

[3]Correspondence: mgulianuss@unr.edu

Cite this introduction as *Cold Spring Harb Protoc;* doi:10.1101/pdb.top107656

tidases (Noriega et al. 2002) and carboxypeptidases (Edwards et al. 2000; Isoe et al. 2009a). Serine peptidases are a class of endopeptidases that are characterized by the presence of the following three invariant residues at the active site: serine, aspartic acid, and histidine. Trypsins are the most extensively studied group of digestive serine proteases in insects and are responsible for the bulk of endoproteolytic activity found in the midguts of blood-fed mosquitoes and other hematophagous insects (Terra and Ferreira 1994; Lemos et al. 1996; Noriega and Wells 1999; Okuda et al. 2002). These proteases are released in a biphasic manner, with an initial "early" phase that begins at ~2 h after a blood meal and continues for 6 h post–blood meal (PBM), followed by the "late" phase beginning ~12 h PBM and reaching maximal activity at 24–30 h PBM.

The *Ae. aegypti* genome has 378 serine protease genes and 172 trypsin-like protease genes (Matthews et al. 2018); however, the expression of only a few genes changes in response to a blood meal. Serine proteases are known to mediate immune responses (Gorman and Paskewitz 2001) and the majority might play a role in immunity, not in blood digestion. The gene for a serine protease named *early trypsin* was cloned and characterized and proposed to be responsible for trypsin-like activity in the early phase of blood digestion (Kalhok et al. 1993; Noriega et al. 1996). Two other genes, *late trypsin* (Barillas-Mury et al. 1991) and *5G1* (Kalhok et al. 1993) were explored because of their high expression in blood-fed mosquitoes, and one or both were hypothesized to contribute toward the late-phase trypsin activity. RNA interference knockdown studies (Brackney et al. 2008; Isoe et al. 2009b) suggest that a late-phase protease (*AaSPVI*, formerly known as *5G1*) is the major trypsin in mosquito midguts, not the *late trypsin* gene identified earlier (Barillas-Mury et al. 1991). At the peak of protease activity (20–30 h PBM), late-phase trypsin concentrations reach ~4–6 µg/midgut (Graf et al. 1988). Neither the early nor the late forms were detected in homogenates of unfed females, suggesting that they are translated de novo after feeding (Graf and Briegel 1989). Transcription is dependent on both the quality and quantity of the protein in the meal (Noriega et al. 1994). Similarly, seven trypsin genes (*Antryp 1–7*) exist in tandem in the genome *of Anopheles gambiae* (Müller et al. 1993); however, only two of them (*Antryp 1* and *Antryp 2*), have been shown to be expressed in the midguts in response to a blood meal (Müller et al. 1993, 1995) and are most likely involved in blood digestion.

TRYPSIN ASSAY

The characterization of purified mosquito trypsins indicates that they have a range of properties in common with other known serine proteases. These similarities include molecular weight range, substrate, inhibitor specificity (Graf et al. 1991), and the level of amino acid homology with other known trypsin sequences (Barillas-Mury et al. 1991). Therefore, the trypsin assays, such as the *N*α-benzoyl-*DL*-arginine 4-nitroanilide hydrochloride (BApNA) assay, developed for vertebrate trypsin (Erlanger et al. 1961), also work with mosquito trypsin and are used frequently to determine blood digestion in mosquitoes and other blood-feeding arthropods.

CONCLUSION

Trypsin is one of the most important enzymes for the digestion of the blood meal in the mosquito midgut. Several factors, such as blood quantity, blood quality, and time since the blood meal was acquired by the female, affect trypsin levels in the mosquito midgut. Therefore, determining trypsin levels using a simple and accurate colorimetric assay, such as the BApNA assay (see Protocol 1: Analyzing Blood Digestion: Estimation of Active Trypsin Levels in the Mosquito Midgut [Reyes et al. 2023] for detailed directions) provides an opportunity for researchers to understand the progression of blood digestion.

 Cite this introduction as *Cold Spring Harb Protoc*; doi:10.1101/pdb.top107656

REFERENCES

Barillas-Mury C, Wells MA. 1993. Cloning and sequencing of the blood meal-induced late trypsin gene from the mosquito *Aedes aegypti* and characterization of the upstream regulatory region. *Insect Mol Biol* **2:** 7–12. doi:10.1111/j.1365-2583.1993.tb00119.x

Barillas-Mury C, Graf R, Hagedorn HH, Wells MA. 1991. cDNA and deduced amino acid sequence of the blood meal-induced trypsin from the mosquito, *Aedes aegypti. Insect Biochem* **21:** 825–831. doi:10.1016/0020-1790(91)90089-W

Bian G, Raikhel AS, Zhu J. 2008. Characterization of a juvenile hormone-regulated chymotrypsin-like serine protease gene in *Aedes aegypti* mosquito. *Insect Biochem Mol Biol* **38:** 190–200. doi:10.1016/j.ibmb.2007.10.008

Brackney DE, Foy BD, Olson KE. 2008. The effects of midgut serine proteases on Dengue virus type 2 infectivity of *Aedes aegypti. Am J Trop Med Hyg* **79:** 267–274. doi:10.4269/ajtmh.2008.79.267

Briegel H, Lea AO. 1975. Relationship between protein and proteolytic activity in the midgut of mosquitoes. *J Insect Physiol* **21:** 1597–1604. doi:10.1016/0022-1910(75)90197-3

Edwards MJ, Moskalyk LA, Donelly-Doman M, Vlaskova M, Noriega FG, Walker VK, Jacobs-Lorena M. 2000. Characterization of a carboxypeptidase A gene from the mosquito, *Aedes aegypti. Insect Mol Biol* **9:** 33–38. doi:10.1046/j.1365-2583.2000.00159.x

Erlanger BF, Kokowsky N, Cohen W. 1961. The preparation and properties of two new chromogenic substrates of trypsin. *Arch Biochem Biophys* **95:** 271–278. doi:10.1016/0003-9861(61)90145-X

Felix CR, Betschart B, Billingsley PF, Freyvogel TA. 1991. Post-feeding induction of trypsin in the midgut of *Aedes aegypti* is separable into two cellular phases. *Insect Biochem* **21:** 197–203. doi:10.1016/0020-1790(91)90050-O

Fisk FW. 1950. Studies on proteolytic digestion in adult *Aedes aegypti* mosquitoes. *Ann Entomol Soc Am* **43:** 555–572. doi:10.1093/aesa/43.4.555

Gooding RH. 1966a. Physiological aspects of digestion of the blood meal by *Aedes aegypti* (Linnaeus) and *Culex fatigans* Wiedemann. *J Med Ent* **3:** 53–60. doi:10.1093/jmedent/3.1.53

Gooding RH. 1966b. *In vitro* properties of proteinases in the midgut of adult *Aedes aegypti* L. and *Culex fatigans* (Wiedemann). *Comp Biochem Physiol* **17:** 115–127. doi:10.1016/0010-406X(66)90013-2

Gorman MJ, Paskewitz SM. 2001. Serine proteases as mediators of mosquito immune responses. *Insect Biochem Mol. Biol* **31:** 257–262. doi:10.1016/s0965-1748(00)00145-4

Graf R, Briegel H. 1989. The synthetic pathway of trypsin in the mosquito *Aedes aegypti* L. (Diptera: Culicidae) and *in vitro* stimulation in isolated midguts. *Insect Biochem* **19:** 129–137. doi:10.1016/0020-1790(89)90083-8

Graf R, Binz H, Briegel H. 1988. Monoclonal antibodies as probes for *Aedes aegypti* trypsin. *Insect Biochem* **18:** 463–470. doi:10.1016/0020-1790(88)90063-7

Graf R, Bohlen H, Briegel H. 1991. Structural diversity of trypsin from different mosquito species feeding on vertebrate blood. *Experientia* **47:** 603–609.

Isoe J, Zamora J, Miesfeld RL. 2009a. Molecular analysis of the *Aedes aegypti* carboxypeptidase gene family. *Insect Biochem Mol Biol* **39:** 68–73. doi:10.1016/j.ibmb.2008.09.006

Isoe J, Rascón AA, Kunz S, Miesfeld RL. 2009b. Molecular genetic analysis of midgut serine proteases in *Aedes aegypti* mosquitoes. *Insect Biochem Mol Biol* **39:** 903–912. doi:10.1016/j.ibmb.2009.10.008

Jiang Q, Hall M, Noriega FG, Wells M. 1997. cDNA cloning and pattern of expression of an adult, female-specific chymotrypsin from *Aedes aegypti* midgut. *Insect Biochem Mol Biol* **27:** 283–289. doi:10.1016/S0965-1748(97)00001-5

Kalhok SE, Tabak LM, Prosser DE, Brook W, Downe AE, White BN. 1993. Isolation, sequencing and characterization of two cDNA clones coding for trypsin-like enzymes from the midgut of *Aedes aegypti. Insect Mol Biol* **2:** 71–79. doi:10.1111/j.1365-2583.1993.tb00127.x

Lemos FJA, Cornell AJ, Jacobs-Lorena M. 1996. Trypsin and aminopeptidase gene expression is affected by age and food composition in *Anopheles gambiae. Insect Biochem Mol Biol* **26:** 651–658. doi:10.1016/S0965-1748(96)00014-8

Matthews BJ, Dudchenko O, Kingan SB, Koren S, Antoshechkin I, Crawford JE, Glassford WJ, Herre M, Redmond SN, Rose NH, et al. 2018. Improved reference genome of *Aedes aegypti* informs arbovirus vector control. *Nature* **563:** 501–507. doi:10.1038/s41586-018-0692-z

Müller HM, Crampton JM, della Torre A, Sinden R, Crisanti A. 1993. Members of a trypsin gene family in *Anopheles gambiae* are induced in the gut by blood meal. *EMBO J* **12:** 2891–2900. doi:10.1002/j.1460-2075.1993.tb05951.x

Müller HM, Catteruccia F, Vizioli J, della Torre A, Crisanti A. 1995. Constitutive and blood meal-induced trypsin genes in *Anopheles gambiae. Exp Parasitol* **81:** 371–385. doi:10.1006/expr.1995.1128

Noriega FG, Wells MA. 1999. A molecular view of trypsin synthesis in the midgut of *Aedes aegypti. J Insect Physiol* **45:** 613–620. doi:10.1016/S0022-1910(99)00052-9

Noriega FG, Barillas-Mury C, Wells MA. 1994. Dietary control of late trypsin gene transcription in *Aedes aegypti. Insect Biochem Mol Biol* **24:** 627–631. doi:10.1016/0965-1748(94)90099-X

Noriega FG, Wang XY, Pennington JE, Barillas-Mury CV, Wells MA. 1996. Early trypsin, a female-specific midgut protease in *Aedes aegypti*: isolation, aminoterminal sequence determination, and cloning and sequencing of the gene. *Insect Biochem Mol Biol* **26:** 119–126. doi:10.1016/0965-1748(95)00068-2

Noriega FG, Edgar KA, Bechet R, Wells MA. 2002. Midgut exopeptidase activities in *Aedes aegypti* are induced by blood feeding. *J Insect Physiol* **48:** 205–212. doi:10.1016/S0022-1910(01)00165-2

Okuda K, de Souza A, Ribolla PEM, de Bianchi AG, Bijovsky AT. 2002. Functional morphology of adult female *Culex quinquafasciatus* midgut during blood digestion. *Tissue Cell* **34:** 210–219. doi:10.1016/S0040-8166(02)00032-0

Reyes J, Nuss A, Gulia-Nuss M. 2023. Analyzing blood digestion: estimation of active trypsin levels in the mosquito midgut. *Cold Spring Harb Protoc* doi:10.1101/pdb.prot108224

Terra WR, Ferreira C. 1994. Insect digestive enzymes: properties, compartmentalization and function. *Comp Biochem Physiol B* **109:** 1–62. doi:10.1016/0305-0491(94)90141-4

Protocol 1

Analyzing Blood Digestion: Estimation of Active Trypsin Levels in the Mosquito Midgut

Jeremiah Reyes,[1] Andrew Nuss,[1,2] and Monika Gulia-Nuss[1,3]

[1]*Department of Biochemistry and Molecular Biology,* [2]*Department of Agriculture, Veterinary, and Rangeland Sciences, University of Nevada, Reno, Nevada 89557, USA*

The *N*α-benzoyl-DL-arginine 4-nitroanilide hydrochloride (BApNA) assay is widely used to quantify trypsin in mosquito midguts and is highly sensitive. BApNA is a chromogenic substrate for proteolytic enzymes such as trypsin and amidase. Hydrolysis of BApNA at the bond between the arginine and the p-nitroaniline moieties releases the chromophore p-nitroaniline, which is detected by colorimetric analysis. The intensity of the color is directly proportional to the amount of trypsin in the solution. Here, we present a trypsin measurement assay specifically using the BApNA substrate.

MATERIALS

It is essential that you consult your appropriate Material Safety Data Sheets and your institution's Environmental Health and Safety Office for proper handling of equipment and hazardous materials used in this protocol.

RECIPES: Please see the end of this protocol for recipes indicated by <R>. Additional recipes can be found online at http://cshprotocols.cshlp.org/site/recipes.

Reagents

Adult female mosquitoes in a cage

This protocol is optimized for Aedes aegypti *at 16–24 h post–blood meal (Gulia-Nuss et al. 2011; Nuss et al. 2018)*

BApNA stock (10-nM) <R>
Deionized H_2O
Ice
Phosphate-buffered saline (PBS; pH 7.0; Santa Cruz Biotechnology SC-24946)
Tris-HCl-$CaCl_2$ buffer (0.02-M) <R>
Trypsin stock <R>

Reagents

Aspirator (Bioquip) or Bug vacuum (Nature bound bug catcher)
Centrifuge (refrigerated)
Clear-bottom plates (96 wells)
Dissecting microscope with illumination
Excel software

[3]Correspondence: mgulianuss@unr.edu

Cite this protocol as *Cold Spring Harb Protoc*; doi:10.1101/pdb.prot108224

Forceps
Microscope slides
Rocker
Sonicator (Qsonica Q55)
Spectrophotometer capable of measuring absorbance at 405 or 410 nm
Tubes (1.7-mL and 15-mL)

METHOD

In this assay, the active trypsin concentration is determined by enzymatic cleavage of the BApNA substrate into Nα-benzoyl-DL-arginine and p-nitroaniline. p-Nitroaniline gives a yellow color to the reaction and is measured by the spectrophotometer (Fig. 1).

This protocol is an adaptation of a method developed by Preiser et al. (1975).

1. Take trypsin stocks out of the −20°C freezer and thaw on ice. Take BApNA stocks out of the freezer and thaw in deionized H_2O at room temperature.
2. Make a BApNA standard by diluting the 10-mM stock to a 4-mM concentration using Tris-HCl-$CaCl_2$ buffer (e.g., add 1 mL of 10-mM BApNA to 1.5 mL Tris-HCl-$CaCl_2$ buffer to get 4 mM).
   This standard is used in Step 12. The complete solubilization of BApNA is critical.
3. Transfer mosquitoes with an aspirator from the mosquito cage into 15-mL tubes. Keep the tubes containing mosquitoes on ice for 10–15 min.
4. With forceps, take out the mosquitoes from the tube and keep them in a drop of cold PBS on a microscope slide. Using two forceps, dissect mosquito midguts individually in PBS buffer under a microscope (Fig. 2).
5. Keep a 1.7-mL tube containing 200 µL of Tris-HCl-$CaCl_2$ buffer on ice. Using forceps, transfer midguts immediately to this cold buffer.
   Keep samples on ice until Step 12.
6. Either proceed immediately with Step 7 or store samples in tubes at −80°C until the next steps can be performed.
7. Use the sonicator to fragment midgut samples on ice with 10 sec on and 30 sec off (five times), and then centrifuge samples for 12 min at 13,000*g* at 4°C. Place the supernatant in a new tube and discard the tubes containing the pellet.
   If a sonicator is not available, a microhomogenizer can be used to disrupt the midgut tissue.

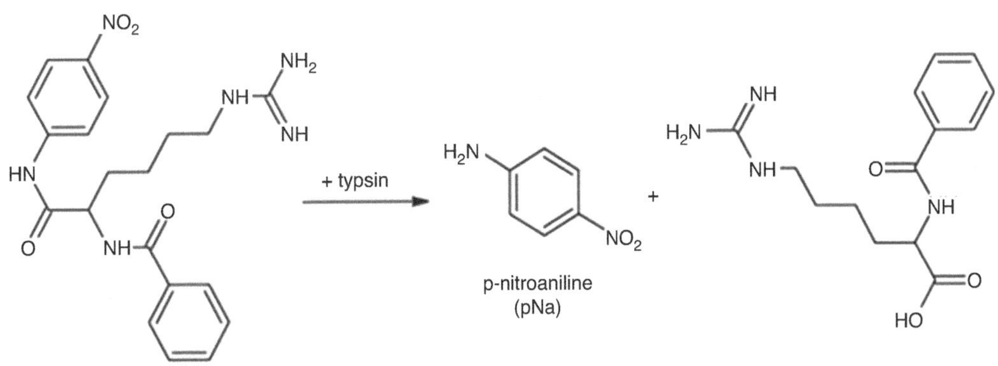

Nα-benzoyl-arginine-p-nitroanilide (BApNA)

FIGURE 1. Trypsin-*N*α-benzoyl-DL-arginine 4-nitroanilide hydrochloride (BApNA) reaction. Hydrolysis of BApNA at the bond between the arginine and the p-nitroaniline moieties by trypsin releases the chromophore p-nitroaniline, resulting in a color detectable at 405–410 nm (Guedidi et al. 2012).

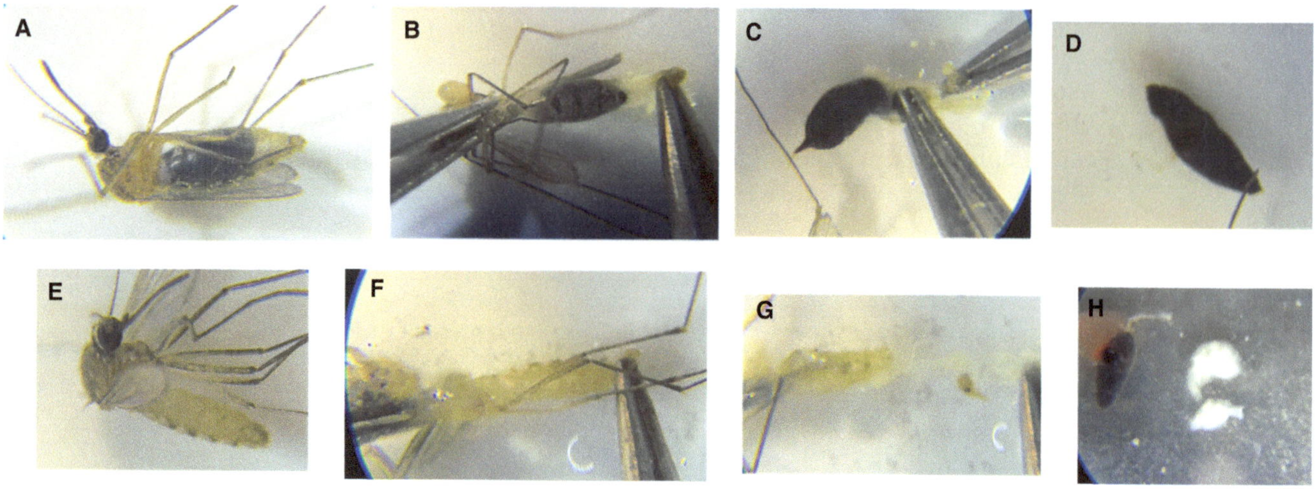

FIGURE 2. Dissection of a mosquito midgut. (*A–D*) The photos show the dissection of blood-fed *Culex pipiens* females. (*A*) Blood-fed female on a microscope slide in a drop of PBS. (*B*) One forceps holds the thorax and the other is used to pull on the abdominal segments. (*C*) After the abdomen is separated, pressure is applied with the forceps to remove the midgut. (*D*) Midgut from a blood-fed female. (*E–H*) Dissection of unfed mosquito midgut. (*E*) An unfed *Cx. pipiens* female. (*F*) Two forceps are used to pull the abdomen away from the thorax. (*G*) The midgut is removed from the abdomen. (*H*) Blood-fed midgut (*left*) and unfed midguts from two females (*right*).

8. While samples are in the centrifuge, dilute trypsin standards according to Table 1.
9. Prepare as many tubes as the number of samples. Label the tubes according to the sample name and number.
10. Add 90 µL of Tris-HCl-$CaCl_2$ buffer to each sample tube.

    See Troubleshooting.

11. Add 10 µL of the sample from Step 7 to the corresponding tube containing 90 µL of Tris-HCl-$CaCl_2$ buffer (from Step 9).

    The total volume will be 100 µL.

    (Optional) When testing the protocol for the first time and with a different mosquito species, take readings using different volumes of the sample. Depending on the time after blood feeding and trypsin activity in the mosquito midgut, a higher or lower sample volume (Step 12) might be needed to keep your sample reading within the range of the standard line. If you use a higher volume of sample, measure and record absorbance at 405 nm for all samples. This is important because at the higher concentration, the blood content of the mosquito midgut might result in a higher absorbance value.

12. Add 200 µL of the BApNA 10-mM stock to all standards and sample tubes.

    The total volume in the tubes is now 300 µL.

13. Incubate samples and standards for 15 min at room temperature on a rocker.

    See Troubleshooting.

14. Add 75 µL of the sample from each tube in triplicate wells of a clear-bottom plate.

TABLE 1. Trypsin standard dilutions for the assay

Standard tube	Trypsin (µL)	Tris-HCl-$CaCl_2$ buffer (µL)	Absolute concentration (µg/75 µL)
1	100	900	2.5
2	75	925	1.87
3	50	950	1.25
4	25	975	0.62
5	10	990	0.25
6	500 µL from tube 5 (1:1)	500	0.125
7	500 µL from tube 6 (1:1)	500	0.062
8	0	1000	0

Cite this protocol as *Cold Spring Harb Protoc*; doi:10.1101/pdb.prot108224

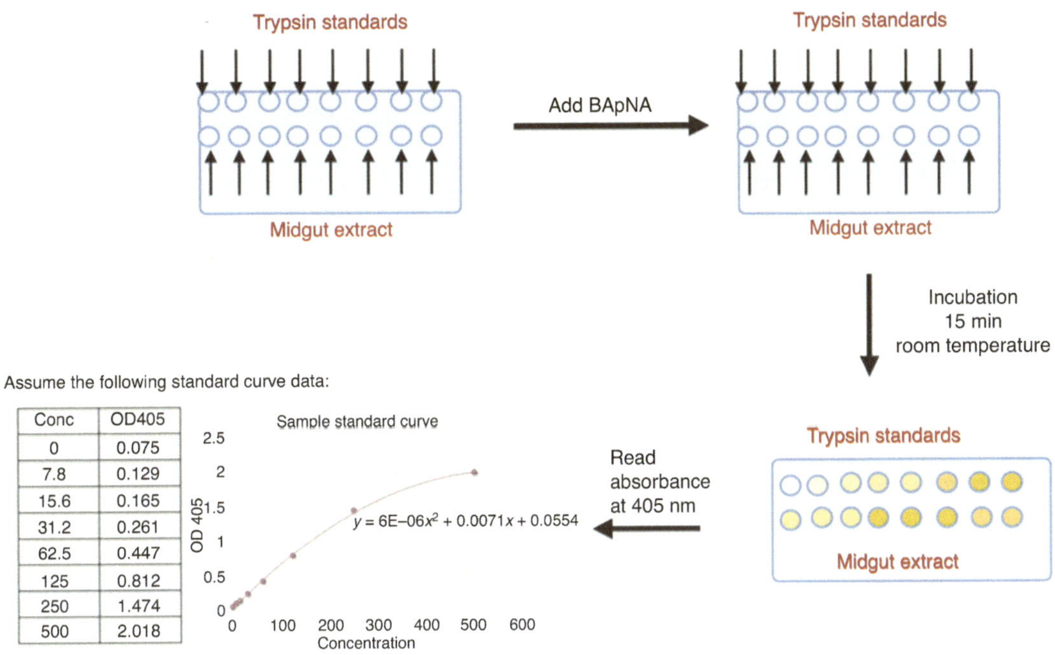

Conc	OD405
0	0.075
7.8	0.129
15.6	0.165
31.2	0.261
62.5	0.447
125	0.812
250	1.474
500	2.018

FIGURE 3. BApNA assay setup.

15. Read absorbance at 405 nm on a spectrophotometer (Fig. 3).

    If the absorbance is greater than the highest standard, the sample is too concentrated.

    See Troubleshooting.

    (Optional) If samples were measured in the optional step mentioned in Step 11, determine the change in absorbance (ΔA) for each sample by subtracting the initial absorbance from the final absorbance (Step 15). This difference should be very small to none.

16. To determine the concentration of the unknown, plot the standards (known concentrations) with absorbance versus concentration in a Microsoft Excel sheet or similar program. Calculate the calibration curve ($y = mx + b$), where y = absorbance, x = concentration, and m = slope of the line. Plot the unknown absorbance and calculate for x, the concentration of the unknown samples. Adjust the value for each sample by multiplying by the sample dilution factor.

TROUBLESHOOTING

Problem (Step 10): The blank sample has a yellow color.
Solution: This indicates contamination of the blank. Make sure to change pipette tips and not add any trypsin to the blank by accident.

Problem (Step 13): Samples were incubated for >15 min.
Solution: If for any reason the sample absorbance cannot be recorded at 15 min, the reaction can be stopped by adding 200 µL of 20% citric acid.

Problem (Step 15): No trypsin activity was detected in samples, or the absorbance values are too low.
Solution: This protocol is optimized for a single midgut of an *Ae. aegypti* blood-fed female at 16–24 h post–blood meal. If a different species or time point is used, increase or decrease the volume of the

midgut extract (Step 11) and accordingly adjust the Tris-HCl-$CaCl_2$ volume (Step 10). If the BApNA solution is hazy, continue to stir over gentle heat (not greater than 65°C) until the solution becomes clear. Do not use the solution if it turns yellow. The yellow color indicates possible chemical decomposition of the substrate due to overheating.

Problem (Step 15): Absorbance is greater than the highest standard.
Solution: If the absorbance is too high, the sample is too concentrated for the standard line. In that case, dilute the sample in Step 11 and use a gradient to identify the best concentration.

DISCUSSION

BApNA is a colorless, chromogenic substrate for trypsin. Hydrolysis of the BApNA at the bond between the arginine and the p-nitroaniline moieties releases the chromophore p-nitroaniline resulting in bright yellow color, which is detected by colorimetric analysis. The higher intensity of color produced in a reaction correlates with the amount of active trypsin. Because this assay requires cleavage of the BApNA substrate by the trypsin enzyme, active trypsin levels in the mosquito gut can be measured rapidly and cost-effectively.

RECIPES

BApNA Stock (10-mM)

Add 0.0174 g of *N*α-benzoyl-DL-arginine 4-nitroanilide hydrochloride (BApNA; ≥99%, TLC; Sigma-Aldrich B3133) to 4 mL of dimethyl sulfoxide (Sigma-Aldrich 276855) in a 5-mL tube to make a 10-mM stock. Store aliquots for up to 2 wk at −20°C.

Tris-HCl-$CaCl_2$ Buffer

Add 1.21 g of Tris and 1.11 g of $CaCl_2$ to 900 mL of deionized H_2O. Adjust the pH to 8.0 by adding concentrated HCl and then bring the volume of the final solution to 1 L with deionized H_2O. Store for up to 6 mo at room temperature.

Trypsin Stock

Weigh 0.002 g of trypsin from the bovine pancreas (Sigma-Aldrich T1426) and dissolve it in 2 mL of Tris-HCl-$CaCl_2$ buffer <R>. This trypsin stock can be aliquoted into eight 250-µL aliquots and stored for up to 2 wk at −20°C.

REFERENCES

Guedidi S, Portugal CA, Innocent C, Janot JM, Deratani A, Crespo JG. 2012. Fluorescence monitoring of trypsin adsorption in layer-by-layer membrane systems. *Enzyme Microb Technol* **51:** 325–333. doi:10.1016/j.enzmictec.2012.07.016

Gulia-Nuss M, Robertson AE, Brown MR, Strand MR. 2011. Insulin-like peptides and the target of rapamycin pathway coordinately regulate blood digestion and egg maturation in the mosquito *Aedes aegypti*. *PLoS ONE* **6:** e20401. doi:10.1371/journal.pone.0020401

Nuss AB, Brown MR, Murty US, Gulia-Nuss M. 2018. Insulin receptor knockdown blocks filarial parasite development and alters egg production in the southern house mosquito, *Culex quinquefasciatus*. *PLoS Negl Trop Dis* **12:** e0006413. doi:10.1371/journal.pntd.0006413

Preiser H, Schmitz J, Maestracci D, Crane RK. 1975. Modification of an assay for trypsin and its application for the estimation of enteropeptidase. *Clin Chim Acta* **59:** 169–175. doi:10.1016/0009-8981(75)90025-X

Cite this protocol as *Cold Spring Harb Protoc*; doi:10.1101/pdb.prot108224

CHAPTER 7

Methods to Assess Blood and Nectar Meals in *Aedes aegypti* Mosquitoes

Krithika Venkataraman,[1,3] Veronica Jové,[1,3] and Laura B. Duvall[2,4]

[1]*The Rockefeller University, New York, New York 10065, USA;* [2]*Department of Biological Sciences, Columbia University in the City of New York, New York, New York 10027, USA*

Male and female *Aedes aegypti* mosquitoes survive by feeding on floral nectar for metabolic energy, but females require blood protein, obtained from biting a host, for egg development. Although males exclusively derive energy from nectar sugars, females must select the meal that best matches their present metabolic and reproductive needs. In females, blood and nectar promote independent feeding behaviors with distinct sensory appendages, meal sizes, digestive tract targets, and metabolic fates. Understanding how male and female mosquitoes recognize, locate, and metabolize nutrients is essential for characterizing the survival and reproductive capabilities of this mosquito. Here, we provide an introduction to blood versus nectar feeding and methods to quantify nectar and blood meal sizes in individual *Ae. aegypti* mosquitoes. Precise quantification of meal size is crucial for ensuring consistency in assays that record events downstream of feeding behavior, including host attraction or fecundity.

BLOOD AND NECTAR MEALS

Although male and female *Aedes aegypti* mosquitoes can sustain energy metabolism with carbohydrates from nectar sugars, females require protein from vertebrate host blood to develop eggs and reproduce (Fig. 1A; Briegel 2003; Duvall et al. 2019). Nectar feeding and blood feeding are linked to behaviorally discrete feeding programs, which are defined by the use of different feeding appendages, meal sizes, and digestive tract targets (Fig. 1B; Gordon and Lumsden 1939; Trembley 1952). Nectar is detected by the labium, an external feeding appendage that ensheathes the stylet (Fig. 1C; Pappas and Larsen 1978; Sanford et al. 2013). Blood is detected by the stylet, which comes into direct contact with blood as a female pierces through her host's skin and retracts the labium (Fig. 1D; Gordon and Lumsden 1939; Trembley 1952; Jové et al. 2020). Females typically take small, frequent nectar meals with an average meal size of 0.87 μL (Jové et al. 2020). In contrast, blood meals approximately double the body weight of a female and have an average size of 3.20 μL (Liesch et al. 2013; Sissoko et al. 2019; Jové et al. 2020). The large blood meal contains enough protein to produce 100–150 eggs and suppress the drive to blood feed for multiple days (Brown et al. 1994; Klowden and Lea 1979; Duvall et al. 2019). The nectar meal is typically routed first to a temporary food storage organ called the crop, leaving the midgut free in case the female encounters an opportunity to feed on blood (Fig. 1B; Gordon and Lumsden 1939; Trembley 1952). Female *Ae. aegypti* mosquitoes thus have mutually exclusive

[3]These authors contributed equally to this work.

[4]Correspondence: lbd2126@columbia.edu

Cite this introduction as *Cold Spring Harb Protoc;* doi:10.1101/pdb.top107657

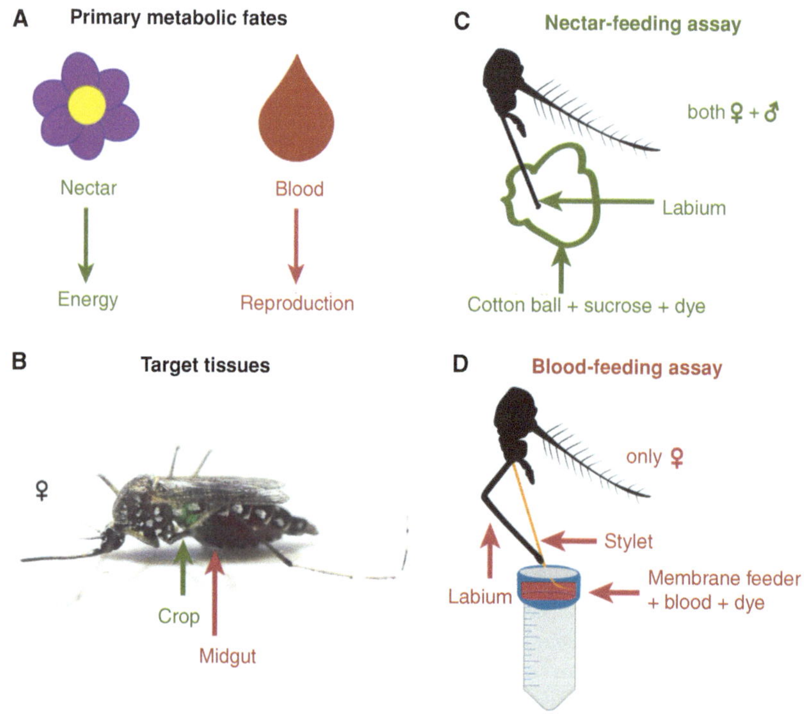

FIGURE 1. Nectar and blood feeding in mosquitoes involve distinct sensory appendages, target tissues, and metabolic fates. (*A*) Nectar and blood meals have distinct metabolic fates in energy and reproduction, respectively. (*B*) Photograph of an *Aedes aegypti* female with a 10% sucrose meal containing food dye in the crop (green), and a sheep blood meal in the midgut (red), illustrating the different target tissues for different meal types. Schematics of example nectar-feeding (*C*) and blood-feeding (*D*) assays that can be used to deliver meals to mosquitoes in the laboratory. (Adapted from Jové et al. 2020.)

feeding programs for blood and nectar meals—from the sensory periphery, to visceral organs, and to metabolic fate (Fig. 1).

Features of each behavioral program can be precisely quantified in the laboratory using nectar- and blood-feeding assays (Costa-da-Silva et al. 2013; Liesch et al. 2013). Room-temperature sucrose meals are presented to females on a cotton ball so that the labium directly contacts the meal upon landing, similar to nectar feeding from flowers (Fig. 1C). In blood-feeding assays, warmed blood meals are offered to females in the presence of attractive human cues like CO_2 and heat to promote landing on an artificial feeder (Fig. 1D; McMeniman et al. 2014; Liu and Vosshall 2019). An artificial membrane prevents the female from directly contacting the blood meal, similar to the way that skin prevents direct blood contact when the female lands on a host. Like with skin, the female must pierce the membrane with the stylet to directly contact the blood meal.

Nectar Feeding

In the wild, female and male *Ae. aegypti* mosquitoes obtain sugars from floral nectar, which is primarily composed of sucrose, fructose, and glucose (Van Handel 1972). In the laboratory, mosquitoes are routinely reared on 10%–20% sucrose solutions as a substitute for floral nectar. When mosquitoes feed on nectar and sucrose solutions, the nectar-feeding appendage, the labium, directly contacts the meal. Sugars are detected by gustatory receptors, and exogenous activation of neurons that express a conserved sweet taste receptor, *Aedes aegypti Gr4*, is sufficient to promote nectar-feeding behavior in the absence of a real nectar meal in these mosquitoes (Jové et al. 2020). It is important to quantify nectar-feeding behavior, because energy derived from sugar metabolism can impact longevity, fertility, and flight time (Harrington et al. 2001; Briegel 2003; Barredo and DeGennaro 2020). In females, nutrients from sugars can also synergize with those from blood to control fecundity. For

Cite this introduction as *Cold Spring Harb Protoc*; doi:10.1101/pdb.top107657

example, isoleucine from human blood enhances egg production only in females that have fed on sugar (Harrington et al. 2001).

Blood Feeding

Female *Ae. aegypti* must ingest vertebrate blood protein to reproduce, and domestic strains of this species preferentially bite human hosts to obtain the required nutrients for egg production (McBride et al. 2014). As a female bites her victim, the needle-like stylet pierces the skin and injects saliva. Compounds in the female's saliva trigger the host's immune response and prevent the blood from clotting while she feeds. As she pumps blood through the stylet, the female's meal is directed to the midgut for digestion (Gordon and Lumsden 1939; Trembley 1952). When allowed to feed to repletion, the volume of blood ingested reliably doubles the female's body weight and takes several days to fully digest (Klowden 1990). Upon ingesting the blood meal, the protein obtained from blood suppresses a female's attraction to human hosts and triggers egg development (Judson 1967; Klowden and Lea 1979; Davis 1984a,b; Klowden 1990; Duvall et al. 2019).

During blood feeding, *Ae. aegypti* are capable of transmitting viral and parasitic pathogens that cause diseases including dengue, lymphatic filariasis, yellow fever, chikungunya, and Zika (World Health Organization 2020). When a mosquito consumes a blood meal from an infected host, she ingests blood-borne pathogens into her midgut, which can then enter the mosquito's circulatory system and migrate to her salivary glands where they are poised to be passed to the next animal that she bites (Rückert and Ebel 2018). After the female lays her clutch of eggs, she is once again motivated to blood feed from her next host, initiating another cycle of host seeking and egg laying (Judson 1967; Brown et al. 1994; Liesch et al. 2013; Duvall et al. 2019). When they feed on subsequent hosts, infected female mosquitoes spread disease by injecting pathogens along with their saliva. Thus, blood-feeding behavior is a crucial step for both reproduction and disease transmission.

Accurate and precise quantification of the blood meal is important for many experimental approaches, because protein content ultimately controls oogenesis and host-seeking suppression behavior (Briegel 1990). If a female fills to repletion on blood, she can develop an entire clutch of more than 100 eggs and will fully suppress her host-seeking drive (Liesch et al. 2013; Duvall et al. 2019). However, partial meals may lead to no or reduced egg production (Briegel 1990) and incomplete host-seeking suppression (Klowden and Lea 1978). Consequently, females will continue to host seek to find an additional source of blood and thereby refeed on multiple hosts. Refeeding on multiple hosts facilitates disease transmission and can also occur if the female experienced poor larval nutrition (Farjana and Tuno 2013).

QUANTIFYING MEAL VOLUME

Selecting a Method

Meal size can be quantified as either an end-point assay by measuring fluorescence (Duvall et al. 2019; Jové et al. 2020) after mosquitoes have consumed meals containing known amounts of fluorescent dye or, dynamically, using video recording and live tracking (Hol et al. 2020). When choosing a measurement method, it is important to consider whether an end-point assay will provide sufficient temporal resolution of the feeding event, and whether live animals will be needed for subsequent experimentation (Fig. 2A). If live mosquitoes are required for downstream experiments, meal size consumed can also be measured by weighing small groups of anesthetized mosquitoes (Fig. 2B). Weights can be benchmarked to known volumes determined by the end-point homogenization assay in which fluorescent dyes are added to the meal (Fig. 2C). To quantify features of blood-feeding behavior prior to engorgement, video recording can be combined with either quantification assay (Fig. 2D; Hol et al. 2020).

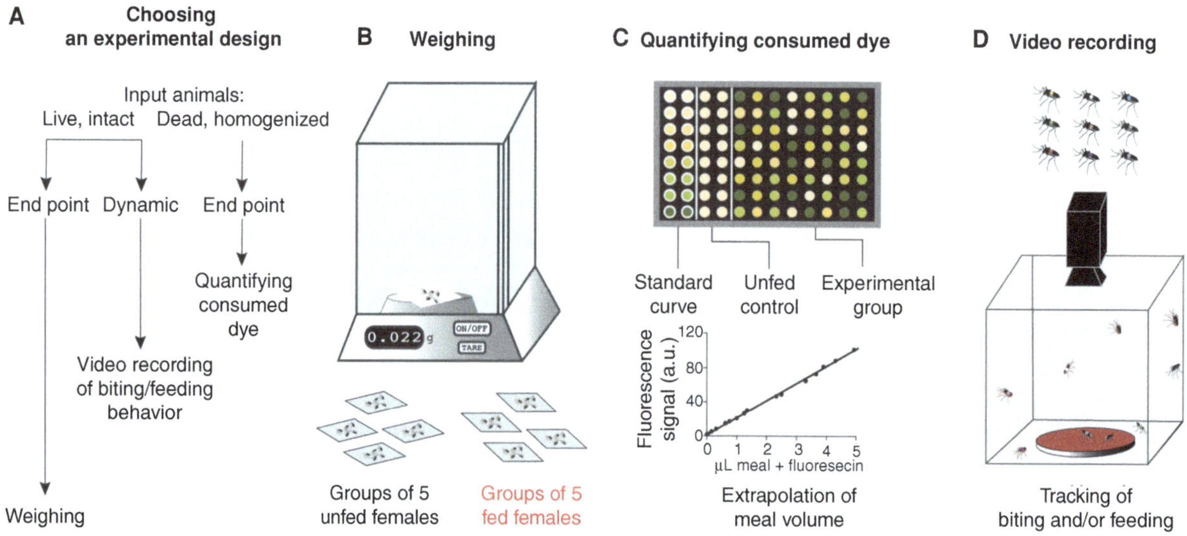

FIGURE 2. Multiple methods can be used to quantify meal volumes consumed by mosquitoes. (*A*) A meal size quantification method should be chosen depending on the ultimate experimental goals and the final state of animals required during/after meal size quantification. (*B*) Mosquitoes can be weighed in small groups if live animals are required following meal size quantification. (*C*) If animals are no longer required after meal size quantification, a more precise end-point method can be used. Meals should contain a quantifiable dye that is released upon homogenization of animals after feeding. Note that the standard curve in *C* depicts a schematic, not real data. (*D*) For dynamic monitoring of biting or feeding behavior, animals can be tracked using video recordings, with or without being individually marked.

In our associated protocol (Protocol 1: Size Quantification of Blood and Sugar Meals in *Aedes aegypti* Mosquitoes [Venkataraman et al. 2022]), we detail a high-throughput, end-point assay to quantify meal volumes consumed by individual mosquitoes.

Applications

Meal size quantification methods are useful for understanding how variables such as the contents of a meal or the genetic background of the mosquito can influence feeding behavior. For example, genetic modification of sweet taste receptors has been shown to modify nectar-feeding behavior in females (Jové et al. 2020), but the analogous receptors for blood-feeding behavior remain unknown. These feeding and quantification assays can also measure changes in feeding vigor observed upon addition of pharmacological compounds like feeding deterrents. Furthermore, dyes of distinct spectra can be added to compare feeding vigor across multiple meal compositions or feeding instances. Because the amount of blood protein consumed determines physiology downstream of feeding—like oogenesis and host-seeking suppression behavior—meal size quantification is important to standardize results in subsequent assays that measure fecundity and host attraction. Finally, these quantification assays can be extended in a time course series to track the meal as it is digested and excreted (Duvall et al. 2019).

CONCLUSIONS

The nectar- and blood-feeding programs in *Ae. aegypti* are linked to nutritional value and reproductive needs. Carbohydrates acquired from floral nectar are sufficient to sustain energy metabolism in females and males, but protein obtained from vertebrate blood is needed for females to produce viable eggs (Briegel 2003; Duvall et al. 2019). If a female ingests blood from an infected host, she can transmit pathogens that cause diseases including Zika, dengue, and chikungunya to her next victim (Rückert and Ebel 2018). Thus, nectar- and blood-feeding behaviors are essential for the survival and repro-

Cite this introduction as *Cold Spring Harb Protoc*; doi:10.1101/pdb.top107657

ductive success of *Ae. aegypti* and underlie their capacity to transmit pathogenic viruses (Rückert and Ebel 2018). Precise measurements of meal size are important to understand how meal content, source, and availability can impact subsequent physiology and behavior, including metabolism, food search motivation, and egg development. These experimental manipulations of meals can be broadly applied to comparisons across mosquito sex, species, and genetic background.

ACKNOWLEDGMENTS

Our work has been supported by a Boehringer Ingelheim Fonds PhD fellowship for funding to K.V., and by funding from the National Institute of General Medical Sciences (R35GM137888), Beckman Young Investigator Award, and a Klingenstein-Simons Fellowship Award in Neuroscience to L.B.D.

REFERENCES

Barredo E, DeGennaro M. 2020. Not just from blood: mosquito nutrient acquisition from nectar sources. *Trends Parasitol* **36:** 473–484. doi:10.1016/j.pt.2020.02.003

Briegel H. 1990. Metabolic relationship between female body size, reserves, and fecundity of *Aedes aegypti*. *J Insect Physiol* **36:** 165–172. doi:10.1016/0022-1910(90)90118-Y

Briegel H. 2003. Physiological bases of mosquito ecology. *J Vector Ecol* **28:** 1–11.

Brown MR, Klowden MJ, Crim JW, Young L, Shrouder LA, Lea AO. 1994. Endogenous regulation of mosquito host-seeking behavior by a neuropeptide. *J Insect Physiol* **40:** 399–406. doi:10.1016/0022-1910(94)90158-9

Costa-da-Silva AL, Navarrete FR, Salvador FS, Karina-Costa M, Ioshino RS, Azevedo DS, Rocha DR, Romano CM, Capurro ML. 2013. Glytube: a conical tube and parafilm M–based method as a simplified device to artificially blood-feed the dengue vector mosquito, *Aedes aegypti*. *PLoS One* **8:** e53816. doi:10.1371/journal.pone.0053816

Davis EE. 1984a. Development of lactic acid-receptor sensitivity and host-seeking behaviour in newly emerged female *Aedes aegypti* mosquitoes. *J Insect Physiol* **30:** 211–215. doi:10.1016/0022-1910(84)90005-2

Davis EE. 1984b. Regulation of sensitivity in the peripheral chemoreceptor systems for host-seeking behavior by a haemolymph-borne factor in *Aedes aegypti*. *J Insect Physiol* **30:** 179–183. doi:10.1016/0022-1910(84)90124-0

Duvall LB, Ramos-Espiritu L, Barsoum KE, Glickman JF, Vosshall LB. 2019. Small-molecule agonists of *Ae. aegypti* neuropeptide Y receptor block mosquito biting. *Cell* **176:** 687–701.e5. doi:10.1016/j.cell.2018.12.004

Farjana T, Tuno N. 2013. Multiple blood feeding and host-seeking behavior in *Aedes aegypti* and *Aedes albopictus* (diptera: Culicidae). *J Med Entomol* **50:** 838–846. doi:10.1603/ME12146

Gordon RM, Lumsden WHR. 1939. A study of the behaviour of the mouth-parts of mosquitoes when taking up blood from living tissue; together with some observations on the ingestion of microfilariae. *Ann Trop Med Parasitol* **33:** 259–278. doi:10.1080/00034983.1939.11685071

Harrington LC, Edman JD, Scott TW. 2001. Why do female *Aedes aegypti* (Diptera: Culicidae) feed preferentially and frequently on human blood? *J Med Entomol* **38:** 411–422. doi:10.1603/0022-2585-38.3.411

Hol FJ, Lambrechts L, Prakash M. 2020. BiteOscope, an open platform to study mosquito biting behavior. *Elife* **9:** 1–24. doi:10.7554/eLife.56829

Jové V, Gong Z, Hol FJH, Zhao Z, Sorrells TR, Carroll TS, Prakash M, Mcbride CS, Vosshall LB. 2020. Sensory discrimination of blood and floral nectar by *Aedes aegypti* mosquitoes. *Neuron* **108:** 1–18. doi:10.1016/j.neuron.2020.09.019

Judson CL. 1967. Feeding and oviposition behavior in the mosquito *Aedes aegypti* (L.). I. Preliminary studies of physiological control mechanisms. *Biol Bull* **133:** 369–377. doi:10.2307/1539832

Klowden MJ. 1990. The endogenous regulation of mosquito reproductive behavior. *Experientia* **46:** 660–670. doi:10.1007/BF01939928

Klowden MJ, Lea AO. 1978. Blood meal size as a factor affecting continued host-seeking by *Aedes Aegypti* (L.). *Am J Trop Med Hyg* **27:** 827–831. doi:10.4269/ajtmh.1978.27.827

Klowden MJ, Lea AO. 1979. Humoral inhibition of host-seeking in *Aedes aegypti* during oocyte maturation. *J Insect Physiol* **25:** 231–235. doi:10.1016/0022-1910(79)90048-9

Liesch J, Bellani LL, Vosshall LB. 2013. Functional and genetic characterization of neuropeptide Y-like receptors in *Aedes aegypti*. *PLoS Negl Trop Dis* **7:** e2486. doi:10.1371/journal.pntd.0002486

Liu MZ, Vosshall LB. 2019. General visual and contingent thermal cues interact to elicit attraction in female *Aedes aegypti* mosquitoes. *Curr Biol* **29:** 2250–2257.e4. doi:10.1016/j.cub.2019.06.001

McBride CS, Baier F, Omondi AB, Spitzer SA, Lutomiah J, Sang R, Ignell R, Vosshall LB. 2014. Evolution of mosquito preference for humans linked to an odorant receptor. *Nature* **515:** 222–227. doi:10.1038/nature13964

McMeniman CJ, Corfas RA, Matthews BJ, Ritchie SA, Vosshall LB. 2014. Multimodal integration of carbon dioxide and other sensory cues drives mosquito attraction to humans. *Cell* **156:** 1060–1071. doi:10.1016/j.cell.2013.12.044

Pappas LG, Larsen JR. 1978. Gustatory mechanisms and sugar-feeding in the mosquito, *Culiseta inornata*. *Physiol Entomol* **3:** 115–119. doi:10.1111/j.1365-3032.1978.tb00141.x

Rückert C, Ebel GD. 2018. How do virus–mosquito interactions lead to viral emergence? *Trends Parasitol* **34:** 310–321. doi:10.1016/j.pt.2017.12.004

Sanford JL, Shields VDC, Dickens JC. 2013. Gustatory receptor neuron responds to DEET and other insect repellents in the yellow-fever mosquito, *Aedes aegypti*. *Naturwissenschaften* **100:** 269–273. doi:10.1007/s00114-013-1021-x

Sissoko F, Junnila A, Traore MM, Traore SF, Doumbia S, Mamadou Dembele S, Schlein Y, Sekou Traore A, Gergely P, Xue R-D, et al. 2019. Frequent sugar feeding behavior by *Aedes aegypti* in Bamako, Mali makes them ideal candidates for control with attractive toxic sugar baits (ATSB). *PLoS One* **14:** e0214170. doi:10.1371/journal.pone.0214170

Trembley HL. 1952. The distribution of certain liquids in the esophageal diverticula and stomach of mosquitoes. *Am J Trop Med Hyg* **1:** 693–710. doi:10.4269/ajtmh.1952.1.693

Van Handel E. 1972. The detection of nectar in mosquitoes. *Mosq News* **32:** 458.

Venkataraman K, Jové V, Duvall LB. 2022. Size quantification of blood and sugar meals in *Aedes aegypti* mosquitoes. *Cold Spring Harb Protoc* doi:10.1101/pdb.prot107862

World Health Organization (WHO). 2020. WHO fact sheet: vector-borne diseases. March 2, 2020. https://www.who.int/news-room/fact-sheets/detail/vector-borne-diseases

Protocol 1

Size Quantification of Blood and Sugar Meals in *Aedes aegypti* Mosquitoes

Krithika Venkataraman,[1,3] Veronica Jové,[1,3] and Laura B. Duvall[2,4]

[1]*The Rockefeller University, New York, New York 10065, USA;* [2]*Department of Biological Sciences, Columbia University in the City of New York, New York, New York 10027, USA*

Both male and female mosquitoes consume sugar-rich nectar meals required for metabolic energy, but only females consume protein-rich blood meals, which are required for egg development. The size of each meal consumed has subsequent effects on behavior and reproduction; therefore, precise quantification is an important aspect of mosquito feeding behavior studies. This protocol describes a high-throughput, end-point assay to quantify meal volumes ingested by individual mosquitoes. The addition of a fluorescent dye to the meal allows for meal size quantification. Individual mosquitoes that have been fed this meal are homogenized in 96-well plates, and the fluorescence levels are measured with a plate reader. This protocol can also be adapted to determine if alteration of meal composition affects the ingested meal volume, if mosquito strain or genotype dictates consumption, or if meals are derived from multiple sources.

MATERIALS

It is essential that you consult the appropriate Material Safety Data Sheets and your institution's Environmental Health and Safety Office for proper handling of equipment and hazardous materials used in this protocol.

Reagents

Adult *Aedes aegypti* mosquitoes

Females should be >3 d old to ensure competence to blood feed, although this can vary by strain. Ae. aegypti *are generally maintained and reared at 25°C–28°C with 70%–80% relative humidity and a 14 h light:10 h dark photoperiod, and animals are allowed to mate freely. Adult mosquitoes are provided constant access to 10% sucrose, which is replaced with a water-soaked cotton ball (to prevent desiccation) ~24 h before performing feeding experiments to standardize feeding rates and meal sizes. Detailed rearing methods are described in DeGennaro et al. (2013) and can be modified to suit experimental needs.*

Feeding deterrents (optional; see Step 1)

Feeding deterrents can be mixed with blood or with other meals delivered through a Glytube (Costa-da-Silva et al. 2013). Known feeding deterrents used with Ae. aegypti *include 1% DEET (N, N-diethyl-meta-toluamide) or bitters like 10 mM denatonium, 1 mM lobeline, and 5 mM quinine, as previously described (Dennis et al. 2019).*

[3]These authors contributed equally to this work.

[4]Correspondence: lbd2126@columbia.edu

Cite this protocol as *Cold Spring Harb Protoc*; doi:10.1101/pdb.prot107862

Fluorescent dye(s) (palatable, nontoxic)

For illustrative purposes, this protocol uses fluorescein as the fluorescent dye. Fluorescein sodium salt (Sigma-Aldrich F6377) must be prepared as 2% aqueous stock. It is light-sensitive and must be kept dark and/or always covered in aluminum foil both as a stock solution and when added to a meal.

If the volume of multiple different meals consumed must be measured, dyes with distinct excitation and emission spectra must be added to each of the meals (see Step 1). For example, fluorescein could be added to one meal, whereas rhodamine B or other water-soluble dye could be added to another meal (Kumar et al. 2021).

Meal(s) appropriate for the experiment

Mammalian blood

Any commercially available defibrinated mammalian blood (i.e., whole blood from which fibrin has been removed to prevent clotting) can be used for laboratory feeding, including blood from sheep, guinea pigs, or humans (Ross et al. 2017). Although optional, feeding rates can be boosted by the addition of 1–2 mM adenosine 5′-triphosphate (ATP) disodium salt hydrate buffered in an aqueous solution of sodium bicarbonate (Galun 1963).

Sucrose (10%)

10% sucrose in aqueous solution is an adequate substitute for nectar in the laboratory and can be delivered to mosquitoes for rearing and meeting metabolic energy needs.

Phosphate-buffered saline (PBS) (1×)

We purchase 10× PBS (Biotium 22020) and dilute 1/10 in DI water.

Equipment

Bead mill homogenizer for use with 96-well plates (e.g., TissueLyser II) (optional; see Step 11)

Another option is to use a pestle grinder (e.g., Kontes Pestle Pellet Grinder) and microcentrifuge tubes to process the samples individually (see Step 11).

Black polystyrene plate (96-well)

Use a plate with a solid bottom if the fluorescence reading is taken from above the plate or a clear bottom if the reading is taken from below.

Borosilicate solid glass beads (3-mm-diameter)

Centrifuge for 96-well plate

Feeder appropriate for the experiment (see Step 4)

Artificial membrane feeder (for blood feeding)

For blood and other meals that trigger engorgement, female mosquitoes must pierce a membrane to consume the meal. Examples of artificial membrane feeders include the commercially available Hemotek feeder or the Glytube (Costa-da-Silva et al. 2013), which can be assembled with standard laboratory equipment.

Cotton balls (for nectar feeding)

Meals ingested via the labium can be delivered easily using cotton balls soaked in the meal solution with added dye.

Fluorescence plate reader

Fluorescence plate readers that can work with a range of excitation and emission wavelengths, as well as those that can take optical density measurements are ideal.

Forceps

Microcentrifuge tubes

Multichannel pipette (optional; see Steps 8 and 14)

PCR plate (96-well)

Sealing film for 96-well plates

Vortex mixer

METHOD

This protocol has been adapted from Liesch et al. (2013), Duvall et al. (2019), and Jové et al. (2020a,b). In these examples, 7- to 21-d-old female mosquitoes are anesthetized at 4°C, sorted into groups of 15 to 20 females, and placed into a 32-oz high-density polyethylene (HDPE) plastic cup (VWR 89009-668) and provided with a water-soaked cotton ball (to prevent desiccation) ~24 h before performing feeding experiments to standardize feeding rates and meal sizes. After at least 15 min of reacclimating to assay room conditions, 1.5–2 mL of fluorescent meal is offered to each cohort of females using an artificial feeder like the Glytube (Costa-da-Silva et al. 2013). Quantification of meal volume is then performed as described below.

Organization is critical for this protocol. Be sure to set aside enough of the meal containing fluorescein at the start of the experiment. This will be used to generate the reference curve and is critical for reliable results. Label all tubes and organize the work area in advance to speed up sample handling. In this protocol, meal volume quantification is performed in a high-throughput manner by using 96-well plates, but it can be scaled up or down as needed. See Figure 1 for recommended plate organization. Carry out all quantification steps at room temperature.

1. For each meal, add fluorescein to a final concentration of 0.002% (e.g., 2 µL of 2% fluorescein stock in a total meal volume of 2 mL).

   If desired, add deterrents to some of the meals at this step. If alternative dyes are to be used, they should also be added at this step.

2. Set aside at least 1 mL of the final meal that has 0.002% fluorescein. To ensure that this sample is comparable to the one delivered to mosquitoes, expose it to the same heating and lighting conditions while preparing the meal for the mosquitoes.

   This sample will be used to generate the reference standard curve for volume calculations.

3. Set aside 32 mosquitoes to use as unfed controls.

   These animals will never be offered a meal and are used to account for tissue autofluorescence. They can be frozen at −20°C until quantification of experimental groups.

4. Deliver the meal to the experimental group of mosquitoes via the appropriate feeding method.
   - Use a saturated cotton ball for nectar feeding.
   - Use an artificial membrane feeder (e.g., Hemotek or Glytube) for blood feeding.

     Detailed methods for feeding labeled or artificial blood meals are included in Costa-da-Silva et al. (2013) and Jové et al. (2020b).

5. Collect the animals after feeding is complete.

   See Troubleshooting.

   To measure meal size immediately after feeding, proceed to Step 6. Alternatively, freeze the animals at −20°C until quantification.

6. Prepare a serial dilution of the reference meal (containing 0.002% fluorescein) that was set aside in Step 2 as follows.

   i. Prepare the first solution by adding 50 µL of meal containing 0.002% fluorescein to 950 µL of 1× PBS. Vortex it so that the fluorescein is uniformly mixed.

   In Step 7, 100 µL of each dilution from the series will be used to generate a standard curve in which fluorescence levels are plotted against the known meal volume (Fig. 1). For this first standard curve solution, the final 100 µL volume contains 5 µL of meal containing 0.002% fluorescein and 95 µL of 1× PBS.

   ii. Prepare the remainder of the standard curve solutions (eight in total) by performing a twofold dilution each time. For each step in the series, take 500 µL from the previous tube and add it to a new tube containing 500 µL of 1× PBS. Vortex the tube well before preparing the next twofold dilution.

   In the final 100-µL volume for each of these eight tubes, the amount of meal containing 0.002% fluorescein will be 5, 2.5, 1.25, 0.625, 0.3125, 0.15625, 0.078125, and 0 µL, respectively.

 Cite this protocol as *Cold Spring Harb Protoc*; doi:10.1101/pdb.prot107862

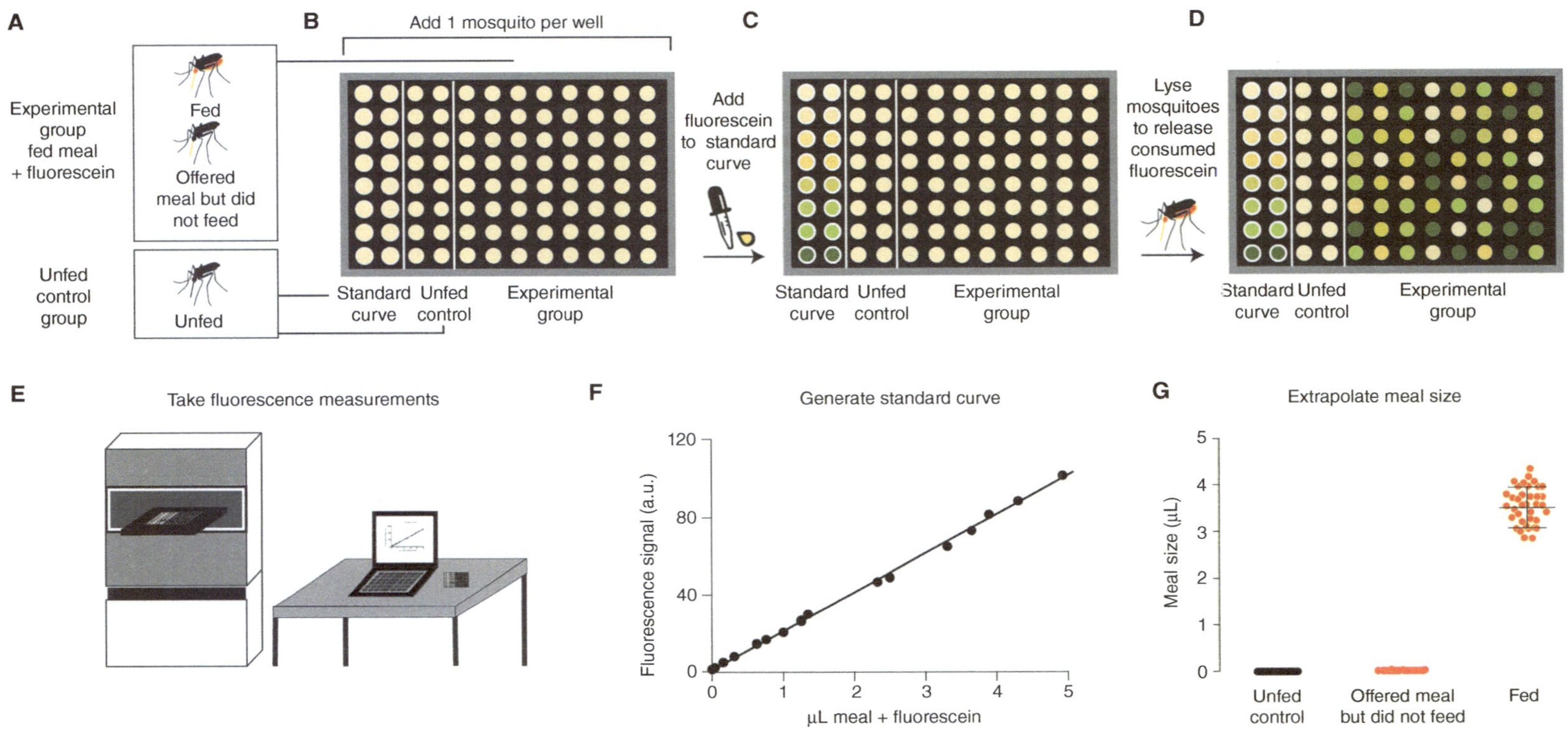

FIGURE 1. Schematic of steps to quantify meal size following nectar- or blood-feeding procedures. (*A*) Mosquitoes must be divided into two groups: an experimental group offered a meal containing a dye, such as fluorescein (*top*) and an unfed negative control group given no meal (*bottom*). (*B*) Mosquitoes are individually transferred to the respective wells in a 96-well plate once feeding is completed. (*C*) Reference standard curve solutions with known amounts of meal containing 0.002% fluorescein are added in replicate to the 96-well plate. (*D*) Mosquitoes are lysed to extract consumed fluorescein. (*E*) A plate reader is used to quantify fluorescence levels in each well of the 96-well plate. (*F*) A reference standard curve is generated using the fluorescence readings for the known standard curve volumes of meal containing 0.002% fluorescein. (*G*) Meal volumes consumed by the experimental group are extrapolated from the standard curve generated, correcting for the baseline autofluorescence of unfed mosquitoes. The graphs in *F* and *G* depict schematics, not real data. This fluorescence quantification method is modified from Liesch et al. (2013). Portions of this figure are adapted from Jové et al. (2020b).

7. Pipette 100 µL of each solution from Step 6 into the first column of a 96-well plate to make the reference curve. Then, use forceps to place one unfed mosquito into each well in the first column of the plate to account for any tissue autofluorescence.

   As shown in Figure 1C, we recommend adding a second column of unfed mosquitoes to generate a replicate curve and account for any edge effects of the plate reader.

   If experimental groups were offered different meals (e.g., blood ± drug), prepare a separate reference standard curve for each meal. Small differences in fluorescein addition can generate big differences in fluorescence levels and thus calculated volume, so each experimental group must be matched to its own reference curve. Calculate R^2 values to assess goodness-of-fit for linear regression to determine the reliability of the standard curve.

8. Add 100 µL of 1× PBS to each of the remaining wells using a multichannel pipette if available. In addition, add a borosilicate solid glass bead to each well.

   A multichannel pipette is not an absolute requirement but improves accuracy and speed.

9. Prepare negative control columns by adding one unfed mosquito using forceps to each well in the next two columns of the plate.

   The volume of fluorescein measured for these mosquitoes sets a baseline fluorescence measurement cutoff, to account for tissue autofluorescence. Use this cutoff to determine whether or not a mosquito in the experimental group is scored as fed.

10. Use forceps to place one mosquito from the experimental groups that were offered a meal into each of the remaining wells. Seal the plate tightly to avoid any leaks.
11. Use one of the following methods to thoroughly homogenize the abdomen and release the meal.
    - Use a TissueLyser II with a borosilicate solid glass bead in each well at 30 Hz for 30 sec at room temperature.
    - Use a vortex mixer with a borosilicate solid glass bead in each well. Vortex using a high setting for 15–30 sec at room temperature or until the tissue and blood meal are well-homogenized with no large pieces visible.

      As an alternative, the tissue in each sample can be manually homogenized in a microcentrifuge tube using a disposable pestle grinder (without beads) until the consistency is uniform and no large pieces are visible. The samples must then be returned to the 96-well plate.

12. Centrifuge the plate for 1–2 min at 2000 rpm and collect the cleared lysate for use in subsequent steps.
13. Add 180 µL of 1× PBS to each well of a new black 96-well plate.
14. Transfer 20 µL of each lysate to the appropriate well of the black 96-well plate (which contains 180 µL of 1× PBS) for a final total volume of 200 µL. Mix using a multichannel pipette if available.
15. Using a plate reader, measure the fluorescence intensity of each well.

    For fluorescein, the 485/520 excitation/emission channel is optimal, but this will vary depending on the dye used.

16. Plot the known meal volumes of the standard curve solutions against their corresponding fluorescence intensity measurements to establish a reference curve.

    If applicable, generate the reference curve by averaging the known concentrations in the two reference rows.

17. Calculate the meal volume consumed by each experimental mosquito. To correct for tissue autofluorescence, average the fluorescence intensity reading among unfed mosquitoes, and subtract this average from the fluorescence intensity reading of each experimental mosquito.

    Because a full blood meal approximately doubles the weight of the female, the volume varies depending on the size of the female; in previous studies (Duvall et al. 2019; Jové et al. 2020a) blood meal volumes from 2–5 µL were typically observed. Nectar meals are generally <2 µL.

    See Troubleshooting.

Cite this protocol as *Cold Spring Harb Protoc*; doi:10.1101/pdb.prot107862

TROUBLESHOOTING

Problem (Step 5): Females do not consume any of the meals offered.

Solution: Confirm that the dye does not render the meal unpalatable. This can be done by offering blood to a positive control cohort alongside the dye-fed cohorts and ensuring that the dye-fed cohorts feed at similar rates and consume similar volumes to the blood-fed cohort. Additionally, consider using a lower concentration of the dye in the meal. Addition of fluorescein has no effect on meal size, but this may not be true for all additives.

Problem (Step 17): The meal volumes calculated are wildly variable or unreliable.

Solution: Small differences in fluorescein addition can generate big differences in fluorescence levels and thus calculated volume. It is critical to confirm that each experimental group is matched to its own reference curve. If meal volumes are likely to be larger or smaller than the bounds of the standard curve suggested, then adjust the standard curve volumes prepared in Step 6 to the range of possible meal volumes. R^2 values can be calculated to determine goodness-of-fit for linear regression and to assess the reliability of the standard curve.

DISCUSSION

The size of each meal consumed by a mosquito has effects on its subsequent behavior and reproduction (Klowden and Lea 1978; Briegel 1990), so methods for the quantification of meal volume in mosquitoes are important. This protocol describes a way to precisely quantify the volume of a meal ingested by a single mosquito by adding a fluorescent dye to the meal.

The protocol can be adapted for nectar or blood meals, depending on the goals of the experiment. Both male and female mosquitoes consume sugar-rich nectar meals required for metabolic energy, but only females consume protein-rich blood meals required for egg development (Briegel 2003). This protocol can also be used to measure changes in feeding vigor in response to variables such as meal composition or genetic background (Jové et al. 2020a). For example, pharmacological compounds and/or feeding deterrents like DEET (*N, N*-diethyl-*meta*-toluamide) or bitters (such as denatonium, lobeline, and quinine) can be added to the meal to determine the impact on meal size (Duvall et al. 2019). This assay can also be modified to track the kinetics of meal digestion and clearance in individual mosquitoes. If dyes of distinct fluorescent spectra are added to each meal, researchers can compare the volume consumed across different meals (e.g., nectar and blood) or across multiple feeding instances (e.g., refeeding after a partial blood meal) (Klowden and Briegel 1994; Ponlawat and Harrington 2005).

Finally, it is important to note that this protocol describes a sensitive end-point assay to precisely measure the volume of meals consumed by individual mosquitoes. However, if mosquitoes will be used for subsequent behavioral testing or experimentation, then, instead, calculate blood meal size by weighing anesthetized mosquitoes in small groups and comparing prefed versus postfed weights as previously described (Duvall et al. 2019; Jové et al. 2020b). Weights can be benchmarked to the volumes calculated in this assay to determine volume consumed, although small nectar meals would be difficult to detect using weight measurements alone. Meal size measurements allow researchers to appropriately select animals for downstream in vivo assays that measure the effect of the meal on processes including fecundity or host attraction. This protocol can also be coupled to versatile feeding assays, such as those to measure feeding dynamics using real-time video recording (Hol et al. 2020), or to examine the effects of interrupted feeding.

ACKNOWLEDGMENTS

Work in our laboratory has been supported by a Boehringer Ingelheim Fonds PhD fellowship for funding to K.V., and by funding from the National Institute of General Medical Sciences (R35GM137888), Beckman Young Investigator Award, and Klingenstein-Simons Fellowship Award in Neuroscience to L.B.D.

REFERENCES

Briegel H. 1990. Metabolic relationship between female body size, reserves, and fecundity of *Aedes aegypti*. *J Insect Physiol* **36:** 165–172. doi:10.1016/0022-1910(90)90118-Y

Briegel H. 2003. Physiological bases of mosquito ecology. *J Vector Ecol* **28:** 1–11.

Costa-da-Silva AL, Navarrete FR, Salvador FS, Karina-Costa M, Ioshino RS, Azevedo DS, Rocha DR, Romano CM, Capurro ML. 2013. Glytube: a conical tube and parafilm M-based method as a simplified device to artificially blood-feed the dengue vector mosquito, *Aedes aegypti*. *PLoS ONE* **8:** e53816. doi:10.1371/journal.pone.0053816

DeGennaro M, McBride CS, Seeholzer L, Nakagawa T, Dennis EJ, Goldman C, Jasinskiene N, James AA, Vosshall LB, Degennaro M, et al. 2013. Orco mutant mosquitoes lose strong preference for humans and are not repelled by volatile DEET. *Nature* **498:** 487–491. doi:10.1038/nature12206

Dennis EJ, Goldman O V, Vosshall LB. 2019. *Aedes aegypti* mosquitoes use their legs to sense DEET on contact. *Curr Biol* **29:** 1551–1556.e5. doi:10.1016/j.cub.2019.04.004

Duvall LB, Ramos-Espiritu L, Barsoum KE, Glickman JF, Vosshall LB. 2019. Small-molecule agonists of *Ae. aegypti* neuropeptide Y receptor block mosquito biting. *Cell* **176:** 687–701.e5. doi:10.1016/j.cell.2018.12.004

Galun R. 1963. Feeding response in *Aedes aegypti:* stimulation by adenosine triphosphate. *Science* **142:** 1674–1675. doi:10.1126/science.142.3600.1674

Hol FJ, Lambrechts L, Prakash M. 2020. BiteOscope, an open platform to study mosquito biting behavior. *Elife* **9:** e56829. doi:10.7554/eLife.56829

Jové V, Gong Z, Hol FJH, Zhao Z, Sorrells TR, Carroll TS, Prakash M, Mcbride CS, Vosshall LB. 2020a. Sensory discrimination of blood and floral nectar by *Aedes aegypti* mosquitoes. *Neuron* **108:** 1–18. doi:10.1016/j.neuron.2020.09.019

Jové V, Venkataraman K, Gabel TM, Duvall LB. 2020b. Feeding and quantifying animal-derived blood and artificial meals in *Aedes aegypti* mosquitoes. *J Vis Exp* **2020:** 1–22.

Klowden MJ, Lea AO. 1978. Blood meal size as a factor affecting continued host-seeking by *Aedes aegypti* (L.). *Am J Trop Med Hyg* **27:** 827–831. doi:10.4269/ajtmh.1978.27.827

Klowden MJ, Briegel H. 1994. Mosquito gonotrophic cycle and multiple feeding potential: contrasts between *Anopheles* and *Aedes* (Diptera: Culicidae). *J Med Entomol* **31:** 618–622. doi:10.1093/jmedent/31.4.618

Kumar S, Hol FJ, Pujhari S, Ellington C, Narayanan HV, Li H, Rasgon JL, Prakash M. 2021. A microfluidic platform for highly parallel bite by bite profiling of mosquito-borne pathogen transmission. *Nat Commun* **12:** 6018. doi:10.1038/s41467-021-26300-0

Liesch J, Bellani LL, Vosshall LB. 2013. Functional and genetic characterization of neuropeptide Y-like receptors in *Aedes aegypti*. *PLoS Negl Trop Dis* **7:** e2486. doi:10.1371/journal.pntd.0002486

Ponlawat A, Harrington LC. 2005. Blood feeding patterns of *Aedes aegypti* and *Aedes albopictus* in Thailand. *J Med Entomol* **42:** 844–849. doi:10.1093/jmedent/42.5.844

Ross PA, Axford JK, Richardson KM, Endersby-Harshman NM, Hoffmann AA. 2017. Maintaining *Aedes aegypti* mosquitoes infected with *Wolbachia*. *J Vis Exp* **2017:** 1–8.

CHAPTER 8

Characterizing Mosquito Biting Behavior at High Resolution

Gregory P.D. Murray,[1,2] Emilie Giraud,[1,2] and Felix J.H. Hol[1,2,3,4]

[1]*Insect-Virus Interactions Unit, Institut Pasteur, UMR2000, CNRS, Paris, France;* [2]*Center for research and Interdisciplinarity, Université de Paris, Paris, France;* [3]*Department of Medical Microbiology, Radboud University Medical Center, Nijmegen 6500 HB, The Netherlands*

Blood feeding is a critical event in the life cycle of female mosquitoes. In addition to providing nutrients to the mosquito, blood feeding facilitates the transmission of parasites and viruses to hosts, potentially having devastating health consequences. Our understanding of these short, yet important, bouts of behavior is incomplete. How and where a mosquito decides to bite and the success of feeding can influence the transmission of pathogens. A more thorough understanding of these processes may allow the development of interventions that reduce or prevent infections. Here, we present an overview of strategies for studying mosquito biting behavior and introduce the biteOscope, which provides an opportunity to observe and understand this behavior at unprecedented spatial and temporal resolution under tightly controlled conditions. The biteOscope combines recent advances in computer vision and automated tracking with designs for behavioral arenas and controllable artificial host cues that use low-cost and readily available materials.

BACKGROUND

The bite of a mosquito is one of the most critical steps in the transmission of mosquito-borne pathogens. The bite and subsequent blood meal are preceded by a complex behavioral trajectory consisting of a mosquito flying toward and landing on a host, selecting a bite site, and engorging a full blood meal (Cardé 2015). Research efforts have focused on understanding the behavior of mosquitoes during host-seeking flight (Dekker et al. 2005; McMeniman et al. 2014; van Breugel et al. 2015; Spitzen and Takken 2018; Coutinho-Abreu et al. 2022) and when interacting with vector control tools such as bed nets and traps (Cribellier et al. 2018; Murray et al. 2020). A critical stage for both the reproductive success of the mosquito and the transmission of pathogens is during the act of blood feeding and the associated movement on the host, with this being a high risk, infrequent event in the mosquito life cycle (Walker and Edman 1985; Reid et al. 2014; Vinauger et al. 2018; Cribellier et al. 2022). Despite the vital role blood feeding plays in the life cycle of the mosquito and the epidemiological implications to their hosts, very few methods can monitor and record biting behavior at high spatial and temporal resolution. Although this task is challenging, ongoing improvements in imaging technology and computational approaches are enabling the development of novel methods to algorithmically characterize mosquito blood feeding at high spatiotemporal resolution (Hol et al. 2020).

To record the biting behavior of mosquitoes, several discrete elements must come together: a lure to induce landing behavior, a bite substrate that mosquitoes will engorge on, an arena to contain the mosquitoes under experimental conditions, and a method to record and track the activity of the mosquitoes. Most studies quantifying blood-feeding behavior have relied on a human or rodent to

[4]Correspondence: felix.hol@radboudumc.nl

Cite this introduction as *Cold Spring Harb Protoc*; doi:10.1101/pdb.top107658

act as both the lure and blood source (Choumet et al. 2012; Hughes et al. 2020; League et al. 2021; Xiang et al. 2022). This practice, however, constrains the type of measurements that can be made (e.g., providing limited opportunity for imaging), may introduce variability between hosts/replicates, limits possibilities to modify the skin surface, and requires behavioral arenas to be designed around access to the host subject. In addition to these experimental limitations, previous studies generally relied on the direct observation and recording of blood-feeding behavior by a human observer, limiting throughput and preventing automated analysis (Gordon and Lumsden 1939; Griffiths and Gordon 1952; Choumet et al. 2012; Dennis et al. 2019; League et al. 2021). More recently, event-logging software (e.g., BORIS by Friard and Gamba 2016) has streamlined and digitized this process; however, these approaches still require a significant time investment due to their requirements for manually scoring behavior and may suffer from observer biases.

To date, there have been few examples of high-resolution recording and in-depth analysis of mosquito feeding behavior. Intravital video microscopy has been used to visualize the mouthparts of *Anopheles gambiae* feeding on anesthetized mice (Choumet et al. 2012). These striking videos revealed that the stylets are highly mobile during probing, prior to feeding. Other studies used videography to examine the late-stage host-seeking behavior of *An. gambiae* (Hughes et al. 2020) and *Aedes aegypti* (Xiang et al. 2022). Analyses in all these studies relied on human observers to manually score and analyze behaviors using real-time playback of the recorded videos. Consequently, throughput in these studies was relatively low, resulting in small sample sizes.

The relatively low cost of video capture equipment means that recording freely behaving animals is now practical with exceptional levels of accuracy. Furthermore, advances in deep learning methods for image analysis enable automated tracking of several key points (e.g., leg joints, proboscis tip, and head) of multiple unmarked animals in a single field of view (Mathis et al. 2018; Graving et al. 2019; Pereira et al. 2019). Such approaches allow accurate pose estimation of individual animals through time, and the obtained data can serve as a natural starting point for the extraction and analysis of behavioral metrics from video data. Automated tracking enables behavioral analysis with sufficient efficiency to produce replicates for hypothesis testing and robust analysis of subtle behavioral phenomena.

Another development is the use of open hardware tools to construct behavioral arenas tailored to address specific experimental questions. The availability of low-cost microcomputers (e.g., the Raspberry Pi), sensors, laser cutters, 3D printers, and other fabrication tools enables researchers to create experimental environments designed to mimic key aspects of the natural world and record the behavior of animals exploring these environments. These cross-disciplinary developments now enable the characterization of a variety of mosquito behaviors at unprecedented levels of detail.

THE BITEOSCOPE

A recent method combining these advances is the biteOscope (Hol et al. 2020). The biteOscope is an open platform that makes use of a transparent behavioral arena and artificial bite substrate to allow the recording and automated detection of mosquitoes during landing, exploration, and feeding. The biteOscope combines high-resolution imaging of mosquito probing and feeding with a machine learning workflow to extract detailed behavioral information on the locomotion, pose, biting, and feeding dynamics of the subject mosquitoes.

In the accompanying protocol, we present a suite of techniques that harness the technology of the biteOscope and combine it with a bespoke behavioral arena and host cues to unlock the potential of high-resolution recordings using inexpensive and non-specialist materials (see Protocol 1: Characterizing Mosquito Biting Behavior Using the BiteOscope [Murray et al. 2023]). When these techniques are paired with automated quantitative image analysis, we are able to record and explain the nuanced, high-resolution biting behavior of large numbers of mosquitoes in high throughput under strictly controlled conditions. In the next sections, we outline several aspects that are critical to an experiment characterizing blood-feeding behavior through quantitative imaging.

 Cite this introduction as *Cold Spring Harb Protoc*; doi:10.1101/pdb.top107658

Bite Substrate and Cage

To observe feeding behavior, sufficient cues need to be present to induce feeding. Traditionally, a live host is used in blood-feeding studies; however, more recent studies have used artificial bite substrates. Although artificial substrates may present a less rich repertoire of stimuli than a live host, they offer several advantages: They can be tailored to present a user-defined set of cues, can be designed to facilitate imaging, and increase reproducibility. In the absence of a live host, carbon dioxide is the most common stimulus to activate mosquitoes and initiate host-seeking behavior and, together with heat, forms a minimal set of cues to induce a host-seeking and probing response. Depending on the experimental design, additional cues, including odorants and visual stimuli, may be presented. The blood meal can come directly from a host, but biting and engorgement can also be promoted using an artificial meal. Previous work demonstrated a robust feeding response on a solution containing 1 mM adenosine triphosphate (ATP), sodium ions, and an osmotic pressure close to that of blood (Galun et al. 1963; Duvall et al. 2019). The use of an artificial meal has the benefit of being transparent, allowing the visualization of the mosquito mouthparts probing and interacting with the feeding liquid and quantification of feeding activity.

The bite substrate is the central component of a behavioral arena that contains the mosquitoes and should be sufficiently large to allow the mosquitoes to interact with the bite substrate. Mounting a transparent bite substrate in the floor, wall, or ceiling of a transparent cage has the advantage that the camera and other equipment can be located outside the cage, whereas the mosquitoes can access the substrate from the inside. Laser cutting sheets of acrylic provides a versatile and easy way to construct custom cages suited for behavioral experiments.

Imaging Considerations

When choosing an imaging setup, one should consider several trade-offs. Choosing a camera and lens combination depends on the level of detail that needs to be resolved (more detail, smaller field of view) and the number of individuals to be imaged (more individuals, larger field of view). Another trade-off exists between camera resolution and frame rate, against file size and the associated storage space and processing time. Striking the right balance depends on the experimental question to be addressed; for example, is it necessary to resolve small body parts such as the labrum? What are the fastest dynamics to be characterized? Our current studies show that imaging an 8-cm × 8-cm bite substrate using a 12-megapixel camera (1/1.7″ sensor) and a 35-mm c-mount lens provides an image quality enabling the reliable detection of the fast movement of slender body parts like legs and proboscis and facilitates the visualization of the labrum when inserted into the bite substrate, while allowing up to 20 mosquitoes to simultaneously interact with the bite substrate. Acquiring frames at a minimum of 25 frames per second is required to temporally resolve distinct behaviors such as biting, grooming, and walking (the stride frequency of mosquitoes is ~10 Hz).

Correct illumination is critical to obtaining high-quality images and, consequently, the amount of information that can be extracted from those images. Light-emitting diode (LED) arrays are a cost-effective means to achieve even illumination, as they have consistent intensity and well-defined spectral properties. Experiments with day-biting mosquitoes (e.g., *Aedes*) are usually conducted using visible wavelengths (e.g., white light), whereas nocturnal or crepuscular species, such as *Anopheles*, are usually recorded under dark conditions using infrared (IR) illumination. Previous work has demonstrated an inability to detect IR light in both *Anopheles* (Gibson 1995) and *Aedes* (Muir et al. 1992) mosquitoes.

Recording and Detecting Behavior

Traditionally, mosquito behavior has been characterized by either direct observation or watching a playback of recorded images, possibly assisted by event logging software to manually log behaviors based on predefined classes. An alternative to the labor-intensive observer-based detection is to use machine vision for detecting and tracking mosquito movement (van Breugel et al. 2015;

Angarita-Jaimes et al. 2016; Hol et al. 2020). Automated image analysis methods may be based on the use of markers to detect and track individuals, but recent advances in deep learning enable markerless tracking and pose estimation of multiple individuals in a single field of view. A recent study characterizing mosquitoes interacting with an artificial bite substrate used the DeepLabCut framework (Mathis et al. 2018) for the markerless tracking of the movement of all legs, head, proboscis, and abdomen of a variety of mosquito species (Hol et al. 2020). The output of body part–tracking algorithms can be used to obtain behavioral statistics (e.g., locomotion, activity, and feeding status) and form the basis for downstream computational pipelines, including behavioral classification. Machine learning tools to automate the extraction of behavioral statistics from large quantities of video data enable high-throughput behavioral characterization and remove biases associated with behavioral classification by human observers. We anticipate that novel tools enabling the high-resolution characterization of mosquito biting and feeding behavior will advance our understanding of the internal and external drivers of bite-site selection and engorgement.

ACKNOWLEDGMENTS

We thank Clayton Ellington, Louis Lambrechts, Ariel Lindner, and all members of the Insect-Virus Interactions Unit and the Center for Research and Interdisciplinarity for valuable discussions. Work in the laboratory of F.J.H.H. was supported by a CRI Long-Term Fellowship, a Marie Curie Fellowship, and a Career Award at the Scientific Interface for the Burroughs Wellcome Fund.

REFERENCES

Angarita-Jaimes N, Parker J, Abe M, Mashauri F, Martine J, Towers C, McCall P, Towers D. 2016. A novel video-tracking system to quantify the behaviour of nocturnal mosquitoes attacking human hosts in the field. *J R Soc Interface* **13:** 20150974. doi:10.1098/rsif.2015.0974

Cardé RT. 2015. Multi-cue integration: how female mosquitoes locate a human host. *Curr Biol* **25:** R793–R795. doi:10.1016/j.cub.2015.07.057

Choumet V, Attout T, Chartier L, Khun H, Sautereau J, Robbe-Vincent A, Brey P, Huerre M, Bain O. 2012. Visualizing non infectious and infectious *Anopheles gambiae* blood feedings in naive and saliva-immunized mice. *PLoS ONE* **7:** e50464. doi:10.1371/journal.pone.0050464

Coutinho-Abreu IV, Riffell JA, Akbari OS. 2022. Human attractive cues and mosquito host-seeking behavior. *Trends Parasitol* **38:** 246–264. doi:10.1016/j.pt.2021.09.012

Cribellier A, van Erp JA, Hiscox A, Lankheet MJ, van Leeuwen JL, Spitzen J, Muijres FT. 2018. Flight behaviour of malaria mosquitoes around odour-baited traps: capture and escape dynamics. *R Soc Open Sci* **5:** 180246. doi:10.1098/rsos.180246

Cribellier A, Straw AD, Spitzen J, Pieters RPM, van Leeuwen JL, Muijres FT. 2022. Diurnal and nocturnal mosquitoes escape looming threats using distinct flight strategies. *Curr Biol* **32:** 1232–1246.e5. doi:10.1016/j.cub.2022.01.036

Dekker T, Geier M, Cardé RT. 2005. Carbon dioxide instantly sensitizes female yellow fever mosquitoes to human skin odours. *J Exp Biol* **208:** 2963–2972. doi:10.1242/jeb.01736

Dennis EJ, Goldman OV, Vosshall LB. 2019. *Aedes aegypti* mosquitoes use their legs to sense DEET on contact. *Curr Biol* **29:** 1551–1556. doi:10.1016/j.cub.2019.04.004

Duvall LB, Ramos-Espiritu L, Barsoum KE, Glickman JF, Vosshall LB. 2019. Small-molecule agonists of *Ae. Aegypti* neuropeptide Y receptor block mosquito biting. *Cell* **176:** 687–701.e5. doi:10.1016/j.cell.2018.12.004

Friard O, Gamba M. 2016. BORIS: a free, versatile open-source event-logging software for video/audio coding and live observations. *Methods Ecol Evol* **7:** 1325–1330. doi:10.1111/2041-210X.12584

Galun R, Avi-Dor Y, Bar-Zeev M. 1963. Feeding response in *Aedes aegypti*: stimulation by adenosine triphosphate. *Science* **142:** 1674–1675. doi:10.1126/science.142.3600.1674

Gibson G. 1995. A behavioural test of the sensitivity of a nocturnal mosquito, *Anopheles gambiae*, to dim white, red and infra-red light. *Physiol. Entomol* **20:** 224–228. doi:10.1111/j.1365-3032.1995.tb00005.x

Graving JM, Chae D, Naik H, Li L, Koger B, Costelloe BR, Couzin ID. 2019. DeepPoseKit, a software toolkit for fast and robust animal pose estimation using deep learning. *eLife* **8:** e47994. doi:10.7554/eLife.47994

Griffiths RB, Gordon RM. 1952. An apparatus which enables the process of feeding by mosquitoes to be observed in the tissues of a live rodent; together with an account of the ejection of saliva and its significance in malaria. *Ann Trop Med Parasitol* **46:** 311–319. doi:10.1080/00034983.1952.11685536

Gordon RM, Lumsden WRH. 1939. A study of the behaviour of the mouthparts of mosquitoes when taking up blood from living tissue; together with some observations on the ingestion of microfilariae. *Ann Trop Med Parasitol* **33:** 259–278. doi:10.1080/00034983.1939.11685071

Hol FJH, Lambrechts L, Prakash M. 2020. BiteOscope: an open platform to study mosquito blood-feeding behavior. *eLife* **9:** e56829. doi:10.7554/eLife.56829

Hughes A, Foster GM, Guy A, Matope A, Abe M, Towers D, McCall PJ. 2020. Quantifying late-stage host-seeking behaviour of *Anopheles gambiae* at the insecticidal net interface using a baited-box bioassay. *Malar J* **19:** 140. doi:10.1186/s12936-020-03213-9

League GP, Degner EC, Pitcher SA, Hafezi Y, Tennant E, Cruz PC, et al. 2021. The impact of mating and sugar feeding on blood-feeding physiology and behavior in the arbovirus vector mosquito *Aedes aegypti*. *PLoS Negl Trop Dis* **15:** e0009815. doi:10.1371/journal.pntd.0009815

Mathis A, Mamidanna P, Cury KM, Abe T, Murthy VN, Mathis MW, Bethge M. 2018. DeepLabCut: markerless pose estimation of user-defined body parts with deep learning. *Nat Neurosci* **21:** 1281–1289. doi:10.1038/s41593018-0209-y

McMeniman CJ, Corfas RA, Matthews BJ, Ritchie SA, Vosshall LB. 2014. Multimodal integration of carbon dioxide and other sensory cues drives mosquito attraction to humans. *Cell* **156:** 5. doi:10.1016/j.cell.2013.12.044

Muir LE, Thorne MJ, Kay BH. 1992. *Aedes aegypti (Diptera: Culicidae)* vision: spectral sensitivity and other perceptual parameters of the female eye. *J Med Entomol* **29:** 278–281. doi:10.1093/jmedent/29.2.278

Murray GPD, Lissenden N, Jones J, Voloshin V, Toé KH, Sherrard-Smith E, Foster GM, Churcher TS, Parker JEA, Towers CE, et al. 2020. Barrier bednets target malaria vectors and expand the range of usable insecticides. *Nat Microbiol* **5:** 40–47. doi:10.1038/s41564019-0607-2

Murray GPD, Giraud E, Hol FJH. 2023. Characterizing mosquito biting behavior using the biteOscope. *Cold Spring Harbor Protoc* doi:10.1101/pdb.prot108176

Pereira TD, Aldarondo DE, Willmore L, Kislin M, Wang SSH, Murthy M, Shaevitz JW. 2019. Fast animal pose estimation using deep neural networks. *Nat Methods* **16:** 117–125. doi:10.1038/s41592-018-0234-5

Reid JN, Hoffmeister TS, Hoi AG, Roitberg BD. 2014. Bite or flight: the response of mosquitoes to disturbance while feeding on a defensive host. *Entomol Exp Appl* **153:** 240–245. doi:10.1111/eea.12245

Spitzen J, Takken W. 2018. Keeping track of mosquitoes: a review of tools to track, record and analyse mosquito flight. *Parasit Vectors* **11:** 123. doi:10.1186/s13071-018-2735-6

van Breugel F, Riffell J, Fairhall A, Dickinson MH. 2015. Mosquitoes use vision to associate odor plumes with thermal targets. *Curr Biol* **25:** 2123–2129. doi:10.1016/j.cub.2015.06.046

Vinauger C, Lahondère C, Wolff GH, Locke LT, Liaw JE, Parrish JZ, Akbari OS, Dickinson MH, Riffell JA. 2018. Modulation of host learning in *Aedes aegypti* mosquitoes. *Curr Biol* **28:** 333–344.e8. doi:10.1016/j.cub.2017.12.015

Walker ED, Edman JD. 1985. Feeding-site selection and blood-feeding behavior of *Aedes triseriatus (Diptera: Culicidae)* on rodent (*Sciuridae*) hosts. *J Med Entomol* **22:** 287–294. doi:10.1093/jmedent/22.3.287

Xiang B, Saron WA, Stewart JC, Hain A, Walvekar V, Missé D, Thomas FR, Kini M, Roche B, Claridge-Chang A, et al. 2022. Dengue virus infection modifies mosquito blood-feeding behavior to increase transmission to the host. *Proc Natl Acad* **119:** e2117589119. doi:10.1073/pnas.2117589119

Protocol 1

Characterizing Mosquito Biting Behavior Using the BiteOscope

Gregory P.D. Murray,[1,2] Emilie Giraud,[1,2] and Felix J.H. Hol[1,2,3,4]

[1]*Insect-Virus Interactions Unit, Institut Pasteur, UMR2000, CNRS Paris, France;* [2]*Center for Research and Interdisciplinarity, Université de Paris, Paris, France;* [3]*Department of Medical Microbiology, Radboud University Medical Center, Nijmegen 6500 HB, The Netherlands*

The biteOscope enables the high-resolution monitoring and video recording of blood-feeding mosquitoes. Mosquito biting is induced by combining host cues, an artificial bloodmeal, a membrane, and a transparent heater in a transparent behavioral arena. Machine vision techniques enable the tracking and pose estimation of individual mosquitoes to discern behavior and resolve individual feeding events. The workflow allows multiple replicates and large amounts of imaging data to be generated rapidly. These data are suitable for downstream analysis using machine learning tools for behavioral analysis, allowing subtle behavioral effects to be characterized.

MATERIALS

It is essential that you consult the appropriate Material Safety Data Sheets and your institution's Environmental Health and Safety Office for proper handling of equipment and hazardous materials used in this protocol.

RECIPES: Please see the end of this protocol for recipes indicated by <R>. Additional recipes can be found online at http://cshprotocols.cshlp.org/site/recipes.

Reagents

Acrylic adhesive (e.g., ACROdis CL [Sunclear France 029019] or any dichloromethane-based adhesive for acrylic) (optional; see Step 2)

Artificial meal <R>

CO_2 source (e.g., Benchtop Flowbuddy; Dutscher 789067)

Alternatively, human breath can be used by blowing in the cage (see Step 28).

Mosquitoes (female, mated, and unfed)

The biteOscope has been tested with a variety of mosquito species, including Aedes aegypti, Aedes albopictus, Anopheles coluzzii, Anopheles stephensi, *and* Culex tarsalis.

Sucrose solution (10%, w/w) <R>

Water (sterile for artificial meal; nonsterile for use in the bite substrate's flask)

Equipment

Adhesive tape (preferably nontoxic and odorless; i.e., transparent, polypropylene-based, pressure-sensitive tape [e.g., Scotch tape, 3M])

---

[4]Correspondence: felix.hol@radboudumc.nl

Supplemental material is available at cshprotocols.cshlp.org.

Cite this protocol as *Cold Spring Harb Protoc*; doi:10.1101/pdb.prot108176

Adhesive tape, waterproof (e.g., Gorilla All Weather Outdoor Waterproof Duct Tape) (optional; see Step 14)
Alligator clips
Aspirator
Behavioral arena (see Steps 1–7)
  Acrylic (clear-cast, 3-mm-thick; e.g., Sunclear France 010042 or McMaster Carr 8560K171)
  Dowel (wooden, 3-mm-thick, ~ 6-cm-long [Mr Bricolage]) (optional; see Step 5)
  Laser cutter (e.g., a Trotec speedy 500 or epilog mini)
  Netting (white polyester; e.g., BioQuip 7250A)

Alternatively, any non-insecticide-treated bed net (~10-cm x 10-cm) can be used.

  Thread (~15-cm of a preferably strong thread; e.g., button thread [Singer])

Larger sizes may be used when looped several times.

Bite substrate (see Steps 11–19)
  Clamp stands

They consist of 14-cm x 21-cm clamp stand bases with 40-cm steel rods and are adjustable to match the focal length of the camera and combined with standard retort clamp stands at 2–90 mm (e.g., Thermo Fisher).

  Knife or scalpel (see Step 11)
  Parafilm
  Tissue culture flask (250-mL Falcon flask; Thermo Fisher Scientific 353136)
  Tube (50-mL; Falcon)
  Vacuum chamber (e.g., a Scienceware vacuum dessicator [Sigma-Aldrich Z119024] with house vacuum or pump)
Hot glue gun with glue (e.g., Dweyka mini hot melt glue gun)
Imaging setup (select one):
  A desktop computer with a USB3 port (e.g., an Intel NUC with i7 processor and SSD), a 12-megapixel USB3 camera (e.g., Basler acA4024-29 μm; 4024-pix × 3036-pix, 31 frames per sec [fps]), and a 35-mm c-mount lens (e.g., Thorlabs MVL35M23)

This configuration can be built for <1000 euros and allows data acquisition of up to 20 animals in an 8-cm × 8-cm field of view, at a high resolution compatible with resolving slender body parts (e.g., the legs or mouthparts) and fast movements associated with walking, feeding, and grooming.

  The relatively cheap, yet powerful, Jetson Nano (NVIDIA) computer, combined with the 12-megapixel Raspberry Pi HQ camera and a 16-mm c-mount lens (OKDO SKU 2012854)

This is a lower-cost configuration (~200 euros). The components used in this configuration will result in slightly lower image quality (e.g., making it harder to resolve the mosquito's mouthparts), and the image acquisition software of this configuration is less user-friendly. However, this configuration allows a user to record locomotion behavior and engorgement data at much reduced cost. We hope that this alternative will make the biteOscope accessible for a broader audience. Although this has not been currently tested, the Jetson Nano can simultaneously be used for imaging and temperature control, removing the need for the Raspberry Pi.

Longpass filter (850-nm; Thorlabs FGL850M with adapter SM1A36 for mounting on lens [only for filming in the dark])
Rubber bands (15-m-diameter, 1-mm-thick)
Syringe (5-mL) (optional; see Steps 2 and 16)
Temperature controller (see Steps 8–10)
  Jumper wires
  Peltier heating elements (two elements; e.g., 12-V 5A Peltier, Adafruit 1330)
  Power source (e.g., USB charger [two chargers; 1.5- to 2.5-amp, voltage corresponding to or lower than voltage of Peltier element] and USB cable [male–male, cut in half, wires stripped])

Alternatively, the ready-made part USB-to-Serial cable, Adafruit 954, can be used. Also, the USB cable and charger can be replaced by stripping the wires of any unused power adapter having a DC output that is compatible with the Peltier elements.

Raspberry Pi (tested with models 4B, 3B+, and 3B; compatible with Pi Zero and Pico [not tested])
Relay switch (e.g., Seeed Grove 2 channel SPDT relay or MCIGICM 2 Channel DC 5 V relay)
Resistor (4.7-kΩ; required for temperature sensor)
Temperature sensor (DS18b20)
Wire strippers

White light-emitting diode (LED) array (e.g., VidPro LED 312) or 840- to 950-nm infrared (IR) LED array (Taobao) (only for filming in the dark)

This procedure can be conducted under light conditions (using white lighting) and dark conditions (using IR illumination) to accommodate the temporal bite preferences of day- versus night-biting mosquitoes.

METHOD

Constructing the biteOscope

See Figure 1 for an overview of the setup.

Behavioral Arena Fabrication

1. Using a laser cutter, cut acrylic into the shapes provided in the supplied cage design files (see Supplemental File S1).
2. Slot together the pieces by hand to form a single transparent cube.

   See Figure 2 for cage assembly details.

   If subsequent disassembly is not required, and for added security, the seams may be permanently joined by application of a dichloromethane-based adhesive (e.g., ACROdis). To do so, simply apply a thin layer of the solution to joining surfaces by beading off a needled syringe and leave to dry.

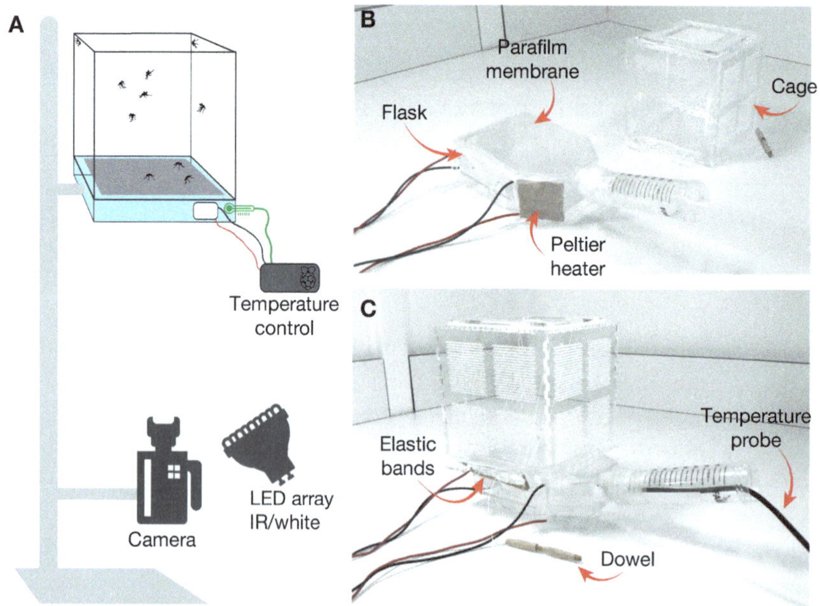

FIGURE 1. The biteOscope setup. (*A*) Schematic of the setup. A camera is mounted on a clamp stand under the bite substrate and cage to provide an abdominal view of mosquitoes interacting with the bite substrate. (*B*) Picture showing the bite substrate (including modified flask, Peltier elements, and Parafilm membrane) and cage separately. (*C*) Picture showing the bite substrate attached to the imaging window (bottom) of the cage using elastic bands.

 Cite this protocol as *Cold Spring Harb Protoc*; doi:10.1101/pdb.prot108176

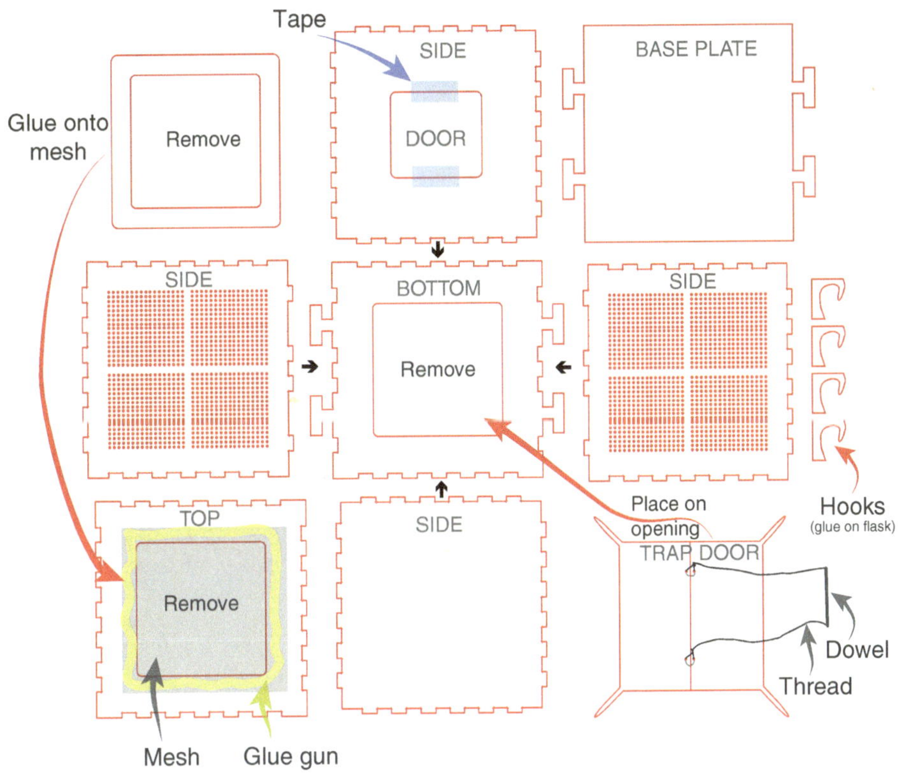

FIGURE 2. Cage assembly. All outlines in red are cut in acrylic using a laser cutter (design in Supplemental File S1). Parts labeled "remove" should be removed before assembly. Crenulated edges of the bottom, sides, and top of the cage slot into each other to form a cube. A mesh covers the window in the upper surface (*bottom left*, "TOP") and is secured by another layer of acrylic (*top left*) placed on top and glued in place with a glue gun. Thread is looped through the holes in the trap door (*bottom right*), threaded through the mesh in the upper surface, and attached to a dowel. If desired, the trap door can be constructed from thinner acrylic (e.g., 2-mm).

3. Cover the upper surface of the cage with netting, and glue the netting in place before placing another layer of acrylic on top and then gluing and sealing around the edges.
4. Loop together the two pieces of acrylic that make up the trap door by inserting strong thread through the precut holes.
5. Thread the 15-cm lengths of thread through the netting in the upper surface.

   If desired, tie and attach the thread to a small dowel with adhesive tape, to act as a handle for pulling open the trap door.
6. Below the closed trap door, place the single-piece acrylic baseplate and secure it using rubber bands around corresponding T-shaped tabs on the side of the cage and baseplate.

   This step is for added security against mosquitoes escaping; the base plate should be removed before filming (Step 26).
7. Once the arena is constructed, transfer mosquitoes by aspiration into the cage via the small door on the side of the cage.

   The small door may then be secured with adhesive tape.

Creating the Temperature Controller

Assemble the temperature controller as explained below and shown in Figure 3. This item maintains the bite substrate at the desired temperature.

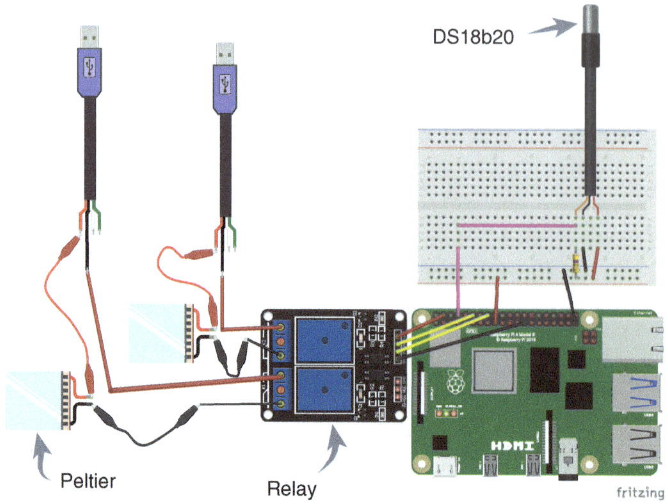

FIGURE 3. Wiring scheme for the temperature controller. The temperature controller consists of a Raspberry Pi (green board), a relay switch (black), a waterproof DS18b20 temperature sensor, and two Peltier elements. Two USB plugs provide power to the Peltier elements and are themselves powered by mains-supplied USB adapters. During the experiment, the Peltier elements are attached to the flask. It is therefore easiest to connect them to the circuit using alligator clips after they have been taped to the flask. When a current runs through the Peltier element, one side will cool, whereas the other side will heat. The heating side should be attached to the flask.

8. Use a standard USB power adapter (e.g., a phone charger) to provide power to the Peltier heating elements as follows.
   i. With wire strippers, strip the wires from the cut USB cables or power adapter.
   ii. Connect the temperature sensor, Raspberry Pi, and relay switch using jumper wires and resistor as shown in the wiring scheme in Figure 3.
   iii. Connect and power the Raspberry Pi.
   iv. Connect the power sources (e.g., USB) to the Peltier elements and relay as shown in Figure 3 using alligator clips.

      These operations only have to be performed once.

      Always unplug the power sources powering the Peltier element when the device is not in use or left unattended.

9. To use the temperature sensor as an input source for the Raspberry Pi, activate the 1-wire interface by running `sudoraspi-config`, following the instructions, and rebooting the Pi.

   This operation only has to be performed once.

10. Transfer the `temperatureControl.py` script available on the biteOscope GitHub repository to the Raspberry Pi.

    Make sure that the Python script addresses the correct GPIO pins that are connected to the relay input (the yellow wires in Fig. 3).

Bite Substrate Assembly and Setup

The bottom opening of the cage is designed to match the dimensions of the top surface of a 250-mL (75-cm^2) culture flask (see Fig. 1). Flasks may be reused after a thorough cleaning.

11. Use knife or scalpel to cut a Falcon tube open on the bottom and attach it to the culture flask opening, using a nontoxic and odorless adhesive (e.g., a hot glue gun), thus creating an elongated neck through which the temperature sensor can be added.
12. Use odorless glue (e.g., glue gun or super glue) to attach the four acrylic hooks to the sides of the culture flask, in such a way as to compliment the T-shaped tabs on the lower section of the cage.

Cite this protocol as *Cold Spring Harb Protoc*; doi:10.1101/pdb.prot108176

13. Fill the modified flask with hot water and degas in a vacuum chamber.

    Ideally, the water temperature should be ~37°C when filming starts. The filled flask will act as a transparent heating surface for the artificial blood meal. Degassing the water prevents bubble formation, improving image quality.

14. Attach the Peltier elements (disconnected from the circuit) to the side of the flask using tape (see Fig. 1). Ensure that the heating side of the element faces the flask.

    Waterproof tape may be used for convenience in case some of the artificial meal leaks.

15. Stretch a rectangular section of Parafilm (~8-cm × 8-cm) to approximately twice its original size in both directions.

    Wear gloves while manipulating the Parafilm and flask to avoid contamination.

16. Apply the artificial meal and stretched Parafilm membrane to the flask in one of the following ways.
    - Apply ~5 mL of the artificial meal to the rectangular outer surface of the culture flask. Carefully stretch the Parafilm membrane over the rectangular section of the tissue culture flask containing the artificial meal. The Parafilm and the outer surface of the flask create a thin fluid cell containing the artificial meal.
    - Carefully stretch the Parafilm membrane over the rectangular section (outer surface) of the tissue culture flask. Inject ~5 mL of the artificial meal under the Parafilm with a syringe. The site of injection should be toward the neck of the flask to avoid damaging the parafilm within the area accessible to mosquitoes.

      Ensure there are no air bubbles in the artificial meal.

17. Support the flask by attaching clamp stands horizontally across the width of the flask at the front and rear of the flask.
18. Connect the wires of the Peltier elements to the circuit using alligator clips.
19. Insert the temperature probe into the flask (outside the field of view) and start the temperature control Python script `tempControlDual.py` on the Raspberry Pi to maintain the water temperature inside the flask at 37°C.

    When current runs through the Peltier element, one side will cool, whereas the other side will heat; ensure the heating side faces the flask.

Experimental Procedure

Depending on the question, assay female mosquitoes of between 6- and 25-d-old. Perform experiments using day-biting mosquitoes (e.g., Aedes*) during daylight hours and using nocturnal species (e.g.,* Anopheles*) in the dark during hours of peak biting activity. A total of one to 35 mosquitoes can be assayed per cage, but depending on the response rate, more than 35 mosquitoes may lead to overcrowding of the field of view. Triplicate cages (at minimum) are typically used for each condition. Ideally, experiments are performed under insectary conditions (e.g., 28°C and 70% relative humidity).*

20. Prior to filming, keep mosquitoes under controlled insectary conditions (e.g., 28°C and 70% relative humidity) and provide a 10% sucrose solution ad libitum.
21. Deprive female mosquitoes of the 10% sucrose solution for >4 h before the recording.

    If using many cages, the duration of sucrose deprivation and the time of day should be considered. Depending on the species of mosquito, the starvation period may need to be adjusted.

22. Mount the camera.

    The camera may be secured directly to the clamp stand to position it under the behavioral arena so that the entire exposed bite surface is within the field of view.

23. Connect the camera to the recording computer, and start the system-specific recording software to preview the image.

    For the recommended Basler camera, the manufacturer-provided Pylon Viewer may be used.

24. Position the LED arrays to provide illumination even across the bite surface. Adjust the exposure time of the camera and/or illumination intensity to maximize the contrast of mosquitoes against the background.

    Ensure that the bite substrate is in focus. If using infrared LEDs to record in the dark, attach the 850-nm longpass filter to the camera lens.

25. Set the appropriate frame rate (typically 25 fps), exposure time, and file write location.
26. Transfer the cage with mosquitoes to the bite substrate.
    i. Place the cage on the bite substrate by positioning the window in the bottom surface over the artificial meal and Parafilm membrane.
    ii. Secure the cage to the flask by attaching the rubber bands to the matching hooks and tabs on the flask and cage, respectively.
    iii. Connect the wires of the Peltier elements to the power supplies.
27. Start the camera recording.
28. Provide a 20-sec puff of CO_2 in the cage.

    Depending on the experimental design, additional puffs of CO_2 may be applied in 10-min intervals.

29. Open the trap door to provide the mosquitoes access to the bite substrate.

    Check the focus and illumination of the recording. Ideally, individual features of each mosquito should be readily visible. Activity is typically highest in the first 15 to 30 min of an experiment, and depending on the research question, multiple short experiments may yield more useful data than a single long experiment.

30. Once the trial is complete, stop the recording.

    The mosquitoes may be aspirated out of the cage through the same entry port, or the entire assembly can be chilled to knockdown and kill the mosquitoes before removal. Both the filming duration and nature of mosquito collection will depend on the research question.

Data Analysis

Imaging data can be acquired as a sequence of images (e.g., .tiff or .png files) or as a video (e.g., .mp4), and these formats can be easily converted in both directions (e.g., with FFmpeg). The acquired data can be analyzed using a variety of methods. We provide two distinct computational pipelines: (1) a deep learning–based approach for mosquito and body part detection using a convolutional neural network (CNN) trained through the DeepLabCut (DLC) framework (Mathis et al. 2018) (Steps 31–35) and (2) a less extensive approach using conventional image processing techniques (Steps 36–40). The native format for the DLC framework is video, whereas image sequences are the most convenient input for the conventional approach.

Deep Learning–Based Processing

A pretrained CNN to detect 38 key points on the body of Ae. aegypti *is available in the biteOscope GitHub repository (snapshot 1030000 in the DLC_aedes directory; see Fig. 4). This ResNet50-based CNN was trained using DLC and can be used for body part detection and tracking of up to approximately 18* Ae. aegypti *freely behaving in an 8-cm × 8-cm field of view (3000-pixels × 3000-pixels using the recommended Basler camera). By labeling a modest number of images, this CNN can be retrained to detect the body parts of mosquito species having a different morphology. This code is written in Python 3 and has been tested with DLC 2.2b8 on Ubuntu 18.04 using a NVIDIA Titan RTX GPU.*

31. Download the biteOscope project from GitHub, and change the `project_path` variable in the config.yaml file to point to your local copy of the project.

    The biteOscope DLC project is set up to analyze mp4 videos using DLC functions as described in Step 32. Running the `/biteOscope/DLC/DLCprocessBiteOscopeVidoes.py` script as outlined in Steps 32 and 33 will detect all body parts per frame and create a validation video showing all detections.

Cite this protocol as *Cold Spring Harb Protoc*; doi:10.1101/pdb.prot108176

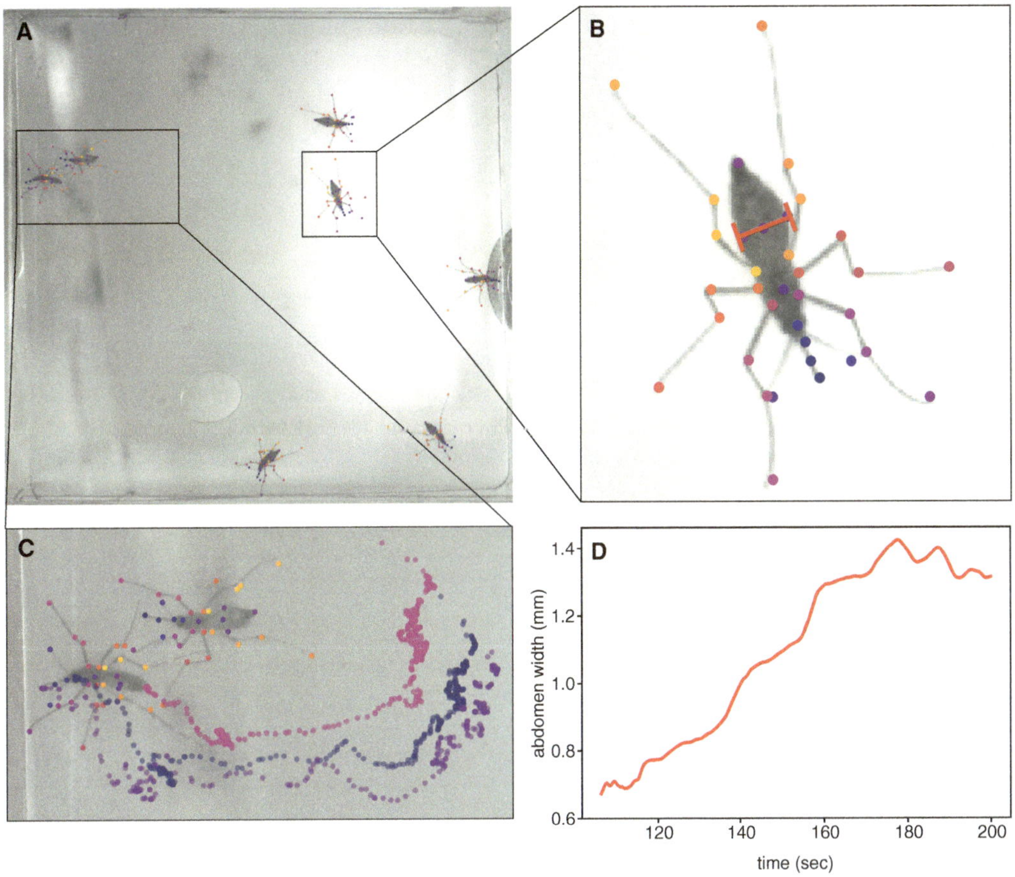

FIGURE 4. Analysis of biteOscope data. (*A*) Frame showing an 8-cm- × 8-cm field of view in which several *Aedes aegypti* mosquitoes are visible. Dots indicate DeepLabCut (DLC)-based detection of 38 key points on the mosquito body. (*B*) Magnification of an engorged mosquito shown in *A*; dots show accurate detection of all key points. (*C*) Magnification of area shown in *A*. Dots on the mosquito bodies show key point detection of a single frame, and trails show the motion of the proboscis (purple), head (blue), and abdominal tip (magenta) of the left mosquito for the preceding 28 sec. (*D*) The width of the abdomen (red bar in *B*) can be used to assess engorgement status. During engorgement, the width of the abdomen approximately doubles.

32. In the DLCprocessBiteOscopeVidoes.py, change "PATH_TO_CONFIG" to the path where the config.yaml file is stored, and change "PATH_TO_VIDEOS" to where the videos to be processed are saved. Run the following script:

```
import os

import glob

import deeplabcut

config_path = /PATH_TO_CONFIG/config.yaml'

videos = glob.glob('/PATH_TO_VIDEOS/*.mp4')

deeplabcut.analyse_videos(config_path, videos, shuffle=1, batchsize=8, dynamic=
(True,0.5, 100))

deeplabcut.create_video_with_all_detections(config_path,
videos,'DLC_resnet50_aedesNov16shuffle1_80000')
```

Use the generated video to check if the accuracy of detected body parts is satisfactory, i.e., the labels match their respective body parts, are well placed, and do not swap or disappear over the duration of the detection.

33. Use the next section of the `DLCprocessBiteOscopeVidoes.py` script to assemble body parts into individuals and track individuals across frames (resulting in "tracklets").

    This step creates a multilevel dataframe holding the predicted body part x,y coordinates of individuals and their associated likelihood scores.

```
deeplabcut.convert_detections2tracklets(config_path, videos, shuffle=1,
videotype='mp4', trainingsetindex=0, track_method='box')
trackletPickles=glob.glob('/PATH_TO_VIDEOS/*_bx.pickle')
for tracklet in trackletPickles:
deeplabcut.convert_raw_tracks_to_h5(config_path, tracklet)
```

34. Use the `processDLCtracks.py` script to assign a new identifier (ID) when individuals enter the field of view and save the resulting tracklet as a dataframe.

    In addition to the DLC-generated body part coordinates (x, y for each body part/frame), this dataframe contains per frame values, as follows:

    - *xSm, x coordinate smoothed using a moving average for each body part;*
    - *ySm, y coordinate smoothed using a moving average for each body part;*
    - *distance, the distance a body part moves between frames for each body part;*
    - *distanceSm, distance smoothed using a moving average for each body part;*
    - *bellyWSm, the distance between the left and right edge of the abdomen, stored with body part abdomenC; and*
    - *engorged, a Boolean indicating if the mosquito is engorged based on observed abdominal width, stored with body part abdomenC.*

      During an experiment, mosquitoes may move in and out of the field of view. For example, when mosquito 1 moves out of the field of view, ID1 becomes vacant and DLC may reassign "1" to the next mosquito entering the field of view. This reassignment is unwanted, as individual identities cannot be correctly assigned when individuals enter and leave the field of view. The processDLCtracks.py script assigns DLC-generated tracklets to unique IDs and saves the tracklets to individual dataframes that can be used for downstream analysis.

35. Use the `processDLCtracks.py` script to generate a dataframe containing behavioral summary statistics per tracklet.

    They can, for example., be used to compare different cohorts of mosquitoes:

    - *ID, a unique ID for the tracklet;*
    - *totDist, total distance (length) of the tracklet;*
    - *totDistSm, total distance smoothed using a moving average;*
    - *totTime, total duration (length) of a tracklet;*
    - *bellyMax09, maximum width of the abdomen during tracklet, calculated as the 90th percentile of observed abdominal widths;*
    - *bellyMin01, minimum width of the abdomen during tracklet, calculated as the 10th percentile of observed abdominal widths;*
    - *engorged, a Boolean indicating if the mosquito engorged;*
    - *timeToEngorge, the tracklet length (in time) before becoming engorged;*
    - *movingTime, total time (in frames) a mosquito moved; and*
    - *movingFrac, fraction of the tracklet that the mosquito moved.*

Data Processing Using Conventional Image Analysis

Alternatively, images can be processed using conventional image processing techniques consisting of background subtraction, morphological operations, and blob detection. The biteOscope GitHub repository contains code to

Cite this protocol as *Cold Spring Harb Protoc*; doi:10.1101/pdb.prot108176

perform these operations resulting in a dataframe holding the x,y coordinates of the center of mass of all mosquitoes. These data can be used to further analyze bouts of behavior of individual mosquitoes.

36. Set all parameters in the trackMosqNB.ipynb.

    Several parameters, including mTreshold and searchRadius may need to be adjusted. Each operation has a preset value and may be adjusted if false-positive or -negative detection occurs.

37. Create a BG image used for background subtraction.

    Make sure inactive mosquitoes do not appear in the image.

38. Test mosquito detection with several frames, and if detection fails to capture all the mosquitoes present or erroneously detects a mosquito, adjust the mTreshold value or the BG image.
39. Detect centroids of mosquitoes in all frames and track detections.
40. Use the output for downstream processing, e.g., using the cropTracks_features.py script provided on the GitHub repository.

DISCUSSION

Building on prior work (Hol et al. 2020), here we describe a device (the biteOscope) to image mosquitoes biting a host mimic and feeding on an artificial blood meal in high resolution. This method takes advantage of a transparent bite substrate to facilitate imaging and uses a custom-built cage to provide freely behaving mosquitoes access to the bite substrate. By using custom-built components, the method provides flexibility, as the cage geometry can be easily adjusted to fit specific experimental needs, for instance using a smaller or larger bite substrate or by offering two different bite substrates side by side. The bite substrate itself can be easily modified, by changing the composition of the artificial meal (as done in Jové et al. 2020), changing the meal's texture by adding a gelling agent (e.g., low-melt 1% agarose), changing its appearance by adding food coloring (note, imaging from the bottom is not compatible with opaque liquids), or by modifying the substrate's surface humidity or coating with attractant or repellent compounds (as in Hol et al. 2020). We anticipate that such modifications will be useful for studying how surface-associated cues govern bite-site selection and induce engorgement.

In this protocol, we present two computational approaches to extract behavioral data from the obtained imaging data—namely, one using deep learning techniques and one based on conventional image processing techniques. The deep learning–based method typically results in more accurate tracking and provides richer data and yet is computationally more intensive. Although not strictly necessary, a dedicated graphics processing unit (GPU) is recommended for the deep learning-based approach. Cloud-based computing resources are an attractive option when a GPU workstation is not available in-house. The computational pipelines enable automatic behavioral characterization, including locomotion and engorgement statistics, and derived activity and feeding metrics. With some modifications, the obtained data can be used as input for novel machine learning tools enabling behavioral characterization, such as B-SOID (Hsu and Yttri 2021) or VAME (Luxem et al. 2020). Algorithmic behavioral classification directly from video data enables the high-throughput comparison of mosquito cohorts undergoing different experimental treatments and provides a powerful approach to explore how internal physiological states shape behavior.

RECIPES

Artificial Meal

Reagent	Final concentration
NaCl (500 mM, stored at room temperature)	110 mM
$NaHCO_3$ (100 mM, stored at room temperature)	20 mM
ATP (20 mM in 25 mM $NaHCO_3$, stored at −20°C)	1 mM

Prepare the artificial meal fresh before each experiment, and use it within 1 d. This formulation was used in Jové et al. (2020) and Hol et al. (2020). See Duvall et al. (2019). for artificial meal formulations that include a variety of common proteins present in blood.

Sucrose Solution (10%, w/w)

Make a 10% (w/w) sucrose solution by dissolving granulated sugar in deionized water. Autoclave in sealed Pyrex containers using standard conditions (e.g., for 15 min at 15 psi and 121°C).

Store at 4°C.

ACKNOWLEDGMENTS

We thank Clayton Ellington, Louis Lambrechts, Ariel Lindner, and all members of the Insect-Virus Interactions Unit and the Center for Research and Interdisciplinarity for valuable discussions. Work in the laboratory of F.J.H.H. was supported by a CRI Long-Term Fellowship, a Marie Curie Fellowship, and a Career Award at the Scientific Interface for the Burroughs Wellcome Fund.

REFERENCES

biteOscope GitHub repository. https://github.com/felixhol/biteOscope [26 February 2022].

Duvall LB, Ramos-Espiritu L, Barsoum KE, Glickman JF, Vosshall LB. 2019. Small-molecule agonists of *Ae. Aegypti* neuropeptide Y receptor block mosquito biting. *Cell* **176:** 687–701.e5. doi:10.1016/j.cell.2018.12.004

Hol FJH, Lambrechts L, Prakash M. 2020. BiteOscope: an open platform to study mosquito blood-feeding behavior. *eLife* **9:** e56829. doi:10.7554/eLife.56829

Hsu AI, Yttri EA. 2021. B-SOiD, an open-source unsupervised algorithm for identification and fast prediction of behaviors. *Nat Commun* **12:** 5188. doi:10.1038/s41467-021-25420-x

Jové V, Gong Z, Hol FJH, Zhao Z, Sorrells TR, Carroll TS, Prakash M, McBride CS, Vosshall LB. 2020. Sensory discrimination of blood and floral nectar by *Aedes aegypti* mosquitoes. *Neuron* **108:** 1163–1180.e12. doi:10.1016/j.neuron.2020.09.019

Luxem K, Fuhrmann F, Kürsch J, Remy S, Bauer P. 2020. Identifying behavioral structure from deep variational embeddings of animal motion. bioRxiv doi:10.1101/2020.05.14.095430

Mathis A, Mamidanna P, Cury KM, Abe T, Murthy VN, Mathis MW, Bethge M. 2018. DeepLabCut: markerless pose estimation of user-defined body parts with deep learning. *Nat Neurosci* **21:** 1281–1289. doi:10.1038/s41593018-0209-y

CHAPTER 9

Determination of Mosquito Probing and Feeding Time to Evaluate Mosquito Blood Feeding

Ines Martin-Martin,[1] Adeline E. Williams,[1,2] and Eric Calvo[1,3]

[1]*Laboratory of Malaria and Vector Research, National Institute of Allergy and Infectious Diseases, National Institutes of Health, Rockville, Maryland 20852, USA;* [2]*Department of Microbiology, Immunology, and Pathology, Colorado State University, Fort Collins, Colorado 80523, USA*

Hematophagous arthropods are animals that feed on vertebrate blood for egg production. Mosquitoes must pierce the host skin, locate blood vessels, and extract blood without being noticed. Mosquito stylets lacerate host tissues, triggering the activation of the three branches of hemostasis, or stopping of blood flow: vasoconstriction, platelet aggregation, and coagulation. Mosquitoes inject saliva into the host skin during their intradermal search for blood (also called probing), and salivary proteins counteract hemostasis. Blood feeding dynamics have been traditionally described by observational studies, in which researchers using magnifying glasses watched mosquitoes in the act of blood feeding. These studies provided the foundation for protocols to evaluate mosquito blood feeding in a more quantitative manner. Here, we introduce mosquito blood feeding biology with a focus on the feeding steps, which include penetration, probing, and feeding. Understanding blood feeding dynamics is crucial for evaluating probing time and other relevant parameters derived from blood feeding, such as blood meal size, fecundity, and fertility. Other considerations, including the relationship between probing and pathogen transmission and novel technologies to address blood feeding, are also discussed.

BLOOD FEEDING PHASES

Mosquitoes require ingestion of vertebrate blood for egg production. For successful blood feeding to occur, mosquitoes must locate a suitable host, penetrate the host skin with their stylets, and inject saliva to counteract host hemostasis. These phases are known as penetration and probing time, respectively. Locating blood vessels is followed by a phase of active ingestion and engorgement referred to as feeding time. Once engorged, mosquitoes withdraw their stylets and retreat from the host. A mosquito must complete blood feeding before the host kills or disturbs it, which would impair blood uptake and therefore egg production. The entire blood feeding process must be completed within the "safety period," or the time between the initial salivary injection and the onset of irritation in host skin (Gillett 1967).

Below is an overview of the feeding steps, which include penetration, probing, and feeding. See Protocol 2: Evaluation of *Aedes aegypti* Penetration, Probing, and Feeding Times on Mice (Martin-Martin et al. 2022a) for additional information on feeding. For anatomy of mosquito mouthparts, see Fig. 1 and Krenn and Aspock (2012).

[3]Correspondence: ecalvo@niaid.nih.gov

Cite this introduction as *Cold Spring Harb Protoc*; doi:10.1101/pdb.top107659

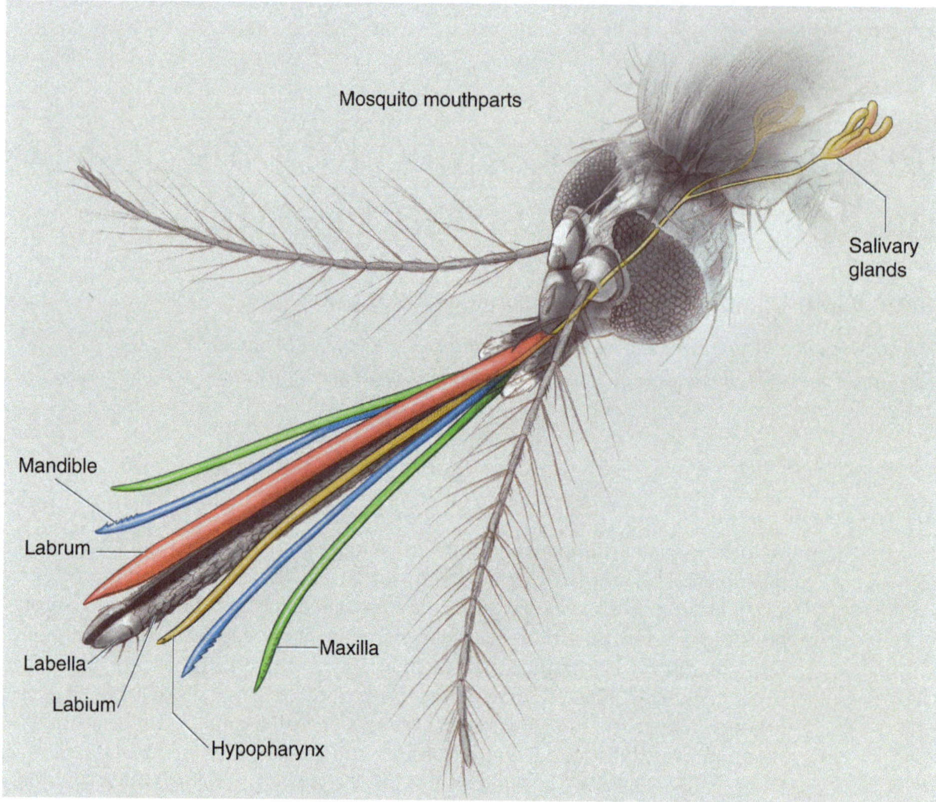

FIGURE 1. Mosquito mouthpart diagram. Credit: Mr. Ryan Kissinger, NIAID.

Penetration Time

Penetration time is the time it takes for the mosquito stylets to penetrate the host skin. Once the mosquito finds a suitable spot for probing, it lowers its metathoracic legs and starts to push its proboscis into the skin. The fascicle is formed by an intricate arrangement of various fine feeding stylets that are ensheathed by a large labium (see Fig. 1; Krenn and Aspock 2012). As soon as the fascicle penetrates the skin, the maxillary palps show a characteristic tremor (Gillett 1967; Jones and Pilitt 1973). The tip of the labrum, which is one of the stylets, first anchors down into the top layer of the skin, whereas the labium retracts and remains folded outside. The fascicle is usually inserted up to about half its length (Kong and Wu 2009). Penetration time is short (<10 sec) and is normally included in the probing time as part of the exploratory phase for blood (Gillett 1967; Ribeiro 2000).

Probing Time

Probing time is recorded as the duration of time between stylet penetration and the appearance of blood in the fascicle and gut as observed through the abdominal pleura (Gillett 1967; Ribeiro 2000). A few seconds after the stylet penetrates the skin, the mosquito uses the toothed maxillae of the fascicle to saw its way into the skin tissue, observed as mosquito head movements (Kong and Wu 2009). The mosquito also bends its flexible fascicle tip to search for blood capillaries. The fascicle acts as both a piercing mechanism and a food canal. During stylet penetration and probing time, mosquitoes release saliva that travels from the salivary glands through the salivary channel, which is an independent conduit from the hypopharynx allowing for simultaneous salivation and blood ingestion (Krenn and Aspock 2012).

Probing time varies from a few seconds, if the mosquito lands right on a blood vessel, to several minutes. Probing times vary by mosquito species, host factors such as temperature, and the level of skin vascularization (Ribeiro et al. 1985; Ribeiro 2000). Slow probing times may also pose fitness costs

 Cite this introduction as *Cold Spring Harb Protoc*; doi:10.1101/pdb.top107659

for mosquitoes. Indeed, mosquitoes that are reared in the absence of selective pressure against slow feeding behavior show longer probing times (Gillett 1967).

Mosquitoes can withdraw their mouthparts before ingesting blood. They may resume probing by reinserting the stylets into the skin, or they can desist. When measuring probing, in the event of reinsertion of the stylets, the timer should be stopped at mouthpart withdrawal and resumed following a subsequent penetration. Time in between probes is measured as "interprobing time" and is excluded from probing time. Accordingly, the total duration of probing time refers to the sum of all probes leading to blood location. In contrast, the duration of an unsuccessful probe is designated as desistance time (Ribeiro et al. 1985). For more detailed guidance, see Protocol 2: Evaluation of *Aedes aegypti* Penetration, Probing, and Feeding Times on Mice (Martin-Martin et al. 2022a)

Feeding Time

Feeding time starts with the sudden appearance of blood in the fascicle and abdomen and ends with a complete distension of the abdomen, followed by the withdrawal of the fascicle. As soon as blood is visualized in the fascicle, the palps abruptly stop rocking, indicating that blood imbibition has begun (Jones and Pilitt 1973). For *Aedes aegypti* and *Ae. albopictus*, feeding time might take 2–3 min on average, and mosquitoes may ingest blood equal to twice their body weight (Kong and Wu 2009). Because female mosquitoes depend on a blood meal for egg production, it is important to evaluate the blood meal size. Blood meal size can be evaluated in a qualitative manner—visually inspecting the abdomen for blood under the stereoscope—or in a quantitative manner—analyzing midgut weight increase or hemoglobin levels. See Protocol 1: Evaluation of *Aedes aegypti* Feeding Status (Martin-Martin et al. 2022b) for details.

Mosquito fecundity and fertility directly depend on the amount of imbibed blood and are relevant parameters to measure fitness. An improved method for evaluating fecundity and fertility was recently published by Tsujimoto and Adelman (2021).

PROBING AND PATHOGEN TRANSMISSION

Saliva is the route of transmission for most mosquito-borne pathogens. Mosquitoes salivate before ingesting blood and can transmit pathogens to the host during this phase of feeding. Notably, blood ingestion is not required for pathogen transmission because mosquitoes that only probe are still able to transmit pathogens (Boyd and Stratman-Thomas 1934; Gillett 1967; Ribeiro et al. 1985). Indeed, virus transmission correlates with the duration of probing and not with the duration of feeding (Mellink 1982). Probing time, as opposed to biting, is actually the important metric for pathogen transmission because salivary gland excretion and blood feeding are independent processes. Therefore, the study of mosquito probing can be applied to determine factors that affect pathogen transmission risks.

Although technically more difficult, feeding parameters can be applied to pool feeders too. For instance, feeding success experiments have proven to be useful in determining the role of certain salivary proteins in sand fly blood feeding (Martin-Martin et al. 2018).

FUTURE STEPS IN BLOOD FEEDING EVALUATION

Studying mosquito feeding mechanisms is becoming increasingly interdisciplinary. For example, microparticle image velocimetry and synchrotron X-ray technologies have been used to analyze the functional relationships between the cibarial and pharyngeal pumps, which are the two muscular organs in the mosquito head whose synchronized movement results in blood extraction (Kim et al. 2011). Furthermore, prediction models of insertion force for the mosquito fascicle into human skin have also been created (Kong and Wu 2009). Indeed, newer technologies complement previous observations made by vector biologists. Electropenetrography is a technique that allows the recording of stylet penetration and probing in which the host and insect become part of an electrical circuit. A small conductive wire is attached to the insect while an electrode and a small electrical current are applied to the host. When the insect starts feeding, a circuit forms between the insect and host and emits a series of waveforms that highlight feeding behaviors such as penetrating the host tissue, salivating, or imbibing plant juices or blood. This technique has mostly been used to study plant-feeding insects or insect pests (Ebert 2019; Shugart et al. 2019). Recent work with *Ae. aegypti* has detailed deep and quantifiable information underlying several phases of feeding: preprobing behavior, shallow penetration of the skin, deeper penetration to reach a blood vessel, and blood ingestion (Wayadande et al. 2020). Additionally, new feeding devices like the biteOscope have been designed to study blood feeding. The biteOscope is a device that attracts mosquitoes to a transparent host mimic that allows imaging of the feeding process (Hol et al. 2020).

In conclusion, mosquito blood feeding studies provide important information that can be applied to prevent mosquito probing and thus pathogen transmission. Feeding studies are also essential for assessing fitness costs of genetically modified insects.

ACKNOWLEDGMENTS

Our research was supported by the Intramural Research Program of the National Institutes of Health (NIH)/National Institute of Allergy and Infectious Diseases (NIAID) (AI001246).

REFERENCES

Boyd MF, Stratman-Thomas WK. 1934. Studies on benign tertian malaria. 7. Some observations on inoculation and onset. *Am J Epidemiol* **20:** 488–495. doi:10.1093/oxfordjournals.aje.a118087

Ebert TA. 2019. The probing behavior component of disease transmission in insect-transmitted bacterial plant pathogens. *Insects* **10:** 212. doi:10.3390/insects10070212

Gillett JD. 1967. Natural selection and feeding speed in a blood-sucking insect. *Proc R Soc Lond B Biol Sci* **167:** 316–329. doi:10.1098/rspb.1967.0029

Hol FJH, Lambrechts L, Prakash M. 2020. BiteOscope, an open platform to study mosquito biting behavior. *eLife* **9:** e56829. doi:10.7554/eLife.56829

Jones JC, Pilitt DR. 1973. Blood-feeding behavior of adult *Aedes aegypti* mosquitoes. *Biol Bull* **145:** 127–139. doi:10.2307/1540353

Kim BH, Kim HK, Lee SJ. 2011. Experimental analysis of the blood-sucking mechanism of female mosquitoes. *J Exp Biol* **214:** 1163–1169. doi:10.1242/jeb.048793

Kong XQ, Wu CW. 2009. Measurement and prediction of insertion force for the mosquito fascicle penetrating into human skin. *J Bionic Eng* **6:** 143–152. doi:10.1016/S1672-6529(08)60111-0

Krenn HW, Aspock H. 2012. Form, function and evolution of the mouthparts of blood-feeding Arthropoda. *Arthropod Struct Dev* **41:** 101–118. doi:10.1016/j.asd.2011.12.001

Martin-Martin I, Chagas AC, Guimaraes-Costa AB, Amo L, Oliveira F, Moore IN, DeSouza-Vieira TS, Sanchez EE, Suntravat M, Valenzuela JG, et al. 2018. Immunity to LuloHya and Lundep, the salivary spreading factors from *Lutzomyia longipalpis*, protects against *Leishmania major* infection. *PLoS Pathog* **14:** e1007006. doi:10.1371/journal.ppat.1007006

Martin-Martin I, Williams AE, Calvo E. 2022a. Evaluation of *Aedes aegypti* penetration, probing, and feeding times on mice. *Cold Spring Harb Protoc* doi:10.1101/pdb.prot108024

Martin-Martin I, Williams AE, Calvo E. 2022b. Evaluation of *Aedes aegypti* feeding status. *Cold Spring Harb Protoc* doi:10.1101/pdb.prot108023

Mellink JJ. 1982. Estimation of the amount of Venezuelan equine encephalomyelitis virus transmitted by a single infected *Aedes aegypti* (Diptera: Culicidae). *J Med Entomol* **19:** 275–280. doi:10.1093/jmedent/19.3.275

Ribeiro JM. 2000. Blood-feeding in mosquitoes: probing time and salivary gland anti-haemostatic activities in representatives of three genera (*Aedes, Anopheles, Culex*). *Med Vet Entomol* **14:** 142–148. doi:10.1046/j.1365-2915.2000.00227.x

Ribeiro JMC, Rossignol PA, Spielman A. 1985. *Aedes aegypti*: Model for blood finding strategy and prediction of parasite manipulation. *Exp Parasitol* **60:** 118–132. doi:10.1016/S0014-4894(85)80029-1

Shugart H, Ebert T, Gmitter F, Rogers M. 2019. The power of electropenetrography in enhancing our understanding of host plant-vector interactions. *Insects* **10:** 407. doi:10.3390/insects10110407

Tsujimoto H, Adelman ZN. 2021. Improved fecundity and fertility assay for *Aedes aegypti* using 24 well tissue culture plates (EAgaL Plates). *JoVE* e61232. doi:10.3791/61232

Wayadande AC, Backus EA, Noden BH, Ebert T. 2020. Waveforms from stylet probing of the mosquito *Aedes aegypti* (Diptera: Culicidae) measured by AC–DC electropenetrography. *J Med Entomol* **57:** 353–368. doi:10.1093/jme/tjz188

Cite this introduction as *Cold Spring Harb Protoc*; doi:10.1101/pdb.top107659

Protocol 1

Evaluation of *Aedes aegypti* Feeding Status

Ines Martin-Martin,[1] Adeline E. Williams,[1,2] and Eric Calvo[1,3]

[1]*Laboratory of Malaria and Vector Research, National Institute of Allergy and Infectious Diseases, National Institutes of Health, Rockville, Maryland 20852, USA;* [2]*Department of Microbiology, Immunology, and Pathology, Colorado State University, Fort Collins, Colorado 80523, USA*

Female mosquitoes need vertebrate blood for egg development. Evaluating mosquito behavior is essential for determining the ability of a mosquito to blood feed. Blood feeding experiments are often performed using artificial membrane feeders; however, such experiments do not represent realistic scenarios in which a mosquito injects saliva into the host to prevent host hemostatic responses. Vertebrate animal models are therefore more representative of a natural blood feeding event. Here, we describe a methodology to evaluate mosquito blood feeding success that can be used to compare blood feeding between mosquito groups—for instance, wild-type versus transgenic mosquitoes lacking salivary proteins or field-collected versus laboratory-reared mosquitoes. We also include a simple procedure to measure blood meal size, allowing for a more quantitative assessment of feeding status. The volume of ingested blood directly affects mosquito fecundity and fertility, important markers of fitness. The methods described herein can be used to evaluate transmission-blocking vaccines, insecticides, or fitness costs associated with transgenic mosquitoes.

MATERIALS

It is essential that you consult the appropriate Material Safety Data Sheets and your institution's Environmental Health and Safety Office for proper handling of equipment and hazardous materials used in this protocol.

RECIPES: Please see the end of this protocol for recipes indicated by <R>. Additional recipes can be found online at http://cshprotocols.cshlp.org/site/recipes.

Reagents

Aedes aegypti female mosquitoes, 5–7 d old (BEI Resources)

Female mosquitoes need to be at least 5 d old to have the whole salivary protein repertoire (Mellink and Von Zeben 1976). Use mosquitoes of the same age and size for comparisons.

Brij-supplemented Drabkin's reagent <R>

Drabkin's reagent is used to measure hemoglobin from blood ingested by insects (Chagas et al. 2014; Martin-Martin et al. 2018; Kojin et al. 2021).

CO_2 for anesthetizing mosquitoes

Mice, 8- to 12-wk-old

Studying both sexes should be a guiding principle to aid in experimental design and hypothesis generation.

Mouse anesthesia (a mixture of ketamine [90 mg/kg] and xylazine [10 mg/kg])

[3]Correspondence: ecalvo@niaid.nih.gov

Cite this protocol as *Cold Spring Harb Protoc*; doi:10.1101/pdb.prot108023

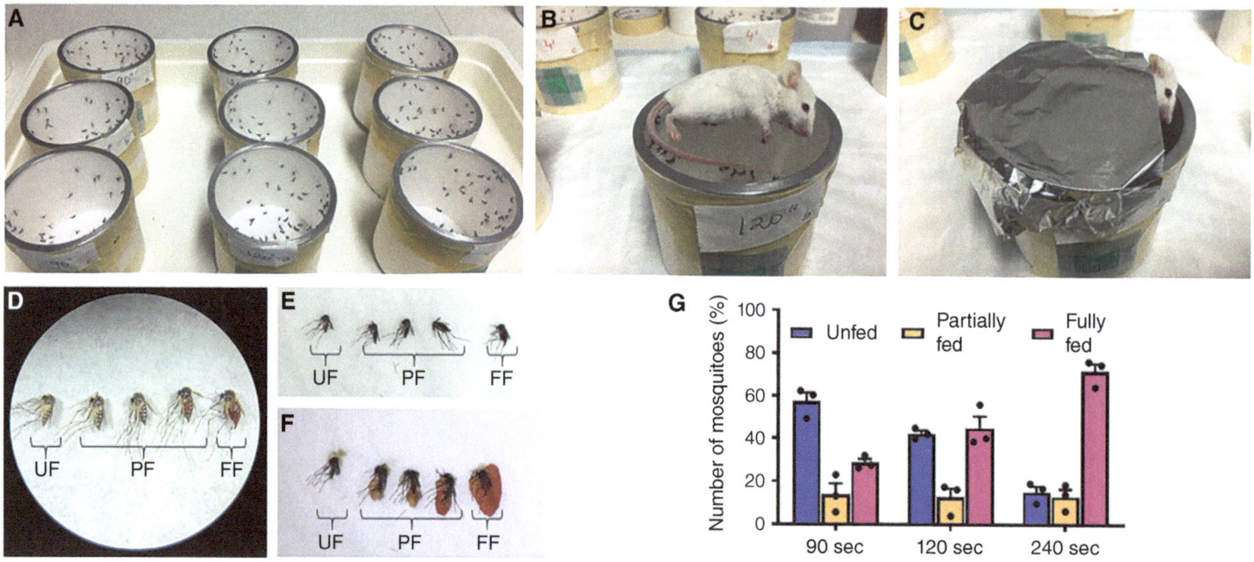

FIGURE 1. Feeding success experiments. (*A*) Cardboard box pints containing starved mosquitoes. (*B,C*) Anesthetized mouse is placed on top of mosquito container and covered with foil to retain body heat. (*D*) Blood feeding assessment under a stereo microscope. (*E,F*) Confirmation of blood feeding status by crushing the mosquitoes onto a tissue paper. (*G*) Feeding success experiment results expressed as the mean ± SEM of biological triplicates. (UF) Unfed, (PF) partially fed, (FF) fully fed.

Equipment

Aluminum sealing foil

Flat-bottom 96-well microtiter plates (Corning, Costar)

Forceps, straight with sharp edges (Dumont #5 preferred)

Glass or plastic Petri dishes

Microcentrifuge tubes (1.5-mL)

Mosquito aspirator (electronic; The John W. Hock Company)

Mosquito cardboard pints (Neptune Paper Products)

Replace the cardboard lid insert with 0.04″ × 0.04″ mesh as shown in Figure 1. Any handmade mosquito container can be used, as long as there is a meshed surface to facilitate contact between the mouse and the mosquito.

Needles, 25G

Self-regulating temperature-control heating pad (PhysioSuite, Kent Scientific)

Stereo microscope (Stemi 508, Zeiss)

Syringes (1-mL)

Tabletop centrifuge

Timer

Tissue homogenizer, such as a Bullet Blender Storm (Next Advance) with microcentrifuge tubes containing disruption beads (Zirconia/Silica, 2.3-mm; Research Product International 9838) or blue polypropylene pestles (Thomas Scientific 3411D62) with pestle motor mixer (Thomas Scientific 1226W45) (see Step 9)

VersaMax microplate reader (Molecular Devices)

METHOD

When we carry out this protocol, we follow Public Health Service Animal Welfare Assurance #A4149-01 guidelines from the National Institute of Allergy and Infectious Diseases (NIAID), National Institutes of Health (NIH) Animal Office of Animal Care and Use (OACU). We carry out our studies according to the NIAID/NIH animal study protocol

 Cite this protocol as *Cold Spring Harb Protoc*; doi:10.1101/pdb.prot108023

(ASP) approved by the NIH Office of Animal Care and Use Committee (OACUC), with approval ID ASP-LMVR3. We use mice that are housed in one of the animal facilities from the NIAID/NIH and are humanely treated according to OACU regulations.

Feeding Success Experiments

1. Use an aspirator to transfer female mosquitoes to cardboard box pints (Fig. 1A).

   At least 30 mosquitoes per pint should be used to account for individual variability. Use two or three replicates, in which each replicate corresponds to mosquitoes fed on one animal.

2. Starve the mosquitoes for 16 h by removing sugar and water cotton pads usually used for feeding.

   Starving the mosquitoes will increase their will to initiate blood feeding.

   For general reference of mosquito rearing, see Das et al. (2007).

3. Anesthetize a mouse with an intraperitoneal injection of ketamine (90 mg/kg)/xylazine (10 mg/kg) per your IACUC protocol.

4. Once the animal is sedated, place it on top of the meshed surface of the mosquito container in a prone position. Cover the animal with the heating pad or aluminum foil and start the timer (Fig. 1B,C).

   According to our protocols, right after feeding and before the animal wakes up from the anesthesia, the mouse is euthanized. For feeding success experiments, anesthesia boosters are not applied and feeding may be extended to a maximum period of time of 15 min.

   It is important to establish a constant time for all mosquito feedings. Comparisons among treatment versus control groups (wild-type versus knockout mosquitoes or naive versus vaccinated mice) are best performed at short times such as 90 or 120 sec. Alternatively, a preliminary test can be done to determine a time at which 60%–80% of the mosquitoes in the control group are fully blood-fed.

   If for some experiments maximal complete feeding needs to be ensured, the feeding time can be extended up to 15 min.

   See Troubleshooting.

5. After the desired amount of time and using the aspirator, collect the mosquitoes from each pint, anesthetize the mosquitoes by applying CO_2, and transfer them to a Petri dish set on ice. Using forceps, place the anesthetized mosquitoes under the stereo microscope and evaluate their feeding status. Visually catalog the animals as unfed (no blood in the midgut), partially fed (some blood in the midgut), or fully fed (engorged mosquitoes with fully distended midgut) as shown in Figure 1D.

   Determining feeding status might not be obviously evident, especially when differentiating between unfed and partially fed mosquitoes. Mosquitoes can be crushed manually onto a tissue paper to confirm engorgement status (Fig. 1E,F).

6. Analyze the data (Fig. 1G).

   If the number of total mosquitoes is equal for all groups, contingency table analyses such as χ^2 of Fisher's exact test can be used to compare values of unfed, partially fed, and fully fed. Otherwise, convert values to percentages and analyze using a one-way ANOVA.

   When increasing the experimental time, the number of unfed mosquitoes decreases, whereas the fully blood-fed numbers increase. The analysis can also be done grouping partially and fully fed together in one group as fed mosquitoes.

Quantitative Assessment of Ingested Blood

7. Randomly select mosquitoes from each group (unfed, partially fed, and fully fed).

   Analyze at least 15 mosquitoes per group. If not analyzed immediately, mosquitoes can be stored for up to 6 mo at −20°C (Kojin et al. 2021).

8. Using forceps, place individual mosquitoes in microcentrifuge tubes containing two to four disruption beads unless using a pestle and pestle motor mixer (see Step 9).

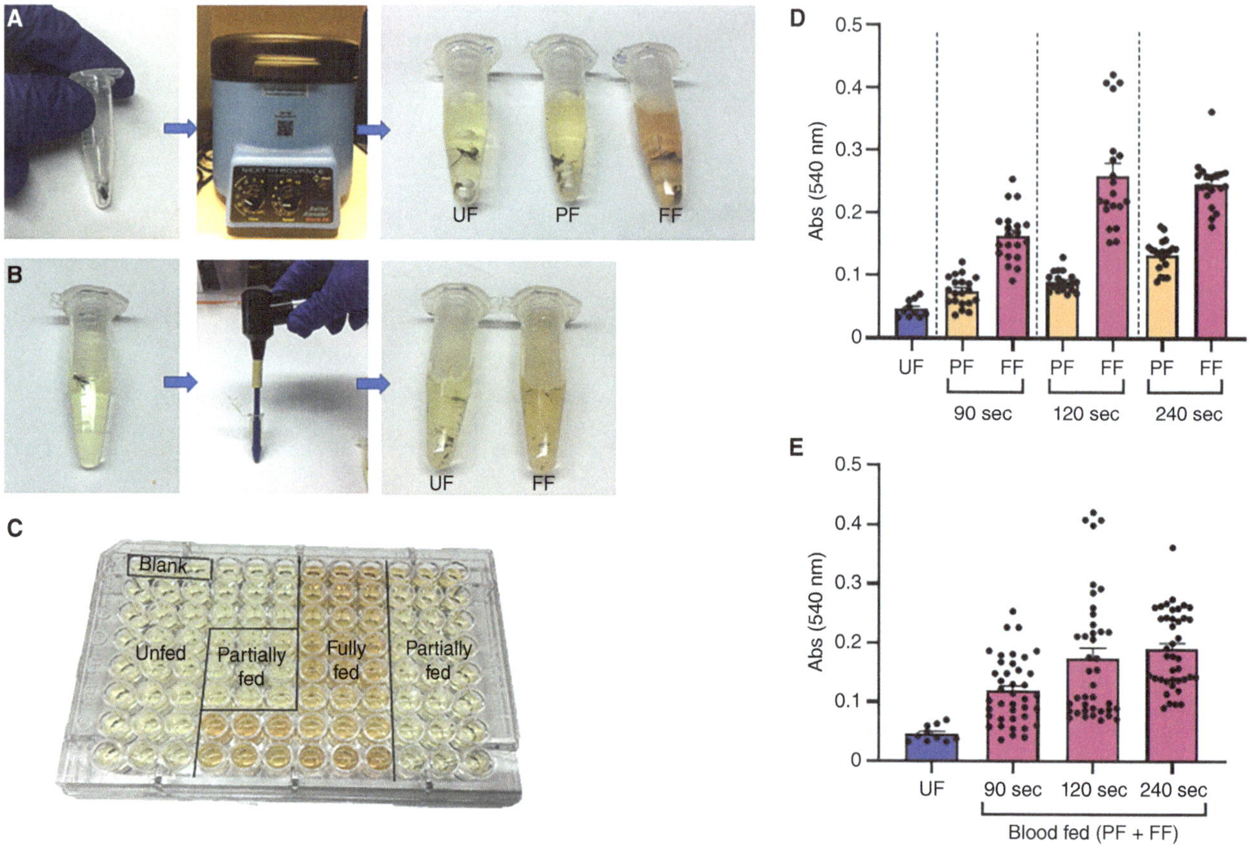

FIGURE 2. Blood meal size. Mosquito homogenization by Bullet Blender Storm (*A*) or polypropylene pestles with pestle motor mixer (*B*). (C) A 96-well plate containing supernatants from homogenized mosquitoes (in technical triplicates) for absorbance reading at 540 nm. Brij-supplemented Drabkin's reagent is used as a blank. (*D,E*) Results expressed as the mean ± SEM of the absorbance of mosquitoes fed for 90, 120, or 240 sec. Data can be separated as PF and FF (*D*) or analyzed together (*E*). (UF) Unfed, (PF) partially fed, (FF) fully fed.

9. Add 1 mL of Brij-supplemented Drabkin's reagent to each tube. Homogenize for 1 min at 4000 rpm in a Bullet Blender Storm (Next Advance) (Fig. 2A).

   Alternatively, homogenize individual mosquitoes in 200 µL of Brij-supplemented Drabkin's reagent with polypropylene pestles and a pestle motor mixer. Then add 800 µL of Brij-supplemented Drabkin's reagent to each tube (Fig. 2B). In this case, beads are not used.

10. Incubate for 10 min at room temperature and clarify by centrifugation for 10 min at 13,500 × *g* at room temperature.

11. Transfer 200 µL of each supernatant to a flat-bottom 96-well microtiter plate and measure the absorbance at 540 nm in a plate reader (Fig. 2C). Read technical triplicates on the same plate.

    Use Brij-supplemented Drabkin's reagent alone as a blank. After mosquito blood feeding, mouse blood can be drawn per your IACUC protocol and used as a standard to extrapolate the volume of ingested blood in the mosquitoes. A standard curve can be performed by adding 1–10 µL of blood instead of a mosquito in Step 8.

12. Analyze data.

    Compare absorbances or blood volumes between experimental and control groups (Fig. 2D). The analysis can also be done grouping partially and fully fed together in one group as fed mosquitoes (Fig. 2E).

TROUBLESHOOTING

Problem (Step 4): Mosquitoes are not interested in blood feeding on the animal.
Solution: Encourage mosquito blood feeding as follows.

Cite this protocol as *Cold Spring Harb Protoc*; doi:10.1101/pdb.prot108023

- Ensure that the temperature of the mouse did not drop after anesthesia administration. Use heating pads and cover the mouse with aluminum foil when possible.
- Breathing near the mosquito vial may encourage blood feeding because of CO_2 exhalation.
- Replace sugar cotton balls with cotton balls impregnated in water 2 d before the experiment. The night before the experiment, remove the water cotton balls.

RECIPE

Brij-Supplemented Drabkin's Reagent

Resuspend one vial of Drabkin's reagent (Sigma-Aldrich D5941) in 1 L of distilled and deionized H_2O (MilliQ grade water). Add 0.5 mL of 30% Brij L23 solution (Sigma-Aldrich B4184). Mix and pass through a 0.45-µm filter if insoluble particles are observed. Cover the bottle with aluminum foil to protect it from the light and store the Brij-supplemented Drabkin's reagent for up to 6 mo at room temperature.

ACKNOWLEDGMENTS

The authors thank Brian Bonilla and Karina Sewell for excellent mosquito rearing. Our research was supported by the Intramural Research Program of National Institutes of Health/National Institute of Allergy and Infectious Diseases (AI001246).

REFERENCES

Chagas AC, Ramirez JL, Jasinskiene N, James AA, Ribeiro JM, Marinotti O, Calvo E. 2014. Collagen-binding protein, Aegyptin, regulates probing time and blood feeding success in the dengue vector mosquito, *Aedes aegypti*. *Proc Natl Acad Sci* **111:** 6946–6951. doi:10.1073/pnas.1404179111

Das S, Garver L, Dimopoulos G. 2007. Protocol for mosquito rearing (*A. gambiae*). *J Vis Exp* **5:** 221. doi:10.3791/221

Kojin BB, Martin-Martin I, Araújo HRC, Bonilla B, Molina-Cruz A, Calvo E, Capurro ML, Adelman ZN. 2021. *Aedes aegypti* SGS1 is critical for *Plasmodium gallinaceum* infection of both the mosquito midgut and salivary glands. *Malar J* **20:** 11. doi:10.1186/s12936-020-03537-6

Martin-Martin I, Chagas AC, Guimaraes-Costa AB, Amo L, Oliveira F, Moore IN, DeSouza-Vieira TS, Sanchez EE, Suntravat M, Valenzuela JG, et al. 2018. Immunity to LuloHya and Lundep, the salivary spreading factors from *Lutzomyia longipalpis*, protects against *Leishmania major* infection. *PLoS Pathog* **14:** e1007006. doi:10.1371/journal.ppat.1007006

Mellink JJ, Von Zeben MS. 1976. Age related differences of saliva composition in *Aedes aegypti*. *Mosquito News* **36:** 247–250.

Protocol 2

Evaluation of *Aedes aegypti* Penetration, Probing, and Feeding Times on Mice

Ines Martin-Martin,[1] Adeline E. Williams,[1,2] and Eric Calvo[1,3]

[1]*Laboratory of Malaria and Vector Research, National Institute of Allergy and Infectious Diseases, National Institutes of Health, Rockville, Maryland 20852, USA;* [2]*Department of Microbiology, Immunology, and Pathology, Colorado State University, Fort Collins, Colorado 80523, USA*

In mosquitoes, the intradermal search for vertebrate blood (probing time) corresponds to the time taken from initial insertion of the mouthparts in the skin until visualization of the initial engorgement of blood in the midgut. Probing time evaluation provides useful information on the ability of a mosquito to initiate successful blood feeding. In this protocol, we describe how to determine feeding parameters in *Aedes aegypti*, a widely distributed mosquito that transmits several deadly pathogens, including yellow fever, dengue, Zika, and Chikungunya viruses. We focus on the different steps of a blood feeding event, including penetration, probing, interprobing, and feeding time. Penetration time corresponds to the insertion of the stylets into the host skin and usually lasts <10 sec. Probing time or intradermal search for blood involves saliva secretion into the skin. Some researchers group penetration and probing time as the exploratory phase for blood. Feeding time is an active phase of blood ingestion and engorgement. Feeding parameters depend on mosquito behaviors and these measurements are visually taken by the investigator. We include a video that provides a close look at a mosquito feeding event in which penetration, probing, and feeding times can be observed. To record these experimental times, one must closely watch the mosquito feeding behavior including stylet penetration in the host skin, visualization of the first traces of blood in the midgut, engorgement of the midgut, and removal of stylets from the skin.

MATERIALS

It is essential that you consult the appropriate Material Safety Data Sheets and your institution's Environmental Health and Safety Office for proper handling of equipment and hazardous materials used in this protocol.

Reagents

Aedes aegypti mosquitoes, 5- to 7-d-old (BEI Resources)

Female mosquitoes must be at least 5 d old to have the whole salivary protein repertoire (Mellink and Von Zeben 1976).

Depilatory cream or clippers (see Step 1)

[3]Correspondence: ecalvo@niaid.nih.gov

Supplemental material is available at cshprotocols.cshlp.org.

Cite this protocol as *Cold Spring Harb Protoc*; doi:10.1101/pdb.prot108024

Mice, 8- to 12-wk-old

Studying both sexes should be a guiding principle to aid in experimental design and hypothesis generation.

Mouse anesthesia (a mixture of ketamine [90 mg/kg] and xylazine [10 mg/kg])

Equipment

Magnifying glass (optional; see Step 7)
Mosquito aspirator (electronic, The John W. Hock Company)
Mosquito meshed cages (2.56″ × 1.7″ × 0.79″)

To make these cages, glue mesh with a size of 0.04″ × 0.04″ to transparent acrylic (Precision Plastics Inc.) so that mesh covers the upper and lower parts of the cages. Cages can be custom built or purchased from Precision Plastics Inc. See Figure 1A for reference.

Needles, 25G
Self-regulating temperature-controlled heating pad (PhysioSuite, Kent Scientific)
Syringes, 1-mL
Timer

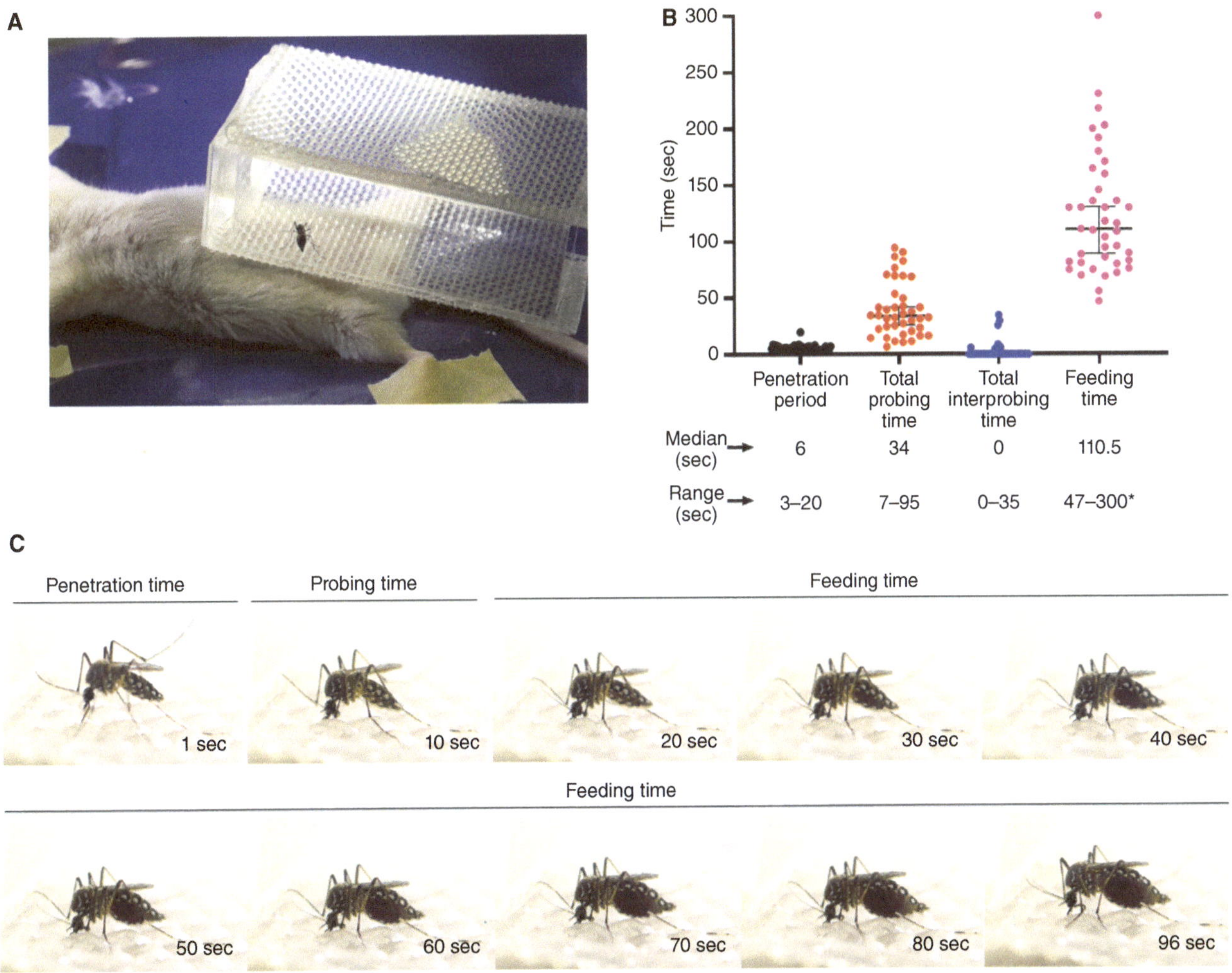

FIGURE 1. Probing time experiments. (*A*) Female mosquito is placed in a meshed cage on an anesthetized mouse. (*B*) Penetration (black), probing (red), interprobing (blue), and feeding (pink) times for wild-type *Aedes aegypti* ($n = 40$). Bars show medians with 95% confidence interval (CI). For plotting and analysis purposes, the maximum experimental time was designated at 300 sec (*). (*C*) Extracted frames from Supplemental Movie S1 showing penetration, probing, and feeding times.

METHOD

When we carry out this protocol, we follow Public Health Service Animal Welfare Assurance #A4149-01 guidelines from the National Institute of Allergy and Infectious Diseases (NIAID), National Institutes of Health (NIH) Animal Office of Animal Care and Use (OACU). We carry out our studies according to the NIAID/NIH animal study protocol (ASP) approved by the NIH Office of Animal Care and Use Committee (OACUC), with approval ID ASP-LMVR3. We use mice that are housed in one of the animal facilities from the NIAID/NIH and are humanely treated according to OACU regulations.

See Supplemental Movie S1, in which an adult female Ae. aegypti *was captured feeding on an anesthetized mouse. The mosquito was caged individually and placed on top of the mouse abdomen. The video shows a successful feeding event without any interprobing time. Penetration, probing, and feeding times are shown.*

1. At least 1 d before the probing experiment, remove hair from the mouse abdomen using clippers or depilatory cream.

   For probing time experiments, removing hair from the mouse abdomen will help to visualize the mosquito proboscis. Because shaving can cause skin irritation, it should be done at least 1 d before the experiment.

2. Using the aspirator, transfer individual mosquitoes to meshed cages (Fig. 1A).

   Label the containers according to the mosquito groups and cover the label to allow for a blinded experiment.

3. Starve the mosquitoes for 16 h by removing sugar and water cotton pads that are usually used for feeding.

   For general reference of mosquito rearing, see (Das et al. 2007).

4. Anesthetize a mouse with an intraperitoneal injection of ketamine (90 mg/kg)/xylazine (10 mg/kg) per your IACUC protocol.
5. Once the animal is sedated, place it on a warm pad at 37°C in the supine position.
6. Place the meshed cage containing the mosquito on top of the mouse such that it covers the whole abdomen of the mouse.

   Different tested mosquitoes will land and probe on different skin areas. If the mosquito does not land on the mouse within 5 min, discard the mosquito and replace it with another one.

7. Using a timer, manually record penetration time as the amount of time it takes for the stylet to be inserted into the mouse skin. Start recording as soon as the mosquito initiates penetration of the mouthparts into the skin. Stop recording when the stylet is completely inserted.

   Penetration time is normally fast and lasts for an average of 6 sec (3–20 sec), as seen in Figure 1B,C and Supplemental Movie S1. Penetration time concludes when the process of stylet insertion has finished. It is determined when the mosquito does not penetrate the skin any further and it stays still on the skin with the stylet inserted.

   A magnifying glass can be used to watch the feeding process.

   See Troubleshooting.

8. Record probing time as the duration between stylet penetration and the appearance of blood in the gut as observed through the thin abdominal pleura (Fig. 1B,C; Supplemental Movie S1).

   Probing time is variable and influenced by, for example, the access to host skin, the anesthesia, host body temperature, and external factors impacting mosquito behavior. Probing time range should be established for each specific experimental setup.

   If mosquitoes withdraw their mouthparts before taking blood, stop timing and resume counting probing time after a subsequent penetration. Total probing time duration refers to the sum of all probes leading to blood engorgement. The time between probing times is called interprobing time. When mosquito probing leads to an unsuccessful feeding, the time spent to locate the blood vessels is referred to as desistance time instead of probing time.

9. Record feeding time as the time elapsed from beginning of blood visualization in the midgut to full engorgement (Fig. 1B,C; Supplemental Movie S1). Full engorgement is assumed when the mosquito stops ingesting blood and its midgut appears distended and full of blood.

Cite this protocol as *Cold Spring Harb Protoc*; doi:10.1101/pdb.prot108024

We terminate all measurements at 300 sec, and mosquitoes that do not start to imbibe blood at this cumulative time are recorded as 300 sec. In this case, probing time can be referred to as desistance time.

Probing and feeding times in mosquitoes are highly variable. A minimum of 30 individual mosquitoes in individual cages per group is recommended for statistical comparisons.

We sequentially perform probing time experiments on the same mouse while the animal remains anesthetized. Additionally, one boost of anesthesia (ketamine/xylazine) can be applied. The total time a mouse can be used for probing time experiments is 30–40 min. According to our protocols, right after feeding and before the animal wakes up from the anesthesia, the mouse is euthanized. Several animals must be used to record data from 30 feeding mosquitoes.

10. Analyze data by comparing probing and feeding time data between control and treatment groups.

    Results can also be expressed as cumulative curves.

TROUBLESHOOTING

Problem (Step 7): Mosquitoes are not interested in blood feeding on the animal.
Solution: Encourage mosquito blood feeding by the following methods.

- Ensure that the temperature of the mouse did not drop after anesthesia administration. Perform probing and feeding experiments in a room at 24°C–26°C. Use heating pads and cover the mouse with aluminum foil when possible.
- Breathing near the mosquito vial may encourage blood feeding because of CO_2 exhalation.
- Replace sugar cotton balls with cotton balls impregnated in water 2 d before the experiment. The night before the experiment, remove the water cotton balls.

DISCUSSION

Mosquito probing and feeding times are highly variable. These parameters depend on mosquito behavior, the host, and the biting area. For instance, *Ae. aegypti* and *Anopheles albimanus* display faster feeding behaviors than *Culex quinquefasciatus* (Ribeiro 2000). *Ae. aegypti* and *An. albimanus* showed shorter probing times when feeding on humans than on mice, whereas *Cx. quinquefasciatus* displayed the opposite behavior. In fact, *Cx. quinquefasciatus* showed fast feeding behaviors on chickens as opposed to feeding on mammals (Ribeiro 2000). The temperature of the host also influences probing and feeding times; therefore, extra care is required to maintain host body temperature at 37°C for probing time experiments. Another factor that affects probing time is the biting area. For example, mosquitoes probe on skin areas that are highly vascularized, such as ears, for shorter times than on areas with a lower degree of vascularization, for instance on the back of an animal (Ribeiro et al. 1984). Feeding parameter measurements are visually taken by the investigator and a certain level of operator subjectivity must be accounted for. Sets of experiments performed by different operators and experiment blinding are crucial requirements for these experiments.

ACKNOWLEDGMENTS

The authors thank Brian Bonilla and Karina Sewell for excellent mosquito rearing. The authors also thank Joy Jackson Farrar, Chia-Chi Charlie Chang, Akura Marshall, and Edward E. Woodhouse Jr., and the National Institutes of Health (NIH) Events Management, for video editing assistance. Our research was supported by the Intramural Research Program of NIH/National Institute of Allergy and Infectious Diseases (NIAID) (AI001246).

REFERENCES

Das S, Garver L, Dimopoulos G. 2007. Protocol for mosquito rearing (*A. gambiae*). *J Vis Exp* **5:** 221. doi:10.3791/221

Mellink JJ, Von Zeben MS. 1976. Age related differences of saliva composition in *Aedes aegypti*. *Mosquito News* **36:** 247–250.

Ribeiro JM. 2000. Blood-feeding in mosquitoes: probing time and salivary gland anti-haemostatic activities in representatives of three genera (*Aedes, Anopheles, Culex*). *Med Vet Entomol* **14:** 142–148. doi:10.1046/j.1365-2915.2000.00227.x

Ribeiro JM, Rossignol PA, Spielman A. 1984. Role of mosquito saliva in blood vessel location. *J Exp Biol* **108:** 1–7. doi:10.1242/jeb.108.1.1

 Cite this protocol as *Cold Spring Harb Protoc*; doi:10.1101/pdb.prot108024

CHAPTER 10

Mosquito Pollination and Sugar Detection Methods: An Overview

Daniel A.H. Peach[1,2]

[1]*Department of Zoology, University of British Columbia, Vancouver, British Columbia, V6T 1Z4, Canada*

The interactions between mosquitoes and plants and, particularly, between mosquitoes and plant sugars from flowers and other structures are often overlooked and are vastly underresearched compared to mosquito–vertebrate or mosquito–pathogen interactions. Given the importance of mosquito nectar-feeding behavior, as well as its impact on vectorial capacity and its implications for vector suppression, a better understanding of mosquito–plant interactions is needed. Direct observation of mosquitoes visiting plants to obtain sugar and other nutrients can be difficult because females may leave flowers to seek a blood meal from the observer, but this can be overcome with the right experimental procedures. This chapter discusses methods for the detection of sugar in mosquitoes and for assessing mosquito pollination.

FLOWERS AS RESOURCES FOR MOSQUITOES

Male and female mosquitoes consume floral nectar as the principal source of their primary food: sugary plant liquids (Foster 1995). These liquids are imbibed using the proboscis—with the fascicle remaining sheathed in the labium—and are primarily diverted to the dorsal diverticula and the ventral diverticulum, or crop (Foster 1995). Sugary plant liquids form the fundamental food of adult mosquitoes, providing sugar and other nutrients for survival, dispersal, and reproduction. In females, sugar enhances egg production and alters biting frequency; in males, it enhances mating performance, which, in turn, alters female biting frequency and oviposition (see reviews by Stone and Foster 2013; Barredo and DeGennaro 2020). Although populations of some highly anthropophilic mosquitoes may achieve high fecundity when fed on human blood alone in the laboratory—a setting in which flight is restricted (Braks et al. 2006), these species are typically found to have fed on sugar when collected from the field (Martinez-Ibarra et al. 1997; Olson et al. 2020), indicating that field populations of these mosquitoes may require plant sugar to persist (Stone et al. 2009). Ultimately, a detailed understanding of mosquito ecology requires a knowledge of their interactions with plants, which play a vital role in combatting mosquito–borne pathogens (Ferguson et al. 2010).

DETECTION OF SUGAR CONSUMPTION

Levels of plant sugar consumption in mosquito populations vary in space and time (Fikrig et al. 2020), and observing nectar-feeding and similar behaviors directly in real time is not always feasible. Fur-

[2]Correspondence: daniel.peach@uga.edu

Cite this introduction as *Cold Spring Harb Protoc*; doi:10.1101/pdb.top107666

thermore, direct determination of the presence and relative amounts of sugar consumed by different species, sexes, or ages, and the frequency of its consumption, is challenging.

A major obstacle is determining the rate of sugar digestion, to establish the time frame after which sugar is undetectable. That rate depends on the sugar type and the physiological state of the mosquito. Sugar presence in field-collected mosquitoes can range from uncommon (Martinez-Ibarra et al. 1997) to very common (Müller et al. 2010; Fikrig et al. 2020), and sugar has been detected in up to 75% of mosquitoes in some populations (Haramis and Foster 1990; Müller et al. 2010).

The hot and cold anthrone tests are based on the dehydration of carbohydrates in concentrated sulfuric acid to form furfural, which reacts with anthrone to give a bluish green color. These tests have been popularized by Van Handel (1972, 1985) for the detection of carbohydrates in mosquitoes (see Protocol 1: Sugar Detection in Mosquitoes via Anthrone Tests [Peach 2023a]). Anthrone tests allow direct measurement of the presence and amount of sugar in a mosquito for up to 24 h after consumption (Fikrig et al. 2020; Olson et al. 2020) and can be used for a variety of purposes. Given the short window during which undigested sugars are present in a mosquito, it is important to field-collect specimens shortly after periods of peak nectar-feeding activity, which can vary by species (Clements 1999).

FLORAL LOCATION

Mosquitoes use a variety of cues to locate floral nectar, among which olfactory cues have been the most studied (Foster 2008; Nyasembe and Torto 2014; Peach and Gries 2020). Mosquitoes respond to floral semiochemicals (Peach et al. 2019a; Lahondère et al. 2020) and volatile emissions from microbes living in floral nectar (Peach et al. 2021). These cues can be exploited for mosquito management technologies, such as phytochemical lures (Nyasembe and Torto 2014; Wooding et al. 2020). Several of the volatiles present in floral semiochemical blends have also been commonly reported in vertebrate odor (Peach et al. 2019a; Nikbakhtzadeh et al. 2014; Ignell and Hill 2020; Lahondère et al. 2020; Hill and Ignell 2021), and vegetative CO_2 emission may play a role in attraction as well (Peach et al. 2019a). This overlap may suggest that the evolutionary origins of mosquito haematophagy lie in nectar-feeding (Peach and Gries 2020; Peach and Matthews 2022; although see Waage 1979 for an alternate theory of origin).

Visual cues also play a role in mosquito attraction to flowers. Demonstration of the importance of visual cues, however, requires behavioral experiments that isolate the effect of olfactory cues alone. Early field studies indicated that mosquitoes preferred flowers with pale colors, such as white or yellow (Foster 1995). However, such evidence was reliant entirely on the subjective experience of human observers and was, thus, biased to the human visual spectrum. Recent work has shown that the visual components of mosquito attraction to flowers likely exist in the ultraviolet (UV) portion of the spectrum and, particularly, that UV-dark patterns on flowers, even when these flowers appear light to humans, are responsible for the visual component of mosquito attraction to flowers (Peach et al. 2019b). It is interesting to note that dark coloration, in combination with olfactory cues, is also responsible for the visual attraction of mosquitoes to vertebrates (van Breugel et al. 2015). UV floral patterns are used by many insect pollinators to find nectar (Koski and Ashman 2014; van der Kooi et al. 2019), and it seems that this pattern includes mosquitoes.

POLLINATION

Although in some cases mosquitoes are observed acting as floral larcenists that obtain nectar without pollinating (Smith and Gadawski 1994; Otienoburu et al. 2012; Pansarin and Pansarin 2017), many studied mosquito–flower associations show at least some pollination (Dexter 1913; Stoutamire 1968; Brantjes and Leemans 1976; Bartareau and Jackes 1994; Peach and Gries 2016). An appreciation for

 Cite this introduction as *Cold Spring Harb Protoc*; doi:10.1101/pdb.top107666

these beneficial interactions requires the use of selective methods when mosquito population management is required. Because of declining pollinator populations and the need to assess environmental effects of mosquito suppression techniques, mosquito pollination is an important topic to explore.

To determine the pollination status of a mosquito for a flower species, one must establish whether those flowers are visited by mosquitoes in the field, whether pollen is picked up, and whether it is effectively transferred to receptive flowers (Ne'eman et al. 2010; Peach and Gries 2016; see Protocol 2: Assessment of Mosquito Pollination [Peach 2023b]). These measures are difficult to ascertain rigorously without designing specific studies to explore them, and a lack of investigation into the topic should not be interpreted as evidence that mosquito pollination in a given system does not occur. After determining whether a plant is pollinated by mosquitoes, measures can be taken to investigate the contribution of mosquito pollination relative to pollination by other insects. However, great care must be taken when trying to design studies and interpret the resulting data (Ne'eman et al. 2010), and special consideration should be given to aspects such as (a) mosquito abundance, (b) density-dependent effects, (c) the order in which mosquitoes visit flowers (e.g., a pollinator visiting species A three times in a row and then species B twice is a row may be more valuable than one alternating back and forth), (d) temporal- and site-specific variation in plant and insect communities, (e) the presence of other flowers at field sites, (f) interactions with other pollinators, and (g) the value of redundant pollinators.

It is currently inappropriate to generate any broad conclusions about mosquito pollination, as this phenomenon remains largely overlooked, even by researchers investigating mosquito–plant interactions (Peach and Gries 2019). In some cases, mosquitoes have been implicated as specialist pollinators (Stoutamire 1968; Brantjes and Leemans 1976; Bartareau and Jackes 1994; Lahondère et al. 2020). Among these close mosquito–plant relationships, the role of mosquitoes as pollinators is likely to be important, if not critical. However, among mosquitoes (and many Diptera) and their generalist nectar hosts (Peach and Gries 2016), the relative contribution of mosquitoes to the pollination of any particular plant species remains poorly investigated and may be limited to times and locations where mosquitoes are abundant or where the majority of mosquito visits occur on the same species of flower. Future studies involving mosquito–flower interactions should include direct field and laboratory pollination experiments, particularly in the tropics, where such studies have been rare.

CONCLUSION

Despite their importance, mosquito–flower interactions are a relatively underresearched aspect of mosquito ecology. Moreover, investigating this behavior requires paying careful attention to the biology of mosquitoes and its unique challenges. Further exploration of mosquito–flower interactions may provide deep insights into mosquito ecology, leading us to reconceptualize aspects of mosquito biology, and may point to novel avenues for mosquito and pathogen control or ways to improve existing mosquito control technologies.

ACKNOWLEDGMENTS

I thank Woodbridge Foster for constructive feedback on this manuscript.

REFERENCES

Barredo E, DeGennaro M. 2020. Not just from blood: mosquito nutrient acquisition from nectar sources. *Trends Parasitol* **36:** 473–484. doi:10.1016/j.pt.2020.02.003

Bartareau T, Jackes B. 1994. Some observations on the flowers and pollination of *Pterostylis procera* M. Celements and D. Jones in Northeast Queensland. *Orchadian* **11:** 198–201.

Braks MAH, Juliano SA, Lounibos LP. 2006. Superior reproductive success on human blood without sugar is not limited to highly anthropophilic mosquito species. *Med Vet Entomol* **20:** 53–59. doi:10.1111/j.1365-2915.2006.00612.x

Brantjes NBM, Leemans JAAM. 1976. *Silene otites* (Caryophyllaceae) pollinated by nocturnal lepidoptera and mosquitoes. *Acta Bot Neerlandica* **25:** 281–295. doi:10.1111/j.1438-8677.1976.tb00240.x

Clements A. 1999. *The biology of mosquitoes, volume 2: sensory reception and behaviour*. CABI, Wallingford, UK.

Dexter J. 1913. Mosquitoes pollinating orchids. *Science* **37:** 867. doi:10.1126/science.37.962.867.a

Ferguson HM, Dornhaus A, Beeche A, Borgemeister C, Gottlieb M, Mulla MS, Gimnig JE, Fish D, Killeen GF. 2010. Ecology: a prerequisite for malaria elimination and eradication. *PLoS Med* **7:** 1–7. doi:10.1371/journal.pmed.1000303

Fikrig K, Peck S, Deckerman P, Dang S, St Fleur K, Goldsmith H, Qu S, Rosenthal H, Harrington LC. 2020. Sugar feeding patterns of New York *Aedes albopictus* mosquitoes are affected by saturation deficit, flowers, and host seeking. *PLoS Negl Trop Dis* **14:** e0008244. doi:10.1371/journal.pntd.0008244

Foster WA. 1995. Mosquito sugar feeding and reproductive energetics. *Annu Rev Entomol* **40:** 443–474. doi:10.1146/annurev.en.40.010195.002303

Foster WA. 2008. Phytochemicals as population sampling lures. *J Am Mosq Control Assoc* **24:** 138–146. doi:10.2987/8756-971X(2008)24[138:PAPSL]2.0.CO;2

Haramis LD, Foster WA. 1990. Gonotrophic cycle duration, population age structure, and onset of sugar feeding and insemination of *Aedes triseriatus* (Diptera: Culicidae). *J. Med. Entomol* **27:** 421–428. doi:10.1093/jmedent/27.4.421

Hill SR, Ignell R. 2021. Modulation of odour-guided behaviour in mosquitoes. *Cell Tissue Res* **383:** 195–206. doi:10.1007/s00441-020-03368-6

Ignell R, Hill SR. 2020. Malaria mosquito chemical ecology. *Curr Opin Insect Sci* **40:** 6–10. doi:10.1016/j.cois.2020.03.008

Koski MH, Ashman TL. 2014. Dissecting pollinator responses to a ubiquitous ultraviolet floral pattern in the wild. *Funct Ecol* **28:** 868–877. doi:10.1111/1365-2435.12242

Lahondère C, Vinauger C, Okubo RP, Wolff GH, Chan JK, Akbari OS, Riffell JA. 2020. The olfactory basis of orchid pollination by mosquitoes. *Proc Natl Acad Sci* **117:** 708–716. doi:10.1073/pnas.1910589117

Martinez-Ibarra JA, Rodriguez MH, Arredondo-Jimenez JI, Yuval B. 1997. Influence of plant abundance on nectar feeding by *Aedes aegypti* (Diptera: Culicidae) in southern Mexico. *J Med Entomol* **34:** 589–593. doi:10.1093/jmedent/34.6.589

Müller GC, Xue R-D, Schlein Y. 2010. Seed pods of the carob tree *Ceratonia siliqua* are a favored sugar source for the mosquito *Aedes albopictus* in coastal Israel. *Acta Trop* **116:** 235–239. doi:10.1016/j.actatropica.2010.08.014

Ne'eman G, Jurgens A, Newstrom-Lloyd L, Potts SG, Dafni A. 2010. A framework for comparing pollinator performance: effectiveness and efficiency. *Biol Rev* **85:** 435–451.

Nikbakhtzadeh MR, Foster WA, Otienoburu PE, Terbot JW. 2014. Olfactory basis of floral preference of the malaria vector *Anopheles gambiae* (Diptera: Culicidae) among common African plants. *J Vector Ecol* **39:** 372–383. doi:10.1111/jvec.12113

Nyasembe V, Torto B. 2014. Volatile phytochemicals as mosquito semiochemicals. *Phytochem Lett* **8:** 196–201. doi:10.1016/j.phytol.2013.10.003

Olson MF, Garcia-Luna S, Juarez JG, Martin E, Harrington LC, Eubanks MD, Badillo-Vargas IE, Hamer GL. 2020. Sugar feeding patterns for *Aedes aegypti* and *Culex quinquefasciatus* (Diptera: Culicidae) mosquitoes in south Texas. *J Med Entomol* **57:** 1111–1119. doi:10.1093/jme/tjaa005

Otienoburu PE, Ebrahimi B, Phelan PL, Foster WA. 2012. Analysis and optimization of a synthetic milkweed floral attractant for mosquitoes. *J Chem Ecol* **38:** 873–81. doi:10.1007/s10886-012-0150-6

Pansarin ER, Pansarin LM. 2017. Crane flies and microlepidoptera also function as pollinators in *Epidendrum* (Orchidaceae: Laeliinae): the reproductive biology of *E. avicula*. *Plant Species Biol* **32:** 200–209. doi:10.1111/1442-1984.12140

Peach DAH. 2023a. Sugar detection in mosquitoes via anthrone tests. *Cold Spring Harb Protoc* doi:10.1101/pdb.prot108301

Peach DAH. 2023b. Assessment of mosquito pollination. *Cold Spring Harb Protoc* doi:10.1101/pdb.prot108302

Peach DAH, Gries G. 2016. Nectar thieves or invited pollinators? A case study of tansy flowers and common house mosquitoes. *Arthropod-Plant Interact* **10:** 497–506. doi:10.1007/s11829-016-9445-9

Peach D, Gries G. 2019. Supplementary data for: Mosquito phytophagy—sources exploited, ecological function, and evolutionary transition to haematophagy. 49427 bytes. http://datadryad.org/stash/dataset/doi:10.5061/dryad.63xsj3tz5 (Accessed December 24, 2021).

Peach DAH, Gries G. 2020. Mosquito phytophagy: sources exploited, ecological function, and evolutionary transition to haematophagy. *Entomol Exp Appl* **168:** 120–136. doi:10.1111/eea.12852

Peach DAH, Matthews BJ. 2022. Sensory mechanisms for the shift from phytophagy to haematophagy in mosquitoes. *Curr Op Insect Sci* **52:** 100930. doi:10.1016/j.cois.2022.100930

Peach DAH, Gries R, Zhai H, Young N, Gries G. 2019a. Multimodal floral cues guide mosquitoes to tansy inflorescences. *Sci Rep* **9:** 3908. doi:10.1038/s41598-019-39748-4

Peach DAH, Ko E, Blake AJ, Gries G. 2019b. Ultraviolet inflorescence cues enhance attractiveness of inflorescence odour to *Culex pipiens* mosquitoes. *PLoS One* **14:** e0217484. doi:10.1371/journal.pone.0217484

Peach DAH, Carroll C, Meraj S, Gomes S, Galloway E, Balcita A, Coatsworth H, Young N, Uriel Y, Gries R, et al. 2021. Nectar-dwelling microbes of common tansy are attractive to its mosquito pollinator, *Culex pipiens* L. *BMC Ecol Evol* **21:** 29. doi:10.1186/s12862-021-01761-5

Smith SM, Gadawski RM. 1994. Nectar feeding by the early-spring mosquito *Aedes provocans*. *Med Vet Entomol* **8:** 201–13. doi:10.1111/j.1365-2915.1994.tb00499.x

Stone CM, Foster WA. 2013. Plant-sugar feeding and vectorial capacity. In *Ecology of parasite–vector interactions* (ed. Takken W, Koenraadt CJM), pp. 35–79. Springer, Wageningen, The Netherlands.

Stone CM, Taylor RM, Roitberg BD, Foster WA. 2009. Sugar deprivation reduces insemination of *Anopheles gambiae* (Diptera: Culicidae), despite daily recruitment of adults, and predicts decline in model populations. *J Med Entomol* **46:** 1327–1337. doi:10.1603/033.046.0611

Stoutamire W. 1968. Mosquito pollination of *Habenaria obtusata*. *Mich Bot* **7:** 203–212.

van Breugel F, Riffell J, Fairhall A, Dickinson MH. 2015. Mosquitoes use vision to associate odor plumes with thermal targets. *Curr Biol* **25:** 2123–2129. doi:10.1016/j.cub.2015.06.046

van der Kooi CJ, Dyer AG, Kevan PG, Lunau K. 2019. Functional significance of the optical properties of flowers for visual signalling. *Ann Bot* **123:** 263–276. doi:10.1093/aob/mcy119

Van Handel E. 1972. The detection of nectar in mosquitoes. *Mosq News* **32:** 458.

Van Handel E. 1985. Rapid determination of glycogen and sugars in mosquitoes. *J Am Mosq Control Assoc* **1:** 299–301.

Waage JK. 1979. The evolution of insect/vertebrate associations. *Biol J Linn Soc* **12:** 187–224. doi:10.1111/j.1095-8312.1979.tb0005.x

Wooding M, Naudé Y, Rohwer E, Bouwer M. 2020. Controlling mosquitoes with semiochemicals: a review. *Parasit Vectors* **13:** 80. doi:10.1186/s13071-020-3960-3

Cite this introduction as *Cold Spring Harb Protoc*; doi:10.1101/pdb.top107666

Protocol 1

Sugar Detection in Mosquitoes via Anthrone Tests

Daniel A.H. Peach[1,2]

[1]*Department of Zoology, University of British Columbia, Vancouver, British Columbia V6T 1Z4, Canada*

Adult mosquitoes feed on plant sugar, frequently in the form of floral nectar. However, because of spatial and temporal variation in this behavior, as well as to the predilection of most mosquitoes to alter their behavior in the presence of a researcher, observing mosquito nectar feeding and similar behaviors directly in real time is not always feasible. In this protocol, I describe methods for hot and cold anthrone tests, which can be used to quantify the degree to which mosquito sugar feeding occurs in nature.

MATERIALS

It is essential that you consult the appropriate Material Safety Data Sheets and your institution's Environmental Health and Safety Office for proper handling of equipment and hazardous materials used in this protocol.

Reagents

2% (w/v) sodium sulfate solution (Sigma-Aldrich 238597) in distilled H_2O
Anthrone (Sigma-Aldrich 52445)
Ethanol, 25% (v/v) in H_2O
Fructose (Sigma-Aldrich 1286504)
Glucose
Ice
Methanol
Sulfuric acid (Sigma-Aldrich 339741)

Equipment

Erlenmeyer flask, 1-L
Glass culture tubes (polypropylene or other acid-stable material), 17-mm × 100-mm
Glass rod
Microcentrifuge
Spectrophotometer and appropriate cuvettes
Tube heater (dry bath or block heater)
Vortex mixer

[2]Correspondence: daniel.peach@uga.edu

Cite this protocol as *Cold Spring Harb Protoc*; doi:10.1101/pdb.prot108301

METHOD

Here, I provide a protocol, synthesized from many sources, for cold and hot anthrone tests to analyze the sugar content of mosquitoes. To carry out this protocol, the user will first collect and chill mosquitoes of interest from the field, prepare the anthrone reagent and optical standards, create a calibration curve from these optical standards, and finally process the mosquitoes and compare them to the calibration curve to determine the amount of sugar present. Whether the cold or hot anthrone test is used depends on whether the user seeks to analyze the presence of fructose or total sugars.

Mosquito Collection

1. Collect mosquitoes by your preferred method.

   Mosquitoes may be collected by a variety of methods, including sweep nets and aspirators at resting sites, variously baited traps, or human landing catches (European Centre for Disease Prevention and Control; European Food Safety Authority 2018).

   Collections made shortly after the time of maximum mosquito foraging activity are best. Resting sites are the best locations for obtaining representative samples of the entire adult population.

2. Chill or freeze specimens by placing them on ice immediately after collection, to minimize crop-emptying and sugar digestion.

Preparation of the Anthrone Reagent

3. Add 150 mL of distilled H_2O to a 1-L Erlenmeyer flask.
4. Cool the flask in ice, and slowly and carefully add 380 mL of concentrated sulfuric acid.

   Mixing sulfuric acid with water results in an exothermic reaction, so follow appropriate safety protocols.

5. While following appropriate safety protocols and using a chemical fume hood, add 750 mg of anthrone to the diluted sulfuric acid and mix.

Cold Anthrone Test (adapted from Van Handel 1968, 1972, 1985)

Preparation of Optical Standards

6. Prepare a 1 mg/mL fructose solution in 25% (v/v) ethanol.
7. Create optical standards by adding 0 (blank), 1, 5, 25, 50, 100, 150, and 200 µL of the fructose solution to separate culture tubes (these volumes correspond to 0, 1, 5, 25, 50, 100, 150, and 200 µg of fructose, respectively). Add the anthrone reagent from Step 5 to the 5-mL level.
8. Allow the solution to sit for 5 h at 20°C (or 90 min at 25°C, depending on the local room temperature) and then determine the optical density at 625 nm. If the optical density is >1, dilute the solution and take another reading. Plot micrograms of fructose against the optical density values obtained to form a calibration curve.

Determination of Fructose Content

9. Prepare a culture tube containing 200 µL of 2% (w/v) sodium sulfate solution. Add a mosquito to the solution and homogenize it using a glass rod.

   When processing a large number of samples simultaneously, 96-well microtiter plates can be used (see Discussion).

10. Add 1 mL of methanol, vortex, and then centrifuge for 1 min. Decant the supernatant (the sugar-containing fraction) to a second tube.
11. Add anthrone reagent to the 5-mL level to the tube with the supernatant. Allow the solution to sit for 5 h at 20°C (or 90 min at 25°C), stir it with a glass rod, and determine optical density at

Cite this protocol as *Cold Spring Harb Protoc*; doi:10.1101/pdb.prot108301

625 nm. If the optical density is >1, dilute the solution and adjust your final values accordingly. Compare the final values for fructose to the calibration curve from Step 8.

12. Determine baseline fructose concentrations for non-sugar-fed mosquitoes by repeating Steps 9–11 on pupae that eclose in the lab. Subtract this value from the results from your field-caught population to control for small amounts of sugar that may be present before nectar feeding.

Hot Anthrone Tests (Adapted from Van Handel 1968, 1972, 1985)

Preparation of Optical Standards

13. Mix glucose at 1 mg/mL in 25% (v/v) ethanol
14. Create optical standards by adding 0, 1, 5, 25, 50, 100, 150, 200, and 500 µL of the glucose or glycogen solution to separate culture tubes. Add anthrone reagent from Step 5 to the 5-mL level, then mix and heat for 17 min to 90°C using the tube heater.
15. Let the solution cool to room temperature and then determine the optical density at 625 nm. If the optical density is >1, dilute the solution and take another reading. Plot micrograms of glucose against the optical density values to form a calibration curve.

Determination of Total Sugar Content

16. Prepare a culture tube containing 200 µL of 2% (w/v) sodium sulfate. Add a mosquito to the solution and homogenize it using a glass rod.
17. Add 1 mL of methanol, vortex, and then centrifuge for 1 min. Decant the supernatant (the sugar-containing fraction) to a second tube and allow the solvent to evaporate to a level of 200 µL.

    The glycogen will stay adsorbed on the precipitated sodium sulfate in the original tube and can be analyzed by performing Steps 17–19 on this precipitate. The mosquito tissue will also remain in the original tube, crushed on the bottom.
18. Add anthrone reagent to the tube with the sugar-containing fraction to the 5-mL level, mix, and heat for 17 min at 90°C using the tube heater.
19. Allow the solution to cool to room temperature, stir it with a glass rod, and determine the optical density at 625 nm. If the optical density is >1, dilute the solution and adjust your final values accordingly. Compare the final values to the calibration curve prepared in Step 15.
20. Determine baseline concentrations for non-sugar fed mosquitoes by repeating Steps 16–19 on pupae that eclose in the lab. Subtract this value from the results from your field-caught population to control for small amounts of sugar which may be present before nectar feeding.

DISCUSSION

Plant sugar is an important resource for adult mosquitoes and is their primary food source. Its consumption prolongs mosquito life span (Foster 1995), affects biting frequency, and plays an important role in pathogen transmission dynamics (Gary and Foster 2001). Anthrone is a tricyclic aromatic ketone used for a common cellulose assay and in the colorimetric determination of carbohydrates (Yemm and Willies 1954). Anthrone tests have been used for decades for evaluating the presence of sugar in a mosquito (Van Handel 1968, 1985). Cold anthrone tests detect fructose, which is the only sugar that is uniquely derived from plants rather than generated within the mosquito. Hot anthrone tests, on the other hand, detect the presence of glycogen and several other sugars, such as glucose and trehalose (Van Handel 1985), which may be derived metabolically either from sugar meals or from blood meals and, hence, are not accurate indications of plant-sugar feeding. Mosquitoes rapidly digest sugars, however, and thus the cold anthrone test is limited by the fact that it must be applied shortly after sugar consumption (Fikrig et al. 2020; Olson et al. 2020), before the fructose can be altered metabolically.

Here, I have described a cold anthrone test approach that uses single culture tubes, which allows the processing of a relatively low number of mosquitoes. An alternative reaction vessel for measuring amounts of fructose by the cold anthrone test is a 96-well microtiter plate. The advantages of using these plates are that large numbers of specimens can be processed simultaneously with small amounts of reagents and that colorimetric measures can be read automatically with a spectrometric plate reader. However, the volumes of the test reactants and reagents, and of their standards (controls), must be adjusted to the small capacity of the wells, perhaps by testing only a portion of each sample. For examples of the application of this method, see Gary (2005), Sissoko et al. (2019), and Swan et al. (2021).

Anthrone tests that differ from the procedure described in this protocol and that require additional chemicals for detecting fructose have been developed and have been shown to exhibit increased sensitivity (see Somani et al. 1987; Nunes et al. 2008). Alternative methods to detect undigested sugar meals involve the use of chromatography (Laarman 1968; Watanabe et al. 1973; Magnarelli and Anderson 1977; Magnarelli 1978, 1979, 1980; Nayar 1978), provided that one can distinguish between plant-derived and metabolically generated sugars. Thin-layer and gas chromatography can also be useful for identifying sugars different from honeydew (Burgin and Hunter 1997a,b,c; Burkett et al. 1998, 1999; Janzen and Hunter 1998; Hunter and Ossowski 1999; Wanjiku et al. 2021) or to help determine candidate plant–host species on the basis of their characteristic sugar ratios (Hamilton and El Naiem 2000; Manda et al. 2007). Although many alternative methods exist, the straightforward anthrone tests described in this protocol allow researchers to determine the proportion of mosquitoes that have recently sugar-fed and, if performed under controlled settings, to compare amounts of sugar consumed.

ACKNOWLEDGMENTS

I thank Woodbridge Foster for constructive feedback on this manuscript.

REFERENCES

Burgin SG, Hunter FF. 1997a. Evidence of honeydew feeding in black flies (Diptera: Simuliidae). *Can Entomol* **129:** 859–869. doi:10.4039/Ent129859-5

Burgin SG, Hunter FF. 1997b. Nectar versus honeydew as sources of sugar for male and female black flies (Diptera: Simuliidae). *J Med Entomol* **34:** 605–608. doi:10.1093/jmedent/34.6.605

Burgin SG, Hunter FF. 1997c. Sugar-meal sources used by female black flies (Diptera: Simuliidae): a four-habitat study. *Can J Zool* **75:** 1066–1072. doi:10.1139/z97-128

Burkett DA, Butler JE, Kline DL. 1998. Field evaluation of colored light-emitting diodes as attractants for woodland mosquitoes and other Diptera in north central Florida. *J Am Mosq Control Assoc* **14:** 186–195.

Burkett DA, Kline DL, Carlson DA. 1999. Sugar meal composition of five north central Florida mosquito species (Diptera: Culicidae) as determined by gas chromatography. *J Med Entomol* **36:** 462–467. doi:10.1093/jmedent/36.4.462

European Centre for Disease Prevention and Control; European Food Safety Authority. 2018. *Field sampling methods for mosquitoes, sandflies, biting midges and ticks—VectorNet project 2014–2018*. ECDC and EFSA, Stockholm and Parma.

Fikrig K, Peck S, Deckerman P, Dang S, St Fleur K, Goldsmith H, Qu S, Rosenthal H, Harrington LC. 2020. Sugar feeding patterns of New York *Aedes albopictus* mosquitoes are affected by saturation deficit, flowers, and host seeking. *PLoS Negl Trop Dis* **14:** e0008244. doi:10.1371/journal.pntd.0008244

Foster WA. 1995. Mosquito sugar feeding and reproductive energetics. *Annu Rev Entomol* **40:** 443–474. doi:10.1146/annurev.en.40.010195.002303

Gary RE. 2005. "Biology of the malaria vector *Anopheles gambiae*: behavioral and reproductive components of sugar feeding." *Doctoral Dissertation*, The Ohio State University, Columbus. ProQuest Dissertations Publishing Accession No. 3182728.

Gary RE, Foster WA. 2001. Effects of available sugar on the reproductive fitness and vectorial capacity of the malaria vector *Anopheles gambiae* (Diptera: Culicidae). *J Med Entomol* **38:** 22–28. doi:10.1603/0022-2585-38.1.22

Hamilton JGC, El Naiem D. 2000. Sugars in the gut of the sandfly *Phlebotomus orientalis* from Dinder National Park, eastern Sudan. *Med Vet Entomol* **14:** 64–70. doi:10.1046/j.1365-2915.2000.00208.x

Hunter FF, Ossowski AM. 1999. Honeydew sugars in wild-caught female horse flies (Diptera: Tabanidae). *J Med Entomol* **36:** 896–899. doi:10.1093/jmedent/36.6.896

Janzen TA, Hunter FF. 1998. Honeydew sugars in wild-caught female deer flies (Diptera: Tabanidae). *J Med Entomol* **35:** 685–689. doi:10.1093/jmedent/35.5.685

Laarman JJ. 1968. The intake of sugars by females of wild *Anopheles gambiae* and *Anopheles funestus*. *Acta Leidensia* **36:** 137–144.

Magnarelli LA. 1978. Nectar-feeding by female mosquitoes and its relation to follicular development and parity. *J Med Entomol* **14:** 527–530. doi:10.1093/jmedent/14.5.527

Magnarelli LA. 1979. Diurnal nectar-feeding of *Aedes cantator* and *Ae. sollicitans* (Diptera: Culicidae). *Environ Entomol* **8:** 949–955. doi:10.1093/ee/8.5.949

Magnarelli LA. 1980. Bionomics of *Psorophora ferox* (Diptera: Culicidae): seasonal occurrence and acquisition of sugars. *J Med Entomol* **17:** 328–332. doi:10.1093/jmedent/17.4.328

Magnarelli LA, Anderson JF. 1977. Follicular development in salt marsh Tabanidae (Diptera) and incidence of nectar feeding with relation to gonotrophic activity. *Ann Entomol Soc Am* **70:** 529–533. doi:10.1093/aesa/70.4.529

Manda H, Gouagna LC, Foster WA, Jackson RR, Beier JC, Githure JI, Hassanali A. 2007. Effect of discriminative plant-sugar feeding on the survival and fecundity of *Anopheles gambiae. Malar J* **6:** 113. doi:10.1186/1475-2875-6-113

Nayar JK. 1978. The detection of nectar sugars in field-collected *Culex nigripalpus* and its application. *Ann Entomol Soc Am* **71:** 55–59. doi:10.1093/aesa/71.1.55

Nunes RD, Oliviera RL, Braz GRC. 2008. A novel method for measuring fructose ingestion by mosquitoes. *J Vector Ecol* **33:** 225–231. doi:10.3376/1081-1710-33.2.225

Olson MF, Garcia-Luna S, Juarez JG, Martin E, Harrington LC, Eubanks MD, Badillo-Vargas IE, Hamer GL. 2020. Sugar feeding patterns for *Aedes aegypti* and *Culex quinquefasciatus* (Diptera: Culicidae) mosquitoes in South Texas. *J Med Entomol* **57:** 1111–1119. doi:10.1093/jme/tjaa005

Sissoko F, Junnila A, Traore M, Traore SF, Doumbia S, Dembele SM, Schlein Y, Traore AS, Gergely P, Xue RD, et al. 2019. Frequent sugar feeding behavior by *Aedes aegypti* in Bamako, Mali makes them ideal candidates for control with attractive toxic sugar baits (ATSB). *PLoS ONE* **14:** e021417. doi:10.1371/journal.pone.0214170

Somani BL, Khanade J, Sinha R. 1987. A modified anthrone-sulfuric acid method for the determination of fructose in the presence of certain proteins. *Anal Biochem* **167:** 327–330. doi:10.1016/0003-2697(87)90172-2

Swan T, Ritmejerytė E, Sebayang B, Jones R, Devine G, Graham M, Zich FA, Staunton KM, Russel TL, Burkot TR. 2021. Sugar prevalence in *Aedes albopictus* differs by habitat, sex and time of day on Masig Island, Torres Strait, Australia. *Parasit Vectors* **14:** 520. doi:10.1186/s13071-021-05020-w

Van Handel E. 1968. Direct microdetermination of sucrose. *Anal Biochem* **22:** 280–283. doi:10.1016/0003-2697(68)90317-5

Van Handel E. 1972. The detection of nectar in mosquitoes. *Mosq News* **32:** 458.

Van Handel E. 1985. Rapid determination of glycogen and sugars in mosquitoes. *J Am Mosq Control Assoc* **1:** 299 301.

Wanjiku C, Tchouassi DP, Sole CL, Pirk C, Torto B. 2021. Plant sugar feeding patterns of wildcaught *Aedes aegypti* from Dengue endemic and non-endemic areas of Kenya. *Med Vet Entomol* **35:** 417–425. doi:10.1111/mve.12514

Watanabe M, Kamimura K, Kubota K. 1973. A simplified method for the analysis of various sugars fed the mosquito, *Culex pipiens sl jap. J Sanit Zool* **24:** 17–21. doi:10.7601/mez.24.17

Yemm EW, Willis AJ. 1954. The estimation of carbohydrates in plant extracts by anthrone. *Biochem J* **57:** 508–514. doi:10.1042/bj0570508

Protocol 2

Assessment of Mosquito Pollination

Daniel A.H. Peach[1,2]

[1]*Department of Zoology, University of British Columbia, Vancouver, British Columbia, V6T 1Z4, Canada*

Adult mosquitoes, sometimes in vast numbers, visit flowers in their search for floral nectar. However, the ability of mosquitoes to pollinate the flowers they visit is often neglected and sometimes even presumptively dismissed. Despite this, mosquito pollination has been reported in many instances, although many questions linger about its extent, importance, and the breadth of floral species or mosquito species which may be involved. In this protocol, I describe methodology to assess whether mosquitoes pollinate a flowering plant they visit, which can create a foundation for future investigations into this topic.

MATERIALS

It is essential that you consult the appropriate Material Safety Data Sheets and your institution's Environmental Health and Safety Office for proper handling of equipment and hazardous materials used in this protocol.

Reagents

Flowering plants for trials (see Step 1)

Lab-reared mosquitoes, starved of sugar for 24 h

Mosquito rearing methods will differ based on the mosquito of interest, but, for examples, see Chapter 1, Protocol 1: Batch Rearing Aedes aegypti *(Wohl and McMeniman 2022), Parker (2020), or Kauffman et al. (2017).*

Potted greenhouse-grown plants with flowers attractive to mosquitoes of interest (see Step 1)

Equipment

Aspirator (1135A Aspirator, BioQuip Products)

Collecting vial, 15-mL (4-dram) or 1.5-mL microcentrifuge tubes or similar

Dissecting microscope (M5 Microscope, Wild Heerbrugg Co.)

Dry ice

Forceps

Freezer

Glass vials, 4-mL (Thomas Scientific 1231B91)

Mesh cages (BugDorm 6M610)

Parafilm (Parafilm M, Bemis Company Inc.)

Pollinator exclusion bags (BugDorm DC3225)

Soft wax

String

[2]Correspondence: daniel.peach@uga.edu

Cite this protocol as *Cold Spring Harb Protoc*; doi:10.1101/pdb.prot108302

METHOD

The method has multiple sections. First, mosquito visitation of flowers is assessed, which will inform the user of which flower species to grow for their experiments. Then, users will obtain an estimate of the maximum theoretical pollen load from the mosquito–flower pair of interest. Last, with the information derived from the previous sections, users will assess pollination.

Assessing Mosquito Visitation of Flowers

1. Plan to conduct field surveys during the period of peak nectar feeding for the mosquito species of interest. If this is at dusk or in the dark, use a headlamp, ideally with a red light to reduce disruption. This will tell users which floral species to use in subsequent experiments.

   The presence of human hosts will disrupt some nectar-feeding mosquitoes; therefore, it is best to sample for a short period of time, leave the site so that mosquitoes can settle again, and return later to sample from the site again.

2. Individually collect mosquitoes from the flowers of interest using a mouth aspirator, and aspirate single mosquitoes directly into separate 15-mL (4-dram) collecting vials. Immediately place the vials on dry ice to minimize mosquito movement and pollen dislocation.

   If difficulty is experienced in directly observing mosquitoes on flowers due to host seeking, or no observations are recorded but visitation is suspected, individually collect host-seeking mosquitoes in the vicinity of the flowers instead.

3. Identify the mosquito species collected and species of flower the pollen comes from (as a mosquito may have visited other flowers before the one it was collected on) under a dissecting microscope.

   This will inform users as to which mosquito species and which floral species to use in subsequent experiments.

Confirming Pollen Accumulation

4. Grow flowers of the species identified in Steps 1–3 from seed in pots in a greenhouse.
5. At an appropriate time of day (chosen so that a time point 2 h later will be within the period of peak nectar feeding for the mosquito species of interest), add a potted greenhouse-grown flower from Step 4 (or a cut flower) to a mesh cage, making sure that no pollen adheres to the cage or falls to the cage floor. Then, using an aspirator, gently transfer to the cage 50 lab-reared mosquitoes that have been starved of sugar for 24 h.
6. After 2 h, at a time overlapping with the mosquitoes' peak feeding period, place the entire cage in a freezer and leave it there overnight.
7. Remove dead mosquitoes individually with forceps, place them under a microscope, and count the number of pollen grains present on the mosquitoes. This will provide an estimate of the maximum theoretical pollen load from the mosquito–flower pair of interest and can be used to inform the number of mosquitoes to use in Step 9.

Testing Seed Set

8. Enclose the flowers in a pollinator exclusion bag tied off with string around the stem (Fig. 1). Place a barrier of soft wax between the stem and the bag where it is tied off, so that the string can be drawn tight enough to prevent aphids or other insects from accessing the flowers without damaging the stem.
9. If the investigator is interested in evaluating cross-pollination, cut off flowers from another plant at the stem 5–10 cm below the flower. Fill a 4-mL glass vial with tap H_2O and cover it with Parafilm, and then insert the stem of the flower through the Parafilm to the base of the vial. Insert these in the mesh bags and reseal the bags.

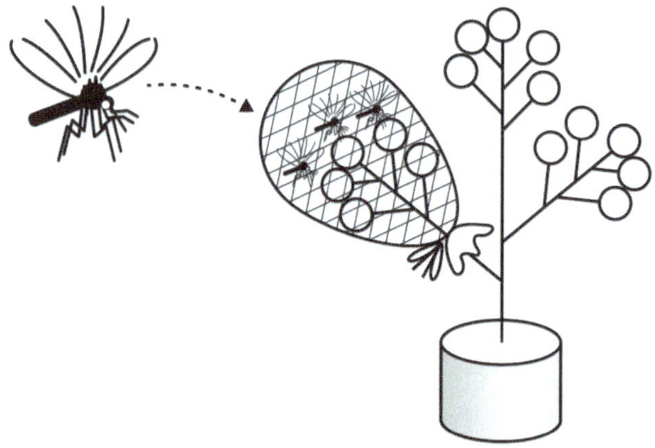

FIGURE 1. Graphical illustration of the experimental setup used for testing seed set. Figure is adapted from Peach and Gries (2016), with permission.

10. Add lab-reared mosquitoes to treatment/flower bags via aspiration and leave control flower bags without mosquitoes. We recommend 50 mosquitoes as a starting point, but this can be altered based on the size of the species of interest, pollen load information from the field or Step 7, or other factors.
11. After a period of time appropriate for seed set to occur in the flower being tested has passed, remove the mosquitoes. Maintain the plants and count the resulting seed set, which indicates successful deposition of pollen and, therefore, pollination by the mosquitoes.

DISCUSSION

Mosquito pollination is a very underresearched aspect of mosquito biology, in part because the presence of an observer may disrupt nectar-feeding mosquitoes by causing them to host seek. Although mosquito pollination of some specialist flowers has been observed, and its existence has been well-known for some time (Müller 1873; Dexter 1913; Stoutamire 1968; Brantjes and Leemans 1976; Lehondère et al. 2020), recent studies and reviews of existing literature suggests that pollination by mosquitoes (Fig. 2) may be more common than previously believed and might exist at a level comparable to pollination by other Diptera (Peach and Gries 2016, 2019, 2020). However, there are also reports of mosquitoes acting as nectar thieves (Smith and Gadawski 1994; Otienoburu et al. 2012), indicating that pollination by mosquitoes is likely flower-specific. Confirmation of successful deposition of pollen is the most widely used measurement to determine pollination (Ne'eman et al. 2010), and is sufficient to assess whether mosquitoes function as pollinators for a plant of interest. The protocol described here details methods to experimentally collect mosquito

FIGURE 2. A female *Culex pipiens* covered in pollen after visiting a composite flower of *Tanacetum vulgare*. Photo adapted from Peach and Gries (2016), with permission.

Cite this protocol as *Cold Spring Harb Protoc*; doi:10.1101/pdb.prot108302

pollination data and test their status as a pollinator of a candidate plant. Ideally, future studies involving mosquito–flower interactions will, as a standard procedure, include an assessment for mosquito pollination to help address the deficit of direct data on the topic. Further steps can be taken to assess the relative contribution of mosquitoes as pollinators (compared to other insects) in different ecosystems and varying field conditions (Ne'eman et al. 2010), but such tests are not covered here as they require much more in-depth experiments and nuanced interpretations of data, as well as multiple seasons of dedicated field studies.

ACKNOWLEDGMENTS

The pertinent methods were developed in Gerhard Gries's lab at Simon Fraser University (SFU), supported by a Natural Sciences and Engineering Research Council of Canada (NSERC) industrial research chair to Gerhard Gries with support from sponsors such as Scotts Canada Ltd. and BASF Canada Inc. I thank Woodbridge Foster for constructive feedback on this manuscript.

REFERENCES

Brantjes NBM, Leemans JAAM. 1976. *Silene otites* (Caryophyllaceae) pollinated by nocturnal lepidoptera and mosquitoes. *Acta Botanica Neerlandica* **25:** 281–295. doi:10.1111/j.1438-8677.1976.tb00240.x

Dexter J. 1913. Mosquitoes pollinating orchids. *Science* **37:** 867. doi:10.1126/science.37.962.867.a

Kauffman E, Payne A, Franke MA, Schmid MA, Harris E, Kramer LD. 2017. The rearing of *Culex* spp. and *Aedes* spp. mosquitoes. *Bio Protocol* **7:** e2542. doi:10.21769/BioProtoc.2542

Lahondère C, Vinauger C, Okubo RP, Wolff GH, Chan JK, Akbari OS, Riffell JA. 2020. The olfactory basis of orchid pollination by mosquitoes. *Proc Natl Acad Sci* **117:** 708–716. doi:10.1073/pnas.1910589117

Müller H. 1873. *Die Befruchtung der Blumen durch Insekten.* Wilhelm Engelmann, Leipzig.

Ne'eman G, Jurgens A, Newstrom-Lloyd L, Potts SG, Dafni A. 2010. A framework for comparing pollinator performance: effectiveness and efficiency. *Biol Rev* **85:** 435–451.

Otienoburu PE, Ebrahimi B, Phelan PL, Foster WA. 2012. Analysis and optimization of a synthetic milkweed floral attractant for mosquitoes. *J Chem Ecol* **38:** 873–881. doi:10.1007/s10886-012-0150-6

Parker C. 2020. Collection and rearing of container mosquitoes and a 24-h addition to the CDC bottle bioassay. *J Insect Sci* **20:** 13. doi:10.1093/jisesa/ieaa059

Peach DAH, Gries G. 2016. Nectar thieves or invited pollinators? A case study of tansy flowers and common house mosquitoes. *Arthropod Plant Interact* **10:** 497–506. doi:10.1007/s11829-016-9445-9

Peach D, Gries G. 2019. Supplementary data for: Mosquito phytophagy—sources exploited, ecological function, and evolutionary transition to haematophagy. 49427 bytes. http://datadryad.org/stash/dataset/doi:10.5061/dryad.63xsj3tz5 (Accessed December 24, 2021).

Peach DAH, Gries G. 2020. Mosquito phytophagy: sources exploited, ecological function, and evolutionary transition to haematophagy. *Entomol Exp Appl* **168:** 120–136. doi:10.1111/eea.12852

Smith SM, Gadawski RM. 1994. Nectar feeding by the early-spring mosquito *Aedes provocans. Med Vet Entomol* **8:** 201–213. doi:10.1111/j.1365-2915.1994.tb00499.x

Stoutamire W. 1968. Mosquito pollination of *Habenaria obtusata. Mich Bot* **7:** 203–212.

Wohl MP, McMeniman CJ. 2022. Batch rearing *Aedes aegypti. Cold Spring Harb Protoc* doi:10.1101/pdb.prot108017

CHAPTER 11

Mosquito Blood Meal Analysis

Lawrence E. Reeves[1] and Nathan D. Burkett-Cadena

University of Florida, Institute of Food and Agricultural Sciences, Department of Entomology and Nematology, Florida Medical Entomology Laboratory, Vero Beach, Florida 32962, USA

The host associations of mosquitoes vary by species, with some species being relative generalists, whereas others specialize, to varying extents, on a particular subset of the available host community. These host associations are driving factors in transmission dynamics of mosquito-vectored pathogens. For this reason, characterizing the host associations of mosquito species is critical for understanding the epidemiology of mosquito-vectored pathogens. Diverse methods have been used to associate mosquito species with their hosts. These typically include collecting mosquitoes that bite a restrained host (bait) or collecting wild blood-engorged mosquitoes and matching their blood meal to reference samples (blood meal analysis). Blood meal analysis refers to a collection of molecular techniques for determining the taxonomic identity of the source of a mosquito blood meal using cytological, serological, or DNA-based characteristics of the blood meal. Blood meal analyses that are based on DNA markers have advantages over cytological and serological methods and are effective for determining species-level identities of hosts from a broad range of potential host taxa. Here, we discuss effective techniques for analyzing blood meals.

INTRODUCTION

As the vectors of diverse viral, protozoal, and nematode pathogens, mosquitoes have had a tremendous impact on human society, influencing our history (Brès 1986; Marr and Cathey 2013), behavior (Pai et al. 2006; Yasuoka et al. 2006), and evolution (Allison 1954). However, mosquito feeding and pathogen transmission is not unique to our species: Many mosquito-vectored pathogens of public health importance, including the West Nile virus, eastern equine encephalitis virus, and the filarial nematodes that cause dog heartworm, are zoonotic (pathogens that typically circulate in nonhuman animals). Zoonotic mosquito-vectored pathogens often have complex transmission systems involving multiple species of vectors or vertebrate hosts. The host associations of mosquitoes drive both the timing and intensity of zoonotic disease outbreaks (Kilpatrick et al. 2006). For this reason, characterizing the host associations of mosquito species is a critical element of understanding the epidemiology of mosquito-vectored pathogens.

The adult female mosquitoes of most species must feed from blood to obtain proteins and nutrients that are necessary for the completion of egg development (Koller and Raikhel 1991). Mosquitoes take blood meals from a wide range of animals, including all vertebrate classes (amphibians, birds, fishes, mammals, and reptiles) and annelids (Tempelis 1975; Reeves et al. 2018a). The host associations of mosquitoes vary by species. Some mosquito species are relative generalists that opportunistically take blood meals from multiple vertebrate classes, whereas many other species are con-

[1]Correspondence: lereeves@ufl.edu

Cite this introduction as *Cold Spring Harb Protoc*; doi:10.1101/pdb.top107706

sidered specialists that take a preponderance of blood meals from only one or two host classes (Tempelis 1975). Specialization on either endothermic (birds and mammals) or ectothermic (reptiles, amphibians, and fish) hosts is a common pattern of host association among mosquito species. Narrower specializations also occur, such as the strict anuran association of *Uranotaenia lowii* (Reeves et al. 2018a).

Some mosquito species, even common and widespread species, feed occasionally or even specialize on unexpected or atypical hosts. For example, *Uranotaenia lateralis* and *Aedes baisasi* specialize on fishes, including mudskippers, other gobies, and eels (Tempelis 1975; Miyake et al. 2019). Marine mammals are also fed upon by mosquitoes: Mosquitoes are known to land upon and bite West Indian manatees (Reeves and Gillett-Kaufman 2020), and there is serological evidence that demonstrates that wild Atlantic bottlenose dolphins have been exposed to mosquito-vectored viruses (Schaefer et al. 2009). Apart from a single historic field report (Howard 1901) and laboratory inducements (Harris et al. 1969; George et al. 2014; Hribar 2014) of mosquitoes feeding from other insects, most studied mosquito species are known to feed from vertebrate animals in nature. Only one species, *Uranotaenia sapphirina* of eastern North America, is known to specialize on nonvertebrate hosts (earthworms and leeches) (Reeves et al. 2018a). Research on the host associations of mosquitoes has largely focused on known or potential pathogen vectors, and the associations of most of Earth's approximately 3600 described mosquito species have yet to be characterized. It is possible that additional invertebrate specialists exist, particularly among the diverse and poorly known mosquito faunas of the tropics.

The patterns of host associations of mosquito species are driving factors in the transmission of mosquito-vectored pathogens. Understanding the epidemiology of mosquito-borne disease relies, in part, on our ability to characterize the host associations of known or potential mosquito vectors. The methods used to characterize mosquito–host associations can be broadly grouped as observational, cytological, serological, and DNA-based (Washino and Tempelis 1983; Kent 2009). The earliest and most basic methods of identifying mosquito hosts involved directly observing interactions between mosquitoes and their hosts (McClelland and Weitz 1960), in some cases, using caged hosts that served as bait for wild host-seeking mosquitoes (Weitz 1960), or through laboratory-based investigations that provided potential hosts to caged mosquitoes (e.g., Vantaux et al. 2014). Direct observation of mosquito–host interactions is still used to study mosquito–host associations (e.g., Reeves and Gillett-Kaufman 2020), but it is not a feasible approach for determining the broader patterns of mosquito host associations because of the potential for bias.

METHODS FOR BLOOD MEAL ANALYSIS

Examining the blood meals of field-collected mosquitoes provides a more informative signal of the host associations of mosquitoes in nature. For nearly a century (King and Bull 1923), cells, proteins, or DNA from host animals ingested by mosquitoes have been examined and used to determine patterns of mosquito host association (Tempelis 1975). Collectively, the techniques that use cytological, serological, or DNA evidence from mosquito blood meals to taxonomically identify the host animal are known as blood meal analysis.

Cytological Methods

Differences in the size of erythrocytes, the presence of nuclei in erythrocytes, and other characteristics of blood cells can provide some, albeit limited, information on the identity of a mosquito host group (Weitz 1956). Such methods are limited in their ability to distinguish host taxa and lack the specificity to reliably differentiate host taxa beyond the vertebrate classes Amphibia, Aves, Mammalia, and Reptilia. Another disadvantage of cytological methods is that cellular components degrade rapidly following ingestion by the mosquito. For these reasons, cytological methods for broadly classifying hosts from mosquito blood meals were largely used before the availability of serological and DNA-

 Cite this introduction as *Cold Spring Harb Protoc*; doi:10.1101/pdb.top107706

based methods of blood meal analysis. The use of cytological methods has diminished as more specific techniques have become widely available.

Serological Methods

Serological tests, including precipitin tests, enzyme-linked immunosorbent assays (ELISAs), fluorescent antibody techniques, and passive hemagglutination-inhibition techniques, have been widely applied to blood meal analysis (Washino and Tempelis 1983). These methods have been in use since the 1920s (Bull and King 1923) and continue to be useful, although blood meal analyses are now predominantly DNA-based. Initial serological blood meal analysis methods were largely variations of precipitin tests, which are based on antigen–antibody reactions. To perform a precipitin test, vertebrate group-specific antisera is produced by injecting blood from various animal groups into other laboratory animals (often rabbits) and then collecting blood from the laboratory animals to obtain the blood serum with the group-specific antisera. A diluted aliquot of a wild mosquito blood meal (of undetermined host origin) is mixed with antisera from potential hosts, in which the production of precipitate indicates a positive reaction. Because of cross-reactivity among related host species, precipitin tests lack specificity beyond the family level, such that determining the host species is rare.

Precipitin tests were the standard method of blood meal analysis until the introduction of ELISAs (Burkot et al. 1981), which improved the specificity of host identification through serological methods. ELISAs also test for the presence of homologous antibodies but determine cross-reactivity using host-specific enzymes (Beier et al. 1988), thereby substantially improving taxonomic sensitivity by identifying host species (Washino and Tempelis 1983). However, a major limitation of ELISAs and other serological methods is the need for specific antisera from a range of potential hosts. For this reason, serological methods are limited by the range of host animals with which a blood meal can be compared. Thus, although serological methods can be useful for identifying higher levels of taxonomic classification or identifying blood meals from a restricted range of taxa, because of these limitations, research questions must be carefully framed when applying serological techniques to blood meal analysis.

DNA-Based Methods

The application of polymerase chain reaction (PCR) and DNA-based tools to blood meal analysis addresses the limitations of the specificity of serological methods, substantially expanding the range of mosquito hosts that can be identified to the species level. When PCR was initially applied to blood meal analysis, variation in the length of amplified DNA fragments (numbers of base pairs) was used to distinguish small groups of potential host species or groups. Over time, the cost of DNA sequencing and molecular reagents declined, and the taxonomic representation of publicly accessible DNA sequences improved. These changes made it increasingly feasible to identify mosquito hosts based on DNA sequences, largely removing constraints on the number of host species with which a blood meal could be compared (Alcaide et al. 2009). Various PCR-based methods have been developed and applied to mosquito blood meal analysis, including multiplexed PCR (Kent and Norris 2005) and restriction fragment-length polymorphism analyses (Oshaghi et al. 2006), and, more recently, next-generation sequencing (Logue et al. 2016; Muturi et al. 2021).

Currently, the most widely used DNA-based method of blood meal analysis involves selectively amplifying a host DNA marker by PCR with vertebrate-specific primers, sequencing the amplicon using the Sanger method (Sanger et al. 1977), and comparing the resulting sequences against publicly accessible sequence reference databases, such as the National Center for Biotechnology Information GenBank database or the Barcode of Life Data System (Ratnasingham and Hebert 2007). Markers from several mitochondrial and ribosomal RNA genes have been used in PCR-based blood meal analyses to identify host species (Kent 2009), and microsatellites have been used to identify individuals of a species (Burkett-Cadena et al. 2010; Martínez-de la Puente et al. 2015). Nuclear genes have also been used in blood meal analysis (Haouas et al. 2007), although infrequently, because of low copy number (one copy per cell) relative to mitochondrial and ribosomal RNA genes (up to thousands of copies per cell). When amplifying DNA from small, degraded samples such as mosquito blood meals, having multiple copies of a gene per cell is advantageous. Mitochondrial genes such as the genes for cytochrome *b* (*MT-CYB*) and cytochrome *c* oxidase subunit I (*COI*) are frequently used in blood meal analysis and provide reliable results (Kent 2009).

Over the past two decades, a 5′ fragment of *COI* has become the standard DNA barcoding marker for animals (Ratnasingham and Hebert 2007). Consequently, the taxonomic representation of reference sequences for *COI* is greater than that for any other gene. This makes *COI* a highly advantageous gene for blood meal analysis as it has the potential to provide the highest likelihood of species-level identification of mosquito hosts. Until recently, there have been few suitable options for the vertebrate-specific *COI* primers required for analyzing blood meals, and *COI* had been underutilized in blood meal analysis (Kent 2009). However, this has changed with the development of novel primer sets intended for DNA barcoding-based mosquito blood meal analysis (Alcaide et al. 2009; Reeves et al. 2018b).

CONCLUSIONS

Mosquito–host association is a driving factor in the transmission dynamics of mosquito-vectored pathogens. The methodologies used to study the host associations of mosquitoes have evolved over time, from direct observation of mosquito–host interactions and cytological and serological methods, to the PCR- and DNA-based blood meal analyses that are currently in use. In particular, the establishment of *COI* as the standardized DNA barcoding gene for animals, the rapid expansion of *COI* taxonomic reference materials in publicly accessible databases, and the availability of effective vertebrate-specific primer sets has made DNA barcoding a convenient and reliable method for blood meal analysis. In the accompanying protocols, we introduce methods for preserving and preparing field-collected mosquito blood meals destined for DNA-based blood meal analysis (see Protocol 1: Preservation of Field-Collected Mosquito Blood Meals [Reeves and Burkett-Cadena 2023a] and Protocol 2: Extracting DNA from Preserved Mosquito Blood Meals [Reeves and Burkett-Cadena 2023b]), and a protocol for identifying mosquito host DNA through DNA barcoding (see Protocol 3: Amplification and Identification of Vertebrate Host Cytochrome *c* Oxidase Subunit I (*COI*) DNA Barcoding Templates from Mosquito Blood Meals [Reeves and Burkett-Cadena 2023c]).

REFERENCES

Alcaide M, Rico C, Ruiz S, Soriguer R, Muñoz J, Figuerola J. 2009. Disentangling vector-borne transmission networks: a universal DNA barcoding method to identify vertebrate hosts from arthropod bloodmeals. *PLoS ONE* **4:** e7092. doi:10.1371/journal.pone.0007092

Allison AC. 1954. The distribution of the sickle-cell trait in East Africa and elsewhere, and its apparent relationship to the incidence of subtertian malaria. *Trans R Soc Trop Med Hyg* **48:** 312–318. doi:10.1016/0035-9203(54)90101-7

Beier JC, Perkins PV, Wirtz RA, Koros J, Diggs D, Gargan TP, Koech DK. 1988. Bloodmeal identification by direct enzyme-linked immunosorbent assay (ELISA), tested on *Anopheles* (Diptera: Culicidae) in Kenya. *J Med Entomol* **25:** 9–16. doi:10.1093/jmedent/25.1.9

Brès PLJ. 1986. A century of progress in combating yellow fever. *Bull World Health Organ* **64:** 775–786.

Bull CG, King WV. 1923. The identification of the blood meal of mosquitoes by means of the precipitin test. *Am J Epidemiol* **3:** 491–496. doi:10.1093/oxfordjournals.aje.a118951

Burkett-Cadena ND, Ligon RA, Liu M, Hassan HK, Hill GE, Eubanks MD, Unnasch TR. 2010. Vector–host interactions in avian nests: do mosquitoes prefer nestlings over adults? *Am J Trop Med Hyg* **83:** 395–399. doi:10.4269/ajtmh.2010.10-0048

Burkot TR, Goodman WG, DeFoliart GR. 1981. Identification of mosquito blood meals by enzyme-linked immunosorbent assay. *Am J Trop Med Hyg* **30:** 1336–1341. doi:10.4269/ajtmh.1981.30.1336

George J, Blanford S, Thomas MB, Baker TC. 2014. Malaria mosquitoes host-locate and feed on caterpillars. *PLoS ONE* **9:** e108894. doi:10.1371/journal.pone.0108894

Haouas N, Passon B, Boudabous R, Dedet J-P, Babba H, Ravel C. 2007. Development of a molecular tool for the identification of *Leishmania* reservoir hosts by blood meal analysis in the insect vectors. *Am J Tropl Med Hyg* **77:** 1054–1059. doi:10.4269/ajtmh.2007.77.1054

Harris P, Riordan DF, Cooke D. 1969. Mosquitoes feeding on insect larvae. *Science* **164:** 184–185. doi:10.1126/science.164.3876.184

Howard LO. 1901. *Mosquitoes. How they live; how they carry disease; how they are classified; how they may be destroyed.* McClure, Phillips, New York.

Hribar LJ. 2014. Mosquitoes feeding on caterpillars of the Common Buckeye butterfly, *Junonia coenia* (Lepidoptera: Nymphalidae). *J Res Lepid* **47:** 45–48. doi:10.5962/p.332197

Kent RJ. 2009. Molecular methods for arthropod bloodmeal identification and applications to ecological and vector-borne disease studies. *Mol Ecol Resour* **9:** 4–18. doi:10.1111/j.1755-0998.2008.02469.x

Cite this introduction as *Cold Spring Harb Protoc*; doi:10.1101/pdb.top107706

Kent RJ, Norris DE. 2005. Identification of mammalian blood meals in mosquitoes by a multiplexed polymerase chain reaction targeting cytochrome *b*. *Am J Trop Med Hyg* **73:** 336–343. doi:10.4269/ajtmh.2005.73.336

Kilpatrick AM, Kramer LD, Jones MJ, Marra PP, Daszak P. 2006. West Nile virus epidemics in North America are driven by shifts in mosquito feeding behavior. *PLoS Biol* **4:** e82. doi:10.1371/journal.pbio.0040082

King WV, Bull CG. 1923. The blood feeding habits of malaria-carrying mosquitoes. *Amer J Hyg* **3:** 497–513. doi:10.1093/oxfordjournals.aje.a118952

Koller CN, Raikhel AS. 1991. Initiation of vitellogenin uptake and protein synthesis in the mosquito (*Aedes aegypti*) ovary in response to a blood meal. *J Insect Physiol* **37:** 703–711. doi:10.1016/0022-1910(91)90048-5

Logue K, Keven JB, Cannon MV, Reimer L, Siba P, Walker ED, Zimmerman PA, Serre D. 2016. Unbiased characterization of *Anopheles* mosquito blood meals by targeted high-throughput sequencing. *PLoS Negl Trop Dis* **10:** e0004512. doi:10.1371/journal.pntd.0004512

Marr JS, Cathey JT. 2013. The 1802 Saint-Domingue yellow fever epidemic and Louisiana Purchase. *J Public Health Man* **19:** 77–82. doi:10.1097/PHH.0b013e318252eea8

Martínez-de la Puente J, Méndez M, Ruiz S, Godoy JA, Soriguer RC, Figuerola J. 2015. Individual identification of endangered species using mosquito blood meals: a proof-of-concept study in Iberian lynx. *Parasitol Res* **114:** 1607–1610. doi:10.1007/s00436-015-4343-0

McClelland GAH, Weitz B. 1960. Further observations on the natural hosts of three species of *Mansonia* Blanchard (Diptera, Culicidae) in Uganda. *Ann Trop Med Parasitol* **54:** 300–304. doi:10.1080/00034983.1960.11685990

Miyake T, Aihara N, Maeda K, Shinzato C, Koyanagi R, Kobayashi H, Yamahira K. 2019. Bloodmeal host identification with inferences to feeding habits of a fish-fed mosquito, *Aedes baisasi*. *Sci Rep* **9:** 4002. doi:10.1038/s41598-019-40509-6

Muturi EJ, Dunlap C, Tchouassi DP, Swanson J. 2021. Next generation sequencing approach for simultaneous identification of mosquitoes and their blood-meal hosts. *J Vector Ecol* **46:** 116–121. doi:10.52707/1081-1710-46.1.116

Oshaghi MA, Chavshin AR, Vatandoost H, Yaaghoobi F, Mohtarami F, Noorjah N. 2006. Effects of post-ingestion and physical conditions on PCR amplification of host blood meal DNA in mosquitoes. *Exp Parasitol* **112:** 232–236. doi:10.1016/j.exppara.2005.11.008

Pai HH, Hong YJ, Hsu EL. 2006. Impact of short-term community-based cleanliness campaign on the sources of dengue vectors: an entomological and human behavior study. *J Environ Health* **88:** 35–39.

Ratnasingham S, Hebert PDN. 2007. BOLD: the barcode of life data system (http://www.barcodinglife.org). *Mol Ecol Notes* **7:** 355–364. doi:10.1111/j.1471-8286.2007.01678.x

Reeves LE, Burkett-Cadena ND. 2023a. Preservation of field-collected mosquito blood meals. *Cold Spring Harb Protoc* doi:10.1101/pdb.prot108290

Reeves LE, Burkett-Cadena ND. 2023b. Extracting DNA from preserved mosquito blood meals. *Cold Spring Harb Protoc* doi:10.1101/pdb.prot108291

Reeves LE, Burkett-Cadena ND. 2023c. Amplification and identification of vertebrate host cytochrome *c* oxidase subunit I (*COI*) DNA barcoding templates from mosquito blood meals. *Cold Spring Harb Protoc* doi:10.1101/pdb.prot108292

Reeves LE, Gillett-Kaufman JL. 2020. Interactions between the imperiled West Indian manatee, *Trichechus manatus*, and mosquitoes (Diptera: Culicidae) in Everglades National Park, Florida. USA. *Sci Rep* **10:** 12971. doi:10.1038/s41598-020-69942-8

Reeves LE, Holderman CJ, Blosser EM, Gillett-Kaufman JL, Kawahara AY, Kaufman PE, Burkett-Cadena ND. 2018a. Identification of *Uranotaenia sapphirina* as a specialist of annelids broadens known mosquito host use patterns. *Commun Biol* **1:** 92. doi:10.1038/s42003-018-0096-5

Reeves LE, Gillett-Kaufman JL, Kawahara AY, Kaufman PE. 2018b. Barcoding blood meals: new vertebrate-specific primer sets for assigning taxonomic identities to host DNA from mosquito blood meals. *PLoS Negl Trop Dis* **12:** e0006767. doi:10.1371/journal.pntd.0006767

Sanger F, Nicklen S, Coulson AR. 1977. DNA sequencing with chain terminating inhibitors. *Proc Natl Acad Sci* **74:** 5463–5467. doi:10.1073/pnas.74.12.5463

Schaefer AM, Reif JS, Goldstein JD, Ryan CN, Fair PA. 2009. Serological evidence of exposure to selected viral, bacterial, and protozoal pathogens in free-ranging Atlantic bottlenose dolphins (*Tursiops truncatus*) from the Indian River Lagoon, Florida, and Charleston, South Carolina. *Aquat Mamm* **35:** 163–170. doi:10.1578/AM.35.2.2009.163

Tempelis CH. 1975. Host-feeding patterns of mosquitoes, with a review of advances in analysis of blood meals by serology. *J Med Entomol* **11:** 635–653. doi:10.1093/jmedent/11.6.635

Vantaux A, Lefèvre T, Dabiré KR, Cohuet A. 2014. Individual experience affects host choice in malaria vector mosquitoes. *Parasit Vectors* **7:** 249. doi:10.1186/1756-3305-7-249

Washino RK, Tempelis CH. 1983. Mosquito host bloodmeal identification. *Annu Rev Entomol* **28:** 179–201. doi:10.1146/annurev.en.28.010183.001143

Weitz B. 1956. Identification of blood meals of blood-sucking arthropods. *Bull World Health Organ* **15:** 473–490.

Weitz B. 1960. Feeding habits of bloodsucking arthropods. *Exp Parasitol* **9:** 63–82. doi:10.1016/0014-4894(60)90011-4

Yasuoka J, Mangione TW, Spielman A, Levins R. 2006. Impact of education on knowledge, agricultural practices, and community actions for mosquito control and mosquito-borne disease prevention in rice ecosystems in Sri Lanka. *Am J Trop Med Hyg* **74:** 1034–1042. doi:10.4269/ajtmh.2006.74.1034

Protocol 1

Preservation of Field-Collected Mosquito Blood Meals

Lawrence E. Reeves[1] and Nathan D. Burkett-Cadena

University of Florida, Institute of Food and Agricultural Sciences, Department of Entomology and Nematology, Florida Medical Entomology Laboratory, Vero Beach, Florida 32962, USA

All PCR- and DNA-based blood meal analyses require host DNA from a mosquito blood meal to be effectively preserved between the time when the specimen is collected and the extraction of DNA. As soon as a mosquito ingests blood from a host animal, digestion of host cells and cellular components within the blood meal by enzymes in the mosquito midgut begins to degrade the host DNA templates that are the targets of polymerase chain reaction (PCR) amplification. Without effective preservation, host DNA is typically undetectable by PCR 48 h after feeding, because of digestion. Preservation methods for mosquito blood meals vary in their efficacy, and the logistics of fieldwork can limit the options for preservation of blood meals and maintenance of the integrity of host DNA. This protocol describes a method of blood meal preservation that is effective, convenient, and amenable to fieldwork in remote locations where cryopreservation at −20°C or −80°C may not be feasible. It uses a Flinders Technology Associates (FTA) card, which is a chemically treated card that lyses cells and allows nucleic acids to be preserved. This method is also expected to preserve the DNA or RNA of pathogens present within the engorged mosquito abdomen, including RNA viruses.

MATERIALS

It is essential that you consult the appropriate Material Safety Data Sheets and your institution's Environmental Health and Safety Office for proper handling of equipment and hazardous materials used in this protocol.

Reagents

Collected mosquitoes
Ethyl acetate (optional; see Step 2)

Equipment

Field notebook or datasheets (see Steps 4.ii and 10.i)
Forceps (fine-point)
Freezer at −20°C or cooler with dry ice (optional; see Step 2)
Freezer at −80°C (optional; see Step 2)
Insect killing jar (optional; see Step 2; Carolina Biological Supply Company 654052)
Nitrile or latex gloves (optional)
Petri dishes or other shallow containers for holding mosquito specimens
Pipette tips (sterile), one per blood meal (20-µL or 200-µL preferred)

[1]Correspondence: lereeves@ufl.edu

Cite this protocol as *Cold Spring Harb Protoc*; doi:10.1101/pdb.prot108290

Plastic bags (zipper-sealed)
Plastic container (see Step 13)
QIAcard Flinders Technology Associates (FTA) Classic Cards (FTA cards; QIAGEN WB120126)

These were formerly called Whatman FTA Classic Cards.

Silica gel (optional; see note below Step 13)
Stereomicroscope

METHOD

This protocol describes a workflow for preserving samples of blood-fed mosquitoes of various species but can be easily adapted to better suit the specific objectives of the research. The protocol describes a process in which 16 blood meals are preserved per card (four per sampling area). With experience, the number of blood meals preserved per card can be increased, though care must be taken to avoid cross-contamination. Perform the entire protocol at room temperature.

1. Collect blood-fed mosquitoes from the field.

   Although small numbers of blood-fed mosquitoes may be incidentally collected in traps that target host-seeking or gravid mosquitoes, blood-fed mosquitoes are more reliably collected using resting shelters that mimic the resting sites of cavity-resting mosquito species (Burkett-Cadena et al. 2008; Burkett-Cadena 2011) or by aspirating natural resting sites (Nasci 1981; Sloyer et al. 2022). See Silver (2008) for information on collection methods that best target blood-fed females of various species.

2. Kill collected mosquitoes in one of two ways:
   - Expose mosquitoes to cold temperatures by placing them in a freezer at −20°C or into a cooler with dry ice for 20–30 min.
   - Expose mosquitoes to 100% ethyl acetate in insect killing jar at room temperature for ~10 min.

     Other methods of killing field-collected mosquitoes may be used, as long as the DNA is not damaged. Cold temperatures and ethyl acetate are known to maintain the integrity of host DNA (Reeves et al. 2019; Reeves and Burkett-Cadena 2022). If blood meal preservation cannot be performed promptly after collection, samples of blood-fed females destined for blood meal analysis can be stored in 95% ethanol such that all specimens are submerged (although ethanol-preserved blood meals cannot be subsequently preserved onto FTA cards) or at −80°C until preservation onto FTA cards.

     Critical: Avoid or minimize storage of such samples for >48 h at −20°C as host DNA will continue to degrade (Reeves et al. 2016).

3. Using a stereomicroscope, sort blood-fed female mosquitoes from male and unfed female mosquitoes (Carpenter and LaCasse 1955). Examine the abdomen of each collected female mosquito for the presence of blood.
   i. Closely inspect the intersegmental membranes between the dorsal tergites and ventral sternites for red or dark coloration indicating the presence of a blood meal (Fig. 1).
   ii. With forceps, transfer all blood-fed females to a Petri dish to separate them from unfed specimens.

   Female mosquitoes that have taken a full blood meal relatively recently should be obvious when viewed under the stereomicroscope, with the abdomen engorged and filled with a scarlet-to-dark-red substance. In some mosquito genera, the presence of a blood meal is readily visible and apparent, but in others, including some species of Aedes, Psorophora, Coquillettidia, *and* Mansonia, *scales or coloration of the integument may obscure the blood meal, particularly if the blood meal is small or partially digested.*

4. Prepare specimens and workspace for blood meal preservation.
   i. Using location-appropriate resources, morphologically identify all blood-fed mosquito specimens and sort each species into unique areas of the dish or into their own dishes so that blood meals from the same species can be consecutively preserved (for convenience).

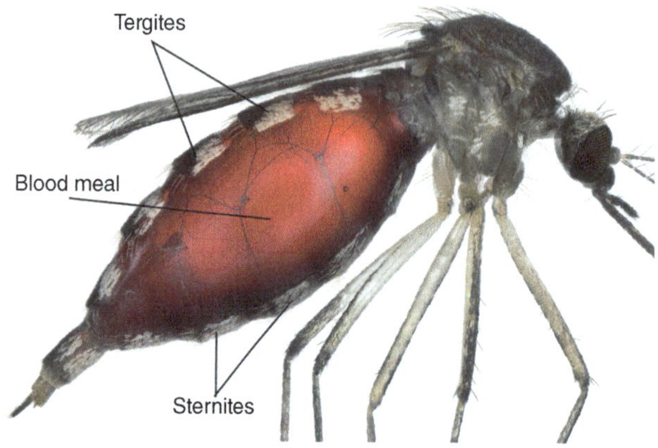

FIGURE 1. Blood-fed adult female *Aedes pertinax* with the locations of the blood meal, dorsal tergites, and ventral sternites indicated. Note that this individual is freshly blood-fed. Over time, the blood meal is digested and as this happens, the color changes from scarlet red to brown or black, and the size of the blood within the abdomen decreases as oocytes develop (see Fig. 3). For some species, the blood meal may not be as apparent as in this example. In some cases, a blood-fed mosquito may be partially blood-fed and in these, the abdomen will contain red coloration, but will not be as distended as in this example.

ii. Prepare table, counter, or benchtop workspace. Have sterile pipette tips and a notebook or datasheet and pen for recording specimen data within reach.

iii. Remove one FTA card from the packaging and open the card, being careful not to touch the sampling area with ungloved hands.

Wearing gloves is recommended to avoid contaminating the FTA card with human DNA.

5. Starting with the most numerous species, use forceps to transfer one specimen to the upper left edge of the top left sampling area of the FTA card.

   The sampling areas are the spaces within each of the four circles of the open card (Fig. 2).

6. Position the specimen so that it lies on its side, with the apex of the abdomen pointed at the upper left corner of the top left sampling area, where the contents of the abdomen are expected to be smeared in Step 8.

7. Place the point of a sterile pipette tip at the base of the mosquito abdomen, where the abdomen meets the thorax, and gently roll the pipette tip starting at the basal area of the abdomen toward the apical end of the abdomen, forcing the contents of the abdomen onto the surface of the card (Fig. 2). Do not touch the sampling areas of the card with ungloved hands.

8. Using the end of the pipette tip, smear the blood meal onto the surface of the card. As much as possible, avoid the head, thorax, wings, and legs of the mosquito when smearing the blood meal onto the card. Create a compact smear that is approximately circular in shape and avoid an elongated smear. Smear onto the card until all viscous droplets are absorbed.

   Host DNA will largely be present only in the abdomen.

9. Use the pipette tip to collect the head, thorax, and wings of the mosquito, which should stick to the end of the pipette tip. If convenient, remove any legs that remain on the card with the pipette tip, being careful to avoid contamination. Discard the pipette tip and remaining mosquito body parts. Take care to avoid cross-contamination during this step. Do not let any of the smeared blood meal or mosquito body come into contact with other blood meals on the card.

10. Estimate the extent of blood meal digestion and record specimen information.

    i. Give each mosquito an identifying number. In the notebook or on a datasheet, record this number along with the mosquito species, collection data (collection locality and date, other relevant information), and estimation of digestion extent. Before the blood meal dries,

Cite this protocol as *Cold Spring Harb Protoc*; doi:10.1101/pdb.prot108290

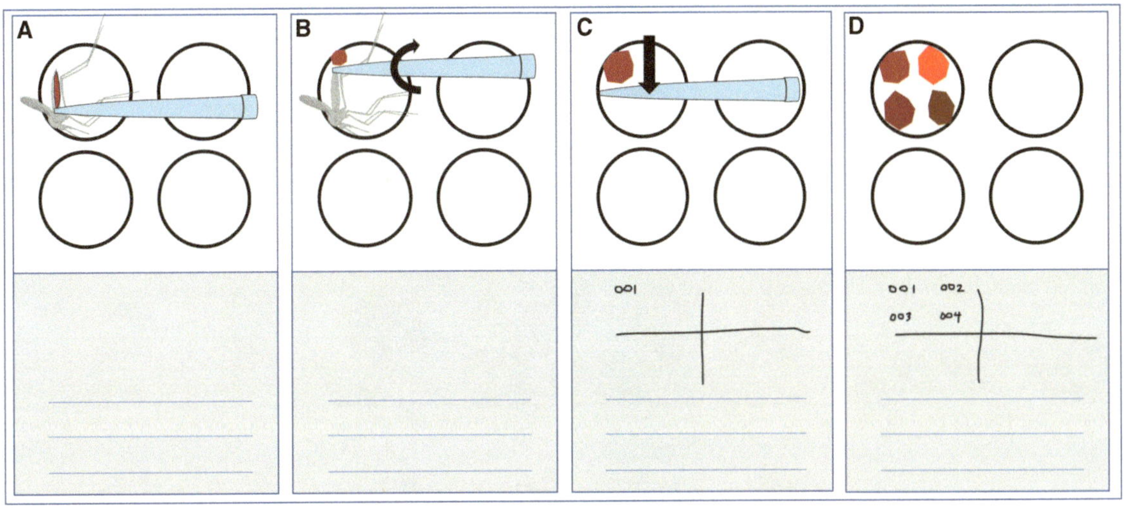

FIGURE 2. Method for smearing mosquito blood meals onto FTA Classic Cards. A mosquito is placed onto the sampling area (the area within the circles) on its side, with the apex of the abdomen pointed at the approximate area where the blood meal is expected to be spread. The point of a sterile pipette top is placed between the thorax and the abdomen (*A*), and a gentle rolling motion is used to force the contents of the abdomen onto the surface of the card (*B*). The mosquito body is discarded, and the blood meal is smeared onto the card until all viscous droplets have been absorbed (*C*). Multiple blood meals can be preserved within each sampling area, working from the *top left* to *bottom right* (*D*). Below the sampling area, the card is labeled so that associated data (collection data, taxonomic identification, etc.) can be matched to each blood meal.

estimate the extent of digestion on a scale of 1–3 (Fig. 3), with 1 being the freshest blood meals that have a scarlet red coloration; 2 being a somewhat digested blood meal, beginning to turn brown, but still with some scarlet or dark red coloration; and 3 being well-digested, with the blood meal brown in color, developing eggs (pale yellow in color when smeared onto card), and no red color.

Estimating the extent of digestion provides information in case troubleshooting is required during blood meal analysis. As a blood meal is digested by the mosquito, the host DNA degrades until it is ultimately undetectable by polymerase chain reaction (PCR), usually 48–72 h after feeding, depending on the PCR target and primers (Ngo and Kramer 2003; Oshaghi et al. 2006). Degradation of host DNA is a common cause of amplification failure in blood meal analysis, and an estimation of the extent of digestion can provide guidance if troubleshooting is needed downstream, or it could suggest that a mosquito's host species cannot be detected using the selected blood meal analysis method.

ii. On the lower half of the card, below the sampling area, record the identifying number of the specimen in such a way to indicate the position of its corresponding preserved blood meal on the card (Fig. 2).

11. Repeat Steps 5–10 for all other blood-fed mosquito specimens. Work from left to right and top to bottom within each sampling area, starting first with the top left sampling area, followed by the lower left, upper right, and lower right areas.

    It can be convenient to preserve all individuals of the same species from a sample consecutively, if samples of blood-fed females consist of more than one species.

12. Prior to closing a card for storage or when the card is full, air-dry preserved blood meals for ~10 min.
13. Store FTA card–preserved blood meals until DNA extraction.
    - If blood meals have been preserved in the field, store the FTA card–preserved blood meals inside a zip-sealed plastic bag, placed inside a plastic container to prevent bending or physical damage.
    - For samples prepared in the laboratory, store FTA cards with preserved blood meals together in zip-sealed plastic bags, placed inside a drawer or other dark, dry location.

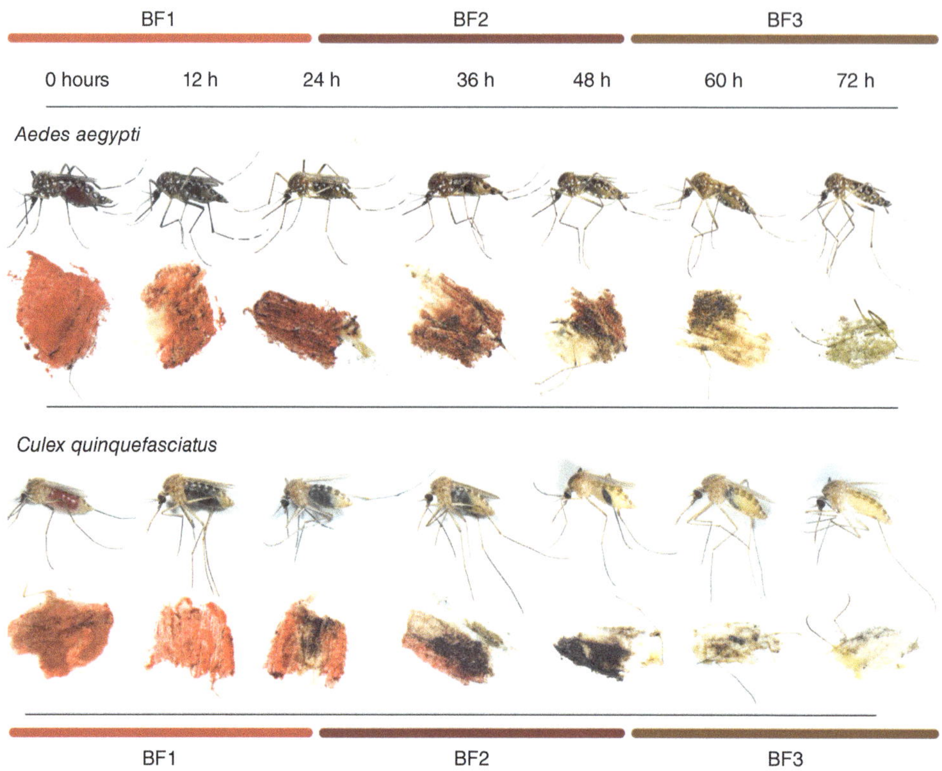

FIGURE 3. Estimation of the extent of digestion of mosquito blood meals. As soon as a mosquito imbibes a blood meal, enzymes in the mosquito midgut begin to degrade the host's DNA until it is rendered undetectable by polymerase chain reaction (PCR) after 48–72 h. Estimating the extent of digestion on a blood-fed (BF) scale of 1–3 can be helpful downstream. Fresh blood meals, with scarlet red coloration and no brown or yellowish coloration, are scored as BF1, blood meals that have some brown coloration but retain some red color are scored as BF2, and blood meals that are brown with some yellowish coloration (developing eggs) are scored as BF3. (Adapted from Reeves et al. 2018.)

Silica gel can be added to serve as a desiccant, and may be helpful when in the field in tropical or humid conditions (Grubaugh et al. 2015).

Host DNA preserved on blood cards is stable at room temperature for extended time periods, probably years (Reeves et al. 2018).

Proceed to Protocol 2: Extracting DNA from Preserved Mosquito Blood Meals (Reeves and Burkett-Cadena 2023).

DISCUSSION

Preservation of mosquito blood meals on FTA cards is a robust and convenient method that effectively preserves the integrity of host DNA at ambient temperatures for downstream PCR-based blood meal analysis, without the need for refrigeration, cryopreservation, or specialized reagents, other than the FTA cards (Reeves et al. 2016). This method is more effective than other methods including preservation of whole-mosquito specimens in ethanol, by desiccation in silica gel, or by cold storage at −20°C (Reeves et al. 2016). Although the intent of this protocol is the maintenance of host DNA integrity between mosquito collection and downstream DNA extraction, preservation of mosquito blood meals on FTA cards has been shown to be effective at preserving viral RNA (Yang et al. 2015), even though there may be variation in the efficacy of preservation or recovery of viral RNA among virus species (Grubaugh et al. 2015).

Although FTA card preservation of host DNA is robust, a potential pitfall of this method is that it is rather time consuming compared to, for example, placing bulk specimens into ethanol or into a −80°C freezer. It can be challenging to promptly preserve blood meals from field-collected samples when large numbers of mosquito blood meals are collected or when field logistics inhibit prompt preservation. This is especially true when large-barrel aspirators are used to collect blood meals, requiring time consuming sorting of mosquito specimens from debris and bycatch. In instances in which samples cannot be promptly preserved onto FTA cards, collected mosquitoes should be stored cold, at −20°C or, preferably, on dry ice or at −80°C until they can be smeared onto the cards. Storing mosquito blood meals at −20°C is not a long-term solution for preservation of host DNA. Storage for >48 h is not recommended and can lead to reduced amplification success of host DNA targets (Reeves et al. 2016). As much as possible, avoid or reduce the time of storage of mosquito blood meals at −20°C.

In some instances, particularly when mosquitoes are collected with an aspirator, mosquito specimens may be damaged and missing diagnostic morphological characteristics, which can inhibit morphological identification. Because mosquito blood meals preserved on FTA cards preserve not only the host DNA but DNA from the mosquito as well, in such instances, it is often possible to identify the mosquito species by DNA barcoding using DNA extracted from the blood meals. Following DNA extraction, diagnostic mosquito DNA targets such as the barcoding region of the cytochrome *c* oxidase subunit I gene (*COI*) can be amplified and sequenced to determine the mosquito species. The LepF1 and LepR1 primers (Hebert et al. 2004) and LCO1490 and HCO2198 (Folmer et al. 1994) can be effective at amplifying mosquito DNA from mixed samples such as blood meals (see the PCR technique in Reeves et al. 2021).

Preservation of host DNA from mosquito blood meals on FTA cards is a reliable method of maintaining such samples at room temperature over long time periods. This technique is also useful because DNA can be extracted from a small piece of the FTA card–preserved blood meal, leaving additional sample on the card should DNA need to be re-extracted or if further assays need to be performed after host identification. Storage of FTA cards can be space-effective. They can be stored in binders within baseball card–style sleeve pages, making it possible to store large numbers (hundreds to thousands) of mosquito blood meal samples in a very limited space.

REFERENCES

Burkett-Cadena ND. 2011. A wire-frame shelter for collecting resting mosquitoes. *J Am Mosquito Contr* **27:** 153–155. doi:10.2987/10-6076.1

Burkett-Cadena ND, Eubanks MD, Unnasch TR. 2008. Preference of female mosquitoes for natural and artificial resting sites. *J Am Mosquito Contr* **24:** 228–235. doi:10.2987/5662.1

Carpenter SJ, LaCasse WJ. 1955. *Mosquitoes of North America*. University of California Press, Berkeley, CA.

Folmer O, Black M, Hoeh W, Lutz R, Vrijenhoek R. 1994. DNA primers for amplification of mitochondrial cytochrome *c* oxidase subunit I from diverse metazoan invertebrates. *Mol Mar Biol Biotechnol* **3:** 294–299.

Grubaugh ND, Sharma S, Krajacich BJ, Fakoli LS III, Bolay FK, Diclaro JW II, Johnson WE, Ebel GD, Foy BD, Brackney DE. 2015. Xenosurveillance: a novel mosquito-based approach for examining the human–pathogen landscape. *PLoS Negl Trop Dis* **9:** e0003628. doi:10.1371/journal.pntd.0003628

Hebert PD, Penton EH, Burns JM, Janzen DH, Hallwachs W. 2004. Ten species in one: DNA barcoding reveals cryptic species in the neotropical skipper butterfly *Astraptes fulgerator*. *Proc Natl Acad Sci* **101:** 14812–14817. doi:10.1073/pnas.0406166101

Nasci RS. 1981. A lightweight battery-powered aspirator for collecting resting mosquitoes in the field. *Mosquito News* **41:** 808–811.

Ngo KA, Kramer LD. 2003. Identification of mosquito blood meals using polymerase chain reaction (PCR) with order-specific primers. *J Med Entomol* **40:** 215–222. doi:10.1603/0022-2585-40.2.215

Oshaghi MA, Chavshin AR, Vatandoost H, Yaaghoobi F, Mohtarami F, Noorjah N. 2006. Effects of post-ingestion and physical conditions on PCR amplification of host blood meal DNA in mosquitoes. *Exp Parasitol* **112:** 232–236. doi:10.1016/j.exppara.2005.11.008

Reeves LE, Burkett-Cadena ND. 2022. Lizards are important hosts for zoonotic Flavivirus vectors, subgenus *Culex*, in the southern USA. *Front Trop Dis* **11:** 842543. doi:10.3389/fitd.2022.842523

Reeves LE, Burkett-Cadena ND. 2023. Extracting DNA from preserved mosquito blood meals. *Cold Spring Harb Protoc* doi:10.1101/pdb.prot108291

Reeves LE, Holderman CJ, Gillett-Kaufman JL, Kawahara AY, Kaufman PE. 2016. Maintenance of host DNA integrity in field-preserved mosquito (Diptera: Culicidae) blood meals for identification by DNA barcoding. *Parasit Vector* **9:** 503. doi:10.1186/s13071-016-1791-z

Reeves LE, Gillett-Kaufman JL, Kawahara AY, Kaufman PE. 2018. Barcoding blood meals: new vertebrate-specific primer sets for assigning taxonomic identities to host DNA from mosquito blood meals. *PLoS Negl Trop Dis* **12:** e0006767. doi:10.1371/journal.pntd.0006767

Reeves LE, Hoyer I, Acevedo C, Burkett-Cadena ND. 2019. Host associations of *Culex (Melanoconion) atratus* (Diptera: Culicidae) and *Culex (Melanoconion) pilosus* from Florida, USA. *Insects* **10:** 239. doi:10.3390/insects10080239

Reeves LE, Medina J, Miqueli E, Sloyer KE, Petrie W, Vasquez C, Burkett-Cadena ND. 2021. Establishment of *Aedes (Ochlerotatus) scapularis* (Diptera: Culicidae) in mainland Florida, with notes on the Ochlerotatus Group in the United States. *J Med Entomol* **58:** 717–729. doi:10.1093/jme/tjaa250

Silver JB. 2008. *Mosquito ecology, field sampling methods*, 3rd ed., p. 1494. Springer, Dordrecht, The Netherlands.

Sloyer KE, Santos M, Riviera E, Reeves LE, Carrera JP, Valderrama A, Burkett-Cadena ND. 2022. Evaluating sampling strategies for enzootic Venezuelan equine encephalitis virus vectors in Florida and Panama. *PLoS Negl Trop Dis* **16:** e0010329. doi:10.1371/journal.pntd.0010329

Yang Y, Garver LS, Bingham KM, Hang J, Jochim RC, Davidson SA, Richardson JH, Jarman RG. 2015. Feasibility of using the mosquito blood meal for rapid and efficient human and animal virus surveillance and discovery. *Am J Trop Med Hyg* **93:** 1377–1382. doi:10.4269/ajtmh.15-0440

Protocol 2

Extracting DNA from Preserved Mosquito Blood Meals

Lawrence E. Reeves[1] and Nathan D. Burkett-Cadena

University of Florida, Institute of Food and Agricultural Sciences, Department of Entomology and Nematology, Florida Medical Entomology Laboratory, Vero Beach, Florida 32962, USA

Mosquito species vary in their host associations. Although some species are relative generalists, most specialize, to varying extents, on particular types of host animals. Mosquito host associations are among the most important factors that influence the transmission dynamics of mosquito-vectored pathogens, and understanding these associations can provide insight on how such pathogens move within ecosystems. Characterization of the host associations of mosquito species requires applying blood meal analysis to the largest possible sample size of mosquito blood meals. Processing large samples of mosquito blood meals can be time-consuming, especially when chain-termination sequencing is used, necessitating individual processing of each specimen. Various methods and commercially available kits and products are available for extracting DNA from mosquito blood meals. The hot sodium hydroxide and Tris (HotSHOT) method is a rapid and inexpensive method of DNA extraction that is compatible with the recovery of DNA from mosquito blood meals preserved on QIAcard Flinders Technology Associates (FTA) Classic Cards (FTA cards). FTA cards allow nucleic acids found in blood meals to be preserved easily, even in field conditions. DNA prepared using this method is suitable for polymerase chain reaction (PCR)-based blood meal analysis.

MATERIALS

It is essential that you consult the appropriate Material Safety Data Sheets and your institution's Environmental Health and Safety Office for proper handling of equipment and hazardous materials used in this protocol.

Reagents

Lysis solution for mosquitoes (25 mM NaOH, 0.2 mM disodium EDTA, pH 12)

Prepare lysis solution beforehand and store for up to 2 mo at room temperature. See Truett et al. (2000) for preparation information. The solution will have a pH of 12 when prepared and should not need to be adjusted.

Mosquito blood meals (preserved on FTA cards) from Protocol 1: Preservation of Field-Collected Mosquito Blood Meals (Reeves and Burkett-Cadena 2023a)

Neutralization solution (40 mM Tris-HCl, pH 5)

Prepare neutralization solution beforehand and store for up to 2 mo at room temperature. See Truett et al. (2000) for preparation information. The solution will have a pH of 5 when prepared and should not need to be adjusted.

Equipment

Aluminum foil

Centrifuge (any model with a rotor that accommodates 96-well plates or eight-well 0.2-mL PCR tube strips as appropriate)

[1]Correspondence: lereeves@ufl.edu

Cite this protocol as *Cold Spring Harb Protoc*; doi:10.1101/pdb.prot108291

Ethanol lamp, Bunsen burner, or other flame source (see Steps 2.v and 6)

Use a flame to sterilize forceps after they have handled blood meals. Ethanol lamps or Bunsen burners are most convenient, but other flame sources can be used.

Forceps (four)
Freezers at −20°C or −80°C (see Step 14)
Hole puncher (1-mm)

Alternatively, use sterile scalpel blades or snap blades to cut a small section of blood meal. A 1-mm hole puncher can often be purchased locally at hobby or craft retailers.

Laboratory notebook or datasheet (see Step 1)
Micropipette and tips (50- to 200-µL)

If available, an eight-channel multichannel pipette can be used.

Microplate (96-well, 0.2-mL) with strip caps (12, or fewer, as needed)

If fewer than 96 samples are to be extracted, this can be replaced with the appropriate number of eight-well PCR strips.

Thermocycler
Vortex

METHOD

The workflow described here involves DNA extraction from a series of 96 mosquito blood meals, although the protocol can be modified to accommodate smaller numbers of samples. Eight-well 0.2-mL PCR tube strips can be used in place of a 96-well plate if convenient. Record keeping is essential to ensure the extracted DNA can be matched to the mosquito specimen from which it is derived and its collection data. DNA will be extracted from hole punches taken from each blood meal and placed into separate wells of a 96-well plate. Perform the entire protocol at room temperature.

1. Plan the order in which wells will be filled, and the mosquito blood meals that will be placed in each. For each well of a 96-well plate, record in a notebook or datasheet the blood meal–identifying number and the well number in which the sample will be placed. Do this for all 96 blood meals included in the extraction before beginning. Label the plate with an identifying number.

   This ensures that all taxonomic and collection data will be associated with the extracted DNA. No additional labeling of the wells of the plate is necessary if this information is appropriately recorded in a notebook or datasheet.

2. Prepare workspace.
   i. Place a rectangular piece of aluminum foil ~45-cm × 35-cm on a flat workspace such as a laboratory bench or tabletop.
   ii. On the foil, place the 96-well plate, 12 strip caps, and four forceps, arranged parallel to one another.
   iii. Within reach, on the foil or the benchtop, place an ethanol lamp, a hole puncher, and FTA cards with the blood meals for extraction.
   iv. Ensure that the 96-well plate is oriented correctly, with well A1 at the upper left corner.
   v. Leave a space clear at the front or side of the foil. Ignite the ethanol lamp or Bunsen burner.
3. Over the foil, in the cleared space, hold the blood card and remove two 1-mm punches from the first blood meal if the blood meal is variable in color (i.e., if there are multiple colors in the blood meal), remove the punches from areas that are most red in color. Take care to position the hole puncher and FTA card in a way that the punches will fall onto an expected area on the foil.
4. Use forceps to transfer both punches individually or together, if convenient, to the bottom of the appropriate well of the plate.

   Static electricity can cause the punches to stick to the sides of the well. Placement of the punches at the bottom of the well is ideal, but on the side is acceptable as long the punches cannot easily be dislodged from the well.

5. Clean the hole puncher by taking two punches from the FTA card outside of the sampling areas (a clean area of the top or sides of the card). Hold the card over a trash can or other surface so that the punches are easily discarded and will not be confused with blood meal punches.

   The punches should not be taken from within the inside of the circles (the sampling areas) where blood meals are preserved.

6. Sterilize the apical tips of the forceps that handled the blood meal punches by exposing the tips to the top of the flame for ∼5 sec. Pay attention to which of the four forceps were used. Allow time (∼3 min) for the sterilized forceps to cool and do not use the same forceps for the subsequent blood meal.

7. Repeat Steps 3–6 for the next seven blood meals so that there are two punches per blood meal in all wells of the column, wells A through H.

8. Add 50 µL of lysis solution to each of the eight wells containing punches and firmly seal the wells with a strip of well caps.

9. Repeat Steps 3–8 for the subsequent well columns until each well of the plate contains two punches and 50 µL of lysis solution, and the wells of all columns are firmly sealed with strip caps.

10. Briefly centrifuge the plate at low speed to draw all punches and lysis solution to the bottom of the wells.

    If available, use the quick spin option of the centrifuge.

11. In the thermocycler, incubate the plate for 60 min at 95°C, followed by 5 min at 4°C.

    The incubation step can be extended to 3 h total if DNA recovery is a concern (e.g., specimens were improperly stored before extraction, small sample size).

12. Briefly centrifuge the plate to draw any condensation droplets to the bottom of the wells.

    If available, use the quick spin option of the centrifuge.

13. To avoid cross-contamination, remove strip caps gently and carefully, and take note of the orientation of the strip caps, so that they can be replaced in the same orientation. Add 50 µL of neutralization solution to each well of the plate, one column at a time. Seal each column in the original orientation before opening the next.

14. Briefly vortex the plate, then briefly centrifuge the plate at low speed to draw all solution to bottom of wells. Promptly store at −20°C or −80°C until ready to perform PCR.

    Avoid storage at room temperature, if possible. Minimizing exposure of extracted DNA to room temperature will better maintain the integrity of extracted host DNA. If storage at −20°C or −80°C is not feasible, perform the extraction on the same day as the PCR.

    Proceed to Protocol 3: Amplification and Identification of Vertebrate Host Cytochrome c *Oxidase Subunit I (COI) DNA Barcoding Templates from Mosquito Blood Meals (Reeves and Burkett-Cadena 2023b).*

DISCUSSION

Here, we describe the application of the HotSHOT method (Truett et al. 2000) to DNA extraction from FTA card–preserved mosquito blood meals. This protocol works well for both FTA card–preserved blood meals (Reeves et al. 2016) and fresh blood meals from which DNA can be extracted shortly after collection (Alcaide et al. 2009). The described workflow involves DNA extraction from 96 FTA card–preserved blood meals using a 96-well plate, but can be adjusted to accommodate a smaller number of blood meals using eight-well PCR strips in place of a 96-well microplate. One benefit of this technique is that, in general, the entire blood meal is not consumed in the extraction. If the extraction fails or more DNA is needed, additional sample remains preserved on the FTA card for future use, if needed. For small samples (e.g., the blood meals of *Uranotaenia* mosquitoes, corethrelid or ceratopogonid midges), DNA can be extracted from a single 1-mm card punch, rather than two.

For PCR-based blood meal–analysis applications, the HotSHOT method is effective, low in cost, and rapid, especially when processing large numbers of samples. However, DNA extracted using this

Cite this protocol as *Cold Spring Harb Protoc*; doi:10.1101/pdb.prot108291

technique can be sensitive to multiple freeze–defrost cycles over time. This sensitivity will not impact the ability to identify the host, but, for some applications, such as preparation of DNA for next-generation sequencing or further molecular analyses beyond PCR-based host identification, other extraction methods, particularly commercially available DNA-extraction kits, may improve the recovery and isolation of DNA, although these should be tested for compatibility with FTA card–preserved blood meals. An alternative to the HotSHOT method that is compatible with FTA card–preserved blood meals is InstaGene Matrix (Bio-Rad), which has been used in multiple studies (McGregor et al. 2019; West et al. 2020; Burkett-Cadena et al. 2021) to extract DNA for PCR-based blood meal analysis of mosquitoes and biting midges (Diptera: Ceratopogonidae: *Culicoides*). The InstaGene extraction can be completed using the same equipment listed in this protocol, following manufacturer-provided instructions. However, a benefit of the HotSHOT method, especially when applied to large numbers of samples, is its cost. Although InstaGene Matrix is an effective means of extracting DNA from FTA cards, and likely better for long-term (e.g., >1 yr) storage of extracted DNA, it costs ~$0.53 USD per sample. Comparatively, the HotSHOT technique costs <$0.01 USD per sample.

REFERENCES

Alcaide M, Rico C, Ruiz S, Soriguer R, Muñoz J, Figuerola J. 2009. Disentangling vector-borne transmission networks: a universal DNA barcoding method to identify vertebrate hosts from arthropod bloodmeals. *PLoS ONE* **4:** e7092. doi:10.1371/journal.pone.0007092

Burkett-Cadena ND, Blosser EM, Loggins AA, Valente MC, Long MT, Campbell LP, Reeves LE, Bargielowski I, McCleery RA. 2021. Invasive Burmese pythons alter host use and virus infection in the vector of a zoonotic virus. *Commun Biol* **28:** 4. doi:10.1038/s42003-021-02347-z

McGregor BL, Stenn T, Sayler KA, Blosser EM, Blackburn JK, Wisely SM, Burkett-Cadena ND. 2019. Host use patterns of *Culicoides* spp. biting midges at a big game preserve in Florida, USA, and implications for the transmission of orbiviruses. *Med Vet Entomol* **33:** 110–120. doi:10.1111/mve.12331

Reeves LE, Burkett-Cadena ND. 2023a. Preservation of field-collected mosquito blood meals. *Cold Spring Harb Protoc* doi:10.1101/pdb.prot108290

Reeves LE, Burkett-Cadena ND. 2023b. Amplification and identification of vertebrate host cytochrome *c* oxidase subunit I (*COI*) DNA barcoding templates from mosquito blood meals. *Cold Spring Harb Protoc* doi:10.1101/pdb.prot108292

Reeves LE, Holderman CJ, Gillett-Kaufman JL, Kawahara AY, Kaufman PE. 2016. Maintenance of host DNA integrity in field-preserved mosquito (Diptera: Culicidae) blood meals for identification by DNA barcoding. *Parasit Vector* **9:** 503. doi:10.1186/s13071-016-1791-z

Truett GE, Heeger P, Mynatt RL, Truett AA, Walker JA, Warman ML. 2000. Preparation of PCR quality mouse genomic DNA with sodium hydroxide and Tris (HotSHOT). *BioTechniques* **29:** 52–54. doi:10.2144/00291bm09

West RG, Mathias DK, Day JF, Acevedo C, Unnasch TR, Burkett-Cadena ND. 2020. Seasonal changes of host use by *Culiseta melanura* (Diptera: Culicidae) in central Florida. *J Med Entomol* **57:** 1627–1634. doi:10.1093/jme/tjaa067

Protocol 3

Amplification and Identification of Vertebrate Host Cytochrome *c* Oxidase Subunit I (*COI*) DNA Barcoding Templates from Mosquito Blood Meals

Lawrence E. Reeves[1] and Nathan D. Burkett-Cadena

University of Florida, Institute of Food and Agricultural Sciences, Department of Entomology and Nematology, Florida Medical Entomology Laboratory, Vero Beach, Florida 32962, USA

Mosquitoes take blood meals from a diverse range of host animals and their host associations vary by species. Characterizing these associations is an important element of the transmission dynamics of mosquito-vectored pathogens. To characterize mosquito host associations, various molecular techniques have been developed, which are collectively referred to as blood meal analysis. DNA barcoding has diverse biological applications and is well-suited to mosquito blood meal analysis. The standard DNA barcoding marker for animals is a 5′ fragment of the cytochrome *c* oxidase I (*COI*) gene. A major advantage of this marker is its taxonomic coverage in DNA sequence reference databases, making it feasible to identify a wider range of mosquito host species than with any other gene. However, the *COI* gene contains high sequence variation at potential priming sites between vertebrate orders. Coupled with the need for primer sequences to be mismatched with mosquito priming sites so that annealing to mosquito DNA is inhibited, it can be difficult to design primers suitable for blood meal analysis applications. Several primers are available that perform well in mosquito blood meal analysis, annealing to priming sites for most vertebrate host taxa, but not to those of mosquitoes. Because priming site sequence variation among vertebrate taxa can cause amplification to fail, a hierarchical approach to DNA barcoding-based blood meal analysis can be applied. In such an approach, no single primer set is expected to be effective for 100% of potential host species. If amplification fails in the initial reaction, a subsequent reaction is attempted with primers that anneal to different priming sites, and so on, until amplification is successful.

MATERIALS

It is essential that you consult the appropriate Material Safety Data Sheets and your institution's Environmental Health and Safety Office for proper handling of equipment and hazardous materials used in this protocol.

Reagents

Agarose gel containing ethidium bromide made with a buffer of choice

If preferred, similar methods of visualizing polymerase chain reaction (PCR) products and amplification success can be used.

DNA ladder

[1]Correspondence: lereeves@ufl.edu

Cite this protocol as *Cold Spring Harb Protoc*; doi:10.1101/pdb.prot108292

TABLE 1. Combinations of vertebrate-specific primers for use in a hierarchical approach to polymerase chain reaction (PCR)-based mosquito blood meal analysis

PCR	Primer combination	Primer sequence	Amplicon size
PCR A	VertCOI_7194_F	CGM ATR AAY AAY ATR AGC TTC TGA Y	395 bp
	Mod_RepCOI_R	TTC DGG RTG NCC RAA RAA TCA	
PCR B	Mod_RepCOI_F	TNT TYT CMA CYA ACC ACA AAG A	244 bp
	VertCOI_7216_R	CAR AAG CTY ATG TTR TTY ATD CG	
PCR C	VertCOI_10096_F	CHC AAT ACC AAA CNC CHY TNT TYG	152 bp
	Mod_RepCOI_R	TTC DGG RTG NCC RAA RAA TCA	

Amplicon size is the same regardless of vertebrate host taxon.

Extracted DNA from mosquito blood meals from Protocol 2: Extracting DNA from Preserved Mosquito Blood Meals (Reeves and Burkett-Cadena 2023)

PCR master mix (dNTPs, $MgCl_2$, PCR buffer, Taq polymerase)

Ready-to-use Apex Taq RED Master Mix (Apex Bioresearch Products) works well in these reactions. Other master mixes may require optimization.

Primers (see Table 1 for primer sequences)
- Primer Mod_RepCOI_F (10 µM)
- Primer Mod_RepCOI_R (10 µM)
- Primer VertCOI_10096_F (10 µM)
- Primer VertCOI_7194_F (10 µM)
- Primer VertCOI_7216_R (10 µM)

Ultrapure H_2O

Equipment

Computer with internet access

DNA electrophoresis equipment

Microcentrifuge tubes (1.2-mL)

Mini-centrifuge with rotor that holds PCR tubes

PCR strips (eight-well)

The number of eight-well strips depends on the number of samples to be analyzed. A 96-well plate can be used for large numbers of samples.

Software for visualizing electropherogram files (e.g., Geneious, UGENE)

Thermocycler

UV light or similar system for visualizing PCR products

Vortex

METHOD

Here, we describe a hierarchical approach to DNA barcoding–based blood meal analysis. Because of priming site sequence variation among vertebrate taxa, in some cases, a primer set may not be compatible with a host species, resulting in failed amplification. To solve this issue, we implement a hierarchical approach to PCR: DNA from mosquito blood meals is amplified in an initial reaction; any DNA that does not amplify is used in a second reaction with another primer set targeting a differing priming site; and any extracts that do not amplify in that second reaction are used in a third reaction, again using a different primer set targeting a different priming site. In the process described here, the order in which primer sets are attempted is based on the size of the amplicon produced, descending from the largest amplicon to the smallest one (Table 1), under the premise that larger amplicons have greater resolving power in distinguishing closely related species. Concomitantly, primer sets targeting small amplicons are expected to perform better than those that target larger amplicons when host DNA templates have been degraded by digestion. Perform the entire protocol at room temperature, but keep PCR master mixes and reagents on ice.

PCR A

1. Prepare a master mix with the VertCOI_7194_F and Mod_RepCOI_R primers as described below.
   i. 10 µL of 2.0× Apex Taq RED Master Mix
   ii. 0.75 µL of forward primer VertCOI_7194_F (10 µM)
   iii. 0.75 µL of reverse primer Mod_RepCOI_R (10 µM)
   iv. 7.5 µL of ultrapure H_2O

   The master mix volume is determined by the number of samples (20 µL × (number of samples + 2)), and consists of the components i–iv, above. Prepare the master mix on ice in 1.2-mL tubes (if number of samples is >48, two tubes will be needed).

2. After all components have been added, close the tube and vortex for 15 sec. On ice, add 19 µL of master mix to each 0.2-mL well of the PCR strips. After the master mix has been added to all tubes, add 1 µL of extracted DNA and close the tube. Using a mini-centrifuge with a rotor that holds PCR strips, centrifuge briefly to pull all liquid to the bottom of the tube, and immediately transfer to the thermocycler.
3. Run each reaction in a thermocycler, with heated lid, under the following conditions:

1 cycle	3 min	95°C
5 cycles	40 sec	95°C
	30 sec	45°C
	3 min	72°C
35 cycles	40 sec	95°C
	30 sec	49°C
	3 min	72°C
1 cycle	7 min	72°C

4. Electrophorese PCR products to determine whether amplification was successful.

   The process described here involves ethidium bromide staining, but any standard gel electrophoresis protocol for PCR amplicons can be used.

   i. Prepare a 1.5% agarose gel with ethidium bromide.
   ii. Load PCR products alongside a DNA ladder and electrophorese them using standard techniques.
   iii. Visualize under UV light.

   Successful reactions will have an amplicon ~395 bp in length. Note which DNA extracts did and did not produce an amplicon of the expected size. Although the proportion of successful reactions varies between sets of samples, in general, the majority of DNA extracts should amplify in the initial reaction.

   See Troubleshooting.

PCR B

5. Using only DNA extracts that did not amplify in PCR A, run a second PCR using the primer set Mod_RepCOI_F and VertCOI_7216_R in PCR strips in a final volume of 20 µL consisting of:
   i. 10 µL of 2.0× Apex Taq RED Master Mix
   ii. 0.75 µL of forward primer Mod_RepCOI_F (10 µM)
   iii. 0.75 µL of reverse primer VertCOI_7216_R (10 µM)
   iv. 7.5 µL of ultrapure H_2O
   v. 1 µL of extracted DNA

   Step 1 notes on preparing the master mix apply here as well. DNA extracts that did not amplify optimally (e.g., dim bands, nonspecific amplification) in PCR A can also be included.

Cite this protocol as *Cold Spring Harb Protoc*; doi:10.1101/pdb.prot108292

6. Run each reaction in a thermocycler, with heated lid, under the following conditions:

1 cycle	3 min	95°C
5 cycles	40 sec	95°C
	30 sec	45°C
	3 min	72°C
35 cycles	40 sec	95°C
	30 sec	46°C
	3 min	72°C
1 cycle	7 min	72°C

7. Electrophorese PCR products alongside a DNA ladder to determine whether amplification was successful as described in Step 4.

Successful reactions will have an amplicon ~244 bp in length. Note the DNA extracts that did and did not produce an amplicon of the expected size.

See Troubleshooting.

PCR C

8. Using only DNA extracts that did not amplify in PCR A or PCR B, run a third PCR using the primer set VertCOI_10096_F and Mod_RepCOI_R in PCR strips in a final volume of 20 µL consisting of:
   i. 10 µL of 2.0× Apex Taq RED Master Mix
   ii. 0.75 µL of forward primer VertCOI_10096_F (10 µM)
   iii. 0.75 µL of reverse primer Mod_RepCOI_R (10 µM)
   iv. 7.5 µL of ultrapure H_2O
   v. 1 µL of extracted DNA

   Step 1 notes on preparing the master mix apply here as well. DNA extracts that did not amplify optimally (e.g., dim bands, nonspecific amplification) in PCR A or B can also be included.

9. Run each reaction in a thermocycler, with heated lid, under the following conditions:

1 cycle	3 min	95°C
40 cycles	40 sec	95°C
	30 sec	48°C
	3 min	72°C
1 cycle	7 min	72°C

Note that these conditions differ from those of PCR A and B.

10. Electrophorese PCR products alongside a DNA ladder to determine whether amplification was successful as described in Step 4.

Successful reactions will have an amplicon ~152 bp in length.

See Troubleshooting.

Sequencing and Analysis

11. Use the chain-termination method (Sanger et al. 1977) to sequence all PCR products that produced an amplicon of the expected size using the forward primer.

Sequencing is most often performed by an external company, with the raw or cleaned PCR products sent to the sequencing company, and digital electropherogram trace files for each sequencing read sent back. If raw PCR products are sent for sequencing, a PCR cleanup step should be performed by the provider of the sequencing service before sequencing. Amplicons can be sequenced in one direction, using the forward primer, or in both directions using the forward and reverse primers. Sequencing in both directions can help to overcome noise in sequencing reads, but if high quality host sequence from a unidirectional read is obtained, bidirectional sequencing is not needed to confidently identify the host of a mosquito blood meal. Amplicons, visualized on an agarose gel, will vary in brightness. Sequence all samples, including those that produce faint bands, as sequences from these are often of sufficient quality to be identified.

12. On the computer, inspect the resulting sequences using software for visualizing and editing electropherogram files (e.g., Geneious, UGENE), examine the Phred scores (Ewing and Green 1998) and overall quality of the sequence, and trim ambiguous stretches from the ends of the sequence.

    See Troubleshooting.

13. Identify sequences using the Barcode of Life Data System (BOLD) Identification Engine (Ratnasingham and Hebert 2007).

    i. For each sequence that includes a stretch of unambiguous nucleotide calls, copy the nucleotide sequence of the unambiguous stretch to the clipboard.

       Some ambiguous nucleotide calls interspersed within the sequence are not necessarily problematic, but avoid stretches of calls with poor Phred scores.

    ii. Navigate to the BOLD (www.boldsystems.org) system and click on "IDENTIFICATION" to get to the Identification Engine.

    iii. Paste and submit the copied sequence into the search field of the "ANIMAL IDENTIFICATION [COI]" tab, with "Species Level Barcode Records" selected.

       A results summary and summary table of matching reference sequences will appear on the following page. In general, if a blood meal sequence is >98% similar to a reference sequence, the host species is conspecific with the corresponding sequence.

       See Troubleshooting.

TROUBLESHOOTING

Problem (Steps 4, 7, and 10): A DNA extract from a mosquito blood meal did not amplify in PCR A, B, or C.

Solution: If the extent of digestion was estimated during blood meal preservation, check its score. If the blood meal was estimated to be highly digested, it is likely that the host DNA templates are too degraded for amplification by PCR, and additional steps are unlikely to lead to identification. If the blood meal was estimated to be fresh, it is possible that the DNA extraction failed, that the priming sites of the host species are mismatched with all primers, that the blood meal was stored for more than 48 h at −20°C, or that the mosquito fed from an invertebrate host animal. To test the former possibility, re-extract DNA from the blood meal if enough preserved sample remains. If priming site mismatches are suspected, additional primer sets (Alcaide et al. 2009; Kent 2009), potentially for other markers such as cytochrome *b*, can be attempted. Cold storage for >48 h at −20°C is not an effective method for preserving blood fed mosquitoes (Reeves et al. 2016), although often the logistics of field or laboratory work do not permit prompt preservation of blood meals on FTA cards. If blood meals estimated to be fresh do not amplify, the amount of time between field collection and FTA card preservation should be considered, and if this is the cause, amplification of a host template is unlikely. Invertebrate hosts are unlikely to be detected using the methods described here. Attempts to use this protocol to identify an invertebrate-derived blood meal may produce no amplification, amplification of mosquito templates, or nonspecific amplification with noisy sequences. Currently, only *Uranotaenia sapphirina* is known to feed from invertebrate hosts (annelids) in nature. Blood meals derived from annelids are red and not visibly different from those of vertebrates. Consider the possibility that the mosquito may have fed from an invertebrate, particularly if all blood meals of a particular mosquito species cannot be identified, if the host associations of the mosquito species are poorly known, or if the mosquito belongs to the genus *Uranotaenia*. Reeves et al. (2018b) include methods for detecting annelid hosts, but the detection of other invertebrate taxa from mosquito blood meals may require development of novel methods.

Problem (Step 12): The sequence is messy or overly noisy or did not prime during sequencing.

Solution: This can be caused by nonspecific amplification of templates other than the host or by blood meals derived from more than one host. In some cases, the signal from the host can be stronger than the noise and will still provide a clear identification at Step 13. In other cases, the quality of the sequence is too poor to be identified. In these cases, reamplify the DNA with a different primer combination. If a sample consistently results in a noisy sequence, the PCR product can be sequenced in both directions and the resulting sequences can be aligned to bolster confidence in base calls. To align forward and reverse sequences, import them into a bioinformatic software (e.g., Geneious) and use the pairwise align or similar function to generate a consensus sequence.

Problem (Step 13): A high-quality sequence submitted to the BOLD Identification Engine returns no match ("Unable to match any records in the selected database").

Solution: This suggests that reference sequences from the host species are not yet included in the database. Rerun the query with the "All Barcode Records on BOLD" option selected to provide information on animal species closely related to the host. Use the results of this search, combined with knowledge of potential host species known to be present at the collection site of the mosquito, to narrow the potential species from which the blood meal may be derived. Using the "TAXONOMY" feature of BOLD, search the names of the potential species to determine whether reference sequences are included in the database. If one or more species is suspected based on this, if possible, collect tissue samples from these species, ideally from museum specimens, extract DNA, and amplify using PCR A. Alternatively, the Mod_RepCOI_F and Mod_RepCOI_R primers can be used for a larger, 664-bp amplicon, with the same reaction components and thermocycler conditions as those in PCR A. Compare the resulting sequence with the unidentified blood meal sequences. If similarity is high (>98%), the host species is likely conspecific.

Problem (Step 13): A high-quality sequence submitted to the BOLD Identification Engine returns a poor-quality (<98%) match to one or more species.

Solution: This can indicate either that the host species is not yet represented in the database or that the host species has higher-than-expected intraspecific *COI* variation. For example, *COI* sequences from green anoles, *Anolis carolinensis*, in Florida, particularly the southern peninsula, are highly divergent from reference green anole sequences. In such cases, collect tissue samples from the suspected host species locally and generate reference sequences as described above for host species that are missing from the BOLD database. Extracted DNA can be amplified and sequenced using the protocols of PCR A or PCR B or the primers LCO1490 and HCO2198 can be used (see Folmer et al. 1994 for PCR protocols). The amplicons produced by PCR A, PCR B, and PCR C fall within the fragment amplified by LCO1490 and HCO2198.

Problem (Step 13): A high-quality sequence submitted to the BOLD Identification Engine is a >98% match to more than one species.

Solution: Products from PCR C (and, more rarely, PCR A or B) may be too short to clearly distinguish closely related, usually congeneric species (e.g., white-tailed deer and mule deer). Use knowledge of the geographic distributions of vertebrate species and local host communities to distinguish between the potential host species, if they are not sympatric. Use other primers producing a longer amplicon to add sequence data and better resolving power. The Mod_RepCOI_F and Mod_RepCOI_R primers can be used to produce a larger, 664-bp amplicon, which may improve the ability to distinguish between closely related species. Reaction reagents and concentration and thermocycler conditions are as described in PCR A.

Problem (Step 13): A sequence submitted to the BOLD Identification Engine matches to mosquito species.

Solution: In rare cases, primers will anneal to and amplify mosquito templates. This can be more of an issue for particular mosquito species than others, especially some aedines. Reamplify the DNA extract with a different primer set.

DISCUSSION

This protocol describes a hierarchical approach for amplifying a short fragment of the DNA barcoding region of the vertebrate host *COI* gene from DNA extracted from mosquito blood meals. A hierarchical approach is applied here to overcome the issue of priming site sequence variation among vertebrate species. Each primer combination is expected to work for the majority of vertebrate host species, but should one fail to anneal because of priming site mismatches, the use of a second or, more rarely, a third primer combination is likely to result in amplification. Another common issue in DNA-based blood meal analysis that can cause amplification to fail is degradation of host templates through digestion in the mosquito midgut. Host DNA is digested over time, and degradation of host DNA templates can cause PCR to fail if digestion is not inhibited within 48 h of feeding (Ngo and Kramer 2003; Oshaghi et al. 2006). As a blood meal is digested, short fragments of template DNA are expected to persist for a greater duration of time than longer fragments. For this reason, primer combinations that target short amplicons are expected to be better suited for amplifying degraded templates. Concomitantly, shorter fragments have less resolving power than longer fragments (Meusnier et al. 2008). This is relevant to Step 13, in which it can be difficult to differentiate between closely related species. In such cases, careful interpretation is required (see Troubleshooting).

For some time, *COI* barcoding had been underutilized in mosquito blood meal analysis applications (Kent 2009), in part because of the lack of primers that work well across vertebrate taxa, while excluding amplification of mosquito templates. More recently, several primer sets have been identified that are compatible with the requirements of mosquito blood meal analysis. In addition to the hierarchical approach described here (modified from Reeves et al. 2018a), primers have been developed for amplification of host *COI* templates in a nested PCR (Alcaide et al. 2009). Primers initially developed as universal barcoding primers for fish (Ivanova et al. 2007) have also been useful in mosquito blood meal analyses. Additional markers including cytochrome *b* (*MT-CYB*), internal transcribed spacer 2 (*ITS2*) and 16S ribosomal RNA gene (*16S*) have been widely used in mosquito blood meal analysis, but, for these markers, reference sequences of comparatively fewer species are available for comparison and identification (Kent 2009). At the same time, identification of host species with *COI* is not perfect. Closely related species separated by short divergence times or species that contain introgressed genes from other species cannot always be resolved (e.g., *Odocoileus hemionus* and *Odocoileus virginianus*, *Rallus elegans* and *Rallus crepitans*, *Lepus alleni*, and *Lepus californicus*), and some species may have higher than expected variation within the *COI* barcoding region sequence (e.g., *Anolis carolinensis*, *Anolis sagrei*, *Osteopilus septentrionalis*). In these cases, it may be advantageous to pair *COI* with another marker (e.g., *MT-CYB*, *ITS2*, *16S*), sequencing amplicon from both markers when finer resolution is required by the research questions. Alternatively, next-generation sequencing–based approaches can be applied that target multiple loci to identify specimens (Liu et al. 2017). However, for most applications, including those in which the objective is the characterization of mosquito host associations, it may not be necessary to distinguish closely related host species

 Cite this protocol as *Cold Spring Harb Protoc*; doi:10.1101/pdb.prot108292

or to verify an amplicon sequence that imperfectly matches a reference sequence when no other closely related hosts are sympatric with the putatively identified species.

REFERENCES

Alcaide M, Rico C, Ruiz S, Soriguer R, Muñoz J, Figuerola J. 2009. Disentangling vector-borne transmission networks: a universal DNA barcoding method to identify vertebrate hosts from arthropod bloodmeals. *PLoS ONE* **4:** e7092. doi:10.1371/journal.pone.0007092

Ewing B, Green P. 1998. Base-calling of automated sequencer traces using *Phred.* II. Error probabilities. *Genome Res* **8:** 186–194. doi:10.1101/gr.8.3.186

Folmer O, Black M, Hoeh W, Lutz R, Vrijenhoek R. 1994. DNA primers for amplification of mitochondrial cytochrome *c* oxidase subunit I from diverse metazoan invertebrates. *Mol Mar Biol Biotechnol* **3:** 294–299.

Ivanova NV, Zemlak TS, Hanner RH, Hebert PDN. 2007. Universal primer cocktails for fish DNA barcoding. *Mol Ecol Notes* **7:** 544–548. doi:10.1111/j.1471-8286.2007.01748.x

Kent RJ. 2009. Molecular methods for arthropod bloodmeal identification and applications to ecological and vector-borne disease studies. *Mol Ecol Resour* **9:** 4–18. doi:10.1111/j.1755-0998.2008.02469.x

Liu J, Jiang J, Song S, Tornabene L, Chabarria R, Naylor GJP, Le C. 2017. Multilocus DNA barcoding: species identification with multilocus data. *Sci Rep* **7:** 16601. doi:10.1038/s41598-017-16920-2

Meusnier I, Singer GAC, Landry J-F, Hickey DA, Hebert PDN, Hajibabaei M. 2008. A universal DNA mini-barcode for biodiversity analysis. *BMC Genomics* **9:** 214. doi:10.1186/1471-2164-9-214

Ngo KA, Kramer LD. 2003. Identification of mosquito blood meals using polymerase chain reaction (PCR) with order-specific primers. *J Med Entomol* **40:** 215–222. doi:10.1603/0022-2585-40.2.215

Oshaghi MA, Chavshin AR, Vatandoost H, Yaaghoobi F, Mohtarami F, Noorjah N. 2006. Effects of post-ingestion and physical conditions on PCR amplification of host blood meal DNA in mosquitoes. *Exp Parasitol* **112:** 232–236. doi:10.1016/j.exppara.2005.11.008

Ratnasingham S, Hebert PDN. 2007. BOLD: the barcode of life data system (http://www.barcodinglife.org). *Mol Ecol Notes* **7:** 355–364. doi:10.1111/j.1471-8286.2007.01678.x

Reeves LE, Burkett-Cadena ND. 2023. Extracting DNA from preserved mosquito blood meals. *Cold Spring Harb Protoc* doi:10.1101/pdb.prot108291

Reeves LE, Holderman CJ, Gillett-Kaufman JL, Kawahara AY, Kaufman PE. 2016. Maintenance of host DNA integrity in field-preserved mosquito (Diptera: Culicidae) blood meals for identification by DNA barcoding. *Parasit Vector* **9:** 503. doi:10.1186/s13071-016-1791-z

Reeves LE, Gillett-Kaufman JL, Kawahara AY, Kaufman PE. 2018a. Barcoding blood meals: new vertebrate-specific primer sets for assigning taxonomic identities to host DNA from mosquito blood meals. *PLoS Negl Trop Dis* **12:** e0006767. doi:10.1371/journal.pntd.0006767

Reeves LE, Holderman CJ, Blosser EM, Gillett-Kaufman JL, Kawahara AY, Kaufman PE, Burkett-Cadena ND. 2018b. Identification of *Uranotaenia sapphirina* as a specialist of annelids broadens known mosquito host use patterns. *Commun Biol* **1:** 92. doi:10.1038/s42003-018-0096-5

Sanger F, Nicklen S, Coulson AR. 1977. DNA-sequencing with chain-terminating inhibitors. *Proc Natl Acad Sci* **74:** 5463–5467.

CHAPTER 12

Quantifying Mosquito Attraction Behavior Using Olfactometry

John S. Castillo, Anthony J. Bellantuono, and Matthew DeGennaro[1]

Department of Biological Sciences and Biomolecular Sciences Institute, Florida International University, Miami, Florida 33199, USA

When blood feeding from human hosts, female mosquitoes can transmit life-threatening pathogens to humans, including dengue virus, chikungunya virus, and Zika virus. Olfaction is the primary sense mosquitoes use to locate and differentiate hosts and studying it can lead to new strategies to reduce the risk of disease. To effectively study host-seeking behavior in mosquitoes, a repeatable, quantitative assay that isolates olfaction from other cues is critical for interpreting mosquito behavior. Here, we contribute an overview of methods and best practices for the study of mosquito attraction (or lack thereof) by using olfactometry to quantify behavior. In the accompanying protocols, we present an olfactory-based behavioral assay using a uniport olfactometer that measures mosquito attraction rate to specific stimuli. We include construction details, setup of the uniport olfactometer, details of the behavioral assay, and data analysis guidelines, as well as how to prepare the mosquitoes before their introduction into the olfactometer. This uniport olfactometer behavioral assay is currently one of the most reliable methods to study mosquito attraction to a single olfactory stimulus.

INTRODUCTION TO MOSQUITO OLFACTION AND OLFACTOMETERS

Olfaction is central to many stages of the mosquito life cycle, particularly in reproduction and blood feeding. This sensory system allows the mosquito to detect different aspects of their environment, including food sources and possible oviposition destinations. For anthropophilic mosquitoes such as *Aedes aegypti*, the human host is a critical part of the life cycle. Anthropophilic mosquitoes are highly attracted to human odors, but the human scent is a complex odor space consisting of more than 1000 different volatile organic compounds (VOCs) (Drabińska et al. 2021). This complexity of the odor plume makes it challenging to identify which components are driving mosquito attraction (Amann et al. 2014).

Olfactometry has been extensively used to study mosquito behavior in different ecological contexts (Willis 1947; Acree et al. 1968; Feinsod and Spellman 1979; Eiras and Jepson 1994; Klowden 1996; Bernier et al. 2000; DeGennaro et al. 2013; Liesch et al. 2013; McBride et al. 2014; McMeniman et al. 2014; Basrur et al. 2020). Most research on mosquito attraction and repulsion has been performed using olfactometers, which are purpose-built devices to introduce odor stimuli in a controlled manner and assess a behavioral response. The Y-tube and dual-port olfactometers are similar in that they allow researchers to simultaneously compare mosquito attraction between two separate stimuli such as carbon dioxide (CO_2), lactic acid, or human hosts (Acree et al. 1968; Braks and Takken 1999; Geier and Boeckh 1999). Y-tube and dual-port olfactometers differ in their geometries, which may affect the distribution of odors and presentation of stimuli in these two assay types (Geier and Boeckh

[1]Correspondence: mdegenna@fiu.edu

Cite this introduction as *Cold Spring Harb Protoc;* doi:10.1101/pdb.top107660

1999). Early olfactometer experiments were critical for the establishment of mosquito olfaction as a field (Willis 1947; Acree et al. 1968; Feinsod and Spellman 1979; Klowden 1996) but involved highly complex setups that imposed challenges for experimental replication. Modern single-choice uniport olfactometers have improvements in usability and are constructed using acrylic material that makes them highly transparent for the detection of insect location, easy to fabricate, and simple to assemble (Liesch et al. 2013). The single-choice olfactometer assay that we introduce here and detail in accompanying protocols (see Protocol 1: Building a Uniport Olfactometer to Assess Mosquito Responses to Odors [Castillo et al. 2023a] and Protocol 2: Quantifying Mosquito Attraction Using a Uniport Olfactometer [Castillo et al. 2023b]) can be used to measure mosquito attraction to isolated volatile compounds or blends of compounds that are associated with human odor (Raji et al. 2019). Careful consideration of the concentration of these compounds is important to replicate the odor stimuli that mosquitoes experience in nature. We have made improvements upon previous uniport olfactometer designs used to study mosquito attraction (Raji et al. 2019).

FEMALE MOSQUITO OLFACTION

Our uniport olfactometer assesses the attraction rate of female mosquitoes to human hosts. Female mosquitoes need a vertebrate blood meal to obtain nutrients for the proper development of their eggs (Detinova 1949). The completion of this sequence of alternate feeding and attraction allows mosquitoes to search for a new human host, enabling the transmission of pathogens such as dengue virus, chikungunya virus, and Zika virus, among others (Patterson et al. 2016). To find these hosts, female mosquitoes use a combination of cues. Body odor (produced by the microbiota on our skin [Shelley et al. 1953; Verhulst et al. 2011]) and CO_2 are key attractants, and olfactometers have been useful tools to document mosquito responses to them. The loss of olfactory receptor function in female mosquitoes leads to a decrease in host attraction, particularly in the absence of CO_2 (DeGennaro et al. 2013; Raji et al. 2019). CO_2 is an important cue because it telegraphs that a vertebrate may be nearby; it activates the female mosquito to begin vertebrate host-seeking, shifting from nectar-seeking behavior (Dekker et al. 2005). Approximately 10–20 msec after exposure, female mosquitoes sense CO_2 emitted by a vertebrate host using gustatory receptors (GRs) Gr1, Gr2, and Gr3 (McMeniman et al. 2014; van Breugel et al. 2015). Once the odor cues have been perceived, females approach the host and sense heat using ionotropic receptors (IRs) Ir21a and Ir25a, and the transient receptor potential channel A1 (TRPA1) (Corfas and Vosshall, 2015; Greppi et al. 2020). By understanding which components of human host odor impact mosquito attraction, we can identify novel repellents and attractants to manipulate mosquito behavior and reduce vector-borne disease transmission.

Mosquitoes seek not only vertebrate hosts but also sugar sources provided by plants (Barredo and DeGennaro 2020). Mosquitoes can distinguish between nutrient-rich or -scarce sugar sources when picking plant hosts and prefer sources that are higher in glycogen, lipid, and protein content (Yu et al. 2018). Critically, these preferences have impacts on mosquito fitness; female mosquitoes that choose to feed on plants with high nutrients live longer and lay more eggs after blood feeding than mosquitoes that select lower nutrition plants (Manda et al. 2007). Additionally, female conspecifics have consistent preferences regarding plants (Otienoburu et al. 2012), indicating that nectar-seeking behavior is a conserved and necessary aspect of the mosquito life cycle. Studies showing attraction to blends of human odor and plant volatiles have been performed in mesocosms, enclosed midsized natural environments that allow for mosquitoes to be observed in a controlled setting (Otienoburu et al. 2012; Busula et al. 2015; van Loon et al. 2015). These mesocosm studies are valuable as they have shown that mosquitoes can sense low concentrations of compounds when performing nectar and host seeking (Torr et al. 2008; van Loon et al. 2015), but mesocosm experiments operate at longer distances than olfactometer experiments. The application of uniport olfactometry to the study of ecologically relevant plants and plant-associated VOCs provides a powerful means to quantitatively test attraction at a range of up to 1 m and to assess olfactory cues in isolation.

 Cite this introduction as *Cold Spring Harb Protoc*; doi:10.1101/pdb.top107660

MALE MOSQUITO OLFACTION

Male mosquitoes exclusively feed on plant nectar, and the acquisition of this meal affects behaviors linked to survival, such as swarming, mating, and nectar seeking (Yu et al. 2018). The attainment of a proper nectar meal is an ecologically important behavior for male mosquitoes that has been previously studied using olfactometers (Gouagna et al. 2010). Due to the importance of a nectar meal in the male mosquito life cycle and their ability to pollinate certain plants, it is plausible that mosquitoes could have coevolved a mutualistic relationship with specific plants (Lahondère et al. 2020; Peach and Gries 2020). Although less studied due to their inability to bite hosts, male mosquito behavior has recently been in the spotlight due to the discovery that males gain attraction to human odor in olfactometers after the genetic ablation of the *fruitless* gene (Basrur et al. 2020). Additionally, male mosquitoes potentially sense human hosts at a distance because it signals possible mating opportunities (Oliva et al. 2014). We expect that our uniport olfactometer design will work to assay male mosquito responses to odors.

MOSQUITO REPELLENTS

Identifying which compounds act as mosquito attractants or repellents can aid in the management of mosquito populations and support interventions to reduce the spread of infectious diseases. Spatial repellents, also known as deterrents, are chemical compounds that cause mosquitoes to move away without having come into physical contact with the compound (Gouck et al. 1967). Olfactometers can be used to study candidate spatial repellents under controlled settings and at a short distance. In single-choice olfactometer studies, repellents are quantified in the presence of a known attractant to assess the repellent effect. Repellency is calculated as the delta between capture with attractant stimuli compared with captured rates of the attractant stimuli in the presence of the potential repellent (Kline et al. 2003; Achee et al. 2009). One of the most used spatial and contact repellents to date, *N-N*-dimethyl toluamide (DEET), has been extensively studied using olfactometers (Dogan and Rossignol et al. 1999; Bernier et al. 2003; Kline et al. 2003; Achee et al. 2009; DeGennaro et al. 2013). Uniport olfactometers can be used to assess repellent activity without the need for host contact and to assess whether the repellent uses an olfactory mechanism to prevent mosquito bites.

REFERENCES

Achee NL, Sardelis MR, Dusfour I, Chauhan KR, Grieco JP. 2009. Characterization of spatial repellent, contact irritant, and toxicant chemical actions of standard vector control compounds1. *J Am Mosquito Control Assoc* **25:** 156–167.

Acree F Jr, Turner RB, Gouck HK, Beroza M, Smith N. 1968. L-Lactic acid: a mosquito attractant isolated from humans. *Science* **161:** 1346–1347. doi:10.1126/science.161.3848.1346

Amann A, de Lacy Costello B, Miekisch W, Schubert J, Buszewski B, Pleil J, Ratcliffe N, Risby T. 2014. The human volatilome: volatile organic compounds (VOCs) in exhaled breath, skin emanations, urine, feces and saliva. *J Breath Res* **8:** 034001. doi:10.1088/1752-7155/8/3/034001

Barredo E, DeGennaro M. 2020. Not just from blood: mosquito nutrient acquisition from nectar sources. *Trends Parasitol* **36:** 473–484. doi:10.1016/j.pt.2020.02.003

Basrur NS, De Obaldia ME, Morita T, Herre M, von Heynitz RK, Tsitohay YN, Vosshall LB. 2020. *Fruitless* mutant male mosquitoes gain attraction to human odor. *eLife* **9:** e63982. doi:10.7554/eLife.63982

Bernier UR, Kline DL, Barnard DR, Schreck CE, Yost RA. 2000. Analysis of human skin emanations by gas chromatography/mass spectrometry. 2. Identification of volatile compounds that are candidate attractants for the yellow fever mosquito (*Aedes aegypti*). *Analyt Chem* **72:** 747–756.

Bernier UR, Kline DL, Posey KH, Booth MM, Yost RA, Barnard DR. 2003. Synergistic attraction of *Aedes aegypti* (L.) to binary blends of L-lactic acid and acetone, dichloromethane, or dimethyl disulfide. *J Med Entomol* **40:** 653–656.

Braks MA, Takken W. 1999. Incubated human sweat but not fresh sweat attracts the malaria mosquito *Anopheles gambiae* sensu stricto. *J Chem Ecol* **25:** 663–672.

Busula AO, Takken W, Loy DE, Hahn BH, Mukabana WR, Verhulst NO. 2015. Mosquito host preferences affect their response to synthetic and natural odour blends. *Malaria J* **14:** 1–9. doi:10.1186/s12936-015-0635-1

Castillo JS, Bellantuono AJ, DeGennaro M. 2023a. Building a uniport olfactometer to assess mosquito responses to odors. *Cold Spring Harb Protoc* doi:10.1101/pdb.prot108174

Castillo JS, Bellantuono AJ, DeGennaro M. 2023b. Quantifying mosquito attraction using a uniport olfactometer. *Cold Spring Harb Protoc* doi:10.1101/pdb.prot108175

Corfas RA, Vosshall LB. 2015. The cation channel TRPA1 tunes mosquito thermotaxis to host temperatures. *eLife* **4:** e11750. doi:10.7554/eLife.11750

DeGennaro M, McBride CS, Seeholzer L, Nakagawa T, Dennis EJ, Goldman C, Jasinskiene N, James AA, Vosshall LB. 2013. *orco* mutant mosquitoes lose strong preference for humans and are not repelled by volatile DEET. *Nature* **498:** 487–491. doi:10.1038/nature12206

Dekker T, Geier M, Cardé RT. 2005. Carbon dioxide instantly sensitizes female yellow fever mosquitoes to human skin odours. *J Exp Biol* **208:** 2963–2972. doi:10.1242/jeb.01736

Detinova TS. 1949. Physiological changes of ovaries in females of *Anopheles maculipennis*. *Med Parazitol Parazitarn Bolezni* **18:** 410.

Dogan EB, Rossignol PA. 1999. An olfactometer for discriminating between attraction, inhibition, and repellency in mosquitoes (Diptera: Culicidae). *J Med Entomol* **36:** 788–793.

Drabińska N, Flynn C, Ratcliffe N, Belluomo I, Myridakis A, Gould O, Fois M, Smart A, Devine T, Costello BD. 2021. A literature survey of all volatiles from healthy human breath and bodily fluids: the human volatilome. *J Breath Res* **15:** 034001. doi:10.1088/1752-7163/abf1d0

Eiras Alvaro E, Jepson PC. 1994. Responses of female *Aedes aegypti* (Diptera: Culicidae) to host odours and convection currents using an olfactometer bioassay. *Bull Entomol Res* **84:** 207–211. doi:10.1017/S0007485300039705

Feinsod FM, Spielman A. 1979. An olfactometer for measuring host-seeking behavior of female *Aedes aegypti* (Diptera: Culicidae). *J Med Entomol* **15:** 282–285. doi:10.1093/jmedent/15.3.282

Geier M, Boeckh J. 1999. A new Y-tube olfactometer for mosquitoes to measure the attractiveness of host odours. *Entomol Exp Appl* **92:** 9–19. doi:10.1046/j.1570-7458.1999.00519.x

Gouagna LC, Poueme RS, Dabiré KR, Ouédraogo JB, Fontenille D, Simard F. 2010. Patterns of sugar feeding and host plant preferences in adult males of *An. gambiae* (Diptera: Culicidae). *J Vector Ecol* **35:** 267–276. doi:10.1111/j.1948-7134.2010.00082.x

Gouck H, McGovern TP, Beroza M. 1967. Chemicals tested as space repellents against yellow-fever mosquitoes I. Esters. *J Econ Entomol* **60:** 1587–1590.

Greppi C, Laursen WJ, Budelli G, Chang EC, Daniels AM, van Giesen L, Smidler AL, Catteruccia F, Garrity PA. 2020. Mosquito heat seeking is driven by an ancestral cooling receptor. *Science* **367:** 681–684. doi:10.1126/science.aay9847

Kline DL, Bernier UR, Posey KH, Barnard DR. 2003. Olfactometric evaluation of spatial repellents for *Aedes aegypti*. *J Med Entomol* **40:** 463–467.

Klowden MJ. 1996. Endogenous factors regulating mosquito host-seeking behaviour. *Ciba Found Symp* **200:** 212–223; discussion 223–232. doi:10.1002/9780470514948.ch16

Lahondère C, Vinauger C, Okubo RP, Wolff GH, Chan JK, Akbari OS, Riffell JA. 2020. The olfactory basis of orchid pollination by mosquitoes. *Proc Natl Acad Sci* **117:** 708–716. doi:10.1073/pnas.1910589117

Liesch J, Bellani LL, Vosshall LB. 2013. Functional and genetic characterization of neuropeptide Y-like receptors in *Aedes aegypti*. *PLoS Negl Trop Dis* **7:** e2486. doi:10.1371/journal.pntd.0002486

Manda H, Gouagna L, Nyandat E, Kabiru E, Jackson R, Foster WA, Githure JI, Beier JC, Hassanali A. 2007. Discriminative feeding behaviour of *Anopheles gambiae s.s.* on endemic plants in western Kenya. *Med Vet Entomol* **21:** 103–111.

McBride CS, Baier F, Omondi AB, Spitzer SA, Lutomiah J, Sang , Ignell R, Vosshall LB. 2014. Evolution of mosquito preference for humans linked to an odorant receptor. *Nature* **515:** 222–227.

McMeniman CJ, Corfas RA, Matthews BJ, Ritchie SA, Vosshall LB. 2014. Multimodal integration of carbon dioxide and other sensory cues drives mosquito attraction to humans. *Cell* **156:** 1060–1071. doi:10.1016/j.cell.2013.12.044

Oliva CF, Damiens D, Benedict MQ. 2014. Male reproductive biology of *Aedes* mosquitoes. *Acta Tropica* **132:** S12–S19.

Otienoburu PE, Ebrahimi B, Phelan PL, Foster WA. 2012. Analysis and optimization of a synthetic milkweed floral attractant for mosquitoes. *J Chem Ecol* **38:** 873–881. doi:10.1007/s10886-012-0150-6

Patterson J, Sammon M, Garg M. 2016. Dengue, Zika and chikungunya: emerging arboviruses in the New World. *West J Emerg Med* **17:** 671.

Peach DAH, Gries G. 2016. Nectar thieves or invited pollinators? A case study of tansy flowers and common house mosquitoes. *Arthropod-Plant Interact* **10:** 497–506. doi:10.1007/s11829-016-9445-9

Peach DAH, Gries G. 2020. Mosquito phytophagy—sources exploited, ecological function, and evolutionary transition to haematophagy. *Entomol Exper Appl* **168:** 120–136.

Raji JI, Melo N, Castillo JS, Gonzalez S, Saldana V, Stensmyr MC, DeGennaro M. 2019. *Aedes aegypti* mosquitoes detect acidic volatiles found in human odor using the IR8a pathway. *Curr Biol* **29:** 1253–1262. doi:10.1016/j.cub.2019.02.045

Shelley WB, Hurley HJ, Nichols AC. 1953. Axillary odor: experimental study of the role of bacteria, apocrine sweat, and deodorants. *AMA Arch Dermatol Syphilol* **68:** 430–446.

Torr SJ, Della Torre A, Calzetta M, Costantini C, Vale GA. 2008. Towards a fuller understanding of mosquito behaviour: use of electrocuting grids to compare the odour-orientated responses of *Anopheles arabiensis* and *An. quadriannulatus* in the field. *Med Vet Entomol* **22:** 93–108. doi:10.1111/j.1365-2915.2008.00723.x

van Breugel F, Riffell J, Fairhall A, Dickinson MH. 2015. Mosquitoes use vision to associate odor plumes with thermal targets. *Curr Biol* **25:** 2123–2129. doi:10.1016/j.cub.2015.06.046

van Loon JJ, Smallegange RC, Bukovinszkiné-Kiss G, Jacobs F, De Rijk M, Mukabana WR, Verhulst NO, Menger DJ, Takken W. 2015. Mosquito attraction: crucial role of carbon dioxide in formulation of a five-component blend of human-derived volatiles. *J Chem Ecol* **41:** 567–573. doi:10.1007/s10886-015-0587-5

Verhulst NO, Mukabana WR, Takken W, Smallegange RC. 2011. Human skin microbiota and their volatiles as odour baits for the malaria mosquito Anopheles gambiae ss. *Entomol Exper Appl* **139:** 170–179.

Willis ER. 1947. The olfactory responses of female mosquitoes. *J Econ Etomol* **40:** 769–778.

Yu B, Hu Y, Ding Y, Tian J, Mo J. 2018. Feeding on different attractive flowering plants affects the energy reserves of *Culex pipiens pallens* adults. *Parasitol Res* **117:** 67–73.

Protocol 1

Building a Uniport Olfactometer to Assess Mosquito Responses to Odors

John S. Castillo, Anthony J. Bellantuono, and Matthew DeGennaro[1]

Department of Biological Sciences and Biomolecular Sciences Institute, Florida International University, Miami, Florida 33199, USA

The uniport olfactometer behavioral assay is currently one of the most reliable single-choice methods to use to study mosquito attraction to olfactory stimuli. It allows for the reproducible calculation of mosquito attraction rate to human hosts or to other olfactory stimuli. Here, we present the design of our modified uniport olfactometer. Consistent carbon-filtered air flows through the assay, creating positive pressure that reduces odor contamination from the room. It includes a precision-milled white acrylic base to facilitate easy setup and consistent placement of the component parts. Our design can be made by a commercial acrylic fabricator or an academic machine shop. This olfactometer is designed to assess the responses of mosquitoes but could be applied to other insects that fly upwind toward an odor stimulus. In a companion protocol, we detail how to perform the experiments with mosquitoes by using the uniport olfactometer.

MATERIALS

It is essential that you consult the appropriate Material Safety Data Sheets and your institution's Environmental Health and Safety Office for proper handling of equipment and hazardous material used in this protocol.

Reagents

Deionized H_2O

Unscented liquid dish soap (Free & Clear Seventh Generation)

Equipment

1/8″ National Pipe Thread (NPT) to 1/4″ barbed fittings (two total)

Activated charcoal (CAS 7440-44-0) packed in a medium canister with barbed fittings (Drierite gas drying unit; W.A. Hammond)

CO_2 cylinder (USP medical grade, high-pressure steel, ≥99.0%; Airgas UN1013)

CO_2 regulator

Environmentally controlled room (25°C–27°C with relative humidity of 50%–70%)

Erlenmeyer flask (2000-mL)

Fiberglass window screen (18 × 16 mesh)

Flowmeter for air (2–24 standard cubic feet per h [SCFH]; King Instrument Company 7520 series)

---

[1]Correspondence: mdegenna@fiu.edu

Cite this protocol as *Cold Spring Harb Protoc;* doi:10.1101/pdb.prot108174

Flowmeter for CO_2 (0.1–1 SCFH; Dwyer Instruments, Inc. VFA-1)
Hot melt adhesive (Gorilla 3033002)
Latex gloves
Stainless steel tube (25-cm)
Sturdy table or benchtop
Tygon tubing (1/4″ internal diameter, vinyl)
Uniport olfactometer and base

Multiple custom parts are constructed from cast acrylic sheets (see Fig. 1)

Vacuum pump (oil-free, Gast DOA-P704-AA high-capacity vacuum pump, with gauge, regulator, and relief; 1.1 cfm, 25.5″ Hg, 115 VAC)

METHOD

Fabricating the Olfactometer

1. Contact a local acrylic fabricator to assist you in the construction of the uniport olfactometer parts and base according to the plans outlined in Figure 1.

   It is recommended that multiple uniports are fabricated at the same time to reduce costs and increase experimental output. Multiple uniport olfactometers can be used at the same time if you have sufficient space.

2. After fabrication, ensure that the parts fit together correctly and sit properly in the computer numerical control (CNC)-milled base.
3. Complete the attraction chamber by gluing (with hot melt adhesive) a fiberglass window screen in the shape of a cone that is 1.500-in (38.100-mm) long with a base diameter of 3.250 in (82.550 mm) and the tip of the cone no larger than 0.750 in (19.050 mm).

Assembling the Olfactometer

Assemble the olfactometer in a climate-controlled room atop a sturdy table or benchtop. The climate of the room should resemble rearing conditions (25°C–27°C with relative humidity of 50%–70% is appropriate if using Aedes aegypti, Aedes albopictus, *and* Culex quinquefasciatus *mosquitoes).*

Additionally, thoroughly clean olfactometer parts before and after each use with unscented dish soap. Rinse thoroughly, do a final rinse with deionized H_2O, and then dry.

4. Place the olfactometer base on a sturdy table or benchtop (Fig. 2A).
5. Place the activated chamber on the olfactometer base so that it is held in place by slots on the base.
6. Fit the activated chamber circular entry (Fig. 2B) and exit plates (Fig. 2C) to the respective ends of the activated chamber.
7. Attach the attraction trap to circular exit plate with the gate closed.
8. Attach the stimulus box to the attraction trap. Seal the large circular arm entry ports with latex gloves (Fig. 2D).
9. Confirm the olfactometer is sealed by lightly pulling on the left latex glove.

   The right latex glove should move, indicating the olfactometer is airtight.

10. Place the vacuum pump on the floor near the olfactometer.
11. Using Tygon tubing, connect the positive pressure outlet of the vacuum pump to the medium canister filled with activated charcoal.
12. Connect the charcoal-filtered air via Tygon tubing to a 25-cm stainless steel tube in a stoppered 2000-mL Erlenmeyer flask containing 1000 mL of deionized H_2O.

Cite this protocol as *Cold Spring Harb Protoc*; doi:10.1101/pdb.prot108174

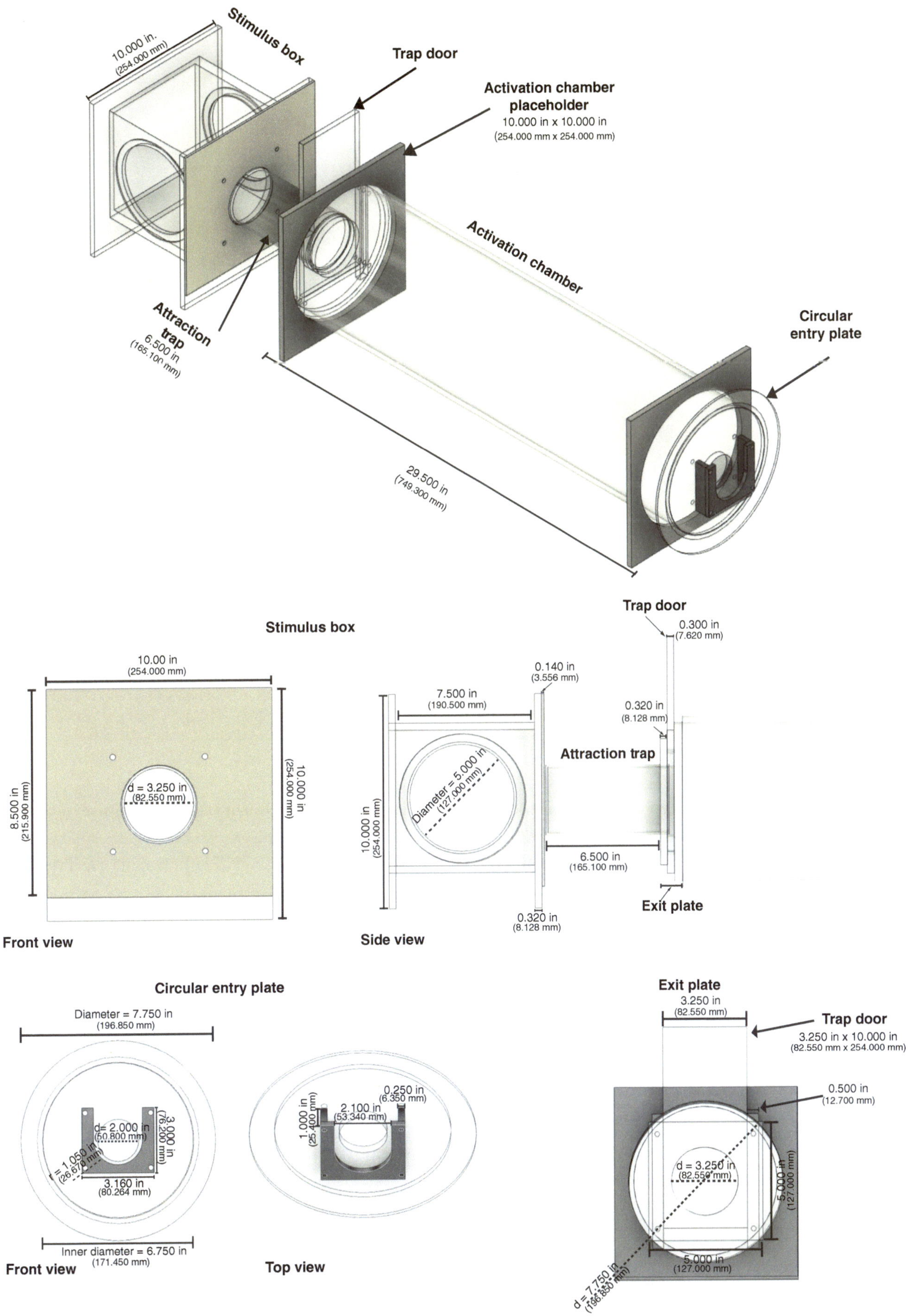

FIGURE 1. Plans for fabricating the uniport olfactometer. All components are made of clear acrylic except for a white acrylic face plate for the stimulus chamber. This removable component of the stimulus chamber was designed to obscure mosquitoes from being able to view items placed in the stimulus chamber from a distance.

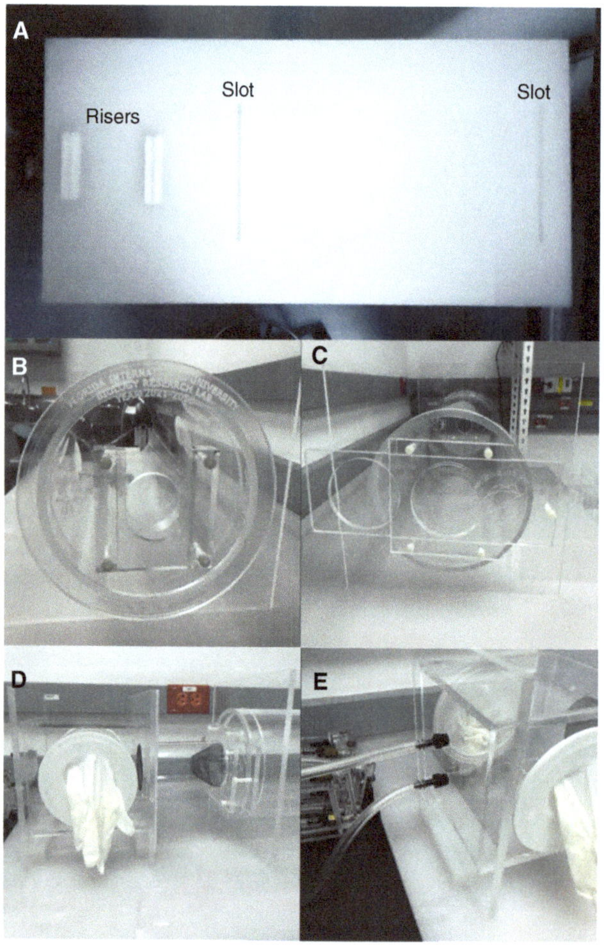

FIGURE 2. The uniport olfactometer placed in custom designed base. (*A*) The olfactometer base is made from computer numerical control (CNC)-milled white acrylic (24-in width × 48-in length × 0.75-in thick) with 12-in slots milled to hold the olfactometer component in fixed positions. The 6-in risers hold the stimulus chamber in position but allow for sliding the chamber for easy disassembly. They are designed to hold the olfactometer components securely together, preventing any changes to the olfactometer component alignment to maintain consistency between trials. It is best to bring the olfactometer to the CNC service provider to assure a good fit. (*B*) Circular entry plate. (*C*) Exit plate. (*D*) Sealed stimulus box attached to attraction trap. (*E*) Tygon tubing attached to barb fittings screwed into the stimulus box.

It is important that the 25-cm stainless steel tube is below the H_2O level in the Erlenmeyer flask to ensure humidity remains in the system.

13. Connect the outlet of the Erlenmeyer flask to the 2- to 24-SCFH flowmeter via Tygon tubing.
14. Connect the CO_2 cylinder by connecting the CO_2 regulator to a 0.1- to 1-SCFH flowmeter.
15. Screw the 1/8″ NPT to 1/4″ NPT barbed fittings into the stimulus box and connect the Tygon tubing from the CO_2 and airstream flowmeters (Fig. 2E).
16. Operate the olfactometer by following the accompanying protocol (Protocol 2: Quantifying Mosquito Attraction Using a Uniport Olfactometer [Castillo et al. 2023]).

ACKNOWLEDGMENTS

This material is based on work supported by the Defense Advanced Research Projects Agency (DARPA) Biological Technologies Office (BTO) under Contract No. HR0011-20-C-0073.

REFERENCES

Castillo JS, Bellantuono AJ, DeGennaro M. 2023. Quantifying mosquito attraction using a uniport olfactometer. *Cold Spring Harb Protoc* doi:10.1101/pdb.prot108175

Cite this protocol as *Cold Spring Harb Protoc*; doi:10.1101/pdb.prot108174

Protocol 2

Quantifying Mosquito Attraction Using a Uniport Olfactometer

John S. Castillo, Anthony J. Bellantuono, and Matthew DeGennaro[1]

Department of Biological Sciences and Biomolecular Sciences Institute, Florida International University, Miami, Florida 33199, USA

Female mosquitoes respond to the world around them by using chemosensory organs, such as their antennae, to detect volatile compounds emitted from a vertebrate host. These chemosensory systems facilitate the interpretation of external stimuli from the periphery by connecting to the central nervous system and eliciting behaviors necessary for survival, such as obtaining a blood meal. This innate behavior leads to the transmission of pathogens, including dengue virus, chikungunya virus, and Zika virus. Olfaction is a primary sense mosquitoes use to differentiate between vertebrate hosts, and studying it can lead to novel strategies to reduce the risk of disease. In this protocol, we present an olfactory-driven behavioral assay using a uniport olfactometer that measures mosquito attraction rate to a specific stimulus. We include details of the behavioral assay and data analysis as well as how to prepare the mosquitoes before their introduction into the olfactometer. This uniport olfactometer behavioral assay is currently one of the most reliable methods to study mosquito attraction to a single stimulus.

MATERIALS

It is essential that you consult the appropriate Material Safety Data Sheets and your institution's Environmental Health and Safety Office for proper handling of equipment and hazardous material used in this protocol.

Reagents

Aedes aegypti mosquitoes (newly hatched larvae)
H_2O (deionized)
Odor stimulus (see Steps 10–16)
Sucrose solution (10% [w/v] autoclaved in deionized H_2O)
Unscented liquid dish soap (Free & Clear Seventh Generation)

Equipment

Aspirator with HEPA filter (John W. Hock Company)
Cotton balls
Environmentally controlled room (25°C–27°C with relative humidity of 50%–70%)
Polycarbonate pan (12.75-in length × 10.38-in width × 4.00-in height; Carlisle 10221B)
Portable vacuum with disposable bags (see Step 21)
Rearing cage (BugDorm1; MegaView Science Co., Ltd.)

[1]Correspondence: mdegenna@fiu.edu

Cite this protocol as *Cold Spring Harb Protoc*; doi:10.1101/pdb.prot108175

Scissors
Small paintbrush
Timer
Uniport olfactometer (fully assembled) setup

For the uniport olfactometer design and assembly, please see our companion protocol (Protocol 1: Building a Uniport Olfactometer to Assess Mosquito Responses to Odors [Castillo et al. 2023]).

WHO tubes (length = 13 cm, diameter = 5 cm; World Health Organization Vector Control Research Unit)

METHOD

Preparing Mosquitoes for Behavior Trials

Maintain mosquitoes in an environmentally controlled room at 25°C–27°C with 50%–70% relative humidity on a 14:10 light–dark cycle. These conditions and other rearing information apply to Ae. aegypti *mosquitoes. If another species is used, different conditions may be needed.*

1. To ensure you have healthy and active female *Ae. aegypti* mosquitoes for the assay, rear larvae under standard conditions in polycarbonate pans at a density of 250 larvae per 1200 mL of H_2O.
2. Place adults in rearing cages and provide ad libitum access to cotton balls soaked with a 10% sucrose solution.
3. Maintain adult females and males together for a minimum of 7 d posteclosion so that they are given sufficient time to mate.

   Females no older than 21 d should be used in the assay. Ideally, they should be between 7 and 14 d old.
4. At 12–16 h before the assay, cold anesthetize female mosquitoes (4°C) until they are knocked down (~5 min), resting motionless on the floor of the cage. Sort them gently using a small paintbrush into groups of 30. Collect 30 mosquitoes with an aspirator and place them into WHO tubes.

   Cold anesthetization makes mosquitoes less active and facilitates their transfer into the WHO tubes. Mosquitoes may perish due to stress from cold anesthetization. Gently graze the anesthetized mosquitoes with small paintbrush to assure they are alive. If mosquitoes do not retract their legs to their original position, they are likely dead and should be discarded. Mosquitoes should not be kept for longer than 1 h at 4°C. Allow mosquitoes to recover for at least 12–16 h before experimentation.
5. Immediately after placing mosquitoes into the WHO tubes (Step 4), provide mosquitoes with access to a cotton ball soaked in deionized H_2O so that they do not desiccate. Store these mosquitoes in an environmentally controlled room at 25°C–27°C with 50%–70% relative humidity with a 14:10 light–dark cycle.

   This 12- to 16-h H_2O-only starvation period is intended to increase mosquito activity.

Performing the Host-Seeking Experiment with the Uniport Olfactometer

Perform the experiment in an environmentally controlled room at 25°C–27°C with 50%–75% relative humidity on a 14:10 light–dark cycle. These conditions apply to Ae. aegypti *mosquitoes. If another species is used, different conditions may be needed.*

6. Turn on the air pump and open the CO_2 regulator valve connected to the uniport olfactometer setup.
7. Check flowmeters and the regulator and adjust air flow to 18–24 standard cubic feet per h (SCFH) and CO_2 flow to 0.2–0.4 SCFH.
8. Open the sliding gate on the WHO tube to allow CO_2 and air to pass through the attraction trap and into the activated chamber.

   CO_2 exhalation is characteristic of living vertebrates and can be modeled by releasing an airstream of CO_2 into the uniport that creates an attractive plume (Dekker et al. 2005; van Breugel et al. 2015). CO_2 along

Cite this protocol as *Cold Spring Harb Protoc*; doi:10.1101/pdb.prot108175

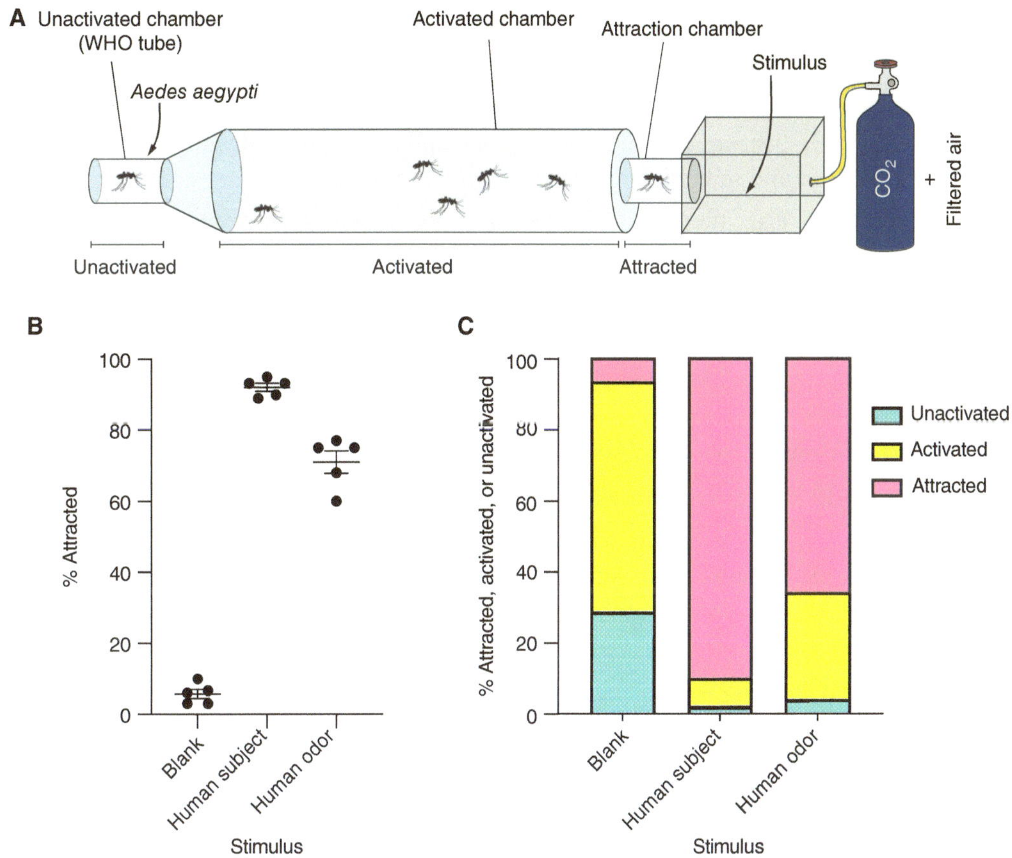

FIGURE 1. *Aedes aegypti* responses to attractive stimuli in the uniport olfactometer. (*A*) Schematic of the uniport olfactometer. Mosquitoes are released from the unactivated chamber (WHO tube) and fly upwind, entering the activation chamber. Mosquitoes may remain in the activated chamber or may continue on to the attraction chamber. After 8 min have elapsed, the assay is scored by counting mosquitoes in each chamber to identify unactivated, activated, and attracted counts and then dividing by the total number of mosquitoes assayed in *B* and *C*. (*B*) Shows *Ae. aegypti* responses to a blank trial (CO_2 + air), a human subject (forearm in the olfactometer stimulus box), and human odor (nylon sleeve in olfactometer stimulus box). Dots represent trials with mean and SEM plotted for each condition. (*C*) Shows the average percentage of female mosquitoes attracted (magenta), activated (yellow), or unactivated (cyan) of the five trials in *B*. Each trial was performed with 30 female mosquitoes.

with filtered air act as a stimulus for the mosquito to leave the unactivated chamber, but both cues alone are not sufficient to trap more than 20% of mosquitoes in the activated chamber (Fig. 1C).

9. Before running an experiment, perform a blank trial with the latex gloves sealing the entry circles of the stimulus box. If attraction is higher than 20%, clean the stimulus box and attraction trap with unscented dish soap and rinse thoroughly, with a final rinse with deionized H_2O, and dry thoroughly. Confirm the CO_2 level is 2250–2750 ppm and then perform another blank trial.

   This step ensures that the system was thoroughly cleaned before the experiment.

   Attraction higher than 20% indicates that contaminating odors are present within the system or that the CO_2 concentration may be erroneously high.

For Inanimate Objects

10. Immediately after completing a successful blank trial (with ≤20% attraction), remove the latex glove and insert the object (i.e., odor stimulus on small platform) into the stimulus box.
11. Place the inanimate object in the center of the stimulus box, elevated sufficiently so that it is in the path of the CO_2 and filtered airstream. Proceed to Step 14.

For Human Subjects

12. Immediately after completing a successful blank trial (with ≤20% attraction), cut one of the latex gloves with sharp scissors to remove four fingers, leaving the thumb intact.
13. Instruct the subject to sit in a comfortable position with their fist and forearm resting on the far and near edges of the entry ports, respectively.
14. Place the WHO tube with mosquitoes onto the activated chamber circular entry plate.
15. Release mosquitoes from the WHO tube by gently sliding open the door.
16. Set a timer for 8 min.

    Eight minutes provides enough time for mosquitoes to respond to the new environment and make a decision.

Calculating the Response Rate in the Assay

17. Once the 8 min allotted for the experiment is concluded, close the gate at the exit plate and at the entrance of the WHO tube.
18. Perform counts of all unactivated mosquitoes in the WHO tube release chamber, in the activated chamber, and in the attraction trap (Fig. 1A).

    Mosquitoes that do not leave the WHO tube release chamber during the experiment are scored as unactivated. Those that leave the WHO tube but never enter the attraction trap are scored as activated, and those that enter the attraction trap are scored as attracted.

    Occasionally, mosquitoes will be killed during handling and will remain in the WHO tube. Any dead mosquitoes should be subtracted from the number of mosquitoes from the total number assayed. Frequently observing dead mosquitoes (i.e., >10% in an assay) is cause for concern and signals the need for careful review of rearing practices and handling procedures to ensure that mosquitoes are in optimal condition for behavioral assessment. Normally, airflow and CO_2 are enough to get most mosquitoes activated. Additional odor cues may be required to get mosquitoes into the attraction trap.

19. To calculate the percentage of mosquitoes attracted, divide the total number of mosquitoes used in the assay trial by the number of mosquitoes in attraction trap.
20. To calculate the percentage of mosquitoes activated, add the total number of mosquitoes in the activated chamber and the attraction trap and divide by the total number of mosquitoes in the assay.

Cleaning the Olfactometer

21. Clean the olfactometer immediately after an experiment by using a portable vacuum. Begin by slowly inserting the vacuum head into the activated chamber, using suction to remove mosquitoes that are in the chamber. Repeat this process for mosquitoes in the WHO tube. Freeze and autoclave the vacuum bags containing mosquitoes for disposal.
22. To clean the olfactometer between subjects, isolate the stimulus box and wash with unscented dish soap and deionized H_2O, rinsing thoroughly with deionized H_2O.

    Avoid using solvent-based cleaning products such as those containing ethanol, as they will damage the acrylic.

23. To determine whether the assay equipment is clean, perform a blank trial (no stimulus) with only CO_2 and air. If mosquito attraction is higher than 20%, clean the stimulus box again with soap and H_2O and perform another blank test.

DISCUSSION

Mosquitoes use their sense of smell to find a suitable host, and the proper physiological conditions must be met for them to initiate host-seeking and blood-feeding behaviors. Initially, a female mos-

Cite this protocol as *Cold Spring Harb Protoc*; doi:10.1101/pdb.prot108175

quito nectar seeks to obtain energy stores that can be critical for her survival and fecundity (Briegel 1990, Foster 1995). The female mosquito is not attracted to a human arm for 24 h after eclosure, and it is therefore crucial that she obtain a sugar meal during this period (Jones and Pilitt 1973). The quality of this sugar meal influences her ability to host seek (Jones and Pilitt 1973). Additionally, when mosquitoes are given sugar ad libitum before a blood meal, they show a significant decline in blood-feeding and host-seeking compared to mosquitoes given access to only H_2O (Jones and Madhukar 1976). Females are also reduced in their ability to host seek for at least 48 h once they obtain a blood meal (Klowden and Blackmer 1987). Female mosquitoes have a refractory period where they cannot be inseminated for 48 h posteclosion (Lea 1968). When working with different mosquito species, it is important to adjust the room temperature, humidity, and light-cycle conditions to be ecologically relevant and ensure increased activity. Randomizing trials by running controls throughout the experiment when using different stimuli in the olfactometer is a good practice to reduce extraneous results. Starving mosquitoes on H_2O and providing a well humidified room can create the proper environment for mosquito participation. If mosquito participation is low, make certain that CO_2 levels in the olfactometer do not exceed 5000 ppm, as overexposure to CO_2 can anesthetize mosquitoes (Williams 1946). These factors have influenced the design of our protocol and, using our methods, we have achieved high response rates to both humans and human odor with *Ae. aegypti* females (Fig. 1B). We have also achieved positive results with other mosquito species such as *Ae. albopictus*, *An. gambiae*, and *Cx. quinquefasciatus*.

Additionally, our uniport olfactometer has a stimulus chamber that allows for the inclusion of a larger region of the human body, increasing the stimulus signal. Our design uses positive air pressure and carbon filtration to remove extraneous odors that may be present in the room from the stimulus being studied, providing the ability to focus on mosquito host-seeking attractants and repellents that are placed in the assay. To occlude visual cues from the stimulus chamber, a white acrylic face plate was placed on its front. The dimensions of the activation chamber provide sufficient space so that the assay does not force a choice. Additionally, to increase the structural integrity of the olfactometer, our design includes a full-length baseplate into which the assay components fit, eliminating leaks due to assembly irregularities and movement of olfactometer components when removing mosquitoes between trials.

CONCLUSION

A repeatable, quantitative method to assess attraction to human hosts, host analogs, and other attractants is critical for the study of mosquito host-seeking behavior. This updated uniport olfactometer assay simplifies the collection of quantitative attraction data. Using the experimental setup described here, researchers can reduce variables when assaying mosquito attraction. A key advantage of this uniport design lies in its exclusion of extraneous odors from the assay itself, as the system is under constant positive pressure provided by filtered air and bottled CO_2. The single-choice, host-seeking assay presented has been used to analyze female mosquito responses to human subjects, to human odor trapped on nylon sleeves, and to individual compounds found on human skin.

ACKNOWLEDGMENTS

This material is based on work supported by the Defense Advanced Research Projects Agency (DARPA) Biological Technologies Office (BTO) under Contract No. HR0011-20-C-0073.

REFERENCES

Briegel H. 1990. Metabolic relationship between female body size, reserves, and fecundity of *Aedes aegypti*. *J Insect Physiol* **36:** 165–172. doi:10.1016/0022-1910(90)90118-Y

Castillo JS, Bellantuono AJ, DeGennaro M. 2023. Building a uniport olfactometer to assess mosquito responses to odors. *Cold Spring Harb Protoc* doi:10.1101/pdb.prot108174

Dekker T, Geier M, Cardé RT. 2005. Carbon dioxide instantly sensitizes female yellow fever mosquitoes to human skin odours. *J Exp Biol* **208:** 2963–2972. doi:10.1242/jeb.01736

Foster WA. 1995. Mosquito sugar feeding and reproductive energetics. *Annu Rev Entomol* **40:** 443–474. doi:10.1146/annurev.en.40.010195.002303

Jones JC, Madhukar BV. 1976. Effects of sucrose in blood avidity in mosquitoes. *J Insect Physiol* **22:** 357–360. doi:10.1016/0022-1910(76)90001-9

Jones JC, Pilitt DR. 1973. Blood-feeding behavior of adult *Aedes aegypti* mosquitoes. *Biol Bull* **145:** 127–139. doi:10.2307/1540353

Klowden MJ, Blackmer JL. 1987. Humoral control of pre-oviposition behaviour in the mosquito, *Aedes aegypti. J Insect Physiol* **33:** 689–692. doi:10.1016/0022-1910(87)90052-7

Lea AO. 1968. Mating without insemination in virgin *Aedes aegypti. J Insect Physiol* **14:** 305–308. doi:10.1016/0022-1910(68)90072-3

van Breugel F, Riffell J, Fairhall A, Dickinson MH. 2015. Mosquitoes use vision to associate odor plumes with thermal targets. *Curr Biol* **25:** 2123–2129. doi:10.1016/j.cub.2015.06.046

Williams CM. 1946. Continuous anesthesia for insects. *Science* **103:** 57. doi:10.1126/science.103.2663.57

Cite this protocol as *Cold Spring Harb Protoc;* doi:10.1101/pdb.prot108175

CHAPTER 13

Quantifying *Aedes aegypti* Host Odor Preference Using a Two-Port Olfactometer

Hillery C. Metz, Jessica L. Zung, and Carolyn S. McBride[1]

Department of Ecology and Evolutionary Biology, and Princeton Neuroscience Institute, Princeton University, Princeton, New Jersey 08540, USA

Blood-feeding mosquitoes are a leading threat to global public health—vectoring dangerous infections including Zika, dengue, and malaria. Mosquitoes identify and target hosts for blood meals by using visual, thermal, and chemical cues. Here we describe an assay for measuring odor-based host-preference behavior—that is, the preferential approach toward one host over another based on differences in the volatile compounds they emit. The assay can be adapted for use with diverse odor sources, from live animals and their breath to odor-scented sleeves with controlled amounts of CO_2. Mosquitoes in this assay fly upwind to within 30 cm of the odor source and then enter a small trap. We therefore believe this assay best replicates medium- to short-range host-seeking, when females approach and are preparing to land on a host animal. We also find that relative response in a two-choice test shows less trial-to-trial variation than the absolute number of responsive mosquitoes, which appears more sensitive to exogenous factors such as rearing conditions. This assay has been used to better understand mosquito host-seeking decisions, which can provide fundamental insight into the brain and behavior as well as information useful for the design of novel vector control strategies.

OVERVIEW

Female mosquitoes must locate hosts to reproduce. In the wild, host cues may include exhaled breath, volatile skin odors, visual cues, heat, and humidity, with different cues being important at different ranges. For example, carbon dioxide (a component of breath) is thought to act at longer ranges (up to ~10 m), whereas heat and humidity are likely short-range cues (<1 m), and volatile odorants guide behavior at intermediate distances or at multiple distances (Lacey et al. 2014; Cardé 2015). Here, we describe a "two-port olfactometer" behavioral assay that allows mosquitoes to fly upwind and choose to enter one of two odor-baited traps (Gouck and Schreck 1965; Gouck 1972; see Protocol 1: An Assay for Quantifying *Aedes aegypti* Host Odor Preference Using a Two-Port Olfactometer [Metz et al. 2023]). The traps are inches away from the odor sources, where volatile odorants may be quite concentrated. When using live host animals, heat and humidity may also be detectable at the entrances to the traps. We therefore believe this assay measures medium- to close-range host attraction and preference (i.e., when mosquitoes are approaching and preparing to land on a host animal) rather than long-range host-seeking, which may be better assayed using a wind tunnel. Consistent with this idea, *Ae. aegypti* do not respond strongly to CO_2 or human breath in this assay when presented alone (DeGennaro et al. 2013; McMeniman et al. 2014).

[1]Correspondence: csm7@princeton.edu
Supplemental material is available for this chapter at cshprotocols.cshlp.org.

Cite this introduction as *Cold Spring Harb Protoc*; doi:10.1101/pdb.top107661

This assay is ideal for characterizing the behavior of large groups of mosquitoes (~25–100 per assay), and therefore can be used to quickly characterize the preference behavior of a colony, treatment group, or mutant strain. Because the starting positions of individual animals vary widely in this assay, it is less suitable for testing smaller numbers.

We find that the proportion of mosquitoes responding to host odor (i.e., the absolute response rate) is highly variable and may be influenced by factors such as rearing and handling, room lighting, air speed, and even the weather (HC Metz et al., unpubl.). In contrast, preference for one host over another appears less sensitive to such factors.

EXPERIMENTAL DESIGN

Animal behavior is intrinsically variable, making it challenging to measure and study. Below, we outline experimental design considerations for minimizing confounds and unwanted environmental variation.

Interleaved or Randomized Trial Structure

All genotypes/treatments should be tested on the same days and at adjacent times of day.

Host Position

Alternate or randomize the side occupied by alternative hosts (i.e., left or right side) to separate the effect of host from the effect of side (i.e., side bias). The side used should be recorded and considered as a factor in analysis.

Testing Mixed Groups

When the different mosquito groups of interest can be distinguished by visible markers (e.g., morphology or fluorescent markers), we recommend testing mixed groups and separating them ex post facto. This design controls for trial-to-trial variation (e.g., due to the exact host odor emitted during a specific trial; paired statistics must be applied), which can be large and may otherwise obscure or confound results. Testing mixed groups also increases assay throughput, as multiple groups are tested simultaneously.

Common-Garden Rearing

In addition to testing mixed groups, it is sometimes possible to rear different genotypes or strains together in the same pans/cages. This provides the added benefit of controlling for variation in environment (e.g., food abundance, crowding), which we find can significantly affect absolute response rates. For example, siblings with three distinct genotypes resulting from an F_1 intercross (i.e., *ab* × *ab* cross producing *aa*, *ab*, and *bb* offspring) may be reared in a common garden, assayed side by side in the same behavioral trials, and identified/genotyped after behavioral results have been gathered.

ANIMAL REARING

Mosquitoes used for behavioral assays should always be reared simultaneously, in the same locations, and handled with consistency and care to minimize environmental sources of variation (Milinski 1997). See Munstermann (1997) or Clemons et al. (2010) for general culturing protocols. Here, we outline additional considerations when rearing mosquitoes for behavior.

Cite this introduction as *Cold Spring Harb Protoc*; doi:10.1101/pdb.top107661

- Eggs should be hatched using "hatch broth" (deoxygenated water mixed with food) to elicit a synchronous hatch and thereby produce cohorts of age-matched animals.
- Tested animals should be age-matched (e.g., by date of eclosion).
- Carefully avoid overcrowding and over- or underfeeding larvae. Overfeeding can lead to excessive bacterial growth that may sicken or kill larvae and pupae. Underfeeding can stunt growth and may have unwanted effects on behavior.
- We typically test host-odor responses in mated, non-blood-fed females and thus house experimental animals in mixed-sex groups with access to 10% (w/v) sucrose solution only.
- Males of *Ae. aegypti* are also attracted to host cues (Amos et al. 2022) and have been successfully tested in this assay (McKenna 1973). They hover around trap entrances but rarely enter them, leading to substantially lower response rates.

ASSAY CONDITIONS

Supplemental Movie S1 shows *Ae. aegypti* females entering the traps of a two-port olfactometer in response to human (*right*) and guinea pig (*left*) odor. Video is taken using the setup described in Protocol 1: An Assay for Quantifying *Aedes aegypti* Host Odor Preference Using a Two-Port Olfactometer (Metz et al. 2023) and replayed at 4× speed. Movie credit: Noah H. Rose.

Host Odor

Human subjects may be instructed to avoid disrupting their natural odor. We typically provide research volunteers with unscented soap and instruct them to bathe with it at a designated time (e.g., 24–72 h before assays) and then abstain from further bathing, use of cosmetic products, and contact with smelly items (e.g., garlic, oranges) before participating in behavioral tests. Likewise, animal hosts should not be bathed, treated with any products, or fed odorous foods for as long as possible before use in behavioral trials, and gloves should be worn during handling.

Individual human and nonhuman animals may have idiosyncratic odor profiles (Zhao et al. 2022), and multiple studies have shown that different individuals evoke different responses from mosquitoes (e.g., Fernández-Grandon et al. 2015; de Obaldia et al. 2022). Therefore, it may be desirable to include multiple individuals of each host/species type (e.g., DeGennaro et al. 2013) or to use one primary host but check whether major conclusions extend to others (e.g., Rose et al. 2020).

Materials

Mosquitoes may be sensitive to volatile organic compounds (VOCs) that off-gas from materials and reagents. Plastics and glues may be especially problematic and can even extinguish host-seeking behavior. We recommend use of low-VOC materials (e.g., glass, metal, PTFE). Unwrapping and washing new materials and allowing them to sit out and off-gas for several days may be helpful. Likewise, always ensure cleaning solvents (e.g., ethanol) have fully evaporated.

Cleaning and Handling of Equipment

All behavioral equipment should be kept free of exogenous odors. Thoroughly wash equipment before use and handle with clean gloves to prevent transfer of human skin odorants. Long gloves may help prevent accidental contact with skin. We wash the inside of the main olfactometer box weekly (or after every 4–5 trial days) and the host chambers/traps daily.

Air and Airspeed

Carrier air should be conditioned to match normal housing conditions and have little exogenous odor content. Correct airspeed is important for mosquito response and can be measured using an anemometer (wind-speed meter). In the two-port olfactometer, airspeed will be highest at the entrances of the traps (i.e., where the cross-sectional area is smallest), and should be in the range of 10–50 cm/sec (ideally ~30 cm/sec). Prior to running behavioral trials, olfactometer airspeed should be calibrated and tested with live mosquitoes. Most box fans will need to be adjusted below their lowest setting using a variable fan meter. An acceptable airspeed will allow mosquitoes to easily enter the traps—that is, although animals may sometimes approach and then retreat from trap openings, they will be able to fly steadily forward when they choose to. Mosquitoes may still respond at higher airspeeds (up to ~1 m/sec), but their flight into the traps will appear jerky and erratic. If the airspeed is too low or too high, mosquitoes will stop responding.

Lighting

Our lights are typically set to ~200 lux for *Ae. aegypti aegypti*. However, some types of mosquitoes may be more responsive in low-light conditions (e.g., Rose et al. 2020 used ~20 lux to test *Ae. aegypti formosus*) or in red light (e.g., Wondwosen et al. 2016).

VARIATIONS ON THE ASSAY

Odor Source

Nylon knee-high stockings may be used in place of live hosts (e.g., McBride et al. 2014). The toe should be cut off to make a "sleeve," which is then worn on a human arm or the body of an animal for 6–24 h depending on goals. To preserve odor, sleeves should be stored in a sealed, odor-free container at −20°C for up to 2–3 mo. Since odor emission from sleeves will decrease over time and with warmer temperatures, we limit use to five trials per sleeve and store them in containers placed on ice between uses. The number of trials across which one observes consistent responses should be tested a priori. The addition of CO_2, heat, and/or dark fabric covers over the stimulus chambers can improve response rates. Pure odorants and multicomponent blends may also be assayed (e.g., Wondwosen et al. 2016), although we noted very low response rates to these stimuli in our laboratory (HC Metz and Z Zhao, unpubl.).

Carbon Dioxide

Exposure to CO_2 gas (a component of host breath) "activates" mosquitoes to seek hosts (i.e., it induces a change in behavioral state from resting to host-seeking) and may increase mosquito responses to other cues (e.g., Dekker et al. 2005; McMeniman et al. 2014; Cardé 2015; van Breugel et al. 2015; Sorrells et al. 2022). If used in this assay, it should be delivered intermittently or in pulses to mimic the turbulent or filamentous plume structure of normal breath (Geier et al. 1999). Note that excessive CO_2 may be repellent in this behavioral context, as it is likely a long-range cue, not a close-range landing cue (Cardé 2015, and references therein).

When using live hosts, the CO_2 in breath will naturally be delivered intermittently. Small animals such as rodents and birds breathe freely inside the host chamber. Human subjects may exhale intermittently into the host chamber (e.g., one nasal exhale every 30 sec throughout the trial as in Rose et al. 2020). When using nylon sleeves or synthetic stimuli, a synthetic air/CO_2 mixture (90%/10%) can be added. For the latter, we typically apply a 1 sec pulse of the mixture every 5 sec using a solenoid valve, aiming for 1200 ppm CO_2 as measured at the mouth of the trap during pulses (Fig. 1I–J of the accompanying Protocol 1: An Assay for Quantifying *Aedes aegypti* Host Odor Preference Using a Two-Port Olfactometer [Metz et al. 2023]). Note that lower amounts of CO_2 may be preferable depending on the mosquito species being tested (Majeed et al. 2017).

Cite this introduction as *Cold Spring Harb Protoc*; doi:10.1101/pdb.top107661

Duration

We run trials for 6–10 min. Females from domestic laboratory strains of *Ae. aegypti* typically respond within the first few minutes, but some mosquitoes/colonies may require a longer trial duration.

Open versus Closed Air System

At its simplest, this assay uses carrier air pulled into the olfactometer from the room, that is, an open-air system; alternatively, clean, conditioned, carbon-filtered air can be pumped into the olfactometer through a "snorkel," creating a closed-air system (Fig. 1A and F, respectively, of Protocol 1: An Assay for Quantifying *Aedes aegypti* Host Odor Preference Using a Two-Port Olfactometer [Metz et al. 2023]). Although the latter is more controlled and prevents experimenter odor from contaminating carrier air, we have found the open system to be reliable for documenting behavioral differences of moderate to large effect. When using an open system, one should be careful to avoid unnecessary use of the room between trials and limit the number of trials per day (see Troubleshooting).

Other Cues

Nonolfactory cues can be manipulated to improve response rates without biasing preference—for example, by adding heat (e.g., microwaveable heating pads) equally to both sides. We strongly recommend wrapping host chambers with dark cloth (Fig. 1A of Protocol 1: An Assay for Quantifying *Aedes aegypti* Host Odor Preference Using a Two-Port Olfactometer [Metz et al. 2023]) to generate visual contrast from the mosquito perspective.

When using live hosts, dark chamber covers can also serve to calm the animal inside and prevent mosquitoes from seeing host size, shape, and movement.

DATA ANALYSIS

The assay measures preference across a group of individuals (e.g., $n=50$ or $n=100$). We suggest specific analyses below.

Analyses

The simplest approach treats each trial as an independent replicate (e.g., McBride et al. 2014). A trial-specific preference index can be calculated as follows: the number of mosquitoes entering the trap for host A minus those that entered the trap for host B, all divided by the total number of mosquitoes entering either trap. The significance of the difference between groups can be assessed using two-sided, two-sample *t*-tests. Paired *t*-tests can be used when two different genotypes or colonies are tested in mixed groups (leading to a pair of genotype-specific preference indices for each trial). Overall response rate is the fraction of mosquitoes that entered either trap and can be analyzed in a similar fashion. This approach is straightforward and makes no assumptions about potential interactions among the individual mosquitoes within a given trial (see below). However, its power is limited when trial-to-trial variability is high.

Alternatively, beta-binomial generalized linear models (or mixed-effect models) can be fit using the R package *glmmTMB* (Brooks et al. 2017) and compared using likelihood ratio tests (e.g., Rose et al. 2020). This statistical approach assumes individual females behave independently while accounting for trial structure and trial-to-trial variability (overdispersion) that may arise from a variety of uncontrollable random sources (e.g., exact starting position of females within the main olfactometer chamber at the start of a trial or trial-to-trial and day-to-day differences in live-host odor, etc.).

Example R code is given below for a data frame in which each row corresponds to a single trial of a single test group (e.g., a strain or genotype; "group" in our code) with the number of females responding to each host given in columns designated as count_host1 and count_host2. In this example, trial_ID is included as a random factor when testing mixed groups of mosquitoes.

```
model1 <- glmmTMB(cbind(count_host1, count_host2) ~ group +
(1|trial_ID), data=MyData, family='betabinomial')
```

Covariates

We suggest recording the following covariates and potentially including them in analyses: trial number (e.g., first, second, third of the day), time of day, age of mosquitoes, barometric pressure (an uncontrolled climate variable), and any unusual occurrences (e.g., urination by an animal host).

TROUBLESHOOTING

Accidental Cues

Large and/or dark objects present in the experimental room (e.g., a visible, moving human researcher) can confound results and should be hidden from host-seeking mosquitoes. The fabric covers and visual blind (see Fig. 1A of Protocol 1: An Assay for Quantifying *Aedes aegypti* Host Odor Preference Using a Two-Port Olfactometer [Metz et al. 2023]) help alleviate this concern.

Odor Habituation

Mosquitoes that have recently been exposed to host odor may become habituated and unresponsive during behavior trials. To prevent habituation: (1) keep mosquito holding cups away from humans in a separate, odor-free room or chamber; and (2) when human proximity to mosquitoes is unavoidable, seal mosquitoes in their holding cups using lids (first removing wet cotton).

Odor Contamination of Carrier Air

Human odor generated by the researcher will accumulate in the room with occupancy. To mitigate this complication, one may wish to limit the number of trials completed per day, use a well-ventilated room, or use fresh air sourced from outdoors (i.e., a closed-air system).

Unhappy Mosquitoes

Mistreatment of mosquitoes can lead to poor response. Avoid overcrowding as larvae and rough handling/stressors as adults (e.g., extended cold exposure). For each group of 50–100 animals set aside in a holding chamber the night before trials, very few (if any) should be found dead. A razor blade can be used to score the smooth, inner walls of holding chambers to provide mosquitoes with ample resting spots.

Odor Plume Structure

To create well-structured host odor plumes (Murlis et al. 1992) and prevent mixing of the two types of odor, air should be pulled through the olfactometer chamber by the terminal box fan—not pushed by incoming air, as may occur with a closed-air system. When using a closed system, airflow rates should be carefully adjusted to equalize inflow and outflow (i.e., the air source produces only as much air as the fan pulls out). A large, low-odor plastic bag (e.g., roasting bags for poultry) may be installed between the air source/snorkel and T-tube (see the accompanying Protocol 1: An Assay for Quantifying *Aedes aegypti* Host Odor Preference Using a Two-Port Olfactometer [Metz et al. 2023]) to visualize air dynamics. Visible inflation or deflation of the bag over time indicates a surplus or deficit of source air, respectively. Alternatively, we prefer a system where inflow is just slightly higher than outflow, but the fitting between the air snorkel and T-tube is loose, allowing extra air to diffuse into the room rather than being forced into the chambers. A smoke pen can be used to visualize airflow and identify problems with plume structure.

Cite this introduction as *Cold Spring Harb Protoc*; doi:10.1101/pdb.top107661

Uncontrolled Weather Variables

Barometric pressure can influence insect behavior (e.g., Austin et al. 2014), and we have noted low response rates when air pressure is low or dropping (HC Metz, unpubl.).

ACKNOWLEDGMENTS

We thank members of the McBride Lab, Rickard Ignell, and Willem Takken for helpful discussions and advice over the years. Work in our lab has been supported by the Pew Scholars Program, the Searle Scholars Program, the Gruber Foundation, and the National Institutes of Health (NIDCD R00DC012069, NIAID DP2AI144246) and the New York Stem Cell Foundation. C.S.M. is a New York Stem Cell Foundation—Robertson Investigator.

REFERENCES

Amos BA, Hoffmann AA, Staunton KM, Lau MJ, Burkot TR, Ross PA. 2022. Long-range but not short-range attraction of male *Aedes aegypti* (Diptera: Culicidae) mosquitoes to humans. *J Med Entomol* **59:** 83–88. doi:10.1093/jme/tjab164

Austin CJ, Guglielmo CG, Moehring AJ. 2014. A direct test of the effects of changing atmospheric pressure on the mating behavior of *Drosophila melanogaster*. *Evol Ecol* **28:** 535–544. doi:10.1007/s10682-014-9689-8

Brooks ME, Kristensen K, Van Benthem KJ, Magnusson A, Berg CW, Nielsen A, Skaug HJ, Mächler M, Bolker BM. 2017. glmmTMB balances speed and flexibility among packages for zero-inflated generalized linear mixed modeling. *R J* **9:** 378–400. doi:10.32614/RJ-2017-066

Cardé RT. 2015. Multi-cue integration: how female mosquitoes locate a human host. *Curr Biol* **25:** R793–R795. doi:10.1016/j.cub.2015.07.057

Clemons A, Mori A, Haugen M, Severson DW, Duman-Scheel M. 2010. Culturing and egg collection of *Aedes aegypti*. *Cold Spring Harb Protoc* doi10.1101/pdb.prot5507

DeGennaro M, McBride CS, Seeholzer L, Nakagawa T, Dennis EJ, Goldman C, Jasinskiene N, James AA, Vosshall LB. 2013. *orco* mutant mosquitoes lose strong preference for humans and are not repelled by volatile DEET. *Nature* **498:** 487–491. doi:10.1038/nature12206

Dekker T, Geier M, Cardé RT. 2005. Carbon dioxide instantly sensitizes female yellow fever mosquitoes to human skin odours. *J Exp Biol* **208:** 2963–2972. doi:10.1242/jeb.01736

de Obaldia ME, Morita T, Dedmon LC, Boehmler DJ, Jiang CS, Zeledon EV, Cross JR, Vosshall LB. 2022. Differential mosquito attraction to humans is associated with skin-derived carboxylic acid levels. *Cell* **185:** 4099–4116. doi:10.1016/j.cell.2022.09.034

Fernández-Grandon GM, Gezan SA, Armour JAL, Pickett JA, Logan JG. 2015. Heritability of attractiveness to mosquitoes. *PLoS ONE* **10:** e0122716. doi:10.1371/journal.pone.0122716

Geier M, Bosch OJ, Boeckh J. 1999. Influence of odour plume structure on upwind flight of mosquitoes towards hosts. *J Exp Biol* **202:** 1639–1648. doi:10.1242/jeb.202.12.1639

Gouck HK. 1972. Host preferences of various strains of *Aedes aegypti* and *Aedes simpsoni* as determined by an olfactometer. *Bull World Health Organ* **47:** 680–683.

Gouck HK, Schreck CE. 1965. An olfactometer for use in the study of mosquito attractants. *J Econ Entomol* **58:** 589–590. doi:10.1093/jee/58.3.589

Lacey ES, Ray A, Cardé RT. 2014. Close encounters: contributions of carbon dioxide and human skin odour to finding and landing on a host in *Aedes aegypti*. *Physiol Entomol* **39:** 60–68. doi:10.1111/phen.12048

Majeed S, Hill SR, Dekker T, Ignell R. 2017. Detection and perception of generic host volatiles by mosquitoes: responses to CO_2 constrains host-seeking behaviour. *Roy Soc Open Sci* **4:** 170189. doi:10.1098/rsos.170189

McBride CS, Baier F, Omondi AB, Spitzer SA, Lutomiah J, Sang R, Ignell R, Vosshall LB. 2014. Evolution of mosquito preference for humans linked to an odorant receptor. *Nature* **515:** 222–227. doi:10.1038/nature13964

McKenna J. 1973. "*Attraction of seven strains of Aedes aegypti to man and guinea pig in the laboratory*." PhD thesis, University of California, Davis.

McMeniman CJ, Corfas RA, Matthew BJ, Ritchie SA, Vosshall LB. 2014. Multimodal integration of carbon dioxide and other sensory cues drives mosquito attraction to humans. *Cell* **156:** 1060–1071. doi:10.1016/j.cell.2013.12.044

Metz HC, Zung JL, McBride CS. 2023. An assay for quantifying *Aedes aegypti* host odor preference in a two-port olfactometer. *Cold Spring Harb Protoc* doi:10.1101/pdb.prot108089

Milinski M. 1997. How to avoid seven deadly sins in the study. *Adv Study Behav* **26:** 159–180. doi:10.1016/S0065-3454(08)60379-4

Munstermann LE. 1997. Care and maintenance of *Aedes* mosquito colonies. In *The molecular biology of insect disease vectors: a methods manual*. Chapman & Hall, London.

Murlis J, Elkinton JS, Cardé RT. 1992. Odor plumes and how insects use them. *Ann Rev Entomol* **37:** 505–532. doi:10.1146/annurev.en.37.010192.002445

Rose NH, Sylla M, Badolo A, Lutomiah J, Ayala D, Aribodor OB, Ibe N, Akorli J, Otoo S, Mutebi JP, et al. 2020. Climate and urbanization drive mosquito preference for humans. *Curr Biol* **30:** 3570–3579. doi:10.1016/j.cub.2020.06.092

Sorrells TR, Pandey A, Rosas-Villegas A, Vosshall LB. 2022. A persistent behavioral state enables sustained predation of humans by mosquitoes. *eLife* **11:** e76663. doi:10.7554/eLife.76663

van Breugel F, Riffell J, Fairhall A, Dickinson MH. 2015. Mosquitoes use vision to associate odor plumes with thermal targets. *Curr Biol* **25:** 2123–2129. doi:10.1016/j.cub.2015.06.046

Wondwosen B, Birgersson G, Seyoum E, Tekie H, Torto B, Fillinger U, Hill SR, Ignell R. 2016. Rice volatiles lure gravid malaria mosquitoes, *Anopheles arabiensis*. *Sci Rep* **6:** 1–8. doi:10.1038/srep37930

Zhao Z, Zung JL, Hinze A, Kriete AL, Iqbal A, Younger MA, Matthews BJ, Merhof D, Thiberge S, Ignell R, et al. 2022. Mosquito brains encode unique features of human odour to drive host seeking. *Nature* **605:** 706–712. doi:10.1038/s41586-022-04675-4

Protocol 1

An Assay for Quantifying *Aedes aegypti* Host Odor Preference Using a Two-Port Olfactometer

Hillery C. Metz, Jessica L. Zung, and Carolyn S. McBride[1]

Department of Ecology and Evolutionary Biology, and Princeton Neuroscience Institute, Princeton University, Princeton, New Jersey 08540, USA

Female mosquitoes use odor cues to locate hosts for blood meals and are often more likely to approach the odor of certain species or individuals over others. Here, we describe an assay for measuring such odor-based host preference. This assay uses a two-port olfactometer and can be adapted to study a wide variety of odor sources including live hosts, host-scented nylon sleeves or host hair samples, and single odorants or odorant blends.

MATERIALS

It is essential that you consult the appropriate Material Safety Data Sheets and your institution's Environmental Health and Safety Office for proper handling of equipment and hazardous materials used in this protocol.

Reagents

Carbogen (10% CO_2/90% air) (optional; see Fig. 1I)
Ethanol (70%)
Mosquitoes; typically mated (see Introduction: Quantifying *Aedes aegypti* Host Odor Preference Using a Two-Port Olfactometer [Metz et al. 2023])
Odor source (e.g., live hosts, odor-impregnated stockings, or single-molecule odorants)
Outdoor air, carbon filtered (~60 ft^3/min, conditioned to 26°C and 75% relative humidity) (for use with closed-air system; Fig. 1F)
Sucrose (10% w/v) in deionized H_2O
Unscented soap

Equipment

Anemometer (optional; e.g., Schiltknecht ThermoAir3, 0–1 m/sec)

This is used to calibrate airspeed ahead of behavioral trials and to troubleshoot problems, for example, low response rate.

Bug vacuum (e.g., InsectaVac, Bioquip)
Cold room (4°C)
Cotton balls

[1]Correspondence: csm7@princeton.edu

Cite this protocol as *Cold Spring Harb Protoc*; doi:10.1101/pdb.prot108089

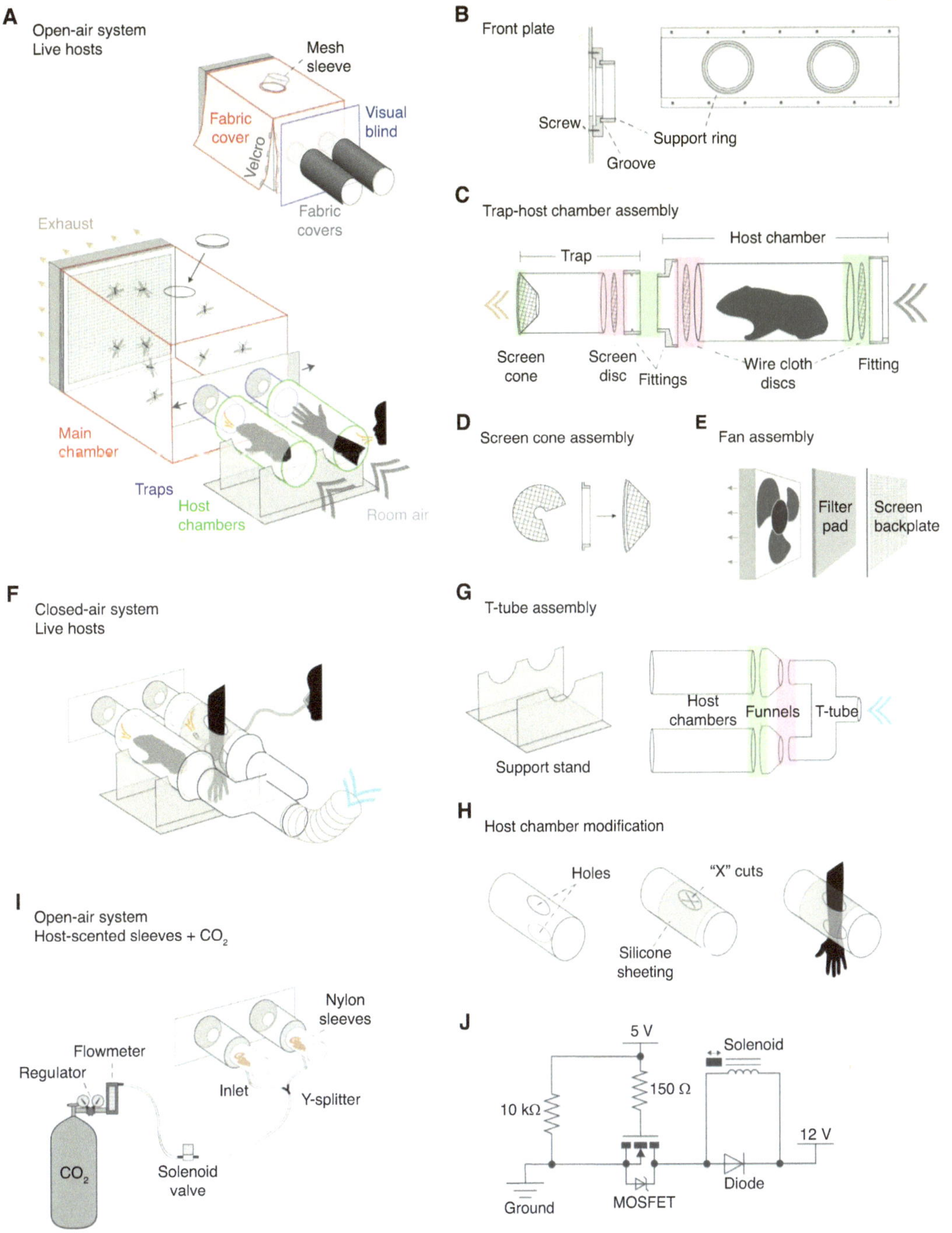

FIGURE 1. Two-port olfactometer assay for measuring host-odor preference. (*A*) Basic olfactometer setup with live hosts and an open-air system. A fan gently pulls air from the room over both hosts, through the traps, and into the main box where acclimated mosquitoes are waiting. A sliding door is opened to allow access to traps at the start of trials. Mosquitoes enter traps as they approach hosts. *Inset* shows setup with recommended fabric cover, visual blind, and live-host chamber covers. (*B–E*) Details of the apparatus. (*B*) Front plate block, which mounts on the main olfactometer box to house the sliding door. (*C*) Trap and live-host chamber with fittings and screens. Pink boxes mark joints that remain fixed throughout the day, whereas green ones mark those that are opened and closed between trials. (*D*) Screen cone assembly for traps. (*E*) Fan assembly. The filter pad helps even airflow generated by the fan. (*F–H*) Modifications for a closed-air system with live hosts. (*F*) Overall setup. An independent source of carbon-filtered, conditioned air is fed into the T-tube (e.g., from a snorkel), which splits the air and feeds it into the host chambers. Inflow from snorkel and outflow from fan must be carefully equalized (see Introduction: Quantifying *Aedes aegypti* Host Odor Preference Using a Two-Port Olfactometer [Metz et al. 2023]). (*G*) Detail of T-tube assembly and support stand. T-tube stand not pictured in *F*. Green and pink boxes as in *C*. (*H*) Detail of host chamber modification to accommodate a human arm in the closed-air system. Human breath can also be introduced via nasal mask and tubing as pictured in *F*. (*I–J*) Modifications for testing inanimate odor sources such as host-scented nylon sleeves. The sleeve chambers are smaller than those used for live hosts. A solenoid valve can be used to introduce intermittent pulses of synthetic CO_2, which can be measured using a CO_2 gas analyzer. (*I*) Overall setup. (*J*) Circuit diagram for solenoid valve.

Environmentally conditioned room, 75% relative humidity, 26°C
Feather or paint brush
Kimwipes and/or paper towels
Mesh fabric fine enough to contain adult mosquitoes, cut into squares (for covering the openings of plastic cups)
Microscope or dissecting scope (optional; for identifying mosquitoes by morphology or fluorescent marker)
Nylon knee-high pantyhose (optional; see Fig. 1I)

Cut off the toe with scissors to create "sleeves" to be worn by hosts, which are thereby impregnated with host odor.

Odor-free air snorkel (optional; see Fig. 1F)
Olfactometer, two-port Gouck-style (Fig. 1A; built as described in Box 1)
Plastic cups with lids (e.g., VWR 89009-668; multipurpose HDPE 32-oz cups)

Use cups for holding mosquitoes overnight ahead of behavior trials.

Plastic trays, white (for contrast with mosquitoes; e.g., HDPE 13 1/4″ × 11 1/4″ × 1 7/8″, Thomas Scientific), precooled to 4°C
Razor blade or rough sandpaper
Rubber bands sized to fit snugly around mouth of 32-oz plastic cups
Scissors (optional; for cutting mesh fabric and other equipment as needed)
Solenoid valve CO_2 delivery system with controller (optional; see Fig. 1I) (built as described in Box 2)

For "activating" mosquitoes, for example, when using host-scented stockings/hair or synthetic odorants rather than live hosts.

Timer
T-tube for splitting air (optional; for use with closed-air system; see Fig. 1F,G) (built as described in Box 3)
Underpads for use with live animals (optional; see Step 12)
Vacuum cleaner, canister type with flexible hose

This is used to remove mosquitoes from the olfactometer main box after completion of each trial.

METHOD

See Introduction: Quantifying Aedes aegypti *Host Odor Preference Using a Two-Port Olfactometer (Metz et al. 2023) for additional experimental considerations.*

Washing the Olfactometer

Wash the olfactometer before use to ensure that it is free of contaminating odors. We wash traps and host chambers daily and the main olfactometer box weekly (or after every 4–5 trial days).

1. Disassemble traps and host chambers and prepare a clean drying area.
2. Wash each individual piece including screens. More specifically, rinse with H_2O, use a gloved hand to gently scrub each piece with unscented, mild soap, and rinse again thoroughly with H_2O. If using tap H_2O, conduct a final rinse with distilled, deionized H_2O. Lay pieces on a clean surface to air-dry.
3. When dry, thoroughly spray each piece with 70% ethanol and dry with Kimwipes.
4. Once completely dry, reassemble each component.
5. The interior of the olfactometer main box can be cleaned from the back end after removing the fan, filter pad, and screened plate. Spray the inside thoroughly with 70% ethanol and dry with Kimwipes. Take care to avoid contact with bare skin.

 Cite this protocol as *Cold Spring Harb Protoc*; doi:10.1101/pdb.prot108089

BOX 1. GOUCK-STYLE TWO-PORT OLFACTOMETER

Note that dimensions are given in metric units, except when components were sourced from U.S. vendors and therefore are listed and sold in imperial units by those vendors.

Materials and Construction

The olfactometer box is made of clear cast acrylic (e.g., plexiglass) sheets assembled with small screws or acrylic glue. Traps and host chambers are made of clear cast acrylic tubing connected by Delrin acetal resin fittings that are meant to fit snugly without glue or other fastenings. Care must be taken to achieve the appropriate fit: tighter fits are desired for connections that are disassembled only for cleaning (Fig. 1C, green boxes), looser fits are desired for connections that are assembled/disassembled at the start/end of each trial (Fig. 1C, pink boxes). Abrasion with sandpaper (coarse grain followed by fine grain) can be used to loosen a fitting that is too tight after original construction. The front plate for the main box (Fig. 1B), as well as fittings for traps/host chambers (Fig. 1C), can be crafted from raw materials in a machine shop. Fiberglass screen discs for traps and wire cloth discs for host chambers are cut to size and then held in place between components (e.g., between tubes and fittings, Fig. 1C). The screen cones of the traps are cut to shape, hand-sewn into cones using fine metal wire, and then carefully affixed to the inside surface of the acrylic trap rings using superglue (Fig. 1D; allow glue to off-gas for 1 wk before use). The front plate block with sliding door and support rings to receive the traps is mounted on the main box. The traps are inserted into the support rings, whereas the host chambers and T-tube are held aloft with plexiglass support stands. A 1″ filter pad inserted between the box fan and screened backplate evens the airflow it generates (Fig. 1E). A variable fan control is necessary to reduce the speed of the fan below its lowest setting, and an extension cord with foot pedal on/off switch allows the experimenter to conveniently activate the precalibrated fan at the start of a trial, even when standing at the "business" end of the olfactometer. Materials for construction are listed in Table 1 and most components are shown in Figure 1.

Components and Dimensions

Main box (Fig. 1A, five sides with separate backplate): 50-cm × 50-cm × 80-cm (H × W × L), ¼″ plexiglass. Holes for traps are round (9.2-cm-diameter), spaced 21.9 cm apart (center to center), and are centered in both directions on the front face of the box. Top access door (14-cm-diameter) has a raised rim (~2-cm-height, made separately and glued in place) for attaching a mesh sleeve using a zip tie.

Front plate block with support rings and sliding door (Fig. 1B): This piece serves to house the sliding door and provide a place to mount the traps. A 48.5-cm × 16.5-cm × ¾″ acrylic plate is mounted to the front of the main box (i.e., over the trap holes) using several small screws. The back side of the plate has a ¼″ deep, 12.7-cm-tall groove running its full length that, when mounted on the main box, forms the housing for the sliding door.

The front plate also has two holes of the exact size and spacing of the trap holes on the main box (9.2-cm-diameter, spaced 21.9 cm apart), which must be aligned during assembly. The external (i.e., "front") side of the front plate has two round grooves centered on each hole to hold rings that support the traps. Support rings (4.5″ OD, 4″ ID, 2.9-cm-length, cast acrylic tubing) are inserted into the grooves and glued in place.

Sliding door: 12.7-cm × 51.3-cm, ¼″ acrylic piece with two 9.2-cm (diameter) holes, spaced 21.9 cm apart (center to center), and centered in both dimensions.

Screen backplate: 50-cm × 50-cm, fine plastic screen sandwiched between two identical plexiglass frames (1/8″-thick, inner dimensions ~48 cm × 48 cm). Screen should be light-colored to avoid attracting mosquitoes to the back of the box.

Extension cord *with foot pedal on/off switch*

Fiberglass screening: Used for screen discs for traps, and for construction of trap cones.

Fabric covers: Hand-sewn from light-colored (for main chamber) and black (for host chambers) fabric and completed with Velcro fasteners. We have used cotton twill (for the main chamber cover) and polyester felt (for host chamber covers).

Fan: Ordinary box fan, 20″ × 20″

Filter pad: 1″-thick reusable polyester plastic filter pad. White or other light color to avoid attracting mosquitoes to the back of the box.

Traps (two or more): Trap body 4″ OD tubing, 3/16″ cast acrylic, 18.4-cm length. Screen discs cut to match OD of trap body. Delrin acetal resin fitting machined to fit snugly and connect trap to host chamber (Fig. 1C). Trap screen cones (two or more): Ring 4″ OD, 3 3/8″ ID, ~1.2-cm length cast acrylic, machined to create a flange (Fig. 1D) that slots snugly into trap body. Fiberglass screen cut in partial donut shape (Fig. 1D),

Continued

BOX 1. *Continued.*

outer circle 16.8-cm-diameter, inner circle 5.1-cm-diameter. After sewing cut ends of donut together with fine wire, the hole at the center of the cone should be ~3-cm-diameter.

Live-host chambers (optional, two or more): 6″ OD, ¼″ cast acrylic tubing, 38.1-cm length. Wire cloth discs cut to match OD of chamber. Delrin acetal resin fittings machined to fit snugly at both ends (Fig. 1C); downwind fitting holds a wire cloth disc in place and connects the host chamber to the trap, upwind fitting holds a wire cloth disc in place. Figure 1A shows a support stand for the host chambers, constructed using acrylic sheets.

Silicone sheets (optional): Used to cover and seal holes in host chambers modified for human arms in a closed-air system (Fig. 1H).

Sleeve chambers (optional, two or more): 3 ¼″ OD tubing, 5/32″ cast acrylic, 12.7-cm length (Fig. 1I).

Lid for access door: Round 14-cm-diameter × ¼″-deep acrylic disc (the plug portion) attached to a wider acrylic disc (the topper) that holds the smaller disc in place and by which it can be easily grasped (Fig. 1A).

Variable fan speed controller

Visual blind: Large opaque plastic or matboard sheet with two holes cut to fit over the Delrin fittings at the upstream end of the traps (Fig. 1A, inset).

Preparing Adult Mosquitoes for Testing

In each behavioral trial, test a group of 50–100 animals, typically mated, non-blood-fed females. Set aside these groups in holding cups 12–20 h before testing (i.e., the previous day). Sugar-starving females during this time may increase response rates.

6. Before the first use of any given holding cup, scratch the interior surface to create edges for mosquitoes to grasp with their tarsi and thus minimize activity within the cup.

   We use a razor blade to etch a diagonal 1″ grid pattern (approximate) across the entire inside surface. Rough sandpaper also works well.

TABLE 1. Components for building a two-port olfactometer

Material	Example supplier	Example product number	Subcomponent of olfactometer
Box fan	Any	Lasko 3-speed	Back
Cast acrylic sheeting, 1/8″, 1/4″, 1/2″, 3/4″	McMaster-Carr	8560K series	Main box, front plate block, frame for screened backplate, support stands
Cast acrylic tubing	McMaster-Carr	8486K481	Host chambers, traps, trap support rings
CPAP nasal mask (optional)	Any (e.g., Amazon.com)		For addition of human breath in closed-air system, may require additional tubing and fittings (Fig. 1F)
Delrin acetal resin	McMaster-Carr	8573K26	Fittings
Extension cord with foot pedal on/off switch	GE	42803	
Fabric, light and dark colors	Any		Covers for main chamber (light) and host chambers (dark)
Fiberglass mesh, commonly used for home windows	Any		Trap cones and screen discs (Fig. 1C,D)
Filter pad, reusable polyester plastic, 1″ thick	McMaster-Carr	2182K14	Box fan assembly (Fig. 1E)
Fine metal wire	Any		For sewing fiberglass screening into trap cones
Funnels (optional)	Winco plastic funnel, 6 ¼″ diameter		Snorkel to T-tube connection (Fig. 1G)
Mesh sleeve with zip tie	Bioquip	1452A	Access door at top of main chamber
Plastic mesh sheet	McMaster-Carr	9318T13	Screen backplate (Fig. 1E)
Rigid plastic sheet or matboard, opaque white	Any		Visual blind (Fig. 1A)
Silicone sheeting, 1/32″	McMaster-Carr	86465K91	Human-arm host chamber used in closed-air system (Fig. 1H)
Small screws	Any		Main box and front plate block
Stainless steel wire cloth	McMaster-Carr	85385T24	Wire cloth discs for host chambers (Fig. 1C)
Superglue	Any		Trap cones
Variable fan controller	MistWorks	VSC5-1	
Velcro fasteners	Any		Fabric covers

 Cite this protocol as *Cold Spring Harb Protoc*; doi:10.1101/pdb.prot108089

BOX 2. SOLENOID VALVE CO_2 DELIVERY SYSTEM WITH CONTROLLER

Materials and Construction

Low, intermittent flow of CO_2 into the assay is achieved by modulating flow from a tank of 10% CO_2/90% air using a pressure regulator, flowmeter, and solenoid valve. Odorless tubing and fittings (e.g., Teflon and metal, respectively) should be used wherever possible.

The solenoid valve is controlled by an Arduino microcontroller. Some simple circuitry is required to allow the 5V-output from the Arduino to drive the higher-voltage solenoid circuit (Fig. 1J). Be sure to connect components in the right orientation. In particular, the 5V-Arduino output should be connected to the "gate" pin of the MOSFET, ground should be connected to the "source" pin, and the negative lead on the solenoid should be connected to the "drain" pin. The diode (used to prevent transient voltage spikes from the solenoid) should be mounted such that it only allows current to flow from the negative to the positive solenoid lead (i.e., end with gray stripe connected to positive lead). All materials are listed in Table 2.

Control Script

The valve can be programmed to open and close at chosen intervals with Arduino software. In the example script below, the valve is open for 1 sec and closed for 4 sec. This script was used with a normally closed valve, which opens when voltage is applied. Change the script accordingly if using a normally open valve.

```
int solPin=13; void setup() {
pinMode(solPin, OUTPUT);
}
void loop() {
digitalWrite(solPin, HIGH);
delay(1000); digitalWrite(solPin, LOW); delay(4000);
}
```

7. One day (12–20 h) before planned trials, anesthetize mosquitoes by cooling them to 4°C in a cold room for ~5 min.

   The duration of this cold stress should be minimized, for example, by working efficiently with small batches of animals.

TABLE 2. Components for building a CO_2 delivery system

Material	Example supplier	Example product number	Notes
Arduino microcontroller	Arduino	Uno R3 A000066	
Breadboard	Arduino	ProtoShield	
Diode	Mouser electronics	637-1N4001	
N-channel MOSFET	Mouser electronics	512-FQP30N06L	
Power adaptors for solenoid valve and Arduino microcontroller	Mouser electronics	992-12V1A-25- POSWALL	Required voltage depends on solenoid valve used.
Resistors	Mouser electronics	660- CF1/4CT52R151J, 660-CF1/4C103J	
Solenoid valve	Cole-Parmer	EW-01540-01	
Flowmeter and gas-tank regulator	Airgas	Y11FM244B590-AG with flowmeter Y21B652-AG	Can be ordered together as a single unit. Regulator should be chosen to fit gas tank.
Low-odor (e.g., Teflon) tubing	McMaster-Carr	52355K54	
Fittings for solenoid valve (2)	McMaster-Carr	51025K774	Example part number given for 1/4″-28 male UNF to push-to-connect.
Fitting for flowmeter	McMaster-Carr	5779K104	Example part number given for 1/8″ male NPT to push-to- connect.
Y-splitter	McMaster-Carr	5779K42	
CO_2 monitoring system	LI-COR	LI-830 CO_2 gas analyzer	May require additional tubing and fittings.

BOX 3. T-TUBE FOR SPLITTING AIR SUPPLY

Materials and Construction

A professional glass blower can construct this component, or it may be constructed from cast acrylic tubing using solvent-welding. Two funnels (6 ¼″/15.875-cm-diameter) are used to direct air from the T-tube into the host chambers. Use a utility razor to remove the narrow end of each funnel, leaving holes just large enough to slide over the T-tube outlets (Fig. 1G).

Dimensions

Total L × W = 41.9 cm × 32 cm. The stem where air enters (16.5-cm-long) and the section perpendicular to it (32-cm-across) are made from tubing with a diameter of 9.5 cm. Air outlets (two) are spaced 20.3 cm apart (center to center), 15.9-cm-long, each with 6.35-cm diameter (Fig. 1G).

8. Transfer anesthetized animals to a precooled white plastic tray and use a soft feather or paint brush to gently separate sexes (if cohoused) and count females into groups of 50–100. Combine equal numbers of females from different genotypes/strains if testing mixed groups.
9. Transfer each group of counted females into a plastic holding cup and seal with fine mesh fabric (held by rubber bands or the original plastic tops with center carved out). Ensure that the fabric is taut.
10. Promptly return cups of mosquitoes to 75% relative humidity and 26°C to allow them to recover from anesthesia. Place two cotton balls presoaked in H_2O or 10% (w/v) sucrose solution on the mesh top of each cup to sustain animals overnight.

Performing Behavioral Assays

Be sure to handle equipment only with freshly gloved hands to avoid contamination with human skin residue.

11. On the day of testing, ensure test mosquitoes are not exposed to host odors while they await their trial lest they become habituated. The holding cups may be kept in a separate, odor-free compartment until use or sealed with plastic lids upon first entry of the experimenter into the testing room.
12. Fully assemble the olfactometer and all (precalibrated) equipment, including traps, visual blind (at upwind end of traps), and host chambers. If using live, non-human animals as odor source, line both host chambers with disposable underpads to absorb urine and feces during trials.
13. Check that the olfactometer's sliding door is closed, and release one holding cup of animals into the main box by lowering it through the sleeved access door and removing the rubber bands and mesh with one hand. Gently invert and tap the cup on the floor of the chamber to release all females and then remove the cup, mesh, and bands from the olfactometer.

    With practice this can be accomplished quickly, before mosquitoes are able to land and probe on the experimenter's arm.
14. Use plug to seal the access door.
15. If using a fabric cover for the olfactometer, secure it around the main box, leaving a small gap for operation of the sliding door.
16. Acclimate females to the olfactometer for a defined period of time that is consistent across all trials of a given experiment.

    Laboratory strains of Ae. aegypti aegypti *(e.g., Orlando) can be acclimated for as few as 5 min, whereas recently field-derived colonies and/or* Ae. aegypti formosus *colonies may behave more consistently with 10–15 min acclimation.*
17. At the conclusion of the acclimation period, load hosts into host chambers. If using a closed-air system, attach air source to host chambers (e.g., by sliding the snorkel-fed T-tube and funnels into place; Fig. 1F).

 Cite this protocol as *Cold Spring Harb Protoc*; doi:10.1101/pdb.prot108089

18. Turn on the box fan, open the sliding door, and start the trial timer.

    The precalibrated fan speed controller can be connected to a foot pedal on/off switch for convenience. Human hosts should sit quietly for the duration of the trial, and may provide breath (e.g., one exhale every 30 sec), if desired.

19. After a specified trial duration (e.g., 6 min), close the sliding door and turn off the fan to conclude the trial.

20. Remove hosts from host chambers and return non-human animals (if used) to their home cages.

21. Remove and count females in each trap.

    Traps may be removed, capped with a plastic lid across the open end, and transferred to a cold room for counting. When testing mixed groups, cold-anesthetized animals may be transferred to labeled microcentrifuge tubes, frozen at −20°C, and sorted/counted using an appropriate microscope at a later time.

22. Remove and count any animals remaining in the main chamber (i.e., nonresponders and dead animals).

    For precise counts, nonresponders may be live-captured from the main chamber using a battery-powered "bug vacuum" or similar device, transferred to the cold room, and counted. Alternatively, the hose attachment of a canister vacuum cleaner may be used to remove nonresponders and the number estimated ex post facto by subtracting the number of responders from the total number prepared (minus any animals that died in holding cups overnight). With either method, care must be taken to spot and remove difficult-to-see animals perched on screws or resting in corners of the box. Be sure to count all dead animals, which may have remained in the holding cup or may have fallen onto the floor of the olfactometer when live mosquitoes were released.

23. Allow cooled traps to return to 26°C before using them in additional trials.

    If two sets of traps are available, they can be used in alternating trials such that one set warms while the other is being used.

24. Clean host chambers as needed before use in additional trials.

    We provide each host with their own, dedicated host chamber to use throughout a day of behavior trials, and only wash equipment between trials when an animal defecates or urinates in their chamber.

RELATED INFORMATION

The use of a two-port olfactometer to study mosquito attractants is described in Gouck and Schreck (1965) and Gouck (1972). Potential odor sources include live hosts (e.g., DeGennaro et al. 2013; McBride et al. 2014; Rose et al. 2020; Zhao et al. 2022), host-scented nylon sleeves or host hair samples (e.g., DeGennaro et al. 2013; McBride et al. 2014; McMeniman et al. 2014; Zhao et al. 2022), and single odorants or odorant blends (e.g., McMeniman et al. 2014; Wondwosen et al. 2016).

REFERENCES

DeGennaro M, McBride CS, Seeholzer L, Nakagawa T, Dennis EJ, Goldman C, Jasinskiene N, James AA, Vosshall LB. 2013. *orco* mutant mosquitoes lose strong preference for humans and are not repelled by volatile DEET. *Nature* **498:** 487–491. doi:10.1038/nature12206

Gouck HK. 1972. Host preferences of various strains of *Aedes aegypti* and *Aedes simpsoni* as determined by an olfactometer. *Bull World Health Organ* **47:** 680–683.

Gouck HK, Schreck CE. 1965. An olfactometer for use in the study of mosquito attractants. *J Econ Entomol* **58:** 589–590. doi:10.1093/jee/58.3.589

McBride CS, Baier F, Omondi AB, Spitzer SA, Lutomiah J, Sang R, Ignell R, Vosshall LB. 2014. Evolution of mosquito preference for humans linked to an odorant receptor. *Nature* **515:** 222–227. doi:10.1038/nature13964

McMeniman CJ, Corfas RA, Matthew BJ, Ritchie SA, Vosshall LB. 2014. Multimodal integration of carbon dioxide and other sensory cues drives mosquito attraction to humans. *Cell* **156:** 1060–1071. doi:10.1016/j.cell.2013.12.044

Metz HC, Zung JL, McBride CS. 2023. Quantifying *Aedes aegypti* host odor preference using a two-port olfactometer. *Cold Spring Harb Protoc* doi:10.1101/pdb.top107661

Rose NH, Sylla M, Badolo A, Lutomiah J, Ayala D, Aribodor OB, Ibe N, Akorli J, Otoo S, Mutebi JP, et al. 2020. Climate and urbanization drive mosquito preference for humans. *Curr Biol* **30:** 3570–3579. doi:10.1016/j.cub.2020.06.092

Wondwosen B, Birgersson G, Seyoum E, Tekie H, Torto B, Fillinger U, Hill SR, Ignell R. 2016. Rice volatiles lure gravid malaria mosquitoes, *Anopheles arabiensis*. *Sci Rep* **6:** 1–8. doi:10.1038/srep37930

Zhao Z, Zung JL, Hinze A, Kriete AL, Iqbal A, Younger MA, Matthews BJ, Merhof D, Thiberge S, Ignell R, et al. 2022. Mosquito brains encode unique features of human odour to drive host seeking. *Nature* **605:** 706–712. doi:10.1038/s41586-022-04675-4

CHAPTER 14

Quantifying Mosquito Host Preference

Diego Giraldo[1] and Conor J. McMeniman[1,2,3]

[1]*W. Harry Feinstone Department of Molecular Microbiology and Immunology, Johns Hopkins Malaria Research Institute, Johns Hopkins Bloomberg School of Public Health, Johns Hopkins University, Baltimore, Maryland 21205, USA;* [2]*The Solomon H. Snyder Department of Neuroscience, Johns Hopkins University School of Medicine, Baltimore, Maryland 21205, USA*

The most dangerous mosquito species for human health are those that blood feed preferentially and frequently on humans (anthropophilic mosquitoes). These include prolific disease vectors such as the African malaria mosquito *Anopheles gambiae* and yellow fever mosquito *Aedes aegypti.* The chemosensory basis for anthropophilic behavior exhibited by these disease vectors, as well as the factors that drive interindividual differences in human attractiveness to mosquitoes, remain largely uncharacterized. Here, we concisely review established methods to quantify mosquito interspecific and intraspecific host preference in the laboratory, as well as semi-field and field environments. Experimental variables for investigator consideration during assays of mosquito host preference across these settings are highlighted.

INTRODUCTION

Several mosquito species exhibit an innate preference to blood feed on specific vertebrate hosts at multiple times during their life span (Takken and Verhulst 2013), with important implications for epidemiological risk (Macdonald 1957; Garrett-Jones 1964). For example, the most dangerous mosquito species for human health are those that have evolved a strong innate drive to preferentially and frequently blood feed on humans (Besansky et al. 2004; Scott and Takken 2012; Takken and Verhulst 2013). Such mosquito species include major vectors of human disease including the African malaria mosquito *Anopheles gambiae,* which obtains the vast majority of its blood meals from humans (Garrett-Jones et al. 1980; Costantini et al. 1999; Dekker et al. 2001; Pates et al. 2001; Mwangangi et al. 2003), and the yellow fever mosquito *Aedes aegypti,* which feeds almost exclusively on human blood often at multiple times during a single gonotrophic cycle (Scott et al. 1993, 2000; Ponlawat and Harrington 2005; Harrington et al. 2014).

Mosquito host preference is classically defined as the tendency of a mosquito species to blood feed on a specific vertebrate host species, rather than other hosts that are equally available in the environment (Takken and Verhulst 2013). Host preference and feeding frequency combine to influence the daily probability that a mosquito species feeds on a particular host species (Macdonald 1952, 1957; Garrett-Jones 1964), and therefore these feeding-related factors are key behavioral components that modulate vectorial capacity of mosquito populations for pathogen transmission (Brady et al. 2016). Anthropophilic species are those with a tendency to blood feed on humans (e.g., *An. gambiae,*

[3]Correspondence: cmcmeni1@jhu.edu

Cite this introduction as *Cold Spring Harb Protoc;* doi:10.1101/pdb.top107663

Anopheles coluzzii, *Anopheles funestus*, and *Ae. aegypti*), whereas zoophilic species are defined as those with a preference to blood feed on vertebrates excluding humans (e.g., *Anopheles quadriannulatus*) (Takken and Verhulst 2013). Such host preferences have been determined to have a heritable genetic basis in many of these mosquito species (Gouck 1972; Trpis and Hausermann 1978; Pates et al. 2014), although the evolutionary (Rose et al. 2020) and associated molecular and cellular mechanisms dictating such preferences are only starting to be uncovered (Main et al. 2016; Matthews et al. 2018).

Within a host species, mosquitoes may also exhibit preferences for certain individuals relative to others. For instance, field studies have documented substantial heterogeneity in mosquito biting among humans occupying the same household or community (Harrington et al. 2014; Guelbéogo et al. 2018). Likewise, such behavioral preferences of mosquitoes for certain humans over others has been documented in a plethora of studies (Khan et al. 1965; Lindsay et al. 1993; Brady et al. 1997; Bernier et al. 2002; Mukabana et al. 2002; Qiu et al. 2006; De Obaldia et al. 2022). Of note, it has been observed that *An. gambiae* may exhibit olfactory preferences for humans carrying infectious stages of the human malaria parasite *Plasmodium falciparum* (Lacroix et al. 2005; Busula et al. 2017; Robinson et al. 2018), as well as for pregnant women who are at risk of severe clinical outcomes related to malaria infection (Lindsay et al. 2000; Ansell et al. 2002; Himeidan et al. 2004). Such behavioral preferences of malaria vectors for certain individuals over others in disease endemic communities may have strong implications for personal risk, intensifying malaria transmission and exacerbating the basic reproduction number of malaria (Smith et al. 2007; Guelbéogo et al. 2018). Currently, the chemosensory basis for interindividual differences in human attractiveness to mosquitoes during health and disease remains poorly understood, although promising advances have recently been made in this area of research (De Obaldia et al. 2022; Zhang et al. 2022).

To detect a vertebrate host, mosquitoes use host-seeking behavior—a collective series of behaviors that involve orientation of a mosquito toward a vertebrate host to blood feed. Host-seeking behavior is a multisensory process that involves detection and integration of visual, olfactory, thermosensory, and gustatory cues by the arthropod peripheral nervous system and brain (McMeniman et al. 2014; van Breugel et al. 2015; Alonso San Alberto et al. 2022). Although the sensory cues that dictate mosquito host preference are still not completely understood, responses of the mosquito olfactory system toward species-specific chemical signatures in host body odor (McBride et al. 2014) are hypothesized to be the primary driver of these behavioral responses to one host species over another.

It is important to recognize that host preferences are indeed a preference, and that shifts in mosquito feeding patterns toward other animal species can readily be observed in some instances, dependent on local availability of the preferred host species (Takken and Verhulst 2013; Yan et al. 2021). Likewise, plasticity in host preference may be observed within particular mosquito species across different geographical locations (e.g., *Anopheles arabiensis* and *Anopheles darlingi*) (Main et al. 2016; Nagaki et al. 2021).

Methods to Quantify Mosquito Host Preference

Across inter- and intraspecific assays of host preference, the attraction of mosquitoes to a point source of host odor is evaluated, often by scoring the number of released mosquitoes that fly into competing entry traps situated inside of an assay flight chamber (Gouck 1972; McBride et al. 2014; De Obaldia et al. 2022), or by employing video tracking to document spatial preferences toward competing scent plumes within the assay (Hinze et al. 2021). Laboratory assays such as uniport olfactometers (Raji et al. 2019), Y-tube olfactometers (Geier and Boeckh 1999), dual port olfactometers (Gouck 1972; De Obaldia et al. 2022), short-range probing measurements (Khan et al. 1965), and wind tunnels (Hinze et al. 2021) have been fundamental toward exploring the chemosensory basis of mosquito host preference, as well as documenting interindividual differences in human attractiveness to mosquitoes.

Mosquito behavioral studies in semi-field environments (large, screened greenhouse enclosures positioned outside) and field environments also have the potential to yield insight into mosquito olfactory preferences in a context in which mosquitoes are exposed to natural variations in environ-

Cite this introduction as *Cold Spring Harb Protoc*; doi:10.1101/pdb.top107663

mental conditions such as temperature, humidity, wind, and associated plume dynamics of whole body odor dissipating from different vertebrate host species. One method to test the behavioral preference of a mosquito species in field settings involves examining blood meals taken by mosquitoes caught in households or surrounding animal pens and matching them back to the inhabitants of the house and/or animal species found in close proximity (Garrett-Jones et al. 1980; Takken and Verhulst 2013; Harrington et al. 2014; Guelbéogo et al. 2018). Alternatively, similar assays may also be performed in a contained fashion under semi-field environments (Harrington et al. 2014).

Another commonly used method to quantify mosquito host preference in field contexts is human landing catches (HLCs), in which participants expose parts of their bodies and collect mosquitoes as they land on them (Lima et al. 2014). However, HLCs may expose participants to mosquito bites and put them at risk of infection, and this technique can be susceptible to collector bias with regard to their efficiency of mosquito capture (Lima et al. 2014; Kenea et al. 2017).

Alternative methods used for assessing mosquito olfactory preference in field-based environments include multichoice tests in which volunteers occupy bed nets or tents, and mosquitoes follow human or animal scent emissions to be collected on walls and ceilings of these housing structures or in associated entry or exit traps (Knols et al. 1995; Lindsay et al. 2000; Torr et al. 2008; Finda et al. 2019). Many of these approaches may also risk investigator exposure to errant biting during manual mosquito collection for quantification (Staunton et al. 2022).

When using methods in which investigators collect or trap field-derived mosquitoes to quantify mosquito host preference, appropriate standard operating procedures for personal protective measures are highly recommended. This minimizes risk associated with transmission of mosquito-borne diseases during these procedures. In addition, institutional approval must be sought before use of live human or animal baits in such assays to adhere to guidelines for human subjects research and animal care and use.

Promising recent advances that mitigate direct exposure of humans and animals to mosquito bites during field-based assays of olfactory preference include the use of decoy traps and electrocution traps (Hawkes et al. 2017; Meza et al. 2019). In these assays, whole body odor from the target vertebrate host species is enclosed within separate tents or screened chambers and is used to lure mosquitoes to their death via adhesion to a sticky warmed surface (e.g., a decoy trap filled with warm water and coated on its exterior with adhesive glue panels) or electrocution grids for quantification, respectively.

Considerations for Controlled Assays of Mosquito Host Preference

Source of Mosquitoes for Assays

A key consideration for mosquito olfactory preference assays is the source of mosquitoes that will be used. For laboratory and semi-field assays, standardized rearing protocols (e.g., Chapter 1, Protocol 1: Batch Rearing *Aedes aegypti* [Wohl and McMeniman 2023]) can be applied to mass rear extant laboratory strains and provide large numbers of pathogen-free mosquitoes of known age and physiological status for behavioral experimentation. In the field, where these parameters cannot be controlled, it is important to recognize that target mosquito species may also vary locally in population density and as such performing sampling at multiple geographic locations in a region is highly recommended.

Environmental Conditions

In the laboratory, environmental parameters such as temperature, humidity, and lighting can be controlled according to user specifications. However, under semi-field and field conditions, these factors can vary dynamically according to local weather conditions. Across all settings, it is highly recommended to record environmental parameters such as temperature, humidity, rainfall, barometric pressure, and lighting levels using portable loggers, meters, or weather stations. These can be placed immediately proximal to each experimental setup to retrospectively correlate recorded metrics for each environmental condition with observed mosquito behaviors.

Mosquito Internal Physiological State

Mosquito host-seeking behavior is strongly modulated by mosquito internal physiological state, which can be influenced by factors such as age (Alto et al. 2003), sugar and water deprivation (Hagan et al. 2018; Dittmer et al. 2019), stage of the gonotrophic cycle (Klowden 1990), and time of day (Rund et al. 2013). When possible, efforts should be made to standardize these factors according to experimental parameters. For instance, we routinely perform semi-field olfactory preference experiments with mated *An. gambiae* Kisumu strain females (3- to 5-d-old) that are nulliparous (have never laid eggs) and maintained with a 6% glucose solution as adults. Prior to the start of behavioral experimentation, mosquitoes are starved of glucose for 12 h and provided with access to a water source (on a soaked cotton ball) to prevent dehydration. Behavioral responses of mosquitoes can be intrinsically variable, even when controlling for internal physiological status, and thus experimental repetition is highly recommended.

Assay Configuration

In the laboratory, olfactory host preference assays with mosquitoes are typically performed in a two-choice configuration, whereby single or small groups of female mosquitoes choose between competing odor stimuli derived from two different hosts, such as a human versus another vertebrate animal (e.g., chicken or guinea pig) (Gouck 1972; McBride et al. 2014), or two competing humans (Busula et al. 2017; De Obaldia et al. 2022). Semi-field assays of mosquito host preference have to date been limited to at most three-choice formats using linked tents with sleeping humans and a central mosquito choice chamber (Mukabana et al. 2002; Lacroix et al. 2005) or have involved contained assays of host preference with small cohorts of humans using blood meal analysis to document feeding preferences (Harrington et al. 2014). Historically, multichoice assay configurations for mosquito olfactory preference within field contexts have occurred by comparing mosquito capture rates of individuals that are geographically segregated over large distances using tent or hut-based collections at several sites within villages in randomized block designs (Lindsay et al. 1993; Knols et al. 1995; Brady et al. 1997). These assay formats provide great flexibility to study mosquito olfactory preference according to experimental parameters and available resources and can be optimized as described below to standardize or document potential sources of confounding variation.

Airspeed and Plume Dynamics

Important considerations for mosquito olfactory preference tests typically performed within laboratory olfactometers or wind tunnels are airspeed and plume dynamics, which will be used to separate the competing odor stimuli inside each laboratory assay (Geier et al. 1999). Careful control of these parameters is crucial to facilitate successful assays of mosquito olfactory preference. Gentle airspeeds that facilitate upwind mosquito flight, while ensuring effective separation of competing odor stimuli in these assays, typically fall within the range of $\sim 0.5 \pm 0.3$ m/sec (Hinze et al. 2021). In semi-field assays that similarly involve directional application of olfactory cues via air flowthrough ducting paths, airspeed can be dynamically controlled by in-line ducting fans and pulse width modulators (see Protocol 1: A Behavioral Assay to Quantify Odor-Guided Thermotaxis with *Anopheles gambiae* under Semi-Field Conditions [Giraldo and McMeniman, 2023a]). In semi-field and field-based assays wind directionality and speed can also be recorded using weather stations. Across these settings, plume dynamics may also be readily visualized using sources of smoke to ensure that olfactory stimuli in carrier airflows are visibly separated (Dekker and Cardé 2011), especially near the release point of each host stimulus.

Carrier Air Source

Within a laboratory context, buffering assay conditions from human observers and animal bystanders is critical to ensure that behavioral preferences are due to the actual host species being tested, and that

 Cite this introduction as *Cold Spring Harb Protoc*; doi:10.1101/pdb.top107663

they are not clouded by background contamination from other competing sensory stimuli emitted by observers in the assay room. Charcoal-filtered room air that is scrubbed of background volatile organic compounds and then humidified, clean synthetic air from pressurized tanks that is then humidified, or air from a custom air handling system with an external air intake, in-line charcoal filter, and temperature and humidity control are all commonly used as carrier air supplies for laboratory assays. In semi-field and field-based assays of mosquito olfactory preference that are subject to field environmental conditions and open air, efforts should be made to ensure that other human or animal hosts not involved in these assays are not immediately proximal to the test subjects or sources of odor to avoid interference with measures of olfactory preference.

Equipment Cleaning and Mitigation of Odorant Contamination

Equipment used for laboratory, semi-field, or field-based assays of olfactory preference (e.g., surfaces of olfactometers, wind tunnels, and tents) should be decontaminated when possible between blocks of trials to remove host-related odor by washing and air drying. Subsequent flushing of this equipment with clean air before the next assays will facilitate evaporation of contaminating volatile organic compounds. Within semi-field and field-based contexts this may also be achieved by placing washed equipment outside in direct sunlight for 24–48 h to off-gas before their next use. Use of a diluted and unscented detergent solution and multiple rinses using warm water is recommended during equipment washing to facilitate removal of contaminating odorants. When handling clean equipment, as well as animal hosts or test odor sources (e.g., host-scented nylon sleeves) during the setup and execution of assays, use of gloves is also required to prevent exogenous contamination from the hands of the experimenter that may confound proper interpretation of assay results.

Interindividual Variability in Host Attractiveness

When testing mosquito olfactory preference for different humans and/or other animals across laboratory, semi-field, and field-based settings, it is important to recognize that basal attractiveness may significantly vary between individuals of the target species being compared (Qiu et al. 2006; Raji et al. 2019; De Obaldia et al. 2022). Given this heterogeneity, use of replicate individuals from each host species in interspecific and intraspecific assays of mosquito host preference is highly recommended. When desirable from an experimental standpoint, control of exogeneous factors such as application of cosmetics, fragrances, and insect repellents and control of dietary items and alcohol consumption that may influence host odor and associated behavioral phenotypes observed are highly recommended (Lefèvre et al. 2010; Rodriguez et al. 2015; Paskewitz et al. 2018).

CONCLUSION

Host preference is a critical feeding-related parameter of mosquito vectorial capacity, which couples with feeding frequency to dictate the daily probability that a vector species feeds on a particular vertebrate host. Anthropophilic disease vectors such as *An. gambiae* and *Ae. aegypti* pose a major threat to public health. An array of methods has been established for quantifying inter- and intraspecific mosquito host preference ranging from blood meal analysis to a battery of different olfactory choice assays under laboratory, semi-field, and field conditions. The majority of host preference assays developed to date are inherently by design end point assays involving capture or death of mosquitoes for quantification of host preference. Our associated protocols describe establishment of a novel multichoice assay under semi-field conditions to study *An. gambiae* olfactory preferences using infrared video tracking (see Protocol 1: A Behavioral Assay to Quantify Odor-Guided Thermotaxis with *Anopheles gambiae* under Semi-Field Conditions [Giraldo and McMeniman 2023a]) as well as procedures for executing preference assays using this system (see Protocol 2: Quantification of *Anopheles gambiae* Olfactory Preferences under Semi-Field Conditions [Giraldo and McMeniman 2023b]).

ACKNOWLEDGMENTS

D.G. was supported by postdoctoral fellowships from the Human Frontier Science Program (LT000310/2019-L) and Johns Hopkins Malaria Research Institute (JHMRI). C.J.M. acknowledges generous support from JHMRI and Bloomberg Philanthropies.

REFERENCES

Alonso San Alberto D, Rusch C, Zhan Y, Straw AD, Montell C, Riffell JA. 2022. The olfactory gating of visual preferences to human skin and visible spectra in mosquitoes. *Nat Commun* **13:** 555. doi:10.1038/s41467-022-28195-x

Alto BW, Lounibos LP, Juliano SA. 2003. Age-dependent blood-feeding of *Aedes aegypti* and *Aedes albopictus* on artificial and living hosts. *J Am Mosq Control Assoc* **19:** 347–352.

Ansell J, Hamilton KA, Pinder M, Walraven GE, Lindsay SW. 2002. Short-range attractiveness of pregnant women to *Anopheles gambiae* mosquitoes. *Trans R Soc Trop Med Hyg* **96:** 113–116. doi:10.1016/S0035-9203(02)90271-3

Bernier UR, Kline DL, Schreck CE, Yost RA, Barnard DR. 2002. Chemical analysis of human skin emanations: comparison of volatiles from humans that differ in attraction of *Aedes aegypti* (Diptera: Culicidae). *J Am Mosq Control Assoc* **18:** 186–195.

Besansky NJ, Hill CA, Costantini C. 2004. No accounting for taste: host preference in malaria vectors. *Trends Parasitol* **20:** 249–251. doi:10.1016/j.pt.2004.03.007

Brady J, Costantini C, Sagnon N, Gibson G, Coluzzi M. 1997. The role of body odours in the relative attractiveness of different men to malarial vectors in Burkina Faso. *Ann Trop Med Parasitol* **91:** S121–S122. doi:10.1080/00034983.1997.11813252

Brady OJ, Godfray HCJ, Tatem AJ, Gething PW, Cohen JM, McKenzie FE, Perkins TA, Reiner RC, Tusting LS, Sinka ME, et al. 2016. Vectorial capacity and vector control: reconsidering sensitivity to parameters for malaria elimination. *Trans R Soc Trop Med Hyg* **110:** 107–117. doi:10.1093/trstmh/trv113

Busula AO, Bousema T, Mweresa CK, Masiga D, Logan JG, Sauerwein RW, Verhulst NO, Takken W, de Boer JG. 2017. Gametocytemia and attractiveness of *Plasmodium falciparum*-infected Kenyan children to *Anopheles gambiae* mosquitoes. *J Infect Dis* **216:** 291–295. doi:10.1093/infdis/jix214

Costantini C, Sagnon N, della Torre A, Coluzzi M. 1999. Mosquito behavioural aspects of vector–human interactions in the *Anopheles gambiae complex. Parassitologia* **41:** 209–217.

Dekker T, Cardé RT. 2011. Moment-to-moment flight manoeuvres of the female yellow fever mosquito (*Aedes aegypti* L.) in response to plumes of carbon dioxide and human skin odour. *J Exp Biol* **214:** 3480–3494. doi:10.1242/jeb.055186

Dekker T, Takken W, Braks MA. 2001. Innate preference for host-odor blends modulates degree of anthropophagy of *Anopheles gambiae sensu lato* (Diptera: Culicidae). *J Med Entomol* **38:** 868–871. doi:10.1603/0022-2585-38.6.868

De Obaldia ME, Morita T, Dedmon LC, Boehmler DJ, Jiang CS, Zeledon EV, Cross JR, Vosshall LB. 2022. Differential mosquito attraction to humans is associated with skin-derived carboxylic acid levels. *Cell* **185:** 4099–4116.e4013. doi:10.1016/j.cell.2022.09.034

Dittmer J, Alafndi A, Gabrieli P. 2019. Fat body–specific vitellogenin expression regulates host-seeking behaviour in the mosquito *Aedes albopictus*. *PLoS Biol* **17:** e3000238. doi:10.1371/journal.pbio.3000238

Finda MF, Moshi IR, Monroe A, Limwagu AJ, Nyoni AP, Swai JK, Ngowo HS, Minja EG, Toe LP, Kaindoa EW, et al. 2019. Linking human behaviours and malaria vector biting risk in south-eastern Tanzania. *PLoS ONE* **14:** e0217414. doi:10.1371/journal.pone.0217414

Garrett-Jones C. 1964. Prognosis for interruption of malaria transmission through assessment of the mosquito's vectorial capacity. *Nature* **204:** 1173–1175. doi:10.1038/2041173a0

Garrett-Jones C, Boreham PFL, Pant CP. 1980. Feeding habits of anophelines (Diptera: Culicidae) in 1971–78, with reference to the human blood index: a review. *Bull Entomol Res* **70:** 165–185. doi:10.1017/S0007485300007422

Geier M, Boeckh J. 1999. A new Y-tube olfactometer for mosquitoes to measure the attractiveness of host odours. *Entomol Exp Appl* **92:** 9–19. doi:10.1046/j.1570-7458.1999.00519.x

Geier M, Bosch OJ, Boeckh J. 1999. Influence of odour plume structure on upwind flight of mosquitoes towards hosts. *J Exp Biol* **202:** 1639–1648. doi:10.1242/jeb.202.12.1639

Giraldo D, McMeniman CJ. 2023a. A behavioral assay to quantify odor-guided thermotaxis with *Anopheles gambiae* under semi-field conditions. *Cold Spring Harb Protoc* doi:10.1101/pdb.prot108303

Giraldo D, McMeniman CJ. 2023b. Quantification of *Anopheles gambiae* olfactory preferences under semi-field conditions. *Cold Spring Harb Protoc* doi:10.1101/pdb.prot108304

Gouck HK. 1972. Host preferences of various strains of *Aedes aegypti* and *A. simpsoni* as determined by an olfactometer. *Bull World Health Organ* **47:** 680–683.

Guelbéogo WM, Gonçalves BP, Grignard L, Bradley J, Serme SS, Hellewell J, Lanke K, Zongo S, Sepúlveda N, Soulama I, et al. 2018. Variation in natural exposure to *Anopheles* mosquitoes and its effects on malaria transmission. *eLife* **7:** e32625. doi:10.7554/eLife.32625

Hagan RW, Didion EM, Rosselot AE, Holmes CJ, Siler SC, Rosendale AJ, Hendershot JM, Elliot KSB, Jennings EC, Nine GA, et al. 2018. Dehydration prompts increased activity and blood feeding by mosquitoes. *Sci Rep* **8:** 6804. doi:10.1038/s41598-018-24893-z

Harrington LC, Fleisher A, Ruiz-Moreno D, Vermeylen F, Wa CV, Poulson RL, Edman JD, Clark JM, Jones JW, Kitthawee S, et al. 2014. Heterogeneous feeding patterns of the dengue vector, *Aedes aegypti*, on individual human hosts in rural Thailand. *PLoS Negl Trop Dis* **8:** e3048. doi:10.1371/journal.pntd.0003048

Hawkes FM, Dabiré RK, Sawadogo SP, Torr SJ, Gibson G. 2017. Exploiting *Anopheles* responses to thermal, odour and visual stimuli to improve surveillance and control of malaria. *Sci Rep* **7:** 17283. doi:10.1038/s41598-017-17632-3

Himeidan YE, Elbashir MI, Adam I. 2004. Attractiveness of pregnant women to the malaria vector, *Anopheles arabiensis*, in Sudan. *Ann Trop Med Parasitol* **98:** 631–633. doi:10.1179/000349804225021307

Hinze A, Lantz J, Hill SR, Ignell R. 2021. Mosquito host seeking in 3D using a versatile climate-controlled wind tunnel system. *Front Behav Neurosci* **15:** 643693. doi:10.3389/fnbeh.2021.643693

Kenea O, Balkew M, Tekie H, Gebre-Michael T, Deressa W, Loha E, Lindtjørn B, Overgaard HJ. 2017. Comparison of two adult mosquito sampling methods with human landing catches in south-central Ethiopia. *Malar J* **16:** 30. doi:10.1186/s12936-016-1668-9

Khan AA, Maibach HI, Strauss WG, Fenley WR. 1965. Screening humans for degrees of attractiveness of mosquitoes. *J Econ Entomol* **58:** 694–697. doi:10.1093/jee/58.4.694

Klowden MJ. 1990. The endogenous regulation of mosquito reproductive behavior. *Experientia* **46:** 660–670. doi:10.1007/BF01939928

Knols BG, de Jong R, Takken W. 1995. Differential attractiveness of isolated humans to mosquitoes in Tanzania. *Trans R Soc Trop Med Hyg* **89:** 604–606. doi:10.1016/0035-9203(95)90406-9

Lacroix R, Mukabana WR, Gouagna LC, Koella JC. 2005. Malaria infection increases attractiveness of humans to mosquitoes. *PLoS Biol* **3:** e298. doi:10.1371/journal.pbio.0030298

Lefèvre T, Gouagna LC, Dabiré KR, Elguero E, Fontenille D, Renaud F, Costantini C, Thomas F. 2010. Beer consumption increases human attractiveness to malaria mosquitoes. *PLoS ONE* **5:** e9546. doi:10.1371/journal.pone.0009546

Lima JB, Rosa-Freitas MG, Rodovalho CM, Santos F, Lourenço-de-Oliveira R. 2014. Is there an efficient trap or collection method for sampling *Anopheles darlingi* and other malaria vectors that can describe the essential parameters affecting transmission dynamics as effectively as human landing catches?: a review. *Mem Inst Oswaldo Cruz* **109:** 685–705. doi:10.1590/0074-0276140134

Lindsay SW, Adiamah JH, Miller JE, Pleass RJ, Armstrong JR. 1993. Variation in attractiveness of human subjects to malaria mosquitoes (Diptera: Culicidae) in The Gambia. *J Med Entomol* **30:** 368–373. doi:10.1093/jmedent/30.2.368

Lindsay S, Ansell J, Selman C, Cox V, Hamilton K, Walraven G. 2000. Effect of pregnancy on exposure to malaria mosquitoes. *Lancet* **355:** 1972. doi:10.1016/S0140-6736(00)02334-5

Macdonald G. 1952. The analysis of equilibrium in malaria. *Trop Dis Bull* **49:** 813–829.

Macdonald G. 1957. *The epidemiology and control of malaria.* Oxford University Press, London.

Main BJ, Lee Y, Ferguson HM, Kreppel KS, Kihonda A, Govella NJ, Collier TC, Cornel AJ, Eskin E, Kang EY, et al. 2016. The genetic basis of host preference and resting behavior in the major African malaria vector, *Anopheles arabiensis*. *PLoS Genet* **12:** e1006303. doi:10.1371/journal.pgen.1006303

Matthews BJ, Dudchenko O, Kingan SB, Koren S, Antoshechkin I, Crawford JE, Glassford WJ, Herre M, Redmond SN, Rose NH, et al. 2018. Improved reference genome of *Aedes aegypti* informs arbovirus vector control. *Nature* **563:** 501–507. doi:10.1038/s41586-018-0692-z

McBride CS, Baier F, Omondi AB, Spitzer SA, Lutomiah J, Sang R, Ignell R, Vosshall LB. 2014. Evolution of mosquito preference for humans linked to an odorant receptor. *Nature* **515:** 222–227. doi:10.1038/nature13964

McMeniman CJ, Corfas RA, Matthews BJ, Ritchie SA, Vosshall LB. 2014. Multimodal integration of carbon dioxide and other sensory cues drives mosquito attraction to humans. *Cell* **156:** 1060–1071. doi:10.1016/j.cell.2013.12.044

Meza FC, Kreppel KS, Maliti DF, Mlwale AT, Mirzai N, Killeen GF, Ferguson HM, Govella NJ. 2019. Mosquito electrocuting traps for directly measuring biting rates and host-preferences of *Anopheles arabiensis* and *Anopheles funestus* outdoors. *Malar J* **18:** 83. doi:10.1186/s12936-019-2726-x

Mukabana WR, Takken W, Coe R, Knols BG. 2002. Host-specific cues cause differential attractiveness of Kenyan men to the African malaria vector *Anopheles gambiae*. *Malar J* **1:** 17. doi:10.1186/1475-2875-1-17

Mwangangi JM, Mbogo CM, Nzovu JG, Githure JI, Yan G, Beier JC. 2003. Blood-meal analysis for anopheline mosquitoes sampled along the Kenyan coast. *J Am Mosq Control Assoc* **19:** 371–375.

Nagaki SS, Chaves LSM, López RVM, Bergo ES, Laporta GZ, Conn JE, Sallum MAM. 2021. Host feeding patterns of *Nyssorhynchus darlingi* (Diptera: Culicidae) in the Brazilian Amazon. *Acta Trop* **213:** 105751. doi:10.1016/j.actatropica.2020.105751

Paskewitz S, Irwin P, Konwinski N, Larson S. 2018. Impact of consumption of bananas on attraction of *Anopheles stephensi* to humans. *Insects* **9:** 129. doi:10.3390/insects9040129

Pates HV, Takken W, Stuke K, Curtis CF. 2001. Differential behaviour of *Anopheles gambiae* sensu stricto (Diptera: Culicidae) to human and cow odours in the laboratory. *Bull Entomol Res* **91:** 289–296. doi:10.1079/BER200198

Pates HV, Curtis CF, Takken W. 2014. Hybridization studies to modify the host preference of *Anopheles gambiae*. *Med Vet Entomol* **28:** 68–74. doi:10.1111/mve.12070

Ponlawat A, Harrington LC. 2005. Blood feeding patterns of *Aedes aegypti* and *Aedes albopictus* in Thailand. *J Med Entomol* **42:** 844–849. doi:10.1093/jmedent/42.5.844

Qiu YT, Smallegange RC, Van Loon JJ, Ter Braak CJ, Takken W. 2006. Interindividual variation in the attractiveness of human odours to the malaria mosquito *Anopheles gambiae s. s. Med Vet Entomol* **20:** 280–287. doi:10.1111/j.1365-2915.2006.00627.x

Raji JI, Melo N, Castillo JS, Gonzalez S, Saldana V, Stensmyr MC, DeGennaro M. 2019. *Aedes aegypti* mosquitoes detect acidic volatiles found in human odor using the *IR8a* pathway. *Curr Biol* **29:** 1253–1262.e1257. doi:10.1016/j.cub.2019.02.045

Robinson A, Busula AO, Voets MA, Beshir KB, Caulfield JC, Powers SJ, Verhulst NO, Winskill P, Muwanguzi J, Birkett MA, et al. 2018. *Plasmodium*-associated changes in human odor attract mosquitoes. *Proc Natl Acad Sci* **115:** E4209–E4218. doi:10.1073/pnas.1721610115

Rodriguez SD, Drake LL, Price DP, Hammond JI, Hansen IA. 2015. The efficacy of some commercially available insect repellents for *Aedes aegypti* (Diptera: Culicidae) and *Aedes albopictus* (Diptera: Culicidae). *J Insect Sci* **15:** 140. doi:10.1093/jisesa/iev125

Rose NH, Sylla M, Badolo A, Lutomiah J, Ayala D, Aribodor OB, Ibe N, Akorli J, Otoo S, Mutebi JP, et al. 2020. Climate and urbanization drive mosquito preference for humans. *Curr Biol* **30:** 3570–3579. doi:10.1016/j.cub.2020.06.092

Rund SS, Bonar NA, Champion MM, Ghazi JP, Houk CM, Leming MT, Syed Z, Duffield GE. 2013. Daily rhythms in antennal protein and olfactory sensitivity in the malaria mosquito *Anopheles gambiae. Sci Rep* **3:** 2494. doi:10.1038/srep02494

Scott TW, Takken W. 2012. Feeding strategies of anthropophilic mosquitoes result in increased risk of pathogen transmission. *Trends Parasitol* **28:** 114–121. doi:10.1016/j.pt.2012.01.001

Scott TW, Chow E, Strickman D, Kittayapong P, Wirtz RA, Lorenz LH, Edman JD. 1993. Blood-feeding patterns of *Aedes aegypti* (Diptera: Culicidae) collected in a rural Thai village. *J Med Entomol* **30:** 922–927. doi:10.1093/jmedent/30.5.922

Scott TW, Amerasinghe PH, Morrison AC, Lorenz LH, Clark GG, Strickman D, Kittayapong P, Edman JD. 2000. Longitudinal studies of *Aedes aegypti* (Diptera: Culicidae) in Thailand and Puerto Rico: blood feeding frequency. *J Med Entomol* **37:** 89–101. doi:10.1603/0022-2585-37.1.89

Smith DL, McKenzie FE, Snow RW, Hay SI. 2007. Revisiting the basic reproductive number for malaria and its implications for malaria control. *PLoS Biol* **5:** e42. doi:10.1371/journal.pbio.0050042

Staunton K, Russell TL, Burkot TR. 2022. Standard operating procedure for human baited tent traps to sample host-seeking mosquitoes. *protocols.io* doi:10.17504/protocols.io.n92ldzq99v5b/v1. [Accessed May 19, 2022].

Takken W, Verhulst NO. 2013. Host preferences of blood-feeding mosquitoes. *Annu Rev Entomol* **58:** 433–453. doi:10.1146/annurev-ento-120811-153618

Torr SJ, della Torre A, Calzetta M, Costantini C, Vale GA. 2008. Towards a fuller understanding of mosquito behaviour: use of electrocuting grids to compare the odour-orientated responses of *Anopheles arabiensis* and *An. quadriannulatus* in the field. *Med Vet Entomol* **22:** 93–108. doi:10.1111/j.1365-2915.2008.00723.x

Trpis M, Hausermann W. 1978. Genetics of house-entering behaviour in East African populations of *Aedes aegypti* (L.) (Diptera: Culicidae) and its relevance to speciation. *Bull Entomol Res* **68:** 521–532. doi:10.1017/S0007485300009494

van Breugel F, Riffell J, Fairhall A, Dickinson MH. 2015. Mosquitoes use vision to associate odor plumes with thermal targets. *Curr Biol* **25:** 2123–2129. doi:10.1016/j.cub.2015.06.046

Wohl MP, McMeniman CJ. 2023. Batch rearing *Aedes aegypti. Cold Spring Harb Protoc* doi:10.1101/pdb.prot108017

Yan J, Gangoso L, Ruiz S, Soriguer R, Figuerola J, Martínez-de la Puente J. 2021. Understanding host utilization by mosquitoes: determinants, challenges and future directions. *Biol Rev Camb Philos Soc* **96:** 1367–1385. doi:10.1111/brv.12706

Zhang H, Zhu Y, Liu Z, Peng Y, Peng W, Tong L, Wang J, Liu Q, Wang P, Cheng G. 2022. A volatile from the skin microbiota of flavivirus-infected hosts promotes mosquito attractiveness. *Cell* **185:** 2510–2522. doi:10.1016/j.cell.2022.05.016

Protocol 1

A Behavioral Assay to Quantify Odor-Guided Thermotaxis with *Anopheles gambiae* under Semi-Field Conditions

Diego Giraldo[1] and Conor J. McMeniman[1,2,3]

[1]*W. Harry Feinstone Department of Molecular Microbiology and Immunology, Johns Hopkins Malaria Research Institute, Johns Hopkins Bloomberg School of Public Health, Johns Hopkins University, Baltimore, Maryland 21205, USA;* [2]*The Solomon H. Snyder Department of Neuroscience, Johns Hopkins University School of Medicine, Baltimore, Maryland 21205, USA*

The African malaria mosquito *Anopheles gambiae* is strongly attracted to human body odor and skin temperature. Quantitative behavioral assays suitable for use in semi-field environments with this nocturnal mosquito species are essential to gain improved insights into *An. gambiae* sensory biology, the mechanistic basis of mosquito attraction to humans, and host preference. In this protocol, we describe steps for engineering equipment for a novel behavioral assay for *An. gambiae*, which we have termed the odor-guided thermotaxis assay (OGTA). The OGTA uses infrared videography to quantify landings of female *An. gambiae* on an aluminum platform heated to human skin temperature that can be baited with volatile odorants such as carbon dioxide or human whole body odor. The OGTA facilitates high-content recordings of *An. gambiae* landing behavior during odor-guided thermotaxis under naturalistic semi-field conditions without the requirement for domestic power.

MATERIALS

It is essential that you consult the appropriate Material Safety Data Sheets and your institution's Environmental Health and Safety Office for proper handling of equipment and hazardous material used in this protocol.

Reagents

3D printing resin, black (e.g., Formlabs RS-F2-GPBK-04)
3D printing resin, clear (e.g., Formlabs RS-F2-GPCL-04)

Equipment

3D printer (e.g., Formlabs Form3)
Alligator clips with wires
Aluminum flashing, 10-cm-diameter circle
Aluminum flex ducting, 4″-diameter (e.g., Lambro 650/AF450UL)
Arduino UNO (Arduino A000066)

Arduino, Raspberry Pi codes, models for 3D printing and a photographic assembly sequence for assay circuitry related to Figure 1 can be downloaded from https://github.com/McMenimanLab/Behavioral-Assay-to-Quantify-Mosquito-Thermotaxis.

[3]Correspondence: cmcmeni1@jhu.edu

Cite this protocol as *Cold Spring Harb Protoc*; doi:10.1101/pdb.prot108303

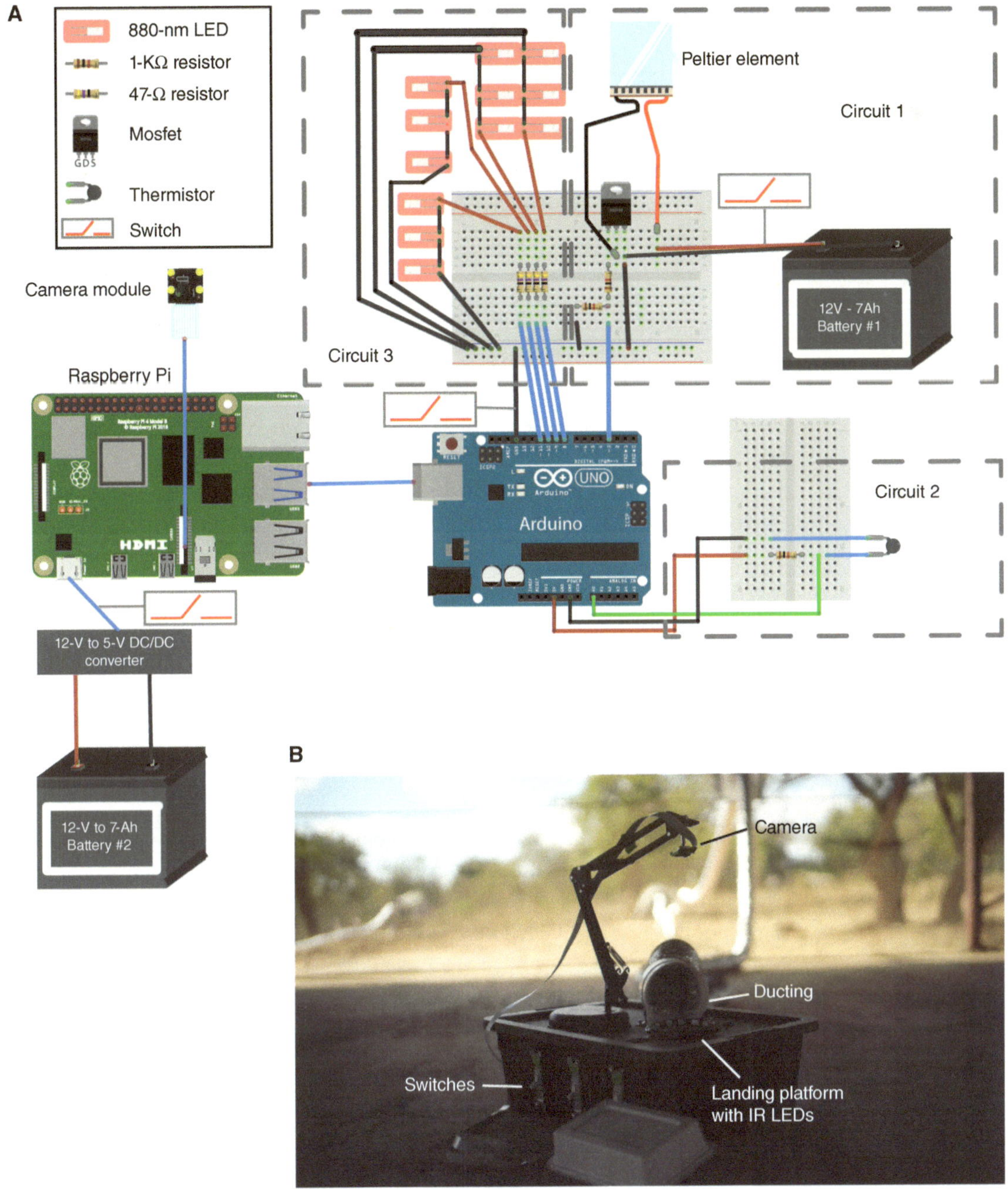

FIGURE 1. Odor-guided thermotaxis assay (OGTA) for quantifying *Anopheles gambiae* attraction to olfactory stimuli. (*A*) Diagram for circuits 1–3 and their components for OGTA assembly. Circuit 1: Peltier element; Circuit 2: thermistor; and Circuit 3: LED array. (*B*) Assembled OGTA positioned within a semi-field flight cage arena with attached ducting for odor delivery, heated landing platform surrounded by a ring of infrared (IR) LEDs and IR-sensitive camera positioned over the platform using a stand to quantify mosquito landing behavior. The three circuits are placed inside a black container, and switches are added to the circuits to turn on and off the Raspberry Pi, LED array, and Peltier element.

Batteries, 12-V–7-Ah (two each)
Camera stand with C-mount (e.g., Myrin Korea B00AUJHKC2)
Circuit breadboard (e.g., Elegoo EL-CP003)
Computer keyboard with trackpad (e.g., Macally Peripherals BTTVKEY)

Container, black (e.g., 19-qt. Rolling Plastic Storage Tote Black, dimensions 21 3/8″ L × 15 1/4″ W × 7 1/8″ H)
DC/DC 5-V power supply adapter (micro-USB connection)
Double-sided tape
Drill bit (1/4″)
Drill, electrical (e.g., Ryobi HP108L 8-V Lithium Ion Cordless)
Duct couplers (e.g., 4″-diameter dryer lint clean out connector; Everbilt 595002)
Electrical tape
HDMI cable
Infrared 850-nm 20-mA LED (e.g., Chanzon 100F5T-IR-FS-850NM) (12 each)
Jumper wires (e.g., EDGELEC 0611355169887)
N-channel MOSFET IRFZ44N
Peltier element, TEC1-12706, 40-mm × 40-mm (e.g., HiLetgo 3-01-0553)
Portable HDMI monitor (e.g., 7″ screen; Toguardgo WR730P)
Raspberry Pi 3 Model B+ (Raspberry Pi 3775)
Raspberry Pi camera module 2 NoIR (Raspberry Pi 3100)
Resistor, 1-KΩ (four each)
Resistor, 47-Ω (four each)
Soldering iron
Solder wire for electrical soldering
Spray paint, flat black (e.g., Rust-Oleum 249127)
Switches (optional; see Steps 21 and 25)
Tape, general purpose laboratory (e.g., VWR 89097-918)
Thermal conductive glue (e.g., Gennel B07TSH6MNT)
Thermistor (e.g., Sensor Scientific Inc. WM222C)
Thumb screws (e.g., #10–24 × 1/2″)
U bracket pipe strap (4″-diameter, two-hole)
USB A to B cable
Wing nuts (e.g., #10–24)
Wire spool (e.g., BNTECHGO PVC18AWG1007SOF20C10)
Wire strippers
Woven wire mesh, stainless steel, 0.41-mm-aperture (e.g., Inoxia, 40 mesh)

METHOD

The odor-guided thermotaxis assay (OGTA) detailed in this protocol has been specifically designed for use with the African malaria mosquito Anopheles gambiae *during nighttime hours and under semi-field conditions. For further details on its composition, see the Discussion. Electrical connections for circuitry in this protocol (Fig. 1A, Circuits 1–3) are performed using circuit breadboards, jumper wires, or wire spool with wire strippers in instances whereby wiring is not already associated with assay components.*

Assembly of Circuits for Heated Landing Platform

The following steps describe the assembly of two circuits required to heat and control the temperature of the landing platform.

Circuit 1: Peltier Element

This circuit controls the heating of an aluminum landing platform for mosquitoes set around human skin temperature (35°C) with a Peltier element (PE). The temperature of the platform is measured by a thermistor and reported to the Arduino microprocessor, which sends a 5-V signal to the Mosfet field effect transistor to heat the PE. The Mosfet has three pins: the gate, source, and drain. Upon receiving a signal at the gate pin, current can flow from the drain to the source, which will result in current flow from the 12-V battery through the PE, thus heating the platform.

1. Connect digital pin 3 of the Arduino to a 1-KΩ resistor and the resistor to the gate of the Mosfet.
2. Connect a second 1-KΩ resistor to the previous resistor and then to ground.

   This resistor prevents the Mosfet from staying open when no signal is coming out of the Arduino.
3. Connect the black wire of the PE to the drain pin of the Mosfet.
4. Connect the red wire of the PE to the positive terminal of the 12-V battery (Battery #1; Fig. 1A) using an alligator clip.
5. Connect the negative terminal of the battery to the source of the Mosfet.
6. Connect the source of the Mosfet to the ground of the Arduino.

Circuit 2: Thermistor

A thermistor changes its resistance with temperature. This resistance value can be measured to calculate the temperature of the landing platform.

7. Connect the 5-V pin of the Arduino to the 1-KΩ reference resistor.
8. Connect the resistor to the analog pin A0 of the Arduino and one pin of the thermistor.
9. Connect the other pin of the thermistor to a ground pin on the Arduino.

Assembly of LED Circuit to Illuminate Landing Platform

This circuit powers the LEDs to illuminate the landing platform with infrared (IR) light using four Arduino pins. IR light facilitates tracking of An. gambiae *in the dark.*

Circuit 3: LED Circuit

10. Solder three IR LEDs in series with a 47-Ω resistor. The LEDs are inserted into a circular 3D-printed black frame (download frame printing instructions from https://github.com/McMenimanLab/Behavioral-Assay-to-Quantify-Mosquito-Thermotaxis). Use the LED frame as a reference for the spacing between LEDs. Repeat the array three more times, for a total of four LED arrays.
11. Secure the LED arrays onto the LED frame with electrical tape.
12. Place the clear 3D-printed light diffuser (download frame printing instructions from https://github.com/McMenimanLab/Behavioral-Assay-to-Quantify-Mosquito-Thermotaxis) over the internal margin of the LED frame.
13. Connect the anodes from LED arrays 1–4 to pins 8–12 of the Arduino.
14. Connect the cathode to ground.

    The cathodes from LED arrays 1–4 may be soldered together onto a single wire connected to a single Arduino ground pin.

Assembly of Heated Landing Platform

The following steps describe integration of Circuits 1–3 to build the heated and IR-illuminated platform.

15. Paint one side of a 10-cm-diameter circle of aluminum flashing with black flat spray paint.
16. Glue the unpainted underside of aluminum circle onto the designated hot side of the PE using thermal conductive glue.

17. Tape the thermistor to the underside of the aluminum sheet using general purpose laboratory tape as close to the margins of the PE as possible.

    Make sure the thermistor is in direct contact with the aluminum.

18. Slot the assembled aluminum disc and PE into the assembled LED frame.

    The LED frame has designated holes for the PE and thermistor wires to be threaded through.

19. Place the assembled landing platform centered on the right side of the lid of the assay base.

    The assay base consists of a rectangular black plastic container with a snap lock lid that serves the dual purpose of housing the two 12-V batteries used to the power the OGTA and concealing the wires associated with the circuits. This protects the electrical components of the assay against inclement weather.

20. Once in the correct position, affix the base of the landing platform to the container lid using double-sided tape and drill two 1/4″ holes on either side.

21. Pass all the LED, PE, and thermistor wires through the holes in the lid so they are directed into the container and connect them to the Arduino and batteries.

    Place the circuit breadboard and batteries inside the container box under the platform.

    Optional: The LED and PE circuits can be connected to switches so that they can be strategically turned on and off. Switches can be positioned on the exterior of the container to facilitate easily turning them on and off, without having to open the lid to the container to access these circuits. The ground of the LED circuit and the negative terminal of the battery of the PE element circuit are suggested positions for installation of switches.

Installation of IR-Sensitive Camera Adjacent to Landing Platform

22. Place the IR-sensitive Raspberry-Pi camera on the camera stand and connect it to the camera port on the Raspberry Pi.

23. Position the camera stand on the container lid adjacent to the landing platform.

24. Connect the Arduino to the Raspberry Pi with a USB A to B cable.

25. Connect the Raspberry Pi to a 12-V battery (Battery #2; Fig. 1A) using a 12-V to 5-V DC/DC convertor.

    This will power the Raspberry Pi. An additional switch can be placed between the DC/DC convertor and the negative terminal of this 12-V battery to conveniently switch the Raspberry Pi on and off.

26. Connect a keyboard with a trackpad to the Raspberry Pi using the USB port, and the monitor to the Raspberry Pi using the HDMI port.

27. Download the "Arduino_scripts" and "Python_scripts" folders (https://github.com/McMenimanLab/Behavioral-Assay-to-Quantify-Mosquito-Thermotaxis) and transfer to the desktop of the Raspberry Pi.

    These codes are used to run the camera using the Raspberry Pi and control the temperature of the platform with the Arduino.

28. Install the Arduino Integrated Development Environment (IDE) (https://www.arduino.cc/en/software) on Raspberry Pi.

29. Open the code "Thermistor_Peltier_control.ino" with the Arduino IDE and load the code onto the Arduino.

    This code measures V_{out} *from the thermistor to calculate the temperature of the platform and turn the PE on for heating.*

    If a constant temperature cannot be achieved, a code that uses a PID (proportional–integral–derivative) controller can be uploaded to the Arduino instead (code Peltier_PID.ino). In this code, coefficients kp, ki, and kd must be adjusted until a stable temperature is obtained. We recommend using the PID code only if a stable temperature cannot be achieved with code "Thermistor_Peltier_control.ino".

Cite this protocol as *Cold Spring Harb Protoc*; doi:10.1101/pdb.prot108303

Test Functionality of Hardware

30. Perform a PE functionality test as follows: Open a terminal on the Raspberry Pi and change the directory to the folder containing "Temperature_arduino.py". Run the script by typing: "`python Temperature_arduino.py`". The temperature of the platform will be displayed and updated on the terminal.

    If the PE is correctly connected to the battery the temperature displayed should be stable at ~35°C. This code will store the temperature of the platform over time onto a .csv file.

31. Perform a camera functionality test by opening a second terminal, changing directory to the folder containing the scripts, and typing: "`python run_preview.py`".

    This code will show a live view of the camera for 50 sec to facilitate adjusting the camera view such that the landing platform is in frame.

Installation of Odor Stimulus

An odor stimulus to the OGTA is supplied via 4″ aluminum ducting that pipes a target odor stimulus (e.g., CO_2 or human whole body odor) from a tent positioned at a predefined distance (e.g., 15 m) outside the semi-field cage. The following steps detail installation of the terminal section of ducting adjacent to the heated landing platform on the lid of the OGTA container.

32. On the end of the 4″ aluminum ducting, screw a 4″ duct coupler.
33. Insert a 4″ circular disc of stainless steel woven wire mesh on the interior rim of the duct coupler.

    This prevents mosquitoes from flying up the ducting.

34. Secure the screened ducting to the container lid immediately adjacent to the landing platform using a 4″ U-bracket pipe strap over the duct coupler with wing nuts and thumb screws.

    This ensures the odor stimulus is directionally applied over the heated landing platform.

    Once assembled, the OGTA (Fig. 1B) is ready for use in semi-field assays as described in Protocol 2: Quantification of Anopheles gambiae *Olfactory Preferences under Semi-Field Conditions (Giraldo and McMeniman 2023).*

DISCUSSION

The OGTA described in this protocol consists of (i) a circular black aluminum platform heated to human skin temperature (35°C) that constitutes the landing target, (ii) a 3D-printed frame, encircling the platform, that houses a ring of IR LEDs to illuminate the mosquitoes landing on this target at night, (iii) an IR-sensitive camera to record mosquito landings on the platform, (iv) associated hardware (Arduino microprocessor and Raspberry Pi computer), circuitry, software, and code to integrate assay components and run the assay, and (v) an odor source such as carbon dioxide or human whole body odor (described in Protocol 2: Quantification of *Anopheles gambiae* Olfactory Preferences under Semi-Field Conditions [Giraldo and McMeniman 2023]) that is directionally applied at a set flow rate via screened ducting over the heated platform to elicit mosquito landings (odor-guided thermotaxis).

Single or multiple OGTAs may be arrayed within a contained semi-field facility (e.g., screened greenhouse or large screened cage). Each OGTA and its components are powered by two 12-V batteries, facilitating its use under semi-field conditions without the requirement for domestic power. An additional 12-V battery per OGTA is required to power a fan assembly to apply the odor stimulus onto the landing platform as outlined in Protocol 2: Quantification of *Anopheles gambiae* Olfactory Preferences under Semi-Field Conditions (Giraldo and McMeniman 2023). Each OGTA has been validated to run for 8 h using 12-V–7-Ah batteries.

In this protocol, we describe the design and assembly of a custom assay to quantify *An. gambiae* odor-guided thermotaxis behavior under semi-field conditions. It is composed of three simple circuits integrated by an Arduino microprocessor that regulate the temperature of an aluminum landing platform (Circuit 1 and 2) and illuminate this platform with IR light to facilitate night-time analysis (Circuit 3). Recording of mosquito behavior is accomplished by an IR-sensitive camera compatible with a Raspberry Pi computer for data storage. The Raspberry Pi is also coupled to the Arduino to store temperature data from the heated landing platform. Use of these easily sourced computational (D'Ausilio 2012; Jolles 2021) and camera components allows this assay for the quantification of *An. gambiae* landing behavior in the dark to be built without specialized engineering expertise and at low cost. We provide codes that can be easily edited to modify experimental parameters such as platform temperature and experiment duration according to user specifications. The ease of construction of this assay allows users to build multiple units for high-throughput data acquisition. This assay in its current configuration has been optimized for use with *An. gambiae* in semi-field conditions, and with modification has the potential to be used with *Aedes aegypti* or other mosquito species across both laboratory, semi-field, and field settings. User-specified manipulation of visual and olfactory cues associated with the temperature stimulus may further be used to investigate multisensory integration in various mosquito species. Finally, multiple OGTAs may be used in competition to test olfactory and thermosensory preferences.

ACKNOWLEDGMENTS

We thank Ben Matthews for initial advice regarding hardware for IR videography and Genevieve Tauxe for contributions to OGTA design. We acknowledge generous funding from Johns Hopkins Malaria Research Institute and Bloomberg Philanthropies to C.J.M. that enabled development of this protocol. D.G. was generously supported by postdoctoral fellowships from the Human Frontier Science Program (LT000310/2019-L) and Johns Hopkins Malaria Research Institute.

REFERENCES

D'Ausilio A. 2012. Arduino: a low-cost multipurpose lab equipment. *Behav Res Methods* **44:** 305–313. doi:10.3758/s13428-011-0163-z

Giraldo D, McMeniman CJ. 2023. Quantification of *Anopheles gambiae* olfactory preferences under semi-field conditions. *Cold Spring Harb Protoc* doi:10.1101/pdb.prot108304

Jolles JW. 2021. Broad-scale applications of the Raspberry Pi: a review and guide for biologists. *Methods Ecol Evol* **12:** 1–18. doi:10.1111/2041-210X.13652

 Cite this protocol as *Cold Spring Harb Protoc*; doi:10.1101/pdb.prot108303

Protocol 2

Quantification of *Anopheles gambiae* Olfactory Preferences under Semi-Field Conditions

Diego Giraldo[1] and Conor J. McMeniman[1,2,3]

[1]*W. Harry Feinstone Department of Molecular Microbiology and Immunology, Johns Hopkins Malaria Research Institute, Johns Hopkins Bloomberg School of Public Health, Johns Hopkins University, Baltimore, Maryland 21205, USA;* [2]*The Solomon H. Snyder Department of Neuroscience, Johns Hopkins University School of Medicine, Baltimore, Maryland 21205, USA*

Anopheles gambiae is a highly anthropophilic (human-preferring) malaria vector that prefers to blood feed frequently and selectively on humans. This mosquito species exhibits a strong innate olfactory preference to seek out human scent over other animals, and certain humans over others—key behavioral traits with the potential to drive heterogeneity in biting risk and malaria transmission. Here, we describe the application of a large-scale, semi-field system in Zambia for the quantification of *An. gambiae* olfactory preferences toward whole body odor sourced from individual humans. We detail steps for modifying one-person canvas tents to duct odor from sleeping humans into a central, semi-field flight cage arena that is securely screened. Using this system, we describe a protocol to perform two-choice olfactory preference assays with two human volunteers using laboratory-reared *An. gambiae* and odor-guided thermotaxis assays that leverage infrared video tracking to quantify mosquito landings on heated targets baited with each body odor sample. This multichoice olfactory preference assay has the potential to be applied with expanded cohorts of humans to define the chemosensory basis of *An. gambiae* host preference and interindividual differences in human attractiveness to mosquitoes and to be used to quantify the effects of protective measures such as personal and spatial repellents on mosquito landing behavior.

MATERIALS

It is essential that you consult the appropriate Material Safety Data Sheets and your institution's Environmental Health and Safety Office for proper handling of equipment and hazardous material used in this protocol.

Reagents

Anopheles gambiae mosquitoes
H_2O (distilled)

Equipment

Adjustable metal drive clamp, 4″-diameter
Alligator clips
Aluminum flex ducting, 4″-diameter (e.g., Lambro 650/AF450UL)
Anemometer (e.g., Testo 405i)

[3]Correspondence: cmcmeni1@jhu.edu

Cite this protocol as *Cold Spring Harb Protoc*; doi:10.1101/pdb.prot108304

Aspirator, mechanical suction (e.g., John W. Hock Company Model 419)
Awl punch
Batteries, 12-V, 9- to 12-Amp Hr.
Battery clamp cable with 7.5-A fuse
Camping mattresses (e.g., Therm-a-rest trail scout self-inflating mattress, large; optional; see Step 9)
Camping stretchers (e.g., Teton sports Outfitter XXL camping cot)
Computer keyboard with trackpad (e.g., Macally Peripherals BTTVKEY)
Cotton balls, medium (e.g., Fisher Scientific 22-456-880)
Cotton camping sheets (optional; see Step 9)
Cups (e.g., 12-oz, Solo 412SIN-0041)
Cutting pliers
Duct couplers (e.g., 4″-diameter dryer lint clean out connector; Everbilt 595002)
Duct flange, 4″ diameter (e.g., DL Wholesale 409004)
Duct to wall connector (e.g., 4″-diameter dryer duct to wall connector; Everbilt 596498)
Dry bags, waterproof
Fencing wire
HDMI cable
In-line marine fans with 4″-diameter connections (e.g., Rule 240 in-line blower, 12-V)
Marker pen
Netting with a small mesh size (<1-mm; preferably polyester, e.g., white tulle)
Odor-Guided Thermotaxis Assay (two each)

Assembly outlined in Protocol 1: A Behavioral Assay to Quantify Odor-Guided Thermotaxis with Anopheles gambiae *under Semi-Field Conditions (Giraldo and McMeniman 2023).*

One-person canvas tents (e.g., Kodiak 8101)
Plastic containers (e.g., Tupperware)
Polypropylene sheeting, white corrugated twinwall
Portable HDMI monitor (e.g., 7″ screen; Toguardgo WR730P)
Prokopack aspirator (e.g., John W. Hock Company 1419)
Pulse width modulator (PWM; e.g., RioRand RR-6-90V-15A-PSC)
Rubber bands
Scissors
Semi-field cage facility, screened (see Step 1)
Tape measure
Thumb screws (e.g., #10–24 × 1/2″)
U bracket pipe strap (4″-diameter, two-hole)
Upholstery needle and heavy gauge thread
USB flash drive (1 per setup, e.g., 256-GB SanDisk SDCZ48-256G-U46)
Utility knife
Velcro, 2″-wide, adhesive backing
Velcro, 2″-wide, soft
Video tracking program (e.g., ivTrace)
Washers (e.g., fender washer 3/16″)
Wing nuts (e.g., #10–24)
Woven wire mesh, stainless steel, 0.41-mm-aperture (e.g., Inoxia, 40-mesh)

METHOD

In this protocol, we describe a semi-field system that we specifically engineered for multichoice olfactory preference assays with laboratory-reared An. gambiae *at the Macha Research Trust campus in Choma District, Zambia. We describe steps for establishment and execution of two-choice olfactory preference assays with* An. gambiae *under*

Cite this protocol as *Cold Spring Harb Protoc;* doi:10.1101/pdb.prot108304

semi-field conditions using the odor-guided thermotaxis assay (OGTA) (see Protocol 1: A Behavioral Assay to Quantify Odor-Guided Thermotaxis with Anopheles gambiae *under Semi-Field Conditions [Giraldo and McMeniman 2023]). In the experimental configuration outlined in this protocol, two OGTAs are baited with human whole body odor derived from two humans sleeping in separate one-person tents that are positioned adjacent to the semi-field facility. We describe steps for modifying one-person canvas tents to connect them individually to aluminum ducting and a fan assembly that draws human whole body odor out of each tent and onto concordant OGTAs positioned inside the semi-field cage. We also describe steps for positioning the OGTAs in the semi-field cage for competition assays, preparing laboratory-reared* An. gambiae *females for use in these assays, executing the two-choice OGTA assay during nighttime trials, and retrieving and analyzing data to quantify mosquito landing on each OGTA.*

Consult with institutional as well as local, state, and federal governmental agencies regarding regulatory requirements for engineering semi-field facilities to ensure contained experimentation.

Setting up a Semi-Field Cage

1. Construct a semi-field cage. To perform semi-field analyses of *An. gambiae* olfactory preferences, a secure and screened enclosure that is exposed to natural weather conditions (i.e., ambient temperature, humidity, wind, and barometric pressure) is required. Typically, semi-field cages are constructed within greenhouse frames or custom-built structures (Knols et al. 2002; Ferguson et al. 2008; Ritchie et al. 2011; Niang et al. 2019). Architecture of the semi-field cage may be established according to user specifications to facilitate experimentation in a habitat reflective of the local region (e.g., to contain local housing structures or local vegetation) or alternatively customized to allow the directional application of olfactory cues (e.g., through ducting cues into the semi-field cage as specified in this protocol) or introduction of live hosts (e.g., humans and other animals for biting assays).

Modifying One-Person Tents for Ducting Human Whole Body Odor to OGTAs

These steps describe the modification of commercially available one-person canvas tents to couple them with a 4″-diameter aluminum flex ducting and an in-line fan to deliver human body odor to OGTAs that are positioned inside the semi-field cage (Fig. 1A). In the two-person assay configuration described in this protocol, each one-person canvas tent is positioned 15 m away from the perimeter of the semi-field cage on a separate concrete slab. For participant comfort and protection from groundwater during storms, each tent is set up on a camping stretcher on top of each slab. One-person canvas tents typically consist of an inner screened chamber where the human sleeps, as well as an outer canvas shell with venting ports at the ends of the tent for air circulation. For this assay, it is important that the canvas flap that constitutes the smaller vent at the end of the tent (i.e., where the feet of the sleeping human are positioned) is modified for installation of a ducting connector flange, which allows the 4″ aluminum flex ducting to be attached to the tent. After installation of the ducting connector flange, the base of this canvas flap is sewn shut to maximize directional flow of human body odor into the semi-field cage via the ducting. These steps can take up to 2 h (1 h per tent).

2. Set up one tent according to the manufacturer's instructions on top of the camping stretcher at a distance of 15 m outside of the semi-field cage.
3. Position the 4″ duct connector flange at the center of the canvas flap that constitutes the tent foot vent.
4. Using a marker pen, mark the positions on the canvas flap where the six wing nuts and thumb screws that will be used to secure the connector flange will go.
5. Using a sharp awl, punch a pilot hole at each screw position to facilitate installation of the wing nuts and thumb screws that will secure the connector flange in place.
6. Install the duct connector flange on the canvas flap and secure it with six wing nuts and thumb screws.

   Once the duct connector flange is secured, the canvas on the interior of the ducting connector should be taut.

7. Using a sharp utility knife, cut the taut canvas on the interior of the ducting connector flange to create a 4″ hole, such that air can flow out of the tent.

   Important: Do not disrupt the interior screened mesh of the sleeping chamber. This ensures that the screened sleeping chambers maintains structural integrity and the human participant is always protected from mosquitoes in the surrounding environment.

FIGURE 1. Key assembly parts to configure the semi-field olfactory preference assay for *Anopheles gambiae*. (*A*) Modified one-person tents, each with an installed ducting flange connected to a fan assembly and stretch of aluminum ducting to pipe human body odor into the flight cage arena. (*B*) Fan assembly mounted on a stand with attached wire guidelines for supporting ducting attachment to the window access port. (*C*) Exterior view of the window access port with wire guidelines attached to the top of the port frame. (*D*) Interior view of the window access port within the flight cage arena.

8. Sew the bottom margin of the canvas flap with the duct connector installed to the base of the tent using an upholstery needle and heavy gauge thread to seal it shut.

   This ensures the whole body odor flowing out of the tent is concentrated through the ducting to be affixed to the ducting connector and ultimately onto the OGTA.

9. Optional: For participant comfort, a camping mattress and washable cotton camping sheets can be placed inside the tent during assays.

Installation of an In-Line Fan Assembly and Ducting Connectors

To duct human body odor from a tent into the semi-field cage, the 4″-diameter ducting connector flange installed in Steps 2–8 onto the tent is first connected via a 30-cm stretch of 4″-diameter aluminum flex ducting onto the inlet of an in-line marine fan with 4″ connections. The in-line fan is fixed onto a steel stand, adjacent to the ducting connector flange at the foot of the tent, such that the airflow from the fan points directly toward the semi-field cage (Fig. 1B). A 15-m stretch of aluminum ducting is then connected to the outlet of the in-line fan and attached to a window access port on the wall of the semi-field cage that facilitates the secure connection of stretches of ducting through the semi-field cage wall to the OGTA (Fig. 1C,D). These steps can be completed in ~2 h per assembly.

Cite this protocol as *Cold Spring Harb Protoc*; doi:10.1101/pdb.prot108304

10. Fix an in-line marine fan with 4″-diameter connections onto a stand adjacent to the foot of the tent with the installed ducting flange connector.

    Use of a marine fan is recommended as they are highly water-resistant in the event of rain. Fix the fan in place onto the stand using self-tapping screws, making sure that the outlet of the fan is pointing toward the semi-field cage.

11. Connect the positive and negative wires from the in-line fan to the corresponding "Motor" screw terminals on the PWM.

12. Connect the "Power" screw terminals on the PWM to positive and negative wires of a battery clamp cable equipped with a 7.5-A fuse and terminal alligator clips. To power the fan, attach the alligator clips to the corresponding terminals on a 12-V battery.

    The PWM has a DC motor pump speed controller on a dial interface that facilitates adjustment of fan speed. Modulation of fan speed allows calibration and standardization of the amount of odor-laden air delivered to each OGTA.

    To protect circuitry on the PWM from water, the unit can be placed inside a Tupperware container with drilled holes to facilitate accessibility of wire connections to the 12-V battery and speed controller. These both can be placed in a dry bag positioned under the tent stretcher to keep them protected from rain.

13. Connect a 30-cm stretch of 4″-diameter aluminum flex ducting from the ducting connector flange on the tent to the fan inlet.

    Tighten each connection with a 4″-diameter adjustable metal drive clamp.

14. Using cutting pliers, cut a separate 15-m segment of 4″-diameter aluminum flex ducting.

15. Affix one end of 15 m ducting to the fan outlet and secure with a 4″-diameter adjustable metal drive clamp.

16. Attach a 4″ duct coupler by manually screwing it into place on the other end of the ducting.

    A series of grooves on the interior walls of the duct coupler accepts the ribbed surface of 4″ aluminum flex ducting.

    Strategic use of commercially available "duct couplers" and "duct to wall connectors" in series conveniently allows the secure connection of segments of flexible aluminum ducting running between the tent and OGTA through an engineered "window access port" in the side of the semi-field cage.

    The margins of each window access port consist of a 30-cm × 30-cm window frame made from steel rebar that is part of the semi-field cage structure. The interior side of the window access port frame within the semi-field cage is covered with soft 2″-wide Velcro backing secured in place using an upholstery needle and heavy gauge thread. A window access port panel is then constructed by centering a 4″-diameter circular hole in a 30-cm × 30-cm square of white corrugated twinwall polypropylene sheeting using a sharp utility knife, and then affixing a 4″ duct to wall connector to this sheet using washer/wing nut/thumb screw assemblies in the four base plate holes on this unit. The interior diameter of the 4″ hole on the duct to wall connector is then covered with a 4″-diameter circle of stainless steel woven wire mesh that slots into position on an internal lip to prevent accidental mosquito exit from the cage in the event that connecting ducts become disconnected. Finally, 4-cm × 30-cm strips of adhesive Velcro are added to the panel margins on the side of the window access port panel that will face the exterior of the cage. Using this configuration, the Velcro surfaces between the window access port frame and panel are then adhered to form a secure connection.

17. Using two 15-m lengths of fencing wire, run two "guidelines" from the fan stand to the frame of the window access port on the semi-field cage (Fig. 1A–C). Place a series of five equidistantly spaced wire hoops (~50 cm in diameter) around these wire lines, through which the 15-m segment of 4″-diameter flexible ducting can be fed.

    Use of fencing wire guidelines and hoops helps to keep the aluminum ducting off the ground and relieves tension from the ducting connection made at the window access port.

18. Snap the male fittings of the duct coupler into the designated female positions on the duct to wall connector of the window access port.

19. For the segment of ducting that runs from the window access port to the OGTA in the interior of the cage, cut a 7.5-m segment of 4″ aluminum flex ducting.

20. Affix one end of this segment of ducting to the interior flange of the duct to wall connector in the window access port using a 4″-diameter adjustable metal drive clamp.

21. Repeat Step 16 to cap the other end of the interior cage ducting with ducting couplers.
22. Using scissors, cut a 4″-diameter circle of stainless steel woven wire mesh and place it on the interior lip of the ducting coupler to securely screen the end of the ducting.

    This prevents mosquitoes from flying up the ducting to the window access port. Make sure there are no gaps.

Position the OGTAs in the Semi-Field Cage

23. For two-choice assays (Fig. 2A,B), strategically space two OGTAs (Fig. 2C,D) inside the semi-field cage at a set distance from each other (e.g., 5.5 m apart) using a tape measure.

    The final positioning of tents and OGTAs can be modified according to user specifications and experimental design for multichoice assays of mosquito olfactory preference.
24. For each OGTA, secure the aluminum flex ducting with a screened duct coupler onto the OGTA container immediately adjacent to the landing platform using a U bracket and wing nuts and thumb screws (Fig. 2D).

    This ensures that human body odor is directionally applied over the warmed OGTA landing platform.
25. Place the camera stand on the OGTA container and center the camera over the platform.

    See Troubleshooting.

Preparing *An. gambiae* for Use in Assays

An. gambiae *used in semi-field assays are typically laboratory strains (e.g.,* An. gambiae *Kisumu strain) that are reared in insectaries adjacent to the semi-field facility according to standard operating procedures. This allows users to control the age and physiological status of mosquitoes and obtain large numbers of mosquitoes for use in assays.*

26. At 3- to 5-d postemergence at 8 a.m., aspirate 200 female *An. gambiae* mosquitoes into a 12-oz. Solo cup or other holding container gently using a mechanical suction aspirator.

    Female An. gambiae *should be maintained with males in a standard cage prior to collection for experimentation to allow them to mate. Secure netting on holding cups with two rubber bands.*
27. Place a large cotton ball with distilled H_2O on the netting of the cup.
28. Maintain the holding container under insectary conditions for 12 h.

Executing the Assay, Retrieving Data, and Analysis

The scripts required to run the experiment can be downloaded from https://github.com/McMenimanLab/Behavioral-Assay-to-Quantify-Mosquito-Thermotaxis.

29. Turn on the fan by connecting the PWM to the 12-V battery using alligator clips.
30. At the output of the OGTA ducting, calibrate the airflow using an anemometer. Using the PWM, adjust the speed of the fan to yield an output of 0.8 m/sec.

    0.8 m/sec is a recommended standardized airflow speed that provides a gentle stimulus from the output of the ducting that is consistent with wind speeds typically used in mosquito laboratory assays. This value can be modified according to different experimental needs.
31. Turn on the Peltier element of the OGTA by connecting it to a second 12-V battery in the semi-field cage.
32. Connect the monitor to the Raspberry Pi HDMI port using the HDMI cable, the keyboard to the Raspberry Pi USB port using the keyboard USB cable, and Raspberry Pi to a third 12-V battery to switch it on.
33. Update the time and date on the Raspberry Pi from the terminal by typing:

    "`sudo date -s 'yyyy-mm-dd hh:mm:ss'`"

    example: "`sudo date -s '2022-08-15 20:00:00'`"

    Alternatively, connect the Raspberry Pi to the internet if available (e.g., by mobile hotspot) and this will automatically update the date and time.

Cite this protocol as *Cold Spring Harb Protoc*; doi:10.1101/pdb.prot108304

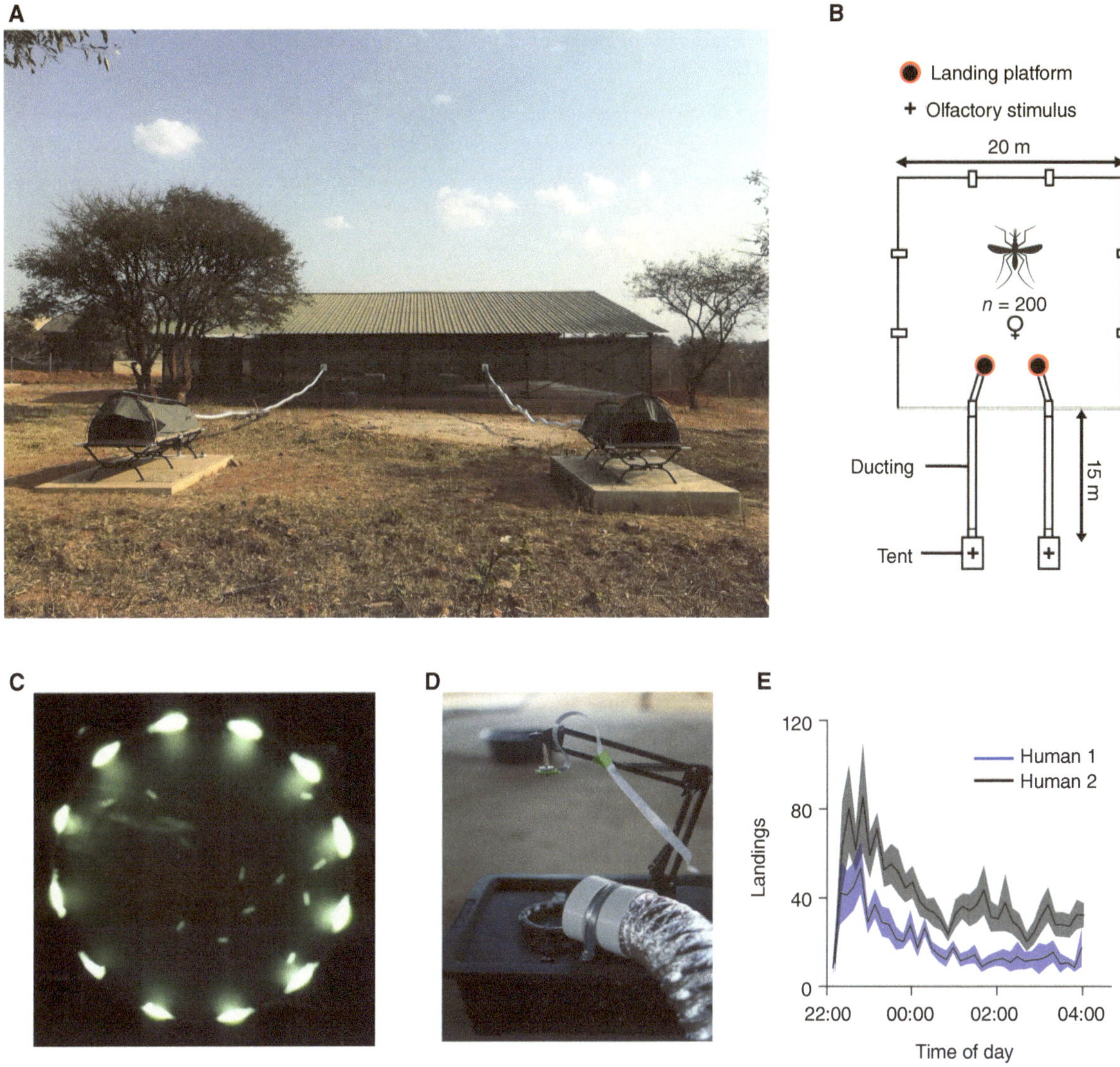

FIGURE 2. Semi-field system configuration for two-choice assay to quantify olfactory preference for the scent of one human over another. (*A*) Two one-person tents connected to the flight cage arena via aluminum ducting. (*B*) Schematic of the multichoice olfactory preference assay. Tents (up to two per side) are individually connected to deliver each olfactory stimulus via ducting fan assemblies to associated landing platforms in the flight cage arena. The output of the ducting is securely screened with stainless steel woven wire mesh to prevent mosquitoes from flying up it. $n = 200$ mosquitoes per nightly trial. (*C*) Example movie frame with mosquitoes on the heated landing platform illuminated by infrared (IR) light. (*D*) Odor-guided thermotaxis assay (OGTA). *Anopheles gambiae* landings on the heated and IR-illuminated landing platform are monitored under low-light conditions at night using an IR-sensitive camera positioned directly above it. Ducting delivering the olfactory stimulus to the landing platform is positioned directly adjacent to it on the assay container lid using a U bracket affixed with wing nuts and thumb screws. (*E*) Olfactory preference of *An. gambiae* females for two individuals. Mean landing on OGTAs baited with "Human 1" versus "Human 2" over replicate 6-h assay periods at night. $n = 7$, mean ± standard error of the mean.

34. Test the camera position by running the script "*run_preview.py*" on the terminal.
35. Create a folder on the desktop called "Data."
36. Create two subfolders in the Data folder called "Temperature" and "Movie."
37. Open a new terminal and run the script "*Temperature_arduino.py*" to test that the landing platform is heated to the desired temperature and that the assay stores temperature data.

    Do not close the terminal. Let the code run for the duration of the experiment so that the data is stored in the "Temperature" folder.

38. Open a new terminal and run the script "*video.py*".

    Make sure the start time of the recording on the script is set to 2 h after the planned release of mosquitoes (e.g., 22:00:00).

    For recordings of multiple hours (e.g., 6 h) it is recommended to record in bouts of 1-h videos. The hours of recording can be set in the video.py script.

    Optional: If desired, baseline landing activity can be recorded by commencing data capture 30 min prior to the introduction of odor stimuli by modifying the code.

    The recorded videos will be stored in the "Movie" folder.

39. Repeat Steps 29–38 for each OGTA unit being used.
40. After all OGTAs have been primed and initiated, remove the monitor and keyboard, as well any other extraneous equipment (e.g., anemometers, notebooks), from the cage.
41. Release the mosquitoes from their holding container into the center of the cage, 2 h prior to the start of the experiment, and leave the cage as quickly as possible.

    In this protocol mosquitoes are released into the cage at 20:00 to acclimate to the semi-field conditions for 2 h.

42. At 22:00, introduce the desired olfactory stimuli into the tents (i.e., one person in each tent) for the duration of the experiment until 4:00.

    To control for position effects over replicate nightly trials the position of the stimuli should be randomized each night.

43. Retrieve the video data stored in the Raspberry Pi using a USB flash drive the morning after the experiment has ended. For a 6-h recording each 1-h video will have a size of ~500 MB.
44. Clear all the mosquitoes from the cage using Prokopack aspirators.
45. Analyze video files with a video tracking program to quantify the number of landings and target occupancy on each OGTA over the 6-h assay period (e.g., ivTrace).

    The number of landings and target occupancy can be calculated as a measure of attractiveness of the stimulus.

    See Troubleshooting.

Shutting Down Raspberry Pi (Optional)

To save battery power, the Raspberry Pi can be set to automatically shut down by setting up a cron job that runs a "shutdown.txt" script. This script will stop all python scripts that are running (i.e., "Temperature_arduino.py") and shut down the Raspberry Pi. By shutting down the Raspberry Pi, the Arduino and Peltier Element are also automatically turned off.

46. Copy the 'shutdown.txt' file onto a folder on the desktop.
47. Open a new terminal and type "`crontab -e.`"
48. When asked to select an editor, type "`1`" and hit enter.
49. Once the editor opens, set the time, date and script to be run:

    `minute/ hour/ day of month/ month/ day of week/ command to execute`

    e.g., "`05 04 * * * bash /home/pi/Desktop/Scripts/shutdown.txt`"

    The example above will run the shutdown.txt script stored in the "Scripts" folder of the Desktop at 4:05 a.m. every day (i.e., 5 min after the end of the assay). The text sequence " * *" in this script for: "day of month/ month/day of week" specifies the script will run every day.*

50. Type "`ctrl x`" to exit the cron job editor.

    Once the cron job is set, it does not have to be set again unless the end time of the assay needs to be modified.

 Cite this protocol as *Cold Spring Harb Protoc*; doi:10.1101/pdb.prot108304

TROUBLESHOOTING

Problem (Step 25): The Peltier element is not heating the platform.

Solution: Ensure the "hot" side of the Peltier element is glued to the bottom of the platform. Alternatively, check for loose connections or wrong connections and make sure the battery connected to the PE is charged. Make sure that the appropriate Arduino file (Thermistor_Peltier_control .ino) has been uploaded. Refer to Protocol 1: A Behavioral Assay to Quantify Odor-Guided Thermotaxis with *Anopheles gambiae* under Semi-Field Conditions (Giraldo and McMeniman 2023) for details.

Problem (Step 25): The landing platform is overheating above the set temperature.

Solution: Ensure the thermistor is in direct contact (i.e., securely taped) to the aluminum platform and that the Arduino code (i.e., Temperature_arduino.py) specifies the correct desired temperature.

Problem (Step 45): Mosquito activity in the assay is low.

Solution: This will be dependent on the semi-field conditions. Ensure that the experiments are carried out during optimal seasonal conditions for host-seeking behavior, and that the timing of the experiment happens around peak activity of the mosquito species. Large containers with water can be placed in the corners or center of the cage to locally increase the humidity in the semi-field environment and mosquito performance. Installation of a weather station (e.g., HOBO U30 USB weather station) adjacent to the semi-field facility is highly recommended to record environmental parameters such as temperature, humidity, and wind direction throughout experimental trials.

DISCUSSION

In this protocol, we describe engineering and establishment of a novel multichoice olfactory preference assay in Zambia for the quantification of *An. gambiae* attraction to whole body odor derived individually from two sleeping humans. By competing two OGTAs (see Protocol 1: A Behavioral Assay to Quantify Odor-Guided Thermotaxis with *Anopheles gambiae* under Semi-Field Conditions [Giraldo and McMeniman 2023]) against each other, it is possible to quantify the number and frequency of *An. gambiae* landings on two competing platforms baited with whole body odor from each volunteer over a 6-h period (Fig. 2E). We have demonstrated the use of IR OGTA illumination and IR-sensitive cameras facilitates performing these assays during nighttime hours (22:00–4:00) which constitutes the peak period of *An. gambiae* host-seeking activity. Replicate trials can be performed over consecutive nights to document olfactory preferences and position effects controlled for by randomizing the location of the two competing tents around the perimeter of the semi-field cage facility. Important prerequisites for running these assays under semi-field conditions away from domestic power include having a ready supply of 12-V batteries and charging mechanism (e.g., solar charging bank) on hand. For each assay night, three 12-V batteries are required per OGTA to power its constituent fan assembly, Peltier element, and Raspberry Pi. Other important considerations include the calibration of fan speeds prior to the start of the assay to standardize the amount of whole body odor flowing over each OGTA and the downstream tracking algorithm that will be used to quantify mosquito landings on each OGTA (e.g., background subtraction, machine learning, and neural networks). The format of this assay can be further modified to increase the number of competing OGTAs for multichoice olfactory preference assays with expanded cohorts of humans. Alternatively, other animal species (e.g., cows or goats) may be introduced into modified tents to quantify *An. gambiae* host preferences versus humans, or repellents applied to human volunteers or the OGTA itself for mosquito exposure-free assays of the impact of these interventions of *An. gambiae* host-seeking behavior and landing.

ACKNOWLEDGMENTS

We thank Johns Hopkins Malaria Research Institute and Bloomberg Philanthropies for generous pilot funding to C.J.M. that enabled development of the semi-field facility and multichoice olfactory preference assay at Macha Research Trust, Zambia detailed in this protocol. D.G. was supported by postdoctoral fellowships from the Human Frontier Science Program (LT000310/2019-L) and Johns Hopkins Malaria Research Institute. A special thanks to Chris Book and the MRT workshop for their expert engineering and construction of the semi-field cage, Genevieve Tauxe for contributions to OGTA design, Dorian Jackson for assistance with data analysis, and Jennifer Stevenson, Phil Thuma, Monicah Mburu, Edgar Simulundu, Limonty Simubali, and the entire MRT entomology team for their support. Human subject research performed in this protocol was approved by the Johns Hopkins Bloomberg School of Public Health Institutional Review Board (IRB Approval No. 00018691) and Macha Research Trust IRB (IRB Approval No. E2021.06).

REFERENCES

Ferguson HM, Ng'habi KR, Walder T, Kadungula D, Moore SJ, Lyimo I, Russell TL, Urassa H, Mshinda H, Killeen GF, et al. 2008. Establishment of a large semi-field system for experimental study of African malaria vector ecology and control in Tanzania. *Malar J* 7: 158. doi:10.1186/1475-2875-7-158

Giraldo D, McMeniman CJ. 2023. A behavioral assay to quantify odor-guided thermotaxis with *Anopheles gambiae* under semi-field conditions. *Cold Spring Harb Protoc* doi:10.1101/pdb.prot108303

Knols BG, Njiru BN, Mathenge EM, Mukabana WR, Beier JC, Killeen GF. 2002. MalariaSphere: a greenhouse-enclosed simulation of a natural *Anopheles gambiae* (Diptera: Culicidae) ecosystem in western Kenya. *Malar J* 1: 19. doi:10.1186/1475-2875-1-19

Niang A, Nignan C, Serge Poda B, Sawadogo SP, Roch Dabiré K, Gnankiné O, Tripet F, Roux O, Diabaté A. 2019. Semi-field and indoor setups to study malaria mosquito swarming behavior. *Parasit Vectors* 12: 446. doi:10.1186/s13071-019-3688-0

Ritchie SA, Johnson PH, Freeman AJ, Odell RG, Graham N, Dejong PA, Standfield GW, Sale RW, O'Neill SL. 2011. A secure semi-field system for the study of *Aedes aegypti*. *PLoS Negl Trop Dis* 5: e988. doi:10.1371/journal.pntd.0000988

CHAPTER 15

Key Features and Considerations for Using Automated Long-Term Monitoring of Heat-Seeking Behavior of Mosquitoes

Chisako Sakuma,[1,2,3] Emi Maekawa,[1,2,4] and Hirotaka Kanuka[1,2,5]

[1]*Department of Tropical Medicine,* [2]*Center for Medical Entomology, The Jikei University School of Medicine, Tokyo 105-8461, Japan;* [3]*Laboratory for Nutritional Biology, RIKEN Center for Biosystems Dynamics Research, Hyogo 650-0047, Japan;* [4]*Group of Genetics, Graduate School of Science, Nagoya University, Aichi 464-8602, Japan*

Mosquitoes are vectors of various deadly infectious diseases, including malaria and dengue fever. Because these pathogens are transmitted by mosquito blood-feeding behavior, it is important to understand how mosquitoes are attracted to their hosts and how they feed on blood. The simplest method is to observe their behavior with the naked eye or video recording. Furthermore, various devices have been invented to assay mosquito behavior, such as olfactometers. Although each method has distinctive advantages, all have disadvantages, such as limitations in the number of individuals that can be assayed at one time and in observable durations, objective quantification methods, and others. To solve these problems, we have developed an automated device to quantify the carbon dioxide–activated heat-seeking behavior of *Anopheles stephensi* and *Aedes aegypti* with continuous monitoring for up to 1 week. This device, which is detailed in an accompanying protocol, can be applied to search for substances and molecules that affect heat-seeking behavior. It may also be applicable to other hematophagous insects.

BACKGROUND

To find ways to control and regulate the behavior of mosquitoes that transmit various diseases, including malaria, yellow fever, and dengue fever, as well as Zika fever, it is important to understand and observe every detail of mosquito behavior. The easiest and most convenient method, used for many years including the present day, is observing mosquito movement with the naked eye. For example, to observe host-seeking behavior, the number of mosquitoes that landed on the host animal or human volunteers within a given period was counted in a mesh cage or a transparent container (DeGennaro et al. 2013; Duvall et al. 2019). However, as long as we use humans in these experiments, there will be an experimental limitation—namely, that pathogen-infected mosquitoes cannot be used because of the risk of infecting the volunteers. Because the above method requires intervals and is limited to the number of observations that can be made with the eye, video recording has recently become mainstream.

[5]Correspondence: kanuka@jikei.ac.jp

Cite this introduction as *Cold Spring Harb Protoc;* doi:10.1101/pdb.top107665

VIDEO RECORDING

Thanks to technological development, high-speed cameras with good pixel counts are now available at relatively low cost, and high-resolution behavioral videos can be obtained with only one click. However, quantifying the behaviors has been a challenge. Tracking the movement of individuals without bias and extracting and analyzing information from the enormous amount of video data require a high-end personal computer, knowledge of bioinformatics, and engineering skills. Recently, an open platform called biteOscope was developed that allows the high-resolution and high-throughput characterization of surface exploration, probing, and engorgement by blood-sucking mosquitoes in two dimensions (Hol et al. 2020). Similar tracking of behavior using machine learning could be applied to other behaviors. However, the analysis of host-seeking behavior in three dimensions will be more difficult. Indeed, the flying activity of groups of female mosquitoes was once tracked using the SciTrackS three-dimensional multi-insect tracking system, which requires custom-built software and further processing, filtering, and analysis in MATLAB (McMeniman et al. 2014). There are limited environments where this equipment and technology can be used. Furthermore, the analysis of even more extended periods, larger spaces, and larger numbers of individuals would also be demanding.

OLFACTOMETERS

To make the quantification process simple and accessible to all, various types of devices were developed. One solution is to allow observers to simply count the number of mosquitoes that have entered a given space after a certain time. For this purpose, an olfactometer, used to determine whether mosquitoes are attracted by odors, is used (Willis 1947). Mosquitoes perceiving an attractant odor placed at one end of the acrylic tube fly toward the stimulus source and start probing, so only mosquitoes approaching the stimulus source need to be counted. Researchers have improved the structures of olfactometers by setting sliding doors and traps. Uniport (Liesch et al. 2013; Raji et al. 2019), two-port (Bohbot et al. 2013; McBride et al. 2014), three-port (Mukabana et al. 2002), Quattroport (Basrur et al. 2020), and even miniport (Duvall et al. 2019) olfactometers are used for different purposes. In addition to two-port olfactometers, Y-shaped olfactometers are also used for dual olfactory choice assays (Emami et al. 2017). Wind tunnels are another type of olfactometer, which are often used to analyze the flight behavior of mosquitoes in response to the plume of human odors or carbon dioxide (Dekker et al. 2005). The versatile climate-controlled chamber and wind tunnel system was recently developed, facilitating tracking insect flight in three dimensions (Hinze et al. 2021). The difficulty in performing these assays is the regulation of airflow; without proper airflow, mosquitoes will not be attracted to the odor source. In addition, there is also concern that the shape of the olfactometer may affect the results. For example, tubes that are too fine generate static electricity, which prevents mosquitoes from flying freely. Also, this analysis does not lend itself to behavioral analysis in the dark, such as host-seeking behavior in nocturnal mosquitoes. An extensive semifield system (SFS) is used as an intermediate step between laboratory and field trials that offers a safe, controlled environment that reproduces field conditions (Ferguson et al. 2008; Ritchie et al. 2011; McMeniman et al. 2014). However, in particular, the high construction costs and the requirement of a large open area are challenging when using SFS.

Previous devices evaluated only the amount of movement in mosquito behavior or focused on a distinct part of host-seeking behavior. Thus, they did not clearly distinguish the actual period of the mosquito's blood-sucking behavior or the successive behavior process. In other words, it is difficult for the above experimental setups to ascertain the whole process of host-seeking behavior, including (1) target recognition from a distance, (2) orientation, (3) landing on the target, and (4) leaving from the target, which is observed in mosquito blood-sucking behavior.

Cite this introduction as *Cold Spring Harb Protoc*; doi:10.1101/pdb.top107665

AUTOMATED DEVICE USING THE INFRARED BEAM BREAK METHOD

With these considerations in mind, we developed an automated long-term monitoring device for host-seeking behavior using the infrared beam break method (Maekawa et al. 2011; see Protocol 1: Automated Long-Term Monitoring of the Heat-Seeking Behavior of Mosquitoes [Sakuma et al. 2023]). We used carbon dioxide and heat sources as possible attractants in the device to elicit selected host-seeking behavior because both factors are commonly observed in all warm-blooded animals. It can simultaneously monitor three independent mosquito behaviors—namely, touchdown on a heated target, sugar feeding, and locomotion activity for up to a week, by using several pairs of infrared laser sensors. The observer can wait for the counts for each sensor to come up as an Excel table from a remote location, thus separating the influence of potential background contamination by the presence of an experimenter. A prototype of this device is reported to use black targets heated by a water bath, in addition to using carbon dioxide to activate mosquito flight (Kawada and Takagi 2004). Because the Peltier element allows accurate and quick control of temperature, in our device, we used a Peltier element as a heated target and an unheated Peltier element as a control. Furthermore, laser sensors were added to simultaneously monitor the food (sugar)-seeking behavior and the locomotion of mosquitoes as they moved freely through the large mesh cage. This device will enable the noninvasive evaluation of mosquito host-seeking behavior and the efficacy of mosquito repellents.

REFERENCES

Basrur NS, De Obaldia ME, Morita T, Herre M, von Heynitz RK, Tsitohay YN, Vosshall LB. 2020. *fruitless* mutant male mosquitoes gain attraction to human odor. *eLife* **9:** e63982. doi:10.7554/eLife.63982

Bohbot JD, Durand NF, Vinyard BT, Dickens JC. 2013. Functional development of the octenol response in *Aedes aegypti*. *Front Physiol* **4:** 39. doi:10.3389/fphys.2013.00039

DeGennaro M, McBride CS, Seeholzer L, Nakagawa T, Dennis EJ, Goldman C, Jasinskiene N, James AA, Vosshall LB. 2013. *orco* mutant mosquitoes lose strong preference for humans and are not repelled by volatile DEET. *Nature* **498:** 487–491. doi:10.1038/nature12206

Dekker T, Geier M, Carde RT. 2005. Carbon dioxide instantly sensitizes female yellow fever mosquitoes to human skin odours. *J Exp Biol* **208:** 2963–2972. doi:10.1242/jeb.01736

Duvall LB, Ramos-Espiritu L, Barsoum KE, Glickman JF, Vosshall LB. 2019. Small-molecule agonists of *Ae. aegypti* neuropeptide Y receptor block mosquito biting. *Cell* **176:** 687–701.e685. doi:10.1016/j.cell.2018.12.004

Emami SN, Lindberg BG, Hua S, Hill SR, Mozuraitis R, Lehmann P, Birgersson G, Borg-Karlson AK, Ignell R, Faye I. 2017. A key malaria metabolite modulates vector blood seeking, feeding, and susceptibility to infection. *Science* **355:** 1076–1080. doi:10.1126/science.aah4563

Ferguson HM, Ng'habi KR, Walder T, Kadungula D, Moore SJ, Lyimo I, Russell TL, Urassa H, Mshinda H, Killeen GF, et al. 2008. Establishment of a large semi-field system for experimental study of African malaria vector ecology and control in Tanzania. *Malar J* **7:** 158. doi:10.1186/1475-2875-7-158

Hinze A, Lantz J, Hill SR, Ignell R. 2021. Mosquito host seeking in 3D using a versatile climate-controlled wind tunnel system. *Front Behav Neurosci* **15:** 643693. doi:10.3389/fnbeh.2021.643693

Hol FJ, Lambrechts L, Prakash M. 2020. BiteOscope, an open platform to study mosquito biting behavior. *eLife* **9:** e56829. doi:10.7554/eLife.56829

Kawada H, Takagi M. 2004. Photoelectric sensing device for recording mosquito host-seeking behavior in the laboratory. *J Med Entomol* **41:** 873–881. doi:10.1603/0022-2585-41.5.873

Liesch J, Bellani LL, Vosshall LB. 2013. Functional and genetic characterization of neuropeptide Y-like receptors in *Aedes aegypti*. *PLoS Negl Trop Dis* **7:** e2486. doi:10.1371/journal.pntd.0002486

Maekawa E, Aonuma H, Nelson B, Yoshimura A, Tokunaga F, Fukumoto S, Kanuka H. 2011. The role of proboscis of the malaria vector mosquito *Anopheles stephensi* in host-seeking behavior. *Parasit Vectors* **4:** 10. doi:10.1186/1756-3305-4-10

McBride CS, Baier F, Omondi AB, Spitzer SA, Lutomiah J, Sang R, Ignell R, Vosshall LB. 2014. Evolution of mosquito preference for humans linked to an odorant receptor. *Nature* **515:** 222–227. doi:10.1038/nature13964

McMeniman CJ, Corfas RA, Matthews BJ, Ritchie SA, Vosshall LB. 2014. Multimodal integration of carbon dioxide and other sensory cues drives mosquito attraction to humans. *Cell* **156:** 1060–1071. doi:10.1016/j.cell.2013.12.044

Mukabana WR, Takken W, Coe R, Knols BG. 2002. Host-specific cues cause differential attractiveness of Kenyan men to the African malaria vector *Anopheles gambiae*. *Malar J* **1:** 17. doi:10.1186/1475-2875-1-17

Raji JI, Melo N, Castillo JS, Gonzalez S, Saldana V, Stensmyr MC, DeGennaro M. 2019. *Aedes aegypti* mosquitoes detect acidic volatiles found in human odor using the IR8a pathway. *Curr Biol* **29:** 1253–1262.e1257. doi:10.1016/j.cub.2019.02.045

Ritchie SA, Johnson PH, Freeman AJ, Odell RG, Graham N, Dejong PA, Standfield GW, Sale RW, O'Neill SL. 2011. A secure semi-field system for the study of *Aedes aegypti*. *PLoS Negl Trop Dis* **5:** e988. doi:10.1371/journal.pntd.0000988

Sakuma C, Maekawa E, Kanuka H. 2023. Automated long-term monitoring of the heat-seeking behavior of mosquitoes. *Cold Spring Harb Protoc* doi:10.1101/pdb.prot108173

Willis ER. 1947. The olfactory responses of female mosquitoes. *J Econ Entomol* **40:** 769–778. doi:10.1093/jee/40.6.769

Cite this introduction as *Cold Spring Harb Protoc*; doi:10.1101/pdb.top107665

Protocol 1

Automated Long-Term Monitoring of the Heat-Seeking Behavior of Mosquitoes

Chisako Sakuma,[1,2,3] Emi Maekawa,[1,2,4] and Hirotaka Kanuka[1,2,5]

[1]*Department of Tropical Medicine,* [2]*Center for Medical Entomology, The Jikei University School of Medicine, Tokyo 105-8461, Japan;* [3]*Laboratory for Nutritional Biology, RIKEN Center for Biosystems Dynamics Research, Hyogo 650-0047, Japan;* [4]*Group of Genetics, Graduate School of Science, Nagoya University, Aichi 464-8602, Japan*

Heat is one of the critical cues female mosquitoes use to find vertebrate hosts, particularly in the final step of attraction (touching down on hosts) before the initiation of blood-sucking. To prevent vector-borne diseases such as malaria and dengue fever, which are transmitted via mosquito blood-sucking, it is essential to understand the dynamics and mechanisms of the heat-seeking behavior of mosquitoes. An automated device to quantify CO_2-activated heat-seeking behavior with continuous monitoring for up to a week was established. The device is based on the infrared beam break method; it can simultaneously monitor three independent mosquito behaviors (touchdown on a heated target, feeding, and locomotion activity) by using several pairs of infrared laser sensors. This protocol describes brief instructions for the device's construction, directions for use, and possible issues and solutions.

MATERIALS

It is essential that you consult the appropriate Material Safety Data Sheets and your institution's Environmental Health and Safety Office for proper handling of equipment and hazardous material used in this protocol.

Reagents

Adult mosquitoes (females, at 4 d after eclosion)
Distilled H_2O
Sucrose solution (10%, w/v)

Equipment

Amplifiers (four total; Keyence Corp. LV-51M)
Aspirator
CO_2 cylinder
CO_2 detector (Testo 535)
CO_2 nozzle (silicone tube, internal diameter, 0.8 cm; outer diameter, 1.2 cm)
Cotton puff (5-cm × 6-cm, medical cotton puff no. 6; Hakujuji Co., Ltd. 11373)
Disposable sink strainer mesh bag (Bonstar Sales Co., Ltd. M-298)

[5]Correspondence: kanuka@jikei.ac.jp

Cite this protocol as *Cold Spring Harb Protoc*; doi:10.1101/pdb.prot108173

Electric fan (Toshiba VFP-8CS3)
Filter paper (Whatman 1001-125)
Glass flask (50-mL)
Incubator (Panasonic MIR-253)
Infrared laser sensors (seven pairs; three pairs on target box and four pairs for locomotion measurement, composed of a laser releaser and acceptor; Keyence Corp. LV-H300)
Intermittent timer (TGK FT-022)
Monitoring software (Data Builder; Keyence Corp.)
Mosquito cage (70-cm-width × 58-cm-length × 161.8-cm-height for the MIR-253 incubator)

We make ours using pipes with pipe joints (metal or polyvinyl chloride) and a cube-shaped nylon mesh cover (Sansho Co., Ltd. 110S, 91-1196) that fits the cage. The nylon mesh cover can be created by sewing nylon mesh fabric, with a sleeve made of the same fabric on the left side, where tubes and cables pass through. It is recommended to sew a two-way zipper at the front side of the cover.

Noncontact infrared thermometer
Paper cup (top diameter, 97.2 mm; height, 100 mm; bottom diameter, 74.3 mm; Tokan Kogyo Co., Ltd. 31-476-001)
Peltier controllers (two total; VICS VPE-10)
Peltier elements (two total; VICS LVPU-30)
Programmable controller unit (Keyence Corp. KV-700)
Shield (optional, see Step 3; e.g., glass protection black plate, Panasonic 6240282596)
Solenoid valve (Flon Industry Co., Ltd. FSD-0408C)
Target boxes (three total; 7-cm-width × 41-cm-length × 12.7-cm-height each)

Our boxes were made by an acrylic processing company. Acrylic boards (3-mm-thick) were cut into the shape of each part and glued together in three dimensions. See Figure 1C for exact dimensions. The box has a space (7-cm × 10-cm × 11.1-cm) with an open top in the center. The space accommodates a target, such as a Peltier element or a conical flask with sucrose solution.

Time switch (Panasonic TB17101)
Trays (two total; plastic photo tray for 6.5 × 8.5 photos, 29-cm × 23-cm × 4.5-cm; As One Corp. 5-182-02)
White paper
Windows personal computer (PC)
Wooden sticks (optional; two total; see Step 10)

METHOD

Construction of the Device

1. Make a hole in the lower side of an incubator and set up an electric fan (Fig. 1A).

   The fan draws air out from the incubator and keeps the air constant. The size of the hole depends on the fan. The fan is always on during the experiment.

   We asked a sales representative to drill this hole (and the one in Step 2) for us.

2. Make another hole (10- to 15-cm-diameter) at the side of the incubator for tubes and cables.
3. Install an external time switch to the incubator (or use a built-in module) to operate the light and dark conditions (e.g., 16 h light and 8 h dark).

   If the incubator has any view window, shade it using a shield to prevent light from entering from outside. A built-in module can be an alternative to the external time switch.

4. Place a mosquito cage inside the incubator.
5. Set the three target boxes parallel to one another at the bottom of the cage and place a pair of infrared laser sensors at the top of each box, ~30 cm apart (Fig. 1B). Wire each laser sensor to a

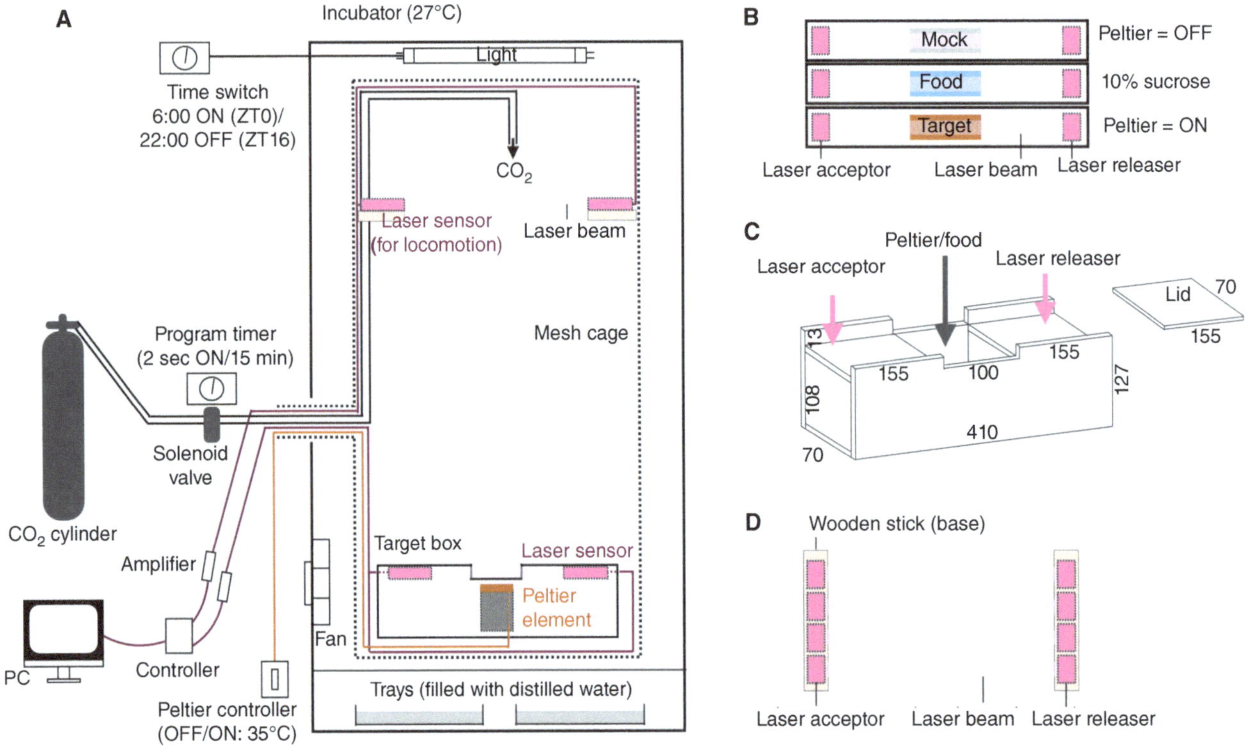

FIGURE 1. Schematic view of the device for monitoring the heat-seeking behavior of mosquitoes. (*A*) The front view of the device. An incubator contains four pairs of laser sensors (pink; the *upper* one for locomotion detection and the others at the *bottom*) connected to amplifiers and a programmable controller unit. To observe the heat-seeking behavior, a heated Peltier element ("target"; orange) is placed in a target box at the *bottom* of the incubator. Two other boxes are also placed in line with the target box ("mock" and "food"). A nozzle for injecting CO_2 into the incubator is fixed in the upper part of a mosquito cage and provides CO_2 at desired intervals using a solenoid valve equipped with a programmable timer. A time switch controls the device's light and dark cycle, and the air is ventilated with an electric fan. Trays filled with distilled H_2O at the bottom of the incubator maintain humidity in the device. (*B*) The overhead view of the three target boxes. A nonheated Peltier element ("mock"; gray), a glass flask containing a 10% (w/v) sucrose solution with a filter paper inserted ("food"; blue), and a heated Peltier element of which the surface is set at 35°C ("target"; orange) are placed in the center of each target box. Laser beams (emitted from the laser releasors and accepted by respective acceptors) pass just above the mock, food, or target, which quantify the heat-seeking and sugar-feeding behaviors by counting the number of beam breaks caused by the movement of mosquitoes. (*C*) A drawing of a target box. The unit for the numbers listed is mm. On the *left* and *right* sides are areas where the laser releasor or acceptor and acrylic plates for the lid are placed on top of the laser acceptor or releasor. (*D*) The overhead view of laser sensors lined up in the upper one-third of the cage for the locomotion measurement. On either side of the cage, a base is built with wooden sticks, with four laser acceptors lined up on one side and four receptors on the other.

respective amplifier connected to a programmable controller unit. Direct the cables of the laser sensors of the incubator through the hole.

6. Install monitoring software (Data Builder) on a Windows PC and connect the PC to the KV-700 unit. Set a program using the software to measure the beam-break counts of each pair of laser sensors in accordance with the manufacturer's instructions.
7. Prepare two Peltier elements and place each in the center of a target box.

   We recommend setting these Peltier elements in the first and third rows.

8. Connect the Peltier elements to the Peltier controllers placed outside of the incubator.
9. Prepare a silicone tube for the CO_2 cylinder and fix the cut end of the tube to the top of the cage. Slide the tube through the hole of the incubator and connect it to a solenoid valve equipped with an intermittent timer and a CO_2 cylinder.
10. (Optional) To monitor the free-flying locomotion of mosquitoes, make a base with a wooden stick on each side of the upper one-third of the cage and line up four pairs of sensors (Fig. 1D).

 Cite this protocol as *Cold Spring Harb Protoc*; doi:10.1101/pdb.prot108173

Instructions for Use

11. At 24 h before the start of the experiment, turn the incubator on and set the temperature inside to 27°C.
12. Confirm the timer switch is set to manage light and dark conditions correctly (e.g., on at 6:00 and off at 22:00).
13. Place trays filled with distilled H_2O at the bottom of the incubator to maintain humidity inside.
14. Turn the electric fan on to ventilate the air in the incubator.
15. Adjust the intermittent timer to open the solenoid valve every 15 min for 60 sec. Open the stopcock of the CO_2 cylinder and subsequently measure the concentration of CO_2 in the incubator every 10 sec using a CO_2 detector. Confirm that the concentration of CO_2 goes back to the basal level within 15 min. Close the stopcock of the cylinder until the experiment starts.
16. Pull out the top of the lid of a paper cup and cover it with a disposable sink strainer mesh bag.
17. Soak cotton puffs in a 10% (w/v) sucrose solution and place them on the cup.
18. Collect 60 female adult mosquitoes (4 d after eclosion) using an aspirator and transfer them to the paper cup.
19. Place the paper cup with cotton puffs at the bottom of the incubator and allow the mosquitoes to become accustomed to the environment overnight (15–18 h).
20. On the first day of the experiment, turn on the PC and start the monitoring software (Data Builder program) to connect the PC to the KV-700 unit.
21. Check whether the device can appropriately count the number of times mosquitoes pass in front of the sensor by dropping a dead mosquito to pass through the middle of each pair of laser sensors. If the count does not increase correctly, confirm the direction and adjust the sensitivity of the laser sensor.

    See Troubleshooting.
22. Place 30 mL of a 10% (w/v) sucrose solution in a 50-mL glass flask and insert a rounded filter paper. Place the flask in the second row of the target boxes.

    Mosquitoes will ingest the sucrose at the top portion of the wet filter paper protruding from the flask.
23. Tune on a Peltier element ("target") to raise the surface temperature to 35°C. Use a noncontact infrared thermometer to adjust the temperature. Keep another Peltier element ("mock") off throughout the experiment. Cover each surface of the Peltier element with white paper to make it visually imperceptible to mosquitoes.
24. Check whether most mosquitoes are alive in the paper cup and release the mosquitoes into the cage by removing the mesh bag on a paper cup with a gloved hand through the slightly opened zipper on the cage.
25. Open the stopcock of the CO_2 cylinder and adjust the intermittent timer to inject the appropriate amount of CO_2 into the incubator (for example, 2 sec at 15-min intervals).
26. Start the experiment and carry it out for up to 1 wk. Check that the counts are working correctly every 12 or 24 h.

    See Troubleshooting.

    In our experience, responses decrease if the behavior is monitored for more than 1 wk.
27. Replace the sucrose solution with a new one by opening the two-way zipper a little at 48–72 h after the experiment begins. Place other substances, such as attractants or repellents, similarly, as desired for your experimental setup.
28. On the last day of the experiment, disconnect the KV-700 unit from Data Builder.
29. Save the data (recorded as an Excel file format).

30. Turn off the electric fan and the Peltier element and close the stopcock of the CO_2 cylinder.
31. For up to 24 h, set the incubator's temperature to 0°C to euthanize the mosquitoes if these mosquitoes are not needed for further experiments.
32. Discard dead mosquitoes, distilled H_2O in the tray, and sucrose solution in the glass flask.
33. Analyze the data to quantify the heat-seeking behavior of the mosquitoes examined.

    Locomotion is calculated as the sum of counts measured by four sensors set at the top of the cage.

TROUBLESHOOTING

Problem (Steps 21 and 26): There are few or no counts at the heated target.
Solution: Check whether the distance of the two laser sensors (releaser and acceptor) is appropriate.

- Check the direction and sensitivity of the laser.
- Check whether the mosquitoes are alive, are active, and have not escaped the cage.
- Check the humidity inside the incubator.
- Check whether the sucrose solution is appropriately supplied (not running out) and whether the filter paper in the solution has not dried up.

Problem (Step 26): There are too many counts at the heated target.
Solution:

- Check whether any vibration is occurring around the incubator. Vibration can be caused by surrounding machines and the movement of people.
- Adjust the laser sensor's sensitivity to not count the resting mosquitoes in front of the laser sensor, which occasionally generates overcounting.

Problem (Step 26): There are too many counts at the mock.
Solution: Check whether the heat does not warm the target boxes from the Peltier element on the opposite side. The heat generated by the Peltier element should be dissipated.

Problem (Step 26): There are no counts at the food treatment.
Solution: Check whether the sucrose solution is appropriately supplied (not running out) and whether the filter paper in the solution has not dried up.

Cite this protocol as *Cold Spring Harb Protoc;* doi:10.1101/pdb.prot108173

CHAPTER 16

Acoustic-Related Mating Behavior in Tethered and Free-Flying Mosquitoes

Garrett P. League,[1,6] Catalina Alfonso-Parra,[2,3] Hoover Pantoja-Sánchez,[4,5] and Laura C. Harrington[1,6]

[1]*Department of Entomology, Cornell University, Ithaca, New York 14853, USA;* [2]*Instituto Colombiano de Medicina Tropical, Universidad CES, Sabaneta, Antioquia, Colombia, 055413;* [3]*Max Planck Tandem Group in Mosquito Reproductive Biology,* [4]*Departamento de Ingeniería Electrónica, SISTEMIC,* [5]*Programa de Estudio y Control de Enfermedades Tropicales, PECET, Universidad de Antioquia, Medellín, Antioquia, Colombia, 050010*

Acoustics play an essential role in mosquito communication, particularly during courtship and mating. Mosquito mating occurs in flight and is coordinated by the perception of wingbeat tones. Flight tone frequencies have been shown to mediate sex recognition in *Aedes*, *Anopheles*, *Culex*, and *Toxorhynchites* genera and are thus a conserved feature of mating across the mosquito family (Culicidae). Upon recognizing a flying female, males respond phonotactically by lunging toward the female and initiating a precopulatory courtship flight interaction. During this interaction, males and females often harmonize their flight tones in a behavior known as harmonic convergence, and male acoustics display rapid frequency modulation. These acoustic phenomena have been characterized both in tethered and free-flying mosquitoes using similar audio recording and analysis methods. Further, the manipulation of mosquito acoustic-related mating behavior shows great promise as a tool for reproductive control strategies. In this brief methodological introduction, we provide an overview of the biological and technical concepts necessary for understanding the recording and analysis of mosquito mating acoustics.

MOSQUITO MATING ACOUSTICS

Mosquito mating occurs across a wide range of behavioral and ecological contexts. For example, mosquito mating can involve intricate courtship movements, small aggregations around hosts indoors, or mass outdoor aerial swarming events (Cator et al. 2021). Mosquitoes in the genus *Aedes* can mate throughout the day, with peaks at dusk and dawn, in relatively small (tens of individuals) male-skewed swarms (Cator et al. 2011, 2021). Mating in *Aedes* tends to be closely associated with human hosts and thus often occurs indoors, with males intercepting and mating with females as they approach the host to blood feed (Hartberg 1971; Yuval 2006). In contrast, in many African species of the genus *Anopheles*, swarms occur primarily at dusk (Sawadogo et al. 2013) and form relative to contrasting ground markers (Sawadogo et al. 2014). Although *Anopheles* swarms often form near human habitations (Diabaté et al. 2011), mating occurs outdoors in large swarms (tens, hundreds, or thousands of individuals) comprised almost entirely of males (Charlwood and Jones 1980; Diabaté et al. 2011). Common to *Aedes* and *Anopheles* species studied to date, mating occurs while both partners are in flight and thus takes place rapidly. Mating typically lasts from 10 to 60 sec (Roth

[6]Correspondence: lch27@cornell.edu; gpl47@cornell.edu

Cite this introduction as *Cold Spring Harb Protoc*; doi:10.1101/pdb.top107667

1948; Charlwood and Jones 1979; Clements 1999), give or take several seconds depending on the species and conditions (e.g., temperature) at the time of mating. As such, mosquito flight acoustics represent a critical form of communication both before and during the act of mating.

Perception of sound in mosquitoes was first documented by Johnston (1855). Air particles displaced by mosquito wingbeats are detected by the many fine hairs (fibrillae) of the antennae. The movement of these hairs is transmitted down the main stalk (flagellum) of the antenna, triggering mechanosensory neurons in the hearing organ at the base of the antenna that now bears Johnston's name: the Johnston's organ (Göpfert et al. 1999; Göpfert and Robert 2001). Males are known for their exquisite hearing abilities thanks to their many Johnston's organ neurons and plumose (i.e., bushy) antennae, which in some species (e.g., *Anopheles* spp.) become erect during swarming periods to enhance detection of females (Pennetier et al. 2010). Until recently, however, the role of female hearing had not been appreciated in acoustic mating interactions, particularly in *Aedes aegypti*, in which females display a particularly active role in the outcome of mating interactions (Cator et al. 2009; Aldersley and Cator 2019). Although major contributions have been made to our understanding of mosquito hearing in recent decades, there are still many lingering questions regarding acoustic-mediated mating behaviors and their implications for operational mosquito control (Andrés et al. 2020; Cator et al. 2021).

The analysis of acoustic mating behavior in mosquitoes focuses on flight activity that begins with swarming and courtship flight and ends with the completion of copulation. Although our accompanying protocol does not directly apply acoustic analysis techniques to basal mosquito flight, host-seeking flight, or oviposition-related flight, the methods we present could be applied to these behavioral applications, as well as others, with some adjustment (see Protocol 1: Recording and Analysis of Mosquito Acoustic-Related Mating Behavior [Pantoja-Sánchez et al. 2022]). Mosquito mating acoustics begin with male swarming or "patrolling" flight, in which males fly in a distinct up-and-down bobbing "dance" (in the case of *Anopheles* spp.) (Butail et al. 2013) or "figure eight" flight pattern (*Aedes* spp.) (Hartberg 1971) in anticipation of mating. Upon detecting a female that flies into their hearing range (Menda et al. 2019), males initiate the courtship phase of a mating interaction (Aldersley and Cator 2019). Although initially overlooked because of its brevity (Yuval 2006), courtship represents the most well-studied portion of mosquito acoustic mating interactions. Defined generally as "the behavior used to obtain copulation with a partner" (Quigley and Fusani 2018), courting in mosquitoes begins with a largely male-driven pursuit of an available female, which in a swarm can involve intense scramble mating competition (Diabaté et al. 2011; Cator et al. 2021; Garcia Castillo et al. 2021). Once a male suitor is in close contact with a female, he lunges toward her to seize her by the legs in flight. After female capture, the male quickly repositions himself to enable genital contact (Charlwood and Jones 1979; Aldersley and Cator 2019). If successful, copulation ensues, during which the male quickly transfers the ejaculate (sperm and seminal fluid) to the female. Although the couple can drop out of flight in copula (Charlwood et al. 2002; Maïga et al. 2014), copulation often occurs partially or completely in flight or on a surface after landing, depending on the species and circumstances (Charlwood and Jones 1980; Maïga et al. 2014). Thus, mating acoustics can persist through copulation, ending only once the male and female genitalia are no longer connected.

One behavior of particular importance to acoustic mating interactions is the harmonization of male and female wingbeat frequencies during courtship and mating in a phenomenon known as harmonic convergence (Cator et al. 2009). Harmonic convergence was initially documented in tethered individuals of numerous mosquito genera, including *Aedes* (Cator et al. 2009), *Anopheles* (Cator et al. 2010; Pennetier et al. 2010), *Culex* (Warren et al. 2009), and *Toxorhynchites* (Gibson and Russell 2006). Recent studies show that harmonic convergence plays essential roles in sex recognition (Gibson and Russell 2006), mate preference and sexual selection (League et al. 2021; Reitmayer et al. 2021), and mating success (Aldersley and Cator 2019). Other important acoustic mating behaviors include the rapid succession of male flight tone frequency increases and decreases during mating interactions, known as rapid frequency modulation (Simões et al. 2016), and phonotactic flight responses to audio stimuli (Dou et al. 2021), often quantified as response latency time (Römer 2020). In the accompanying protocol, we focus on the recording and quantification of mosquito

Cite this introduction as *Cold Spring Harb Protoc*; doi:10.1101/pdb.top107667

flight tone parameters with application to these and closely related acoustic mating behaviors (see Protocol 1: Recording and Analysis of Mosquito Acoustic-Related Mating Behavior [Pantoja-Sánchez et al. 2022]).

Mating acoustics are a fundamental and conserved aspect of reproduction in mosquitoes. Thus, despite arising mainly from work conducted in two genera—namely, *Aedes* and *Anopheles*—the methods presented here are widely applicable across mosquitoes.

ANALYSIS OF MOSQUITO MATING ACOUSTICS

Analysis of mating acoustics has been conducted on data from experiments on tethered mosquitoes (e.g., Gibson and Russell 2006), free-flying mosquitoes (e.g., Cator et al. 2011; Garcia Castillo et al. 2021), and a combination of tethered and free-flying mosquitoes (e.g., League et al. 2019; Pantoja-Sánchez et al. 2019; Montoya et al. 2021). Because of the ongoing utility of both techniques and the similarity between their recording and analysis methods, the accompanying protocol integrates both tethered and free-flight experimental approaches for examining mosquito mating acoustics (see Protocol 1: Recording and Analysis of Mosquito Acoustic-Related Mating Behavior [Pantoja-Sánchez et al. 2022]). The technical introductions that follow provide brief summaries of the main components of mosquito mating acoustics analysis, including mosquito tethering and free-flight techniques, flight audio recording equipment, flight audio recording, and flight audio analysis.

Mosquito Tethering and Free-Flight Techniques

Mosquito flight audio can be measured by tethering mosquitoes to an insect pin either directly (Charlwood and Jones 1979; Cator et al. 2009) or via a human hair (Cator et al. 2010), which is referred to as "semitethering" because it permits greater flight mobility (Cator and Harrington 2011). Tethers are attached to the mosquito thorax using small amounts of adhesive. Tethering has three primary advantages: (1) It permits adult flight while maintaining proximity to a recording device (microphone); (2) it reduces complex flight tone patterns arising from physical contact, dramatic changes in flight position, and interactions from other mosquitoes into simplified spectra, allowing for greater clarity and ease of analysis; and (3) it enables measurements of responses to live or artificial flight tones that have discernible, precise acoustic behavior outcomes. However, tethering has two major disadvantages: (1) it alters natural flight tone frequencies (usually by lowering them), with the degree of the effect dependent on the mosquito species, sex, and size (Pantoja-Sánchez et al. 2019; Montoya et al. 2021); and (2) it restricts natural mosquito mating flight patterns, including male swarming and rapid frequency modulation (Simões et al. 2016), as well as female mate rejection behaviors (Charlwood and Jones 1979; Aldersley and Cator 2019).

As the name suggests, free-flight techniques are those that record audio from free-flying, rather than tethered, mosquitoes. Because free-flight techniques do not require restraining or manipulating mosquitoes before audio recording, they typically involve placing mosquitoes, along with one or more microphones, inside a recording arena to restrict flight to a confined area. Flight audio recording arenas typically take the form of plastic terrariums or mosquito cages lined with netting, with the size of the arena dependent on the number of mosquitoes and specific behaviors under investigation. Free-flying techniques have the advantage of permitting the full range of natural flight pattern and mating interaction acoustics in the recorded mosquitoes, with the disadvantage of making audio capture and analysis of flight acoustics more technically challenging.

Tethering and free-flight techniques are frequently used in conjunction to capitalize on the strengths of both methods. Most often when this is the case, free-flying males are permitted to acoustically interact with tethered females placed near a microphone. This allows for the ideal capture of mating flight tones by keeping mating interactions near a microphone while permitting dynamic male flight behaviors and introducing only minimal tethering artifacts to female acoustics, which are generally less impacted by tethering.

Flight Audio Recording Equipment

Capturing mosquito flight audio requires several pieces of equipment whose use varies by application. Generally, flight audio recording setups consist of the following: (1) a recording arena in which tethered or free-flying mosquitoes are placed to limit the flight area and reduce background noise; (2) a microphone and amplifier/digitizer connected to a computer via a USB adapter to record flight tone audio; and (3) a computer with audio recording and analysis software for digitally capturing and analyzing flight audio spectrograms. Other optional components include microphone accessories such as parabolic reflectors or funnels to improve audio capture of free-flying individuals in larger spaces (Cator et al. 2011; Simões et al. 2016), ground or horizon markers for stimulating swarm behavior in some species (i.e., *Anopheles* spp.), red or infrared spectrum lighting to visualize mosquitoes while minimizing light interference during recordings performed in the dark, heat and humidity sources to control for impacts on flight tones (Villarreal et al. 2017), acoustic foam to line the recording room/arena, various host stimuli (i.e., heat, odors, CO_2) to provide sensory cues for species that mate near hosts (i.e., *Aedes* spp.), and video cameras to simultaneously record flight patterns with audio recordings (although we do not cover the analysis of flight video here).

Flight Audio Recording

Audio recording of mosquito mating acoustics is performed with microphones placed near the flying individual(s). Near-field, unidirectional particle velocity microphones are often used for capturing only the flight tones of tethered or free-flying individuals in close proximity to the microphone from a single direction. Far-field omnidirectional pressure microphones are ideal for capturing both near and distant free-flying mosquito sounds from any direction. Microphones are typically attached to individual or integrated amplifiers and digitizers that in turn interface with computer programs like Audacity (Audacity Team 2021) via USB attachment. Recording typically starts with the introduction of mosquitoes into the recording arena or at the onset of swarming and/or mating behavior. Recording usually ends once flying male and female pairs have acoustically interacted in a mating attempt or once swarm and/or mating activity ceases and males no longer interact with females.

Flight Audio Analysis

Mosquito flight tones are comprised of a fundamental frequency, or the first harmonic, and its harmonic overtones, or the second, third, etc., harmonics. Analysis of flight tone audio is generally performed by spectral analysis using software like Raven (Center for Conservation Bioacoustics 2014). Audio spectrograms are the most common format for analyzing flight sounds, as they permit visual representation of audio parameters such as frequency, in hertz (Hz) or kilohertz (kHz), and amplitude, in decibels (dB), over time. Analyses can either be manual, involving manual sampling of flight audio parameters at desired interactions or time points (e.g., League et al. 2019), or automated, using software algorithms to automatically extract audio parameters (e.g., Aldersley et al. 2014). Analyses focus primarily on the detection and quantification of male and female fundamental flight tone and harmonic overtone frequencies and amplitudes. Once male and female flight tones are identified, their frequency overlap (generally, differences of <5 Hz are considered converging) and overlap duration (typically ≥ 1 sec) at various harmonic ratios can be calculated and visually assessed to test for the presence of harmonic convergence (Cator and Harrington 2011). Further, the characteristics of flight tone modulations can be assessed in detail to detect and quantify swarming flight (Garcia Castillo et al. 2021), rapid frequency modulation (Simões et al. 2016), flight response latency (Dou et al. 2021), and other acoustic-related mating behaviors.

Cite this introduction as *Cold Spring Harb Protoc*; doi:10.1101/pdb.top107667

CONCLUSION

The biological importance of flight tone acoustics in mediating mosquito mating interactions has become increasingly clear over recent decades because of the core methodologies detailed here. Because of the relatively inexpensive equipment and software required for these analyses, both tethered and free-flight recordings of mosquito flight audio are amenable to a range of research environments, including laboratory (discussed here), semifield, field, and resource-limited settings. Further, the potential efficacy of mosquito flight tones for use in surveillance and trapping techniques has been repeatedly shown, and manipulation of mating acoustics shows great promise for use in reproductive control methods (Sawadogo et al. 2017; Andrés et al. 2020; Cator et al. 2021). Future fine-scale studies of mosquito mating acoustics using the techniques described here, particularly in natural free-flying swarms, will be of great utility for advancing our understanding of mosquito reproductive biology and will pave the way for breakthroughs in mosquito-borne disease control.

ACKNOWLEDGMENTS

We thank Dr. Frank W. Avila and Dr. Lauren J. Cator for technical feedback and scientific content review.

REFERENCES

Aldersley A, Cator LJ. 2019. Female resistance and harmonic convergence influence male mating success in *Aedes aegypti*. *Sci Rep* **9:** 2145. doi:10.1038/s41598-019-38599-3

Aldersley A, Champneys A, Homer M, Robert D. 2014. Time-frequency composition of mosquito flight tones obtained using Hilbert spectral analysis. *J Acoust Soc Am* **136:** 1982–1989. doi:10.1121/1.4895689

Andrés M, Su MP, Albert J, Cator LJ. 2020. Buzzkill: targeting the mosquito auditory system. *Curr Opin Insect Sci* **40:** 11–17. doi:10.1016/j.cois.2020.04.003

Audacity Team. 2021. Audacity®: Free Audio Editor and Recorder [Computer application]. httpss://www.audacityteam.org/

Butail S, Manoukis NC, Diallo M, Ribeiro JMC, Paley DA. 2013. The dance of male *Anopheles gambiae* in wild mating swarms. *J Med Entomol* **50:** 552–559. doi:10.1603/ME12251

Cator LJ, Harrington LC. 2011. The harmonic convergence of fathers predicts the mating success of sons in *Aedes aegypti*. *Anim Behav* **82:** 627–633. doi:10.1016/j.anbehav.2011.07.013

Cator LJ, Arthur BJ, Harrington LC, Hoy RR. 2009. Harmonic convergence in the love songs of the dengue vector mosquito. *Science* **323:** 1077–1079. doi:10.1126/science.1166541

Cator LJ, Ng'Habi KR, Hoy RR, Harrington LC. 2010. Sizing up a mate: variation in production and response to acoustic signals in *Anopheles gambiae*. *Behav Ecol* **21:** 1033–1039. doi:10.1093/beheco/arq087

Cator LJ, Arthur BJ, Ponlawat A, Harrington LC. 2011. Behavioral observations and sound recordings of free-flight mating swarms of *Ae. aegypti* (Diptera: Culicidae) in Thailand. *J Med Entomol* **48:** 941–946. doi:10.1603/ME11019

Cator LJ, Wyer CAS, Harrington LC. 2021. Mosquito sexual selection and reproductive control programs. *Trends Parasitol* **37:** 330–339. doi:10.1016/j.pt.2020.11.009

Center for Conservation Bioacoustics. 2014. Raven Pro: Interactive sound analysis software version 1.5 [computer software]. http:// ravensoundsoftware.com/

Charlwood JD, Jones MDR. 1979. Mating behaviour in the mosquito, *Anopheles gambiae s.l.* I. Close range and contact behaviour. *Physiol Entomol* **4:** 111–120. doi:10.1111/j.1365-3032.1979.tb00185.x

Charlwood JD, Jones MDR. 1980. Mating in the mosquito, *Anopheles gambiae s.l.* II. Swarming behaviour. *Physiol Entomol* **5:** 315–320. doi:10.1111/j.1365-3032.1980.tb00241.x

Charlwood JD, Pinto J, Sousa CA, Madsen H, Ferreira C, do Rosario VE. 2002. The swarming and mating behaviour of *Anopheles gambiae s.s.* (Diptera: Culicidae) from São Tomé Island. *J Vector Ecol* **27:** 178–183.

Clements AN. 1999. *The biology of mosquitoes: sensory, reception, and behavior,* Vol. 2. CABI, Wallingford, United Kingdom.

Diabaté A, Yaro AS, Dao A, Diallo M, Huestis DL, Lehmann T. 2011. Spatial distribution and male mating success of *Anopheles gambiae* swarms. *BMC Evol Biol* **11:** 184. doi:10.1186/1471-2148-11-184

Dou Z, Madan A, Carlson JS, Chung J, Spoleti T, Dimopoulos G, Cammarato A, Mittal R. 2021. Acoustotactic response of mosquitoes in untethered flight to incidental sound. *Sci Rep* **11:** 1884. doi:10.1038/s41598-021-81456-5

Garcia Castillo SS, Pritts KS, Krishnan RS, Harrington LC, League GP. 2021. Harmonic convergence coordinates swarm mating by enhancing mate detection in the malaria mosquito *Anopheles gambiae*. *Sci Rep* **11:** 24102. doi:10.1038/s41598-021-03236-5

Gibson G, Russell I. 2006. Flying in tune: sexual recognition in mosquitoes. *Curr Biol* **16:** 1311–1316. doi:10.1016/j.cub.2006.05.053

Göpfert MC, Robert D. 2001. Active auditory mechanics in mosquitoes. *Proc Biol Sci* **268:** 333–339. doi:10.1098/rspb.2000.1376

Göpfert MC, Briegel H, Robert D. 1999. Mosquito hearing: sound-induced antennal vibrations in male and female *Aedes aegypti*. *J Exp Biol* **202:** 2727–2738. doi:10.1242/jeb.202.20.2727

Hartberg WK. 1971. Observations on the mating behaviour of *Aedes aegypti* in nature. *Bull World Health Organ* **45:** 847–850.

Johnston C. 1855. Original communications: auditory apparatus of the *Culex* mosquito. *J Cell Sci* **s1–3:** 97–102. doi:10.1242/jcs.s1-3.10.97

League GP, Baxter LL, Wolfner MF, Harrington LC. 2019. Male accessory gland molecules inhibit harmonic convergence in the mosquito *Aedes aegypti*. *Curr Biol* **29:** R196–R197. doi:10.1016/j.cub.2019.02.005

League GP, Harrington LC, Pitcher SA, Geyer JK, Baxter LL, Montijo J, Rowland JG, Johnson LM, Murdock CC, Cator LJ. 2021. Sexual selection theory meets disease vector control: testing harmonic convergence as a "good genes" signal in *Aedes aegypti* mosquitoes. *PLoS Negl Trop Dis* **15:** e0009540. doi:10.1371/journal.pntd.0009540

Maïga H, Niang A, Sawadogo SP, Dabiré RK, Lees RS, Gilles JRL, Tripet F, Diabaté A. 2014. Role of nutritional reserves and body size in *Anopheles gambiae* males mating success. *Acta Trop* **132:** S102–S107. doi:10.1016/j.actatropica.2013.08.018

Menda G, Nitzany EI, Shamble PS, Wells A, Harrington LC, Miles RN, Hoy RR. 2019. The long and short of hearing in the mosquito *Aedes aegypti*. *Curr Biol* **29:** 709–714. doi:10.1016/j.cub.2019.01.026

Montoya JP, Pantoja-Sánchez H, Gomez S, Avila FW, Alfonso-Parra C. 2021. Flight tone characterisation of the South American malaria vector *Anopheles darlingi* (Diptera: Culicidae). *Mem Inst Oswaldo Cruz* **116:** e200497. doi:10.1590/0074-02760200497

Pantoja-Sánchez H, Gomez S, Velez V, Avila FW, Alfonso-Parra C. 2019. Precopulatory acoustic interactions of the New World malaria vector *Anopheles albimanus* (Diptera: Culicidae). *Parasit Vectors* **12:** 386. doi:10.1186/s13071-019-3648-8

Pantoja-Sánchez H, League GP, Harrington LC, Alfonso-Parra C. 2022. Recording and analysis of mosquito acoustic-related mating behavior. *Cold Spring Harb Protoc* doi:10.1101/pdbprot107989

Pennetier C, Warren B, Dabiré KR, Russell IJ, Gibson G. 2010. "Singing on the wing" as a mechanism for species recognition in the malarial mosquito *Anopheles gambiae*. *Curr Biol* **20:** 131–136. doi:10.1016/j.cub.2009.11.040

Quigley C, Fusani L. 2018. Courtship and mating. In *Encyclopedia of reproduction* (ed. Skinner MK), pp. 67–71. Elsevier, New York.

Reitmayer CM, Pathak AK, Harrington LC, Brindley MA, Cator LJ, Murdock CC. 2021. Sex, age, and parental harmonic convergence behavior affect the immune performance of *Aedes aegypti* offspring. *Commun Biol* **4:** 723. doi:10.1038/s42003-021-02236-5

Römer H. 2020. Directional hearing in insects: biophysical, physiological and ecological challenges. *J Exp Biol* **223:** jeb203224. doi:10.1242/jeb.203224

Roth LM. 1948. A study of mosquito behavior. An experimental laboratory study of the sexual behavior of *Aedes aegypti* (Linnaeus). *Am Midl Nat* **40:** 265–352. doi:10.2307/2421604

Sawadogo SP, Costantini C, Pennetier C, Diabaté A, Gibson G, Dabiré RK. 2013. Differences in timing of mating swarms in sympatric populations of *Anopheles coluzzii* and *Anopheles gambiae s.s.* (formerly *An. gambiae* M and S molecular forms) in Burkina Faso, West Africa. *Parasit Vectors* **6:** 275. doi:10.1186/1756-3305-6-275

Sawadogo PS, Namountougou M, Toé KH, Rouamba J, Maïga H, Ouédraogo KR, Baldet T, Gouagna LC, Kengne P, Simard F, et al. 2014. Swarming behaviour in natural populations of *Anopheles gambiae* and *An. coluzzii*: review of 4 years survey in rural areas of sympatry, Burkina Faso (West Africa). *Acta Trop* **132:** S42–S52. doi:10.1016/j.actatropica.2013.12.011

Sawadogo SP, Niang A, Bilgo E, Millogo A, Maïga H, Dabire RK, Tripet F, Diabaté A. 2017. Targeting male mosquito swarms to control malaria vector density. *PLoS One* **12:** e0173273. doi:10.1371/journal.pone.0173273

Simões PMV, Ingham RA, Gibson G, Russell IJ. 2016. A role for acoustic distortion in novel rapid frequency modulation behaviour in free-flying male mosquitoes. *J Exp Biol* **219:** 2039–2047.

Villarreal SM, Winokur O, Harrington L. 2017. The impact of temperature and body size on fundamental flight tone variation in the mosquito vector *Aedes aegypti* (Diptera: Culicidae): implications for acoustic lures. *J Med Entomol* **54:** 1116–1121. doi:10.1093/jme/tjx079

Warren B, Gibson G, Russell IJ. 2009. Sex recognition through midflight mating duets in *Culex* mosquitoes is mediated by acoustic distortion. *Curr Biol* **19:** 485–491. doi:10.1016/j.cub.2009.01.059

Yuval B. 2006. Mating systems of blood-feeding flies. *Annu Rev Entomol* **51:** 413–440. doi:10.1146/annurev.ento.51.110104.151058

Cite this introduction as *Cold Spring Harb Protoc*; doi:10.1101/pdb.top107667

Protocol 1

Recording and Analysis of Mosquito Acoustic-Related Mating Behavior

Hoover Pantoja-Sánchez,[1,2] Garrett P. League,[3] Laura C. Harrington,[3] and Catalina Alfonso-Parra[4,5,6]

[1]*Departamento de Ingeniería Electrónica, SISTEMIC,* [2]*Programa de Estudio y Control de Enfermedades Tropicales, PECET, Universidad de Antioquia, Medellín, Antioquia, Colombia, 050010;* [3]*Department of Entomology, Cornell University, Ithaca, New York 14853, USA;* [4]*Instituto Colombiano de Medicina Tropical, Universidad CES, Sabaneta, Antioquia, Colombia, 055413;* [5]*Max Planck Tandem Group in Mosquito Reproductive Biology, Universidad de Antioquia, Medellín, Antioquia, Colombia, 050010*

In mosquitoes, courtship and mating sounds are produced by the movement of the wings during flight. These sounds, usually referred to as flight tones, have been studied using tethered and free-flying individuals. Here, we describe a general approach for recording and analyzing mosquito acoustic-related mating behaviors that can be broadly adapted to a variety of experimental designs.

MATERIALS

It is essential that you consult the appropriate Material Safety Data Sheets and your institution's Environmental Health and Safety Office for proper handling of equipment and hazardous materials used in this protocol.

Reagents

Cyanoacrylate-based adhesive (e.g., superglue, Boss 7.05), polyvinyl acetate glue (e.g., Elmer's glue), or nail glue (e.g., L.A. Colors) (see Step 5.ii)

Cyanoacrylate adhesives are recommended because of their strength, rapid drying, and lack of odor. Such adhesives minimize the amount of glue applied to the mosquito and reduce interference from chemical sensory stimuli (Boiteau et al. 2009).

Human hair (~1-cm-length)

Ice or another anesthetic (see Step 5.i)

Male and female mosquitoes (see Step 1)

Equipment

See Figure 1A for a schematic overview of the equipment required for the recording setup.

Amplifier and digitizer

These components can be individual (e.g., amplifier: INA128 Texas Instruments; digitizer: 779676-01 National Instruments) or integrated (e.g., amplifier + digitizer: M-Track Quad Four Channel Audio, M-Audio).

[6]Correspondence: catalfonso@gmail.com

Cite this protocol as *Cold Spring Harb Protoc*; doi:10.1101/pdb.prot107989

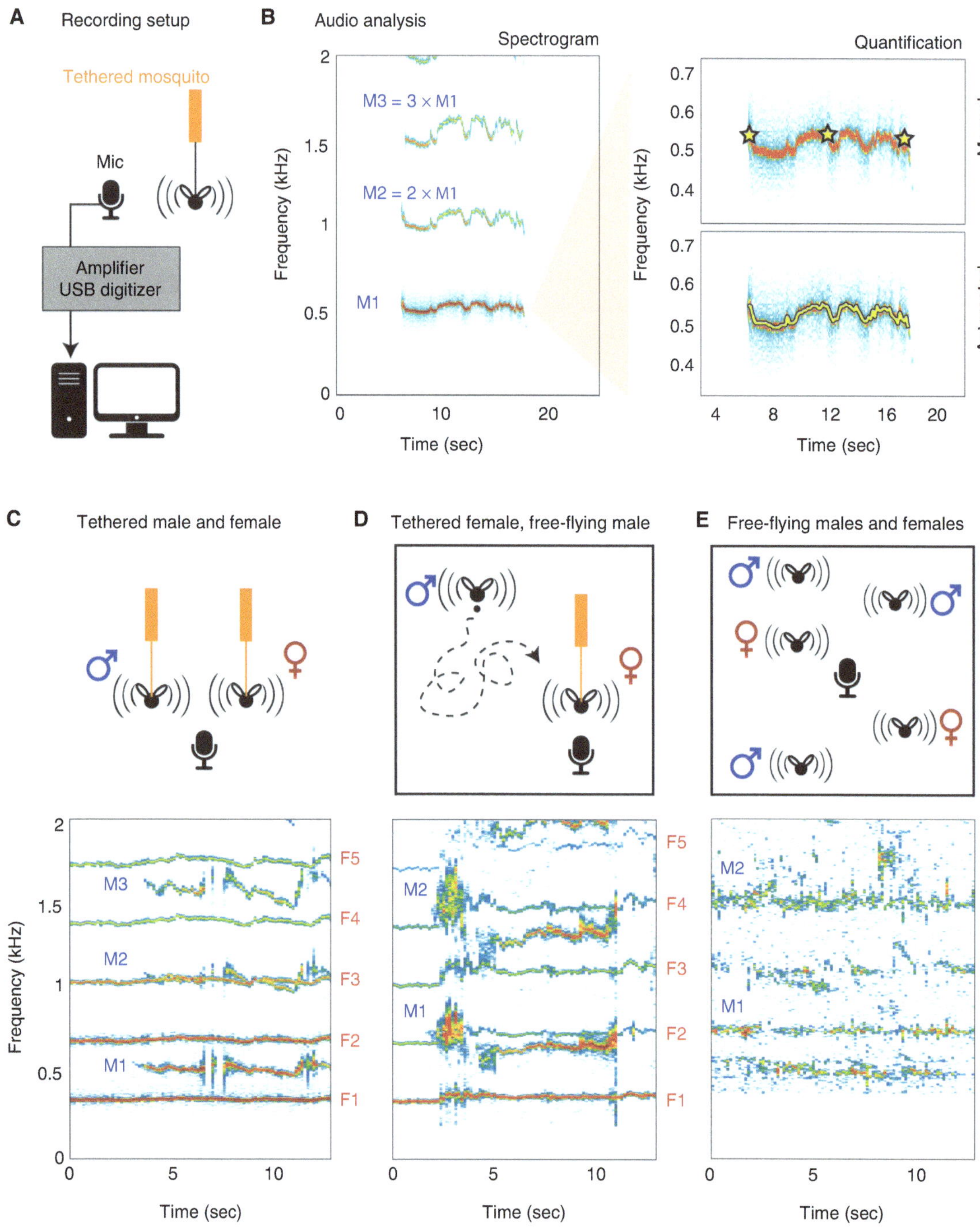

FIGURE 1. Setup and techniques for recording and analyzing mosquito acoustic-related mating behaviors. (*A*) Diagram of a single microphone recording setup. Several microphones can be used in a similar manner to record from multiple channels. (*B*) Spectrogram and quantification of a representative male flight tone, its fundamental frequency (M1) and harmonic overtones (M2 to M3). Colors indicate amplitude of frequency components, red being the largest, blue the smallest, and white indicating the noise floor. To extract quantitative information from spectrograms, manual analysis measures frequency and amplitude by manually selecting specific time points (yellow stars), whereas automated analysis uses algorithms to track the flight tone fundamental frequency (yellow line) throughout an entire recording. Tethered (*C*,*D*) and free-flying (*E*) techniques for recording acoustic mating behaviors with representative recording examples. The frequency components of males (M1 to M3) and females (F1 to F5) are indicated in blue and red, respectively.

 Cite this protocol as *Cold Spring Harb Protoc*; doi:10.1101/pdb.prot107989

Brush (optional; see Step 5.vii)

Flight audio recording arena

Arenas can take the form of a plastic terrarium or a Faraday cage to block electromagnetic fields in the recording area. Recording arenas limit flight to the immediate vicinity of microphones and limit background noise. Arenas are essential for recording free-flying mosquitoes and are optional, although still recommended, for recording tethered mosquitoes. Some free-flight recordings have been conducted in small glass vials or plexiglass tubes (Wekesa et al. 1998), although larger arenas that range from ~1000 to 10,000 cm^3 (Villarreal et al. 2017; Pantoja-Sánchez et al. 2019; League et al. 2021) are encouraged to allow for a fuller range of movement during flight.

Insect pins (steel; 0.3–0.5-mm-diameter) (e.g., Austerlitz or Phusis)

Omnidirectional pressure microphones

Omnidirectional microphones record sounds from all directions with similar gain and are capable of recording sounds broadcast by distant far-field sources (e.g., Knowles FG-23329-C05). Microphones can also be mounted at the focal point of parabolic reflectors (Simões et al. 2016) to focus recordings on a specific area for sound detection.

Paper, cardboard, or pipe cleaners (see Step 5.iv)

Particle velocity microphones

These acoustic sensors record sounds from near-field sources and significantly reduce background noise (e.g., Knowles NR-23158-000).

Software

Specialized software to record and process audio (e.g., Raven [K. Lisa Yang Center for Conservation 2016], Avisoft, Audacity [Audacity Team 2021]) or software including audio processing packages (e.g., MATLAB—Signal processing toolbox [The MathWorks, Inc. 2019]; R—Seewave [Sueur et al. 2008]).

Soundproof chamber

Ideally, recordings are conducted in an anechoic soundproofed room. However, as access to specialized chambers is rare, acoustic foam and heavy blankets can be used to minimize echoes and ambient noise amplification. When such arrangements are infeasible (i.e., recording outdoors), background noise can be reduced using near-field microphones and/or software noise reduction procedures (see Steps 4 and 8).

Temperature and humidity data logger

HOBO U23 Pro v2 Temperature/Relative Humidity Data Logger, Onset, or similar product.

METHOD

This protocol is intended to promote research on mosquito acoustics in laboratory conditions. However, procedures can be adapted to perform recordings in larger spaces, or in semifield or field conditions, by using microphone accessories like parabolic reflectors (Mankin 1994; Simões et al. 2016) or funnels (Cator et al. 2011).

Recording Audio

1. Prepare mosquitoes for experiments.

   Because size and age affect flight tones (Cator et al. 2010; Staunton et al. 2019), these factors should be controlled by synchronously hatching and rearing larvae (Das et al. 2007; Clemons et al. 2010). Body sizes that are field-relevant are recommended. Physiological parameters (e.g., nutritional status) should also be controlled to reduce variability stemming from differences in energy reserves. Virgin females should be used for mating interactions, as mated females are not sexually receptive (Helinski et al. 2012; Gabrieli et al. 2014) and differ from virgins in their acoustic interactions with males (League et al. 2019).

2. Control and track environmental conditions, such as temperature and humidity, within the recording room.

   Flight tone fundamental frequency increases proportionally with temperature (Villarreal et al. 2017). Ambient temperature (typically 27°C) and humidity (70%–80% relative humidity) during recordings can be tracked using data loggers (e.g., HOBO U23 Pro v2). Further, as swarming and/or mating activity for some species is dependent on circadian rhythms, light intensity, or host cues (e.g., odors, CO_2), variables such as time of day, ambient lighting, and chemosensory cues should be accounted for (Howell and Knols 2009; Oliva et al. 2014; Montell and Zwiebel 2016; Wang et al. 2021).

3. Prepare recording equipment.

   i. Introduce the microphone(s) into the recording arena.

   ii. Turn on the amplifier and digitizer and ensure the microphones are connected (see Fig. 1A).

   Most microphones require batteries. Verify that batteries are fully charged before each recording session.

   iii. Use audio software to record and verify that the digitizer is being used for recording and not the computer's built-in microphone.

   Most audio software record from the computer's built-in microphone by default unless directed to record from another device.

   iv. Adjust, if possible, the recording sample rate to values ≥10 kHz/16 bits to analyze frequencies up to 5 kHz, which is sufficient to characterize flight tones.

   Low sample rates (i.e., <8 kHz) might fail to record relevant information. High sample rates (i.e., >45 kHz) do not provide relevant information, increase the size of the audio file, and consequently reduce processing efficiency.

   v. Verify that the digitizer is obtaining signals from each microphone and adjust the gain to obtain recordings with an optimal signal-to-noise ratio.

   An ideal signal-to-noise ratio facilitates the amplification of sounds of interest without exceeding the maximum allowed limit. Exceeding the limit leads to distortion and lowers the quality of the audio.

   vi. If background noise is too high, consider reducing airborne and/or electrical noise sources.

   Low- and mid-frequency noise (<2 kHz) can interfere with flight tone recordings. To reduce airborne noise, cover the recording arena with heavy materials such as heavy blankets or bass traps (i.e., SonoLite Bass trap). Low-frequency noise might be produced by vibrations of nearby equipment. Place the recording arena on top of an antivibration platform (i.e., Speirs Robertson EW-11500-04) or a heavy table. To reduce electromagnetic noise, set the recording arena in a Faraday cage and cover cables and electric power sources with aluminum foil.

4. Record background noise samples to further reduce ambient noise interference after recordings are completed.

   Use a 5-sec recording from within the recording arena for software noise reduction (see Step 9).

5. Record acoustic behaviors, preferably in uncompressed audio formats such as WAV, AIFF, or PCM.

Tethered Mosquitoes (see Fig. 1C)

   i. Anesthetize mosquitoes with ice.

   For most species, anesthetization for 30–60 sec provides ample time to immobilize and tether a mosquito without negatively impacting subsequent flight performance. However, use only the minimum time needed for anesthetizing and tethering, as sensitivity to temperature varies depending on the species. Plastic Petri dishes or vials are typically used to hold mosquitoes on ice during tethering as placing mosquitoes directly on ice is discouraged. Although ice is preferred, FlyNap (triethylamine) and CO_2 can also be used as alternative anesthetics.

   ii. While anesthetized, tether the mosquito to an insect pin either directly or preferably via a human hair (referred to as "semitethering") to the dorsal thorax using a cyanoacrylate-based adhesive.

   For tethering with a human hair, use adhesive to attach one end of the hair to an insect pin and the opposite end to the mosquito's thorax. This technique allows for additional flight mobility compared to directly tethering mosquitoes to an insect pin. Using a new hair for each recording is recommended. If reapplying fresh glue to the same tether containing hardened glue, take care not to let the glue accumulate, as this will weigh down the mosquito in flight.

   iii. Verify that the tether allows for natural wing movement.

   To allow for optimal tether tension and ideal mosquito distance from the insect pin, the tether should be oriented in a 45° arc pointing outward from the pin and cut to ~1 cm in length: shorter than this, and the mosquito will touch the pin, preventing flight; longer than this, and the mosquito's flight will be hampered as it bends the tether downward under its own weight, resulting in an unnatural flight

Cite this protocol as *Cold Spring Harb Protoc*; doi:10.1101/pdb.prot107989

orientation and interactions with the pin. When gluing the hair to the thorax, ensure it does not restrain wing movement.

iv. Use a removable surface support to inhibit leg movement.

Avoid unnecessary movement and flight activity before recording by supporting the legs of the mosquito temporarily using a piece of paper, cardboard, or pipe cleaners.

v. Place one or more particle velocity microphones 1–5 cm from the mosquito.

Set an individual microphone below or beside (depending on mic and mosquito orientation) each mosquito or mosquito group(s) as needed.

vi. When ready to record, initiate flight by removing the leg support and releasing tarsal inhibition.

The length of sustained periods of flight tends to decline over time as the mosquito habituates to tethering and depletes its energy reserves. Thus, initial bouts of flight after tethering are typically the most desirable for recording.

vii. If necessary, promote flight and/or mating by gently touching individual mosquitoes' legs with a brush or by blowing on them.

Keep in mind that blowing on the mosquitoes introduces other chemosensory cues (CO_2, odors, etc.) that may introduce behavioral variability. If access to the arena is not possible, an opening into which flight stimuli may be introduced can be added. Tapping the recording arena might also elicit flight. However, this option can introduce significant noise interference during audio recording.

viii. Record the flight tones of individual mosquitoes or mosquito group(s).

To record acoustic mating interactions between individuals, move two or more mosquitoes into (and subsequently out of, if desired) close auditory range (typically >2 cm). Such interactions can, but need not, include physical touch. See Troubleshooting.

Free-Flying Mosquitoes (see Fig. 1E)

ix. Place one or more omnidirectional pressure microphones in the center of the recording arena or on its walls, depending on the experiment.

x. Introduce individual mosquitoes to the recording arena using an aspirator and, after a sufficient acclimation period, start recording.

Acclimation to the arena should last for at least 5–10 min before audio recording begins. If analyzing male–female interactions, add males to the arena several minutes before carefully adding females. This will allow males to land on arena walls and minimize mating interactions with females during the acclimation period.

xi. If desired, note the time of a specific behavior, including flight or mating activity. Record male–female interactions in their entirety—that is, until the couple has mated or the female rejects the male by pushing him away with her legs. If recording swarm mating, record until the desired number of mating attempts are obtained or until swarm activity ceases.

Free-flying mating attempts involve a male approaching a female in flight and rapid flight tone frequency modulations performed by males and females (Simões et al. 2016; Pantoja-Sánchez et al. 2019; Garcia Castillo et al. 2021). If a male reaches a female, the couple will drop out of flight, often while in copula. Copulation can occur partially or completely in flight or on a surface. If the couple forms a mating copula, allow them to remain together long enough to ensure insemination (10–60 sec). Successful mating can be verified by dissecting females' spermatheca(e) to visually assess the presence or absence of sperm. See Troubleshooting.

Tethered Female and Free-Flying Males (see Fig. 1D)

xii. Introduce a tethered female to the recording arena and set a particle velocity microphone at 1–5 cm from the mosquito.

xiii. Initiate female flight if necessary (see Step 5.vii).

xiv. Add free-flying male(s) into the arena using an aspirator.

xv. Note the time a male approaches/contacts the female and record the interaction in its entirety until the couple has mated or the female finally rejects the male by pushing him away with her legs (note that some initial female kicking is common to both successful and unsuccessful mating attempts). If recording swarming behavior, record until the desired number of mating attempts are obtained or until swarm activity ceases (Garcia Castillo et al. 2021; Montoya et al. 2021).

Mating attempts between tethered females and free-flying males involve a male approaching a tethered female, rapid flight tone frequency modulation by the male, the male landing on the female, increased male and female flight tone frequencies and amplitudes, and the couple interacting until either copulation or male rejection. If the couple forms a mating copula, allow them to remain together long enough to ensure insemination (10–60 sec). Successful mating can be verified by dissecting females' spermatheca(e) to visually assess the presence or absence of sperm. See Troubleshooting.

6. Stop recording and save the audio file, preferably in WAV or mp3 format.

   Include metadata in the file description with specifications about the experiment and technical data.

Audio Analysis

7. Use preferred audio software to visualize the flight tone components of audio recordings using spectrograms (see Fig. 1B).

   The Fourier transform is commonly used to generate time-frequency representations called spectrograms. To generate a spectrogram, recordings are broken up into small time windows and transformed to calculate the frequency spectrum for each window. Each spectrum corresponds to a vertical section in the spectrogram and constitutes a measurement of amplitude for frequency components at a specific moment in time. Individual spectrums are organized side by side to form a spectrogram. Spectrograms are thus often "blurred" because of selection of inadequate parameters. Adjust time-frequency resolution as needed to obtain accurate information (Nisar et al. 2016).

8. If recordings are too noisy, perform a noise reduction procedure like spectral subtraction (Zavarehei 2021) or Audacity's "Noise Reduction" effect using a background noise sample (see Step 4).
9. Identify flight tones of individual mosquitoes or mosquito groups.

   Flight tones consist of a fundamental frequency, or first harmonic, and its harmonic overtones, or second, third, etc., harmonics (see Fig. 1B). The fundamental frequency (F0 in general or M1 and F1 when referring to male and female fundamentals, respectively) is equivalent to the wingbeat frequency, and harmonic overtones are multiple integers of the fundamental (2 × F0, 3 × F0, 4 × F0, etc.). Males typically fly at higher (1.5×) frequencies than females.

10. Use manual or automated approaches to quantify flight tone parameters such as frequency and amplitude. (Each approach can quantify spectral information of acoustic mating interactions.)

Manual Approaches

i. Characterize flight tones by sampling frequencies (Hz) and amplitudes (dB) of fundamental frequencies and/or harmonic overtones from several time points during the acoustic behavior of interest (see Fig. 1B).

   Typically, fundamental frequencies are sampled directly, as this is the most reliable signal, with harmonic overtone frequencies calculated as multiples of the fundamental. For manual analysis, Raven and Audacity software are recommended.

Automated Approaches

ii. Quantify the frequency components of flight tones throughout an entire recording from the acoustic behavior of interest (see Fig. 1B).

   Custom algorithms (i.e., MATLAB scripts) based on sinusoidal models (Klingbeil 2005), high-order spectral analysis (Arthur et al. 2014), and Hilbert spectral analysis (Aldersley et al. 2014) have been used to quantify flight acoustics.

11. Characterize acoustic behaviors using extracted flight tone parameters.

Flight parameters are used to characterize common acoustic behaviors during mating interactions, including (1) harmonic convergence—that is, the maintenance over time (i.e., ≥1 sec) of overlapping male and female flight tone frequencies (i.e., <5 Hz difference) at one of four to five commonly converging male: female harmonic ratios (Aldersley et al. 2016; Garcia Castillo et al. 2021); (2) rapid frequency modulation—that is, changes in frequency per time unit (Hz/sec) (Simões et al. 2016); and (3) flight response latency time—that is, the lapse between audio stimulus presentation and the consequent flight response (Römer 2020).

TROUBLESHOOTING

Problem (Step 5.viii, Step 5.xi, and Step 5.xv) Mosquitoes do not engage in mating interactions after 5 min.

Solution: The simplest solutions are to increase the male:female ratio, as most mosquito species mate in highly male-skewed swarms or to replace the reluctant mosquitoes with a new set of mosquitoes. If these measures also fail, it is most likely that the lack of mating is due to patterns in circadian swarm activity (Garcia Castillo et al. 2021; Somers et al. 2022), in which case the experiments should be optimized to occur primarily during periods of high mosquito mating activity (typically dusk and dawn, but this not clear or known for all species). Alternatively, resource-starved and/or stressed mosquitoes that have experienced adverse conditions may be energy-depleted and less likely to fly and mate. Thus, care should be taken to ensure proper rearing and maintenance conditions. Finally, laboratory generation of F_1 mosquitoes hatched from field-collected F_0 parent eggs can display highly variable mating behavior, and thus when conducting studies on low-generation (F_1–F_5) field-derived colonies, it is recommended that experiments be conducted on mosquitoes no earlier than laboratory generation F_2 when possible (Aldersley and Cator 2019; League et al. 2021).

ACKNOWLEDGMENTS

We thank Dr. Frank W. Avila and Dr. Lauren J. Cator for technical feedback and scientific content review.

REFERENCES

Aldersley A, Cator LJ. 2019. Female resistance and harmonic convergence influence male mating success in *Aedes aegypti*. *Sci Rep* **9:** 2145. doi:10.1038/s41598-019-38599-3

Aldersley A, Champneys A, Homer M, Robert D. 2014. Time-frequency composition of mosquito flight tones obtained using Hilbert spectral analysis. *J Acoust Soc Am* **136:** 1982–1989. doi:10.1121/1.4895689

Aldersley A, Champneys A, Homer M, Robert D. 2016. Quantitative analysis of harmonic convergence in mosquito auditory interactions. *J R Soc Interface* **13:** 20151007. doi:10.1098/rsif.2015.1007

Arthur BJ, Emr KS, Wyttenbach RA, Hoy RR. 2014. Mosquito (*Aedes aegypti*) flight tones: frequency, harmonicity, spherical spreading, and phase relationships. *J Acoust Soc Am* **135:** 933–941. doi:10.1121/1.4861233

Audacity Team. 2021. Audacity^R: Free Audio Editor and Recorder [Computer application]. https://www.audacityteam.org/

Boiteau G, Meloche F, Vincent C, Leskey TC. 2009. Effectiveness of glues used for harmonic radar tag attachment and impact on survival and behavior of three insect pests. *Environ Entomol* **38:** 168–175. doi:10.1603/022.038.0121

Cator LJ, Ng'Habi KR, Hoy RR, Harrington LC. 2010. Sizing up a mate: variation in production and response to acoustic signals in *Anopheles gambiae*. *Behav Ecol* **21:** 1033–1039. doi:10.1093/beheco/arq087

Cator LJ, Arthur BJ, Ponlawat A, Harrington LC. 2011. Behavioral observations and sound recordings of free-flight mating swarms of *Ae. aegypti* (Diptera: Culicidae) in Thailand. *J Med Entomol* **48:** 941–946. doi:10.1603/ME11019

Clemons A, Mori A, Haugen M, Severson DW, Duman-Scheel M. 2010. Culturing and egg collection of *Aedes aegypti*. *Cold Spring Harb Protoc* **2010:** pdb.prot5507. doi:10.1101/pdb.prot5507

Das S, Garver L, Dimopoulos G. 2007. Protocol for mosquito rearing (*A. gambiae*). *J Vis Exp* 221. doi:10.3791/221.

Gabrieli P, Kakani EG, Mitchell SN, Mameli E, Want EJ, Mariezcurrena Anton A, Serrao A, Baldini F, Catteruccia F. 2014. Sexual transfer of the steroid hormone 20E induces the postmating switch in *Anopheles gambiae*. *Proc Natl Acad Sci* **111:** 16353–16358. doi:10.1073/pnas.1410488111

Garcia Castillo SS, Pritts KS, Krishnan RS, Harrington LC, League GP. 2021. Harmonic convergence coordinates swarm mating by enhancing mate detection in the malaria mosquito *Anopheles gambiae*. *Sci Rep* **11:** 24102. doi:10.1038/s41598-021-03236-5

Helinski MEH, Deewatthanawong P, Sirot LK, Wolfner MF, Harrington LC. 2012. Duration and dose-dependency of female sexual receptivity responses to seminal fluid proteins in *Aedes albopictus* and *Ae. aegypti* mosquitoes. *J Insect Physiol* **58:** 1307–1313. doi:10.1016/j.jinsphys.2012.07.003

Howell PI, Knols BGJ. 2009. Male mating biology. *Malar J* **8:** 1–10. doi:10.1186/1475-2875-8-S2-S8

Klingbeil M. 2005. Software for spectral analysis, editing, and synthesis. In *Proceedings of the International Computer Music Conference*, pp. 107–110.

K. Lisa Yang Center for Conservation. 2016. Raven Lite: Interactive Sound Analysis Software. http://ravensoundsoftware.com/.

League GP, Baxter LL, Wolfner MF, Harrington LC. 2019. Male accessory gland molecules inhibit harmonic convergence in the mosquito *Aedes aegypti*. *Curr Biol* **29:** R196–R197. doi:10.1016/j.cub.2019.02.005

League GP, Harrington LC, Pitcher SA, Geyer JK, Baxter LL, Montijo J, Rowland JG, Johnson LM, Murdock CC, Cator LJ. 2021. Sexual selection theory meets disease vector control: testing harmonic convergence as a "good genes" signal in *Aedes aegypti* mosquitoes. *PLoS Negl Trop Dis* **15:** e0009540. doi:10.1371/journal.pntd.0009540

Mankin RW. 1994. Acoustical detection of *Aedes taeniorhynchus* swarms and emergence exoduses in remote salt marshes. *J Am Mosq Control Assoc* **10:** 302–308.

Montell C, Zwiebel LJ. 2016. Chapter Ten: Mosquito sensory systems. In *Progress in mosquito research* (ed. Raikhel AS), Vol. **51**, pp. 293–328. Academic, London.

Montoya JP, Pantoja-Sánchez H, Gomez S, Avila FW, Alfonso-Parra C. 2021. Flight tone characterisation of the South American malaria vector *Anopheles darlingi* (Diptera: Culicidae). *Mem Inst Oswaldo Cruz* **116:** e200497. doi:10.1590/0074-02760200497

Nisar S, Khan OU, Tariq M. 2016. An efficient adaptive window size selection method for improving spectrogram visualization. *Comput Intell Neurosci* **2016:** 6172453. doi:10.1155/2016/6172453

Oliva CF, Damiens D, Benedict MQ. 2014. Male reproductive biology of Aedes mosquitoes. *Acta Trop* **132:** S12–S19. doi:10.1016/j.actatropica.2013.11.021

Pantoja-Sánchez H, Gomez S, Velez V, Avila FW, Alfonso-Parra C. 2019. Precopulatory acoustic interactions of the New World malaria vector *Anopheles albimanus* (Diptera: Culicidae). *Parasit Vectors* **12:** 386. doi:10.1186/s13071-019-3648-8

Römer H. 2020. Directional hearing in insects: biophysical, physiological and ecological challenges. *J Exp Biol* **223:** jeb203224. doi:10.1242/jeb.203224

Simões PMV, Ingham RA, Gibson G, Russell IJ. 2016. A role for acoustic distortion in novel rapid frequency modulation behaviour in free-flying male mosquitoes. *J Exp Biol* **219:** 2039.

Somers J, Georgiades M, Su MP, Bagi J, Andrés M, Alampounti A, Mills G, Ntabaliba W, Moore SJ, Spaccapelo R, et al. 2022. Hitting the right note at the right time: circadian control of audibility in *Anopheles* mosquito mating swarms is mediated by flight tones. *Sci Adv* **8:** eabl4844. doi:10.1126/sciadv.abl4844

Staunton KM, Usher L, Prachar T, Ritchie SA, Snoad N, Johnson BJ. 2019. A novel methodology for recording wing beat frequencies of untethered male and female *Aedes aegypti*. *J Am Mosq Control Assoc* **35:** 169–177. doi:10.2987/18-6799.1

Sueur J, Aubin T, Simonis C. 2008. Seewave, a free modular tool for sound analysis and synthesis. *Bioacoustic* **18:** 213–226. doi:10.1080/09524622.2008.9753600

The MathWorks, Inc. 2019. MATLAB and signal processing toolbox. https://www.mathworks.com/

Villarreal SM, Winokur O, Harrington L. 2017. The impact of temperature and body size on fundamental flight tone variation in the mosquito vector *Aedes aegypti* (Diptera: Culicidae): implications for acoustic lures. *J Med Entomol* **54:** 1116–1121. doi:10.1093/jme/tjx079

Wang G, Vega-Rodríguez J, Diabate A, Liu J, Cui C, Nignan C, Dong L, Li F, Ouedrago CO, Bandaogo AM, et al. 2021. Clock genes and environmental cues coordinate *Anopheles* pheromone synthesis, swarming, and mating. *Science* **371:** 411–415. doi:10.1126/science.abd4359

Wekesa JW, Brogdon WG, Hawley WA, Besansky NJ. 1998. Flight tone of field-collected populations of *Anopheles gambiae* and *An. arabiensis* (Diptera: Culicidae). *Physiol Entomol* **23:** 289–294. doi:10.1046/j.1365-3032.1998.233087.x

Zavarehei E. 2021. Boll Spectral Subtraction. (https://www.mathworks.com/matlabcentral/fileexchange/7675-boll-spectral-subtraction), MATLAB Central File Exchange. Retrieved June 25, 2021.

Cite this protocol as *Cold Spring Harb Protoc*; doi:10.1101/pdb.prot107989

CHAPTER 17

Insemination in *Aedes aegypti* and *Aedes albopictus*

Dhwani Parsana,[1] Ferdinand Nanfack-Minkeu,[1] and Laura K. Sirot[2]

[1]*Department of Biology, The College of Wooster, Wooster, Ohio 44691, USA*

Aedes mosquitoes are the vectors of several arboviruses that cause human disease. A better understanding of their reproduction helps to improve their management and contributes insights into the fundamental biology of mosquitoes. During mating, inseminated mosquito females receive seminal fluids and sperm from males that they then store in the spermathecae. In *Aedes aegypti* and *Aedes albopictus*, most mated females become resistant to further insemination within 2 h of initial insemination. Although the male seminal fluids are known to be involved in initiating the resistance of inseminated females to further insemination, the mechanism underlying this resistance is not well-understood. The determination of insemination status is a key step in investigating the behavioral and molecular interactions between males and females and for exploring the proximate influences and evolutionary implications of interspecific copulations. Several methods exist for determining insemination status, as discussed here. The choice of method depends on the research question and the availability of resources.

BACKGROUND

The reproductive biology of *Aedes* mosquitoes has been of particular interest ever since their role in the transmission of arboviruses that cause several human diseases was first recognized (Reed et al. 1900; Lounibos 2002; Gratz 2004). Their reproductive behavior has been carefully described in several studies performed both in the wild and in the laboratory (e.g., Roth 1948; Downes 1969; Oliva et al. 2014). In some *Aedes* species, including *Aedes aegypti* and *Aedes albopictus*, most females become resistant to insemination (i.e., receipt and/or retention of sperm) within 2 h after an initial insemination (Craig 1967; Boyer et al. 2012; Degner and Harrington 2016b), although they may be inseminated by additional males during the initial 2-h period (Oliva et al. 2013; Richardson et al. 2015). Even after they become refractory to insemination, the females often continue to engage in apparent copulations (i.e., physical coupling of the female and male's genitalia) throughout their lives, at least under laboratory conditions. These copulations may result in the transfer of seminal fluid and even sperm, but the sperm are rarely retained and stored (Spielman et al. 1967). The factors contributing to female refractoriness to insemination include effects of seminal fluid proteins produced in the male reproductive accessory glands (Fuchs et al. 1969), although the identity of these contributors and how they interact with the female reproductive system have yet to be determined (but see Duvall et al. 2017).

Insemination occurs quickly in *Ae. aegypti* (<20 sec on average) and *Ae. alboctius* (<2 min on average) once the insects are properly positioned for copulation, although achieving that positioning can take some time. Once a pair is physically in contact, the male grasps the female dorsally and then

[2]Correspondence: lsirot@wooster.edu

Cite this introduction as *Cold Spring Harb Protoc*; doi:10.1101/pdb.top107668

attempts to maneuver rapidly into a venter-to-venter position. Females may resist these attempts through body and leg movements. Males that are successful in assuming the venter-to-venter position grasp the female cerci by their claspers, stimulating the extension of the male aedeagus, which widens the female vagina to enlarge the opening of the bursa (Roth 1948; Spielman et al. 1967; Oliva et al. 2014). Once the opening is enlarged, the males transfer the semen, containing a mixture of sperm and accessory gland secretions, to the females. When sperm enter the female reproductive tract, they are temporarily stored in the bursa (Barr 1974; Oliva et al. 2014; Degner and Harrington 2016a). By about 5 minutes after copulation, the majority of sperm are completely transferred to the spermathecae (Jones and Wheeler 1965). Female *Aedes* have three spermathecae, one bigger medial spermatheca and two smaller lateral spermathecae, and how many are filled after copulation can vary (Carrasquilla et al. 2019). Multiple studies have shown that at least two of the three spermathecae will be filled after insemination, and in 99.95% of cases, the medial and one of the lateral spermathecae were filled (Jones and Wheeler 1965; Carrasquilla et al. 2019). Sperm that enter the spermathecae can be stored there for >3 mo (Clements and Potter 1967). If the female is dead, or the sperm are dead, the spermathecae of *Ae. aegypti* will not be filled, suggesting that both female action and sperm locomotion are necessary for sperm to move to the spermathecae for storage (Jones and Wheeler 1965).

DETERMINING INSEMINATION STATUS

Determining the insemination status of female *Aedes* mosquitoes is useful for many reasons, including (i) understanding the basic biology of mosquito reproduction (e.g., Helinski et al. 2012a; Oliva et al. 2014; Degner and Harrington 2016a; Carrasquilla et al. 2019); (ii) investigating behavioral and molecular interactions between the sexes (e.g., Sirot et al. 2008, 2011; Boes et al. 2014; Aldersley and Cator 2019); (iii) exploring proximate influences on and evolutionary implications of interspecific copulations (e.g., Bargielowski et al. 2013; Carrasquilla and Lounibos 2015b; Honório et al. 2018); and (iv) improving current mosquito control methods and providing novel approaches to mosquito control (e.g., Duvall et al. 2017; Aldersley et al. 2019; Chen et al. 2021). Several factors influence insemination success, including the prior mating history, age, and body size of both the female and the male (Ponlawat and Harrington 2009; Helinski and Harrington 2011).

Determining insemination status (inseminated or not) of female *Ae. aegypti* and *Ae. albopictus* in cases in which the researcher does not need the female for further studies is quick and easy (~3–5 min per female once you are practiced in dissection (see Protocol 1: Determining the Insemination Status of *Aedes* Mosquitoes [Parsana et al. 2022]). When the females are to be used for further study, the protocol described by Carrasquilla and Lounibos (2015a) for detecting the insemination status of live *Aedes* females can be used. If it is desirable to determine the number of sperm transferred, adaptations of the protocol described by Ponlawat and Harrington (2009) to count sperm or Gutzler et al. (2020)'s molecular approach to estimate sperm numbers by quantifying DNA may be used. For assessing sperm motility, readers are encouraged to consult Noble et al. (2019)'s description of a simple, quick assay based on latency to sperm exit from opened spermathecae. Sperm viability may be assessed using commercially available live/dead sperm assays; however, the reliability of these methods has been brought into question (Holman 2009; Eckel et al. 2017). An alternative approach based on relative sperm viability in a stressor medium has been proposed by Eckel et al. (2017).

Because some *Ae. aegypti* females copulate with multiple males before becoming refractory, it may be desirable to score the occurrence of multiple inseminations or determine offspring paternity. In controlled lab or semifield enclosure studies, one can use transgenic strains with markers detectable in either the sperm (e.g., DsRed; Smith et al. 2007; Aldersley et al. 2019; Degner et al. 2019) or the resulting larvae (e.g., white- or fluorescence-expressing eyes; Duvall et al. 2017) to determine whether a female has been inseminated by multiple males and the proportion of offspring from each mate. If transgenic strains are not available or not amenable to the research question, other methods to identify multiple pairings include labeling different groups of males with different fluorescent powders (Rojas-Araya et al. 2020) or stable isotopes (Helinski et al. 2012b), although care should be taken in inter-

Cite this introduction as *Cold Spring Harb Protoc*; doi:10.1101/pdb.top107668

preting results, as pairings and even seminal fluid transfer are not always indicative of sperm transfer, and transferred sperm may be released from the female reproductive tract without being transferred to the spermathecae (Spielman et al. 1967). To investigate sperm use from multiple males in field-caught females, multiple inseminations can also be detected through microsatellite analysis of the offspring produced by individual females (Boyer et al. 2012; Wong et al. 2012; Richardson et al. 2015). In combination, these techniques will allow detailed analyses into the causes and consequences of variation in insemination timing and frequency in *Aedes* mosquitoes and aid in the improvement or development of mosquito control approaches.

ACKNOWLEDGMENTS

We appreciated the helpful suggestions of Dr. Ethan Degner on an earlier version of this manuscript. D.P. was supported in part by the College of Wooster Biochemistry and Molecular Biology Program. All authors were supported by a U.S. National Institutes of Health grant to L.K.S. (R15AI140223).

REFERENCES

Aldersley A, Cator LJ. 2019. Female resistance and harmonic convergence influence male mating success in *Aedes aegypti*. *Sci Rep* **9:** 2145. doi:10.1038/s41598-019-38599-3

Aldersley A, Pongsiri A, Bunmee K, Kijchalao U, Chittham W, Fansiri T, Pathawong N, Qureshi A, Harrington L, Ponlawat A, et al. 2019. Too "sexy" for the field? Paired measures of laboratory and semi-field performance highlight variability in the apparent mating fitness of *Aedes aegypti* transgenic strains. *Parasit Vectors* **12:** 357. doi:10.1186/s13071-019-3617-2

Bargielowski IE, Lounibos LP, Carrasquilla MC. 2013. Evolution of resistance to satyrization through reproductive character displacement in populations of invasive dengue vectors. *Proc Natl Acad Sci* **110:** 2888–2892. doi:10.1073/pnas.1219599110

Barr A. 1974. Symposium on reproduction of arthropods of medical and veterinary importance: V. Reproduction in Diptera of medical importance with special reference to mosquitoes. *J Med Entomol* **11:** 35–40. doi:10.1093/jmedent/11.1.35

Boes K, Ribeiro J, Wong A, Harrington L, Wolfner M, Sirot L. 2014. Identification and characterization of seminal fluid proteins in the Asian tiger mosquito, *Aedes albopictus*. *PLoS Negl Trop Dis* **8:** e2946. doi:10.1371/journal.pntd.0002946

Boyer S, Toty C, Jacquet M, Lempérière G, Fontenille D. 2012. Evidence of multiple inseminations in the field in *Aedes albopictus*. *PLoS One* **7:** e42040. doi:10.1371/journal.pone.0042040

Carrasquilla M, Lounibos P. 2015a. Detection of insemination status in live *Aedes aegypti* females. *J Insect Physiol* **75:** 1–4. doi:10.1016/j.jinsphys.2015.01.015

Carrasquilla MC, Lounibos LP. 2015b. Satyrization without evidence of successful insemination from interspecific mating between invasive mosquitoes. *Biol Lett* **11:** 20150527. doi:10.1098/rsbl.2015.0527

Carrasquilla MC, Lounibos LP, Honorio NA, Murr S. 2019. Spermathecal filling in *Aedes aegypti* and *Aedes albopictus*: effects of female and male body sizes and species. *J Med Entomol* **56:** 334–340. doi:10.1093/jme/tjy158

Chen J, Luo J, Wang Y, Gurav AS, Li M, Akbari OS, Montell C. 2021. Suppression of female fertility in *Aedes aegypti* with a CRISPR-targeted male-sterile mutation. *Proc Natl Acad Sci* **118:** e2105075118. doi:10.1073/pnas.2105075118

Clements A, Potter S. 1967. The fine structure of the spermathecae and their ducts in the mosquito *Aedes aegypti*. *J Insect Physiol* **13:** 1825–1836. doi:10.1016/0022-1910(67)90018-2

Craig G. 1967. Mosquitoes: female monogamy induced by male accessory gland substance. *Science* **156:** 1499–1501. doi:10.1126/science.156.3781.1499

Degner E, Harrington L. 2016a. A mosquito sperm's journey from male ejaculate to egg: mechanisms, molecules, and methods for exploration. *Mol Reprod Dev* **83:** 897–911. doi:10.1002/mrd.22653

Degner E, Harrington L. 2016b. Polyandry depends on postmating time interval in the dengue vector *Aedes aegypti*. *Am J Trop Med Hyg* **94:** 780–785. doi:10.4269/ajtmh.15-0893

Degner E, Ahmed-Braimah Y, Borziak K, Wolfner M, Harrington L, Dorus S. 2019. Proteins, transcripts, and genetic architecture of seminal fluid and sperm in the mosquito *Aedes aegypti*. *Mol Cell Proteomics* **18:** S6–S22. doi:10.1074/mcp.RA118.001067

Downes JA. 1969. The swarming and mating flight of Diptera. *Annu Rev Entomol* **14:** 271–298. doi:10.1146/annurev.en.14.010169.001415

Duvall L, Basrur N, Molina H, McMeniman C, Vosshall L. 2017. A peptide signaling system that rapidly enforces paternity in the *Aedes aegypti* mosquito. *Curr Biol* **27:** 3734–3742.e5. doi:10.1016/j.cub.2017.10.074

Eckel B, Guo R, Reinhardt K. 2017. More pitfalls with sperm viability staining and a viability-based stress test to characterize sperm quality. *Front Ecol Evol* **5:** 165. doi:10.3389/fevo.2017.00165

Fuchs MS, Craig GB, Despommier DD. 1969. The protein nature of the substance inducing female monogamy in *Aedes aegypti*. *J Insect Physiol* **15:** 701–709. doi:10.1016/0022-1910(69)90265-0

Gratz NG. 2004. Critical review of the vector status of *Aedes albopictus*. *Med Vet Entomol* **18:** 215–227. doi:10.1111/j.0269-283X.2004.00513.x

Gutzler B, Pugh T, Watson W. 2020. A novel method of quantifying sperm to determine the potential for sperm depletion in male American lobsters *Homarus americanus* (H. Milne Edwards, 1837) (Decapoda: Astacidea: Nephropidae). *J Crustac Biol* **40:** 471–478. doi:10.1093/jcbiol/ruaa030

Helinski MEH, Harrington LC. 2011. Male mating history and body size influence female fecundity and longevity of the dengue vector *Aedes aegypti*. *J Med Entomol* **48:** 202–211. doi:10.1603/ME10071

Helinski MEH, Deewatthanawong P, Sirot L, Wolfner M, Harrington L. 2012a. Duration and dose-dependency of female sexual receptivity responses to seminal fluid proteins in *Aedes albopictus* and *Ae. aegypti* mosquitoes. *J Insect Physiol* **58:** 1307–1313. doi:10.1016/j.jinsphys.2012.07.003

Helinski MEH, Valerio L, Facchinelli L, Scott T, Ramsey J, Harrington L. 2012b. Evidence of polyandry for *Aedes aegypti* in semifield enclosures. *Am J Trop Med Hyg* **86:** 635–641. doi:10.4269/ajtmh.2012.11-0225

Holman MEH. 2009. Sperm viability staining in ecology and evolution: potential pitfalls. *Behav Ecol Sociobiol* **63:** 1679–1688. doi:10.1007/s00265-009-0816-4

Honório N, María CC, Irka EB, Naoya N, Tom S LPL. 2018. Male origin determines satyrization potential of *Aedes aegypti* by invasive *Aedes albopictus*. *Biol Invasions* **20:** 653–664. doi:10.1007/s10530-017-1565-3

Jones J, Wheeler R. 1965. Studies on spermathecal filling in *Aedes aegypti* (linnaeus). *Biol Bull* **129:** 532–545. doi:10.2307/1539731

Lounibos P. 2002. Invasions by insect vectors of human disease. *Annu Rev Entomol* **47:** 233–266. doi:10.1146/annurev.ento.47.091201.145206

Noble J, Degner E, Harrington L, Kourkoutis L. 2019. Cryo-electron microscopy reveals that sperm modification coincides with female fertility in the mosquito *Aedes aegypti*. *Sci Rep* **9:** 18537. doi:10.1038/s41598-019-54920-6

Oliva CF, Damiens D, Vreysen MJB, Lemperière G, Gilles J. 2013. Reproductive strategies of *Aedes albopictus* (Diptera: Culicidae) and implications for the sterile insect technique. *PLoS ONE* **8:** e78884. doi:10.1371/journal.pone.0078884

Oliva CF, Damiens D, Benedict M. 2014. Male reproductive biology of *Aedes* mosquitoes. *Acta Trop* **132:** S12–S19. doi:10.1016/j.actatropica.2013.11.021

Parsana D, Nanfack-Minkeu F, Sirot LK. 2022. Determining the insemination status of *Aedes* mosquitoes. *Cold Spring Harb Protoc* doi:10.1101/pdb.prot107954

Ponlawat A, Harrington LC. 2009. Factors associated with male mating success of the dengue vector mosquito, *Aedes aegypti*. *Am J Trop Med Hyg* **80:** 395–400. doi:10.4269/ajtmh.2009.80.395

Reed W, Carroll J, Agramonte A, Lazear JW. 1900. The etiology of yellow fever: a preliminary note. *Public Health Pap Rep* **26:** 37–53.

Richardson JB, Jameson SB, Gloria-Soria A, Wesson DM, Powell J. 2015. Evidence of limited polyandry in a natural population of *Aedes aegypti*. *Am J Trop Med Hyg* **93:** 189–193. doi:10.4269/ajtmh.14-0718

Rojas-Araya D, Alto B, Cummings D, Burkett-Cadena N. 2020. Differentiation of multiple fluorescent powders, powder transfer, and effect on mating in *Aedes aegypti* (Diptera: Culicidae). *Insects* **11:** E727. doi:10.3390/insects11110727

Roth L. 1948. A study of mosquito behavior. An experimental laboratory study of the sexual behavior of *Aedes aegypti* (Linnaeus). *Am Midl Nat* **40:** 265–352. doi:10.2307/2421604

Sirot L, Poulson R, McKenna C, Girnary H, Wolfner M, Harrington L. 2008. Identity and transfer of male reproductive gland proteins of the dengue vector mosquito, *Aedes aegypti*: potential tools for control of female feeding and reproduction. *Insect Biochem Mol Biol* **38:** 176–189. doi:10.1016/j.ibmb.2007.10.007

Sirot L, Hardstone M, Helinski M, Ribeiro J, Kimura M, Deewatthanawong P, Wolfner M, Harrington L. 2011. Towards a semen proteome of the dengue vector mosquito: protein identification and potential functions. *PLoS Negl Trop Dis* **5:** e989. doi:10.1371/journal.pntd.0000989

Smith D, Arrigo N, Leal G, Muehlberger L, Weaver S. 2007. Infection and dissemination of Venezuelan equine encephalitis virus in the epidemic mosquito vector, *Aedes taeniorhynchus*. *Am J Trop Med Hyg* **77:** 176–187. doi:10.4269/ajtmh.2007.77.176

Spielman A, Leahy MG, Skaff V. 1967. Seminal loss in repeatedly mated female *Aedes aegypti*. *Biol Bull* **132:** 404–412. doi:10.2307/1539644

Wong J, Chu YY, Stoddard S, Lee Y, Morrison A, Scott T. 2012. Microsatellite-based parentage analysis of *Aedes aegypti* (Diptera: Culicidae) using nonlethal DNA sampling. *J Med Entomol* **49:** 85–93. doi:10.1603/ME11152

Protocol 1

Determining the Insemination Status of *Aedes* Mosquitoes

Dhwani Parsana,[1] Ferdinand Nanfack-Minkeu,[1] and Laura K. Sirot[1,2]

[1]*Department of Biology, The College of Wooster, Wooster, Ohio 44691, USA*

Within the genus *Aedes*, the reproductive biology of two species has been most thoroughly studied: *Aedes aegypti* and *Aedes albopictus*. In these species, females tend to copulate with one or more males once sexually mature. Within a few hours after an initial insemination, most females become refractory to insemination for the rest of their lives. *Aedes* females store sperm in three sclerotized spherical structures called spermathecae, where they can remain viable for >3 mo after copulation. This protocol outlines a quick (~3–5 min per female once you are practiced in dissection) and effective method for dissecting the female spermathecae from *Ae. aegypti* and *Ae. albopictus* mosquitoes and scoring them for the presence or absence of sperm in cases in which the researcher does not need the female for further studies.

MATERIALS

It is essential that you consult the appropriate Material Safety Data Sheets and your institution's Environmental Health and Safety Office for proper handling of the equipment and hazardous materials used in this protocol.

RECIPES: Please see the end of this protocol for recipes indicated by <R>. Additional recipes can be found online at http://cshprotocols.cshlp.org/site/recipes.

Reagents

Aedes aegypti adult females previously housed with males

Copulations occur most quickly when females are placed with a group of males, in low light conditions, relative humidity >50%, and temperature ~25°C, during dawn or dusk hours according to the mosquitoes' light cycle. Females can also be left for multiple days singly or in groups with males. A ratio of two to three males per female generally ensures high levels of insemination when mosquitoes are left together for 48 h.

Ephrussi–Beadle Ringer's solution <R>

Ringer's solution is used in this protocol, but 1× phosphate buffered saline or an alternative saline solution would also be suitable.

Ice or alternative source of low temperature, such as a chill table or cold block (for keeping the females anesthetized; see Step 2)

Equipment

Compound microscope with phase contrast or dark field, with 10× eyepiece and 4×, 10×, and 40× objectives

---

[2]Correspondence: lsirot@wooster.edu

Cite this protocol as *Cold Spring Harb Protoc*; doi:10.1101/pdb.prot107954

Coverslip (22-mm × 22-mm)
Dissecting microscope with 10× eyepiece and at least 4× objective
Glass microscope slides (75-mm × 25-mm)
Fine forceps (two pairs; e.g., Dumont 55)
Petri dishes (100-mm × 15-mm) with a strip of black tape down the center of the top to provide contrast
Sharpie marker or equivalent waterproof marker pen

METHOD

Dissection of Females (Fig. 1)

It is critical that the tissues remain in Ringer's solution throughout the dissection so that they do not desiccate or stick to the slide.

1. Place females for ~5 min at −20°C or on ice to anesthetize them.

   If females are frozen for a longer time, sperm should still be visible, but will be more difficult to detect as they may no longer be moving. Additionally, after more than a few hours of freezing, tissues may become brittle and more difficult to dissect. Alternatively, females can be anesthetized with carbon dioxide.

2. Fill a Petri dish with ice or use another source of low temperature. Place a clean glass slide over the black tape on the Petri dish.

   Change the ice frequently to maintain cold temperature (see Troubleshooting).

3. Using a micropipette, place an ~3-µL droplet of Ringer's solution on the glass slide.
4. Place the anesthetized female laterally on the glass slide with the posterior end of her abdomen just touching the edge of the droplet (Fig. 1A).
5. Place the Petri dish under a dissecting microscope.
6. Hold the female's thorax or upper abdomen with one pair of forceps. Squeeze these forceps gently.

   The membrane between the terminal segments of the female's abdomen should extend, making it easier to grasp the cerci in the next step.

7. Pinch the female's cerci with the second forceps (Fig. 1B).
8. Gently but firmly pull the cerci away from the rest of the body.

   At this point the three spermathecae should be visible as distinct, dark brown, sclerotized spheres. If they are not visible, you may need to look for them within the upper abdomen or the exoskeleton of the terminal segments.

9. If the spermathecae (Fig. 1C) are not yet visible, carefully remove the exoskeleton from the most distal segment of the abdomen. To do so, pinch the cerci with one pair of forceps, and use the second pair of forceps to grasp the outside and inside of the exoskeleton of the distal abdominal segment. Pull the forceps away from each other gently but firmly to pull off the ring of exoskeleton.

   See Troubleshooting.

10. If the remaining tissue of the female is not needed, kill the female on ice or by pinching in a paper towel so that she does not awaken from the anesthesia.
11. Remove excess tissue surrounding the spermathecae using the forceps.

    It is important to clear off most of the surrounding tissues before mounting the sample. Otherwise, it can be difficult to visualize the sperm or to distinguish sperm from other structures. During cleaning, try not to disrupt the spermathecal ducts, which form a connection between the spermathecae and make it easier to move them together to a new droplet for mounting. At this point, with experience, the female can be scored as inseminated or not by looking for sperm through the walls of the spermathecae. Spermathecae with sperm present are darker than those with no sperm. With proper lighting, movement of sperm may be visible.

 Cite this protocol as *Cold Spring Harb Protoc*; doi:10.1101/pdb.prot107954

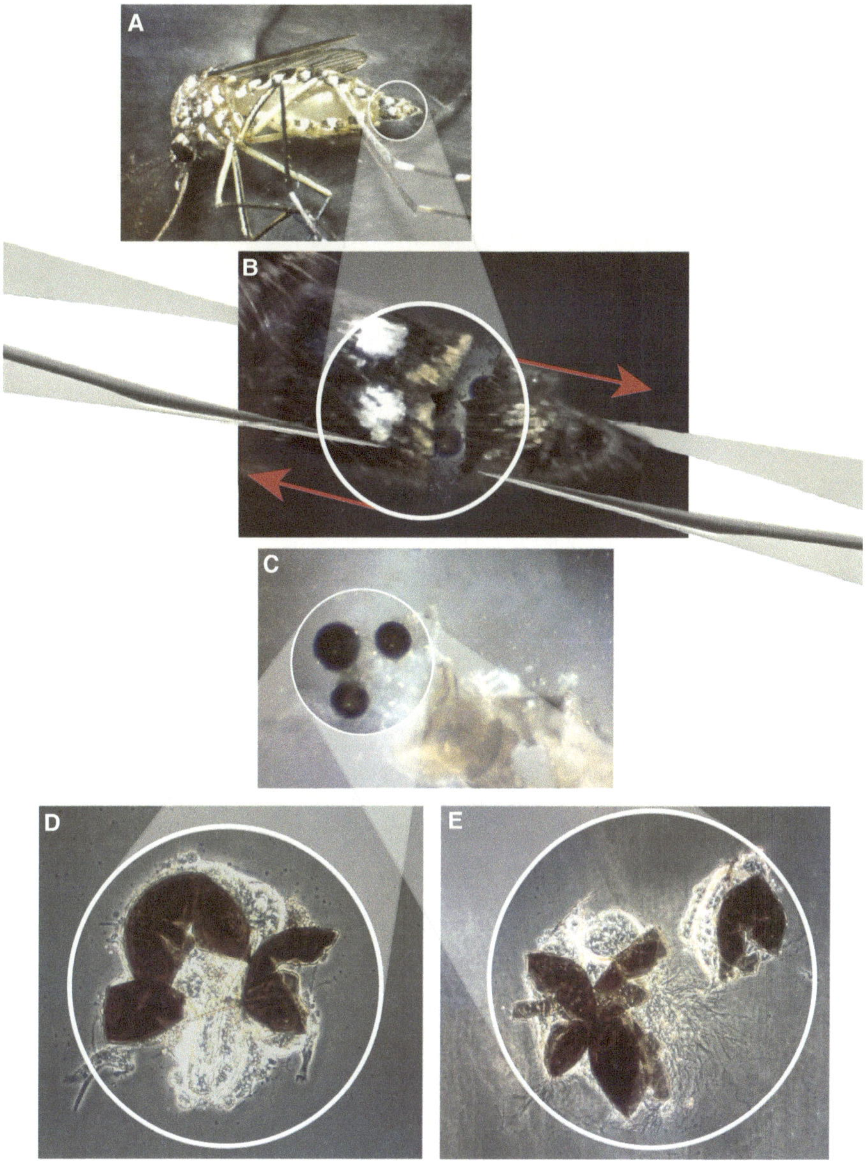

FIGURE 1. Steps of spermathecae dissection for determining insemination status of *Aedes* mosquitoes. (*A*) Anesthetized female placed laterally in a drop of saline buffer on glass slide. (*B*) Female's cerci and terminal abdominal segments pulled apart with forceps, containing spermathecae that are visible but not yet exposed. (*C*) Female's cerci and terminal abdominal segment covered by exoskeleton, with spermathecae exposed. (*D*) Opened spermathecae without sperm. (*E*) Opened spermathecae with sperm.

Sample Mounting

The spermathecal samples should be maintained in Ringer's solution to avoid tissue dehydration. The spermathecae can remain in the Ringer's solution for 2–3 min, but it is advisable to mount the sample as soon as possible.

12. Using a micropipette, place a 5- to 10-µL drop of Ringer's solution on a separate microscope slide or a different part of the same slide.
13. Use the fine forceps to transfer the cleaned spermathecae to the new Ringer's solution droplet.

    Be sure that all three spermathecae are transferred (see Troubleshooting). Try to have the three spermathecae in close proximity but not on top of each other if it is important to score which or how many spermathecae contain sperm.

14. Place a coverslip on the spermathecae and give it a gentle push to open the spermathecae (Fig. 1D,E).

Make sure that the spermathecae are cracked open (Fig. 1D,E). Failure to open up the spermathecae can result in inaccurate scoring. To facilitate finding the spermathecae quickly on the compound microscope, use a fine-tipped Sharpie marker (or equivalent fine-tipped waterproof marker pen) to draw a circle around the spermathecae either on the coverslip or on the bottom of the slide. The marker ink can be removed later by wiping with ethanol.

Spermatheca Scoring

Observe the spermathecae within 5 min of sample mounting. Delay may result in dehydration of the tissue and sperm and inaccurate scoring of insemination status (see Troubleshooting). If the sample begins to dry, a drop of Ringer's solution can be added to the edge of the coverslip to rehydrate the sample.

15. Place the microscope slide in the compound microscope.
16. Use the 4× objective to locate the spermathecae on the slide.
17. Use the 10× objective with the appropriate phase contrast setting to observe the three spermathecae and surrounding fluid for presence (Fig. 1E) or absence (Fig. 1D) of sperm.

Sperm will appear as long, thin strands and are generally moving within the spermathecae and/or the surrounding fluid.

18. Use the 40× objective with the appropriate phase contrast setting to observe each spermatheca individually.
19. Record the presence or absence of sperm.

If the presence or absence of sperm in each spermatheca is important, data can be recorded for each spermatheca individually. Note that, in some cases, only a few sperm will be present in a spermatheca, or some but not all spermathecae will contain sperm. Therefore, to confidently conclude that a female is not inseminated, each spermatheca must be examined carefully.

TROUBLESHOOTING

Problem (Step 2): During dissection, if the ice in the Petri dish starts melting, the female may awaken and start moving.

Solution: Change the ice in the Petri dish frequently to keep the female anesthetized.

Problem (Step 9): During the removal of the exoskeleton surrounding the spermathecae, the spermathecae can remain attached to the abdomen and may not come out attached to the cerci.

Solution: If this happens, use the forceps to gently open the abdominal exoskeleton to retrieve the spermathecae. To open the exoskeleton, carefully pinch the abdomen on opposite sides with two pairs of forceps, being careful not to pierce the spermathecae. Pull the forceps away from each other to reveal the spermathecae.

Problem (Step 13): During the transfer of the spermathecae to a new slide, all three spermathecae might not be attached to the spermathecal ducts, causing one or two spermathecae to be left behind on the dissection slide.

Solution: To transfer the individual spermatheca, touch the loose spermatheca with one tine of the forceps. Generally, the spermatheca will adhere to the forceps. If that does not happen, you can use the forceps to gently grab the spermatheca.

Problem (Steps 15–19): During the recording of sperm presence or absence, sample dehydration may occur, resulting in improper interpretation of whether the sample contains sperm.

Cite this protocol as *Cold Spring Harb Protoc*; doi:10.1101/pdb.prot107954

Solution: To avoid sample dehydration, the samples should be scored within 5 min of mounting. The sperm will still be alive and moving, allowing the observer to record their presence or absence in the spermatheca. If the sample does become dehydrated, a drop of Ringer's solution can be added to the sides of the coverslip to rehydrate the sample, which may make the sperm visible.

RECIPE

Ephrussi–Beadle Ringer's Solution

Reagent	Amount (for 1× solution)	Amount (for 10× stock)
NaCl	7.5 g	75 g
KCl	0.35 g	3.5 g
$CaCl_2$	0.21 g	2.1 g

Ringer's solution can be made as a 1× solution or as a 10× stock. To prepare 1 L of either 1× or 10× Ringer's solution, dissolve the reagents listed above in 1 L of deionized H_2O and adjust the pH to 7 ±0.5. Store for up to 12 mo at 4°C.

This recipe is derived from Ephrussi and Beadle (1936).

ACKNOWLEDGMENTS

We thank Gemma Briggs, Dr. Ethan Degner, Rachel Greer, Dr. Laura Harrington, Nosherwan Mughal, and Jodi Robison for suggestions for improving this manuscript. L.K.S. thanks Dr. Laura Harrington for training in *Aedes* dissection techniques. D.P. was supported in part by the College of Wooster Biochemistry and Molecular Biology Program. All authors were supported by a U.S. National Institutes of Health grant to L.K.S. (R15AI140223).

REFERENCES

Ephrussi B, Beadle G. 1936. A technique of transplantation for *Drosophila*. *Am Nat* **70:** 218–225. doi:10.1086/280658

CHAPTER 18

Mosquito Egg Development and Eggshell Formation

Jun Isoe,[1,3] Michael A. Riehle,[1] and Roger L. Miesfeld[2]

[1]*Department of Entomology,* [2]*Department of Chemistry and Biochemistry, The University of Arizona, Tucson, Arizona 85721, USA*

Anautogenous female mosquitoes, which ingest a blood meal from warm-blooded vertebrates to produce eggs, have become a valuable model organism for investigating signaling pathways and physiological processes that occur during egg development. Different molecular pathways tightly regulate the initiation of egg development and are governed by a balance among different insect hormones. Gravid (mature egg-carrying) females deposit fully developed eggs at the end of each gonotrophic cycle, which is defined as the time interval between the ingestion of a blood meal to oviposition. An intact eggshell protects the oocyte and embryo inside from external factors such as desiccation, physical damage, etc., and the various eggshell proteins are spatially and temporary deposited during oogenesis. Additionally, follicle resorption (oosorption) during blood meal–induced mosquito ovarian follicle development is an adapted physiological process that optimizes reproductive fitness. Mosquito oocytes grow and mature synchronously throughout oogenesis; however, during the later stages of oogenesis, some oocytes may undergo oosorption if sufficient nutrients are unavailable. This introduction highlights how mosquito egg development can be used to investigate follicular resorption and identify proteins involved in eggshell formation in *Aedes aegypti* mosquitoes.

BACKGROUND

Anautogenous female mosquitoes consume discrete, intermittent blood meals, unlike many other insect species. The *Drosophila melanogaster* fruit fly, which feeds continuously to obtain nutrients for reproduction, has 16 ovarioles in each ovary, and each ovariole is composed of six follicles at different development stages, allowing *D. melanogaster* females to lay dozens of eggs daily. In contrast, once anautogenous female mosquitoes such as *Aedes aegypti* acquire a sufficient amount of vertebrate blood, they allocate their newly acquired nutrient reserves to facilitate follicle growth and maturation over the next 48 to 60 h (Fig. 1; Clements 1992; Airs et al. 2023). *Ae. aegypti* females have a pair of ovaries, and each ovary contains approximately 100 discrete ovarioles. Each ovariole contains a primary and a secondary follicle, plus a germarium. The primary follicle within the ovariole will become the mature egg, and the secondary follicle will mature into the primary follicle following oviposition. The germarium contains undifferentiated stem cells that mature into secondary follicles. After a blood meal, the primary follicles develop and mature by receiving vitellogenin yolk and other nutrients, whereas the secondary follicles become quiescent during the first gonotrophic cycle. The vital roles of insect hormones, including 20-hydroxyecdysone, juvenile hormone, ovarian ecdysteroidogenic hormone 1, insulin-like peptides, and others during mosquito ovarian follicle development have been studied

[3]Correspondence: jisoe@ag.arizona.edu

Cite this introduction as *Cold Spring Harb Protoc*; doi:10.1101/pdb.top107669

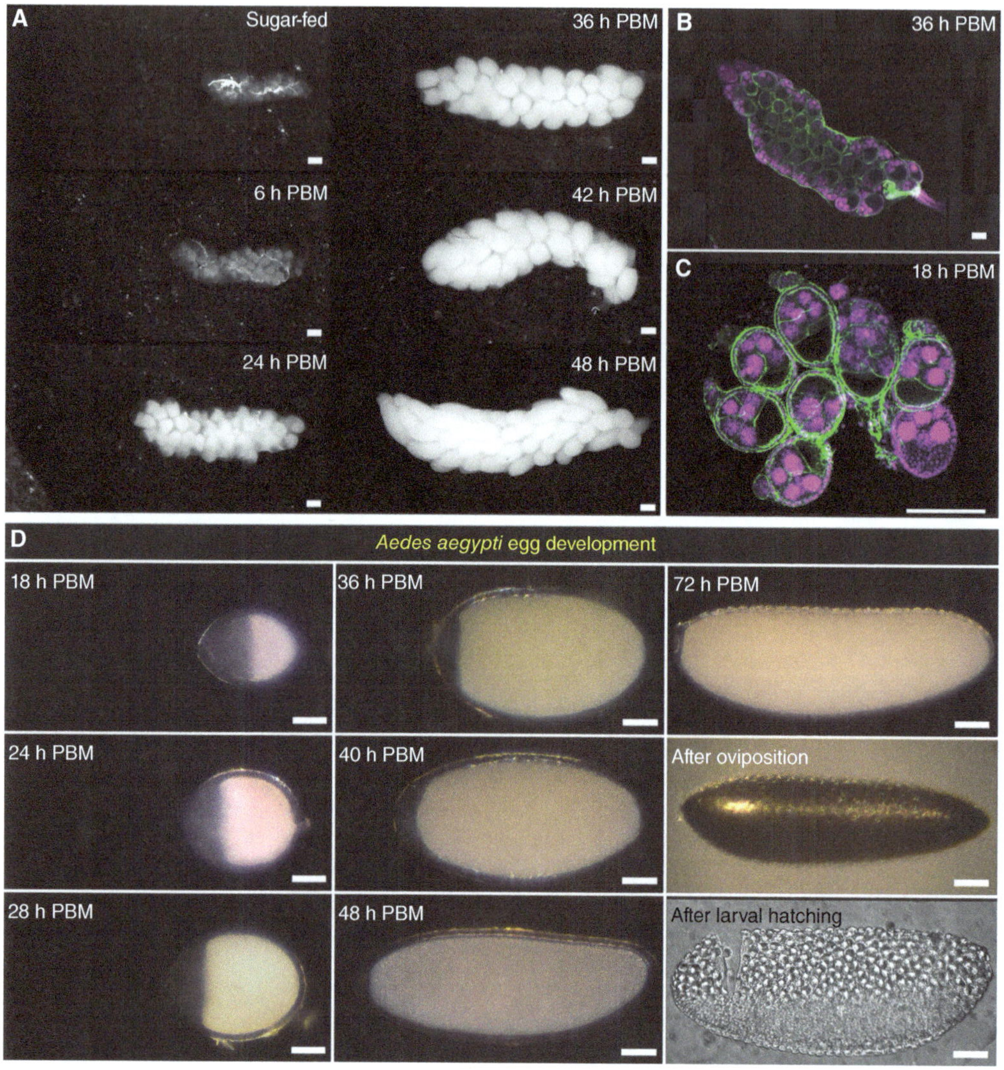

FIGURE 1. Representative images of dissected ovaries and individual eggs during *Aedes aegypti* vitellogenesis. (*A*) Light microscopy images of ovaries are shown from mosquitoes that were sugar-fed to those at 6, 24, 36, 42, and 48 h post-blood meal (PBM). Bars, 150 µm. (*B,C*) Ovaries dissected at 36 and 18 h post-blood feeding were fixed with 4% paraformaldehyde and stained with phalloidin for actin cytoskeleton (green) and 4′,6-diamidino-2-phenylindole (DAPI; purple), showing a synchronous primary follicle development within the ovary (C1si confocal laser scanning microscopy, Nikon). Bars, 150 µm. (*D*) Light microscopy images of eggs that were dissected from *Ae. aegypti* females at 18, 24, 28, 36, 40, 48, and 72 h PBM. Melanized eggs after oviposition and an eggshell after larval hatching are also shown. Bars, 50 µm.

and reviewed (Raikhel et al. 2002; Attardo et al. 2005; Brown et al. 2008; Smykal and Raikhel 2015; Roy et al. 2018; Allman et al. 2020). Mosquito ovarian primary follicles develop synchronously throughout oogenesis, and gravid female mosquitoes oviposit the entire egg clutch at the end of each gonotrophic cycle (Fig. 1; Clements 1992; Airs et al. 2023). When *Ae. aegypti* female mosquitoes feed to repletion on blood or an optimized concentration of artificial protein diet such as bovine serum albumin (BSA) under laboratory conditions, only about half of the primary follicles in the mosquito complete follicular maturation. Although all primary follicles initiate development following a blood meal, if insufficient nutritional resources are available, nutrients can be reclaimed from partially developed oocytes through a process of oocyte resorption (oosorption), also known as follicular atresia. This allows the mosquito to maximize reproductive fitness by ensuring that the greatest number of oocytes are sufficiently provisioned.

Cite this introduction as *Cold Spring Harb Protoc*; doi:10.1101/pdb.top107669

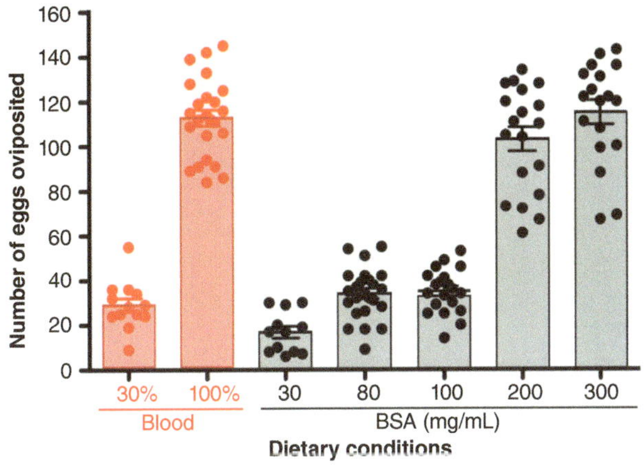

FIGURE 2. Dietary protein levels affect *Aedes aegypti* egg production during the first gonotrophic cycle. Phosphate-buffered saline was used to prepare specific concentrations of bovine blood or bovine serum albumin (BSA) (see *x*-axis). Only fully fed females were allowed to lay eggs in individual containers 3 days after feeding.

Mosquitoes are ideal models for studying egg maturation because their reproductive cycle can be easily initiated via blood feeding. This maturation is tightly regulated by a series of hormonal signals, resulting in the follicles developing synchronously. The number of fully maturing follicles within a pair of mosquito ovaries is largely influenced by the total protein concentration ingested by the female mosquitoes (Attardo et al. 2005; Isoe et al. 2009; Gonzales et al. 2015), whereas sucrose levels and juvenile hormone titers in female mosquitoes during the previtellogenenic stage (from adult eclosion until the blood meal) have been shown to be important for the degree of follicle resorption during a reproductive cycle (Clifton and Noriega 2012). Mosquito reproductive performance, as measured by fecundity, is significantly altered by dietary protein levels during the first gonotrophic cycle (Fig. 2; Isoe et al. 2009; Gonzales et al. 2015). Females that fed to repletion on whole blood oviposited a significantly higher number of eggs than those provisioned with a 30% diluted blood meal during the first gonotrophic cycle (Isoe et al. 2009). Similarly, mosquitoes that fed on a high-protein BSA diet had a significantly higher number of eggs than those fed on a low-BSA protein diet, suggesting that female mosquitoes on the low protein diet had significantly more follicles undergoing resorption because of increased follicle–follicle competition for vitellogenin yolk protein uptake. These studies collectively show that the number of developing ovarian follicles in female mosquitoes is strongly affected by the amount of protein in their diet.

The precise timing of synthesis and secretion of eggshell building blocks and construction of intact eggshell during oocyte development is critical for fully supporting embryonic development after fertilization and oviposition. Follicular epithelial cells, which surround the oocyte and 15 nurse cells, are the cells responsible for synthesizing and secreting most eggshell components between 30 and 72 h after blood meals (Fig. 1; Clements 1992; Airs et al. 2023). Once eggshell formation is completed, the follicular epithelial cells go through programmed cell death in each ovariole before oviposition. Scanning electron microscope images show the complex structure of the mosquito eggshells (Fig. 3; Dos Santos-Mallet et al. 2010; Suman et al. 2011; Isoe et al. 2019). When mature mosquito eggs are laid, the protective eggshell immediately undergoes maturation by hardening and tanning processes catalyzed by multiple enzymes, of which several have been identified and biochemically characterized (Li 1994; Ferdig et al. 1996; Han et al. 2000; Johnson et al. 2001; Fang et al. 2002; Kim et al. 2005; Li and Li 2005). Furthermore, liquid chromatography with tandem mass spectrometry (LC-MS/MS) proteomic studies on purified mosquito eggshells have been conducted to identify the abundant eggshell proteins in *Anopheles gambiae* (Amenya et al. 2010) and in *Ae. aegypti*, where 130 unique eggshell proteins have been identified (Marinotti et al. 2014). In summary, mosquitoes are ideal model organisms in which to study ovarian follicle development and eggshell formation because these physiologies are precisely controlled in response to ingestion of a blood meal and the follicles mature synchronously throughout egg development.

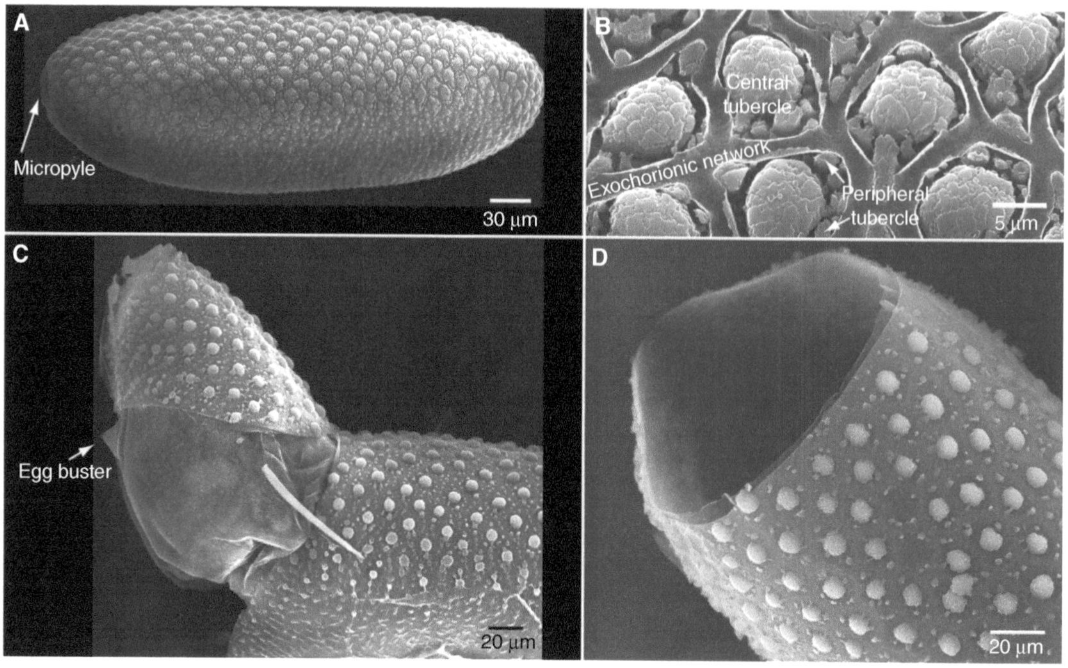

FIGURE 3. Ultrastructural characteristics of the *Aedes aegypti* eggshell. (*A*) An image of an entire egg. (*B*) A high-resolution image of eggshell surface that shows central tubercles, exochorionic network, and peripheral tubercles. (*C*) An image of the first-instar larva emerging from the eggshell. The larvae use an egg buster to break open the eggshell. (*D*) An image of an anterior eggshell after hatching. Samples were prepared as in Isoe et al. (2019) and analyzed with a scanning electron microscope (Inspect-S; FEI).

BIOCHEMICAL ASSAYS TO STUDY MOSQUITO EGGSHELLS

Using Fluorescent-Labeled Inhibitor of Caspases to Study Caspase-Mediated Apoptosis in Mosquito Follicles

Apoptotic markers are useful for understanding the timing of primary follicle apoptosis and resorption in mosquitoes. One such indicator is the activation of caspase enzymes during the initial phases of apoptosis (Uchida et al. 2004; Ahmed and Hurd 2006; Isoe et al. 2019). Additional apoptotic markers, such as the TUNEL assay or acridine orange staining, have been successfully used to monitor apoptotic events in mosquito ovaries (Hopwood et al. 2001; Uchida et al. 2004; Almeida and Suesdek 2017). The distribution and occurrence of caspase-positive follicles within an ovary can be determined under different dietary conditions using a fluorescent-labeled inhibitor of caspases (FLICA) under a confocal laser scanning microscope. In this assay, the FLICA reagent enters cells and binds covalently to active caspases within the cell (Bedner et al. 2000; Darzynkiewicz et al. 2011). This leads to retention of the fluorescent signal only in apoptotic cells. Our associated Protocol 1: Visualization of Apoptotic Ovarian Follicles during *Aedes aegypti* Mosquito Egg Maturation by Fluorescent Imaging Studies (Isoe et al. 2023a) provides detailed procedures for carrying out a FLICA assay to assess ovarian follicular apoptosis in *Ae. aegypti* mosquitoes. The synchronous development of mosquito ovarian primary follicles under different nutrient levels coupled with our FLICA assay provides an ideal system with which to study oosorption initiation, timing, and extent in mosquito ovaries. We have observed that primary follicles undergoing resorption exhibit high caspase activity based on fluorescent inhibitor probe SR-VAD-FMK fluorescence. Furthermore, the protocol can also be applied to other tissues in diverse organisms that undergo caspase-mediated apoptosis.

Cite this introduction as *Cold Spring Harb Protoc*; doi:10.1101/pdb.top107669

USING LC-MS/MS ANALYSIS TO IDENTIFY AND CHARACTERIZE MOSQUITO EGGSHELL PROTEINS

Following oviposition, the soft eggshell hardens over several hours, as described above, to produce the final protective eggshell. During this hardening process the newly deposited eggs are sensitive to desiccation until the formation of a serosal cuticle that surrounds the developing embryo (Rezende et al. 2008; Vargas et al. 2014). Because eggshell components are transported into the extracellular space between the oocyte and the surrounding follicular epithelial layer, intimate communication between these participating cells within the ovariole likely is present throughout the maturation of follicles, leading to shedding of follicular epithelial layer, ovulation, fertilization, and oviposition. Our associated Protocol 2: Identification of Mosquito Eggshell Proteins from *Aedes aegypti* by Liquid Chromatography with Tandem Mass Spectrometry (LC-MS/MS) Proteomic Analysis (Isoe et al. 2023b) describes how to enrich eggshells from *Ae. aegypti* mosquito ovaries for proteomic analysis. The protocol was developed to ensure a relatively fast, accurate, and efficient physical separation of eggshell from dissected *Ae. aegypti* mosquito ovaries for subsequent use to detect eggshell proteins by liquid chromatography-mass spectrometry analysis (Isoe et al. 2023c) Important features of this approach include (1) an enrichment of primary follicles from the ovaries that contain different cell types, (2) physical separation of partially cross-linked extracellular eggshell from the oocytes by Dounce homogenization, (3) extraction of eggshell proteins with a guanidine hydrochloride and ethanol protein precipitation, and (4) LC-MS/MS analysis. Although this current approach was developed for examining mosquito eggshells, the method may be applicable to highly melanized and cross-linked extracellular structures such as the mosquito larval head capsule.

In summary, female mosquitoes are an ideal organism in which to study egg development, because ingestion of vertebrate blood tightly induces follicle development and eggshell formation. Two assays described here, FLICA-based apoptotic follicle detection and eggshell proteomics, will enable elucidation of the dynamic processes controlling ovarian follicle development and eggshell formation and melanization in mosquitoes. An understanding of these events may lead to the discovery of mosquito-specific molecular mechanisms and development of novel biosafe mosquitocides.

ACKNOWLEDGMENTS

We thank Alyssa J. Peterson at the University of Arizona for technical assistance and Dr. Brooke B. Massani of the University of Arizona W.M. Keck Center for Surface and Interface Imaging Facility for assistance with scanning electron and confocal microscopy work. We thank reviewers for their valuable and thoughtful comments.

REFERENCES

Ahmed AM, Hurd H. 2006. Immune stimulation and malaria infection impose reproductive costs in *Anopheles gambiae* via follicular apoptosis. *Microbes Infect* **8:** 308–315. doi:10.1016/j.micinf.2005.06.026

Airs PM, Nazarchyk MJ, Tucker BJ, Bartholomay LC. 2023. Characterizing oogenesis and programmed cell death in the eastern tree hole mosquito *Aedes* (*Protomacleaya*) *triseriatus*. *Front Insect Sci* **2:** 1073308. doi:10.3389/finsc.2022.1073308

Allman MJ, Fraser JE, Ritchie SA, Joubert DA, Simmons CP, Flores HA. 2020. *Wolbachia*'s deleterious impact on *Aedes aegypti* egg development: the potential role of nutritional parasitism. *Insects* **11:** 735. doi:10.3390/insects11110735

Almeida F, Suesdek L. 2017. Effects of *Wolbachia* on ovarian apoptosis in *Culex quinquefasciatus* (Say, 1823) during the previtellogenic and vitellogenic periods. *Parasit Vectors* **10:** 398. doi:10.1186/s13071-017-2332-0

Amenya DA, Chou W, Li J, Yan G, Gershon PD, James AA, Marinotti OJ. 2010. Proteomics reveals novel components of the *Anopheles gambiae* eggshell. *J Insect Physiol* **56:** 1414–1419. doi:10.1016/j.jinsphys.2010.04.013

Attardo GM, Hansen IA, Raikhel AS. 2005. Nutritional regulation of vitellogenesis in mosquitoes: implications for anautogeny. *Insect Biochem Mol Biol* **35:** 661–675. doi:10.1016/j.ibmb.2005.02.013

Bedner E, Smolewski P, Amstad P, Darzynkiewicz Z. 2000. Activation of caspases measured in situ by binding of fluorochrome-labeled inhibitors of caspases (FLICA): correlation with DNA fragmentation. *Exp Cell Res* **259:** 308–313. doi:10.1006/excr.2000.4955

Brown MR, Clark KD, Gulia M, Zhao Z, Garczynski SF, Crim JW, Suderman RJ, Strand MR. 2008. An insulin-like peptide regulates egg maturation and metabolism in the mosquito *Aedes aegypti*. *Proc Natl Acad Sci* **105:** 5716–5721. doi:10.1073/pnas.0800478105

Clements AN. 1992. *The biology of mosquitoes: development, nutrition and reproduction*. Chapman & Hall, London.

Clifton ME, Noriega FG. 2012. The fate of follicles after a blood meal is dependent on previtellogenic nutrition and juvenile hormone in *Aedes aegypti*. *J Insect Physiol* **58:** 1007–1019. doi:10.1016/j.jinsphys.2012.05.005

Darzynkiewicz Z, Pozarowski P, Lee BW, Johnson GL. 2011. Fluorochrome-labeled inhibitors of caspases: convenient *in vitro* and *in vivo* markers of apoptotic cells for cytometric analysis. *Methods Mol Biol* **682:** 103–114. doi:10.1007/978-1-60327-409-8_9

Dos Santos-Mallet JR, Müller GA, Gleiser RM, Alencar J, Marques Wde A, Sarmento JS, Marcondes CB. 2010. Scanning electron microscopy of the eggs of *Aedes scapularis* from southern South America. *J Am Mosq Control Assoc* **26:** 205–209. doi:10.2987/09-5959.1

Fang J, Han Q, Johnson JK, Christensen BM, Li J. 2002. Functional expression and characterization of *Aedes aegypti* dopachrome conversion enzyme. *Biochem Biophys Res Commun* **290:** 287–293. doi:10.1006/bbrc.2001.6200

Ferdig MT, Li J, Severson DW, Christensen BM. 1996. Mosquito dopa decarboxylase cDNA characterization and blood-meal-induced ovarian expression. *Insect Mol Biol* **5:** 119–126. doi:10.1111/j.1365-2583.1996.tb00046.x

Gonzales KK, Tsujimoto H, Hansen IA. 2015. Blood serum and BSA, but neither red blood cells nor hemoglobin can support vitellogenesis and egg production in the dengue vector *Aedes aegypti*. *PeerJ* **3:** e938. doi:10.7717/peerj.938

Han Q, Li G, Li J. 2000. Purification and characterization of chorion peroxidase from *Aedes aegypti* eggs. *Arch Biochem Biophys* **378:** 107–115. doi:10.1006/abbi.2000.1821

Hopwood JA, Ahmed AM, Polwart A, Williams GT, Hurd H. 2001. Malaria-induced apoptosis in mosquito ovaries: a mechanism to control vector egg production. *J Exp Biol* **204:** 2773–2780. doi:10.1242/jeb.204.16.2773

Isoe J, Rascón AA Jr, Kunz S, Miesfeld RL. 2009. Molecular genetic analysis of midgut serine proteases in *Aedes aegypti* mosquitoes. *Insect Biochem Mol Biol* **39:** 903–912. doi:10.1016/j.ibmb.2009.10.008

Isoe J, Koch LE, Isoe YE, Rascón AA Jr, Brown HE, Massani BB, Miesfeld RL. 2019. Identification and characterization of a mosquito-specific eggshell organizing factor in *Aedes aegypti* mosquitoes. *PLoS Biol* **17:** e3000068. doi:10.1371/journal.pbio.3000068

Isoe J, Miesfeld RL, Riehle MA. 2023a. Visualization of apoptotic ovarian follicles during *Aedes aegypti* mosquito egg maturation by fluorescent imaging studies. *Cold Spring Harb Protoc* doi: 10.1101/pdb.prot108226

Isoe J, Parsawar K, Davis CL, Riehle MA, Miesfeld RL. 2023b. Identification of mosquito eggshell proteins from *Aedes aegypti* by liquid chromatography with tandem mass spectrometry (LC-MS/MS) proteomic analysis. *Cold Spring Harb Protoc* doi: 10.1101/pdb.prot108227

Isoe J, Simington CJ, Oscherwitz ME, Peterson AJ, Rascón AA Jr, Massani BB, Miesfeld RL, Riehle MA. 2023c. Characterization of essential eggshell proteins from *Aedes aegypti* mosquitoes. *BMC Biol* **21:** 214. doi:10.1186/s12915-023-01721-z

Johnson JK, Li J, Christensen BM. 2001. Cloning and characterization of a dopachrome conversion enzyme from the yellow fever mosquito, *Aedes aegypti*. *Insect Biochem Mol Biol* **31:** 1125–1135. doi:10.1016/S0965-1748(01)00072-8

Kim SR, Yao R, Han Q, Christensen BM, Li J. 2005. Identification and molecular characterization of a prophenoloxidase involved in *Aedes aegypti* chorion melanization. *Insect Mol Biol* **14:** 185–194. doi:10.1111/j.1365-2583.2004.00547.x

Li J. 1994. Egg chorion tanning in *Aedes aegypti* mosquito. *Comp Biochem Physiol* **109:** 835–843. doi:10.1016/0300-9629(94)90231-3

Li JS, Li J. 2005. Characterization of *N*-linked oligosaccharides in chorion peroxidase of *Aedes aegypti* mosquito. *Protein Sci* **14:** 2370–2386. doi:10.1110/ps.051419105

Marinotti O, Ngo T, Kojin BB, Chou SP, Nguyen B, Juhn J, Carballar-Lejarazú R, Marinotti PN, Jiang X, Walter MF, et al. 2014. Integrated proteomic and transcriptomic analysis of the *Aedes aegypti* eggshell. *BMC Dev Biol* **14:** 15. doi:10.1186/1471-213X-14-15

Raikhel AS, Kokoza VA, Zhu J, Martin D, Wang SF, Li C, Sun G, Ahmed A, Dittmer N, Attardo G. 2002. Molecular biology of mosquito vitellogenesis: from basic studies to genetic engineering of antipathogen immunity. *Insect Biochem Mol Biol* **32:** 1275–1286. doi:10.1016/S0965-1748(02)00090-5

Rezende GL, Martins AJ, Gentile C, Farnesi LC, Pelajo-Machado M, Peixoto AA, Valle D. 2008. Embryonic desiccation resistance in *Aedes aegypti*: presumptive role of the chitinized serosal cuticle. *BMC Dev Biol* **8:** 82. doi:10.1186/1471-213X-8-82

Roy S, Saha TT, Zou Z, Raikhel AS. 2018. Regulatory pathways controlling female insect reproduction. *Annu Rev Entomol* **63:** 489–511. doi:10.1146/annurev-ento-020117-043258

Smykal V, Raikhel AS. 2015. Nutritional control of insect reproduction. *Curr Opin Insect Sci* **11:** 31–38. doi:10.1016/j.cois.2015.08.003

Suman DS, Shrivastava AR, Pant SC, Parashar BD. 2011. Differentiation of *Aedes aegypti* and *Aedes albopictus* (Diptera: Culicidae) with egg surface morphology and morphometrics using scanning electron microscopy. *Arthropod Struct Dev* **40:** 479–483. doi:10.1016/j.asd.2011.04.003

Uchida K, Nishizuka M, Ohmori D, Ueno T, Eshita Y, Fukunaga A. 2004. Follicular epithelial cell apoptosis of atretic follicles within developing ovaries of the mosquito *Culex pipiens pallens*. *J Insect Physiol* **50:** 903–912. doi:10.1016/j.jinsphys.2004.07.002

Vargas HC, Farnesi LC, Martins AJ, Valle D, Rezende GL. 2014. Serosal cuticle formation and distinct degrees of desiccation resistance in embryos of the mosquito vectors *Aedes aegypti*, *Anopheles aquasalis* and *Culex quinquefasciatus*. *J Insect Physiol* **62:** 54–60. doi:10.1016/j.jinsphys.2014.02.001

Protocol 1

Visualization of Apoptotic Ovarian Follicles during *Aedes aegypti* Mosquito Egg Maturation by Fluorescent Imaging Studies

Jun Isoe,[1,3] Roger L. Miesfeld,[2] and Michael A. Riehle[1]

[1]*Department of Entomology,* [2]*Department of Chemistry and Biochemistry, The University of Arizona, Tucson, Arizona 85721, USA*

In insects, oocyte resorption (oosorption) or follicular atresia is one of the key physiological processes and evolutionary strategies used to optimize reproductive fitness. Mosquitoes are ideal model organisms for studying egg maturation in arthropods, as their follicle development is initiated only following the ingestion of a blood meal, followed by a carefully orchestrated series of hormonally regulated events leading to egg maturation. A cohort of approximately 100 follicles per mosquito ovary begin developing synchronously. However, a significant fraction of follicles ultimately undergo apoptosis and oosorption, especially when available resources from the blood meal are limited. Therefore, simple, rapid, and reliable techniques to accurately evaluate follicular atresia are necessary to understand mechanisms underlying follicle development in insects. This protocol describes how to detect apoptotic follicle cells within the *Aedes aegypti* mosquito ovaries using a commercially available fluorescent-labeled inhibitor of caspases (FLICA). Caspases are key players in animal apoptosis. In this assay, the FLICA reagent enters the intracellular compartment of follicles in dissected mosquito ovaries and covalently binds to active caspases. The bound reagent remains within the cell and its fluorescent signal can be observed by confocal microscopy. Although this method was specifically developed for visualizing apoptotic ovarian follicles during *Ae. aegypti* mosquito egg development, it should be applicable to other mosquito tissues that undergo caspase-mediated program cell death in a time-dependent manner.

MATERIALS

It is essential that you consult the appropriate Material Safety Data Sheets and your institution's Environmental Health and Safety Office for proper handling of equipment and hazardous material used in this protocol.

RECIPES: Please see the end of this protocol for recipes indicated by <R>. Additional recipes can be found online at http://cshprotocols.cshlp.org/site/recipes.

Reagents

Acti-stain 488 phalloidin (Cytoskeleton, Inc.)

Adult *Aedes aegypti* mosquitoes (e.g., Rockefeller, Liverpool, UGAl, or other widely used strain) in a cage

Rear mosquitoes under standard laboratory conditions as previously described (Imam et al. 2014) (e.g., 28°C, 72% relative humidity, with a photoperiod of 16 h light and 8 h dark cycle).

[3]Correspondence: jisoe@ag.arizona.edu

Cite this protocol as *Cold Spring Harb Protoc*; doi:10.1101/pdb.prot108226

Adult mosquito artificial protein diet <R>

Prepare both full-protein and protein-deficient versions of the recipe.

Dimethyl sulfoxide (DMSO)
Glycine (25 mM)
Ice
Medium 199 (Thermo Fisher Scientific 11150059)
Nail polish
Paraformaldehyde (4%, w/v, solution in 1× PBS [pH 6.9])
Phosphate-buffered saline (PBS), pH 6.9 <R>
ProLong Gold Antifade reagent (Thermo Fisher Scientific)
SR-FLICA Poly Caspase Assay Kit (ImmunoChemistry Technologies)
Sucrose (10%, w/v, solution)
Triton X-100 (0.5%, v/v, solution)

Equipment

Dark chamber or room for fluorescence visualization
Fluorescent or confocal microscope
Forceps (e.g., Dumont Biology style 5)
Glass blood feeder (e.g., Chemglass CG-1836-75)

A Hemotek membrane feeding system can also be used.

Glass Pasteur pipette
Glass scintillation vial (20-mL)
Incubator
Lab clamps
Light microscope
Microscope slides and coverslips
Mosquito cage (e.g., BugDorm-1 or similar)
Parafilm
Petri dish (glass, 10-mm × 15-mm)
Refrigerator (4°C)
Water bath

METHOD

Perform all procedures at room temperature unless indicated otherwise.

Preparation of Mosquito Ovaries

1. Set up glass blood feeders wrapped in stretched Parafilm and connected to a water bath (40°C) according to the manufacturer's instructions.
2. Place a glass feeder on the screened top of mosquito cage, secure with a lab clamp, and add 1 mL of artificial protein diet (either full-protein or protein deficient) prewarmed to 37°C.

   The type of diet used will depend on the experimental question.

3. Allow approximately 100 4-d-old adult female mosquitoes to feed on the adult mosquito artificial protein diet for 1 h. Feeding success with the *Aedes aegypti* UGAl strain is typically 70%–90%, but

Cite this protocol as *Cold Spring Harb Protoc*; doi:10.1101/pdb.prot108226

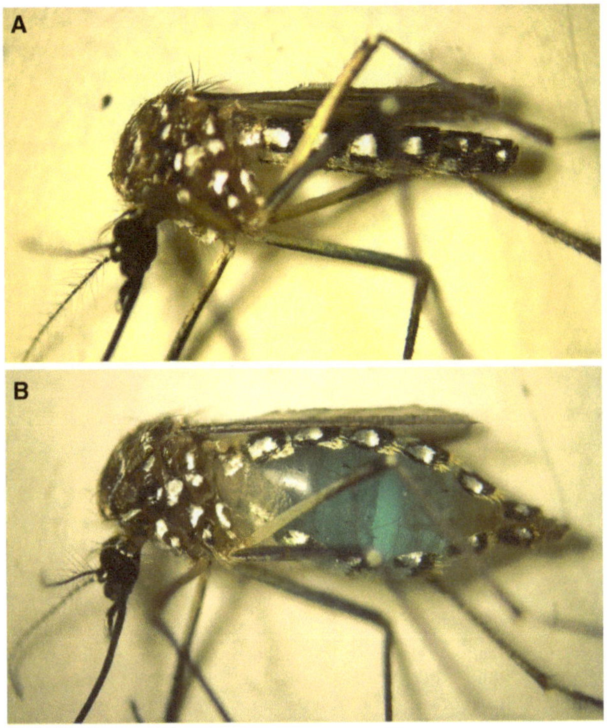

FIGURE 1. A feeding status of individual mosquitoes observed under a light microscope. (*A*) An image of a 3-d-old sugar-fed mosquito. (*B*) An image of fully engorged mosquito taken 20 min after feeding bovine serum albumin (300 mg/mL) with a color dye.

this will vary based on mosquito species and strain. Engorgement rates can be increased by removing 10% sucrose-soaked cotton and water sources for 2 h before feeding the mosquitoes.

4. Cold-anesthetize mosquitoes by placing the cage in a refrigerator for 5 min. Transfer anesthetized mosquitoes to a chilled glass Petri dish on ice using fine forceps.
5. Examine the feeding status of individual mosquitoes on ice under a light microscope. Discard nonfed or partially fed mosquitoes. Transfer fully engorged mosquitoes, as indicated by a fully distended abdomen, to a new cage using forceps (Fig. 1).
6. Follow standard rearing protocols (see Chapter 1, Protocol 1: Batch Rearing *Aedes aegypti* [Wohl and McMeniman 2023]; Clemons et al. 2010; Imam et al. 2014). Maintain mosquitoes with 10% sucrose via soaked cotton balls.

Preparation of Caspase Inhibitor

7. Prepare the caspase inhibitor (SR-FLICA) from the Poly Caspase Assay Kit according to the manufacturer's instruction.
   i. Briefly, reconstitute SR-FLICA in 50 µL of DMSO and store it in the dark at −20°C.
      The stock remains effective for at least 1 yr at −20°C in the dark.
   ii. Dilute the stock SR-FLICA solution 1:5 by adding 200 µL of 1× PBS to the stock tube before each experiment. Maintain in the dark at room temperature.
      This solution will be used in Step 12.

Incubation of the Ovaries with a Caspase Inhibitor

8. Cold-anesthetize mosquitoes as described in Step 4 at 12, 24, and 36 h after feeding on the artificial protein diet.

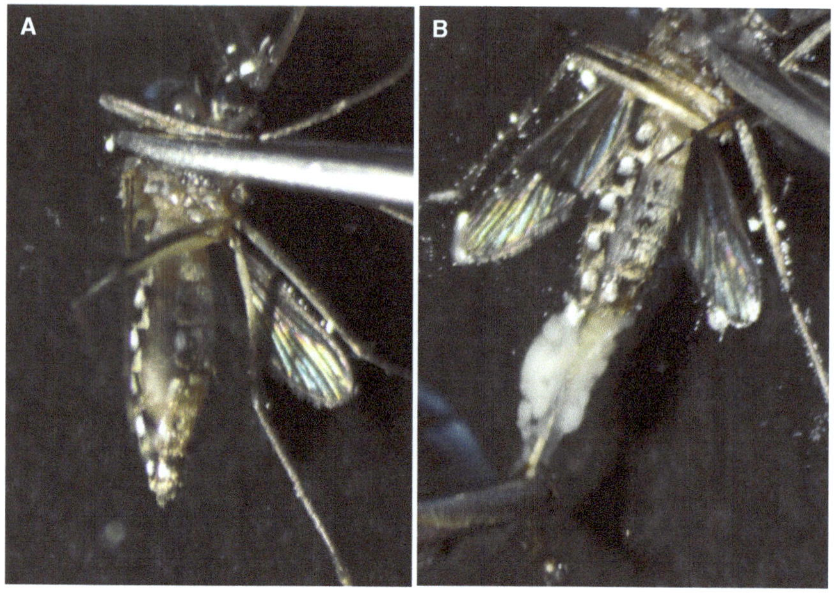

FIGURE 2. A dissection step to remove intact ovaries from a blood-fed *Aedes aegypti* mosquito. (*A*) Cold-anesthetized mosquito was placed on a drop of ice-cold 1× PBS buffer onto a microscope slide. (*B*) Ovaries were gently dissected from the mosquito abdomen by grasping the last two abdominal segments of the female mosquito with forceps and gently pulling until the developing paired ovaries emerged.

9. Place a drop of ice-cold 1× PBS buffer onto a glass microscope slide and transfer a cold anesthetized mosquito to the slide using forceps. Gently dissect ovaries from the mosquito abdomen by grasping the last two abdominal segments of the female mosquito with forceps and gently pulling until the developing paired ovaries emerge (Fig. 2).
10. Immediately transfer 10 pairs of intact ovaries into a glass scintillation vial containing 1× PBS.
11. Remove the 1× PBS with a glass Pasteur pipette and wash ovaries once with 1 mL of medium 199 at room temperature. Discard the medium and replace with 580 µL of medium 199.
12. Store the vial for 5 min at room temperature to acclimate the intact ovary tissues, and then add 20 µL of diluted SR-FLICA (from Step 7.ii) to the 199 medium (1:30). Gently swirl the vial.
13. Store the vial in a dark chamber for 1 h at 28°C.

Preparation of the Ovaries for Microscopic Analysis

14. Following the incubation period, remove the medium and rinse the ovaries five times with 200 µL of wash buffer provided in the Poly Caspase Assay Kit, followed by five additional washes with 1 mL of 1× PBS in a dark room. Remove the final wash with a glass Pasteur pipette.
15. Fix the ovaries with 1 mL 4% paraformaldehyde solution for 30 min in the dark.
16. Remove the paraformaldehyde solution and wash the ovaries with 1 mL of 1× PBS five times.
17. Quench the residual paraformaldehyde in the ovary vial with 1 mL of 25 mM glycine for 5 min at room temperature in the dark. Wash the ovaries with 1 mL of 1× PBS five times.
18. Permeabilize the ovaries with 1 mL of 0.5% Triton X-100 for 5 min. Remove the Triton and wash the ovaries with 1 mL of 1× PBS five times.
19. Stain the filamentous actin cytoskeleton of the ovaries with 300 µL Acti-stain 488 phalloidin (100 nM in PBS) for 16 h at 4°C.

Cite this protocol as *Cold Spring Harb Protoc*; doi:10.1101/pdb.prot108226

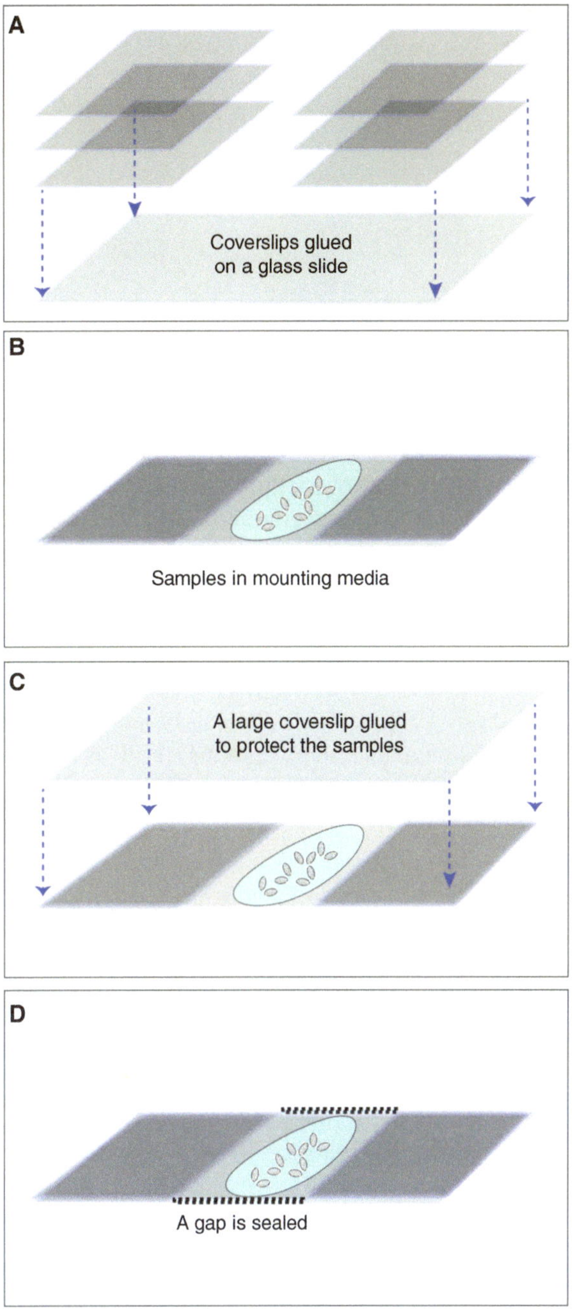

FIGURE 3. A preparation of a bridged-glass slide. (*A*) Three square coverslips on both edges of the glass slide were glued with clear nail polish. (*B*) Fixed and stained ovaries were placed on the center of the glass slide. Three drops of ProLong Gold Antifade reagent were applied to the ovaries. (*C*) A large coverslip was gently placed on top of the ovary samples and secured by gluing on the top coverslip. (*D*) The side gaps are closed with clear nail polish.

20. Prepare a bridged-glass slide by gluing three square coverslips on both edges of the slide with clear nail polish to protect ovaries from being crushed. Transfer the ovaries using a glass Pasteur pipette onto the center of the bridged-glass slide (Fig. 3).
21. Remove residual 1× PBS and cover the ovaries with three drops of ProLong Gold Antifade reagent.
22. Place an overlaying coverslip on the sample and seal the sides with clear nail polish.

23. Visualize the ovaries using a fluorescence or confocal microscope at 550- to 580-nm excitation and at 590- to 600-nm emission for SR-FLICA. Detect Acti-stain 488 phalloidin at 485 nm excitation and at 535 nm emission.
24. Determine the percentage of follicles undergoing apoptosis by counting the number of stained follicles.

DISCUSSION

In the present protocol, we use a commercially available fluorescent-labeled inhibitor of caspases (FLICA) technique (Bedner et al. 2000) to identify ovarian follicles that are undergoing apoptosis in *Ae. aegypti* mosquitoes. The FLICA methods use a cell-permeable caspase inhibitor tripeptide (e.g., VAD) conjugated to a fluorochrome (e.g., sulforhodamine [SR], red; and carboxyfluorescein [FAM], green) as in vivo markers of apoptotic cells. Once the inhibitor interacts with active caspase in the cells, it irreversibly binds to the enzyme active site. After washing away unbound inhibitor, the localization of active caspases can therefore be detected by a confocal microscope. Advantages of the FLICA-labeling approach are that the labeling reaction is fairly fast and readily detected by most fluorescence and confocal microscopes (Darzynkiewicz et al. 2011). After FLICA treatment, the cells can be subsequently fixed and counterstained with phalloidin and 4′,6-diamidino-2-phenylindole (DAPI) for further microscopic studies. FLICA techniques have been applied to assess apoptotic mosquito ovarian follicles (Uchida et al. 2004; Ahmed and Hurd 2006; Isoe et al. 2019). Other methods including acridine orange staining and TUNEL assays have been also used to study ovarian follicular apoptosis in different mosquito species (Hopwood et al. 2001; Uchida et al. 2004; Almeida and Suesdek 2017).

Anautogenous mosquitoes ingest a blood meal to begin a gonotrophic cycle and develop a clutch of eggs, and ovarian follicle development is strictly regulated by blood feeding. Female mosquitoes can also consume an artificial protein meal such as bovine serum albumin (BSA) to develop ovarian follicles (Gonzales et al. 2015). Once female mosquitoes acquire blood or an artificial meal, hormones, including ovarian ecdysteroidogenic hormone 1 (OEH1), insulin-like peptides, and 20-hydroxyecdysone, orchestrate the synthesis of three vitellogenin yolk proteins in the fat body and their uptake via receptor-mediated accumulation by oocytes, gradually transforming the spherical primary follicles into the mature ellipsoid form by 48 h after blood feeding (Chen et al. 1994; Romans et al. 1995; Raikhel et al. 2002; Attardo et al. 2005; Isoe and Hagedorn 2007; Brown et al. 2008; Smykal and Raikhel 2015; Roy et al. 2018; Allman et al. 2020). Each mosquito ovary contains approximately 100 ovarioles, which are composed of primary and secondary follicles and a germarium, with the primary follicles developing synchronously throughout oogenesis. The number of fully mature follicles within the pair of mosquito ovaries at the end of vitellogenesis is strongly affected by the quantity of protein found in the blood meal or artificial diet such as BSA that female mosquitoes consume (Attardo et al. 2005; Isoe et al. 2009; Gonzales et al. 2015). Some females may acquire a blood meal and become fully engorged, in turn producing a large clutch of eggs (e.g., 130 eggs), whereas others may acquire a smaller meal than the former, yielding a reduced batch of eggs (e.g., 30 eggs). In the latter adverse instances, instead of equally allocating digested and absorbed nutrients to all developing oocytes, leading to the incomplete development of nutritionally deprived embryos, some or most primary follicles undergo apoptosis. Limited nutritional resources are then provided to the remaining cohort of primary follicles that develop synchronously to reach full maturity (Isoe et al. 2009; Gonzales et al. 2015). The levels of both sucrose and juvenile hormone in female mosquitoes before blood feeding may also influence the extent of follicle resorption after blood feeding (Clifton and Noriega 2012). Similarly, female mosquitoes that fed on a high-BSA protein diet displayed a significantly higher number of mature follicles than those fed on low-BSA protein diet during the first gonotrophic cycle, indicating that a low-protein diet is associated with higher rates of resorption (Fig. 4; Isoe et al. 2009; Gonzales et al. 2015). Thus, timely activation of

Cite this protocol as *Cold Spring Harb Protoc*; doi:10.1101/pdb.prot108226

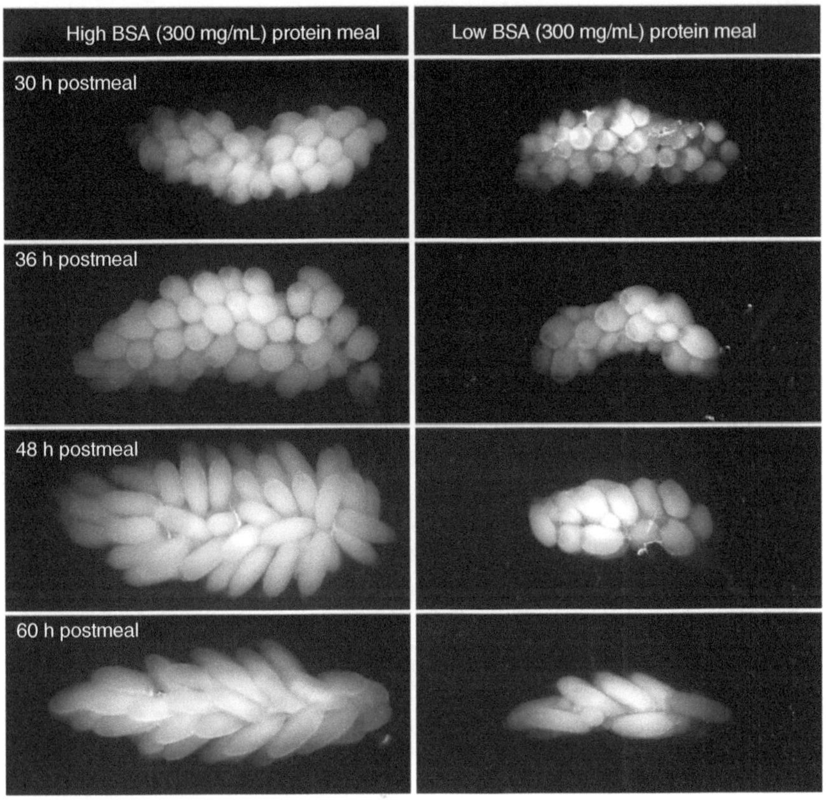

FIGURE 4. Dietary protein level effects on *Aedes aegypti* follicle development during the first gonotrophic cycle. Female mosquitoes were fed either a high-BSA protein meal (300 mg/mL; *left* images) or low-BSA protein meal (30 mg/mL; *right* images). Representative light microscopy images of ovaries are shown from 30 to 60 h post-bovine serum albumin (BSA) protein meal.

caspase-mediated cell death pathways in the mosquito ovarian follicles in response to blood feeding or an artificial protein meal is an irreversible event, strictly defining the number of eggs resulting from each gonotrophic cycle.

Our results indicate that the FLICA method followed by confocal microscopy analysis reliably identifies FLICA-labeled follicles (Fig. 5). Apoptotic cells were visualized by their strong caspase activity (red), while the actin cytoskeleton (green) was counterstained. FLICA-labeled follicles undergoing follicular atresia are most frequently observed in ovaries dissected from females 36 h after being provisioned with a low-BSA protein meal (Gonzales et al. 2015; Isoe et al. 2019). These ovaries from females fed a low-BSA diet show an increased number of primary follicle cells undergoing caspase-mediated apoptosis compared to ovaries from mosquitoes provisioned with a high-BSA diet (Gonzales et al. 2015). Because female mosquitoes are discontinuous blood feeders, the level of nutrient intake during each individual meal significantly influences egg production, although it does not affect egg developmental time. This suggests that mosquitoes evolved to synchronously and consistently develop a clutch of eggs independent of nutritional intake but at the expense of other primary follicles, which then undergo caspase-mediated apoptosis. This allows the surviving primary follicles to fully mature within 48–60 h in preparation for fertilization and oviposition when a suitable oviposition site is located. The FLICA methods described here are suitable for studying caspase-mediated apoptosis of mosquito ovarian follicle cells during egg maturation. Also, a combination of the FLICA approach with other apoptosis assays, such as acridine orange staining for chromatin condensation and TUNEL assays for detecting fragmented DNA, could further confirm apoptotic follicle cells in mosquitoes.

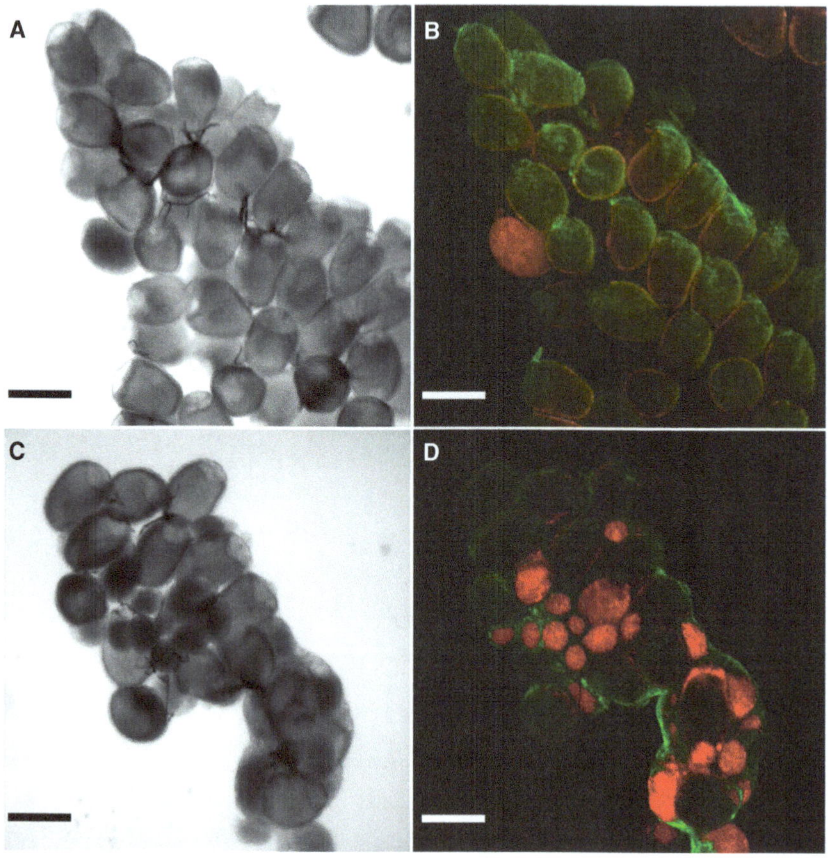

FIGURE 5. Caspase-mediated apoptosis in ovarian follicles of *Aedes aegypti* mosquitoes induced by low–bovine serum albumin (BSA) protein meal. Light microscopic images of representative mosquito ovaries from high-BSA protein meal (*A*; 300 mg/mL) and low-BSA protein meal (*C*; 30 mg/mL). Confocal images of the corresponding ovaries (*B*,*D*). Mosquito ovaries were dissected at 36 h after BSA protein meal feeding. Intact dissected ovaries were incubated with SR-FLICA (red), rinsed with 1× PBS in a dark room, fixed with 4% paraformaldehyde, rinsed with 1× PBS, and stained with phalloidin for actin cytoskeleton (green). Bars, 300 µm.

RECIPES

Adult Mosquito Artificial Protein Diet

Completely dissolve the desired amount of bovine serum albumin (BSA) in 1× Phosphate-buffered saline (PBS) <R>, pH 7.4 by incubating the mixture for 10 min in a 37°C water bath and vigorously vortexing. Use concentrations of 30 mg of BSA/mL of 1× PBS for protein-deficient conditions and 300 mg of BSA/mL of 1× PBS for full-protein conditions. Add 5.0 µL of food dye (Food Colors and Egg Dye from McCormick & Company, Inc.; any color can be used) to the mixture to easily identify fed mosquitoes. Supplement the mixture with adenosine triphosphate (5 mM final concentration) to facilitate the movement of the meal into the mosquito midgut. Freshly prepare.

 Cite this protocol as *Cold Spring Harb Protoc*; doi:10.1101/pdb.prot108226

Phosphate-Buffered Saline (PBS)

Reagent	Amount to add (for 1× solution)	Final concentration (1×)	Amount to add (for 10× stock)	Final concentration (10×)
NaCl	8 g	137 mM	80 g	1.37 M
KCl	0.2 g	2.7 mM	2 g	27 mM
Na_2HPO_4	1.44 g	10 mM	14.4 g	100 mM
KH_2PO_4	0.24 g	1.8 mM	2.4 g	18 mM
If necessary, PBS may be supplemented with the following:				
$CaCl_2 \cdot 2H_2O$	0.133 g	1 mM	1.33 g	10 mM
$MgCl_2 \cdot 6H_2O$	0.10 g	0.5 mM	1.0 g	5 mM

PBS can be made as a 1× solution or as a 10× stock. To prepare 1 L of either 1× or 10× PBS, dissolve the reagents listed above in 800 mL of H_2O. Adjust the pH to 7.4 (or 7.2, if required) with HCl, and then add H_2O to 1 L. Dispense the solution into aliquots and sterilize them by autoclaving for 20 min at 15 psi (1.05 kg/cm^2) on liquid cycle or by filter sterilization. Store PBS at room temperature.

ACKNOWLEDGMENTS

We thank Yurika E. Isoe at the University of Arizona for technical assistance and Dr. Brooke B. Massani of the University of Arizona W. M. Keck Center for Surface and Interface Imaging Facility for assistance with confocal microscopic work. We thank reviewers for their valuable and thoughtful comments.

REFERENCES

Ahmed AM, Hurd H. 2006. Immune stimulation and malaria infection impose reproductive costs in *Anopheles gambiae* via follicular apoptosis. *Microbes Infect* **8:** 308–315. doi:10.1016/j.micinf.2005.06.026

Allman MJ, Fraser JE, Ritchie SA, Joubert DA, Simmons CP, Flores HA. 2020. *Wolbachia*'s deleterious impact on *Aedes aegypti* egg development: the potential role of nutritional parasitism. *Insects* **11:** 735. doi:10.3390/insects11110735

Almeida F, Suesdek L. 2017. Effects of *Wolbachia* on ovarian apoptosis in *Culex quinquefasciatus* (Say, 1823) during the previtellogenic and vitellogenic periods. *Parasit Vectors* **10:** 398. doi:10.1186/s13071-017-2332-0

Attardo GM, Hansen IA, Raikhel AS. 2005. Nutritional regulation of vitellogenesis in mosquitoes: implications for anautogeny. *Insect Biochem Mol Biol* **35:** 661–675. doi:10.1016/j.ibmb.2005.02.013

Bedner E, Smolewski P, Amstad P, Darzynkiewicz Z. 2000. Activation of caspases measured *in situ* by binding of fluorochrome-labeled inhibitors of caspases (FLICA): correlation with DNA fragmentation. *Exp Cell Res* **259:** 308–313. doi:10.1006/excr.2000.4955

Brown MR, Clark KD, Gulia M, Zhao Z, Garczynski SF, Crim JW, Suderman RJ, Strand MR. 2008. An insulin-like peptide regulates egg maturation and metabolism in the mosquito *Aedes aegypti*. *Proc Natl Acad Sci* **105:** 5716–5721. doi:10.1073/pnas.0800478105

Chen JS, Cho WL, Raikhel AS. 1994. Analysis of mosquito vitellogenin cDNA. Similarity with vertebrate phosvitins and arthropod serum proteins. *J Mol Biol* **237:** 641–647. doi:10.1006/jmbi.1994.1261

Clemons A, Mori A, Haugen M, Severson D, Duman-Scheel M. 2010. *Aedes aegypti* culturing and egg collection. *Cold Spring Harb Protoc* doi:10.1101/pdb.prot5507

Clifton ME, Noriega FG. 2012. The fate of follicles after a blood meal is dependent on previtellogenic nutrition and juvenile hormone in *Aedes aegypti*. *J Insect Physiol* **58:** 1007–1019. doi:10.1016/j.jinsphys.2012.05.005

Darzynkiewicz Z, Pozarowski P, Lee BW, Johnson GL. 2011. Fluorochrome-labeled inhibitors of caspases: convenient *in vitro* and *in vivo* markers of apoptotic cells for cytometric analysis. *Methods Mol Biol* **682:** 103–114. doi:10.1007/978-1-60327-409-8_9

Gonzales KK, Tsujimoto H, Hansen IA. 2015. Blood serum and BSA, but neither red blood cells nor hemoglobin can support vitellogenesis and egg production in the dengue vector *Aedes aegypti*. *PeerJ* **3:** e938. doi:10.7717/peerj.938

Hopwood JA, Ahmed AM, Polwart A, Williams GT, Hurd H. 2001. Malaria-induced apoptosis in mosquito ovaries: a mechanism to control vector egg production. *J Exp Biol* **204:** 2773–2780. doi:10.1242/jeb.204.16.2773

Imam H, Zarnigar, Sofi G, Seikh A. 2014. The basic rules and methods of mosquito rearing (*Aedes aegypti*). *Trop Parasitol* **4:** 53–55. doi:10.4103/2229-5070.129167

Isoe J, Hagedorn HH. 2007. Mosquito vitellogenin genes: comparative sequence analysis, gene duplication, and the role of rare synonymous codon usage in regulating expression. *J Insect Sci* **7:** 1–49. doi:10.1673/031.007.0101

Isoe J, Rascón AA Jr, Kunz S, Miesfeld RL. 2009. Molecular genetic analysis of midgut serine proteases in *Aedes aegypti* mosquitoes. *Insect Biochem Mol Biol* **39:** 903–912. doi:10.1016/j.ibmb.2009.10.008

Isoe J, Koch LE, Isoe YE, Rascón AA Jr, Brown HE, Massani BB, Miesfeld RL. 2019. Identification and characterization of a mosquito-specific eggshell organizing factor in *Aedes aegypti* mosquitoes. *PLoS Biol* **17:** e3000068. doi:10.1371/journal.pbio.3000068

Raikhel AS, Kokoza VA, Zhu J, Martin D, Wang SF, Li C, Sun G, Ahmed A, Dittmer N, Attardo G. 2002. Molecular biology of mosquito vitellogenesis: from basic studies to genetic engineering of antipathogen immunity. *Insect Biochem Mol Biol* **32:** 1275–1286. doi:10.1016/S0965-1748(02)00090-5

Romans R, Tu Z, Ke Z, Hagedorn HH. 1995. Analysis of a vitellogenin gene of the mosquito, *Aedes aegypti* and comparisons to vitellogenins from other organisms. *Insect Biochem Mol Biol* **25:** 939–958. doi:10.1016/0965-1748(95)00037-V

Roy S, Saha TT, Zou Z, Raikhel AS. 2018. Regulatory pathways controlling female insect reproduction. *Annu Rev Entomol* **63:** 489–511. doi:10.1146/annurev-ento-020117-043258

Smykal V, Raikhel AS. 2015. Nutritional control of insect reproduction. *Curr Opin Insect Sci* **11:** 31–38. doi:10.1016/j.cois.2015.08.003

Uchida K, Nishizuka M, Ohmori D, Ueno T, Eshita Y, Fukunaga A. 2004. Follicular epithelial cell apoptosis of atretic follicles within developing ovaries of the mosquito *Culex pipiens pallens. J Insect Physiol* **50:** 903–912. doi:10.1016/j.jinsphys.2004.07.002

Wohl MP, McMeniman CJ. 2023. Batch rearing *Aedes aegypti. Cold Spring Harb Protoc* doi:10.1101/pdb.prot108017

Cite this protocol as *Cold Spring Harb Protoc*; doi:10.1101/pdb.prot108226

Protocol 2

Identification of Mosquito Eggshell Proteins from *Aedes aegypti* by Liquid Chromatography with Tandem Mass Spectrometry (LC–MS/MS) Proteomic Analysis

Jun Isoe,[1,4] Krishna Parsawar,[2] Cynthia L. David,[2] Michael A. Riehle,[1] and Roger L. Miesfeld[3]

[1]*Department of Entomology,* [2]*Analytical and Biological Mass Spectrometry Core Facility,* [3]*Department of Chemistry and Biochemistry, The University of Arizona, Tucson, Arizona 85721, USA*

The insect eggshell is a multifunctional structure with several important roles, including generating an entry point for sperm via the micropyle before oviposition, serving as an oviposition substrate attachment surface, and functioning as a protective layer during embryo development. Eggshell proteins play major roles in eggshell tanning and hardening following oviposition and provide molecular cues that define dorsal–ventral axis formation. Precise eggshell formation during ovarian follicle maturation is critical for normal embryo development and the synthesis of a defective eggshell often gives rise to inviable embryos. Therefore, simple and accurate methods for identifying eggshell proteins will facilitate our understanding of the molecular pathways regulating eggshell formation and the mechanisms underlying normal embryo development. This protocol describes how to isolate and enrich eggshells from mature oocytes of *Aedes aegypti* mosquitoes and how to extract their eggshell proteins for liquid chromatography with tandem mass spectrometry (LC–MS/MS) proteomic analysis. Although this methodology was developed for studying mosquito eggshells, it may be applicable to eggs from a variety of insects. Mosquitoes are ideal model organisms for this study as their ovarian follicle development and eggshell formation are meticulously regulated by blood feeding and their follicles develop synchronously throughout oogenesis in a time-dependent manner.

MATERIALS

It is essential that you consult the appropriate Material Safety Data Sheets and your institution's Environmental Health and Safety Office for proper handling of equipment and hazardous material used in this protocol.

RECIPES: Please see the end of this protocol for recipes indicated by <R>. Additional recipes can be found online at http://cshprotocols.cshlp.org/site/recipes.

Reagents

Acetonitrile
Acrylamide SDS–PAGE gels (12%; Bio Rad)
Adenosine triphosphate (100 mM stock)
Adult *Aedes aegypti* mosquitoes (e.g., Rockefeller, widely used strain) in a cage

[4]Correspondence: jisoe@ag.arizona.edu

Cite this protocol as *Cold Spring Harb Protoc;* doi:10.1101/pdb.prot108227

Rear mosquitoes under standard laboratory conditions as previously described (Imam et al. 2014) (e.g., 28°C, 72% relative humidity, with a photoperiod of 16 h light and 8 h dark cycle)

Ammonium bicarbonate (50 mM, 100 mM)
Dithiothreitol (DTT; 25 mM) in 50 mM ammonium bicarbonate
Eggshell lysis buffer <R>
Eggshell protease cocktail <R>
Ethanol (90%, v/v, and 100%)
Formic acid
GelCode Blue Stain Reagent (Thermo Scientific)
Glacial acetic acid
Guanidine hydrochloride (6.0 M, pH 9.0)
H_2O (boiling and room temperature)
Ice
Iodoacetamide (55 mM) in 50 mM ammonium bicarbonate
Methanol
Phosphate-buffered saline (PBS), pH 7.4 <R>
Pierce 660 nm protein assay reagent containing Ionic Detergent Compatibility Reagent (Thermo Scientific)
Protein sample buffer <R>
Solvent mixture for peptide extraction <R>
Sucrose (10%, w/v, solution)
Tris–Glycine–SDS protein electrophoresis buffer <R>
Ultrapure H_2O
Whole vertebrate blood (e.g., whole bovine blood with citrate anticoagulant; HemoStat laboratories)

Equipment

Acclaim PepMap 100 trap column (75-µm inner diameter [ID] × 2-cm)
Acclaim PepMap rapid separation liquid chromatography (RSLC) analytical column (75-µm ID × 25-cm)
Bath sonicator
Centrifuge
Cotton balls
Dionex Ultimate 3000 RSLCnano System
Dounce homogenizer (5-mL-volume) with pestles A and B
Forceps
Gel staining box or large Petri dish (see Step 19)
Glass blood feeder (e.g., Chemglass CG-1836-75)

A Hemotek system (Hemotek Ltd.) could be used instead.

Glass Pasteur pipettes
Glass plate
Glass scintillation vials (20-mL)
Glass slide
Incubator
Lab clamp
Light microscope
Mass spectrometer equipped with a nano-flow liquid chromatography system
Mesh strainer (40-µm)
Microcentrifuge tubes (1.7-mL)
Microscissors (optional, see Step 7)

Cite this protocol as *Cold Spring Harb Protoc*; doi:10.1101/pdb.prot108227

Mini gel electrophoresis unit
Mosquito cage
Parafilm
Petri dish, glass (100-mm × 15-mm)
Proteome Discoverer software
Razor blades
Refrigerator (4°C)
Scaffold Q + S software (Proteome Software Inc.)
Shaker table
SpeedVac concentrator
Surgical scalpels
Vortexer
Vortex tube adapter
Water bath

METHOD

This protocol requires access to a mass spectrometer and extensive expertise with proteomic techniques, so users new to these approaches are encouraged to consult with relevant experts or a core facility.

Mosquito Eggshell Protein Extract Preparation

1. Set up glass blood feeders wrapped in stretched Parafilm, connected to a water bath (40°C), and place the glass feeders on the screened top of the mosquito cage, secured with a lab clamp. Prewarm vertebrate blood (e.g., cow) to 37°C.
2. Load 1 mL of prewarmed vertebrate blood supplemented with 5 mM adenosine triphosphate into the feeders according to the manufacturer's instructions.
3. Allow 3- to 4-d-old adult female mosquitoes in a cage to feed on the blood source for 1 h.
4. Cold-anesthetize mosquitoes by placing the cage for 5 min in a refrigerator (4°C).
5. Place 100 cold-anesthetized mosquitoes onto a glass Petri dish on ice under a light microscope to facilitate sorting of engorged mosquitoes based on abdominal distension (Fig. 1). Use forceps to transfer fully engorged mosquitoes to a new cage. Follow standard rearing protocols (see Chapter 1, Protocol 1: Batch Rearing *Aedes aegypti* [Wohl and McMeniman 2023]; Clemons et al. 2010; Imam et al. 2014). Maintain blood-fed mosquitoes with 10% sucrose (via soaked cotton balls) for several days.

   Although in general eggshell formation for each oocyte is completed ∼48 h after blood feeding, we recommend harvesting fully mature ovarian follicles at 4 d post–blood meal, which can be accomplished by depriving the gravid females of an oviposition substrate. Place the 10% sucrose-soaked cotton on the cage for only 1 h per day during this period to avoid mosquitoes ovipositing on the cotton.

6. On the day of dissection, cold-anesthetize mosquitoes as described in Step 4.
7. Place a drop of ice-cold PBS onto a glass slide. Place an anesthetized mosquito onto the drop of PBS. Under a light microscope, gently dissect ovaries from the mosquitoes by grasping the last two abdominal segments with forceps and pulling until the ovaries emerge (see Fig. 2 in Protocol 1: Visualization of Apoptotic Ovarian Follicles during *Aedes aegypti* Mosquito Egg Maturation by Fluorescent Imaging Studies (Isoe et al. 2023a).

   If this technique is not successful because of the size of the fully developed ovaries, the dorsal side of the abdomen can be cut along the longitudinal axis with microscissors and the ovaries gently lifted out.

8. Using a glass Pasteur pipette, transfer ovaries into a glass scintillation vial containing 5 mL of ice-cold PBS (Fig. 2).

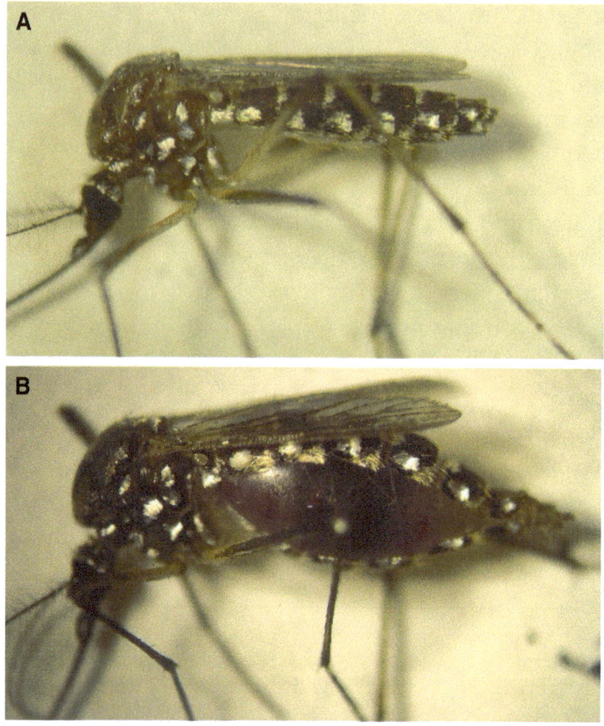

FIGURE 1. A feeding status of individual mosquitoes observed under a light microscope. (*A*) An image of a 3-d-old sugar-fed mosquito. (*B*) An image of fully engorged mosquito taken 20 min after feeding bovine blood.

9. Once approximately 100 pairs of ovaries are collected in the vial, pipette up and down 20 times. As soon as the mature elongated oocytes settle to the bottom of the vial, use a glass Pasteur pipette to remove all floating tissues and cells including inner, lateral, and common oviducts; trachea; muscles; germarium; secondary follicles; and follicular epithelial cells. Add 5 mL of ice-cold PBS and repeat Step 9 10 times to ensure a minimum contamination of nontarget cell types.

   Oocytes quickly descend because of gravity, whereas other ovarian tissues float or descend slowly.

10. Transfer the purified oocytes in 1 mL of ice-cold PBS into a Dounce homogenizer using a glass Pasteur pipette. Homogenize the oocytes on ice using 40 strokes in a Dounce homogenizer with pestle B. Let the insoluble eggshells form a pellet by gravity at the bottom of the glass homogenization tube. Remove the cloudy portion that contains vitellogenin yolk proteins and other soluble proteins with a glass Pasteur pipette. Add 1 mL of ice-cold PBS to resuspend the eggshell

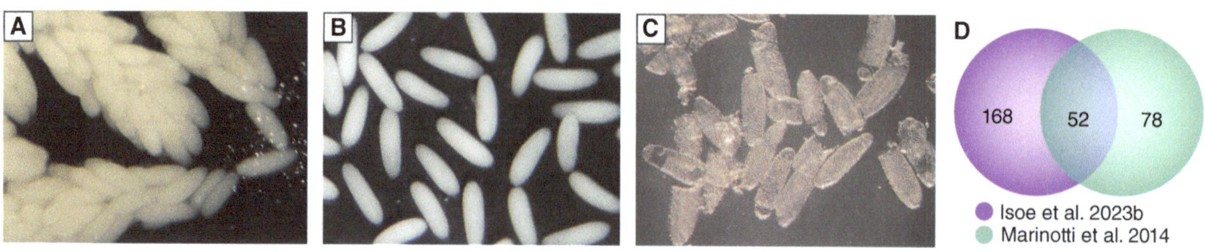

FIGURE 2. Representative images of dissected ovaries (*A*), enriched primary follicles (*B*), and isolated eggshell without cytosolic contents (*C*). A comparison of number of *Ae aegypti* eggshell proteins identified (*D*) between the method described here and in Isoe et al. (2023b) and the previous liquid chromatography with tandem mass spectrometry (LC–MS/MS) studies Marinotti et al. (2014).

without homogenization, remove the cloudy portion, and repeat this process until the cloudiness disappears.

11. Transfer the homogenates containing eggshells onto a nylon mesh (40-µm) using a glass Pasteur pipette and wash thoroughly with 5 mL of PBS to remove oocyte cytosolic components. Repeat this washing process two additional times by pouring 5 mL of PBS onto the mesh to ensure a minimum contamination of cytosolic and membrane proteins (Fig. 2).
12. Add 1 mL of ice-cold PBS to the enriched eggshells on the mesh and quickly remove the PBS and eggshells with a glass Pasteur pipette. Transfer the enriched eggshells in PBS into a Dounce homogenizer using a glass Pasteur pipette. Remove the PBS and homogenize the eggshells in 5 mL of 6 M guanidine hydrochloride, pH 9 in a Dounce homogenizer with pestle A for 30 strokes. Place the homogenates in an incubator for 16 h at 37°C.
13. Precipitate the proteins by adding 45 mL of 100% ethanol and incubate the samples for 16 h at −20°C.
14. Pellet the eggshell proteins by centrifugation at 16,000*g* for 15 min at 4°C. Remove the supernatant carefully and wash the pellet with 90% ice-cold ethanol. Repeat the 90% ethanol wash two additional times to remove residual guanidine hydrochloride. Pellet the eggshell proteins by centrifugation at 16,000*g* for 15 min at 4°C after each wash.
15. Air-dry the pellets for 10 min to completely eliminate the ethanol.
16. Resuspend the eggshell proteins in 50 µL eggshell lysis buffer and use 2 µL of the resuspended protein to assay the protein concentration using the Pierce 660-nm reagent containing the Ionic Detergent Compatibility Reagent according to the manufacturer's instructions.
17. Add an equal volume of protein sample buffer to up to 20 µg of protein sample from Step 16 for a maximum of 50 µL and then denature the purified eggshell proteins in boiling H_2O for 5 min.

Protein Sample Preparation for Mass Spectrometry

18. Load the mosquito eggshell protein samples onto an sodium dodecyl-sulfate polyacrylamide gel electrophoresis (SDS–PAGE) gel in Tris-Glycine-SDS protein electrophoresis buffer. Run the protein gels via a mini gel electrophoresis unit using standard techniques for ~5 min until the proteins have entered the resolving gel (~1 cm). This step is used to remove any nonprotein impurities.
19. Transfer the gel to a suitably sized gel staining box or large Petri dish and fix the gels with a 25-mL solution of 10% (v/v) glacial acetic acid and 50% (v/v) methanol for 30 min on a shaker table at room temperature. Stain the gel with 20 mL GelCode Blue Stain Reagent with shaking for 16 h at room temperature. Destain the gel using 25 mL ultrapure H_2O three times for ~2 h at room temperature on a benchtop orbital shaker. This step enhances stain sensitivity.
20. Excise gel bands (~1.0 cm^2) on a clean glass plate using a clean surgical scalpel and transfer the gel bands into 1.7-mL microcentrifuge tubes. Destain bands in 1 mL of 50% (v/v) methanol in 50 mM ammonium bicarbonate at room temperature on a vortexer with a tube adapter at low speed for 1 h or until the stain is removed; change the solution several times if necessary. Remove the destain.
21. To reduce disulfide bonds in the protein sample, add 100 µL of 25 mM DTT in 50 mM ammonium bicarbonate to the 1.7-mL microcentrifuge tube containing the gel bands for 45 min at 60°C. Quickly vortex the tube. Remove the DTT and immediately add 100 µL of 55 mM iodoacetamide in 50 mM ammonium bicarbonate for 30 min at room temperature to alkylate-free cysteines.
22. Remove the solution and wash the bands in 1 mL of 50 mM ammonium bicarbonate on a vortexer with a tube adapter or shaker for 1 h at room temperature. Repeat this wash two times.

23. Cut the bands into small pieces with a razor blade and transfer the bands into a 1.7-mL microcentrifuge tubes. Dehydrate the bands by adding 500 µL of 100% acetonitrile and incubating for 10 min at room temperature. Remove as much acetonitrile as possible. Dry the samples in a SpeedVac concentrator to remove the remaining acetonitrile.
24. Add 30 µL of eggshell protease cocktail to the samples. Incubate for 16 h at 37°C.
25. Centrifuge the gel pieces for 2 min at 4000*g* at room temperature, and transfer the supernatant to a fresh 1.7-mL microcentrifuge tube. Add 50 µL (or just enough to cover the gel pieces) of solvent mixture for peptide extraction to the remaining gel pieces. Sonicate for 10 min at room temperature. Centrifuge for 2 min. Transfer the supernatant to combine with the original supernatant.
26. Repeat the extraction in Step 25, transfer the supernatant to a new 1.7-mL microcentrifuge tube, pool the supernatants, and completely dry the supernatants in a SpeedVac concentrator for 30 min at room temperature.

LC–MS/MS and Protein Identification

27. Identify eggshell proteins by detecting tryptic/lysC peptides using a mass spectrometer equipped with a nano-flow liquid chromatography system followed by matching them to databases containing theoretical protein digestions.

    In our laboratory, we use Thermo Scientific instruments, columns, and software. We have a Q Exactive Plus mass spectrometer incorporated with an EASY-Spray nanoESI source.

    i. Resuspend peptides in 10 µL of 0.1% formic acid (1.0 µg/µL).
    ii. Inject peptides (500 ng in 0.1% formic acid) onto an Acclaim PepMap 100 trap C18 Reversed Phase HPLC column, and elute peptides onto an Acclaim PepMap RSLC analytical column (separation column), using an H_2O (0.1% formic acid) and acetonitrile (0.1% formic acid) gradient over 120 min.
    iii. For MS, set the flow rate at 300 nL/min using a Dionex Ultimate 3000 RSLCnano System. Analyze peptides using a data-dependent method.
    iv. For MS/MS, use a normalized higher collision energy set at 27 and an isolation width of 1.5 m/z.
    v. Search MS and MS/MS raw data against the *Ae. aegypti* NCBI protein database and a common human-derived contaminant protein database using Proteome Discoverer software using a SEQUEST based search engine.

        MS/MS spectra matches consider fully tryptic peptides with at most two missed tryptic cleavage sites. Variable post-translational peptide modifications are also considered (e.g., methionine oxidation and cysteine carbamidomethylation). To identify proteins at 95% confidence, use XCorr score cutoffs as determined by a reversed database search.

    vi. Further analyze the peptide and protein identification outputs with proteomics software (e.g., Scaffold Q + S), a program that uses Bayesian statistics to detect additional spectra.
    vii. Accept protein identifications that have at least two peptides at a 0.1% peptide false discovery rate with 95% protein confidence.
    viii. Subject the identified eggshell proteins to BLASTP protein searches against the nonredundant NCBI protein database to find possible functions.

DISCUSSION

Eggshell proteins from *Ae. aegypti* mosquitoes were first identified and described more than 28 years ago (Lin et al. 1993; Edwards et al. 1998). When mosquito eggs are laid, a protective extracellular eggshell immediately undergoes final maturation by melanization and protein cross-linking reactions catalyzed by several eggshell enzymes. The key enzymes involved in these reactions for *Ae. aegypti* have been well identified and biochemically studied (Li 1994; Ferdig et al. 1996; Han et al. 2000; Johnson et al. 2001; Fang et al. 2002; Kim et al. 2005; Li and Li 2005). In addition, proteomic studies on purified mosquito eggshells have been conducted to identify novel eggshell protein components. The first proteomic study on the mosquito eggshell was performed in *Anopheles gambiae* in 2010 (Amenya et al. 2010). In *Ae. aegypti*, eggshell proteomic analysis identified 130 unique proteins (Marinotti et al. 2014). Traditionally, in these studies, a meticulous hand sorting of endochorion and exochorion after homogenization was performed under a light microscope to enrich and purify the eggshell before the extraction of eggshell proteins. Mosquito eggshell proteomics using the protocol described here have identified an additional 168 proteins in *Ae. aegypti*, indicating that a synthesis of mosquito eggshell is far more complex process than previously thought (Fig. 2; Isoe et al. 2023b). Some of these extracellular eggshell proteins may function together to rapidly engage in cross-linking and melanization steps upon oviposition to protect and support the embryonic development and the first larval instar during the extended periods of time these embryos may encounter before hatching.

The present protocol was developed to ensure a fast, accurate, and efficient separation of eggshells from dissected *Ae. aegypti* mosquito ovaries to subsequently identify eggshell proteins by LC–MS/MS analysis. Important features of this approach involve (1) an enrichment of primary oocytes that are surrounded by a protective eggshell from other components of the ovaries, (2) the physical separation of partially cross-linked extracellular eggshell from the oocytes by Dounce homogenization, (3) the extraction of eggshell proteins with the chaotrope guanidine hydrochloride, and (4) the subsequent LC–MS/MS analysis of mosquito eggshell proteins. The first step allows a thorough physical separation of the fully matured primary oocytes from other ovarian tissues and cell types including the inner, lateral, and common oviducts; trachea; longitudinal and circular muscles; germarium; secondary follicles; and follicular epithelial cells, without compromising the eggshell structure. We strongly recommend an enrichment of the primary oocytes as described herein, otherwise proteins from contaminating tissues will be identified as noise during downstream proteomic analysis. Thus, Step 9 should be thoroughly repeated until no unwanted tissues are visible in the vial under a light microscope, exclusively recovering primary oocytes covered with eggshell that can be used immediately in the next step. We chose Dounce homogenization in PBS on ice with pestle B in Step 10 to physically disrupt the primary oocytes to separate the eggshells from oocytes. Oocyte intracellular contents including vitellogenin yolk proteins become readily soluble and cloudy in buffer after homogenization, whereas insoluble eggshells quickly sediment to form a pellet at the bottom of the glass homogenization tube. This process is repeated following a removal of the top phase that includes a soluble fraction and nonsedimented cellular debris with a glass Pasteur pipette until the cloudiness disappears. The eggshell samples are filtered on nylon mesh (40-µm) and washed several times to remove remaining particulate and protein contaminants. The eggshell samples are then further homogenized with pestle A and treated with guanidine hydrochloride overnight to extract eggshell proteins. In summary, this isolation and sequencing approach can be used to quickly identify eggshell proteins from a variety of mosquito species, allowing us to greatly expand our understanding of eggshell biology in these important vectors of human disease (Fig. 2).

RECIPES

Eggshell Lysis Buffer

1× Phosphate-buffered saline (PBS) <R>, pH 7.4
1.0% (v/v) Triton X-100
12 mM Sodium deoxycholate
2.0% (w/v) Sodium dodecyl sulfate (SDS)

Store for up to 1 yr at 4°C.

Eggshell Protease Cocktail

20 ng/μL of Trypsin/LysC protease (Promega)
0.1% (w/v) ProteaseMAX (Promega)
50 mM ammonium bicarbonate

Freshly prepare.

Phosphate-Buffered Saline (PBS)

Reagent	Amount to add (for 1× solution)	Final concentration (1×)	Amount to add (for 10× stock)	Final concentration (10×)
NaCl	8 g	137 mM	80 g	1.37 M
KCl	0.2 g	2.7 mM	2 g	27 mM
Na_2HPO_4	1.44 g	10 mM	14.4 g	100 mM
KH_2PO_4	0.24 g	1.8 mM	2.4 g	18 mM
If necessary, PBS may be supplemented with the following:				
$CaCl_2 \cdot 2H_2O$	0.133 g	1 mM	1.33 g	10 mM
$MgCl_2 \cdot 6H_2O$	0.10 g	0.5 mM	1.0 g	5 mM

PBS can be made as a 1× solution or as a 10× stock. To prepare 1 L of either 1× or 10× PBS, dissolve the reagents listed above in 800 mL of H_2O. Adjust the pH to 7.4 (or 7.2, if required) with HCl, and then add H_2O to 1 L. Dispense the solution into aliquots and sterilize them by autoclaving for 20 min at 15 psi (1.05 kg/cm^2) on liquid cycle or by filter sterilization. Store PBS at room temperature.

Protein Sample Buffer

Reagent	Final concentration
Tris (pH 6.8)	50 mM
Bromophenol blue	0.05% (w/v)
Sodium dodecyl sulfate (SDS)	2% (w/v)
Glycerol	10% (v/v)
Dithiothreitol (DTT)	50 mM

Prepare protein sample buffer without DTT and store for up to 1 yr at room temperature. Freshly add DTT before use.

Solvent Mixture for Peptide Extraction

45% (v/v) acetonitrile
5% (v/v) isopropanol
0.2% (v/v) formic acid
Dilute in H_2O.

Freshly prepare.

 Cite this protocol as *Cold Spring Harb Protoc*; doi:10.1101/pdb.prot108227

Tris–Glycine–SDS Protein Electrophoresis Buffer

Reagent	Final concentration
Tris	25 mM
Glycine	192 mM
Sodium dodecyl sulfate (SDS)	1% (w/v)

Store for up to 1 yr at room temperature.

ACKNOWLEDGMENTS

We thank Carter J. Simington, Max E. Oscherwitz, and Alyssa J. Peterson at the University of Arizona for technical assistance. We thank Dr. Ning Zhang and reviewers for their valuable and thoughtful comments. Proteomic studies like those described are performed at the Analytical and Biological Mass Spectrometry Core Facility at the University of Arizona BIO5 Institute. Our research is funded internally through a Core Facilities Pilot Program provided by the Research, Innovation & Impact at the University of Arizona (to J.I.).

REFERENCES

Amenya DA, Chou W, Li J, Yan G, Gershon PD, James AA, Marinotti OJ. 2010. Proteomics reveals novel components of the *Anopheles gambiae* eggshell. *J Insect Physiol* **56:** 1414–1419. doi:10.1016/j.jinsphys.2010.04.013

Clemons A, Mori A, Haugen M, Severson D, Duman-Scheel M. 2010. *Aedes aegypti* culturing and egg collection. *Cold Spring Harb Protoc* doi:10.1101/pdb.prot5507

Edwards MJ, Severson DW, Hagedorn HH. 1998. Vitelline envelope genes of the yellow fever mosquito, *Aedes aegypti*. *Insect Biochem Mol Biol* **28:** 915–925. doi:10.1016/S0965-1748(98)00083-6

Fang J, Han Q, Johnson JK, Christensen BM, Li J. 2002. Functional expression and characterization of *Aedes aegypti* dopachrome conversion enzyme. *Biochem Biophys Res Commun* **290:** 287–293. doi:10.1006/bbrc.2001.6200

Ferdig MT, Li J, Severson DW, Christensen BM. 1996. Mosquito dopa decarboxylase cDNA characterization and blood-meal-induced ovarian expression. *Insect Mol Biol* **5:** 119–126. doi:10.1111/j.1365-2583.1996.tb00046.x

Han Q, Li G, Li J. 2000. Purification and characterization of chorion peroxidase from *Aedes aegypti* eggs. *Arch Biochem Biophys* **378:** 107–115. doi:10.1006/abbi.2000.1821

Imam H, Zarnigar, Sofi G, Seikh A. 2014. The basic rules and methods of mosquito rearing (*Aedes aegypti*). *Trop Parasitol* **4:** 53–55. doi:10.4103/2229-5070.129167

Isoe J, Miesfeld RL, Riehle MA. 2023a. Visualization of apoptotic ovarian follicles during *Aedes aegypti* mosquito egg maturation by fluorescent imaging studies. *Cold Spring Harb Protoc* doi:10.1101/pdb.prot108226

Isoe J, Simington CJ, Oscherwitz ME, Peterson AJ, Rascón AA Jr, Massani BB, Miesfeld RL, Riehle MA. 2023b. Characterization of essential eggshell proteins from *Aedes aegypti* mosquitoes. *BMC Biol* **21:** 214. doi:10.1186/s12915-023-01721-z

Johnson JK, Li J, Christensen BM. 2001. Cloning and characterization of a dopachrome conversion enzyme from the yellow fever mosquito, *Aedes aegypti*. *Insect Biochem Mol Biol* **31:** 1125–1135. doi:10.1016/S0965-1748(01)00072-8

Kim SR, Yao R, Han Q, Christensen BM, Li J. 2005. Identification and molecular characterization of a prophenoloxidase involved in *Aedes aegypti* chorion melanization. *Insect Mol Biol* **14:** 185–194. doi:10.1111/j.1365-2583.2004.00547.x

Li J. 1994. Egg chorion tanning in *Aedes aegypti* mosquito. *Comp Biochem Physiol* **109:** 835–843. doi:10.1016/0300-9629(94)90231-3

Li JS, Li J. 2005. Characterization of *N*-linked oligosaccharides in chorion peroxidase of *Aedes aegypti* mosquito. *Protein Sci* **14:** 2370–2386. doi:10.1110/ps.051419105

Lin Y, Hamblin MT, Edwards MJ, Barillas-Mury C, Kanost MR, Knipple DC, Wolfner MF, Hagedorn HH. 1993. Structure, expression, and hormonal control of genes from the mosquito, *Aedes aegypti*, which encode proteins similar to the vitelline membrane proteins of *Drosophila melanogaster*. *Dev Biol* **155:** 558–568. doi:10.1006/dbio.1993.1052

Marinotti O, Ngo T, Kojin BB, Chou SP, Nguyen B, Juhn J, Carballar-Lejarazú R, Marinotti PN, Jiang X, Walter MF, et al. 2014. Integrated proteomic and transcriptomic analysis of the *Aedes aegypti* eggshell. *BMC Dev Biol* **14:** 15. doi:10.1186/1471-213X-14-15

Wohl MP, McMeniman CJ. 2023. Batch rearing *Aedes aegypti*. *Cold Spring Harb Protoc* doi:10.1101/pdb.prot108017

CHAPTER 19

Measuring Oviposition Preference in *Aedes aegypti* Mosquitoes

Yunusa Garba Mohammed, Nicholas Kristoff Tochor, and Benjamin J. Matthews[1]

Department of Zoology, The University of British Columbia, Vancouver, British Columbia V6T 1Z4, Canada

The female mosquito is endowed with the responsibility of gene transfer from one generation to another through careful delivery of her eggs to an appropriate environment. To guarantee a successful reproduction process, female mosquitoes use an evaluative approach to select an oviposition site that will ensure that their progeny successfully hatch and develop from larvae to pupa to adult. Specific conditions must be met during oviposition site selection before gravid females deposit eggs, including but not limited to appropriate temperature, salinity, and pH; the presence of nutrition; and a low risk of predation. Mosquito species exhibit a remarkable diversity in their oviposition site selection behaviors. For instance, *Aedes aegypti*, which is a vector for Zika, dengue, yellow fever, and chikungunya viruses, prefers to lay its eggs in natural depressions (e.g., tree holes) or artificial containers. In contrast, other mosquito species, like *Anopheles*, gravitate toward more open water bodies. Understanding the oviposition preference behavior of the *Ae. aegypti* mosquito will facilitate the development of surveillance and control efforts to hinder the successful progression of this vector. Here, we present information on some of the known behaviors and preferences of *Ae. aegypti* during oviposition site selection, techniques to measure these preferences, and some open questions.

BACKGROUND

The reproductive success of a mosquito depends on the ability of the aquatic immature life stages to thrive under the specific biotic and abiotic conditions in breeding sites chosen by egg-carrying females. Thus, oviposition preference is a critically important behavior determining mosquito distribution and abundance. In mosquitoes like *Aedes aegypti*, which need a blood meal to produce eggs, one feeding can mature a batch of oocytes, leading to the laying of up to 100 eggs in a single gonotrophic cycle (Christophers 2009). A gravid mosquito is a female mosquito who has completed a blood meal and mated and now contains mature eggs. She needs to navigate her environment and evaluate potential breeding sites before deciding where to lay these eggs. This process is influenced by weather, wind, the presence of water, the size and shape of potential water repositories, water quality, the presence of food or potential predators, the suitability of the water surface, and many other environmental factors (O'Gower 1963). To use this information, the mosquito relies on long-range sensory modalities, including olfaction and vision, to determine a viable site during flight. Once she is near a potential oviposition location, short-range sensory modalities including mechanosensation and chemosensation provide a final opportunity for her to evaluate the site before laying eggs.

Mosquito species use different behaviors to avoid ovipositing in a suboptimal environment and to ensure successful larval development. These behaviors include laying eggs in a small hole to limit access by predators, skip oviposition in which they spread the eggs across different oviposition sites,

[1]Correspondence: ben.matthews@zoology.ubc.ca

Cite this introduction as *Cold Spring Harb Protoc*; doi:10.1101/pdb.top107670

and egg brooding in which females guard their eggs (Bentley and Day 1989; Day 2016; Swan et al. 2018). *Ae. aegypti* oviposits both in natural and artificial containers that include rock holes, discarded tires, tree holes, discarded water bottles, and many garbage sites with water-holding containers (Kraemer et al. 2015). Unlike other species that lay eggs directly on water, *Ae. aegypti* mosquitoes lay their eggs above the waterline. The eggs will lay dormant until submerged in water (e.g., the next rainfall) and then will hatch as first-instar larvae. A detailed understanding of mosquito oviposition biology is important for designing vector control protocols for a specific species or strain (Day 2016). An extensive review (Bentley and Day 1989) is recommended reading for those interested in oviposition behavior of different mosquito species; here, we focus primarily on *Ae. aegypti*.

LONG-RANGE OVIPOSITION SITE SELECTION BEHAVIOR

Ae. aegypti females have different sets of cues they use for oviposition site selection versus host seeking. After obtaining a blood meal, gravid *Ae. aegypti* stop host-seeking behavior (Judson 1967) and instead search for an oviposition site. This behavior is initiated by a nondirected upwind flight (Hancock and Foster 1993). During this flight, a gravid *Ae. aegypti* female uses long-range senses including olfactory, visual, thermal, and/or humidity cues to locate potential oviposition sites (O'Gower 1963). Semiochemicals, which include conspecific pheromones, infochemicals, toxins, and nutrients, act as long-distance signaling cues for potential egg-laying sites (Bentley and Day 1989). As shown in other insects, she follows a chemical concentration gradient to determine the direction and estimated distance, as odors are carried by convection through air (Hopfield 1999). *Ae. aegypti* also use vision as a long-range cue to orient themselves for oviposition site selection, similar to other species. The oviposition site water must be visible to gravid mosquitoes so that they can evaluate the site from a distance (Bernáth et al. 2012). During visual site selection, gravid *Ae. aegypti* tend to choose wider containers over smaller ones (Harrington et al. 2008). Conspecific larvae and eggs have been proposed as a positive cue in oviposition site selection (Wong et al. 2011), but poor conditions or overcrowding might result as a repellent cue. In one experiment, gravid females showed a preference for sites under a shed in warmer regions, whereas in cooler regions, they favored sites exposed to the sun (Wong et al. 2011). Surtees (1967) reviewed the influence of light on *Ae. aegypti* oviposition, discussing that the predominant cue for egg laying is light, with most eggs being laid in the late afternoon under typical light conditions (Gillett et al. 1959). Brief exposure to light, even as short as 5 sec, followed by darkness, can establish a 24-h oviposition cycle for females previously kept in darkness. Continuous light disrupts this regularity (Corbet 1964). It appears that the beginning of darkness serves as a prompt for oviposition, aligning egg laying right before sunset irrespective of seasonal day length variations (Gillett 1962).

SHORT-RANGE SELECTION BEHAVIOR

As the gravid female reaches a potential oviposition site, she uses her chemotactile receptors to evaluate the location. *Ae. aegypti* mosquitoes require a wet surface to lay and use tarsal and labral receptors to evaluate taste cues in the water (Kennedy 1942; Wallis 1954; Bentley and Day 1989; Afify and Galizia 2015; Matthews et al. 2019; Baik and Carlson 2020). They use contact sensation as a final opportunity to evaluate the water in which their offspring will live. At this point, this "go/no-go" decision can be important for population fitness, and results of studies aimed at understanding the specific cues involved in accepting or rejecting individual oviposition sites can have important implications for mosquito surveillance or control methods that use oviposition traps to monitor or kill mosquitoes (Day 2016).

Cite this introduction as *Cold Spring Harb Protoc*; doi:10.1101/pdb.top107670

TECHNIQUES USED TO MEASURE OVIPOSITION PREFERENCES

Oviposition assays can be used to test the egg-laying preference of gravid *Ae. aegypti* for different sensory cues. These assays are generally performed by providing single or groups of gravid mosquitoes multiple options of oviposition sites and counting the number of eggs or quantifying the density of eggs deposited during each trial. These assays can be performed under controlled laboratory conditions, in semifield environments, or directly in the field by placing traps in mosquito-infested areas. The choice of oviposition preference assay also depends on the scientific questions being asked and is perhaps most importantly determined by whether the cue(s) in question is related to long-range or short-range oviposition. We describe two examples of oviposition assays, although many others can be found in the literature.

Technique for Measuring Long-Range Site Selection

A study by Harrington et al. (2008) investigated the influence of container size and location on oviposition site selection in a semifield environment. They constructed a 10-m × 10-m × 4-m arena, covered with a 1.2-mm² mesh net, to ensure the maintenance of external conditions inside the study area.

They placed water containers of different sizes and colors in various locations within the arena; the sizes ranged from 5 to 45 cm in diameter, and colors included black, gray, and brown. The container inner surfaces were covered with egg-laying papers (here, seed germination paper), allowing for water retention and a coarse surface to which eggs could be affixed. This allowed researchers to manually count the eggs and determine that the surface area, container volume, and container diameter was positively correlated with egg number. This protocol is an elegant means of testing long-range oviposition cues with some level of control while allowing for ambient environmental factors like temperature, rainfall, humidity, and wind to reflect real-world conditions. In addition to measuring egg numbers, "sticky traps" can be deployed to measure site selection by capturing gravid female mosquitoes (de Santos et al. 2012). This work concluded that the behavior of *Ae. aegypti* females indicates a selectivity in choosing oviposition sites, with preferences potentially influenced by environmental factors and trap designs tailored to exploit these behaviors. The goal and scope of each experiment will determine the exact design when using this method.

Techniques for Measuring Short-Range Selection

When considering oviposition preference by egg numbers, it is important to note that close-range cues still influence the final choice. To measure oviposition site selection based on short-range cues, small containers can be used to evaluate egg-laying preference from female mosquitoes that can "sample" multiple potential oviposition sites. To understand how *Ae. aegypti* of different genotypes carry out site selection within a prospective egg-laying site, a high-throughput, single-animal egg-laying assay was developed in which a single gravid female can walk around and have access to two potential oviposition sites. These sites are each constructed from a small Petri dish containing 2 mL of liquid, one containing deionized water and the other containing varying concentrations of seawater, with filter paper acting as an egg-laying surface (Matthews et al. 2019). The small size of the chamber reduces any effects of long-range cues, and the white filter paper in this instance facilitates photography and an automated analysis of egg-laying preference. Technical details of this assay can be found in the accompanying protocols (see Protocol 1: Evaluating Egg-Laying Preference of Individual *Aedes aegypti* Mosquitoes [Tochor et al. 2023a] and Protocol 2: Automated Egg-Counting Approaches for *Aedes aegypti* Oviposition Experiments [Tochor et al. 2023b]).

Experiments with a Single Animal versus Groups of Animals

An additional parameter to consider when developing oviposition preference assays is the number of gravid females to use. To reflect real-world conditions, the experimental design should include the

TABLE 1. Information about using a single mosquito or a group for oviposition assays

	Pros	Cons
Single	Can measure individual variability	More replicates required
	No social cues	Time-consuming
	Easier to automate egg counting	Limited throughput
Group	Reduced variability for group measurements	Masks individual variability
	Better reflects real-world conditions	Social behavior/cues are a possible confounder
		High egg numbers can be more difficult to automatically count

presence of conspecifics and/or heterospecific larvae and eggs in potential oviposition sites. However, this can complicate the analysis, as these cues can have both positive and negative effects on egg laying by *Ae. aegypti* depending on density, feeding state, or infection state of larvae (Allan and Kline 1998; Zahiri and Rau 1998; Gonzalez et al. 2016). In addition, the possibility of social signals acting between gravid females cannot be ruled out. To pinpoint the steps involved in egg-laying behaviors in an individual mosquito, the signals affecting this mosquito need to be confined to what is set by the experimenter to avoid unexplainable behavior resulting from unknown sources. Table 1 includes some pros and cons of either using a single mosquito or a group of mosquitoes to quantify egg-laying behaviors. Recent examples of laboratory oviposition behavior assays for *Ae. aegypti* with groups or individual animals can be found in Melo et al. (2020) and Matthews et al. (2019), respectively. In short, there is no single oviposition assay that can address all scientific questions that might be asked in relation to oviposition preference in *Ae. aegypti*. However, with meticulous design and tailored approaches, researchers can effectively address their specific questions of interest.

ACKNOWLEDGMENTS

We thank members of the (Ben) Matthews Laboratory for discussions and camaraderie.

REFERENCES

Afify A, Galizia CG. 2015. Chemosensory cues for mosquito oviposition site selection. *J Med Entomol* **52:** 120–130. doi:10.1093/jme/tju024

Allan SA, Kline DL. 1998. Larval rearing water and preexisting eggs influence oviposition by *Aedes aegypti* and *Ae. albopictus* (Diptera: Culicidae). *J Med Entomol* **35:** 943–947. doi:10.1093/jmedent/35.6.943

Baik LS, Carlson JR. 2020. The mosquito taste system and disease control. *Proc Natl Acad Sci* **117:** 32848–32856. doi:10.1073/pnas.2013076117

Bentley MD, Day JF. 1989. Chemical ecology and behavioral aspects of mosquito oviposition. *Annu Rev Entomol* **34:** 401–421. doi:10.1146/annurev.en.34.010189.002153

Bernáth B, Horváth G, Meyer-Rochow VB. 2012. Polarotaxis in egg-laying yellow fever mosquitoes *Aedes* (Stegomyia) *aegypti* is masked due to infochemicals. *J Insect Physiol* **58:** 1000–1006. doi:10.1016/j.jinsphys.2012.05.004

Christophers SR. 2009. Aedes aegypti *(L.) the yellow fever mosquito: its life history, bionomics and structure.* Cambridge University Press, Cambridge.

Corbet PS. 1964. Observations on mosquitoes ovipositing in small containers in Zika Forest, Uganda. *J Anim Ecol* **33:** 141–164. doi:10.2307/2354

Day JF. 2016. Mosquito oviposition behavior and vector control. *Insects* **7:** 65. doi:10.3390/insects7040065

de Santos EMM, de Melo-Santos MAV, de Oliveira CMF, Correia JC, de Albuquerque CMR. 2012. Evaluation of a sticky trap (AedesTraP), made from disposable plastic bottles, as a monitoring tool for *Aedes aegypti* populations. *Parasit Vectors* **5:** 195. doi:10.1186/1756-3305-5-195

Gillett JD, Corbet PS, Haddow AJ. 1959. Observations on the oviposition-cycle of *Aëdes (Stegomyia) aegypti* (Linnaeus), III. *Ann Tropic Med Parasitol* **53:** 132–136. doi:10.1080/00034983.1959.11685910

Gillett JD, Haddow AJ, Corbet PS. 1962. The sugar-feeding-cycle in a cage-population of mosquitoes. *Entomol Exper Applic* **5:** 223–232. doi:10.1111/j.1570-7458.1962.tb00584.x

Gonzalez PV, González Audino PA, Masuh HM. 2016. Oviposition behavior in *Aedes aegypti* and *Aedes albopictus* (Diptera: Culicidae) in response to the presence of heterospecific and conspecific larvae. *J Med Entomol* **53:** 268–272. doi:10.1093/jme/tjv189

Hancock RG, Foster WA. 1993. Effects of preblood-meal sugar on sugar seeking and upwind flight by gravid and parous *Aedes aegypti* (Diptera: Culicidae). *J Med Entomol* **30:** 353–359. doi:10.1093/jmedent/30.2.353

Harrington LC, Ponlawat A, Edman JD, Scott TW, Vermeylen F. 2008. Influence of container size, location, and time of day on oviposition patterns of the dengue vector, *Aedes aegypti*, in Thailand. *Vector Borne Zoonotic Dis Larchmt N* **8:** 415–423. doi:10.1089/vbz.2007.0203

Hopfield JJ. 1999. Odor space and olfactory processing: collective algorithms and neural implementation. *Proc Natl Acad Sci* **96:** 12506–12511. doi:10.1073/pnas.96.22.12506

Judson CL. 1967. Feeding and oviposition behavior in the mosquito *Aedes aegypti* (L.). I. Preliminary studies of physiological control mechanisms. *Biol Bull* **133:** 369–378. doi:10.2307/1539832

Kennedy JS. 1942. On water-finding and oviposition by captive mosquitoes. *Bull Entomol Res* **32:** 279–301. doi:10.1017/S0007485300017235

Kraemer MUG, Sinka ME, Duda KA, Mylne AQN, Shearer FM, Barker CM, Moore CG, Carvalho RG, Coelho GE, Van Bortel W, et al. 2015. The global distribution of the arbovirus vectors *Aedes aegypti* and *Ae. albopictus*. *eLife* **4:** e08347. doi:10.7554/eLife.08347

Matthews BJ, Younger MA, Vosshall LB. 2019. The ion channel *ppk301* controls freshwater egg-laying in the mosquito *Aedes aegypti*. *eLife* **8:** e43963. doi:10.7554/eLife.43963

Melo N, Wolff GH, Costa-da-Silva AL, Arribas R, Triana MF, Gugger M, Riffell JA, DeGennaro M, Stensmyr MC. 2020. Geosmin attracts *Aedes aegypti* mosquitoes to oviposition sites. *Curr Biol* **30:** 127–134.e5. doi:10.1016/j.cub.2019.11.002

O'Gower AK. 1963. Environmental stimuli and the oviposition behaviour of *Aedes aegypti* var. *queenslandis* Theobald (Diptera, Culicidae). *Anim Behav* **11:** 189–197. doi:10.1016/0003-3472(63)90028-9

Surtees G. 1967. Factors affecting the oviposition of *Aedes aegypti*. *Bull World Health Organ* **36:** 594–596.

Swan T, Lounibos LP, Nishimura N. 2018. Comparative oviposition site selection in containers by *Aedes aegypti* and *Aedes albopictus* (Diptera: Culicidae) from Florida. *J Med Entomol* **55:** 795–800. doi:10.1093/jme/tjy034

Tochor NK, Mohammed YG, Matthews BJ. 2023a. Evaluating egg-laying preference of individual *Aedes aegypti* mosquitoes. *Cold Spring Harb Protoc* doi:10.1101/pdb.prot108341

Tochor NK, Mohammed YG, Matthews BJ. 2023b. Automated egg-counting approaches for *Aedes aegypti* oviposition experiments. *Cold Spring Harb Protoc* doi:10.1101/pdb.prot108342

Wallis RC. 1954. A study of oviposition activity of mosquitoes. *Am J Hyg* **60:** 135–168.

Wong J, Astete H, Morrison AC, Scott TW. 2011. Sampling considerations for designing *Aedes aegypti* (Diptera: Culicidae) oviposition studies in Iquitos, Peru: substrate preference, diurnal periodicity, and gonotrophic cycle length. *J Med Entomol* **48:** 45. doi:10.1603/ME10149

Wong J, Stoddard ST, Astete H, Morrison AC, Scott TW. 2011. Oviposition site selection by the dengue vector *Aedes aegypti* and its implications for dengue control. *PLoS Negl Trop Dis* **5:** e1015. doi:10.1371/journal.pntd.0001015

Zahiri N, Rau ME. 1998. Oviposition attraction and repellency of *Aedes aegypti* (Diptera: Culicidae) to waters from conspecific larvae subjected to crowding, confinement, starvation, or infection. *J Med Entomol* **35:** 782–787. doi:10.1093/jmedent/35.5.782

Protocol 1

Evaluating Egg-Laying Preference of Individual *Aedes aegypti* Mosquitoes

Nicholas Kristoff Tochor, Yunusa Garba Mohammed, and Benjamin J. Matthews[1]

Department of Zoology, The University of British Columbia, Vancouver, British Columbia V6T 1Z4, Canada

Egg-laying behavior in insects has a dramatic impact on their fitness, controlling their extent and distribution across ecosystems. Thus, understanding the cues that guide mosquito egg laying is an important goal. Measuring mosquito egg-laying preference is critically important for understanding the sensory and neural mechanisms guiding adaptive breeding decisions. Here, we present a method for ascertaining the egg-laying preference of individual *Aedes aegypti* mosquitoes in small chambers with two distinct oviposition substrate choices. This assay is best suited for medium- to high-throughput analysis of egg-laying preference to liquids containing distinct contact chemosensory cues, such as differing salinities. We discuss all steps of the process from assay construction and setup to data acquisition and analysis.

MATERIALS

It is essential that you consult the appropriate Material Safety Data Sheets and your institution's Environmental Health and Safety Office for proper handling of equipment and hazardous material used in this protocol.

Reagents

Aedes aegypti mosquitoes, adult (from lab colony or wild-caught)

Defibrinated sheep blood (Hemostat DSB100; Cedarlane Labs) or alternative method of feeding (e.g., artificial blood meal, anesthetized animal, etc.; see Chapter 1, Protocol 1: Batch Rearing *Aedes aegypti* [Wohl and McMeniman 2023])

Ethanol solution (80%, v/v)

Ice

Mild dish detergent

Sucrose solution (10%, w/v)

Test and control solution

A test solution could be dechlorinated H_2O, as used for egg collection during standard rearing; test solutions could include NaCl or other salt solutions with varying concentrations.

Equipment

Braided cotton wick for feeding sugar solution (Richmond Dental 201208)

Cell phone camera or equivalent device

[1]Correspondence: ben.matthews@zoology.ubc.ca

Cite this protocol as *Cold Spring Harb Protoc*; doi:10.1101/pdb.prot108341

Cold (4°C) room or large refrigerator
Drill bit (3/16″- or 5/32″-diameter)
Equipment for image acquisition (see Steps 19–30)
- LED light panel (e.g., A4 size)
- MDF board (1/8″ × 24″ × 24″)
- MicroSD card (>16 GB)
- Optical posts (2× Ø1/2″ 6″; Thorlabs TR4)
- Post-mountable component clamp (Ø1/2″; Thorlabs PMTR)
- Raspberry Pi CSI ribbon cable for camera (30-cm)
- Raspberry Pi HQ camera
- Raspberry Pi (model 3+) with power supply, keyboard, mouse, and monitor
- Retort stand (Ø1/2″; Fisher Scientific 12000103)
- Socket head bolt (alloy steel, 2″ × size 6–32; eight each; McMaster Carr 91251A159)
- Wide-angle C-mount lens (12-mm)
- Wing nuts (18-8 stainless steel, size 6–32; eight each; McMaster Carr 92001A281)

Equipment for tray construction (see Steps 1 and 7–14)
- Bakers trays and racks (optional; see Step 14; e.g., Chef Equipment SKU 5000/404010 and 016/58182640)
- Cast acrylic, black (12″ × 24″ × 0.125″; one each; McMaster Carr 8505K12)
- Cast acrylic, white (12″ × 24″ × 0.25″; five each; McMaster Carr 8505K92)
- Cast acrylic, white (12″ × 24″ × 0.125″; two each; McMaster Carr 8505K12)
- Laser cutter with bed size of >12″ × 24″

  Alternatively, cutting can be outsourced to fabrication facilities. Also, a smaller bed size can be used with modifications to the parts.
- Socket head bolt (alloy steel, 2″ × size 6–32; 12 each; McMaster Carr 91251A159)
- Wing nuts (18-8 stainless steel, size 6–32; 12 each; McMaster Carr 92001A281)

Filter paper (Whatman grade 1 qualitative, circle, 47-mm; Fisher Scientific 09-805-342)
Fine paint brush (e.g., size 00 or 0)
Forceps
Insectary room or environmental chamber (ideally set to ~27°C)
Laboratory tape (1/2″; Fisher Scientific 15958)
Mechanical mosquito aspirator (John W. Hock Company)
Membrane feeders or other method for blood-feeding (Chemglass CG-1836)
Paper cup with a mesh lid
Paper towels
Petri dishes (60-mm-diameter, 15-mm-height; Fisher Scientific AS-4050)
Power drill
Rearing cages (BugDorm 4M-3030)
Sealable container for mosquito disposal
Serological pipette gun and 10-mL pipettes
Stereomicroscope (optional; see Step 17)

METHOD

Assay Construction

For details on the original assay construction, see the Github repository online at https://github.com/VosshallLab/MatthewsYoungerVosshall2018/ (Matthews et al. 2019).

Each tray has 14 chambers, each designed for a single female mosquito with two distinct choices of liquids, each with a filter paper substrate.

Modifications to this design to accommodate smaller trays (e.g., fewer single-animal chambers and fewer replicates) have been performed successfully. In addition, novel designs for enclosures that contain multiple egg-laying substrates will likely be successful, but for simplicity's sake, this protocol assumes the use of 60-mm Petri dishes and 47-mm Whatman filter paper.

1. Using available laser-cutting facilities, cut eight sheets of 12″ × 24″ acrylic of the specified thickness and color from the following design files online at *https://github.com/bnmtthws/CSHL-Protocols-Mosquito-Egg-Counting/*:
   Main Assay Parts (one sheet of each per tray, seven sheets in total)
   1. oviposition_assay_top_and_bottom.pdf (3-mm or 0.125″; white)
   2. floor-1×_6 mm_white.pdf (6-mm or 0.25″; white)
   3. wells-3×_6 mm_white.pdf (6-mm or 0.25″; white)
   4. wells-ledge-1×_6 mm_white.pdf (6-mm or 0.25″; white)
   5. wells-3×_6 mm_white.pdf (6-mm or 0.25″; white)
   6. wells-3×_6 mm_white.pdf (6-mm or 0.25″; white)
   7. oviposition_assay_top_and_bottom.pdf (3-mm or 0.125″; black)

   Petri Dish Holders (each sheet results in 36 holders, each tray needs 28, 1 sheet in total)
   - Petri_dish_holders_3 mm_white.pdf (3-mm or 0.125″; white)
2. Wash parts before first use by soaking them in mild dish detergent and rinsing thoroughly.

Mosquito Rearing

3. Rear *Ae. aegypti* mosquitoes in mixed-sex rearing cages following best rearing practices (e.g., see Chapter 1, Protocol 1: Batch Rearing *Aedes aegypti* [Wohl and McMeniman 2023]), with the following exception: When picking pupae, divide them into up to six separate cages (each to be used on a single day of experimentation) with ad libitum access to 10% sucrose saturating a cotton dental wick.
4. At 5–10 d after emergence and 4 d before behavioral assays, blood-feed a cage of mosquitoes with a membrane feeder or other feeding strategy (see Chapter 1, Protocol 1: Batch Rearing *Aedes aegypti* [Wohl and McMeniman 2023]).

   Keep as close to ~96 h as possible between blood feeding and assay setup. Record feeding time on the cage for future reference. This time difference was empirically determined for the Liverpool strain of Ae. aegypti *to maximize the number of gravid mosquitoes that will participate in the experiment by laying eggs, and may need to be adjusted for additional strains or species of interest.*
5. Within 48 h after blood feeding, visually inspect the abdomens of each female mosquito and separate blood-fed from unfed mosquitoes by aspiration or cold anesthesia followed by manual manipulation with a paintbrush, only retaining fully engorged females. Record the number of animals available for the behavioral assay. Retain nonfed or partially fed mosquitoes for other uses, or alternatively, place into any sealed container before freezing for at least 12 h at <−20° before disposal.

   Separation of fully fed mosquitoes immediately after a blood meal is possible, but the risk of damage ("popping") to the mosquitoes is higher. However, waiting until after 48 h reduces the experimenter's ability to discern fully engorged females from those who might have partially fed. We have found that ~24 h after a blood meal is an optimum time for separation.
6. Store fed mosquitoes in a rearing cage or a small paper cup, ensuring ad libitum access to >10% sucrose until needed in the assay setup.

   Gravid females will begin to lay eggs on any H_2O that they have access to by the end of the holding period. Ensure that there is no water condensing on the floors or mesh of the cage, and if you notice eggs being laid on the sugar wick, increase the concentration of sucrose to prevent egg laying.

 Cite this protocol as *Cold Spring Harb Protoc*; doi:10.1101/pdb.prot108341

Assay Setup

7. Assemble the laser-cut parts by stacking them in the appropriate order and using at least eight socket cap bolts (four on each long edge) with the threaded part pointing up (Fig. 1A).
8. Perform a "dry setup" by placing one laser-cut Petri dish holder, one Petri dish bottom, and one 47-mm filter paper into each half of each chamber (two per chamber, 28 per tray).

   This step can be completed in advance of loading animals. If done in advance, place the lid on the assay to prevent dust accumulation or other contamination. Wear gloves during all phases of setup to prevent contamination of assay by human secretions. We typically load animals in the late afternoon and allow them to lay eggs overnight (~14–18 h), although researchers interested in circadian regulation of mosquito behavior may choose to use different times of day in their experiments.
9. Immediately before loading the animals, complete a "wet setup." Using a serological pipette, load 2 mL of the test or control solution to each Petri dish (Fig. 1B). If performing a two-choice assay, alternate the position of the test solution and control solution between chambers, recording the identity of the solution in each well on a data sheet that contains all pertinent experimental information.

   A template data sheet can be found online at https://github.com/bnmtthws/CSHL-Protocols-Mosquito-Egg-Counting/blob/main/assay_setup/datasheet.pdf and can be modified according to the needs of an individual experiment.
10. Using a mechanical aspirator, transfer 14 blood-fed mosquitoes per tray from Step 6 into a paper cup with a mesh lid.
11. Place the cup on wet ice until the animals are cold-anesthetized (~5 min).
12. Working quickly, use forceps to gently transfer a single anesthetized female mosquito into the center of each chamber. Alternatively, use a mechanical aspirator to transfer them.
13. Place the lid on top and secure it with the eight bolts from Step 7 with wingnuts.

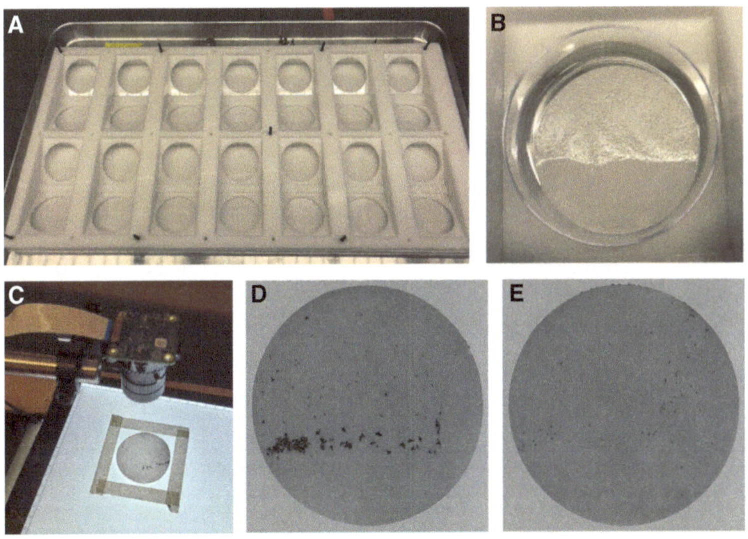

FIGURE 1. Photographs of the fully assembled assay setup (*A*) and a single Petri dish loaded with filter paper and 2 mL of a test or control solution (*B*). (*C*) Imaging setup using Raspberry Pi with a high-quality camera and manual-focus lens. The camera is fixed to a retort stand and is kept in place for the duration of data collection. Light-colored tape should be used to allow for consistent placement of egg papers. (*D,E*) Images taken with the Raspberry Pi camera using a standardized imaging setup. Backlighting provides consistent illumination between images, and camera angle and field of view remain unchanging. *D* shows a heavily clustered egg paper from an assay with multiple animals. *E* shows an egg paper with minimal clumping from a single animal assay.

14. Transfer the tray to an insectary room or other environmental chamber for 18 h, taking care not to spill the liquid in the dishes by jolting, tilting, or otherwise disturbing the tray.

    If multiple trays are being used, a wheeled bakers rack with full-sized sheet trays is a space-efficient method of storing and transporting assay setups.

    See Troubleshooting.

Assay Deconstruction

15. Transfer the tray to a cold room or refrigerator, leaving it for at least 45 min to anesthetize the adult mosquitoes.
16. Unscrew the wingnuts and open the lid; remove each female mosquito with forceps and place it into any sealed container before freezing for at least 12 h at <−20° before disposal.

    Record any escapees or chambers without any eggs on the data sheet.

17. Remove Petri dishes from the tray and proceed immediately to imaging before the egg papers dry out.

    Manual egg counting can be performed using a stereoscope if no suitable imaging setup is available. However, we highly recommend taking an image of each Petri dish even if just with a cell phone camera or equivalent device. The image becomes a record of the raw data for each assay and can be used for subsequent egg counting. See Discussion.

18. Clean the assay setup by spraying it with an 80% ethanol solution and wipe it down with paper towels. Every week, or if the assay setup will not be used for extended periods of time, disassemble it and wash all pieces with mild detergent, rinse well, and allow to air-dry.

    Assay setups can be stored assembled.

Image Acquisition

Relevant files can be found on the Github repository associated with this protocol (https://github.com/bnmtthws/CSHL-Protocols-Mosquito-Egg-Counting).

19. Assemble the imaging setup as shown in Figure 1C. Attach the wide angle lens to the Raspberry Pi HQ Camera and connect the camera to the Raspberry Pi using the CSI ribbon connector. Using the sheet of MDF as a work surface, set up the camera over the LED tracing panel using a component clamp, optical posts, and a retort stand. Use laboratory tape to mark a box to aid in consistently positioning egg papers. Care must be taken to ensure that the camera and tracing panel will not move for the duration of data collection.
20. Turn on the Raspberry Pi and illuminate the LED tracing panel.

    If setting up the Raspberry Pi for the first time, set up the Pi with a microSD card with a suitable operating system loaded. Documentation is available (https://www.raspberrypi.com/documentation/computers/getting-started.html) for detailed instructions for initializing the system.

21. Download and open the "ovi_paper_imaging.py" script from the Github repository.

    This is a Python script that allows researchers to efficiently collect images of egg papers using a Raspberry Pi computer and camera.

22. Change parameters in the script to suit the requirements of the experiment (e.g., replicates and number of treatments).

    Changing these parameters will ensure that all replicates are efficiently imaged in a single session, and that images are saved with appropriate file names.

    See Troubleshooting.

23. Using forceps, carefully remove the egg paper from the Petri dish taking care not to drop any eggs. Any eggs that remain in the dish can be transferred to the egg paper using a fine paint brush.

Cite this protocol as *Cold Spring Harb Protoc*; doi:10.1101/pdb.prot108341

24. Run the script, using the automatic preview to align the egg paper in the center of the image. Ensure that the camera is clearly focused on the eggs. Figure 1D and E provide examples of egg papers that are from multiple animal and single animal assays, respectively.

    The script also has settings for defining the region of interest and for adjusting brightness and contrast. Adjust them for a new imaging setup, but then keep them consistent for the duration of data collection.

25. Once the camera is in an ideal position and angle, tighten all clamps attaching the camera to the retort stand to secure the position of the camera.
26. Mark the position of the retort stand and tracing panel (one mark per side, for a total of at least eight marks) and drill holes using a power drill with a 5/32″ or 3/16″ drill bit.
27. Push the bolts up so that the threads are sticking out of the MDF sheet. Thread eight wingnuts onto the bolts to secure them in place. Return the retort stand and tracing panel to their positions; the bolts should rest against the sides of each component in order to keep them in place.
28. When the egg paper is aligned in the field of view, press "enter" to take the image.
29. Wipe down the tracing light with 80% ethanol solution to remove any loose eggs and destroy any potentially viable eggs that have been left behind.
30. Repeat the image acquisition steps as needed for the remaining replicates of the experiment.

TROUBLESHOOTING

Problem (Step 14): Animals are not laying eggs.

Solution: Animals must be blood-fed, mated, prevented from laying eggs while in the mass rearing cage, and given enough time to lay in the assay chambers. Double-check each of these steps to ensure that appropriate guidelines are followed. If it is suspected that animals are laying eggs ahead of the assay (e.g., on sugar wick or droplets of water found in the cage) the sucrose concentration can be raised to ~30%. Also ensure that there is no standing water available for laying.

Problem (Step 14): Animals are "escaping" the assay or moving between chambers.

Solution: Ensure that the layers of acrylic used to construct each tray are properly aligned without any gaps between layers. Add more bolts to ensure that there is no "bowing" in the middle of the assay setup.

Problem (Step 22): Images taken with the Raspberry Pi are being cropped incorrectly (either under- or overcropped).

Solution: The cropping extent or region of interest is set in the line of code that captures the image using the "picamera" package. The "camera.zoom" argument sets the cropping extent, with its units as proportion of the camera sensor. The first two values define the top left boundary of the image extent, and the next two values set the width and height of the image. Adjust these values as needed to set a region of interest that crops the image to include the egg paper and exclude the rest of the image's background.

DISCUSSION

Preference Assays

The assay described here is best suited for evaluating oviposition preference in large numbers of single *Ae. aegypti* females in response to contact chemosensory or tactile cues. Because of the small size of the chamber, any volatile cues will quickly saturate the headspace of the chamber, providing little to no directional information for the mosquito to select one potential site over the other. Therefore, we do not suggest using it to study volatiles.

Imaging Considerations

The strategy used for imaging egg papers will depend on the downstream methods used for counting eggs. If researchers plan to use automated egg-counting methods, they will need collect consistent, standardized images. Here, we briefly outline imaging methods in increasing complexity and describe best practices for collecting sufficiently standardized images of egg papers to facilitate automated egg counting.

In the case of studies with few replicates or few test conditions, researchers may prefer to count eggs manually using a low-power (10×–30× magnification) microscope. Even if eggs will be counted by hand, we still recommend quickly taking an image of each egg paper in case a recount is required and as a way of retaining raw data without storing physical egg papers. In this case, images can be captured with a cell phone or other camera. Another option for manually counting eggs is using a counting software such as the Cell Counter plug-in (https://imagej.net/ij/plugins/cell-counter.html) for Fiji (Schindelin et al. 2012) or ImageJ (Schneider et al. 2012). This option is preferable to manual tallies and can also save an overlay image of counted eggs, reducing the likelihood that the researcher will "lose their place" while counting. Although counting is done manually in this case, we still recommend taking steps to standardize images to enable efficient counting. Researchers may opt to use a fixed camera position and more complex lighting setups, but for manually counting eggs with Fiji/ImageJ, ensuring even lighting and clear focus will be sufficient. If an experiment has a significantly large number of replicates, researchers may also wish to modify the imaging setup to take pictures of multiple egg papers at a time. Although our workflow is designed for single paper images, the imaging scripts could be easily modified to update file names for multiple papers.

Camera Considerations

Maintaining a stable camera position is essential for collecting standardized images to achieve the best automated counts, regardless of the counting method used. Consistent camera position is best achieved by fixing a camera in place, ideally on a wall or a bench, where it can remain until all data collection is complete, as any minor changes in camera angle can compromise count accuracy. We use Raspberry Pi single-board computers, as there are several affordable camera options available and their lenses can be manually focused, preventing minor differences between images that may occur when using autofocus. As egg paper images will serve as the raw data for this type of assay, it is highly valuable to collect consistently captured images. Regardless of the counting method initially used, raw images can be counted and recounted using any method, including more complex methods than those described here.

It is also important to use consistent lighting when collecting images for automated counting (e.g., see Protocol 2: Automated Egg-Counting Approaches for *Aedes aegypti* Oviposition Experiments [Tochor et al. 2023]). The algorithms we describe use pixel-value thresholding to convert images into binary matrices, so any variations in illumination or image brightness will prevent these thresholds from being generalized between images and experiments. The easiest way to achieve consistent lighting is by using a backlit LED tracing panel. By backlighting the images with a bright light source, small differences in room lighting (e.g., from overhead lights or time of day) are negated and images will have consistent illumination. Although we used an A4-sized LED light panel, any size can be used so long as the lit surface is large enough for the egg papers used in the experiment. Ensure that there is nothing dark present in the background so that eggs are the only dark features in each image. Background features, including edges of the light panel, dark-colored lines for delineating paper placement, or even loose eggs from previous papers, can negatively affect the accuracy of automated egg counts.

 Cite this protocol as *Cold Spring Harb Protoc*; doi:10.1101/pdb.prot108341

ACKNOWLEDGMENTS

We thank members of the (Ben) Matthews Laboratory for discussions and camaraderie, as well as Krithika Venkataraman, Stephanie Marcus, and Solomon Dworkin for early work on developing previous iterations of the assay.

REFERENCES

Matthews BJ, Younger MA, Vosshall LB. 2019. The ion channel *ppk301* controls freshwater egg-laying in the mosquito *Aedes aegypti*. *eLife* **8:** 143. doi:10.7554/eLife.43963

Schindelin J, Arganda-Carreras I, Frise E, Kaynig V, Longair M, Pietzsch T, Preibisch S, Rueden C, Saalfeld S, Schmid B, et al. 2012. Fiji: an open-source platform for biological-image analysis. *Nat Methods* **9:** 676–682. doi:10.1038/nmeth.2019

Schneider CA, Rasband WS, Eliceiri KW. 2012. NIH Image to ImageJ: 25 years of image analysis. *Nat Methods* **9:** 671–675. doi:10.1038/nmeth.2089

Tochor NK, Mohammed YG, Matthews BJ. 2023. Automated egg-counting approaches for *Aedes aegypti* oviposition experiments. *Cold Spring Harb Protoc* doi:10.1101/pdb.prot108342

Wohl MP, McMeniman CJ. 2023. Batch rearing *Aedes aegypti*. *Cold Spring Harb Protoc* doi:10.1101/pdb.prot108017

Protocol 2

Automated Egg-Counting Approaches for *Aedes aegypti* Oviposition Experiments

Nicholas Kristoff Tochor, Yunusa Garba Mohammed, and Benjamin J. Matthews[1]

Department of Zoology, The University of British Columbia, Vancouver, British Columbia V6T 1Z4, Canada

Egg-laying preferences of mosquitoes can reveal key neurosensory mechanisms informing the decision-making process for this critical behavior. A single blood meal results in a gravid female *Aedes aegypti* mosquito laying more than 100 eggs. Therefore, egg counting represents a potentially time-consuming and laborious component to behavioral assays, such as those that measure oviposition preference or fecundity. Automated algorithms that count eggs from images can dramatically reduce the time required for this step of data processing and analysis and increase reproducibility associated with having multiple human observers count the eggs. Here, we present two distinct approaches for counting melanized *Ae. aegypti* eggs laid on white filter paper.

MATERIALS

It is essential that you consult the appropriate Material Safety Data Sheets and your institution's Environmental Health and Safety Office for proper handling of equipment and hazardous material used in this protocol.

Equipment

Anaconda virtual environment software (version 3 and onward)

Images of *Aedes aegypti* eggs on filter paper (e.g., from Protocol 1: Evaluating Egg-Laying Preference of Individual *Aedes aegypti* Mosquitoes [Tochor et al. 2023])

OpenCV software (version 4.1.x and onward)

Personal computer (Windows, MacOS, or Linux) capable of running Python, Anaconda, and OpenCV

Image analysis can be performed on a Raspberry Pi if following the protocol for image acquisition in Protocol 1: Evaluating Egg-Laying Preference of Individual Aedes aegypti *Mosquitoes (Tochor et al. 2023).*

Python-integrated development environment (IDE; e.g., Spyder)

Python software (version 3.6.x and onward)

METHOD

This protocol outlines the steps for counting eggs from previously collected images of egg papers from an oviposition assay. Eggs can be counted manually using a stereomicroscope immediately after conducting an end point assay (e.g., after following Protocol 1: Evaluating Egg-Laying Preference of Individual Aedes aegypti *Mosquitoes [Tochor et al.*

[1]Correspondence: ben.matthews@zoology.ubc.ca

Cite this protocol as *Cold Spring Harb Protoc*; doi:10.1101/pdb.prot108342

2023]) or from a computer screen if imaging was performed; however, for consistency, reproducibility, and high throughput, we recommend automated egg-counting approaches. Two methods, contour-based counting (Fig. 1A) and pixel-based counting (Fig. 1B) are outlined here. Both automated egg-counting methods function by detecting distinct contours in binarized images of egg papers.

However, the pixel-based method calculates an average egg area from example egg papers and can result in more accurate measurements with an appropriate training set containing well-separated eggs.

More detailed documentation is available in a Github Repository associated with this protocol (https://github.com/bnmtthws/CSHL-Protocols-Mosquito-Egg-Counting).

Software Installation

We highly recommend using a virtual environment software such as Anaconda. This allows for a much simpler installation of OpenCV.

1. Install Anaconda virtual environment software (https://docs.anaconda.com/anaconda/install/index.html) on a personal computer.
2. Set up a new virtual environment for your OpenCV installation and install Python software.
3. Install OpenCV software (Bradski 2000) within the newly created virtual environment (https://anaconda.org/conda-forge/opencv).
4. Using a Python-IDE, test the installation of OpenCV by importing it into the console.
5. Access the scripts from the Github page for this protocol, and change parameters to define the directories where the images to be analyzed are saved.

Setting a Binarization Threshold

These steps are important for either counting method, so a suitable threshold should be determined before starting them. Figure 1C shows an example of the thresholding process.

6. Open the "set-threshold.py" script and read in the first image of egg papers to the Python environment.

   It will be converted to grayscale to simplify the thresholding process.
7. Set the threshold at the midpoint (128) and check the output image.

   Overall, pixels are represented as values from 0 (black) to 255 (white).
8. Change the threshold based on the binarized image.

   See Troubleshooting.

   - If nonegg pixels are being converted to white pixels, lower the threshold and check the output again.
   - If some eggs are lost or if eggs are being fragmented into smaller white patches (more than one per egg), raise the threshold and check the binary output.
9. Repeat Step 8, as needed, to determine a threshold that effectively binarizes the images.

   See Troubleshooting and Figure 1D–G.
10. Make note of this threshold and proceed to the egg-counting method of choice. Steps 11–16 describe the contour-based method and Steps 17–21 describe the pixel-based method.

Option 1: Simple Contour-Based Method

11. Access the OpenCV environment in Anaconda and open the "egg-count-polygon.py" script for contour-based egg counting.

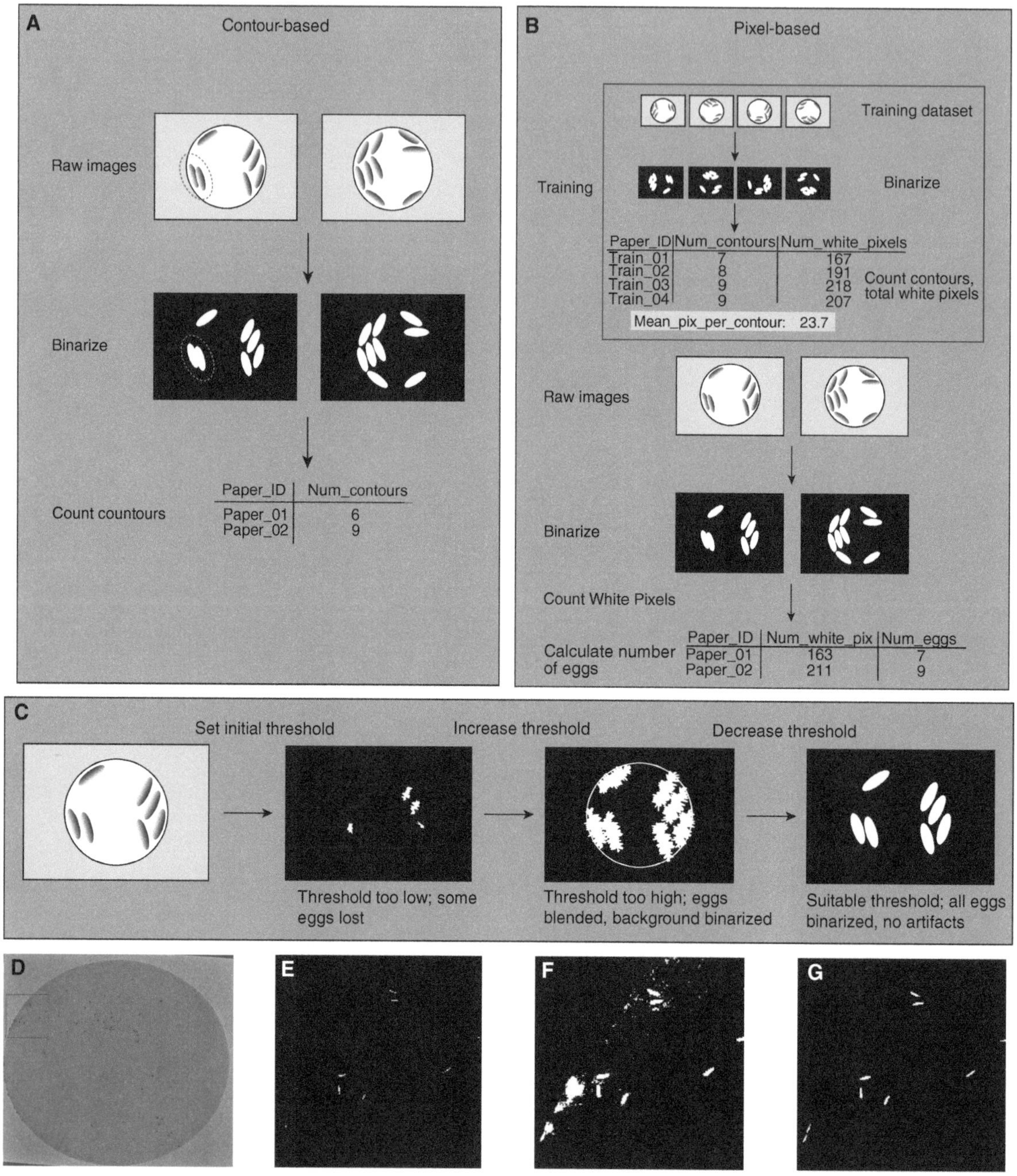

FIGURE 1. Two workflows for automated egg counting. (*A*) Outline of the simpler contour-based counting method. This method binarizes images of egg papers and then simply counts the number of distinct white polygons as eggs. Note the red dotted oval, highlighting an overlapping pair of eggs in the first egg paper. The contour-based method will make an error when counting this egg paper, as the pair of eggs appears as a single contour. (*B*) The workflow for pixel-based automated egg counting, which uses a training data set to determine the average egg size in pixels. This method then counts the total number of white pixels in the binarized images and divides them by the average egg size to estimate the number of eggs in each image. The pixel-based method will likely be the better performing method for images of egg papers with highly clustered eggs. (*C*) A schematic of the process of determining a suitable threshold for image binarization. (*D*) An example image with the inset region for panels *E*, *F*, and *G* highlighted. The tape for aligning egg papers is visible in the background of the image, and attention should be given to ensure that this background feature is not included as a highlighted feature. (*E*) A close-up view of a binarized image with an unsuitably low threshold, causing fragmentation of individual eggs. (*F*) A close-up view of a binarized image with an unsuitably high threshold, resulting in artifacts from the egg paper and other shaded regions of the raw image. Using a threshold this low will result in falsely inflated automated counts. (*G*) A close-up view of eggs that have been suitably binarized. In this image, eggs are well-differentiated and there are no artifacts that will impact the accuracy of downstream counting processes.

Cite this protocol as *Cold Spring Harb Protoc*; doi:10.1101/pdb.prot108342

12. Set the working directory to the folder containing the egg paper images from the desired experiment.
13. Batch read the images, converting them to grayscale in the process.
14. Set the threshold to your optimized threshold for image binarization.
15. Run the remaining section of the script to count the egg paper images using contour-based counts. The script will output a .csv with egg counts for each image by file name.
16. Compare the performance of count data with manual counts for a subset of the egg paper images. Ensure that the egg counts produced by this method are acceptably close to the manual counts. The amount of acceptable error between automated and manual counting methods is at the researcher's discretion and will depend on the number of replicates and the number of eggs per paper (which may vary between single- and multiple-animal experiments).

    See Troubleshooting.

Option 2: Pixel-Based Method with Training Data

17. Access the OpenCV environment in Anaconda and open the "egg-count-pixels.py" script for pixel-based egg counting. The script is divided into annotated sections that can be run on their own within the console of a Python IDE while working through the remaining steps.
18. Assign a directory for training images (use at least 20 images to calculate representative average egg size).

    The training images should be images with minimally clumped eggs (see Fig. 1 in Protocol 1: Evaluating Egg-Laying Preference of Individual Aedes aegypti *Mosquitoes [Tochor et al. 2023]) and must have been taken with an identical imaging setup to the one used for collecting experimental data. The contour-based method used to determine mean egg area in pixels does not handle clustered eggs well, so selecting sparsely covered egg papers will help produce more accurate pixels-per-egg values.*

19. Run the section of the code for determining mean egg size from the training data set.

    This step will generate a table of contour-based egg counts along with the total area of pixels that comprise the eggs on the paper. The script will determine the mean egg size, which will be used to count eggs for experimental papers.

20. Check the performance of the contour-based counts for the training data set by comparing them to manual counts. If satisfied with the count accuracy for the training data, proceed to running the final section of code to produce pixel-based egg counts.

    See Troubleshooting.

21. Export pixel-based egg counts for data analysis.

TROUBLESHOOTING

Problem (Steps 8 and 9): A threshold that produces a suitable binary image for automated counting cannot be identified.

Solution: If a suitable threshold cannot be found for egg paper images, the physical imaging setup will need to be adjusted. If it is not already used, ensure that bright, even lighting is used during data collection. If egg papers are not evenly lit while imaging, locally bright or dim regions of the image will be harder to binarize with a single threshold. Alternatively, the "ovi_paper_imaging.py" script can be edited to increase brightness, exposure, digital gain, or other settings during egg imaging (as in Protocol 1: Evaluating Egg-Laying Preference of Individual *Aedes aegypti* Mosquitoes [Tochor et al. 2023]). Improving the contrast between eggs and their background will make threshold selection much easier.

Problem (Steps 8 and 9): Egg papers are highly clumped and automatic counting is performing poorly.

Solution: Oviposition between different strains of *Ae. aegypti* shows a high degree of variation. We developed these initial scripts for single-animal assays using the *Ae. aegypti* Liverpool IB12 strain, which does not oviposit with a high degree of clumping. Most small clumps can be counted reliably by adjusting the threshold to a higher value, which will reduce the size of "egg" shapes in the binarized image. If the image has been backlit, this is often sufficient to discriminate between individual eggs within clusters. More complex counting methods can be used, and several features within OpenCV can help identify and parameterize clusters of eggs that cannot be accurately counted using our methods, as discussed in the Further Considerations section of this protocol.

Problem (Steps 8 and 9): Automated counts from either method are consistently severely overcounting eggs.

Solution: If either counting method is significantly overcounting eggs compared to manual counts, the threshold is likely too low. If the threshold is too conservative, individual eggs will be fragmented into several small clusters of pixels (see Fig. 1E). Because both counting methods rely on contour-based egg counts (the contour-based method uses this as its main approximation, the pixel-based method uses contours to estimate the number of eggs in the training data set), a low threshold will create artificially inflated egg counts regardless of the method used. Check the binarized image from the threshold testing script and zoom in to see several eggs up close. If single eggs are fragmented, increase the threshold as much as possible without producing background artifacts in the binary output image.

Problem (Steps 16 and/or 20): A previously well-performing method is no longer producing accurate counts.

Solution: Compare the images of accurately counted egg papers with the images of egg papers that are being counted inaccurately. If any difference in framing, angle, or lighting is present, it is likely that the camera has moved since the beginning of data collection. Another way that the system can be checked is by comparing the pixels-per-egg values of the training data sets. These values should be relatively consistent over time, and there should not be large differences in the mean egg size between imaging sessions. If there are large differences, it is also likely that the physical imaging setup has been altered. As a solution, attempt to reproduce the camera angle as closely as possible or try setting a new binarization threshold.

Problem (Step 20): The pixel-based counting method is performing poorly by over- or undercounting eggs.

Solution: Ensure that imaging conditions between the training data set and the experimental data set are identical. Any changes in camera angle or position will change the relative size of eggs, rendering the pixels-per-egg values determined in the training data useless for counting eggs from other imaging sessions. If multiple imaging setups are used, training data must exist for each setup.

Problem (Step 20): Automatic count methods are reliably counting at least one more egg than manual counts.

Solution: Ensure that the backlight or background surface is cleaned well between trials. Eggs can easily fall off of egg papers and cling to the background surface. Any eggs that are in-frame will be counted, including eggs left behind from other papers. One way to confirm that background eggs are causing the counting mistakes is by examining the binarized images produced during count-

Cite this protocol as *Cold Spring Harb Protoc*; doi:10.1101/pdb.prot108342

ing. If white clumps beyond the boundaries of the egg paper are visible in the binary image, it is likely that eggs from previous papers remain in the background.

DISCUSSION

Automated Egg Counting

As discussed, eggs in small-scale studies can be counted manually or with manual image-counting software, but when standardized images are captured, eggs can be quickly counted using a variety of automated methods. Our egg-counting methods use the open-access image analysis software OpenCV. OpenCV facilitates the analysis of thousands of images and can process large data sets in under a minute, depending on workstation specifications. We highlight two methods for automatically counting eggs and recommend starting with the simpler method and assessing its performance before using the more complicated workflow.

After consistent image collection, the next most important step for producing accurate automated counts is selecting a suitable binarization threshold. Carefully selecting a threshold that effectively captures the geometries of eggs without highlighting other regions of the image (e.g., paper boundaries) is essential for producing accurate automated counts. This step can be tedious, but its accuracy will determine the performance of counting for either method used. It is important to check the output of the thresholded binary image (Fig. 1) to ensure that eggs are the only objects converted to white pixels and that the threshold is not set so conservatively that binarized eggs are being split into multiple small contours. See the Troubleshooting section for more information on the threshold-setting process.

The most basic counting method functions using contour detection. Our method binarizes an image at a predetermined threshold and counts the number of distinct contours in the binarized image, using that as a proxy for the number of eggs. When eggs are not excessively clumped and images are consistently captured and brightly illuminated, this method will often perform within an acceptable margin of error to manual counting. A slightly more complex counting method that is more robust to clumped patches of eggs uses a training data set to determine the average egg size, measured in pixels. For this method, we count the number of contours for the training data set, as was done for the first counting method. Next, the total number of white "egg" pixels are counted for each training image to calculate the mean egg size for each training image. When repeated for many egg papers (we recommend training with at least 20 images), the mean egg size should be sufficiently representative to facilitate accurate size-based counts. Our pixel-based counting method determines the average area of each binarized contour and will calculate a much more accurate mean egg size if the training papers contain well-separated eggs. As the training step still uses contour-based counting methods, we recommend selecting a subset of egg papers with minimally clumped eggs for the training data set. Additionally, it is important to select egg papers with various amounts of eggs (e.g., some sparsely covered, some densely covered) to produce a representative mean pixels-per-egg value. Once the mean egg size has been calculated using the training data, the pixel-based method is robust, even for papers with highly clustered eggs.

Performance Evaluation

When initially using an automated egg count method, researchers need to validate its accuracy before conducting further analyses. Both egg-counting methods we use are fundamentally simple and will produce counts from images of any objects as long as the threshold is set in a way that creates clusters of white pixels. It is important to ensure that the imaging setup, camera settings, and threshold selection are all configured appropriately and are producing reasonable count data before using the data from either counting method.

We recommend manually counting eggs using a software such as Fiji to build a reference data set to compare to automated egg counts. If doing this, count the eggs for a reasonably large number of egg papers (at least 25, but more is better) and record these count data along with the file names for each image. With this data set made, the counts produced by either counting method in this protocol can be quickly compared to determine the accuracy of the method. Because both methods generate a .csv with image file names and the number of eggs counted for each image, it is easiest to merge them to the data set of manual counts and quickly calculate the counting error for each method. We recommend calculating the relative error between either method and the manual counts and averaging the error over the number of images in the manual data set. Some judgment is required for determining an acceptable margin of error, but we have found both counting methods perform with relatively high degrees of accuracy for a variety of imaging setups.

Further Considerations

After a sufficiently well-performing counting method has been selected and egg papers have been analyzed, we still recommend periodically validating the automated count data against manual counts. Be sure to save the initial counting performance results so that they can be compared to future validations to ensure consistent results. If a previously well-performing method begins performing poorly, proceed to the Troubleshooting section to identify the cause of the performance issues.

If neither method is able to produce satisfactory egg counts after troubleshooting, other tools included in OpenCV can be used to create more sophisticated analyses. Before binarizing an image, repeated passes of dilation and erosion can be performed to improve separation between clustered eggs. When repeated, these steps can remove noise from the perimeters of eggs and can potentially result in better-defined boundaries between clustered eggs that were previously binarized into single polygons. Additionally, conditions based on concavity and convexity can help with egg counting on highly clustered papers. Single eggs will have exclusively convex geometries, whereas clusters will have concave regions where eggs contact one another. These parameters can help identify clustered eggs so that they can be excluded from the eggs used for calculating mean egg size.

ACKNOWLEDGMENTS

We thank members of the (Ben) Matthews Laboratory for discussions and camaraderie. In particular, Orna Phelan provided valuable feedback during the development of the imaging and counting scripts and shared raw images that were highly useful for testing both counting methods.

REFERENCE

Bradski G. 2000. The OpenCV library. *Dr Dobb's J Softw Tools* **120:** 122–125.

Tochor NK, Mohammed YG, Matthews BJ. 2023. Evaluating egg-laying preference of individual *Aedes aegypti* mosquitoes. *Cold Spring Harb Protoc* doi:10.1101/pdb.prot108341

CHAPTER 20

Retrospective Review of and Introduction to the Analysis of Mosquito Optomotor Responses

Clément Vinauger[1,3] and Jeffrey A. Riffell[2]

[1]*Department of Biochemistry, Virginia Polytechnic Institute and State University, Blacksburg, Virginia 24061, USA;* [2]*Department of Biology, University of Washington, Seattle, Washington 98195, USA*

Adult hematophagous female mosquitoes require nutrients and proteins from vertebrate blood to produce progeny. To find these hosts, mosquitoes rely on olfactory, thermal, and visual cues. Among these sensory modalities, vision has received far less attention than olfaction, in part because of a lack of experimental tools providing sufficient control on the delivery of visual stimuli and the recording of mosquito responses. Although free-flight experiments (e.g., wind tunnel and cage) ensure higher ecological relevance and allow the observation of more natural flight dynamics, tethered flight assays offer a greater level of control on the suite of sensory stimuli experienced by mosquitoes. In addition, these tethered assays provide a stepping stone toward understanding the neural underpinnings of mosquito optomotor behavior. Advances in computer vision tracking systems and programmable light-emitting diode displays have permitted significant discoveries in models such as the fly *Drosophila melanogaster*. Here, we introduce the use of these methods with mosquitoes.

BACKGROUND

To locate their hosts, mosquitoes rely on multiple sensory modalities including olfaction, thermosensation, and vision. Because hosts become visible to mosquitoes at 2–15 m, visual cues play a critical role by coupling long-range odor tracking (>10 m) to behaviors responsive to short-range cues (heat and skin odors; ~0.1–1 m distance). Although substantial efforts have been made to unravel the behavioral, physiological, neural, and molecular underpinnings of mosquitoes' sense of smell, their vision has been historically overlooked.

A BRIEF HISTORY OF THE STUDY OF MOSQUITO VISION

From the foundational work of Brett in the late 1930s (Brett 1938), Kennedy in the early 1940s (Kennedy 1940), and Brown in the 1950s (Brown 1951), it was clear that vision was implicated in *Aedes*' host-seeking behavior, as evidenced both in field and laboratory assays. Two decades later, the spectral sensitivity of the ocelli of *Aedes aegypti* larvae was characterized (Brown and White 1972; Seldin et al. 1972) followed, another 20 years later, by the characterization of the spectral sensitivity of the adult female's compound eye and the biochemical demonstration that rhodopsin was a key visual pigment involved in mosquito photosensitivity (Muir et al. 1992a). Behaviorally, this species' respons-

[3]Correspondence: vinauger@vt.edu

Cite this introduction as *Cold Spring Harb Protoc*; doi:10.1101/pdb.top107672

es to visual objects were suggested to depend on the spectral reflectance of a target (Muir et al. 1992b), and with the progress in molecular biology and immunohistochemistry, rhodopsin and ultraviolet photoreceptors were shown to be present in species-specific expression and coexpression patterns across different regions of the retina (Hu et al. 2014). However, to this point, the neuroethological role of vision in guiding mosquito navigation to the host remained to be determined.

Recently, work in *Ae. aegypti* showed that the knockout of the genes encoding two opsins, namely, *Op1* and *Op2*, using CRISPR–Cas9, led to a disruption of the CO_2-induced target-seeking behavior of adult females but not of their phototactic and odor-tracking behaviors (Zhan et al. 2021). These genes code for the two most abundant rhodopsins of the five rhodopsins present in the eyes of this species and may be key molecular players in the CO_2-induced host tracking. By using real-time tracking wind-tunnel assays where visual and olfactory stimuli can be delivered, researchers showed that this CO_2-induced target tracking behavior was further shown to be spectral-band specific (Alonso San Alberto et al. 2022). Remarkably, the most pronounced attraction by mosquitoes was to wavelength bands that humans perceive as cyan, orange, and red, which dominate the color spectrum of human skin.

ANALYSIS OF MOSQUITO VISUAL RESPONSES

Despite the potential importance of vision in mediating mosquito flight and biting behaviors, surprisingly, details regarding the visual cues that attract mosquitoes remain unclear. Visual spectra of human skin, flowers, and oviposition sites can be diverse and dominated by short- to long-wavelength bands (325–700 nm). Unfortunately, prior studies often neglected to control for the visual stimuli's reflectance, intensity, and contrast with the background, thus often making the results from these studies contradictory. For instance, studies have shown specific preferences but for different bands of the color spectrum; *Ae. aegypti* mosquitoes were attracted to blue in one study (Smart and Brown 1956) and only to green-yellow in another (Snow 1971). In addition, mosquitoes have been shown to be as attracted to red as to black (Brett 1938; Fay and Prince 1968; Muir et al. 1992a).

Bioassays that allow careful control of the visual environment and the tested stimuli are needed to assess the visual behaviors of mosquitoes. Controlling for light intensity contrast and perceptual color contrast, Alonso San Alberto et al. (2022) showed that *Ae. aegypti* mosquitoes were highly attracted to medium-long wavelengths of the visual spectrum (>600 nm; orange to red to the human observer). Additional studies, using a combination of free-flight and tethered assays, will be needed to determine whether other mosquito species show similar spectral preferences and the neural bases of these behaviors. The two experimental paradigms mentioned above—that is, free flight and tethered flight—each present a series of advantages and limitations. On the one hand, free flight provides a greater ecological relevance as mosquito flight dynamics and, by extension, optomotor feedbacks are more realistic. A range of aerodynamic conditions can be simulated and other sensory cues (e.g., warm objects, odor plumes) can be introduced in a controlled manner by using setups such as wind tunnels (van Breugel et al. 2015). However, the size and volume of the wind tunnel are often limiting factors because of its footprint and associated accessories that require a large-sized room. Because of their fast maneuvers, the three-dimensional video tracking of mosquitoes in free flight requires multiple cameras and a dedicated high-performance computer with ad hoc software (Straw et al. 2011). Another aspect to consider is that, in most cases, free-flight optomotor responses are bound to be analyzed a posteriori, from recorded trajectories, which limits options for closed-loop experiments (but see Straw et al. 2011).

On the other hand, tethered flight preparations are, by definition, restrained in their behavioral and biomechanical responses, and only a subset of variables can be extracted from the mosquito, as follows: changes in wing stroke frequency or amplitude; turning attempts; or head, leg, and abdomen movements. Relating changes in these variables to free-flight behaviors is not straightforward, as insects tend to overcompensate their actions during tethered flight to correct for the lack of motor and aerodynamic feedback (Vinauger et al. 2019). However, because tethered individuals are exposed to programmable visual displays, the subjective space in which they fly is virtually unlimited. This feature allows for longer flight durations, repeated presentations of stimuli, and closed-loop conditions where aspects of the virtual visual environment are under the control of the mosquito's flight dynamics (Vinauger et al. 2018, 2019; Liu and Vosshall 2019).

Vision is a critical modality for insect vectors and is integrally tied to other sensory modalities. Without vision, a mosquito's ability to fly, locate hosts, or find oviposition sites is severely affected. Moreover, vision is tied to olfaction. After encountering an odor, a mosquito uses its vision to stabilize its flight and move upwind toward the host (termed odor-conditioned optomotor anemotaxis) (van Breugel et al. 2015; Cardé 2021). It is hypothesized that if the visual object is attractive (appropriate color) and has attractive near-field cues (temperature and skin odors), mosquitoes may land on and bite the object's surface. Thus, mosquito behaviors are multimodal and integrally tied with vision. Nonetheless, how these different sensory cues interact to influence biting behavior remains unclear. Another critical gap is determining how the mosquito can identify a potential host in a complex visual scene and what visual features it uses (specific bands of the color spectrum, size, or intensity). Finally, how visual information is processed in the mosquito brain and integrated with other sensory cues remains an open question that requires new behavioral and electrophysiological tools.

TOOLS AVAILABLE FOR THE STUDY OF INSECT OPTOMOTOR RESPONSES

Here, we focus on systems that control the visual and sensory environment in tethered preparations and allow the repeatable presentation of stimuli under closed-loop control.

Projector-Based Systems

Beyond light-emitting diode (LED) panels for displaying panoramic visual stimuli (discussed in the subsection LED-Based Systems) or the older rotating drums (Hassenstein and Reichardt 1956), researchers have also used short-throw video projectors as a method to display complex scenes and visual stimuli in both open-loop and closed-loop experiments (Creamer et al. 2019; Palavalli-Nettimi and Theobald 2020; Rusch et al. 2021). Because of the flicker-fusion frequency of the insect visual system, high-frame-rate projectors (>360 Hz) are necessary to prevent motion artifacts and scene transitions that are invisible to the human eye but can be easily detected by the mosquito. Projectors can be used to create panoramic stimuli by projecting scenes onto spheres or mirrors, although image distortion must be considered. Some of the additional complications associated with using projectors include cost, footprint of the rig, the lag in update rates, and limitations in the visual spectrum (e.g., RGB). The mosquito can be tethered while surrounded by the screens or within a sphere, and its wingbeat dynamics and behavior can be read out via a computer vision system or a pair of photodiodes.

LED-Based Systems

Although limited in spatial resolution, LED-based systems offer some advantages over projector-based systems, as follows: (1) the narrower spectral tuning of LEDs and the subtle control over the brightness of each individual LED allow a greater control over the color and contrast of stimuli presented to the mosquitoes; (2) rapid refresh rates enable closed-loop experiments with minimal lag between the insect's behavior and the resulting visual feedback; and (3) because of their modular nature and small footprint, LED displays can be easily adapted for simultaneous tethered flight and brain imaging or electrophysiological recordings (Vinauger et al. 2019). A commonly used setup, originally developed for *Drosophila melanogaster* (Reiser and Dickinson 2008), has been successfully adapted to mosquito studies for both behavioral (Vinauger et al. 2018; Liu and Vosshall 2019) and neurobiological experiments (Vinauger et al. 2019). Other LED panels, including multicolored RGB boards, are commercially available and can be programmed with Raspberry Pi (Adafruit Industries) for additional portability and reduced cost. In the accompanying protocol (see Protocol 1: Tethered Preparation for the Analysis of Mosquito Visual-Motor Responses Using Modular Visual Displays [Vinauger and Riffell 2023]), we focus on the modular LED display (Reiser and Dickinson 2008) because of its ease of use and growing popularity among mosquito studies. This common platform for the study of mosquito visual (optomotor) responses allows comparisons with other fly species and across studies performed in different laboratories, hereby further supporting the reproducibility of the results.

REFERENCES

Alonso San Alberto D, Rusch C, Zhan Y, Straw AD, Montell C, Riffell JA. 2022. The olfactory gating of visual preferences to human skin and visible spectra in mosquitoes. *Nat Commun* **13:** 555. doi:10.1038/s41467-022-28195-x

Brett GA. 1938. On the relative attractiveness to *Aëdes aegypti* of certain coloured cloths. *Trans R Soc Trop Med Hyg* **32:** 113–124. doi:10.1016/S0035-9203(38)90101-4

Brown AWA. 1951. Studies of the responses of the female *Aëdes* mosquito. Part IV. Field experiments on Canadian species. *Bull Entomol Res* **42:** 575–582. doi:10.1017/S0007485300028984

Brown PK, White RH. 1972. Rhodopsin of the larval mosquito. *J Gen Physiol* **59:** 401–414. doi:10.1085/jgp.59.4.401

Cardé RT. 2021. Navigation along windborne plumes of pheromone and resource-linked odors. *Annu Rev Entomol* **66:** 317–336. doi:10.1146/annurev-ento-011019-024932

Creamer MS, Mano O, Tanaka R, Clark DA. 2019. A flexible geometry for panoramic visual and optogenetic stimulation during behavior and physiology. *J Neurosci Methods* **323:** 48–55. doi:10.1016/j.jneumeth.2019.05.005

Fay RW, Prince WH. 1968. A trap based on visual responses of adult mosquitoes. *Mosq News* **28:** 1–7.

Hassenstein B, Reichardt W. 1956. Systemtheoretische Analyse der Zeit-, Reihenfolgen- und Vorzeichenauswertung bei der Bewegungsperzeption des Rüsselkäfers Chlorophanus. *Z Naturforschung B* **11:** 513–524. doi:10.1515/znb-1956-9-1004

Hu X, Leming MT, Whaley MA, O'Tousa JE. 2014. Rhodopsin coexpression in UV photoreceptors of *Aedes aegypti* and *Anopheles gambiae* mosquitoes. *J Exp Biol* **217:** 1003–1008. doi:10.1242/jeb.096347

Kennedy JS. 1940. The visual responses of flying mosquitoes. *Proc Zool Soc Ser A* **109A:** 221–242. doi:10.1111/j.1096-3642.1940.tb00831.x

Liu MZ, Vosshall LB. 2019. General visual and contingent thermal cues interact to elicit attraction in female *Aedes aegypti* mosquitoes. *Curr Biol* **29:** 2250–2257.e4. doi:10.1016/j.cub.2019.06.001

Muir LE, Thorne MJ, Kay BH. 1992a. *Aedes aegypti* (Diptera: Culicidae) vision: spectral sensitivity and other perceptual parameters of the female eye. *J Med Entomol* **29:** 278–281. doi:10.1093/jmedent/29.2.278

Muir LE, Kay BH, Thorne MJ. 1992b. *Aedes aegypti* (Diptera: Culicidae) vision: response to stimuli from the optical environment. *J Med Entomol* **29:** 445–450. doi:10.1093/jmedent/29.3.445

Palavalli-Nettimi R, Theobald JC. 2020. Small eyes in dim light: implications to spatio-temporal visual abilities in *Drosophila melanogaster. Vision Res* **169:** 33–40. doi:10.1016/j.visres.2020.02.007

Reiser MB, Dickinson MH. 2008. A modular display system for insect behavioral neuroscience. *J Neurosci Methods* **167:** 127–139. doi:10.1016/j.jneumeth.2007.07.019

Rusch C, Alonso San Alberto D, Riffell JA. 2021. Visuo-motor feedback modulates neural activities in the medulla of the honeybee, *Apis mellifera. J Neurosci* **41:** 3192–3203. doi:10.1523/JNEUROSCI.1824-20.2021

Seldin EB, White RH, Brown PK. 1972. Spectral sensitivity of larval mosquito ocelli. *J Gen Physiol* **59:** 415–420. doi:10.1085/jgp.59.4.415

Smart MR, Brown AWA. 1956. Studies on the responses of the female *Aëdes* mosquito. Part VII. The effect of skin temperature, hue and moisture on the attractiveness of the human hand. *Bull Entomol Res* **47:** 89–100. doi:10.1017/S000748530004654X

Snow WF. 1971. The spectral sensitivity of *Aedes aegypti* (L.) at oviposition. *Bull Entomol Res* **60:** 683–696. doi:10.1017/S0007485300042437

Straw AD, Branson K, Neumann TR, Dickinson MH. 2011. Multi-camera real-time three-dimensional tracking of multiple flying animals. *J R Soc Interface* **8:** 395–409. doi:10.1098/rsif.2010.0230

van Breugel F, Riffell J, Fairhall A, Dickinson MH. 2015. Mosquitoes use vision to associate odor plumes with thermal targets. *Curr Biol* **25:** 2123–2129. doi:10.1016/j.cub.2015.06.046

Vinauger C, Riffell JA. 2023. Tethered preparation for the analysis of mosquito visual-motor responses using modular visual displays. *Cold Spring Harb Protoc* doi:10.1101/pdb.prot108179

Vinauger C, Lahondère C, Wolff GH, Locke LT, Liaw JE, Parrish JZ, Akbari OS, Dickinson MH, Riffell JA. 2018. Modulation of host learning in *Aedes aegypti* mosquitoes. *Curr Biol* **28:** 333–344.e8. doi:10.1016/j.cub.2017.12.015

Vinauger C, van Breugel F, Locke LT, Tobin KKS, Dickinson MH, Fairhall AL, Akbari OS, Riffell JA. 2019. Visual-olfactory integration in the human disease vector mosquito, *Aedes aegypti. Curr Biol* **29:** 2509–2516.e5. doi:10.1016/j.cub.2019.06.043

Zhan Y, San Alberto DA, Rusch C, Riffell JA, Montell C. 2021. Elimination of vision-guided target attraction in *Aedes aegypti* using CRISPR. *Curr Biol* **31:** 4180–4187. doi:10.1016/j.cub.2021.07.003

Cite this introduction as *Cold Spring Harb Protoc*; doi:10.1101/pdb.top107672

Protocol 1

Tethered Preparation for the Analysis of Mosquito Visual-Motor Responses Using Modular Visual Displays

Clément Vinauger[1,3] and Jeffrey A. Riffell[2]

[1]*Department of Biochemistry, Virginia Polytechnic Institute and State University, Blacksburg, Virginia 24061, USA;* [2]*Department of Biology, University of Washington, Seattle, Washington 98195, USA*

In the present protocol, we describe methods to assess mosquito visual-motor responses using the Reiser–Dickinson light-emitting diode (LED) panels arranged in a cylindrical arena and fixed-tethered preparations where the insect cannot adjust its orientation relative to the visual display. Variations around this approach might be better adapted for the specific requirements of each research project and must be considered by the investigators. Other types of displays may provide other stimulation possibilities (e.g., color range, refresh rate, field of view). Also, other types of preparations, such as rotating (magneto-tethered) preparations where the insect can rotate around a vertical axis and reorient relative to the visual display, may reveal other aspects of mosquito optomotor responses. Finally, the methods described here are applicable to multiple species and were used to produce data published previously using 6-d-old *Aedes aegypti* females.

MATERIALS

It is essential that you consult the appropriate Material Safety Data Sheets and your institution's Environmental Health and Safety Office for proper handling of equipment and hazardous materials used in this protocol.

Reagents

Glue (optional; see Step 4)
Ice
Mosquitoes, *Aedes aegypti*, 6-d-old female
UV-curing glue (e.g., Bondic)

Equipment

Air line (building air or medical air tank connected to an adjustable flowmeter [e.g., SP Bel-Hart, Riteflow H40407-0125]; optional; see Step 18)

CO_2, temperature, and humidity sensor

Any affordable air quality sensor can be used for this purpose (e.g., Amazon, ASIN: B01966BZDY).

Computer(s)

[3]Correspondence: vinauger@vt.edu

Cite this protocol as *Cold Spring Harb Protoc*; doi:10.1101/pdb.prot108179

One computer compatible with MATLAB is required to control the LED modular display. If the experimenters decide to rely on Kinefly or Flytrax to track the motion of the mosquito's wings, we recommend running Kinefly on a separate Linux-based computer.

Controller

The panels are controlled by a USB interface that can be acquired and assembled by following the instructions provided on the modular display repository (https://reiserlab.github.io/Modular-LED-Display/).

Cooling station and micromanipulator setup

The cooling station consists of a cooling platform aligned with a micromanipulator that allows the precise positioning of tethers relative to the thorax of mosquitoes. Aluminum blocks placed on ice can be used or, preferably, a temperature-controlled cooling stage (Duistermars and Frye 2008) with a set point at 5°C.

Data acquisition software

Multiple options are available, and the authors have used WinEDR (University of Strathclyde, Glasgow, Scotland; https://spider.science.strath.ac.uk/sipbs/software_ses.htm).

Data acquisition system

Multiple options are commercially available for data acquisition systems (e.g., National Instruments BNC-2099A [used by the authors] or Molecular Devices 1550B). Technical requirements are low noise, enough ports, and high sampling rates. The resulting data can be read and recorded from MATLAB (The MathWorks, Inc.) or other software systems (e.g., WinEDR, Strathclyde University; or Axon PCLAMP Software Suite, Molecular Devices).

Experiment script

The experiment script will be designed to answer the biological question of interest, and although there are scripts of previously published work available on GitHub (described below in the MATLAB entry), there is no universal script that one can download and use. We recommend, however, to consider the following when designing your experiment: (1) allow for a sufficient number of stimuli presentations and sufficient numbers of conditions and controls to capture intraindividual variability; (2) allow time between discreet stimuli for mosquitoes to return to their baseline wingbeat kinematics; and (3) avoid long experiments as this will limit your ability to test multiple individuals within mosquitoes' peak activity time. Finally, note that depending on the nature of the experiment, it may become difficult to maintain mosquitoes in flight for long periods of time (e.g., if continuously stimulated with aversive looming stimuli). As a benchmark, in tasks of bar fixation assays, mosquitoes sustained flight for ~45 min.

Fixed tethers and tether holders (see Steps 1–4)

- Aluminum rods (e.g., MacMaster Carr 2655N11)
- Connectors (TE Connectivity 205089-1)
- Minutien pins (optional, see Step 4)
- Solder gun
- Soldering wire
- Tungsten rod (254-µm-diameter, one cut to a length of 2.5 cm; A-M Systems 717000)

Forceps

Kimwipes

LED displays

Modular LED displays can be set up following the instructions available at https://reiserlab.github.io/Modular-LED-Display/. At the time this protocol was written, no preassembled and preprogrammed system (i.e., "off-the-shelf" solution) was commercially available. To acquire a new setup, one must therefore follow instructions provided on the modular display repository (https://reiserlab.github.io/Modular-LED-Display/G4/). Although originally designed for work on fruit flies, the version described on the repository is usable with mosquitoes.

The 8 × 8 LED panels (40-mm × 40 mm) can be fitted with LEDs of various emission wavelengths to stimulate with different colors. The latest versions of the panels can be fitted with two different color LEDs for multicolor stimuli presentations. An estimate of the total cost of a full system (LED arena + controller) is $2000. We also estimate a full day of work by someone with a basic knowledge of MATLAB and programming to assemble a complete rig.

MATLAB

Although the modular LED display can be controlled by custom-made codes and scripts, the most straightforward way to get started with the LED arena is to use the MATLAB code and start from example patterns and functions made available by the Reiser lab (https://github.com/reiserlab/Panel-G3-Software).

Mosquito container (see Step 7)

Plastic syringe (e.g., 20- to 50-mL)

Cite this protocol as *Cold Spring Harb Protoc*; doi:10.1101/pdb.prot108179

Temporary container (e.g., an empty pipette tip box in which either a piece of Styrofoam, a band of Play-Doh, or dental wax has been placed; simply stick the tether pins into it; for use in Step 14)

UV light source

An affordable solution is the Bondic starter kit (Amazon ASIN: B01MR7J5ZX), which provides UV-curing glue and a UV LED light.

Wing-tracking system

To record the optomotor response of mosquitoes, the motion of the wings needs to be captured. Key parameters are the amplitude and frequency of wing strokes, as those variables can be used to estimate forces exerted by wings and to quantify changes in the fictive flight dynamics of mosquitoes (e.g., behaviors that could be analogous to surges, or saccades, although in a tethered preparation the behavior can be quite different from free-flight responses). Multiple options can be used for this purpose, including analog optical tracking solutions (Lehmann and Dickinson 1997). Computer vision tools (e.g., Kinefly for fixed tethers or Flytrax for rotating tethers) can also be used, although the specifications of the camera (such as the frame rate) must be considered to accurately quantify the wingbeat frequency. The tracked variables can be either recorded digitally or converted to analog voltage signals (e.g., PhidgetAnalog 4-Output, 1002_0B, Phidget) that can be fed back to the display controller to allow for closed-loop experiments where aspects of the visual display are modulated by the behavior of the insect. These options are equally capable, and the choice for one over the other should be informed by the biological question being asked. However, although camera-based systems may be more cost efficient than optical solutions to track wing stroke amplitude (e.g., the wingbeat analyzer used in Lehmann and Dickinson, 1997), capturing mosquitoes' high wingbeat frequency (400–500 Hz) with a camera will be costly and computationally intensive. If frequency is an important parameter for your specific experiment, opting for an optical tachometer should be considered an alternative.

XYZ linear stages (Newport Corporation MT-XYZ)

METHOD

Before mosquito optomotor responses are assessed, individual mosquitoes must be prepared and isolated for tethering. The choice of the sex, age, or physiological state (e.g., mated vs. virgins, fed vs. starved) of the individuals to be tested should be guided by the biological question(s) asked by the experimenters. Similarly, external factors such as time of day, room temperature, and relative humidity will affect mosquito flight and must be determined according to the biological question being investigated. As an example, previous work using tethered-flight preparations used nongravid female Ae. aegypti *between 5 and 7 d postemergence. These mosquitoes were maintained on a 10% sucrose diet until the day of the experiments and were starved for 6 h before starting the experiment at a time corresponding to the peak in spontaneous locomotor activity for this species (Zeitgeber times: 10–12).*

Set Up the Equipment

Prepare Fixed Tethers and Tether Holders

1. Insert a 254-µm-diameter tungsten rod cut at a length of 2.5 cm into a male connector pin and solder it with a soldering gun in a vertical position using melted soldering wire.
2. To hold the tethers during the tethering of mosquitoes and to maintain them in position in the arena, prepare two tether holders using the female connector pin soldered into 1/4-inch aluminum rods.
3. Use XYZ linear stages as tether holders.

   These items are easier to manipulate and position when held in three-axis micromanipulators.
4. (Optional) For magnetic tether experiments (*sensu* Bender and Dickinson 2006), glue the mosquito to minutien pins.

   The magnets and assembly of the rotating preparation are described in Duistermars and Frye (2008) and Bender and Dickinson (2006).

Environmental Monitoring

5. Deploy the CO_2, temperature, and humidity sensor near the arena.

6. Acquire the resulting data streams during the experiment.

   This step ensures that background CO_2 or other environmental variables (temperature and humidity) do not affect the experiments.

Cold-Anesthetize Mosquitoes

7. Switch on the cooling station and place a mosquito container (e.g., cup or dish used to isolate females of the desired age and physiological condition) on ice while the platform reaches the temperature setpoint (5°C).
8. Once mosquitoes are immobilized, use forceps to transfer one mosquito onto the cooling platform.

   To reduce the accumulation of H_2O on the mosquito wings due to potential condensation on the cooling station, wipe the surface of the cooling block with a Kimwipe before transferring the mosquito.

9. Position the mosquito on its side, close to the edge of the platform and at an angle of 60° from the side of the platform (i.e., 150° from the axis of the tether; Fig. 1).

   The thorax and wings of the mosquito should be facing the experimenter. Note that this angle is well-adapted for Ae. aegypti *females but may need to be adjusted when other species or males are used.*

Tether Mosquitoes

10. Place a tether on the tether holder and use the micromanipulator to get within 5 mm of the thorax.
11. Apply a small amount of UV-curing glue to the tip of the tether using a tungsten rod.
12. Using the micromanipulator, bring the tether as close to the mosquito as necessary for the droplet of glue to contact the thorax. Let the glue spread to the back of the thorax by capillary action.
13. Shine a UV light on the glue for ~15 sec on each side.
14. Let mosquitoes recover from cold anesthesia. Move tethered mosquitoes to a temporary container where they can be held upside down until the time of the experiment.

    Positioning them in the same order as they were tethered allows consistency in the recovery time across mosquitoes. A recovery time of 30–60 min has led to robust sustained flight bouts in our studies.

15. Tether 15–25 mosquitoes by repeating Steps 7–14.

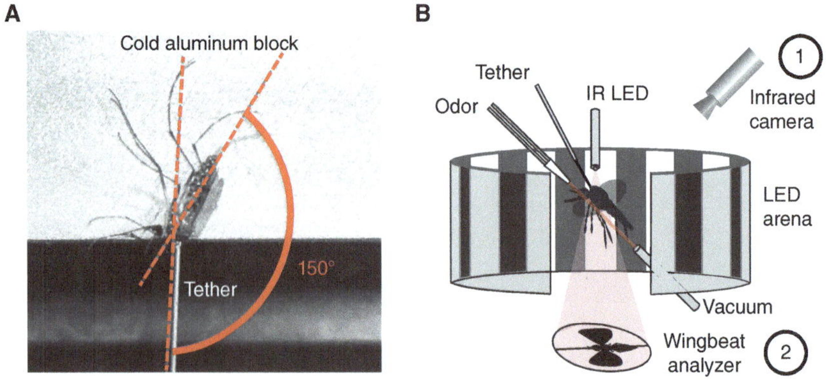

FIGURE 1. (*A*) Illustration of the tethering position and angle of an adult *Aedes aegypti* female. The mosquito is cold-anesthetized and maintained on a cold aluminum block during tethering. Note that the tungsten rod (tether) and the mosquito's main body axis are on the same plane. (*B*) Schematic of the modular LED display with a mosquito (not to scale) positioned at the center of a cylinder of LED panels (adapted from Vinauger et al. 2019). This setup can be used with machine vision solutions using infrared sensitive cameras (1) or an infrared (IR) emitter can be positioned above the mosquito to cast the shadows of its wings onto optical sensors (2).

Prepare the Arena

16. Switch on the computer and start MATLAB.
17. Switch on the LED display. Once the panels' address numbers are all displayed, switch on the arena controller and start PControl, the MATLAB tool that allows communication between the display system and MATLAB (https://reiserlab.github.io/Modular-LED-Display/G4/#pcontrol). Verify that PControl can access the controller's SD card and displays "Current working mode: Controller mode."
18. (Optional) To increase the mosquito's motivation to fly, add an air line.

    This step is particularly important in situations where experimenters want to combine olfactory and visual stimulation (see Vinauger et al. 2019). In such cases, a vacuum line (e.g., from the building vacuum or from a vacuum pump) should be positioned diametrically opposed to the airflow inlet to ensure that odorants do not accumulate in the arena. If such a setup is in place, open the air flow and vacuum lines to generate a bulk air flow of about 50 cm sec^{-1}.
19. Position one mosquito in at the center of the arena, by inserting its tether in the tether holder.

    Start with the first mosquito that was tethered and follow the tethering order.

    It is critical that the mosquito is positioned at the center of the LED drum, at midheight of the LED panels, and at a body angle of about 30° (pitch). The plane in which wings beat (roll) should be parallel to the ground, and when looked at from above, the mosquito's main body axis should be pointing forward (i.e., a yaw angle of 0°).

    Verify that the mosquito is well-positioned in between the air inlet and the vacuum lines, which should be aligned with each other.
20. Adjust the wing-tracking system.

    Regardless of whether an optical (e.g., tachometer or wingbeat analyzer) or a camera-based (Kinefly) system is used, adjust the mosquito position via the micromanipulator so that the average amplitudes of the left and right wing are approximately equal during flight. This step is critical, and if there is any pronounced asymmetry, then the mosquito should be replaced. During this point, a stripe fixation task or a closed-loop starfield pattern can help keep the mosquito engaged and flying forward.

Conduct and Stop the Experiment

21. Once the mosquito's position is adjusted, start recording the data with the National Instruments BNC-2090A system in WinEDR and launch the experiment script you designed.

    If a mosquito stops flying before the experiment can be started, deliver a gentle puff of air using a plastic syringe. If the mosquito does not sustain flight before the experiment is started, swap with another mosquito, following the sequence in which they have been tethered.
22. Once the experiment script has run to completion, stop the experiment and stop the recording. Check that a new filename is entered in WinEDR.
23. Repeat Steps 16–22 with the next mosquito until all mosquitoes have been tested.
24. Use the PControl or the command line to turn all panels off and switch off the controller.
25. Switch off the LED arena. Close PControl and exit MATLAB.
26. In WinEDR, export the saved data by using File > Export > ASCII text.
27. Import the .txt files in Python or R for subsequent analysis.

    The data are formatted as one column per recorded variable, and the first column indicates the recording time. If using MATLAB, a custom script is available (https://gist.github.com/ptweir/3984712) to load the .edr files directly into MATLAB.

DISCUSSION

Mosquitoes rely on multimodal sensory inputs to navigate in their environment and locate their hosts. Olfactory cues play an important role in gating (van Breugel et al. 2015) or modulating (Vinauger et al.

2019) their responses to visual objects. As such, the uncontrolled presence of odor contamination and dynamic changes in environmental humidity and temperature can all dramatically affect the outcome of the experiments. A slight contamination from host odors can modify visual responses. Investigators should always be monitoring and reporting these environmental variables. Finally, the interindividual variability in mosquitoes' flight kinematics and optomotor responses seem to be higher than those in *Drosophila*. To satisfy statistical power requirements, larger sample sizes must therefore be anticipated. As an example, using means and standard deviations observed in Vinauger et al. (2019) and two-sample t-test power calculations result in 24 mosquitoes being required in each group to reach a statistical power of 0.8.

ACKNOWLEDGMENTS

We thank Drs. Peter Weir, Floris van Breugel, Michael Dickinson, and Michael Reiser for their help adapting the LED arena to mosquitoes and for fruitful discussions on insect vision and neuroethology. Support for our laboratory was funded by Air Force Office of Scientific Research under grant FA9550-20-1-0422 (J.A.R.), the National Institutes of Health under grants R01-AI148300 (J.A.R.) and R21-AI137947 (J.A.R.), and the USDA National Institute of Food and Agriculture Hatch project 1017860 (C.V.).

REFERENCES

Bender JA, Dickinson MH. 2006. Visual stimulation of saccades in magnetically tethered *Drosophila*. *J Exp Biol* **209:** 3170–3182. doi:10.1242/jeb.02369

Duistermars BJ, Frye M. 2008. A magnetic tether system to investigate visual and olfactory mediated flight control in *Drosophila*. *J Vis Exp* doi:10.3791/1063

LED display. https://reiserlab.github.io/Modular-LED-Display/G4/ [accessed September 19, 2022].

Lehmann FO, Dickinson MH. 1997. The changes in power requirements and muscle efficiency during elevated force production in the fruit fly *Drosophila melanogaster*. *J Exp Biol* **200:** 1133–1143. doi:10.1242/jeb.200.7.1133

MATLAB code. https://github.com/reiserlab/Panel-G3-Software [accessed September 19, 2022].

Modular display repository. https://reiserlab.github.io/Modular-LED-Display/ [accessed September 19, 2022].

PControl. https://reiserlab.github.io/Modular-LED-Display/G4/#pcontrol [accessed September 19, 2022].

Reiser MB, Dickinson MH. 2008. A modular display system for insect behavioral neuroscience. *J Neurosci Methods* **167:** 127–139. doi:10.1016/j.jneumeth.2007.07.019

Script for loading .edr files into MATLAB. https://gist.github.com/ptweir/3984712 [accessed September 19, 2022].

van Breugel F, Riffell J, Fairhall A, Dickinson MH. 2015. Mosquitoes use vision to associate odor plumes with thermal targets. *Curr Biol* **25:** 2123–2129. doi:10.1016/j.cub.2015.06.046

Vinauger C, Lahondère C, Wolff GH, Locke LT, Liaw JE, Parrish JZ, Akbari OS, Dickinson MH, Riffell JA. 2018. Modulation of host learning in *Aedes aegypti* mosquitoes. *Curr Biol* **28:** 333–344.e8. doi:10.1016/j.cub.2017.12.015

Vinauger C, van Breugel F, Locke LT, Tobin KKS, Dickinson MH, Fairhall AL, Akbari OS, Riffell JA. 2019. Visual-olfactory integration in the human disease vector mosquito, *Aedes aegypti*. *Curr Biol* **29:** 2509–2516.e5. doi:10.1016/j.cub.2019.06.043

WinEDR. University of Strathclyde, Glasgow, Scotland; https://spider.science.strath.ac.uk/sipbs/software_ses.htm [Accessed September 22, 2022].

 Cite this protocol as *Cold Spring Harb Protoc*; doi:10.1101/pdb.prot108179

CHAPTER 21

Experiments and Analysis of Mosquito Flight Behaviors in a Wind Tunnel: An Introduction

Diego Alonso San Alberto,[1] Claire Rusch,[1] and Jeffrey A. Riffell[1,2]

[1]*Department of Biology, University of Washington, Seattle, Washington 98195, USA*

Mosquitoes detect and navigate to important resources, like a host, using combinations of different sensory stimuli. The relative importance of the sensory cues can change as the mosquito gets closer to their target. Other factors, both internal and external, can also influence the mosquito behavior. A mechanistic understanding of these sensory stimuli, and how they impact mosquito navigation, can now be readily studied using wind tunnels and associated computer vision systems. In this introduction, we present a behavioral paradigm using a wind tunnel for flight behavior analysis. The wind tunnel's large size with its associated cameras and software system for analysis of the mosquito flight tracks can be sophisticated and sometimes cost-prohibitive. Nevertheless, the wind tunnel's flexibility in allowing the testing of multimodal stimuli and scaling of environmental stimuli makes it possible to reproduce conditions from the field and test them in the laboratory, while also allowing the observation of natural flight kinematics.

INTRODUCTION

Mosquitoes navigate to hosts, sources of nectar, and oviposition sites and locate mates using multiple sensory inputs during flight (McMeniman et al. 2014; van Breugel et al. 2015; Montell and Zwiebel 2016; Peach et al. 2019). For instance, CO_2 and host odors can mediate long-distance flight (>10 m), whereas at intermediate distances (2–5 m) the integration of odor and visual cues allows the mosquito to locate visual objects that may appear like a host and initiate landing and biting based on the presence of heat, humidity, gustatory and olfactory cues. The integration of these sensory streams also provides the mosquito a robust ability to navigate within an uncertain and time-varying environment, where perturbations from wind gusts or moving and defensive hosts (to name a few examples) must be rapidly processed during flight. In the field, mosquitoes will often navigate over 10 sec of meters distance. Under these conditions the aerodynamic environment can be quite complex; for instance, turbulent air motion causes an odor plume to grow and meander. Can all aerodynamic conditions be replicated in the laboratory? Laboratory wind tunnels provide a controlled environment to test different environmental and experimental conditions, including those that mosquitoes experience in the natural environment. Wind tunnels can simulate the physical environment, such as the wind, temperature, and humidity, as well as the chemical and biological environment. Unfortunately, the size of a typical wind tunnel (~0.25–1 m height and width; and 1–5 m length) (Visser 1976; Miller and Roelofs 1978) is much smaller than the turbulent eddies in the field (1–100 m); therefore, an odor

[2]Correspondence: jriffell@uw.edu

Cite this introduction as *Cold Spring Harb Protoc*; doi:10.1101/pdb.top107674

plume in a wind tunnel may not exhibit the exact characteristics seen in the field. However, because of modern wind tunnel design, it is now possible to recreate specific physical environmental conditions, including turbulent fluctuations, allowing an experimenter to test specific sensory and physical features in a reproducible manner. Wind tunnel design characteristics, in concert with computer vision methods for the tracking of mosquito flight, allow fine-scale control and evaluation of mosquito behavior under simulated natural conditions.

Wind Tunnels and Mosquito Behavior

Historically, low-speed wind tunnels have been used with a variety of insect species to examine their odor tracking behaviors, but some of the earliest such experiments involved the optomotor anemotaxis responses of mosquitoes, a form of navigation that involves visual and wind cues to orient and move toward a goal (Kennedy 1939). As mentioned, wind tunnels cannot replicate the turbulent airflow characteristics and low odor concentrations of typical odor plumes from long distances away (>100 m) (Visser 1976). They can, however, be used to simulate the physical environmental characteristics of the field (airflow, temperature, humidity) and experimentally manipulate them one at a time over long periods. Moreover, sustained flights from distances >2 m can be replicated in the wind tunnel by simulating optic flow with a moving floor pattern beneath the mosquitoes as they fly. This method was used by Kennedy (1939) to examine visual feedback by mosquitoes while orienting to host odors. Since that time, wind tunnels have been designed to examine mosquito attraction behavior (Kellogg and Wright 1962; Colvin et al. 1989; Lacey and Cardé 2011) or the effects of repellents on flight trajectories (Wright and Rayner 1960).

There are several design features of wind tunnels that enable them to provide fluid dynamical conditions that can be replicated between laboratories. The first is a contract zone for air once entering the tunnel intake. This streamlines and forces the air to move downwind and prevents any physical perturbations within the working section (e.g., jet actuators) from moving upwind. Air movement through honeycombs and screens further minimizes any turbulent fluctuations and allows laminar flow within the tunnel. Although mosquitoes and other insects track turbulent odor plumes (Cardé and Willis 2008; van Breugel et al. 2015), laminar airflow within the working section allows the experimenter to simulate the turbulent statistics of the flow to recreate the desired field conditions. For instance, using grid-generated turbulence, cylinder-generated vortices, or other techniques allows a controlled method for creating the dominant timescales and length scales of turbulent momentum fluctuations within the tunnel. Wind tunnels also allow for fine-scale climate control and air filtration within the working section. This provides reproducible environmental conditions while minimizing the potential for odor contamination (Hinze et al. 2021).

Sensory Environment and the Tracking of Mosquito Flight Behavior

The tracking of mosquito behavior in the tunnel has allowed researchers better understanding of the sensory stimuli used by navigating mosquitoes and how the stimuli are integrated. Real-time, low-latency tracking permits the testing of different stimuli under automated control based on the mosquito flight behavior (closed-loop) or sequentially through time (open-loop). For example, similar to rotating floor patterns used by Kennedy and others (Kennedy 1939; Baker and Linn 1984), short-throw projectors can project vertical stripes or more complex scenes onto the floor and sides of the tunnel. These can be used under closed-loop conditions based on the mosquito position. Computer controlled mass flow controllers and solenoid systems can create dynamic olfactory plumes within the working section (van Breugel et al. 2015), and pistons can create gusts of wind to perturb the flight trajectories (Fuller et al. 2014).

Using these systems within a working section of a wind tunnel, researchers have begun to examine the principles by which mosquitoes respond to different odors during flight (Dekker et al. 2005), and how the optic flow environment and airspeed affect mosquito flight and navigation (Kennedy 1939; van Breugel et al. 2015). Moreover, the controlled environment within the working section allows

 Cite this introduction as *Cold Spring Harb Protoc*; doi:10.1101/pdb.top107674

extreme reproducibility between experiments. This is an important point, especially for comparative studies testing different mosquito species or subtle differences in an odor (Hinze et al. 2021).

Another important feature is the lighting within the tunnel and the camera systems and associated software for recording the mosquito behavior. Infrared (IR) LEDs provide lighting that reflects from the body and wings of the flying mosquitoes, and multiple cameras (>60 frames per seconds [fps]) allow tracking and digitizing of the 3D flight trajectories. The video recordings can be synchronized and analyzed offline or analyzed in real time via automated tracking software.

Tools Available for the Study of Mosquito Responses

Here, we focus on wind tunnel–like systems that allow control of the sensory environment for the mosquito in free-flight.

Y-Tube and Multiport Olfactometers

Beyond the wind tunnels described above and discussed below, there are several other types of chambers with flowing air (olfactometers) that we and others have used to characterize mosquito behaviors. The strength of these systems is that the limited footprint of the olfactometers makes it accessible and easy for setup and running of the experiments. Moreover, Y-mazes have the advantage of allowing the presentation of non-olfactory stimuli (i.e., tactile, visual) in a sequential manner (entrance vs. arm). Y-mazes permit an increased probability of encounter with the stimuli presented as they are smaller in size. On the other hand, wind tunnels provide a more naturalistic environment where the animal can freely explore a bigger environment and the probability of encountering the boundary of the enclosure is reduced in comparison with the Y-maze. Another advantage of wind tunnels is the possibility to release and track groups of animals.

However, in the case of the "Y-tube" olfactometer, the region where the two tubes join causes a difference in pressure and flow disturbance that can impact the odor plume and the mosquito flight. Other olfactometers that use a single tube can limit the flight kinematics of the mosquito, which can impact the behavioral response (van Breugel et al. 2015; Vinauger et al. 2019; Amos et al. 2021).

Multiport olfactometers are another system to examine mosquito behavior. In multichamber olfactometers, a cage containing the mosquitoes is connected to ports that can emit the stimuli while trapping the attracted mosquitoes (Posey et al. 1998; McBride et al. 2014; Amos et al. 2021). The strength of this system is its ease of setup and trapping of mosquitoes within the odor ports. However, care must be taken using these olfactometers. Multiport olfactometers have limited and uncharacterized airflow where the odor plume and turbulent dynamics within the cage often are uncharacterized, making it difficult to accurately replicate the conditions between laboratories. Thus, subtle design changes in the olfactometer have unintended consequences for behavior.

Classic Wind Tunnel Systems

It should be noted that in contrast with the Y-tube and other olfactometers already described, wind tunnels often take up a much larger footprint than olfactometers, and with that larger volume they require additional cameras and software calibration to characterize flight behaviors. Nonetheless, wind tunnels provide an intermediate between bioassays in arenas or small cages and those that occur in the field. Generally, the working section of the low-speed wind tunnel should be ~0.25–1 m in height and width, and 1–2 m in length, and capable of air velocities of 0.1 to 2 m/sec. These dimensions still enable the mosquito to exhibit a crosswind search behavior (called "casting") and enable characterization of its upwind odor-guided behavior (called "surges"). During odor-guided flight, mosquitoes navigate robustly in low air velocities (0.2–0.5 m/sec) (Clements 1999; van Breugel et al. 2015). In our associated protocol, we focus on the wind tunnel because similar designs have been used in a growing number of studies on mosquito behavior (see Protocol 1: Conducting an Analysis of Mosquito Flight Behaviors in a Wind Tunnel [Alonso San Alberto et al. 2023]). The advantage of this system lies in its ability to simulate environmental characteristics and, knowing the turbulent conditions within the working section, allowing comparisons between other mosquito species and results from other laboratories. Moreover, similar wind tunnels, both in dimensions and qualities, can be used for a wide variety of insect species (e.g., *Manduca sexta*, Bomphrey et al. 2005; bees, Crall et al. 2017; *Drosophila*, van Breugel et al. 2018). Importantly, the wind tunnel airflow speed and the ambient illumination may need to be adjusted to the species-specific preference.

REFERENCES

Amos BA, Hoffmann AA, Staunton KM, Lau MJ, Burkot TR, Ross PA. 2022. Long-range but not short-range attraction of male *Aedes aegypti* (Diptera: Culicidae) mosquitoes to humans. *J Med Entomol* **59:** 83–88.

Alonso San Alberto D, Rusch C, Riffell JA. 2024. Conducting and analysis of mosquito flight behaviors in a wind tunnel. *Cold Spring Harb Protoc* doi:10.1101/pdb.prot108257

Baker T, Linn C. 1984. Wind tunnels in pheromone research. In *Techniques in pheromone research*, pp. 75–110. Springer, New York.

Bomphrey RJ, Lawson NJ, Harding NJ, Taylor GK, Thomas AL. 2005. The aerodynamics of *Manduca sexta*: digital particle image velocimetry analysis of the leading-edge vortex. *J Exp Biol* **208:** 1079–1094.

Cardé RT, Willis MA. 2008. Navigational strategies used by insects to find distant, wind-borne sources of odor. *J Chem Ecol* **34:** 854–866.

Clements AN. 1999. *The biology of mosquitoes. Volume 2: sensory reception and behaviour.* CABI, Wallingford, United Kingdom.

Colvin J, Brady J, Gibson G. 1989. Visually-guided, upwind turning behaviour of free-flying tsetse flies in odour-laden wind: a wind-tunnel study. *Physiol Entomol* **14:** 31–39.

Crall JD, Chang JJ, Oppenheimer RL, Combes SA. 2017. Foraging in an unsteady world: bumblebee flight performance in field-realistic turbulence. *Interface Focus* **7:** 20160086.

Dekker T, Geier M, Cardé RT. 2005. Carbon dioxide instantly sensitizes female yellow fever mosquitoes to human skin odours. *J Exp Biol* **208:** 2963–2972.

Fuller SB, Straw AD, Peek MY, Murray RM, Dickinson MH. 2014. Flying *Drosophila* stabilize their vision-based velocity controller by sensing wind with their antennae. *Proc Natl Acad Sci* **111:** E1182–E1191.

Hinze A, Lantz J, Hill SR, Ignell R. 2021. Mosquito host seeking in 3D using a versatile climate-controlled wind tunnel system. *Front Behav Neurosci* **15:** 33.

Kellogg F, Wright R. 1962. The guidance of flying insects. V. Mosquito attraction. *Can Entomol* **94:** 1009–1016.

Kennedy JS. 1939. The visual responses of flying mosquitoes. *Proc Zool Soc Lond A* **109:** 221–242.

Lacey E, Cardé R. 2011. Activation, orientation and landing of female *Culex quinquefasciatus* in response to carbon dioxide and odour from human feet: 3D flight analysis in a wind tunnel. *Med Vet Entomol* **25:** 94–103.

McBride CS, Baier F, Omondi AB, Spitzer SA, Lutomiah J,, Sang R, Ignell R, Vosshall LB. 2014. Evolution of mosquito preference for humans linked to an odorant receptor. *Nature* **515:** 222–227.

McMeniman CJ, Corfas RA, Matthews BJ, Ritchie SA, Vosshall LB. 2014. Multimodal integration of carbon dioxide and other sensory cues drives mosquito attraction to humans. *Cell* **156:** 1060–1071.

Miller J, Roelofs W. 1978. Sustained-flight tunnel for measuring insect responses to wind-borne sex pheromones. *J Chem Ecol* **4:** 187–198.

Montell C, Zwiebel L. 2016. Mosquito sensory systems. *Adv Insect Physiol* **51:** 293–328.

Peach DA, Ko E, Blake AJ, Gries G. 2019. Ultraviolet inflorescence cues enhance attractiveness of inflorescence odour to *Culex pipiens* mosquitoes. *PLoS ONE* **14:** e0217484.

Posey KH, Barnard DR, Schreck AC. 1998. Triple cage olfactometer for evaluating mosquito (Diptera: Culicidae) attraction responses. *J Med Entomol* **35:** 330–334.

van Breugel F, Riffell JA, Fairhall A, Dickinson MH. 2015. Mosquitoes use vision to associate odor plumes with thermal targets. *Curr Biol* **25:** 2123–2129.

van Breugel F, Huda A, Dickinson MH. 2018. Distinct activity-gated pathways mediate attraction and aversion to CO_2 in *Drosophila*. *Nature* **564:** 420–424.

Vinauger C, Van Breugel F, Locke LT, Tobin KK, Dickinson MH, Fairhall AL, Akbari OS, Riffell JA. 2019. Visual-olfactory integration in the human disease vector mosquito *Aedes aegypti*. *Curr Biol* **29:** 2509–2516.

Visser J. 1976. The design of a low-speed wind tunnel as an instrument for the study of olfactory orientation in the Colorado beetle (*Leptinotarsa decemlineata*). *Entomol Exp Appl* **20:** 275–288.

Wright R, Rayner H. 1960. The olfactory guidance of flying insects. II. Mosquito repulsion. *Can Entomol* **92:** 812–817.

Cite this introduction as *Cold Spring Harb Protoc*; doi:10.1101/pdb.top107674

Protocol 1

Conducting an Analysis of Mosquito Flight Behaviors in a Wind Tunnel

Diego Alonso San Alberto,[1] Claire Rusch,[1] and Jeffrey A. Riffell[1,2]

[1]*Department of Biology, University of Washington, Seattle, Washington 98195, USA*

Prior to conducting wind tunnel experiments, mosquitoes must be prepared for testing. Important factors and state-dependent processes of the mosquito—like the sex, age, infection status, reproductive status, or nutritional status—should be evaluated and motivated by questions and hypotheses one seeks to address. Other critical external factors that can impact the mosquitoes' behavior and should be controlled for both in the colony and in the room where the wind tunnel experiments take place include the circadian rhythm, room temperature, light intensity, and relative humidity. Together, the internal and external factors, and wind tunnel design, ultimately control the behavior of the mosquito and, hence, the success of the experiments. In the present protocol, we describe methods using a standard wind tunnel design in which the fan pulls the air through the working section of the wind tunnel and the mosquito behavior is recorded by a multicamera system. Variations around the camera tracking system can be adapted according to the research questions being asked and include real-time tracking for both closed-loop and open-loop control of the stimulus environment or recording of the videos for off-line digitization and analysis. Within the working section, the sensory environment (odor, visual, wind) can be controlled to test the mosquito responses to different stimuli, and below we include different equipment and tools for modifying the stimuli the mosquito experiences during flight. Finally, the methods described here are applicable to multiple mosquito species, although the experiment parameters may need to be changed (e.g., ambient luminosity).

MATERIALS

It is essential that you consult the appropriate Material Safety Data Sheets and your institution's Environmental Health and Safety Office for proper handling of equipment and hazardous materials used in this protocol.

Reagents

Aedes aegypti mosquitoes

Allow 6 d after emergence for mating to occur for the study of host-seeking behavior in females (Vinauger et al. 2014).

Equipment

Cameras and lighting for 3D tracking of mosquito flight

Infrared (IR)-sensitive cameras should be used to record the mosquito trajectories (e.g., Basler AC640gm). To minimize the effect of other lighting on mosquito tracking, cameras should include opaque IR filters (e.g., Kodak

[2]Correspondence: jriffell@uw.edu

Cite this protocol as *Cold Spring Harb Protoc*; doi:10.1101/pdb.prot108257

89B). IR backlights (e.g., LedLightsWorld HK-F3528IR30-X) should be installed below and at the sides of the wind tunnel to provide constant illumination beyond the visual sensitivity of the mosquitoes. Rear projection screens that allow IR transmission are placed along the sides and bottom of the tunnel, allowing visual scenes and patterns from the short-throw projectors to be projected into the working section.

CO_2 Flypad (optional; see Step 1)

Cotton pad

Environmental chamber

For Ae. aegypti *mosquitoes, maintain at 25°C, >70% relative humidity (RH), and under a 12 h–12 h light–dark cycle.*

Fishing line

Mass flow controllers

Use two mass flow controllers (MFCs; e.g., Alicat Scientific MC-200SCCM-D) to control odor input into the working section: one 200-sccm (standard cubic cm per min) MFC to provide filtered air into the odor outlet, and the other 100-sccm for the odor input (CO_2, or similar).

Petri dish (e.g., 90-mm-diameter, 14.2-mm-height)

Programming language for analysis (e.g., Python or Matlab)

Small acrylic box with the top covered by mesh (rearing container)

Small acrylic box with a sliding lateral door that can be connected to a fishing line

This is the testing container. The size should be adjusted to the size of the mosquito group used—for example, for 50+ individuals, 10-cm x 10-cm x 10-cm.

Software for data acquisition and tracking for multicamera systems (including commercial and open-source software; e.g., Braid [see https://github.com/strawlab])

Short-throw projectors for showing optic flow patterns or visual stimuli

High intensity (>500 lux) short-throw projectors with fast refresh rates (>100 frames per sec) are ideal for both open-loop and closed-loop experiments and provide bright patterns that project on the screens (e.g., LG PH450U). Additionally, visual cues can be placed on the floor of the wind tunnel to explore visual search behaviors.

Stereomicroscope

Wind tunnel

There are many different design options for wind tunnels, including those that are custom-built and commercially available (e.g., low-speed tunnel, ELD Inc.) (Alonso San Alberto et al. 2022) (Fig. 1). Consider several key design principles: (1) The air intake system and exhaust system should be evaluated to ensure that clean air reaches the working section, and that odor-contaminated air from the tunnel does not recirculate in the room. (2) The aerodynamic conditions within the tunnel are well-characterized and correspond with the optimal flight behaviors of the mosquito and the turbulent statistics that will impact mosquito flight and the odor plume. (3) The wind tunnel intake should go through a 6:1 contraction zone and honeycombs and screens for the airflow entering the working section to be laminar. Once entering the working section, the experimenter can use grids, cylinders, or actuators to create realistic turbulent intensities and coefficients to simulate field conditions. An alternate method to control the temporal dynamics of an odor include using a solenoid or MFC. (4) The working section should be made from clear acrylic with a length that is three-times or more the width and height of the tunnel.

METHOD

During the entirety of this experiment, it is critical that the mosquitoes do not experience any odor or CO_2 from the experimenter or the environment. Accidental odor exposure greatly impacts mosquito behavior and lasts ~10 min or more in duration.

1. One day before the experiment, cold-anesthetize mosquitoes either by placing their rearing container on ice or in a 5°C cold room.

   Typically, experiments are performed with group sizes of 50 individuals. Larger group size may impact the quality of the tracking.

2. After 3 min, transfer the mosquitoes by turning the container over a Petri dish under a dissecting microscope and sort according to sex. Place sorted mosquitoes within a small acrylic box with the top covered by mesh.

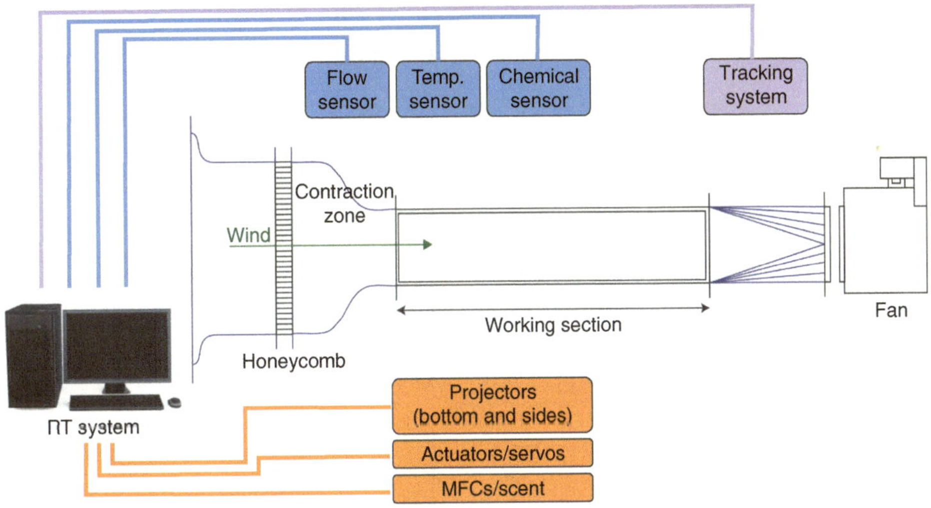

FIGURE 1. Schematic of low-speed wind tunnel. The modular construction and design permit a wide range of environmental conditions while enabling feedback from the real time–tracking and sensor-based systems. Real time–tracking systems can modify visual (projectors), chemical (MFCs), and mechanosensory (actuators, servos) stimuli, allowing open- and closed-loop control of the environment within the wind tunnel.

A CO_2 Flypad can be used if the species is cold-sensitive. Use a CO_2 gun to anesthetize mosquitoes inside their rearing container before transferring onto the Flypad.

3. After sorting, place the mosquitoes into the testing container and back within an environmental chamber where the humidity, temperature, and light cycle can be controlled and correspond with the optimal time period for running the wind tunnel experiments. In the case of the diurnal *Ae. aegypti* mosquitoes, maintain at 25°C, >70% RH, and a light cycle that allows testing within the wind tunnel at 3 h before ZT12 (i.e., 3 h before turning off the lights).
4. Allow the mosquitoes to feed on a water-soaked cotton pad. During this time, ensure that mosquitoes do not experience odors, such as CO_2.

   Mosquitoes are starved for 24 h because feeding on a 10% sucrose pad suppresses the mosquitoes' odor-tracking and visual responses.
5. Let mosquitoes recover from cold anesthesia for 24 h.

   To prevent accidental contamination, the experimenter should wear gloves at all times while manipulating the wind tunnel, odor source, or mosquito box.
6. Before the experiment, ensure that the multicamera system is properly calibrated.

   When using multiple cameras, calibrating the system is critical for proper mosquito trajectory reconstruction. Camera calibration is often conducted before the performed experiments, and the calibration system and procedures are customized to the software being used for the trajectory digitization and analysis.
7. One hour before the experiment, prepare the wind tunnel.

   Turn on the IR flights, short-throw projectors, and computer vision system (cameras, computers, and associated hardware). Make sure the wind tunnel air velocity is adjusted to the preferred speed: In our experience, 20–60 cm/sec is an optimal velocity for studying odor-tracking and visual behavior. If running continuous experiments over the course of several days, consider leaving the wind tunnel running between experiments. This allows the wind tunnel fans to maintain a steady state flow within the working section and helps clear any odor contaminants.
8. Place the mosquitoes in the back and center of the working section.

   Position the mosquito testing box in the working section and let the mosquitoes adjust to the environment for 3 min. Attach a fishing line connector to the sliding door of the box. The fishing line extends outside of the tunnel, allowing the experimenter a low friction method to open the lid without entering the working section.
9. After 3 min, gently open the box and let mosquitoes fly within the working section. Let them fly for 10 min.

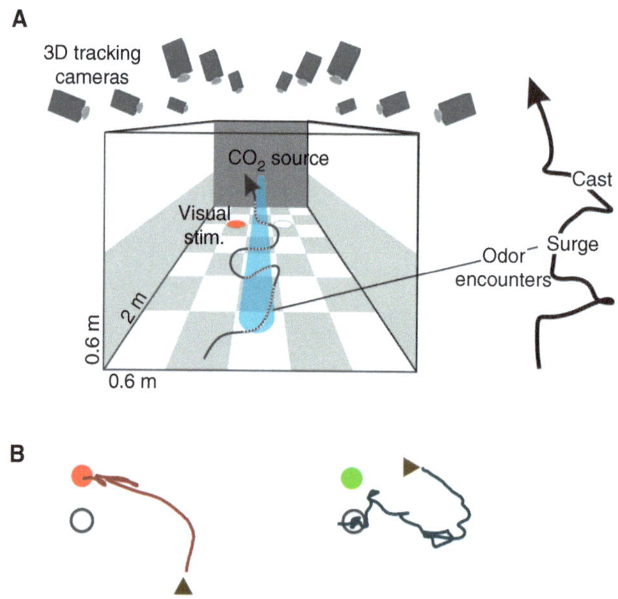

FIGURE 2. (*A*) Schematic of the wind tunnel depicting odor (CO_2) and visual stimuli. A checkerboard and horizon displays are projected onto the bottom and sides of the working section. The 3D tracking system allows determination of when and where the mosquitoes encounter the CO_2 plume (trajectory depicted on the *right* of wind tunnel) and how the CO_2 modifies the mosquito behavior by eliciting upwind surges (when encountering the odor) and casts (when leaving the plume). (*B*) Mosquito trajectories to different visual stimuli on the floor of the wind tunnel working section (modified from Alonso San Alberto et al. 2022).

Mosquitoes are very sensitive to mechanical perturbations and prior olfactory experiences which can impact their behaviors for prolonged time periods. The 10-min flight time in filtered air lets them adjust to the wind tunnel and "reset" their behavioral state.

10. Conduct your experiment as follows:
    - i. Turn on the data acquisition system and scripts controlling the sensory stimuli.
    - ii. Once the experiment is initiated, either leave the room or move to a different area that is well ventilated and is immediately near an outflow duct.
    - iii. At the end of the experiment, check that a new file is stored on the drive, and data was successfully recorded. Ensure that the timestamps for the onset and offset of the sensory stimuli (odor delivery, visual stimuli, or wind actuators) were recorded.
11. Collect the mosquitoes from the working section. Turn off the projectors, IR lights, and cameras.
12. Import the data into a programming language for analysis (e.g., python or Matlab).

    Output from the FLYDRA software and other computer vision systems (e.g., Noldus, Photonic Sentry) include a trajectory identifier in the collected data, allowing processing and analysis of the data on a per trajectory basis (Fig. 2). An important note is that, when releasing multiple mosquitoes for an experiment, multiple trajectories in the same data set may be from one individual. But the number of trajectories per experimental replicate and ease of data collection circumvents the problems associated with pseudoreplication.

DISCUSSION

During free flight, mosquitoes rely on multimodal sensory inputs to navigate in their environment and locate their hosts. Olfactory cues play an important role in gating (van Breugel et al. 2015) or modulating their responses to visual objects (Vinauger et al. 2019; Zhan et al. 2022). As such, the uncontrolled presence of odor contamination and dynamic changes in environmental humidity and temperature can all dramatically impact the outcome of the experiments. Even slight contamination from host odors can modify visual responses. Investigators should monitor and report these environmental variables at all times.

ACKNOWLEDGMENTS

Support for this project was funded by Air Force Office of Scientific Research under grants FA9550-20-1-0422 (J.A.R.) and by the National Institutes of Health under grants R01-AI148300 and R21-AI137947 (J.A.R.).

REFERENCES

Alonso San Alberto D, Rusch C, Zhan Y, Straw AD, Montell C, Riffell JA. 2022. The olfactory gating of visual preferences to human skin and colors visible spectra in mosquitoes. *Nat Commun* **13:** 555. doi:10.1038/s41467-022-28195-x

van Breugel F, Riffell JA, Fairhall A, Dickinson MH. 2015. Mosquitoes use vision to associate odor plumes with thermal targets. *Curr Biol* **25:** 2123–2129. doi:10.1016/j.cub.2015.06.046

Vinauger C, Lutz EK, Riffell JA. 2014. Olfactory learning and memory in the disease vector mosquito *Aedes aegypti*. *J Exp Biol* **217:** 2321–2330. doi:10.1242/jeb.101279

Vinauger C, van Breugel F, Locke LT, Tobin KK, Dickinson MH, Fairhall AL, Akbari OS, Riffell JA. 2019. Visual-olfactory integration in the human disease vector mosquito *Aedes aegypti*. *Curr Biol* **29:** 2509–2516.e5. doi:10.1016/j.cub.2019.06.043

Zhan Y, Alonso San Alberto D, Rusch C, Riffell JA, Montell C. 2021. Elimination of vision-guided target attraction in *Aedes aegypti* using CRISPR. *Curr Biol* **31:** 4180–4187. doi:10.1016/j.cub.2021.07.003

CHAPTER 22

Wind Tunnels and Airflow-Driven Assays: Methods for Establishing the Cues and Orientation Mechanisms That Modulate Female Mosquito Attraction to Human Hosts

Ring T. Cardé[1]

Department of Entomology, University of California, Riverside, California 92521, USA

Understanding how female mosquitoes find a prospective host is crucial to developing means that can interfere with this process. Many methods are available to researchers studying cues and orientation mechanisms that modulate female mosquito attraction to hosts. Behaviors that can be monitored with these assays include activation, taking flight, upwind flight along an odor plume (optomotor anemotaxis), close approach to the stimulus (including hovering), and landing. Video recording can three-dimensionally document flight tracks and can correlate overall distribution patterns and moment-to-moment movements with odor contact and the presence of nearby cues such as a visual target. Here, we introduce mosquito host-seeking behaviors and methods to study them: wind tunnels (which allow orientation in free-flight), airflow-driven assays (using either tethered mosquitoes or small assay chambers that permit flight but also often dictate walking orientation), and still-air assays (wherein in odor concentration and spatial distribution are the orientation cues). We also describe factors that affect the assays and provide assay design considerations.

INTRODUCTION

Female anthropophilic (human-seeking) mosquitoes are generally thought to be attracted to humans over distances of 10 meters or more by sensing an above-ambient concentration of CO_2, after which they head upwind. As they near the host, other cues such as host-specific odors, visual cues, heat, and perhaps humidity can further guide orientation. Ultimately, if a mosquito is to blood-feed, she must land on the host, and heat and an elevation in humidity may play significant roles in these late-stage behaviors. Precisely how these cues are integrated in upwind flight to an odor source and then landing remains unclear (Cardé and Gibson 2010; McMeniman et al. 2014; Cardé 2015; Sumner and Cardé 2022).

Laboratory-based bioassays including wind tunnels and other wind-driven and still-air assays can be used to disentangle how different stimuli contribute to these behaviors. Besides studies of mosquito orientation, there is a wealth of parallel work with other insects, especially moths and *Drosophila*, that can inform experimental designs. Wind tunnels are ideal for allowing upwind orientation followed by landing on an odor-imbued stimulus. Presumably, the skin-released odor or odor blend induces both reactions, although this assumption requires verification in each study species.

[1]Correspondence: ring.carde@ucr.edu

Cite this overview as *Cold Spring Harb Protoc*; doi:10.1101/pdb.over107675

An airborne organism does not sense the direction of ambient airflow by mechanoreception. Using female *Aedes aegypti*, John Kennedy (1940) was the first researcher to establish how a flying insect detects ambient wind direction. His simple but ingenious wind tunnel had a projected floor pattern that could be moved either with or opposite the direction of airflow. Kennedy's breath and CO_2 induced mosquitoes to initiate flight. A pattern projected on the tunnel's floor allowed Kennedy to show that the mosquito sped upwind when the pattern was moved in the direction of airflow; when the pattern was moved in the downwind direction, upwind progress was slowed. It is even possible to entice a mosquito to drop downwind by using the floor's movement to simulate upwind displacement. If a flying insect is not heading directly upwind, it experiences transverse (not front-to-rear) image flow and uses this feedback to adjust its trajectory. It is possible to fly an insect "in place" by matching the speed of a floor pattern that is moved in the downwind direction with the insect's "preferred" retinal velocity. The use of visual surroundings to gauge the direction of airflow while heading upwind is termed positive optomotor anemotaxis (Cardé and Gibson 2010; Cardé 2021). A thorough consideration of this mechanism, zigzag flight, and downwind movements is provided in David (1984).

STRENGTHS AND APPLICATIONS OF WIND TUNNELS

Wind tunnels have been used to decipher the role of the cues used in orientation and their relative valences (whether inducing a positive or negative response, in this case movement toward or way from an odorant). Wind tunnels allow the simulation of most conditions that a mosquito experiences during navigation in wind toward a host or to host cues. Landing platforms or substrates containing odors and visual cues also can be positioned upwind, on the tunnel's floor (see Fig. 1) or at the height of the CO_2 release tube (Healy and Copland 2000; Healy et al. 2002; Lacey and Cardé 2011; Lacey et al. 2014; Sumner et al. 2022). In such arrangements, both attraction and landing can be monitored.

A wind tunnel enables the systematic manipulation of cues and an analysis of flight tracks recorded on video. Moment-to-moment maneuvers can be linked to current or very recent contact with odor filaments (swirls of odor that comprise a turbulent plume). Movements also can be correlated with flight to, flight near, or contact with visual or heat cues. Video analyses typically use two video cameras with software designed to pinpoint the precise *x*, *y*, and *z* position of the mosquito in each video frame (25–30/sec) (e.g., Dekker et al. 2005; Cooperband et al. 2008; Dekker and Cardé 2011; Lacey and Cardé 2011; Hawkes and Gibson 2016; Hinze et al. 2021; Zhao et al. 2022). van Breugel et al. (2015) and San Alberto et al. (2022) used multiple video cameras to document simultaneous flights of multiple *Ae. aegypti* clustered near visual cues.

Rather than documenting the *x*, *y*, *z* positions sequentially, another method of quantification of maneuvers over time is to use heat maps to represent a mosquito's position in an assay chamber. Such graphical representations of density, however, tell us nothing about directions of flight and high values may indicate that the stimulus of interest does not evoke close-range orientation, but rather arrestment of orientation nearby.

Wind tunnels (Baker and Linn 1984) are widely used to verify that a moth pheromone has been correctly and completely identified and therefore induces flight to and landing on the pheromone source. Wind tunnels serve a similar role in identifying odors or odor blends that are attractive for mosquitoes. Using *Anopheles gambiae*, Healy and Copland (2000) established that 2-pentanoic acid, a component of human sweat, evokes upwind flight and landings on its odor source. Subsequently, Healy et al. (2002) determined that C4–C6, 2-oxocarboxylic acids also elicit attraction and landing. In these assays, quantifications of response were enabled by video recording (Healy and Copland 1995). Other studies in which researchers used wind tunnels and choice assays to identify attractive odorants include Geier et al. (2002), Cooperband et al. (2008), and Ghaninia et al. (2019).

Many recent efforts to identify odorants of interest, however, have bypassed behavioral bioassays, relying for example on electrophysiological monitoring of either single-sensillum or whole-antennal

 Cite this overview as *Cold Spring Harb Protoc*; doi:10.1101/pdb.over107675

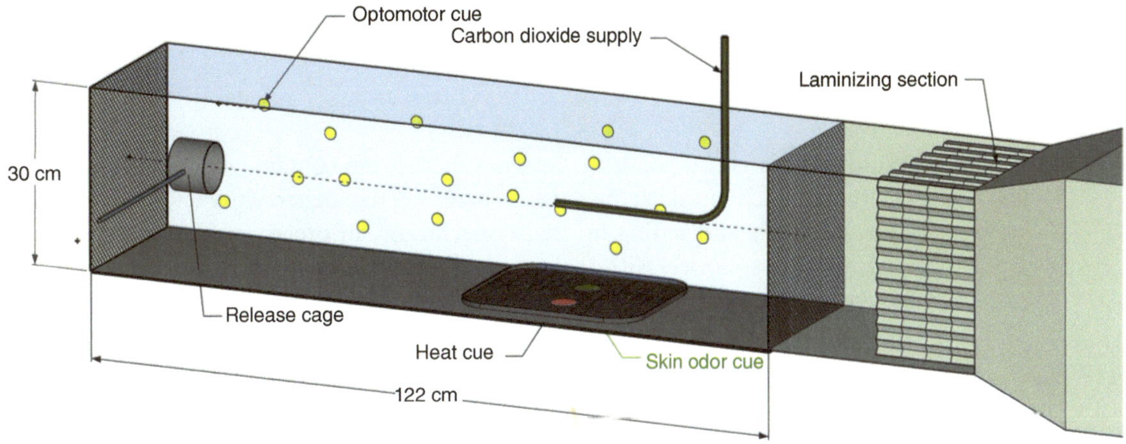

FIGURE 1. Wind tunnel used in Sumner and Cardé (2022). Here, the optomotor cues are dots and only those on the far side of the tunnel are shown. Many tunnels use visual feedback patterns that have straight edges (e.g., LED arrays, alternating white and black squares, or alternating vertical stripes or stripes on the tunnel's floor). David (1982) cautioned that, unlike dots, these patterns provide directional feedback that could influence trajectory, especially in assays that use a moving pattern. In the configuration shown here, the host-associated cues (human foot odor, a visual cue, and heat) were presented in a tray on the tunnel's floor, such that the CO_2 plume would engulf the downwind release cage, thereby stimulating activation and flight. Because the CO_2 plume is vertically separated from the cues presented on the tray below, the CO_2 plume could influence the propensity to land upon arrival over the tray, but the CO_2 plume would not be sensed concurrent with landing. In other configurations, the upwind odor cue could be positioned on a stand at the same height as the CO_2 plume, and then both plumes would be sensed simultaneously (Lacey et al. 2014). When starting an assay, the researcher should open the release cage or, if it has an open end, turn it so that the open end is facing upwind; mechanical movement may cause inadvertent activation or taking flight of mosquitoes and therefore this process should induce as little vibration as feasible. (Adapted from Sumner and Cardé [2022], published under a Creative Commons Attribution 4.0 International License [http://creativecommons.org/licenses/by/4.0/].)

responses to identify compounds passing through gas chromatographic columns as potentially important to a behavioral response. Using an electrophysiological response as a bio monitor is an extraordinarily valuable technique, but if a compound does not pass through such columns, it will be missed (Bello and Cardé 2022). Compounds can be identified from, for example, human skin head space (odor released from the skin's surface) and then tested for behavioral response. Given that there are hundreds of identified odorants in human head space (Bernier et al. 2002), screening of combinations and ratios for behavioral activity is a logistical conundrum. Diagnostic bioassays therefore are needed to verify the behavioral activity of odorants and to establish how such compounds mediate responses.

WIND-TUNNEL ASSAYS

General Bioassay Design Considerations and Standardizations

In a bioassay, researchers must consider and standardize many conditions to ensure that mosquito responses are consistent. In a wind-tunnel assay, of particular interest are temperature, relative humidity, light levels, time of assays within the daily rhythm of responsiveness, mosquito age, prior access of females to males for mating, number of mosquitoes per holding cage, and access to water or sugar water. It is also important to ensure that the rearing conditions (particularly avoidance of larval crowding) produce mosquitoes of a consistent and "normal" size, as small mosquitoes can be less responsive to host stimuli (Takken et al. 1998; Farjana and Tumo 2013). Some studies have tested the same cohort of mosquitoes more than once over a short time interval. Such a protocol could cause habituation of response and, if possible, should be avoided. Odorant purity is a little appreciated issue and should be specified. Impurities in odorants may give false positives in studies of ligands, receptors, and behavioral output (Schorkopf et al. 2019). Finally, the origin of the mosquito should be considered carefully, as behavioral responses may differ among strains (Williams et al. 2006).

Contamination

Wind tunnels and other assay devices can become contaminated with human skin odor and test odorants, which can affect the results. Therefore, researchers should wear nitrile gloves while handling assay materials and setting up trials. Specifically, in pretest protocols, experimenters should clean surfaces of assay devices with a solvent or other methods appropriate to their construction materials. A periodic "airing out" of assay systems also can mitigate odorant contamination. Wind tunnels should have a cross-sectional area large enough to have the odorant plume remain isolated from the tunnel's walls. Small assay tubes are particularly prone to having their surface contaminated by odorants in the airflow and need to be cleaned between each treatment.

Appropriate humidity levels are also important in ensuring a high level of responsiveness. The air that is drawn into the tunnel typically must be humified to a level appropriate for the species being studied. One strategy is to have an unoccupied room that is humified and temperature-regulated as the source of air for the tunnel. This eliminates human odors and fluctuations in background levels of CO_2 in the wind tunnel. Many tunnels use an upwind charcoal filter to further ensure that the air is free of background odors.

Another contamination concern is the inadvertent exposure of mosquitoes to CO_2 (and possibly other odorants) from human breath. We simply hold our breath while in the assay room while setting up mosquitoes for assay. A typical assay room will experience an increase in CO_2 levels from the presence of an observer (Webster et al. 2015). This is not an issue in tunnels or other assay devices that have a flowthrough system drawing air from an outside source (e.g., Sumner and Cardé 2022).

Available Protocols

Many published assays exist describing wind-tunnel assays. A detailed wind-tunnel protocol for testing the attractiveness of a natural extract of human skin versus heat and/or visual cues for *Ae. aegypti* is provided in Sumner and Cardé (2022). A similar wind-tunnel protocol establishing the comparative attractiveness of human skin and/or odorants from human skin that also induce attraction is given in Sumner et al. (2022). See also Chapter 21, Introduction: Experiments and Analysis of Mosquito Flight Behaviors in a Wind Tunnel: An Introduction (San Alberto et al. 2023a) and Chapter 21, Protocol 1: Conducting an Analysis of Mosquito Flight Behaviors in a Wind Tunnel (San Alberto et al. 2023b) for general information and a protocol on running wind-tunnel experiments. A still-air, small-cage bioassay for *Anopheles coluzzii* that uses proportion landing and residence time on the odorant substrate as metrics is described in Webster et al. (2015). This method was used by Bello and Cardé (2022) in the identification of a new odorant blend for *Ae. aegypti*. A landing assay can serve as a proxy for attraction in the more complicated and tedious wind-tunnel assay (Bello and Cardé 2022).

Assay Design Considerations

Wind-Tunnel Size

The size of wind tunnels varies considerably. At a minimum, its crosswind expanse should be sufficient to enable flight maneuvers without permitting the mosquito to have incidental contact with the tunnel's walls. A 0.3-m by 0.3-m cross section (e.g., Sumner and Cardé 2022) is adequate, provided there are cues that induce upwind movement or sustained orientation along the direction of wind flow. The working section of the tunnel can be ~1 m or more long, which is sufficient to allow takeoff and upwind flight to the area of cue presentation. Much larger tunnels are useful for documenting large-scale maneuvers such as persistent flight near a prominent visual cue (van Breugel et al. 2015). Hawkes and Gibson (2016) devised a tunnel large enough (1.2-m-wide and 1.6-m-high) to permit a human to sit at its upwind end. Cooperband and Cardé (2006) used a field wind tunnel (2.5-m-wide, 1.85-m-high, and 6.2-m-long) to determine the capture efficacy of traps for

Culex mosquitoes. McMeniman et al. (2014) found that differences in the attraction of various genotypes of *Ae. aegypti* were modulated by the size of the assay system, which serves as a cautionary example of how assay setup and mosquito strain differences can influence results.

Assay Duration

Kennedy (1977) cautioned that the total duration of an assay could influence its outcome. For example, *Ae. aegypti* is considered an opportunistic or "quick to bite" species, and therefore, if the appropriate stimuli are offered, rapid attraction should be expected if the mosquito is in a responsive state. However, assays are often run for more than several minutes. In a still air-port bioassay, Dekker et al. (2001) tested six treatments that elicited similar very high capture levels of *Ae. aegypti* at the 15-min mark (the standard time for this assay). However, counts over 2 to 6 min were lower for two of the six treatments, meaning that attraction levels would differ depending on the time that the assay's end point was set. Therefore, care should be taken when selecting bioassay duration, as the time to produce the highest capture/attraction values may not be diagnostic of cue valence. A lengthy (several-minute-long) time to odor source location may be indicative of the stimulus not being a good representation of the natural host odor. Timing could also be influenced by whether the assay is performed in a conventional wind tunnel. Positioning the odor stimulus directly downwind of the mosquitoes could induce immediate orientation. Conversely, a large still-air flight chamber (Dekker et al. 2011) allows prolonged flight that is not oriented to the odor source, although the odor's presence may stimulate flight; upwind navigation along a stream of odor only occurs near the entrance port to the trapping device and the mosquito may only contact this cue via a random flight.

Initiation of Flight

Oriented flight to a cue can begin from quiescence. In a wind-tunnel assay, typically one or more mosquitoes are released downwind from a small cage, either by removing a barrier to exit upwind or by turning an open-ended cage toward the upwind direction. Such techniques can introduce a mechanical disturbance that itself may trigger flight. Latencies to flight thus could have an artifactual component (Sumner et al. 2022). Some researchers use mosquitoes that either are presented a cue in-flight or encounter a cue given their flight path. One unresolved question is whether freely flying and quiescent mosquitoes have the same threshold for response to odors or other host cues. Work with the moth *Lymantria dispar* suggests that the male's threshold of response for upwind navigation along a pheromone plume is lower than that for flight initiation (Charlton et al. 1983). This finding is potentially of relevance in a laboratory assay if the odor threshold for flight initiation from quiescence is coupled with plume diffusion models to estimate the maximum distances of odor attraction in the field. An in-flight mosquito may have a lower threshold for response than a quiescent one.

Plume Structure

The fine-scale structure of the odor plume modulates the upwind orientation of mosquitoes (Geier et al. 1999; Dekker et al. 2001; Dekker and Cardé 2011), like its effect on male moth orientation to pheromone (Mafra-Neto and Cardé 1994; Vickers and Baker 1994) and on an attractive ethanol plume by *Drosophila* (van Breugel and Dickinson 2014; Pang et al. 2018). Upwind surges are dictated by contact with odor filaments generated by turbulent flow, whereas crosswind casting follows the loss of odor when encountering either an odor gap within the turbulent plume or exiting into clean air outside of the plume's envelope.

The direct influence of a filamentous plume structure in evoking upwind flight has been verified by comparing orientation along turbulent versus relatively homogenous odor plumes (Geier et al. 1999), by using an electric net to tally orientation near the entrance ports of odor sources that were filamentous or relatively homogeneous (Dekker et al. 2001), and by observations of individual mosquitoes in a wind tunnel in which the movements relative to a thin odor plume could be established (Dekker and Cardé 2011).

To provide an airflow free of turbulence, tunnels can use an upwind panel comprised of hexagonal aluminum cells ~2 cm in diameter and ~10 cm deep. Turbulent structure can then be introduced as desired. Justus et al. (2002) used a 2.5-cm-diameter disk just upwind of the odor source to generate a turbulent odor plume. Similarly, a stimulus generator can produce odor pulses timed for duration and interpulse interval (Mafra-Neto and Cardé 1994; Vickers and Baker 1994; Justus et al. 2002). Plume structure can be visualized by using a small source of titanium tetrachloride, which upon contact with moist air generates a visible structure comprised of titanium dioxide and hydrochloric acid. Sometimes such "smoke" in the plume, however, can coalesce into droplets that fall from the plume; these artifacts are not part of the plume's turbulent structure. Alternatively, plume structure can be measured by use of a tracer gas, such as propene, and the plume's fine-scale structure can be measured with a photo-ionization detector (PID) (Justus et al. 2002). These tracers need to be exhausted from the tunnel to the outside. Care should be taken to ensure that airflow velocity in the assay system is set to a level that avoids distortion of the PID signal (see True and Crimaldi 2022).

In many assays CO_2 plumes are a common stimulus, either alone or combined with other cues. As vertebrates exhale at ~4% CO_2, flowmeters are used to blend air and pure CO_2 to this level. When CO_2 or air is taken from a tank, their temperatures will be below the wind tunnel's temperature and consequently the plume will slowly sink as it is carried downwind. It is therefore necessary to verify its position relative to the mosquitoes' release cage using a CO_2 gas detector. The CO_2 flow can become isothermal by using a gas bag or a long delivery tube, and then its position will not drop as the plume is carried downwind. Regardless, it is important to verify that flow from the CO_2 tube is isothermal with air in the wind tunnel.

Visual Cues

As mentioned during our discussion of optomotor anemotaxis, upwind navigation requires the perception of the visual surround. This is best accomplished by providing a defined set of visual cues. Most often these cues are placed on the wind tunnel's floor, but based on studies with moths, they also can be positioned on the wind tunnel's sides or even on the ceiling. To avoid directional cues, David (1982) advocated using dots, rather than stripes or other visual features with straight sides. Undoubtedly, some structural features of the wind tunnel and other objects outside of the tunnel provide additional visual feedback. However, unless the objective of a study is to explore how the placement, size, and other features of such cues influence orientation, using a floor or sidewall pattern of randomly placed dots provides sufficient optomotor feedback for flight orientated along an odorant plume. Clutter in the visual surround can be reduced by using white sheets placed over a tunnel.

Assay Caveats

A particular concern in all odor assays is ensuring that odors are presented at physiologically and ecologically relevant concentrations. Positive orientation to a naturally present odor cue needs to be at a naturally occurring concentration if it is to be meaningful in establishing that this cue governs natural orientation. Unfortunately, measuring naturally occurring airborne concentrations and those generated by test stimuli is methodologically complicated and is rarely considered. A piezo-electric atomizer delivering odorants from solvent mixture is one method for precise manipulation of concentrations and ratios of airborne odorants (Girling and Cardé 2007). Mosquitoes can be attracted to many odors (such as floral and oviposition site volatiles) that are presumably not part of natural odor cues used in orientation to a host. Majeed et al. (2016), for example, found that (*R*)-1-octen-3-ol induced landing in several mosquito species but at concentrations from a host that were above those that would occur naturally.

Another consideration is the light levels in many assays vary from those naturally occurring in the field. Video recordings often add infrared light to enhance the mosquitos' image. Generally, it is desirable to conduct assays during the mosquitoes' normal rhythm of host seeking (Cardé and Gibson 2010).

 Cite this overview as *Cold Spring Harb Protoc*; doi:10.1101/pdb.over107675

ALTERNATIVE BEHAVIORAL ASSAYS

Besides wind tunnels, many other kinds of air-driven assays have been used to determine the cues used by mosquitoes in orientation. A common method for documenting the attractiveness of odor cues is the use of a choice between a candidate odor and an odorless control or a choice between two odor-laden streams. Often these cues are presented with airflow with a Y-tube choice chamber (Geier and Boeckh 2003) in which responders must choose a right or left upwind path. Steib et al. (2001) and Geier et al. (2002) used this system to identify odors attractive to *Ae. aegypti*. This bioassay has also been used to characterize human-released odors that interfere with attraction to humans (Logan et al. 2008). To identify odors attractive to *Culex quinquefasciatus*, Cooperband et al. (2008) used a similar assay in which mosquitoes enter a small Plexiglas box and then choose between two 10-cm-diameter tubes. Mosquitoes in such devices often proceed upwind with their flight punctuated by periodic contact with the walls of the apparatus or even landing and walking upwind. Although these kinds of assay systems allow the testing of individual mosquitoes, typically a cohort of five or more individuals is tested per replicate. Testing a cohort speeds up the evaluation of different odor stimuli, but it may not permit the fine-scale evaluation of behavioral responses of individuals, such as the time to reach the stimulus source or total time in orientation.

Alternatively, odors can simply be presented in a no-choice assay. For example, Ghaninia et al. (2019) used a relatively small, 9.5-cm-diameter clear plastic tube to show that acetone plus CO_2 induced upwind attraction in *Ae. aegypti* but decreased attraction of *An. gambiae*. Such an assay system essentially functions as a free-flight wind tunnel, provided its diameter is sufficient to prevent incidental contact with the tube's wall. A limitation of this sort of assay system is that it is not as useful for recording flight tracks.

Another approach uses a still-air flight chamber. Acree et al. (1968) devised an assay comprised of a flight box leading to a small capture port from which CO_2 and test compounds or air flowing over a human arm could be introduced. This bioassay led to the identification of lactic acid as an attractive component of human skin odor for female *Ae. aegypti*, although lactic acid, even when offered with CO_2, is generally only weakly attractive to this mosquito (Bernier et al. 2002; Dekker et al. 2002; Sumner et al. 2022). Dekker et al. (2001) used a large flight arena with two 2-cm-diameter entrance ports at one end, each leading to a capture container. The plume emanating from these ports retained some structural identity for ∼20 cm into the flight chamber. Because there is no airflow in the flight chamber, mosquitoes presumably did not orient directly to an odor source via anemotaxis unless they encountered a plume from one of two entrance ports. Because the assay time totaled 15 min, mosquitoes presumably encountered the test odors in the flight chamber, which influenced flight activity before capture.

Although it is possible to correlate a single flight maneuver with contact with an individual odor filament (e.g., Dekker and Cardé 2011), an insect freely flying along an odor plume will experience intermittent but unknown patterns of odor contact. However, mosquitoes can be tethered by a wire attached to their thorax and their heading in an airflow toward odor cues monitored continually, as has been done with *Drosophila* (Cheng and Frye 2021). The advantage of using this method when studying odors is that they can be presented in various patterns and concentrations to determine how they modulate upwind heading. Tethering also can be used to establish how visual feedback enables orientation (Kaushik et al. 2020). However, because a tethered insect is constrained, it experiences a different mechanosensory feedback than it would in free flight (Cardé 2021).

Assessments of Landing

Besides attraction to an odor source or other host-associated cues, another diagnostic behavior is landing, which is a prelude to biting. Landing generally involves preparing the legs for source contact, orienting the body to be parallel to the landing surface, and allowing expansion of the frontal image, in which the object ahead takes up an increasing proportion of the visual field. Smith et al. (2020), however, found that *Ae. aegypti* females generally had one or two "bounces" before attaching to a substrate; the assay in their study did not provide host-odor cues and so it is unclear whether the bouncing maneuvers mimic natural alighting on a human host.

Surrogate Assays for In-Flight Orientation or Port Entrance to Odor Sources

Webster et al. (2015) bypassed attraction in a wind tunnel by simply scoring the proportion of *An. gambiae* landing on fabric imbued with a human skin rubbing in a 30-cm by 30-cm by 30-cm screen cage without a defined airflow. Landing counts were tallied every 30 sec over a 6-min interval. This provided a measure of presumed attraction and tendency to remain on the cue substrate. The 6-min assay time also allowed an approximation of latencies. To identify compounds attractive to *Ae. aegypti* females, Bello and Cardé (2022) used this cage bioassay to monitor in isolated fractions the propensity of a mosquito to land on glass beads treated with human foot odor (this assay also allows the monitoring of proboscis probing). This method bypasses upwind attraction as a diagnostic assay, but nonetheless, the odors identified evoked attraction in a wind tunnel (Sumner et al. 2022). This assay also permits a relatively rapid throughput of odors for evaluation, compared to a testing of individual mosquitoes in a wind tunnel. Examples of using a nonflight bioassay to identify attractive odors are well-established with bark beetles, which perform poorly in a wind tunnel. Instead, pheromones and host-tree volatiles have been identified using a quite simple upwind walking assay as a substitute for flight (Wood and Bushing 1963).

Orientation in Still Air

Although long-distance orientation via navigation along an odor plume presumes the detection of wind direction, klinotaxis (here, orientation that is guided by sensing spatial differences in odor concentrations) can mediate odor-directed movement that is not wind-directed (Baker and Kuenen 1982; Cardé 2021). Very close to a host, airflow may not be adequate for orientation. There the odor plume may be relatively homogeneous in structure, which can enhance a positive response (Dekker and Cardé 2011).

Creating a defined plume in still air is possible with a wind tunnel that has louvers at the upwind and downwind ends. Once the plume is created in a wind flow, the louvers are quickly closed; the plume retains its structure for many seconds until dissipated by molecular diffusion and residual air movement (Lacey and Cardé 2012; Wisniewska and Cardé 2012). The general structural features of such plumes are best understood by use of a visual tracer such as titanium tetrachloride. Odor orientation in still air would be via a klinotactic mechanism. However, if the insect has already initiated upwind flight along the odor plume before stoppage of the airflow, the insect could continue its set trajectory by using visual cues. To verify unambiguous klinotaxis, an insect would need either to enter the odor plume or initiate flight after the cessation of airflow.

Additionally, several anthropophilic species have "house-entering" behaviors, wherein entry to human dwellings precedes biting, and these species may "lie in wait" (preattack resting) until their normal time of biting and the appearance of a host (Trpis and Hausermann 1978; Dekker and Cardé 2011). Such orientation to and within a dwelling also may involve displacement that is not governed by upwind movement. Bioassay protocols that measure house entering and preattack resting have not been developed.

MISINTERPRETING THE VALENCE OF ODOR CUES IN ATTRACTION AND LANDING ASSAYS

It is tempting to interpret a positive, statistically supported, response to an odor source as verification that this odor or odor blend will induce natural orientation behaviors at physiologically and ecologically appropriate concentrations and natural plume structures. Few behavioral studies, however, have attempted to use odor presentations that mimic airborne concentrations of human skin odors (but see Zhao et al. 2022). Instead, the dispenser dose offered usually is largely arbitrary, often just a single dose is tested, and there is little consideration of a compound's volatility. Also, as noted earlier, an assay's duration and use of an unnaturally high dose could contribute to an interpretation that the odor cues tested are evoking "natural" attraction.

A cautionary viewpoint recognizes that the well-studied mosquitoes (*Ae. aegypti*, *An. gambiae* s.l., and *Culex pipiens/Cx. quinquefasciatus*) have numerous compounds identified to be attractive, to induce landing, or both (Coutinho-Abreu et al. 2022). One issue is that an attractive odor may mediate finding flowers or oviposition sites and not hosts. Floral odors could add to the attractiveness of a skin odor blend, even though the orientation responses would be governed by different proximate selective forces.

The seemingly large number of odors that are reported to induce attraction and/or landing may be in part artifactual. The Coutinho-Abreu et al. (2022) review, for example, lists for *Ae. aegypti* more than two dozen compounds that induce attraction when presented, usually in combination with CO_2 and lactic acid. Some of these compounds are not a natural constituent of the head space of human skin.

Additionally, some human emitted odors have repellent properties and humans can differ substantially in their intrinsic attractiveness (e.g., see Logan et al. 2008 and De Obaldia et al. 2022 for data on *Ae. aegypti*). Dosage, missing odors, off-ratios, and repellent compounds all could mediate differential attractiveness. The bioassay techniques used to substantiate attraction also can be used to establish how an odor interferes with this process.

It seems improbable that the human odor signature for mosquitoes is comprised of so many compounds, and this idea should offer a salutary perspective on the interpretation of what odors are attractive in nature. However, it also may be that some compounds that have the same moiety and are very similar structurally can be redundant, that is, they activate the same odor receptor. Furthermore, Herre et al. (2022) established that odor coding in *Ae. aegypti* involves many neurons that coexpress multiple receptor genes with broad sensitivity to many ligands. Both redundancy mechanisms could account for the broad behavioral responses of *Ae. aegypti* and other mosquitoes to many odors. Such a flexibility in odor response also could be adaptive in facilitating the finding and recognition of the variable odor profile of humans.

Finally, odors used in field trapping devices may use odor release rates that well exceed the rates from natural human emitters. In such applications, the goal is to trap the maximum number of mosquitoes, and, consequently, unnatural rates and ratios and possibly nonhuman odors may be optimal for lures.

CONCLUSION

Wind tunnels and other air-driven assays enable the determination of cues used by female mosquitoes in navigation to a prospective host. Wind tunnels are particularly suited for use in establishing how cues determine maneuvers moment to moment. Assays that allow mosquitoes to choose between cues can establish relative cue valence. Bioassays also are a mainstay in efforts to identify new cues, particularly odors. A difficulty in interpretation will be to identify odors or odor blends that are attractive only because they are offered at unnaturally high concentrations. Other major considerations with odors are recreating plume structures and airborne concentrations that closely mimic those encountered in nature and accounting for the possibility that some odors are redundant. A continuing challenge will be to devise assays that allow the expression of the full repertory of responses and presentation of all cue modalities.

ACKNOWLEDGMENTS

Funding from the Pacific Southwest Regional Center of Excellence for Vector-Borne Diseases funded by the U.S. Centers for Disease Control and Prevention (Cooperative Agreement 1U01CK000516) is acknowledged. I thank Dr. Genevieve Tauxe for valuable suggestions and review.

REFERENCES

Acree F Jr, Turner RB, Gouck HK, Beroza M, Smith N. 1968. L-lactic acid—a mosquito attractant isolated from humans. *Science* **161:** 1346–1347. doi:10.1126/science.161.3848.1346

Baker TC, Kuenen LPS. 1982. Pheromone source location be flying moths: a supplementary non-anemotactic mechanism. *Science* **216:** 424–427. doi:10.1126/science.216.4544.424

Baker TC, Linn CE Jr. 1984. Wind tunnels in pheromone research. In *Techniques in pheromone research* (ed. Hummel HE, Miller TA), pp. 75–110. Springer-Verlag, New York.

Bello JE, Cardé RT. 2022. Compounds from human odor that induce landing and attraction in female yellow fever mosquitoes (*Aedes aegypti*). *Sci Rep* **12:** 15638. doi:10.1038/s41598-022-19254-w

Bernier U, Kline D, Schreck C, Yost R, Barnard D. 2002. Chemical analysis of human skin emanations: comparison of volatiles from humans that differ in attraction of *Aedes aegypti*. *J Am Mosq Control Assoc* **18:** 186–195.

Cardé RT. 2015. Multi-cue integration: how female mosquitoes locate a human host. *Curr Biol* **18:** 21–25.

Cardé RT. 2021. Navigation along windborne plumes of pheromone and resource-linked odors. *Annu Rev Entomol* **66:** 317–336. doi:10.1146/annurev-ento-011019-024932

Cardé RT, Gibson G. 2010. Long-distance orientation of mosquitoes to host odours and other host-related cues. In *Ecology of vector-borne diseases. Vol. 2. Olfaction in vector–host interactions* (ed. Takken W, Knols BGF), pp. 115–141. Wageningen Academic, Wageningen, The Netherlands.

Charlton RE, Kanno H, Collins RD, Cardé RT. 1983. Influence of pheromone concentration and ambient temperature on flight of the gypsy moth, *Lymantria dispar* (L.), in a sustained-flight wind tunnel. *Physiol Entomol* **18:** 349–362. doi:10.1111/j.1365-3032.1993.tb00608.x

Cheng KY, Frye MA. 2021. Odour boosts visual object approach in flies. *Biol Lett* **17:** 20200770. doi:10.1098/rsbl.2020.0770

Cooperband MF, Cardé RT. 2006. Orientation of *Culex* mosquitoes to carbon dioxide-baited traps: flight manoeuvres and trapping efficiency. *Med Vet Entomol* **26:** 11–26. doi:10.1111/j.1365-2915.2006.00613.x

Cooperband MF, McElfresh JS, Millar JG, Cardé RT. 2008. Attraction of *Culex quinquefasciatus* say (Diptera: Culicidae) to odors from chicken feces. *J Insect Physiol* **53:** 1184–1192. doi:10.1016/j.jinsphys.2008.05.003

Coutinho-Abreu A, Riffell JA, Akbari OS. 2022. Human attractive cues and host-seeking behavior. *Trends Parasitol* **38:** 246–264. doi:10.1016/j.pt.2021.09.012

David CT. 1982. Competition between fixed and moving stripes in the control of orientation in flying *Drosophila*. *Physiol Entomol* **7:** 151–156. doi:10.1111/j.1365-3032.1982.tb00283.x

David CT. 1984. Mechanisms of directional flight in wind. In *Mechanisms of insect olfaction* (ed. Payne T, Birch M, Kennedy K), pp. 49–57. Oxford University Press, Oxford.

Dekker T, Cardé RT. 2011. Moment-to-moment flight manoeuvres of the female yellow fever mosquito (*Aedes aegypti* L.) in response to plumes of carbon dioxide and human skin odour. *J Exp Biol* **214:** 3480–3494. doi:10.1242/jeb.055186

Dekker T, Takken W, Cardé RT. 2001. Structure of host-odour plumes influences catch of *Anopheles gambiae* s.s. and *Aedes aegypti* in a dual-choice olfactometer. *Physiol Entomol* **26:** 124–134. doi:10.1046/j.1365-3032.2001.00225.x

Dekker T, Steib B, Cardé RT, Geier M. 2002. L-lactic acid: a human-signifying host cue for the anthropophilic mosquito *Anopheles gambiae*. *Med Vet Entomol* **16:** 91–98. doi:10.1046/j.0269-283x.2002.00345.x

Dekker T, Geier M, Cardé RT. 2005. Carbon dioxide instantly sensitizes female yellow fever mosquitoes to human skin odours. *J Exp Biol* **208:** 2963–2972. doi:10.1242/jeb.01736

De Obaldia ME, Morita T, Dedmon LC, Boehmler DJ, Jiang CS, Zeledon EV, Cross JR, Vosshall LB. 2022. Differential mosquito attraction to humans is associated with skin-derived carboxylic acid levels. *Cell* **85:** 4099–4116. doi:10.1016/j.cell.2022.09.034

Farjana T, Tuno N. 2013. Blood feeding and host-seeking behavior in *Aedes aegypti* and *Aedes albopictus* (Diptera: Culicidae). *J Med Entomol* **50:** 838–846. doi:10.1603/ME12146

Geier M, Boeckh J. 2003. A new Y-tube olfactometer for mosquitoes to measure the attractiveness of host odours. *Entomol Exp Appl* **92:** 9–19. doi:10.1046/j.1570-7458.1999.00519.x

Geier M, Bosch OJ, Boeckh J. 1999. Influence of odour plume structure on upwind flight of mosquitoes towards hosts. *J Exp Biol* **202:** 1639–1648. doi:10.1242/jeb.202.12.1639

Geier M, Bosch O, Steib B, Rose A, Boeckh J. 2002. Odour guides host finding mosquitoes: identification of new attractants on human skin. In Proc Int Conf Urban Pests, Regensburg, Germany, pp. 37–46.

Ghaninia M, Majeed S, Dekker T, Hill SR, Ingell R. 2019. Hold your breath: differential behavioral and sensory acuity of mosquitoes to acetone and carbon dioxide. *PLoS ONE* **14:** e0226815. doi:10.1371/journal.pone.0226815

Girling RD, Cardé RT. 2007. Analysis and manipulation of odor plumes from a piezo-electric release system and measurements of upwind flight of male almond moths, *Cadra cautella*, to pheromone plumes. *J Chem Ecol* **33:** 1927–1945. doi:10.1007/s10886-007-9349-3

Hawkes F, Gibson G. 2016. Seeing is believing: the nocturnal malarial mosquito *Anopheles coluzzii* responds to visual host-cues when odour indicates a host is nearby. *Parasites Vectors* **9:** 320. doi:10.1186/s13071-016-1609-z

Healy TP, Copland MJW. 1995. Activation of *Anopheles gambiae* mosquitoes by carbon dioxide and human breath. *Med Vet Entomol* **9:** 331–336. doi:10.1111/j.1365-2915.1995.tb00143.x

Healy T, Copland M. 2000. Human sweat and 2-oxopentanoic acid elicit a landing response from *Anopheles gambiae*. *Med Vet Entomol* **14:** 195–200. doi:10.1046/j.1365-2915.2000.00238.x

Healy T, Copland M, Cork A, Przyborowska A, Halket J. 2002. Landing responses of *Anopheles gambiae* elicited by oxocarboxylic acids. *Med Vet Entomol* **16:** 126–132. doi:10.1046/j.1365-2915.2002.00353.x

Herre M, Goldman OV, Lu TC, Caballero-Vidal G, Qi Y, Gilbert ZN, Gong Z, Morita T, Rahiel S, Ghaninia M, et al. 2022. Non-canonical odor coding in the mosquito. *Cell* **85:** 3104–3123. doi:10.1016/j.cell.2022.07.024

Hinze A, Lantz J, Hill S, Ingell R. 2021. Mosquito host seeking in 3D using a versatile climate controlled wind tunnel system. *Front Behav Neurosci* **15:** 643693. doi:10.3389/fnbeh.2021.643693

Justus KA, Murlis J, Jones C, Cardé RT. 2002. Measurement of odor-plume structure in a wind tunnel using a photoionization detector and a tracer gas. *Environ Fluid Mech* **2:** 115–142. doi:10.1023/A:1016227601019

Kaushik PK, Renez M, Olsson SB. 2020. Characterizing long-range search behavior in Diptera using complex 3D virtual environments. *Proc Natl Acad Sci* **117:** 12201–12207. doi:10.1073/pnas.1912124117

Kennedy JS. 1940. The visual responses of flying mosquitoes. *Proc Zool Soc Lond Ser A* **109:** 221–242. doi:10.1111/j.1096-3642.1940.tb00831.x

Kennedy JS. 1977. Behaviorally discriminating assays of attractants and repellents. In *Chemical control of insect behavior* (ed. Shorey HH, McKelvey JJ). pp. 215–229. Wiley, New York.

Lacey ES, Cardé RT. 2011. Activation, orientation and landing of female *Culex quinquefasciatus* in response to carbon dioxide and odour from human feet: 3D flight analysis in a wind tunnel. *Med Vet Entomol* **25:** 94–103. doi:10.1111/j.1365-2915.2010.00921.x

Lacey ES, Cardé RT. 2012. Location of and landing on a source of human body odour by female *Culex quinquefasciatus* in still and moving air. *Physiol Entomol* **37:** 153–159. doi:10.1111/j.1365-3032.2012.00829.x

Lacey ES, Ray A, Cardé RT. 2014. Close encounters: contributions of carbon dioxide and human skin odour to finding and landing on a host in *Aedes aegypti*. *Physiol Entomol* **39:** 60–68. doi:10.1111/phen.12048

Logan JG, Birkett MA, Clark SJ, Powers S, Seal NJ, Wadhams LJ, Mordue Luntz AJ, Pickett JA. 2008. Identification of human-derived volatile chemicals that interfere with attraction of *Aedes aegypti* mosquitoes. *J Chem Ecol* **34:** 308. doi:10.1007/s10886-008-9436-0

Mafra-Neto A, Cardé RT. 1994. Fine-scale structure of pheromone plumes modulates upwind orientation of flying moths. *Nature* **369:** 142–144. doi:10.1038/369142a0

Majeed S, Hill SR, Birgersson G, Ingell R. 2016. Detection and perception of generic host volatiles by mosquitoes modulate host preference: context dependence of (*R*)-1-octen-3-ol. *R Soc Open Sci* **3:** 160467. doi:10.1098/rsos.160467

McMeniman CJ, Corfas RA, Matthews BJ, Ritchie SA, Vosshall LB. 2014. Multimodal integration of carbon dioxide and other sensory cues drives mosquito attraction to humans. *Cell* **156:** 1060–1071. doi:10.1016/j.cell.2013.12.044

Pang R, van Breugel F, Dickinson M, Riffell JA, Fairhall A. 2018. History dependence in insect flight decisions during odor tracking. *PLoS Comput Biol* **14:** e1005969. doi:10.1371/journal.pcbi.1005969

San Alberto AD, Rusch C, Zhan Y, Straw AD, Montell C, Riffell JA. 2022. The olfactory gating of visual preferences to human skin and visible spectra in mosquitoes. *Nat Commun* **13:** 555. doi:10.1038/s41467-022-28195-x

San Alberto DA, Rusch C, Riffell JA. 2023a. Experiments and analysis of mosquito flight behaviors in a wind tunnel: an introduction. *Cold Spring Harb Protoc* doi:10.1101/pdb.top107674

San Alberto DA, Rusch C, Riffell JA. 2023b. Conducting an analysis of mosquito flight behaviors in a wind tunnel. *Cold Spring Harb Protoc* doi:10.1101/pdb.prot108257

Schorkopf DLP, Molnár BP, Solum M, Larsson MC, Millar JG, Kárpáti Z, Dekker T. 2019. False positives from impurities result in incorrect functional characterization of receptors in chemosensory studies. *Prog Neurobiol* **181:** 101661. doi:10.1016/j.pneurobio.2019.101661

Smith NB, Balsalobre JB, Doshi M, Willenberg BJ, Dickerson AK. 2020. Landing mosquitoes bounce when engaging a substrate. *Sci Rep* **10:** 15744. doi:10.1038/s41598-020-72462-0

Steib BM, Geier M, Boeckh J. 2001. The effect of lactic acid on odour-related host preference of yellow fever mosquitoes. *Chem Senses* **26:** 523–528. doi:10.1093/chemse/26.5.523

Sumner BD, Cardé RT. 2022. Primacy of human-odors over visual and heat cues in inducing landing in female *Aedes aegypti* mosquitoes. *J Insect Behav* **35:** 31–43. doi:10.1007/s10905-022-09796-2

Sumner BD, Amos BA, Bello JE, Cardé RT. 2022. L-lactic and 2-ketoglutaric acids, odors from human skin, govern attraction and landing in host-seeking female *Aedes aegypti* mosquitoes. *J Insect Behav* **5:** 213–226. doi:10.1007/s10905-022-09812-5

Takken W, Klowden MJ, Chambers GM. 1998. Effect of body size on host seeking and blood meal utilization in *Anopheles gambiae* sensu stricto (Diptera: Culicidae): the disadvantage of being small. *J Med Entomol* **35:** 639–645. doi:10.1093/jmedent/35.5.639

Trpis M, Hausermann W. 1978. Genetics of house-entering behaviour in East African populations of *Aedes aegypti* (L.) (Diptera: Culicidae) and its relevance to speciation. *Bull Entomol Res* **68:** 521–532. doi:10.1017/S0007485300009494

True CC, Crimaldi JP. 2022. Distortion of passive scalar structure during suction-based plume sampling. *Sens Actuat B Chem* **367:** 132018. doi:10.1016/j.snb.2022.132018

van Breugel F, Dickinson MH. 2014. Plume-tracking behavior of flying *Drosophila* emerges from a set of distinct sensory-motor reflexes. *Curr Biol* **24:** 274–286. doi:10.1016/j.cub.2013.12.023

van Breugel F, Riffell J, Fairhall A, Dickinson MH. 2015. Mosquitoes use vision to associate odor plumes with thermal targets. *Curr Biol* **25:** 2123–2129. doi:10.1016/j.cub.2015.06.046

Vickers NJ, Baker TC. 1994. Reiterative responses to single strands of odor promote sustained upwind flight and odor source location by moths. *Proc Natl Acad Sci* **91:** 5756–5760. doi:10.1073/pnas.91.13.5756

Webster B, Lacey ES, Cardé RT. 2015. Waiting with bated breath: opportunistic orientation to human odor in the malaria mosquito, *Anopheles gambiae*, is modulated by minute changes in carbon dioxide concentration. *J Chem Ecol* **41:** 59–66. doi:10.1007/s10886-014-0542-x

Williams CR, Ritchie SC, Russell RC, Eiras AE, Kline DL, Geier M. 2006. Geographic variation in attraction to human odor compounds by *Aedes aegypti* mosquitoes (Diptera: Culicidae): a laboratory study. *J Chem Ecol* **32:** 1625–1634. doi:10.1007/s10886-006-9097-9

Wisniewska J, Cardé RT. 2012. Visual cues collimate the trajectories of almond moth, *Cadra cautella*, males flying in wind and still air within a wind-formed plume of pheromone. *Physiol Entomol* **37:** 42–52. doi:10.1111/j.1365-3032.2011.00807.x

Wood DL, Bushing RW. 1963. The olfactory response of *Ips confusus* (LeConte) (Coleoptera: Scolytidae) to the secondary attraction in the laboratory. *Can Entomol* **95:** 1066–1078. doi:10.4039/Ent951066-10

Zhao Z, Zung JL, Hinze A, Kriete AL, Iqbal A, Younger MA, Matthews BJ, Merhof D, Thiberge S, Ignell R, et al. 2022. Mosquito brains encode unique features of human odour to drive host seeking. *Nature* **605:** 706–712. doi:10.1038/s41586-022-04675-4

CHAPTER 23

Using Videography to Study the Biomechanics and Behavior of Freely Moving Mosquitoes

Andrew K. Dickerson,[1,3] Florian T. Muijres,[2] and Remco Pieters[2]

[1]*Department of Mechanical, Aerospace, and Biomedical Engineering, University of Tennessee, Tennessee 37996, USA;* [2]*Department of Experimental Zoology, Wageningen University, 6708 PB Wageningen, the Netherlands*

Female mosquitoes of most species require a blood meal for egg development. When biting a human host to collect this blood meal, they can spread dangerous diseases such as malaria, yellow fever, or dengue. Researchers use videography to study many aspects of mosquito behavior, including in-flight host-seeking, takeoff, and landing behaviors, as well as probing and blood feeding, and more. Here, we introduce protocols on how to use videography to capture and analyze mosquito movements at high spatial and temporal resolution, in two and three dimensions.

STUDYING MOSQUITO BEHAVIOR USING VIDEOGRAPHY

Mosquitoes are dipteran insects with a slender body, long legs, and relatively high-aspect ratio wings. When flying, mosquitoes beat their wings back and forth at exceptionally high frequencies and low amplitudes. Female mosquitoes of most species need protein found in vertebrate blood meal for egg development. To acquire this blood meal, female mosquitoes bite a blood host, such as a human. Not only are mosquito bites annoying, but they can result in the contraction of several dangerous infectious diseases, including malaria, yellow fever, dengue, chikungunya, and Zika virus (Lee et al. 2018). As a result, the behavior of this deadly disease-carrying insect is studied extensively using videography systems (Dickerson et al. 2012, 2015a,b, 2018; Dickerson and Hu 2014; Bomphrey et al. 2017; Sutcliffe et al. 2017).

Using video analysis, one can capture and quantify mosquito behavior at temporal resolutions that cannot be captured by other means. By using multiple synchronized cameras in a single video system, one can reconstruct the full three-dimensional (3D) kinematics of moving mosquitoes. Videography system complexity can be tuned to study a large range of mosquito behaviors, including feeding, egg laying, biting of blood hosts, and landing on and taking off from a range of complex surfaces, including H_2O and human hosts (Muijres et al. 2017; Smith et al. 2018, 2020). Videography also makes it possible to capture various in-flight behaviors, such as searching for food, blood hosts, and mates, as well as in-flight escape maneuvers (Cribellier et al. 2022). Such mosquito behavior studies often aim to contribute to the development and improvement of vector control methods and techniques (Cribellier et al. 2018, 2020; Batista et al. 2019). For example, based on knowledge of the in-flight search and escape dynamics of host-seeking mosquitoes, one can improve repellents and mosquito traps and create pathways for new control tactic development.

[3]Correspondence: ad@utk.edu

Cite this introduction as *Cold Spring Harb Protoc;* doi:10.1101/pdb.top107676

A mosquito's minute stature and uncommonly high wingbeat frequencies present a considerable challenge for in vivo biomechanical measurements. Therefore, dedicated high-speed cameras are needed for high-fidelity visualization. Recent developments in high-speed camera technology and computing power have made the required hardware more available and affordable. For all videography-based studies on moving mosquitoes, the choice of video system depends directly on the goal of the current study (see Fig. 1). Therefore, to effectively design a videography experiment, researchers should first clearly define their research question, and then determine hardware requirements. With the research question and hardware in mind, researchers can then select an appropriate software tool for video analysis. However, the process of designing a videography-based mosquito behavior study is an ongoing process based on continual feedback and is thus nonlinear.

In the accompanying protocols, we present the videography experiment design process in detail, and provide best practices, guidelines, and alternatives for experimental and analysis tools for studying mosquito behavior using videography. For guidelines and points of consideration for designing a generic videography experiment for studying mosquito behavior, see Protocol 1: Designing a Generic Videography Experiment for Studying Mosquito Behavior (Muijres et al. 2022). For a discussion on how to use a real-time tracking system, called Braid, to study moving mosquitoes, see Protocol 2: Real-Time Tracking of Multiple Moving Mosquitoes (Straw et al. 2022). The Braid tracking system facilitates fully automated location tracking of multiple mosquitoes moving freely about an arena without the need of storing video data. Moreover, how to use the software tool DLTdv to track the body, wing, and leg kinematics of moving mosquitoes in 3D and at high spatial and temporal resolution is detailed in Protocol 3: Tracking the Body, Wing, and Leg Kinematics of Moving Mosquitoes (Hedrick et al. 2022). DLTdv can be used to track mosquito movements both manually and automatically. Finally, how to quantify the body, leg, and wingbeat kinematics of moving mosquitoes post–data collection is discussed in Protocol 4: Quantifying and Analyzing Mosquito Movement from Video Tracking Results (Muijres 2022).

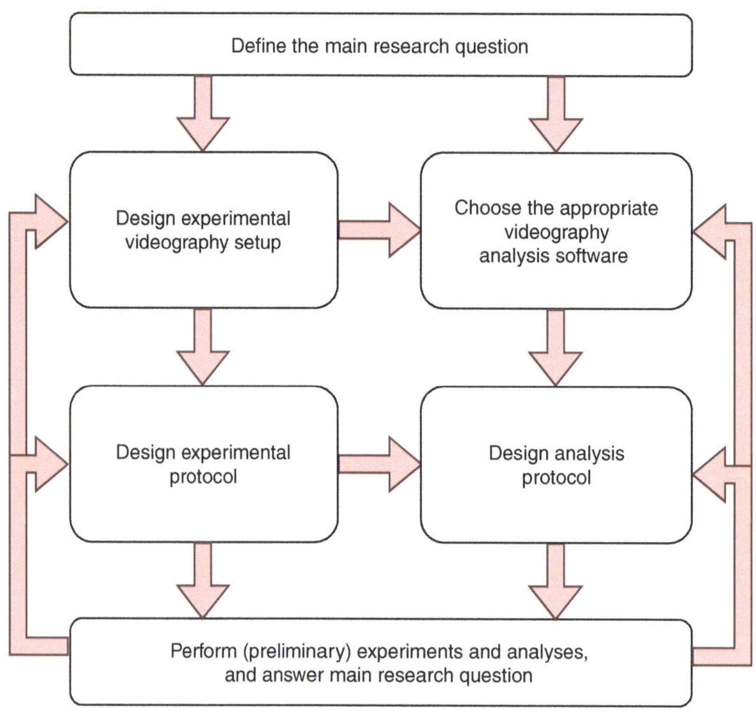

FIGURE 1. The workflow for designing a videography-based mosquito behavior study.

 Cite this introduction as *Cold Spring Harb Protoc*; doi:10.1101/pdb.top107676

DESIGNING YOUR VIDEOGRAPHY EXPERIMENT

As with any research project, it is crucial to align the experimental design and procedure with the main research question. The workflow for experimental videography design targeting moving mosquitoes is summarized in Figure 1. This plan consists of four steps.

1. Define your main research question.
2. Based on the main research question, determine the hardware requirements for the experimental setup. The most important questions to answer are thus what size the experimental arena should be, how many cameras are needed, and what camera lighting conditions are required. For details, see the section on hardware requirements below and Protocol 1: Designing a Generic Videography Experiment for Studying Mosquito Behavior (Muijres et al. 2022).
3. Based on the research question and experimental design, choose the appropriate analysis software and tools for processing and analyzing the video data. Here, make certain the chosen software permits the extraction of the type(s) of data produced by the videography system, and that the analysis output can be used to answer the main research question. For details, see the section on analysis software requirements below, and Protocol 2: Real-Time Tracking of Multiple Moving Mosquitoes (Straw et al. 2022) and Protocol 3: Tracking the Body, Wing, and Leg Kinematics of Moving Mosquitoes (Hedrick et al. 2022).
4. Following videography design and software selection, you should design your experimental and analysis methods, such that they are aligned with the research question and the used hardware and software. For details, see the "analysis software requirements" section below and Protocol 4: Quantifying and Analyzing Mosquito Movement from Video Tracking Results (Muijres 2022).
5. Finally, build the experimental setup, perform your research, and answer your research question.

Note that here we describe this plan as a single set of steps, but the experimental setup may often need to be redesigned and optimized based on preliminary experiments and analyses, as indicated by the return arrows in Figure 1.

Hardware Requirements

As stated above, the nature of the desired data determines the number of cameras needed in a particular experiment. The most important consideration here is whether 3D movements need to be measured. A single-camera setup can only track movements accurately within the viewing plane of the camera. With a multicamera setup, whereby multiple (synchronized) cameras simultaneously film an event, 3D movements can be reconstructed. As a general rule, using more cameras increases precision measurement and mitigates view obstruction. However, multicamera systems are more complex to set up, calibrate, and operate, and data analysis is more labor-intensive. Therefore, a multicamera setup should only be used if needed. Consult Protocol 1: Designing a Generic Videography Experiment for Studying Mosquito Behavior (Muijres et al. 2022) to determine the level of system complexity appropriate for your experiments.

When using a single-camera setup, ensure that movements occur mostly within the camera viewing plane. By doing so, single-camera setups have been used to determine, in high temporal and spatial resolution, the center of mass of mosquitoes impacted by falling H_2O drops (Dickerson et al. 2012), mosquitoes falling through dense gas (Dickerson et al. 2015b), taking off (Smith et al. 2018), in short flight sequences (Ortega-Jiménez and Combes 2018), and probing for holes in a net (Dickerson et al. 2018). Single-camera setups have also been used to track moving body parts such as tarsi (Smith et al. 2018), wings (Dickerson and Hu 2014), and halteres (Dickerson et al. 2015b).

Many movements of mosquitoes are inherently 3D, and when 3D motion is of interest a multicamera system is needed. Depending on the spatial and temporal resolution of the multicamera system, one can study the 3D kinematics of moving mosquitoes at varying levels of detail. A low-resolution system allows for location tracking of mosquitoes within a relatively large area, up to meters in size. Tracking over a relatively large volume allows, for example, the study of flying mosquito interaction with large objects such as mosquito traps (Cribellier et al. 2018, 2020; Batista et al. 2019). Multicamera systems with both high spatial and temporal resolutions can be used to track the body and wingbeat kinematics of mosquitoes in detail. This facilitates the observation of the bio(fluid) mechanics of mosquito flight (Bomphrey et al. 2017; Nakata et al. 2020) during takeoff (Muijres et al. 2017) and landing (Smith et al. 2020).

Analysis Software Requirements

There are several software tools available for analyzing videography data of moving animals. These include both open-source and commercial software. The choice of analysis tool is driven by the research question, the resulting required measurement complexity and analysis volume, and the researcher's experience. Below, we list seven analysis packages ranging from basic to complex, but this is by no means a comprehensive list.

1. Tracker Video Analysis and Modeling Tool is likely the simplest and most accessible two-dimensional tracking software available. It runs in Java and was developed by Douglas Brown at Cabrillo College (Brown 2007). The software allows for digitizing single-camera video data and is user-friendly but has limited options.
2. Direct Linear Transformation data viewer (DLTdv) is a MATLAB package for analyzing video data of moving animals (Hedrick 2008), packaged for use in MATLAB or the freely available MATLAB runtime environment. It is developed and maintained by Tyson Hedrick at the University of North Carolina at Chapel Hill. It is user-friendly, has extensive tutorials and support, and has many features. It can be used to digitize animal movements in a single camera or simultaneously in multiple cameras. It includes sophisticated camera calibration routines, including a DLT calibration for 3D reconstruction, and lens distortion corrections (using the accompanying pose and position-based "Wand calibration tools"). The analysis tool allows for both manual and automatic tracking of movements. The automatic tracking can be done using classic machine vision techniques and using a deep learning algorithm (Breslav et al. 2017). We discuss this system in detail in Protocol 3: Tracking the Body, Wing, and Leg Kinematics of Moving Mosquitoes (Hedrick et al. 2022).
3. Braid is an open-source real-time tracking software for animal movement research developed by Andrew Straw at Freiburg University (https://strawlab.org/braid) and is a rewrite of a similar software called Flydra (Straw et al. 2011). Because it tracks moving animals in real time, it is fully integrated in the experimental videography system and thus sets specific requirements for the hardware. This software can track multiple mosquitoes in real time based on video streams, and therefore does not require saving and postprocessing of video data. We discuss the Braid system in detail in Protocol 2: Real-Time Tracking of Multiple Moving Mosquitoes (Straw et al. 2022).
4. Track3D is a commercial 3D videography-based tracking software developed by the company Noldus Information Technology in collaboration with professor Willem Takken (Laboratory of Entomology) and professor Johan van Leeuwen (Experimental Zoology Group) of Wageningen University (Noldus et al. 2011; Spitzen et al. 2013). It is a user-friendly software package that allows automatic 3D tracking of the position of a moving animal such as a mosquito and has good postprocessing and visualization tools.
5. Hull Reconstruction Motion Tracking (HRMT) is an automated insect tracking algorithm developed by the Itai Cohen Group (Ristroph et al. 2009). HRMT allows for automatic tracking of detailed wing and body kinematics of moving mosquitoes and is therefore particularly useful for studying mosquito flight in detail. That said, HRMT requires the use of several high-speed cameras and expert knowledge of machine vision algorithms (Ristroph et al. 2009; Walker et al. 2012; Bomphrey et al. 2017).

 Cite this introduction as *Cold Spring Harb Protoc*; doi:10.1101/pdb.top107676

6. Model-based 3D insect tracker was originally developed by Ebraheem Fontaine at the California Institute of Technology (Fontaine et al. 2009). It has been used to track the movements of swimming fish, moving worms, and flying insects, including mosquitoes (Fontaine et al. 2009; Muijres et al. 2014, 2017). Similar to the HRMT method, these algorithms require expert knowledge of machine vision. However, model-based tracking tends to require fewer cameras (typically three) than HRMT (four or more).
7. Deep Lab Cut (DLC) is a Python-based program that uses sophisticated deep-learning algorithms to track moving animals (Mathis et al. 2018). It was developed by Alexander and Mackenzie Mathis at École Polytechnique Fédérale de Lausanne and shows great potential for behavioral research on mosquitoes. It has an active community of developers and good tutorials and tech support. To effectively use deep-learning algorithms, a dedicated workstation with a high-end graphics card is needed, although simple artificial neural networks can be developed even on a laptop. DLC supports single-camera analysis and two-camera stereoscopic setups for 3D reconstruction.

In the accompanying protocols, we present high-level instructions for tracking moving mosquitoes in real time and in detail. For the real-time tracking of multiple moving mosquitoes, we present Braid (see Protocol 2: Real-Time Tracking of Multiple Moving Mosquitoes [Straw et al. 2022]). For detailed tracking of body, wing, and leg movements of mosquitoes, we highlight DLTdv because of its long history of use, number of users, available documentation, ease of use, and overall capability (see Protocol 3: Tracking the Body, Wing, and Leg Kinematics of Moving Mosquitoes [Hedrick et al. 2022]). DLTdv can also perform both manual and automatic tracking, whereby automatic tracking can be done using a classic machine vision algorithm and a deep-learning algorithm.

Within particular dimensionality constraints, each software tool listed above can ultimately provide the user with the same types of data. For example, Tracker, the simplest package, can produce the same types of tracks as DLC, but perhaps at a higher manual labor cost. Thus, the choice which software is best for your project may not be clear. There is in fact no correct or incorrect choice. Instead, consider the experience of the user, your short- and long-term goals, the video quality, and available computational power when approaching the various software packages. Tracker requires the least initial investment but lacks high-throughput abilities. In contrast, DLC requires much greater computing capabilities, familiarity with Python, and considerable algorithm training before any data can be realized. However, the payoff is the ability to track points in multiple videos without much user input. If videos are inconsistent one to the next, or otherwise demand a high degree of human subjectivity, a manual tracker with the ability to switch to (semi)automatic tracking such as DLTdv will be the natural choice. Finally, if one is only interested in the position of the mosquito, Braid might be the best option. When set up and well-calibrated, Braid provides location data of mosquitoes without any postprocessing, allowing you to track a large number of mosquitoes at low labor costs.

Regardless of the analysis software, every analysis must include a calibration routine to transform video data (pixel-based coordinates) to a measurement with physical units. For single-camera systems, a scale reference such as an object of known dimension often suffices. Once a pixels/length metric is determined, multiple videos can be captured in sequence without providing a new scale, so long as no camera adjustments are made. Multicamera systems often use more sophisticated calibration routines, where an object with known dimensions is placed in the camera's field of view. Depending on the calibration routine, this can be a 3D object with many reference points, a checkerboard placed at different locations, or even a moving stick or set of light-emitting diodes. These calibration routines are often part of the tracking software. Camera calibrations are specific to the precise relative position of the cameras, lens settings, and pixel resolution. Changes to any of these parameters, even a small inadvertent nudge given to one of the cameras, will reduce calibration accuracy. Because the accuracy and quality of the videography experiments depends directly on the quality of the calibration, calibration routines need to be performed very carefully and regularly.

REFERENCES

Batista EP, Mapua SA, Ngowo H, Matowo NS, Melo EF, Paixão KS, Eiras AE, Okumu FO. 2019. Videographic analysis of flight behaviours of host-seeking *Anopheles arabiensis* towards BG-malaria trap. *PLoS ONE* **14:** e0220563. doi:10.1371/journal.pone.0220563

Bomphrey RJ, Nakata T, Phillips N, Walker SM. 2017. Smart wing rotation and trailing-edge vortices enable high frequency mosquito flight. *Nature* **544:** 92–95. doi:10.1038/nature21727

Breslav M, Hedrick TL, Sclaroff S, Betke M. 2017. Automating image analysis by annotating landmarks with deep neural networks. *arXiv* doi:10.48550/arXiv.1702.00583

Brown D. 2007. *Combining computational physics with video analysis in tracker. Proceedings of the American Association of Physics Teachers AAPT Summer Meeting, Greensboro, NC, August 2007.* American Institute of Physics (AIP), College Park, MD.

Cribellier A, van Erp JA, Hiscox A, Lankheet MJ, van Leeuwen JL, Spitzen J, Muijres FT. 2018. Flight behaviour of malaria mosquitoes around odour-baited traps: capture and escape dynamics. *R Soc Open Sci* **5:** 180246. doi:10.1098/rsos.180246

Cribellier A, Spitzen J, Fairbairn H, van de Geer C, van Leeuwen JL, Muijres FT. 2020. Lure, retain, and catch malaria mosquitoes. How heat and humidity improve odour-baited trap performance. *Malar J* **19:** 1–16. doi:10.1186/s12936-020-03403-5

Cribellier A, Straw AD, Spitzen J, Pieters RPM, van Leeuwen JL, Muijres FT. 2022. Diurnal and nocturnal mosquitoes escape looming threats using distinct flight strategies. *Curr Biol* **32:** 1–15. doi:10.1016/j.cub.2022.01.036

Dickerson AK, Hu DL. 2014. Mosquitoes actively remove drops deposited by fog and dew. *Integr Comp Biol* **54:** 1008–1013. doi:10.1093/icb/icu042

Dickerson AK, Shankles PG, Madhavan NM, Hu DL. 2012. Mosquitoes survive raindrop collisions by virtue of their low mass. *Proc Natl Acad Sci* **109:** 9822–9827. doi:10.1073/pnas.1205446109

Dickerson AK, Liu X, Zhu T, Hu DL. 2015a. Fog spontaneously folds mosquito wings. *Physics Fluids* **27:** 021901. doi:10.1063/1.4908261

Dickerson AK, Shankles PG, Berry BE, Hu DL. 2015b. Fog and dense gas disrupt mosquito flight due to increased aerodynamic drag on halteres. *J Fluids Struct* **55:** 451–462. doi:10.1016/j.jfluidstructs.2015.03.016

Dickerson AK, Olvera A, Luc Y. 2018. Void entry by *aedes aegypti* (diptera: Culicidae) mosquitoes is lower than would be expected by a randomized search. *J Insect Sci* **18:** 9. doi:10.1093/jisesa/iey115

Fontaine EI, Zabala F, Dickinson MH, Burdick JW. 2009. Wing and body motion during flight initiation in *Drosophila* revealed by automated visual tracking. *J Exp Biol* **212:** 1307–1323. doi:10.1242/jeb.025379

Hedrick TL. 2008. Software techniques for two- and three-dimensional kinematic measurements of biological and biomimetic systems. *Bioinspir Biomim* **3:** 034001. doi:10.1088/1748-3182/3/3/034001

Hedrick TL, Dickerson AK, Muijres FT, Pieters R. 2022. Tracking the body, wing, and leg kinematics of moving mosquitoes. *Cold Spring Harbor Protoc* doi:10.1101/pdb.prot107928

Lee H, Halverson S, Ezinwa N. 2018. Mosquito-borne diseases. *Prim Care* **45:** 393–407. doi:10.1016/j.pop.2018.05.001

Mathis A, Mamidanna P, Cury KM, Abe T, Murthy VN, Mathis MW, Bethge M. 2018. Deeplabcut: markerless pose estimation of user-defined body parts with deep learning. *Nat Neurosci* **21:** 1281–1289. doi:10.1038/s41593-018-0209-y

Muijres FT. 2022. Quantifying and analyzing mosquito movement from video tracking results. *Cold Spring Harbor Protoc* doi:10.1101/pdb.prot107929

Muijres FT, Elzinga MJ, Melis JM, Dickinson MH. 2014. Flies evade looming targets by executing rapid visually directed banked turns. *Science* **344:** 172–177. doi:10.1126/science.1248955

Muijres F, Chang S, van Veen W, Spitzen J, Biemans B, Koehl M, Dudley R. 2017. Escaping blood-fed malaria mosquitoes minimize tactile detection without compromising on take-off speed. *J Exp Biol* **220:** 3751–3762. doi:10.1242/jeb.163402

Muijres FT, Dickerson AK, Pieters R. 2022. Designing a generic videography experiment for studying mosquito behavior. *Cold Spring Harbor Protoc* doi:10.1101/pdb.prot107926

Nakata T, Phillips N, Simões P, Russell IJ, Cheney JA, Walker SM, Bomphrey RJ. 2020. Aerodynamic imaging by mosquitoes inspires a surface detector for autonomous flying vehicles. *Science* **368:** 634–637. doi:10.1126/science.aaz9634

Noldus L, Grieco F, Spoor C, van Leeuwen J, Spitzen J, Takken W. 2011. Track3D: A new system for tracking, visualization and analysis of insect flight behavior. In *Proceedings of the 6th Asia-Pacific Conference on Chemical Ecology; Chemical Ecology Future: Green Chemoecology in 21st Century*, pp. 112. Beijing, China. Wageningen University, Wageningen, Netherlands.

Ortega-Jiménez VM, Combes SA. 2018. Living in a trash can: turbulent convective flows impair *Drosophila* flight performance. *J R Soc Interface* **15:** 20180636. doi:10.1098/rsif.2018.0636

Ristroph L, Berman GJ, Bergou AJ, Wang ZJ, Cohen I. 2009. Automated hull reconstruction motion tracking (HRMT) applied to sideways maneuvers of free-flying insects. *J Exp Biol* **212:** 1324–1335. doi:10.1242/jeb.025502

Smith NM, Clayton GV, Khan HA, Dickerson AK. 2018. Mosquitoes modulate leg dynamics at takeoff to accommodate surface roughness. *Bioinspir Biomim* **14:** 016007. doi:10.1088/1748-3190/aaed87

Smith NM, Balsalobre JB, Doshi M, Willenberg BJ, Dickerson AK. 2020. Landing mosquitoes bounce when engaging a substrate. *Sci Rep* **10:** 15744. doi:10.1038/s41598-020-72462-0

Spitzen J, Spoor CW, Grieco F, ter Braak C, Beeuwkes J, van Brugge SP, Kranenbarg S, Noldus LPJJ, van Leeuwen JL, Takken W. 2013. A 3D analysis of flight behavior of *Anopheles gambiae* sensu stricto malaria mosquitoes in response to human odor and heat. *PLoS ONE* **8:** e62995. doi:10.1371/journal.pone.0062995

Straw AD, Branson K, Neumann TR, Dickinson MH. 2011. Multi-camera real-time three-dimensional tracking of multiple flying animals. *J R Soc Interface* **8:** 395–409. doi:10.1098/rsif.2010.0230

Straw AD, Pieters R, Muijres FT. 2022. Real-time tracking of multiple moving mosquitoes. *Cold Spring Harbor Protoc* doi:10.1101/pdb.prot107927

Sutcliffe J, Ji X, Yin S. 2017. How many holes is too many? A prototype tool for estimating mosquito entry risk into damaged bed nets. *Malar J* **16:** 304. doi:10.1186/s12936-017-1951-4

Walker SM, Thomas AL, Taylor GK. 2012. Operation of the alula as an indicator of gear change in hoverflies. *J R Soc Interface* **9:** 1194–1207. doi:10.1098/rsif.2011.0617

Protocol 1

Designing a Generic Videography Experiment for Studying Mosquito Behavior

Florian T. Muijres,[1,3] Andrew K. Dickerson,[2] and Remco Pieters[1]

[1]*Department of Experimental Zoology, Wageningen University, 6708 PB Wageningen, the Netherlands;*
[2]*Department of Mechanical, Aerospace, and Biomedical Engineering, University of Tennessee, Tennessee 37996, USA*

In this protocol, we describe the basic design considerations and general method to set up a videography system to study mosquito behavior. A basic videography system to study mosquito behavior requires one or more cameras with an optical lens, camera lighting, a calibration setup, and a system to record the video data or otherwise control the camera. Here, we define two types of systems: (1) a real-time videography-based tracking system for determining the position of multiple moving (flying) mosquitoes, and (2) a high-fidelity videography system that can track the detailed movements of body, wings, and legs of a single mosquito at high spatial and temporal resolutions. These high-fidelity trackers are divided into single-camera systems for studying two-dimensional (2D) movements, and multicamera systems that can reconstruct three-dimensional (3D) movements of the mosquito.

MATERIALS

It is essential that you consult the appropriate Material Safety Data Sheets and your institution's Environmental Health and Safety Office for proper handling of equipment and hazardous materials used in this protocol.

Reagents

Mosquitoes

Equipment

Calibration objects, such as a simple ruler (see Step 6)

All videography systems require a calibration to convert the videography measurements in pixel coordinates into spatial units such as millimeters. To perform this calibration, a calibration object is needed. For single-camera systems, this calibration object can consist of a simple ruler with an easily distinguishable scale. The ruler should be placed in the same plane of focus as the mosquito, spanning most of the camera frame. For multicamera systems, more complex calibration objects are needed to determine how combined pixel coordinates in the set of cameras can be converted into 3D spatial coordinates (see Step 6).

Camera(s), either a one-, two-, or three-camera system

3D tracking requires two or more cameras (see Step 2).

Flight arena

[3]Correspondence: florian.muijres@wur.nl

Cite this protocol as *Cold Spring Harb Protoc*; doi:10.1101/pdb.prot107926

Lights for set illumination

Backlighting provides sharp silhouettes but hides on-body features (see Step 4).

Tracking software

For more information on choosing a tracking software, see Introduction: Using Videography to Study the Biomechanics and Behavior of Freely Moving Mosquitoes (Dickerson et al. 2022).

Video recording system (see Steps 7 and 8)

METHOD

1. Design and build the flight arena.

   The flight arena should permit optical access and allow mosquitoes to fly unencumbered but constrain the volume for easy filming. Tethered experiments require no enclosure.

2. Determine the required number of cameras.

   The most important design decision when developing a videography system to study freely moving mosquitoes is the number of cameras used. This decision is primarily driven by the aim of the study.

Tracking Basic Behavioral Measurements

If one is interested in basic behavioral measurements, a simple single-camera system, like that shown in Figure 1A, is often sufficient. Using such a system, one can track 2D positional data, wingbeat frequency, feet and proboscis placement on a substrate, and some basic behavioral metrics. However, a single camera cannot be used to accurately quantify movements out of the image plane, and only one side of the mosquito is visible. Thus, if one needs to quantify 3D movement and wants to reconstruct the position and orientation of the complete animal, a multicamera system is needed.

Tracking the 3D Position of the Mosquito

If one is only interested in tracking the 3D position of the mosquito, only two cameras are needed (see Fig. 1B), although using more cameras will generate higher precision positional estimates or encompass a larger volume-of-interest. In a multicamera system, it is imperative that at least two cameras capture each point within the volume of interest. Systems that are used to determine the 3D position of a mosquito need to run at frame rates of at least 50–100 fps, depending on the speed of movement. Thus, flying tends to require higher frame rates than walking or probing.

Tracking the 3D Movements of Body Parts

When one needs to quantify the 3D movements of body parts, such as head, thorax, abdomen, legs, wings, and proboscis, more sophisticated multicamera systems are needed. To prevent temporarily obscured body parts, a system with at least three cameras is required to be positioned around the mosquito. That way, all body parts are almost always visible from at least two cameras. Frame rate requirements are driven by the movement frequency or the number of pixels traversed by the body part of interest. This is particularly problematic when studying mosquito flight, as flying mosquitoes beat their wings at exceptionally high wingbeat frequencies (up to 600 Hz). To reconstruct the 3D wingbeat kinematics, one should film the wing movement at 30–40 images per wingbeat, resulting in a required frame rate up to 24,000 fps.

Regardless of the type of multicamera system used, there are a couple of critical design rules that should always be considered to accurately determine 3D movements, including camera synchronization (see Step 5) and viewing angles.

To minimize 3D reconstruction errors, the stereoscopic cameras should view the object from different angles. A difference in viewing angle of 90° is best. For viewing angle differences of 0° (cameras in the same position) and 180° (cameras opposite each other), 3D reconstruction is not possible. Therefore, it is best to aim for angles of 90° between the cameras, but angles of >45° and <135° are most often sufficient.

If one wants to reconstruct the 3D kinematics of different body parts, it is often best to use a three-camera system with one camera viewing the mosquito from the top, one from the side, and one from the front. This system allows viewing the mosquito from all sides, minimizing occlusions, and the viewing angle between all cameras is exactly 90° (see Fig. 1C).

3. Align camera and focus the lens and set the aperture for the desired depth of field (DOF). Place an alignment object in the center of the volume of view. Set the lens aperture to fully open. Focus the

Cite this protocol as *Cold Spring Harb Protoc*; doi:10.1101/pdb.prot107926

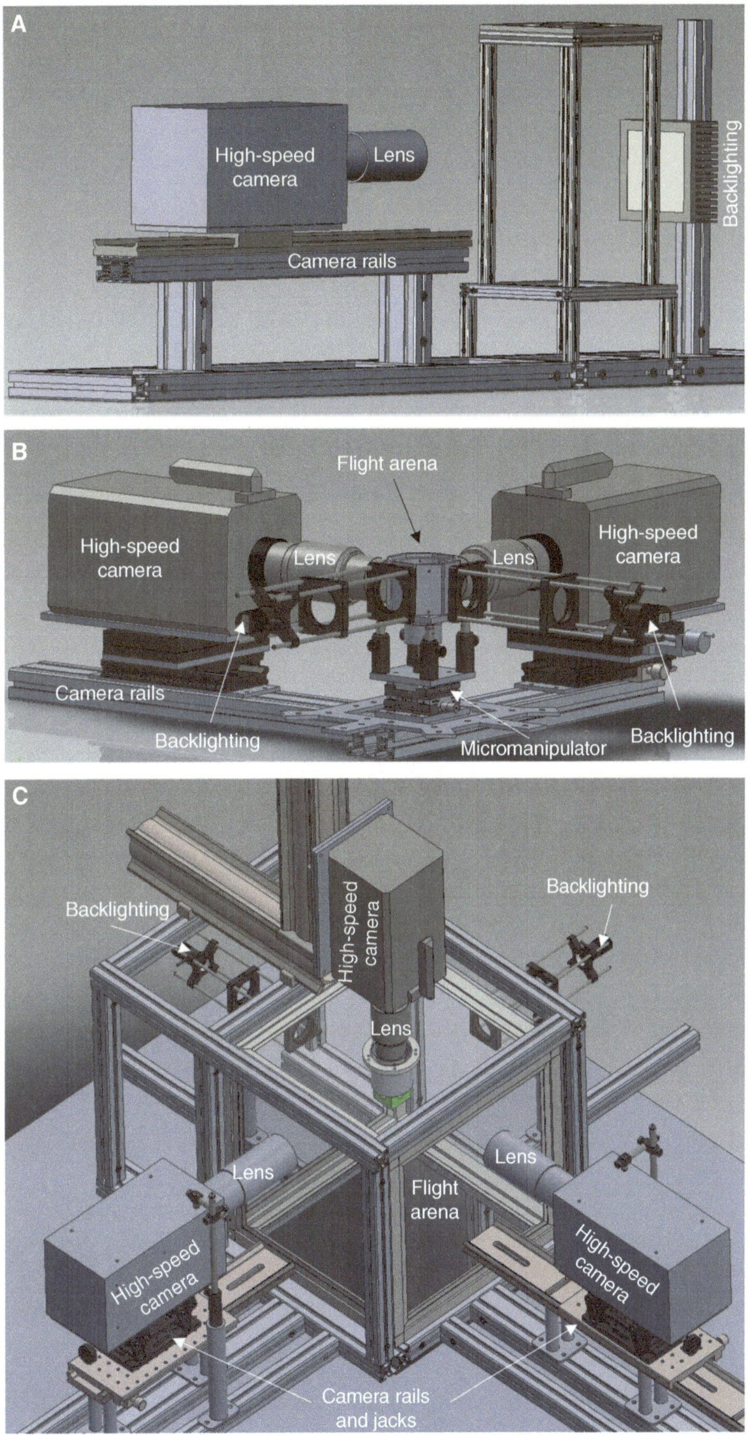

FIGURE 1. Example of videography setups with (*A*) one, (*B*) two, and (*C*) three cameras.

lens. Close the aperture to the desired DOF—a piece of printed text is an easy-to-use object to focus the lens. Frame rate is not important for Step 3.

Any videography system consists of either a single or multiple cameras that are positioned such that the viewing area of the camera(s) cover the volume of interest. In addition, the object of interest should be filmed at high enough spatial and temporal resolution. To achieve all three requirements (sufficient spatial and temporal resolution and coverage of the complete volume of interest), one should carefully design the videography setup. This includes the choice of camera and lenses, the placement of the camera(s), and the camera and lens settings. The most important trade-offs include the following.

Monochrome Versus Color Cameras

Full color recordings show the natural environment of the mosquito. But this color sampling comes at a cost: Color cameras use a color filter array (often a Bayer pattern) to capture the primary colors. For each pixel, only one color is measured directly. To get the full color representation of each pixel, it must be combined with the neighboring pixels. Therefore, monochrome cameras are more sensitive (three times more light per pixel), have more detail (~20%), and can record video at higher frame rates than equivalent color cameras. Furthermore, monochrome cameras are in general sensitive to near-infrared light and can therefore be used in combination with infrared lights to simulate nighttime conditions for the mosquito. Thus, monochrome camera sensors have many advantages compared to color sensors, so when color information is not needed, a monochrome camera sensor is highly recommended.

Spatial Resolution and Volume of Interest

The pixel resolution of the video camera is limited, and thus for a given volume of interest, the maximum achievable spatial resolution of the filmed mosquito is also limited. Thus, when designing the camera setup, one should determine the minimum required field of view and spatial resolution, and thus choose a camera with a corresponding pixel resolution.

Spatial Resolution and Temporal Resolution

Most professional video cameras can record video at various combinations of frame rate (temporal resolution) and pixel (spatial) resolution, whereby the pixel resolution decreases with increasing frame rate. Thus, based on the research question, one should select the camera type that can best achieve the required combination of frame rate and pixel resolution.

Image Brightness and Motion Blur

When filming rapidly moving mosquitoes, one should be careful to prevent motion blur in the video. This is done by reducing the camera exposure time. However, reducing the exposure time reduces image brightness. Therefore, a balance needs to be found between image brightness and motion blur. It is again important to base such a balance on the research question: If one is interested in the rapid wingbeat movements of flying mosquitoes, then exposure time should be very short and images sharp; but if wing motion does not need to be digitized, blurred wing images are not problematic and exposure time can be increased to produce higher quality (brighter) images. The problem of image brightness and motion blur can also be solved by increasing the camera illumination (see next section) or increasing the sensor gain.

Image Brightness and DOF

When filming small animals such as mosquitoes, the DOF or focal depth of the camera system can be relatively small, especially when using macro lenses. In other words, one can only accurately digitize the movements of the mosquito within a very thin focal plane, and sometimes only portions of a complete mosquito are in focus. Thin focal planes are particularly problematic for multicamera systems, as the DOF should cover the complete volume of interest. One can increase the DOF by reducing the aperture of the lens, but doing so reduces the brightness of the image. Thus, similarly to motion blur, there is a trade-off between maximizing the DOF and image brightness, a problem that can also be reduced by increasing the camera image illumination.

4. Arrange the lights around the arena such that the cameras are viewing mosquitoes lit from either the front or back.

   Illumination of the camera setup is crucial for producing good quality video data. This is particularly relevant for high-speed stereoscopic videography because of the short exposure time and the small lens aperture required for maximizing the DOF.

Recording Nocturnal Mosquito Species

In contrast, high-intensity lights can disturb the behavior of mosquitoes, especially nocturnal species. To solve this problem, one can use near-infrared LED lights. Most insects cannot visually detect the near-infrared spectrum, minimizing the disturbance to mosquitoes and allowing one to recreate nighttime conditions while providing illumination for the cameras. There are a wide range of LED light sources available with wavelengths of 850 nm and 940 nm, generally designed for nighttime CCTV security applications. Also, most monochrome camera systems for machine vision application have good sensitivity in the 850- to 940-nm range. There are also cameras available that are specially designed to operate in the near-infrared range. Color cameras tend to have infrared filters in front of their sensor and should be avoided if using near-infrared illumination. In general, monochrome cameras tend to be more sensitive to light, so if color vision is not needed, we suggest using monochrome cameras.

High-Contrast Lighting to Analyze 3D Body Kinematics

A high contrast between the mosquito and background is advantageous to detect the mosquito and analyze its 3D body kinematics. Very high contrast is easily achievable with backlighting by placing an array of LEDs in combination with a diffusing sheet behind the mosquito, thus shining the light straight into the camera. The mosquito will then be visible as a dark shadow in front of a bright background. This illumination setup has the added benefit that the light intensity is relatively high, resulting in sharp video images.

The downside of shadow images produced by backlighting is that detailed information about the body surface is often not visible. If this information is desired, one can use a conventional illumination setup with lights positioned next to the cameras. Thus, the cameras will record the light reflected from the mosquito, which will be seen as a light object with relatively high surface contrast in front of a dark background.

A less conventional camera illumination setup is to place a single LED in combination with a long focal collimator lens opposite the camera (Walker and Taylor 2021; Voesenek et al. 2019). This setup also results in backlit video images, but due to the collimated light beam, the DOF becomes relatively high. In this case, the downside is that the field of view becomes maximally equal to the size of the image sensor, or the maximum aperture of the lens.

5. When using more than one camera, synchronize the cameras using a (TTL) signal from a digital pulse generator or from a transmitting camera to the receiving camera(s).

   All cameras should be synchronized in time, which is best achieved using a synchronization signal. A synchronization signal can be produced by a digital pulse generator, or by a transmitting camera that sends the signal to all other receiving cameras. If synchronization is not possible because of hardware limitations, video data must later be aligned (e.g., using a filmed action such as an LED flash at a random time interval during the whole recording), but this method will inherently result in larger 3D reconstruction errors.

6. Calibrate the camera(s) using a calibration object.

   Camera calibration should be done regularly, especially after the camera or lens settings are changed or a camera is (slightly) moved or rotated.

   The calibration routine is often included in the analysis software, and each routine requires its dedicated calibration object. This can be a 3D object with many points with known 3D locations, or a checkerboard or stick that should be moved throughout the field of view of all cameras. The calibration object should be made very precisely and cover as much of the experimental volume as possible to minimize calibration error. Calibration routines using an object of known 3D location typically use Direct Linear Transformation (DLT) to convert pixel coordinates of multiple cameras into 3D spatial world coordinates; the resulting spatial coordinate orientation matches that of the calibration object. Calibration routines using a moving object typically first solve for the relative position and orientation of the cameras along with their lens properties; these may later be converted to DLT coefficients for calculating 3D spatial world coordinates. When the optics of the camera lenses are inferior, lens distortion corrections are needed. The most sophisticated calibration routines include such distortion corrections.

7. Perform your experiment and record videos of the moving mosquitoes inside the arena with your camera setup.

   Video recordings at low frame rates can be streamed to permanent storage in real time, but at high frame rates, the recording should be triggered manually or automatically.

8. Store video files on a data storage device.

   Most machine vision cameras do not have their own controller or data storage system. These cameras must therefore be connected to a video recording system, typically consisting of one or more personal computers. The video recording system sets the acquisition parameters: frame rate, exposure time, and sensor gain, and stores the video data. At frame rates on the order of 100 fps, video can often be streamed directly to the computer, where it will be stored or processed. For recordings at a higher frame rate, the camera footage must often first be stored on local camera RAM, after which it can be downloaded to the computer (typically via a gigabit ETHERNET or USB3 connection).

9. Extract kinematic data from videos using an appropriate tracking software.

   See Protocol 2: Real-Time Tracking of Multiple Moving Mosquitoes (Straw et al. 2022) and Protocol 3: Tracking the Body, Wing, and Leg Kinematics of Moving Mosquitoes (Hedrick et al. 2022) for details.

10. Analyze the kinematic data.

See Protocol 4: Quantifying and Analyzing Mosquito Movement from Video Tracking Results (Muijres 2022) for details.

DISCUSSION

We present three examples of generic videography systems in Figure 1, which represent nearly idealized systems. These setups incorporate rail sliders upon which cameras rest, micromanipulators to position flight arenas, and lights affixed to the rig. Such levels of build-out and fine adjustment allow for the precise placement and adjustment of the camera, lights, flight arena, and calibration object. The sophistication shown in Figure 1 is not required. Cameras can be mounted to tripods and lights affixed to arms. However, although more rudimentary setups have been the basis of successful studies, they are more prone to misalignment, calibration errors, and inconsistency over the course of experiments. The cameras in Figure 1 are positioned with mutually orthogonal viewing angles. This is ideal for reconstructing 3D position, but is not strictly required, and any pair of cameras, excluding the special case of cameras with viewing angles separated by 0° or 180°, can be used to collect 3D measurements.

REFERENCES

Dickerson AK, Muijres FT, Pieters R. 2022. Using videography to study the biomechanics and behavior of freely moving mosquitoes. *Cold Spring Harbor Protoc* doi:10.1101/pdb.top107676

Hedrick TL, Dickerson AK, Muijres FT, Pieters R. 2022. Tracking the body, wing, and leg kinematics of moving mosquitoes. *Cold Spring Harbor Protoc* doi:10.1101/pdb.prot107928

Muijres FT. 2022. Quantifying and analyzing mosquito movement from video tracking results. *Cold Spring Harbor Protoc* doi:10.1101/pdb.prot107929

Straw AD, Pieters R, Muijres FT. 2022. Real-time tracking of multiple moving mosquitoes. *Cold Spring Harbor Protoc* doi:10.1101/pdb.prot107927

Voesenek CJ, Pieters RP, Muijres FT, van Leeuwen JL. 2019. Reorientation and propulsion in fast-starting zebrafish larvae: an inverse dynamics analysis. *J Exp Biol* **222:** jeb203091. doi:10.1242/jeb.203091

Walker SM, Taylor GK. 2021. A semi-empirical model of the aerodynamics of manoeuvring insect flight. *J R Soc Interface* **18:** 20210103. doi:10.1098/rsif.2021.0103

Cite this protocol as *Cold Spring Harb Protoc*; doi:10.1101/pdb.prot107926

Protocol 2

Real-Time Tracking of Multiple Moving Mosquitoes

Andrew D. Straw,[1,3] Remco Pieters,[2] and Florian T. Muijres[2]

[1]*Institute of Biology I & Bernstein Center Freiburg, Faculty of Biology, Albert-Ludwigs-Universität Freiburg, 79085 Freiburg im Breisgau, Germany;* [2]*Department of Experimental Zoology, Wageningen University, 6708 PB Wageningen, the Netherlands*

Tracking mosquitoes in real time, as opposed to recording video files and performing the tracking step later, is useful for two reasons. The first is efficiency. Real-time tracking requires less storage because video images do not need to be saved and followed by a tracking step. The second is that tracking data can be used to interact with the animal in some way, such as triggering the approach of a looming object. In this protocol, we discuss the use of Braid, free software for performing real-time, multi-camera, multianimal tracking. We describe a setup with four cameras capable of tracking the three-dimensional (3D) position of mosquitoes at 100 frames per second in a volume of 30 cm × 30 cm × 60 cm with millimeter accuracy. The specific hardware configuration is flexible and can be substituted using different or additional components to adjust the tracking parameters as needed.

MATERIALS

It is essential that you consult the appropriate Material Safety Data Sheets and your institution's Environmental Health and Safety Office for proper handling of equipment and hazardous materials used in this protocol.

Equipment

Arduino Nano to generate camera synchronization pulses

Download, compile, and flash the firmware to the Nano using the instructions at https://github.com/strawlab/triggerbox.

Braid

Available at https://strawlab.org/braid/.

Cameras (four), C-mount, global shutter, external triggered, 2.3-MP, monochrome, 160-fps (Basler a2A1920-160umBAS)

Braid version 0.10.1 supports only Basler cameras.

Each camera should have a USB3 cable and 1/4″-20 tripod mount.

Camera lenses (four), varifocal 4–13-mm, f/1.5, near-infrared corrected, C-mount (Tamron M118VM413IR)

Checkerboard with flat and rigid backing for camera calibration

Computer, operating system, and camera software

[3]Correspondence: straw@bio.uni-freiburg.de

Cite this protocol as *Cold Spring Harb Protoc;* doi:10.1101/pdb.prot107927

Minimum specifications: Intel 64-bit CPU, 3-GHz or higher; 8-GB RAM or more; four USB3 ports on two separate USB3 busses (e.g., two USB3 ports on motherboard, two additional USB3 ports using PCIe card); Ubuntu version 20.04 desktop operating system (https://ubuntu.com); Braid version 0.10.1 (https://strawlab.org/braid).

Hardware trigger cables with M8 six-pin connector for camera synchronization

Available with cameras. Solder cable pin 4 (black) to Nano pin GND and cable pin 6 (pink) to Nano pin D9.

Near-infrared lights for illumination (850-nm)

White or red light-emitting diode (LED) for camera calibration (small, handheld, battery-powered)

METHOD

Expanded documentation is available online for free in User's Guide for Braid and Strand Camera *(Straw et al. 2011; Straw 2021). The overall workflow is shown in Figure 1. This protocol has been used, for example, to track* Anopheles *malaria mosquitoes and* Aedes *yellow-fever mosquitoes in laboratory and semifield studies (Cribellier et al. 2020; Cribellier et al. 2022).*

Lighting, Camera, and Lens Setup

Proper alignment/setup of all cameras is important from experimental inception (see Protocol 1: Designing a Generic Videography Experiment for Studying Mosquito Behavior [Muijres et al. 2022]). When the camera orientation, zoom, aperture, or focus position is changed, one must reconsider the camera calibration for both lens distortions and stereo calibration of the multiple cameras in the tracking system. The pylon Viewer provided by Basler (2021) provides some useful features for camera alignment such as a grid, histogram, and focus assistant.

1. Launch pylon Viewer.
2. Connect all cameras and start the video stream for each. Set camera trigger mode to "off" and use free run.
3. Arrange infrared lights such that the entire tracking volume is well lit for all cameras.

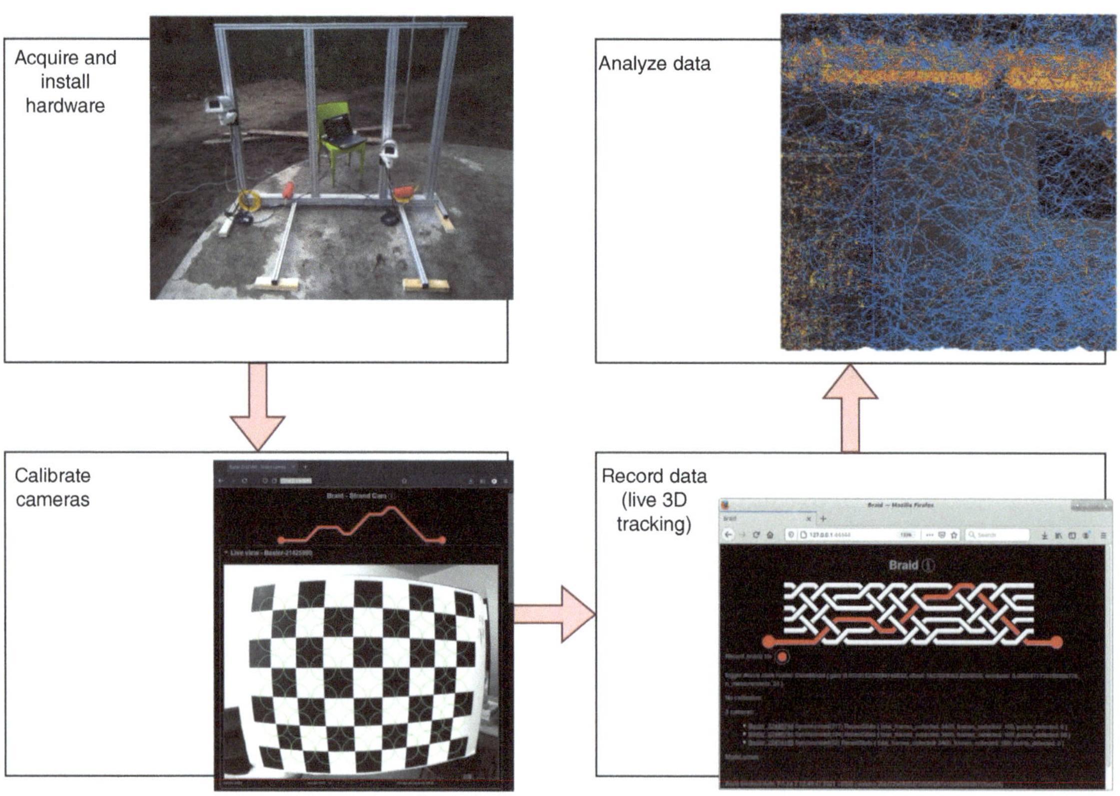

FIGURE 1. The workflow for real-time tracking experiments and analysis.

 Cite this protocol as *Cold Spring Harb Protoc*; doi:10.1101/pdb.prot107927

4. Fully open the lens aperture.
5. Adjust the exposure time to get a well-lit scene and so that the reference points in the setup are clearly visible; adjust the focus if needed. Frame rate is not yet important for this step.

   Use the camera's histogram to check whether the entire dynamic range is used. Sensor gain or pixel binning is optional to increase image brightness.

6. Adjust the camera orientation and lens zoom such that as much of the tracking volume (or part of it) is in view.

   Stop down the lens aperture to increase the depth of field.

7. Repeat Steps 1–6 for each camera. Ensure that each point in the tracking volume is visible by at least two camera views. A 90° angle between cameras maximizes stereo information. More cameras are helpful to improve tracking performance.

Synchronization Cables and Trigger Box Setup

8. Connect each camera with the trigger cable to the Arduino Nano.

Configuring Braid to Connect to Your Cameras

9. Prior to running Braid, generate a configuration file with the identity of the cameras, a calibration, if available, and tracking parameters. Any values not explicitly set in the configuration file will default to reasonable values.
10. To see all possible configuration items and their default settings, type "braid default-config" at the command line. Save the output of this program to a TOML file like "config.toml" then edit the file name "config.toml" to include the individual names of your cameras. Remove all values for which the defaults are OK.
11. Ensure your configuration is correct by running "braid show-config config.toml."

Camera Intrinsic Parameter Calibration

12. Start Braid with "braid run config.toml."

    Two stages of camera calibration are required, and they must be performed in sequential order. The first stage calibrates the intrinsic parameters of the camera. These are the focal length, optical center pixel, and any image distortion but do not include the position of the camera.

13. From the Braid main page, launch Strand Cam for the first camera to open the Strand Cam user interface for each camera.
14. Disable object detection by unchecking the "object detection" checkbox and go to the "checkerboard calibration" panel.
15. Enter the checkerboard size in the user interface.
16. Click "enable checkerboard calibration."
17. Collect at least 10 checkerboard images by holding the checkerboard in front of the camera. To accurately calibrate the camera's distortion, which is highest near the corners, ensure the checkerboard is visible in the corners of the image.
18. To perform and save the calibration, click the "perform and save calibration" button. Repeat this procedure for all cameras.

Multicamera Calibration

Here, the MultiCamSelfCal calibration routine (Svoboda et al. 2005) is used to create a single calibration for all cameras.

19. Reduce camera exposure duration and set the object detection parameters in Strand Camera so that the LED is detected when turned on, but no other points are detected.

    Turn off the infrared lights used for illumination.

20. Begin recording a data file in the Braid user interface.
21. Wave the LED smoothly through the tracking volume and ensure it tracked from all cameras, with at least some frames where it is visible from three or more cameras.

    Turn the LED on and off intermittently while moving through the tracking volume to validate synchronization.

22. Stop saving the data.
23. Run the MultiCamSelfCal calibration code to generate an initial multicamera calibration that is not yet "aligned."
24. Align the new calibration to the correct coordinate frame with the "flydra analysis calibration align GUI" tool.

    Use an LED to track a path along a known path in the coordinate frame.

Data Collection

25. In the Braid user interface, begin recording a data file.
26. Perform your experiment.
27. Stop saving the data. Braidz files can be viewed using the viewer at https://braidz.strawlab.org/ and analyzed starting with example scripts available in the *User's Guide for Braid and Strand Camera* (Straw et al. 2011; Straw 2021).

ACKNOWLEDGMENTS

This work was supported by a grant from the Momentum Programme of the Volkswagen Foundation to A.D.S.

REFERENCES

Basler AG. 2021. pylon Camera Software Suite. https://docs.baslerweb.com/pylon-camera-software-suite#documentation.

Cribellier A, Spitzen J, Fairbairn H, van de Geer C, van Leeuwen JL, Muijres FT. 2020. Lure, retain, and catch malaria mosquitoes. How heat and humidity improve odour-baited trap performance. *Malar J* **19:** 357. doi:10.1186/s12936-020-03403-5

Cribellier A, Straw AD, Spitzen J, Pieters RPM, van Leeuwen JL, Muijres FT. 2022. Diurnal and nocturnal mosquitoes escape looming threats using distinct flight strategies. *Curr Biol* **32:** 1–15. doi:10.1016/j.cub.2022.01.036

Muijres FT, Dickerson AK, Pieters R. 2022. Designing a generic videography experiment for studying mosquito behavior. *Cold Spring Harbor Protoc* doi:10.1101/pdb.prot107926

Straw AD. 2021. User's guide for braid and strand camera. https://strawlab.github.io/strand-braid/braid-and-strand-camera.html.

Straw AD, Branson K, Neumann TR, Dickinson MH. 2011. Multi-camera real-time three-dimensional tracking of multiple flying animals. *J R Soc Interface* **8:** 395–409. doi:10.1098/rsif.2010.0230

Svoboda T, Martinec D, Pajdla T. 2005. A convenient multi-camera self-calibration for virtual environments. *Presence Teleoperators Virtual Environ* **14:** 407–422. doi:10.1162/105474605774785325

Protocol 3

Tracking the Body, Wing, and Leg Kinematics of Moving Mosquitoes

Tyson L. Hedrick,[1,4] Andrew K. Dickerson,[2] Florian T. Muijres,[3] and Remco Pieters[3]

[1]*Department of Biology, University of North Carolina at Chapel Hill, North Carolina 27599, USA;* [2]*Department of Mechanical, Aerospace, and Biomedical Engineering, University of Tennessee, Knoxville, Tennessee 37996, USA;* [3]*Experimental Zoology Group, Wageningen University, 6708 PB Wageningen, the Netherlands*

In this protocol, we discuss general techniques for tracking the three-dimensional (3D) locations of the mosquito body, wings, legs, or other features of interest using videos. Tracking data must be acquired to produce detailed kinematics of moving mosquitoes. The software of focus for this protocol, DLTdv, was chosen for its widespread use and excellent support and because it is open-source. In addition, DLTdv allows both manual and automatic tracking. The automatic tracking can be done using a classic machine vision or machine-learning algorithm. The software supports both single-camera analysis and multicamera systems and can take advantage of sophisticated calibration algorithms, both for intrinsic lens distortion correction and for 3D DLT-based reconstruction. For this protocol, we assume all kinematic data is acquired *post hoc* through video analysis.

MATERIALS

It is essential that you consult the appropriate Material Safety Data Sheets and your institution's Environmental Health and Safety Office for proper handling of equipment and hazardous materials used in this protocol.

Equipment

Computer with Windows, Mac OS, or Linux operating system

DLTdv

DLTdv is written in MATLAB and runs either in the freely available MATLAB runtime environment or as an app or source code in a licensed MATLAB installation.

Here, we describe the use of DLTdv version 8.2.3 (i.e., DLTdv8) for digitizing mosquito movements from videography data (Hedrick 2008). The open-source software is available at https://biomech.web.unc.edu/dltdv/. An in-depth online manual for DLTdv8, including several video tutorials, can be found at https://biomech.web.unc.edu/dltdv8_manual/. This includes video tutorials of how to use manual tracking and automatic tracking of both two-dimensional (2D) and 3D video data, how to calibrate the multicamera system, and tutorials for using the deep learning tracking algorithm.

DLTdv8 includes an interface for performing a basic multicamera calibration, based on the Direct Linear Transformation (DLT) routine. If a more sophisticated calibration is needed, one can use the easyWand calibration tool (Theriault et al. 2014) available at https://biomech.web.unc.edu/wand-calibration-tools/, which produces optimized camera calibrations from a wide variety of possible inputs.

[4]Correspondence: thedrick@bio.unc.edu

Cite this protocol as *Cold Spring Harb Protoc*; doi:10.1101/pdb.prot107928

DLTdv can directly read the "raw" save format of a variety of high-speed camera systems, including those from Phantom (Vision Research Inc.) and Photron (Photron USA, Inc.) along with videos stored in AVI, MOV, or MP4 containers using the underlying operating system libraries.

METHOD

Below, we provide a list of user instructions for analyzing the 3D movements of mosquitoes from multicamera videography, including the DLT camera system calibration. This list is intended to be a broad overview; thus, all the steps must not always be performed nor performed in the given order. For example, when using a single-camera system, camera calibration is not required. Moreover, tutorials that cover more options and functionality are available online for experienced users. The list of steps below fits into the broader workflow for tracking detailed mosquito kinematics, as shown in Figure 1.

Point Tracking in DLTdv

1. Open DLTdv8 if installed as a MATLAB app or packaged runtime application or run "DLTdv8a" in MATLAB's command window if installed as source code to open the DLTdv8 main window.
2. Click on Project menu → New. Select the number of cameras in the videography system.
3. Use the file browser to select video files, equal to the number of cameras chosen in Step 2.
4. When prompted, select the DLT calibration coefficients file.

   To generate the DLT calibration coefficients file, see the DLT calibration routine beginning with Step 13. Projects with only one camera cannot use DLT, and multicamera projects may choose not to, at which point the resulting output is only the (u,v) pixel coordinate time series instead of an (x,y,z) spatial coordinate time series.
5. Set the Frame Number to the first frame to be tracked and adjust Video Gamma such that there is high contrast between the mosquito(es) to be tracked and the background.
6. Add points using the "add point" button according to the number of points to be tracked.

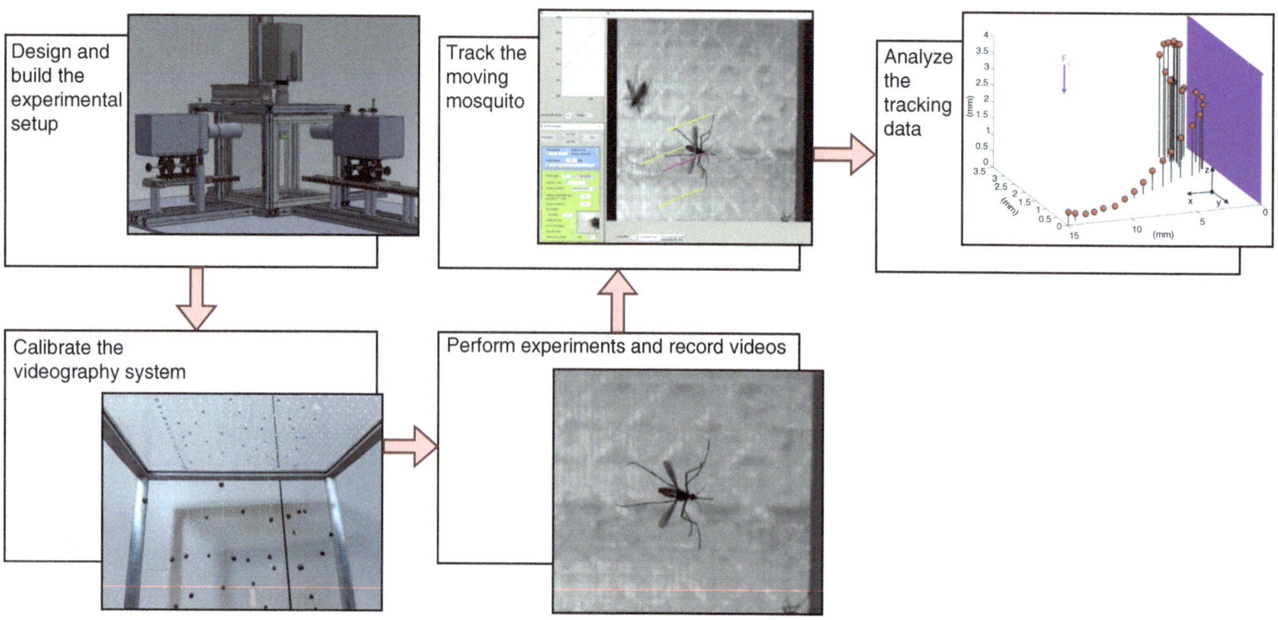

FIGURE 1. The workflow for tracking detailed mosquito kinematics.

Cite this protocol as *Cold Spring Harb Protoc*; doi:10.1101/pdb.prot107928

7. For manual tracking, set autotrack mode to "off" or "auto-advance." For automated tracking, set the mode to "automatic" and adjust Tracker Threshold, DLT Residual Threshold, and Marker Centroid Search to suit video needs.

   Begin with defaults. Further experience with DLTdv8 will inform optimum settings for individual videos.

8. Select a marker location in one video and press the "f" key to advance the frame if autotrack mode is set to "off." If DLT coefficients are present, once a marker location is defined in one video, DLTdv displays the "epipolar line," or known marker location from the first camera projected into the other camera views. Once the marker location is defined in two or more cameras, the 3D location is calculated and shown in all camera views.

   DLTdv also has autotracking capabilities that can be used in place of Step 8.

9. Adjust the interface settings to control what information is shown in the video and time series windows.

   Video windows can show none, all, or selected 2D tracks as the user progresses through tracking. The time-series window can display position data for individual videos or global coordinates if multiple videos have been tracked. The user can also specify the time-series window width in frames.

10. Track all named points in each loaded video.

11. Click on Project menu → Save.

    This saves all the project metadata and tracked points as a MATLAB binary file but does not save the movie files themselves, which must be present to reload the saved project.

12. Click on Points menu → Export points.

    This creates text format outputs of the pixel and spatial coordinates for import and processing in the data analysis platform of your choice.

    Upon export, DLTdv asks the user to select various export options with sequential pop-up windows. The first asks the user to specify a prefix, which should be some unique identifier to the current project; the second asks the user to select export format (sparse or flat); and the third asks if the user would like to create 95% confidence intervals. The recommended choices for most projects are flat format with no confidence intervals.

Calibrating Cameras in DLTdv

Here, we briefly describe how to calibrate a camera using Direct Linear Transformation in DLTdv. This method is suitable for cameras with negligible optical distortion.

13. Prepare a calibration object, such as a 3D object of known dimensions with six or more landmarks that have their (*x,y,z*) location described in a comma-separated value spreadsheet.

    The landmarks must not lie in a single 2D plane (even if this plane passes through a 3D volume). Ideally, the calibration object should cover one-third or more of the camera view. Lego bricks and other construction toys are often suitable for creating small custom calibration objects.

14. Position the calibration object in the field of view of the cameras then record and save a brief movie.

15. *The camera frame rate and exposure time need not be the same used for mosquito recordings, but all other camera, lighting, and lens parameters should be held constant. Each camera must see at least six calibration landmarks that do not lie in a 2D plane, but these need not be the same landmarks in each camera. Additional calibration points beyond the minimum of six will improve calibration stability and accuracy; 15 to 20 points are recommended.* Open DLTdv8 if installed as a MATLAB app or packaged runtime application or run "DLTdv8a" in MATLAB's command window if installed as source code to open the DLTdv8 main window.

16. Click on Project menu → New. Create a multicamera project and select some videos. When asked if there is a DLT calibration, select "no."

    Any videos may be used for Step 16.

17. Launch DLTcal8, a specialized interface for collecting and computing calibrations, by selecting "3D menu" and then "start calibration tool." Once launched, DLTcal8 is independent of DLTdv.

18. Click on the DLTcal8 "initialize" button.
19. Use the file browser to select the calibration frame specification, that is, the comma-separated value spreadsheet with the (x,y,z) locations of the landmarks.
20. Select the calibration movie file for your first camera; DLTcal8 will load and display the image.
21. Using your mouse, mark the location of the first calibration point, then advance to the next point and mark it as well. Mark all points, skipping those that are not visible.
22. Click on the "compute coefficients" button.

    This computes coefficients for the camera in question, reporting quality as the "calibration residual." This number is the disagreement between the ideal and observed landmark locations, measured in pixels. Lower values are better, and values <1 are high-quality. Larger values may be acceptable, especially for high-resolution cameras.

23. Click on the "error analysis" button to verify that no single point contributes excessively to the calibration residual. Reexamine any points thus identified.
24. Click on the "add a camera" button and select the image or movie for your next camera.
25. Mark the landmarks and compute the calibration for this camera.
26. Once all the cameras are calibrated, click on the "save data" button and select the folder where you want to save the output. Specify a prefix for the output files.

    Saving will produce a prefix DLTcoefs.csv file (calibration output) and a prefix xypts.csv file (calibration input).

27. Close DLTcal8 by closing the DLTcal8 window.
28. Validate your calibration by restarting DLTdv, creating a project with the calibration videos, selecting the new DLT coefficients, and clicking the calibration landmarks or other locations.

DISCUSSION

Here are some additional tips for operating DLTdv8.

- The DLTdv8 program has a large range of keyboard shortcuts. When used correctly, these shortcuts allow you to increase your productivity when analyzing videos. To get familiar with these keyboard shortcuts, check the "list of mouse and keyboard commands" via the Help menu or look for the shortcut cheat sheet "DLTdv keyboard map.pdf" if you have downloaded the source code.
- If manually digitizing videos, it is typically fastest to work through all the frames marking one point in a first video that best shows the landmark of interest, then going back to the first frame and switching to a second video.
- DLTdv8 includes an image cross-correlation based autotracker for interactive use. For best results set the search area width to about twice the width of the feature in question and watch the "tracker score" in the interface to determine a good Tracker Threshold value. If the tracker makes an error press the letter "x" on your keyboard to rapidly exit autotracking.
- DLTdv8 includes a variety of deep learning-based autotracking tools; these are more suitable for interactive use than most other deep learning-based automated video analysis tools but are less well suited for large-scale batch processing. See https://biomech.web.unc.edu/dltdv/ for more details.
- When digitizing single-camera data, one can calibrate the image based on objects of known dimensions that are visible in the video. For correct calibration, ensure that the calibration object is located the same distance from the camera as the mosquito.
- See https://biomech.web.unc.edu/wand-calibration-tools/ for an alternative calibration tool compatible with DLTdv.

 Cite this protocol as *Cold Spring Harb Protoc*; doi:10.1101/pdb.prot107928

REFERENCES

Hedrick TL. 2008. Software techniques for two- and three-dimensional kinematic measurements of biological and biomimetic systems. *Bioinspir Biomim* 3: 034001. doi:10.1088/1748-3182/3/3/034001

Theriault DH, Fuller NW, Jackson BE, Bluhm E, Evangelista D, Wu Z, Betke M, Hedrick TL. 2014. A protocol and calibration method for accurate multi-camera field videography. *J Exp Biol* 217: 1843–1848. doi:10.1242/jeb.100529

Protocol 4

Quantifying and Analyzing Mosquito Movement from Video Tracking Results

Florian T. Muijres[1,2]

[1]*Department of Experimental Zoology, Wageningen University, 6708 PB Wageningen, the Netherlands*

After tracking the kinematics of a moving mosquito from video, one needs to convert the tracking output into the kinematics data needed to answer the research question. Here, we provide general guidelines for how to do this by discussing how to quantify body position and orientation in the world reference frame and wing and leg orientation in the mosquito body reference frame. These guidelines should be adapted based on the goal of your research. To answer your research question, the resulting kinematics data must then be further analyzed. Because the nature of this analysis depends strongly on your specific research question, we refer to literature for designing these postprocessing routines.

MATERIALS

It is essential that you consult the appropriate Material Safety Data Sheets and your institution's Environmental Health and Safety Office for proper handling of equipment and hazardous materials used in this protocol.

Equipment

Computer with Windows, Mac OS, or Linux operating system

Programming software such as Python, MATLAB, or R

METHOD

Here, we describe how to quantify the movement kinematics of mosquitoes from video tracking data. We first describe how to define a time–space reference frame for this. Then, we describe how to define the temporal dynamics of mosquito body position and orientation. Finally, we explain how to describe wing and leg kinematics.

Define the Coordinate System for Your Analysis

1. Define a world reference frame in two dimensions (2D) ($\mathbf{X} = \{x, y\}$) or three dimensions (3D) ($\mathbf{X} = \{x, y, z\}$) (see Fig. 1).

   The coordinate system of this world reference frame should be right-handed, and the coordinate axes should be explicitly defined. Position the center of the coordinate system in an appropriate location (e.g., the center of a landing platform), and orient the axes (e.g., set the y-axis or z-axis vertically up in 2D and 3D, respectively).

[2]Correspondence: florian.muijres@wur.nl

Cite this protocol as *Cold Spring Harb Protoc*; doi:10.1101/pdb.prot107929

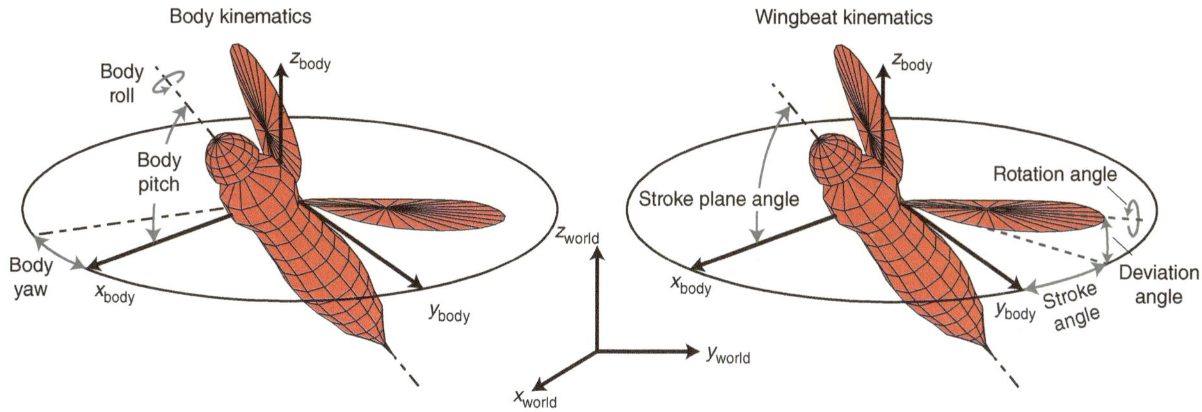

FIGURE 1. Definitions of the body and wingbeat kinematic parameters of flying mosquitoes. The *left* panel shows body orientation parameters: the three consecutive Euler angles body yaw, body pitch, and body roll. The *right* panel shows the wingbeat kinematics parameters: the three consecutive Euler angles wing stroke angle, wing deviation angle, and wing rotation angle. The body location and orientation are defined relative to the world reference frame with axes (*x*, *y*, *z*); The wingbeat kinematics angles are defined in the body-fixed stroke-plane reference frame, with a stroke plane angle of ~45° and with axes (x_{body}, y_{body}, z_{body}).

2. Define time *t*, where $t = 0$ sec is a suitable moment in time (e.g., the start of the experiment or the moment of touchdown during a landing experiment).
3. Define all mosquito movement kinematics parameters within a video relative to this time–space coordinate system (t, **X**).

Quantify the Position of a Moving Mosquito

For many studies, only the position of the moving mosquito is of relevance (e.g., Cribellier et al. 2020, 2022). Here, we explain how to define the temporal dynamics of this positional data throughout a video.

4. Define the position of a mosquito in each video frame as (t, **X**).

   For a 2D single-camera study, this positional data will consist of the 2D array $\mathbf{X} = (x, y)$*, where x is the horizontal position, and y is the equivalent vertical position. For 3D studies with multiple cameras, the positional data will consist of the 3D data* $\mathbf{X} = (x, y, z)$.

5. Compile the positional data of all digitized video frames into a time–space array ($\mathrm{t}(n)$, $\mathbf{X}(n)$), where the video frame number n ranges from 1 to N from the first to the last digitized video frame.

   Here, the positional data thus consists of a set of two or three arrays for a 2D or 3D study, respectively: ($\mathbf{X} = (\mathbf{x}(\mathrm{n}), \mathbf{y}(\mathrm{n}))$ *and* $\mathbf{X} = (\mathbf{x}(\mathrm{n}), \mathbf{y}(\mathrm{n}), \mathbf{z}(\mathrm{n}))$*, respectively).*

Quantify the Body Orientation of a Moving Mosquito

When tracking body orientation in addition to body location, these data need to be added to the kinematics time–space array using the following steps.

6. Define the orientation of the mosquito in each video frame using the body angle array $\mathbf{\Gamma}$.

   For 3D kinematics analyses, the body orientation array is defined as $\boldsymbol{\Gamma} = (\gamma, \beta, \theta)$*, which consists of the three consecutive Euler angles, body yaw angle* γ*, body pitch angle* β*, and body roll angle* θ *(see Fig. 1).*

   For 2D analyses with a single sideview camera, only body pitch β *can be estimated, and thus the body angle array reduces to* $\boldsymbol{\Gamma} = \beta$*. Similarly, for studies with only a top view camera, only body yaw* γ *can be estimated, and thus the body angle array reduces to* $\boldsymbol{\Gamma} = \gamma$.

7. Add the body kinematics array for all digitized video frames to the time–space kinematics array ($\mathrm{t}(n)$, $\mathbf{X}(n)$, $\mathbf{\Gamma}(n)$).

   Here, the body orientation angle array thus consists of three body angle arrays as $\boldsymbol{\Gamma}(\mathrm{n}) = (\boldsymbol{\gamma}(\mathrm{n}), \boldsymbol{\beta}(\mathrm{n}), \boldsymbol{\theta}(\mathrm{n}))$.

Quantify the Wingbeat Kinematics of Flying Mosquitoes

When also tracking the wingbeat kinematics of a flying mosquito, the body kinematics time–space array should be augmented with a wing kinematics array, according to the following steps.

8. Define a wing stroke reference frame that is fixed to the mosquito body (see Fig. 1).

   This wing stroke reference frame should be located in a plane in which most wing stroke movements occur. For mosquitoes, this stroke plane is oriented at a stroke plane angle of ~45° relative to the long axis of the body (see Fig. 1; Muijres et al. 2017).

9. Define the orientation data of the left and right wing in this wing stroke reference frame using the wing angle arrays $\mathbf{\Phi}_L$, $\mathbf{\Phi}_R$, respectively. Each wing angle array $\mathbf{\Phi} = (\phi, \eta, \alpha)$ consists of the three consecutive wing kinematics Euler angles: the wing stroke angle ϕ, the wing deviation angle η, and the wing rotation angle α (see Fig. 1; Muijres et al. 2017).

   The stroke angle describes the wing rotation within the stroke plane; the deviation angle defines rotations out of the stroke plane; and the rotation angle defines wing pitch rotation around the wingspan axis.

10. Add the wing kinematics array of both wings to the complete time–space kinematics array for the digitized video $(\mathbf{t}(n), \mathbf{X}(n), \mathbf{\Gamma}(n), \mathbf{\Phi}_L(n), \mathbf{\Phi}_R(n))$.

    The wing kinematics array thus consists of data for both wings and each combination of video frame and kinematics angle. For example, the kinematics of the left wing is defined using a set of three arrays as $\boldsymbol{\phi}_L(n) = (\boldsymbol{\phi}_L(n), \boldsymbol{\eta}_L(n), \boldsymbol{\alpha}_L(n))$.

Quantify the Leg Kinematics of Moving Mosquitoes

When also tracking leg kinematics, these data need to be added to the kinematics time–space array using the following steps:

11. Define the leg orientation using the coordinates of the leg joints and tip of each foot.

    These coordinates can be given in both the world reference frame $\mathbf{X} = \{x, y, z\}$, *or similar to the wings in the body reference frame (i.e., relative to the center of the body* $\mathbf{X}_{body} = \{x_{body}, y_{body}, z_{body}\}$). *The specific research question will drive the choice of kinematic reference frame for the legs.*

12. Add the kinematics array of each digitized leg to the complete time–space kinematics array for the digitized video $(\mathbf{t}(n), \mathbf{X}(n), \mathbf{\Gamma}(n), \mathbf{X}_{leg}(n))$.

    With six legs in total and multiple tracked points per leg (several leg joints and foot tip), the leg kinematics array can become quite large. It is therefore important to precisely and uniquely define each leg kinematics parameter.

REFERENCES

Cribellier A, Spitzen J, Fairbairn H, van de Geer C, van Leeuwen JL, Muijres FT. 2020. Lure, retain, and catch malaria mosquitoes. How heat and humidity improve odour-baited trap performance. *Malar J* **19:** 1–16. doi:10.1186/s12936-020-03403-5

Cribellier A, Straw AD, Spitzen J, Pieters RPM, van Leeuwen JL, Muijres FT. 2022. Diurnal and nocturnal mosquitoes escape looming threats using distinct flight strategies. *Curr Biol* **32:** 1–15. doi:10.1016/j.cub.2022.01.036

Muijres F, Chang S, van Veen W, Spitzen J, Biemans B, Koehl M, Dudley R. 2017. Escaping blood-fed malaria mosquitoes minimize tactile detection without compromising on take-off speed. *J Exp Biol* **220:** 3751–3762. doi:10.1242/jeb.163402

CHAPTER 24

Mosquito (Diptera: Culicidae) Vision and Associated Electrophysiological Techniques

Daniel A.H. Peach[1,2,5] and Adam J. Blake[3,4]

[1]*Department of Zoology, University of British Columbia, Vancouver, British Columbia, V6T 1Z4, Canada;* [2]*Savannah River Ecology Laboratory, Department of Infectious Diseases, The University of Georgia, Aiken, South Carolina 29802, USA;* [3]*Department of Biological Sciences, Simon Fraser University, Burnaby, British Columbia, V5A 1S6, Canada;* [4]*Department of Biology, University of Washington, Seattle, Washington 98195, USA*

Mosquitoes are considered the world's deadliest animal because of the pathogens they spread. Additionally, they are an unbearable nuisance in many areas. Visual stimuli play an important role in the mosquito life cycle, helping them find vertebrate hosts, floral nectar, and oviposition sites. Here, we review mosquito vision, including its influences on mosquito behavior, the photoreceptors involved, and mosquito spectral sensitivity, as well as provide an overview of techniques used for the analysis of mosquito vision, including electroretinograms, single-cell recordings, and the use of opsin-deficient mutants. We anticipate that this information will be useful for researchers studying mosquito physiology, evolution, ecology, and management.

REVIEW OF MOSQUITO VISION

Vision is an important sense mosquitoes use to gather information about their surroundings and locate resources, including vertebrate hosts (van Breugel et al. 2015), floral nectar (Peach et al. 2019), and oviposition sites (Beehler et al. 1993). From a mosquito control perspective, visual cues are important elements in mosquito trap design and trap capture rates (Bidlingmayer 1994). Additionally, knowledge about the visual cues used by mosquitoes is used to provide advice to the public regarding how to reduce their exposure to mosquito bites (i.e., to avoid wearing dark-colored clothing) (CBC News 2021).

Behavior

Humans become visible to mosquitoes at ~5–15 m (Bidlingmayer and Hem 1980), and visual cues are important to bridge the gap between long-range cues, such as CO_2, and short-range cues, such as volatiles and thermal cues (van Breugel et al. 2015). Host-seeking mosquitoes are attracted to darker colors (Kennedy 2009), preferring blacks, browns, and blues to lighter colors like white or yellow (Brett 1938; Sippell and Brown 1953; Fay 1968; Muir et al. 1992a; Wen et al. 1997). This attraction is gated by CO_2 (van Breugel et al. 2015). In its absence, mosquitoes ignore dark objects, but once exposed to a plume of CO_2, a mosquito will investigate nearby dark objects, even if this plume does not directly emanate from the object (van Breugel et al. 2015). CO_2 also induces attraction to objects

[5]Correspondence: daniel.peach@uga.edu

Cite this introduction as *Cold Spring Harb Protoc;* doi:10.1101/pdb.top107671

reflecting wavelengths of >600 nm (San Alberto et al. 2021). Female mosquitoes also prefer dark-colored oviposition sites (Hoel et al. 2011). Until recently, this attraction to dark colors conflicted with mosquito attraction to light-colored flowers; however, Peach et al. (2019) found that ultraviolet (UV)-dark floral cues, which are invisible to humans, were involved in the visual attraction of mosquitoes to such flowers.

Photoreceptors

Similar to higher flies (Lunau 2014), the hundreds of ommatidia in mosquito compound eyes each contain eight photoreceptor cells (R1–R8), most with their own rhabdomere (Fig. 1). Unlike higher flies, mosquitoes have a semifused rhabdom, with the rhabdomeres of the six outer photoreceptors (R1–R6) fused in a ring-like structure surrounding the central photoreceptors (R7 and R8). The outer photoreceptors express similar opsins, with the two central photoreceptors, which share a rhabdomere, expressing various opsins (Hu et al. 2009, 2011, 2012), and R7 contributes to the rhabdom only in the most distal part of the ommatidium (Brammer 1970).

Rather than opsin expression varying among ommatidial types across the entire eye, opsin expression varies between eye regions (Hu et al. 2009, 2011). *Aedes aegypti* possess 10 putative opsins, of which five are expressed in the compound eye (Giraldo-Calderón et al. 2017). The R1–R6 photoreceptors, and most R8 photoreceptors, express the long-wave-sensitive Aaop1 opsin (Hu et al. 2012). Aaop2, a long-wavelength opsin, and Aaop8, a UV-sensitive opsin, are expressed in nonoverlapping patterns in R7 cells (Hu et al. 2009). The Aaop2-expressing R7 cells also coexpress the Aaop9 opsin, broadening their range of sensitivity and enhancing functionality under low-light conditions (Hu et al. 2011). Aaop9 is further expressed in R8 cells in the dorsal portion of the eye. Last, the medium-wavelength-sensitive opsin Aaop10 is found coexpressed with Aaop8 in a subset of R7 cells (Hu et al. 2014).

Genomic studies show that *Anopheles gambiae* and *Culex quinquefasciatus* have a similar set of opsins as *Ae. aegypti* (Giraldo-Calderón et al. 2017). These three species show many long-wave opsins, suggesting gene duplication and active selection. The expression of these rhodopsins in *An. gambiae* is less studied than that in *Ae. aegypti* but is largely similar, with some regional differences in the

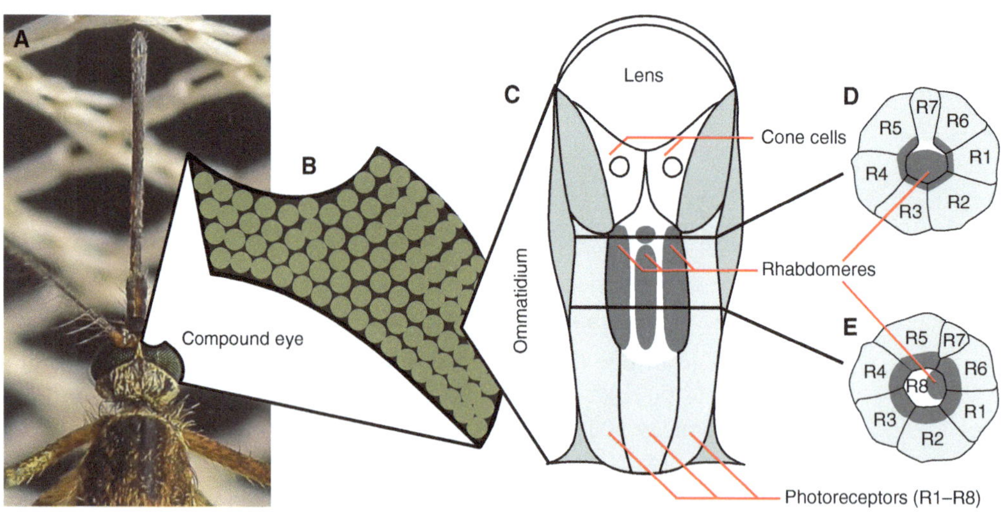

FIGURE 1. Anatomy of the mosquito compound eye. (*A*) Photograph of a mosquito eye showing the two compound eyes made of many ommatidia. (*B*) Enlarged diagram of a single compound eye with each ommatidium represented by a green circle. (*C*) A diagram of a longitudinal section of a single generalized mosquito ommatidium. (*D,E*) Diagrams of transverse sections of mosquito ommatidium at the level indicated by the thick black lines. Photoreceptor R7 is positioned above photoreceptor R8. The section in *D* is more distal, showing the rhabdomere of R7, whereas the section in *E* is more proximal, showing the rhabdomere of R8. Diagrams are based on the drawings of Sato (1957) and electron microscopy of Hu et al. (2009) and Zhan et al. (2021).

expression pattern (Hu et al. 2009, 2014). Opsin expression outside these two species remains unexamined.

Mosquito Spectral Sensitivity

Like many insects and other flies (Brown and Anderson 1996), mosquitoes show sensitivity peaks in the UV range (300–400 nm) and the green range (495–570 nm), although interspecific variability occurs in the relative sensitivity of these peaks. In one study, *Ae. aegypti* compound eyes had a primary sensitivity peak in the green at ~520 nm with a secondary peak in the UV at ~340 nm (Muir et al. 1992b). In a subsequent study, the primary peak was in the violet at ~410 nm with a secondary peak still at ~520 nm (San Alberto et al. 2021). Compound eyes of *Culex pipiens*, *Cx. quinquefasciatus*, and *Anopheles stephensi* show the greatest sensitivity in the UV, with peaks occurring at 335–360 nm (Peach et al. 2019; San Alberto et al. 2021). These three species also show a secondary green peak, at ~540 nm for *Cx. pipiens* and ~520 nm for *Cx. quinquefasciatus* and *An. stephensi*.

TECHNIQUES USED TO ANALYZE MOSQUITO VISION

Electroretinograms

Many techniques are used to characterize the spectral sensitivity of different insect species. However, only electroretinograms (ERGs) have been used with mosquitoes thus far. ERGs use a single electrode to record data from the intercellular space and capture the response of many photoreceptors as a negative deflection in the recorded voltage (Goldsmith and Bernard 1974). These recordings can reveal the extent and relative strength of the spectral sensitivity of the insect compound eye (or ocelli). In traditional ERG methods, like those described in Protocol 1: Measuring Mosquito Spectral Sensitivity Using Electroretinograms (Peach and Blake 2023), a series of flashes of varying spectral content but with constant intensity or photon flux are presented to the eye (DeVoe et al. 1997). The eye responses to flashes of increasing intensity are then recorded, and a sigmoid function is used to describe the relationship between the response and the logarithm of the stimulus intensity (VLogI). Sensitivity is then calculated using this relationship by estimating the intensity of light necessary to generate a particular criterion response value at each measured wavelength. Constant-response ERG methods use constant light rather than flashes, which can offer several advantages over traditional methods, but they require more elaborate and complicated equipment (DeVoe et al. 1997).

Recordings are typically performed after an insect is dark adapted or held in a dark environment, which maximizes the compound eye's electrical response (Goldsmith and Bernard 1974). As this method records from multiple photoreceptors, this response is dominated by the more common photoreceptor types. In Diptera, the response is dominated by the outer R1–R6 photoreceptors (Minke et al. 1975), which express long-wave opsins but often have their sensitivity broadened with UV-sensitizing pigments (Stavenga et al. 2017). Color-adaptation lights can reduce the contribution of some photoreceptors, such as the R1–R6 photoreceptors, making the contributions of others, such as the central R7 and R8 receptors, more apparent in the ERG response (Minke et al. 1975; Kirchner et al. 2005; Telles et al. 2014; Pirih et al. 2018). For example, using a green-adaptation light can reduce the ERG response (relative to dark adaptation) to green light while increasing the response to UV light, thereby implying the presence of photoreceptors primarily sensitive to both green and UV light, as was shown in *Cx. pipiens* (Peach et al. 2019).

Intracellular Recordings

As with other small insects (Sharkey et al. 2020), the potential for intracellular recording from single cells in mosquitoes is limited, both due to the small size and relative rarity of central R7 and R8 receptors. However, these recordings can reveal the spectral sensitivity of individual photoreceptors, which are often shaped by filtering effects of pigments or other photoreceptors, and can be difficult to detect with ERGs (Pirih et al. 2018). Intracellular recordings are also necessary to characterize the polarization sensitivity of receptors, as photoreceptors with similar spectral sensitivities will often have differing polarization sensitivities (i.e., R1–R6 photoreceptors in Diptera; Wernet et al. 2012).

Recording from Opsin-Deficient Mutants

Genetic methods, such as single-opsin rescue, can allow for the isolation of the response of single spectral classes of photoreceptors without the use of intracellular recordings, a technique that has been successfully used in *Drosophila melanogaster* (Sharkey et al. 2020). Although *Ae. aegypti* lines deficient in long-wave opsins (Aaop1 and Aaop2) have been created (San Alberto et al. 2021), the expression of these opsins in multiple photoreceptors in a single ommatidia (i.e., Aaop1, expressed in both R1–R6 and R8 in some regions of the *Ae. aegypti* compound eye) complicates the use of this technique in mosquitoes.

CONCLUSION

As the spectral sensitivities of additional mosquitoes are measured, we may be able to glean additional information about correlations between sensitivities and ecological niches. The use of techniques beyond simple dark-adapted ERGs in such investigations should also allow for a better estimation of the spectral sensitivities of individual photoreceptors. These sensitivities are a prerequisite for a more mechanistic understanding of mosquito vision. Such an understanding would allow us to model photoreceptor catch in response to different stimuli and provide information on the input being provided to downstream processes with the mosquito neural system.

REFERENCES

Beehler JW, Millar JG, Mulla MS. 1993. Synergism between chemical attractants and visual cues influencing oviposition of the mosquito, *Culex quinquefasciatus* (Diptera: Culicidae). *J Chem Ecol* **19:** 635–644. doi:10.1007/BF00984998

Bidlingmayer WL. 1994. How mosquitoes see traps: role of visual responses. *J Am Mosq Control Assoc* **10:** 272–279.

Bidlingmayer WL, Hem DG. 1980. The range of visual attraction and the effect of competitive visual attractants upon mosquito (Diptera: Culicidae) flight. *Bull Entomol Res* **70:** 321–342. doi:10.1017/S0007485300007604

Brammer JD. 1970. The ultrastructure of the compound eye of a mosquito *Aedes aegypti* L. *J Exp Zool* **175:** 181–195. doi:10.1002/jez.1401750207

Brett GA. 1938. On the relative attractiveness to *Aëdes aegypti* of certain coloured cloths. *Trans R Soc Trop Med Hyg* **32:** 113–124. doi:10.1016/S0035-9203(38)90101-4

Brown PE, Anderson M. 1996. Spectral sensitivity of the compound eye of the cabbage root fly, *Delia radicum* (Diptera: Anthomyiidae). *Bull Entomol Res* **86:** 337–342. doi:10.1017/S0007485300034908

CBC News. 2021. The invisible reasons mosquitoes prefer some people over others | CBC News. CBC. https://www.cbc.ca/news/canada/north/mosquitoes-prefer-some-people-yukon-1.6113232 [Accessed August 14, 2021].

DeVoe RD, de-Souza JM, Ventura DF. 1997. Electrophysiological measurements of spectral sensitivities: a review. *Braz J Med Biol Res* **30:** 169–177. doi:10.1590/S0100-879X1997000200003

Fay R. 1968. A trap based on visual responses of adult mosquitoes. *Mosq News* **28:** 1–7.

Giraldo-Calderón GI, Zanis MJ, Hill CA. 2017. Retention of duplicated long-wavelength opsins in mosquito lineages by positive selection and differential expression. *BMC Evol Biol* **17:** 1–20. doi:10.1186/s12862-017-0910-6

Goldsmith TH, Bernard GD. 1974. The visual system of insects. In *The physiology of insecta*, pp. 165–272. Elsevier. https://linkinghub.elsevier.com/retrieve/pii/B9780125916028500126 [Accessed August 12, 2021].

Hoel DF, Obenauer PJ, Clark M, Smith R, Hughes TH, Larson RT, Diclaro JW, Allan SA. 2011. Efficacy of ovitrap colors and patterns for attracting *Aedes albopictus* at suburban field sites in north-central Florida. *J Am Mosq Control Assoc* **27:** 245–251. doi:10.2987/11-6121.1

Hu X, England JH, Lani AC, Tung JJ, Ward NJ, Adams SM, Barber KA, Whaley MA, O'Tousa JE. 2009. Patterned rhodopsin expression in R7 photoreceptors of mosquito retina: implications for species-specific behavior. *J Comp Neurol* **516:** 334–342. doi:10.1002/cne.22114

Hu X, Whaley MA, Stein MM, Mitchell BE, O'Tousa JE. 2011. Coexpression of spectrally distinct rhodopsins in *Aedes aegypti* R7 photoreceptors. *PLoS One* **6:** e23121. doi:10.1371/journal.pone.0023121

Hu X, Leming MT, Metoxen AJ, Whaley MA, O'Tousa JE. 2012. Light-mediated control of rhodopsin movement in mosquito photoreceptors. *J Neurosci* **32:** 13661–13667. doi:10.1523/JNEUROSCI.1816-12.2012

Hu X, Leming MT, Whaley MA, O'Tousa JE. 2014. Rhodopsin coexpression in UV photoreceptors of *Aedes aegypti* and *Anopheles gambiae* mosquitoes. *J Exp Biol* **217:** 1003–1008. doi:10.1242/jeb.096347

Kennedy JS. 2009. The visual responses of flying mosquitoes. *Proc Zool Soc Lond A* **109:** 221–242. doi:10.1111/j.1096-3642.1940.tb00831.x

Kirchner SM, Döring TF, Saucke H. 2005. Evidence for trichromacy in the green peach aphid, *Myzus persicae* (Sulz.) (Hemiptera: Aphididae). *J Insect Physiol* **51:** 1255–1260. doi:10.1016/j.jinsphys.2005.07.002

Lunau K. 2014. Visual ecology of flies with particular reference to colour vision and colour preferences. *J Comp Physiol A* **200:** 497–512. doi:10.1007/s00359-014-0895-1

Minke B, Wu C-F, Pak WL. 1975. Isolation of light-induced response of the central retinula cells from the electroretinogram of *Drosophila*. *J Comp Physiol A* **98:** 345–355. doi:10.1007/BF00709805

Muir LE, Kay BH, Thorne MJ. 1992a. *Aedes aegypti* (Diptera: Culicidae) vision: response to stimuli from the optical environment. *J Med Entomol* **29:** 445–450. doi:10.1093/jmedent/29.3.445

Muir LE, Thorne MJ, Kay BH. 1992b. *Aedes aegypti* (Diptera: Culicidae) vision: spectral sensitivity and other perceptual parameters of the female eye. *J Med Entomol* **29:** 278–281. doi:10.1093/jmedent/29.2.278

 Cite this introduction as *Cold Spring Harb Protoc*; doi:10.1101/pdb.top107671

Peach DAH, Blake AJ. 2023. Measuring mosquito spectral sensitivity using electroretinograms. *Cold Spring Harb Protoc* doi:10.1101/pdb.prot108190

Peach DAH, Ko E, Blake AJ, Gries G. 2019. Ultraviolet inflorescence cues enhance attractiveness of inflorescence odour to *Culex pipiens* mosquitoes. *PLoS ONE* **14:** e0217484. doi:10.1371/journal.pone.0217484

Pirih P, Ilić M, Rudolf J, Arikawa K, Stavenga DG, Belušič G. 2018. The giant butterfly-moth *Paysandisia archon* has spectrally rich apposition eyes with unique light-dependent photoreceptor dynamics. *J Comp Physiol A* **204:** 639–651. doi:10.1007/s00359-018-1267-z

San Alberto DA, Rusch C, Zhan Y, Straw AD, Montell C, Riffell JA. 2021. The olfactory gating of visual preferences to human skin and colors in mosquitoes. *Nat Commun* **13:** 555. doi:10.1038/s41467-022-28195-x

Sato S. 1957. On the dimensional characters of the compound eye of *Culex pipiens var. pallens* Coquillett (Morphological studies of the compound eye in the mosquito, No. V). *Sci Rep Tohoku Imp Univ Ser 4 Biol* **23:** 83–90.

Sharkey CR, Blanco J, Leibowitz MM, Pinto-Benito D, Wardill TJ. 2020. The spectral sensitivity of *Drosophila* photoreceptors. *Sci Rep* **10:** 18242. doi:10.1038/s41598-020-74742-1

Sippell WL, Brown AWA. 1953. Studies on the response of the female *Aedes* mosquito. V. The role of visual factors. *Bull Entomol Res* **43:** 567–574. doi:10.1017/S0007485300026651

Stavenga DG, Wehling MF, Belušič G. 2017. Functional interplay of visual, sensitizing and screening pigments in the eyes of *Drosophila* and other red-eyed dipteran flies. *J Physiol* **595:** 5481–5494. doi:10.1113/JP273674

Telles FJ, Lind O, Henze MJ, Rodríguez-Gironés MA, Goyret J, Kelber A. 2014. Out of the blue: the spectral sensitivity of hummingbird hawkmoths. *J Comp Physiol A* **200:** 537–546. doi:10.1007/s00359-014-0888-0

van Breugel F, Riffell J, Dickinson MH, Breugel FV, Riffell J, Fairhall A, Dickinson MH. 2015. Mosquitoes use vision to associate odor plumes with thermal targets. *Curr Biol* **25:** 2123–2129. doi:10.1016/j.cub.2015.06.046

Wen Y, Muir LE, Kay BH. 1997. Response of *Culex quinquefasciatus* to visual stimuli. *J Am Mosq Control Assoc* **13:** 150–152.

Wernet MF, Velez MM, Clark DA, Baumann-Klausener F, Brown JR, Klovstad M, Labhart T, Clandinin TR. 2012. Genetic dissection reveals two separate retinal substrates for polarization vision in *Drosophila*. *Curr Biol* **22:** 12–20. doi:10.1016/j.cub.2011.11.028

Zhan Y, Alonso San Alberto D, Rusch C, Riffell JA, Montell C. 2021. Elimination of vision-guided target attraction in *Aedes aegypti* using CRISPR. *Curr Biol* **31:** 4180–4187.e6. doi:10.1016/j.cub.2021.07.003

Protocol 1

Measuring Mosquito Spectral Sensitivity Using Electroretinograms

Daniel A.H. Peach[1,2,5] and Adam J. Blake[3,4]

[1]Department of Zoology, University of British Columbia, Vancouver, British Columbia, V6T 1Z4, Canada; [2]Savannah River Ecology Laboratory, Department of Infectious Diseases, The University of Georgia, Aiken, South Carolina 29802, USA; [3]Department of Biological Sciences, Simon Fraser University, Burnaby, British Columbia, V5A 1S6, Canada; [4]Department of Biology, University of Washington, Seattle, Washington 98195, USA

Mosquitoes use a variety of cues to locate resources in the world around them, including olfactory, thermal, and visual stimuli. Understanding how mosquitoes perceive these stimuli is important for understanding and investigating mosquito behaviors and mosquito ecology. Specifically, mosquito vision can be studied in many ways, including by using electrophysiological recordings from their compound eyes. Electroretinograms can be used to characterize the spectral sensitivity of a mosquito species, revealing the wavelengths of light they can perceive. Here, we provide details on how to perform and analyze these recordings.

MATERIALS

It is essential that you consult the appropriate Material Safety Data Sheets and your institution's Environmental Health and Safety Office for proper handling of equipment and hazardous material used in this protocol.

RECIPES: Please see the end of this protocol for recipes indicated by <R>. Additional recipes can be found online at http://cshprotocols.cshlp.org/site/recipes.

Reagents

Mosquitoes (adults)
Ringer's solution for mosquitoes <R>

Equipment

Data acquisition system (Lab-Trax-4/16; World Precision Instruments)
Equipment for assembling the electroretinogram (ERG) optical components (see Steps 1–5)
Adaptation light-emitting diodes (LEDs)

The peak wavelength of the LEDs should overlap with the spectral sensitivity of the photoreceptors being adapted. For mosquitoes this would typically include LED with peak wavelengths (λ_{max}) in the green (520–570 nm) and UV (330–370 nm). For example, Peach et al. (2019) used green ($\lambda_{max} = 528$ nm; Roithner Laser-Technik B5B-433-B25) and UV LEDs ($\lambda_{max} = 351$ nm, Roithner LaserTechnik RLT350-0.3-15).

Beam splitter (Optometrics; "polka dot" 4–2001)
Circular variable neutral density wheel (fused silica, 200- to 2500-nm; Reynard Corp.)

[5]Correspondence: daniel.peach@uga.edu

Cite this protocol as *Cold Spring Harb Protoc*; doi:10.1101/pdb.prot108190

Collimator lenses (a total of two; Multimode Fiber Optics LC-4U-THD)
Cosine corrector (Ocean Optics CC-3-UV-S)
Custom-built LED driver (Science Technical Centre, Simon Fraser University) or an Arduino UNO microcontroller
Electronic shutter

A custom-built programmable shutter (R. Holland, Science Technical Centre, Simon Fraser University) was used in Eby et al. (2013) and Peach et al. (2019) but any shutter with an opening/closing time of <10 msec would be appropriate (ex. Thorlabs SHB05T).

Faraday cage
Fiber-optic cable with a subminiature-A (SMA) terminus (Multimode Fiber Optic PCU-1000-2-SS)
Fiber-optic scanning monochromator (MonoScan 2000; Mikropak GmbH)
Optical breadboard
Optical fiber (600-µm; Ocean Optics QP600-1-SR-B X)
Optical mounts, posts, and post mounts (e.g., Thor Labs, LMR2, UPH1, TR1)
Polycarbonate sheet (0.3175-cm-thick; Lexan, SABIC, Riyadh, Saudi Arabia)
Spectrometer (Ocean Optics HR-4000)
Xenon arc light source (35-W; Ocean Optics HPX-2000)

Equipment for preparing the specimens (see Steps 11–16)

Borosilicate glass capillary tubes (1.2/0.69 mm outer/inner diameter) tapered with a micropipette puller (Sutter Instrument Co. P-1000)
Container (see Step 11)
Forceps
Kite manual micromanipulator (World Precision Instruments)
Microscope slide
Microscope (Wild M10; Leica Microsystems)
Micromanipulators (Leitz)
Magnetic mounts (see Fig. 1 legend)
Silver wire electrode
Sticky tack

Equipment for taking the measurements (see Steps 17–22)

Amplifier (Duo 773 electrometer; World Precision Instruments)
Recording software (LabScribe v4; iWorx Systems Inc.)

METHOD

These methods were partially adapted from Peach et al. (2019) and Eby et al. (2013).

Assembling the ERG Optical Components

1. Place the optical components in the appropriate optical mounts and fix these mounts to optical posts. Arrange the positions of the posts of these components on the breadboard using post mounts (see Fig. 1A).
2. Connect the xenon arc light source to the monochromator and light from the monochromator to the optical breadboard via an optical fiber (600-µm).
3. Create and direct a collimated light beam to optical components on the breadboard via a lens mounted on the fiber-optic cable from the monochromator.

   The beam passes through a neutral density wheel, followed by a beam splitter, which directs a portion of light toward a fiber-optic cable fitted with cosine corrector, which carries that light to a spectrometer (Fig. 1A). The main portion of the beam then passes through a shutter before being directed down a fiber-optic cable into the Faraday cage to the mosquito.

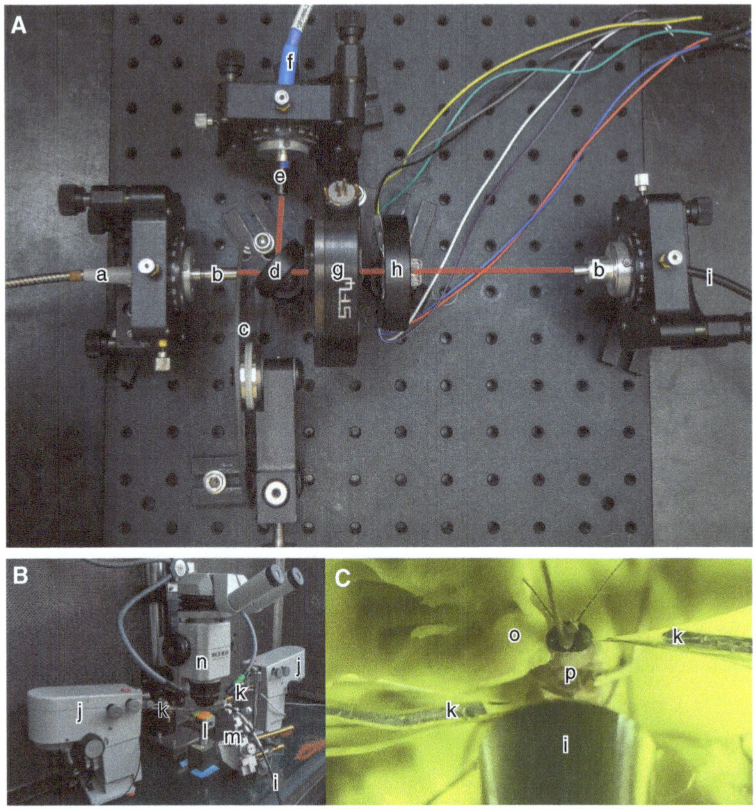

FIGURE 1. Arrangement of electroretinogram (ERG) materials on the optical breadboard (*A*), within the Faraday cage (*B*), and in the vicinity of the prepared insect (*C*). (*C*, courtesy Elton Ko and Adam J. Blake.) The red line in *A* indicates the path of light through the components. (a) Fiber-optic cable from the monochromator, (b) collimator lenses, (c) neutral density wheel, (d) beam splitter, (e) cosine corrector, (f) fiber-optic cable to spectrometer, (g) shutter, (h) light-emitting diode (LED) mount with wires leading to LED driver, (i) fiber-optic cable to the Faraday cage, (j) micromanipulators (Lietz) holding the (k) electrodes, (l) magnetic mount where microscope slide is mounted, (m) KITE micromanipulator holding the fiber-optic cable on a magnetic mount, (n) microscope, and (o) sticky tack brace for (p) restrained mosquito.

4. Fix adaptation LEDs to the fiber-optic cable terminus in the Faraday cage (Kirchner et al. 2005) or place them between the shutter and second collimating lens in a mount that holds them immediately adjacent to the beam.

   In our setup, the LEDs are inserted through small slits in a 50-mm circle of cardstock with a central hole for the beam. This circle is then held in a mount for 50 mm or 2-in optics (e.g., LMR2).

   Placing the LEDs immediately adjacent to beam requires very careful positioning of the LEDs to allow their emitted light to be carried by the fiber, whereas mounting the LEDs on the fiber terminus is much more forgiving. However, this positioning has several drawbacks as the fiber and each LED will stimulate a different set of ommatidia. The fiber-mounted LEDs also limit the minimum achievable distance between the fiber terminus and the compound eye.

5. (Optional) Connect the shutter to the data acquisition system.

   This step allows the digital TTL signal to be recorded alongside the electrical response of the insect eye and allows the stimulus flashes to be synchronized with the eye response. Electrical responses can still be recorded without this synchronization, but the eye responses to stimuli must be manually identified. This identification is trivial when responses are large and noise is low but can be difficult when responses are small relative to the noise.

Calibrating the Setup

With this setup, the stimulus intensity is measured indirectly via the beam splitter, so the wavelength specific ratio of intensity between the portion of the light reaching the spectrometer and the portion of the light being delivered to fiber-

optic terminus in the Faraday cage must be determined to produce stimuli at the target intensity (5.0×10^{13} photons/cm^2/sec).

6. With the neutral density wheel transmitting in a position to maximally transmit light and with the cosine corrector at its normal position on the optical breadboard, use the spectrometer to determine the intensity of light in photons/cm^2/second at 10-nm intervals across the spectral range (310–700 nm).
7. To avoid leakage of UV light from the monochromator at wavelengths >600 nm (both during calibration and during measurements), insert a small sheet of polycarbonate (Lexan) between the shutter and adaptation LEDs (Fig. 1Ag,h) in order to block the secondary peak that appears 300 nm below the main wavelength peak.
8. Take similar intensity measurements with the cosine corrector of the spectrometer positioned directly in front of the terminus of the fiber-optic cable in the Faraday cage.
9. By comparing these two measurements, calculate wavelength-specific targets for spectrometer measurements to achieve the target intensity.
10. With the cosine corrector again positioned in front of the fiber-optic terminus, adjust the intensity of the adaptation LEDs to an intensity of 1.0×10^{14} photons/cm^2/sec by either varying the resistance of the circuit or through pulse width modulation (frequency $\approx$ 980 Hz).

Preparing the Specimens

11. Cold anesthetize the mosquito, in a container of your choice, for 1–4 min at −4°C depending on the species and size.
12. Use forceps to place the anesthetized mosquito on a glass microscope slide, ventral side up. Immobilize the mosquito by rolling sticky tack into thin strips and placing them, with ends tacked to the glass slide, over the legs, abdomen, and wings of the specimen. Restrict the mosquito's head movement by bracing the head with a piece of sticky tack on one side (Fig. 1B,C).

    As any motion of the animal can generate noise or affect electrode placement, proper immobilization of the insect and especially of the head is critical for a good recording.
13. Add Ringer's solution for mosquitoes to tapered glass capillaries and fit the silver wire electrodes into them. Ensure the electrodes are submerged in the solution. Then attach these glass capillary electrodes to the micromanipulators and the amplifier.
14. While looking under the microscope, use the micromanipulators (Lietz) to insert one glass capillary electrode into the mosquito's eye and a second electrode into the thorax (Fig. 1B,C).

    If the voltage is wildly varying or drifting, check the electrodes' connection and reinsert them if necessary.
15. Check that the connection is working by first lining up the fiber-optic terminus with the mosquito's eyes using the KITE micromanipulator. Then look for a corresponding voltage peak by opening and closing the shutters.
16. Dark adapt the mosquito by shutting off the lights and allowing the room to sit in darkness for 45 min.

Taking the Measurements

17. For all stimulus flashes, set the shutter exposure length to 0.5 sec.

    There should be ~10 sec between each measurement.

    As the electrical signal passes through the electrometer, it is low-pass filtered at 1 kHz and amplified by 10× before being digitized by the data acquisition system at 1 kHz and recorded in LabScribe.
18. Perform a set of vLogI measurements before each spectral sweep.

These measurements consist of a series of stimulus flashes ramping up in intensity by $\log_{10}$ steps of 0.125 and are typically performed at the wavelength with peak sensitivity for that species under those adaptation conditions (typically in the UV for most mosquito species after dark adaptation). This would give a series of intensity targets for the spectrometer of 3.20×10^{12}, 4.20×10^{12}, 5.60×10^{12}, 7.50×10^{12}, 1.00×10^{13}, 1.33×10^{13}, 1.77×10^{13}, 2.37×10^{13}, 3.16×10^{13}, and 4.22×10^{13} photons/cm^2/sec

19. Begin the recording and complete a spectral sweep of the spectral range (310–700 nm) increasing in 10-nm intervals from the lowest wavelength, followed by another set of vLogI measurements and another spectral sweep starting with highest wavelength and decreasing in 10-nm intervals. Following the same procedure as during calibration, insert the polycarbonate sheet in front of the shutter when recording the response of wavelengths of 600 nm and above.
20. Adjust the stimulus intensity between stimulus flashes by turning the neutral density wheel to hit the target intensity.
21. To adapt the mosquito to specific wavelengths of light (e.g., UV or green), turn on the adaptation LEDs, allowing the light to shine into the mosquito's eyes for 60 sec, before repeating the vLogI measurements and spectral sweeps described in Steps 19 and 20.
22. If the specimen dies during the process, restart the process with a fresh specimen.
    Specimen death is determined by a lack of movement.

Analyzing the Data

23. After the recording is completed, export it from LabScribe as a csv file and produce a separate csv file for each subexperiment (e.g., Vlogl, upward spectral sweep, downward spectral sweep).
24. Use the R code (Blake et al. 2022) to process the recording and estimate the spectral sensitivity (Steps 24–27). First import the subexperiment recording and automatically estimate and graph the height of the peaks relative to the background in each subexperiment.
25. Estimate a sigmoid function from the VLogI measurements, which can then be used to estimate sensitivity from the spectral sweeps.
26. Normalize the spectral sensitivities by the 97.5% quantile value of their sensitivity, and average between the up- and downsweep to give final sensitivity values for each insect under dark and color light adaptation.
27. Average the sensitivities across recordings from multiple insects, allowing for both a mean and standard error or sensitivity to be calculated for the across the entire spectral range.
    Comparing the averaged sensitivities between dark and colored adaptations can reveal changes in the relative contribution of the photoreceptor classes making up the ERG response (e.g., green adaptation light reducing sensitivity in the green spectral range).

DISCUSSION

By examining the averaged spectral sensitives under dark adaptation, one is easily able to characterize the range of wavelengths that the insect is capable of perceiving. The use of colored adaptations can better characterize individual photoreceptors; however, the filtering effects of pigments or other photoreceptors can be difficult to detect. It should be noted that an electrophysiological response to visual stimuli does not necessarily imply that the mosquito will behaviorally respond to those stimuli. Complementary behavioral tests under a variety of conditions are necessary to determine the behavioral response to those stimuli. Despite these caveats, mosquito spectral sensitivities can inform mosquito population management, such as through the design of improved traps, as well as yield a more thorough understanding of mosquito biology.

 Cite this protocol as *Cold Spring Harb Protoc*; doi:10.1101/pdb.prot108190

RECIPE

Ringer's Solution for Mosquitoes

Reagent	Amount
NaH_2PO_4	69 mg
Na_2SO_4	137 mg
$KHCO_3$	45 mg
$MgCl_2$	369 mg
$CaCl_2$	42 mg
KCl	46 mg

Dissolve all reagents in 100 mL of deionized H_2O and adjust the solution to pH 6.8. Store for up to 1 mo at 4°C.

ACKNOWLEDGMENTS

The pertinent technology was developed in Gerhard Gries's laboratory at Simon Fraser University (SFU), which was supported by a Natural Sciences and Engineering Research Council of Canada (NSERC) industrial research chair to Gerhard Gries with support from sponsors such as Scotts Canada Ltd and BASF Canada Inc.

REFERENCES

Blake AJ, Hung E, Peach DAH. 2022. Electroretinogram Analysis Code from: Measuring Mosquito Spectral Sensitivity Using Electroretinograms, Mendeley Data, v1. doi:10.17632/tzfyx575m6.1

Eby C, Weis M, Gardiner MGT, Judd GJR, Gries G. 2013. Spectral efficiency and microstructure of the compound eyes of *Synanthedon myopaeformis* (Lepidoptera: Sesiidae). *Can Entomol* **145:** 529–538. doi:10.4039/tce.2013.25

Kirchner SM, Döring TF, Saucke H. 2005. Evidence for trichromacy in the green peach aphid, *Myzus persicae* (Sulz.) (Hemiptera: Aphididae). *J Insect Physiol* **51:** 1255–1260. doi:10.1016/j.jinsphys.2005.07.002

Peach DAH, Ko E, Blake AJ, Gries G. 2019. Ultraviolet inflorescence cues enhance attractiveness of inflorescence odour to *Culex pipiens* mosquitoes. *PLoS ONE* **14:** e0217484. doi:10.1371/journal.pone.0217484

CHAPTER 25

Mosquito Electroantennography

Chloé Lahondère[1,2,3,4,5,6]

[1]*Department of Biochemistry,* [2]*The Fralin Life Science Institute,* [3]*The Global Change Center,* [4]*Department of Entomology,* [5]*Center of Emerging, Zoonotic and Arthropod-borne Pathogens, Virginia Polytechnic Institute and State University, Blacksburg, Virginia 24061, USA*

Mosquitoes use multiple cues to locate food sources (animal and plants), mates, and oviposition sites. The sense of smell plays an important role in these behaviors, and olfactory cues are detected primarily by the appendages on the head—in particular, the antennae. Thus, mosquito olfaction can be studied by conducting electroantennogram (EAG) recordings. EAGs have emerged as a reliable technique to screen for bioactive compounds and are useful in the development of attractants and repellents for mosquito population control. Here, we focus on comparing the two main experimental approaches used in the literature (whole-body and whole-head mounting) and highlight key considerations for conducting EAGs in multiple species and for obtaining reliable and reproducible data.

BACKGROUND

Mosquitoes use a wide range of cues (i.e., thermal, visual, auditory, and olfactory) to locate and identify food sources (plants and hosts), for mating and oviposition, and to avoid predators (Montell and Zwiebel 2016). Olfaction plays a key role in mosquito behavior for the detection of short, medium, and long-range chemical cues (Takken 1991; Zwiebel and Takken 2004). These cues are primarily detected by the peripheral olfactory system composed of the antennae, maxillary palps, and proboscis located on the mosquito head (Fig. 1; Kwon et al. 2006). Receptors have also been identified on the mosquito tarsi and wings (Owen 1963; Dennis et al. 2019; Yang et al. 2021).

Three primary classes of olfactory receptors are implicated in the detection of a wide range of chemicals: olfactory receptors (ORs), gustatory receptors (GRs), and ionotropic receptors (IRs) (Wolff and Riffell 2018; Raji et al. 2019; Wheelwright et al. 2021). Proteins (e.g., olfactory binding proteins [OBPs]) located in the sensillar lymph also play an important role in chemical detection by allowing the transport of hydrophobic molecules to the receptors' binding sites (Leal 2013). Integration of the signal occurs at the level of the antennal lobes (ALs) and subesophageal zone (SEZ) in the brain (Riabinina et al. 2016; Wheelwright et al. 2021).

Because olfaction is implicated in host-seeking behavior, it has been widely studied, as a better understanding of mosquito olfaction allows for the development of tools to potentially disrupt host-seeking behavior and limit pathogen transmission through mosquito bites, as well as the development of baits and repellents for population control (Paluch et al. 2010; Fiorenzano et al. 2017; Wooding et al. 2020).

[6]Correspondence: lahonder@vt.edu

Cite this introduction as *Cold Spring Harb Protoc*; doi:10.1101/pdb.top107679

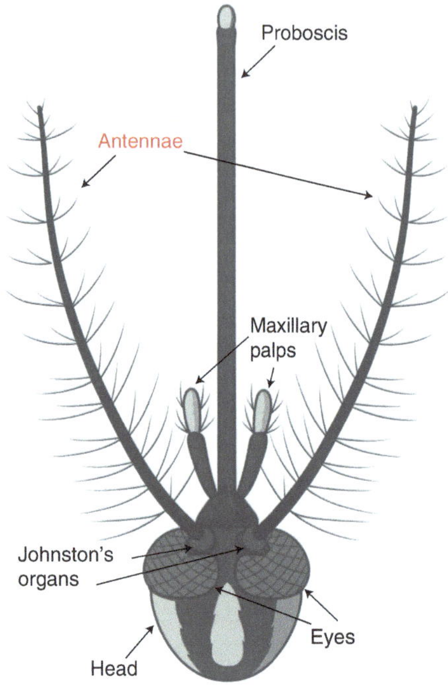

FIGURE 1. Schematic of the mosquito head highlighting the sensory tissues and organs. The antennae that are used for electroantennograms (EAGs) are highlighted in red. (Created with BioRender.com.)

ELECTROANTENNOGRAPHY

Electroantennography (EAG) is a powerful, yet simple technique to determine chemical compounds detected by the mosquito antennae (Roelofs 1984). Originally developed by Schneider (1957) in the late 1950s for pheromone research in moths, this method consists of measuring the voltage difference between a recording electrode and an indifferent electrode placed over an insect antenna(e) preparation after the presentation of an odorant stimulus. This technique has since allowed researchers to screen for bioactive compounds, perform dose curve response experiments to determine thresholds of detection (sensitivity), and to determine possible interactions between chemical compounds (e.g., Afify et al. 2019; Lahondère 2021). The negative deflections (in mV) in the signal observed when a given chemical compound is detected in an EAG correspond to the receptors' activation and cell depolarization/repolarization. The EAG signal's amplitude and overall shape depend on the concentration and structure of the chemical compound to which the insect's antennae have been exposed.

A diversity of EAG procedures can be found in the literature but two main techniques are widely used in mosquitoes: (1) The whole mosquito is mounted on a holder, or (2) the whole head is used for the preparation. These differences lead to a large variability in results and may sometimes render comparisons difficult. Below is a discussion of key factors in both techniques.

Mosquito Preparation

Whole-Body Mounting

This technique consists of mounting a whole mosquito on a substrate or holder to prevent movement. Lacher (1971) developed a holder onto which a legless mosquito was fixed, wings were glued, and the antennae were immobilized using tape. Meijerink et al. (2000) also used this immobilization technique and placed the recording electrode over the cut tip of the antenna while the reference electrode was inserted in the mosquito eye. Plasticine has also been used to immobilize mosquitoes for EAGs (Cork and Park 1996). In this study, the recording electrode was inserted in a segment of the antenna and the reference electrode inserted in either the insect abdomen or at the base of the antenna.

 Cite this introduction as *Cold Spring Harb Protoc*; doi:10.1101/pdb.top107679

Whole-Head Mounting

This technique, as described in Protocol 1: Mosquito Electroantennogram Recordings (Lahondère 2022), consists of excising the mosquito head and mounting it on the reference electrode. The tips of the antennae are generally clipped to increase contact and inserted in the recording electrode. In some studies, polyvinylpyrrolidone (PVP) is added to the saline solution filling the electrodes to minimize evaporation (Rund et al. 2013). Capillaries are generally filled with mosquito saline solution (e.g., Collins and Blackwell 1998; Wolff et al. 2019; Lahondère et al. 2020; Melo et al. 2020; Lahondère 2021) or KCl solution (e.g., Qiu et al. 2004; Rund et al. 2013; Meza et al. 2020).

Important Biological Factors to Consider

Many factors influence mosquito olfactory behavior (Klowden 1994; Hill et al. 2021) and it is thus essential to control for these factors to ensure replicability and reliability of EAG data. Before conducting experiments, minimizing the time the insects spend on ice for anesthesia can also reduce cold stress. If working with a cold-tolerant species, it is possible to use a CO_2 fly pad equipped with a humidifier to anesthetize mosquitoes. Other factors modulating olfactory behavior and thus antennal responses include physiological status, feeding status, and time of day, as described below.

Physiological Status

The mosquito age and overall physiological status (e.g., mating status) need to be considered when designing experiments. Indeed, aging and mating each induce a cascade of physiological events that lead to a drastic modification of the host-seeking behavior (Klowden 1994; Tallon et al. 2019; Hill et al. 2021).

Feeding Status

The mosquito feeding status is another factor to control for in EAG experiments, given that ingestion of a blood meal can affect sensitivity to some compounds (Klowden 1994; Takken et al. 2001; Siju et al. 2010; Duvall 2019). Similarly, sugar feeding can influence mosquito behavior, affecting responses to both host and plant cues (Klowden 1986). Consequently, sugar depriving non-blood-fed females for several hours before EAG recordings usually increases the insect's motivation, which leads to higher responses to both host and plant volatiles.

Time of Day

The time of day greatly influences mosquito behavior and consequently their sensitivity to given odorants as shown by Eilerts et al. (2018) and Rund et al. (2013) in two different mosquito species. It is thus critical to conduct experiments at a time of day that is ecologically relevant for the species that is studied (e.g., nocturnal vs. diurnal). Depending on the mosquito species and sex, feeding behavior varies throughout the day and should also be taken into account for EAG experiments to obtain coherent results.

Technical Details to Maintain Preparation Responsiveness

Several factors influence the responsiveness of the EAG preparation and allow the experimenter to determine the detection threshold of compounds at low concentrations.

Gradual Decrease in Responsiveness

If conducting whole-head mounting EAGs, supplying the main airflow bathing the antennae with humidity minimizes the desiccation of the neck tissues. It is especially important when conducting gas chromatography (GC)-EAGs as the transfer line is usually heated up to >200°C, which can lead to a rapid decrease in mosquito antennae responsiveness through time because of overheating and desiccation of the preparation.

Noise-to-Signal Ratio

Mosquito antennae can sometimes produce very small EAG responses, which can easily be missed in the background noise, especially for compounds at low concentrations. A low noise baseline signal is thus essential. Filtering the signal *a posteriori* is recommended and can help reduce some electrical noise. Ensuring a proper grounding of the system, working under a Faraday cage with minimal electrical instruments inside, grounding electrodes with alligator clips, and adding a noise reducer to the amplifier will lead to better results.

Limitations of the Technique

Although EAG is a simple and reliable technique, it can only be used to determine whether a compound, or a blend of compounds, at a given concentration, is detected by the mosquito antennae. To assess the valence (i.e., attractant, repellent, neutral) of this compound or blend, it is necessary to conduct subsequent behavioral experiments such as detection with an olfactometer, landing assays, biting assays, etc. (Qiu et al. 2004; Wolff et al. 2019; Lahondère et al. 2020; Meza et al. 2020). Moreover, EAG recordings do not inform on single-neuron responses, receptor type and location, or specificity for a given compound. Complementary experiments such as single-sensillum recording, two-photon microscopy, or the use of the empty neuron system are necessary to obtain fine-scale understanding of a response to a given chemical.

CONCLUSION

EAG should be seen as a useful technique for rapid compound screening and used in conjunction with other behavioral assays and electrophysiological recordings to obtain a complete picture on the valence of a specific compound. Ensuring that biological and ecological factors (which are species-specific) are taken into account will allow more relevant data to be collected. Our EAG protocol (see Protocol 1: Mosquito Electroantennogram Recordings [Lahondère 2022]) is applicable to both females and males of multiple genera including *Aedes*, *Anopheles*, *Culex*, and *Toxorhynchites* (Lahondère 2021). It allows for long-duration recordings, which are important when conducting both EAGs and GC-EAGs.

REFERENCES

Afify A, Betz JF, Riabinina O, Lahondère C, Potter CJ. 2019. Commonly used insect repellents hide human odors from *Anopheles* mosquitoes. *Curr Biol* **29:** 3669–3680. doi:10.1016/j.cub.2019.09.007

Collins LE, Blackwell A. 1998. Electroantennogram studies of potential oviposition attractants for *Toxorhynchites moctezuma* and *T. amboinensis* mosquitoes. *Physiol Entomol* **23:** 214–219. doi:10.1046/j.1365-3032.1998.233091.x

Cork A, Park KC. 1996. Identification of electrophysiologically-active compounds for the malaria mosquito, *Anopheles gambiae*, in human sweat extracts. *Med Vet Entomol* **10:** 269–276. doi:10.1111/j.1365-2915.1996.tb00742.x

Dennis EJ, Goldman OV, Vosshall LB. 2019. *Aedes aegypti* mosquitoes use their legs to sense DEET on contact. *Curr Biol* **29:** 1551–1556. doi:10.1016/j.cub.2019.04.004

Duvall LB. 2019. Mosquito host-seeking regulation: targets for behavioral control. *Trends Parasitol* **35:** 704–714. doi:10.1016/j.pt.2019.06.010

Eilerts DF, VanderGiessen M, Bose EA, Broxton K, Vinauger C. 2018. Odor-specific daily rhythms in the olfactory sensitivity and behavior of *Aedes aegypti* mosquitoes. *Insects* **9:** 147. doi:10.3390/insects9040147

Fiorenzano JM, Koehler PG, Xue RD. 2017. Attractive toxic sugar bait (ATSB) for control of mosquitoes and its impact on non-target organisms: a review. *Int J Environ Res Public Health* **14:** 398. doi:10.3390/ijerph14040398

Hill SR, Taparia T, Ignell R. 2021. Regulation of the antennal transcriptome of the dengue vector, *Aedes aegypti*, during the first gonotrophic cycle. *BMC Genomics* **22:** 1–19. doi:10.1186/s12864-020-07336-w

Klowden MJ. 1986. Effects of sugar deprivation on the host-seeking behaviour of gravid *Aedes aegypti* mosquitoes. *J Insect Physiol* **32:** 479–483. doi:10.1016/0022-1910(86)90009-0

Klowden MJ. 1994. Endogenous regulation of the attraction of *Aedes aegypti* mosquitoes. *J Am Mosq Control Assoc* **10:** 326–332.

Kwon HW, Lu T, Rützler M, Zwiebel LJ. 2006. Olfactory responses in a gustatory organ of the malaria vector mosquito *Anopheles gambiae*. *Proc Natl Acad Sci* **103:** 13526–13531. doi:10.1073/pnas.0601107103

Lacher V. 1971. Electrophysiological equipment for measuring the activity of single olfactory nerve cells on the antennae of mosquitoes. *J Econ Entomol* **64:** 313–314. doi:10.1093/jee/64.1.313

Lahondère C. 2021. A step-by-step guide to mosquito electroantennography. *J Visual Exp* doi:10.3791/62042

Lahondère C. 2022. Mosquito electroantennogram recordings. *Cold Spring Harb Protoc* doi: 10.1101/pdb.prot107871

Lahondère C, Vinauger C, Okubo RP, Wolff GH, Chan JK, Akbari OS, Riffell JA. 2020. The olfactory basis of orchid pollination by mosquitoes. *Proc Natl Acad Sci* **117:** 708–716. doi:10.1073/pnas.1910589117

Leal WS. 2013. Odorant reception in insects: roles of receptors, binding proteins, and degrading enzymes. *Annu Rev Entomol* **58:** 373–391. doi:10.1146/annurev-ento-120811-153635

Meijerink J, Braks MAH, Brack AA, Adam W, Dekker T, Posthumus MA, van Beek TA, van Loon JJA. 2000. Identification of olfactory stimulants for *Anopheles gambiae* from human sweat samples. *J Chem Ecol* **26:** 1367–1382. doi:10.1023/A:1005475422978

Melo N, Wolff GH, Costa-da-Silva AL, Arribas R, Triana MF, Gugger M, Riffell JA, DeGennaro M, Stensmyr MC. 2020. Geosmin attracts *Aedes aegypti* mosquitoes to oviposition sites. *Curr Biol* **30:** 127–134. doi:10.1016/j.cub.2019.11.002

Meza FC, Roberts JM, Sobhy IS, Okumu FO, Tripet F, Bruce TJ. 2020. Behavioural and electrophysiological responses of female *Anopheles gambiae* mosquitoes to volatiles from a mango bait. *J Chem Ecol* **46:** 387–396. doi:10.1007/s10886-020-01172-8

Montell C, Zwiebel LJ. 2016. Mosquito sensory systems. *Adv Insect Physiol* **51:** 293–328. doi:10.1016/bs.aiip.2016.04.007

Owen WB. 1963. The contact chemoreceptor organs of the mosquito and their function in feeding behaviour. *J Insect Physiol* **9:** 73–87. doi:10.1016/0022-1910(63)90085-4

Paluch G, Bartholomay L, Coats J. 2010. Mosquito repellents: a review of chemical structure diversity and olfaction. *Pest Manag Sci* **66:** 925–935. doi:10.1002/ps.1974

Raji JI, Melo N, Castillo JS, Gonzalez S, Saldana V, Stensmyr MC, DeGennaro M. 2019. *Aedes aegypti* mosquitoes detect acidic volatiles found in human odor using the IR8a pathway. *Curr Biol* **29:** 1253–1262. doi:10.1016/j.cub.2019.02.045

Riabinina O, Task D, Marr E, Lin CC, Alford R, O'Brochta DA, Potter CJ. 2016. Organization of olfactory centres in the malaria mosquito *Anopheles gambiae. Nat Commun* **7:** 13010. doi:10.1038/ncomms13010

Roelofs WL. 1984. Electroantennogram assays: rapid and convenient screening procedures for pheromones. In *Techniques in pheromone research* (ed. Hummel HE, Miller TA), pp. 131–159. Springer, New York.

Rund SS, Bonar NA, Champion MM, Ghazi JP, Houk CM, Leming MT, Syed Z, Duffield GE. 2013. Daily rhythms in antennal protein and olfactory sensitivity in the malaria mosquito *Anopheles gambiae. Sci Rep* **3:** 2494. doi:10.1038/srep02494

Schneider D. 1957. Elektrophysiologische Untersuchungen von Chemo-und Mechanorezeptoren der Antenne des Seidenspinners *Bombyx mori* L. *Z Vergleichende Phys* **40:** 8–41. doi:10.1007/BF00298148

Siju KP, Hill SR, Hansson BS, Ignell R. 2010. Influence of blood meal on the responsiveness of olfactory receptor neurons in antennal sensilla trichodea of the yellow fever mosquito, *Aedes aegypti. J Insect Physiol* **56:** 659–665. doi:10.1016/j.jinsphys.2010.02.002

Takken W. 1991. The role of olfaction in host-seeking of mosquitoes: a review. *Int J Tropical Insect Sci* **12:** 287–295. doi:10.1017/S1742758400020816

Takken W, van Loon JJ, Adam W. 2001. Inhibition of host-seeking response and olfactory responsiveness in *Anopheles gambiae* following blood feeding. *J Insect Physiol* **47:** 303–310. doi:10.1016/S0022-1910(00)00107-4

Tallon AK, Hill SR, Ignell R. 2019. Sex and age modulate antennal chemosensory-related genes linked to the onset of host seeking in the yellow-fever mosquito, *Aedes aegypti. Sci Rep* **9:** 43. doi:10.1038/s41598-018-36550-6

Qiu YT, Smallegange RC, Hoppe S, van Loon JJA, Bakker EJ, Takken W. 2004. Behavioural and electrophysiological responses of the malaria mosquito *Anopheles gambiae* Giles *sensu stricto* (Diptera: Culicidae) to human skin emanations. *Med Vet Entomol* **18:** 429–438. doi:10.1111/j.0269-283X.2004.00534.x

Wheelwright M, Whittle CR, Riabinina O. 2021. Olfactory systems across mosquito species. *Cell Tissue Res* **383:** 75–90. doi:10.1007/s00441-020-03407-2

Wolff GH, Riffell JA. 2018. Olfaction, experience and neural mechanisms underlying mosquito host preference. *J Exp Biol* **221:** jeb157131. doi:10.1242/jeb.157131

Wolff GH, Lahondère C, Vinauger C, Riffell JA. 2019. Neuromodulation and differential learning across mosquito species. bioRxiv doi:10.1101/755017

Wooding M, Naudé Y, Rohwer E, Bouwer M. 2020. Controlling mosquitoes with semiochemicals: a review. *Parasit Vectors* **13:** 80. doi:10.1186/s13071-020-3960-3

Yang L, Agramonte N, Linthicum KJ, Bloomquist JR. 2021. A survey of chemoreceptive responses on different mosquito appendages. *J Med Entomol* **58:** 475–479. doi:10.1093/jme/tjaa208

Zwiebel LJ, Takken W. 2004. Olfactory regulation of mosquito–host interactions. *Insect Biochem Mol Biol* **34:** 645–652. doi:10.1016/j.ibmb.2004.03.017

Protocol 1

Mosquito Electroantennogram Recordings

Chloé Lahondère[1,2,3,4,5,6]

[1]*Department of Biochemistry,* [2]*The Fralin Life Science Institute,* [3]*The Global Change Center,* [4]*Department of Entomology,* [5]*Center of Emerging, Zoonotic and Arthropod-borne Pathogens, Virginia Polytechnic Institute and State University, Blacksburg, Virginia 24061, USA*

Electroantennography (EAG) is a simple, yet powerful technique that can be used to indicate that chemical compounds are detected by mosquito antennae. When coupled with gas chromatography (GC), it allows for the identification of detectable compounds within complex mixtures, including those in plant and animal headspaces that might influence mosquito behavior. EAGs and GC-EAGs reveal compounds that can be tested in behavioral assays with the goal of developing new repellents or attractants for mosquito control.

MATERIALS

It is essential that you consult the appropriate Material Safety Data Sheets and your institution's Environmental Health and Safety Office for proper handling of equipment and hazardous materials used in this protocol.

RECIPES: Please see the end of this protocol for recipes indicated by <R>. Additional recipes can be found online at http://cshprotocols.cshlp.org/site/recipes.

Reagents

Bleach solution (100%)

Carbon dioxide (gas tank) (optional; see Step 18)

Chemical odorants and controls

Dilute chemicals in mineral oil, paraffin oil, water, hexane, etc., depending on the solubility of the compound. Concentrations will vary depending on the experiment to be conducted. Prepare dilutions in gas chromatography–mass spectrometry (GC-MS) vials. Use 0.1% or 1% benzaldehyde as a positive control and the solvent used for diluting the chemicals as the negative control. See Lahondère (2021) for more details regarding the preparation of the chemicals.

Deionized H_2O

Electrode gel, salt-free (e.g., Parkerlabs 12-08-Spectra-360)

Medical air (gas tank)

Mosquitoes

This protocol can be used in multiple mosquito genera, including Culex, Anopheles, Aedes, *and* Toxorhynchites, *and in both females and males. Gerberg (1970) describes rearing of laboratory mosquito species, and Moll et al. (2013) describes rearing of malaria mosquitoes.*

[6]Correspondence: lahonder@vt.edu

Cite this protocol as *Cold Spring Harb Protoc*; doi:10.1101/pdb.prot107871

Mosquito saline <R>

Around 50 mL of saline is needed per experiment. Add 5.0 mM sucrose to the saline right before running an electroantennogram.

Equipment

30V/5A DC Power Supply (e.g., Dr. Meter PS-305DM)
Air table (e.g., TMC)
Alligator clips (optional; see Step 29)
Aluminum block (4 in × 4 in × 1 in)
Analog-to-digital board (National Instruments BNC-2090A)
Borosilicate glass capillaries with filament (Sutter Instrument B100-78-10)

Pull capillaries based on the equipment available to the experimenter and store in a Petri dish on unscented modeling clay or wax. Programs to design capillaries will depend on the pipette puller available (e.g., P-1000 or P-2000). Using capillaries with filaments will reduce the risk of having bubbles inside the holder during the mounting process.

Bubbler (e.g., Genesee Scientific 59-181)
Carbon dioxide anesthesia system (Genesee Scientific 59-122BC) (optional; see Step 18)
Cold light source (e.g., Volpi NCL-150)
Cotton tip applicator
Disposable syringes (2-mL)
Electrode cables (World Precision Instruments 5371)
Electrode holders (A-M Systems 675748) (two; see Step 12)
Faraday cage (e.g., TMC)
Filter paper (e.g., Whatman, Cole Parmer UX-06648-03)
Flowmeters (65-mm [Bel-Art H40406-0010]/150-mm [Bel-Art H40407-0075])
Forceps (e.g., #5, Dumostar Biology)
Gas chromatograph coupled with flame ionization detection (GC-FID) (e.g., Trace, Thermo Fisher Scientific) (optional; see Step 10)
Gas chromatography–mass spectrometry (GC–MS) vials and caps (Thermo Fisher Scientific 2-SVWKA8-CPK)
Glass syringes (2-mL; Sigma-Aldrich Z314307) or Pasteur pipettes (Fisher Scientific 13-678-6A)
Magnetic bases × 2 (e.g., Kanetec MB-FX) (two; see note above Step 6)
Microelectrode amplifier with headstage (Model 1800, A-M Systems)
Microscissors (e.g., F.S.T. 15000-003)
Microscope (e.g., Nikkon SMZ-800N)
Needles, 25G (BD 305127) and 21G (BD 305165)
Noise reducer for electrophysiology recordings (e.g., HUMBUG, Quest Scientific)
Polytetrafluoroethylene (PTFE) tubing of different diameters appropriate to connect the flowmeters, air table, etc. (e.g., McMaster-Carr)
Recording software for EAGs (e.g., WinEDR)
Relay for solenoid valve (custom made as in Lahondère 2021)
Silver wire 0.01" (A-M Systems 782500)
Solenoid valve (three-way) (The Lee Company LHDA0533115H)
Statistical analysis software (e.g., R)
Task wipers (e.g., Kimwipes)
Three-Axis Coarse/Fine Compact Micromanipulator (Narishige MHW-3) (two; see Steps 24 and 25)
Tubes (15-mL; e.g., Millipore Sigma-Aldrich CLS430791)
Vortex

METHOD

Conduct this experiment at room temperature except where noted.

Mosquitoes

For information on how physiological status, feeding status, and time of day can affect electroantennograms (EAGs), see Introduction: Mosquito Electroantennography (Lahondère 2022).

1. Starve mosquitoes for up to 12 h before the experiment to increase their motivation.
2. Isolate mosquitoes in individual 15-mL tubes 1 h before EAGs start as follows.
   i. Place a container or rearing jar full of mosquitoes for 5–10 min at 4°C to prevent mosquitoes from flying away.
   ii. Select individual mosquitoes with forceps, placing one individual per 15-mL tube.
   iii. To avoid desiccation, keep mosquitoes in the rearing incubator until the experiments start.

Chemicals

3. Remove GC-MS vials containing odorant dilutions from the freezer/refrigerator 30 min before conducting EAGs and let them reach room temperature.

   Along with chemicals to be tested, always include positive (e.g., 0.1% or 1% benzaldehyde) and negative (i.e., solvent used for diluting the chemicals) controls. This is critical for data analyses as well as to ensure that no carryover occurs.

4. Vortex each vial before pipetting 10 µL of each odorant dilution onto a piece of 0.5-cm × 2-cm filter paper. Using forceps, place the filter paper in a glass syringe equipped with a 21G needle or in a Pasteur pipette.

   Use one syringe per odorant to avoid contamination. Load the syringes at least 10 min before starting to allow for diffusion of the chemical inside the syringes. Keep syringes closed at all times to minimize evaporation. The connector that sends air into the syringe should not touch the filter paper to avoid contamination. Replace the filter paper after each EAG session. Replace the needle every 10 sessions to minimize risk of clogging.

5. Plan the order in which the chemicals will be presented to the mosquito antennae.
   - If conducting a dose–response curve experiment of a given chemical, present the mosquito antennae with the most diluted solution first.
   - If conducting experiments with a set of chemicals, consider randomizing the odorants' order.

     The presentation of an odorant can impact the response to the following one. It is thus essential to randomize odorant presentation to ensure reliable data collection.

EAG Rig Checkup and Preparation

The EAG rig consists of an air table with a Faraday cage, a microscope, electrode holders, magnetic bases, computer(s), air line, flowmeters, bubbler, power supply, amplifier, noise reducer, cold light source, microelectrode amplifier with headstage, analog-to-digital board, and solenoid valve, micromanipulators, and associated tubing. See Lahondère (2021) for more details and a picture of this rig.

6. Turn on the EAG computer and load any specific script written for the experiment for odorant pulse delivery. See Lahondère (2021) for details on writing a script for odorant pulse delivery. Turn on the software for the recordings and set up the needed parameters (e.g., duration of recording; see Step 31 for details).
7. Turn on the air line and check the flowmeters. Ensure that the bead inside each flowmeter rises and stabilizes. Set the airflow at 140 mL/min, and the pulse at 15 mL/min. Check that the air table is up/on to prevent vibrations and noise in your signal. Check the level of water in the bubbler to ensure that the airstream is humidified.

Cite this protocol as *Cold Spring Harb Protoc*; doi:10.1101/pdb.prot107871

8. Turn on the power supply and set at 5.2 V. Test the solenoid valve by delivering a test pulse.
9. Verify the parameters of the amplifier: low cutoff = 0.1 Hz; high cutoff = 500 Hz; gain = ×100.

GC-FID Rig Checkup and Preparation

See Lahondère et al. (2020) for information on using GC-FID in these experiments.

10. If conducting GC-EAGs, turn on the gas tanks, the GC-FID machine, transfer line, and associated computer and load the appropriate GC method. Initiate the GC-FID flame.

Electrode Holders and Capillaries

11. Gently break the tips of two pulled capillaries with forceps under the microscope.

    The reference electrode that will fit the mosquito head should be slightly larger than the recording electrode into which the tips of the antennae will be inserted. These capillaries can be reused with multiple head preparations if not broken.

12. Inspect both electrode holders (i.e., reference and recording) and wires. Remove any glass debris or salt traces present in the electrodes before starting.

    Debris and salt residues can break the bottom part of the capillary and create a glass plug that will accumulate inside the holder. This can eventually break the electrode wire.

13. Soak the tip (1-cm) of two pieces of wire cut to the length of the capillary in 100% bleach solution for 5 min. Rinse with deionized H_2O to remove excess bleach.

    Chloridization causes the silver wires to turn matte dark gray.

14. Loosen the rubber stopper of each holder and fill both capillaries and holders with mosquito saline solution using a syringe with a 21G needle, making sure to avoid inserting bubbles.

    To prevent bubbles from entering the holders and capillaries, keep pushing saline inside the compartment as the needle is gently removed.

15. Insert the silver wire into the capillary with the silver wire <1 mm from the capillary tip. Gently insert each capillary into an electrode holder. Tighten the rubber stoppers and verify that no air bubbles are present.
16. Leave the two mounted electrodes on a piece of wet task wiper while you move on to the next step.
17. After the experiments, if the capillaries are still intact, rinse them with deionized H_2O, dry them by applying a cleaning wipe at the tip to remove excess water inside, and place them back in their storage box until future use.

Mosquito Head Mounting

18. Place a tube containing a mosquito on ice to anesthetize it (~3 min).

    Some species are cold-resistant and may need to be anesthetized using a carbon dioxide pad. Minimizing the time spent on ice/carbon dioxide pad will increase the preparation longevity.

19. Prepare an aluminum block on ice and place a slightly wet task wiper on top. Using a syringe and a 25G needle, place a small dollop of electrode gel in the middle. Place the block under a microscope.
20. Once the mosquito is anesthetized, place it delicately on the task wiper using forceps. Remove the last segment of both antennae with micro scissors and dip each tip in the electrode gel, bringing them closely together.

    Avoid dipping the antennae in the gel too much to prevent blockage of the sensilla, which would lead to smaller EAG responses.

21. Once the antennae tips are close, gently pull the mosquito by a leg using forceps while supporting the antennae so when they come out of the gel, the tips remain in contact.

    Maintain the antennae tips away from the wet task wiper surface or they will separate.

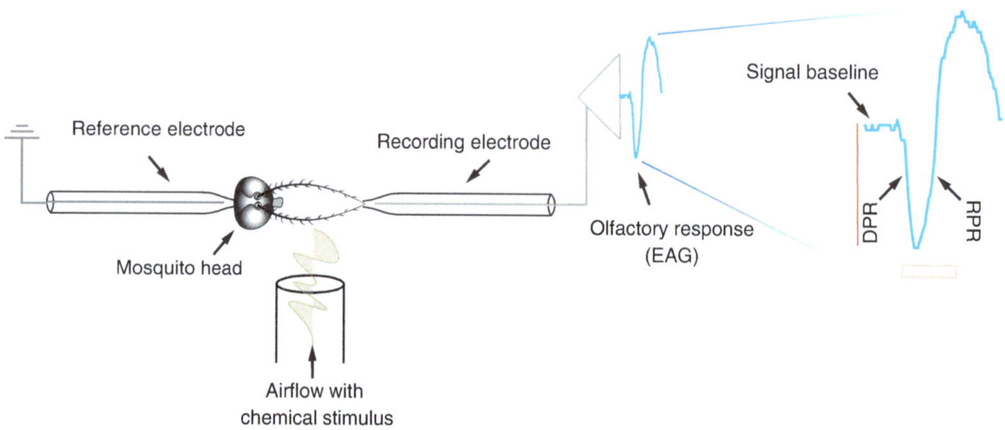

FIGURE 1. Schematic representation of the mosquito electroantennogram (EAG) preparation and details of the olfactory response. The mosquito head is excised and mounted between the reference electrode and the recording electrode connected to an amplifier and noise reducer. A constant humid airflow bathes the antennae and also carries the odorant plume during stimulation (light brown square). Detection of a tested chemical leads to cell depolarization (DPR) followed by repolarization (RPR). The amplitude of the signal (in mV) (red bar) is measured and compared across preparation and stimuli. (Created with BioRender.com.)

22. Place the mosquito on its side on the wet task wiper and remove the mosquito head using micro scissors.

    When cutting the head off, make sure to target the soft tissues of the neck and leave the head cuticle intact. If the brain is exposed to air, it will dry out faster and the preparation will not be responsive for more than a few minutes.

23. Dip the reference electrode in electrode gel and apply the tip of the capillary at the base of the mosquito head.

    This will create a suction effect and the head will stick to the capillary and will be movable.

    The position of the head is critical here and should be planned based on the position of the jack connector to avoid pulling on the wire when recording. Ideally, the head and the jack should be in the same plane.

24. Bring both electrodes to the EAG rig and first place the reference electrode with the mounted head on the micromanipulator. Center the head in the middle using the microscope.
25. Place the recording electrode in front of the reference electrode and bring it as close as possible to the reference electrode using the micromanipulator. (See Fig. 1.)
26. Once both electrodes are close, plug both into the amplifier.

    The amplifier should still be off at this point.

27. Insert the antennae tips into the recording electrode.

    The antennae tips should be "sucked in" and enter the capillary easily. The tips should be visible through the capillary and placed as closed as possible to the electrode silver wire.

28. Position the vacuum line behind the mosquito preparation to remove chemicals after the pulses (distance: 20 cm). Position the main airline in front of the preparation (distance: 1 cm).

Recordings

29. Turn the amplifier and noise reducer on and observe the baseline signal.

    If the baseline is noisy (>0.01 mV), adjust the position of the head and use alligator clips to ground things that might be introducing the noise within the Faraday cage (e.g., lamp, head stage). The Faraday cage should remain closed during the recording.

30. Once the signal is stable, introduce the first odor cartridge to test in the airline.
31. On the EAG software, click on "record" and deliver the odor pulse.

Cite this protocol as *Cold Spring Harb Protoc*; doi:10.1101/pdb.prot107871

The number and duration of the odorant pulses are at the discretion of the experimenter based on the research questions and experimental conditions. 500-msec to 1-sec pulses are commonly used, separated by 30–45 sec between odorant stimulations.

32. Observe the response of the mosquito antennae and note whether a clear deflection in the signal is observed, which indicates that the chemical is detected.
33. Proceed with the next odorants to be tested.

    A positive (e.g., 0.1% or 1% benzaldehyde) and negative (e.g., solvent) control should be used as well to ensure that the preparation is responsive and that responses are not due to potential noise.

34. Once the odorant panel has been presented, turn off the amplifier and noise reducer before removing the electrodes from the EAG rig and start with the next preparation.

    Refill both electrodes with fresh saline solution and remove any extra electrode gel or neck tissue that could be clogging the capillaries.

35. If conducting GC-EAGs (e.g., for testing a complex mixture of chemicals), inject your sample in the GC-FID machine and start the recording on the GC-FID computer.

    See Lahondère et al. (2020) for information on using GC-FID in these experiments.

Cleaning

36. After the experiment is completed, turn off all equipment, and return the chemical vials to the freezer/fridge. Remove the filter papers inside the syringes and wipe excess liquid away using a cotton tip applicator.
37. Flush the electrode holders with deionized H_2O using a designated syringe and needle.

    If intact, capillaries can also be rinsed, dried, and placed in their container for future use.

Data Analysis

See Lahondère (2021) for additional information on data analysis.

38. Manually or automatically measure the EAG responses. Average responses in mV between preparations.
39. Conduct data analysis using the recording software or in R.

RECIPE

Mosquito Saline

Reagent	Final concentration
NaCl	150.0 mM
HEPES	25.0 mM
KCl	3.4 mM
$NaHCO_3$	1.8 mM
$CaCl_2$	1.7 mM
$MgCl_2$	1.0 mM

Combine the ingredients in deionized H_2O for a total volume of 2 L, and adjust the pH to 7.1 with 1 M NaOH. Store indefinitely at 4°C.

ACKNOWLEDGMENTS

The author thanks Dr. Clément Vinauger and Dr. Jeffrey Riffell for helpful discussions.

REFERENCES

Gerberg EJ. 1970. Manual for mosquito rearing and experimental techniques. *Am Mosq Control Assoc Bull* **5:** 72–73.

Lahondère C. 2021. A step-by-step guide to mosquito electroantennography. *J Vis Exp* doi:10.3791/62042

Lahondère C. 2022. Mosquito electroantennogram recordings. *Cold Spring Harb Protoc* doi:10.1101/pdb.top107679

Lahondère C, Vinauger C, Okubo RP, Wolff GH, Chan JK, Akbari OS, Riffell JA. 2020. The olfactory basis of orchid pollination by mosquitoes. *Proc Natl Acad Sci* **117:** 708–716. doi:10.1073/pnas.1910589117

Moll K, Ljungström I, Perlmann H, Scherf A, Wahlgren M, Manassas V. 2013. *Methods in malaria research.* https://www.beiresources.org/portals/2/MR4/Methods_In_Malaria_Research-6th_edition.pdf (accessed January 27, 2022).

CHAPTER 26

Taste Sensory Responses in Mosquitoes

Adriana Medina Lomelí[1] and Anupama Arun Dahanukar[2,3]

[1]*Entomology Graduate Program,* [2]*Department of Molecular, Cell and Systems Biology, University of California, Riverside, California 92521, USA*

Analysis of taste sensory responses has been a powerful approach for understanding principles of taste detection and coding. The shared architecture of external taste sensing units, called sensilla, in insects opened up the study of tastant-evoked responses in any model of choice using a single-sensillum tip recording method that was developed in the mid-1900s. Early studies in blowflies were instrumental for identifying distinct taste neurons based on their responses to specific categories of chemicals. Broader system-wide analyses of whole organs have since been performed in the genetic model insect *Drosophila melanogaster*, revealing principles of stereotypical organization and function that appear to be evolutionarily conserved. Although limited in scope, investigations of taste sensory responses in mosquitoes showcase conservation in sensillar organization, as well as in groupings of functionally distinct taste neurons in each sensillum. The field is now poised for more thorough dissections of mosquito taste function, which should be of immense value in understanding close-range chemosensory interactions of mosquitoes with their hosts and environment. Here, we provide an introduction to the basic structure of a taste sensillum and functional analysis of the chemosensory neurons within it.

THE MOSQUITO TASTE SYSTEM

Taste sensitivity in insects has been a topic of investigation for about 100 years, beginning with observations of feeding responses upon stimulation of distal leg segments (tarsi) in the nymphalid butterflies *Pyrameis atlanta Linn.* and *Vanessa antiopa Linn.* (Minnich 1921). Insects rely on taste, or contact chemoreception, for behaviors such as feeding, mate recognition and courtship, and oviposition, which are essential for their survival and reproductive success. Some of the very same behaviors are drivers of an enormous economic and health burden caused by pests and disease vectors, including anthropophilic, or human-seeking, mosquitoes. Studies of taste cues and how they are sensed provide important insight into the biology and evolution of insect taste-guided behaviors.

Insects possess taste sensory units, called sensilla, in many different locations across the body surface: on mouthparts, tarsi, wing margins, ovipositors (female-specific structures used to lay eggs), and, in some cases, the antennae and maxillary palps. The external mouthparts of blood-feeding mosquitoes include a labellum on the distal part of the labium that acts as a sheath to a set of six needle-like structures or stylets that the mosquito uses to obtain a blood meal from its host (Fig. 1A). Of the six stylets, the labrum is the one that is used to pierce the host skin and locate blood via chemosensory sensilla. Two types of taste sensilla can be found in mosquitoes: apical/subapical sensilla, which are small, straight, hair-like structures that lie flat against the tip of the labrum of

[3]Correspondence: anupama.dahanukar@ucr.edu

Cite this introduction as *Cold Spring Harb Protoc;* doi:10.1101/pdb.top107680

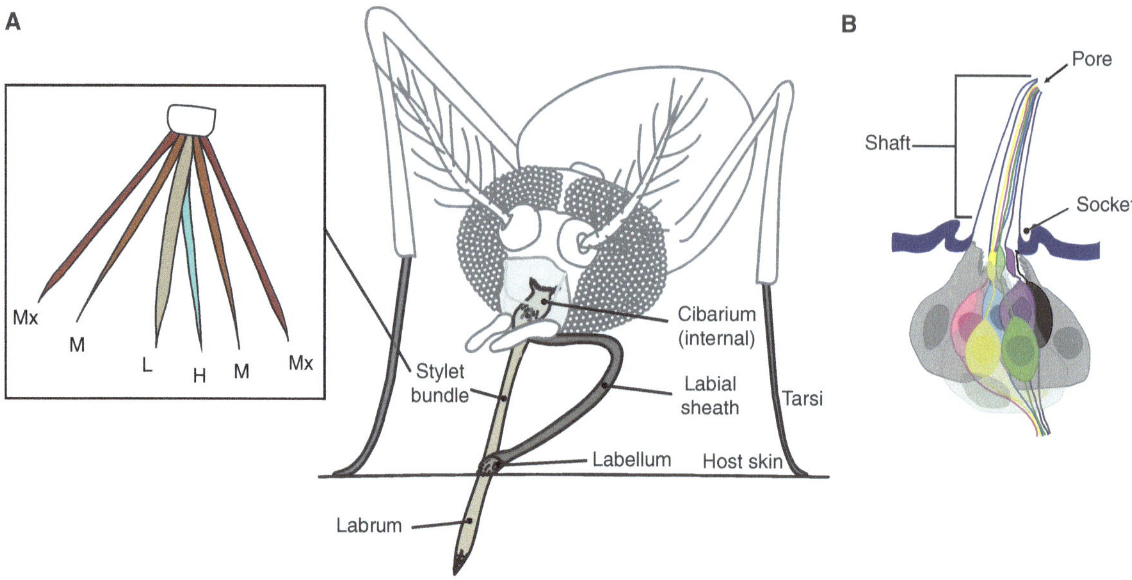

FIGURE 1. (*A*) Schematic showing the mosquito's taste organs (colored in) while probing for a blood meal. Upon landing on a host, the mosquito will use its tarsi and labellum to survey the surface of the host skin. Once a suitable bite site is found, the female mosquito will apply pressure to the skin, causing the labial sheath to retract and expose the stylets. Although the stylets have the appearance of a single feeding tube structure, there are in fact six stylets (L, labrum; H, hypopharynx; M, mandibles; Mx, maxillae) that stack together neatly to form it. The mosquito uses its collection of stylets to cut through the host skin, but the labrum is the only one inserted through the skin to locate host blood. Once imbibed, the blood passes through the feeding tube formed by the stylets and through the cibarium, an internal structure, which is positioned to be a last taste checkpoint before the blood meal enters the midgut. Image credits: (Inset) Teodros Hailye/KQED based on research by Young-Moo Choo and colleagues. (*B*) Schematic showing the basic structure of a uniporous trichoid sensillum. In mosquitoes, the trichoid sensilla can be innervated by up to five gustatory neurons (colored) and a single mechanosensory neuron (black).

hematophagous (blood-feeding) mosquito species, and longer trichoid sensilla that are located throughout external taste organs. Olfactory sensilla are also located on the mosquito labellum and are easily distinguished from the larger and more numerous trichoid taste sensilla. The mosquito cibarium, an internal taste organ that is likely to serve as a checkpoint for evaluating meal quality before ingestion, is coated with small taste papillae in addition to a few trichoid taste sensilla and mechanosensory campaniform sensilla (Lee and Craig 1983). Comparative studies suggest that numbers of cibarial papillae may be lower in non-blood-feeding species (Lee and Davies 1978). For this reason, they are of great interest in terms of investigating blood-sensing function; however, they are difficult to access for electrophysiological analysis.

Despite the morphological differences between the trichoid and apical sensilla, these, along with the papillae in the cibarium, are categorized as taste sensilla because of the presence of a single pore that allows the dendrites of sensory neurons to contact tastants in the environment. The basic architecture of a trichoid taste sensillum, also known as a taste bristle or taste hair, is conserved in arthropods and consists of an external cuticular shaft that is socketed at the base and contains two compartments, one of which is innervated by the dendrites of a number of taste sensory neurons housed in the sensillum (Fig. 1B). Each sensillum also includes a single mechanosensory neuron with a dendrite that terminates below the shaft. The cell bodies of all sensory neurons lie below the socket and are enveloped by accessory cells that serve to electrically insulate the sensory unit and to supply components of the sensillar lymph that bathes the dendrites. The single pore at the tip of the cuticular hair serves as a conduit between the taste sensors and soluble chemicals in the environment, making the hairs easily accessible for electrophysiological analysis.

The arrangement and morphology of external taste sensilla has been mapped for only a handful of the thousands of mosquito species. As described in *Drosophila* and other flies, sensilla are found

 Cite this introduction as *Cold Spring Harb Protoc*; doi:10.1101/pdb.top107680

in a stereotypical organization that is characteristic for a species (Hill and Berry Smith 1999). In *Aedes aegypti* and *Anopheles gambiae*, topographic maps of the labellum have been generated with unique names for each of the trichoid sensilla identified by morphology and location (Hill and Berry Smith 1999; Kessler et al. 2015). Trichoid sensilla are also found throughout the tarsi with higher numbers on the forelegs and midlegs as compared to hindlegs, in *Ae. aegypti* and possibly other species as well (Baik and Carlson 2020). Although tarsal sensilla have not been mapped to the same degree of completeness as those in the labellum, it is clear that differently sexed individuals of the same species show consistent patterns of sensilla (Seenivasagan et al. 2009). In addition, sexual dimorphism has been observed in the number of trichoid sensilla in some organs of some species. Females of *Ae. aegypti*, for example, have many more trichoid sensilla on their legs than males, but the numbers on the labellum are the same in the two sexes (McIver and Siemicki 1978).

ANALYSIS OF TASTE SENSORY RESPONSES

The responsiveness of taste neurons within a sensillum can be assessed using the extracellular tip recording method, developed in blowflies, *Phormia regina*, by Hodgson et al. (1955). This method simultaneously captures the activity of all neurons housed in the sensillum, visualized as train(s) of action potentials. In this method the mosquito, or indeed any insect, is immobilized and then its body is pierced with a reference electrode. With this live restrained preparation, any water-soluble stimulus can be delivered to the open pore at the tip of a selected external taste sensillum via a glass micropipette electrode that also records neural responses. The activity of individual neurons can often, but not always, be parsed by amplitude of the action potential and sensitivity to "diagnostic" tastants. The strength of a neuron's response is then quantified by measuring action potential frequency, typically in the first second or part thereof after contact of the sensillum tip with the tastant solution. In our associated protocol, we introduce such a method (see Protocol 1: Single-Sensillum Taste Recordings in Mosquitoes [Lomelí and Dahanukar 2023]).

Selected labellar and tarsal sensilla in a few different mosquito species, including both hematophagous and nonhematophagous species, have been tested for responsiveness to sugars, bitter compounds, salts, amino acids, and water. Compounds in each of these categories are found in plant nectar, which is a food source for both sexes of hematophagous and nonhematophagous mosquito species (Lomelí and Dahanukar 2022). Additionally, anthropophilic mosquitoes encounter sugars, salts, and free amino acids as human sweat and skin-associated tastants, as well as various components in blood. In *Culex pipiens*, hematophagous female-specific apical and subapical labral sensilla have been tested with tip recordings for responses to ATP, which is a blood component, and related nucleotides such as ADP, AMP, adenylyl-imidodiphosphate (AMP-PNP), and GTP (Liscia et al. 1993). In *Ae. aegypti*, calcium imaging has been used to examine neurons innervating these same sensilla for responses to blood, as well as individual components such as ATP, glucose, salt (NaCl), and sodium bicarbonate ($NaHCO_3$) (Jové et al. 2020). In the labellum, taste neuron groupings similar to those described in *Drosophila* were found in *Culiseta inornata*, with neurons responding to water, sugars, and salts at low and high concentrations (Pappas and Larsen 1976). Bitter sensitivity has also been explored in *Ae. aegypti* and *Anopheles quadrimaculatus*, and a taste neuron activated by bitter compounds is indeed found in labellar sensilla (Sanford et al. 2013; Sparks and Dickens 2016). Moreover, as found in *Drosophila* (Meunier et al. 2009; Zhang et al. 2013; French et al. 2015), bitter compounds and salts inhibit sugar and water taste responses in mosquitoes (Kessler et al. 2013). The observed inhibition of neuronal activity is not a permanent effect, and the dampened activity of the neuron recovers when the inhibitory stimulus is removed and the activating stimulus is reintroduced to the tip of the sensillum. These neuronal properties can be correlated with behavior, exemplifying the importance of understanding how not only pure compounds but also mixtures are encoded by the insect taste system.

CONCLUSION

The external distribution of taste sensilla in insects makes them easily accessible for functional analyses, and, if body size is not a limiting factor, the physiological and behavioral activity of even a single-sensillum can be measured (Minnich 1926; Dethier 1976). This feature presents a tremendous advantage for performing high-resolution system-wide analyses of peripheral taste sensitivity. Responses to diverse sets of chemicals have been mapped to taste neurons in individual sensilla in entire taste organs in *Drosophila*, which allows comparison of how chemicals are distributed in the "tastant space" of different organs. Clearly, similar analyses in anthropophilic mosquitoes would offer a glimpse into the similarities and disparities in taste coding in a blood-feeding dipteran as compared to the extensively studied phytophagous, or plant-feeding, *Drosophila*. Such analyses would also afford a comprehensive view of taste sensory input that can be evaluated for correlation with behavior.

ACKNOWLEDGMENTS

We thank Manali Dey and Vaibhav Menon for helpful comments. Research in A.A.D.'s laboratory is supported by grants from the National Institutes of Health (R01DC017390 and R21AI140065) and the University of California, Riverside Agricultural Experimental Station and the National Institute of Food and Agriculture—U.S. Department of Agriculture (Hatch Project Grant 1011543). A.M.L. is a UC-Hispanic Serving Institutions Doctoral Diversity Initiative (UC-HSI DDI) fellow.

REFERENCES

Baik L, Carlson JR. 2020. The mosquito taste system and disease control. *Proc Natl Acad Sci* **117:** 32848–32856. doi:10.1073/pnas.2013076117

Dethier VG. 1976. *The hungry fly: a physiological study of the behavior associated with feeding*. Harvard University Press, Cambridge, MA.

French A, Sellier MJ, Agha MA, Guigue A, Chabaud MA, Reeb PD, Mitra A, Grau Y, Soustelle L, Marion-Poll F. 2015. Dual mechanism for bitter avoidance in *Drosophila*. *J Neurosci* **35:** 3990–4004. doi:10.1523/JNEUROSCI.1312-14.2015

Hill SR, Berry Smith J. 1999. Consistent pattern in the placement of taste sensilla on the labellar lobes of *Aedes aegypti*. *Int J Insect Morphol Embryol* **28:** 281–290. doi:10.1016/S0020-7322(99)00031-8

Hodgson E, Lettvin JY, Roeder KD. 1955. Physiology of a primary chemoreceptor unit. *Science* **122:** 417–418. doi:10.1126/science.122.3166.417.b

Jové V, Gong Z, Hol FJH, Zhao Z, Sorrells TR, Carroll TS, Prakash M, McBride CS, Vosshall LB. 2020. Sensory discrimination of blood and floral nectar by *Aedes aegypti* in mosquitoes. *Neuron* **108:** 1163–1180 .e12. doi:10.1016/j.neuron.2020.09.019

Kessler S, Vilmant M, Guerin PM. 2013. The sugar meal of the African malaria mosquito *Anopheles gambiae* and how deterrent compounds interfere with it: a behavioral and neurophysiological study. *J Exp Biol* **216:** 1292–1306. doi:10.1242/jeb.076588

Kessler S, Vilmant M, Guerin PM. 2015. Sugar-sensitive neurone responses and sugar feeding preferences influence lifespan and biting behaviors of the Afrotropical malaria mosquito, *Anopheles gambiae*. *J Comp Physiol* **201:** 317–329. doi: 10.1007/s00359-015-0978-7

Lee R, Craig DA. 1983. Cibarial sensilla and armature in mosquito adults (Diptera:Culicidae). *Can J Zool* **61:** 1568–79. doi:10.1139/z83-211

Lee R, Davies DM. 1978. Cibarial sensilla of *Toxorhynchites* mosquitoes (Diptera: Culicidae). *Int J Insect Morphol Embryol* **7:** 189–194. doi:10.1016/0020-7322(78)90001-6

Liscia A, Crnjar R, Barbarossa IT, Esu S, Muroni P, Galun R. 1993. Electrophysiological responses of labral apical chemoreceptors to adenine nucleotides in *Culex pipiens*. *J Insect Physiol* **39:** 261–265. doi:10.1016/0022-1910(93)90097-B

Lomelí AM, Dahanukar AA. 2022. Host-plant feeding in mosquitoes. Ch. 16. In *Sensory ecology of disease vectors* (ed. Ignell R, Lazzari CR, Lorenzo MG, et al.), pp. 489–509. Wageningen Academic Publishers, Wageningen, the Netherlands.

Lomelí AM, Dahanukar AA. 2023. Single-sensillum taste recordings in mosquitoes. *Cold Spring Harb Protoc* doi:10.1101/pdb.prot108195

McIver S, Siemicki R. 1978. Fine structure of tarsal sensilla of *Aedes aegypti* (L.) (Diptera: Culicidae). *J Morphol* **155:** 137–156. doi:10.1002/jmor.1051550202

Meunier N, Marion-Poll F, Lucas P. 2009. Water taste transduction pathway is calcium dependent in *Drosophila*. *Chem Senses* **34:** 441–449. doi:10.1093/chemse/bjp019

Minnich DE. 1921. An experimental study of the tarsal chemoreceptors of two nymphalid butterflies. *J Exp Zool* **33:** 172–203. doi:10.1002/jez.1400330105

Minnich DE. 1926. The chemical sensitivity of the tarsi of certain muscid flies. *Biol Bull* 166–178. doi:10.2307/1536476

Pappas LG, Larsen JR. 1976. Gustatory hairs on the mosquito, *Culiseta inornata*. *J Exp Zool* **196:** 351–360. doi:10.1002/jez.1401960309

Sanford JL, Shields VDC, Dickens JC. 2013. Gustatory receptor neuron responds to DEET and other insect repellants in the yellow-fever mosquito, *Aedes aegypti*. *Naturwissenschaften* **100:** 269–273. doi:10.1007/s00114-013-1021-x

Seenivasagan T, Sharma KR, Shrivastava A, Parahar BD, Pant SC, Prakash S. 2009. Surface morphology and morphometric analysis of sensilla of Asian tiger mosquito, *Aedes albopictus* (Skuse): an SEM investigation. *J Vector Borne Dis* **46:** 125–135.

Sparks JT, Dickens J. 2016. Electrophysiological responses of gustatory receptor neurons on the labella of the common malaria mosquito, *Anopheles quadrimaculatus* (Diptera: Culicidae). *J Med Entomol* **53:** 1148–1155. doi:10.1093/jme/tjw073

Zhang YV, Ni J, Montell C. 2013. The molecular basis for attractive salt-taste coding in *Drosophila*. *Science* **340:** 1334–1338. doi:10.1126/science.1234133

Protocol 1

Single-Sensillum Taste Recordings in Mosquitoes

Adriana Medina Lomelí[1] and Anupama Arun Dahanukar[2,3]

[1]*Entomology Graduate Program,* [2]*Department of Molecular, Cell and Systems Biology, University of California, Riverside, California 92521, USA*

In insects, gustatory neurons sense chemicals upon contact and directly inform many behaviors critical for survival and reproduction, including biting, feeding, mating, and egg laying. However, the taste sensory system is underexplored in many anthropophilic disease vectors such as mosquitoes, which acquire and transmit human pathogens during blood feeding from human hosts. This results in a big gap in vector biology—the study of organisms that spread disease by transmitting pathogens—because insect vectors closely interact with humans while selecting suitable individuals and appropriate bite sites for blood meals. Human sweat and skin-associated chemistries are rich in nonvolatile compounds that can be sensed by the mosquito's taste system when she lands on the skin. Taste sensory units, called sensilla, are distributed in many organs across the mosquito body, including the mouthparts, legs, and ovipositors (female-specific structures used to lay eggs). Each sensillum is innervated by as many as five taste neurons, which allow detection and discrimination between various tastants such as water, sugars, salts, amino acids, and plant-derived compounds that taste bitter to humans. Single-sensillum recordings provide a robust way to survey taste responsiveness of individual sensilla to various diagnostic and ecologically relevant chemicals. Such analyses are of immense value for understanding links between mosquito taste responses and behaviors to specific chemical cues and can provide insights into why mosquitoes prefer certain hosts. The results can also aid development of strategies to disrupt close-range mosquito–human interactions to control disease transmission. Here we describe a protocol that is curated for electrophysiological recordings from taste sensilla in mosquitoes and sure to yield exciting results for the field.

MATERIALS

It is essential that you consult the appropriate Material Safety Data Sheets and your institution's Environmental Health and Safety Office for proper handling of equipment and hazardous material used in this protocol.

RECIPES: Please see the end of this protocol for recipes indicated by <R>. Additional recipes can be found online at http://cshprotocols.cshlp.org/site/recipes.

Reagents

Aedes aegypti in a rearing cage (DeGennaro et al. 2013)
Beadle–Ephrussi Ringer solution <R>
Recording electrolyte of choice
  Potassium chloride (KCl; Sigma-Aldrich P3911)
  Tricholine citrate (TCC; Sigma-Aldrich T0252)

[3]Correspondence: anupama.dahanukar@ucr.edu

Cite this protocol as *Cold Spring Harb Protoc;* doi:10.1101/pdb.prot108195

A recording electrolyte is necessary for conductance. Include it in all tastant stimulus solutions that are prepared. Stock solutions of electrolytes can be prepared in autoclaved deionized H_2O and stored at −20°C.

Within a sensillum, there may be a neuron that is sensitive to H_2O, which will be activated when any H_2O-soluble stimulus is brought to the tip of a sensillum. Depending on the experiment, one may choose to silence the activity of this neuron using 30 mM TCC as the recording electrolyte (see Dahanukar and Benton 2023; also Dethier 1976; Wieczorek and Wolff 1989). If the presence of H_2O neuron activity is not an issue for the experiment or if there are problems with electrical conductivity, then 1 mM KCl may be used as the recording electrolyte.

Taste stimuli

Examples:

100 mM sucrose in 30 mM TCC

10 mM caffeine in 30 mM TCC

10 mM methionine in 30 mM TCC

100 mM NaCl in 30 mM TCC or 1 mM KCl

1 mM KCl for stimulation of water-sensing neurons

Any H_2O-soluble tastant can be used, but always prepare it with either 30 mM TCC or 1 mM KCl. Store stock solutions prepared with autoclaved deionized H_2O for weeks to months at −20°C depending on the stability of the specific tastant. Prepare 1- to 1.5-mL aliquots of working solutions in 1.5-mL microcentrifuge tubes and store them for up to 1 wk at 4°C when not in use. Carefully consider those stimuli that do not properly dissolve in H_2O. For tastants that do not dissolve in H_2O, explore alternatives such as base recordings with tungsten electrodes (see Dahanukar and Benton 2023; also Lacaille et al. 2007; Dweck et al. 2022).

Equipment

Equipment for mosquito preparation

Borosilicate glass capillaries (World Precision Instruments 1B100F-4; see Steps 1 and 4)

Box cutter blade

Bunsen burner

Clay (optional; see note below Step 34)

Cup (small, paper) with a mesh lid

Foam tape (double-sided; Scotch)

Forceps (two pairs, fine, straight-tip, carbon steel, Dumont #5; FST 11251-10)

These are ideal for manipulating the anesthetized mosquito.

Humidified chamber (e.g., a clear plastic box that allows exposure to ambient light with wet paper towels inside)

Ice in a small bucket

Microscope slide (75-mm × 25-mm × 1-mm; VWR 16004-368)

Modified 1-mL syringe with microloading tip (see Step 2)

Multistep electrode puller (flaming/brown micropipette puller; Sutter Instrument P-1000 with a trough or box filament)

Petri dish (polystyrene, 100-mm × 15-mm)

Preparation station with a stereomicroscope (Olympus SZ51, 0.67–4.5×; Olympus SZ61-P) and a gooseneck fiber optic illumination source (AmScope HL250-AY)

Scissors

Serological pipette (10-mL)

Squeeze-bulb pump

Stereo dissecting microscope with fiber-optic light source (e.g., Olympus SZ51)

Tape for mosquito preparation (double-sided; Scotch)

Tape for sealing (VWR 89097-990)

Tubing (plastic, 0.5″ outer diameter × 0.375″ inner diameter)

Woven nylon mesh (fine)

Cite this protocol as *Cold Spring Harb Protoc*; doi:10.1101/pdb.prot108195

Equipment for the electrophysiological recording rig

Aquarium pump (e.g., API, Rena Air 200, model number 702C)

Autospike program (Syntech)

Bicycle air pump

Computer with Windows and USB ports for IDAC-4, and monitor

Electrode holders, two (Syntech)

Faraday cage (optional; see Step 8)

Gooseneck fiber optic light source (e.g., AmScope HL250-AY)

Grounding metal plate

IDAC-4 signal acquisition system (Syntech)

Insulated wires with a banana plug on one end and an alligator clip on the other for grounding the Faraday cage and other connections

Magnetic stand with clamps for a glass slide

Manual manipulator on a magnetic stand for a reference electrode holder (e.g., Narishige YOU-1)

Manual manipulator on a magnetic stand for holding a glass tube (e.g., Narishige UM-3C with GJ-1)

Metal mesh (~1-mM mesh opening)

Micromanipulator for the TasteProbe with recording electrode (three-axis fine-movement manipulator with minimal drift; e.g., Narishige NMN-21)

Serological pipette (2-mL)

Sidearm filtering flask (500-mL) with rubber stopper

Silver wire (0.3- to 0.5-mm diameter; World Precision Instruments AGW1510)

Soldering iron

Stereomicroscope with adequate zoom (e.g., 0.8×–8×; Nikon SMZ1000, with a Nikon stereomicroscope objective lens [HR Plan Apo 1.6×; working distance, 24; Nikon])

Tape for sealing (e.g., VWR 89097-990)

TasteProbe (Syntech)

This is an amplifier designed to obtain reliable DC recordings with the single-sensillum tip recording technique described in this protocol. The amplifier performs contact detection, as well as DC compensation and amplification (Marion-Poll and van der Pers 1996).

Tubing (various sizes, plastic)

Vibration platform or table (air table; BenchMate vibration-free platform; Kinetic Systems 2210-01-22)

Window screen frames or block strips of wood for Faraday cage (see Step 8)

METHOD

Prepare Glass Micropipette Recording and Reference Electrodes

Prepare the Recording Electrode and Micropipette Washer

1. On an electrode puller, use a multistep program that yields electrodes with open tips that have a ~3- to 5-mm taper. A specific size is not required for the tip diameter of the recording electrode, but one that is ~5–10 µm allows for comfortable contact with most sensilla of interest without disturbing any neighboring ones. For the recommended glass capillaries, a good pull program to begin using on a multistep puller with a 2.5-mm × 2.5-mm box filament is Heat RAMP ± 20, Pull 0, Velocity 14–16, Time 250, Pressure 500, looped twice.

   Although the size, distribution, and orientation of taste sensilla vary across mosquito species, most external trichoid taste sensilla can be stimulated with micropipette electrodes with these parameters. For sensilla that are smaller or more difficult to access, micropipettes with a smaller tip diameter can be made by increasing heat, velocity, and pull settings in the multistep program.

   Taste recordings can also be performed using a base recording method with a tungsten electrode (see Dahanukar and Benton 2023; also Lacaille et al. 2007; Dweck et al. 2022), although this method is technically more challenging and not as commonly used.

2. To prepare a syringe micropipette loader/washer, take a 200-µL pipette tip and use a box cutter blade to trim the wide end so that it fits onto a 1-mL syringe tip. Use the blade to trim the narrow end of the pipette tip and attach it to a 20-µL microloading tip. Both ends of the pipette tip adaptor should fit snugly (Fig. 1A).
3. Use the plunger of the syringe to aspirate tastant solution into the microloader tip and dispense it into the pulled glass recording micropipette. Gently tap the glass micropipette to dislodge any air bubbles and blot the open end with a Kimwipe. Fill the glass electrode to no more than 85%–90% of its length.

   This prevents overflow when it is placed on the TasteProbe electrode holder.

   If testing multiple tastants, prepare a separate microloader syringe for each tastant for a more streamlined process.

   Multiple recording micropipettes can be preloaded with different stimulus solutions at the start of an experiment and arranged in a labeled order on a strip of modeling clay in a Petri dish. With this strategy, stimuli can be efficiently switched for sequential recordings from the same mosquito preparation. If any crystallization or air bubbles are noticed in the micropipettes, avoid preloading those solutions and prepare the micropipettes just before use. If using the same micropipette for testing different stimuli, rinse 3×–4× with H_2O with the syringe micropipette washer between stimuli.

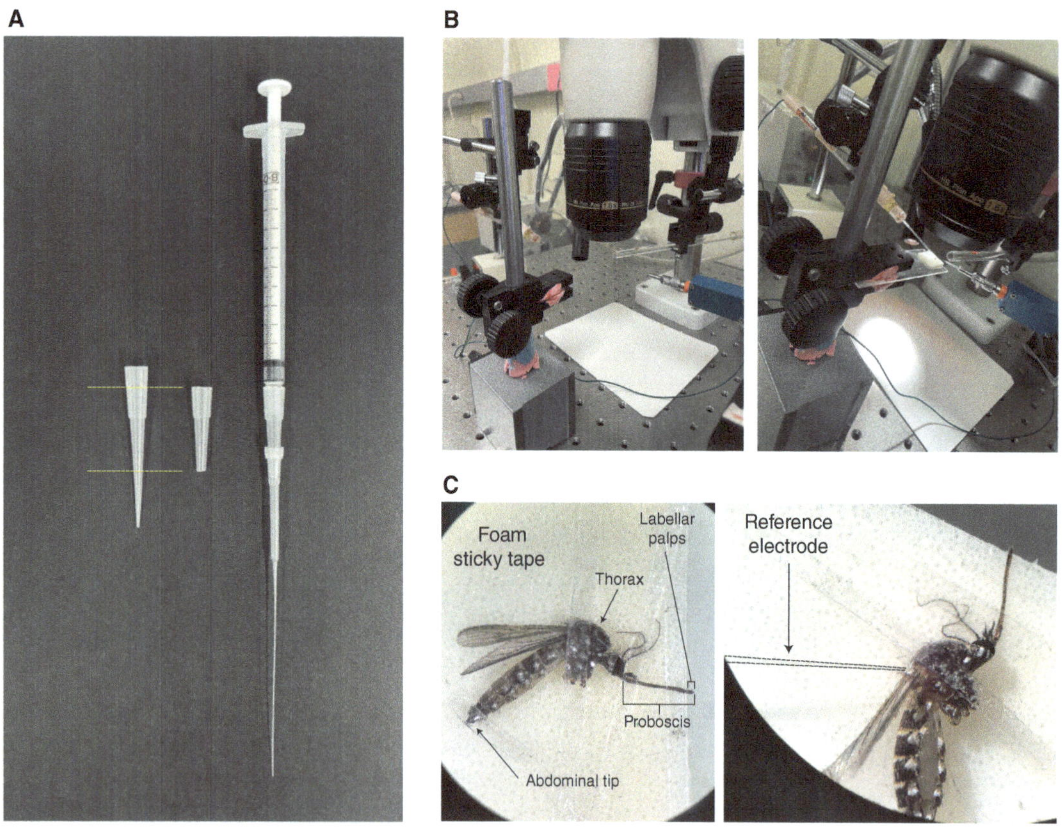

FIGURE 1. (*A*) From *left* to *right*, image showing modifications to a 200-µL natural pipette tip to fit a 1-mL syringe. The modified pipette tip can then be inserted into a 20-µL microloading tip. (*B*) Image on the *left* shows the base and clamp on the air table. Clay can be used between the clamps to provide a tighter grip on the glass slide. Image on the *right* shows the glass slide in place. Once a suitable location on the table is identified, the reference electrode, airstream, and micromanipulator that holds the TasteProbe can be adjusted to reach the mosquito at the end of the glass slide. Note that the thick double-sided foam tape is placed at the short edge of the glass slide. (*C*) Image on the *left* shows a mosquito positioned on the thick double-sided foam sticky tape after her legs have been removed. Strips of Scotch double-sided sticky tape are placed across the base of the labellum, the base of the proboscis, the thorax, and tip of the abdomen. Image on the *right* shows the tip of the glass reference electrode (arrow) inserted into the thorax. With this configuration, the TasteProbe will access the mosquito labellar sensilla from the right.

Cite this protocol as *Cold Spring Harb Protoc*; doi:10.1101/pdb.prot108195

Prepare the Reference Electrode

4. Use a single-step program on an electrode puller to create a glass micropipette with a long thin taper. For the recommended glass capillaries, a typical single-step pull program is Heat RAMP + 70, Pull 50, Velocity 70, Time 240, Pressure 500.

   The precise shape and length of the taper is not critical, because this micropipette will be used to pierce the mosquito's abdomen.

Prepare the Mosquito Aspirator

Make aspirators in the laboratory with a piece of plastic tubing attached to a serological pipette that is modified to aspirate mosquitoes.

5. Cut the tip of a 10-mL serological pipette using a box cutter to allow for a wider opening and use a Bunsen burner to create a slight bend ~4″–5″ from one end of the pipette.

   This will be the part that is inserted into the mosquito cages to easily reach mosquitoes.

6. Take a small piece of fine woven mesh and place it over the opening on the longer side of the pipette and then insert the pipette end into a piece of plastic tubing (0.5″ outer diameter × 0.375″ inner diameter), making sure that it fits snugly.
7. Use tape to secure the junction where the tubing and pipette meet and create an airtight seal.

Assembly of the Recording Rig

Detailed reports describing the assembly of electrophysiology rigs for single-sensillum tip recordings in Drosophila melanogaster *are published elsewhere. As the setup is essentially the same for taste recordings from any insect, we direct readers to these other sources (see Dahanukar and Benton 2023; also Delventhal et al. 2014) rather than repeat the description here. However, differences in the anatomy of flies and mosquitoes do call for modifications in preparing a mosquito for recordings, for which an additional magnetic stand with a clamp is needed (Fig. 1B). Use the magnetic stand in place of the microscope stage, which facilitates a wider range of movement, translating to a bigger range in field of view and more access to manipulate the preparation.*

Faraday Cage

8. The use of a Faraday cage is optional, but it will minimize electrical noise. If used, customize the cage to fit the size of the setup and ensure that three sides (left, right, and back) surround the air table with the microscope and the front is open. The height of the cage should allow sufficient room for the experimenter to look through the microscope eyepieces comfortably.

   Faraday cages can be constructed with four appropriately sized, framed wire mesh panels. Panels can be made using wood block strips or metal window frames. The dimensions of the three sides and the top panel should be customized to fit the rig.

9. Prepare insulated grounding wires with banana plugs and alligator clips according to the needs of the setup. Use a soldering iron to seal wires to the appropriate components. For best conductivity results, use a light coating of solder and try to minimize clumps.
10. Use a grounding metal plate to ground all wires, including any grounding wires that are clipped to the inside of the Faraday cage.

Humidified Airstream

11. Set up a humidified airstream by filling a sidearm filtering flask about midway with distilled H_2O.
12. Take a small 2-mL serological pipette and push it through a rubber stopper inserted in the mouth of the flask until its tip is in the H_2O.
13. Using thin plastic tubing, attach an aquarium pump to the top of the serological pipette and seal with tape to create a humidified airstream.

14. Attach appropriately sized plastic tubing to the sidearm of the flask.

The humidified airstream will be expelled from the loose end of the tubing that is attached to the sidearm of the flask.

Vibration Platform or Table

An air table will make the preparation more stable by minimizing any vibrations caused by movement in the room.

15. Use a bicycle air pump to inflate the air table and make sure that it is level and lifted off the tabletop on which it sits.

Mosquito Preparation

Laboratory-reared Ae. aegypti *mosquitoes are typically maintained in temperature- (27°C), humidity- (80%), light cycle- (14 h light:10 h dark), and density-controlled conditions. With the objective of the experiment in mind, select mosquitoes of the appropriate age, sex, and rearing conditions for recordings.*

16. On the day of recordings, aspirate four to six mosquitoes from a rearing cage and transfer them into a small paper cup with a mesh lid placed in a humidified chamber. Allow mosquitoes to remain in the humidified chamber (2–3 h) while preparing for recordings.
17. Fill a reference micropipette (Step 4) with Beadle–Ephrussi Ringer solution using a previously prepared syringe and microloader tip (Step 2). Be sure that the solution is filled to the tip of the glass electrode and tap to dislodge any air bubbles near the tip.
18. While viewing the electrode under a stereomicroscope using fiber-optic illumination, use a pair of forceps to break off the tip of the electrode. Once the electrode tip is the desired length, slide it on the silver wire attached to an electrode holder and secure the holder on a manual manipulator attached to a magnetic stand.
19. Chill a 100-mm × 15-mm polystyrene Petri dish on ice.
20. Prepare a plain microscope slide with a small square of double-sided sticky foam tape on one end of the slide.

    The piece of tape can either be aligned to the edge of the slide or overhang slightly (Fig. 1C).
21. Using scissors, cut four thin strips of double-sided Scotch tape (not the foam tape) and set them aside where they will not be damaged but are still easily accessible.

    These thin strips of tape will be used to immobilize the mosquito.
22. Place the microscope slide under the stereomicroscope and adjust the zoom and focus so that the sticky foam tape is in clear view.
23. Using a mosquito aspirator attached to a squeeze bulb, draw a single mosquito from the container.
24. Open the Petri lid slightly and place the tip of the aspirator between the plate and lid.
25. Gently push the mosquito into the dish and close the lid quickly. Let the mosquito lay on ice for a few seconds until it stops moving.
26. Remove the Petri lid. While the mosquito is still on ice, use forceps to remove all six legs.
27. Once the legs are removed, use forceps to pick up the mosquito by the wings. Place the mosquito on its side on the square of double-sided sticky foam tape (Fig. 1C), with the mosquito proboscis facing the hanging edge of the tape.
28. Adjust the head of the mosquito so that only the labellar palps are positioned off the edge of the sticky tape.

    The recording electrode is typically located on the right side of the microscope.

Cite this protocol as *Cold Spring Harb Protoc*; doi:10.1101/pdb.prot108195

29. Adjust the head and labellar palps of the mosquito so that the sensilla of interest are easily accessible to the TasteProbe electrode. Adjust/rotate the labellum carefully with forceps or with the thin strip of tape that will be placed at its base.
30. Take one of the thin strips of double-sided sticky tape and place it across the base of the labellum.
31. Place the next strip at the base of the proboscis, followed by another placed across the thorax.
32. If the wings are not held down by these tape strips, use forceps to press the wings into the sticky foam square for added stability and prevention of any wing movement.
33. Place the last tape strip across the tip of the abdomen.

    This should hold the mosquito securely in place.
34. Place the glass slide between the clamps of the magnetic stand near the microscope. Be sure that the slide is inserted in an orientation so that sensilla of interest are easily accessible to the recording electrode located on the right side.

    If the clamps do not hold the glass slide properly, add clay to create a more secure hold.
35. Using the lowest magnification on the microscope, bring the mosquito labellum into view. Increase magnification to the highest setting and ensure that the sensilla of interest are visible and accessible.
36. Return the magnification to the lowest setting (0.8×).

    The thorax, head, and proboscis should be in view.
37. Use the manual manipulator to guide the reference electrode to the center of the thorax. Proceed to pierce the thorax of the mosquito without perturbing the slide. Return to the highest magnification and focus on the sensilla.

Performing Recordings

Compare tastant responses with those observed using the recording electrolyte alone, which will reveal baseline activity.

38. Turn on the aquarium pump and place the humidified air stream as close to the mosquito as possible. Use a holder to aim the open end of the tube directly at the mosquito during taste recordings.

    Doing so will prevent the mosquito from desiccating during recordings.
39. Slide the filled recording micropipette (see Step 3) on the silver wire of an electrode holder attached to the TasteProbe head stage. Use the three-axis micromanipulator to bring the tip of the recording electrode into the field of view and focus the stereomicroscope.
40. Focus on the desired sensillum and position the recording electrode so that it faces the tip of the sensillum at a distance of ~5 µm. Enable the trigger using the pedal and move the electrode forward using the micromanipulator so that it envelops the tip of the sensillum.

    The sensillum should make contact with the tastant/electrolyte solution. Upon contact, a beep will be heard and recording will begin automatically if the trigger has been armed. Alternatively, recordings can be started and stopped manually using control buttons in the Autospike software.
41. Maintain contact between the sensillum and the electrode for the duration of the recording, for which the sampling rate and length of time can be adjusted in the AutoSpike program on the computer.
42. If the preparation is stable, continue recording with other tastants. Wipe the silver wire with a Kimwipe each time after a stimulus-filled micropipette is switched.

DISCUSSION

This single-sensillum recording protocol can easily be adapted to investigate responsiveness of external trichoid taste sensilla. No transgenesis or genome editing is required for this technique; therefore, any insect or arthropod of choice could be investigated, although consideration should be given to the immobilization and preparation of the animal for taste recordings, which will need adjustment according to the insect's size and activity and the location and orientation of its taste organs. Appropriate numbers of recordings from each sensillum are necessary for statistical power, and especially when testing an animal species that has not been well-investigated. In most cases, bilaterally symmetrical sensilla can be treated as the same; focusing on one side of the animal reduces survey time and extends the number of stimuli that can be tested on a single preparation.

Data Analysis

Responses to tastants are quantified by counting the number of action potentials within a given window of stimulus application in Autospike (see Dahanukar and Benton 2023). Typically, firing rates are reported as the number of spikes per second. In single-sensillum tip recordings, it is important to bear in mind that the method captures the activity of all neurons innervating the sensillum. It is sometimes possible, although not always, to identify the neuron from which a response originates by the amplitude of the action potentials. Sample traces from trichoid taste sensilla of *Aedes albopictus* mosquitoes show differences in the amplitudes of neurons responding to different tastants (Baik and Carlson 2020). If a tastant elicits responses from multiple neurons within the sensillum, spikes may be assigned to separate neurons during analysis by virtue of their amplitude.

Tastant responses should be compared with those observed with recording electrolyte alone, which will reveal baseline activity. It is also important to use a positive control that yields a robust response, such as 100 mM sucrose, to be sure that recordings are being obtained from a responsive individual and that the sensillum is not damaged in any way.

For electrophysiological surveys in which the number of chemosensory neurons housed in specific sensillum is unknown, mixtures of different categories of compounds (e.g., sugar and bitter) can be tested as a strategy to assign responses to identified neurons within a sensillum. Hierarchical clustering analyses can then be used for functional analyses following acquisition of single-sensillum response data.

RECIPE

Beadle–Ephrussi Ringer Solution

Reagent	Amount
NaCl	7.5 g
KCl	0.35 g
$CaCl_2 \cdot 2H_2O$	0.279 g

Dissolve the ingredients in 1 L of H_2O. Aliquot and store at –20°C.

ACKNOWLEDGMENTS

We thank Manali Dey and Vaibhav Menon for helpful comments. Research in A.A.D.'s laboratory is supported by grants from the National Institutes of Health (R01DC017390 and R21AI140065) and the

University of California, Riverside Agricultural Experimental Station and the National Institute of Food and Agriculture—U.S. Department of Agriculture (Hatch Project Grant 1011543). A.M.L. is a UC-Hispanic Serving Institutions Doctoral Diversity Initiative (UC-HSI DDI) fellow.

REFERENCES

Baik L, Carlson JR. 2020. The mosquito taste system and disease control. *Proc Natl Acad Sci* **117:** 32848–32856. doi:10.1073/pnas.2013076117

Dahanukar A, Benton R. 2023. Recording from fly taste sensilla. *Cold Spring Harb Protoc* doi:10.1101/pdb.prot108064

DeGennaro M, McBride CS, Seeholzer L, Nakagawa T, Dennis EJ, Goldman C, Jasinkskiene N, James AA, Vosshall LB. 2013. *orco* mutant mosquitoes lose strong preference for humans and are not repelled by volatile DEET. *Nature* **498:** 487–491. doi:10.1038/nature12206

Delventhal R, Kiely A, Carlson JR. 2014. Electrophysiological recording from *Drosophila* labellar taste sensilla. *J Vis Exp* **84:** e51355. doi:10.3791/51355

Dethier VG. 1976. *The hungry fly.* Harvard University Press, Cambridge, MA.

Dweck HKM, Talross GJS, Luo Y, Ebrahim SAM, Carlson JR. 2022. Ir56b is an atypical ionotropic receptor that underlies appetitive salt response in *Drosophila. Curr Biol* **32:** 1776–1787. doi:10.1016/j.cub.2022.02.063

Lacaille F, Hiroi M, Twele R, Inoshita T, Umemoto D, Maniere G, Marion-Poll F, Ozaki M, Francke W, Cobb M, et al. 2007. An inhibitory sex pheromone tastes bitter for *Drosophila* males. *PLoS ONE* **2:** e661. doi:10.1371/journal.pone.0000661

Marion-Poll F, van der Pers J. 1996. Un-filtered recordings from insect taste sensilla. *Entomol Exp Appl* **80:** 113–115. doi:10.1111/j.1570-7458.1996.tb00899.x

Wieczorek H, Wolff G. 1989. The labellar sugar receptor of *Drosophila. J Comp Physiol A* **164:** 825–834. doi:10.1007/BF00616754

CHAPTER 27

Identification of Bioactive Volatile Organic Compounds Using Combined Gas Chromatography and Single-Sensillum Recording

Rickard Ignell,[1,4] Majid Ghaninia,[2,3] and Sharon Rose Hill[1]

[1]*Disease Vector Group, Unit of Chemical Ecology, Department of Plant Protection Biology, Swedish University of Agricultural Sciences, 23422 Alnarp, Sweden;* [2]*School of Life Sciences, Arizona State University, Tempe, Arizona 85281, USA;* [3]*Division of Entomology, Department of Plant Protection, Gorgan University of Agricultural Sciences and Natural Resources, Gorgan, Iran*

Most arthropods rely heavily on their sense of smell (i.e., olfaction) to locate and discriminate among mates, food, and egg laying sites. The odors emanating from these resources are composed of blends of volatile compounds that are detected by olfactory sensory neurons (OSNs) that are housed in hair-like structures, called sensilla, on the olfactory organs of arthropods. By inserting an electrode into a single-sensillum and recording the activity of the OSNs while stimulating with volatile compounds eluting from a gas chromatograph, combined gas chromatography and single-sensillum recording (GC-SSR) provides a high-resolution tool to identify bioactive compounds and to functionally characterize the peripheral olfactory system of arthropods.

INVESTIGATING CHEMOSENSATION IN INSECTS

Insects detect volatile chemicals (i.e., odors) in their environment to locate and discriminate among resources, such as mates, food, and egg laying sites. These odors are composed of blends of volatile compounds, of which only some are detected by the insect "nose," the antennae, or other secondary appendages (e.g., the mouthparts, legs, and wing margins). These olfactory organs are covered in hair-like structures, called sensilla, which house the olfactory sensory neurons (OSNs) responsible for the detection and relay of olfactory signals to higher brain centers. For researchers to identify which chemical compounds are detected by the insect, a method to separate the volatile odor blends into component parts and then survey the insect for their response to each compound is needed.

Moorhouse et al. (1969) developed a method called gas chromatography–electroantennographic detection (GC-EAD), by which individual bioactive volatile compounds in complex blends (e.g., natural odor collections) could be identified through combining extracellular electrophysiology with the separation of volatile compounds over a column by gas chromatography. Although GC-EAD provides a valuable screening tool for the identification of salient volatile compounds for most arthropods, a subsequent refinement of the method, from detecting field potentials of a large population of neurons to the focused extracellular recording from the few neurons housed in individual sensilla, known as combined gas chromatography and single-sensillum recording (GC-SSR) was developed. This technique allows for the definition of the volatile detection profiles of individual

[4]Correspondence: rickard.ignell@slu.se

Cite this introduction as *Cold Spring Harb Protoc;* doi:10.1101/pdb.top107681

OSNs and identification of functional neuronal classes based on natural ligands with ecological relevance to the insect under investigation (Ghaninia et al. 2008). We have refined this technique and adapted it for mosquitoes (see Protocol 1: Combined Gas Chromatography and Single-Sensillum Recording for Mosquitoes [Ignell et al. 2022]).

USING GC-SSR

Individual tuning profiles of OSNs using GC-SSR are attained by injecting ecologically relevant odor collections (see Kunert et al. 2009) onto a gas chromatograph fitted with a splitter (Fig. 1). The splitter directs half of the airstream to the flame ionization detector (FID), which detects the relative time and quantity of the compounds emanating from the column, and the other half of the airstream over the sensillum from which the neuronal activity is recorded, using SSR (Fig. 1). The relative timing of the appearance of the volatile compound on the FID and the OSN response during SSR provides the initial evidence indicating that a compound is a natural ligand for a given OSN. Comparing the Kovats' retention indices between the GC-SSR and the subsequent injection of the odor collection on a GC-mass spectrometer (MS), coupled with a comprehensive mass spectral library, will putatively identify the ligand (for more information on fundamental GS-MS analysis of volatile compounds, please see Karasek and Clement 2012). Cooccurrence of the synthetic compound with the putative ligand in the odor collection when coinjected on the GC-MS, and a neuronal response to the injection of the synthetic compound on the GC-SSR provide the final evidence of the identity of the ligand.

The role of the gas chromatograph in GC-SSR is to separate volatile compounds within a complex odor blend, without modification or degradation, for delivery to the FID to provide a reference for qualitative identification and relative quantification, and for delivery to potentially activate the OSN(s). The gas chromatograph consists of an injection port, a temperature-controlled chamber into which the sample is injected through a rubber septum leading to a narrow, lined, flowthrough

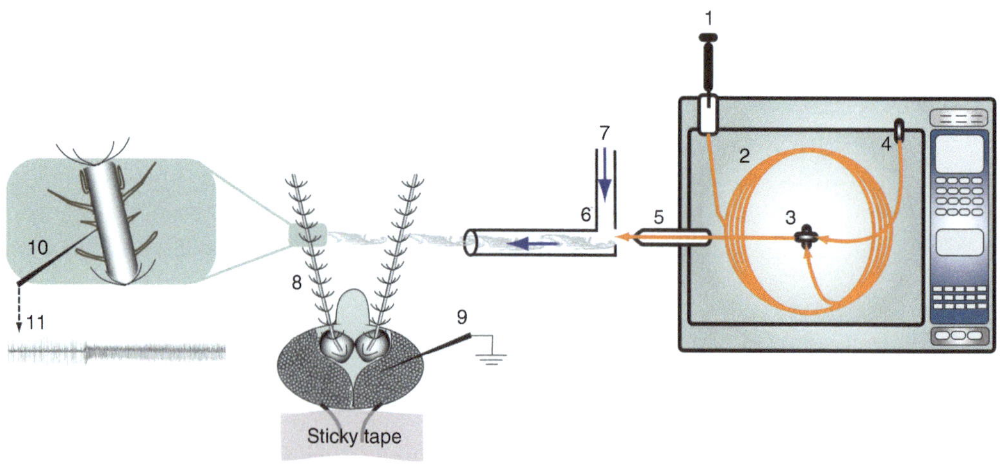

FIGURE 1. The gas chromatography–coupled single-sensillum recording (GC-SSR) technique in arthropods. For GC, headspace extracts are injected using a microsyringe (*1*) onto a GC column (*2*). The column is situated in an oven. As the oven temperature increases, the components of the extract are separated, travel through the column, and reach a split (*3*) from which half of the effluent goes to a flame ionization detector (FID) (*4*). The other half leaves the column and passes through a transfer line (*5*) to a glass tube (*6*) where a continuous humidified–purified airflow (*7*) blows the separated components of the extract over the arthropod antenna (*8*). For SSRs, two tungsten electrodes, a ground and a recording electrode (*9* and *10*), are placed into the eye and at the base of a single-sensillum, respectively. Action potentials of the OSNs housed in a sensillum and their responses to the odor components are recorded (*11*). (Reproduced from Ghaninia et al. 2008, with permission, © Company of Biologists Ltd.)

 Cite this introduction as *Cold Spring Harb Protoc*; doi:10.1101/pdb.top107681

capillary tube called a column, through which the sample and an inert carrier gas (e.g., helium or nitrogen), known as the mobile phase, pass (Fig. 1). The rate at which a sample moves through the column is determined by its interaction with the lining of the column, known as the stationary phase, as well as the temperature, length of column, and the flow rate of the carrier gas. As each compound elutes from the column, it is sent to one or more detectors, which in GC-SSR is usually an FID and an insect antennal sensillum. The choice of column is highly dependent on the polarity of the sample, and the stationary phase should match the polarity of the sample closely to maximize separation and minimize the time of the run. When analyzing complex odor blends, as many headspace extracts from natural odor sources are, it is good practice to run the GC-SSR using two columns with differing physiochemical properties (e.g., nonpolar HP-5 and polar DB-wax columns) in order to increase the likelihood of all compounds separating and eluting from the column during the run.

Extracellular recordings from OSNs in a single-sensillum provide not only a physiological detector for determining the biological activity of volatiles from natural sources, similar to GC-EAD, but also a technique for functionally characterizing OSNs and, by proxy, olfactory receptors. Under common conditions, it is rare to be able to record extracellularly from a restricted number of neurons. A sensillum provides a sealed chamber for OSN dendrites, insulated from other neural activity, into which the placement of a sharpened tungsten recording electrode can record both the passive dendritic activity and the active activity of the axon hillock, resulting in extracellular recordings from which the frequency of OSN action potentials can be inferred. These discrete peaks of neural activity are referred to as spikes. The relative diameter of the individual OSNs in each functional class of sensillum are consistent, and thus identifiable, resulting in the ability to classify the OSNs by the amplitude of the spikes, with the highest amplitude designated "A," the second highest "B," and so forth.

Factors affecting the resolution of GC-SSR include the placement of both the recording and ground electrodes, the ability to move the recording electrode in fast, discrete steps in a fine scale, and the stability of the contact with the sensillum, which is dependent on the stability of the tissue mounting. The recording electrode should be placed in the shaft of the sensillum for the best signal-to-noise ratio. Placement of the recording electrode at the base of the sensillum, although usually easier to achieve, may reduce the resolution of the signal with the proximity to the antennal nerve. To accurately place the recording electrode in the shaft of the sensillum, a precision piezoelectric micromanipulator is required. Such a micromanipulator provides the fine motor control and ability to rapidly move the electrode forward stepwise a small distance, in order to pierce the cuticular shaft, and not simply have it bend out of the way of the electrode. It is recommended that the ground electrode be placed contralaterally from the recording electrode, often in the eye, to reduce the risk of damaging associated nerves, neuropils, and ganglia. Because GC-SSR requires stability of the contact for the duration of the gas chromatograph run, often 25–30 min, the stability of the mounting is paramount.

CONCLUSIONS

Although GC-EAD remains the best method for identifying biologically active volatile compounds from ecologically relevant resources, GC-SSR provides a high-resolution tool to functionally characterize the peripheral olfactory system of arthropods. Moreover, GC-SSR can be used in conjunction with other techniques to broaden its functional range. A common pairing is with the so-called *Drosophila* empty neuron system (Hallem and Carlson 2004), in which odorant receptors (Ors) are expressed ectopically in an OSN (e.g., ab3A) lacking a tuning Or but still expressing the odorant receptor co-receptor (Orco) required for a functional Or complex (see Benton and Dahanukar 2011; Gonzalez et al. 2016). With this technique, individual tuning profiles for known Ors can be generated to ecologically relevant ligands.

REFERENCES

Benton R, Dahanukar A. 2011. Electrophysiological recording from *Drosophila* olfactory sensilla. *Cold Spring Harb Protoc* doi:10.1101/pdb.prot5630

Gonzalez F, Witzgall P, Walker WB III. 2016. Protocol for heterologous expression of insect odourant receptors in *Drosophila*. *Front Ecol Evol* **4:** 24. doi:10.3389/fevo.2016.00024

Ghaninia M, Larsson M, Hansson BS, Ignell R. 2008. Natural odor ligands for olfactory receptor neurons of the female mosquito *Aedes aegypti*: use of gas chromatography-linked single sensillum recordings. *J Exp Biol* **211:** 3020–3027. doi:10.1242/jeb.016360

Hallem EA, Carlson JR. 2004. The odor coding system of *Drosophila*. *Trends Genet* **20:** 453–459. doi:10.1016/j.tig.2004.06.015

Ignell R, Ghaninia M, Hill SR. 2022. Combined gas-chromatography and single sensillum recording for mosquitoes. *Cold Spring Harb Protoc* doi:10.1101/pdb.prot107869

Karasek FW, Clement RE. 2012. *Basic gas chromatography–mass spectrometry: principles and techniques*. Elsevier, Amsterdam.

Kunert M, David A, Becher J, Boland W. 2009. Volatile sampling from biological sources by the closed-loop-stripping technique. *Cold Spring Harb Protoc* doi:10.1101/pdb.prot5233

Moorhouse JE, Yeadon R, Beevor PS, Nesbitt BF. 1969. Method for use in studies of insect chemical communication. *Nature* **223:** 1174–1175. doi:10.1038/2231174a0

Cite this introduction as *Cold Spring Harb Protoc*; doi:10.1101/pdb.top107681

Protocol 1

Combined Gas Chromatography and Single-Sensillum Recording for Mosquitoes

Rickard Ignell,[1,4] Majid Ghaninia,[2,3] and Sharon Rose Hill[1]

[1]*Disease Vector Group, Unit of Chemical Ecology, Department of Plant Protection Biology, Swedish University of Agricultural Sciences, 23422 Alnarp, Sweden;* [2]*School of Life Sciences, Arizona State University, Tempe, Arizona 85281, USA,* [3]*Division of Entomology, Department of Plant Protection, Gorgan University of Agricultural Sciences and Natural Resources, Gorgan, Iran*

Combined gas chromatography and single-sensillum recording (GC-SSR) joins together the established volatile compound separation and relative quantification techniques of GC with the ability to functionally screen olfactory sensory neurons (OSNs) for volatile selectivity and sensitivity using SSR. With minimal equipment modification, including a splitter and a heated transfer line, half of the effluent from the gas chromatograph column is directed to the flame ionization detector and half to the mosquito antennae. This GC-SSR combination provides a bioassay capable of determining the salient natural ligands for individual OSNs and of providing accurate and comparable dose–response curves irrespective of differences in the volatility of the compounds.

MATERIALS

It is essential that you consult the appropriate Material Safety Data Sheets and your institution's Environmental Health and Safety Office for proper handling of equipment and hazardous materials used in this protocol.

Reagents

Hexane (GC grade; Sigma-Aldrich 650552)

Hydrogen (LAB LINE, Strandmöllen A/S)

Mosquitoes of the species of choice

For rearing protocols of common laboratory mosquito species, see Gerberg (1970), and for malaria mosquitoes, Moll et al. (2013).

The age (days postemergence), sex, and physiological state (e.g., starved, sugar-fed, blood-fed, time following a blood meal, and parity) of the mosquito, as well as the time of day, can affect the sensitivity and selectivity of the peripheral olfactory system. Choose these parameters carefully to match the hypothesis tested and be consistent in the implementation.

Nitrogen (LAB LINE, Strandmöllen A/S)

Saturated potassium hydroxide (KOH; Sigma-Aldrich 1310-58-3)

Signature volatile or set of volatiles (optional; see Step 21)

Source of volatiles

[4]Correspondence: rickard.ignell@slu.se

Cite this protocol as *Cold Spring Harb Protoc*; doi:10.1101/pdb.prot107869

These volatiles can be derived from many sources, including natural odor headspace extracts and commercial or synthesized volatile compounds in solvent, odors collected by solid phase microextraction (SPME), and volatiles generated by the thermally induced vaporization of a sample (e.g., an entire mosquito). In this protocol, we describe an example of using a dynamic headspace odor extract; headspace odor extracts are not commercially available. There are several solvent and solvent-less headspace techniques available for use (e.g., Fäldt et al. 2000). Use caution when selecting the headspace technique, as the methods used will affect which odors are collected. We commonly opt for dynamic headspace collections using solvent extraction to obtain sufficient amounts of the extracts for subsequent behavioral and physiological assays.

Equipment

Electrodes

Tungsten electrode etching device (Ockenfels Syntech GmbH) or a custom-made etching device (see Step 7)

Instructions for building a custom-made etching device are included in the legend for Figure 1. Required materials include a carbon grounding electrode (LxWxH, 4.3 × 2.4 × 0.4 in, Aileen SS Gph5pcs), light source, micromanipulator (World Precision Instruments, M3301L), power unit, DC power supplier (e.g., Hanmatek model HM310), retort stand, stainless-steel shaft (Ockenfels Syntech GmbH), stereomicroscope, and syringe (60-mL).

Tungsten wire (World Precision Instruments TGW0325)

Tungsten wire electrode holders (two; see Step 8) (Ockenfels Syntech GmbH)

Equipment for mounting the mosquito

Container for anesthetizing mosquitoes (e.g., *Drosophila*-rearing vials or 15-mL Falcon tubes)

Coverslips (VWR 631-1569)

Dental wax (Surgident Periphery Wax, Kulzer 50092189)

Double-sided scotch tape (3M)

Featherweight forceps (VWR 470315-238)

Microscope slides (VWR 631-1553)

Small arts and crafts paint brush

Standard aspirator (John W. Hock PN 419)

Standard glass capillaries, no filament (World Precision Instruments 1B100F), pulled (optional; see Step 6)

The settings for pulling these capillaries and the final diameter of the tip are not important because they are used solely as a nonconductive restraint for the antennae.

Equipment for odor stimulation

Automatic liquid sampler (Agilent Technologies 7693A) (optional; see Step 23)

Controlled air delivery system, Stimulus Controller CS 55 (Ockenfels Syntech GmbH)

Cylinder bubble diffuser air stone (Pawfly ASC025)

DB-WAX column (30-m × 0.25-mm inner diameter and 0.25-µm film thickness; Agilent Technologies 122-7032)

Filter paper

Gerstel Graphpack 3D/2 low dead volume four-way cross (Gerstel)

Glass odor delivery tube (custom-made)

Prepare a glass tube (10-mm diameter, 250-mm length), angled at 90° (arm 1, 150 mm, and arm 2,100 mm) with two small diameter holes, one hole 100–120 mm from the mouth, and one hole at the 90° angle directing the effluent from the gas chromatograph into the glass tube in the direction of the air flow.

Glass wash bottle (Thermo Fisher Scientific 11836104)

Glass pipettes

Heated transfer line, Effluent conditioner assembly type EC-05 (Ockenfels Syntech GmbH)

HP 6890 gas chromatograph (Agilent Technologies)

HP-5 column (30-m × 0.25-mm inner diameter and 0.25-µm film thickness, Agilent Technologies 19091J-433E)

Porapak Q cartridge (50/80 mesh; Waters Associates) (optional; see Step 22)

 Cite this protocol as *Cold Spring Harb Protoc*; doi:10.1101/pdb.prot107869

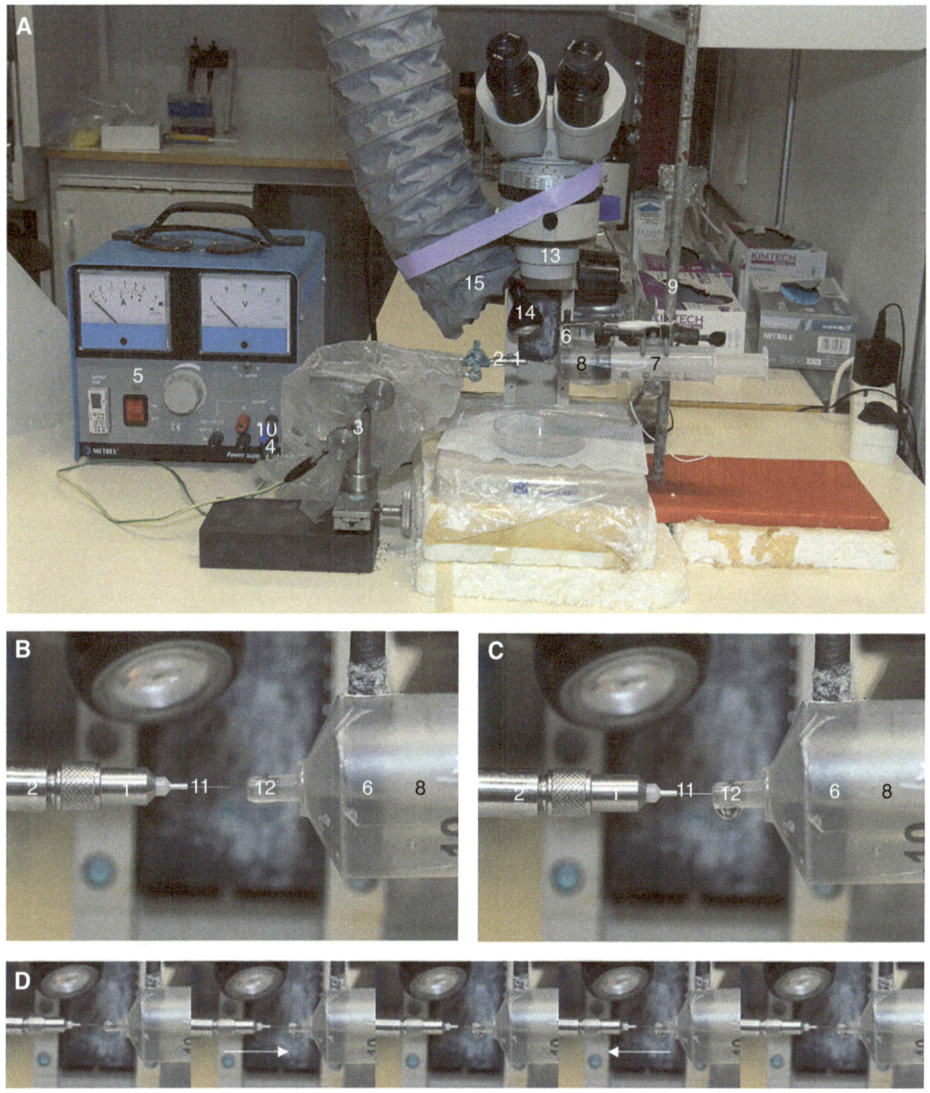

FIGURE 1. Custom-made electrode etching device. (*A*) A tungsten wire is inserted into the electrode holder (*1*) and mounted to the stainless-steel shaft (*2*) on the etching micromanipulator (*3*), which is connected to the positive pole (*4*) of the power unit (*5*). The carbon grounding electrode (*6*), placed inside a 60-cc syringe (*7*) filled with potassium hydroxide (1 M) (*8*) mounted on a retort stand (*9*), is connected to the negative pole (*10*) of the power unit (*5*). The tungsten wire (*B–D*) (*11*) and the mouth of the syringe (*B–D*) (*12*) are positioned in the center of the field of view of a dissecting microscope (*13*) with a light source (*14*). With the power unit on and adjusted to ~1–4 V, move the tungsten wire (*B–D*) (*11*) into and out of the potassium hydroxide solution (*8*) at an even rate, with the goal to produce a tungsten electrode with an even, gradual taper to a sharp tip (see Fig. 3F). The fumes from the potassium hydroxide are removed through an exhaust pipe (*15*). Close-up view of the tungsten electrode and the mouth of the syringe containing the etching solution (potassium hydroxide) (*B–D*) demonstrating the sharpening technique. Arrows indicate the direction of movement of the tungsten wire relative to the potassium hydroxide (*D*).

Silicone tubing (10-mm outer diameter × 7-mm inner diameter; VWR 228-1455)

Equipment for recording

Antenna Signal Simulator (Ockenfels Syntech GmbH)

This is a device meant to test the recording system before initiating experiments. This device will help to isolate and control for electrical noise in the setup by providing a clean set of artificial "OSN" signals, with varying preparation qualities.

Autospike software for SSR (Ockenfels Syntech GmbH)

Desktop PC computer

Four-channel USB acquisition controller (Ockenfels Syntech GmbH IDAC-4)
Grounding wires (e.g., World Precision Instruments 504714)
High-impedance universal single-ended probe (Ockenfels Syntech GmbH)
High-precision motorized micromanipulator (World Precision Instruments DC3001) (see Step 11)
Light microscope (e.g., Olympus BX51W1)
Magnetic stands (World Precision Instruments M10L) (two; see Step 11)
Objectives for long working distance (e.g., Olympus UMPLFLN10XW, LUMPLFLN50XW)
Piezo translator (World Precision Instruments SYS-MPM-10)
Standard electrical wires (12- or 14-gauge)
Vibration dampening table and Faraday cage (e.g., Sutter Instrument Company AT-3036-M6-F)

A dampening table is crucial for the success of GC-SSR because the preparation is sensitive to the most minor movements. The Faraday cage prevents electrical noise from the environment (e.g., computers, odor stimulator, microscope light source, mobile telephones).

METHOD

The following method is based on the method by Karasek and Clement (2012), pioneered in mosquitoes by Ghaninia et al. (2008), and then refined in the intervening years by the authors.

Carry out the entire protocol at room temperature, except as noted. If climatic factors are to be explored, vary the temperature accordingly.

Mounting the Mosquito

1. Aspirate the mosquitoes into any appropriate container for this task (e.g., *Drosophila*-rearing vials, 15-mL Falcon tubes) using a standard aspirator.

   If interaction among mosquitoes is hypothesized to affect the peripheral olfactory system under particular investigation, be sure to individualize the mosquitoes. Likewise, if interactions between the sexes may have an effect, divide the sexes.

2. Anesthetize the mosquitoes for 30 sec to 2 min in a −20°C freezer, depending on the species. Transfer the container containing the mosquitoes to ice before mounting.

   Do not use CO_2 to anesthetize the mosquitoes, as CO_2 is a salient semiochemical for mosquitoes, which has the potential to modulate the olfactory response under investigation. The anophelines are usually more sensitive to the cold and need only ~30 sec in the freezer, whereas Culex *and* Aedes *species generally need more time to anesthetize.*

3. Using featherweight forceps, mount and restrain the mosquito in one of two ways (Fig. 2).
   - Gently sandwich the mosquito between two layers of double-sided scotch tape affixed to a microscope slide, dorsal side up and wings splayed, with the edge of the top layer of tape running perpendicularly across the space between the head and thorax and abutting the head (Fig. 2A,C).
   - Place the mosquito in a 200-µL pipette tip, cut so that the anterior one-third of the head protrudes from the tip. Cut the back end of the pipette tip just after the abdomen of the mosquito. Fix the cut section of the pipette tip containing the mosquito to a microscope slide with dental wax (Fig. 2B,D).

     Tissues remain viable longer when the individual mosquito is kept intact. The most common tissues to use for GC-SSR in mosquitoes are the primary olfactory organs, the antennae, and the other secondary olfactory organ associated with the mouthparts, the maxillary palps, located on mosquitoes as a pair of appendages protruding from the head on either side of the proboscis. The mounting procedure is similar for recording from either organ.

4. Very gently affix the protruding antennae or maxillary palps to a double-sided scotch tape covered coverslip mounted on a ball of dental wax (~5-mm diameter) to allow for ease of positioning, and to provide sufficient stability to prevent movement during recording (Fig. 2). If recordings are to

Cite this protocol as *Cold Spring Harb Protoc*; doi:10.1101/pdb.prot107869

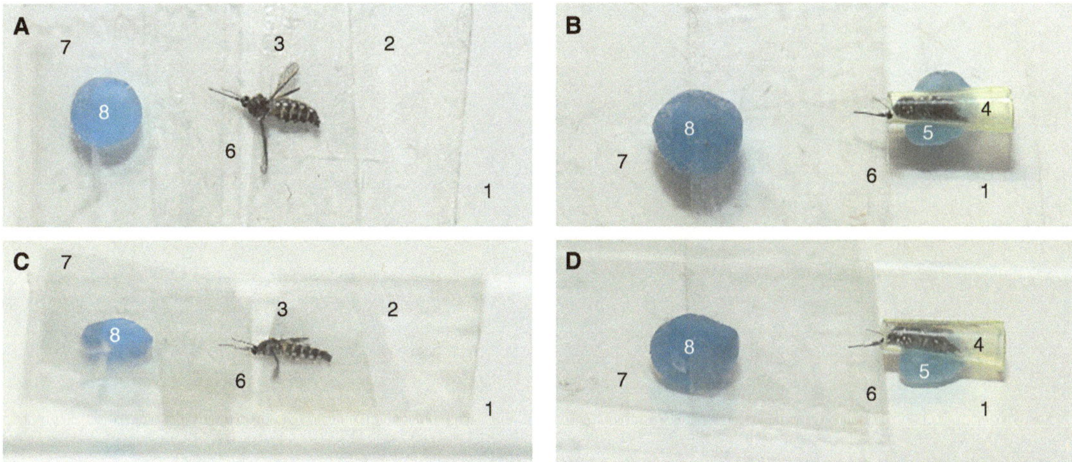

FIGURE 2. Mounting the mosquito for antennal single-sensillum recordings. An *Aedes aegypti* female is mounted on a microscope slide (*1*), by either gently sandwiching the mosquito between two layers of double-sided tape, one flat on the microscope slide (*2*), and the other covering the thorax and abdomen like a blanket (*3*) (*A*,*C*), or inserting the mosquito head-first into a trimmed 200-µL pipette tip (*4*), and secured to the microscope slide by a small ball of dental wax (*5*) (*B*,*D*), viewed from above (*A*,*B*) and the side (*C*,*D*). Antennae are fixed to double-sided tape protruding slightly (~1–2 mm) over the edge (*6*) of a coverslip (*7*) placed on a small ball of dental wax (*8*) on the microscope slide.

be made from the maxillary palps, gently remove the scales using a paint brush or other similar soft appliance.

5. On a plane above and parallel to the microscope slide, position the edge of the coverslip under the antennae or maxillary palps, as close as possible to the base of the tissue and, using a flexible fiber similar to a paint brush bristle, gently press the tissue onto the sticky surface from the base out to the tip (Fig. 2).
6. (Optional) Place the narrowed end of a pulled glass capillary across the base of the antenna or between the second and third segment of the maxillary palp to further immobilize the tissue, which can improve the stability of recordings.

Sharpening the Electrodes

7. Sharpen the electrodes using a commercially available etching device or a custom-made setup (Fig. 1).
8. Cut tungsten wire to a length of ~6–8 cm and thread it through an electrode holder with a minor kink in the wire that helps to hold the wire in place in the holder when inverted.

   For more information see the manual for the tungsten wire electrode holders.

9. Bend the wire protruding from the base of the holder at 90°, so that the hook does not extend beyond the white Teflon insulation (Fig. 3D).
10. Place the electrode holder in the tungsten electrode etching device (Fig. 1). Repeatedly dip the tip of the tungsten wire in and out of a solution of saturated KOH connected through a carbon electrode to an adjustable power source set at ~1–4 V that will electrolytically carve the sharpened tip of the electrode. Sharpen both recording and reference electrodes in this manner. Take care to make the electrode both sharp enough to penetrate the sensillum shaft and strong enough not to bend upon contact with the cuticle (Fig. 3F).

    Optimizing this process is a matter of trial and error. The shape of an optimal recording electrode is a long taper from the shaft to the tip, without obvious steps (Fig. 3F). If you find that the tip is bending into a hook on contact with the sensillum shaft, try to extend the taper further back on the shaft, making it more gradual.

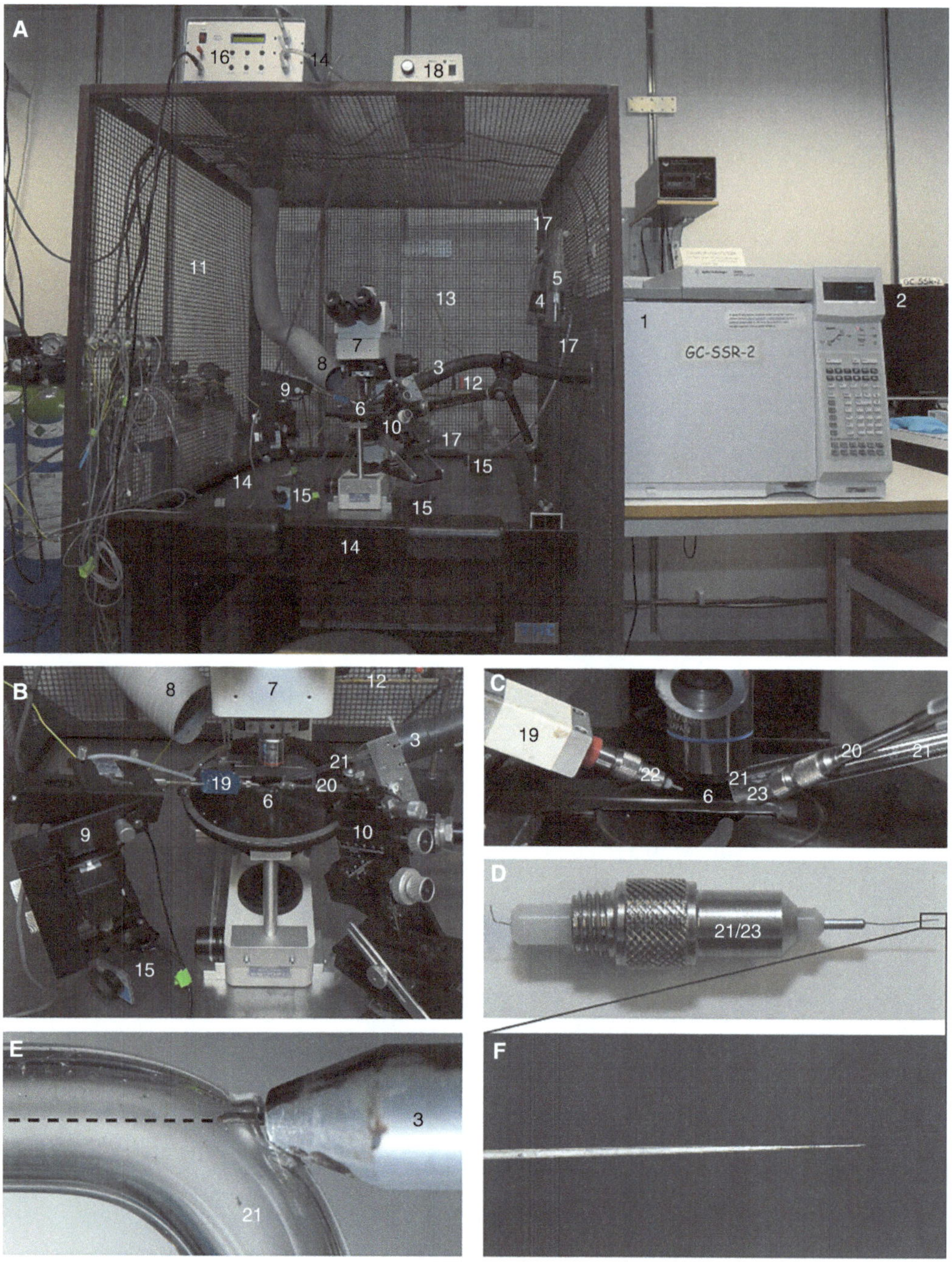

FIGURE 3. The gas chromatography–coupled single-sensillum recording (GC-SSR) setup for mosquitoes (*A*). Headspace extracts are injected into a gas chromatograph (*1*), connected to a computer (*2*). The headspace extract is separated through a GC column as the temperature increases, and delivered, through a transfer line (*3*), into the center of the glass odor delivery tube (*B,C,E*) (*21*; dotted line), in which a continuous, charcoal (*4*) filtered air stream, humidified by a cylinder bubble diffuser air stone (*5*), is directed over the antennal preparation area (*6*) on the microscope (*7*) stage, and actively removed by an exhaust (*8*). To record from the olfactory sensory neurons, a tungsten recording electrode (*F*), in a holder (*B–D*) (*22*), connected to a high-impedance universal single-ended probe (*B,C*) (*19*) mounted on a high-precision motorized micromanipulator (*9*) is positioned at the base of a sensillum using the piezo translator. A tungsten reference electrode, in its holder (*B,C*) (*23*), connected to the stainless-steel shaft (*B–D*) (*20*) and mounted on the manually controlled micromanipulator (*10*) is placed in the contralateral eye and connected to the Faraday cage (*11*) through a ground hub (*12*) by 12- or 14-gauge wire (*13*). Both micromanipulators are mounted on the vibration dampening table (*14*) by magnetic stands (*15*). The airflow is controlled through an air delivery system and stimulus controller (*16*) and directed through the glass wash bottles to the glass tube via silicone tubing (*17*). The microscope light source (*18*) is mounted outside the Faraday cage to minimize the electrical noise detected by this highly sensitive electrophysiological system.

Cite this protocol as *Cold Spring Harb Protoc*; doi:10.1101/pdb.prot107869

Recording

The parts of the GC-SSR setup are depicted in Figure 3.

11. Place the microscope in the middle of the vibration dampening table and within the Faraday cage, with the two micromanipulators mounted on magnetic stands on either side of the microscope stage (Fig. 3A,B). Ground the microscope and micromanipulators to a single grounding point on the Faraday cage with standard electrical wire (12- or 14-gauge) attached by the grounding screws provided on the equipment (Fig. 3A,B). Place the glass odor-delivery tube to one side, out of the way of the micromanipulators, with its mouth terminating within 0.5–1 cm of the antennal preparation.

    Do not place the odor delivery tube where it is likely to be "bumped," as such movement can damage the end of the GC column inserted within this tube (Fig. 3).

12. Place the reference electrode holder into the stainless steel shaft mounted in one micromanipulator and place the recording electrode into the high-impedance universal single-ended probe mounted on the high-precision motorized micromanipulator with the piezo translator (Fig. 3A,B).
13. Place the mosquito mounted on the slide under the microscope and, using a low magnifying objective, insert the sharpened tungsten reference electrode into the eye contralateral to the tissue from which recordings will be taken, minimizing the inadvertent severing of the antennal nerve by poor placement of the reference electrode.
14. Position the segment of the antennae or maxillary palp from which recordings will be made in the center of the field of view.
15. Carefully move the sharpened tungsten recording electrode into close proximity with the segment manually with the micromanipulator.
16. Switch the objective to the high magnification (50×).
17. With the motorized controller, use the micromanipulator to position the tip of the recording electrode against the shaft, just above the base of the sensillum.
18. Penetrate the sensillum with the recording electrode by taking a rapid, short step forward with the piezo function of the micromanipulator.
19. Once a contact is established, if needed, make small movements (in and out) of the recording electrode using the motorized micromanipulator to attempt to strengthen the signal.

    Ideally, the spikes of individual olfactory sensory neurons (OSNs) should be visible above the background noise in a ratio of about 3:1 to 5:1. Note that most OSNs show regular spontaneous activity, which can be both seen and heard (a function provided by most software for SSR). In cases in which the recording electrode is inserted too abruptly, or too deep, this may lead to the death of the OSN(s), which will be evident by the rapid firing rate that diminishes to none over a short period of time (1–5 sec; producing a sound like "wheeeee"). In this case, try to establish a contact with a neighboring sensillum. More than one sensillum can be recorded from on a single preparation in succession.

20. Monitor OSN activity via software for SSR (e.g., Autospike).

    The OSN activity is recorded extracellularly and reported in units of either spikes per second (spikes s^{-1}), or Hertz (Hz). Spikes are classified by shape and amplitude, with those demonstrating the largest amplitude denoted as A, and the next largest as B, and so on.

(Optional) Sensillum Functional Type Identification

The purpose of this step is to link natural bioactive compound(s) to an OSN, which have previously been functionally characterized using a panel of odorants, which often have an unknown behavioral role.

21. Once a contact is established, confirm the sensillum functional type, if such information is available from previous functional characterization studies (e.g., Ghaninia et al. 2007). Deliver a known amount of a signature volatile or set of volatiles (10 µL) emitted from a filter paper inside

a glass pipette into the airstream directed over the antennae. To do so, trigger a 0.5-sec programmed pulse through a controlled air delivery system (e.g., Stimulus Controller CS55) that injects the volatiles into the odor delivery glass tube.

Odor Stimulation

Odor stimulation using a gas chromatograph can deliver defined quantities of well-separated volatile compounds into the airstream and over the sensillum. These volatiles can be derived from many sources, including natural odor headspace extracts and commercial or synthesized volatile compounds in solvent, odors collected by SPME, and volatiles generated by the thermally induced vaporization of a sample (e.g., an entire mosquito). A gas chromatograph can also be used to deliver accurate and repeatable doses of volatile compounds providing the means to generate clear, comparable dose–response curves of OSN activity that is not confounded by differences in the volatility among compounds. An example protocol for eluting headspace odor extracts from a cartridge and separating the individual components within it is described below.

22. Elute a headspace odor extract off of a Porapak Q (50/80 mesh) cartridge in 500-µL hexane.

    There are several options for both the solid phase (e.g., Porapak Q) and solvents (e.g., n-hexane, pentane) to be used in this step. In our experience, we have found that Porapak Q along with redistilled n-hexane and pentane (highest-purity) have proven to be sufficient to identify the bioactive compounds in resources (e.g., human body headspace extracts, floral extracts, oviposition site extracts). For more options, please see Fäldt et al. (2000). If an alternate odor source (i.e., not a headspace odor extract) is being used, this step can be skipped.

23. Inject the eluted extract from Step 22 onto an HP 6890 GC at 220°C fitted with a splitless injector (30 sec) and flame ionization detector (FID, 220°C).

    Samples can be injected onto the GC column either manually or using an automated injector (e.g., Automatic Liquid Sampler). For manual injections, inject volumes between 0.5 and 2 µL, depending on the polarity of the sample, and the solvent (mobile phase) and column (stationary phase) used, splitless onto the injector, programmed at between 220°C and 300°C, depending on the thermal stability of the samples. We regularly inject odor collections from both SPME and 1- to 2-µL natural headspace extracts of animals and plants on a HP6890 gas chromatograph at 220°C.

24. Separate compounds on a polar DB-Wax or a nonpolar HP-5 column (30-m × 0.25-mm inner diameter and 0.25-µm film thickness) using hydrogen as the mobile phase (45-cm sec^{-1}).

    It is highly recommended that odor collection samples be tested using at least two GC columns of differing polarity (e.g., HP-5 and DB-WAX columns) to accurately identify the bioactive compounds eliciting OSN responses. Please note that when we refer to columns, we are referring to the GC columns (that use solid and mobile phases) installed in the gas chromatograph. Do not confuse this with the Porapak Q cartridge in steps above. For good separation of the volatile compounds, while retaining a narrow distribution along the column, we often use 30-m × 0.25-mm inner diameter columns, and programs of ~30 min.

25. Hold the oven temperature for 3 min at 35°C. Ramp up to 300°C at 10°C min^{-1} and hold for 10 min at 300°C. An alternate program setting of the GC may be required to allow optimal separation of compounds in some cases.

26. At the GC effluent splitter, add nitrogen (4-psi) and split 1:1 in a Gerstel 3D/2 low dead volume four-way cross, which directs half of the effluent to the FID for compound peak detection, and the other half through a heated transfer line (tracking the oven temperature +5°C) into the humidified airstream (1.5-L min^{-1}) passing over the sensilla.

    Each recording from an individual mosquito is considered to be a single replicate, and at least three consistent replicates are required to confirm bioactivity for each condition tested.

REFERENCES

Fäldt J, Eriksson M, Valterová I, Borg-Karlson AK. 2000. Comparison of head-space techniques for sampling volatile natural products in a dynamic system. *Z Naturforsch C* **55:** 180–188. doi:10.1515/znc-2000-3-408

Gerberg EJ. 1970. Manual for mosquito rearing and experimental techniques. *Am Mosq Control Assoc Bull* 5: 72–73.

Ghaninia M, Ignell R, Hansson BS. 2007. Functional classification and central nervous projections of olfactory receptor neurons housed in antennal trichoid sensilla of female yellow fever mosquitoes, *Aedes aegypti*. *Eur J Neurosci* **26:** 1611–1623. doi:10.1111/j.1460-9568.2007.05786.x

Cite this protocol as *Cold Spring Harb Protoc*; doi:10.1101/pdb.prot107869

Ghaninia M, Larsson M, Hansson BS, Ignell R. 2008. Natural odor ligands for olfactory receptor neurons of the female mosquito *Aedes aegypti*: use of gas chromatography-linked single sensillum recordings. *J Exp Biol* **211:** 3020–3027. doi:10.1242/jeb.016360

Karasek FW, Clement RE. 2012. *Basic gas chromatography–mass spectrometry: principles and techniques.* Elsevier, Amsterdam.

Moll K, Ljungström I, Perlmann H, Scherf A, Wahlgren M, Manassas V. 2013. *Methods in malaria research,* 6th ed. https://www.beiresources.org/portals/2/MR4/Methods_In_Malaria_Research-6th_edition.pdf (accessed on 2021-12-06).

CHAPTER 28

GCaMP Imaging in Mosquitoes: Central Nervous System

Zhilei Zhao,[1,2,3] Lukas Weiss,[1,2] and Carolyn S. McBride[1,2,4]

[1]*Princeton Neuroscience Institute,* [2]*Department of Ecology and Evolutionary Biology, Princeton University, Princeton, New Jersey 08544, USA*

Mosquitoes spread dengue, Zika, malaria, and other pathogens to hundreds of millions of people every year. A better understanding of mosquito behavior and its underlying neural mechanisms may lead to new control strategies, but such an understanding requires the development of tools and approaches for exploring the nervous system of key vector species. For example, we can now image neural activity in mosquito brains using genetically encoded calcium sensors like GCaMP. Compared with other types of neural recording, GCaMP imaging has the advantage of allowing one to record from many neurons simultaneously and/or to record from specific neuronal types. Successful implementation requires careful consideration of many factors, including the choice of microscope and how to make the brains of experimental animals visible and stable while minimizing damage. Here, we elaborate on these points and provide a concise introduction to GCaMP imaging in the mosquito central nervous system.

BACKGROUND

GCaMP Imaging versus Other Neural Recording Methods

The development and continuous improvement of genetically encoded calcium indicators (GECIs), combined with microscopes that allow imaging in large areas and/or deep regions of the brain, have revolutionized neuroscience (Knot et al. 2005; Grienberger and Konnerth 2012; Jercog et al. 2016; Lin and Schnitzer 2016). When neurons are activated, intracellular calcium concentrations usually increase (Baker et al. 1971; Tank et al. 1988; Grienberger and Konnerth 2012). Visualization of these chemical fluctuations as changes in fluorescence gives researchers an indirect readout of neural activity and forms the basis of calcium imaging. Early efforts relied on small, hydrophobic calcium-sensitive dyes (e.g., Fura-2 or Fluo-3) that passively diffuse into cells through the membrane (Knot et al. 2005). Although broadly applicable, including in organisms for which genetic tools are not available, dye-based calcium imaging suffers from low signal-to-noise ratios and limited ability to distinguish among different neuronal types. However, these two issues were largely resolved by the rise of GECIs, especially the GCaMP family of molecules—modified forms of green fluorescent protein (GFP) engineered to change conformation under the control of a fused calmodulin domain (Lin and Schnitzer 2016). GCaMP6, in particular, was developed through systematic mutagenesis and screening and is sensitive enough to detect single action potentials (Chen et al. 2013). Furthermore, because GECIs are introduced into cells via endogenous gene expression, they can be targeted to specific types of neurons or brain regions (Luo et al. 2018). For example, in the insect antennal lobe, whereas classic work in honeybees using calcium-sensitive dyes recorded mixed signals from all neuronal types

[3]Present address: Department of Neurobiology and Behavior, Cornell University, Ithaca, New York 14853, USA
[4]Correspondence: csm7@princeton.edu

Cite this introduction as *Cold Spring Harb Protoc;* doi:10.1101/pdb.top107682

(Galizia et al. 1999), more recent work in mosquitoes takes advantage of GECIs to record specifically from incoming sensory neurons (Shankar et al. 2021; Herre et al. 2022; Zhao et al. 2022).

GCaMP imaging has both strengths and weaknesses relative to direct electrophysiological recording. As discussed above, imaging allows one to monitor more neurons simultaneously and/or to monitor only specific neuronal types. It is also unbiased with respect to the baseline spontaneous activity of focal neurons (whereas electrophysiological approaches sometimes rely on this baseline activity to locate focal neurons). On the other hand, imaging suffers from low temporal resolution. Even the fastest version of GECIs, such as jGCaMP7f, cannot fully capture a neuron's electrical events that unfold at the millisecond scale. Furthermore, increases in intracellular calcium concentration do not scale linearly with a neuron's firing rate, making calcium imaging an indirect measure of neural activity (Pologruto et al. 2004; Wei et al. 2020). The novel genetically encoded voltage sensors are promising candidates for applications that require fast dynamics and direct readout of neural activity (Yang and St-Pierre 2016; Bando et al. 2019).

Current Availability of Transgenic Reagents for Imaging

More and more genetic tools are becoming available in mosquitoes thanks to the successful development of CRISPR–Cas9 genome editing (Gantz et al. 2015; Kistler et al. 2015; Itokawa et al. 2016). Our group has developed pan-neuronal driver lines that label all neurons in the brain of the arbovirus vector *Aedes aegypti* (Zhao et al. 2021). In theory, these lines can be used to image activity in any brain region. Lines that target specific neuronal types are also widely available for peripheral olfactory circuits, including genetic drivers for sensory neurons that make up the entire odorant receptor (OR) pathway or ionotropic receptor (IR) pathways (Riabinina et al. 2016; Shankar et al. 2021; Herre et al. 2022; Zhao et al. 2022). Although drivers for neurons that express specific neurotransmitters are not yet available, they are under active development. Many versions of GCaMP effectors are also available in mosquitoes, including GCaMP6s (Matthews et al. 2019) and jGCaMP7s (Zhao et al. 2022). By crossing these GCaMP effector lines with various driver lines, mosquito researchers now have the opportunity to test different combinations and find the best one for their specific application.

Key Considerations for Imaging

One important consideration for designing GCaMP imaging experiments is the choice of microscope. In theory, any fluorescence microscope can be used, including epifluorescence, confocal, two-photon, or lightsheet (Grienberger and Konnerth 2012). Each setup has advantages and disadvantages in terms of the potential imaging area, depth, speed, resolution, and laser power. Two-photon (2P) microscopes are particularly popular, as they allow high-resolution imaging in deep brain tissues with relatively low laser power (Svoboda and Yasuda 2006). When coupled with high-speed scanning, 2P microscopes have been used to carry out whole-brain GCaMP imaging in flies (Mann et al. 2017; Pacheco et al. 2021). In mosquitoes, researchers have used epifluorescence, confocal, and two-photon microscopes (Riabinina et al. 2016; Afify et al. 2019; Bui et al. 2019; Matthews et al. 2019; Jové et al. 2020; Lahondère et al. 2020; Melo et al. 2020; Shankar et al. 2021; Herre et al. 2022; Zhao et al. 2022).

Another key consideration is the brain preparation. The goal is to achieve optimal visibility and stability, with minimal side effects on physiology and behavior. To make the regions of interest visible, one must remove part of the cuticle covering the head. Moreover, because much of a mosquito's head is covered by its eyes, parts of the eyes may also need to be removed. Although 2P microscopes make it possible to image deep within the brain, it is best to minimize the length of the light path (and associated light scatter) by optimizing the mounting angle. For example, ventral brain regions may be imaged by mounting the animal upside down (Matthews et al. 2019). Stability of the prep is another important factor, because large brain movements cause artifacts that are difficult to correct post hoc. In cases in which stability is the priority, certain muscles that control the head or proboscis may be severed (Pacheco et al. 2021). Finally, it is important to keep the open brain tissue healthy by bathing it in a slow but continuous flow of saline buffer bubbled with carbogen.

 Cite this introduction as *Cold Spring Harb Protoc*; doi:10.1101/pdb.top107682

CONCLUSION

Uncovering the neural basis of mosquito behaviors promises to inform the development of more effective vector control strategies and to provide fundamental insight into animal brains and behavior. GCaMP imaging is a powerful method to record neural activity, allowing researchers to record from specific subsets of neurons with high signal-to-noise ratio. The tool kit of neuronal drivers and GCaMP effectors available for imaging in mosquitoes is growing rapidly, and recent efforts have already begun to reveal key mechanisms underlying interesting aspects of mosquito behavior. We expect these revelations to continue as imaging approaches are further developed and improved in this fascinating and important group of insects.

REFERENCES

Afify A, Betz JF, Riabinina O, Lahondère C, Potter CJ. 2019. Commonly used insect repellents hide human odors from *Anopheles* mosquitoes. *Curr Biol* **29:** 3669–3680.e5. doi:10.1016/j.cub.2019.09.007

Baker PF, Hodgkin AL, Ridgway EB. 1971. Depolarization and calcium entry in squid giant axons. *J Physiol* **218:** 709–755. doi:10.1113/jphysiol.1971.sp009641

Bando Y, Grimm C, Cornejo VH, Yuste R. 2019. Genetic voltage indicators. *BMC Biol* **17:** 71. doi:10.1186/s12915-019-0682-0

Bui M, Shyong J, Lutz EK, Yang T, Li M, Truong K, Arvidson R, Buchman A, Riffell JA, Akbari OS. 2019. Live calcium imaging of *Aedes aegypti* neuronal tissues reveals differential importance of chemosensory systems for life-history-specific foraging strategies. *BMC Neurosci* **20:** 27. doi:10.1186/s12868-019-0511-y

Chen T-W, Wardill TJ, Sun Y, Pulver SR, Renninger SL, Baohan A, Schreiter ER, Kerr RA, Orger MB, Jayaraman V, et al. 2013. Ultrasensitive fluorescent proteins for imaging neuronal activity. *Nature* **499:** 295–300. doi:10.1038/nature12354

Galizia CG, Sachse S, Rappert A, Menzel R. 1999. The glomerular code for odor representation is species specific in the honeybee *Apis mellifera*. *Nat Neurosci* **2:** 473–478. doi:10.1038/8144

Gantz VM, Jasinskiene N, Tatarenkova O, Fazekas A, Macias VM, Bier E, James AA. 2015. Highly efficient Cas9-mediated gene drive for population modification of the malaria vector mosquito *Anopheles stephensi*. *Proc Natl Acad Sci* **112:** E6736–E6743. doi:10.1073/pnas.1521077112

Grienberger C, Konnerth A. 2012. Imaging calcium in neurons. *Neuron* **73:** 862–85. doi:10.1016/j.neuron.2012.02.011

Herre M, Goldman OV, Lu T-C, Caballero-Vidal G, Qi Y, Gilbert ZN, Gong Z, Morita T, Rahiel S, Ghaninia M, et al. 2022. Non-canonical odor coding in the mosquito. *Cell* **185:** 3104–3123.e28. doi:10.1016/j.cell.2022.07.024

Itokawa K, Komagata O, Kasai S, Ogawa K, Tomita T. 2016. Testing the causality between CYP9M10 and pyrethroid resistance using the TALEN and CRISPR/Cas9 technologies. *Sci Rep* **6:** 24652. doi:10.1038/srep24652

Jercog P, Rogerson T, Schnitzer MJ. 2016. Large-scale fluorescence calcium-imaging methods for studies of long-term memory in behaving mammals. *Cold Spring Harb Perspect Biol* **8:** a021824. doi:10.1101/cshperspect.a021824

Jové V, Gong Z, Hol FJH, Zhao Z, Sorrells TR, Carroll TS, Prakash M, McBride CS, Vosshall LB. 2020. Sensory discrimination of blood and floral nectar by *Aedes aegypti* mosquitoes. *Neuron* **108:** 1163–1180. doi:10.1016/j.neuron.2020.09.019

Kistler KE, Vosshall LB, Matthews BJ. 2015. Genome engineering with CRISPR–Cas9 in the mosquito *Aedes aegypti*. *Cell Rep* **11:** 51–60. doi:10.1016/j.celrep.2015.03.009

Knot HJ, Laher I, Sobie EA, Guatimosim S, Gomez-Viquez L, Hartmann H, Song L-S, Lederer WJ, Graier WF, Malli R, et al. 2005. Twenty years of calcium imaging: cell physiology to dye for. *Mol Interv* **5:** 112–127. doi:10.1124/mi.5.2.8

Lahondère C, Vinauger C, Okubo RP, Wolff GH, Chan JK, Akbari OS, Riffell JA. 2020. The olfactory basis of orchid pollination by mosquitoes. *Proc Natl Acad Sci* **117:** 708–716. doi:10.1073/pnas.1910589117

Lin MZ, Schnitzer MJ. 2016. Genetically encoded indicators of neuronal activity. *Nat Neurosci* **19:** 1142–1153. doi:10.1038/nn.4359

Luo L, Callaway EM, Svoboda K. 2018. Genetic dissection of neural circuits: a decade of progress. *Neuron* **98:** 256–281. doi:10.1016/j.neuron.2018.03.040

Mann K, Gallen CL, Clandinin TR. 2017. Whole-brain calcium imaging reveals an intrinsic functional network in *Drosophila*. *Curr Biol* **27:** 2389–2396.e4. doi:10.1016/j.cub.2017.06.076

Matthews BJ, Younger MA, Vosshall LB. 2019. The ion channel *ppk301* controls freshwater egg-laying in the mosquito *Aedes Aegypti*. *eLife* **8:** e43963. doi:10.7554/eLife.43963

Melo N, Wolff GH, Costa-da-Silva AL, Arribas R, Fernandez Triana M, Gugger M, Riffell JA, DeGennaro M, Stensmyr MC. 2020. Geosmin attracts *Aedes aegypti* mosquitoes to oviposition sites. *Curr Biol* **30:** 127–134.e5. doi:10.1016/j.cub.2019.11.002

Pacheco DA, Thiberge SY, Pnevmatikakis E, Murthy M. 2021. Auditory activity is diverse and widespread throughout the central brain of *Drosophila*. *Nat Neurosci* **24:** 93–104. doi:10.1038/s41593-020-00743-y

Pologruto TA, Yasuda R, Svoboda K. 2004. Monitoring neural activity and $[Ca^{2+}]$ with genetically encoded Ca^{2+} indicators. *J Neurosci* **24:** 9572–9579. doi:10.1523/JNEUROSCI.2854-04.2004

Riabinina O, Task D, Marr E, Lin C-C, Alford R, O'Brochta DA, Potter CJ. 2016. Organization of olfactory centres in the malaria mosquito *Anopheles gambiae*. *Nat Commun* **7:** 13010. doi:10.1038/ncomms13010

Shankar S, Tauxe GM, Spikol ED, Li M, Akbari OS, Giraldo D, McMeniman CJ. 2021. Synergistic coding of carbon dioxide and a human sweat odorant in the mosquito brain. bioRxiv doi:10.1101/2020.11.02.365916

Svoboda K, Yasuda R. 2006. Principles of two-photon excitation microscopy and its applications to neuroscience. *Neuron* **50:** 823–839. doi:10.1016/j.neuron.2006.05.019

Tank DW, Sugimori M, Connor JA, Llinás RR. 1988. Spatially resolved calcium dynamics of mammalian Purkinje cells in cerebellar slice. *Science* **242:** 773–777. doi:10.1126/science.2847315

Wei Z, Lin B-J, Chen T-W, Daie K, Svoboda K, Druckmann S. 2020. A comparison of neuronal population dynamics measured with calcium imaging and electrophysiology. *PLoS Comp Biol* **16:** e1008198. doi:10.1371/journal.pcbi.1008198

Yang HH, St-Pierre F. 2016. Genetically encoded voltage indicators: opportunities and challenges. *J Neurosci* **36:** 9977–9989. doi:10.1523/JNEUROSCI.1095-16.2016

Zhao Z, Tian D, McBride CS. 2021. Development of a pan-neuronal genetic driver in *Aedes aegypti* mosquitoes. *Cell Rep Methods* **1:** 100042. doi:10.1016/j.crmeth.2021.100042

Zhao Z, Zung JL, Hinze A, Kriete AL, Iqbal A, Younger MA, Matthews BJ, Merhof D, Thiberge S, Ignell R, et al. 2022. Mosquito brains encode unique features of human odour to drive host seeking. *Nature* **605:** 706–712. doi:10.1038/s41586-022-04675-4

Protocol 1

Two-Photon Calcium Imaging in the Brain of *Aedes aegypti* Mosquitoes

Zhilei Zhao,[1,2,3] Lukas Weiss,[1,2] and Carolyn S. McBride[1,2,4]

[1]*Princeton Neuroscience Institute,* [2]*Department of Ecology and Evolutionary Biology, Princeton University, Princeton, New Jersey 08544, USA*

Understanding the neural basis of mosquito behavior is critical for designing effective vector control strategies and can potentially shed new light on basic nervous system function. Because mosquitoes are a non-model species, however, functional studies of mosquito nervous systems have long been restricted to electrophysiological recording from peripheral sensory organs such as the antenna. This is now changing with the advent of CRISPR–Cas9 gene editing and the development of other powerful new genetic tools. Transgenic mosquitoes that carry genetically encoded calcium sensors, for example, open the door to optical recording of neural activity with two-photon calcium imaging. Compared with electrophysiology, calcium imaging permits continuous monitoring of neural activity from large populations of neurons, even deep in the brain. When combined with selective neural drivers, it also allows targeted recording from specific neuronal types. Here, we describe a calcium imaging protocol we use in our laboratory to study neural activity in the brain of *Aedes aegypti* mosquitoes.

MATERIALS

It is essential that you consult the appropriate Material Safety Data Sheets and your institution's Environmental Health and Safety Office for proper handling of equipment and hazardous materials used in this protocol.

RECIPES: Please see the end of this protocol for recipes indicated by <R>. Additional recipes can be found online at http://cshprotocols.cshlp.org/site/recipes.

Reagents

Aedes aegypti saline solution for imaging <R>
Carbogen (5% CO_2 and 95% oxygen)
Epoxy glue (Eurotool GLU-735.90)
TES (Sigma-Aldrich T5691-100G)
Transgenic *Ae. aegypti* mosquitoes that express GCaMP in neurons (e.g., Zhao et al. 2022)
UV light curing glue (RapidFix 6121830ES)
Wet ice

Equipment

Blunt tip dispensing needle, 22-gauge (Grainger 5FVK6)
Dissecting microscope

[3]Present address: Department of Neurobiology and Behavior, Cornell University, Ithaca, New York 14853, USA
[4]Correspondence: csm7@princeton.edu

Cite this protocol as *Cold Spring Harb Protoc*; doi:10.1101/pdb.prot108070

Epifluorescence microscope
Fine wire for applying small quantities of UV light curing glue
Flowmeter
Glass Petri dish (15-cm-diameter)
Ice bucket
Mosquito aspirator (John Hock Company)
Mosquito imaging holder (see Method, Step 3)
Plastic shell vials with cotton plug (Fisher Scientific AS-519)
Plastic transfer pipettes, 3-mL (Fisher Scientific 137119D)
Sharp forceps (Dumont #5)
Software for imaging and data collection (e.g., ScanImage software [Pologruto et al. 2003])
Stimulus delivery device (depending on the research question)
Two-photon microscope
UV flashlight
Vacuum air line terminating in a 1000-µL pipette tip (for manipulating mosquito)
Vacuum air line with 1/16-in inner diameter tubing (for perfusing saline)
Volumetric flasks, 500-mL and 2-L

METHOD

Conduct experiments at room temperature unless otherwise indicated.

Preparing Mosquitoes and Imaging Holder

1. Hatch transgenic *Ae. aegypti* mosquitoes.

   Avoid overcrowding or underfeeding. We typically record from adults that are ~7 d old, so mosquito eggs need to be hatched ~2 wk in advance.

2. Screen third- or fourth-instar larvae under an epifluorescence microscope for any markers associated with GCaMP expression in the brain.

   Skip this step if all individuals are expected to carry the expression constructs (e.g., if the mosquitoes are homozygous). We avoid bulk screening at the adult stage since it involves anesthetizing mosquitoes on ice for an extended period of time, which may reduce their fitness and increase the likelihood of a mosquito dying during an imaging session.

3. Prepare the imaging holder, which consists of a plastic body and metal plate (Fig. 1A). The body is 3D-printed with fine detail plastic and has a hole in it for odor exhaust. The metal plate is a rectangular piece of stainless steel (thickness 0.001 in) with a tiny hole in the shape of a mosquito head photochemically etched in the center (Fig. 1A). The plate is folded down the center at the level of the hole. Align the holder and plate and glue them together with epoxy (Fig. 1A). Ensure the rectangular portion of the hole (Fig. 1C) is facing away from the main body of the plastic holder.

   The imaging holder can be reused many times. The hole in the metal plate (Fig. 1A,C) is designed to fit the head of adult female Ae. aegypti, *but can be photochemically etched in other sizes and shapes to fit male* Ae. aegypti *or different species. Design files for 3D printing and photochemical etching are available at https://github.com/mcbridelab/Zhao_2020_HumanOdorRepresentation/tree/master/TwoPhotonMosquitoHolder.*

Dissection

4. On the day of imaging, remove an appropriately sized aliquot of *Ae. aegypti* saline solution from the 4°C refrigerator and allow it to warm to room temperature in a 500-mL volumetric flask. Experiments are conducted at room temperature.

   The saline will be exchanged continuously during imaging at a flow rate of 125 mL/h. An imaging session that lasts 4 h will therefore require ~500 mL of saline.

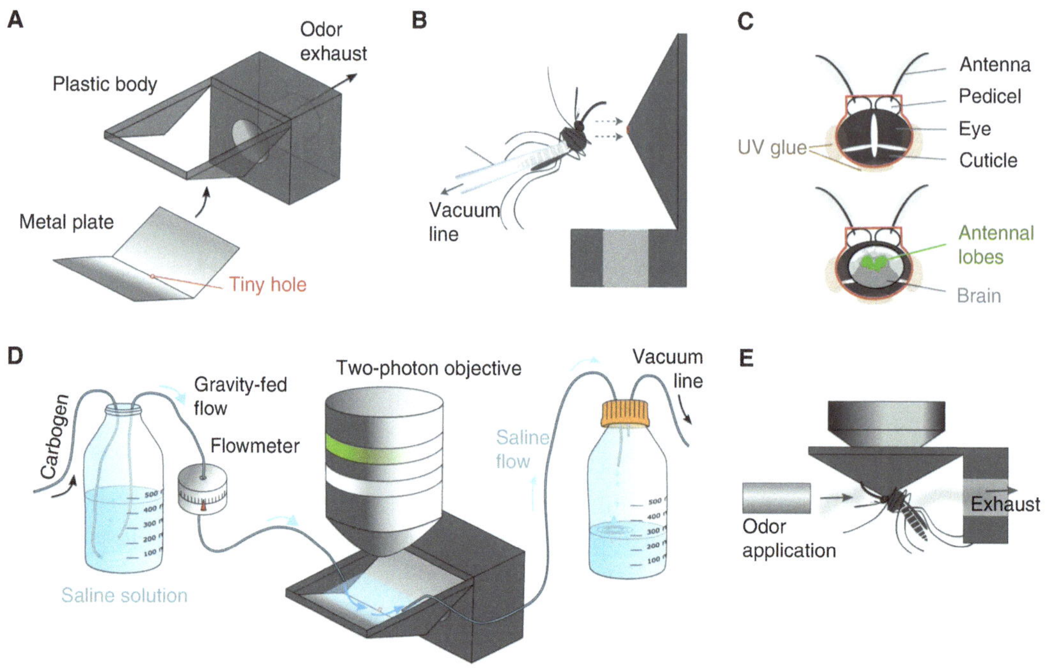

FIGURE 1. Imaging of odor-evoked responses in the antennal lobe of *Aedes aegypti* mosquitoes. (*A*) Imaging holder, comprising a plastic body plus a metal plate. A tiny hole (roughly the shape of a mosquito head) is photochemically etched in the center of the plate (red; see also *C*). (*B*) The mosquito head is pushed into the hole in the metal plate. We use a vacuum line with slight negative pressure to move the mosquito up to the imaging holder for mounting. (*C*, *Top*) The head is kept in position using UV-cured glue (brown), with the pedicels in the rectangular portion of the hole. (*Bottom*) Part of the eyes and cuticle are removed with fine forceps to expose the antennal lobes (green). (*D*) Schematic showing the saline exchange line. Saline is bubbled with carbogen and constantly flowing into the imaging holder from one side, then removed via suction from the other side. (*E*) Imaging setup with odors flowing from the outlet of the odor delivery system through the exhaust hole in the plastic holder.

5. Place the mosquito holder in an upright position under a dissecting microscope (Fig. 1B). Use a fine wire to spread UV glue along the edge of the hole in the metal plate.

   If recording olfactory responses, be careful to apply glue only to the circular part of the hole, avoiding the rectangular part (which will eventually contact the base of the mosquito antenna) (Fig. 1C).

6. Aspirate one mosquito out of the cage and place in a plastic shell vial. Place the vial on ice (half submerged) for ~1 min to anesthetize the mosquito.
7. Fill the glass Petri dish with ice so that both the bottom and top are in contact with the ice. Place the anesthetized mosquito on top of the closed Petri dish.
8. Gently push the abdomen of the mosquito into a pipette tip connected to a vacuum line under slight negative pressure (Fig. 1B). This allows one to gently hold and move the mosquito without risking the damage that can be incurred by use of forceps.
9. Move the mosquito close to the mosquito holder under the dissecting scope and gently push its head into the hole in the metal plate. The anterior–dorsal part of the head should be above the plate, while the antennae and proboscis are below the plate. The two pedicels should fill the square portion of the hole (Fig. 1C).
10. Turn on a UV flashlight for ~15 sec to cure the glue and thus cement the head to the circular part of the hole. Add and cure more glue at the cuticle–plate and eye–plate junctions to further fix the head and reduce brain movements during imaging (see Fig. 1C for glue locations).

    See Troubleshooting.

11. Flip the mosquito holder to the normal position (e.g., Fig. 1D) and fill the metal plate with ~2 mL of saline.

 Cite this protocol as *Cold Spring Harb Protoc*; doi:10.1101/pdb.prot108070

12. Under the dissecting scope, use sharp forceps to gently remove the cuticle and any part of the eyes necessary to uncover the brain region of interest.

    See bottom of Figure 1C for the location of the window we expose for antennal lobe imaging, which involves partial removal of both cuticle and eyes. We start by cutting open the cuticle at the eye–eye and eye–pedicel junction using the forceps tip, then gently peel it off to both sides. Hold forceps as flat/horizontal as possible to avoid damaging the underlying tissue

13. Exchange the saline in the metal plate several times with a plastic transfer pipette to remove cuticle debris and lipid droplets.

    To check whether the mosquito has survived the dissection procedure, we gently exhale on it. The mosquito usually wakes from anesthesia during the dissection and will be active when sensing CO_2 in exhaled breath. Only proceed if the mosquito is responsive.

Two-Photon Imaging

14. Place the mosquito holder under the two-photon microscope objective and fix it to the microscope stage. We use a screw for this purpose, but magnets may also work.
15. Set up the saline exchange line (Fig. 1D). Position the 500-mL volumetric flask containing the saline solution higher than the imaging setup to facilitate gravity-fed flow through the tubing positioned on one side of the imaging holder. Position the suction syringe connected to the 2-L volumetric flask and vacuum line at the other side of the imaging holder.

    We continuously bubble the saline with carbogen (5% CO_2/95% oxygen) and exchange it using the vacuum line at a flow rate of 125 mL/h, controlled by a flowmeter (Fig. 1D). Failing to add carbogen to the saline will decrease gas exchange in the brain and lead to accelerated tissue damage/death.

    See Troubleshooting.

16. Move the two-photon stage with a micromanipulator to center the brain region of interest under the objective.

    We use a water immersion objective with a numerical aperture of 0.8 for our experiments (Olympus LUMPLFL 40× water, NA 0.8). Objectives with different optical elements can be used, as long as the diameter does not exceed the dimensions of the metal plate/imaging holder.

17. Set up the stimulation device.

    For olfactory stimuli, we bring the output port of our odor-delivery system within ~3 mm of the mosquito head. The output port is wide enough to cover all peripheral olfactory organs (1/4 in diameter).

18. Start imaging and data collection.

    This step depends heavily on the specific two-photon microscope and data acquisition software used. In our laboratory, we use a custom two-photon microscope controlled by the ScanImage software (Pologruto et al. 2003) to acquire data. Besides recording responses to stimuli, we also record a high-resolution movie of the imaged part of the brain at the end of each session to obtain more detailed anatomical information useful for offline registration.

    See Troubleshooting.

19. After the experiment, withdraw the mosquito holder from the microscope stage, and remove the mosquito and saline from the metal plate. The hardened glue can be carefully scraped off the plate with forceps, and the plate can be rinsed with water or saline before reuse.

TROUBLESHOOTING

Problem (Steps 10 and 18): The mosquito head is moving during imaging.
Solution: To prevent the mosquito head from moving too much, make sure to apply enough glue to the circular portion of the hole in the metal plate. If the mosquito is already mounted but still moves too much, more glue can be applied at the borders between the eyes/cuticle and the metal plate to increase stability of the prep.

Problem (Steps 15 and 18): Saline leakage occurs.

Solution: When recording olfactory responses, we avoid placing any UV glue along the rectangular portion of the hole in the metal plate adjacent to the mosquito pedicels (Fig. 1C). However, saline may sometimes leak from the holder at this position, lowering the quality of the recordings or damaging the microscope. In this case, attempt to create a tighter fit between the mosquito head and the edge of the hole in this area, minimizing any gaps.

Problem (Step 18): Mosquito neurons are unresponsive.

Solution: In our olfactory imaging experiments, before collecting data, we first check the responsiveness of a prep using standard odorant dilutions. In our hands, roughly one out of two preps have decent olfactory responses. Potential causes of low responsiveness include, but are not limited to, (1) damage to the peripheral nerve or central brain during dissection; (2) poor mounting angle of the prep, such that fluorescence is scattered by brain tissue before reaching the objective; (3) excessive laser power, causing tissue damage during imaging (we never use >20 mW); (4) saline not correctly formulated, not yet warmed to ambient temperature, or not bubbled with carbogen; and (5) malfunction of the stimulus-delivery device.

DISCUSSION

With this protocol, we can reliably record from the same brain for 4–5 h (Zhao et al. 2022). This is not an upper limit since typically the mosquito is still alive after imaging and we rarely observe a significant decrease in neural response. Since *Aedes aegypti* is a tropical and subtropical mosquito species, the optimal rearing condition in the laboratory is 26°C with 75% humidity. In our two-photon recording chamber, we do not condition the temperature or humidity of the air. The imaging is therefore conducted at room temperature and humidity. We found that this does not cause obvious problems for laboratory strains of the globally invasive subspecies *Ae. aegypti*. However, it may be an issue for field strains that have not fully adapted to laboratory environments or for the ancestral African subspecies *Ae. aegypti* ssp. *formosus*. It may be beneficial to add temperature and humidity control to the two-photon recording chamber in these situations.

The research in our laboratory focuses mostly on the antennal lobe (AL) for olfactory processing. The mosquito mounting angle described in this protocol is therefore optimized for visualizing activity in the AL. If the region of interest is located in the ventral or posterior brain, the mounting orientation (and shape of the hole in the metal plate) may need to be adjusted in order to minimize the scattering and decay of fluorescence signals with imaging depth. Note that our prep involves damage to the eyes. Researchers interested in visual responses will likely need to find a way to access the brain from the back without damaging the eyes—a potentially tricky feat given that mosquito eyes cover most of the surface area of the head.

 Cite this protocol as *Cold Spring Harb Protoc*; doi:10.1101/pdb.prot108070

RECIPES

Aedes aegypti Saline Solution for Imaging

Reagent	Final concentration
NaCl	103 mM
KCl	3 mM
TES	5 mM
$NaHCO_3$	26 mM
NaH_2PO_4	1 mM
$CaCl_2$	1.5 mM
$MgCl_2$	4 mM
Trehalose	10 mM
Glucose	10 mM

We typically make a 2-L stock in a volumetric flask, adding water and commercial solutions of $CaCl_2$ and $MgCl_2$ before other ingredients to avoid calcium precipitation. Adjust the pH to 7.1 with HCl or NaOH solution using a pH meter (Beyenbach and Masia 2002). Store at 4°C for up to 2 wk. Note that this recipe is based on a standard recipe used for *Drosophila melanogaster*, with the pH adjusted to a published value for *Ae. aegypti* (Beyenbach and Masia 2002). It is also likely to be compatible with other mosquitoes.

REFERENCES

Beyenbach KW, Masia R. 2002. Membrane conductances of principal cells in Malpighian tubules of *Aedes aegypti*. *J Insect Physiol* **48:** 375–386. doi:10.1016/S0022-1910(02)00057-4

Design files for 3D printing and photochemical etching. 2020. https://github.com/mcbridelab/Zhao_2020_HumanOdorRepresentation/tree/master/TwoPhotonMosquitoHolder [Accessed October 24, 2022]

Pologruto TA, Sabatini BL, Svoboda K. 2003. ScanImage: flexible software for operating laser scanning microscopes. *Biomed Engineer Online* **2:** 13. doi:10.1186/1475-925X-2-13

Zhao Z, Jessica L, Zung JL, Hinze A, Kriete AL, Iqbal A, Younger MA, Matthews BJ, Merhof D, Thiberge S, et al. 2022. Mosquito brains encode unique features of human odour to drive host seeking. *Nature* **605:** 706–712. doi:10.1038/s41586-022-04675-4

CHAPTER 29

Genetically Encoded Calcium Indicators for Functional Imaging of Mosquito Olfactory Neurons

Ali Afify[1] and Christopher J. Potter[1,2]

[1]*The Solomon H. Snyder Department of Neuroscience, Johns Hopkins University School of Medicine, Baltimore, Maryland 21205, USA*

Mosquitoes transmit a multitude of diseases to humans and animals through biting and blood feeding. To locate their hosts, mosquitoes primarily use their sense of smell. Therefore, an understanding of mosquito olfaction will help develop strategies to control the diseases they transmit. A mosquito's sense of smell is determined by the response of olfactory neurons on its peripheral olfactory organs. Traditionally, mosquito olfactory neuron activity has been examined using electrophysiological techniques such as electroantennography and single-sensillum recordings. Electroantennography examines if an odorant is detectable by the ensemble of all antennal neurons. In contrast, single-sensillum electrophysiology allows detailed recordings of the activity of two to three neurons at a time. However, single-sensillum recording of olfactory neurons is difficult, laborious, and typically allows examination of only a few neurons on the antenna. A promising new approach is to use optical imaging techniques to provide a way to visualize the global response of olfactory organs to an odor, as well as the specific responses of several olfactory neurons to that odor. In particular, calcium imaging has progressed significantly, from the use of chemical calcium indicators to the development of genetically encoded calcium sensors. These advances have opened the way to study the mode of action of known mosquito attractants and repellents as well as a way to screen potential new attractants and repellents. Here, we provide an introduction to the different types of calcium indicators and their uses for investigating the function of mosquito sensory neurons.

MOSQUITO OLFACTION: AN INTRODUCTION

Mosquitoes transmit several diseases to humans such as West Nile virus (Farajollahi et al. 2011), yellow fever (Christophers 1960), and malaria (Russell 1959). Every year, malaria alone infects more than 200 million people and kills more than 400,000 people worldwide (WHO 2020). Female mosquitoes transmit diseases through blood feeding, which is required for egg production. Some mosquitoes, such as the *Anopheles* malaria mosquitoes (Costantini et al. 1998), prefer to blood-feed on humans (anthropophilic), making them particularly dangerous as vectors for human diseases. Understanding how such mosquitoes find humans and distinguish them from other animals is an important step in the effort to develop methods for preventing mosquito bites and the spread of diseases. The important role of olfaction in mosquito host seeking has been appreciated for decades (Rudolfs 1922). Recent studies have further highlighted the primary role olfaction plays, along with other senses, to find hosts for blood feeding (Potter 2014). In fact, most mosquito behaviors are guided by their sense of smell

[2]Correspondence: cpotter@jhmi.edu

Cite this introduction as *Cold Spring Harb Protoc*; doi:10.1101/pdb.top107683

(Konopka et al. 2021), which suggests that understanding how a mosquito smells its environment can be used to control its behaviors.

The mosquito olfactory system consists of three sensory organs: the antennae, the maxillary palps, and the labella (Potter 2014; Lombardo et al. 2017; Konopka et al. 2021). These peripheral organs are covered with sensory hairs called sensilla, with the antenna containing the most sensilla. For example, the antenna of an *Anopheles gambiae* female is covered by more than 700 sensilla, while the maxillary palp contains only 67 sensilla (McIver 1982). Each of these sensilla contains sensory neurons that express different types of odorant receptors (ORs), gustatory receptors (GRs), and/or ionotropic receptors (IRs) (Potter 2014; Lombardo et al. 2017; Konopka et al. 2021). Insect ORs are odor-gated ion channels that function as a complex of two proteins: the first is the "tuning" odorant receptor (ORx) that binds the odorant to induce activation, and the second component is an obligate odorant receptor coreceptor (Orco) that traffics the ORx to the membrane and helps form the receptor complex (Carey and Carlson 2011; Potter 2014).

ORx/Orco activation by an odorant leads to an influx of ions into the neuron and results in action potentials and spiking of that neuron. The electrical activity of the neurons can be monitored by electrophysiological approaches such as electroantennography (EAG) and single-sensillum recording (SSR). EAG is a global yet low-resolution approach that records the summed responses of the antenna to an odor stimulus. In contrast, SSR is a higher-resolution approach, in which an electrode is directly inserted into a sensillum and used to record the extracellular potentials of the neurons within that sensillum. SSR has been used in some mosquitoes to survey responses to panels of odorants (Lu et al. 2007; Syed and Leal 2008, 2009), but it is a difficult and challenging technique, and a full survey of the antenna requires the testing of a large number of sensilla, one at a time.

CALCIUM IMAGING TO STUDY MOSQUITO OLFACTION

Neural activity leads to a significant increase in intracellular calcium (Baker et al. 1971). Calcium imaging depends on the use of calcium-sensitive molecules to serve as "sensors" for detecting changes in intracellular calcium, and hence neural activity. These calcium indicators are weakly fluorescent at rest conditions ("baseline activity"), and the intensity of their emitted fluorescence increases as intracellular calcium levels increase (Hendel et al. 2008). Calcium imaging is less sensitive at detecting weak neural activity, but unlike EAG and SSR, the use of calcium imaging allows for visualizing the activity of large numbers of individual neurons at once. In addition, the mosquito preparation for calcium imaging is simpler than that required for EAG or SSR.

Two types of calcium indicators are typically used. The first of these are chemical indicators, which are fluorescent calcium chelators that bind strongly to calcium ions to increase their fluorescence properties and can be used to measure changes in intracellular calcium concentrations. These chemicals are generally membrane impermeable but can be combined with acetoxymethyl ester (AM) to make them membrane permeable, allowing them to be added to the solution bathing the neuronal sample and thereby label many neurons at once (e.g., the Calcium Green AM dye). Once inside the cell, such indicators are freed from the acetoxymethyl ester by the action of cytosolic esterases (Galizia and Vetter 2004). Chemical indicators can also be used in their membrane impermeable form, without acetoxymethyl ester. In this case, they are directly injected into neurons, allowing more confidence in the identity of the neurons being imaged (e.g., staining projection neurons with Fura dextran dye). These chemical indicators have been successfully used to study olfactory responses in the brain in different insects such as honeybees, *Drosophila*, and moths (Galizia and Vetter 2004). However, they require loading of dye into neurons before each imaging session.

The second type of calcium sensors are genetically encoded calcium indicators (GECIs). These sensors are calcium-sensitive proteins that can be genetically engineered to be expressed in targeted neurons. The most commonly used GECI is GCaMP, a GFP-based calcium indicator that results in recordings with high signal-to-noise ratio (Nakai et al. 2001). GCaMP contains a circularly permuted

 Cite this introduction as *Cold Spring Harb Protoc*; doi:10.1101/pdb.top107683

eGFP molecule in the center, connected to the M13 fragment of myosin light chain kinase at the amino terminus, and calmodulin (CaM) at the carboxyl terminus. When the calmodulin subunit binds calcium, it goes through a conformational change that enables binding to M13. This leads the carboxyl and amino termini of the enhanced green fluorescent protein (eGFP) to come together and helps optimize the fluorescent activity of eGFP (Nakai et al. 2001). Improved versions of GCaMP continue to be developed, such the three variants of GCaMP6, named GCaMP6f, GCaMP6m, and GCaMP6s, that have fast, medium, and slow kinetics, respectively (Chen et al. 2013).

GECIS FOR MOSQUITO OLFACTION

GECIs can be expressed in a population of neurons, such as Orco or IR expressing neurons, using a tissue-specific promoter. This allows control over which tissues will be examined by calcium imaging. Binary expression systems such as the Q-system (Riabinina and Potter 2016) can also be used to provide versatility in expressing GECIs in different populations of neurons. In brief, this is accomplished by combining two genetic components in the same animal: a QF transcription factor driver line expressed in a tissue-specific population, and a QF-responsive *QUAS-GCaMP* line that will express the calcium reporter in the QF-expressing population of neurons. Binary expression systems also allow for transcriptional amplification of GCaMP, which can boost the fluorescent signal. GECIs have been used for years in *Drosophila* to monitor sensory or olfactory functions (Simpson and Looger 2018) but were only recently introduced into mosquitoes. In one such experiment, the Orco promoter was used to drive expression of GCaMP6f in olfactory neurons of *Anopheles coluzzii* mosquitoes (*Orco-QF2 + QUAS-GCaMP6f*) (Afify et al. 2019). The method by which this was done is described in detail in our associated protocol (see Protocol 1: Calcium Imaging of *Anopheles coluzzii* Mosquito Antennae Expressing the Calcium Indicator GCaMP6f [Afify and Potter 2022]). This allowed direct monitoring of the odor-induced neuronal responses for the majority of olfactory neurons on the peripheral organs of *Anopheles* mosquitoes and was used to perform calcium imaging to test *Anopheles* antennal olfactory neuron responses toward mosquito repellents. Natural repellents (e.g., lemongrass oil and eugenol) were found to activate discrete odorant receptor neurons on the antenna while synthetic repellents (DEET, IR3535, and picaridin) did not robustly activate *Anopheles* olfactory neurons, but instead acted to decrease the amount of human odorants that reached the mosquito antenna (Afify et al. 2019). Calcium imaging with these transgenic mosquitoes was also used to test the response of *Anopheles* mosquitoes to repellent blends and showed that mixing repellents does not necessarily produce a better repellent (Afify and Potter 2020). These studies, which involved testing the activity of hundreds of neurons on the mosquito antennae, maxillary palps, and proboscis, would have been exceedingly difficult to perform using electrophysiology approaches.

In *Aedes aegypti*, the *Bruchpilot* (*Brp*) promoter has been used to drive the expression of GCaMP6s in all neurons (Jové et al. 2020; Zhao et al. 2021). This transgenic line was used to examine the response of the mosquito stylet to sugar and blood. "Stylet" is a common term used for the piercing tubular "needle" inserted into the skin during blood feeding. In this experiment, stylet neurons were imaged ex vivo while ligands were delivered using a microfluidics device, an experiment that would have been very difficult to perform using a technique like SSR that is more sensitive to movements.

GECIs can also be used to perform calcium imaging of neurons in the brain. For example, GCaMP6s was expressed in all cells in *Ae. aegypti* mosquitoes using the polyubiquitin (PUb) promoter (Bui et al. 2019; Lahondère et al. 2019). This enabled the use of calcium imaging to study responses to floral odors in the antennal lobe, the brain region targeted by olfactory neurons. Another example is the expression of GCaMP6f in Orco$^+$ neurons in *Ae. aegypti* and examining odor-induced responses in the antennal lobes. Although human, animal, and nectar odors consistently activated a "universal glomerulus," human odors alone activated a "human-sensitive" glomerulus, suggesting a potential neural mechanism by which *Ae. aegypti* mosquitoes might be able to distinguish human hosts (Zhao et al. 2020). Calcium imaging of brain regions can help identify how odors are represented by the activities of neurons within the brain, a task that would be far more challenging to accomplish using electrophysiological approaches.

Calcium imaging of peripheral organs can be a powerful approach to characterize the response properties of sensory neurons and to potentially isolate and identify the sensory receptors expressed by these neurons that enable these responses. This could be applied, for example, to identify the olfactory receptors that are required for the response to known mosquito attractant and repellent odors. In addition, calcium imaging of peripheral organs can be used as a quick and easy screening method for the development or optimization of new odor-based attractants and repellents.

REFERENCES

Afify A, Potter CJ. 2020. Insect repellents mediate species-specific olfactory behaviours in mosquitoes. *Malar J* **19:** 127. doi:10.1186/s12936-020-03206-8

Afify A, Potter CJ. 2022. Calcium imaging of *Anopheles coluzzii* mosquito antennae expressing the calcium indicator GCaMP6f. *Cold Spring Harb Protoc* doi: 10.1101/pdb.prot107918

Afify A, Betz JF, Riabinina O, Lahondère C, Potter CJ. 2019. Commonly used insect repellents hide human odors from *Anopheles* mosquitoes. *Curr Biol* **29:** 3669–3680.e3665. doi:10.1016/j.cub.2019.09.007

Baker PF, Hodgkin AL, Ridgway EB. 1971. Depolarization and calcium entry in squid giant axons. *J Physiol* **218:** 709–755. doi:10.1113/jphysiol.1971.sp009641

Bui M, Shyong J, Lutz EK, Yang T, Li M, Truong K, Arvidson R, Buchman A, Riffell JA, Akbari OS. 2019. Live calcium imaging of *Aedes aegypti* neuronal tissues reveals differential importance of chemosensory systems for life-history-specific foraging strategies. *BMC Neurosci* **20:** 27. doi:10.1186/s12868-019-0511-y

Carey AF, Carlson JR. 2011. Insect olfaction from model systems to disease control. *Proc Natl Acad Sci* **108:** 12987–12995. doi:10.1073/pnas.1103472108

Chen T-W, Wardill TJ, Sun Y, Pulver SR, Renninger SL, Baohan A, Schreiter ER, Kerr RA, Orger MB, Jayaraman V, et al. 2013. Ultra-sensitive fluorescent proteins for imaging neuronal activity. *Nature* **499:** 295–300. doi:10.1038/nature12354

Christophers SR. 1960. *Aedes aegypti. The yellow fever mosquito: its life history, bionomics, and structure.* Cambridge University Press, Cambridge.

Costantini C, Sagnon N, Ad T, Diallo M, Brady J, Gibson G, Coluzzi M. 1998. Odor-mediated host preferences of West African mosquitoes, with particular reference to malaria vectors. *Am J Trop Med Hyg* **58:** 56–63. doi:10.4269/ajtmh.1998.58.56

Farajollahi A, Fonseca DM, Kramer LD, Marm Kilpatrick A. 2011. "Bird biting" mosquitoes and human disease: a review of the role of *Culex pipiens* complex mosquitoes in epidemiology. *Infect Genet Evol* **11:** 1577–1585. doi:10.1016/j.meegid.2011.08.013

Galizia CG, Vetter RS. 2004. Optical methods for analyzing odor-evoked activity in the insect brain. In *Advances in insect sensory neuroscience* (ed. Christensen TA), pp. 349–392. CRC Press, Boca Raton, FL.

Hendel T, Mank M, Schnell B, Griesbeck O, Borst A, Reiff DF. 2008. Fluorescence changes of genetic calcium indicators and OGB-1 correlated with neural activity and calcium in vivo and in vitro. *J Neurosci* **28:** 7399–7411. doi:10.1523/JNEUROSCI.1038-08.2008

Jové V, Gong Z, Hol FJH, Zhao Z, Sorrells TR, Carroll TS, Prakash M, McBride CS, Vosshall LB. 2020. Sensory discrimination of blood and floral nectar by *Aedes aegypti* mosquitoes. *Neuron* **108:** 1163–1180. doi:10.1016/j.neuron.2020.09.019

Konopka JK, Task D, Afify A, Raji J, Deibel K, Maguire S, Lawrence R, Potter CJ. 2021. Olfaction in *Anopheles* mosquitoes. *Chem Senses* **46:** bjab021. doi:10.1093/chemse/bjab021

Lahondère C, Vinauger C, Okubo RP, Wolff GH, Chan JK, Akbari OS, Riffell JA. 2019. The olfactory basis of orchid pollination by mosquitoes. *Proc Natl Acad Sci* **117:** 708–716. doi:10.1073/pnas.1910589117

Lombardo F, Salvemini M, Fiorillo C, Nolan T, Zwiebel LJ, Ribeiro JM, Arcà B. 2017. Deciphering the olfactory repertoire of the tiger mosquito *Aedes albopictus. BMC Genomics* **18:** 770. doi:10.1186/s12864-017-4144-1

Lu T, Qiu YT, Wang G, Kwon Jae Y, Rutzler M, Kwon H-W, Pitts RJ, van Loon JJA, Takken W, Carlson JR, et al. 2007. Odor coding in the maxillary palp of the malaria vector mosquito *Anopheles gambiae. Curr Biol* **17:** 1533–1544. doi:10.1016/j.cub.2007.07.062

McIver SB. 1982. Sensilla of mosquitoes (Diptera: Culicidae). *J Med Entomol* **19:** 489–535. doi:10.1093/jmedent/19.5.489

Nakai J, Ohkura M, Imoto K. 2001. A high signal-to-noise Ca^{2+} probe composed of a single green fluorescent protein. *Nat Biotechnol* **19:** 137–141. doi:10.1038/84397

Potter CJ. 2014. Stop the biting: targeting a mosquito's sense of smell. *Cell* **156:** 878–881. doi:10.1016/j.cell.2014.02.003

Riabinina O, Potter CJ. 2016. The Q-system: a versatile expression system for *Drosophila. Methods Mol Biol* **1478:** 53–78. doi:10.1007/978-1-4939-6371-3_3

Rudolfs W. 1922. Chemotropism of mosquitoes. *Bull N J Agric Exp Stn* **367:** 4–23.

Russell PF. 1959. Insects and the epidemiology of malaria. *Annu Rev Entomol* **4:** 415–434. doi:10.1146/annurev.en.04.010159.002215

Simpson JH, Looger LL. 2018. Functional imaging and optogenetics in *Drosophila. Genetics* **208:** 1291–1309. doi:10.1534/genetics.117.300228

Syed Z, Leal WS. 2008. Mosquitoes smell and avoid the insect repellent DEET. *Proc Natl Acad Sci* **105:** 13598–13603. doi:10.1073/pnas.0805312105

Syed Z, Leal W. 2009. Acute olfactory response of *Culex* mosquitoes to a human- and bird-derived attractant. *Proc Natl Acad Sci* **106:** 18803–18808. doi:10.1073/pnas.0906932106

WHO. 2020. *World Malaria Report 2020: 20 years of global progress and challenges.* WHO Press–World Health Organization, Geneva, Switzerland.

Zhao Z, Zung JL, Kriete AL, Iqbal A, Younger MA, Matthews BJ, Merhof D, Thiberge S, Strauch M, McBride CS. 2020. Chemical signatures of human odour generate a unique neural code in the brain of *Aedes aegypti* mosquitoes. bioRxiv doi:10.1101/2020.11.01.363861

Zhao Z, Tian D, McBride CS. 2021. Development of a pan-neuronal genetic driver in *Aedes aegypti* mosquitoes. *Cell Rep Methods* **1:** 100042. doi:10.1016/j.crmeth.2021.100042

Cite this introduction as *Cold Spring Harb Protoc*; doi:10.1101/pdb.top107683

Protocol 1

Calcium Imaging of *Anopheles coluzzii* Mosquito Antennae Expressing the Calcium Indicator GCaMP6f

Ali Afify[1] and Christopher J. Potter[1,2]

[1]*The Solomon H. Snyder Department of Neuroscience, Johns Hopkins University School of Medicine, Baltimore, Maryland 21205, USA*

Calcium imaging is a technique used to measure functional neuronal activities in response to stimuli. It has been used for years to study odorant-induced responses in insects (i.e., honeybees, *Drosophila*, and moths) and was recently introduced into mosquitoes. Traditionally, calcium imaging in mosquitoes was performed using nonspecific calcium indicator dyes to examine neuronal responses in whole insect brain regions, but the development of genetically encoded calcium indicators (GECIs) has facilitated the ability to perform functional calcium imaging on specific tissues. For example, by specifically expressing a GECI in olfactory neurons, the odor-induced responses of these neurons in peripheral organs can be examined. Calcium imaging of mosquito antennae further provides an advantageous method for simultaneously visualizing the activity of several antennal neurons in a single experiment. In this protocol, we describe a calcium imaging method to study odor-evoked responses in *Anopheles coluzzii* antennae expressing the calcium indicator GCaMP6f. This method requires imaging equipment (compound microscope, light sources, and camera), an odorant delivery system, and image acquisition software. The mosquito preparation is straightforward but requires practice to minimize mosquito movement during imaging. Recorded videos can be analyzed using Fiji software to generate heatmaps and activity traces for odorant-evoked responses. This protocol can also be used, with some modifications, to study other peripheral organs (such as labella, palps, and tarsi).

MATERIALS

It is essential that you consult the appropriate Material Safety Data Sheets and your institution's Environmental Health and Safety Office for proper handling of equipment and hazardous materials used in this protocol.

Reagents

Anopheles coluzzii mosquitoes (3- to 10-d-old non-blood-fed females) that express GCaMP6f in Orco$^+$ neurons

We generated these mosquitoes by crossing AgOrco-QF2 (BEI Resources MRA-1300) and QUAS-GCaMP6f mosquitoes (available upon request to C. Potter) (Afify et al. 2019).

This protocol can also be adapted to other species of mosquitoes in which olfactory neurons are similarly labeled by a calcium sensor. However, it is possible that a darker cuticle, such as that typically found in Aedes *mosquito antennae, might interfere with calcium imaging.*

Pure odorant

In this example, we use 1-octen-3-ol (SAFC W280518).

[2]Correspondence: cpotter@jhmi.edu

Cite this protocol as *Cold Spring Harb Protoc*; doi:10.1101/pdb.prot107918

Solvent for odorant

Choose a solvent that has a weak or no odor and that completely dissolves the test odorants. Mineral and paraffin oils are good solvents, but odorants in solid forms might not dissolve in them and require a solvent like ethanol. In this example, we use paraffin oil (Sigma-Aldrich 18512).

Equipment

Calcium imaging setup

10× objective (Zeiss EC Epiplan-Neofluar 10x/0.25)

This objective is used to image the whole mosquito antenna.

50× objective (LD EC Epiplan-Neofluar 50×/0.55 DIC)

This objective is used to image single segments of the mosquito antenna.

Air isolator table (TMC Gimbal Piston)

Compound microscope (Zeiss Axio Examiner D1)

EMCCD camera (Andor iXon Ultra)

To capture movies/images, we use this camera mounted on top of the microscope and connected to a computer with image acquisition software.

Fluorescence light source (Zeiss Illuminator HXP 200C) with eGFP filter cube (FL Filter Set 38 HE GFP shift free)

NIS Elements Advanced Research software (Nikon instruments) or Andor Solis software

This software is used for image acquisition.

Stimulus controller (Syntech CS-55)

A calcium imaging setup consists of a compound microscope with bright-field and fluorescence light sources, a camera with image acquisition software, and an odorant delivery system. The microscope should be mounted on an air table to reduce movements during imaging. There are many products on the market that offer different solutions and can be chosen based on the type of planned experiments. In this protocol, we used the products listed above.

Conical tubes (50-mL)

Coverslips

Dissecting microscope (Zeiss Stemi 2000)

Dremel rotary tool (Dremel 8000-03)

Fiji software with the Image Stabilizer and HeatMap Histogram plug-ins (Schindelin et al. 2012)

Filter paper (Whatman 1001090)

Forceps (Dumont #3)

Handheld insect aspirator (John W. Hock Company PN 419)

Micromanipulator (Kite Manual Micromanipulator, World Precision Instrument)

Microscope slide

Modeling clay (Crayola)

Pasteur pipettes (14.6-cm; Fisher Scientific 13-678-6A)

Plastic serological pipette (10-mL; Denville Scientific Inc)

Pulled glass capillary tubes (1-mm × 0.5-mm × 100-mm [OD×ID×L]; Harvard Apparatus)

The exact size of the pulled glass capillaries is not crucial. They can be pulled by a filament- or laser-based micropipette puller.

PVC tubing, clear (4.8-mm inner diameter, 15-cm-long) (two; see Step 2)

PVC tubing, clear (4.8-mm inner diameter, ~1.5-m-long)

PVC tubing, clear (8-mm inner diameter, ~2-m-long)

Razor blade

Y-shaped connection (e.g., Wye Connectors, Tube, or McMaster-Carr)

Cite this protocol as *Cold Spring Harb Protoc*; doi:10.1101/pdb.prot107918

METHOD

Setup of Calcium Imaging Equipment

1. Program the A and B output ports on the Syntech stimulus controller to deliver equal amounts of air. Connect a long clear PVC tubing (4.8-mm inner diameter, ∼1.5-m-long) to the A port (Fig. 1A). Next, connect the foot pedal (included with the stimulus controller) to the "Pedal Switch" port on stimulus controller and place that pedal under the air table.

   During imaging, the pedal will be used to divert air between the A and B ports (see the Imaging section below).

2. Connect two short (15-cm-long) clear pieces of PVC tubing (4.8-mm inner diameter) to the B and C output ports on the Syntech stimulus controller. Use a Y-shaped connection to connect the two short lengths of tubing to one long piece of tubing (8-mm inner diameter, ∼2-m-long). Insert a 10-mL plastic serological pipette at the end of the long tubing (Fig. 1A).
3. Make a small hole (3-mm diameter) using a Dremel rotary tool into the long plastic pipette 20 cm from the tip. Mount the pipette on a micromanipulator so it can be directed at the mosquito antenna.

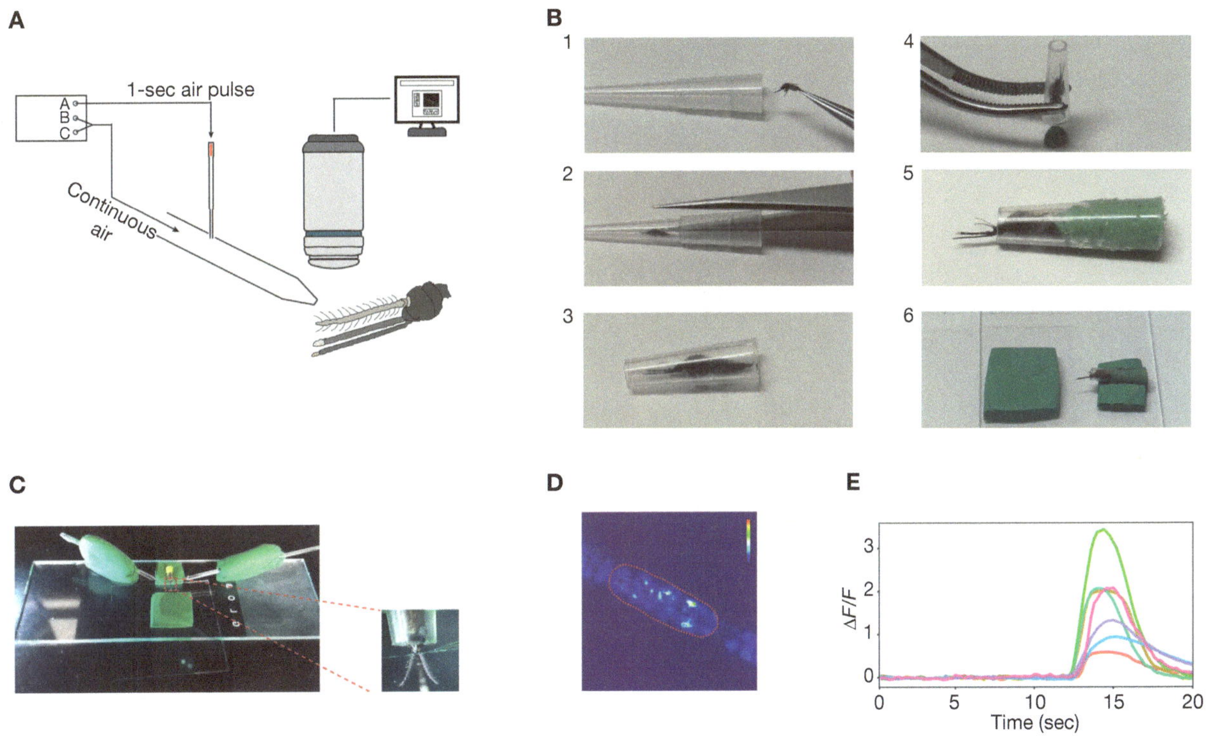

FIGURE 1. Calcium imaging of antennal olfactory neurons in *Anopheles coluzzii* mosquitoes. (*A*) Schematic of the imaging setup. Arrows indicate the direction of airflow. A 1-sec air pulse leaves the A port of the stimulus controller (rectangle), passes through the Pasteur pipette that contains the odorant, and ends in the long pipette directed at the mosquito antenna. Continuous air starts from the B and C ports of the stimulus controller and ends in the long pipette. (*B*) Mosquito preparation for imaging. (*1*) The mosquito is inserted into the pipette tip using forceps. (*2*) Forceps are used to push the mosquito thorax and move the mosquito forward. (*3*) The pipette tip is cut from the front and back around the mosquito. (*4*) A small piece of clay is used to push the mosquito forward. (*5*) The mosquito is trapped in the pipette tip with only the antenna stretching out. (*6*) Two pieces of clay are used to attach the pipette tip and a coverslip to the slide. (*C*) The mosquito is ready for imaging. The mosquito is trapped in a pipette tip with only the antennae extended outside the pipette tip. Two pulled glass capillaries are used to hold the antenna down on a coverslip. (*D*) Example heatmap of the responses toward 1% of 1-octen-3-ol (a human odorant). Dashed red line indicates the borders of the 11th antennal segment where imaging was performed. (*E*) Traces from the calcium imaging recording in *C*. Colors represent different ROIs (neurons), and the *y*-axis shows Δ*F*/*F* values (odor-evoked change in fluorescence from baseline divided by baseline fluorescence).

During imaging, this long pipette will deliver charcoal-filtered continuous air from the stimulus controller to the antenna, and the small hole will be used to inject an airstream containing odorants into the continuous airstream (Fig. 1A; see the Imaging section below).

The stimulus controller can be adjusted to deliver air at a desired airflow. We use 8.3 mL/sec.

Odorant Preparation

4. Prepare odorant stocks by diluting pure odorants in a solvent.

   In this example, we used 1-octen-3-ol, a human skin odorant, diluted to a 1% concentration in paraffin oil.

   The solvent will also be used as the control odorant.

   Diluted odorants should not be kept for extended periods (e.g., >3 mo) because they change over time (degrade or are oxidized). These stocks should be kept in tightly closed and dark containers at the same temperature as recommended for the pure odorant, away from direct sun light, and discarded/reprepared on a regular basis.

5. On the day of the experiment, place a piece of filter paper (1-cm × 2-cm) in a glass Pasteur pipette and then pipette the desired amount of diluted odorant (~20 µL) onto the filter paper. Cover the wide end of the Pasteur pipette with a 1000-µL pipette tip. Prepare pipettes with solvent only (e.g., paraffin oil or mineral oil) as a control.

   Pasteur pipettes containing odorants should be prepared fresh every day, preferably before preparing mosquitoes for imaging and discarded at the end of the experiment or after approximately five odorant deliveries. More volatile odorants should be reprepared more frequently.

Mosquito Preparation

6. Use a handheld aspirator to insert a single mosquito into a 50-mL conical tube. Place the tube on ice for ~1 min to immobilize the mosquito.
7. Under a dissecting microscope, use forceps to carefully hold the mosquito from the tip of its abdomen and insert the mosquito into a 200-µL pipette tip (Fig. 1B1), with the mosquito head facing the tip of the pipette tip. Use forceps to push the mosquito thorax and move the mosquito toward the tip of the pipette tip (Fig. 1B2), while making sure the two antennae are always extended forward, until the mosquito is halfway through the pipette tip. Then, use a sharp razor to cut the front of the pipette tip so the front opening is around the same diameter as the mosquito head (~1-mm in diameter, Fig. 1B3).

   When pushing the mosquito forward, the antennae might bend to the back of the pipette tip. This will lead the fragile antennae to be stuck between the mosquito thorax and the wall of the pipette tip. This can be resolved by pulling the mosquito back and pushing it forward again. Alternatively, carefully insert one branch of the fine forceps from behind the mosquito to push the antennae forward until it is freed from the thorax.

8. Push the mosquito again using forceps, while making sure the antennae are always extended forward, until the antennae reach the front opening of the pipette tip. Cut the back of the pipette tip right behind the mosquito abdomen (Fig. 1B3) and use small pieces of modeling clay to push the mosquito forward until only the antennae are extended outside the pipette tip (Fig. 1B4,5).
9. Attach a small piece of modeling clay to a glass slide and press the pipette tip down into the clay (Fig. 1B6). Use another piece of modeling clay to attach a glass coverslip to the slide in front of the pipette tip, in front of the mosquito antennae (Fig. 1B6). Now, use a finger to push the coverslip toward the pipette tip and place the antennae on the coverslip (Fig. 1C).
10. Put a piece of modeling clay on the slide on each side of the pipette tip, and mount two pulled glass capillaries on the two pieces of clay. Place the pulled tip of one of the capillaries by hand on the third to fourth segment of one antenna to press it down on the coverslip. Place the other capillary on the 12th to 13th antennal segment (the two most distal segments) and make sure the antenna is flattened on the slide (Fig. 1C).

Cite this protocol as *Cold Spring Harb Protoc*; doi:10.1101/pdb.prot107918

Flattening the antenna on the coverslip is the most crucial step in mosquito preparation and requires practice. The fragile antenna can be easily damaged by the capillaries. However, failing to adequately flatten the antenna allows it to move while imaging; this will also cause the antenna to slightly bend which will require imaging at different Z planes.

Now the mosquito is ready for imaging.

Imaging

11. Place the prepared mosquito on the stage of the compound microscope. Turn on the bright-field light source, the fluorescence light source, the stimulus controller, and the camera.
12. Start the image acquisition software, which has many settings to control the length, quality, and format of the output videos. We record 20-sec videos, at a resolution of 512 × 512 pixels, and an exposure time of 200 msec (5 Hz).

    Although the length of videos can be freely adjusted, a lower resolution or slower frame rate could make the analysis of the imaging videos challenging and potentially miss important data.

    The image acquisition software will give an error message if it is started before the camera is turned on.
13. Check the antenna through the microscope eyepieces at low (10×) and high (50×) magnification to make sure it is flattened on the coverslip. Locate the shutter on the microscope, which allows switching between viewing through the eyepieces and viewing with the image acquisition software on the computer screen. Open this shutter to view live images on the image acquisition software.
14. Use fluorescent light to verify that antennal neurons are visible.

    For viewing GCaMP6-labeled neurons, use a GFP filter cube (stimulate with 470-nm light, visualize the 525-nm emitted light). If the antenna is not adequately flattened, only a few neurons will be visible in a single plane under fluorescent light.

    Close the shutter on the fluorescence light source when not using the fluorescent light. This will protect neurons from continuous exposure to fluorescent light and prevent photobleaching of the GCaMP fluorescent protein.
15. Use the micromanipulator to direct the long pipette (the one connected to the B and C ports of the stimulus controller) at the mosquito antenna (~1 cm away from the antenna).

    This will deliver a continuous charcoal-filtered airstream to the antenna (Fig. 1A).
16. To deliver odorants to the antenna, connect the back of the Pasteur pipette that contains odorant-soaked filter paper to the long tubing that is connected to the A port of the stimulus controller. Then, insert the tip of the Pasteur pipette into the hole in the long pipette (Fig. 1A).
17. Start recording the video on the software. Allow a few seconds of recording before puffing the odorant.

    We deliver the odorants on the 10th sec of a 20-sec video. The 9-sec segment at the beginning of each video serves as the "prestimulus" and represents the neurons' baseline activity.
18. To deliver the odorant (on the 10th sec), press the foot pedal connected to the stimulus controller.

    This will divert a 1-sec pulse of air into the Pasteur pipette and deliver the odorant from the filter paper into the continuous airstream, reaching the mosquito antenna at the end.

    Program the stimulus controller to deliver equal amounts of air through the A and B ports at 5 mL/sec. Airstreams from the B and C ports merge at the Y-shaped part of the hose to deliver 8.3-mL/sec continuous airstream to the antenna. When the foot pedal is pressed, the stimulus controller will stop air from the B port and will deliver air through the A port for 1 sec. This will ensure that a continuous undisturbed airstream reaches the mosquito antenna at all times. To identify potential issues caused by odor-induced changes to neuronal activity, randomize the stimulus order of odorants and controls.
19. Save the video in a suitable format.

    In Solis software, save videos in AVI format using the "Export As" function. This function uses different codecs to compress videos. Use the "raw video" codec which allows the videos to be imported by Fiji for data analysis. The "raw" (uncompressed format) is also recommended to avoid potential compression artifacts being introduced into the images.

Analysis of Calcium Imaging Recordings

There are different ways to extract data from calcium imaging videos. We use Fiji (version 1.53c), an open-source software, to produce two types of data outputs: (1) antennal heatmaps—these heatmaps assign colors to different parts of the mosquito antenna based on their response level (Fig. 1D) and (2) intensity time traces—these are the mean intensity values of different regions of interests (ROIs) on the antenna across time (Fig. 1E).

Antennal Heatmaps

20. The first time Fiji is needed, install the latest version of Fiji and the Image Stabilizer plug-in into Fiji (https://imagej.net/plugins/image-stabilizer) (Li and Kang 2008).
21. Import the calcium imaging video "x.avi" in Fiji. File → Open, then choose the video to be imported. Be sure to uncheck the "Use Virtual Stack" box.

    Note that the imported video will show some information like the number of slices (frames) and duration of the video. The number of frames will depend on the recording speed and the duration of the video. If the recording was at 5 Hz and for 20 sec, it will be a 100-frame video.
22. Use the Image Stabilizer plug-in to correct for any movements that occurred during recording. Go to Plugins → Image Stabilizer. Check the "Output to a New Stack" box. Be aware that this will produce a new corrected or stabilized video, which will be automatically named "Stabilized x," while keeping the original video "x." Close the original video file to avoid any confusion.
23. Produce the ΔF image, which shows the change in fluorescence intensity after odorant delivery. To do this, first choose the frame with the maximum fluorescence intensity "F_i" by going through the stabilized video, frame by frame, and manually choosing the frame that shows the maximum intensity after odorant delivery. Try to choose the frame that contains the brightest spots; sometimes multiple olfactory neurons reach their maximum responses at different times. In this case, use the frame in which the majority of the neurons have reached their maximum responses. Next, click Image → Stacks → Tools → Slice Keeper and write the number that corresponds to the "F_i" frame in the "First Slice" and in the "Last Slice." Input "1" in the "Increment" window, which will extract the "F_i" image from the stabilized video and automatically name it "Stabilized x kept stack."
24. Use the stabilized video to produce an F_0 image—that is, the image with the mean baseline fluorescence ("prestimulus" fluorescence), which will be representative of the 9 sec before odorant delivery (45 frames). To do this, click Image → Stacks → Z project, and write 1 in the "Start slice," 45 in the "Stop slice," and choose Average Intensity from the "Projection type" drop-down menu.

    The output image will be automatically named "AVG_ Stabilized x."
25. Produce the ΔF image by subtracting the mean baseline fluorescence F_0 from the image of maximum fluorescence F_i. To do this, click Process → Image calculator, and choose "Stabilized x kept stack" in the "Image1" drop-down menu, "subtract" in the "Operation" drop-down menu, and "AVG_ Stabilized x" in the "Image2" drop-down menu. Check the "Create new window" box to produce the ΔF image in a new window automatically named "Result of Stabilized x kept stack."
26. Produce a heatmap from the ΔF image using a plug-in called "HeatMap Histogram." Install the plug-in into Fiji (https://www.researchgate.net/publication/322420028_HeatMap-Histogram_from_Samuel_Pean) (Pean 2018), then click on the ΔF image (Result of Stabilized x kept stack) to select it. Click Plugins → HeatMap Histogram to produce two identical heatmap images, one of which will contain a calibration bar for the color code of the heatmap (Fig. 1D). Save those heatmaps using the "Save As" function from the "File" drop-down menu. Close all windows except the "Stabilized x" video, which will be used to produce intensity time traces.

    Available methods for generating heatmaps differ in the colors they assign to different intensity levels.

 Cite this protocol as *Cold Spring Harb Protoc*; doi:10.1101/pdb.prot107918

Intensity Time Traces

27. Click Analyze → Tools → ROI Manager, which will open the ROI Manager window that can be used to draw ROIs.
28. Select the "Oval" selection tool from the main Fiji window and use it to draw circles or oval outlines around the regions of interest (responding neurons) in the "Stabilized x" video. To do this, go through the video, frame by frame, and draw outlines around the neurons that show increased fluorescence after odorant delivery. The heatmap can be used as a reference to choose the responding neurons.
29. After drawing an outline around each neuron/ROI, click "Add [t]" in the ROI manager window to save it. Otherwise, it will be lost if a new ROI is drawn before saving.

    Check the "Show All" box in the ROI manager to show the ROIs on the video. If the "Labels" box is checked, it will show the number assigned to each ROI in the middle of that ROI.

30. After drawing all ROIs, click "More" in the ROI manager → Multi Measure, and check the "Measure all 100 slices" and the "One row per slice" boxes. This will produce a "Results" window that contains one column numbered 1–100 (the number of each frame), and one column for each ROI (Mean1, Mean2, Mean3, ...). Each of the ROI columns will contain mean intensity values from all pixels in that ROI across time (all frames).

    Before clicking "Multi Measure," make sure not to select any particular ROI in the video. Otherwise, the produced "Results" window will contain intensity values for only that particular ROI.

31. Click "File" in the "Results" window → Save as, to save the intensity values as an Excel spreadsheet.
32. Finally, calculate $\Delta F/F$ values in an Excel spreadsheet using the formula

$$\Delta F/F = F_i - F_0/F_0,$$

where F_i is the fluorescence intensity value at frame i (the frame with maximum intensity) and F_0 is the mean fluorescence intensity before odorant delivery (first 9 sec, 45 frames).

$\Delta F/F$ values can be used to plot time traces (Fig. 1E) in R (R Core Team 2018). Odorant responses or further analysis can be compared using R or other statistical analysis software.

REFERENCES

Afify A, Betz JF, Riabinina O, Lahondère C, Potter CJ. 2019. Commonly used insect repellents hide human odors from *Anopheles* mosquitoes. *Curr Biol* **29:** 3669–3680.e3665. doi:10.1016/j.cub.2019.09.007

Li K, Kang S. 2008. The image stabilizer plugin for ImageJ. Available at https://imagejnet/plugins/image-stabilizer (Accessed on 2/9/2022).

Pean S. 2018. HeatMap-Histogram from Samuel Pean. Available at https://wwwresearchgatenet/publication/322420028_HeatMap-Histogram_from_Samuel_Pean (Accessed on 2/9/2022).

R Core Team. 2018. *R: a language and environment for statistical computing.* R Foundation for Statistical Computing, Vienna.

Schindelin J, Arganda-Carreras I, Frise E, Kaynig V, Longair M, Pietzsch T, Preibisch S, Rueden C, Saalfeld S, Schmid B, et al. 2012. Fiji: an open-source platform for biological-image analysis. *Nat Methods* **9:** 676. doi:10.1038/nmeth.2019

CHAPTER 30

Acoustic Physiology in Mosquitoes

Matthew P. Su,[1,2] Marta Andrés,[3,4] Marcos Georgiades,[3,4] Judit Bagi,[3,4] and Joerg T. Albert[3,4,5]

[1]*Graduate School of Science,* [2]*Institute for Advanced Research, Nagoya University, Nagoya 464-8602, Japan;* [3]*Ear Institute, University College London, London WC1X 8EE, United Kingdom;* [4]*The Francis Crick Institute, London NW1 1AT, United Kingdom*

The acoustic physiology of mosquitoes is perhaps the most complex within the entire insect class. Past research has uncovered several of its—sometimes stunningly unconventional—principles, but many mysteries remain. Their solution necessitates a concerted transdisciplinary effort to successfully link the neuroanatomical and biophysical properties of mosquito flagellar ears to the behavioral ecology of entire mosquito populations. Neuroanatomically, mosquito ears can rival those of humans in both complexity and sheer size. The approximately 16,000 auditory hair cells within the human organ of Corti, for example, are matched by the approximately 16,000 auditory neurons in the Johnston's organ of a male *Anopheles* mosquito. Both human and mosquito ears receive very extensive efferent innervation, which modulates their function in ways that are as yet poorly understood. Different populations of neuronal and nonneuronal cell types divide the labor of the mosquito ear amongst themselves. Yet, what exactly this labor is, and how it is achieved, is at best vaguely known. For the majority of mosquitoes, biologically relevant sounds are inextricably linked to their flight tones. Either these flight tones are (directly) the sounds of interest or they contribute (indirectly) to the production of audible sound through a process called nonlinear distortion. Finally, male ears can generate tones themselves: The generation of an internal "phantom copy" of a female flight tone (or *self-sustained oscillation*) is believed to aid the male hearing process. Here, we introduce protocols that target the mosquitoes' auditory neuroanatomy, electrophysiology, and behavior to help shed light on some of these issues.

HEARING IS A KEY SENSE FOR MOSQUITOES

The sense of hearing is key to mosquito survival; without it, the mid-air copulation between male and female could not take place (Fig. 1A). Males identify—and locate—a female mating partner by listening out for her faint flight tones (Fig. 1B). Within the (acoustically and spatially) crowded environment of mosquito mating swarms (Downes 1969), where hundreds or thousands of males often compete for a few dozen females, this is a challenging sensory task. For a mosquito male, *reproductive fitness* might thus be approximated as *acoustic fitness*; the (positive) selection for genes that improve hearing, as well as the (negative) selection against genes that cause deafness, can be expected to have shaped the evolution of male mosquitoes.

Despite their crucial importance (without their sense of hearing, males could not execute the phonotactic chase of a female mating partner) and seeming simplicity (the only relevant sounds are the pure tone-like "flight tones"), the mechanisms of mosquito hearing have retained many of their secrets to this day. The assumption of simplicity has proven deceptive.

[5]Correspondence: joerg.albert@ucl.ac.uk

Cite this introduction as *Cold Spring Harb Protoc;* doi:10.1101/pdb.top107685

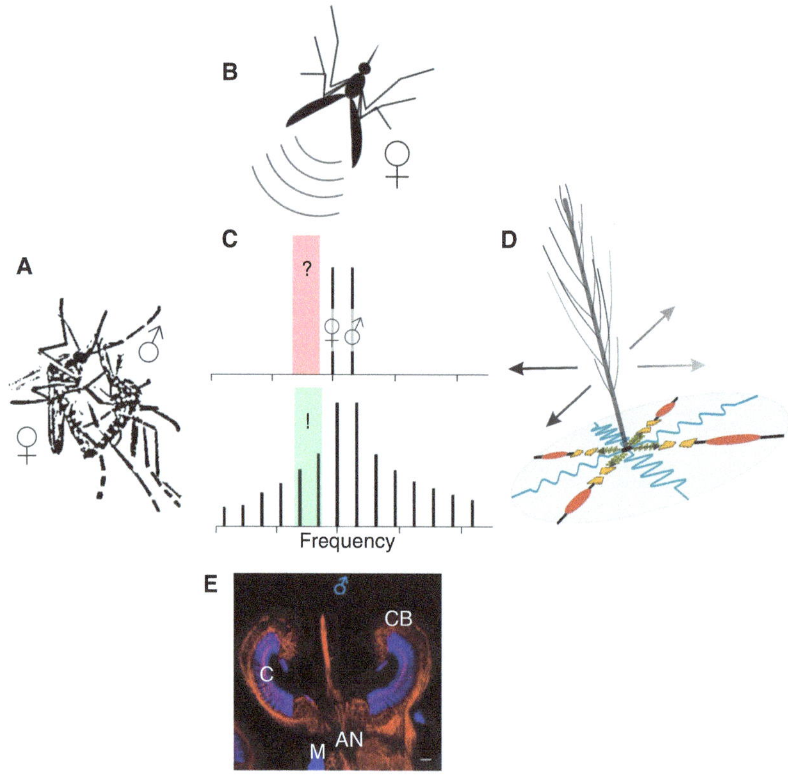

FIGURE 1. The acoustic ecology of mosquitoes. (*A*) Mosquitoes copulate mid-flight, following a precopulatory pursuit (often taking place in large mating swarms) during which the male chases the female. (*B*) Males use the females' flight tones to identify females within the swarm and to track them during the phonotactic chase. (*C*) (*Top*) The mosquitoes' flight tones (bars) are nearly inaudible to both sexes, as they fall outside the mosquitoes' hearing range (red highlighted area); (*bottom*) when male and female flight tones mix nonlinearly within the mosquitoes' ears, however, they produce audible distortion tones (additional bars), which fall partly within the mosquito hearing range (green highlighted area). (*D*) Audibility thus arises from an interaction between (male and female) flight tones and the mosquito flagellar ear (see schematic drawing). Understanding the biophysical operation of the mosquito flagellar ear is key to understanding mosquito hearing and auditory behavior. (*E*) Within the Johnston's organ (JO), different populations of ciliated neurons (neurons red, cilia blue), and nonneuronal cell types, coalesce to mediate, and modulate, the process of mosquito hearing. The antennal nerve carries both afferent and efferent nerve fibers, which transmit, or modulate, sensory signals, respectively. (CB) Cell body, (C) cilia, (AN) antennal nerve, (M) muscle.

A BRIEF BREAKDOWN OF MOSQUITO HEARING

On the receiver side, the mosquitoes' flagellar ears convert airborne vibrations (such as those associated with a flight tone) into neuro-electrical signals, which are then propagated to the mosquito brain. The flagellar shaft is the actual sound receiver: it acts like an inverted pendulum that is set into motion by friction with the surrounding air (Albert and Kozlov 2016). The greater the velocity of the air particles, the larger the flagellar oscillation. The flagellum's plane of oscillation depends on the direction of the velocity vector, and thus the location of the sound source. To boost sound-induced motion, the flagellar receivers of male mosquitoes possess arrays of fibrillar extensions (or "fibrillae"), which increase the flagellum's effective acoustic surface. In Anopheline mosquitoes these fibrillae canalso be collapsed. Mosquito ears have an exquisite sensitivity: In males, angular displacements of <1 mdeg (corresponding to flagellar tip displacements of <20 nm) (Su et al. 2018) are enough to excite the auditory neurons of the Johnston's organ (JO), the mosquitoes' "inner ear."

Mosquitoes, especially males, should thus be well-adapted to detect the faint sounds produced by the wingbeats of a nearby flying mate. However, testing the responses of their ears to naturally occurring flight tone frequencies tells a different story. Mosquito flight tones change with ambient

 Cite this introduction as *Cold Spring Harb Protoc*; doi:10.1101/pdb.top107685

temperatures (Villarreal et al. 2017). Under laboratory conditions (20°C–22°C), females emit flight tones of ~400 Hz and males emit flight tones of ~600 Hz. In both sexes, however, these tones fail to elicit responses in the auditory neurons of JO. Instead, the male JO shows a response plateau centered on frequencies of ~200 Hz (ranging from ~150 Hz to ~250 Hz; see Somers 2022 and highlighted areas in Fig. 1C). No mosquito sound emissions in this frequency range have been reported. The solution to this puzzling mismatch is provided by an essential property of mosquito (and other) ears. As a direct consequence of their mode of operation, sound receivers display characteristic nonlinearities (e.g., gating compliances). The receivers' nonlinearities, in turn, introduce distortions into their response to external sound. Even when stimulated with two pure tones only (such as the mosquitoes' flight tones; see Fig. 1C, top), the receiver produces mathematically predictable distortion products (DPs) (Julicher et al. 2001), some of which will fall into the sensitivity range of the mosquitoes' auditory nerve (Fig. 1C, bottom). The nonlinear mixing of two—in themselves inaudible—flight tones thus can create an audible distortion tone within the mosquito ear (Gibson et al. 2010; Somers et al. 2022). DPs are, however, at least two orders of magnitude smaller than the original flight tones, and thus the exquisite displacement sensitivity of the flagellar ear is crucial to their detection.

AUDIBILITY IS A COMPLEX, MULTIVARIATE PROBLEM IN MOSQUITOES

Two likewise immediate, and far-reaching, conclusions can be drawn from the relations sketched out above: (i) The audibility of females is a key prerequisite for the males' precopulatory chase and thus for reproduction in mosquitoes in general; and (ii) audibility between mosquitoes is a nontrivial, interrelational problem, as it does not simply arise from a match between an emitted signal (e.g., a female flight tone) and a receiver's properties (e.g., a male ear). Uniquely among all other reported hearing systems, audibility in mosquitoes also depends on sounds generated (and possibly controlled) by the receiver (Fig. 1D): For a male to be able to hear a given female, he must himself fly with the correct wingbeat frequency (and thus the correct flight tone), which generates audible distortions when mixing nonlinearly with the female's flight tone in his own ear. Studies have started to shed light on these issues, exploring the mutual relation, and possible optimization, of male and female flight tones within the context of the mating swarm. However, many known (and most certainly many unknown) unknowns remain. The biggest mystery here is arguably the role of the self-sustained oscillations (SSOs) specific to male ears. SSOs represent self-generated flagellar oscillations, which closely mimic those induced by female flight tones. It has been speculated that SSOs act as internal amplifiers of female flight tones, but their specific contributions and ecological relevance—and cellular origin (Fig. 1E)—are still unclear.

FROM BIOPHYSICS TO VECTOR CONTROL

From what we do know, direct lines can be drawn from the in-depth study of mosquito auditory biology to vector control efforts. First, hearing is a key sensory modality that guides male reproductive behavior within the mating swarm. The mating swarm has been identified as a prime target for future vector control efforts (Sawadogo et al. 2017), and current programs aiming at the release of genetically modified mosquitoes have almost exclusively concentrated on males (as the nonbiting sex). A better quantitative understanding of mosquito hearing—and audibility—thus carries the dual promise of (i) providing novel molecular targets and mechanistic routes for mosquito population control efforts and —perhaps even more importantly—(ii) supporting ongoing release programs through improved predictive models of expected reproductive interactions (and compatibility) between populations of genetically modified and native mosquitoes (Su et al. 2020). One simple way that this could be achieved is to "acoustically match" mutant and wild-type mosquitoes in order to optimize their respective mating interactions. This discussion and the three associated protocols have been written in an effort to promote auditory research in mosquitoes on the cellular (see Protocol 1: Immunohistochemical Staining of the Mosquito Ear [Andrés and Albert 2022]), biophysical (see Protocol 2: Electrophysiological Measurements of Compound Action Potential Responses from the Antennal Nerve in Response to Stimulation [Su and Albert 2022], and behavioral (see Protocol 3: Recording and Extraction of Mosquito Flight Tones [Georgiades and Albert 2022a], and Protocol 4: Mosquito Phonotaxis Assay [Georgiades and Albert 2022b]) levels.

ACKNOWLEDGMENTS

Our work benefited from funding through the European Research Council (ERC) under the Horizon research and innovation programme (grant agreement no. 648709, to J.T.A.), a pump-priming award from the BBSRC Vector Borne Disease (VBD) Network ANTI-VeC (AV/PP/0028/1, to J.T.A., Sarah Moore, and M.P.S.), a BBSRC project grant (BB/V007866/1 to J.T.A.), and a UCL Global Challenges Research Fund (GCRF) small grant (to J.T.A.).

REFERENCES

Albert JT, Kozlov AS. 2016. Comparative aspects of hearing in vertebrates and insects with antennal ears. *Curr Biol* **26:** R1050–R1061. doi:10.1016/j.cub.2016.09.017

Andrés M, Albert JT. 2022. Immunohistochemical staining of the mosquito ear. *Cold Spring Harb Protoc* doi:10.1101/pdb.prot108008

Downes JA. 1969. Swarming and mating flight of diptera. *Annu Rev Entomol* **14:** 271–298. doi:10.1146/annurev.en.14.010169.001415

Georgiades M, Albert J. 2022a. Recording and extraction of mosquito flight tones. *Cold Spring Harb Protoc* doi:10.1101/pdb.prot108011

Georgiades M, Albert J. 2022b. Mosquito phonotaxis assay. *Cold Spring Harb Protoc* doi:10.1101/pdb.prot108012

Gibson G, Warren B, Russell IJ. 2010. Humming in tune: sex and species recognition by mosquitoes on the wing. *J Assoc Res Otolaryngol* **11:** 527–540. doi:10.1007/s10162-010-0243-2

Julicher F, Andor D, Duke T. 2001. Physical basis of two-tone interference in hearing. *Proc Natl Acad Sci* **98:** 9080–9085. doi:10.1073/pnas.151257898

Sawadogo SP, Niang A, Bilgo E, Millogo A, Maiga H, Dabire RK, Tripet F, Diabate A. 2017. Targeting male mosquito swarms to control malaria vector density. *PLos ONE* **12:** e0173273. doi:10.1371/journal.pone.0173273

Somers J, Georgiades M, Su MP, Bagi J, Andrés M, Mills G, Ntabaliba W, Moore SJ, Spaccapelo R, Albert JT. 2022. Hitting the right note at the right time: circadian control of audibility in *Anopheles* mosquito mating swarms is mediated by flight tones. *Sci Adv* **8:** eabl4844. doi:10.1126/sciadv.abl4844

Su MP, Albert J. 2022. Electrophysiological measurements of compound action potential responses from the antennal nerve in response to stimulation. *Cold Spring Harb Protoc* doi:10.1101/pdb.prot108010

Su MP, Andrés M, Boyd-Gibbins N, Somers J, Albert JT. 2018. Sex and species specific hearing mechanisms in mosquito flagellar ears. *Nat Commun* **9:** 3911. doi:10.1038/s41467-018-06388-7

Su MP, Georgiades M, Bagi J, Kyrou K, Crisanti A, Albert JT. 2020. Assessing the acoustic behaviour of *Anopheles gambiae* (s.l.) dsxF mutants: implications for vector control. *Parasite Vector* **13:** 507. doi:10.1186/s13071-020-04382-x

Villarreal SM, Winokur O, Harrington L. 2017. The impact of temperature and body size on fundamental flight tone variation in the mosquito vector *Aedes aegypti* (Diptera: Culicidae): implications for acoustic lures. *J Med Entomol* **54:** 1116–1121. doi:10.1093/jme/tjx079

Cite this introduction as *Cold Spring Harb Protoc*; doi:10.1101/pdb.top107685

Protocol 1

Immunohistochemical Staining of the Mosquito Ear

Marta Andrés,[1] Judit Bagi, and Joerg T. Albert

Ear Institute, University College London, London WC1X 8EE, United Kingdom; The Francis Crick Institute, London NW1 1AT, United Kingdom

Immunohistochemistry has played a major role in improving our understanding of the anatomy and function of the nervous system. The use of fluorescent dyes that label different antigens reveals how biological tissues are built and how interactions between cells take place. Obtaining this information is particularly important in the case of the mosquito ear given its highly complex anatomy. This protocol describes an immunohistochemical technique to stain the mosquito ear. The first steps of the procedure include the embedding of the tissue in albumin-gelatin and its sectioning into thin slices to allow antibody penetration. The immunohistochemical procedure can be exploited to detect protein expression and localization by using antibodies specifically raised against the protein of interest or that recognize epitope tags fused to proteins using genome editing methods.

MATERIALS

It is essential that you consult the appropriate Material Safety Data Sheets and your institution's Environmental Health and Safety Office for proper handling of equipment and hazardous materials used in this protocol.

RECIPES: Please see the end of this protocol for recipes indicated by <R>. Additional recipes can be found online at http://cshprotocols.cshlp.org/site/recipes.

Reagents

Albumin-gelatin mounting medium <R>

Antibodies

The selection of antibodies to be used depends on the proteins to be detected. Several primary antibodies can be used together if they are raised in different animals and can be recognized by secondary antibodies conjugated to different fluorochromes. It is important to plan the combination of primary and secondary antibodies in advance to ensure that they work together.

Commonly used antibodies to study the mosquito ear structure are the primary antibody anti-HRP (1:500; Sigma-Aldrich) that stains the sensory neurons and the secondary antibody Alexa Fluor Phalloidin 633 (1:50; ThermoFisher) that binds to actin and labels the actin-based rods in scolopale cells that surround the auditory cilia. As secondary antibodies, we usually use Alexa Flour from ThermoFisher at 1:500 dilution.

Blocking solution (5% bovine albumin serum, 2% normal goat serum, and 0.3% Triton X-100 in PBS)

The solution is aliquoted into 1-mL tubes and stored at −20°C until used.

We generally use a blocking solution containing bovine and goat serum. However, be aware that this will prevent the use of any anti-goat secondary antibody. If it is necessary to use an anti-goat secondary antibody, adapt the blocking solution by omitting the goat serum.

---

[1]Correspondence: marta.andres@ucl.ac.uk

Cite this protocol as *Cold Spring Harb Protoc*; doi:10.1101/pdb.prot108008

Formaldehyde, 4% in PBS with 1% Triton X-100

The solution is divided into aliquots in 1-mL tubes and stored at −20°C until used.

Formaldehyde, 6% in PBS

The solution can be kept in the refrigerator for 1–2 mo. To post-fix the albumin-gelatin blocks, the solution will be poured into a 50-mL Falcon tube where blocks will be placed. The solution can be reused three to four times.

Methanol

Mosquitoes

We recommend using young mosquitoes (3- to 5-d-old).

Mounting medium

We recommend using an antifading mounting medium such as Vectashield or glycerol with DABCO.

Phosphate-buffered saline (PBS)

PBS-T (1× PBS with 0.3% Triton X-100)

We use 100× PBS (Oxoid BR0014G) and dilute it 1:100 in water.

Equipment

Cell culture plate, 24-well
Cold chamber
Disposable glass Pasteur pipettes and rubber pipette bulb
Embedding silicone or rubber molds
Filter paper
Fridge
Hot plate with magnetic stirrer
Microcentrifuge tubes, 1.5-mL
Micro-dissecting tweezers, preferably Dumont #55
Microscope slides and coverslips, 50-mm-long
Orbital shaker
P200 micropipette, with a 200-μL nonfilter tip
Parafilm
Polyethylene Pasteur pipettes, fine-tip
Polypropylene tubes, 50-mL
Razors

Use single-edge steel razor blades for higher security.

Rotating shaker suitable for 1.5-mL microcentrifuge tubes
Stereomicroscope
Tweezers
Vibratome

Different vibratomes can be used. We recommend using a Leica VT1000S.

Water bath

METHOD

Some conditions (fixation and antibody incubation times) may require modification depending on the antibodies used. Although the times for most incubations and washes are relatively flexible, it is important to be accurate with fixation times.

Head Fixation and Vibratome Sectioning

1. Remove mosquito heads from the bodies using tweezers and fix them in 1 mL of 4% formaldehyde in PBS-T for 1 h at room temperature. Place the samples in a rotator shaker.

 Cite this protocol as *Cold Spring Harb Protoc*; doi:10.1101/pdb.prot108008

Use an ice pad to sedate mosquitoes and work under a stereomicroscope. To facilitate the diffusion of the fixative in the mosquito head, remove the mosquito proboscis. Formaldehyde is toxic, so it is essential to work under a fume hood. We recommend using six mosquitoes for each antibody to be tested.

2. While the samples are being fixed, heat the frozen albumin-gelatin aliquot in a water bath to 42°C.

   It can take 30–45 min for the albumin-gelatin medium to reach this temperature. Keep a thermometer in the albumin-gelatin and monitor the temperature carefully, as an incorrect temperature can ruin the embedding process (it curdles the albumin-gelatin).

3. After fixing the samples for 1 h in the rotator, proceed to embed them in the silicone molds filled with albumin-gelatin.

   i. Before embedding, place the materials required (tweezers, glass Pasteur pipettes, and molds) on a hot plate to warm them up.

      The albumin-gelatin medium in the Falcon tube should be kept in the water bath during sample embedding to avoid cooling and solidifying.

   ii. Take a P200 micropipette, with a 200-µL nonfilter tip, and cut the end of the tip. Use the micropipette to collect three heads from the fixative, and place them on a filter paper to dry them and remove any residual fixative.

   iii. Use the glass Pasteur pipette to carefully pour albumin-gelatin into the molds, avoiding the formation of bubbles. Use tweezers to collect the heads and place them on the surface of the albumin-gelatin. Work under a stereomicroscope for better precision. Gently push the heads under the surface, so that they are completely covered by the albumin-gelatin but still stay superficial. Use tweezers to orient them in a frontal plane (Fig. 1A).

   iv. Let the molds chill for 5 min in a fridge before the post-fixation step.

      Any remaining albumin-gelatin can be frozen and reused later.

4. Place the molds in 50-mL Falcon tubes containing 6% formaldehyde and allow them to post-fix overnight at 4°C.

5. Take the molds out of the fixative, briefly dry them on a filter paper, and place them for 10 min in a 50-mL Falcon tube containing 20 mL of methanol.

   Methanol is neurotoxic, so work should be conducted under a hood. Molds can be dabbed onto filter paper to remove residual fixative.

6. Take the molds out of the methanol, briefly dry them on a filter paper, and place them in 50-mL Falcon tubes containing 20 mL of PBS.

   Molds can be sectioned immediately or kept for 1–2 h in PBS at room temperature.

7. Use a vibratome to create 40-µm sections of the embedded heads.

   i. Fill the buffer tray with PBS.

   ii. To section the albumin-gelatin block, cut it in a trapezoidal shape and cut sections in a frontal plane (Fig. 1B).

      For imaging under a confocal microscope, we recommend using a slice thickness of 1 µm.

   iii. Place the sections in a well of a 24-well cell culture plate filled with 1 mL of PBS.

      We recommend collecting sections from two blocks in the same well.

Immunostaining

8. Rinse and wash the sections for 10 min on an orbital shaker in 1 mL of PBS-T at room temperature. Repeat three times.

   PBS-T contains 0.3% Triton X-100. "Rinse" means adding PBS-T to the cell culture plate well and immediately removing it. "Wash" means adding the PBS-T and allowing it to sit for 5 min. Avoid bubble formation. To remove the solutions, use fine-tip polyethylene Pasteur pipettes.

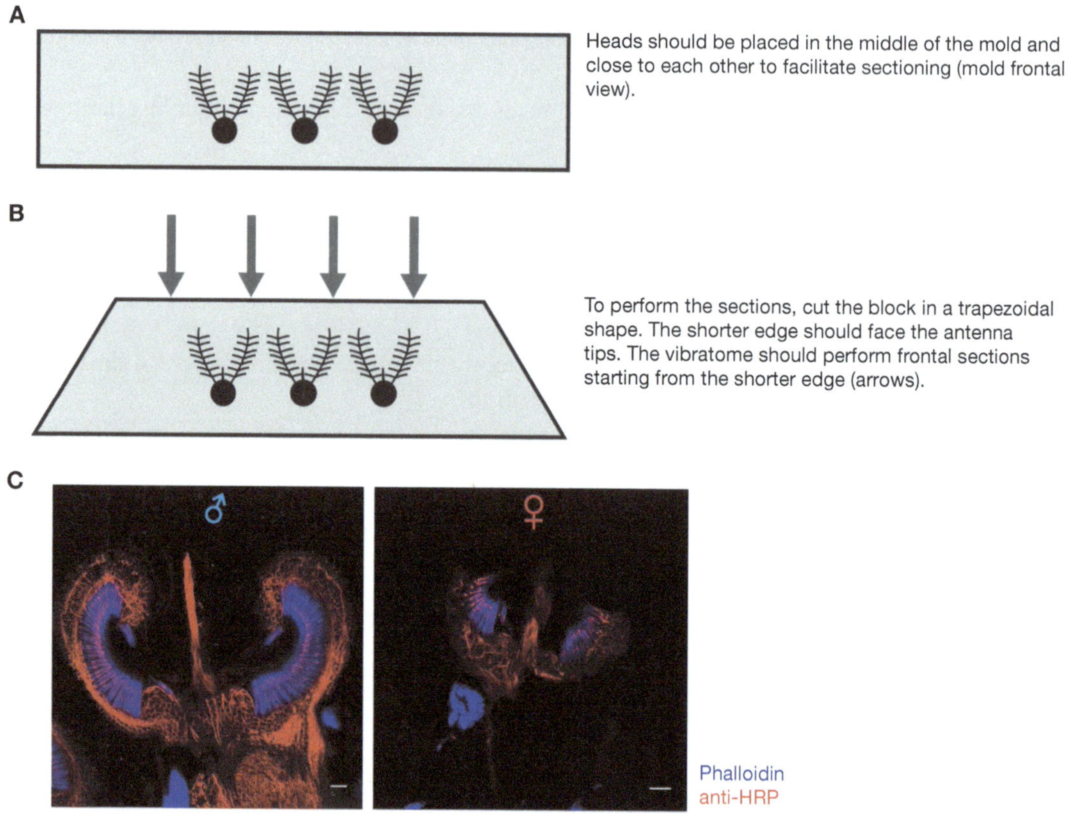

FIGURE 1. Immunostaining of the mosquito ear. (*A*) Position of the mosquito heads in the mold filled with albumin-gelatin. (*B*) The hardened block should be cut into a trapezoidal shape for vibratome sectioning. (*C*) Example of male (*left*) and female (*right*) mosquito ear immunostaining showing labeling with a neuronal marker, anti-HRP, and phalloidin, which binds to actin-rods in scolopale cells that surround the auditory cilia (Su et al. 2018).

9. Incubate the samples on an orbital shaker in 0.5–1 mL of blocking solution for 1 h at room temperature.

    Incubation time can be increased up to 3 h.

10. Incubate the samples on an orbital shaker in 0.3–0.5 mL of primary antibody diluted in blocking solution overnight at 4°C. Use Parafilm to seal the culture plate.

    If a primary antibody conjugated to a fluorochrome is used, add it directly in Step 12. If only a primary antibody conjugated to a fluorochrome is used (without any other primary antibody), Steps 10–12 can be skipped.

11. Rinse and wash the samples on an orbital shaker for 15 min in 1 mL of PBS-T at room temperature. Repeat four times.

12. Incubate the samples on an orbital shaker in the secondary antibody diluted in blocking solution for 1 h at room temperature. Cover the samples to keep them in the dark.

    Incubation time can be increased up to 3 h. If a primary antibody conjugated to a fluorochrome is used, add it in this step. From this point on, exposure of samples to light should be minimized to avoid photobleaching. Aluminum foil or a cardboard microtube storage box can be used to cover the culture plate.

13. Rinse and wash the samples on an orbital shaker for 15 min in 1 mL of PBS-T at room temperature. Repeat four times.

    Keep in the dark.

14. Rinse and wash on an orbital shaker the samples once for 10 min in 1 mL of PBS at room temperature.

    Keep in the dark. This step is performed to avoid any autofluorescence caused by Triton X-100.

15. Mount the sections on a slide and seal them for confocal imaging as follows (Fig. 1C).

Cite this protocol as *Cold Spring Harb Protoc*; doi:10.1101/pdb.prot108008

i. Cut the tip of a 200-µL tip and transfer 30 µL of mounting medium to the slide.

ii. Use a wooden stick to collect the sections and carefully place them in the mounting medium. Using tweezers, carefully arrange the sections in the slide (five to seven per slide).

Mounting medium can be added if the sections stick to each other and cannot be separated.

iii. Carefully use a 50-mm-long coverslip to cover the sections, trying to avoid the formation of bubbles. Place a small, moderately heavy object on top of the coverslip and remove any mounting medium seeping out of the coverslip. Seal the preparation using nail polish

The preparations can be stored at 4°C if imaged in the next days. For long-term storage, we recommend keeping the preparations at −20°C, avoiding frequent thawing and freezing.

RECIPE

Albumin-Gelatin Embedding Medium

1. Carefully dissolve 24.2 g albumin (Sigma-Aldrich A5253) in 66 mL of distilled water. Use a magnetic stirrer to slowly dissolve the albumin in the water, and add the albumin gradually (over the course of 1 h) to avoid the formation of clumps or foam. Try to dissolve any clumps using a spoon. After albumin has been dissolved, stir gently for another 30 min and discard any foam that forms.
2. Warm 25 mL of distilled water to 55°C and add 5.7 g of gelatin (Sigma-Aldrich G2500). Place the solution on the magnetic stirrer, keep it warm, and gently mix until gelatin has been completely dissolved.
3. Add the dissolved gelatin to the albumin solution while slowly stirring on the magnetic stirrer for ~5 min.
4. Divide into 15-mL aliquots in 50-mL Falcon tubes and freeze at −20°C until used.

ACKNOWLEDGMENTS

We thank the European Commission (Marie Curie Individual Fellowship, 752472) and the UKRI (Future Leaders Fellowship MR/S015493/1) for funding. We also thank Stephanie Pauls and Martin Goepfert for their help in developing this protocol.

REFERENCES

Su MP, Andrés M, Boyd-Gibbins N, Somers J, Albert JT. 2018. Sex and species specific hearing mechanisms in mosquito flagellar ears. *Nat Commun* **9:** 3911. doi:10.1038/s41467-018-06388-7

Protocol 2

Electrophysiological Measurements of Compound Action Potential Responses from the Antennal Nerve in Response to Stimulation

Matthew P. Su[1,2] and Joerg T. Albert[3,4,5]

[1]*Graduate School of Science,* [2]*Institute for Advanced Research, Nagoya University, Nagoya 464-8602, Japan;* [3]*Ear Institute, University College London, London WC1X 8EE, United Kingdom;* [4]*The Francis Crick Institute, London NW1 1AT, United Kingdom*

Electrophysiological recordings taken from the antennal nerve can provide essential information on the general auditory condition of the mosquito tested. Furthermore, electrophysiological recordings provide detailed information on what types of stimulation induce the largest nerve responses. When these are used in conjunction with a vibrometer to measure the corresponding movement of the antennal ear during stimulation, a comprehensive overview of hearing function can be obtained. This protocol can be applied to male and female adults from any mosquito strain and can be scaled relative to available resources.

MATERIALS

It is essential that you consult the appropriate Material Safety Data Sheets and your institution's Environmental Health and Safety Office for proper handling of equipment and hazardous materials used in this protocol.

Equipment

Blue-light-cured glue (e.g., DEHP Flowable Composite Syringe; Henry Schein Dental 530017)

Environmental incubator

The mosquitoes' circadian state and environmental conditions at the time of the experiment must be controlled. Incubators that allow environmental parameters (light, temperature, humidity) to be controlled are recommended for entraining mosquitoes (e.g., I-30VL, Percival Scientific). Common entrainment settings are 12 h:12 h light-dark (LD) cycles at constant temperature (ranging from 22°C to 28°C) and 60%–80% relative humidity.

Forceps, regular (for mosquito transfer) and inverted (for glue application)

Ice

Mosquitoes can survive on ice for at least 1 h.

Laser Doppler vibrometer/interferometer (optional; see Step 14)

Although the use of a vibrometer/interferometer is not strictly necessary for obtaining high-quality electrophysiology data, these measurement tools can provide extremely valuable information regarding the actual movement of the antennal ear during stimulus presentation. This allows careful calibration of ear movement in response to a range of stimulus amplitudes, and can allow confirmation of hearing function.

[5]Correspondence: joerg.albert@ucl.ac.uk

Cite this protocol as *Cold Spring Harb Protoc;* doi:10.1101/pdb.prot108010

Microscope

Mosquito gluing and electrode insertion into the base of the ear both need to be done under a dissection microscope.

Micromanipulators

Micromanipulators are necessary to hold the rod onto which the mosquito is glued, to hold the two electrodes that will be inserted into the mosquito, and to mount the stimulation source (actuators or speaker).

Plastic rod, ~5 cm in length

Recording software

The Spike2 software offers a high-end option for electrophysiology recordings; free alternatives are available however. Likewise, the VibSoft software offered by Polytec can be used for vibrometer recordings.

Stimulation source: electrostatic actuators (preferred) or speaker (see Step 11)

Although speakers can provide a range of stimuli to the mounted mosquito, the use of electrostatic actuators provides a number of benefits (not least because required speaker volumes can be inconveniently loud for researchers). Actuators are straightforward and inexpensive to make (for details, see Albert et al. 2007; Duong and Kim 2015; Weinberger et al. 2017; Su et al. 2018).

Temperature-controlled room

Many mosquito behaviors and physiological performances are strongly influenced by changes in temperature, including mosquito antennal frequency tuning. It is vital that recordings of hearing function be made under strict temperature control to ensure that the results obtained are comparable across experiments.

Tungsten electrodes

Electrodes can be reused by sharpening/cleaning through electrochemical etching in between recordings (Duong and Kim 2015). Two electrodes are required per recording: one unbiased/reference electrode inserted in the abdomen, and one biased/recording electrode inserted near the antennal nerve.

METHOD

In principle, the hearing function of female and male mosquitoes of any genotype can be tested. Rearing conditions for individual mosquito lines are dependent on the respective lines and local safety protocols.

Preparation and Entrainment of Mosquitoes

1. Three days before the start of the experiment, transfer mosquitoes from the group of interest into light- and temperature-controlled incubators.
2. Entrain mosquitoes to desired entrainment pattern for 3 d.

   The use of 12 h:12 h light-dark (LD) cycles at constant temperature is suggested as a standard entrainment pattern.
3. Ensure that the circadian time in the incubator matches desired experimental time.

   A circadian/experimental time matching that of dusk is suggested.

Mounting of Mosquitoes

4. Remove mosquitoes from incubator and sedate by placing them on ice.
5. Apply glue to the tip of the plastic rod, and carefully transfer the mosquito onto the glue using forceps.

   It is essential that glue not be spread onto the mosquito ears, as this may inhibit flagellar movements. Only the mosquito body below the head should be placed into the glue at this stage.
6. Carefully apply further glue to the body and head so that only the antennal ear can move freely.

   Spread glue across each part of abdomen and thorax to prevent kicking and general movement. Then apply glue to the mouthparts and neck to keep the head in a stable position. Take care to leave the respiratory openings (spiracles) on thorax and abdomen as open as possible.

7. Direct the blue light source at the glue surrounding the mosquito and hold for ~5–10 sec to cure the glue.

Insertion of Electrodes

8. Place rod in micromanipulator so that the mosquito head is clearly visible under a microscope.
9. Position micromanipulators holding electrodes so that the unbiased electrode can be placed into the mosquito abdomen and the biased/recording electrode can penetrate the cuticle between the mosquito Johnston's organ (JO) and head.
10. Insert electrodes gently, starting with the unbiased/reference electrode, so that the minimum amount of electrode enters the mosquito at first.

    The male JO is substantially larger than the female JO, which makes electrode insertion typically far easier for males than for females. JO size also varies between species, with Anopheles females having perhaps the smallest JO size amongst human-biting species.

    See Troubleshooting.

Stimulus Provision and Measurement

11. Place actuators/speaker directly in front of the mosquito flagellar tip.
12. Start recording, and provide stimulus.

    The stimulus could be a sweep of pure tones covering a distinct relevant frequency range (e.g., 1–1000 Hz) at different amplitudes.

    Action potential magnitude increases nonlinearly with the stimulus amplitude. Carefully measure the minimum stimulus required to evoke the smallest measurable action potential.

13. Record multiple repeats of stimulus (at least 10 per amplitude).

    See Troubleshooting.

(Optional) Measurement of Flagellar Movement

14. If access to a vibrometer/interferometer is available, focus the laser toward the tip of the mosquito flagellum.

    For females, focus on the base of the third flagellomere from the tip; for males, on the base of the second.

15. Record the movement of the flagellum both before and during stimulus provision.

    See Troubleshooting.

Data Analysis

16. Antennal nerve responses can be analyzed for multiple key components, but perhaps the most important are the maximum compound action potential (CAP) amplitudes and the stimulus frequency at which these occur.

    Male and female mosquitoes have well-defined sexual dimorphisms in hearing function—specifically, the peak male action potential response should occur for stimuli between ~250 and 350 Hz, whereas female peak responses occur at ~150 Hz. Male action potential responses are typically greater than those of females (because of the higher number of neurons).

17. First, create averages of nerve responses from the multiple repeats conducted per mosquito.
18. Then smooth the resulting data to reduce the influence of noise on peak extraction.
19. The peak response amplitude can be identified, in addition to the frequency which generated this peak.
20. For analysis of flagellar movements, a harmonic oscillator function can be fit to raw data that enables the extraction of other key parameters (such as the flagellar best frequency and tuning sharpness).

Cite this protocol as *Cold Spring Harb Protoc*; doi:10.1101/pdb.prot108010

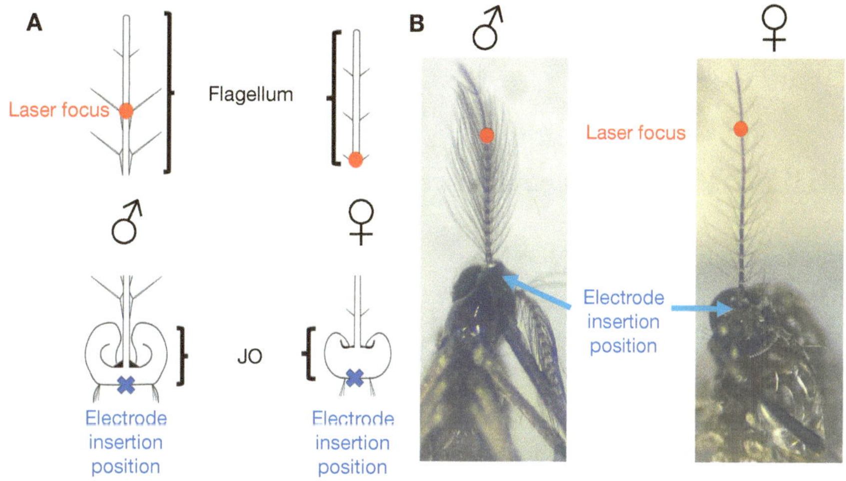

FIGURE 1. Anatomical differences between male and female ears and resulting effects on electrode insertion position and laser focus point. (*A*) Schematic drawing; (*B*) photographs of real anatomy. Male (♂) and female (♀) ears are composed of a Johnston's organ (JO) ("inner ear") and a flagellum ("sound receiver"). Although the electrode insertion position (marked by blue cross) is the same for both sexes (i.e., at the base of the pedicel), the focal position of the laser (marked by red dot) is slightly different due to dimorphism in flagellar anatomy. For males, the laser should be focused at the base of the second flagellomere from the tip, whereas for females, the third flagellomere should serve as the focal point.

TROUBLESHOOTING

Problem (Steps 10, 13, and 15): The mosquito dies or flagellar movement is compromised.
Solution: Consider the following.

- Pay particular attention to the location of glue application on the mosquito—no glue should be applied to the ear that will be tested, as this could alter flagellar movement. The application of too much glue can also shorten mosquito life expectancy significantly.
- Electrodes should be inserted as gently as possible, as deep insertion can cause critical damage. Pay attention to correct positioning of the electrode and laser beam (see Fig. 1 for details).

ACKNOWLEDGMENTS

This protocol benefited from funding through the European Research Council (ERC) under the Horizon research and innovation programme (grant agreement no. 648709, to J.T.A.), a pump-priming award from the BBSRC Vector Borne Disease (VBD) Network ANTI-VeC (AV/PP/0028/1, to J.T.A., Sarah Moore, and M.P.S.), a BBSRC project grant (BB/V007866/1 to J.T.A.), and a UCL Global Challenges Research Fund (GCRF) small grant (to J.T.A.).

REFERENCES

Albert JT, Nadrowski B, Göpfert MC. 2007. Mechanical signatures of transducer gating in the *Drosophila* ear. *Curr Biol* **17:** 1000–1006. doi:10.1016/j.cub.2007.05.004

Duong T-H, Kim H-C. 2015. Electrochemical etching technique for tungsten electrodes with controllable profiles for micro-electrical discharge machining. *Int J Precis Eng Manuf* **16:** 1053–1060. doi:10.1007/s12541-015-0136-8

Su MP, Andrés M, Boyd-Gibbins N, Somers J, Albert JT. 2018. Sex and species specific hearing mechanisms in mosquito flagellar ears. *Nat Commun* **9:** 3911. doi:10.1038/s41467-018-06388-7

Weinberger S, Topping MP, Yan J, Claeys A, De Geest N, Ozbay D, Hassan T, He X, Albert JT, Hassan BA, et al. 2017. Evolutionary changes in transcription factor coding sequence quantitatively alter sensory organ development and function. *eLife* **6:** e26402. doi:10.7554/eLife.26402

Protocol 3

Recording and Extraction of Mosquito Flight Tones

Marcos Georgiades and Joerg T. Albert[1]

Ear Institute, University College London, London WC1X 8EE, United Kingdom; The Francis Crick Institute, London NW1 1AT, United Kingdom

Despite the artificial conditions, flight tone recordings taken from tethered mosquitoes can provide valuable information on the acoustic signals produced by male and female mosquitoes. Although auditory responsiveness appears to be largely (and possibly exclusively) restricted to males, the flight tones of both sexes have sensory-ecological relevance, as it is the mixing of the two tones that produces audibility in males and thereby facilitates reproduction. This protocol describes how to record wing flapping from mounted mosquitoes and how to estimate wingbeat frequencies from those recordings.

MATERIALS

It is essential that you consult the appropriate Material Safety Data Sheets and your institution's Environmental Health and Safety Office for proper handling of equipment and hazardous materials used in this protocol.

Reagents

Mosquitoes

To control for the confounding effects of aging on behavioral performance, all mosquitoes recorded should be of similar age, and averaging across age groups (unless statistically verified) should be avoided. We recommend using 4- to 7-d-old mosquitoes; these have already assumed their final body size but can be expected not to suffer from age-dependent declines of flight performance (Wishart and Riordan 1959).

The mating status of all experimental mosquitoes should be registered and controlled for. Unless the mating status is itself the object of investigation, we recommend separating all mosquitoes by sex during the pupal stage and keeping them separate and virgin throughout adulthood before the experiment.

Equipment

Blue light source

Blue-light-cured glue

Most light-cured dental glues would be suitable (e.g., DEHP Flowable Composite Syringe; Henry Schein Dental 53001).

Compartment for transferring mosquitoes (e.g., glass vial)

Container for gluing (e.g., 3-cm-diameter Petri dish or small plastic lid)

Cotton wool

[1]Correspondence: joerg.albert@ucl.ac.uk

Cite this protocol as *Cold Spring Harb Protoc*; doi:10.1101/pdb.prot108011

Environmental incubator

Because of its multiple effects on neural function and behavior, the mosquitoes' circadian state at the time of the experiment must be controlled. This also applies to environmental conditions, which could alter the circadian state during the experiment. Incubators that allow environmental parameters (light, temperature, humidity) to be controlled are recommended for entraining mosquitoes (e.g., Percival Scientific I-30VL). Common entrainment settings are 12 h:12 h light-dark (LD) cycles at constant temperature (ranging from 22°C to 28°C) and 60%–80% relative humidity.

Ice blocks

Microscope (optional; see Step 7)

Particle velocity microphone and holding device

Microphones sensitive to the air particle velocity changes associated with the mosquitoes' wingbeat should be used (e.g., pressure gradient microphone, Knowles Acoustics NR-3158). Prior calibration of microphones is preferable for insect wingbeat recordings as it makes it possible to extract both spectral (flight tone frequency) and magnitude (air particle velocity) information.

Temperature-controlled room

Mosquito wingbeat frequency is sensitive to temperature fluctuations. Before the experiment, adjust the room (or setup) temperature to the value under study. If the effect of temperature is not being investigated, experimental temperatures should be kept at constant value (within the range 22°C–28°C, with 22°C being the most widely used laboratory temperature).

Temperature logger

Tethered mosquito holding device

The holding device can be custom-made. A simple clamp, which can hold the tungsten wire, is here sufficient. For greater experimental convenience this can be mounted to a conventional micromanipulator.

Tissues

Tungsten wire

Tweezers

METHOD

Microphone and Mosquito Holder Setup

1. Place the particle velocity microphone and mosquito holding devices onto a stable surface.

   Recording microphones should be placed at a fixed distance of 2–4 cm between microphone surface and mosquito and oriented such that their sensitive side (if they have one, as is the case for the particle velocity microphone NR-3158) faces the mosquito. If more than one mosquito is to be recorded (e.g., same- or opposite-sex pairs) simultaneously, a separate microphone should be used for each. In this case, it is important to ensure that each microphone is sufficiently distant from mosquitoes other than the one it is meant to record from to prevent signal cross talk (experimental validation is required).

Mounting of Mosquitoes

2. Transfer the mosquitoes from the incubator into a small compartment (e.g., glass vial), and place in the recording room along with an ice block.
3. Cut a small piece of tungsten wire and leave to the side.
4. Expel a small portion of glue inside a small container, or onto a 3-cm-diameter Petri dish or small plastic lid.
5. Place a piece of tissue on the ice block, and unload the mosquitoes from the compartment on top of this tissue.

   The tissue protects mosquitoes from damage via direct contact with the ice block. Wait a few seconds until the mosquitoes are anesthetized before removing the glass vial. Return excess mosquitoes to the compartment using tweezers, carefully holding them by their legs.

6. With the tweezers, grab the small piece of tungsten wire and dip it in the glue.

7. Attach the wire to the mosquito's thorax and hold it steady while applying blue light to the glue to cure it (typical curing time ~10–20 sec).

   Blue light can be damaging/straining to the eyes; avoid the direction of application during the glue-curing step. The identification of an appropriate position on the mosquito's thorax for attaching the wire should be done under a microscope, if one is available, to facilitate visibility. When attaching the wire to the mosquito, take care to avoid blocking its wings with the wire (or the glue) or damaging or getting glue on the two flagellar protrusions (external compartment of the mosquito ear, located on its head).

8. Transfer the wire-attached mosquito to the holding apparatus and leave for a minute to recover.

Prompting and Recording Flight

9. Start the recording.

   To ensure sufficient resolution, set the sampling frequency reasonably high (e.g., ≥10 kHz). Higher sampling rates allow the application of running averages to the time series data before the analysis and thereby noise removal.

   It is strongly recommended that ambient temperature fluctuations be tracked with a temperature logger throughout the experiment for future reference and analysis.

10. Activate the tarsal reflex to promote wing flapping, by cutting a small piece of cotton wool and transferring it, using tweezers, underneath the mosquito; while still holding the cotton wool, wait for the mosquito to place its legs on the cotton wool and then swiftly remove it. This will activate the tarsal reflex, which provokes the mosquito to initiate flight.

    Alternatively, to trigger wing flapping, try blowing a gentle puff of air at the mosquito's back.

    See Troubleshooting.

11. Aim to record continuous flight over at least 10 sec.

    This will facilitate subsequent necessary data processing steps such as cropping of the flight recording. If the mosquito stops flapping its wings before the 10-sec mark, repeat the aforementioned flight-triggering procedures a few times before resorting to a new animal. Recorded flight lengths of different mosquitoes that you plan to compare should not vary greatly—stop recording before flight cessation by the animal, if necessary.

Extracting Wingbeat Frequency from Recordings

This protocol considers a fast Fourier transform (FFT) procedure for estimation of wingbeat frequencies from tethered flight recordings. The procedure can be coded and processed using a programming language of personal preference (e.g., Python, R, MATLAB). The steps outlined here provide a 5-Hz frequency resolution for the wingbeat frequency estimates, assuming a sampling rate of 10 kHz for the recordings. Frequency resolution F_{res} is given by

$$F_{res} = \frac{F_{sampling}}{F_{data}},$$

where $F_{sampling}$ is the sampling frequency and N_{data} is the number of data points for any given FFT window.

12. Discard the data for the first and last 2 sec of flight before the start of data analysis.

    This step ensures that steep gradients of wingbeat frequency due to flight initiation and cessation are not included in the analysis.

13. Apply a bandpass filter to the cropped data with a low-pass cut-off point at 1 kHz and high-pass cutoff point at 200 Hz.

    Cutoff values assume the recorded mosquito's wingbeat frequency falls well within the 200- to 1000-Hz frequency range; modify accordingly if necessary.

14. Apply a sliding FFT with a 200-msec window (corresponding to 2000 data points, given a 10-kHz sampling rate) over the recording, and shift the FFT window by 100-msec increments for every sliding step until the end of the recording length is reached (Fig. 1).

    The 50% overlap between successive FFTs enhances the visibility of events occurring faster than the processing duration of samples in the FFT window.

 Cite this protocol as *Cold Spring Harb Protoc*; doi:10.1101/pdb.prot108011

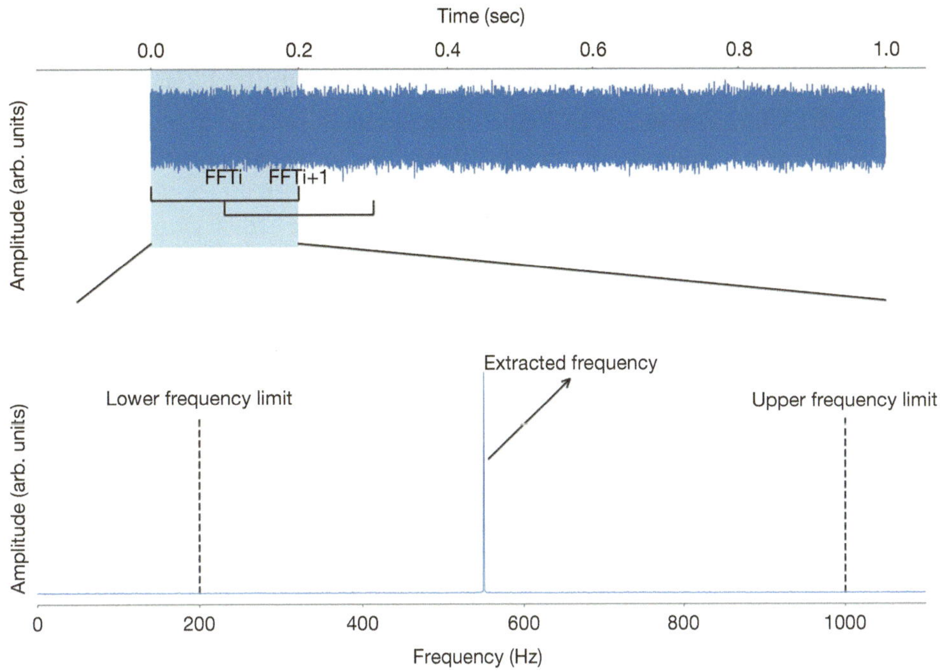

FIGURE 1. Sketch of the paradigm for the wingbeat frequency extraction algorithm. A fast Fourier transform (FFT) sliding window scans through the recording; the dominant frequency is extracted with each sliding step modified from Su et al. (2020).

15. For each step (and thus each FFT), extract and store f_0, the value in the frequency domain (x-value), which gives the maximum FFT amplitude (y-value) (Fig. 1).
16. Append the extracted frequencies together to obtain the wingbeat frequencies of the mosquito over the processed flight recording.
17. Use values according to experimental goals.

    Suggestion: For phonotactic experiments, assign the mean (if data is normally distributed) or median (if not normally distributed) of the extracted frequencies as the mosquito's wingbeat frequency.

TROUBLESHOOTING

Problem (Step 10): It is difficult to stimulate flapping in the animals.

Solution: Some disease-transmitting species (e.g., *Anopheles gambiae*) are prone to inactivity throughout most hours of the day, with the exception of short circadian time slots. The animals' circadian entrainment (and thereby their subjective time of the day) or the environmental conditions of the experimental room (which can influence the circadian state) may therefore influence wing flapping. To stimulate flapping, ensure that (a) the animals' entrainment is known, (b) the experiments are conducted at a time when the animal is expected to be active based on their entrainment, and (c) the light conditions in the experimental room are aligned with those expected by the animals given their circadian state. Use a red-light source to conduct experiments in the dark, if necessary.

RELATED INFORMATION

Flight tone information can supplement parallel phonotaxis experiments to provide an improved picture of the acoustic space mosquitoes inhabit. To carry out a phonotaxis experiment, see Protocol 4: Mosquito Phonotaxis Assay (Georgiades and Albert 2022).

ACKNOWLEDGMENTS

This protocol benefited from funding through the European Research Council (ERC) under the Horizon research and innovation programme (grant agreement no. 648709 to J.T.A.), a pump-priming award from the BBSRC Vector Borne Disease (VBD) Network ANTI-VeC (AV/PP/0028/1, to J.T.A., Sarah Moore, and Mathew P. Su), a BBSRC project grant (BB/V007866/1 to J.T.A.), and a UCL Global Challenges Research Fund (GCRF) small grant (to J.T.A.).

REFERENCES

Georgiades M, Albert JT. 2022. Mosquito phonotaxis assay. *Cold Spring Harb Protoc* doi:10.1101/pdb.prot108012

Su MP, Georgiades M, Bagi J, Kyrou K, Crisanti A, Albert JT. 2020. Assessing the acoustic behaviour of *Anopheles gambiae* (*s.l.*) *dsxF* mutants: implications for vector control. *Parasit Vectors* **13:** 507. doi:10.1186/s13071-020-04382-x

Wishart G, Riordan DF. 1959. Flight responses to various sounds by adult males of *Aedes aegypti* (L.) (Diptera: Culicidae). *Can Entomol* **91:** 181–191. doi:10.4039/Ent91181-3

Cite this protocol as *Cold Spring Harb Protoc*; doi:10.1101/pdb.prot108011

Protocol 4

Mosquito Phonotaxis Assay

Marcos Georgiades and Joerg T. Albert[1]

Ear Institute, University College London, London WC1X 8EE, United Kingdom; The Francis Crick Institute, London NW1 1AT, United Kingdom

Phonotaxis experiments can provide information on the spectrum of sounds relevant to mosquito acoustic behaviors. It is widely known that males of disease-transmitting species are attracted to tones with frequencies resembling the wingbeat frequencies of their conspecific females. Thus, phonotaxis experiments can be coupled with wingbeat frequency measurements to inform the development of vector control tools such as acoustic traps and lures. This protocol describes how to set up and execute a phonotaxis experiment.

MATERIALS

It is essential that you consult the appropriate Material Safety Data Sheets and your institution's Environmental Health and Safety Office for proper handling of equipment and hazardous materials used in this protocol.

Reagents

Mosquitoes

To control for the confounding effects of aging on behavioral performance, all mosquitoes recorded should be of similar age, and averaging across age groups (unless statistically verified) should be avoided. We recommend using 4- to 7-d-old mosquitoes; these have already assumed their final body size but can be expected not to suffer from age-dependent declines of flight performance (Wishart and Riordan 1959).

The mating status of all experimental mosquitoes should be registered and controlled for. Unless the mating status is itself the object of investigation, we recommend separating all mosquitoes by sex during the pupal stage and keeping them separate and virgin throughout adulthood before the experiment.

Equipment

Decibel meter (optional; see Step 6)

Environmental incubator

Because of its effects on neural function and behavior, the mosquitoes' circadian state at the time of the experiment must be controlled. This also applies to environmental conditions, which could alter the circadian state during the experiment. Incubators that allow environmental parameters (light, temperature, humidity) to be controlled are recommended for entraining mosquitoes (e.g., Percival Scientific I-30VL). Common entrainment settings are 12 h:12 h light-dark (LD) cycles at constant temperature (ranging from 22°C to 28°C) and 60%–80% relative humidity.

Insect cage

[1]Correspondence: joerg.albert@ucl.ac.uk

Cite this protocol as *Cold Spring Harb Protoc*; doi:10.1101/pdb.prot108012

An insect cage with sides of woven mesh is used for phonotaxis stimulation of mosquitoes (e.g., BugDorm BD42222F). For 20–30 mosquitoes, we recommend using a cage with dimensions 25-cm × 25-cm × 25-cm.

Particle velocity microphone (e.g., pressure gradient microphone, Knowles Acoustics NR-3158)

This is used for evaluating the frequency content and loudness of the stimuli.

Sound speaker

Temperature- and light-controlled room

Mosquito wingbeat frequency is sensitive to temperature fluctuations. Before the experiment, increase or decrease the room temperature to the value under study. If the effect of temperature is not being investigated, room temperatures should be kept at a constant value (within the range 22°C–28°C, with 22°C being the most widely used laboratory temperature).

The illumination conditions of the experimental room must also be considered. Many disease-transmitting mosquito species, such as Anopheles gambiae, *mate in reproductive swarms that are under circadian regulation. Such species show phonotaxis only during their preferred mating time period, which could be, under natural conditions, at sunset or during darkness (and possibly suppressed by light). For this reason, the mosquitoes' circadian time (i.e., their "subjective" time relative to a prior entrainment) and the illumination conditions must be considered. A red-light source (~620–650-nm) can be used to conduct the phonotaxis experiments under scotopic conditions or effective darkness.*

METHOD

Speaker Setup and Experimental Calibration

1. Set the sound speaker on the surface of a temperature- and light-controlled room where the recordings are to take place.
2. Place an empty mosquito (insect) cage so that one of its meshy sides makes contact with the speaker.
3. Mark the region where the speaker makes contact with the cage with a marker.

   This region constitutes the area that counts as indicating positive phonotaxis, if approached by mosquitoes during sound stimulation.
4. Mark the position of the cage and speaker on the surface where they sit for reference in subsequent phonotaxis experiments.
5. Append a particle velocity microphone to the cage on the side that is most distant from the speaker.
6. Ensure tones emitted from the speaker are sufficiently pure and audible to the mosquitoes.

   Play the tones of interest via the speaker and record with the microphone. Fourier analysis can then be applied to the recording to ensure that the right frequencies were being produced. To get an estimate for the intensity of each tone and ensure similarity of loudness across tones, calculate the root mean square amplitude of the signals recorded for each tone and compare across tones. Alternatively, use a decibel meter. As a reference point, mosquitoes can be expected to respond to tones with a sound amplitude of ≥60 dB sound pressure level (SPL) when assessed within a 25-cm × 25-cm × 25-cm cage volume.

Phonotaxis Experiment

7. Transfer 20–30 mosquitoes from the environmental incubator to the insect cage and place the cage back to the marked spot where calibration was conducted.
8. Control for mosquitoes being attracted to cues other than sound.

   At the beginning of the experiment, when you position the cage and speaker in contact with each other, count how many mosquitoes are attracted to the marked area of contact for 1 min during which no sound is being emitted.
9. Present the tones to the mosquitoes in the desired pattern.

Cite this protocol as *Cold Spring Harb Protoc*; doi:10.1101/pdb.prot108012

A suggested protocol for a single experimental replicate is as follows: Randomize the order of tone presentation, then present the first tone for 1 min, followed by 1 min of silence, followed by the next tone, and repeat until the sequence of tones has reached its end. Repeat for three runs (each with a newly randomized sequence of tones).

10. Count the mosquitoes approaching the area of positive phonotaxis during stimulation with each tone.

The counting can be done by eye. Count mosquitoes that make contact with the area of positive phonotaxis during stimulation.

See Troubleshooting.

TROUBLESHOOTING

Problem (Step 10): You observe a large variance in responses to the same tones across experiments conducted on different days.

Solution: We recommend looking for and controlling for confounding factors. The usual suspects, both of which influence the animals' wingbeat frequencies, are the ambient temperature of the room (Villarreal et al. 2017) and the animals' circadian state (Somers et al. 2022). Note that changes in the mosquitoes' own wingbeats will change the flight tones they can hear.

RELATED INFORMATION

The phonotaxis experiments described here can be coupled with wingbeat frequency measurements as described in Protocol 3: Recording and Extraction of Mosquito Flight Tones (Georgiades and Albert 2022).

ACKNOWLEDGMENTS

This protocol benefited from funding through the European Research Council (ERC) under the Horizon research and innovation programme (grant agreement no. 648709, to J.T.A.), a pump-priming award from the BBSRC Vector Borne Disease (VBD) Network ANTI-VeC (AV/PP/0028/1, to J.T.A., Sarah Moore, and Matthew P. Su), a BBSRC project grant (BB/V007866/1 to J.T.A), and a UCL Global Challenges Research Fund (GCRF) small grant (to J.T.A.).

REFERENCES

Georgiades M, Albert JT. 2022. Recording and extraction of mosquito flight tones. *Cold Spring Harb Protoc* doi:10.1101/pdb.prot108011

Somers J, Georgiades M, Su MP, Bagi J, Andrés M, Alampounti A, Mills G, Ntabaliba W, Moore SJ, Spaccapelo R, et al. 2022. Hitting the right note at the right time: circadian control of audibility in *Anopheles* mosquito mating swarms is mediated by flight tones. *Sci Adv* **8:** eabl4844. doi:10.1126/sciadv.abl4844

Villarreal SM, Winokur O, Harrington L. 2017. The impact of temperature and body size on fundamental flight tone variation in the mosquito vector *Aedes aegypti* (Diptera: Culicidae): implications for acoustic lures. *J Med Entomol* **54:** 1116–1121. doi:10.1093/jme/tjx079

Wishart G, Riordan DF. 1959. Flight responses to various sounds by adult males of *Aedes aegypti* (L.) (Diptera: Culicidae). *Canadian Entomologist* **91:** 181–191.

CHAPTER 31

Physiological and Behavioral Characterization of Larval Olfaction in the Malaria Vector Mosquito *Anopheles coluzzii*

Feng Liu,[1] Huahua Sun,[1] and Laurence J. Zwiebel[2]

Department of Biological Sciences, Vanderbilt University, Nashville, Tennessee 37235, USA

When viewed from both academic and vector-control perspectives, the chemosensory biology of larval-stage anopheline mosquitoes is both enigmatic and paradoxical. As is true for all mosquito species, anopheline larvae are free-swimming organisms that use complex sensory processes to both locate nutrients and avoid predators. Because of their obligatory and therefore restrictive aquatic habitats, mosquito larvae are the most easily sampled and targeted mosquito life stage and as such they have been the focus of the majority of vector control strategies used to date. Although this might reasonably have resulted in the accumulation of a robust body of knowledge of the natural and molecular biology of larval-stage chemosensory processes, there is, instead, a paucity of such information relative to adults. Here, we describe two relatively simple laboratory-based bioassays that allow for the characterization of larval chemosensory-driven behaviors as well as an electrophysiological approach to examine the responses of larval peripheral neurons to volatile odorant stimuli. Taken together, these approaches provide a road map for the study of the chemosensory biology and chemical ecology during this important stage in the life cycle of anophelines that transmit malaria.

THE CRITICAL IMPORTANCE OF LARVAL STAGE OLFACTION

The characterization of olfaction in mosquitoes has thus far, and not unreasonably, largely focused on adult females, which are solely responsible for disease transmission (Montell and Zwiebel 2016). However, because the prevalence and the presence of disease-transmitting adults is entirely dependent on the successful development/survival of larval stages to pupae and ultimately into those adults, this narrow prism reinforces an underappreciation of the importance of mosquito larvae. Furthermore, the focus on adults ignores the central role that larval stages play in numerous aspects of mosquito control/management programs (Tusting et al. 2013).

Larval stage anophelines are typically found in standing (rather than running) water (Amani et al. 2014; Soleimani-Ahmadi et al. 2014) containing a wide range of natural components that include plant debris and microorganisms (Sattler et al. 2005; Nilsson et al. 2018). These constituents, and more importantly their breakdown products, represent a rich source of chemosensory signals that emanate into aqueous larval habitats and the headspace just above the water surface. Here they act as semiochemicals, which are biologically relevant chemical signals released by one organism that affect the behavior of another (Danos et al. 1983; Wotton et al. 1997). Mosquito larvae develop over ~10 d and four morphologically distinctive stages known as instars (Clements 1992). The larval development

[1]These authors contributed equally to this work.

[2]Correspondence: l.zwiebel@vanderbilt.edu

Cite this introduction as *Cold Spring Harb Protoc*; doi:10.1101/pdb.top107677

time of anophelines varies among species and reflects the availability of food sources and the ambient water temperature. In addition, larval mosquitoes typically exist in restricted and often exceedingly dangerous environments, where predators abound and there are a variety of intra- and interspecific competitors for food and, indeed, survival. Success depends upon a robust ability to both find food for survival and avoid becoming food for the survival of others. This tenuous existence contrasts the stereotypic viewpoint that mosquito larvae simply represent an immature, largely inactive preadult life stage with a limited sensory and cognitive capacity. Instead, mosquito larvae are actually highly active aquatic insects that depend on an acutely robust suite of sensory systems to detect and respond to a wide range of chemical and other stimuli that dramatically impact their behavior (Sun et al. 2021).

Larval chemosensory systems mediate both taste (gustation) and smell (olfaction) senses to identify and react to both direct and volatile chemical stimuli associated with potential food sources, predators, and other aspects of their aquatic environments. Unlike culicine larvae, which are bottom feeders and spend most their time at depth, anopheline larvae are surface filter feeders that largely position themselves parallel and just below the water surface. Here they have access to a rich selection of organic animal or plant detritus as well as diverse microorganisms, such as algae and bacteria (Danos et al. 1983; Wotton et al. 1997; Gimmg et al. 2001; Fillinger et al. 2004). From this vantage point, there is an opportunity to extend their antennae above the surface of their aquatic habitats to sample the rich spectrum of volatile odorants that emanate from potential food sources, providing an obvious evolutionary advantage to anopheline larvae.

The limited number of studies performed on larval chemosensory processes have largely been restricted to terrestrial insects. In that context, extracellular electrophysiological studies of odorant receptor neurons (ORNs) from the larval dorsal organ of the vinegar fruit fly *Drosophila melanogaster* uncovered several odorants that elicit excitatory neuronal responses (Kreher et al. 2005). Moreover, functional expression of those fly larva–specific odorant receptors in the "empty" neuron system further revealed their response spectrum to a panel of 27 olfactory stimuli. Thus far, the study of larval chemosensory processes in anopheline mosquitoes has mirrored that of adult stages, focusing first on behavioral responses and subsequently developing peripheral electrophysiology approaches to assess neuronal activity. As such, we initially developed and used a suite of aqueous behavioral bioassays to facilitate the characterization of larval chemosensory responses to soluble odorants that are dispersed in their aquatic environment (Xia et al. 2008; Liu et al. 2010). We also adapted these approaches to provide a high-throughput aversive response bioassay that served as a proxy for repellent activity. This assay was particularly useful in establishing the efficacy of a suite of nonvolatile small molecule modulators of Orco, the insect odorant receptor coreceptor, that comprises the Vanderbilt University Allosteric Agonists (VUAAs) (Taylor et al. 2012). More recently, we have also incorporated direct electrophysiological assessments and behavioral bioassays specifically tailored to examine responses to volatile cues that are available in the headspace just above larval habitats (Sun et al. 2021). Extending traditional insect electrophysiological approaches to mosquito larvae was, at first glance, very challenging as we assumed assessing the peripheral responses of larval chemosensory neurons would require the added technical challenge of performing in vivo recordings under aqueous conditions. However, as noted above, because *An. coluzzii* larvae spend most of their time at the surface of their aqueous habitats, their antennae can sample volatiles in the nonaqueous headspace just above the water. This novel adaptation allowed us to use a nonaqueous-based protocol to carry out in vivo extracellular electrophysiological recordings of peripheral neuronal activity derived from the larval sensory cone in response to those biologically relevant volatiles (Sun et al. 2021).

In addition to the inherent biological significance of larval stages and their importance in mosquito control efforts, when one considers the simplicity of the larval olfactory system, for example, the substantially smaller number of sensory appendages and neurons as well as the significantly less complex central nervous system, mosquito larvae may be considered to represent an attractive "model within a nonmodel" system in which to study the fundamental properties of mosquito olfactory processes and odor coding. The methodologies that we have developed for the behavioral and electrophysiological characterization of anopheline larval chemosensory processes will be presented in associated protocols (see Protocol 1: Examining Peripheral Electrophysiological Responses

 Cite this introduction as *Cold Spring Harb Protoc*; doi:10.1101/pdb.top107677

of the Larval Antennal Sensory Cone of *Anopheles coluzzii* to Volatile Odorants [Liu et al. 2022a] and Protocol 2: Cup and Pan Behavioral Assays for Assessing *Anopheles coluzzii* Larval Volatile Responses [Liu et al. 2022b]). These methods reveal that anopheline larvae have a complex and multifaceted peripheral olfactory apparatus that is highly capable of chemosensation in the aquatic habitat of mosquito larvae. As is the case for adults, in addition to providing valuable insight into mosquito biology, these studies will likely directly inform the development of novel vector control strategies that target the olfactory system of anopheline larvae.

REFERENCES

Amani H, Yaghoobi-Ershadi MR, Kassiri H. 2014. The ecology and larval habitats characteristics of anopheline mosquitoes (Diptera: Culicidae) in Aligudarz county (Luristan province, western Iran). *Asian Pac J Trop Biomed* **4:** S233–S241. doi:10.12980/APJTB.4.2014C186

Clements AN. 1992. *The biology of mosquitoes: development, nutrition and reproduction*, Vol. **1**, pp. 231–234. CABI, Wallingford, UK

Danos SC, Maki JS, Remsen CC. 1983. Stratification of microorganisms and nutrients in the surface microlayer of small freshwater ponds. *Hydrobiologia* **98:** 193–202. doi:10.1007/BF00021021

Fillinger U, Sonye G, Killeen GF, Knols BGJ, Becker N. 2004. The practical importance of permanent and semipermanent habitats for controlling aquatic stages of *Anopheles gambiae sensu lato* mosquitoes: operational observations from a rural town in western Kenya. *Trop Med Int Heal* **9:** 1274–1289. doi:10.1111/j.1365-3156.2004.01335.x

Gimmg JE, Ombok M, Kamau L, Hawley WA. 2001. Characteristics of larval anopheline (Diptera: Culicidae) habitats in western Kenya. *J Med Entomol* **38:** 282–288. doi:10.1603/0022-2585-38.2.282

Kreher SA, Kwon JY, Carlson JR. 2005. The molecular basis of odor coding in the *Drosophila* larva. *Neuron* **46:** 445–456. doi:10.1016/j.neuron.2005.04.007

Liu C, Pitts RJ, Bohbot JD, Jones PL, Wang G, Zwiebel LJ. 2010. Distinct olfactory signaling mechanisms in the malaria vector mosquito *Anopheles gambiae*. *PLoS Biol* **8:** e1000467. doi:10.1371/journal.pbio.1000467

Liu F, Sun H, Zwiebel LJ. 2022a. Examining peripheral electrophysiological responses of the larval antennal sensory cone of *Anopheles coluzzii* to volatile odorants. *Cold Spring Harb Protoc* doi:10.1101/pdb.prot108020

Liu F, Sun H, Zwiebel LJ. 2022b. Cup and pan behavioral assays for assessing *Anopheles coluzzii* larval volatile responses. *Cold Spring Harb Protoc* doi:10.1101/pdb.prot108021

Montell C, Zwiebel LJ. 2016. Mosquito sensory systems. *Adv Insect Physiol* **51:** 293–328. doi:10.1016/bs.aiip.2016.04.007

Nilsson LK, Sharma A, Bhatnagar RK, Bertilsson S, Terenius O. 2018. Presence of *Aedes* and *Anopheles* mosquito larvae is correlated to bacteria found in domestic water-storage containers. *FEMS Microbiol Ecol* **94:** fiy058. doi:10.1093/femsec/fiy058

Sattler MA, Mtasiwa D, Kiama M, Premji Z, Tanner M, Killeen GF, Lengeler C. 2005. Habitat characterization and spatial distribution of *Anopheles* sp. mosquito larvae in Dar es Salaam (Tanzania) during an extended dry period. *Malaria J* **4:** 1–15. doi:10.1186/1475-2875-4-4

Soleimani-Ahmadi M, Vatandoost H, Zare M. 2014. Characterization of larval habitats for anopheline mosquitoes in a malarious area under elimination program in the southeast of Iran. *Asian Pac J Trop Biomed* **4:** S73–S80. doi:10.12980/APJTB.4.2014C899

Sun H, Liu F, Baker AP, Honegger HW, Raiser G, Zwiebel LJ. 2021. Neuronal odor coding in the larval sensory cone of *Anopheles coluzzii*: complex responses from a simple system. *Cell Rep* **36:** 109555. doi:10.1016/j.celrep.2021.109555

Taylor RW, Romaine IM, Liu C, Murthi P, Jones PL, Waterson AG, Sulikowski GA, Zwiebel LJ. 2012. Structure–activity relationship of a broad-spectrum insect odorant receptor agonist. *ACS Chem Biol* **7:** 1647–1652. doi:10.1021/cb300331z

Tusting LS, Thwing J, Sinclair D, Fillinger U, Gimnig J, Bonner KE, Bottomley C, Lindsay SW. 2013. Mosquito larval source management for controlling malaria. *Cochrane Database Syst Rev* **2013:** CD008923. doi:10.1002/14651858.CD008923.pub2

Wotton RS, Chaloner DT, Yardley CA, Merritt RW. 1997. Growth of *Anopheles* mosquito larvae on dietary microbiota in aquatic surface microlayers. *Med Vet Entomol* **11:** 65–70. doi:10.1111/j.1365-2915.1997.tb00291.x

Xia Y, Wang G, Buscariollo D, Pitts RJ, Wenger H, Zwiebel LJ. 2008. The molecular and cellular basis of olfactory-driven behavior in *Anopheles gambiae* larvae. *Proc Natl Acad Sci* **105:** 6433–6438. doi:10.1073/pnas.0801007105

Protocol 1

Examining Peripheral Electrophysiological Responses of the Larval Antennal Sensory Cone of *Anopheles coluzzii* to Volatile Odorants

Feng Liu,[1] Huahua Sun,[1] and Laurence J. Zwiebel[2]

Department of Biological Sciences, Vanderbilt University, Nashville, Tennessee 37235, USA

Anopheline larvae rely on their antennae to respond to a complex suite of stimuli with which they navigate their aquatic environment, search for food, and avoid predators and pollution. Chemosensory signaling initiates on dendrites innervating the sensory peg and sensory cone, which are adjacently located at the distal (apical) end of the larval antennae. These structures are the primary sites for the detection of both soluble and volatile semiochemicals, which are biologically relevant chemical signals (typically unitary or blends of compounds) released by one organism that affect the behavior of another. The sensory neurons housed in the larval antennal sensory cone are responsible for the signal transduction processes that initiate responses to a range of volatile stimuli. To investigate the mosquito larval olfactory neuronal response to volatile odorants, we developed this method to record, extracellularly, the electrophysiological responses of sensory cone neurons to a range of chemical stimuli. This method provides an in vivo demonstration of how mosquito larvae perceive volatile semiochemicals in their environment.

MATERIALS

It is essential that you consult the appropriate Material Safety Data Sheets and your institution's Environmental Health and Safety Office for proper handling of equipment and hazardous materials used in this protocol.

Reagents

Anopheles coluzzii and other anopheline larvae

Diethyl pyrocarbonate (DEPC)-treated double-distilled H_2O, dimethyl sulfoxide (DMSO), or paraffin oil

Distilled H_2O

Fleischmann's Rapid-Rise Instant Yeast (ACH Food Companies, Inc.)

Kaytee Koi's Choice Premium Fish Food

KNO_2 (10%)

Odorants of the highest purity, typically ≥99%, diluted 100-fold (Kreher et al. 2005) (Sigma-Aldrich)

Dilute the compounds in paraffin oil, DMSO, or DEPC-treated double distilled H_2O to make 1:100 solutions at specified concentrations in 1.5-mL centrifuge tubes and vortex well. Store the solutions for no longer than 1 wk before use at 4°C.

---

[1]These authors contributed equally to this work.

[2]Correspondence: l.zwiebel@vanderbilt.edu

Cite this protocol as *Cold Spring Harb Protoc*; doi:10.1101/pdb.prot108020

Equipment

AC power supply (BK Precision)
Analog-to-digital signal converter (e.g., IDAC-4, Ockenfels Syntech GmbH)
Climate-controlled rearing chamber (Percival Scientific)
Custom-made glass tubing (10-mm inner diameter) with a small hole ~10 cm from one end of the tube into which the tip of the Pasteur pipette can be inserted
Dental wax (e.g., Orthomechanics)
Disposable plastic pipette, large-bore
Double-sided tape (3M Corporation)
Electronic balance
Filter paper (3-mm × 10-mm)
Fine forceps with sharp tips (Dumont #5)
Glass container with a lid, 500-mL
Insect pins
Insectary at 27°C and 75% relative humidity
Inverted microscope (e.g., Olympus BX51WI)
KimWipes (Kimberly Clark/Fisher)
Manual micromanipulator (e.g., World Precision Instruments)
Mechanical (bubble-type) variable area air flowmeter (Fisher Scientific)
Microscope slides (76-mm × 26-mm)
Pasteur pipette
Piezo-controlled micromanipulator (e.g., Burleigh Model PCS6000)
Preamplifier (e.g., Syntech Universal AC/DC 10×)
Refrigerator (4°C)
Sterilite plastic rearing pan (35.6-cm × 27.9-cm × 8.3-cm)
Stimulus controller (e.g., CS-55, Ockenfels Syntech GmbH)
Tungsten wire (0.008 in diameter × 3 in length, A-M Systems)
Vortex mixer
Windows-based computer with Syntech (or comparable) electrophysiology recording/analysis software (e.g., AutoSpike)

METHOD

Rearing Healthy *An. coluzzii* Larvae

Colonies of anopheline mosquitoes are typically maintained in dedicated insectaries to provide security and containment. Rear anopheline larvae and adults at 27°C and 75% relative humidity (RH) in a climate-controlled rearing chamber with distilled H_2O under the standard 12 h light/12 h dark cycle (Fox et al. 2001; Suh et al. 2016). To keep the aquatic environment clean, the amount of larval food used is critical in the protocol.

1. Prepare 12 g Kaytee Koi's Choice Premium Fish Food and 6 g yeast using an electronic balance. Pour the powder into a glass container with a lid, then add 100 mL distilled H_2O. Screw the lid on tightly and mix well. Subsequently incubate the mixture overnight at 4°C for fermentation. This is now referred to as larval food, which can be stored for up to 1 wk at 4°C.
2. In the rearing chamber, transfer approximately 300 first-instar larvae using a large-bore disposable plastic pipette into the rearing pan filled with 1000 mL distilled H_2O. Add 0.08 mL larval food into the H_2O and mix well. Repeat the addition of larval food every 24 h until the larvae develop to third instar (7–9 mm).

Mounting *An. coluzzii* Larvae

This protocol should be restricted solely to third-instar larvae as their bodies are soft and fragile, especially the appendages on the head. Carry out the following steps gently at room temperature.

3. Place a bead of dental wax (1 cm diameter) on the microscope slide to be used as a pivot point. Affix double-sided tape to a coverslip and lean that coverslip on the dental wax bead at ~60°–90° relative to where the head will be placed, with the tape side opposite to the dental wax (Fig. 1A).
4. Grasp the abdomen of a larva with fine forceps and transfer it from the rearing pan to a KimWipe for 30 sec to remove excess liquid. Take care to avoid rolling or otherwise damaging the antennae.
5. After drying, use forceps to mount the individual larva on the microscope slide. Keep the dorsal surface of the larval abdomen and thorax in contact with the microscope slide and bend the larval head on the coverslip. Using an insect pin, gently press the larval head onto the double-sided tape to prevent further movement. Adjust larval antennae gently at ~15°–20° relative to the head by manipulating them with the insect pin (Fig. 1A).

Electrophysiological Recording of *An. coluzzii* Larval Antennal Sensory Cone

6. Sharpen tungsten microelectrodes by dipping the tungsten wire (0.008 in diameter ×3 in length) into 10% KNO_2 at 10 V from an AC power supply. Move the wire back and forth by hand for ~100 times until the tip is sharpened.

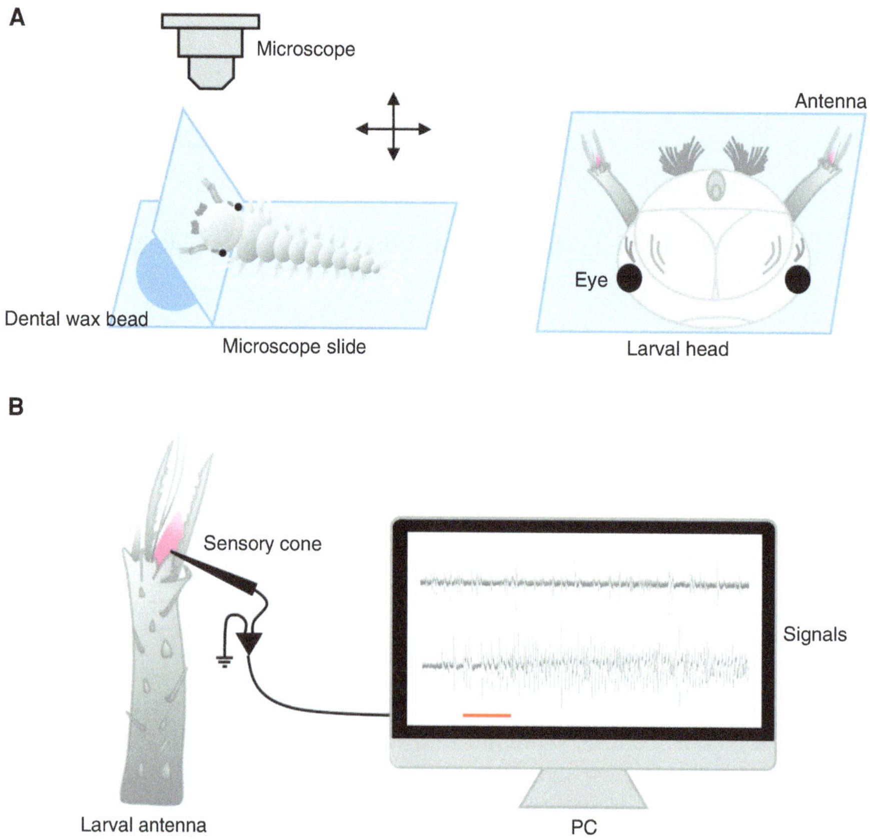

FIGURE 1. Electrophysiological recording of the mosquito larval antennal sensory cone. (*A*) (*Left*) Schematic of larval electrophysiology recording setup. (D) Dorsal, (V) ventral, (L) left, (R) right. (*Right*) The location of the larval head and antennae on the plastic coverslip. (*B*) (*Left*) Schematic of the external morphology of the mosquito larval antenna and the position of the recording electrode in the larval sensory cone. (*Right*) Representative signal traces in the mosquito larval sensory cone.

 Cite this protocol as *Cold Spring Harb Protoc*; doi:10.1101/pdb.prot108020

Sharpened electrodes can be prepared in bulk and stored at room temperature before the recording. Replace electrodes when the tip becomes blunt or curved.

7. Connect the preamplifier to an analog-to-digital signal converter at a sample rate of ~96,000 samples per sec and 10× amplification and connect the signal converter to a PC for signal recording. Connect all electrodes to the preamplifier.
8. Once mounted, place the specimen onto the stage of the inverted microscope. Visualize the larval antenna and its sensory cone at moderate (200×) and high (1000×) magnification to confirm placement of the entire preparation.

   The exact position and shape of larval antenna and sensory cone are presented in the supporting information of Xia et al. (2008).
9. Slowly move the ground (reference) electrode toward the larval compound eye using a micromanipulator at moderate (200×) magnification until the tip has completely inserted into the compound eye.
10. Carefully insert the recording electrode into the base of the antennal sensory cone using a piezo-controlled micromanipulator at high (1000×) magnification to complete the electrical circuit to facilitate extracellular recording of olfactory sensory neuron potentials from the larval sensory cone (Fig. 1B).

Odorant Preparation and Stimulus Application

This assay is closely related to single-sensillum recordings (SSRs) from adult mosquitoes and, in that context, we recommend viewing Pellegrino et al. (2010) for additional information and context on insect SSRs. For larval recordings, we test at least six animals independently for each compound and test no more than 30 compounds for each animal/preparation. Always assay control compounds at the onset of each recording preparation. For concentration response assays of any single compound, always proceed from low to high concentration with at least a 30 sec interval between each stimulation.

11. Allow the odorant solution to equilibrate to room temperature and then aliquot 10 µL onto a small piece (3-mm × 10-mm) of filter paper. Insert the filter paper into a Pasteur pipette to create the stimulus cartridge.
12. Using a bubble-tube flowmeter, ensure that a constant 20 mL/sec airflow is maintained from the stimulus controller through a glass tube that has a small hole (10 cm from its end) into which the stimulus cartridge will be inserted. Once an adequate airflow is established, place the glass tube toward the mosquito larval head such that the end of the glass tube is positioned within 1 cm of the larval head.
13. For stimulus application, insert the Pasteur pipette stimulus cartridge into the small hole of the glass tube. Using the foot pad switch of the stimulus controller, divert a portion of the air stream (0.5 L/min) to flow through the stimulus cartridge for 0.5 sec (Fig. 1B).
14. Set AutoSpike software at a 96,000/sec sampling rate. Record signals were for 10 sec starting 1 sec before odorant stimulation.

DISCUSSION

Although anopheline mosquitoes live in the water for most of their larval stage, they are also known to display modest amphibian characteristics. These can appear when oviposition takes place such that subsequent egg hatching occurs on damp soil rather than in aqueous habitats. Here, larval instars sometimes survive and can show a slow and forward form of caterpillar-like movement toward standing water (Koenraadt et al. 2003; Miller et al. 2007). Furthermore, unlike aedine or culicine mosquito larvae, which usually feed on the bottom of their aqueous habitats, anopheline larvae largely use their mouth brushes to filter feed on algae, bacteria, and other microorganisms or organic compounds in the water surface microlayer (Gimmg et al. 2001; Fillinger et al. 2004). This unique biology suggests that anopheline larvae are likely to position their antennae above the surface of their aqueous habitats, such that their sensory cone is effectively exposed to the air and thus able to sample and detect airborne odors in that headspace. The direct recording of neuronal responses from the larval sensory cone to such volatiles closely approximates the well-established technique of SSR of adult olfactory responses, thereby facilitating the characterization of larval chemosensory neurons' responses to volatile compounds (Sun et al. 2021).

REFERENCES

Fillinger U, Sonye G, Killeen GF, Knols BGJ, Becker N. 2004. The practical importance of permanent and semipermanent habitats for controlling aquatic stages of *Anopheles gambiae sensu lato* mosquitoes: Operational observations from a rural town in western Kenya. *Trop Med Int Heal* **9:** 1274–1289. doi:10.1111/j.1365-3156.2004.01335.x

Fox AN, Pitts RJ, Robertson HM, Carlson JR, Zwiebel LJ. 2001. Candidate odorant receptors from the malaria vector mosquito *Anopheles gambiae* and evidence of down-regulation in response to blood feeding. *Proc Natl Acad Sci* **98:** 14693–14697. doi:10.1073/pnas.261432998

Gimmg JE, Ombok M, Kamau L, Hawley WA. 2001. Characteristics of larval anopheline (Diptera: Culicidae) habitats in western Kenya. *J Med Entomol* **38:** 282–288. doi:10.1603/0022-2585-38.2.282

Koenraadt CJ, Paaijmans KP, Githeko AK, Knols BG, Takken W. 2003. Egg hatching, larval movement and larval survival of the malaria vector *Anopheles gambiae* in desiccating habitats. *Malar J* **2:** 20. doi:10.1186/1475-2875-2-20

Kreher SA, Kwon JY, Carlson JR. 2005. The molecular basis of odor coding in the *Drosophila* larva. *Neuron* **46:** 445–456. doi:10.1016/j.neuron.2005.04.007

Miller JR, Huang J, Vulule J, Walker ED. 2007. Life on the edge: African malaria mosquito (*Anopheles gambiae s.l.*) larvae are amphibious. *Naturwissenschaften* **94:** 195–199. doi:10.1007/s00114-006-0178-y

Pellegrino M, Nakagawa T, Vosshall LB. 2010. Single sensillum recordings in the insects *Drosophila melanogaster* and *Anopheles gambiae*. *J Vis Exp* **2010:** 1–5. doi:10.3791/1725

Suh E, Choe D-H, Ahmed SM, Zwiebel LJ. 2016. Suboptimal larval habitats mediate oviposition of the malaria vector mosquito *Anopheles coluzzii*. *PLoS ONE* **11:** e0149800. doi:10.1371/journal.pone.0149800

Sun H, Liu F, Baker AP, Honegger HW, Raiser G, Zwiebel LJ. 2021. Neuronal odor coding in the larval sensory cone of *Anopheles coluzzii*: complex responses from a simple system. *Cell Rep* **36:** 109555. doi:10.1016/j.celrep.2021.109555

Xia Y, Wang G, Buscariollo D, Pitts RJ, Wenger H, Zwiebel LJ. 2008. The molecular and cellular basis of olfactory-driven behavior in *Anopheles gambiae* larvae. *Proc Natl Acad Sci* **105:** 6433–6438. doi:10.1073/pnas.0801007105

 Cite this protocol as *Cold Spring Harb Protoc*; doi:10.1101/pdb.prot108020

Protocol 2

Cup and Pan Behavioral Assays for Assessing *Anopheles coluzzii* Larval Volatile Responses

Feng Liu,[1] Huahua Sun,[1] and Laurence J. Zwiebel[2]

Department of Biological Sciences, Vanderbilt University, Nashville, Tennessee 37235, USA

Larval stage *Anopheles coluzzii* are highly reliant on their olfactory system to locate food sources and to avoid predators and less advantageous microenvironments within their aqueous habitats. The major larval chemosensory appendage, the antenna, is a complex organ with multiple sensory components that is responsible for both gustation and olfaction, thereby facilitating the detection and of both soluble and volatile compounds of biological relevance. Such compounds include food sources, predators, and a range of environmental toxicants. Unlike other mosquitoes, *Anopheles coluzzii* often position themselves parallel and just under the surface of their aqueous habitats, where they can detect and respond to volatile stimuli. We describe two assays for assessing the behavioral responses of larval anophelines in response to volatile chemicals. The first is a dual-choice, water-surface, inverted-cup assay designed to behaviorally characterize the response valences (attraction, neutral, and repulsion) of anopheline larvae by monitoring and recording the distribution of larvae proximate to chemical volatiles relative to solvent controls. Second, an aqueous-based larval pan behavior assay is designed to assess the responses of mosquito larvae to soluble compounds (as well as potential headspace volatiles) that are released from a point source within larval water. Here, the response valence (attractive, neutral, and repulsive) of mosquito larvae is assessed by quantifying the numbers of larvae in predefined zones proximate to chemical sources.

MATERIALS

It is essential that you consult the appropriate Material Safety Data Sheets and your institution's Environmental Health and Safety Office for proper handling of equipment and hazardous materials used in this protocol.

Reagents

Anopheles coluzzii larvae

Diethyl pyrocarbonate (DEPC)-treated double-distilled H_2O, dimethyl sulfoxide (DMSO), or paraffin oil

Distilled H_2O

Fleischmann's Rapid-Rise Instant Yeast (ACH Food Companies, Inc.)

Kaytee Koi's Choice Premium Fish Food

NuSieve, GTG low-melting temperature agarose (Cambrex Bio Science)

This item is used for the larval pan assay only.

[1]These authors contributed equally to this work.

[2]Correspondence: l.zwiebel@vanderbilt.edu

Cite this protocol as *Cold Spring Harb Protoc*; doi:10.1101/pdb.prot108021

Odorants of the highest purity, typically ≥99% (Sigma-Aldrich)

Many of these compounds are purchased as liquids and can therefore be prepared as direct (v/v) dilutions in appropriate solvents (typically paraffin oil, DMSO, or DEPC-treated double distilled H_2O) to make 1:100 (e.g., 10^{-2} dilution) stimulus solutions. For compounds purchased as solid powders, carefully weigh out the appropriate amount using an electronic balance to prepare a similar m/v dilution (Kreher et al. 2005). Place solutions at specified concentrations in 1.5-mL centrifuge tubes and vortex well. Prepare the odorant solutions immediately before the behavioral assay and place them in the rearing chamber (27°C/70% humidity) where the assay will be conducted to allow them to equilibrate to those conditions.

Equipment

Clear glass dish (e.g., Pyrex; 38.1-cm × 25.4-cm × 5.08-cm)

Computer; Windows PC with GraphPad Prism software that can be used to assess images from digital cameras and/or flash cards

Electronic balance

Equipment needed for the larval pan assay only

- Clear adhesive tape (e.g., Scotch Magic Tape; 1.9-cm-width, 3M Corporation)
- Cotton thread
- Custom-made metal container (5-mL)
- Heat block
- Micropipettes (10-µL, 1000-µL)
- Mini glass vial (1-mL; 1.5-cm (d) × 1.9-cm (h), Qorpak CAP-00545)
- Nylon mesh (0.25-mm opening; Elko, Fisher Scientific)
- Scissors
- Sewing needle (or comparable)

Glass container with a lid, 500-mL

Insect rearing chambers or incubators maintained at 27°C and 75% relative humidity

Large-bore disposable plastic pipette

Metal sieve (6″ diameter, 1-mm opening, Gilson Co., No. 18)

Mobile phone, GoPro camera, or other device capable of taking a digital image

Pasteur pipette

Plastic cups, 5.5-oz with 5-cm-diameter lid (AV Inc.)

Plastic rearing pan (e.g., Sterilite; 35.6-cm × 27.9-cm × 8.3-cm)

Vortex mixer

METHOD

Rearing Healthy *An. coluzzii* Larvae and Preparing Larval Food

Colonies of anopheline mosquitoes are typically maintained in dedicated insectaries to provide security and containment. Anopheline larvae and adults are reared at 27°C and 75% relative humidity (RH) in a climate-controlled rearing chamber with distilled H_2O under the standard 12 h light/12 h dark cycle (Fox et al. 2001; Suh et al. 2016). To keep the aquatic environment clean, strictly limiting the amount of larval food used is critical in the protocol.

1. Prepare 12 g Kaytee Koi's Choice Premium Fish Food and 6 g yeast using an electronic balance. Pour the powder into a glass container with a lid, then add 100 mL distilled H_2O. Screw the lid on tightly and mix well. Subsequently incubate the mixture overnight at 4°C for fermentation. This is now referred to as larval food, which can be stored for up to 1 wk at 4°C.
2. In the rearing chamber, transfer approximately 300 first-instar larvae using a large-bore disposable plastic pipette into the rearing pan filled with 1000 mL distilled H_2O. Add 0.08 mL larval food into the H_2O and mix well. Repeat the addition of larval food every 24 h until the larvae develop to the third instar (7–9 mm) (Jannat and Roitberg 2013).

Cite this protocol as *Cold Spring Harb Protoc*; doi:10.1101/pdb.prot108021

Introducing the Mosquito Larvae into the Arena

3. Gently collect 100 *An. coluzzii* third-instar larvae (7–9 mm) using a Pasteur pipette and place them in 5.5-oz plastic cups containing 27°C distilled H_2O. Gently rinse the larvae by removing and replacing the H_2O five times using a 6-in-diameter metal sieve to retain the larvae. Thereafter maintain them in the cups without larval food for 2 h and leave the cups in the rearing chamber.
4. Fill a glass dish with 500 mL of 27°C distilled H_2O and place it on a level surface in a temperature-controlled walk-in environmental rearing chamber that is maintained at 27°C/75% humidity.
5. Using a wide-bore Pasteur pipette, gently introduce the food-deprived larvae into the center of the dish and allow them to swim freely for 30 min to acclimate to the arena.

   To perform the cup assay, proceed to Step 6. To perform the pan assay, proceed to Step 10.

Preparing the Odorant Stimuli Setup (Cup Assay)

6. Cut a 3-cm-diameter hole in the center of a 5.5-oz plastic cup lid.

   Importantly, leave the raised edge of inner side of the cup lid around the 3-cm-diameter hole intact.

7. Transfer 500 µL of the diluted odorant or solvent to the inner side of the cup lid so that it always remains out of contact with the larval rearing H_2O to provide a volatile odor (treatment) or solvent (control) stimuli within each inverted-cup stimulus chamber.
8. Assemble the 5.5-oz plastic cup with the lid containing solvent or odorant. Place the chambers on the surface of the H_2O within the glass dish that contains larvae. Place the cups (by leaning them on the inner side of the glass container so as to allow the very lightweight cups to float on the water surface) at opposite ends of the behavioral arena (Fig. 1A).
9. Allow mosquito larvae to swim freely in the arena between treatment and control cups for 1 h in the rearing chamber. Thereafter, obtain digital image recording (i.e., with a mobile phone or GoPro camera or other setup equipped with automated time-lapse photography) and manually count the number of larvae directly under each cup lid. The relative position of the odorant cup and control cup should be randomly interchanged among different replicates.

   Proceed to Step 17.

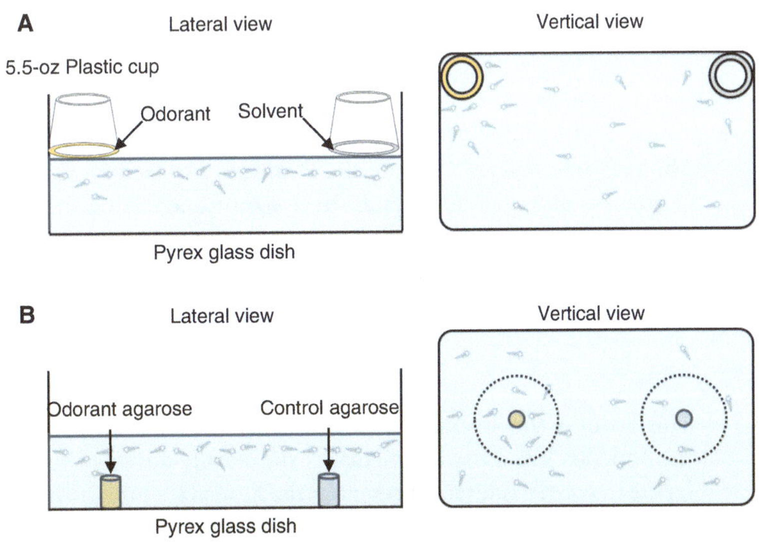

FIGURE 1. Schematic of the mosquito larval behavior assays. (*A*) Schematic of inverted cup behavioral assay for volatile semiochemicals. (*Left*) Lateral view. (*Right*) Vertical view. (*B*) Schematic of pan behavior assay for chemosensory stimuli. (*Left*) Lateral view. (*Right*) Vertical view.

Preparing the Odorant Stimuli Setup (Pan Assay)

10. Cut the nylon mesh into a 0.8-cm × 5.5-cm rectangle and then sew up the short edges of the rectangle mesh using a needle and thread to form a mesh ring. Prepare two rings.
11. Prepare a 2% agarose solution with GTG low-melting temperature agarose and distilled H_2O in the 5-mL mini metal container. Mix well and incubate the mixture for 10–15 min at 70°C on a metal heat block.

    Prepare 1 mL of agarose for each agarose block that will be used.
12. At the same time, turn the mini glass vial upside down and place adhesive tape around the bottom, leaving a 1-cm edge of tape extending beyond the bottom edge of the mini glass vial. This will be the mold for casting the odorant (or solvent)-containing agarose blocks in Step 14. Prepare as many agarose blocks as required for the day's trials.
13. When the solution becomes transparent, place the container containing agarose on the laboratory bench at room temperature. Remove 1 mL of agarose to use in the control block. Add the appropriate amount of odorant stock solution to achieve the desired final concentration to the agarose solution using a 10-µL micropipette and mix well. Add an equal amount of solvent to 1 mL of agarose as a solvent-only control block.
14. Add 1 mL odorant agarose solution into the adhesive tape mold prepared in Step 12 using a 1000-µL micropipette. Add 1 mL agarose solution without odorant to a second mold as a control. When the solution becomes solid at room temperature, remove the tape and insert the solid agarose block into the mesh ring prepared as described in Step 12.
15. Print templates for the test/control circular zones to be visualized on both ends of the glass dish on a sheet of paper sized to match the entire behavioral arena. Place the template directly under the clear glass dish to create an easily visualized behavioral arena with two circular zones (Fig. 1B). Place the odorant/control agarose mesh ring in the center of each test/control zone area.
16. After 1 min, activate a prepositioned digital camera setup to capture a 30-min real-time video of the entire experimental arena.

    Proceed to Step 17.

Quantifying the Preference of Larvae to Odorant or Solvent

17. After the assay is completed calculate the performance index (PI) (Heimbeck et al. 1999).

    For the cup assay, count the number of larvae directly under the cup lids at discrete (typically 1-h) time points. For the pan assay, calculate the PI at a discrete 15 min time point. Calculate the PI as follows: PI = (#odorant − #control)/(#odorant + #control), where the #odorant indicates the number of larvae in the odorant-treated side and the #control indicates the number in the control side.
18. For both assays, compare the PI values for each compound with that of solvent alone control assays and analyze for statistical significance using unpaired, two-tailed Student's *t*-tests with Welch's within the GraphPad Prism software platform.

DISCUSSION

The layout of the inverted cup assay, which prevents the dispersal and solubilization of odorants into the aqueous larval rearing media, is specifically designed to characterize the responses of mosquito larvae to volatile odorants present in the headspace just above the water surface. It has been found that mosquito larvae are more likely to stay in standing water (Amani et al. 2014; Soleimani-Ahmadi et al. 2014) containing potential organically abundant food sources, such as plant or animal debris, rotten vegetation, and microorganisms (Sattler et al. 2005; Nilsson et al. 2018). The ability to detect the odorants in the air may increase larval opportunities to locate nearby food sources, especially rotten or

Cite this protocol as *Cold Spring Harb Protoc*; doi:10.1101/pdb.prot108021

fermented organic debris, which are expected to emanate multiple compounds into both the water and air.

The water-based pan behavior assay allows anopheline larvae to be in both direct as well as potentially volatile-based contact with chemical stimuli released from the agarose plug. Accordingly, this assay cannot distinguish between gustatory and olfactory chemosensory responses (Xia et al. 2008). Instead, it uses low-melting agarose to create a slow-release point source for dispersal of chemical stimuli in larval water, thereby creating gradient responses. As such, although this assay is useful for a range of hydrophilic compounds, it may not be amenable for assessing the behavioral responses of mosquito larvae to most hydrophobic compounds, which do not readily dissolve in melted agarose.

REFERENCES

Amani H, Yaghoobi-Ershadi MR, Kassiri H. 2014. The ecology and larval habitats characteristics of anopheline mosquitoes (Diptera: Culicidae) in Aligudarz County (Luristan province, western Iran). *Asian Pac J Trop Biomed* **4:** S233–S241. doi:10.12980/APJTB.4.2014C186

Fox AN, Pitts RJ, Robertson HM, Carlson JR, Zwiebel LJ. 2001. Candidate odorant receptors from the malaria vector mosquito *Anopheles gambiae* and evidence of down-regulation in response to blood feeding. *Proc Natl Acad Sci* **98:** 14693–14697. doi:10.1073/pnas.261432998

Heimbeck G, Bugnon V, Gendre N, Haberlin C, Stocker RF. 1999. Smell and taste perception in *Drosophila melanogaster* larva: toxin expression studies in chemosensory neurons. *J Neurosci* **19:** 6599–6609. doi:10.1523/JNEUROSCI.19-15-06599.1999

Kreher SA, Kwon JY, Carlson JR. 2005. The molecular basis of odor coding in the *Drosophila* larva. *Neuron* **46:** 445–456. doi:10.1016/j.neuron.2005.04.007

Nilsson LK, Sharma A, Bhatnagar RK, Bertilsson S, Terenius O. 2018. Presence of *Aedes* and *Anopheles* mosquito larvae is correlated to bacteria found in domestic water-storage containers. *FEMS Microbiol Ecol* **94:** fiy058. doi:10.1093/femsec/fiy058

Sattler MA, Mtasiwa D, Kiama M, Premji Z, Tanner M, Killeen GF, Lengeler C. 2005. Habitat characterization and spatial distribution of *Anopheles* sp. mosquito larvae in Dar es Salaam (Tanzania) during an extended dry period. *Malaria J* **4:** 1–15. doi:10.1186/1475-2875-4-4

Soleimani-Ahmadi M, Vatandoost H, Zare M. 2014. Characterization of larval habitats for anopheline mosquitoes in a malarious area under elimination program in the southeast of Iran. *Asian Pac J Trop Biomed* **4:** S73–S80. doi:10.12980/APJTB.4.2014C899

Suh E, Choe D-H, Ahmed SM, Zwiebel LJ. 2016. Suboptimal larval habitats mediate oviposition of the malaria vector mosquito *Anopheles coluzzii*. *PLoS ONE* **11:** e0149800. doi:10.1371/journal.pone.0149800

Xia Y, Wang G, Buscariollo D, Pitts RJ, Wenger H, Zwiebel LJ. 2008. The molecular and cellular basis of olfactory-driven behavior in *Anopheles gambiae* larvae. *Proc Natl Acad Sci* **105:** 6433–6438. doi:10.1073/pnas.0801007105

CHAPTER 32

Mosquito Embryo Microinjection

Robert A. Harrell II[1,2]

[1]*The Institute for Bioscience and Biotechnology Research, University of Maryland Insect Transformation Facility, Rockville, Maryland 20850, USA*

Genetically modified (GM) mosquitoes are an important tool in the fight against mosquito-borne disease, both indirectly through their use in research investigating host–pathogen interaction, mosquito olfaction, and anthropomorphic behavior and in future direct uses for suppression and possibly eradication through sterile insect technique (SIT) and/or gene-drive programs. Successful creation of GM mosquitoes depends on microinjection procedures that precisely deliver injection materials while causing as little damage to mosquito embryos as possible. Genetic modification reagents, such as transposon system components (vector plasmids, helper plasmids, and helper mRNA), and CRISPR–Cas9 components (guide RNAs, Cas9 protein, plasmids expressing Cas9 and/or guide RNAs, and donor plasmids used in homology-directed repair [HDR]), must be delivered into the preblastoderm embryo at the posterior end where the pole cells will form before cellularization occurs. Sharp needles that pierce the embryo easily are important tools in this procedure and work best when the embryos are not desiccated. The two main procedures for mosquito embryo microinjection involve injecting embryos under halocarbon oil or under aqueous solution.

INTRODUCTION

Embryo microinjection has been the "go to" method for delivery of genetic modification reagents such as transposon system components (vector plasmids, helper plasmids, and helper mRNA) and CRISPR–Cas9 components (guide RNAs, Cas9 protein, Cas9- and/or guide-RNA-expressing plasmids, and donor plasmids used in homology-directed repair [HDR]) since Rubin and Spradling (1982) developed embryo injection for *Drosophila* genetic modification. Other processes for delivery have been tried over the years, such as electroporation, biolistics, and adult ovary injection (receptor-mediated ovary transduction of cargo [ReMOT] and direct parental CRISPR [DIPA-CRISPR]), but these methods have proven either inefficient (electroporation and biolistics) or were only demonstrated to work in CRISPR mutation, not in transposon delivery or HDR (Chaverra-Rodriguez et al. 2018; Macias et al. 2020; Shirai et al. 2022). The process of embryo microinjection for genetic modification in mosquitoes is more of an art than a technique, as success requires both optimally reared insects and delivery of injection materials at a precise time and location during development. Although optimal rearing is of equal importance for the successful genetic modification of mosquitoes, this introduction will focus on the delivery and timing aspects of insect embryo microinjection.

INJECTION TIMING AND LOCATION

Successful insect genetic modification requires a basic understanding of early embryo development for the species of insect being modified. Specifically, the relevant period of development for genetic

[2]Correspondence: harrelr@umd.edu

Cite this introduction as *Cold Spring Harb Protoc*; doi:10.1101/pdb.top107686

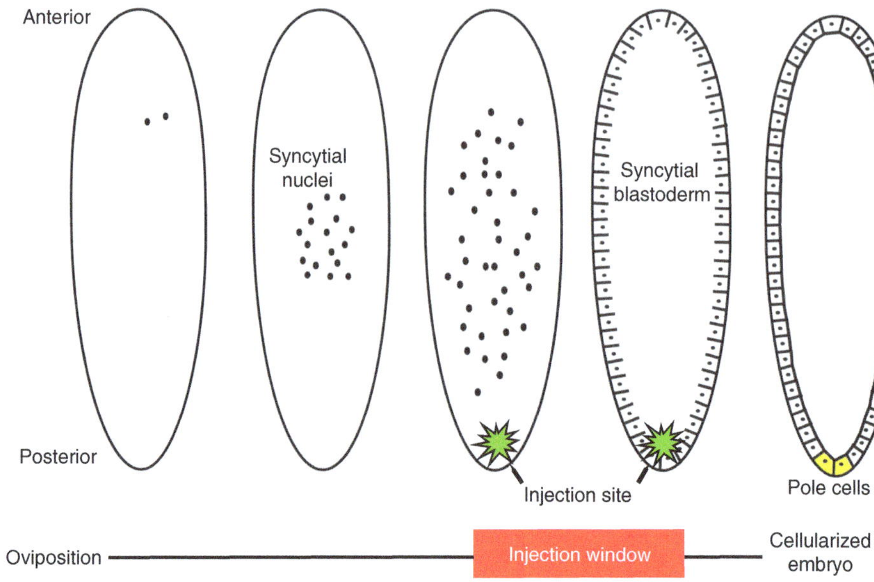

FIGURE 1. Embryo development as related to injection site and timing. Schematic representation of mosquito embryo development from oviposition, time 0, to cellularization at ~2 h postoviposition (*left* to *right*). Green injection icons indicate injection site placement at the posterior polar end of the embryo. Approximate injection time window indicated in red on the timeline between oviposition and cellularization. (Figure based on Gilbert 2000.)

modification is from embryo fertilization to cellularization. At this early stage in development, mosquito embryos develop similarly to those of *Drosophila*, the first insect to be genetically modified (Rubin and Spradling 1982). Like the *Drosophila* embryo, the early mosquito embryo develops as a syncytium, where nuclei divide without cell division (Fig. 1; Clements 1992). Eventually the nuclei migrate toward the periphery of the embryo, with the polar nuclei migrating to the posterior end of the embryo. Pole cells, which are embryonic stem cells that will become the gametes in adults, form in the posterior end of the embryo, and this is the region in which injection is carried out. It is critical for the injection mix to be delivered prior to cellularization so that the modifying agents such as transposon system components or CRISPR–Cas9 components have access to the pole cell nuclei. Once membranes form between the nuclei, injection materials are no longer free to diffuse. Hypothetically, injecting at an earlier point in development may be advantageous, because if the modifying agents have access to dividing nuclei at an earlier stage and are able to modify the DNA, then all cells derived from those modified nuclei will also be modified, leading to a greater proportion of modified offspring. Injecting at earlier stages has been successful in other insects, such as *Danaus plexippus*, the monarch butterfly (Markert et al. 2016) and *Phlebotomus papatasi*, the sand fly (Louradour et al. 2019). Injecting mosquito embryos at an earlier stage, however, markedly decreases survival of injected individuals, therefore lowering the efficiency of the process. In mosquito embryos, injections should therefore be timed to occur during the late stages of the syncytial blastoderm just prior to cellularization (Clements 1992).

IMPORTANCE OF SHARP NEEDLES

Successful embryo microinjection in mosquitoes depends on injections that cause as little damage to the injected embryo as possible. Ideally the injection needle should slip into the embryo, deliver the injection material, and then be removed without losing any embryoplasm, the cytoplasm of the embryo. Embryos that leak embryoplasm will either die from the trauma caused to the embryo during a bad injection or because of fungal growth attracted to the rich nutrients of the leaking

 Cite this introduction as *Cold Spring Harb Protoc*; doi:10.1101/pdb.top107686

embryoplasm. Alternatively, embryos that leak may result in infertile adults because of the loss of pole nuclei in the leaking material. Good injections require sharp needles and good technique for insertion and removal of the needle, thereby avoiding injection site leakage.

Sharp needles are a major key to successful microinjections. Glass needles are fabricated by heating glass microcapillaries until the glass is elastic, at which point a force is exerted on the ends of the microcapillary, drawing the ends away from the center and creating a taper down to a closed point. The type of glass used and the puller used for fabricating injection needles does not ultimately matter, as long as the end product is a sharp needle capable of delivering injection materials without causing embryoplasm to leak after the injection is complete and the needle is withdrawn. Sharp needles can be pulled and used directly or can be further modified by beveling, a process using a diamond abrasive disk to hone the tip of the microneedle. Whether using needles directly after pulling or beveling the needle, the process for developing a sharp needle for injections is the same; it involves multiple rounds of adjusting needle pulling parameters followed by injection of appropriately aged embryos with injection materials. It is important to inject embryos with actual injection material and not just buffer solution because buffer alone has much lower viscosity. The following parameters on a Sutter P-2000 needle puller are what is used by the University of Maryland Insect Transformation Facility for creating needles: quartz needles (heat: 735, fil: 4 vel: 42: del: 126, and pul:160) or borosilicate needles that are beveled (heat: 305, fil: 4, vel: 70, del: 235, and pul:160).

When starting to pull needles for the first time in a laboratory, one option is to start with an example needle from a laboratory successfully microinjecting mosquito embryos, adjusting the parameters of the needle puller until the shape of the needle is similar to the example, and then to keep adjusting and testing needles until the needle is piercing smoothly. Alternatively, start with pulling parameters that are known to produce good mosquito embryo quality needles, and then test and adjust the parameters. Keep in mind that needle pullers are not highly calibrated between machines, meaning parameters on one puller will achieve a similar needle but not the exact same needle; therefore, all needles produced need to be tested and refined. Appropriately sharp needles slip through the membrane of the embryo with little or no resistance. Smooth piercing is evident from the indentation of the membrane from the time the needle first touches the membrane to the point at which it pierces and enters into the embryo. Extremely sharp needles will have barely perceptible indentations of the membrane, whereas less sharp needles will have exaggerated indentations.

EMBRYO DESICCATION

During collection and alignment of mosquito embryos for microinjection, it is important to prevent water loss from the embryos. Mosquito embryos that desiccate during preparation for injections do not develop and die before hatching. It is critical to postinjection survival that the embryos remain properly hydrated during the entire alignment phase of preparing embryos for injection. When the membrane of a properly hydrated embryo is lightly touched with the tip of the needle, it should spring back in place once the needle is removed. Embryos with membranes that do not spring back are overdesiccated. Prevention of desiccation is also important for proper needle piercing of the embryo membrane. For the needle to pierce efficiently and smoothly, the internal pressure on the embryo membrane of a properly hydrated embryo provides resistance against the needle tip. As the tip of a sharp needle touches a properly hydrated embryonic membrane, it will slip through the membrane, whereas the tip of a needle will cause the membrane to bend inward on a desiccated embryo. The lower the internal pressure, the harder it is for the tip of the needle to pierce the embryonic membrane even if it is sharp. If the needle is dull or the internal embryo pressure is somewhat low, there is a tendency for the membrane to tear, which leads to embryoplasm leaking after injection. The key to good injections is, therefore, a combination of appropriately turgid embryos and very sharp needles.

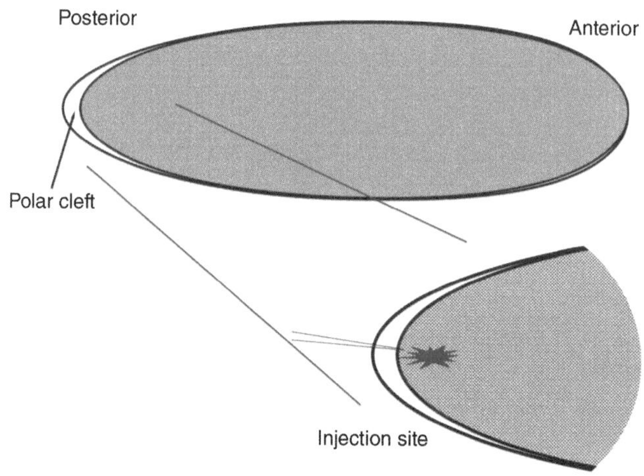

FIGURE 2. Injection site location. A schematic view of a mosquito embryo with *inset* view indicating the target area for injection.

INJECTION, A FOUR-STEP PROCESS

Each injection can be broken into four phases: orientation, insertion, injection, and withdrawal. Orientation is the process of locating the position on the embryo that is appropriate for easy insertion of the needle at the best location for delivery of modification material to the posterior pole plasm. Insertion is the process of sliding the needle through the embryo membrane so that it causes as little damage to the membrane as possible. Avoiding membrane damage relies on needle sharpness, embryo desiccation, membrane location injected, and technique used for insertion. Insertion can be by slow steady pressure until the needle pierces or it may be a slight bounce of the needle against the membrane. Once the needle has pierced, it should be guided slowly to the correct depth into the embryo. This depth is just beyond the clear area of the polar cleft, if visible, and just interior to the inner membrane that divides the clear area of embryoplasm and the light dense region of embryoplasm (Fig. 2). Upon reaching this region the injection should be triggered, delivering enough material such that the injection is just barely perceivable as a slight lightening in this region of embryoplasm. The needle should then be removed, making sure that no embryoplasm leaks from the injection site.

THE TWO MAJOR STRATEGIES FOR EMBRYO ALIGNMENT AND INJECTION

Ideally, embryos should be kept in as native an environment as possible during injection. There are two main strategies currently being used for mosquito embryo microinjection: alignment and injection under oil and alignment and injection in aqueous solution. Both are described in our associated protocol (see Protocol 1: Mosquito Embryo Microinjection under Halocarbon Oil or in Aqueous Solution [Harrell 2023]). Alignment and injection under oil is the older of the two processes. This technique was modified from Rubin and Spradling's (1982) method for injecting *Drosophila* embryos and was first used by Miller et al. (1987) to transform *Anopheles gambiae.* With modifications, this procedure was later used to genetically modify *Aedes aegypti* (Jasinskiene et al. 1998), *Culex quinquefasciatus* (Allen et al. 2001), and *Anopheles albimanus* (Perera et al. 2002). Alignment and injection in aqueous solution was first used in *Anopheles stephensi* (Catteruccia et al. 2000), and with refinements is routinely used for genetic modification of *Ae. gambiae* (Volohonsky et al. 2015). Both of these strategies can be used to successfully produce genetic modifications in mosquitoes. A valid argument can be made that injection in aqueous solution offers a slight advantage for *Anopheles* injections because *Anopheles* are particularly susceptible to desiccation. The choice of method depends mainly on which species is being modified and which method ultimately works best in a researcher's hands.

Cite this introduction as *Cold Spring Harb Protoc*; doi:10.1101/pdb.top107686

REFERENCES

Allen ML, O'Brochta DA, Atkinson PW, Levesque CS. 2001. Stable, germline transformation of *Culex quinquefasciatus* (Diptera: Culicidae). *J Med Entomol* **38:** 701–710. doi:10.1603/0022-2585-38.5.701

Catteruccia F, Nolan T, Loukeris TG, Blass C, Savakis C, Kafatos FC, Crisanti A. 2000. Stable germline transformation of the malaria mosquito *Anopheles stephensi*. *Nature* **405:** 959–962. doi:10.1038/35016096

Chaverra-Rodriguez D, Macias VM, Hughes GL, Pujhari S, Suzuki Y, Peterson DR, Kim D, McKeand S, Rasgon JL. 2018. Targeted delivery of CRISPR–Cas9 ribonucleoprotein into arthropod ovaries for heritable germline gene editing. *Nat Commun* **9:** 3008. doi:10.1038/s41467-018-05425-9

Clements AN. 1992. *The biology of mosquitoes. Development, nutrition and reproduction*. Chapman and Hall, London.

Gilbert SF. 2000. Early *Drosophila* development. In *Developmental biology*, 6th ed. https://www.ncbi.nlm.nih.gov/books/NBK10081/

Harrell RA II. 2023. Mosquito embryo microinjection under halocarbon oil or in aqueous solution. *Cold Spring Harb Protoc* doi:10.1101/pdb.prot108203

Jasinskiene N, Coates CJ, Benedict MQ, Cornel AJ, Rafferty CS, James AA, Collins FH. 1998. Stable transformation of the yellow fever mosquito, *Aedes aegypti*, with the Hermes element from the housefly. *Proc Natl Acad Sci* **95:** 3743–3747. doi:10.1073/pnas.95.7.3743

Louradour I, Ghosh K, Inbar E, Sacks DL. 2019. CRISPR/Cas9 mutagenesis in *Phlebotomus papatasi*: the immune deficiency pathway impacts vector competence for leishmania major. *mBio* **10:** e01941-19. doi:10.1128/mBio.01941-19

Macias VM, McKeand S, Chaverra-Rodriguez D, Hughes GL, Fazekas A, Pujhari S, Jasinskiene N, James AA, Rasgon JL. 2020. Cas9-mediated gene-editing in the malaria mosquito *Anopheles stephensi* by ReMOT control. *G3(Bethesda)* **10:** 1353–1360. doi:10.1534/g3.120.401133

Markert MJ, Zhang Y, Enuameh MS, Reppert SM, Wolfe SA, Merlin C. 2016. Genomic access to monarch migration using TALEN and CRISPR/Cas9-mediated targeted mutagenesis. *G3(Bethesda)* **6:** 905–915. doi:10.1534/g3.116.027029

Miller LH, Sakai RK, Romans P, Gwadz RW, Kantoff P, Coon HG. 1987. Stable integration and expression of a bacterial gene in the mosquito *Anopheles gambiae*. *Science* **237:** 779–781. doi:10.1126/science.3039658

Perera OP, Harrell RA II, Handler AM. 2002. Germ-line transformation of the South American malaria vector, *Anopheles albimanus*, with a PiggyBac/EGFP transposon vector is routine and highly efficient. *Insect Mol Biol* **11:** 291–297. doi:10.1046/j.1365-2583.2002.00336.x

Rubin GM, Spradling AC. 1982. Genetic transformation of *Drosophila* with transposable element vectors. *Science* **218:** 348–353. doi:10.1126/science.6289436

Shirai Y, Piulachs M-D, Belles X, Daimon T. 2022. DIPA-CRISPR is a simple and accessible method for insect gene editing. *Cell Rep Methods* **2:** 100215. doi:10.1016/j.crmeth.2022.100215

Volohonsky G, Terenzi O, Soichot J, Naujoks DA, Nolan T, Windbichler N, Kapps D, Smidler AL, Vittu A, Costa G, et al. 2015. Tools for *Anopheles gambiae* transgenesis. *G3 (Bethesda)* **5:** 1151–1163. doi:10.1534/g3.115.016808

Protocol 1

Mosquito Embryo Microinjection under Halocarbon Oil or in Aqueous Solution

Robert A. Harrell II[1,2]

[1]*The Institute for Bioscience and Biotechnology Research, University of Maryland Insect Transformation Facility, Rockville, Maryland 20850, USA*

The process of genetically modifying mosquitoes requires skilled delivery of reagents for modification. Plasmids, RNA, DNA, and/or protein must be transported into the developing embryo during an appropriate time in development when these agents will have access to the genome. Embryo microinjection has been the main method by which such modifying agents have been delivered. Ideally the microinjection process will deliver these modifying agents in sufficient quantity to effect the genetic modification without severely damaging or killing the injected embryo in the process. As semiaquatic insects, mosquitoes have embryos that are susceptible to desiccation and the degree to which embryos are susceptible is based on species. Two microinjection methods are outlined here. The first method describes embryo microinjections performed under Halocarbon-27 oil. The oil is used to reduce desiccation during the injection process. A second method limits desiccation by injecting the mosquito embryos in water. In both procedures, the embryos are first aligned and then injected before the embryos cellularize, ~1 h and 45 min after oviposition.

MATERIALS

It is essential that you consult the appropriate Material Safety Data Sheets and your institution's Environmental Health and Safety Office for proper handling of equipment and hazardous materials used in this protocol.

Reagents

Bovine whole blood (Lampire Biological Labs 7200801)
Female mosquitoes (~200) in a cage
Halocarbon oil-27 (Sigma-Aldrich H8773) (optional; see Step 10)
Material to be injected (RNA, DNA, or protein; see examples below)
- CRISPR homology-directed repair (HDR) injection mix: 0.3 ng/µL recombinant Cas9 protein; 0.04 ng/µL sgRNA (each) plus 0.2 ng/µL single-stranded oligodeoxynucleotides (ssODN) or 0.5 ng/µL double-stranded plasmid DNA (Kistler et al. 2015)
- CRISPR injection mix: 0.3 ng/µL recombinant Cas9 protein; 0.04 ng/µL single guide RNA (sgRNA) (each) (Kistler et al. 2015)
- piggyBac transposon injection mix: 150 ng/µL vector plasmid; 190 ng/µL piggyBac transposase mRNA (RA Harrell II, unpubl.)
- RNAi injection mix: 3 µg/µL dsRNA (Billingsley et al. 2021)

Reverse osmosis distilled H_2O

[2]Correspondence: harrelr@umd.edu

Cite this protocol as *Cold Spring Harb Protoc*; doi:10.1101/pdb.prot108203

Equipment

Conical tube, 50-mL (Fisher Scientific 12-565-270), modified with a dental dam port for inserting female mosquitoes

Prepare the side port by drilling a 2-cm-diameter hole in the side of the conical tube. Next, cut two squares (2.5-cm × 2.5-cm) of dental dam. Place the squares on top of each other, and then fold them over so that a 1-cm cut can be made in the center of the two squares. Tape one square over the hole in the 50-mL conical tube with Scotch tape. Tape the second square of dental dam over the first but rotated 90° so that the cut in the dental dam is perpendicular to the first dental dam. The dental dam arranged in this manner forms a port that is closed but allows an aspirator to be pushed through the port to insert mosquitoes and closes when the aspirator is withdrawn.

Cup, 6-oz.

Dark cloth or opaque box

Double-sided tape (Scotch Brand permanent double-sided tape) (optional; see Step 6)

Environmental chamber (28°C, 75% relative humidity [RH], with 12 h:12 h light:dark cycle)

Filter papers (1-cm × 3-cm; 3-cm× 6-cm) (optional; see Step 3)

Fine paintbrush

Fine probe made by hot gluing a size "0" insect pin to a wooden applicator (Fisher Scientific 22-363-158)

Food storage buckets (160-oz; Berry Plastics T811166)

Gel loading tips

Glass coverslips (20-mm × 40-mm)

Glass slides

Injection needle, pulled as described in Introduction: Mosquito Embryo Microinjection (Harrell 2023)

Microcentrifuge

Microcentrifuge tubes

Microinjection controller with a reliable working supply pressure between 20 and 80 psi

Micromanipulator

Use a good quality micromanipulator that allows for smooth operation and provides a solid platform for needle manipulation.

Microscope with a mobile stage for microinjection

Microscope can be a stereo-dissecting microscope (minimum useful magnification, 630×), inverted microscope, or a compound microscope.

Motorized aspirator for transferring mosquitoes

Parafilm (optional; see Step 25)

Petri dishes (4.25-cm, 10-cm)

Rearing tray

Stage adapter

A stage adapter is a raised platform so that the injection needle can approach the injection sample at a very shallow angle. This can be made by gluing or taping four to five standard glass slides in a stack.

Stereo-dissecting microscope for use in aligning embryos for injection

Transparent film dressing (e.g., Tegaderm) (optional; see Step 6)

Vacuum chamber (optional; see Step 26)

White filter paper (round, 4.25-cm)

METHOD

Keep cages of mosquitoes and egg collection chambers in the environmental chamber for the entire procedure. All times mentioned in the procedure are based on an injection room temp of 22°C. If the temperature of injection room is lower, these times will increase or if temperature of the injection room is high, times will decrease.

Preparation of Mosquitoes and Injection Mix

1. Four days before injection, blood feed a cage of mosquitoes (Kaufman 2015), containing approximately 200 females. Once fed, hold mosquitoes in an environmental chamber at 28°C, 75% RH, with a 12 h:12 h light:dark cycle.
2. On the day of injection, prepare the injection mix and store it on ice unless it has been prepared ahead of time. If prepared ahead of time, store CRISPR, CRISPR HDR, and mRNA injection mixes at −80°C, and plasmid-based and dsRNA injection mixes at −20°C. Prepare all injection mixes in microcentrifuge tubes, and centrifuge at 12,000*g* for 10 min at 4°C before using for injection.

   This centrifugation step is to pellet any undissolved particulate matter in the injection mix that may cause clogging issues during microinjection.
3. Set up the egg collection chamber as follows.

   i. If collecting *Aedes aegypti* eggs, place two 1-cm × 3-cm filter papers near the top of a 3-cm × 6-cm filter paper with a gap between them, and then add 1 mL of H_2O to the filter papers (Fig. 1). Place the prewet filter papers on the inside wall of the modified 50-mL conical tube with a small amount of paper sticking out of the bottom of the tube.

   Ae. aegypti *like to lay eggs at the edge of the two smaller filter papers, making it easier to collect eggs.* Anopheles *do not need these pieces of filter paper.*

   ii. For both genera, place the 50-mL conical tube into a 4.25-cm Petri dish top lined with 4.25-cm round white filter paper. For *Aedes*, the filter paper in the conical tube should contact the filter paper lining the Petri dish top (Fig. 1).

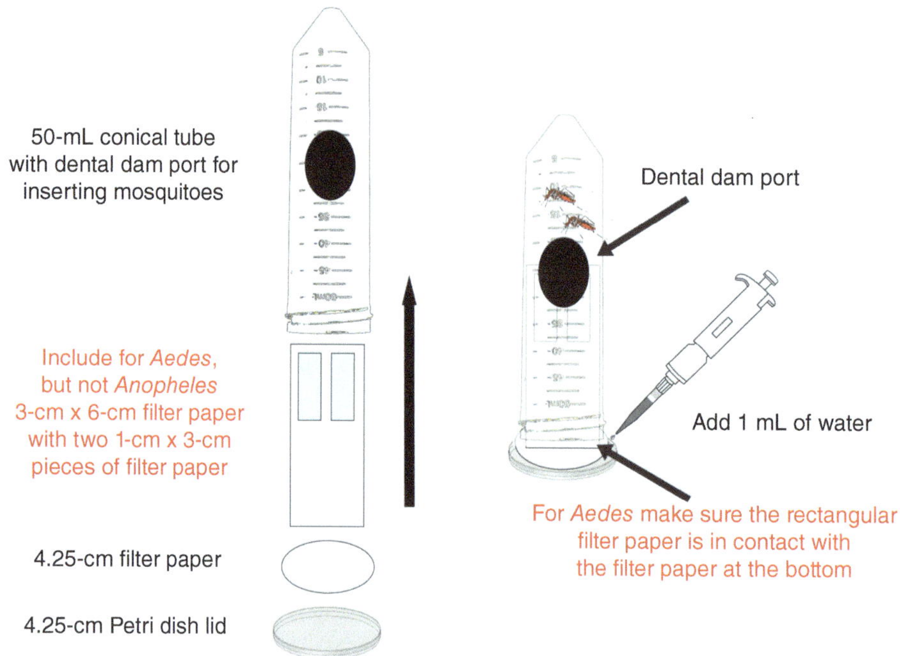

FIGURE 1. Egging chamber assembly. The egging chamber is assembled by cutting a hole in the side of a 50-mL conical centrifuge tube. The hole is then covered with two layers of dental dam with a slit cut into each. Each piece of dental dam is taped into place with the slits perpendicular to each other. For *Aedes*, two pieces of 1-cm × 3-cm filter paper are placed on top of a 3-cm × 6-cm filter paper. The papers should be dampened to hold them in place. The filter papers are slid into the 50-mL conical tube and pressed to the wall of the tube. A small portion of the larger filter paper should slightly hang out of the tube so that when the tube is placed on top of the 4.25-cm circle filter paper in the Petri dish lid, the filter paper disk and the rectangular filter paper are in contact. Contact between the filter papers is critical to facilitate wicking of water into the rectangular pieces of filter paper. *Aedes* lay eggs along the edges of the small filter paper pieces, whereas *Anopheles* lay eggs on the 4.25-cm circle of filter paper therefore the rectangular pieces of filter paper can be omitted when using the chamber for *Anopheles*. Ten to 15 female mosquitoes can be added to the egging chamber through the dental dam port, and then 1 mL of distilled H_2O is added to the filter paper disk. Once water is added, cover the egging chamber to block light and allow the mosquitoes to lay eggs for 25 min.

 Cite this protocol as *Cold Spring Harb Protoc*; doi:10.1101/pdb.prot108203

Anopheles only need the 4.25-cm round white filter paper that lines the Petri dish.

4. Using a mechanical aspirator, add 10 blood-fed female mosquitoes to the egg collection chamber through the side port. Add 1 mL of H_2O to the filter paper lining the Petri dish bottom.
5. Allow the mosquitoes to lay eggs by placing the tube containing mosquitoes into an environmental chamber on a solid surface that will not allow light through from below. Cover the egg collection tube with a dark cloth or opaque box and leave in darkness for 25 min (postoviposition setup, POS).

   Proceed to Step 6 for injection under oil and Step 11 for injection under aqueous solution.

Preparation of Embryos for Injection

Figure 2 shows an overview of the injection process under oil, and Figure 3 shows modifications specific to injecting in an aqueous solution.

Preparation for Injection under Oil

6. Prepare injection coverslips/slides as follows.
   i. Place a strip of double-sided tape, ~2–3 mm in width, at the edge along one of the 20-mm sides of a 20-mm × 40-mm glass coverslip.
   ii. On top of the double-sided tape place a strip of Tegaderm (~2 mm in width), making sure that the adhesive side of the Tegaderm is facing up. Do not remove the Tegaderm backing to expose the adhesive until it is time to pick up embryos.
7. After the 25 min in Step 5 has elapsed, empty the mosquitoes into a collection cage and remove the filter paper with the embryos.
8. Under a dissecting microscope, use a fine paintbrush to transfer embryos from the collection filter paper to a new damp filter paper for alignment.

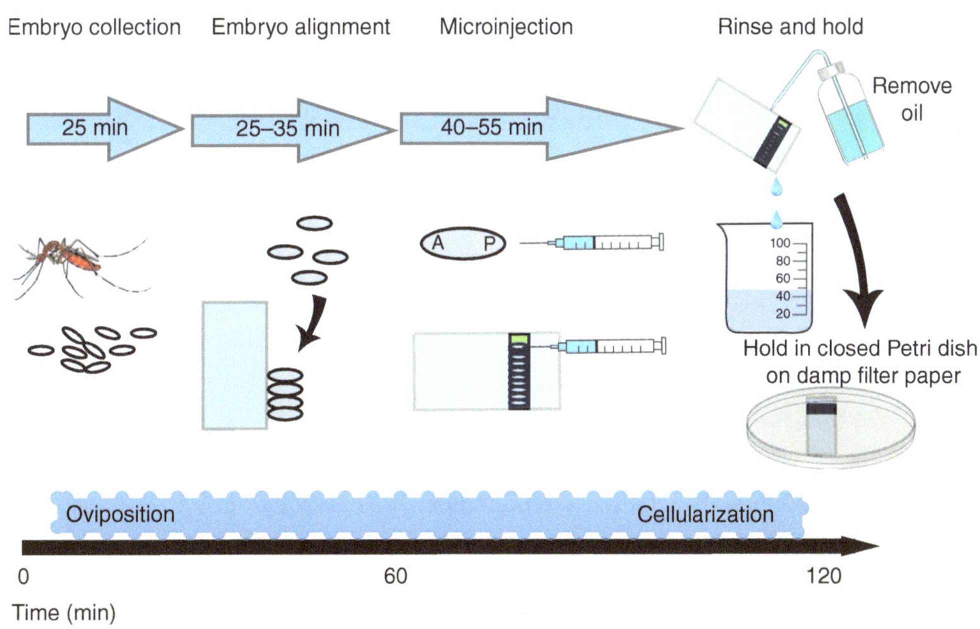

FIGURE 2. Overview of microinjection in oil. Time line of the injection process, starting at time zero with embryo collection, and then proceeding through embryo alignment, microinjection, and rinse and hold.

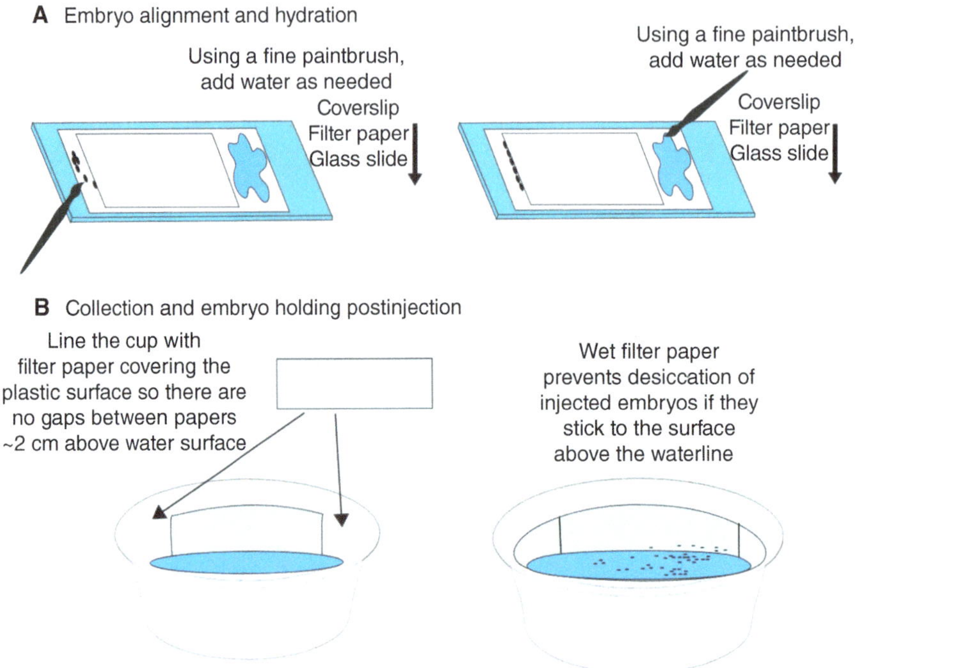

FIGURE 3. Diagram illustrating (*A*) embryo alignment and hydration in aqueous solution and (*B*) postinjection collection and embryo holding.

9. Place a new glass coverslip onto the damp filter paper and use a fine paintbrush to move the embryos over against the edge of the coverslip so that the embryos are perpendicular to the coverslip edge. Place all embryos in the same orientation with the posterior end of each embryo pointing in the same direction.

   In mosquitoes, the more rounded end is the posterior end of the embryo.

10. Once all embryos are aligned or 25 additional minutes has elapsed (50 min POS), attach the embryos to a prepared injection coverslip as follows.
    i. Carefully slide the coverslip away from the line of embryos and wick away water with a piece of filter paper to slightly dry the embryos, making them easier to pick up.
    ii. Remove the backing from the Tegaderm on the prepared injection coverslip, and then lower the coverslip down on to the embryo, making sure that the posterior ends of the embryos are oriented toward the closest edge of the prepared coverslip.
    iii. Gently press down on the coverslip to attach the embryos to the Tegaderm.
    iv. Cover the embryos with a layer of Halocarbon 27.

       A fine probe can be used to gently press the coverslip down on the embryos.

       Proceed to Step 15.

Preparation for Injection under Aqueous Solution

11. Prepare slides for injection under aqueous solution.
    i. Place a piece of filter paper onto a glass slide, and then wet the paper with reverse osmosis distilled H_2O.
    ii. On top of the wet filter paper place a glass coverslip.

12. After the 25 min in Step 5 has elapsed, empty the mosquitoes into a collection cage and remove the filter paper with embryos.

 Cite this protocol as *Cold Spring Harb Protoc*; doi:10.1101/pdb.prot108203

13. Under a dissecting microscope, use a fine paintbrush to transfer embryos from the collection filter paper to the filter paper of the prepared slide (Fig. 3A). Push the embryos in line with the edge of the coverslip. The filter paper must stay damp during alignment but not so damp that the embryos are drawn under the coverslip edge or float away from the edge because of too much water.
14. Once all embryos are aligned or 25 min have elapsed, move to injecting embryos.

Embryo Injections

15. Start embryo injections at 50 to 60 min post start of embryo collection (PSE) (Step 7 or 12) and continue injections until 105 min PSE. For oil injections, place the prepared coverslip of collected embryos on the stage adapter such that the aligned embryos have their posterior ends facing the needle. For aqueous injections, place the slide so that embryos are perpendicular to the injection needle.
16. Set up an injection needle by inserting 0.5 µL or less of injection mix into the injection needle using a gel loading tip to back fill the needle.
17. Start with the injection pressure set at 30 psi. Press and hold the injection trigger to clear air from the needle until injection mix flows from the needle tip.

    See Troubleshooting.
18. Inject embryos by inserting the needle at the posterior end of the embryo. Inject the needle in the posterior end for both oil and aqueous injections; for oil injections the needle approaches straight, whereas for aqueous injections it approaches from the side. Make sure that the needle inserts deep enough that it passes the polar cleft (if present) and is inserted just beyond the inner membrane so that injection material can be delivered to the posterior pole plasm but do not insert too deeply, where it will be beyond where the pole cells develop. Deliver a pulse of injection mix in the area of the posterior pole plasm. Deliver enough injection material that a slight change in the embryoplasm can be observed as a lightening at the tip of the needle. Adjust the injection pressure up or down based on the observed injection volume inserted into the embryo.

    With a sharp needle, insertion should be smooth, causing minimal invagination of the embryo membrane. Dull needles and/or overdesiccated embryos will cause the needle to pierce the embryo very harshly, resulting in the membrane violently snapping back into place as the needle punches through the membrane. Dull needles cause significant damage to embryos which lowers survival and thereby the efficiency of the injections.

    See Figure 2 in Introduction: Mosquito Embryo Microinjection (Harrell 2023) to see where in the embryo the polar cleft is located.

    See Troubleshooting.
19. After material is injected into the posterior pole plasm, gently withdraw the needle tip from the embryo and observe to make sure no embryoplasm leaks from the insertion site.

    Material leaking from the embryo indicates the embryo was damaged and is likely to die. Leaking embryoplasm is also a rich media for mold, which can take over the injected embryos and cause even nearby healthy injected embryos to die.
20. Between each embryo, check to make sure the needle is not clogged and is still delivering the same amount of injection mix when the injection trigger is activated.

    Press the injection trigger to test if the needle is clogged.

    See Troubleshooting for tips on resolving clogging. Once unclogged, readjust the injection pressure and resume injections on the next embryo.
21. Repeat the injection process for each embryo until all embryos are either injected or removed from the coverslip.

    Remove uninjected and damaged embryos from the coverslip by teasing the embryo with a fine probe until the embryo floats off of the Tegaderm for oil injections and by using a brush or fine probe for aqueous injections. Push the free embryos away from the immobilized embryos where they can be removed.

22. Count the number of injected embryos.

    Keeping a running tally of the number of injected embryos for that day's injection count, along with hatch and survival figures, allows you to gauge injection success and improvement.

23. For mosquitoes injected under oil only, remove the coverslip with the injected embryos from the microscope stage and remove the oil.

    Remove the oil from the coverslip by holding the coverslip at a slight angle with the embryo side slightly down over a waste collection beaker. Direct a stream of distilled H_2O onto the coverslip above where the embryos are attached and allow the H_2O to flow down over the embryos. Do not spray H_2O directly on embryos. Spraying directly onto the embryos could crush or remove the embryos.

Hatching the Embryos

Proceed to Step 24 for Aedes *embryos and Step 27 for* Anopheles *embryos.*

Aedes *Embryos*

24. For embryos injected in oil, place the coverslip of embryos on to a wet filter paper in a Petri dish. For embryos injected in aqueous solution, remove the coverslip from the filter paper by gently sliding it away from the injected embryos and place the filter paper in a Petri dish.

    Hold the coverslips or filter papers of injected embryos in a closed Petri dish until injections are complete for that day.

25. Once all injections have been completed for the day, wrap Parafilm around the Petri dish and hold in an environmental chamber for 3 d postinjection.

    See Troubleshooting.

26. On the third day postinjection, hatch the embryos in the afternoon by placing the coverslips or filter papers into a 6-oz. cup with reverse osmosis distilled H_2O. Place the container into a vacuum chamber and hold under vacuum for 2 h. Initial hatchlings should be present upon removal from the vacuum chamber.

    Proceed to Step 29.

Anopheles *Embryos*

27. Place embryos in a cup.

    i. For embryos injected in oil, submerge the coverslip with embryos in a 6-oz. cup with distilled H_2O filled to a depth of 1 cm. Embryos will start to hatch ~2 d postinjection.

    ii. For embryos injected in aqueous solution, prepare a 6-oz. cup containing distilled H_2O filled to a depth of 1 cm. Line the sides of the cup with damp filter paper, making sure that the bottom edge of the filter paper is submerged below the surface of the water so that the filter paper stays wet (Fig. 3B). Remove the coverslip by gently sliding it away from the injected embryos. Remove the filter paper with injected embryos and using a squirt bottle rinse the injected embryos into the cup. Some of the embryos will sink but most will develop normally and hatch.

       The filter paper liner prevents desiccation of any embryos that stick to the side of the cup above the water line.

28. Hold embryo cups in a rearing tray in an environmental chamber for 3 d postinjection.

    See Troubleshooting.

Rearing and Mating Transgenic Mosquitoes

29. Rear larvae according to standard mosquito rearing protocol: For *Anopheles* mosquitoes use Kaufman (2015) or for *Aedes* mosquitoes use Munstermann (1997).

Cite this protocol as *Cold Spring Harb Protoc*; doi:10.1101/pdb.prot108203

30. Sort injected generation (G_0) adults by sex as virgins then mate with uninjected virgin mosquitoes by placing male and female mosquitoes together in 160-oz food storage buckets (Clifton et al. 2014).

    Mate mosquitoes in pools of 15 G_0 individuals at a ratio of one female to five males in 160-oz food storage buckets.

31. Screen generation one (G_1) individuals for the presence of a marker. The screening method used will depend on the genetic modification materials being injected, examples being a transposon vector or HDR repair product marked using a promoter-fluorescent protein expression system (Grossman et al. 2001; O'Brochta et al. 2011; Riabinina et al. 2016) or a PCR screen to identify CRISPR–Cas9 indels (Kistler et al. 2015).

TROUBLESHOOTING

Problem (Step 17): The needle clogs immediately after adding injection mix and clearing air from the needle tip.

Solution: This usually indicates material from the injection mix is clogging the needle. Sometimes some particulate matter is forced into the tip, causing the needle to clog. Using low pressure, sometimes the injection mix will flow past the particulate matter without forcing it into the needle opening. High pressure can quickly force the particulate matter into the tip. There are three possible corrective actions.

- Try increasing the injection pressure, then insert the needle into an embryo without pressing inject, and then wait ~10 sec then press and hold inject. If this clears the needle, lower the injection pressure and continue injecting with the next embryo in line.
- Fill and change to a new needle. Lower the injection pressure to less than 30 psi. Press and hold the injection trigger until injection mix flows.
- If using low pressure causes the needle to clog, then the injection mix may be a little viscous. Change to (and fill) a new needle. Increase the injection pressure to 70 psi, press the injection trigger until air clears and injection mix starts to flow. Once the injection mix starts to flow decrease injection pressure as needed.

Problem (Step 18): The embryo membrane folds in on itself during injection and does not rebound after the needle is withdrawn.

Solution: This occurs because the embryos have become desiccated during collection and alignment and setup. Discard embryos and make sure to keep embryos moister during subsequent collection and alignment of embryos.

Problem (Step 18): Embryo membrane folds in on itself during injection and rebounds after needle is withdrawn.

Solution: This indicates the needle is not sharp enough to pierce the embryo. Improve needle sharpness. Try different needle pull parameters and/or bevel needle.

Problem (Step 20): The needle clogs after injecting some number of embryos.

Solution: This occurs because embryo material has clogged the needle tip. There are two possible corrective actions.

- Try increasing the injection pressure, then insert needle into an embryo without pressing inject, and then wait ~10 sec then press and hold inject. If this clears the needle, lower injection pressure and continue injecting with the next embryos in line. If injection mix does not start flowing after two to three tries, switch to a new needle.

- Alternatively, back the needle away from the embryos, and then lower the needle down to the substrate that the embryos are sitting on. Once touching the substrate, back the needle away from the embryos so that the needle is dragged backward across the substrate while pressing the injection trigger. If this works, injection mix should start flowing from the needle again.

Problem (Step 25): Embryos appear deflated postinjection.
Solution: This occurs when embryos have desiccated after being injected. In future experiments, make sure to add more H_2O to the filter paper where embryos are being held. Also seal the Petri dish holding the embryos with Parafilm.

Problem (Step 28): Injected embryos develop mold.
Solution: This can occur because embryoplasm leaking from injected embryos provides a rich environment for mold growth. In future experiments, improve needle sharpness. Try different needle pull parameters and/or a bevel needle.

DISCUSSION

Prevention of desiccation during mosquito embryo microinjection is crucial to survival of the injected embryos and, in turn, successful genetic modification. The process of mosquito embryo injection proceeds through multiple rounds of embryo collection, alignment, and injection. Each step must be performed in a timely manner such that the injection components are delivered before cellularization occurs, and desiccation must be avoided at each of these stages. A protocol for injecting mosquito embryos in an aqueous environment was first used in Grossman et al. (2001). The researchers modified the *Drosophila* protocol developed by Rubin and Spradling (1982) by covering the embryos in 25 mM NaCl instead of Halocarbon 700. The protocol was further refined by injecting embryos against a saturated membrane (Bossin 2005) or on a saturated membrane as described here.

In *Ae. aegypti*, cellularization occurs at ~2 h POS, therefore each round of injection must be completed by ~1 h 45 min POS to get successful genetic modification. The protocol specifically names *Ae. aegypti* and *Anopheles* throughout and the procedure has been used regularly for the modification of *Anopheles stephensi* (O'Brochta et al. 2011; Hauck et al. 2013; Cheatle Jarvela et al. 2020), *Anopheles coluzzii* (Riabinina et al. 2016), and *Culex pipiens* (Meuti and Harrell 2020).

ACKNOWLEDGMENTS

The author thanks the staff of the Insect Transformation Facility (ITF), Channa Aluvihare, Robert Alford, and Daniel Gay, for their dedication to providing ITF clients with excellent service and without whom there would not be an ITF. Special thanks to ITF staff past and present, Channa Aluvihare, Robert Alford, and Yonas Gebremicale, for helping to develop and perfect the mosquito microinjection protocols used by the ITF.

REFERENCES

Billingsley PF, George KI, Eappen AG, Harrell RA, Alford R, Li T, Chakravarty S, Sim BKL, Hoffman SL, O'Brochta DA. 2021. Transient knockdown of *Anopheles stephensi* LRIM1 using RNAi increases *Plasmodium falciparum* sporozoite salivary gland infections. *Malar J* **20:** 284. doi:10.1186/s12936-021-03818-8

Bossin H. 2005. *Microinjection methods for* Anopheles *embryos. MR4 vector component technical manual.* Malaria Research and Reference Reagent Resource Center, Manassas, VA.

Cheatle Jarvela AM, Trelstad CS, Pick L. 2020. Regulatory gene function handoff allows essential gene loss in mosquitoes. *Commun Biol* **3:** 540. doi:10.1038/s42003-020-01203-w

Clifton ME, Correa S, Rivera-Perez C, Nouzova M, Noriega FG. 2014. Male *Aedes aegypti* mosquitoes use JH III transferred during copulation to influence previtellogenic ovary physiology and affect the reproductive output of female mosquitoes. *J Insect Physiol* **64:** 40–47. doi:10.1016/j.jinsphys.2014.03.006

Grossman GL, Rafferty CS, Clayton JR, Stevens TK, Mukabayire O, Benedict MQ. 2001. Germline transformation of the malaria vector, *Anopheles gambiae*, with the piggyBac transposable element. *Insect Mol Biol* **10:** 597–604. doi:10.1046/j.0962-1075.2001.00299.x

Harrell RA II. 2023. Mosquito embryo microinjection. *Cold Spring Harb Protoc* doi:10.1101/pdb.top107686

Hauck ES, Antonova-Koch Y, Drexler A, Pietri J, Pakpour N, Liu D, Blacutt J, Riehle MA, Luckhart S. 2013. Overexpression of phosphatase and tensin homolog improves fitness and decreases *Plasmodium falciparum* development in *Anopheles stephensi*. *Microbes Infect* **15:** 775–787. doi:10.1016/j.micinf.2013.05.006

Kaufman C. 2015. *Methods in* Anopheles *research manual*. BEI Resources, Washington, DC.

Kistler KE, Vosshall LB, Matthews BJ. 2015. Genome-engineering with CRISPR–Cas9 in the mosquito *Aedes aegypti*. *Cell Rep* **11:** 51–60. doi:10.1016/j.celrep.2015.03.009

Meuti ME, Harrell R. 2020. Preparing and injecting embryos of *Culex* mosquitoes to generate null mutations using CRISPR/Cas9. *J Vis Exp* e61651. doi:10.3791/61651

Munstermann LE. 1997. Care and maintenance of *Aedes* mosquito colonies. In *Molecular biology of insect disease vectors: a methods manual* (ed. Crampton JM, Beard CB, Louis C), pp. 13–20. Chapman and Hall, London.

O'Brochta DA, Alford RT, Pilitt KL, Aluvihare CU, Harrell RA. 2011. piggyBac transposon remobilization and enhancer detection in *Anopheles* mosquitoes. *Proc Natl Acad Sci* **108:** 16339–16344. doi:10.1073/pnas.1110628108

Riabinina O, Task D, Marr E, Lin C-C, Alford R, O'Brochta DA, Potter CJ. 2016. Organization of olfactory centres in the malaria mosquito *Anopheles gambiae*. *Nat Commun* **7:** 13010. doi:10.1038/ncomms13010

Rubin GM, Spradling AC. 1982. Genetic transformation of *Drosophila* with transposable element vectors. *Science* **218:** 348–353. doi:10.1126/science.6289436

CHAPTER 33

Mosquito Transposon-Mediated Transgenesis

Vanessa Bottino-Rojas[1] and Anthony A. James[1,2,3]

[1]*Department of Microbiology and Molecular Genetics, University of California, Irvine, California 92697-4500, USA;* [2]*Department of Molecular Biology and Biochemistry, University of California, Irvine, California 92697-3900, USA*

Transposon-mediated transgenesis of mosquito vectors of disease pathogens followed the early success of transgenesis in the vinegar fly, *Drosophila melanogaster*. The *P* transposable element used in *Drosophila* does not function canonically in mosquitoes, and repeatable, routine transgenesis in mosquitoes was not accomplished until new transposable elements were discovered and validated. A number of distinct transposons were subsequently identified that mediate the introduction of exogenous DNA in a stable and heritable manner in mosquito species, including members of the genera *Aedes*, *Anopheles*, and *Culex*. The most versatile element, *piggyBac*, is functional in all of these mosquito genera, as well as in many other insects in diverse orders, and has been used extensively outside the class. Transposon-mediated transgenesis of recessive and dominant marker genes and reporter systems has been used to define functional fragments of gene control sequences, introduce exogenous DNA encoding products beneficial to medical interests, and act as "enhancer traps" to identify endogenous genes with specific expression characteristics.

BACKGROUND

Following the discovery that the *P* element could be used to efficiently transform *Drosophila melanogaster* (Rubin and Spradling 1982), concerted efforts were made to develop transposon-based transgenic technologies for a wide variety of arthropod species of agricultural and public health significance. Early efforts in mosquitoes focused unproductively on trying to adapt the *P* element to several species (Miller et al. 1987: McGrane et al. 1988; Morris et al. 1989), but it was not until the discovery of a number of new elements that success was finally achieved. "Plasmid mobility assays" were used to determine whether or not a particular element *could* be mobilized (excise and insert) in the germline of a specific insect (O'Brochta and Handler 1988). These assays clearly showed that despite its power in *D. melanogaster*, the *P* element could not transform the vast majority of other, non-drosophilid species in which it was tested. Efforts were made to use the *P* element by identifying host factors necessary for activity in *D. melanogaster* (Rio and Rubin 1988; Handler et al. 1993), but the parallel discovery and testing of new elements ultimately led to their adoption.

Work in mosquitoes has focused on Class II transposable elements, which mobilize through a DNA intermediate and can result in either conservative (no increase in element copy number) or replicative (results in an increase in copy number of the element) transposition (Atkinson and James 2002). Class II elements have variable-length terminal repeat DNA sequences, which are most often inverted (designated inverted terminal repeats [ITRs] or terminal inverted repeats [TIRs]). These

[3]Correspondence: aajames@uci.edu

Cite this introduction as *Cold Spring Harb Protoc*; doi:10.1101/pdb.top107687

repeats flank an open reading frame encoding a transposase that recognizes the ITRs and a conserved target site 2–8 bp in length and catalyzes the mobility of the element from one site in a plasmid or chromosome to another (Fig. 1). Following successful testing in mobility assays, the elements *Mos1 mariner*, *Minos*, and *Hermes* (Franz and Savakis 1991; Medhora et al. 1991; Warren et al. 1994) were adapted to transform mosquito species (Table 1; Sarkar et al. 1997; Coates et al. 1998; Jasinskiene et al. 1998; Catteruccia et al. 2000a,b; Allen et al. 2001). Molecular characterization of the baculovirus insertion element IFP2 using mobility assays in insect cells set the groundwork for the element now known as *piggyBac* to be applied widely across mosquito species (Fraser et al. 1995; Elick et al. 1997). Currently, *piggyBac* is sufficiently versatile to be useful across species from yeast to mammals (Yusa 2015) and is now the element used predominantly for transposon-mediated transgenesis of all the major mosquito vectors (Table 1).

Transposon-mediated mosquito transgenesis experiments have focused primarily on defining functional fragments of gene control sequences (promoters, 5′- and 3′-end DNAs, introns), introducing exogenous DNA encoding products beneficial to mitigating pathogen transmission, modifying expression profiles of endogenously derived genes, and introducing "enhancer traps" to identify endogenous genes with specific expression characteristics. Enhancer traps based on *piggyBac* were used to identify genes with potentially useful expression profiles (O'Brochta et al. 2012; Reid et al. 2018). Functional promoter analysis studies were biased at the outset by looking at genes that were expressed in mosquito tissues important for their role as pathogen vectors. This bias was expected to reduce potential transgenesis-related fitness costs by restricting the expression of transgenes to the infection-relevant sex, developmental stage, and mosquito body compartments in which the pathogens are found. To date, transgenesis technologies have identified gene sequences that allow stage-, tissue-, and sex-specific expression of dominant marker and reporter genes, and DNA sequences whose products interfere with viral and protozoan pathogen development and transmission (Table 1). Future studies could identify effective and robust regulatory sequences that are uniquely activated in response to pathogen infection in order to minimize even further any fitness issues associated with transgenesis. Additionally, reporter strains for binary expression systems (e.g., QF2) can be generated easily through transposon-mediated transgenesis and are useful for the characterization of neural circuits and neuronal activity visualization (Lynd and Lycett 2012; Riabinina et al. 2016).

EXPERIMENTAL DESIGN CONSIDERATIONS

Element choice is significant for transposon-based experiments. Although all elements have a preferred target nucleotide sequence for integration, these vary in size and complexity (Table 1). Both *Minos* and *Mos1* recognize the dinucleotide TA, which occurs frequently enough in the genome to make integrations of these elements appear random. *piggyBac* and *Hermes* recognize longer target sequences, four (TTAA) and eight bases (GTNCAGAC) in length, respectively, but these sequences also occur frequently enough in the genome to appear to be distributed randomly. In our experience with *Aedes aegypti*, *Mos1* tends to insert in fewer copy numbers in the genome than *piggyBac*, making post-transformation characterization of the transgenes easier to manage (V Bottino-Rojas and AA James, unpubl.). However, *Anopheles* transgenesis is almost exclusively performed with *piggyBac*. Elements can exhibit noncanonical and unstable integration and postintegration behavior, making it important to accurately characterize the insertion copy number, structure, and stability in individual transgenic lines (Jasinskiene et al. 2000; Adelman et al. 2004; Palavesam et al. 2013).

Transposon-mediated transgenesis is usually performed with a binary system of donor and helper plasmids (Fig. 1). The donor plasmid contains ITRs flanking the transgene, whereas the helper plasmid transiently expresses the transposase enzyme, which catalyzes element excision from the donor plasmid and insertion into the host genome (Fraser et al. 1996). Our associated protocol provides detailed experimental considerations (see Protocol 1: Generating and Validating Transgenic Mosquitoes with Transposon-Mediated Transgenesis [Bottino-Rojas and James 2023]). Plasmid

 Cite this introduction as *Cold Spring Harb Protoc*; doi:10.1101/pdb.top107687

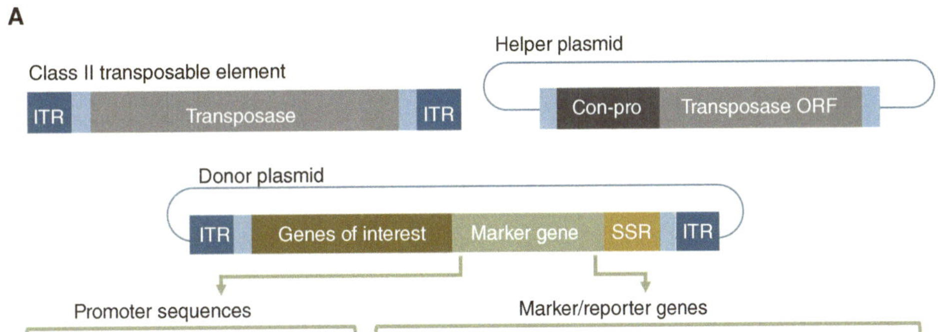

Gene	Visible pattern
3xP3	Eyes/nerve cells
IE1	Whole body
Ubiquitin	Whole body
Actin 5C	Whole body

Name	Early report
Enhanced green fluorescent protein (EGFP)	Pinkerton et al. 2000
Discosoma species red fluorescent protein (DsRed)	Nolan et al. 2002
Cyan fluorescent protein (CFP)	Nirmala et al. 2006
Yellow fluorescent protein (YFP)	Lynd & Lycett, 2012
Tandem-dimer tomato red fluorescent protein (tdTomato)	O'Brochta et al. 2012
Luciferase (luc)	Coates et al. 1999; Johnson et al. 1999

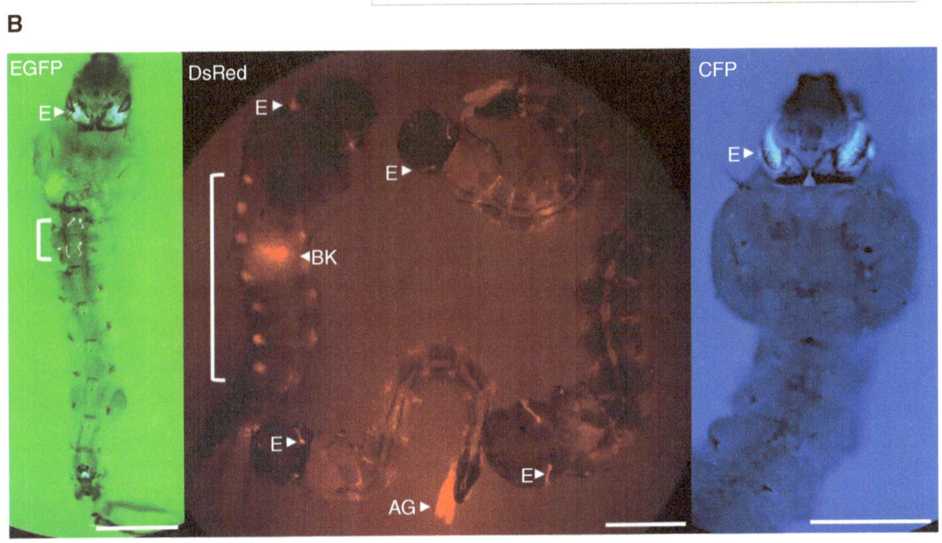

FIGURE 1. General organization and use of Class II transposable elements. (*A*) Schematic representations of a Class II transposable element and engineered helper and donor plasmids. Active Class II transposable elements have inverted terminal repeated (ITR) DNA sequences flanking a functional transposase gene complete with promoter and non-transcribed control DNA (Transposase). The helper plasmid (Helper) has the transposase open reading frame (ORF) under the control of a constitutive promoter (Con-pro). The donor plasmid (Donor) contains ITRs flanking genes of interest, a dominant marker gene along with control sequences to allow visible screening of transgenic animals, and in some cases sites for site-specific recombination (SSR). Both donor and helper DNA sequences are cloned into bacterial plasmids (thin lines) that allow production and purification for use in embryo microinjections or other DNA-introduction technologies. (*Table*) Examples of promoter sequences and marker/reporter genes along with early- or first-use citations in mosquitoes. (*B*) Examples of transposon-mediated transgenic mixed-instar *Anopheles stephensi* mosquito larvae marked with 3xP3-fluorescence genes. All larvae show transgene-mediated fluorescence in their eyes (E). Some larvae exhibit additional transgene-specific fluorescence in the segmented nervous tissue (vertical brackets), and anal gills (AG). The food bolus in the gut can produce background (BK) fluorescence. Horizontal bars, ~1 mm. (EGFP) Enhanced green fluorescent protein, (CFP) cyan fluorescent protein, (DsRed) *Discosoma* species red fluorescent protein.

TABLE 1. Transposable elements (A) and common promoters (B) used in mosquito transgenesis

A

Transposable elements	Origin	Family	Size (kb)[a]	Recognition site	Species[b]	Reference(s)[c]
Hermes	*Musca domestica*	hAT	2.7	GTNCAGAC	*Aeae, Cuqu*	Jasinskiene et al. 1998; Allen et al. 2001
Mos1	*Drosophila mauritiana*	mariner	1.3	TA	*Aeae*	Coates et al. 1998
Minos	*Drosophila hydei*	TC1-like	1.8	TA	*Anst*	Catteruccia et al. 2000b
piggyBac	*Trichoplusia ni* baculovirus	Sleeping beauty	2.5	TTAA	*Aeae, Aeal, Aefl, Anal, Anga, Anst, Ansi*	Kokoza et al. 2001; Grossman et al. 2001; Nolan et al. 2002; Perera et al. 2002; Rodrigues et al. 2006; Labbé et al. 2010; Liu et al. 2021

B

Promoters: Gene	Origin	Expression pattern	Reference(s)[c]
Heat-shock protein 70 (hsp70)	*Drosophila melanogaster; Aedes aegypti*	Ubiquitous; constitutive/heat-inducible	Morris et al. 1991; Carpenetti et al. 2012
Heat-shock protein 82 (Hsp82)	*Drosophila pseudoobscura*	Ubiquitous; constitutive	Coates et al. 1998
Act88f	*D. melanogaster*	Flight muscles	Allen and Christensen 2004
Actin5C	*D. melanogaster*	Ubiquitous; constitutive	Pinkerton et al. 2000
Actin4	*Ae. aegypti*	Female-specific; flight muscles	Fu et al. 2010
Ubiquitin (L40 and PUb)	*Ae. aegypti*	Ubiquitous; constitutive	Anderson et al. 2010
D7-related (D7r)	*Anopheles gambiae*	Salivary glands	Lombardo et al. 2005
Antiplatelet protein (aapp)	*Anopheles stephensi*	Female-specific; salivary glands; blood-meal inducible	Yoshida and Watanabe 2006
Apyrase	*Ae. aegypti; An. gambiae*	Female-specific; salivary glands	Coates et al. 1999; Lombardo et al. 2005
Aegyptin (30k a and 30k b)	*Ae. aegypti*	Female-specific; salivary glands	Mathur et al. 2010
Vitellogenin	*Ae. aegypti; An. stephensi; An. gambiae*	Female-specific; fat body (hemolymph); blood-meal inducible	Kokoza et al. 2000; Nirmala et al. 2006; Chen et al. 2007
Carboxypeptidase	*An. gambiae*	Female-specific; midgut; blood-meal inducible	Moreira et al. 2000;
G12	*An. gambiae*	Female-specific; midgut; blood-meal inducible	Nolan et al. 2011;
Peritrophin	*An. gambiae*	Female-specific; midgut; constitutive	Abraham et al. 2005
Antryp1	*An. gambiae*	Female-specific; midgut; blood-meal inducible	Nolan et al. 2011
Vasa	*An. gambiae*	Gonads	Papathanos et al. 2009
nanos	*Ae. aegypti*	Female-specific; ovaries/embryos	Adelman et al. 2007
exuperantia	*Ae. aegypti*	Female-specific; ovaries; blood-meal inducible	Akbari et al. 2013
bZip1	*Ae. aegypti*	Female-specific; ovaries/embryos	Kojin et al. 2020
β2 tubulin	*An. gambiae; Ae. aegypti*	Male-specific; testes	Catteruccia et al. 2005; Smith et al. 2007

[a]Approximate size in kilobase pairs (kb).
[b]Mosquito species transformed: (*Aeae*) *Ae. aegypti*, (*Aeal*) *Aedes albopictus*, (*Aefl*) *Aedes fluviatilis*, (*Anal*) *Anopheles albimanus*, (*Anga*) *An. gambiae*, (*Ansi*) *Anopheles sinensis*, (*Anst*) *An. stephensi*, (*Cuqu*) *Culex quinquefasciatus*.
[c]First reported use in mosquito species using transposon-mediated transgenesis.

Cite this introduction as *Cold Spring Harb Protoc*; doi:10.1101/pdb.top107687

cloning allows the production and purification of large amounts of DNA before experimental usage. The minimal donor plasmid design contains the ITRs flanking a marker gene and genes of interest. Target sequences for site-specific recombination (SSR) also can be included that allow manipulation of the transgene insert once it is integrated into the genome. DNA-insulating sequences can be added just inside the ITRs to help mitigate insertion-site effects (Carballar-Lejarazú et al. 2013). Fully functional mosquito-derived introns have been used to modify gene-of-interest configurations when needed (Franz et al. 2006). Other components of the genetic toolbox (e.g., Gal4-UAS systems, SSR docking sites) are discussed elsewhere (see Chapter 36: Genetic Toolbox Approaches in Mosquitoes [Riabinina et al. 2022]). The exact order of the genes in the element depends on the purpose of the experiment. The success rate of transposase-mediated mobilization from a plasmid into a mosquito genome is likely to depend on the size of the DNA being inserted, but a meta-analysis of the published work showed that most elements can efficiently insert 10–15 kb of DNA (Gregory et al. 2016).

Helper plasmids contain the transposase open reading frame (ORF) under the control of a constitutive promoter cloned into a plasmid. Heat-shock protein gene promoters (*hsp70* from *D. melanogaster* and *hsp82* from *Drosophila pseudoobscura*; Table 1) are active constitutively in mosquito embryos and have been used successfully in many transgenesis experiments. Helper plasmids will not replicate in the insect embryo and are expected to be lost by dilution as the injected animals progress through development.

Purified transposases also have been used in place of a helper plasmid (Coates et al. 2000). Hyperactive transposase proteins were shown to improve transformation rates in nonmosquito species, and transposase mRNAs can be injected when helper plasmids are not readily available (Otte et al. 2018).

A key contribution to mosquito transgenic technology was the discovery and application of dominant-acting fluorescent proteins that can be assayed in live animals and are visible when expressed from a single copy of their encoding gene (Fig. 1; Handler and Harrell 2001). Genes encoding proteins that fluoresce in green, red, and blue wavelengths were adapted quickly as marker genes in mosquitoes and have been used extensively. These gene products also can be used as reporter sequences in both qualitative and quantitative assays. Indeed, characterization of specific gene control DNA (promoters, 5′- and 3′-end untranslated regions [UTRs]) has been performed successfully for many genes by linking the predicted control elements to one of the fluorescent protein–encoding reading frames and scoring activity in live animals under fluorescence microscopy (Table 1).

DISCUSSION

Transposon-mediated generation of genetically engineered mosquito vectors of disease has revolutionized both basic and applied studies of these insects. The application of CRISPR–Cas technologies to mosquito transgenesis is likely to make obsolete some aspects of transposon-based technologies except in circumstances where quasi-random insertion generation is desirable (enhancer traps, random "docking" sites for SSR, insertion-site impact analyses) (see Chapter 34, Introduction: Design and Validation of Guide RNAs for CRISPR–Cas9 Genome Editing in Mosquitoes [Lo and Matthews 2023]; Amenya et al. 2010; O'Brochta et al. 2012; Carballar-Lejarazú et al. 2013). Therefore, efficient and routine methods for transposon-mediated germline transgenesis and genomic analyses provide tools for investigations where its unique, remobilizable, insertional mutagenesis features are needed.

ACKNOWLEDGMENTS

Funding was provided by the University of California Irvine Malaria Initiative. A.A.J. is a Donald Bren Professor at the University of California, Irvine. We are grateful to Kiona Parker for providing images of transgenic *An. stephensi* larvae.

REFERENCES

Abraham EG, Donnelly-Doman M, Fujioka H, Ghosh A, Moreira L, Jacobs-Lorena M. 2005. Driving midgut-specific expression and secretion of a foreign protein in transgenic mosquitoes with *AgAper1* regulatory elements. *Insect Mol Biol* **14:** 271–279. doi:10.1111/j.1365-2583.2004.00557.x

Adelman ZN, Jasinskiene N, Vally KJ, Peek C, Travanty EA, Olson KE, Brown SE, Stephens JL, Knudson DL, Coates CJ, et al. 2004. Formation and loss of large, unstable tandem arrays of the *piggyBac* transposable element in the yellow fever mosquito, *Aedes aegypti. Transgenic Res* **5:** 411–425. doi:10.1007/s11248-004-6067-2

Adelman ZN, Jasinskiene N, Onal S, Juhn J, Ashikyan A, Salampessy M, MacCauley T, James AA. 2007. *nanos* gene control DNA mediates developmentally regulated transposition in the yellow fever mosquito *Aedes aegypti. Proc Natl Acad Sci* **104:** 9970–9975. doi:10.1073/pnas.0701515104

Akbari OS, Antoshechkin I, Amrhein H, Williams B, Diloreto R, Sandler J, Hay B. 2013. The developmental transcriptome of the mosquito *Aedes aegypti*, an invasive species and major arbovirus vector. *G3 (Bethesda)* **3:** 1493–1509. doi:10.1534/g3.113.006742

Allen ML, Christensen BM. 2004. Flight muscle-specific expression of *act88F*: GFP in transgenic *Culex quinquefasciatus* Say (Diptera: Culicidae). *Parasitol Int* **53:** 307–314. doi:10.1016/j.parint.2004.04.002

Allen ML, O'Bhochta DA, Atkinson PW, Levesque CS. 2001. Stable, germ-line transformation of *Culex quinquefasciatus* (Diptera: Culicidae). *J Med Entomol* **38:** 701–710. doi:10.1603/0022-2585-38.5.701

Amenya DA, Bonizzoni M, Isaacs AT, Jasinskiene N, Chen H, Marinotti O, Yan G, James AA. 2010. Comparative fitness assessment of *Anopheles stephensi* transgenic lines receptive to site-specific integration. *Insect Mol Biol* **19:** 263–269. doi:10.1111/j.1365-2583.2009.00986.x

Anderson MAE, Gross TL, Myles KM, Adelman ZN. 2010. Validation of novel promoter sequences derived from two endogenous ubiquitin genes in transgenic *Aedes aegypti. Insect Mol Biol* **19:** 441–449. doi:10.1111/j.1365-2583.2010.01005.x

Atkinson PW, James AA. 2002. Germline transformants spreading out to many insect species. *Adv Genet* **47:** 49–86. doi:10.1016/S0065-2660(02)47002-2

Bottino-Rojas T, James AA. 2023. Generating and validating transgenic mosquitoes with transposon-mediated transgenesis. *Cold Spring Harb Protoc* doi: 10.1101/pdb.prot108194

Carballar-Lejarazú R, Jasinskiene N, James AA. 2013. Exogenous gypsy insulator sequences modulate transgene expression in the malaria vector mosquito, *Anopheles stephensi. Proc Natl Acad Sci* **110:** 7176–7181. doi:10.1073/pnas.1304722110

Carpenetti TLG, Aryan A, Myles KM, Adelman ZN. 2012. Robust heat-inducible gene expression by two endogenous *hsp70*-derived promoters in transgenic *Aedes aegypti. Insect Mol Biol* **21:** 97–106. doi:10.1111/j.1365-2583.2011.01116.x

Catteruccia F, Nolan T, Blass C, Muller HM, Crisanti A, Kafatos FC, Loukeris TG. 2000a. Toward *Anopheles* transformation: Minos element activity in anopheline cells and embryos. *Proc Natl Acad Sci* **97:** 2157–2162. doi:10.1073/pnas.040568397

Catteruccia F, Nolan T, Loukeris TG, Blass C, Savakis C, Kafatos FC, Crisanti A. 2000b. Stable germline transformation of the malaria mosquito *Anopheles stephensi. Nature* **405:** 959–962. doi:10.1038/35016096

Catteruccia F, Benton JP, Crisanti A. 2005. An *Anopheles* transgenic sexing strain for vector control. *Nat Biotechnol* **23:** 1414–1417. doi:10.1038/nbt1152

Chen X, Marinotti O, Whitman L, Jasinskiene N, Romans P, James AA. 2007. The *Anopheles gambiae* vitellogenin gene (*VGT2*) promoter directs persistent accumulation of a reporter gene product in transgenic *Anopheles stephensi* following multiple blood meals. *Am J Trop Med Hygiene* **76:** 1118–1124. doi:10.4269/ajtmh.2007.76.1118

Coates CJ, Jasinskiene N, Miyashiro L, James AA. 1998. *Mariner* transposition and transformation of the yellow fever mosquito, *Aedes aegypti. Proc Natl Acad Sci* **95:** 3743–3747. doi:10.1073/pnas.95.7.3748

Coates CJ, Jasinskiene N, Pott GB, James AA. 1999. Promoter-directed expression of recombinant fire-fly luciferase in the salivary glands of *Hermes*-transformed *Aedes aegypti. Gene* **226:** 317–325. doi:10.1016/S0378-1119(98)00557-5

Coates CJ, Jasinskiene N, Morgan D, Tosi LR, Beverley SM, James AA. 2000. Purified *mariner* (*Mos1*) transposase catalyzes the integration of marked elements into the germ-line of the yellow fever mosquito, *Aedes aegypti. Insect Biochem Mol Biol* **30:** 1003–1008. doi:10.1016/S0965-1748(00)00110-7

Elick TA, Lobo N, Fraser MJ Jr. 1997. Analysis of the *cis*-acting DNA elements required for *piggyBac* transposable element excision. *Mol Gen Genet* **255:** 605–610. doi:10.1007/s004380050534

Franz G, Savakis C. 1991. Minos, a new transposable element from *Drosophila hydei*, is a member of the Tc1-like family of transposons. *Nucl Acids Res* **19:** 6646. doi:10.1093/nar/19.23.6646

Franz AWE, Sanchez-Vargas I, Adelman ZN, Blair CD, Beaty BJ, James AA, Olson KE. 2006. Engineering RNA interference-based resistance to dengue virus type-2 in genetically-modified *Aedes aegypti. Proc Natl Acad Sci* **103:** 4198–4203. doi:10.1073/pnas.0600479103

Fraser MJ, Cary L, Boonvisudhi K, Wang HG. 1995. Assay for movement of Lepidopteran transposon IFP2 in insect cells using a baculovirus genome as a target DNA. *Virology* **211:** 397–407. doi:10.1006/viro.1995.1422

Fraser MJ, Ciszczon T, Elick T, Bauser C. 1996. Precise excision of TTAA-specific lepidopteran transposons *piggyBac* (IFP2) and *tagalong* (TFP3) from the baculovirus genome in cell lines from two species of Lepidoptera. *Insect Mol Biol* **5:** 141–151. doi:10.1111/j.1365-2583.1996.tb00048.x

Fu G, Lees R, Aw D, Jin L, Gray P, Berendonk TU, White-Cooper H, Scaife S, Phuc HK, Marinotti O, et al. 2010. Female-specific flightless phenotype for mosquito control. *Proc Natl Acad Sci* **107:** 4550–4554. doi:10.1073/pnas.1000251107

Gregory M, Alphey L, Morrison NI, Shimeld SM. 2016. Insect transformation with *piggyBac*: getting the number of injections just right. *Insect Mol Biol* **25:** 259–271. doi:10.1111/imb.12220

Grossman GL, Rafferty CS, Clayton JR, Stevens TK, Mukabayire O, Benedict MQ. 2001. Germline transformation of the malaria vector, *Anopheles gambiae*, with the *piggyBac* transposable element. *Insect Mol Biol* **10:** 597–604. doi:10.1046/j.0962-1075.2001.00299.x

Handler AM, Harrell RA. 2001. Polyubiquitin-regulated DsRed marker for transgenic insects. *Biotechniques* **31:** 820. 824-8.

Handler AM, Gomez SP, O'Brochta DA. 1993. A functional analysis of the *P*-element gene-transfer vector in insects. *Arch Insect Biochem Physiol* **22:** 373–384. doi:10.1002/arch.940220306

Jasinskiene N, Coates CJ, Benedict MQ, Cornel AJ, Rafferty CS, James AA, Collins FH. 1998. Stable transformation of the yellow fever mosquito, *Aedes aegypti*, using the *Hermes* element from the housefly. *Proc Natl Acad Sci* **95:** 3748–3751. doi:10.1073/pnas.95.7.3743

Jasinskiene N, Coates CJ, James AA. 2000. Structure of *Hermes* integrations in the germline of the yellow fever mosquito, *Aedes aegypti. Insect Mol Biol* **9:** 11–18. doi:10.1046/j.1365-2583.2000.00153.x

Johnson BW, Olson KE, Allen-Miura T, Carlson JO, Coates CJ, Jasinskiene N, James AA, Beaty BJ, Higgs S. 1999. Inhibition of luciferase expression in transgenic *Aedes aegypti* mosquitoes by Sindbis virus expression of antisense luciferase RNA. *Proc Natl Acad Sci* **96:** 13399–13403. doi:10.1073/pnas.96.23.13399

Kojin BB, Biedler JK, Tu Z, Adelman ZN. 2020. Characterization of a female germline and early zygote promoter from the transcription factor bZip1 in the dengue mosquito *Aedes aegypti. Parasit Vectors* **13:** 1–11. doi:10.1186/s13071-020-04216-w

Kokoza V, Ahmed A, Cho WL, Jasinskiene N, James AA, Raikhel A. 2000. Engineering blood meal-activated systemic immunity in the yellow fever mosquito, *Aedes aegypti. Proc Natl Acad Sci* **97:** 9144–9149. doi:10.1073/pnas.160258197

Kokoza V, Ahmed A, Wimmer EA, Raikhel AS. 2001. Efficient transformation of the yellow fever mosquito *Aedes aegypti* using the *piggyBac* transposable element vector pBac [3xP3-EGFP afm]. *Insect Biochem Mol Biol* **31:** 1137–1143. doi:10.1016/S0965-1748(01)00120-5

Labbé GM, Nimmo DD, Alphey L. 2010. *piggybac*- and *PhiC31*-mediated genetic transformation of the Asian tiger mosquito, *Aedes albopictus* (Skuse). *PLoS Negl Trop Dis* **4:** e788. doi:10.1371/journal.pntd.0000788

Liu JG, Qiao L, Zhang JJ, Chen B, He ZB. 2021. *piggyBac*-mediated germline transformation of the malaria mosquito *Anopheles sinensis* (Diptera: Culicidae). *Insect Sci* **28:** 1202–1206. doi:10.1111/1744-7917.12836

Lo IHY, Matthews B. 2023. Design and validation of guide RNAs for CRISPR–Cas9 genome editing in mosquitoes. *Cold Spring Harb Protoc* doi:10.1101/pdb.top107688

Lombardo F, Nolan T, Lycett G, Lanfrancotti A, Stich N, Catteruccia F, Louis C, Coluzzi M, Arcà B. 2005. An *Anopheles gambiae* salivary gland promoter analysis in *Drosophila melanogaster* and *Anopheles stephensi*. *Insect Mol Biol* **14:** 207–216. doi:10.1111/j.1365-2583.2004.00549.x

Lynd A, Lycett GJ. 2012. Development of the bi-partite Gal4-UAS system in the African malaria mosquito, *Anopheles gambiae*. *PLoS ONE* **7:** e31552. doi:10.1371/journal.pone.0031552

Mathur G, Sanchez-Vargas I, Alvarez D, Olson KE, Marinotti O, James AA. 2010. Transgene-mediated suppression of dengue viruses in the salivary glands of the yellow fever mosquito, *Aedes aegypti*. *Insect Mol Biol* **19:** 753–763. doi:10.1111/j.1365-2583.2010.01032.x

McGrane V, Carlson JO, Miller BR, Beaty BJ. 1988. Microinjection of DNA into *Aedes triseriatus* ova and detection of integration. *Am J Trop Med Hyg* **39:** 502–510. doi:10.4269/ajtmh.1988.39.502

Medhora M, Maruyama K, Hartl DL. 1991. Molecular and functional analysis of the mariner mutator element Mos1 in *Drosophila*. *Genetics* **128:** 311–318. doi:10.1093/genetics/128.2.311

Miller LH, Sakai RK, Romans P, Gwadz RW, Kantoff P, Coon HG. 1987. Stable integration and expression of a bacterial gene in the mosquito *Anopheles gambiae*. *Science* **237:** 779–781. doi:10.1126/science.3039658

Moreira LA, Edwards MJ, Adhami F, Jasinskiene N, James AA, Jacobs-Lorena M. 2000. Robust gut-specific gene expression in transgenic *Aedes aegypti* mosquitoes. *Proc Natl Acad Sci* **97:** 10895–10898. doi:10.1073/pnas.97.20.10895

Morris AC, Eggelston P, Crampton JM. 1989. Genetic transformation of the mosquito *Aedes aegypti* by micro-injection of DNA. *Med Vet Entomol* **3:** 1–7. doi:10.1111/j.1365-2915.1989.tb00467.x

Morris AC, Schaub TL, James AA. 1991. FLP-mediated recombination in the vector mosquito, *Aedes aegypti*. *Nucl Acids Res* **19:** 5895–5900. doi:10.1093/nar/19.21.5895

Nirmala X, Marinotti O, Sandoval JM, Phin S, Gakhar S, Jasinskiene N, James AA. 2006. Functional characterization of the promoter of the vitellogenin gene, *AsVg1*, of the malaria vector, *Anopheles stephensi*. *Insect Biochem Molec Biol* **36:** 694–700. doi:10.1016/j.ibmb.2006.05.011

Nolan T, Bower TM, Brown AE, Crisanti A, Catteruccia F. 2002. *piggyBac*-mediated germline transformation of the malaria mosquito *Anopheles stephensi* using the red fluorescent protein dsRED as a selectable marker. *J Biol Chem* **277:** 8759–8762. doi:10.1074/jbc.C100766200

Nolan T, Petris E, Müller H-M, Cronin A, Catteruccia F, Crisanti A. 2011. Analysis of two novel midgut-specific promoters driving transgene expression in *Anopheles stephensi* mosquitoes. *PLoS ONE* **6:** e16471. doi:10.1371/journal.pone.0016471

O'Brochta DA, Handler AM. 1988. Mobility of *P* elements in drosophilids and nondrosophilids. *Proc Natl Acad Sci* **85:** 6052–6056. doi:10.1073/pnas.85.16.6052

O'Brochta DA, Pilitt KL, Harrell RA II, Aluvihare C, Alford RT. 2012. Gal4-based enhancer-trapping in the malaria mosquito *Anopheles stephensi*. *G3 (Bethesda)* **2:** 1305–1315. doi:10.1534/g3.112.003582

Otte M, Netschitailo O, Kaftanoglu O, Wang Y, Page RE Jr, Beye M. 2018. Improving genetic transformation rates in honeybees. *Sci Rep* **8:** 16534. doi:10.1038/s41598-018-34724-w

Palavesam A, Esnault C, O'Brochta DA. 2013. Post-integration silencing of *piggyBac* transposable elements in *Aedes aegypti*. *PLoS ONE* **8:** e68454. doi:10.1371/journal.pone.0068454

Papathanos PA, Windbichler N, Menichelli M, Burt A, Crisanti A. 2009. The vasa regulatory region mediates germline expression and maternal transmission of proteins in the malaria mosquito *Anopheles gambiae*: a versatile tool for genetic control strategies. *BMC Mol Biol* **10:** 1–13. doi:10.1186/1471-2199-10-65

Perera OP, Harrell RA II, Handler AM. 2002. Germline transformation of the South American malaria vector, *Anopheles albimanus*, with a *piggyBac/EGFP* transposon vector is routine and highly efficient. *Insect Mol Biol* **11:** 291–297. doi:10.1046/j.1365-2583.2002.00336.x

Pinkerton AC, Michel K, O'Brochta DA, Atkinson PW. 2000. Green fluorescent protein as a genetic marker in transgenic *Aedes aegypti*. *Insect Mol Biol* **9:** 1–10. doi:10.1046/j.1365-2583.2000.00133.x

Reid W, Pilitt K, Alford R, Cervantes-Medina A, Yu H, Aluvihare C, Harrell R, O'Brochta DA. 2018. An *Anopheles stephensi* promoter-trap: augmenting genome annotation and functional genomics. *G3 (Bethesda)* **8:** 3119–3130. doi:10.1534/g3.118.200347

Riabinina O, Task D, Marr E, Lin C-C, Alford R, O'Brochta DA, Potter CJ. 2016. Organization of olfactory centres in the malaria mosquito *Anopheles gambiae*. *Nat Commun* **7:** 13010. doi:10.1038/ncomms13010

Riabinina O, Quinn M, Whitehead JP. 2022. Genetic toolbox approaches in mosquitoes. *Cold Spring Harb Protoc* doi:10.1101/pdb.prot107691

Rio DC, Rubin GM. 1988. Identification and purification of a *Drosophila* protein that binds to the terminal 31-base-pair inverted repeats of the P transposable element. *Proc Natl Acad Sci* **85:** 8929–8933. doi:10.1073/pnas.85.23.8929

Rodrigues FG, Oliveira SB, Rocha BC, Moreira LA. 2006. Germline transformation of *Aedes fluviatilis* (Diptera: Culicidae) with the *piggyBac* transposable element. *Mem Inst Oswaldo Cruz* **101:** 755–777. doi:10.1590/S0074-02762006000700008

Rubin GM, Spradling AC. 1982. Genetic transformation of *Drosophila* with transposable element vectors. *Science* **218:** 348–353. doi:10.1126/science.6289436

Sarkar A, Yardley K, Atkinson PW, James AA, O'Brochta DA. 1997. Transposition of the *Hermes* element in embryos of the vector mosquito, *Aedes aegypti*. *Insect Mol Biol* **27:** 359–363. doi:10.1016/S0965-1748(97)00018-0

Smith RC, Walter MF, Hice RH, O'Brochta DA, Atkinson PW. 2007. Testis-specific expression of the β2 tubulin promoter of *Aedes aegypti* and its application as a genetic sex-separation marker. *Insect Mol Biol* **16:** 61–71. doi:10.1111/j.1365-2583.2006.00701.x

Warren WD, Atkinson PW, O'Brochta DA. 1994. The Hermes transposable element from the house fly, *Musca domestica*, is a short inverted repeat-type element of the *hobo*, *Ac*, and *Tam3* (*hAT*) element family. *Genet Res* **64:** 87–97. doi:10.1017/S0016672300032699

Yoshida S, Watanabe H. 2006. Robust salivary gland-specific transgene expression *in Anopheles stephensi* mosquito. *Insect Mol Biol* **15:** 403–410. doi:10.1111/j.1365-2583.2006.00645.x

Yusa K. 2015. *piggyBac* transposon. *Microbiol Spectr* **3:** MDNA3-0028-2014. doi:10.1128/microbiolspec.MDNA3-0028-2014

Protocol 1

Generating and Validating Transgenic Mosquitoes with Transposon-Mediated Transgenesis

Vanessa Bottino-Rojas[1] and Anthony A. James[1,2,3]

[1]*Department of Microbiology and Molecular Genetics, University of California, Irvine, California 92697-4500, USA;* [2]*Department of Molecular Biology and Biochemistry, University of California, Irvine, California 92697-3900, USA*

Transposon-mediated transgenesis has revolutionized both basic and applied studies of mosquito vectors of disease. Currently, techniques such as enhancer traps and transposon tagging, which rely on remobilizable insertional mutagenesis, are only possible with transposon-based vector systems. Here, we provide general descriptions of methods and applications of transposon-based mosquito transgenesis. The exact procedures must be adapted to each mosquito species and comparisons of some differences among different mosquito species are outlined. A number of excellent publications showing detailed and specific protocols and methods are featured and referenced.

MATERIALS

It is essential that you consult the appropriate Material Safety Data Sheets and your institution's Environmental Health and Safety Office for proper handling of equipment and hazardous materials used in this protocol.

RECIPES: Please see the end of this protocol for recipes indicated by <R>. Additional recipes can be found online at http://cshprotocols.cshlp.org/site/recipes.

Reagents

Agarose

Deionized or double-distilled sterile H_2O

DNeasy Blood & Tissue Kit (QIAGEN 69504)

Donor and helper plasmid DNA for transposon-mediated transgenesis (see Introduction: Mosquito Transposon-Mediated Transgenesis [Bottino-Rojas and James 2023]).

Helper plasmids are available by request from researchers who have published transgenesis studies and some may be available from Addgene (https://www.addgene.org).

Resuspend plasmids in injection buffer for mosquitoes. The concentration will depend on the experiment (see Step 4).

EndoFree Plasmid Maxi kit (QIAGEN 12362)

Ethanol, absolute, molecular-biology-grade (optional; see Step 2)

[3]Correspondence: aajames@uci.edu

Cite this protocol as *Cold Spring Harb Protoc*; doi:10.1101/pdb.prot108194

Gel electrophoresis buffer of choice
Halocarbon oil 27 (Sigma-Aldrich H8773)
Halocarbon oil 700 (Sigma-Aldrich H8898)
Injection buffer for mosquitoes <R>
Isotonic buffer <R>
Mosquitoes (species/strains specific for the study) (Benedict 2015)
Oligonucleotide primers to confirm transgenesis

Design oligonucleotide primers that amplify specific regions of the transposable element to confirm the presence of the construct-genome junctions in the genomes of apparent transgenic organisms.

PCR reagents

Equipment

Coverslips
Dissecting microscope (e.g., Leica MZ12)

This microscope is used for embryo alignment.

Double-sided Scotch tape
Fine paintbrush
Gel electrophoresis apparatus
Glass slides
Insectary maintained with the following conditions: 28°C temperature, 75% humidity, with a 12 h cycle of light and darkness
Inverted microscope (e.g., Leica DM 1000 LED or M165 FC)

This microscope is used for microinjection.

Larval trays (plastic, 33-cm×19-cm×11-cm)
Microinjector and micromanipulator (e.g., Sutter Instrument—XenoWorks)
Microinjection needles/micropipettes (pulled from Quartz glass capillaries [Sutter Instrument QF100-70-10] as described in **Mosquito Embryo Microinjection** [Harrell 2023a])
Microloader pipette tips (2-µL) (Eppendorf)
Micron filter (0.22-µm) or magnetic beads (optional; see Step 2)
Microscope with fluorescence filter set (e.g., Leica MZFLIII)

This microscope is used for larval screening.

Minimum-fiber filter paper (e.g., Fisher Brand 05-714-4)
Oviposition containers (conical; paper cup, *Drosophila* vials, or 50-mL centrifuge tubes)
Petri dishes (plastic, 90-mm×15-mm) or glass containers (125-mm × 65-mm) (Pyrex 3140)
Thermocycler
Transfer pipettes (e.g., Fisher Brand 13-711-7M)

METHOD

Step-by-step instructions for making transgenic Aedes aegypti, Anopheles gambiae, Anopheles stephensi, *and* Culex quinquefasciatus, *some of which include video images that allow replication and troubleshooting, are available online (Allen et al. 2001; Lobo et al. 2006; Jasinskiene et al. 2007; Terenius et al. 2007; Fuchs et al. 2013; Adolfi et al. 2021; Carballar-Lejarazú et al. 2021). Significant differences among species are detailed in Table 1. See the Discussion for further details.*

Plasmid Preparation

1. Design and construct the desired donor plasmid with the following essential components: fluorescent marker (with an appropriate promoter); terminal-repeat sequences from transpos-

TABLE 1. Significant embryo microinjection procedure differences among mosquito species

	Anopheles stephensi	*Anopheles gambiae*	*Aedes aegypti*	*Culex quinquefasciatus*
Preparation for egg laying	Blood-feed mosquitoes twice and collect embryos 2–4 d after the second blood meal.	Blood-feed mosquitoes twice and collect embryos 2–4 d after the second blood meal.	Blood-feed mosquitoes once and collect embryos 3–4 d after feeding.	Blood-feed mosquitoes once and collect embryos 4–5 d after feeding.
Egg laying setup (embryo collection)	*Drosophila* vial tube with wet cotton and filter paper—forced egg laying for 1 h.	50-mL centrifuge tube with filter paper adapted on a Petri dish—forced egg laying for 45 min.	*Drosophila* vial tube with wet cotton and filter paper—forced egg laying for 1 h 15 min.	Egg rafts collected immediately after oviposition in a cup of water—allowed to mature for 30–45 min.
Manipulation solution	Isotonic buffer	Sterile deionized water	Sterile deionized water	Sterile deionized water
Halocarbon oil to prevent desiccation	Yes, 2:1 proportion (700:27)	No	Yes, 1:1 proportion (700:27)	Yes. 1:1 proportion (700:27)
Embryo microinjection angle	30°	15°	30°	45°
Handling of injected embryos	Remove excess oil and keep embryos in a Petri dish covered with isotonic buffer to allow hatching. Transfer hatched larvae to a rearing container.	Transfer eggs into a glass container lined with filter paper and filled with water. Transfer hatched larvae to a rearing container.	Incubate eggs in a humidified chamber, in a vertical, anterior-down, orientation. Hatch 4 d after injection and observe hatching for up to 10 d.	Incubate eggs in a humidifier chamber, in a vertical, anterior-down, orientation. Hatch 24–30 h after injection.
Reference	Terenius et al. 2007	Carballar-Lejarazú et al. 2021	Jasinskiene et al. 2007	Allen et al. 2001

able elements; desired transgene cargo (see Introduction: Mosquito Transposon-Mediated Transgenesis [Bottino-Rojas and James 2023]).

2. Purify donor and helper plasmids using an endotoxin-free plasmid purification kit (e.g., Endo-Free Plasmid Maxi kit).

   Proceed with an additional purification (ethanol precipitation, magnetic beads, or 0.22 µm filtration) to remove any impurities that can poison the embryos and particulate matter that may clog the microinjection needles.

3. Dilute the purified plasmids in injection buffer for mosquitoes into aliquots for microinjection (~1 µg/µL) that can be stored until needed at −70°C.

4. Mix donor and helper plasmids at a desirable ratio and microinject into preblastoderm mosquito embryos as described in the next section.

   The helper:donor ratio is adjusted for plasmid size and overall concentration. Typically, the total DNA concentration for injection should not exceed 1 mg/mL. Standard protocols suggest a higher proportion of donor to helper plasmid (two- to fourfold; e.g., 0.5 mg/mL donor:0.3 mg/mL helper) but successful applications have been reported for equal (1:1) ratios as well (Coates et al. 1998; Jasinskiene et al. 1998; Handler and O'Brochta 2011).

Microinjection Procedures

Detailed microinjections procedures can be found in Chapter 32, Introduction: Mosquito Embryo Microinjection (Harrell 2023a) and Chapter 32, Protocol 1: Mosquito Embryo Microinjection under Halocarbon Oil or in Aqueous Solution [Harrell 2023b].

5. Collect preblastoderm embryos by allowing gravid females (3–5 d after blood feeding) to force-lay eggs on an oviposition container for 45–75 min at insectary conditions.

6. Allow eggs to mature until their color has changed from creamy yellow to gray. This color change is characteristic of the eggshell melanization process and it is important to not let the eggs turn

 Cite this protocol as *Cold Spring Harb Protoc*; doi:10.1101/pdb.prot108194

dark brown or black, as these will be too difficult to inject and break the needles. Timing varies among species and general times are listed in Table 1.

7. Using a fine paintbrush, align individual eggs on a strip of double-sided tape mounted on a glass coverslip. Place the coverslip on a glass slide to elevate the coverslip and allow movement of the embryos on the microscope platform during injection.

   Keep a humid microenvironment to prevent desiccation (e.g., use a piece of moistened filter paper as a support for alignment).

8. Proceed with manual dechorionation of the eggs as needed.

   Some researchers use manual and chemical dechorionation of the eggs to make them easier to inject (Kumar and Puttaraju 2012). However, if the eggs are injected before melanization (Table 1), then dechorionation is not necessary. Forgoing dechorionation greatly increases embryo postinjection survival. Others have used delayed melanization, but this runs the risk of not getting the injected DNA into the germline cells before they develop membranes that would exclude the DNA (Catteruccia et al. 2000).

9. Use a microloader tip to fill the needles with 2 µL of DNA (donor and helper plasmids) mixture. Mount beveled prefilled microcapillary needles in the microinjection apparatus at an appropriate angle and inject solution laterally targeting the posterior pole (region of formation of the future germline) of the embryo.

   The volume injected depends proportionally on the degree of preinjection aging (Step 6). A successful injection will lead to a small movement of the cytoplasm within the egg.

   See Troubleshooting

Postinjection Procedures

10. After microinjection, move the injected eggs undisturbed on the injection slides to insectary conditions (e.g., place the coverslip in a covered Petri dish with moist filter paper or in a glass slide box with water to incubate the embryos in a vertical orientation).

    Hatching should start ∼48 h after injection in some species. Since injection may cause a developmental delay, it is advisable to keep monitoring for late-hatching larvae for at least 3–5 d. We have kept some eggs for 10 d.

11. Isolate surviving injected (G_0) insects as pupae, sort according to sex, and cross resulting adult mosquitoes to an excess of fresh (5–7 d after emergence) wild-type adults (Adolfi et al. 2021; Carballar-Lejarazú et al. 2021).

    Standard protocols call for establishing founder families in separate containers by mating each surviving G_0 male to five to 10 wild-type females of the strain of origin, and mating females in batches (of five to 10) with 20 or more wild-type males. Allow adults to mate for 4–5 d.

12. Blood-feed G_0 families, collect eggs, and rear emerging next generation (G_1) (Adolfi et al. 2021; Carballar-Lejarazú et al. 2021).

13. Screen G_1 larvae (third-/fourth-instar) for inheritance of the transgenes (see Fig. 1 in Introduction: Mosquito Transposon-Mediated Transgenesis [Bottino-Rojas and James 2023]). Verify expression of the visible marker using a stereomicroscope with an attached fluorescence module.

    See Troubleshooting.

14. Transfer transformed G_1 individuals using a transfer pipette into a larval tray and rear to pupae (Benedict 2015). Discard nonfluorescent individuals.

15. Sort marker-expressing pupae by sex and prepare them for mating. Outcross *en masse* transformed G_1 individuals from each G_0 founder family with opposite-sex age-matched wild-type individuals.

    Repeat the outcrossing procedures over the next generations until sufficient progeny numbers allow intercrossing of putative transgenics and establishment of new transgenic pure-breeding colonies.

16. Set aside a subset of next-generation marker-positive (fluorescent) individuals (G_2 onward) for molecular analysis.

17. Use PCR-based methods for transgenesis confirmation.
    i. Isolate genomic DNA from pupa or adult specimens of each colony of transgenic insects using a commercial genomic DNA extraction kit (e.g., DNeasy Blood & Tissue Kit) from animal tissues in accordance with the manufacturer's instructions.
    ii. Perform PCR using oligonucleotide primers that amplify specific regions of the transposable element to confirm the presence of the construct-genome junctions in the genomes of apparent transgenic organisms.
18. Visualize the presence of the expected diagnostic PCR amplicons through agarose gel electrophoresis using standard methods. Sequence PCR products to confirm anticipated sequences.
19. Upon establishment of transgenic pure-breeding (homozygous) colonies, further experiments can be performed to evaluate transgene copy number and chromosomal location of the insertion sites, as well as effects such as insertional mutagenesis and position-dependent differences in gene expression pattern and timing. Furthermore, expression profile studies can be used to link gene function with the sequence information from genomics data (see Discussion).

TROUBLESHOOTING

Problem (Step 9): The DNA solution is too viscous for the needles.
Solution: The purified plasmids must be injected at a defined concentration (<1 mg/mL total DNA). Make sure the appropriate concentration is being used.

Problem (Step 13): Transformation efficiency is low and/or fitness of injected individuals is low.
Solution: In this protocol, the helper plasmid expresses transposase from a constitutive promoter. The ubiquitous presence of the enzyme may lead to transformation of somatic cells if microinjections are not directed precisely to the posterior region of the embryo where the germline forms. Although such transformation events are not heritable, somatic effects can decrease the transformation efficiency and the fitness of injected individuals. If this occurs, repeat the experiment with more precise injections.

DISCUSSION

Space limitations prevent the detailed and specific description of protocols; however, step-by-step instructions for making transgenic *Ae. aegypti*, *An. gambiae*, *An. stephensi*, and *Cx. quinquefasciatus*, some of which include video images that allow replication and troubleshooting, are available online (Allen et al. 2001; Lobo et al. 2006; Jasinskiene et al. 2007; Terenius et al. 2007; Fuchs et al. 2013; Adolfi et al. 2021; Carballar-Lejarazú et al. 2021). These protocols have been applied to other species that are closely related—for example, *Aedes albopictus*, *Aedes triseriatus*, and *Aedes fluviatilis* (Lobo et al. 2001; Rodrigues et al. 2006; Labbé et al. 2010). The procedures include embryo collection and preparation, injection needle specifics, injection microscope variations, injections, postinjection care, and screening (see Chapter 32, Introduction: Mosquito Embryo Microinjection [Harrell 2023a] and Chapter 38, Introduction: Techniques for Identifying and Sorting Transgenic Mosquito Larvae [Marois 2023]). Some of the differences among species are outlined in Table 1. For example, many Culicine mosquitoes must undergo a period of estivation before hatching and this must be accommodated in the postinjection handling procedures. Transgene insert validation or copy number and genomic insertion site and transgene validation by microscopy and functional tests also are provided in the reference materials.

A carefully examined plasmid design is paramount for the success of this experiment (see Introduction: Mosquito Transposon-Mediated Transgenesis [Bottino-Rojas and James 2023]). For example, a larger insert generally correlates negatively with transformation efficiency and transposons most successfully integrate DNA fragments of 10–15 kilobases in length (Volohonsky et al. 2015; Gregory et al. 2016).

Additionally, special care must be paid to the rearing and crossing of individuals derived from injected embryos to maximize the chances of recovering transformants. Lines derived from male founders are used as the numerator in determining transformation efficiencies, and extra care must be taken with female-derived lines as they may represent more than one independent transformation event. Recovery of "clusters" of transformed progeny can be expected if the transposon inserts into a germ cell early in its differentiation to multiple gametes. Once a line has been established with sufficient numbers to ensure its continued propagation, molecular approaches such as inverse polymerase chain reaction (iPCR) (Ochman et al. 1988) can be used to establish transgene copy number and chromosomal location of the insertion sites. Alternately, the confirmation of transgenesis and number of insertions can be verified with Southern blotting (Southern 2006). Depending on the other genes in the construct, fluorescence microscopy can be used to verify the correct expression profiles of DNA control sequences of interest linked to a fluorescent reporter construct (e.g., Nirmala et al. 2006; Chen et al. 2007; Mathur et al. 2010).

RECIPES

Injection Buffer for Mosquitoes

5 mM KCl
0.1 mM $NaPO_4$

Adjust pH to 7.2 and filter sterilize with a 0.22-µm filter. Store for 6 mo at −20°C.

Isotonic Buffer

150 mM NaCl
5 mM KCl
10 mM HEPES
2.5 mM $CaCl_2$

Adjust pH to 7.2 and filter sterilize with a 0.22-µm filter. Store for 1 mo at 4°C.

ACKNOWLEDGMENTS

Funding was provided by the University of California Irvine Malaria Initiative. A.A.J. is a Donald Bren Professor at the University of California, Irvine.

REFERENCES

Adolfi A, Lynd A, Lycett GJ, James AA. 2021. Site-directed φC31-mediated integration and cassette exchange in *Anopheles* vectors of malaria. *J Vis Exp* **168:** e62146. doi:10.3791/62146

Allen ML, O'Brochta DA, Atkinson PW, Levesque CS. 2001. Stable, germ-line transformation of *Culex quinquefasciatus* (Diptera: Culicidae). *J Med Entomol* **38:** 701–710. doi:10.1603/0022-2585-38.5.701

Benedict MQ. 2015. Chapter 2: *Anopheles* laboratory biology and culture 2.4 *Anopheles* culture. In *Methods in* Anopheles *research, 4th ed.* BEI Resources, Manassas, Virginia.

Bottino-Rojas T, James AA. 2023. Mosquito transposon-mediated transgenesis. *Cold Spring Harb Protoc* doi:10.1101/pdb.top107687

Carballar-Lejarazú R, Tushar T, Pham TB, James AA. 2021. Microinjection method for *Anopheles gambiae* embryos. *J Vis Exp* doi:10.3791/62591

Catteruccia F, Nolan T, Loukeris TG, Blass C, Savakis C, Kafatos FC, Crisanti A. 2000. Stable germline transformation of the malaria mosquito *Anopheles stephensi*. *Nature* **405:** 959–962. doi:10.1038/35016096

Chen X, Marinotti O, Whitman L, Jasinskiene N, Romans P, James AA. 2007. The *Anopheles gambiae* vitellogenin gene (*VGT2*) promoter directs persistent accumulation of a reporter gene product in transgenic *Anopheles stephensi* following multiple blood meals. *Am J Trop Med Hygiene* **76:** 1118–1124. doi:10.4269/ajtmh.2007.76.1118

Coates CJ, Jasinskiene N, Miyashiro L, James AA. 1998. *Mariner* transposition and transformation of the yellow fever mosquito, *Aedes aegypti*. *Proc Natl Acad Sci* **95:** 3743–3747. doi:10.1073/pnas.95.7.3748

Fuchs S, Nolan T, Crisanti A. 2013. Mosquito transgenic technologies to reduce *Plasmodium* transmission. *Methods Mol Biol* **923:** 601–622. doi:10.1007/978-1-62703-026-7_41

Gregory M, Alphey L, Morrison NI, Shimeld SM. 2016. Insect transformation with *piggyBac*: getting the number of injections just right. *Insect Mol Biol* **25:** 259–271. doi:10.1111/imb.12220

Handler AM, O'Brochta DA. 2011. Transposable elements for insect transformation. In *Insect molecular biology and biochemistry* (ed. Gilbert LI), pp. 90–133. Elsevier, Amsterdam.

Harrell R II. 2023a. Mosquito embryo microinjection. *Cold Spring Harb Protoc* doi:10.1101/pdb.top107686

Harrell R II. 2023b. Mosquito embryo microinjection under halocarbon oil or in aqueous solution. *Cold Spring Harb Protoc* doi:10.1101/pdb.prot108203

Jasinskiene N, Coates CJ, Benedict MQ, Cornel AJ, Rafferty CS, James AA, Collins FH. 1998. Stable transformation of the yellow fever mosquito, *Aedes aegypti*, using the Hermes element from the housefly. *Proc Natl Acad Sci* **95:** 3748–3751. doi:10.1073/pnas.95.7.3743

Jasinskiene N, Juhn J, James AA. 2007. Microinjection of *Aedes aegypti* embryos to obtain transgenic mosquitoes. *J Vis Exp* **5:** 219. doi:10.3791/219

Kumar SS, Puttaraju HP. 2012. Improvised microinjection technique for mosquito vectors. *Indian J Med Res* **136:** 971–978.

Labbé GM, Nimmo DD, Alphey L. 2010. *piggybac*- and *PhiC31*-mediated genetic transformation of the Asian tiger mosquito, *Aedes albopictus* (Skuse). *PLoS Negl Trop Dis* **4:** e788. doi:10.1371/journal.pntd.0000788

Lobo N, Li X, Hua-Van A, Fraser MJ Jr. 2001. Mobility of the *piggyBac* transposon in embryos of the vectors of Dengue fever (*Aedes albopictus*) and La Crosse encephalitis (*Ae. triseriatus*). *Mol Genet Genomics* **265:** 66–71. doi:10.1007/s004380000388

Lobo NF, Clayton JR, Fraser MJ, Kafatos FC, Collins FH. 2006. High efficiency germ-line transformation of mosquitoes. *Nat Protoc* **1:** 1312–1317. doi:10.1038/nprot.2006.221

Marois E. 2023. Technique for identifying and sorting transgenic mosquito larvae. *Cold Spring Harb Protoc* doi:10.1101/pdb.top107686

Mathur G, Sanchez-Vargas I, Alvarez D, Olson KE, Marinotti O, James AA. 2010. Transgene-mediated suppression of dengue viruses in the salivary glands of the yellow fever mosquito, *Aedes aegypti*. *Insect Mol Biol* **19:** 753–763. doi:10.1111/j.1365-2583.2010.01032.x

Nirmala X, Marinotti O, Sandoval JM, Phin S, Gakhar S, Jasinskiene N, James AA. 2006. Functional characterization of the promoter of the vitellogenin gene, *AsVg1*, of the malaria vector, *Anopheles stephensi*. *Insect Biochem Mol Biol.* **36:** 694–700. doi:10.1016/j.ibmb.2006.05.011

Ochman H, Gerber AS, Hartl DL. 1988. Genetic applications of an inverse polymerase chain reaction. *Genetics* **120:** 621–623. doi:10.1093/genetics/120.3.621

Rodrigues FG, Oliveira SB, Rocha BC, Moreira LA. 2006. Germline transformation of *Aedes fluviatilis* (Diptera: Culicidae) with the *piggyBac* transposable element. *Mem Inst Oswaldo Cruz* **101:** 755–777. doi:10.1590/S0074-02762006000700008

Southern E. 2006. Southern blotting. *Nat Protoc* **1:** 518–525. doi:10.1038/nprot.2006.73

Terenius O, Juhn J, James AA. 2007. Injection of *Anopheles stephensi* embryos to generate malaria-resistant mosquitoes. *J Vis Exp* **5:** 216.

Volohonsky G, Terenzi O, Soichot J, Naujoks DA, Nolan T, Windbichler N, Kapps D, Smidler AL, Vittu A, Costa G, et al. 2015. Tools for *Anopheles gambiae* transgenesis. *G3(Bethesda)* **5:** 1151–1163. doi:10.1534/g3.115.016808

 Cite this protocol as *Cold Spring Harb Protoc*; doi:10.1101/pdb.prot108194

CHAPTER 34

Design and Validation of Guide RNAs for CRISPR–Cas9 Genome Editing in Mosquitoes

Ivan Hok Yin Lo and Benjamin J. Matthews[1]

Department of Zoology, The University of British Columbia, Vancouver, British Columbia V6T 1Z4, Canada

CRISPR–Cas9 has revolutionized gene editing for traditional and nontraditional model organisms alike. This tool has opened the door to new mechanistic studies of basic mosquito biology as well as the development of novel vector control strategies based on CRISPR–Cas9, including gene drives that spread genetic elements in the population. Although the promise of the specificity, flexibility, and ease of deployment CRISPR is real, its implementation still requires empirical optimization for each new species of interest, as well as to each genomic target within a given species. Here, we provide an overview of designing and testing single-guide RNAs for the use of CRISPR-based gene editing tools.

CRISPR IN MOSQUITOES AND OTHER NONTRADITIONAL MODEL ORGANISMS

CRISPR is a core component of the immune response pathway in bacteria and archaea that has been co-opted and developed into a set of precision gene editing tools that enable easier, more flexible, and more scalable means of inducing double-strand DNA (dsDNA) breaks in any organism's genome (Doudna and Charpentier 2014). Repair of these dsDNA breaks by messy processes such as nonhomologous end joining (NHEJ) can render genes nonfunctional, and more precise pathways such as homology-directed repair (HDR) can insert novel sequences into specific parts of the genome by reading from template "donor DNA" provided alongside the single-guide RNA (sgRNA) and Cas9.

In the past 10 years, CRISPR has been transformative for many fields by introducing the power of gene editing to new organisms or supercharging the speed of its application in others, such as mosquitoes, that previously relied on lower throughput protein-based technologies like zinc-finger nucleases. This tool has permitted the development of new nontraditional model species in laboratories across the world and enabled scientists to ask questions that simply cannot be modeled in the standard stable of genetic workhorses (Matthews and Vosshall 2020). In mosquitoes, CRISPR has been deployed through embryo microinjection of various combinations of CRISPR components. sgRNAs made through in vitro transcription and Cas9 provided as recombinant protein or messenger RNA (mRNA) can be injected into the developing embryo; sgRNAs can be injected into transgenic strains that express Cas9 in the developing germline (Basu et al. 2015; Dong et al. 2015; Kistler et al. 2015; Li et al. 2017; Zhu et al. 2021); or plasmids that contain regulatory sequences, which express the necessary components at the appropriate time and place to generate stable germline mutants, can be injected (Kyrou et al. 2018).

[1]Correspondence: ben.matthews@zoology.ubc.ca

Cite this introduction as *Cold Spring Harb Protoc;* doi:10.1101/pdb.top107688

In addition to enabling mechanistic studies of basic mosquito biology, CRISPR is a key component in several genetic strategies of vector-borne disease control, such as gene drives, precision genetic sterile insect technique, and more. These strategies use CRISPR gene editing to generate transgenic mosquitoes that pass on genetic elements, making them sterile or resistant to pathogens (Wang et al. 2021).

sgRNA DESIGN AND SELECTION

The CRISPR–Cas9 system relies on RNA–DNA base-pairing interactions between the sgRNA and a complementary sequence of genomic DNA to guide the Cas9 nuclease to its specific genomic target. The recognition site for the most commonly used *Streptococcus pyogenes* Cas9 takes the following form: N_{17-20} (NGG), with N representing any standard DNA nucleotide (A, T, C, or G). NGG is the protospacer-associated motif (PAM) and is not contained within the sgRNA DNA template or the transcribed RNA but is required to be located immediately adjacent to the target sequence in the genome for Cas9 nuclease activity.

Although one can scan the gene of interest manually when designing sgRNAs by simply looking for "GG" (or "CC"), we highly recommend the use of online sgRNA selection tools to identify potential sgRNAs. These have the benefit of scanning the rest of your organism's genome for predicted off-target activity, as well as designing primers for polymerase chain reaction (PCR) verification of mutagenesis and oligos for sgRNA synthesis. We highly recommend using CHOPCHOP (Labun et al. 2019), which has the latest versions of the common mosquito genomes loaded and ready for use, although countless other tools exist as well. In an accompanying protocol (Protocol 1: Validating Single-Guide RNA for *Aedes aegypti* Gene Editing [Lo and Matthews 2023]), we detail how to design and synthesize sgRNAs as well as how to validate sgRNA activity after injection (which is also discussed in the section sgRNA Testing here).

sgRNA TESTING

The ultimate goal of most gene editing experiments in mosquitoes is to produce stable and heritable mutant lines by editing the germline of a developing embryo. This requires that the Cas9/sgRNA complex generates a double-strand break that is repaired in a developing germ cell so that it can be passed on to the offspring of the founder. sgRNA efficiency is one variable that can influence the rate of success in creating these mutations. Once approximately three potentially suitable sgRNAs have been identified, we recommend checking their efficacy through small-scale embryo injections of sgRNA mixed with recombinant Cas9 protein, which should involve up to 50–100 embryos for each sgRNA. Large-scale injections of >500 embryos can be performed after identifying the sgRNAs that lead to the highest gene editing activities (Kistler et al. 2015). This testing is especially important if knock-ins, which have far lower rates of success in *Aedes aegypti* than loss-of-function NHEJ mutations, are planned (Kistler et al. 2015). Knock-ins generated by HDR are much less common than NHEJ in *Ae. aegypti* no matter the identity of sgRNAs used. Another option that is more commonly being used is to simply pool two to four sgRNAs into a single injection without testing, which is a strategy that has both potential benefits and caveats. On the positive side, multi-sgRNA injections appear to be very mutagenic, particularly in generating deletions between two cut sites. On the negative side, any potential off-target effects are multiplied, and structural rearrangements can occur because of multiple breaks.

ALTERNATIVES TO EMBRYO MICROINJECTION

One exciting new avenue for gene editing in nontraditional model organisms is the advent of alternatives to embryo microinjection, including receptor-mediated ovary transduction of cargo

 Cite this introduction as *Cold Spring Harb Protoc*; doi:10.1101/pdb.top107688

(ReMOT) control (Chaverra-Rodriguez et al. 2018; Terradas et al. 2022). This technique requires the delivery of CRISPR components directly into the ovaries of the model organism, where gene editing can happen directly in the developing embryos. In addition to providing a means of gene editing that bypasses the expensive and sometimes finicky process of embryo microinjection, adult injections can enable gene editing in animals whose eggs are difficult to collect or to inject, such as ticks (Sharma et al. 2022). However, one major limitation with ReMOT control is that there is currently no way to introduce exogenous DNA for HDR, meaning that for the time being, embryo microinjections are the best option for researchers trying to introduce exogenous DNA into their study organism's genome.

REFERENCES

Basu S, Aryan A, Overcash JM, Samuel GH, Anderson MAE, Dahlem TJ, Dahlem TJ, Myles KM, Adelman ZN. 2015. Silencing of end-joining repair for efficient site-specific gene insertion after TALEN/CRISPR mutagenesis in *Aedes aegypti*. *Proc Natl Acad Sci* **112:** 4038–4043. doi:10.1073/pnas.1502370112

Chaverra-Rodriguez D, Macias VM, Hughes GL, Pujhari S, Suzuki Y, Peterson DR, Kim D, McKeand S, Rasgon JL. 2018. Targeted delivery of CRISPR–Cas9 ribonucleoprotein into arthropod ovaries for heritable germline gene editing. *Nat Commun* **9:** 3008. doi:10.1038/s41467-018-05425-9

Dong S, Lin J, Held NL, Clem RJ, Passarelli AL, Franz AWE. 2015. Heritable CRISPR/Cas9-mediated genome editing in the yellow fever mosquito, *Aedes aegypti*. *PLoS ONE* **10:** e0122353. doi:10.1371/journal.pone.0122353

Doudna JA, Charpentier E. 2014. Genome editing. The new frontier of genome engineering with CRISPR–Cas9. *Science* **346:** 1258096. doi:10.1126/science.1258096

Kistler KE, Vosshall LB, Matthews BJ. 2015. Genome engineering with CRISPR–Cas9 in the mosquito *Aedes aegypti*. *Cell Rep* **11:** 51–60. doi:10.1016/j.celrep.2015.03.009

Kyrou K, Hammond AM, Galizi R, Kranjc N, Burt A, Beaghton AK, Nolan T, Crisanti A. 2018. A CRISPR–Cas9 gene drive targeting *doublesex* causes complete population suppression in caged *Anopheles gambiae* mosquitoes. *Nat Biotechnol* **36:** 1062–1066. doi:10.1038/nbt.4245

Labun K, Montague TG, Krause M, Torres Cleuren YN, Tjeldnes H, Valen E. 2019. CHOP-CHOP v3: expanding the CRISPR web toolbox beyond genome editing. *Nucl Acids Res* **47:** W171–W174. doi:10.1093/nar/gkz365

Li M, Bui M, Yang T, Bowman CS, White BJ, Akbari OS. 2017. Germline Cas9 expression yields highly efficient genome engineering in a major worldwide disease vector, *Aedes aegypti*. *Proc Natl Acad Sci* **114:** E10540–E10549. doi:10.1073/pnas.1711538114

Lo IHY, Matthews BJ. 2023. Validating single-guide RNA for *Aedes aegypti* gene editing. *Cold Spring Harb Protoc* doi:10.1101/pdb.prot108340

Matthews BJ, Vosshall LB. 2020. How to turn an organism into a model organism in 10 "easy" steps. *J Exp Biol* **223:** jeb218198. doi:10.1242/jeb.218198

Sharma A, Pham MN, Reyes JB, Chana R, Yim WC, Heu CC, Kim D, Chaverra-Rodriguez D, Rasgon JL, Harrell RA, et al. 2022. Cas9-mediated gene editing in the black-legged tick, *Ixodes scapularis*, by embryo injection and ReMOT control. *iScience* **25:** 103781. doi:10.1016/j.isci.2022.103781

Terradas G, Macias VM, Peterson H, McKeand S, Krawczyk G, Rasgon JL. 2022. Receptor-mediated ovary transduction of cargo—ReMOT control: a comprehensive review and detailed protocol for implementation. In *Transgenic insects*, 2nd ed (ed. Benedict MQ, Scott MJ), pp. 125–148. CABI, Wallingford, UK.

Wang GH, Gamez S, Raban RR, Marshall JM, Alphey L, Li M, Rasgon JL, Akbari OS. 2021. Combating mosquito-borne diseases using genetic control technologies. *Nat Commun* **12:** 4388. doi:10.1038/s41467-021-24654-z

Zhu GH, Albishi NM, Chen X, Brown RL, Palli SR. 2021. Expanding the toolkit for genome editing in a disease vector, *Aedes aegypti*: transgenic lines expressing Cas9 and single guide RNA induce efficient mutagenesis. *CRISPR J* **6:** 846–853. doi:10.1089/crispr.2020.0052

Protocol 1

Validating Single-Guide RNA for *Aedes aegypti* Gene Editing

Ivan Hok Yin Lo and Benjamin J. Matthews[1]

Department of Zoology, The University of British Columbia, Vancouver, British Columbia V6T 1Z4, Canada

Creating transgenic mosquitoes allows for mechanistic studies of basic mosquito biology and the development of novel vector control strategies. CRISPR–Cas9 gene editing has revolutionized gene editing, including in mosquitoes. This protocol details part of the gene editing process of *Aedes aegypti* mosquitoes via CRISPR–Cas9, through testing and validating single-guide RNAs (sgRNAs). Gene editing activity varies depending on the sequence of sgRNAs used, so validation of sgRNA activity should be done before large-scale generation of mutants or transgenics. sgRNA is designed using online tools and synthesized in <1 h. Once mutants or transgenics are generated via embryo microinjection, sgRNA activity is validated by quick genotyping polymerase chain reaction (PCR) and DNA sequencing.

MATERIALS

It is essential that you consult the appropriate Material Safety Data Sheets and your institution's Environmental Health and Safety Office for proper handling of equipment and hazardous materials used in this protocol.

Reagents

Agarose
Buffer of choice for gel electrophoresis
Column purification kit (optional, see Step 20)
DNA ladder
DNA polymerase (e.g., NEBNext master mix M0541 or equivalent)
Engen sgRNA synthesis kit (New England BioLabs E3322)
Ethanol (80%, v/v, freshly made with nuclease-free H_2O)
Ice
Monarch genomic DNA extraction kit (NEB; or equivalent)
Nuclease-free H_2O
Recombinant Cas9 protein (lyophilized; e.g., PNA Bio CP01 or equivalent)

Equipment

Agencourt Ampure RNAclean XP magnetic beads (e.g., Beckman Coulter A63987 or equivalent)
Electric tissue grinder (e.g., Cole Palmer RK-44468-25) and disposable pestles (e.g., Bel-Art F65000-9002)

A bead mill homogenizer can be used in place of a tissue grinder.

[1]Correspondence: ben.matthews@zoology.ubc.ca

Cite this protocol as *Cold Spring Harb Protoc*; doi:10.1101/pdb.prot108340

Gel electrophoresis equipment
Magnetic stand (1.5-mL tube format; e.g., Millipore Sigma-Aldrich Z740155-4EA or equivalent)
Microcentrifuge tubes (1.5-mL, nuclease-free; e.g., Eppendorf)
Nuclease-free tips (P10, P200, P1000)
Polymerase chain reaction (PCR) tube strips (0.1-mL; e.g., Eppendorf)
Spectrophotometer (e.g., NanoDrop)
Vortexer

METHOD

Perform all steps at room temperature unless otherwise indicated.

sgRNA Design

1. Use CHOPCHOP (https://chopchop.cbu.uib.no/) to select potential single-guide RNAs (sgRNAs) for your gene or region of interest.

   Typically, there will be many suitable sgRNAs for your gene and exon of interest. To facilitate sgRNA synthesis, we recommend enabling the parameter that filters results by those containing a 5′ G. This ensures that there is at least one "G" at the 5′ end of the sgRNA sequence, which is required for T7 RNA polymerase function.

2. For each sgRNA, order an oligonucleotide (standard yield and desalting is fine) of the following form: 5′ TTCTAATACGACTCACTATA N_{20} GTTTTAGAGCTAGA 3′.

   Do not include the protospacer-associated motif (PAM) in the middle of the N_{20} sgRNA, where N_{20} refers to the sgRNA sequence designed in the previous step.

   Oligos can be ordered from any synthesis provider (e.g., IDT, ThermoFisher, Genewiz)

3. For each sgRNA, order the top pair of genotyping primers recommended by CHOPCHOP. If designing primers on your own, design primers to generate a 200- to 800-bp amplicon with the sgRNA recognition roughly in the middle of the amplicon (avoiding the first and last 50 bp—this ensures that the Sanger sequencing reaction will be giving clean peaks when it reaches the PAM).

sgRNA Synthesis and Cleanup

4. Using the NEB Engen sgRNA synthesis kit, generate the sgRNA per the manufacturer's directions with the following modifications.
   i. Increase the incubation time from 30 min to 1 h to ensure maximum yield.
   ii. Perform the optional DNase removal step, increasing the final reaction volume to 52 µL.
5. Purify sgRNA using magnetic beads.
   i. Mix the sgRNA synthesis reaction mixture (from Step 4) with a 1.8× volume (94 µL) of Ampure RNAClean XP beads in a nuclease-free 1.5-mL tube.
   ii. Vortex the tube to mix well.
   iii. Incubate the mixture for 5 min at room temperature to allow the RNA to bind to the beads.
   iv. Place the tube on a magnet stand for 5 min to allow beads to collect at the side of the tube.

      The supernatant should be completely clear.

   v. Remove the supernatant and discard it, taking care not to disturb beads.
   vi. Rinse the beads by adding 300 µL of 80% ethanol with a P1000 tip.

vii. Let the tube remain on the magnet stand for 30 sec to 1 min.

viii. Remove the supernatant and discard it, taking care not to disturb the beads.

ix. Perform a second rinse with 300 µL of 80% ethanol.

x. Let the tube remain on the magnet stand for 30 sec to 1 min.

xi. Remove the supernatant carefully, taking care not to disturb beads.

xii. Use a small (P10) tip to remove all remaining ethanol from the bottom of the tube, taking care not to disturb beads.

xiii. Leave the tube on the magnet stand for 10 min until the beads are completely dry.

xiv. Remove the tube from the magnet stand and resuspend the bead pellet in 55 µL of nuclease-free H_2O. Using a P200 tip, pipette up and down 10 times to mix thoroughly.

xv. Incubate the tube at room temperature off the magnet stand for 10 min, and then return it to the magnet stand.

xvi. Let the tube stand for 5 min, until the solution has cleared.

xvii. Transfer 50 µL of the supernatant to a new tube, taking care not to disturb the beads.

If some beads are accidentally transferred to the new tube at this step, simply place the new tube on the magnet stand for 5 min and retransfer 45 µL to a third fresh tube.

This step ensures injection-ready sgRNAs are generated; many of the RNA column cleanup kits will leave residual salts or ethanol that can increase embryonic lethality following microinjection.

6. Measure RNA concentration and purity on a spectrophotometer.

The Engen synthesis kit should yield several micrograms of sgRNA in total. Aliquots should be made at concentrations >200 ng/µL to facilitate dilution for sgRNA injection. RNA should be pure and free of any indication of EtOH or salt contamination via spectrophotometer to prevent lethality to embryos during injection.

7. Divide sgRNAs into 5-µL aliquots in PCR strip tubes and store them at −80°C until use.

Each aliquot is "single use" and will be used for a single round of microinjection.

More aliquots can be made if desired, either by diluting sgRNAs with nuclease-free H_2O (maintaining concentrations of >200 ng/µL to facilitate injection mix dilution) or by making aliquots of smaller volumes.

According to commercial vendors, sgRNAs can be stored for up to 18 mo at −80° without losing activity. If there is any doubt about storage conditions, it is better to remake fresh sgRNAs before a large set of injections.

Reconstitution and Dilution of Cas9

8. Reconstitute the lyophilized Cas9 protein to 900 ng/µL in nuclease-free H_2O.

Other concentrations are fine, this one simplifies the math for injection mix creation. Avoid using any buffers that contain glycerol or other additives as they may interfere with injection survival.

9. Aliquot the reconstituted Cas9 protein in small (2- to 5-µL) aliquots in PCR strip tubes and freeze them until use at −80°C.

Injection Mix Preparation

10. From the freezer, remove aliquots of Cas9 and sgRNA(s) and thaw them on wet ice.

11. Mix each aliquot thoroughly by pipetting the entire volume up and down >10 times.

12. Generate the injection mixture by mixing sgRNA and Cas9 in nuclease-free H_2O to a final concentration of 300 ng/µL of Cas9 and 80 ng/µL of sgRNA.

sgRNA validation can be multiplexed across multiple genes at the same time. In this instance, reduce the concentration of each sgRNA to 40 ng/µL (total of 120 ng/µL if testing three sgRNAs across three genes). Three sgRNAs have been tested before without issue (see e.g., Kistler et al. 2015). Potential issues with multiplexing include increased likelihood of off-target cuts, potential for larger-scale chromosomal rearrangements, and viscosity of injection mix.

 Cite this protocol as *Cold Spring Harb Protoc*; doi:10.1101/pdb.prot108340

13. Place the injection mixture generated in Step 12 on wet ice and use for microinjections the same day. Discard any leftover mix at the end of the day.

    Alternatively, the injection mixture can be shipped on dry ice to an injection facility. The injection mixture can be stored for up to 18 mo at −80°.

Injections

14. Perform embryo microinjections as detailed elsewhere (see Chapter 32, Protocol 1: Mosquito Embryo Microinjection under Halocarbon Oil or in Aqueous Solution [Harrell 2023]).

Molecular Validation of sgRNA Activity

15. Rear surviving injected embryos to pupal stages, and freeze them at −20°C in at least three groups of five individuals. If using a tissue grinder in Step 16, freeze pupae in the specific tubes that fit the disposable pestles.

    Follow standard mosquito rearing procedures, see Chapter 1, Protocol 1: Batch Rearing Aedes aegypti *[Wohl and McMeniman 2022].*

    Animals can be frozen at other life stages as well, including at larval or adult stages; however, differences in the levels of somatic mosaicism may be observed due to different developmental trajectories.

    If few embryos survive, you can reduce the number of animals per replicate to three. Assaying individual animals is not recommended due to variability in injections from egg to egg.

16. Disrupt tissues using an electric tissue grinder and disposable pestles for ∼1 min until no visible large chunks of tissue are left.
17. Perform genomic DNA extractions from pools of pupae using a Monarch genomic DNA extraction kit, following the manufacturer's protocol. Store genomic DNA at −20°C.

    If DNA from wild-type mosquitoes of the injection strain is not readily available for Step 18, be sure to perform DNA extractions on a pool of wild-type mosquitoes during this step.

18. Perform a genotyping PCR (50 µL) with the primers designed by CHOPCHOP and DNA polymerase, using 10 ng of genomic DNA from Step 17 as the template. The amplicon size is provided by CHOPCHOP depending on the sgRNA design chosen. Set PCR conditions according to the type of DNA polymerase used and melting temperatures of the primers.

    Be sure to run a positive control using genomic DNA from wild-type animals of the injection strain to account for any differences from the reference genome.

    See Troubleshooting.

19. Run 2 µL of the reaction product on a 1.5% agarose gel with the appropriate DNA ladder, to check for the presence of a specific band of the expected size.

    The percentage of agarose gel can be altered according to the expected band size.

    See Troubleshooting.

20. Clean the remainder of the PCR product for Sanger sequencing using magnetic beads as in Step 5 or column purification kits, according to the manufacturer's instructions.
21. Send amplicons from each replicate, as well as from an uninjected positive control, for Sanger sequencing using the primer that is furthest away from the putative cut site as the sequencing primer.
22. Analyze sequencing data using online tools, such as ICE by Synthego or TIDE (Brinkman et al. 2014). These tools look for degradation in the major peaks of a chromatograph generated as part of a Sanger sequencing run as evidence of mosaicism in sequence caused by Cas9-induced insertions and deletions.

    Comparing estimated mutation rates for amplicons from injected pools to amplicons generated from a similarly sized pool of unmanipulated animals from the injection strain is essential to account for single nucleotide polymorphisms or other variants that might exist in the injection strain.

    See Troubleshooting.

TROUBLESHOOTING

Problem (Steps 18 and 19): Genotyping PCR does not work.

Solution: Standard PCR troubleshooting steps can be taken (e.g., reduce annealing temperature, alter concentration of template DNA, and perform a positive control for the genomic DNA using a set of known good primers). If these steps fail to resolve the issue, alternate primer pairs as suggested by CHOPCHOP can be ordered.

Problem (Step 22): Sanger sequencing results are not "clean," even for the positive control.

Solution: Ensure that a single band of the right size is seen on the gel (Step 19). If not, redo PCRs (Step 18) with wild-type genomic DNA and increase annealing temperatures until a single band is seen. If the band appears to be specific, try redoing Sanger sequencing with the other primer in the pair. If no troubleshooting steps help, redo genotyping PCRs with alternate primer pairs.

Problem (Step 22): An analysis of the amplicons reveals no evidence of sgRNA activity for any putative guides.

Solution: To distinguish between an issue with the non-sgRNA-related techniques and procedures (e.g., microinjection problems, inactive Cas9, or other problems) and inactive sgRNAs, we recommend including a positive-control sgRNA from previously published work. This is especially important when setting up microinjections in the laboratory for the first time, as it allows you to troubleshoot that process with a known good sgRNA.

DISCUSSION

Why Test sgRNAs?

In our experience, the in vivo efficiency of individual sgRNAs can vary dramatically, even when they meet all design criteria and are located nearby one another on the chromosome (Kistler et al. 2015). Given that large-scale injections are time-consuming and the subsequent rearing and screening for desired mutations or insertions are labor-intensive, we feel that the time and effort spent testing a handful of sgRNAs to identify the most active one (or at least to avoid the least active or inactive sgRNAs) is a valid trade-off.

Why Not Deep Sequence?

Original versions of this protocol used deep sequencing on the Illumina MiSeq platform to quantify the activity of the sgRNA (Kistler et al. 2015). Although the method is highly effective and more precise than those analyzing the degradation of Sanger sequencing traces recommended here, preparing sequencing libraries is a relatively complex procedure that might be inaccessible to some laboratories. However, some commercial sequencing providers now offer amplicon sequencing as a fee-for-service (e.g., Genewiz Amplicon-EZ), which might, over time, represent a better option given trade-offs between cost, time at the bench, and the precision of estimates.

How Active Is "Active Enough"?

It is difficult to pinpoint a target level of activity that is "acceptable" to declare a sgRNA to be viable moving forward. More than likely, experimental design dictates that a sgRNA be chosen that is in a certain region of the genome, as follows: near the 5′ region of a gene within an exon to induce a loss-of-function mutation, perhaps, or near the stop codon of a gene to insert a fusion cassette for functional genetics (Herre et al. 2022). With these limitations in mind, the sgRNA testing process can be seen as a way of avoiding the most low-performing sgRNAs. In other words, if

 Cite this protocol as *Cold Spring Harb Protoc*; doi:10.1101/pdb.prot108340

you must choose among three guides near a genomic region of interest, these test assays can be seen as helping to rank order them (or you can go back and design additional sgRNAs if none shows any activity).

What Next? Types of Mutations and Donors

Once you have identified a guide in your genomic region of interest that shows evidence of activity, congratulations are in order! You can now proceed to a larger-scale experiment, with the goal of knocking out the function of a gene or introducing novel genetic elements into your organism's genome. The type of mutation that you generate will depend largely on the scientific questions you are asking but will also depend on your capacity to maintain and genotype lines through molecular genotyping (necessary for nonhomologous end joining [NHEJ] mutations) or through fluorescence microscopy or other visible markers (Matthews and Vosshall 2020).

ACKNOWLEDGMENTS

We thank Katie Kistler and members of the Vosshall laboratory for help with the initial variants of this protocol and to Robert Harrell II at the University of Maryland/IBBR Insect Transformation Facility for injections and troubleshooting early versions of these methods. Thanks to Cassidy Mark and Zeynab Kohandel for testing and suggesting updates to these protocols in 2022.

REFERENCES

Brinkman EK, Chen T, Amendola M, van Steensel B. 2014. Easy quantitative assessment of genome editing by sequence trace decomposition. *Nucl Acids Res* **42:** e168. doi:10.1093/nar/gku936

Harrell R. 2023. Mosquito embryo microinjection under halocarbon oil or in aqueous solution. *Cold Spring Harb Protoc* doi:10.1101/pdb.prot108203

Herre M, Goldman OV, Lu TC, Caballero-Vidal G, Qi Y, Gilbert ZN, Gong Z, Morita T, Rahiel S, Ghaninia M. 2022. Non-canonical odor coding in the mosquito. *Cell* **185:** 3104–3123.e28. doi:10.1016/j.cell.2022.07.024

Kistler KE, Vosshall LB, Matthews BJ. 2015. Genome engineering with CRISPR–Cas9 in the mosquito *Aedes aegypti*. *Cell Rep* **11:** 51–60. doi:10.1016/j.celrep.2015.03.009

Matthews BJ, Vosshall LB. 2020. How to turn an organism into a model organism in 10 "easy" steps. *J Exp Biol* **223:** jeb218198. doi:10.1242/jeb.218198

Wohl MP, McMeniman CJ. 2022. Batch rearing *Aedes aegypti*. *Cold Spring Harb Protoc* doi:10.1101/pdb.prot108017

CHAPTER 35

Oral RNAi for Gene Silencing in Mosquitoes: From the Bench to the Field

Keshava Mysore,[1,2,3] Limb K. Hapairai,[1,2,3] Jacob S. Realey,[1,2] Longhua Sun,[1,2] Joseph B. Roethele,[1,2] and Molly Duman-Scheel[1,2,4]

[1]*Department of Medical and Molecular Genetics, Indiana University School of Medicine, South Bend, Indiana 46617, USA;* [2]*Eck Institute for Global Health, University of Notre Dame, Notre Dame, Indiana 46556, USA*

RNA interference (RNAi) has played a key role in the field of insect functional genomics, a discipline that has enhanced the study of developmental, evolutionary, physiological, and molecular biological phenomena in a wide variety of insects, including disease vector mosquitoes. Here we introduce a recently optimized RNAi procedure in which adult mosquitoes are fed with a colored sugar bait containing small interfering RNA (siRNA). This procedure effectively and economically leads to gene silencing, is technically straightforward, and has been successfully used to characterize a number of genes in adult mosquitoes. We also discuss how, in addition to laboratory applications, this oral RNAi procedure might one day be used in the field for controlling insect pests.

RNAi FOR GENE SILENCING IN MOSQUITOES

Vector mosquitoes transmit a variety of medically important human disease-causing pathogens (Centers for Disease Control 2021). Vector control is the primary method for preventing mosquito-borne illnesses in humans. However, as a result of the increasing global incidence of insecticide resistance and concerns for potential negative impacts of chemical pesticides on nontarget organisms, current mosquito control methods are reaching the limits of sustainability, necessitating the development and introduction of innovative vector control strategies (Airs and Bartholomay 2017; Centers for Disease Control 2021). Mosquito genome projects (Holt et al. 2002; Nene et al. 2007) have facilitated research in new aspects of mosquito biology, including functional genetic studies in medically important *Aedes* (dengue, Zika, chikungunya, and yellow fever vector) and *Anopheles* (malaria vector) human disease vector mosquitoes (Centers for Disease Control 2021). These advancements have precipitated the development of novel gene-centered vector control strategies. RNA interference (RNAi), which is currently used to functionally characterize mosquito genes in the laboratory, could potentially be applied as a new method of vector control (Wiltshire and Duman-Scheel 2020).

RNAi, an innate genetic regulatory mechanism conserved in many eukaryotes, prevents the expression of targeted mRNA transcripts. These transcripts are silenced in a sequence-specific manner by interfering RNA species, including 100- to 300-bp double-stranded RNAs (dsRNAs), 21- to 25-bp small interfering RNAs (siRNAs), or short hairpin RNAs (shRNAs) (Fire et al. 1998; Timmons et al. 2001). In the laboratory, RNAi has allowed characterization of gene function in a wide variety of

[3]These authors contributed equally to this work.
[4]Correspondence: mscheel@nd.edu

Cite this introduction as *Cold Spring Harb Protoc*; doi:10.1101/pdb.top107690

insects, including mosquitoes (Airs and Bartholomay 2017). Despite the successful use of this technique by many laboratory research programs, the efficiency of RNAi is historically inconsistent and seemingly dependent on a number of variables (Wiltshire and Duman-Scheel 2020), including the interfering RNA species administered (dsRNA, siRNA, or shRNA), the stability of the interfering RNA and the mRNA target, the interfering RNA delivery system used (Whitten 2019), the arthropod species targeted, the life stage at which RNAi is introduced, and the cellular physiology of the tissue(s) targeted (Airs and Bartholomay 2017; Kunte et al. 2020). Such variables can complicate RNAi studies in the laboratory and hamper the development of RNAi as an insect pest control technology. Recent advances in the oral administration of interfering RNA to both larvae (Singh et al. 2013; Whyard et al. 2015; Hapairai et al. 2017) and adult mosquitoes (Hapairai et al. 2020; Mysore et al. 2020) may facilitate translation of this technology to the field (Wiltshire and Duman-Scheel 2020). As a result, orally administered RNAi pesticides may represent a new class of biorational insecticides that could combat the increased global incidence of insecticide resistance and be successfully deployed as a new component of integrated mosquito control programs.

ORAL DELIVERY OF siRNAs FOR GENE SILENCING IN ADULT MOSQUITOES

We recently optimized a procedure for the analysis of gene function in which adult mosquitoes are fed with a colored sugar bait containing siRNA (Hapairai et al. 2020; Mysore et al. 2020). The accompanying protocol presents step-by-step details for carrying out this procedure in both *Aedes aegypti* and *Anopheles gambiae* (see Protocol 1: Sugar-Baited Delivery of Small Interfering RNA for Gene Silencing in Adult Mosquitoes [Mysore et al. 2022]). Adapted from a previously published technique for feeding long dsRNAs to *Ae. aegypti* (Coy et al. 2012), the procedure is less technically challenging and physically disruptive than subcutaneous injection, is designed to enhance sugar feeding and yield reproducible data, and facilitates the analysis of behavioral and morbidity phenotypes resulting from target gene suppression. Concerns about off-site targeting with RNAi can be alleviated by using multiple different short siRNAs corresponding to the gene of interest (rather than long dsRNA) to confirm phenotypes. Moreover, the use of siRNAs involves the design of interfering RNA with short 21- to 25-bp target sites that are conserved in orthologous genes, but that lack target sites in non–target organisms (Hapairai et al. 2020; Mysore et al. 2020), an advantage that may facilitate the design of species-specific attractive targeted sugar baits (ATSBs), which are emerging as a new paradigm for vector control (Fiorenzano et al. 2017). This methodology could be adapted to study other laboratory insects that consume sugar bait and could potentially be applied for the control of mosquitoes or other insect pests in the field.

THE APPLICATION OF ORAL RNAi FOR FUNCTIONAL CHARACTERIZATION OF MOSQUITO GENES

Our optimized oral RNAi strategy was successfully used to characterize the function of the *Shaker (Sh)* and *dopamine-1 receptor (dop1)* genes. These studies, which verified that these genes are required for mosquito viability and adult brain function, have been published (Hapairai et al. 2020; Mysore et al. 2020) and are summarized here to show the efficacy and utility of the sugar-baited siRNA delivery technique. In these studies, adult female mosquitoes were permitted to feed for 4 h on sugar bait alone, sugar bait with control siRNA that lacks a target in mosquitoes, sugar bait with Sh.463 siRNA (which silences *Sh*), or sugar bait with dop1.462 siRNA (which silences *dop1*) in replicate trials conducted using 25 adult females per treatment. Although negligible morbidity was observed in mosquitoes fed with sugar bait alone (10% ± 2%) or control siRNA sugar bait (10% ± 2%), high levels of morbidity were observed in mosquitoes that consumed Sh.463 (89% ± 6%, $P < 0.001$ compared to sugar bait or to control siRNA sugar bait) or dop1.462 siRNA (88% ± 4%, $P < 0.001$ com-

 Cite this introduction as *Cold Spring Harb Protoc*; doi:10.1101/pdb.top107690

pared to sugar bait or to control siRNA sugar bait). Sh.463- and dop1.462-treated adults died over the course of 6 d, with the time of death spread fairly evenly over the 6-d trial period (Hapairai et al. 2020; Mysore et al. 2020). These findings indicated that Sh.463 and dop1.462 siRNAs have insecticidal qualities that could potentially be exploited in the field for mosquito control.

Both Sh.463- and dop1.462-treated mosquitoes (Hapairai et al. 2020; Mysore et al. 2020) failed to fly, displayed uncoordinated walking behavior, often fell, and did not venture beyond the bottom of the vial in which they were housed. Use of the sugar-baited delivery system removed the trauma of siRNA microinjection, ensuring confidence that resulting behavioral phenotypes and morbidity were not secondary to injection trauma. The Sh.463 and dop1.462 target sites are well-conserved in mosquitoes (Hapairai et al. 2020; Mysore et al. 2020), and in future studies oral sugar-baited RNAi can be used to target these genes or other genes of interest in additional *Aedes, Anopheles,* or potentially *Culex* species. It is also likely that the sugar-baited siRNA delivery system could be used to study gene function in many other insects that consume sugar solutions.

POTENTIAL DEPLOYMENT OF RNAi PESTICIDES IN INTEGRATED MOSQUITO CONTROL PROGRAMS

An ultimate goal of our research is to consider whether siRNAs that cause mortality, such as Sh.463 and dop1.462, might represent a new class of biorational pesticides that could be used in integrated human disease vector mosquito control programs (Hapairai et al. 2020; Mysore et al. 2020). The expression and delivery of these pesticides in yeast expression systems shows great promise as a larvicide intervention (Duman-Scheel 2019), and yeast larvicides corresponding to these siRNAs effectively kill multiple species of mosquito larvae (Hapairai et al. 2020; Mysore et al. 2020). In adults, ATSBs, which take advantage of the sugar-feeding behavior of adult male and female mosquitoes that are lured to feed on a sugar bait laced with an insecticide, are emerging as a promising new means of vector control (Fiorenzano et al. 2017). ATSBs can be supplied through bait stations or in sprays and can be used indoors or outdoors (Müller and Galili 2016). Although ATSB technology facilitates targeted delivery of a variety of pesticides, insecticide resistance is nevertheless an ongoing concern (Faraji and Unlu 2016), and the use of RNAi insecticides, a new class of pesticides, is therefore of interest. The species specificity associated with RNAi pesticides would also be advantageous, as insecticides typically used in many ATSBs are not specific to mosquitoes (Faraji and Unlu 2016). Moreover, despite efforts to limit ATSB applications to nonflowering vegetation and the use of protective bait stations that are more selective to mosquitoes (Müller and Galili 2016), it is still difficult to completely eliminate insecticide risks to pollinators and other non–target organisms. RNAi-based pesticides appear to have a highly desirable safety profile (Environmental Protection Agency 2014) particularly with respect to conventional chemical pesticides, and therefore have the potential to enhance existing ATSB technology. Optimized laboratory sugar feeding protocols, which should be further adjusted for field deployment in the future, will facilitate further analysis of the utility of this promising new technology.

CONCLUDING REMARKS

Gene silencing by oral RNAi offers several useful advantages (Wiltshire and Duman-Scheel 2020). The long-term maintenance of genetically modified mosquito strains bearing germline modifications is not required. Researchers can control the timing of RNAi-mediated gene silencing and choose the stage at which gene silencing begins, circumventing challenges such as developmental lethality, which has been observed as a result of *Sh* and *dop1* gene silencing in mosquito larvae (Hapairai et al. 2020; Mysore et al. 2020) and which can hinder the production and maintenance of strains bearing heritable mutations. Finally, despite the significant advancements associated with the advent of CRISPR–Cas9 technology, genetic engineering in nonmodel insects is labor-intensive, requires skill, and is still relatively expensive. For these reasons, RNAi remains a useful strategy for studying gene function in insects (Wiltshire and Duman-Scheel 2020).

ACKNOWLEDGMENTS

Our RNAi pesticide research has been supported by the National Institutes of Health/National Institute of Allergy and Infectious Disease (1 R21 AI128116-01), the United States Agency for International Development (AID-OAA-F-16-00097), the U.S. Department of Defense (W81XWH-17-1-0294), the Innovative Vector Control Consortium, and the Department of the Army, U.S. Army Contracting Command, Aberdeen Proving Ground, Natick Contracting Division, Fort Detrick, MD, USA under a Deployed Warfighter Protection (DWFP) Program Grant (W911QY-17-1-0002). The sponsors of these awards (to M.D.-S.) did not play a role in the writing of this introduction nor the decision to submit it for publication.

REFERENCES

Airs PM, Bartholomay LC. 2017. RNA interference for mosquito and mosquito-borne disease control. *Insects* **8:** 4. doi:10.3390/insects8010004

Centers for Disease Control. 2021. Mosquitoes. Available at https://www.cdc.gov/mosquitoes/index.html. Accessed March 20, 2020.

Coy MR, Sanscrainte ND, Chalaire KC, Inberg A, Maayan I, Glick E, Paldi N, Becnel JJ. 2012. Gene silencing in adult *Aedes aegypti* mosquitoes through oral delivery of double-stranded RNA. *J Applied Ent* **136:** 741–748. doi:10.1111/j.1439-0418.2012.01713.x

Duman-Scheel M. 2019. *Saccharomyces cerevisiae* (baker's yeast) as an interfering RNA expression and delivery system. *Curr Drug Targets* **20:** 942–952. doi:10.2174/1389450120666181126123538

Environmental Protection Agency. 2014. RNAi technology as a pesticide: problem formulation for human health and ecological risk assessment. Docket ID: EPA-HQ-OPP-2013-0485. Available at https://www.regulations.gov/docket?D=EPA-HQ-OPP-2013-0485. Accessed May 5, 2015.

Faraji A, Unlu I. 2016. The eye of the tiger, the thrill of the fight: effective larval and adult control measures against the Asian tiger mosquito, *Aedes albopictus* (Diptera: Culicidae), in North America. *J Med Entomol* **53:** 1029–1047. doi:10.1093/jme/tjw096

Fiorenzano JM, Koehler PG, Xue RD. 2017. Attractive toxic sugar bait (ATSB) for control of mosquitoes and its impact on non–target organisms: a review. *Int J Environ Res Public Health* **2017:** 14. doi:10.3390/ijerph14040398

Fire A, Xu S, Montgomery MK, Kostas SA, Driver SE, Mello CC. 1998. Potent and specific genetic interference by double-stranded RNA in *Caenorhabditis elegans*. *Nature* **391:** 806–811. doi:10.1038/35888

Hapairai LK, Mysore K, Chen Y, Harper EI, Scheel MP, Lesnik AM, Sun L, Severson DW, Wei N, Duman-Scheel M. 2017. Lure-and-kill yeast interfering RNA larvicides targeting neural genes in the human disease vector mosquito *Aedes aegypti*. *Sci Rep* **7:** 13223. doi:10.1038/s41598-017-13566-y

Hapairai LK, Mysore K, Sun L, Li P, Wang C-W, Scheel ND, Lesnik A, Scheel MP, Igiede J, Wei N, et al. 2020. Characterization of an adulticidal and larvicidal interfering RNA pesticide that targets a conserved sequence in mosquito G protein–coupled *dopamine 1 receptor* genes. *Insect Biochem Mol Biol* **120:** 103359. doi:10.1016/j.ibmb.2020.103359

Holt RA, Subramanian GM, Halpern A, Sutton GG, Charlab R, Nusskern DR, Wincker P, Clark AG, Ribeiro JM, Wides R, et al. 2002. The genome sequence of the malaria mosquito *Anopheles gambiae*. *Science* **298:** 129–149. doi:10.1126/science.1076181

Kunte N, McGraw E, Bell S, Held D, Avila LA. 2020. Prospects, challenges and current status of RNAi through insect feeding. *Pest Manag Sci* **76:** 26–41. doi: 10.1002/ps.5588

Müller GC, Galili A. 2016. Attractive toxic sugar baits (ATSB): from basic science to product—a new paradigm for vector control. Available at https://endmalaria.org/sites/default/files/7_Gunter%20Mueller.pdf. Accessed October 20, 2019.

Mysore KM, Hapairai LK, Sun L, Li P, Wang CW, Scheel ND, Lesnik A, Igiede A, Scheel MP, Wei N, et al. 2020. Characterization of a dual-action adulticidal and larvicidal interfering RNA pesticide targeting the *Shaker* gene of multiple disease vector mosquitoes. *PLoS Negl Trop Dis* **14:** e0008479. doi:10.1371/journal.pntd.0008479

Mysore KM, Hapairai LK, Realey JS, Sun L, Roethele JB, Duman-Scheel M. 2022. Sugar-baited delivery of small interfering RNA for gene silencing in adult mosquitoes. *Cold Spring Harb Protoc* doi: 10.1101/pdb.prot107808

Nene V, Wortman JR, Lawson D, Haas B, Kodira C, Tu ZJ, Loftus B, Xi Z, Megy K, Grabherr M, et al. 2007. Genome sequence of *Aedes aegypti*, a major arbovirus vector. *Science* **316:** 1718–1723. doi:10.1126/science.1138878

Singh AD, Wong S, Ryan CP, Whyard S. 2013. Oral delivery of double-stranded RNA in larvae of the yellow fever mosquito, *Aedes aegypti:* implications for pest mosquito control. *J Insect Sci* **13:** 69. doi:10.1673/031.013.6901

Timmons L, Court DL, Fire A. 2001. Ingestion of bacterially expressed dsRNAs can produce specific and potent genetic interference in *Caenorhabditis elegans*. *Gene* **263:** 103–112. doi:10.1016/S0378-1119(00)00579-5

Whitten MM. 2019. Novel RNAi delivery systems in the control of medical and veterinary pests. *Curr Opin Insect Sci* **34:** 1–6. doi:10.1016/j.cois.2019.02.001

Whyard S, Erdelyan CN, Partridge AL, Singh AD, Beebe NW, Capina R. 2015. Silencing the buzz: a new approach to population suppression of mosquitoes by feeding larvae double-stranded RNAs. *Parasit Vectors* **8:** 96. doi:10.1186/s13071-015-0716-6

Wiltshire RM, Duman-Scheel M. 2020. Advances in oral RNAi for disease vector mosquito research and control. *Curr Opin Insect Sci* **40:** 18–23. doi:10.1016/j.cois.2020.05.002

Protocol 1

Sugar-Baited Delivery of Small Interfering RNA for Gene Silencing in Adult Mosquitoes

Keshava Mysore,[1,2,3] Limb K. Hapairai,[1,2,3] Jacob S. Realey,[1,2] Longhua Sun,[1,2] Joseph B. Roethele,[1,2] and Molly Duman-Scheel[1,2,4]

[1]*Department of Medical and Molecular Genetics, Indiana University School of Medicine, South Bend, Indiana 46617, USA;* [2]*Eck Institute for Global Health, University of Notre Dame, Notre Dame, Indiana 46556, USA*

RNA interference (RNAi), an innate regulatory mechanism that is conserved across many eukaryotic species, has been harnessed for experimental gene silencing in many organisms, including mosquitoes. This protocol describes an optimized method for inducing RNAi in adult *Aedes aegypti* and *Anopheles gambiae* mosquitoes that involves feeding them a red-colored sugar bait containing small interfering RNA (siRNA). This oral delivery method is less physically disruptive than delivery by subcutaneous injection, and the use of siRNAs (in contrast to long dsRNAs) for RNAi enables the design of molecules that target conserved sites so that gene function can be studied in multiple species. After feeding, the behavioral and morbidity phenotypes that result from the suppression of target gene expression can then be analyzed.

MATERIALS

It is essential that you consult the appropriate Material Safety Data Sheets and your institution's Environmental Health and Safety Office for proper handling of equipment and hazardous materials used in this protocol.

Reagents

Aedes aegypti or *Anopheles gambiae* mosquitoes

Liver powder (60 g/L slurry prepared in double-distilled H_2O and stored at 4°C) (for *Ae. aegypti* only; see Table 1)

Reagents for quantitative real-time polymerase chain reaction (qRT-PCR) or in situ hybridization (optional; see Step 14)

Reverse osmosis (RO) water

RNase-free sterile water

Small interfering RNAs (siRNAs)

Design an siRNA against a target of interest as described (see Clemons et al. 2010a). We typically use siRNA prepared with standard desalting by Integrated DNA Technologies, which has a useful siRNA design tool (IDT 2021). Control siRNA with the sequence 5′-GAAGAGCACUGAUAGAUGUUAGCGU-3′, which has no known target in mosquitoes, should also be synthesized.

---

[3]These authors contributed equally to this work.

[4]Correspondence: mscheel@nd.edu

Cite this protocol as *Cold Spring Harb Protoc*; doi:10.1101/pdb.prot107808

Staple diet Koi food (for *An. gambiae* only; see Table 1)

Sucrose (10%, dissolved in autoclaved RO water and filter-sterilized with a 0.2-µm filter)

Prepare fresh. For sucrose solutions used in Step 9, add 0.5 µL of red food color (e.g., McCormick) per mL of sugar solution after filter sterilization. The red color serves as an attractant and promotes increased feeding rates. It also facilitates confirmation that the mosquitoes have eaten the bait.

Equipment

Aspirator (Gempler's RHM200) with collecting tube (Gempler's RHM201)

Clear larval tray cover (Bioquip 1426BC)

Coffee filters

Colony cage (30 cm × 30 cm × 30-cm; Bioquip 1452)

Drosophila vials (Fisher Scientific 07-000-482)

Erlenmeyer flask (50-mL)

Feeding cages

Adapt 3.8-L buckets (Berry Plastics T801163-5) for use as feeding cages by cutting a small incision in the top through which feeders can be suspended.

Insect room or incubator that maintains 26.5°C ± 1°C and ~80% relative humidity

Large cotton balls (Fisher Scientific NC9689872)

Microcentrifuge

Microcentrifuge tubes (0.2-mL thin-walled tubes and standard 1.5-mL tubes)

Plastic cups (500-mL)

Razor blade

Syringe (60-mL)

Transfer pipettes

Vacuum incubator (VWR 6291) (for *Ae. aegypti* only; see Step 1)

White larval trays (Bioquip 1426B)

METHOD

This entire procedure should be performed in an insectary maintained at 26.5°C ± 1°C, at ~80% relative humidity, and with a 12-h light/dark cycle that ideally includes 1-h crepuscular periods.

Rearing Larvae and Preparing Adults for siRNA Feeding

A recommended daily schedule for using this protocol in conjunction with Ae. aegypti *and* An. gambiae *is provided in Table 1. Refer to Clemons et al. (2010b) or MR4 (2015) for details on rearing* Ae. aegypti *or* An. gambiae*, respectively.*

Larval Stages

1. Age-synchronize mosquito larvae as follows.
    - To age-synchronize *Ae. aegypti* larvae, rehydrate an egg paper in RO water and place it under a vacuum at ~20–25 psi for 45 min.
    - To age-synchronize *An. gambiae* larvae, see MR4 (2015) and Table 1.
2. Rear age-synchronized larvae in white trays with clear covers (to facilitate visualization) at a density of 200 larvae/L of water.

    Avoid higher larval densities, which can result in less robust adults and poor adult siRNA feedings.
3. Provide larval food as described in Table 1.

    Avoid overfeeding because uneaten food can pollute the water and impact mosquito development/survival.

 Cite this protocol as *Cold Spring Harb Protoc*; doi:10.1101/pdb.prot107808

TABLE 1. An experimental time line for preparing and feeding small interfering RNA (siRNA) to adult female mosquitoes

Day	Day	Stage	Food	Amount (mg/larva)	Activities
Aedes aegypti					
D1	Wed	L1	LP	0.2	Hatch and feed
D2	Thu	L2	LP	0.2	Feed
D3	Fri	L2–L3	LP	0.5	Feed
D4	Sat	L3	LP	0.5	Feed
D5	Sun	L3–L4	LP	0.5	Feed
D6	Mon	L4	LP	0.5	Feed
D7	Tue	L4-pupae	LP	If needed	Discard pupae[a,e]
D8	Wed	L4-pupae	LP	If needed	Pick pupae[b,e]
D9	Thu	Pupae	LP	If needed	Pick pupae[b,e]
D10	Fri	Adult D0	SF	Ad libitum	NA
D11	Sat	Adult D1	SF	Ad libitum	NA
D12	Sun	Adult D2	SF	Ad libitum	NA
D13	Mon	Adult D3	None	NA	Sugar starve
D14	Tue	Adult D4	None	NA	Sugar starve; transfer to feeding cage
D15	Wed	Adult D5	SF + siRNA	At least 65 µL	siRNA treatment[c]
D16	Thu	Adult D6	SF	Ad libitum	Begin to monitor phenotypes
Anopheles gambiae					
D1	Mon	L1	KSD	0.2	Eggs hatch
D2	Tue	L1–L2	KSD	0.2	Feed
D3	Wed	L2	KSD	0.5	Feed
D4	Thu	L2–L3	KSD	0.5	Feed
D5	Fri	L3	KSD	0.5	Feed
D6	Sat	L3–L4	KSD	0.5	Feed
D7	Sun	L4	KSD	0.5	Feed
D8	Mon	L4-pupae	KSD	In needed	Discard pupae[d,e]
D9	Tue	L4-pupae	KSD	In needed	Discard pupae[d,e]
D10	Wed	Pupae	KSD	In needed	Pick pupae[b,e]
D11	Thu	Adult D0	SF	Ad libitum	NA
D12	Fri	Adult D1	SF	Ad libitum	NA
D13	Sat	Adult D2	SF	Ad libitum	NA
D14	Sun	Adult D3	SF	Ad libitum	Transfer to feeding cage
D15	Mon	Adult D4	None	NA	Sugar starve
D16	Tue	Adult D5	SF + siRNA	At least 65 µL	siRNA treatment[c]
D17	Wed	Adult D6	SF	Ad libitum	Begin to monitor phenotypes

This table details the recommended time line of the experimental procedure, from the larval growing regimen to feeding assays, for *Ae. aegypti* (top) and *An. gambiae* (bottom) adult females.

(KSD) Koi staple diet, (LP) liver powder, (SF) sugar feed (10% sucrose).

[a]To permit age synchronization on D8–9.

[b]Rear and use these mosquitoes for adult siRNA feedings.

[c]siRNA treatment for no more than 4 h ensures single feedings; additional feedings on subsequent days can be added to the protocol if desired.

[d]To permit age synchronization on D10–11 (discarded pupae of either species can be used for colony maintenance).

[e]Males, which are smaller in size, typically pupate ahead of females and can be presorted at this stage.

Pupal Stage

4. Add 40 mL of 10% sucrose to an Erlenmeyer flask, which will be sufficient to feed a single colony cage. Roll a coffee filter into a conical tube and insert it into the flask, creating a wick.
5. Select approximately 30 female pupae and approximately five male pupae (permitting a 6 female:1 male ratio for mating) per treatment group and use a transfer pipette to place them in a cup of RO water. Load the cup into the colony cage.

   Feedings are typically conducted on mated females. It may be possible to use unmated females, but this has not been trialed in conjunction with this protocol to date.

   Males, which are smaller in size, typically pupate ahead of females and can be presorted (see schedule in Table 1).

Adults

For best results, adults should be at least 5 d old, and females should be mated before siRNA feedings (Table 1).

6. Once adults eclose, make sure that 10% sucrose is available and mortality is minimal.
7. Sugar-starve adults by removing the 10% sucrose from colony cages 48 h before *Ae. aegypti* female siRNA feedings (which initiate on D15; Table 1) or 12–24 h for *An. gambiae* female feedings (which initiate on D16; Table 1). Using an aspirator, transfer adult mosquitoes of the desired sex to an experimental feeding cage 24 h before siRNA feedings.

   To prevent excessive adult manipulation, use the pulse mode of the aspirator to aspirate mosquitoes into the collecting tube, and gently blow mosquitoes out into the experimental feeding cages.

   Allow mosquitoes to rest for a day following transfer so that they are not stressed at the initiation of the assays.

   See Troubleshooting.

Preparing siRNA Feeders and Bait

8. To create an siRNA feeder, cut off the pointed end of a thin-walled 0.2-mL microcentrifuge tube with a razor blade, creating an ~1-mm opening, and then insert a small piece of cotton (~4-mg) through the opening, creating a wick (Fig. 1A). Prepare as many feeders as there are experimental conditions and label each feeder.

   These can be prepared in advance of the feeding assays.

   See Troubleshooting.

9. On the day of feeding, dissolve control or experimental siRNA in the 10% sugar solution, gently pipetting up and down to ensure adequate mixing. Collect the solution with brief centrifugation at low speed.

   Preparation and use of at least 65 µL of 2.5 µg/µL siRNA in 10% sugar solution prepared using RNase-free water in conjunction with a cage of 25 adult females has yielded good results for several genetic targets (e.g., Hapairai et al. 2020), but dosage should be calibrated for each specific target.

   Suggested controls include sugar bait alone and sugar bait containing siRNA of the sequence 5′-GAAGAG-CACUGAUAGAUGUUAGCGU-3′, which has no known target in mosquitoes.

10. Carefully transfer the siRNA/sugar solutions from Step 9 into individual feeders (from Step 8) and quickly cap each tube.

    Prolonged time without a cap will vent the feeder, causing it to drip. An empty P-1000 pipette tip box can serve as a feeder holder.

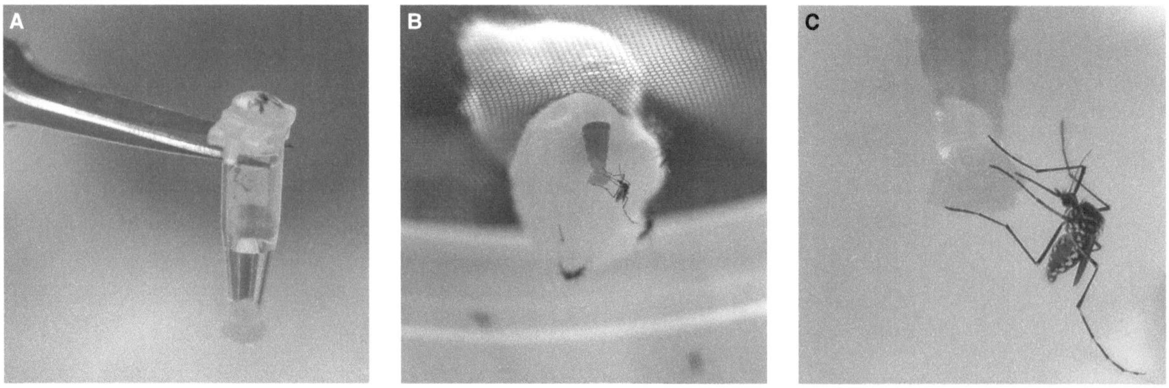

FIGURE 1. Sugar-baited delivery of small interfering RNA (siRNA) to adult mosquitoes. (*A*) Feeder filled with siRNA-sugar solution that is delivered through a cotton wick. (*B*) The feeder is hung within the feeding cage. (*C*) An engorged adult *Aedes aegypti* female is shown.

 Cite this protocol as *Cold Spring Harb Protoc*; doi:10.1101/pdb.prot107808

Feeding Adults with Control and Treatment siRNA

11. Hang the feeder carefully from a 2-cm hole in the feeding cage and secure it with a cotton ball (Fig. 1B).

    If the wick dries, carefully open and close the cap to vent the feeder and rehydrate the wick.

12. Allow 25 5-d-old adult females to feed for 4 h (Fig. 1B).

    For best results, start the feeding at dawn, which can be simulated through manipulation of the light–dark settings in the insectary if desired. Alternatively, covering the feeding cages with a black trash bag before the feeding can help.

    See Troubleshooting.

Monitoring Phenotypes

13. Identify engorged females by visualizing the red tracer dye in the abdomen (Fig. 1C), and separate them from mosquitoes that fail to consume the sugar bait by carefully aspirating and transferring them to a new culturing cage in the insectary where they can be monitored.

    Watch for engorged mosquitoes frequently, as engorgement is not obvious once the mosquitoes defecate (within ~2 h). See Troubleshooting.

    In the event that phenotypes must be monitored closely in individual mosquitoes, carefully aspirate each individual into a separate 10-mL fly rearing vial. Cap the vial with a small cotton ball and store it in the insectary. Isolation of individual mosquitoes may not be necessary depending on which phenotypes are evaluated.

14. Begin to assess phenotypes of interest within several hours of completing the feeding assays and verify gene silencing via qRT-PCR or in situ hybridization (see Mysore et al. 2019 or Hapairai et al. 2020 for suggestions).

    Our analyses of mortality/morbidity typically conclude within 6 d postfeeding. However, the duration of phenotype assessments can be custom-designed for each experiment.

15. Twenty-four hours after completing the siRNA feedings, feed mosquitoes with 10% sucrose using the Erlenmeyer flask/coffee filter wick system placed in the cage as described in Step 4.

    For feeding mosquitoes reared and observed as individuals, add 75 µL of 10% sucrose to the fly rearing vial using a syringe. The syringe can be used to displace a bit of cotton from the cotton plug, and the sucrose can then be added to these displaced cotton fibers, which serve as a wick. Add additional sucrose to individual vials in this manner every 2 d.

TROUBLESHOOTING

Problem (Step 7): The adults die before the experiment.
Solution: It may be helpful to reduce the sugar starving time, as excessive sugar starving can result in death.

Problem (Step 8): The feeders are leaking.
Solution: The cotton may need to be packed more tightly, or the wick may need to be shortened. You may want to prepare trial feeders and fill them with 10% sugar solution as a test (Fig. 1A).

Problem (Step 12): Insects are not feeding.
Solution: Consider the following.

- It may be helpful to reduce the sugar starving time, as excessive sugar starving can result in poor feedings.

- Initiating the feedings at dawn or simulated dawn is also critical. To ensure that mosquitoes drink from the feeder, be sure to completely dry the feeding cages and remove water condensation.
- Improved feeding rates were observed when staff members avoided use of gloves made of latex, which can have repellent qualities (Ramos et al. 2011); nitrile gloves can be substituted.

Problem (Step 13): The adults are not completely engorged after feeding.
Solution: Consider the following.

- Try not to disrupt the mosquitoes during feedings, which should ideally be conducted when the insectary is not heavily accessed.
- Avoid exposing the feeding cages to direct light by placing them on lower shelves of the insectary or hanging a covering above the cage.
- The feeders may need to be vented occasionally to prevent drying of the wick.
- Use of a second feeder may improve feedings, but requires additional siRNA, which increases costs.

RELATED INFORMATION

This protocol was adapted from a previously published technique for feeding long double-stranded RNAs (dsRNAs) to adult *Ae. aegypti* (Coy et al. 2012). For a protocol for gene silencing in larvae, see Mysore et al. (2019).

ACKNOWLEDGMENTS

Protocol optimization was funded by the Innovative Vector Control Consortium and the Department of the Army, U.S. Army Contracting Command, Aberdeen Proving Ground, Natick Contracting Division, Fort Detrick, MD, USA under Deployed Warfighter Protection (DWFP) Program Grant Award W911QY-17-1-0002 to M.D.-S. The award sponsors did not play a role in the writing of this protocol nor in the decision to submit it for publication. Thanks to members of the laboratory for useful discussions.

REFERENCES

Clemons A, Haugen M, Severson DW, Duman-Scheel M. 2010a. Functional analysis of genes in *Aedes aegypti* embryos. *Cold Spring Harb Protoc* doi:10.1101/pdb.prot5511

Clemons A, Mori A, Haugen M, Severson DW, Duman-Scheel M. 2010b. Culturing and egg collection of *Aedes aegypti*. *Cold Spring Harb Protoc* doi:10.1101/pdb.prot5507

Coy MR, Sanscrainte ND, Chalaire KC, Inberg A, Maayan I, Glick E, Paldi N, Becnel JJ. 2012. Gene silencing in adult *Aedes aegypti* mosquitoes through oral delivery of double-stranded RNA. *J Applied Ent* **136:** 741–748. doi:10.1111/j.1439-0418.2012.01713.x

Hapairai LK, Mysore K, Sun L, Li P, Wang C-W, Scheel ND, Lesnik A, Scheel MP, Igiede J, Wei N, et al. 2020. Characterization of an adulticidal and larvicidal interfering RNA pesticide that targets a conserved sequence in mosquito G protein-coupled *dopamine 1* receptor genes. *Insect Biochem Mol Biol* **120:** 103359. doi:10.1016/j.ibmb.2020.103359

IDT. 2021. Custom Dicer-Substrate siRNA (DsiRNA). Available at https://www.idtdna.com/site/order/designtool/index/DSIRNA_CUSTOM. Accessed February 1, 2021.

MR4. 2015. Methods in *Anopheles* research. Available at https://www.beiresources.org/Portals/2/VectorResources/2016%20Methods%20in%20Anopheles%20Research%20full%20manual.pdf. Accessed March 20, 2020.

Mysore K, Hapairai LK, Wei N, Realey JS, Scheel ND, Severson DW, Duman-Scheel M. 2019. Preparation and use of a yeast shRNA delivery system for gene silencing in mosquito larvae. *Methods Mol Biol* **1858:** 213–231. doi:10.1007/978-1-4939-8775-7_15

Ramos MV, Araújo ES, Oliveira RSB, Teixeira FM, Pereira DA, Cavalheiro MG, Souza DP, Oliveira JS, de Freitas CDT. 2011. Latex fluids are endowed with insect repellent activity not specifically related to their proteins or volatile substances. *Braz J Plant Physiol* **23:** 57–66. doi:10.1590/S1677-04202011000100008

CHAPTER 36

Genetic Toolbox Approaches in Mosquitoes

Olena Riabinina,[1,2] Matthew Quinn,[1] and Joe P. Whitehead[1]

[1]*Department of Biosciences, Durham University, Durham, DH1 3LE, United Kingdom*

Mosquitoes are fascinating and diverse study organisms that are rapidly becoming genetically tractable. Here, we review the latest and most commonly used approaches to silence or edit genes and express transgenes in mosquitoes. We also focus on gene drives—a genetic technology that may prove essential for combating mosquito-borne diseases.

INTRODUCTION

Mosquitoes are not genetic model organisms. However, they attracted intense research interest because of their ability to transmit zoonotic diseases, and the first transgenic mosquitoes were created more than 20 years ago (Coates et al. 1998; Jasinskiene et al. 1998; Catteruccia et al. 2000; Grossman et al. 2001; Perera et al. 2002). Since then, the genome editing field has been revolutionized by CRISPR–Cas9, which was quickly and successfully implemented in eight mosquito species (Gantz et al. 2015; Kistler et al. 2015; Dong et al. 2018; Li et al. 2018; Anderson et al. 2019; Duvall et al. 2019; Liu et al. 2019; Raji et al. 2019; Greppi et al. 2020; Li et al. 2020a, 2021; Meuti and Harrell 2020; Redhai et al. 2020; Sun et al. 2020; Feng et al. 2021a,b; Inbar et al. 2021; Purusothaman et al. 2021; Quinn et al. 2021; Zhan et al. 2021). It has also become apparent that although mosquito genetic tools are often instructed by similar tools in *Drosophila*, the *Drosophila* solutions cannot be ported directly into mosquitoes and need to be adjusted for each mosquito species of interest. Efforts to eliminate mosquito species that transmit malaria, dengue, and West Nile virus (*Anopheles spp.*, *Aedes spp.*, and *Culex spp.*, respectively) have fueled the development of gene drives in mosquitoes more than in any other organism to date (Gantz et al. 2015; Hammond et al. 2016, 2017, 2021a; Kyrou et al. 2018; Adolfi et al. 2020; Li et al. 2020b; Hoermann et al. 2021). Here, we review recent progress in gene silencing by RNAi, genome editing and transgenesis, and the development of CRISPR–Cas9 gene drives. We have highlighted the most widely used or promising approaches that may be of use to mosquito researchers, focusing on the genetic techniques rather than on the original research questions.

GENE SILENCING BY RNAi

The suppression of gene expression can be induced by RNA interference (RNAi), as first observed in the nematode worm *Caenorhabditis elegans* (Fire et al. 1998). RNAi has now been used in *Aedes aegypti*

[2]Correspondence: olena.riabinina@durham.ac.uk

Supplemental material is available at www.cshprotocols.cshlp.org.

Cite this introduction as *Cold Spring Harb Protoc*; doi:10.1101/pdb.top107691

(Hapairai et al. 2017), *Aedes albopictus* (Gu et al. 2011), *Culex pipiens* (Lopez-Martinez et al. 2012), *Anopheles gambiae* (Taracena et al. 2019), and *Anopheles stephensi* (Brown et al. 2003) mosquitoes, and may be implemented via delivery of double-stranded RNA (dsRNA), small interfering RNA (siRNA), or short hairpin RNA (shRNA). The RNAi process can be broken into three critical stages. First, a long dsRNA molecule that is introduced into the cell is cut into smaller RNA fragments by a Dicer or Ribonuclease III enzyme (RNase III). These smaller fragments are known as siRNAs and have a complementary sequence to the target mRNA (Zhang et al. 2004). Second, the siRNAs complex with the RNA-induced silencing complex (RISC), and the sense strand of the siRNA is degraded, leaving the antisense strand complexed (Kim and Rossi 2008). Third, the RISC-antisense siRNA strand complex binds to the complimentary target mRNA sequence before translation can occur and degrades the mRNA, preventing the expression of the gene (Fire et al. 1998). There are three primary methods of delivering RNAi into mosquito cells: (1) microinjection (e.g., Liu et al. 2010; Singh et al. 2013); (2) feeding (e.g., Zhang et al. 2010; Mysore et al. 2017); and (3) microbial delivery (Taracena et al. 2019).

Microinjection

Microinjection has been the most common method to induce RNAi in mosquitoes, particularly at the pupal and adult stages of development (Regna et al. 2016; Isoe et al. 2019). The advantage of using microinjection is that it ensures the delivery of dsRNA directly into the tissue of choice, avoiding the cuticle, gut epithelium, and hardy outer exoskeleton. Another benefit of microinjections is knowing precisely how much dsRNA is delivered into the organism, unlike feeding, in which the amount of dsRNA ingested may be unknown (Yu et al. 2013). However, the disadvantage of using microinjections is that they require a high degree of skill and may cause injury, thus affecting survival and experimental readouts. For example, Blitzer et al. (2005) targeted sterol carrier protein-2 (AeSCP-2*)* expression by injecting chemically synthesized and HPLC-purified dsRNA into adult and fourth-instar larvae of *Ae. aegypti.* Knockdown of *AeSCP-2* in fourth instars reduced cholesterol accumulated in pupae by 33% but caused 42% adult mortality in the first 3 d of emergence.

Microinjection is rarely used in the larvae because of the small size of the larvae and increased likelihood of fatal injury to the organism. However, Liu et al. (2010) used this approach to down-regulate *An. gambiae olfactory receptors 7* (*AgOr7*) and *40* (*AgOr40*), and *An. gambiae variant ionotropic receptor 76b* (*AgIr76b*) in fourth-instar larvae. Quantitative analysis showed a significant transcript level reduction of *AgOr7*, *AgOr40,* and *AgIr76b* after siRNA treatment. Larvae injected with the siRNA had a significantly reduced behavioral response to DEET compared to the control.

Whyard et al. (2015) injected *Ae. aegypti* pupae with dsRNA to target 10 genes that regulate fertility in adult males (see Supplemental Table S1). A quantitative reverse transcription polymerase chain reaction (qRT-PCR) confirmed that each of the genes was down-regulated by 70%–95% in the adult males 4 d postinjection and in total, nine out of 10 of the dsRNAs induced sterility in half of the males. Similarly, Meuti et al. (2015) microinjected adult female *Cx. pipiens* with dsRNA to knock down several circadian clock genes (*per, tim, cry2, cyc,* and *pdf*) to measure the effect on egg follicle length and lipid content. Reduction of mRNA expression by RNAi had varied success per gene: *per* (40%), *tim* (25%), *cry2* (45%), *cyc* (15%), and *pdf* (60%).

It appears that the effects of RNAi can be highly variable not only between studies, but also between targets of the same study in the same species of mosquito. The duration of RNAi effects is also unclear. We thus suggest that down-regulation of selected genes should be experimentally validated for each target gene. In summary, delivery of dsRNA by microinjection is a suitable method that has been successfully used in, for example, *An. gambiae* (Michel et al. 2006; Liu et al. 2010; Liu and Zwiebel 2013; Regna et al. 2016), *An. stephensi* (Brown et al. 2003), *Ae. aegypti* (Blitzer et al. 2005; Berois et al. 2012; Erdelyan et al. 2012; Hapairai et al. 2017; Isoe et al. 2019), and *Cx. pipiens* (see Supplemental Table S1; Meuti et al. 2015).

Cite this introduction as *Cold Spring Harb Protoc*; doi:10.1101/pdb.top107691

Feeding or Soaking in a ds/siRNA Solution

Feeding or soaking is the most straightforward delivery method because of its simplicity and the lack of injury caused to the insect. This method is well-established, quick, repeatable, and convenient for larval and pupal stages with effects that can persist into the adult stage. The disadvantage of soaking is that large volumes of dsRNA needed to make it effective. Moreover, a study by Fischer et al. (2017) showed that dsRNA quickly degrades in aquatic environments, giving it limited application to larviciding in the field.

Singh et al. (2013) were the first to perform soaking on mosquito larva. They successfully targeted *β-tubulin, chitin synthase-1* and *-2*, and *heat shock protein 83* by soaking first-instar larvae of *Ae. aegypti* in H_2O containing β-*tub*-dsRNA for 2 h before returning them to fresh H_2O. The authors confirmed 100% uptake of the dsRNA solution by adding a blue dye that could be seen in the gut of the larvae when ingested. The length of time to soak a sample varies between developmental stages. Figueira-Mansur et al. (2013) soaked second-instar *Ae. aegypti* larvae for 24 h to silence the insecticide resistant *AaegP-gp* gene. Larvae soaked in the dsRNA solution had a 30% higher mortality in insecticides than control groups (see Supplemental Table S1).

Inducing RNAi in larvae may have effects that persist into the adult stage. Whyard et al. (2015) soaked larvae in a dsRNA solution for 1 h over 6 d to successfully target genes that regulate male fertility (*bol, fzo, gas8, nht,* and *zpg*). They used dsRNA at two concentrations: 0.02 mg/mL (which rendered 20%–35% of males sterile) and 0.2 mg/mL (which rendered 72%–92% of males sterile). The same method was used by the authors to target the female isoform of the sex determination gene *doublesex,* creating a male-biased population. Similarly, Lopez-Martinez et al. (2012) soaked dehydrated fourth-instar *Cx. pipiens* larvae in H_2O containing dsRNA to down-regulate *heat shock protein 90 (hsp90).* The effects persisted through to the pupal and adult stages, with *hsp90* levels reduced by 77% on the third day of adult life.

There have been many studies aimed at optimizing ds/siRNA delivery by feeding to adult mosquitoes. Mysore et al. (2020) targeted the gene *Shaker,* which when knocked down, produced lethal phenotypes and neural defects that affected coordination and prevented flight behavior in dsRNA-treated mosquitoes. RNAi targeting the *dop1* gene has now been trialed as a pesticide delivered to adults in attractive toxic sugar baits (Hapairai et al. 2020). The study found that 88% of mosquitoes that fed on the sugar bait died and the remaining 12% could not fly. Toxic sugar bait studies thus show promise for the control of mosquito populations in the field (see Supplemental Table S1).

Chitosan Nanoparticle and Agarose Gel Delivery

The instability of dsRNA after delivery is a major difficulty for RNAi applications. The degradation of dsRNA may be delayed by encapsulating the ds/siRNA in chitosan nanoparticles (Howard et al. 2006) that may then be fed to mosquitoes. Mysore et al. (2013) used chitosan/siRNA nanoparticle feeding to investigate the role of the gene *semaphorin-1a* (*sema1a*) in the olfactory development of *Ae. aegypti.* Antennal lobe defects were detected in 44% of treated larvae and 32% of pupae. In the behavioral assay, 48% of treated larvae failed to touch the yeast pellet, a normally attractive food stimulus for the larvae. In another study, researchers disrupted cuticle formation in larvae and induced mortality in adult *Ae. aegypti* by feeding the mosquitoes chitosan/dsRNA nanoparticles to target the gene *3, 4-dihydroxyphenylacetaldehyde* (*DOPAL*) *synthase* (Chen et al. 2019). Furthermore, Ramesh et al. (2016) compared two methods of dsRNA delivery to the third-instar larvae of *Ae. aegypti.* First, they fed the larvae bacterially expressed dsRNA, and second, they fed them chitosan/dsRNA nanoparticles to target the *vestigial gene (vg).*There was no significant difference in the efficiency of dsRNA delivery by either method, and both techniques resulted in high adult deaths and wing abnormalities.

Chitosan nanoparticles may also be used in combination with agarose gel to deliver RNAi through feeding. Zhang et al. (2010) reduced the transcript levels of two chitin synthase genes, *AgCHS1* (by 62.8%) and *AGCH2* (by 33.8%), by feeding third-instar larvae of *Ae. aegypti* with dsRNA coated in a mixture of agarose gel, which was further entrapped in chitosan/*AgCHS* dsRNA-based nanoparticles (see Supplemental Table S1). Incorporating RNAi into chitosan nanoparticles coated with an agarose gel helps protect the dsRNA from breaking down in the environment, as previously shown in mice (Gao et al. 2009). The technique is particularly well-suited for larval feeding because larvae eat almost continuously and because without the gel/particle protection, RNA quickly degrades in H_2O.

Microbial Delivery (Yeast, Bacteria, Viruses, and Algae)

RNAi delivery through microbes has been developed in recent years as another method of oral delivery. Mysore et al. (2017) used engineered *Saccharomyces cerevisiae* that produced shRNAs to silence *suppressor of actin* (*Sac1*), *leukocyte receptor complex member* (*lrc*), and *offtrack* (*Otk*) by feeding the yeast to *An. gambiae* larvae. The knockdown of these genes resulted in a loss of neuronal synapses and 100% of larval mortality in *An. gambiae*.

Bacteria have also been used as an effective method of delivering RNAi to mosquito. Taracena et al. (2019) knocked down expression of the female-specific isoform of the *doublesex* gene *AgdsxF* in *An. gambiae* larvae through oral delivery of bacteria that were producing dsRNA. Feeding *AgdsxF* dsRNA reduced the mRNA transcript by 66% in females, and only half of the females survived to adulthood. Similar to the protection that chitosan nanoparticles give to dsRNA, microbial delivery offers RNAi protection against RNases and from breaking down in the gut of insects (Wiltshire and Duman-Scheel 2020).

Viral delivery has been used as another delivery method through soaking. Gu et al. (2011) developed a recombinant *Ae. aegypti* Densovirus (AeDNV) siRNA expression system that was introduced to *Ae. albopictus* by soaking second-instar larvae in H_2O containing the virus stock. siRNA entered the body through the anal papillae before spreading to the muscle fibers, midgut, salivary glands, and several other tissues, down-regulating V-ATPase expression by up to 70% (see Supplemental Table S1).

Microalgae are one of the primary sources of nutrition for larvae in aquatic environments, therefore using transgenic algae as a vector for delivering dsRNA to the larvae has offered some potential. The advantage of microalgae is that they can be heat-killed, reducing the amount of genetically modified organisms released in the wild. Kumar et al. (2013) fed first-instar larvae of *An. stephensi* transgenic microalgae, targeting 3-hydroxykynurenine transaminase (3-HKT), a gene that catalyzes the transamination of 3-hydroxykynurenine (3-HK) to xanthurenic acid in the tryptophan catabolism pathway, resulting in an increase in reactive oxygen species that can kill the mosquito. The study found that dsRNA-treated larvae had 40% greater mortality than the control in every bioassay. Mortality was first observed in the second-instar stage, whereas treated larvae had significantly reduced levels of 3-HKT transcripts.

In summary, microbial delivery shows great potential because of its ease of manipulation, application to larviciding in the field, and ability to protect dsRNA from degradation in aquatic environments.

Transgenic Delivery

The use of transgenes to express dsRNA is a commonly used approach in *Drosophila melanogaster*, and a large collection of commercially available RNAi lines is currently available (Dietzl et al. 2007). Transgenic delivery has been successfully used in *An. gambiae* to down-regulate *cyp4g16* and *cyp4g17* genes involved in the production of cuticular hydrocarbons (Lynd et al. 2019). In their study, Lynd and colleagues (2019) used an oenocyte-specific driver to express *cyp4g16/17* dsRNA, leading to mortality or high susceptibility of mosquitoes to desiccation during the pupal stage. The advantage of this transgenic approach is the ease of delivery via a genetic cross, which is suitable for research applications. However, establishing an efficient RNAi line may be tricky, and breeding several test lines is labor-intensive. Nevertheless, efficient RNAi lines could potentially be applied in the field by using gene drives (see the section CRISPR–Cas9 Gene Drives) for transgene propagation.

 Cite this introduction as *Cold Spring Harb Protoc*; doi:10.1101/pdb.top107691

GENE EDITING

Gene editing in mosquitoes has so far been used to primarily study mosquito olfaction and the mechanisms of disease transmission (for reviews, see Dong et al. 2021; Konopka et al. 2021; Wheelwright et al. 2021), as well as for population disruption or replacement (see, e.g., Adolfi and Lycett 2018; Raban et al. 2020 and the section CRISPR–Cas9 Gene Drives). All these studies were published after 2012, when CRISPR–Cas9 was introduced as a tool for gene editing (Jinek et al. 2012). Therefore, it is not surprising that only a few studies used early gene editing tools, such as transcription activator–like effector nucleases (TALENs) and zinc-finger nucleases (ZFNs). Notable examples of pre-CRISPR studies are works on chemosensation and thermosensation in *Ae. aegypti* that used ZFNs to mutate olfactory receptor coreceptor *AeORCO* (Degennaro et al. 2013), a CO_2 coreceptor gene *AeGR3* (McMeniman et al. 2014), and a temperature receptor *TRPA1* (Corfas et al. 2015). The use of TALENs in mosquitoes was pioneered in *Ae. aegypti* by a proof-of-principle study that mutated the *kmo* gene, a gene essential for eye pigmentation (Aryan et al. 2013). At almost the same time, TALENs were used to mutate an immunity gene *TEP1* in *An. gambiae* (Smidler et al. 2013). Although TALENs and ZFNs are still an option for creating DNA breaks at a predefined location, the low cost and ease of use of CRISPR–Cas9 has made it the gene editing tool of choice for mosquito researchers.

To date, CRISPR–Cas9 has been used in eight mosquito species: *An. gambiae* (Dong et al. 2018; Greppi et al. 2020; Redhai et al. 2020; Sun et al. 2020), *An. stephensi* (Gantz et al. 2015; Inbar et al. 2021), *Anopheles funestus* (Li et al. 2018; Quinn et al. 2021), *Anopheles albimanus* (Li et al. 2018), *Ae. aegypti* (Kistler et al. 2015; Duvall et al. 2019; Raji et al. 2019; Zhan et al. 2021), *Ae. albopictus* (Liu et al. 2019), *Culex quinquefasciatus* (Anderson et al. 2019; Li et al. 2020a; Feng et al. 2021a,b; Purusothaman et al. 2021), and *Cx. pipiens* (Meuti and Harrell 2020; Li et al. 2021), all of which are vectors of human disease (see Supplemental Table S2). CRISPR–Cas9 was used in these species to create null mutants via nonhomologous end joining (NHEJ) repair of Cas9-induced DNA breaks, and/or to introduce transgenes via homology-directed repair (HDR) of DNA breaks. Kistler et al. (2015) provide an excellent step-by-step method for CRISPR–Cas9 gene editing in *Ae. aegypti*, which, briefly, consists of injecting mosquito embryos with (1) Cas9 protein or mRNA, (2) sgRNA, and (3) ssDNA or dsDNA plasmid donor for HDR. Injections of recombinant Cas9 protein led to the highest rate of mutagenesis, up to 30%, followed by Cas9 mRNA. The authors also tried injecting a plasmid containing a *polyubiquitin_promoter-Cas9* construct but found it ineffective. The rate of HDRs observed was also up to 30%. A somewhat different method was used for *Anopheles* (Hammond et al. 2016; Sun et al. 2020; Quinn et al. 2021) and *Ae. aegypti* (Zhu et al. 2021). The authors used *Vasa* promoter, which is active in the germline, to drive expression of Cas9 at the early embryonic stage. *Vasa_promoter-Cas9* and *U6_promoter-sgRNA* constructs were cloned into the same plasmid that was injected into the mosquito embryos together with a second plasmid containing a homology repair template. This technique led to a 2%–7% rate of HDR insertions in *Anopheles*, indicating that the Vasa promoter can efficiently drive Cas9 expression in the germline.

Chaverra-Rodriguez et al. (2018) have pioneered ReMOT Control (Receptor-Mediated Ovary Transduction of Cargo)—an alternative, high-throughput method of CRISPR component delivery based on the phenomenon known as maternal deposition, whereby protein material is synthesized by the mother and delivered to the embryos. In the ReMOT Control method, a carrier peptide P2C is fused to the Cas9 protein and injected into the abdomen of a gravid female together with sgRNA and an endosomal escape reagent. The P2C peptide ensures that the Cas9 + sgRNA complex is taken up by the embryo from the mother's hemolymph, and the endosomal escape reagent ensures the Cas9 + sgRNA complex is released from the uptake endosome and reaches the embryo's nuclei. This method bypasses one of the main bottlenecks of mosquito genetic modification: the labor-intensive manual injection of individual embryos (Bui et al. 2020; Meuti and Harrell 2020; Carballar-Lejarazú et al. 2021). ReMOT Control has been successfully used in *Ae. aegypti* (Chaverra-Rodriguez et al. 2018), *An. stephensi* (Macias et al. 2020), and *Cx. pipiens* (Li et al. 2021) to produce knockout mutants with success rates as high as 0.3 mutants per injected female. No knock-ins or random transgene insertions were yet reported with ReMOT Control, but when further developed, this method will make in-house mosquito genetic editing and transgenesis available to numerous laboratories.

TRANSGENES AND BINARY EXPRESSION SYSTEMS

Transgenes

The ability to introduce transgenes, instead of only creating mutations, takes genetic editing of mosquitoes to the next level. Interestingly, random insertions of transgenes were successfully achieved in *An. stephensi* (Catteruccia et al. 2000), *An. gambiae* (Grossman et al. 2001), *An. albimanus* (Perera et al. 2002), and *Ae. aegypti* (Coates et al. 1998; Jasinskiene et al. 1998) decades before successful gene editing (see Supplemental Table S2). Random insertions of transgenes into *Aedes fluviatilis* (Rodrigues et al. 2006) and *Ae. albopictus* (Labbé et al. 2010) followed soon after. More recently, *piggyBac* and *Mos* vectors were successfully used for random transgene insertions in *An. gambiae* (Riabinina et al. 2016) and *Ae. aegypti* (Buchman et al. 2019; Bui et al. 2019; Aryan et al. 2020; Shankar et al. 2020). Transgenes may also be integrated into the "docking sites" of known genomic locations, such as p*attB* (Franz et al. 2011; Meredith et al. 2011, 2013). The p*attB* sites are first integrated as random insertions into the genome, and then their position is mapped out and characterized for strength of expression and leakiness. The *pattB* insertion may be preferred to a random insertion when researchers do not need to obtain a number of transgenic lines with different strengths of transgene expression. The transgene often consists of a mosquito-active promoter fused to the reporter of interest, such as green fluorescent protein (GFP). The transgenes also usually carry an easily visible marker for screening. In contrast to *Drosophila*, in which curly wings, number and shape of bristles, eye color and shape, etc., serve as markers, mosquito markers are limited and consist of a fluorescent protein driven by an eye (*3xP3*) or a midgut (*actin5C*) enhancer and promoter (Volohonsky et al. 2015; Bernardini et al. 2018; Webster and Scott 2021). Thus, having good, easily visible markers in mosquitoes is important —genetic crosses of several transgenic or mutant lines would be greatly simplified if each had its own visual marker. Volohonsky et al. (2015) provide a useful compilation of tools that they developed for *Anopheles* transgenesis, including expression patterns of commonly used markers and reporters, and *An. gambiae* codon usage preference.

Binary Expression Systems

Binary systems for transgene expression have proved to be effective and popular in mosquitoes (Driesschaert et al. 2021 offers an excellent recent introduction to transgene expression and overview of binary expression systems). Briefly, binary expression systems have two key components: One determines in which cells and at what times a transgene is expressed; the other carries the coding sequence of the transgene to be expressed. Binary systems provide several advantages over promoter-reporter fusions: (1) The expression of the reporter gene will be greatly amplified; (2) different reporters/effectors will always be expressed in the same pattern if the same driver is used; and (3) fewer transgenic lines need to be maintained as soon as the number of promoters and reporters exceeds two of each. So far, three binary systems have been introduced into mosquitoes: TetOn-Off (Gossen and Bujard 1992; Lycett et al. 2004), GAL4/UAS (Brand and Perrimon 1993; Kokoza and Raikhel 2011), and the Q-system (Potter et al. 2010; Riabinina et al. 2015, 2016).

The GAL4/UAS system was piloted in *Ae. aegypti* (Kokoza and Raikhel 2011), in which the *vitellogenin* (*Vg*) promoter was used to drive expression of a shortened version of GAL4, consisting of its DNA-binding and activation, but not the middle, domains. The *Vg-GAL4* driver was inducible by a blood meal in females, which resulted in GFP expression in the fat bodies of females after blood feeding. In *An. gambiae*, the *carboxypeptidase* promoter was used to drive GAL4 expression in the midgut, visualized via *UAS-eYFP* reporter (Lynd and Lycett 2012). However, to achieve sufficiently high expression levels, the activation domain of GAL4 had to be replaced by the VP16 activation domain, which removed the ability of the modified GAL4 to bind GAL80, the repressor of the GAL4/

UAS system. Another study on *An. gambiae* used the GAL4/UAS system to drive *Cyp4g16/17*-RNAi in oenocytes to affect production of cuticular hydrocarbons (Lynd et al. 2019). Random insertions of GAL4 were also used for enhancer-trapping in *An. stephensi*, resulting in six GAL4 driver lines with larval and/or adult expression in the fat body, midgut, maxillary palps, antennae, muscles, salivary glands, and other tissues (O'Brochta et al. 2012).

The Q-system is another binary expression system that functions similarly to GAL4/UAS but employs the *qa* gene cluster from the bread fungus *Neurospora crassa* (Potter et al. 2010; Riabinina et al. 2015; Riabinina and Potter 2016). It was also originally developed in *Drosophila* but turned out to work well across species from plants to vertebrates. In mosquitoes, it was first used to target *Orco*$^+$ neurons of *An. gambiae* (Riabinina et al. 2016). The 9-kb putative promoter/enhancer region immediately upstream of the ATG translation start codon of *Orco* was polymerase chain reaction (PCR)-amplified and cloned into a *piggyBac* vector, followed by the QF2 transcription factor. The resulting expression pattern of QF2 faithfully labeled *Orco*$^+$ neurons, and the *Orco-QF2* driver has since been used to study the anatomy and function of the olfactory system in *An. gambiae* (Riabinina et al. 2016; Afify et al. 2019; Maguire et al. 2022). However, it proved very challenging to create other driver lines in mosquitoes by PCR-amplifying a putative enhancer/promoter region in the same manner (although see Ye et al. 2020). The intergenic sequences in mosquitoes are very large, and without an accurate method that predicts the location of an enhancer, it is impossible to pinpoint and PCR-amplify the enhancer. Instead, mosquito researchers turned to an alternative strategy that bypasses enhancer search. Matthews et al. (2019) used CRISPR–Cas9 to create a double-stranded break at the stop codon of the *ppk301* gene, and by providing a DNA template for HDR, replaced the stop codon with a *T2A–QF2* construct, creating a *ppk301–QF2* driver line in *Ae. aegypti*. The *QF2* was thus expressed in exactly the same pattern as the *ppk301* gene under the control of endogenous *ppk301* enhancer, the exact location of which is not known. *QF2* and *ppk301* were produced as two separate proteins because of the ribosome skipping the *T2A* peptide. This method has been repeatedly and successfully used in *Ae. aegypti* to study their sense of taste (Jové et al. 2020), olfaction (Shankar et al. 2020; Sorrells et al. 2021; Younger et al. 2020; Zhao et al. 2022), and the anatomy of their nervous system (Zhao et al. 2021). It has also been implemented in *An. gambiae* to make a chemosensory IR76b–QF2 driver line (Ye et al. 2021).

Both GAL4/UAS and the Q-system can efficiently drive transgene expression in various mosquito tissues; however, modifications of GAL4 may be required for stronger expression. The QF2 transactivator of the Q-system has been reliably strong, but some toxicity effects may be present for broad expression patterns (Basrur et al. 2020; Zhao et al. 2021). The weaker version of the QF2, named QF2^{W}, may thus be the best choice for future Q-driver lines (Riabinina et al. 2015). In addition, QF2 or QF2^{W} may be silenced by QS and reactivated by quinic acid—manipulations that work well in *Drosophila* (Riabinina et al. 2015) and *C. elegans* (Wei et al. 2012) but that have not yet been implemented in mosquitoes.

CRISPR–CAS9 GENE DRIVES

Overview of Gene Drives in Mosquitoes

Gene drives increase gene inheritance rates above Mendelian rates (super-Mendelian inheritance), allowing a genetic modification, the "payload," to rapidly spread throughout mosquito populations. The use of gene drives in mosquitoes has attracted interest because of their potential to eradicate serious mosquito-borne diseases, such as malaria.

Gene drives were originally proposed in 2003 using naturally occurring homing endonuclease genes (HEGs) (Burt 2003). Although some HEG-based drives were produced, for example, in *An. gambiae*, *I-SceI* was used to disrupt GFP at an allele conversion ("homing") rate of 56% (Windbichler et al. 2011), their development was limited because of the complexity of endonuclease engineering. The development of CRISPR–Cas9 as a tool for genome editing and the ease of engineering gRNAs

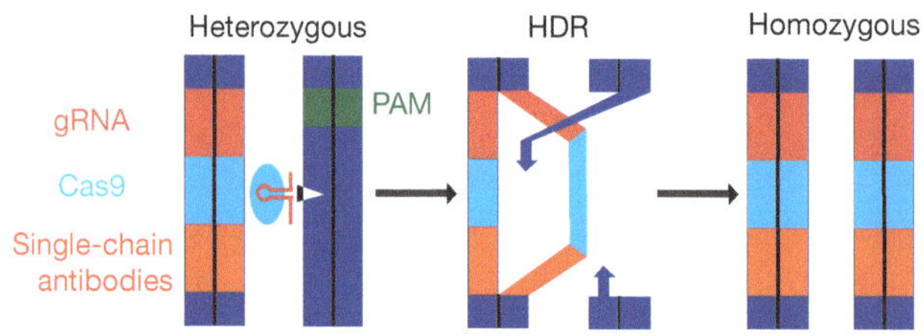

FIGURE 1. Mutagenic chain reaction (MCR) mechanism driving the first mosquito CRISPR–Cas9 gene drive (CGD) (Gantz et al. 2015). (*Left*) The drive construct contains sequences for gRNA (red), Cas9 (cyan), and *Plasmodium falciparum*–targeting single-chain antibodies (orange). Cas9 scans for the PAM sequence, and then gRNA checks for a complementary sequence. When the target sequence is found, the Cas9–gRNA complex cleaves (black and white arrow) the wild-type allele (dark blue). (*Middle*) The double-stranded break is repaired by homology-directed repair (HDR). (*Right*) Following HDR, the mosquito is homozygous for the drive construct.

originally facilitated the development of CRISPR–Cas9 gene drives (CGDs) in *Drosophila* (Gantz and Bier 2015). CGDs convert organisms heterozygous for a gene into homozygotes in an autocatalytic process termed mutagenic chain reaction (MCR) (Fig. 1; Gantz and Bier 2015; Gantz et al. 2015). Interestingly, a recent analysis by Garrood et al. (2021) in *An. gambiae* suggested that although CGDs can produce off-target mutations, these can be reduced to below detectable levels by tightly restricting spatiotemporal *Cas9* expression to the germline and carefully choosing the guide RNA (gRNA) target site. Furthermore, there was no evidence that off-target mutations were capable of driving, even when using the drive *vas-7280*^{CRISPRh}, which shows high rates of off-target mutations (Hammond et al. 2017; Garrood et al. 2021).

Currently, mosquito CGDs are used for two general purposes: population suppression (reducing the number of mosquitoes) and population replacement (replacing a wild-type population with a modified one). For example, the population suppression drive *vas-7280*^{CRISPRh} disrupted the *An. gambiae* female fertility gene *AGAP007280*. At peak drive frequency, this suppressed the egg production of a caged population by 92%. However, functional resistance alleles caused drive frequency to decline, making it unsuitable for field trials (Hammond et al. 2016, 2017). A population replacement drive was used to express single-chain antibodies targeting the malarial parasite *P. falciparum* (m2A10 and m1C3) in *An. stephensi*, showing inheritance rates of ~99.5% (Fig. 1; Gantz et al. 2015). Although the efficacy was not tested, the coexpression of m2A10 and m1C3 has previously been shown to abolish *P. falciparum* development (Isaacs et al. 2012).

Resistance to Gene Drives

A significant challenge for CGDs is resistant alleles. In addition to natural genetic variation, CGDs can produce resistance alleles when double-stranded breaks are repaired by NHEJ pathways, producing indels that resist further cleavage (Esvelt et al. 2014). The protospacer-adjacent motif (PAM) sequence (Fig. 1) is particularly vulnerable to mutations, as a single-nucleotide mutation can prevent Cas9 function (Jinek et al. 2012). The PAM sequence is a short motif near the target sequence and is essential for Cas9 recognition of target DNA. Mathematical modeling suggests that resistant alleles will inevitably arise and spread through the population (Esvelt et al. 2014; Unckless et al. 2017), calling for approaches to reduce resistance development.

Approaches that reduce resistance rates may be combined to increase effectiveness. First, NHEJ rates can be reduced by tightly restricting drive expression to the germline, which shows higher rates of HDR (Gantz and Bier 2015). Previously, the *vasa* promoter has been used to drive Cas9 expression in *An. stephensi* (Gantz et al. 2015). However, leaky expression of Cas9 in female mosquitoes resulted in its maternal deposition into embryos, leading to resistance-generating NHEJ. Hammond et al. (2021a) found the *zero population growth* (*zpg*) promoter reduces leaky Cas9 expression and generates fewer resistance alleles in *An. gambiae*, significantly increasing drive inheritance, although functional resistance alleles were still able to emerge and reduce the drive prevalence (Hammond et al. 2021a).

 Cite this introduction as *Cold Spring Harb Protoc*; doi:10.1101/pdb.top107691

Inhibiting the NHEJ protein ligase IV by coexpressing the adenovirus4 proteins E1B55K and E4orf6 increases HDR efficiency by up to eightfold in mammalian cells (Chu et al. 2015), although this has not been tested in mosquitoes. However, suppressing NHEJ may not entirely prevent resistance-generation mutations, as HDR can produce errors (for review, see Guirouilh-Barbat et al. 2014).

Alternatively, targeting highly conserved sequences of essential genes can prevent the evolution of resistance. This has been showed in *An. gambiae* by targeting the female isoform *dsxF* of the sex determination gene *doublesex* (*dsx*). Any *dsx* mutations that may resist CGD action produce a non-functional *dsxF* transcript and are unable to spread (Kyrou et al. 2018; Simoni et al. 2020). The CGD produced by Kyrou et al. (2018) also recently became the first drive to be tested in larger cages containing age-structured populations with conditions closer to field conditions. Even in this setup, the drive reached 100% prevalence to crash the population (Hammond et al. 2021b). However, a recent drive targeting the *An. gambiae* female fertility gene *AGAP029113* failed to spread in caged populations and was lost in three generations, despite heavy target sequence conservation. This was in part due to the emergence of a functional single-nucleotide substitution (Fuchs et al. 2021), demonstrating that targeting a conserved sequence is not always sufficient to prevent drive resistance.

For population replacement, essential genes could be targeted by a drive containing a rescue sequence that encodes the cognate protein. Mutated resistance alleles lacking this rescue sequence will experience negative selection due to the associated fitness cost, so will be unlikely to spread through the population (Esvelt et al. 2014). This has been showed by disrupting the *kynurenine hydroxylase* gene in *An. stephensi* (Adolfi et al. 2020). Resistant alleles lacking the rescue sequence significantly reduced female fecundity, which led resistant allele prevalence to decrease between generations. The drive reached >95% prevalence in five to 11 generations and remained stable for the following generations (tested up to 20 generations) (Adolfi et al. 2020).

Targeting multiple sites by multiplexing gRNAs will reduce resistance rates as it is unlikely that resistance alleles can form against each individual gRNA (Hammond et al. 2017; Marshall et al. 2017; Unckless et al. 2017). A recent population suppression CGD developed in *Drosophila* using four multiplexed gRNAs produced lower resistance and higher inheritance rates but plateaued at ~63% population frequency and failed to suppress the population because of the drive fitness cost (Yang et al. 2022).

Controlling Gene Drive Spread

CGDs could potentially modify entire mosquito species, causing global effects. As such, it may be desirable to limit CGD spread. One approach to this is the split drive system, which allows a payload gene to be transiently driven to high frequency in a population while preventing global spread. In this system, drive elements are arranged so that each element drives the next, with no element able to drive itself. The top element will contain the payload gene (DiCarlo et al. 2015; Noble et al. 2019). Although this remains to be experimentally verified, the successive loss of each element of the chain over time would cause the payload gene to be diluted and eventually disappear from the population (Noble et al. 2019). Although split drives of three or more elements have yet to be developed, the transient nature of this system could theoretically be useful for field studies. However, mathematical modeling has suggested that split drives are likely to spread to neighboring populations, even at low migration rates. This is particularly true for population replacement approaches, which likely carry a lower fitness cost (Dhole et al. 2018).

Another approach that may reduce the invasive potential of gene drives is to unlink Cas9 from the other drive components. A proof-of-concept drive system using this approach has been developed in *Ae. aegypti* to disrupt the sex-linked *white* gene with the fluorescent marker tdTomato (Li et al. 2020b). Mathematical modeling suggests that this system will probably be reversible over time and less likely to invade neighboring mosquito populations. However, as *Cas9* will not drive, this would require inundating the population with homozygous males, in contrast to other drives that require a single release (Li et al. 2020b).

When developing CGDs, strong biosafety procedures should be followed to prevent accidental release (Gantz and Bier 2015). In the case of an accident, one option could be to release mosquitoes resistant to CGD action, such as those carrying resistant alleles (Burt 2003; Unckless et al. 2017). This approach has not been implemented yet, but the occurrence of resistant alleles in previous drive experiments has decreased drive frequency (Hammond et al. 2017, 2021a; Simoni et al. 2020; Fuchs et al. 2021). This approach relies on a lower fitness cost of resistant alleles, so it could be an effective approach for population suppression drives, although it may be ineffective when targeting a highly constrained sequence or against population replacement drives (Taxiarchi et al. 2021). An alternative approach is releasing mosquitoes expressing anti-CRISPR effectors (e.g., the phage protein AcrIIA4), which inhibits both the PAM-binding site and RuvC endonuclease site of Cas9 (Dong et al. 2017). In caged populations with 50% of the mosquitoes carrying the sex-distortion drive produced by Kyrou et al. (2018), having 20% of the male *An. gambiae* express AcrIIA4 reduced drive inheritance to Mendelian rates. This caused drive frequency in a caged population to plateau at 76.6%–98.1% by the third generation. Egg production was maintained at 27%–43% of wild type by generation 16, preventing population collapse. However, if anti-CRISPR alleles are maintained in the population, this would likely prevent future CGD use (Taxiarchi et al. 2021).

Integral Gene Drives

Integral gene drives (IGDs) could hypothetically be used to test population replacement approaches before spread by a drive. IGDs separately insert Cas9 and the payload gene into essential germline or somatic genes, respectively (Nash et al. 2019; Hoermann et al. 2021). The payload-carrying mosquitoes could be released for field trials. Cas9-carrying mosquitoes could be released later to cross with payload-carrying mosquitoes, allowing the payload to initiate driving (Fig. 2).

Hoermann and colleagues (2021) used a partial IGD strategy to drive an anti-*Plasmodium* peptide *Scorpine (Sco)* in *An. gambiae. Sco* was fused to the native somatic gene *peritrophin1*, expressed in the adult gut, and was driven by the native promoter. A separate *vasa-Cas9* line was generated, in which *Cas9* was driven by a *vasa* promoter but not associated with the native *vasa* gene (Fig. 2; Hoermann

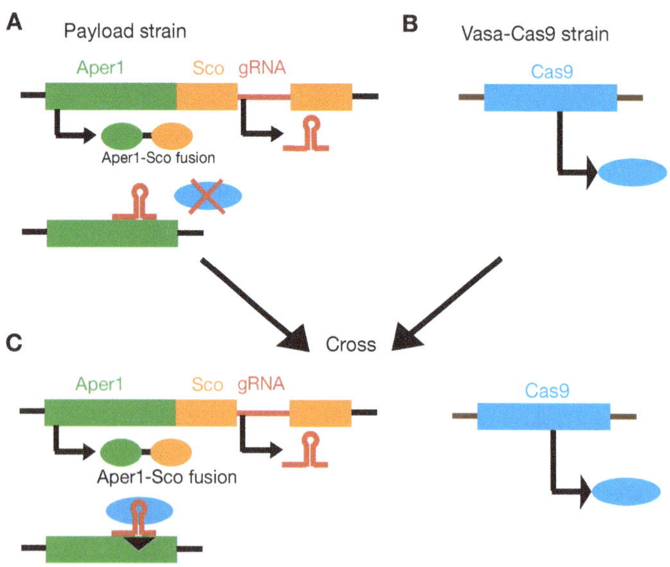

FIGURE 2. An example of a partial integral gene drive (IGD) developed by Hoermann and colleagues (2021). Separate mosquito strains were generated containing sequences for (*A*) the anti-Plasmodium effector peptide Scorpine (Sco, orange) fused to the somatic gene peritrophin1 (Aper1, green). Guide RNA (gRNA, red) is encoded by an artificial intron in Sco. As Cas9 (blue) is not present, the construct is inherited at Mendelian rates, allowing field testing in the absence of drive. (*B*) Cas9 (blue), driven by the vasa promoter (E. Marois, unpubl data). (*C*) Crossing the strains produces offspring carrying both the Sco payload and Cas9, allowing Sco to drive, as the homologous allele can be cleaved (black triangle) to initiate homing.

 Cite this introduction as *Cold Spring Harb Protoc*; doi:10.1101/pdb.top107691

et al. 2021). IGDs can reduce the rate of evolution of functional resistant alleles, as frameshift-generation mutations could disrupt the native gene. However, nonframeshift mutations may still generate resistance alleles (Nash et al. 2019). This IGD gave homing rates of 98.43% and reduced infection intensity in mosquitoes fed *P. falciparum*. However, when *Sco* was linked to the amino-terminal of *zinc carboxypeptidase A1*, *Plasmodium* infectivity increased, demonstrating the importance of careful host gene choice (Hoermann et al. 2021).

FUTURE DIRECTIONS AND OUTLOOK

The development of genetic intervention in mosquitoes is strongly dictated by the practical needs to develop field applications and combat mosquito-borne diseases. RNAi applications and CGDs offer great promise here, although optimizing and rolling out these approaches to a wider range of species is required. For example, the efficiency of RNAi delivery via feeding may be increased by the down-regulation of larval gut RNases (Giesbrecht et al. 2020). To date, CGDs have mostly been produced for *An. gambiae* and *An. stephensi* (see Supplemental Table S3). CGD development has now begun in *Ae. aegypti*, although the drives developed so far have a lower drive efficiency (Li et al. 2020b). No CGDs have been produced for *Culex* mosquitoes, despite their epidemiological importance. However, early steps toward CGD development have begun in *Cx. quinquefasciatus*—expression toolkits for gRNA and Cas9 have been designed (Feng et al. 2021b) and gRNA target sites identified (Feng et al. 2021a). Interestingly, in *Cx. quinequefasciatus*, a gRNA scaffold variant containing a 5-bp loop that better represents the CRISPR–Cas native state increased allele conversion rates and improved drive efficiency when applied to *Drosophila* (Feng et al. 2021b), suggesting this could be applied to other species. Further research in other species could validate the efficacy of this loop variant and provide valuable information about the likely effectiveness of CGDs in eliminating a wide range of mosquito-borne diseases.

In addition to practical applications, recent interest in the basic biology of mosquitoes, such as their sensory neuroscience and sensory ecology, has driven the development of complementary genetic tools. Transgenic knock-ins and binary expression systems (see Supplemental Table S2) allow us to study neuronal morphology and function and to manipulate mosquito behavior. These approaches will also be applicable to mosquito species that do not transmit diseases but that are interesting because of their unusual biology and ecology.

ACKNOWLEDGMENTS

O.R. was funded by the Wellcome Trust (217440/Z/19/Z), the Royal Society (RGS/R2/192005), and the Natural Environment Research Council (NE/W00402X/1). M.Q. was funded by the BBSRC NLD DTP scholarship.

REFERENCES

Adolfi A, Lycett GJ. 2018. Opening the toolkit for genetic analysis and control of *Anopheles* mosquito vectors. *Curr Opin Insect Sci* **30:** 8–18. doi:10.1016/j.cois.2018.07.014

Adolfi A, Gantz VM, Jasinskiene N, Lee H-F, Hwang K, Terradas G, Bulger EA, Ramaiah A, Bennett JB, Emerson JJ, et al. 2020. Efficient population modification gene-drive rescue system in the malaria mosquito *Anopheles stephensi*. *Nat Commun* **11:** 5553. doi:10.1038/s41467-020-19426-0

Afify A, Betz JF, Riabinina O, Lahondère C, Potter CJ. 2019. Commonly used insect repellents hide human odors from *Anopheles* mosquitoes. *Curr Biol* **29:** 3669–3680. doi:10.1016/j.cub.2019.09.007

Anderson ME, Mavica J, Shackleford L, Flis I, Fochler S, Basu S, Alphey L. 2019. CRISPR/Cas9 gene editing in the West Nile Virus vector, *Culex quinquefasciatus* say. *PLoS ONE* **14:** e0224857. doi:10.1371/journal.pone.0224857

Aryan A, Anderson MAE, Myles KM, Adelman ZN. 2013. TALEN-based gene disruption in the dengue vector *Aedes aegypti*. *PLoS ONE* **8:** e60082. doi:10.1371/journal.pone.0060082

Aryan A, Anderson MAE, Biedler JK, Qi Y, Overcash JM, Naumenko AN, Sharakhova M V, Mao C, Adelman ZN, Tu Z. 2020. *Nix* alone is sufficient to convert female *Aedes aegypti* into fertile males and myo-sex is needed for male flight. *Proc Natl Acad Sci* **117:** 17702–17709. doi:10.1073/pnas.2001132117

Basrur NS, Elena De Obaldia M, Morita T, Herre M, von Heynitz RK, Tsitohay YN, Vosshall LB. 2020. Fruitless mutant male mosquitoes gain attraction to human odor. *Elife* **9:** e63982. doi:10.7554/eLife.63982

Bernardini F, Haghighat-Khah RE, Galizi R, Hammond AM, Nolan T, Crisanti A. 2018. Molecular tools and genetic markers for the generation of

transgenic sexing strains in Anopheline mosquitoes. *Parasit Vectors* **11:** 660. doi:10.1186/s13071-018-3207-8
Berois M, Romero-Severson J, Severson DW. 2012. RNAi knock-downs support roles for the mucin-like (*AeIMUC1*) gene and short-chain dehydrogenase/reductase (SDR) gene in *Aedes aegypti* susceptibility to *Plasmodium gallinaceum*. *Med Vet Entomol* **26:** 112–115. doi:10.1111/j.1365-2915.2011.00965.x
Blitzer EJ, Vyazunova I, Lan Q. 2005. Functional analysis of AeSCP-2 using gene expression knockdown in the yellow fever mosquito, *Aedes aegypti*. *Insect Mol Biol* **14:** 301–307. doi:10.1111/j.1365-2583.2005.00560.x
Brand AH, Perrimon N. 1993. Targeted gene expression as a means of altering cell fates and generating dominant phenotypes. *Development* **118:** 401–415. doi:10.1242/dev.118.2.401
Brown AE, Bugeon L, Crisanti A, Catteruccia F. 2003. Stable and heritable gene silencing in the malaria vector *Anopheles stephensi*. *Nucleic Acids Res* **31:** 1–6. doi:10.1093/nar/gkg009
Buchman A, Gamez S, Li M, Antoshechkin I, Li H-H, Wang H-W, Chen C-H, Klein MJ, Duchemin J-B, Paradkar PN, et al. 2019. Engineered resistance to Zika virus in transgenic *Aedes aegypti* expressing a polycistronic cluster of synthetic small RNAs. *Proc Natl Acad Sci* **116:** 3656–3661. doi:10.1073/pnas.1810771116
Bui M, Shyong J, Lutz EK, Yang T, Li M, Truong K, Arvidson R, Buchman A, Riffell JA, Akbari OS. 2019. Live calcium imaging of *Aedes aegypti* neuronal tissues reveals differential importance of chemosensory systems for life-history-specific foraging strategies. *BMC Neurosci* **20:** 27. doi:10.1186/s12868-019-0511-y
Bui M, Li M, Raban RR, Liu N, Akbari OS. 2020. Embryo microinjection techniques for efficient site-specific mutagenesis in *Culex quinquefasciatus*. *J Vis Exp* doi:10.3791/61375.
Burt A. 2003. Site-specific selfish genes as tools for the control and genetic engineering of natural populations. *Proc R Soc B Biol Sci* **270:** 921–928. doi:10.1098/rspb.2002.2319
Carballar-Lejarazú R, Tushar T, Pham T, James A. 2021. Microinjection method for *Anopheles gambiae* embryos. *J Vis Exp* doi:10.3791/62591.
Catteruccia F, Nolan T, Loukeris TG, Blass C, Savakis C, Kafatos FC, Crisanti A. 2000. Stable germline transformation of the malaria mosquito *Anopheles stephensi*. *Nature* **405:** 959–962. doi:10.1038/35016096
Chaverra-Rodriguez D, Macias VM, Hughes GL, Pujhari S, Suzuki Y, Peterson DR, Kim D, McKeand S, Rasgon JL. 2018. Targeted delivery of CRISPR–Cas9 ribonucleoprotein into arthropod ovaries for heritable germline gene editing. *Nat Commun* **9:** 3008. doi:10.1038/s41467-018-05425-9
Chen J, Lu HR, Zhang L, Liao CH, Han Q. 2019. RNA interference-mediated knockdown of 3, 4-dihydroxyphenylacetaldehyde synthase affects larval development and adult survival in the mosquito *Aedes aegypti*. *Parasites and Vectors* **12:** 1–11. doi:10.1186/s13071-018-3256-z
Chu VT, Weber T, Wefers B, Wurst W, Sander S, Rajewsky K, Kühn R. 2015. Increasing the efficiency of homology-directed repair for CRISPR–Cas9-induced precise gene editing in mammalian cells. *Nat Biotechnol* **33:** 543–548. doi:10.1038/nbt.3198
Coates CJ, Jasinskiene N, Miyashiro L, James AA. 1998. Mariner transposition and transformation of the yellow fever mosquito, *Aedes aegypti*. *Proc Natl Acad Sci* **95:** 3748–3751. doi:10.1073/pnas.95.7.3748
Corfas RA, Vosshall LB, Scott K, Corfas RA, Vosshall LB, Scott K, Corfas RA, Vosshall LB. 2015. The cation channel TRPA1 tunes mosquito thermotaxis to host temperatures. *Elife* **4:** 1–16. doi:10.7554/eLife.11750
Degennaro M, McBride CS, Seeholzer L, Nakagawa T, Dennis EJ, Goldman C, Jasinskiene N, James AA, Vosshall LB. 2013. Orco mutant mosquitoes lose strong preference for humans and are not repelled by volatile DEET. *Nature* **498:** 487–491. doi:10.1038/nature12206
Dhole S, Vella MR, Lloyd AL, Gould F. 2018. Invasion and migration of spatially self-limiting gene drives: a comparative analysis. *Evol Appl* **11:** 794–808. doi:10.1111/eva.12583
DiCarlo JE, Chavez A, Dietz SL, Esvelt KM, Church GM. 2015. Safeguarding CRISPR–Cas9 gene drives in yeast. *Nat Biotechnol* **33:** 1250–1255. doi:10.1038/nbt.3412
Dietzl G, Chen D, Schnorrer F, Su KC, Barinova Y, Fellner M, Gasser B, Kinsey K, Oppel S, Scheiblauer S, et al. 2007. A genome-wide transgenic RNAi library for conditional gene inactivation in *Drosophila*. *Nature* **448:** 151–156. doi:10.1038/nature05954
Dong D, Guo M, Wang S, Zhu Y, Wang S, Xiong Z, Yang J, Xu Z, Huang Z. 2017. Structural basis of CRISPR–SpyCas9 inhibition by an anti-CRISPR protein. *Nature* **546:** 436–439. doi:10.1038/nature22377
Dong Y, Simões ML, Marois E, Dimopoulos G. 2018. CRISPR/Cas9 -mediated gene knockout of *Anopheles gambiae* FREP1 suppresses malaria parasite infection. *PLoS Pathog* **14:** e1006898. doi:10.1371/journal.ppat.1006898
Dong S, Dong Y, Simões ML, Dimopoulos G. 2021. Mosquito transgenesis for malaria control. *Trends Parasitol* **38:** 54–66. doi:10.1016/j.pt.2021.08.001
Driesschaert B, Mergan L, Temmerman L. 2021. Conditional gene expression in invertebrate animal models. *J Genet Genomics* **48:** 14–31. doi:10.1016/j.jgg.2021.01.005
Duvall LB, Ramos-Espiritu L, Barsoum KE, Glickman JF, Vosshall LB. 2019. Small-molecule agonists of *Ae. aegypti* neuropeptide Y receptor block mosquito biting. *Cell* **176:** 687–701.e5. doi:10.1016/j.cell.2018.12.004
Erdelyan CNG, Mahood TH, Bader TSY, Whyard S. 2012. Functional validation of the carbon dioxide receptor genes in *Aedes aegypti* mosquitoes using RNA interference. *Insect Mol Biol* **21:** 119–127. doi:10.1111/j.1365-2583.2011.01120.x
Esvelt KM, Smidler AL, Catteruccia F, Church GM. 2014. Concerning RNA-guided gene drives for the alteration of wild populations. *Elife* **3:** e03401. doi:10.7554/eLife.03401
Feng X, Kambic L, Nishimoto JHK, Reed FA, Denton JA, Sutton JT, Gantz VM. 2021a. Evaluation of gene knockouts by CRISPR as potential targets for the genetic engineering of the mosquito *Culex quinquefasciatus*. *Cris J* **4:** 595–608. doi:10.1089/crispr.2021.0028
Feng X, López Del Amo V, Mameli E, Lee M, Bishop AL, Perrimon N, Gantz VM. 2021b. Optimized CRISPR tools and site-directed transgenesis towards gene drive development in *Culex quinquefasciatus* mosquitoes. *Nat Commun* **12:** 2960. doi:10.1038/s41467-021-23239-0
Figueira-Mansur J, Ferreira-Pereira A, Mansur JF, Franco TA, Alvarenga ESL, Sorgine MHF, Neves BC, Melo ACA, Leal WS, Masuda H, et al. 2013. Silencing of P-glycoprotein increases mortality in temephos-treated *Aedes aegypti* larvae. *Insect Mol Biol* **22:** 648–658. doi:10.1111/imb.12052
Fire A, Xu S, Montgomery MK, Kostas SA, Driver SE, Mello CC. 1998. 35888. *Nature* **391:** 806–811. doi:10.1038/35888
Fischer JR, Zapata F, Dubelman S, Mueller GM, Uffman JP, Jiang C, Jensen PD, Levine SL. 2017. Aquatic fate of a double-stranded RNA in a sediment–water system following an over-water application. *Environ Toxicol Chem* **36:** 727–734. doi:10.1002/etc.3585
Franz AWE, Jasinskiene N, Sanchez-Vargas I, Isaacs AT, Smith MR, Khoo CCH, Heersink MS, James AA, Olson KE. 2011. Comparison of transgene expression in *Aedes aegypti* generated by *mariner Mos1* transposition and *ΦC31* site-directed recombination. *Insect Mol Biol* **20:** 587–598. doi:10.1111/j.1365-2583.2011.01089.x
Fuchs S, Garrood WT, Beber A, Hammond A, Galizi R, Gribble M, Morselli G, Hui T-YJ, Willis K, Kranjc N, et al. 2021. Resistance to a CRISPR-based gene drive at an evolutionarily conserved site is revealed by mimicking genotype fixation. *PLoS Genet* **17:** e1009740. doi:10.1371/journal.pgen.1009740
Gantz VM, Bier E. 2015. The mutagenic chain reaction: a method for converting heterozygous to homozygous mutations. *Science* **348:** 442–444. doi:10.1126/science.aaa5945
Gantz VM, Jasinskiene N, Tatarenkova O, Fazekas A, Macias VM, Bier E, James AA. 2015. Highly efficient Cas9-mediated gene drive for population modification of the malaria vector mosquito *Anopheles stephensi*. *Proc Natl Acad Sci* **112:** E6736–E6743. doi:10.1073/pnas.1521077112
Gao S, Dagnaes-Hansen F, Nielsen EJB, Wengel J, Besenbacher F, Howard KA, Kjems J. 2009. The effect of chemical modification and nanoparticle formulation on stability and biodistribution of siRNA in mice. *Mol Ther* **17:** 1225–1233. doi:10.1038/mt.2009.91
Garrood WT, Kranjc N, Petri K, Kim DY, Guo JA, Hammond AM, Morianou I, Pattanayak V, Joung JK, Crisanti A, et al. 2021. Analysis of off-target effects in CRISPR-based gene drives in the human malaria mosquito. *Proc Natl Acad Sci* **118:** e2004838117. doi:10.1073/pnas.2004838117
Giesbrecht D, Heschuk D, Wiens I, Boguski D, Lachance P, Whyard S. 2020. RNA interference is enhanced by knockdown of double-stranded RNases in the yellow fever mosquito *Aedes aegypti*. *Insects* **11:** 327. doi:10.3390/insects11060327

Gossen M, Bujard H. 1992. Tight control of gene expression in mammalian cells by tetracycline-responsive promoters. *Proc Natl Acad Sci* **89:** 5547–5551. doi:10.1073/pnas.89.12.5547

Greppi C, Laursen WJ, Budelli G, Chang EC, Daniels AM, van Giesen L, Smidler AL, Catteruccia F, Garrity PA. 2020. Mosquito heat seeking is driven by an ancestral cooling receptor. *Science* **367:** 681–684. doi:10.1126/science.aay9847

Grossman GL, Rafferty CS, Clayton JR, Stevens TK, Mukabayire O, Benedict MQ. 2001. Germline transformation of the malaria vector, *Anopheles gambiae*, with the piggyBac transposable element. *Insect Mol Biol* **10:** 597–604. doi:10.1046/j.0962-1075.2001.00299.x

Gu J, Liu M, Deng Y, Peng H, Chen X. 2011. Development of an efficient recombinant mosquito densovirus-mediated RNA interference system and its preliminary application in mosquito control. *PLoS ONE* **6:** e21329. doi:10.1371/journal.pone.0021329

Guirouilh-Barbat J, Lambert S, Bertrand P, Lopez BS. 2014. Is homologous recombination really an error-free process? *Front Genet* **5:** 175. doi:10.3389/fgene.2014.00175

Hammond A, Galizi R, Kyrou K, Simoni A, Siniscalchi C, Katsanos D, Gribble M, Baker D, Marois E, Russell S, et al. 2016. A CRISPR–Cas9 gene drive system targeting female reproduction in the malaria mosquito vector *Anopheles gambiae*. *Nat Biotechnol* **34:** 78–83. doi:10.1038/nbt.3439

Hammond AM, Kyrou K, Bruttini M, North A, Galizi R, Karlsson X, Kranjc N, Carpi FM, D'Aurizio R, Crisanti A, et al. 2017. The creation and selection of mutations resistant to a gene drive over multiple generations in the malaria mosquito. *PLoS Genet* **13:** e1007039. doi:10.1371/journal.pgen.1007039

Hammond A, Karlsson X, Morianou I, Kyrou K, Beaghton A, Gribble M, Kranjc N, Galizi R, Burt A, Crisanti A, et al. 2021a. Regulating the expression of gene drives is key to increasing their invasive potential and the mitigation of resistance. *PLoS Genet* **17:** e1009321. doi:10.1371/journal.pgen.1009321

Hammond A, Pollegioni P, Persampieri T, North A, Minuz R, Trusso A, Bucci A, Kyrou K, Morianou I, Simoni A, et al. 2021b. Gene-drive suppression of mosquito populations in large cages as a bridge between lab and field. *Nat Commun* **12:** 4589. doi:10.1038/s41467-021-24790-6

Hapairai LK, Mysore K, Chen Y, Harper EI, Scheel MP, Lesnik AM, Sun L, Severson DW, Wei N, Duman-Scheel M. 2017. Lure-and-kill yeast interfering RNA larvicides targeting neural genes in the human disease vector mosquito *Aedes aegypti*. *Sci Rep* **7:** 13223. doi:10.1038/s41598-017-13566-y

Hapairai LK, Mysore K, Sun L, Li P, Wang CW, Scheel ND, Lesnik A, Scheel MP, Igiede J, Wei N, et al. 2020. Characterization of an adulticidal and larvicidal interfering RNA pesticide that targets a conserved sequence in mosquito G protein-coupled dopamine 1 receptor genes. *Insect Biochem Mol Biol* **120:** 103359. doi:10.1016/j.ibmb.2020.103359

Hoermann A, Tapanelli S, Capriotti P, Del Corsano G, Masters EKG, Habtewold T, Christophides GK, Windbichler N. 2021. Converting endogenous genes of the malaria mosquito into simple non-autonomous gene drives for population replacement. *Elife* **10:** e58791. doi:10.7554/eLife.58791

Howard KA, Rahbek UL, Liu X, Damgaard CK, Glud SZ, Andersen M, Hovgaard MB, Schmitz A, Nyengaard JR, Besenbacher F, et al. 2006. RNA Interference in vitro and in vivo using a novel chitosan/siRNA nanoparticle system. *Mol Ther* **14:** 476–484. doi:10.1016/j.ymthe.2006.04.010

Inbar E, Eappen AG, Alford RT, Reid W, Harrell RA, Hosseini M, Chakravarty S, Li T, Sim KL, Billingsley PF, et al. 2021. Knockout of *Anopheles stephensi* immune gene LRIM1 by CRISPR-Cas9 reveals 2 its unexpected role in reproduction and vector competence. *PLoS Pathog* **17:** e1009770. doi:10.1371/journal.ppat.1009770

Isaacs AT, Jasinskiene N, Tretiakov M, Thiery I, Zettor A, Bourgouin C, James AA. 2012. Transgenic *Anopheles stephensi* coexpressing single-chain antibodies resist *Plasmodium falciparum* development. *Proc Natl Acad Sci* **109:** E1922–E1930. doi:10.1073/pnas.1207738109

Isoe J, Koch LE, Isoe YE, Rascón AA, Brown HE, Massani BB, Miesfeld RL. 2019. Identification and characterization of a mosquito-specific eggshell organizing factor in *Aedes aegypti* mosquitoes. *PLoS Biol* **17:** 1–23. doi:10.1371/journal.pbio.3000068

Jasinskiene N, Coates CJ, Benedict MQ, Cornel AJ, Rafferty CS, James AA, Collins FH. 1998. Stable transformation of the yellow fever mosquito, *Aedes aegypti*, with the Hermes element from the housefly. *Proc Natl Acad Sci* **95:** 3743–3747. doi:10.1073/pnas.95.7.3743

Jinek M, Chylinski K, Fonfara I, Hauer M, Doudna JA, Charpentier E. 2012. A programmable dual-RNA-guided DNA endonuclease in adaptive bacterial immunity. *Science* **337:** 816–821. doi:10.1126/science.1225829

Jové V, Gong Z, Hol FJH, Zhao Z, Sorrells TR, Carroll TS, Prakash M, McBride CS, Vosshall LB. 2020. Sensory discrimination of blood and floral nectar by *Aedes aegypti* mosquitoes. *Neuron* **108:** 1163–1180.e12. doi:10.1016/j.neuron.2020.09.019

Kim DH, Rossi JJ. 2008. RNAi mechanisms and applications. *Biotechniques* **44:** 613–616. doi:10.2144/000112792

Kistler KEE, Vosshall LBB, Matthews BJJ. 2015. Genome engineering with CRISPR–Cas9 in the mosquito *Aedes aegypti*. *Cell Rep* **11:** 51–60. doi:10.1016/j.celrep.2015.03.009

Kokoza VA, Raikhel AS. 2011. Targeted gene expression in the transgenic *Aedes aegypti* using the binary Gal4-UAS system. *Insect Biochem Mol Biol* **41:** 637–644. doi:10.1016/j.ibmb.2011.04.004.

Konopka JK, Task D, Afify A, Raji J, Deibel K, Maguire S, Lawrence R, Potter CJ. 2021. Olfaction in *Anopheles* mosquitoes. *Chem Senses* **46:** bjab021. doi:10.1093/chemse/bjab021

Kumar A, Wang S, Ou R, Samrakandi M, Beerntsen BT, Sayre RT. 2013. Development of an RNAi based microalgal larvicide to control mosquitoes. *MWJ* **4:** 7.

Kyrou K, Hammond AM, Galizi R, Kranjc N, Burt A, Beaghton AK, Nolan T, Crisanti A. 2018. A CRISPR–Cas9 gene drive targeting *doublesex* causes complete population suppression in caged *Anopheles gambiae* mosquitoes. *Nat Biotechnol* **36:** 1062–1071. doi:10.1038/nbt.4245

Labbé GMC, Nimmo DD, Alphey L. 2010. piggybac- and PhiC31-mediated genetic transformation of the Asian tiger mosquito, *Aedes albopictus* (Skuse). *PLoS Negl Trop Dis* **4:** e788. doi:10.1371/journal.pntd.0000788

Li M, Akbari OS, White BJ. 2018. Highly efficient site-specific mutagenesis in malaria mosquitoes using CRISPR. *G3* **8:** 653–658. doi:10.1534/g3.117.1134

Li M, Li T, Liu N, Raban RR, Wang X, Akbari OS. 2020a. Methods for the generation of heritable germline mutations in the disease vector *Culex quinquefasciatus* using clustered regularly interspaced short palindrome repeats-associated protein 9. *Insect Mol Biol* **29:** 214–220. doi:10.1111/imb.12626

Li M, Yang T, Kandul NP, Bui M, Gamez S, Raban R, Bennett J, Sánchez C HM, Lanzaro GC, Schmidt H, et al. 2020b. Development of a confinable gene drive system in the human disease vector *Aedes aegypti*. *Elife* **9:** e51701. doi:10.7554/eLife.51701

Li X, Xu Y, Zhang H, Yin H, Zhou D, Sun Y, Ma L, Shen B, Zhu C. 2021. ReMOT control delivery of CRISPR–Cas9 ribonucleoprotein complex to induce germline mutagenesis in the disease vector mosquitoes *Culex pipiens pallens (Diptera: Culicidae)*. *J Med Entomol* **58:** 1202–1209. doi:10.1093/jme/tjab016

Liu C, Zwiebel LJ. 2013. Molecular characterization of larval peripheral thermosensory responses of the malaria vector mosquito *Anopheles gambiae*. *PLoS ONE* **8:** e72595. doi:10.1371/journal.pone.0072595

Liu C, Pitts RJ, Bohbot JD, Jones PL, Wang G, Zwiebel LJ. 2010. Distinct olfactory signaling mechanisms in the malaria vector mosquito *Anopheles gambiae*. *PLoS Biol* **8:** e72595. doi:10.1371/journal.pbio.1000467

Liu T, Yang W-Q, Xie Y-G, Liu P-W, Xie L-H, Lin F, Li C-Y, Gu J-B, Wu K, Yan G-Y, et al. 2019. Construction of an efficient genomic editing system with CRISPR/Cas9 in the vector mosquito *Aedes albopictus*. *Insect Sci* **26:** 1045–1054. doi:10.1111/1744-7917.12645

Lopez-Martinez G, Meuti M, Denlinger DL. 2012. Rehydration driven RNAi: a novel approach for effectively delivering dsRNA to mosquito larvae. *J Med Entomol* **49:** 215–218. doi:10.1603/ME11122

Lycett GJ, Kafatos FC, Loukeris TG. 2004. Conditional expression in the malaria mosquito *Anopheles stephensi* with Tet-On and Tet-Off systems. *Genetics* **167:** 1781–1790. doi:10.1534/genetics.104.028175

Lynd A, Lycett GJ. 2012. Development of the bi-partite Gal4-UAS system in the African malaria mosquito, *Anopheles gambiae*. *PLoS ONE* **7:** e31552. doi:10.1371/journal.pone.0031552

Lynd A, Balabanidou V, Grosman R, Maas J, Lian L-Y, Vontas J, Lycett GJ. 2019. Development of a functional genetic tool for *Anopheles gambiae* oenocyte characterisation: application to cuticular hydrocarbon synthesis. bioRxiv doi:10.1101/742619

Macias VM, McKeand S, Chaverra-Rodriguez D, Hughes GL, Fazekas A, Pujhari S, Jasinskiene N, James AA, Rasgon JL. 2020. Cas9-mediated gene-editing in the malaria mosquito *Anopheles stephensi* by ReMOT Control. *G3* **10:** g3.401133.2020.

Maguire SE, Afify A, Goff LA, Potter CJ. 2022. Odorant-receptor-mediated regulation of chemosensory gene expression in the malaria mosquito *Anopheles gambiae. Cell Rep* **38:** 110494. doi:10.1016/j.celrep.2022.110494

Marshall JM, Buchman A CHMS, Akbari OS. 2017. Overcoming evolved resistance to population-suppressing homing-based gene drives. *Sci Rep* **7:** 3776. doi:10.1038/s41598-017-02744-7

Matthews BJ, Younger MA, Vosshall LB, Scott K, Matthews BJ, Younger MA, Vosshall LB. 2019. The ion channel *ppk301* controls freshwater egg-laying in the mosquito *Aedes aegypti. Elife* **8:** e43963. doi:10.7554/eLife.43963

McMeniman CJ, Corfas RA, Matthews BJ, Ritchie SA, Vosshall LB. 2014. Multimodal integration of carbon dioxide and other sensory cues drives mosquito attraction to humans. *Cell* **156:** 1060–1071. doi:10.1016/j.cell.2013.12.044

Meredith JM, Basu S, Nimmo DD, Larget-Thiery I, Warr EL, Underhill A, McArthur CC, Carter V, Hurd H, Bourgouin C, et al. 2011. Site-specific integration and expression of an anti-malarial gene in transgenic *Anopheles gambiae* significantly reduces *Plasmodium* infections. *PLoS ONE* **6:** e14587. doi:10.1371/journal.pone.0014587

Meredith JM, Underhill A, McArthur CC, Eggleston P. 2013. Next-generation site-directed transgenesis in the malaria vector mosquito *Anopheles gambiae*: self-docking strains expressing germline-specific phiC31 integrase. *PLoS ONE* **8:** e59264. doi:10.1371/journal.pone.0059264

Meuti ME, Harrell R. 2020. Preparing and injecting embryos of *Culex* mosquitoes to generate null mutations using CRISPR/Cas9. *J Vis Exp* doi:10.3791/61651

Meuti ME, Stone M, Ikeno T, Denlinger DL. 2015. Functional circadian clock genes are essential for the overwintering diapause of the Northern house mosquito, *Culex pipiens. J Exp Biol* **218:** 412–422. doi:10.1242/jeb.113233

Michel K, Suwanchaichinda C, Morlais I, Lambrechts L, Cohuet A, Awono-Ambene PH, Simard F, Fontenille D, Kanost MR, Kafatos FC. 2006. Increased melanizing activity in *Anopheles gambiae* does not affect development of *Plasmodium falciparum. Proc Natl Acad Sci* **103:** 16858–16863. doi:10.1073/pnas.0608033103

Mysore K, Flannery EM, Tomchaney M, Severson DW, Duman-Scheel M. 2013. Disruption of *Aedes aegypti* olfactory system development through chitosan/siRNA nanoparticle targeting of semaphorin-1a. *PLoS Negl Trop Dis* **7:** 1–12. doi:10.1371/journal.pntd.0002215

Mysore K, Hapairai LK, Sun L, Harper EI, Chen Y, Eggleson KK, Realey JS, Scheel ND, Severson DW, Wei N, et al. 2017. Yeast interfering RNA larvicides targeting neural genes induce high rates of *Anopheles* larval mortality. *Malar J* **16:** 461. doi:10.1186/s12936-017-2112-5

Mysore K, Hapairai LK, Sun L, Li P, Wang CW, Scheel ND, Lesnik A, Igiede J, Scheel MP, Wei N, et al. 2020. Characterization of a dual-action adulticidal and larvicidal interfering RNA pesticide targeting the shaker gene of multiple disease vector mosquitoes. *PLoS Negl Trop Dis* **14:** 1–22. doi:10.1371/journal.pntd.0008479

Nash A, Urdaneta GM, Beaghton AK, Hoermann A, Papathanos PA, Christophides GK, Windbichler N. 2019. Integral gene drives for population replacement. *Biol Open* **8:** bio037762.

Noble C, Min J, Olejarz J, Buchthal J, Chavez A, Smidler AL, DeBenedictis EA, Church GM, Nowak MA, Esvelt KM. 2019. Daisy-chain gene drives for the alteration of local populations. *Proc Natl Acad Sci* **116:** 8275–8282. doi:10.1073/pnas.1716358116

O'Brochta DA, Pilitt KL, Harrell RA, Aluvihare C, Alford RT. 2012. Gal4-based enhancer-trapping in the malaria mosquito *Anopheles stephensi. G3* **2:** 1305–1315. doi:10.1534/g3.112.003582

Perera OP, Harrell RA II, Handler AM. 2002. Germ-line transformation of the South American malaria vector, *Anopheles albimanus*, with a piggyBac/EGFP transposon vector is routine and highly efficient. *Insect Mol Biol* **11:** 291–297. doi:10.1046/j.1365-2583.2002.00336.x

Potter CJ, Tasic B, Russler E V, Liang L, Luo L. 2010. The Q system: a repressible binary system for transgene expression, lineage tracing, and mosaic analysis. *Cell* **141:** 536–548. doi:10.1016/j.cell.2010.02.025

Purusothaman D-K, Shackleford L, Anderson MAE, Harvey-Samuel T, Alphey L. 2021. CRISPR/Cas-9 mediated knock-in by homology dependent repair in the West Nile Virus vector *Culex quinquefasciatus* Say. *Sci Rep* **11:** 14964. doi:10.1038/s41598-021-94065-z

Quinn C, Anthousi A, Wondji C, Nolan T. 2021. CRISPR-mediated knock-in of transgenes into the malaria vector *Anopheles funestus. G3* **11:** jkab201. doi:10.1093/g3journal/jkab201

Raban RR, Marshall JM, Akbari OS. 2020. Progress towards engineering gene drives for population control. *J Exp Biol* **223:** jeb208181. doi:10.1242/jeb.208181

Raji JI, Melo N, Castillo JS, Gonzalez S, Saldana V, Stensmyr MC, DeGennaro M. 2019. *Aedes aegypti* mosquitoes detect acidic volatiles found in human odor using the IR8a pathway. *Curr Biol* **29:** 1253–1262.e7. doi:10.1016/j.cub.2019.02.045

Ramesh Kumar D, Saravana Kumar P, Gandhi MR, Al-Dhabi NA, Paulraj MG, Ignacimuthu S. 2016. Delivery of chitosan/dsRNA nanoparticles for silencing of wing development vestigial (vg) gene in *Aedes aegypti* mosquitoes. *Int J Biol Macromol* **86:** 89–95. doi:10.1016/j.ijbiomac.2016.01.030

Redhai S, Pilgrim C, Gaspar P, van Giesen L, Lopes T, Riabinina O, Grenier T, Milona A, Chanana B, Swadling JB, et al. 2020. An intestinal zinc sensor regulates food intake and developmental growth. *Nature* **580:** 263–268. doi:10.1038/s41586-020-2111-5

Regna K, Harrison RM, Heyse SA, Chiles TC, Michel K, Muskavitch MAT. 2016. RNAi trigger delivery into *Anopheles gambiae* pupae. *J Vis Exp* 53738. doi:10.3791/53738.

Riabinina O, Potter CJ. 2016. The Q-system: a versatile expression system for *Drosophila. Methods Mol Biol* **1478:** 53–78. doi:10.1007/978-1-4939-6371-3_3

Riabinina O, Luginbuhl D, Marr E, Liu S, Wu MN, Luo L, Potter CJ. 2015. Improved and expanded Q-system reagents for genetic manipulations. *Nat Methods* **12:** 219–222. doi:10.1038/nmeth.3250

Riabinina O, Task D, Marr E, Lin C-C, Alford R, O'Brochta DA, Potter CJ. 2016. Organization of olfactory centres in the malaria mosquito *Anopheles gambiae. Nat Commun* **7:** 13010. doi:10.1038/ncomms13010

Rodrigues FG, Oliveira SB, Rocha BC, Moreira LA. 2006. Germline transformation of *Aedes fluviatilis (Diptera:Culicidae)* with the piggyBac transposable element. *Mem Inst Oswaldo Cruz* **101:** 755–757. doi:10.1590/S0074-02762006000700008

Shankar S, Tauxe GM, Spikol ED, Li M, Akbari OS, McMeniman CJ, Harry Feinstone W. 2020. Synergistic coding of human odorants in the mosquito brain. bioRxiv doi:10.1101/2020.11.02.365916

Simoni A, Hammond AM, Beaghton AK, Galizi R, Taxiarchi C, Kyrou K, Meacci D, Gribble M, Morselli G, Burt A, et al. 2020. A male-biased sex-distorter gene drive for the human malaria vector *Anopheles gambiae. Nat Biotechnol* **38:** 1054–1060. doi:10.1038/s41587-020-0508-1

Singh AD, Wong S, Ryan CP, Whyard S. 2013. Oral delivery of double-stranded RNA in larvae of the yellow fever mosquito, *Aedes aegypti*: implications for pest mosquito control. *J Insect Sci* **13:** 69. doi:10.1673/031.013.6901

Smidler AL, Terenzi O, Soichot J, Levashina EA, Marois E. 2013. Targeted mutagenesis in the malaria mosquito using TALE nucleases. *PLoS ONE* **8:** e74511. doi:10.1371/journal.pone.0074511

Sorrells TR, Pandey A, Rosas-Villegas A, Vosshall LB. 2021. A persistent behavioral state enables sustained predation of humans by mosquitoes. *Elife* **11:** e76663. doi:10.7554/eLife.76663

Sun H, Liu F, Ye Z, Baker A, Zwiebel LJ. 2020. Mutagenesis of the orco odorant receptor co-receptor impairs olfactory function in the malaria vector *Anopheles coluzzii. Insect Biochem Mol Biol* **127:** 103497. doi:10.1016/j.ibmb.2020.103497

Taracena ML, Hunt CM, Benedict MQ, Pennington PM, Dotson EM. 2019. Downregulation of female *doublesex* expression by oral-mediated RNA interference reduces number and fitness of *Anopheles gambiae* adult females. *Parasit Vectors* **12:** 170. doi:10.1186/s13071-019-3437-4

Taxiarchi C, Beaghton A, Don NI, Kyrou K, Gribble M, Shittu D, Collins SP, Beisel CL, Galizi R, Crisanti A. 2021. A genetically encoded anti-CRISPR protein constrains gene drive spread and prevents population suppression. *Nat Commun* **12:** 3977. doi:10.1038/s41467-021-24214-5

Unckless RL, Clark AG, Messer PW. 2017. Evolution of resistance against CRISPR/Cas9 gene drive. *Genetics* **205:** 827–841. doi:10.1534/genetics.116.197285

Volohonsky G, Terenzi O, Soichot J, Naujoks DA, Nolan T, Windbichler N, Kapps D, Smidler AL, Vittu A, Costa G, et al. 2015. Tools for *Anopheles gambiae* transgenesis. *G3* **5:** 1151–1163. doi:10.1534/g3.115.016808

Webster SH, Scott MJ. 2021. The *Aedes aegypti* (Diptera: Culicidae) *hsp83* gene promoter drives strong ubiquitous DsRed and ZsGreen marker expression in transgenic mosquitoes. *J Med Entomol* **56:** 2533–2537. doi:10.1093/jme/tjab128

Wei X, Potter CJ, Luo L, Shen K. 2012. Controlling gene expression with the Q repressible binary expression system in *Caenorhabditis elegans. Nat Methods* **9:** 391. doi:10.1038/nmeth.1929

Wheelwright M, Whittle CR, Riabinina O. 2021. Olfactory systems across mosquito species. *Cell Tissue Res* **383:** 75–90. doi:10.1007/s00441-020-03407-2

Whyard S, Erdelyan CNG, Partridge AL, Singh AD, Beebe NW, Capina R. 2015. Silencing the buzz: a new approach to population suppression of mosquitoes by feeding larvae double-stranded RNAs. *Parasit Vectors* **8:** 1–11. doi:10.1186/s13071-015-0716-6

Wiltshire RM, Duman-Scheel M. 2020. Advances in oral RNAi for disease vector mosquito research and control. *Curr Opin Insect Sci* **40:** 18–23. doi:10.1016/j.cois.2020.05.002

Windbichler N, Menichelli M, Papathanos PA, Thyme SB, Li H, Ulge UY, Hovde BT, Baker D, Monnat RJ, Burt A, et al. 2011. A synthetic homing endonuclease-based gene drive system in the human malaria mosquito. *Nature* **473:** 212–215. doi:10.1038/nature09937

Yang E, Metzloff M, Langmüller AM, Xu X, Clark AG, Messer PW, Champer J. 2022. A homing suppression gene drive with multiplexed gRNAs maintains high drive conversion efficiency and avoids functional resistance alleles. *G3* **12:** jkac081.

Ye Z, Liu F, Sun H, Barker M, Pitts RJ, Zwiebel LJ. 2020. Heterogeneous expression of the ammonium transporter *AgAmt* in chemosensory appendages of the malaria vector, *Anopheles gambiae. Insect Biochem Mol Biol* **120:** 103360. doi:10.1016/j.ibmb.2020.103360

Ye Z, Liu F, Sun H, Ferguson ST, Baker A, Ochieng SA, Zwiebel LJ. 2021. Discrete roles of the Ir76b Ionotropic co-receptor impact olfaction, blood feeding, and mating in the malaria vector mosquito *Anopheles coluzzii. Proc Natl Acad Sci* **119:** e2112385119. doi:10.1073/pnas.2112385119

Younger MA, Herre M, Ehrlich AR, Gong Z, Gilbert ZN, Rahiel S, Matthews BJ, Vosshall LB. 2020. Non-canonical odor coding ensures unbreakable mosquito attraction to humans. bioRxiv doi:10.1101/2020.11.07.368720

Yu N, Christiaens O, Liu J, Niu J, Cappelle K, Caccia S, Huvenne H, Smagghe G. 2013. Delivery of dsRNA for RNAi in insects: an overview and future directions. *Insect Sci* **20:** 4–14. doi:10.1111/j.1744-7917.2012.01534.x

Zhan Y, Alonso San Alberto D, Rusch C, Riffell JA, Montell C. 2021. Elimination of vision-guided target attraction in *Aedes aegypti* using CRISPR. *Curr Biol* **31:** 4180–4187.e6. doi:10.1016/j.cub.2021.07.003

Zhang H, Kolb FA, Jaskiewicz L, Westhof E, Filipowicz W. 2004. Single processing center models for human Dicer and bacterial RNase III. *Cell* **118:** 57–68. doi:10.1016/j.cell.2004.06.017

Zhang X, Zhang J, Zhu KY. 2010. Chitosan/double-stranded RNA nanoparticle-mediated RNA interference to silence chitin synthase genes through larval feeding in the African malaria mosquito (*Anopheles gambiae*). *Insect Mol Biol* **19:** 683–693. doi:10.1111/j.1365-2583.2010.01029.x

Zhao Z, Tian D, McBride CS. 2021. Development of a pan-neuronal genetic driver in *Aedes aegypti* mosquitoes. *Cell Rep Methods* **1:** 100042. doi:10.1016/j.crmeth.2021.100042

Zhao Z, Zung JL, Hinze A, Kriete AL, Iqbal A, Younger MA, Matthews BJ, Merhof D, Thiberge S, Strauch M, McBride CS. 2022. Mosquito brains encode unique features of human odour to drive host seeking. *Nature* **605:** 706–712.

Zhu G-H, Albishi NM, Chen X, Brown RL, Palli SR. 2021. Expanding the Toolkit for genome editing in a disease vector, *Aedes aegypti*: transgenic lines expressing Cas9 and single guide RNA induce efficient mutagenesis. *CRISPR J* **4:** 846–853.

CHAPTER 37

CRISPR–Cas9 Methods and Key Considerations in the Production of *Aedes aegypti* Mutant Strains

Ruichen Sun, Robyn Raban, and Omar S. Akbari[1]

Division of Biological Sciences, Section of Cell and Developmental Biology, University of California, San Diego, La Jolla, California 92093, USA

Since the characterization of the CRISPR–Cas9 system in prokaryotes, it has become the prime choice in gene editing because of its exceptional flexibility, ease of use, high efficiency, and superior specificity. As a result, CRISPR–Cas9-mediated gene-editing technologies have enabled researchers not only to engineer transgenic animal strains with site-directed insertions more efficiently but also to generate desired mutants for previously intractable species. One such species is the invasive yellow fever mosquito, *Aedes aegypti*, which is notorious for its ability to transmit many blood-borne human pathogens. Methods for developing new transgenic strains of the yellow fever mosquito may aid in the effort to control its populations and provide significant benefits for the public. Here, we provide an overview of injection and noninjection methods for generating transgenic mosquitoes and also highlight important experimental design features.

BACKGROUND

Gene Editing with CRISPR–Cas9

Clustered regularly interspaced short palindromic repeats (CRISPR) is a class of bacteriophage-derived DNA sequences found in the genomes of prokaryotic organisms (Jinek et al. 2012). Prokaryotes use CRISPR sequences coupled with an RNA-guided DNA endonuclease enzyme called CRISPR-associated protein 9 (Cas9) to recognize and cleave foreign DNA fragments complementary to the CRISPR sequences (Jinek et al. 2012). The CRISPR–Cas9 system has enabled a wide range of gene editing applications since its functional characterization. During editing, the endonuclease Cas9 first forms a ribonucleoprotein complex with a single-guide RNA (sgRNA)—an RNA segment complementary to a 20-bp-long genomic DNA sequence. Next, this ribonucleoprotein complex induces a double-stranded DNA break at the targeted genomic locus. The DNA break then activates the endogenous cellular repair machinery, which repairs the break via one of the following two main mechanisms: end-joining (EJ) repair or homology-directed repair (HDR) (Capecchi 1989; Takata et al. 1998; Hefferin and Tomkinson 2005). The EJ repair mechanism is error prone and tends to incur small base pair insertions or deletions (indels), whereas HDR promotes high-fidelity repairs, as it uses DNA templates with homology to the breakpoint to repair breaks, thereby facilitating site-specific insertion of desired exogenous DNA fragments.

[1]Correspondence: oakbari@ucsd.edu

Cite this introduction as *Cold Spring Harb Protoc*; doi:10.1101/pdb.top107693

CRISPR–Cas9-Based Gene Editing in *Aedes aegypti*

The yellow fever mosquito, *Ae. aegypti*, is one of the world's most widespread mosquito species. It is notorious for its ability to transmit many blood-borne human pathogens, such as yellow fever, dengue, Zika, and chikungunya viruses (Powell 2018). Due to the significant public health burden caused by *Ae. aegypti*, effective control of its populations has long been desired. The CRISPR–Cas9 system offers potential genome engineering solutions for reducing *Ae. aegypti* populations, which may prevent the spread of various pathogens (Li et al. 2020a, 2021b).

Several CRISPR–Cas9-based gene-editing methods have been developed for *Ae. aegypti* and several other mosquito species (Kistler et al. 2015; Li et al. 2017, 2018, 2020b, 2021a; Gantz and Akbari 2018; Bui et al. 2020). Based on differences in the Cas9 endonuclease and sgRNA delivery to mosquitoes, these methods can be grouped into the following two categories: injection-based methods and non-injection-based methods.

The injection-based methods involve the direct delivery of the CRISPR–Cas9 system components (e.g., sgRNAs, Cas9, and donor DNA) into mosquito embryos via microinjection (Kistler et al. 2015). During this process, the injection mixture containing all or some of the CRISPR–Cas9 editing components is injected into early mosquito embryos during the presyncytial blastoderm stage while the embryo is one large multinucleated cell. It is worth noting that when wild-type mosquito embryos are used for injection, the injection mixture needs to include all components, whereas if Cas9-expressing mosquito embryos are used for injection, only non-Cas9 components are required (e.g., sgRNAs and donor DNA if HDR-mediated insertion is desired). Importantly, the injected components can be of different forms. Specifically, either recombinant Cas9 protein or purified Cas9 mRNA can be used as the Cas9 component. The sgRNA sequence encoding an in vitro–transcribed single-stranded RNA or double-stranded DNA plasmid (generated from a U6 backbone plasmid) can be used as the sgRNA component (Li et al. 2021a). If HDR repair is desired, donor DNA must be included as the third component for embryonic microinjection. Donor DNA containing sequence homology adjacent to the cut site can be supplied as short single-stranded DNA (ssDNA) oligonucleotides or plasmid DNA.

In addition to embryonic injection, an alternative injection method called receptor-mediated ovary transduction of cargo (ReMOT control) can also induce heritable germline gene editing in mosquitoes without the need to inject embryos (Chaverra-Rodriguez et al. 2018). In the ReMOT control method, an injection mixture containing CRISPR–Cas9 system components is instead injected into the thorax of blood-fed adult mosquito mothers, which results in gene editing in the mothers' ovaries, ultimately producing transgenic progeny. ReMOT control has been applied successfully for generating heritable EJ mutations in several mosquito species; however, it has not yet been adapted for inserting DNA via HDR (Chaverra-Rodriguez et al. 2018).

In the non-injection-based methods, both Cas9 and/or sgRNA are genetically encoded in the mosquito genome. Cas9 is expressed under the control of promoters specific to the germline or soma, and sgRNAs are expressed by small nuclear RNA promoters that are expressed ubiquitously. Editing can be achieved by crossing genetically encoded sgRNA-expressing mosquitoes with Cas9-expressing mates, and their offspring thus inherit both components of the CRISPR–Cas9 system and are capable of very high rates of self-editing (Li et al. 2017, 2020a). The transgenic Cas9-expressing and/or sgRNA-expressing mosquito strains are usually generated beforehand via the direct delivery method. Depending on project needs, some of the transgenic mosquito lines (such as Cas9-expressing mosquito strains) may be readily available to the research community and work quite efficiently (we have several such strains available in house for interested parties, Li et al. 2017), whereas others (such as mosquitos expressing an sgRNA of interest) may need to be generated using the embryonic microinjection method.

Here, we focus on the embryonic injection-based method that uses genetically encoded Cas9 mosquito strains to generate site-directed mutants/HDR-mediated insertions, as it is the foundational method that can accommodate most project requirements about the generation of mutant *Ae. aegypti* lines.

Cite this introduction as *Cold Spring Harb Protoc*; doi:10.1101/pdb.top107693

KEY PROJECT PLANNING DECISIONS

Repair Mechanism

Two predominating DNA mechanisms are used to repair DNA breaks generated by the CRISPR–Cas9 system, namely, the EJ and HDR mechanisms. EJ repair is generally error prone and induces small base-pair repairs that can be a deletion, an imprecise insertion, or point mutations (Hefferin and Tomkinson 2005). This mechanism is suitable when the only desired feature in generated mutant lines is a simple disruption of a target gene in an unspecified way.

The HDR mechanism is required if precise editing, such as a precisely mutated sequence or precise insertion of certain DNA fragments, is desired. In addition, HDR allows for the inclusion of a positive marker (e.g., 3xP3-dsRed), which enables fast postediting genotyping and screening. During HDR-based repair, the cellular repair machinery uses a donor DNA fragment as the template for repairing double-stranded breaks on the genome (Capecchi 1989). The desired sequence provided by the donor DNA template is inserted into the genome at the breakpoint, leading to precise editing of the targeted gene.

Whether DNA is repaired by EJ or HDR is heavily dependent on the cell cycle. For example, the HDR mechanism required for donor DNA insertion is favored in certain stages of meiosis and nearly absent in other stages (Branzei and Foiani 2008). Therefore, the timing of Cas9 expression or delivery is important for increasing HDR efficiency, which is typically very low in *Ae. aegypti.* High HDR editing efficiency has been reported in certain species, such as *Anopheles stephensi*, offering hope for the future improvement of the HDR technique (Gantz et al. 2015). Many research groups are working on methods to improve HDR efficiency to promote more accurate HDR-mediated genome editing by manipulating the cell cycle, adding or stimulating factors that support HDR, or inhibiting factors that promote EJ (Liu et al. 2018; Li and Margolis 2021; Manoj et al. 2021); however, most of these new methods are still proof-of-concept and would need further development and validation in *Ae. aegypti.*

CRISPR–Cas9 Component Forms

Cas9 may be supplied in the injection mixture either in the form of a recombinant protein (enzyme), which offers the freedom to work in arbitrary genetic backgrounds, or in the form of mRNA. Alternatively, genetically encoded Cas9-expressing mosquito embryos can be used for injection, avoiding the need to include Cas9 in the injection mixture altogether. Both the Cas9 protein and Cas9 mRNA can be generated in-house or purchased commercially. Although both forms provide reliable editing efficiency, they are challenging to produce in-house. Also, the Cas9 mRNA is more unstable than the Cas9 protein. Using the genetically encoded Cas9-expressing mosquito embryos directly for injection, in contrast, offers reproducible high editing efficiency compared to using Cas9 mRNA and Cas9 protein directly (Li et al. 2017). The challenge is that generating true-breeding Cas9-expressing mosquito lines is time-consuming and nontrivial. Many of these lines, however, are readily available to the research community, allowing researchers to streamline their efforts. As there is a limited need to customize Cas9 sequences or expression for most applications, using Cas9 mosquito embryos is the recommended method to achieve consistent high editing efficiency.

The sgRNA may be provided in the injection mixture as RNA or as a DNA plasmid. Of the two options, the RNA form tends to deliver higher editing efficiency. The RNA form can also be rapidly synthesized either in-house or via commercial vendors, but it is generally unstable and requires −80°C storage and minimal freeze–thaw handling, which means additional effort, care, and cost for projects. A DNA plasmid of sgRNAs offers good stability but suffers from lower editing efficiency. In addition to these two options, it is possible to use genetically encoded sgRNA-expressing mosquito embryos for injection with Cas9 solutions or by crosses with Cas9-expressing mosquitoes. Yet, generating such transgenic lines is time-consuming and different projects often require different sgRNA designs, such as target sequence and copy number of the selected sgRNAs. It is also common for a project to undergo several rounds of sgRNA design before an optimal version is chosen. Therefore, it is more cost effective to use sgRNA in the form of either an RNA or DNA plasmid.

HDR-mediated insertions require a third component, namely, a donor DNA template. This template can be designed in the form of either ssDNA or DNA plasmids. The ssDNA can be rapidly synthesized, and its size is usually limited to ~200 nt (Bier et al. 2018; Kanca et al. 2019), which can impede the insertion of many genes and markers. In comparison, DNA plasmids do not have size limitations and are more stable, but they take longer to generate. Similar to the choice of sgRNA, donor DNA templates can also be provided via genetically encoded transgenic mosquitoes, although the efforts involved are significant.

Target Site Selection and sgRNA Design

Target site selection to generate mutations in the *Ae. aegypti* genome first requires choosing a target gene region and then identifying putative target sites within the target gene region. The choice of a target gene is usually project specific. The target genes can be selected using the *Ae. aegypti* reference genome and/or reference genomes of other closely related species (other mosquito species or *Drosophila melanogaster*). Confirmation of the selected genes' transcription profile in *Ae. aegypti* is recommended and can be achieved by performing NCBI-BLAST searches against available *Ae. aegypti* transcriptional databases (Vector Caltech, vectorbase.org/vectorbase/app; Akbari et al. 2013). Once target genes are defined, sgRNA target sites within the target gene sequence can be selected by scanning both the sense and antisense strands for 20-nt-long sequences that include an NGG sequence (either inspect visually or use available software) (Xie et al. 2014; Labun et al. 2016). The NGG sequence is called a protospacer-adjacent motif (PAM) sequence, N represents any of the four nucleotides (A, T, C, and G), and G refers to guanine. Online sgRNA design programs are also available, such as CHOPCHOP V3.0.0 and CRISPR software, to aid in the design of effective sgRNAs (Labun et al. 2016).

We recommend using NCBI-BLAST or BLAT to check for any potential off-target effects of the sgRNAs (Bhagwat et al. 2012; Stover and Cavalcanti 2014). Select the most specific sgRNAs within the specified target regions with the least potential off-target binding sites. In addition, to increase the chance of successful editing, it is recommended to design several different sgRNAs for each target gene and coinject these different sgRNAs during microinjection.

Once the target sequences are chosen, one can design the sgRNAs. An sgRNA consists of the following two segments: a scaffold sequence and target sequence. The scaffold sequence is kept constant, whereas the target sequence needs to have the following features: (1) a 20-nt sequence complementary to the selected target site sequence, (2) NGG present at its 3′ end, and (3) guanine padding at its 5′ end if it does not already have two guanines as a T7 RNA polymerase-preferred initiation motif.

Donor DNA Design (for HDR Only)

The donor DNA plasmid and genetically encoded template components (for HDR-mediated insertion) minimally contain the following three elements: user-defined DNA insertion fragment, a marker sequence, and two homology arms of 500–1000 bases flanking the target cleavage site. For gene knockout experiments, it is common to insert a DNA fragment (typically a fluorescent protein) into the open reading frame of the target gene (Basu et al. 2015; Kistler et al. 2015; Li et al. 2017). The marker sequence is used to track transgenesis, which is optional if the DNA insertion fragment already includes a fluorescent marker. For the homology arms, one arm end should be near the cleavage site, and the other arm should exclude the six nucleotides 3′ to this cleavage site (inclusive of the PAM). The sgRNA sequence is often left out of the donor arms. In addition to the basic framework discussed here, more sophisticated tricks in donor plasmid design, such as in-frame fusion, have also been reported using HDR (Fig. 1; Shankar et al. 2021).

The ssDNA templates, sometimes referred to as single-stranded oligo donors, facilitate HDR-mediated insertions of small DNA fragments, such as short peptide tags or base pair edits, but require only short homology arms, typically of the same length (Renaud et al. 2016). These templates are routinely designed so the insertion site and double-stranded break site are in close proximity (<10 nt),

 Cite this introduction as *Cold Spring Harb Protoc*; doi:10.1101/pdb.top107693

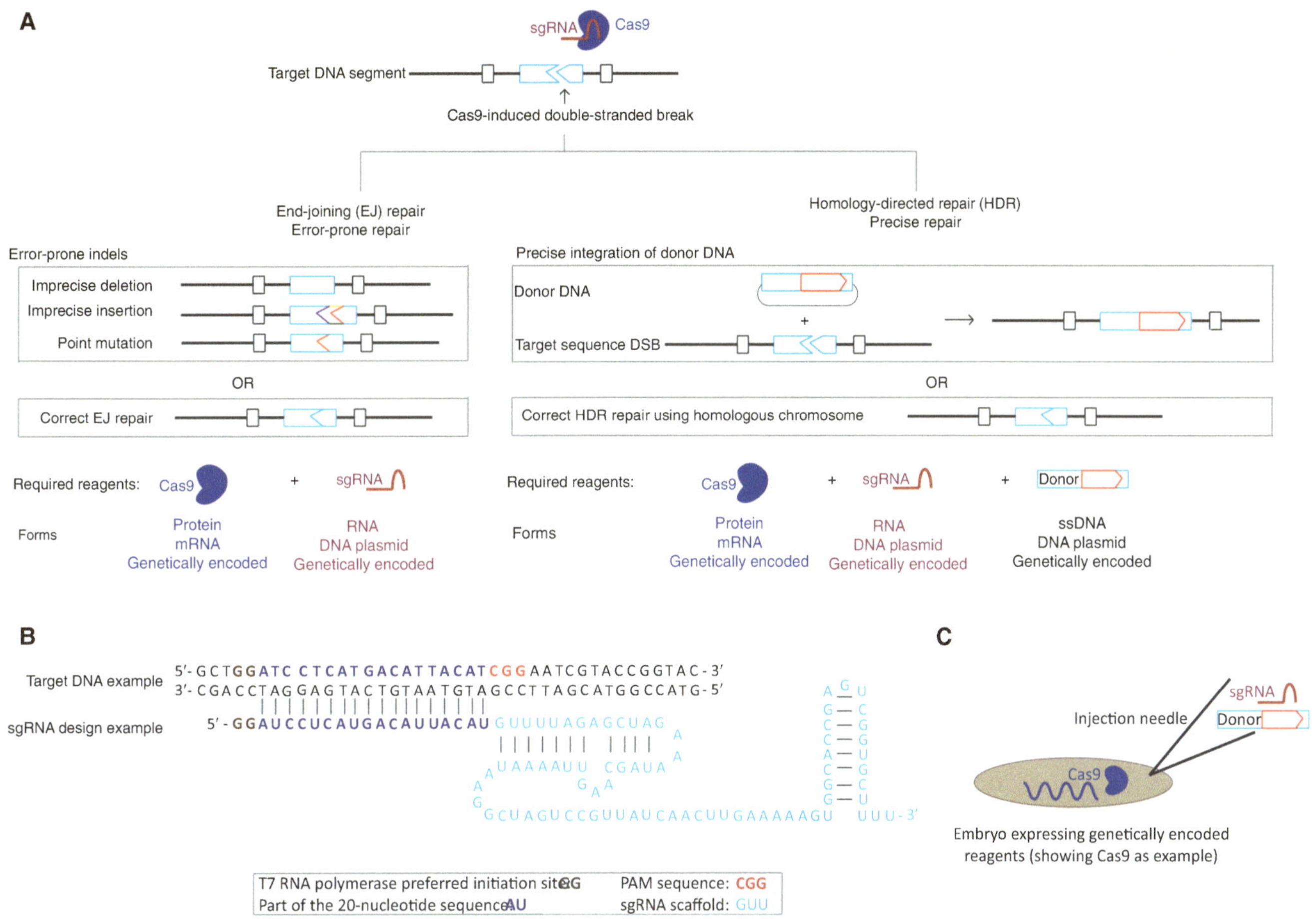

FIGURE 1. Basic principles in using CRISPR–Cas9 to generate mutant mosquitoes. (*A*) The repair mechanisms and their required reagents enabled by the CRISPR–Cas9 system. After Cas9-induced genome cleavage, the double-stranded break is repaired by either end-joining (EJ) or homology-directed repair (HDR). Both repair mechanisms required the Cas9 and single-guide RNA (sgRNA) components to induce and direct cleavage of a specific site in the genome. EJ repair is the more common repair pathway but is error-prone, often resulting in gene knockout because of the generation of deletions, insertions, and point mutations that disrupt gene expression. HDR, on the other hand, can use a donor DNA template to precisely repair the break (knock-in). The HDR mechanism, therefore, requires donor DNA supplied as ssDNA or a DNA plasmid, or it can be genetically encoded to serve as a template for this process. The DNA donors are designed with the desired modification and homology arms flanking the cleavage site. (*B*) An sgRNA design example. (*C*) Schematic of non-genetically encoded component delivery into embryos expressing genetically encoded Cas9 via microinjection. (DSB) Double-stranded break, (PAM) protospacer-adjacent motif.

which was shown to improve HDR efficiency (Elliott et al. 1998; Yang et al. 2013). In *Ae. aegypti*, this approach generated low frequencies of somatic and germline modifications (Kistler et al. 2015) and has also been used to modify the expression of endogenous genes in this same organism (Ling et al. 2017).

CONCLUSION

The CRISPR–Cas9 system offers exceptional flexibility, ease of use, high efficiency, and superior specificity for editing DNA sequences on almost any genomic location. As a result, it has been used in numerous species, of which many were previously intractable for genome engineering, including *Ae. aegypti*. One of the important applications for the CRISPR–Cas9-mediated gene-editing technologies is generating mutant strains. Here, we provided a method overview of CRISPR–Cas9-based

Ae. aegypti mutant strain generation and discussed some of the key design considerations based on our experience. In the accompanying protocol (see Protocol 1: Generating *Aedes aegypti* Mutant Strains with Transgenic Cas9 [Sun et al. 2023]), we present a detailed step-by-step CRISPR–Cas9-based genome-editing procedure for inducing desired heritable edits in *Ae. aegypti.* Notably, this protocol describes a ready-to-use method that accommodates most of the common needs for mutant generation while serving as a starting point for further customization for more sophisticated project requirements. Using this tool in mosquitoes will help researchers unravel the genes important for basic physiological functions and also may prove useful for generating innovative population control strategies that can leverage genetically encoded CRISPR reagents such as gene drives (Champer et al. 2016; Li et al. 2020a; Raban et al. 2020; Verkuijl et al. 2020) and precision-guided sterile insect techniques (Kandul et al. 2019; Li et al. 2021b) to develop transformative technologies for preventing the spread of deadly mosquito-borne pathogens.

ACKNOWLEDGMENTS

Work in our laboratory is supported by funding from a Defense Advanced Research Projects Agency (DARPA) Safe Genes Program Grant (HR0011-17-2-0047) and a National Insitutes of Health (NIH) award (R01AI151004, R01AI148300, R01GM132825, DP2AI152071, and R01AI151004) granted to O.S.A. The views, opinions, and/or findings expressed are those of the authors and should not be interpreted as representing the official views or policies of the U.S. government.

CONFLICT OF INTEREST

O.S.A. is a founder of Agragene, Inc., with equity interest. The terms of this arrangement have been reviewed and approved by the University of California, San Diego in accordance with its conflict of interest policies.

REFERENCES

Akbari OS, Antoshechkin I, Hay BA, Ferree PM. 2013. Transcriptome profiling of *Nasonia vitripennis* testis reveals novel transcripts expressed from the selfish B chromosome, paternal sex ratio. *G3* **3:** 1597–1605. doi:10.1534/g3.113.007583

Basu S, Aryan A, Overcash JM, Samuel GH, Anderson MAE, Dahlem TJ, Myles KM, Adelman ZN. 2015. Silencing of end-joining repair for efficient site-specific gene insertion after TALEN/CRISPR mutagenesis in *Aedes aegypti. Proc Natl Acad Sci* **112:** 4038–4043. doi:10.1073/pnas.1502370112

Bhagwat M, Young L, Robison RR. 2012. Using BLAT to find sequence similarity in closely related genomes. *Curr Protoc Bioinformatics* **Chapter 10:** 10.8.1–10.8.24. doi:10.1002/0471250953.bi10008s37

Bier E, Harrison MM, O'Connor-Giles KM, Wildonger J. 2018. Advances in engineering the fly genome with the CRISPR-Cas system. *Genetics* **208:** 1–18. doi:10.1534/genetics.117.1113

Branzei D, Foiani M. 2008. Regulation of DNA repair throughout the cell cycle. *Nat Rev Mol Cell Biol* **9:** 297–308. doi:10.1038/nrm2351

Bui M, Li M, Raban RR, Liu N, Akbari OS. 2020. Embryo microinjection techniques for efficient site-specific mutagenesis in *Culex quinquefasciatus. J Vis Exp* 159. doi:10.3791/61375

Capecchi MR. 1989. Altering the genome by homologous recombination. *Science* **244:** 1288–1292. doi:10.1126/science.2660260

Champer J, Buchman A, Akbari OS. 2016. Cheating evolution: engineering gene drives to manipulate the fate of wild populations. *Nat Rev Genet* **17:** 146–159. doi:10.1038/nrg.2015.34

Chaverra-Rodriguez D, Macias VM, Hughes GL, Pujhari S, Suzuki Y, Peterson DR, Kim D, McKeand S, Rasgon JL. 2018. Targeted delivery of CRISPR–Cas9 ribonucleoprotein into arthropod ovaries for heritable germline gene editing. *Nat Commun* **9:** 3008. doi:10.1038/s41467-018-05425-9

Elliott B, Richardson C, Winderbaum J, Nickoloff JA, Jasin M. 1998. Gene conversion tracts from double-strand break repair in mammalian cells. *Mol Cell Biol* **18:** 93–101. doi:10.1128/MCB.18.1.93

Gantz VM, Akbari OS. 2018. Gene editing technologies and applications for insects. *Curr Opin Insect Sci* **28:** 66–72. doi:10.1016/j.cois.2018.05.006

Gantz VM, Jasinskiene N, Tatarenkova O, Fazekas A, Macias VM, Bier E, James AA. 2015. Highly efficient Cas9-mediated gene drive for population modification of the malaria vector mosquito *Anopheles stephensi. Proc Natl Acad Sci* **112:** E6736–E6743. doi:10.1073/pnas.1521077112

Hefferin ML, Tomkinson AE. 2005. Mechanism of DNA double-strand break repair by non-homologous end joining. *DNA Repair* **4:** 639–648. doi:10.1016/j.dnarep.2004.12.005

Jinek M, Chylinski K, Fonfara I, Hauer M, Doudna JA, Charpentier E. 2012. A programmable dual-RNA–guided DNA endonuclease in adaptive bacterial immunity. *Science* **337:** 816–821. doi:10.1126/science.1225829

Kanca O, Zirin J, Garcia-Marques J, Knight SM, Yang-Zhou D, Amador G, Chung H, Zuo Z, Ma L, He Y, et al. 2019. An efficient CRISPR-based strategy to insert small and large fragments of DNA using short homology arms. *eLife* **8:** e51539. doi:10.7554/eLife.51539

Kandul NP, Liu J, Sánchez C HM, Wu SL, Marshall JM, Akbari OS. 2019. Transforming insect population control with precision guided sterile males with demonstration in flies. *Nat Commun* **10:** 84. doi:10.1038/s41467-018-07964-7

Kistler KE, Vosshall LB, Matthews BJ. 2015. Genome engineering with CRISPR–Cas9 in the mosquito *Aedes aegypti. Cell Rep* **11:** 51–60. doi:10.1016/j.celrep.2015.03.009

Labun K, Montague TG, Gagnon JA, Thyme SB, Valen E. 2016. CHOP-CHOP v2: a web tool for the next generation of CRISPR genome engineering. *Nucleic Acids Res* **44:** W272–W276. doi:10.1093/nar/gkw398

Li PP, Margolis RL. 2021. Use of single guided Cas9 nickase to facilitate precise and efficient genome editing in human iPSCs. *Sci Rep* **11:** 9865. doi:10.1038/s41598-021-89312-2

Li M, Bui M, Yang T, Bowman CS, White BJ, Akbari OS. 2017. Germline Cas9 expression yields highly efficient genome engineering in a major worldwide disease vector, *Aedes aegypti. Proc Natl Acad Sci* **114:** E10540–E10549. doi:10.1073/pnas.1711538114

Li M, Akbari OS, White BJ. 2018. Highly efficient site-specific mutagenesis in malaria mosquitoes using CRISPR. *G3* **8:** 653–658. doi:10.1534/g3.117.1134

Li M, Yang T, Kandul NP, Bui M, Gamez S, Raban R, Bennett J, Sánchez C HM, Lanzato GC, Schmidt H, et al. 2020a. Development of a confinable gene drive system in the human disease vector *Aedes aegypti. eLife* **9:** e51701. doi:10.7554/eLife.51701

Li M, Li T, Liu N, Raban RR, Wang X, Akbari OS. 2020b. Methods for the generation of heritable germline mutations in the disease vector *Culex quinquefasciatus* using clustered regularly interspaced short palindrome repeats-associated protein 9. *Insect Mol Biol* **29:** 214–220. doi:10.1111/imb.12626

Li H-H, Li J-C, Su MP, Liu K-L, Chen C-H. 2021a. Generating mutant *Aedes aegypti* mosquitoes using the CRISPR/Cas9 system. *STAR Protoc* **2:** 100432. doi:10.1016/j.xpro.2021.100432

Li M, Yang T, Bui M, Gamez S, Wise T, Kandul NP, Liu J, Alcantara L, Lee H, Edula JR, et al. 2021b. Eliminating mosquitoes with precision guided sterile males. bioRxiv doi:10.1101/2021.03.05.434167

Ling L, Kokoza VA, Zhang C, Aksoy E, Raikhel AS. 2017. MicroRNA-277 targets insulin-like peptides 7 and 8 to control lipid metabolism and reproduction in *Aedes Aegypti* mosquitoes. *Proc Natl Acad Sci* **114:** E8017–E8024. doi:10.1073/pnas.1707369114

Liu M, Rehman S, Tang X, Gu K, Fan Q, Chen D, Ma W. 2018. Methodologies for improving HDR efficiency. *Front Genet* **9:** 691. doi:10.3389/fgene.2018.00691

Manoj F, Tai LW, Wang KSM, Kuhlman TE. 2021. Targeted insertion of large genetic payloads using cas directed LINE-1 reverse transcriptase. *Sci Rep* **11:** 23625. doi:10.1038/s41598-021-03130-0

Powell JR. 2018. Mosquito-borne human viral diseases: why *Aedes aegypti*? *Am J Trop Med Hyg* **98:** 1563–1565. doi:10.4269/ajtmh.17-0866

Raban RR, Marshall JM, Akbari OS. 2020. Progress towards engineering gene drives for population control. *J Exp Biol* **223:** jeb208181. doi:10.1242/jeb.208181

Renaud J-B, Boix C, Charpentier M, De Cian A, Cochennec J, Duvernois-Berthet E, Perrouault L, Tesson L, Edouard J, Thinard R, et al. 2016. Improved genome editing efficiency and flexibility using modified oligonucleotides with TALEN and CRISPR–Cas9 nucleases. *Cell Rep* **14:** 2263–2272. doi:10.1016/j.celrep.2016.02.018

Shankar S, Tauxe GM, Spikol ED, Li M, Akbari OS, Giraldo D, McMeniman CJ. 2021. Synergistic coding of human odorants in the mosquito brain. bioRxiv doi:10.1101/2020.11.02.365916

Stover NA, Cavalcanti ARO. 2014. Using NCBI BLAST. *Curr Protoc Essential Lab Tech* 11.1.1–11.1.35. doi:10.1002//9780470089941.et1101s08

Sun R, Raban R, Akbari OS. 2023. Generating *Aedes aegypti* mutant strains with transgenic Cas9. *Cold Spring Harb Protoc* doi:10.1101/pdb.prot108085

Takata M, Sasaki MS, Sonoda E, Morrison C, Hashimoto M, Utsumi H, Yamaguchi-Iwai Y, Shinohara A, Takeda S. 1998. Homologous recombination and non-homologous end-joining pathways of DNA double-strand break repair have overlapping roles in the maintenance of chromosomal integrity in vertebrate cells. *EMBO J* **17:** 5497–5508. doi:10.1093/emboj/17.18.5497

Verkuijl SAN, Gonzalez E, Li M, Ang J, Kandul NP, Anderson MAE, Akbari OS, Bonsall MB, Alphey L. 2020. A CRISPR endonuclease gene drive reveals two distinct mechanisms of inheritance bias. bioRxiv doi:10.1101/2020.12.15.421271

Xie S, Shen B, Zhang C, Huang X, Zhang Y. 2014. sgRNAcas9: a software package for designing CRISPR sgRNA and evaluating potential off-target cleavage sites. *PLoS ONE* **9:** e100448. doi:10.1371/journal.pone.0100448

Yang L, Guell M, Byrne S, Yang JL, De Los Angeles A, Mali P, Aach J, Kim-Kiselak C, Briggs AW, Rios X, et al. 2013. Optimization of scarless human stem cell genome editing. *Nucl Acids Res* **41:** 9049–9061. doi:10.1093/nar/gkt555

Protocol 1

Generating *Aedes aegypti* Mutant Strains with Transgenic Cas9

Ruichen Sun, Robyn Raban, and Omar S. Akbari[1]

Division of Biological Sciences, Section of Cell and Developmental Biology, University of California, San Diego, La Jolla, California 92093, USA

Here, we provide a protocol for generating *Aedes aegypti* mutant strains via end-joining (EJ) or homology-directed repair (HDR) mechanisms using genetically encoded Cas9.

MATERIALS

It is essential that you consult the appropriate Material Safety Data Sheets and your institution's Environmental Health and Safety Office for proper handling of equipment and hazardous materials used in this protocol.

RECIPES: Please see the end of this protocol for recipes indicated by <R>. Additional recipes can be found online at http://cshprotocols.cshlp.org/site/recipes.

Reagents

Aedes aegypti

We use the Liverpool strain as the wild-type (BEI Resources 735) and a genetically encoded Cas9-expressing strain (e.g., expressed using the exuperantia promoter or other; also called Exu-Cas9 strain, marked with Opie2-dsRed [Li et al. 2017]).

Ambion RNaseZap (Sigma-Aldrich R2020)
DNA extraction kit (DNeasy Blood and Tissue Kit; QIAGEN 69506)
Donor DNA template synthesis reagents (see Step 5)
  Gibson assembly master mix (2×; New England BioLabs E2611S)
  piggyBac plasmid pBac-3xP3-dsRed (Addgene 100707)
  ZymoPURE II Plasmid Maxiprep Kit (Zymo Research D4202)
Embryonic microinjection
  Halocarbon oil solution for embryo microinjection <R>
  Nitrogen gas (research purity, 99.9999%; Matheson G2173112)
H_2O (deoxygenated deionized and ultrapure molecular grade [Invitrogen 10977-015])
Ice
Mosquito husbandry reagents
  Larval food (crushed TetraMin Tropical Flakes; Tetra Werke 77101)
  Mice (live, anesthetized intraperitoneally using ketamine/xylazine/sterile saline solution [82.5 mg/kg ketamine; 10 mg/kg xylazine])
  Sucrose (Sigma-Aldrich S0389; crystal form, purity ≥ 99.5%) or glucose (see Step 9)

[1]Correspondence: oakbari@ucsd.edu

Cite this protocol as *Cold Spring Harb Protoc*; doi:10.1101/pdb.prot108085

PCR product purification kit (Zymoclean Gel DNA Recovery Kit; Zymo Research D4007)
Reagents for in-house sgRNA synthesis (optional; see Step 4)
  MegaScript T7 (Ambion AM1334)
  Primers for generating a DNA template of sgRNA using example target sequence (see Step 4)
  Q5 high-fidelity DNA polymerase (New England BioLabs M0491S)
RNA 6000 Nano Kit (Agilent 5067-1511)
Single-guide RNA (sgRNA)
  sgRNA can be obtained via a commercial synthesis service (we use Synthego Corporation) or via in-house synthesis (see Step 4).

Equipment

Bioanalyzer (Agilent Bioanalyzer 2100)
Dissecting microscope with light source (optional; see Step 20)
Embryonic microinjection equipment
  Coverslip (disposable; Fisherbrand S17525C)
  Glass needles (quartz capillary glass tube; Sutter Instrument QF100-70-10)
  Microinjector (FemtoJet 4×; Eppendorf 5252000021)
  Microloader tips (Eppendorf 930001007)
  Micromanipulator (Sutter Instrument MM-33R)
  Micropipette puller (Sutter Instrument P-1000 or P-2000)
  Needle beveler (Sutter BV-10)
  Slides (Fisherbrand 12-550-A3)
  Tape (double-sided; Scotch 378988)
  Vacuum drying oven (ECO 250C 1.9 Cu Ft; Across International AT19e)
Filter paper (Whatman 1825-021)
Fluorescent stereomicroscope (Leica M165FC)
Items for in-house sgRNA synthesis (optional; see Step 4)
  Magnetic beads (AMPure XP; Beckman Coulter)
  MegaClear columns (Life Technologies AM1908)
  Kimwipes
  Microcentrifuge tubes (Fisher Scientific 05-408-130)
Mosquito husbandry equipment
  Cages (24.5-cm × 24.5-cm × 24.5-cm; Bugdorm DP1000)
  Containers (plastic, 34.6-cm × 21-cm × 12.4-cm; Sterilite 1642)
  Cups (plastic; 9-oz [Karat C-KC9] and 16-oz [Amazon B015TK4N30])
  Dehydration chamber (Nalgene transparent polycarbonate; Thermo Scientific 5311-0250) (optional; see Step 20)
  Desiccant (e.g., Drierite granules; Carolina 858963) (optional; see Step 20)
  Membrane feeding system (Hemotek SP6W1-3) (optional; see Step 10)
  Vials (narrow, polystyrene) with plugs (Genesee Scientific 32-116)
Paintbrush (fine-tipped; ZEM 25950)
Paper towels
PCR thermal cycler (T100 Series Touch; Bio-Rad 1861096)
Software to design sgRNA (e.g., CHOPCHOP V3.0.0) (see Step 3)
Spectrophotometer (NanoDrop 2000; Thermo Scientific ND-2000)
Transgenesis screening in G_1 microscopy (see Step 27)
  Camera (Leica DMC2900)
  Fluorescent microscope filters (ECFP, Leica 10447409; EGFP, Leica 10447408; EYFP, Leica 10447410; RFP/dsRed/tdTomato/mCherry, Leica 10447410; ECFP/EGFP/RFP, Leica 10450611)
  LED light source (Lumencor SOLA-III)

METHOD

1. Select regions within the target genes for editing based on project requirements, gene structures, transcripts, protospacer adjacent motif (PAM) requirement (e.g., NGG for Cas9), and exon–intron junction boundaries.

   It is highly recommended to validate the target region sequence using polymerase chain reaction (PCR) amplification (use DNA extraction kit, purification kit, and external primer synthesis service offered by Integrated DNA Technologies, Inc.) and Sanger sequencing (use external sequencing service offered by Integrated DNA Technologies, Inc.) before designing sgRNAs for the target region.

2. Rear *Ae. aegypti* Exu-Cas9 strain and wild-type larvae at an optimal density at 27°C ± 1°C, with 50%–80% humidity, and 12:12-h (light:dark) photoperiod (~200 larvae in 3 L of deoxygenated deionized H_2O) in plastic containers. Feed larvae with a standardized amount of larval food daily until pupation.

   Standardizing the rearing procedure and feeding regimen according to larval density is highly recommended to achieve synchronous pupation and uniform adult size. A standard food density is ~50 mg/200 larvae daily for first-instar (L1) and L2 larvae and ~1300 mg/200 larvae daily for L2–L4 larvae. While going through this protocol, it is important to keep rearing wild-type Ae. aegypti *continuously under the above-mentioned condition. Pupae from these wild-type mosquitoes will be needed for Step 23.*

3. Use CHOPCHOP V3.0.0 or CRISPR software to design the sgRNAs of interest.

4. Prepare sgRNAs in the form of in vitro–transcribed sgRNA or plasmid DNA to be used in the injection mixture.

   i. Thoroughly clean the work area including the microinjection apparatus, gloves, and pipettes with RNaseZap before conducting experiments.

   ii. Prepare in vitro sgRNAs, or sgRNA-expressing plasmid DNA, using ultrapure molecular grade nuclease-free deionized H_2O.

   Option 1: Use a commercial sgRNA synthesis service.

   Option 2: Perform in-house sgRNA synthesis using T7-mediated in vitro transcription. Use the reagents Q5 high-fidelity DNA polymerase, MegaClear columns, magnetic beads, Ambion MegaScript T7, and primers for generating DNA template of sgRNA (Li et al. 2017).

   The primers we use are sgRNA-F (replace underlined sequence with your target sequence, 5′-GAAATTAATACGACTCACTATAGGATCCTCATGACATTACATGTTTTAGAGCTAGAAATAGC-3′) and sgRNA-R (universal; 5′-AAAAGCACCGACTCGGTGCCACTTTTTCAAGTTGATAACGGACTAGCCTTATTTTAACTTGCTATTTCTAGCTCTAAAAC-3′)

   Option 3: Synthesize an sgRNA-expressing plasmid according to Li et al. (2021a).

   iii. Determine the concentration of the sgRNA or plasmid DNA using a spectrophotometer.

   The target size of 5–100 µg is acceptable.

   iv. Verify the size and concentration using an Agilent Bioanalyzer with the RNA 6000 Nano Kit.

   The bioanalyzer's result of its size may not be accurate due to the nature of the sgRNA structure, but it is important to have a single sharp band/peak around the approximate size with little evidence of degradation

   v. Store the final solution in aliquots at 1 µg/µL of a volume 1–10 µL in microcentrifuge tubes at −80°C.

5. If HDR-mediated insertion is the desired editing method, design and assemble a donor DNA plasmid using Gibson assembly (Gibson et al. 2009).

   i. Assemble the donor DNA plasmid using Gibson assembly master mix and the piggyBac plasmid.

   ii. To obtain high-quality donor DNA plasmid for microinjection (ultrapure and endotoxin-free), isolate and purify the assembled HDR plasmid using the ZymoPURE II Plasmid Maxiprep kit.

Cite this protocol as *Cold Spring Harb Protoc*; doi:10.1101/pdb.prot108085

The donor DNA plasmid template generally contains a user-defined DNA segment to be inserted, a transgenic marker sequence (we use 3xP3-dsRed), and two homology arms of 500–1000 bases flanking the target cleavage site. An additional marker can be added outside the homology arms and used as an indicator of proper insertion (e.g., CFP), as one would not expect an HDR-mediated insertion into the correct location to result in expression from both markers.

iii. After obtaining the donor plasmid, confirm the plasmid sequence with Sanger sequencing.

iv. Store plasmids for 1 yr+ at −80°C.

6. Prepare final injection mixtures with nuclease-free deionized H_2O, and store the mixture in aliquots (5–10 µL) in microcentrifuge tubes at −80°C until use (Fig. 1A,B).

   The recommended final concentrations for different components are as follows: sgRNA-expressing plasmid (300 ng/µL), sgRNA (100 ng/µL), and donor DNA plasmid (100 ng/µL, HDR only). Only one source of sgRNA is needed in the injection mixtures.

7. Manually collect 200–300 *Ae. aegypti* pupae from the plastic rearing container as the parental generation to be reared to adults and used for embryo collection.
8. Place the pupae into 9-oz cups containing ~100 mL of ultrapure H_2O.
9. Place the pupal cups inside a holding cage maintained at 27°C ± 1°C, with 50%–80% humidity, and 12:12-h (light:dark) photoperiod. Provide the adults with sucrose or glucose in a 9-oz cup with a filter paper capable of saturation by capillary action for ad libitum adult feeding. Allow adults to mate freely in the cage for 4–5 d.

   Ensure no sources of moisture are present in the cage other than the sugar source, as adult mosquitoes may prematurely lay on this surface, preventing timed oviposition required for injections.

10. Blood feed the females from each cage with anesthetized mice until satiated (for ~15 min daily on two consecutive days).

    Another way of blood feeding is to use the Hemotek membrane feeding system loaded with fresh animal blood. However, blood feeding with live anesthetized mice usually leads to more eggs being laid and higher egg survival rates.

11. Prepare an oviposition cup (9-oz) by filling it with H_2O to 2/3 of its volume, and place paper towels (soaked with H_2O) in the cup. Then, place the cup inside the cage on the day of embryonic microinjections.

    Females are ready to oviposit eggs 72 h after the second blood meal.

    Blowing on, or agitating the cage, in addition to heating the oviposition cup to just above room temperature, may help stimulate egg laying.

12. Collect 50 individual eggs that are white to light gray in coloration (within ~15 min to 1 h of oviposition) and align them side-by-side on moistened filter paper (sufficiently wet with double-distilled H_2O to keep the eggs moist without a meniscus forming) using a paintbrush. Orient the narrow posterior poles (which is the narrower end of the two poles) of the embryos to the same direction for microinjection.

    After embryos are aligned, use fresh filter paper to absorb H_2O from the wet filter paper that is being used as an alignment substrate. This step soaks up residual moisture to help the embryos stick to the double-sided tape in Step 13.

13. Adhere double-sided tape to a glass coverslip and a microscope slide. Transfer aligned embryos to the taped coverslip by gently pressing the coverslip against the surface of the aligned embryos. Flip the glass coverslip so that the eggs are on top and adhere the untaped side of the coverslip to the taped microscope slide to immobilize the coverslip.
14. Immediately pour a little of the ultrapure H_2O-saturated halocarbon oil over the embryos so that they are immersed.
15. Thaw the injection mix on ice.
16. Prepare glass needles for injection.

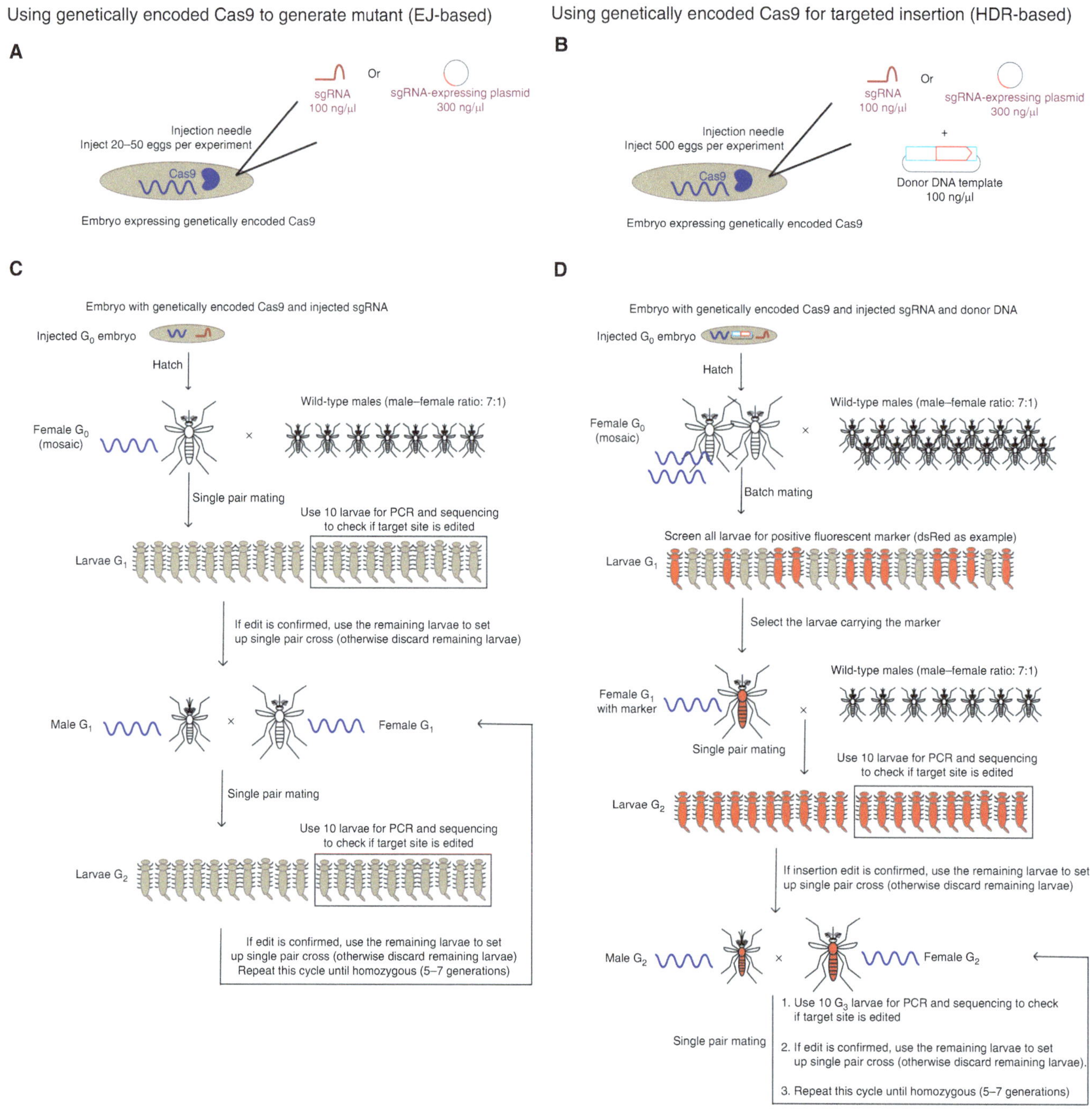

FIGURE 1. Injection and crossing steps when using genetically encoded Cas9 to generate site-directed mutations and insertions. (*A*) Injection mixture for end-joining (EJ)-based editing in mutant generation includes single-guide (sgRNA) or sgRNA-expressing plasmid only. (*B*) Injection mixture for homology-directed repair (HDR)-based editing in mutant generation includes sgRNA or sgRNA-expressing plasmid and donor DNA template. (*C*) Postinjection genetic crosses for obtaining homozygous lines of EJ-based editing. (*D*) Postinjection genetic crosses for obtaining homozygous lines of HDR-based editing include an additional screening and cross step compared to EJ-based editing.

Cite this protocol as *Cold Spring Harb Protoc*; doi:10.1101/pdb.prot108085

We recommend using quartz capillary glass needles pulled by a Sutter P-2000 puller (heat: 750, filament: 4, velocity: 40, delay: 150, pull: 165). The tip of the needle needs to be beveled (using Sutter needle beveler) to create a sharp opening.

17. Set the Femtojet microinjector to a constant compensation pressure (P_c) of ~1 Psi.

    This pressure minimizes capillary backflow/blockage.

18. Load a glass needle with 2 µL of the injection mixture using microloader tips while avoiding bubble formation and then mount it on the microscope micromanipulator stage.

19. Examine embryos under a fluorescent stereomicroscope. If embryos are white/light gray (ideally <1 h old), carefully insert the needle into the posterior pole of the embryo and inject the mixture using an Eppendorf Femtojet 4× with a typical injection pressure (P_I) of 30–50 Psi (using nitrogen gas) and injection time (T_I) of 0.1–0.3 sec, which results in an injection quantity of ~10% of the volume of the embryo.

    For best results, keep the injection needle at an angle slightly greater than 45° relative to the surface of the posterior quarter of the embryo.

20. (Optional) If the injected embryo is leaking fluid, perform light desiccation of the eggs with one of the following options.

    - Place the slide containing embryos briefly under the illumination of a dissecting microscope and constantly monitor the embryos under high magnification until the slightest sign of a dimple or indentation is observed in the side of the embryos.
    - Place the slide containing aligned embryos in a dehydration chamber consisting of a sealed Nalgene jar with desiccant (e.g., Drierite) until liquid is no longer leaking from the embryos.

21. Carefully remove all halocarbon oil within 20 min postinjection using a Kimwipe. Remove the injected G_0 eggs gently from the coverslip with a clean paintbrush and transfer them to a moist filter paper to recover and develop for 4 d at 27°C.

22. After 4 d postinjection, hatch the injected G_0 eggs in deoxygenated deionized H_2O in a vacuum drying oven (27°C, 20 Psi, and 1 h). Follow normal larval rearing procedures as in Step 2, with special care to prevent interlarval competition due to underfeeding or high population density.

    For nonlethal gene targets, typically 40%–80% of injected embryos hatch.

23. Rear larvae to the pupal stage. Sort and separate surviving pupae (G_0) into 16-oz cups based on sex (use sex-specific cages for HDR-G_0). Add the opposite sex of wild-type pupae into each cup (or cage) at 7:1 male-to-female ratios (wild type:G_0) (Fig. 1C).

    If single EJ events need to be isolated, place only one injected G_0 pupa per cup. A 16-oz cup is the smallest cup size in which mosquitoes will reliably mate in single pairs or small-group crosses. Do not mate mosquitoes in smaller containers and vials (e.g., Drosophila *vials). If isolating single EJ events is not needed, batch mate G_0 with excess virgin wild-type mosquitoes in larger cages.*

24. Blood feed adult G_0 females with anesthetized mice until satiated at 4 d posteclosion.

25. (EJ only) Use an empty vial to capture an individual blood-fed adult G_0 female and transfer it directly to another single narrow polystyrene vial with a moist filter paper cone for egg-laying (G_1 eggs) 3 d post-blood meal.

    For HDR editing, put a 9-oz cup in each cage for oviposition.

26. Hatch G_1 eggs in deoxygenated deionized H_2O using a vacuum drying oven (27°C, 20 Psi, and 1 h) after 4 d. Then follow normal larval rearing procedures as in Step 2.

    If a vacuum drying oven is not available, at 2 d postoviposition, condition the moist egg papers for synchronous hatching by removing them individually from their vials and drying them for 5 min at room temperature. Dry the inside of the original oviposition vials with Kimwipes and place the moist oviposition papers that have been drying back into their respective polystyrene vials and plug using cotton stoppers. Allow an additional 2 d to let these egg papers in each vial to desiccate in the insectary. Hatch conditioned eggs in deoxygenated deionized H_2O that has been boiled, capped in a bottle, and cooled to room temperature, and rear mosquitoes as specified in Step 2.

27. (EJ only) Select 10 larvae (G_1) randomly from each isolated colony (out of 40–80 larvae per female) and then pool and extract DNA from these larvae using a DNA extraction kit. Conduct PCR and Sanger sequencing (using commercial PCR primer synthesis and sequencing service) to check for CRISPR–Cas9-induced mutations in the genomic target site (Fig. 1C). Keep colonies with confirmed mutant genotypes and discard other colonies.

    For HDR, screen all G_1 larvae of L3 and older under a fluorescent stereomicroscope (with filters, an LED light source, and camera) for positive fluorescent markers (Fig. 1D). Perform Step 23 on the selected larvae using 16-oz cups. Next, perform Steps 24–26 to obtain G_2 larvae carrying fluorescent markers. Select and pool 10 G_2 larvae carrying fluorescent markers for each sample, perform DNA extraction and then conduct PCR and Sanger sequencing to validate HDR donor DNA insertion. Keep the G_2 colonies with confirmed insertion.

28. (EJ only) Use sexed pupae from each G_1 colony with confirmed CRISPR-induced mutations or insertion to set up multiple single pair crosses of G_1 mosquitoes (mate one female only with one male) (Fig. 1C). Blood-feed mated G_1 females and let them lay eggs. Genotype the candidate G_1 mutant parents using DNA extraction, followed by PCR and Sanger sequencing or restriction fragment analysis after egg laying. Keep G_2 colonies with confirmed mutant genotypes/HDR-mediated insertions and discard other colonies. Continue intercrossing this way until a homozygous mutant/transgenic line is obtained, which can take up to 5–10 generations.

    For HDR, either set up multiple single crosses using G_2 mosquitoes (single intercross) with the confirmed insertion or outcross the G_2 mosquitoes with wild-type mosquitoes (Fig. 1D). For a single intercross, repeat this step until homozygous transgenic lines are obtained. For outcrossing, it is recommended to outcross for five to seven generations to remove any potential off-target editing. After outcrossing, a single intercross is still needed.

    Homozygosity, the presence of two identical copies of the edited gene location, should be determined via sequencing.

    For all PCR and Sanger sequencing steps, it is recommended to sequence for any putative off-target site editing.

DISCUSSION

In the process of generating mutant strains of mosquitoes with CRISPR–Cas9, researchers should keep several key items in mind. Validation of target DNA sequences is important as the reference genome sequence of *Ae. aegypti* is imperfect. Single-nucleotide polymorphisms, insertions, and deletions are common in the genomes of in-house laboratory strains relative to the genome assembly strains. The use of genetically encoded Cas9 strains streamlines the generation of mutant strains and site-directed insertions. Additionally, researchers using this tool in mosquitoes will help unravel the genes important for basic physiological functions and also may prove useful for generating innovative population control strategies that can leverage genetically encoded CRISPR reagents such as gene drives (Champer et al. 2016; Li et al. 2020; Raban et al. 2020; Verkuijl et al. 2022) and precision-guided sterile insect techniques (Kandul et al. 2019; Li et al. 2021b) to develop transformative technologies to prevent the spread of deadly mosquito-borne pathogens.

RECIPE

Halocarbon Oil Solution for Embryo Microinjection

Reagent	Amount
Halocarbon oil 700 (Sigma-Aldrich H8898)	9 mL
Halocarbon oil 27 (Sigma-Aldrich H8773)	1 mL
Deoxygenated deionized H_2O	20 mL

Store in a tightly sealed container in a dry, well-ventilated space at room temperature.

ACKNOWLEDGMENTS

Work in the laboratory of O.S.A. was supported by funding from a Defense Advanced Research Projects Agency (DARPA) Safe Genes Program Grant (HR0011-17-2-0047) and a National Institutes of Health (NIH) award (R01AI151004). The views, opinions, and/or findings expressed are those of the authors and should not be interpreted as representing the official views or policies of the U.S. government

CONFLICT OF INTEREST

O.S.A. is a founder of Agragene, Inc., with equity interest. The terms of this arrangement have been reviewed and approved by the University of California, San Diego in accordance with its conflict of interest policies.

REFERENCES

Champer J, Buchman A, Akbari OS. 2016. Cheating evolution: engineering gene drives to manipulate the fate of wild populations. *Nat Rev Genet* **17:** 146–159. doi:10.1038/nrg.2015.34

Gibson DG, Young L, Chuang R-Y, Venter JC, Hutchison CA III, Smith HO. 2009. Enzymatic assembly of DNA molecules up to several hundred kilobases. *Nat Methods* **6:** 343–345. doi:10.1038/nmeth.1318

Kandul NP, Liu J, Sánchez C HM, Wu SL, Marshall JM, Akbari OS. 2019. Transforming insect population control with precision guided sterile males with demonstration in flies. *Nat Commun* **10:** 84. doi:10.1038/s41467-018-07964-7

Li M, Bui M, Yang T, Bowman CS, White BJ, Akbari OS. 2017. Germline Cas9 expression yields highly efficient genome engineering in a major worldwide disease vector, *Aedes aegypti*. *Proc Natl Acad Sci* **114:** E10540–E10549. doi:10.1073/pnas.1711538114

Li M, Yang T, Kandul NP, Bui M, Gamez S, Raban R, Bennett J, Sánchez C HM, Lanzato GC, Schmidt H, et al. 2020. Development of a confinable gene drive system in the human disease vector *Aedes aegypti*. *eLife* **9:** e51701. doi:10.7554/eLife.51701

Li H-H, Li J-C, Su MP, Liu K-L, Chen C-H. 2021a. Generating mutant *Aedes aegypti* mosquitoes using the CRISPR/Cas9 system. *STAR Protoc* **2:** 100432. doi:10.1016/j.xpro.2021.100432

Li M, Yang T, Bui M, Gamez S, Wise T, Kandul NP, Liu J, Alcantara L, Lee H, Edula JR, et al. 2021b. Suppressing mosquito populations with precision guided sterile males. *Nat Commun* **12:** 5374 doi:10.1038/s41467-021-25421-w

Raban RR, Marshall JM, Akbari OS. 2020. Progress towards engineering gene drives for population control. *J Exp Biol* **223:** jeb208181. doi:10.1242/jeb.208181

Verkuijl SAN, Gonzalez E, Li M, Ang J, Kandul NP, Anderson MAE, Akbari OS, Bonsall MB, Alphey L. 2022. A CRISPR endonuclease gene drive reveals two distinct mechanisms of inheritance bias. *Nat Commun* **13:** 7145. doi:10.1038/s41467-022-34739-y

CHAPTER 38

Techniques for Identifying and Sorting Transgenic Mosquito Larvae

Eric Marois[1]

CNRS UPR9022, INSERM U1257, Institut de Biologie Moléculaire et Cellulaire, Université de Strasbourg, 67000 Strasbourg, France

Transgenic mosquitoes are used in many aspects of mosquito research. First, they can help answer biological questions to advance scientific knowledge—for example, in the fields of mosquito–pathogen interactions, insect immunity, or olfaction. Second, transgenic technologies may be used to develop much needed novel vector control strategies, such as mosquitoes that are unable to transmit disease or transgenes that sterilize mosquito females to suppress vector populations. Here, we introduce how researchers use various selection markers to screen for transgenic mosquito larvae following a transgenesis experiment. Common procedures include using a binocular fluorescence microscope for initial screening. For higher-throughput screening, a flow cytometer known as Complex Object Parametric Analyzer and Sorter (COPAS) can be used to stabilize transgenic lines through the purification of homozygous individuals or to manage transgene frequency in established transgenic lines. In particular, COPAS sorting allows the production of mosquito larval cultures composed of a mixture of genotypes (control and genetically modified larvae) with the goal of raising both groups of mosquitoes under the same environmental conditions in preparation for a controlled phenotype assessment. It can also be used to produce large populations of male mosquitoes, which should facilitate the development of mosquito control intervention strategies similar to the sterile insect technique (SIT), which aims to release large numbers of sterile males that will mate with and sterilize wild females to suppress mosquito populations. Finally, the utilization of a puromycin resistance marker cassette to screen for transgenic *Anopheles* larvae is also introduced.

INTRODUCTION TO TRANSGENIC MOSQUITO PRODUCTION TECHNIQUES

Transgenic mosquitoes have been produced since the late 1990s (Jasinskiene et al. 1998; Catteruccia et al. 2000), mainly by embryo microinjection. Gene gun bombardment is an alternative approach that has yielded transgenic *Aedes aegypti* (Lule-Chávez et al. 2021). The receptor-mediated ovary transduction of cargo (ReMOT) approach (i.e., ovarian uptake of nucleic acid and protein complexes injected in adult females) has allowed CRISPR–Cas9 editing of marker genes in *Aedes* and *Anopheles* (Chaverra-Rodriguez et al. 2018; Macias et al. 2020) and could potentially also be used to insert transgenes in the genome. In all cases, DNA-injected individual mosquitoes must be backcrossed to wild-type mosquitoes, in a cross termed G0, and the G1 progeny, composed of hundreds to tens of thousands of larvae, must be screened to identify transgenic individuals. Depending on the transgenesis approach used (transposon, docking site [which is the integration of foreign DNA into a recombination site previously inserted in the genome], or CRISPR–Cas9-based construct knock-in)

[1]Correspondence: e.marois@unistra.fr

Cite this introduction as *Cold Spring Harb Protoc*; doi:10.1101/pdb.top107694

and on the technical quality of microinjection, the number of transgenic individuals in a G1 population of 10,000 larvae may range from just a few (e.g., following *piggyBac* transposon transformation in *Anopheles*) to a few hundred (CRISPR–Cas9 knock-in of a fluorescent marker gene in *Anopheles* or *piggyBac* transposon transformation in *Aedes*).

The rarity of the desired event underscores the absolute necessity for a robust selection system, in which positive larvae will immediately be distinguishable from wild types without the need for polymerase chain reaction (PCR) genotyping. This is best achieved using fluorescence markers expressed in live mosquitoes, especially during their larval stages, and requires visual examination under a fluorescence microscope as described in an accompanying protocol (Protocol 1: Screening Mosquito Larvae under a Fluorescence Binocular Microscope [Marois 2023a]). Alternatively, selection based on resistance to the antibiotic puromycin has been developed for *Anopheles* larvae and is also described in an accompanying protocol (Protocol 3: Selecting Transgenic Mosquito Larvae with Puromycin [Marois 2023b]). Once a transgenic line is obtained and amplified, it is often desirable to isolate individuals carrying the transgene in the homozygous state to stabilize the line. Alternatively, it may be decided to maintain the transgenic line as a mix of negative, heterozygous, and homozygous individuals. This ensures that control and transgenic mosquitoes to be compared in subsequent experiments continue to share identical genetic backgrounds and microbiota over time and are grown under the same environmental conditions. When an experiment is planned, animals with the correct homozygous transgenic and control genotypes in equal numbers can be extracted from a population of larvae. For all these purposes, higher-throughput screening techniques are helpful and must be able to distinguish between homozygous and heterozygous individuals based on their fluorescence intensity, which correlates with transgene copy number (two in homozygotes, one in heterozygotes). A flow cytometer–like sorting machine (Complex Object Parametric Analyzer and Sorter [COPAS]) provides high-throughput sorting capacity and is described in Protocol 2: Sorting Mosquito Larvae with a COPAS Machine (Marois 2023c). The same technique can be applied to the sorting of large quantities of pure male or female larvae, which may facilitate future mosquito control intervention strategies related to the sterile insect technique (SIT) (Lutrat et al. 2019, 2023).

MARKER GENES AND THE CHOICE OF THEIR PROMOTERS IN MOSQUITOES

Several marker genes have been validated for the selection of transgenic mosquitoes. The popular fluorescent proteins CFP (cyan fluorescent protein), GFP (green fluorescent protein), YFP (yellow fluorescent protein), and DsRed or mCherry (red fluorescent proteins) and their many derivatives are suitable and frequently used to identify transgenic mosquito larvae and adults. Two major promoters are commonly used to express these fluorescence markers in mosquitoes—namely, the synthetic 3xP3 promoter (Berghammer et al. 1999) and the *Ae. aegypti polyubiquitin* promoter (PUb) (Anderson et al. 2010). The 3xP3 promoter appears to function quasiuniversally across insects (Horn et al. 2002) with the possible exception of tephritid flies (Schetelig and Handler 2013; Handler and Schetelig 2014). It drives expression primarily in the eyes, brain, and optic nerve of larvae and adults. In *Aedes* mosquito larvae, pupae, and adults, expression from this promoter can be weak (Fig. 1A–E). A fluorescence binocular microscope with excellent optics ensuring high sensitivity to the chosen fluorophore (CFP is often more difficult to visualize than GFP or YFP, although its mTurquoise2 derivative version [Goedhart et al. 2012] improves fluorescence intensity) and careful optical screening are essential to avoid missing weakly expressing transgenic individuals. The same promoter in *Anopheles* can yield stronger fluorescence than in *Aedes* (Fig. 1F,G), which, depending on the line and therefore on genomic position effects, will often extend to the entire ventral neural chain and anal papillae of the larvae more frequently than in *Aedes*. As a result, the 3xP3 promoter is well-suited for automated sorting of *Anopheles* larvae using a COPAS machine (Marois et al. 2012; Bernardini et al. 2014) but will generally be too weak for this purpose with *Aedes* larvae. In *Aedes*, the PUb promoter yields bright whole-body fluorescence (Fig. 1B–D,H) and makes it ideal for COPAS sorting.

Cite this introduction as *Cold Spring Harb Protoc*; doi:10.1101/pdb.top107694

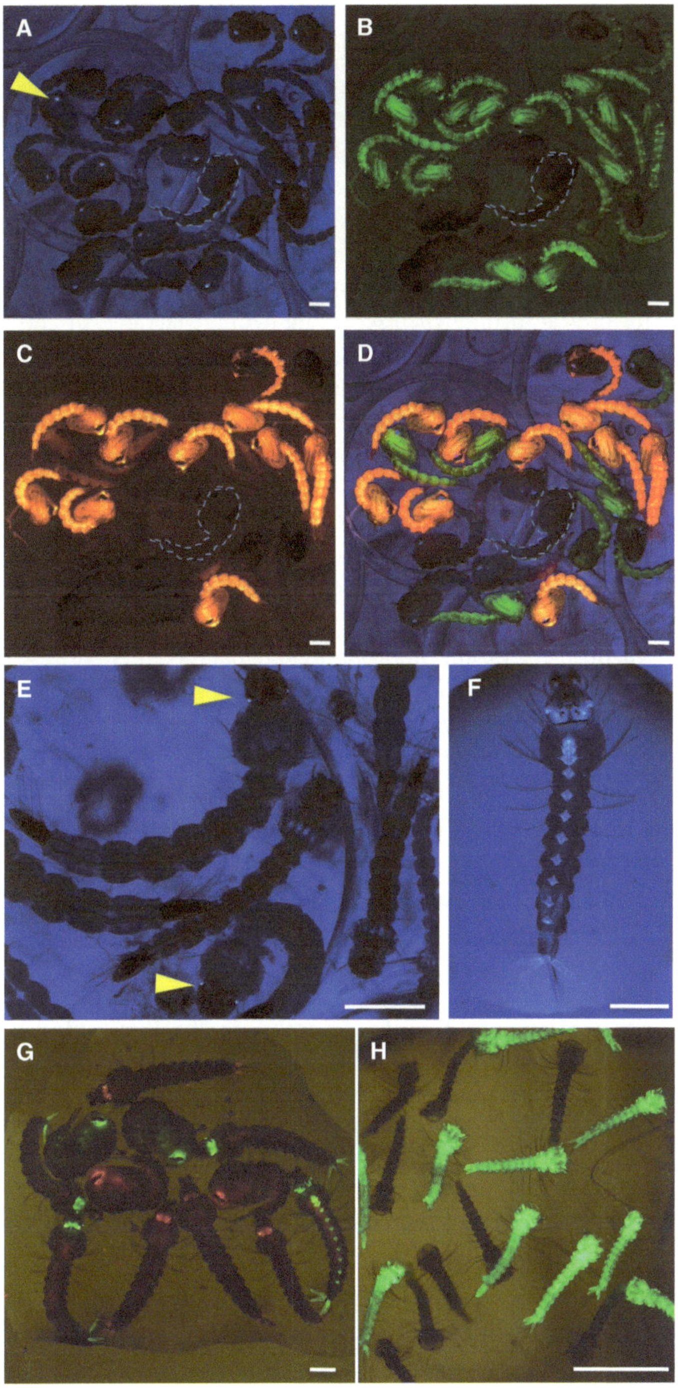

FIGURE 1. Commonly used fluorescence markers for mosquito larvae. (*A–D*) *Aedes albopictus* pupae carrying combinations of three transgenic constructs, marked with 3xP3-mTurquoise (*A*), PUb-YFPvenus (*B*), and PUb-DsRed (*C*). (*D*) Shows the overlay of *A–C*. A negative pupa is indicated with dotted lines, and the arrowhead in *A* shows mTurquoise expression in the eye. (*E*) Fourth-instar *Ae. albopictus* larvae with 3xP3-mTurquoise expression in the eyes (arrowheads) and control larvae. (*F*) Fourth-instar *Anopheles coluzzii* larva with 3xP3-mTurquoise expression extending to the abdominal nerve chain and anal papillae. (*G*) Fourth-instar *An. coluzzii* larvae and pupae expressing 3xP3-EGFP and/or 3xP3-DsRed markers. The rightmost larva is oriented with the ventral side upward, showing GFP expression in the abdominal nerve chain, which is hidden from view in the other larvae. (*H*) Neonate *Aedes* larvae expressing PUb-GFP and wild types. Scale bars, 1 mm.

Unfortunately, the *Aedes* PUb promoter does not work well in *Anopheles*, except for weak expression in developing pupal wing veins, so that marker cassettes that incorporate the *Aedes* PUb promoter cannot be used in both of these mosquito genera. In *Anopheles*, an endogenous *polyubiquitin* promoter for ubiquitous expression has been described (Adolfi et al. 2018) and may be used as an alternative to 3xP3. The baculovirus *OpIE2* promoter (Theilmann and Stewart 1992) can also offer a useful alternative, showing midgut and cardia expression in mosquito larvae, which becomes ubiquitous in late larvae, pupae, and adults. One drawback of *OpIE2* is that, depending on its genomic insertion site, this promoter may become active only during late stages of larval development in some lines, which is problematic for COPAS sorting (optimal at the neonate stage, i.e., before larvae start feeding) as well as for optical sorting when screening neonate larvae is preferred. Several reporter constructs have been described in mosquitoes (e.g., Volohonsky et al. 2015), of which some could serve as alternative transgenesis selection markers, to avoid overlapping marker expressions in tissues of interest. Of note, the purpose of transgenesis can be to reveal the expression pattern of a newly investigated promoter by expressing a fluorescent reporter protein under its control. In this case, it is desirable to associate this reporter construct with a transgenesis marker showing no interference with the reporter (i.e., no overlapping fluorescence colors or fluorescence bleed-through). This can be achieved by choosing a transgenesis marker with weak or highly tissue-restricted expression, with very different fluorescence (e.g., associate a GFP or CFP reporter with a DsRed transgenesis marker, or associate YFP with CFP), or one that is nonfluorescent (puromycin resistance). Finally, if immunostaining experiments using various fluorophores are planned on tissue dissected from transgenic mosquitoes, it may be wise to choose a nonfluorescent transgenesis marker to avoid any interference with fluorescent staining. A puromycin resistance cassette (*puromycin acetyltransferase* gene cloned under control of the *OpIE2* promoter) is efficient as a selection marker in *Anopheles* (Volohonsky et al. 2015). Unfortunately, it does not work for *Aedes*, whose larvae are naturally more resistant to puromycin. In this mosquito genus, a similarly convenient chemical selection marker remains to be developed.

ACKNOWLEDGMENTS

In our laboratory, research on mosquitoes is supported by funding from Inserm, CNRS, the University of Strasbourg, the city of Strasbourg, through contrat triennal "Strasbourg capitale européenne" 2018–2020 and Agence Nationale de la Recherche (ANR) through grants ANR-11-EQPX-0022, ANR-19-CE35-0007, and ANR-11-JSV3-0001.

REFERENCES

Adolfi A, Pondeville E, Lynd A, Bourgouin C, Lycett GJ. 2018. Multi-tissue GAL4-mediated gene expression in all *Anopheles gambiae* life stages using an endogenous polyubiquitin promoter. *Insect Biochem Mol Biol* **96:** 1–9. doi:10.1016/j.ibmb.2018.03.005

Anderson MAE, Gross TL, Myles KM, Adelman ZN. 2010. Validation of novel promoter sequences derived from two endogenous ubiquitin genes in transgenic *Aedes aegypti*. *Insect Mol Biol* **19:** 441–449. doi:10.1111/j.1365-2583.2010.01005.x

Berghammer AJ, Klingler M, Wimmer EA. 1999. A universal marker for transgenic insects. *Nature* **402:** 370–371. doi:10.1038/46463

Bernardini F, Galizi R, Menichelli M, Papathanos P-A, Dritsou V, Marois E, Crisanti A, Windbichler N. 2014. Site-specific genetic engineering of the *Anopheles gambiae* Y chromosome. *Proc Natl Acad Sci* **111:** 7600–7605. doi:10.1073/pnas.1404996111

Catteruccia F, Nolan T, Loukeris TG, Blass C, Savakis C, Kafatos FC, Crisanti A. 2000. Stable germline transformation of the malaria mosquito *Anopheles stephensi*. *Nature* **405:** 959–962. doi:10.1038/35016096

Chaverra-Rodriguez D, Macias VM, Hughes GL, Pujhari S, Suzuki Y, Peterson DR, Kim D, McKeand S, Rasgon JL. 2018. Targeted delivery of CRISPR–Cas9 ribonucleoprotein into arthropod ovaries for heritable germline gene editing. *Nat Commun* **9:** 3008. doi:10.1038/s41467-018-05425-9

Goedhart J, von Stetten D, Noirclerc-Savoye M, Lelimousin M, Joosen L, Hink MA, van Weeren L, Gadella TWJ, Royant A. 2012. Structure-guided evolution of cyan fluorescent proteins towards a quantum yield of 93%. *Nat Commun* **3:** 751. doi:10.1038/ncomms1738

Handler AM, Schetelig MF. 2014. Tephritid fruit fly transgenesis and applications. In *Transgenic insects, CABI biotechnology* (ed. Benedict MQ, Scott MJ), pp. 117–137. CAB International, Wallingford, UK.

Horn C, Schmid BGM, Pogoda FS, Wimmer EA. 2002. Fluorescent transformation markers for insect transgenesis. *Insect Biochem Mol Biol* **32:** 1221–1235. doi:10.1016/s0965-1748(02)00085-1

Jasinskiene N, Coates CJ, Benedict MQ, Cornel AJ, Rafferty CS, James AA, Collins FH. 1998. Stable transformation of the yellow fever mosquito, *Aedes aegypti*, with the Hermes element from the housefly. *Proc Natl Acad Sci* **95:** 3743–3747. doi:10.1073/pnas.95.7.3743

Lule-Chávez AN, Carballar-Lejarazú R, Cabrera-Ponce JL, Lanz-Mendoza H, Ibarra JE. 2021. Genetic transformation of mosquitoes by microparticle bombardment. *Insect Mol Biol* **30:** 30–41. doi:10.1111/imb.12670

Lutrat C, Giesbrecht D, Marois E, Whyard S, Baldet T, Bouyer J. 2019. Sex sorting for pest control: it's raining men! *Trends Parasitol* **35:** 649–662. doi:10.1016/j.pt.2019.06.001

Lutrat C, Burckbuchler M, Olmo RP, Beugnon R, Fontaine A, Akbari OS, Argilés-Herrero R, Baldet T, Bouyer J, Marois E. 2023. Combining two genetic sexing strains allows sorting of non-transgenic males for *Aedes* genetic control. *Commun Biol* **6:** 646. doi:10.1038/s42003-023-05030-7

Macias VM, McKeand S, Chaverra-Rodriguez D, Hughes GL, Fazekas A, Pujhari S, Jasinskiene N, James AA, Rasgon JL. 2020. Cas9-mediated gene-editing in the malaria mosquito *Anopheles stephensi* by ReMOT control. *G3 (Bethesda Md)* **10:** 1353–1360. doi:10.1534/g3.120.401133

Marois E. 2023a. Screening mosquito larvae under a fluorescence binocular microscope. *Cold Spring Harb Protoc* doi:10.1101/pdb.prot108306

Marois E. 2023b. Selecting transgenic mosquito larvae with puromycin. *Cold Spring Harb Protoc* doi:10.1101/pdb.prot108308

Marois E. 2023c. Sorting mosquito larvae with a COPAS machine. *Cold Spring Harb Protoc* doi:10.1101/pdb.prot108307

Marois E, Scali C, Soichot J, Kappler C, Levashina EA, Catteruccia F. 2012. High-throughput sorting of mosquito larvae for laboratory studies and for future vector control interventions. *Malar J* **11:** 302. doi:10.1186/1475-2875-11-302

Schetelig MF, Handler AM. 2013. A functional comparison of the 3xP3 promoter by recombinase-mediated cassette exchange in *Drosophila* and a tephritid fly, *Anastrepha suspensa. G3 (Bethesda Md)* **3:** 687–693. doi:10.1534/g3.112.005488

Theilmann DA, Stewart S. 1992. Molecular analysis of the trans-activating *IE-2* gene of *Orgyia pseudotsugata* multicapsid nuclear polyhedrosis virus. *Virology* **187:** 84–96. doi:10.1016/0042-6822(92)90297-3

Volohonsky G, Terenzi O, Soichot J, Naujoks DA, Nolan T, Windbichler N, Kapps D, Smidler AL, Vittu A, Costa G, et al. 2015. Tools for *Anopheles gambiae* transgenesis. *G3 (Bethesda Md)* **5:** 1151–1163. doi:10.1534/g3.115.016808

Protocol 1

Screening Mosquito Larvae under a Fluorescence Binocular Microscope

Eric Marois[1]

CNRS UPR9022, INSERM U1257, Institut de Biologie Moléculaire et Cellulaire, Université de Strasbourg, 67000 Strasbourg, France

Transgenic mosquitoes used in many aspects of mosquito research and convenient selection markers are crucial to identifying transgenic individuals even among large numbers of wild types. Visual markers, in the form of fluorescent proteins expressed in larval and adult mosquito tissues, are the most commonly used. This requires observing large numbers of mosquitoes under the fluorescence microscope and recognizing positive individuals expressing the fluorescent genetic marker. As research models, mosquito larvae possess the following two advantages over many other insects, greatly facilitating fluorescence screening: (1) Being aquatic, they can be isolated in a drop of clear water, an ideal medium for live observations under the binocular microscope; and (2) synchronous hatching from many eggs is easily obtained, so that large populations of larvae can be screened in batches of several tens of thousands at a time. Screening at the neonate stage, when larvae are ~1-mm-long, allows the concentration of hundreds of larvae in a drop of water that fits in the observation field of the microscope at medium magnification. Thus, many individuals can be screened rapidly. We strongly recommend screening larvae at the neonate stage and before any feeding for two reasons, as follows: (1) Food particles can be strongly autofluorescent, thereby dramatically increasing the fluorescence background noise around larvae, and (2) tissue autofluorescence increases during development, notably in the digestive tract, significantly decreasing the specific signal-to-noise ratio. In this protocol, we guide the experimenter step-by-step for a fast and efficient medium-throughput manual screening for fluorescent larvae.

MATERIALS

It is essential that you consult the appropriate Material Safety Data Sheets and your institution's Environmental Health and Safety Office for proper handling of equipment and hazardous material used in this protocol.

Reagents

Larval food (finely ground TetraMin fish food; Tetra)

Transgenic and wild-type mosquito larvae (recently hatched and unfed)

For detailed protocols for mosquito rearing, see Benedict (2015).

For the detailed process of microinjection of Anopheles *eggs and transgenesis, refer to the following references: Fuchs et al. (2013); Pondeville et al. (2014); Volohonsky et al. (2015); Carballar-Lejarazú et al. (2021).*

[1]Correspondence: e.marois@unistra.fr

Cite this protocol as *Cold Spring Harb Protoc*; doi:10.1101/pdb.prot108306

Equipment

Breeding containers (plastic boxes of various sizes, with lids such as Plexiglas plates)
Glass bowl (optional; see Step 1)
Manually controlled binocular microscope equipped with a good source of fluorescence excitation light and high sensitivity to fluorescence

The sensitivity of fluorescence detection is of paramount importance for efficient visual screening. We have noticed that weakly expressing transgenic mosquito larvae or mosquito larvae expressing low transient fluorescence following their injection as embryos can easily be missed when using suboptimal optics. A good set of filters is also extremely important. Ideally, filters for GFP, DsRed, YFP, and CFP should all be mounted on the system, without bleed-through of the CFP and YFP channels into each other and as little bleed-through as possible between YFP and DsRed, while preserving a high fluorescence signal intensity. An additional filter for bright field and a double filter (GFP + DsRed) may also be very useful, particularly when screening multiple wavelengths at the same time (performing simultaneous transgenesis with multiple plasmids labeled with different fluorescent markers is a considerable labor and time saver, particularly when using docking sites [a defined recombination site previously inserted in the genome]). We mainly use a Nikon SMZ18 binocular microscope with CFP, YFP, DsRed, and GFP/DsRed filters and a Lumencor SOLA Light engine excitation source.

Mosquito-rearing equipment
Petri dishes (100-mm-diameter)
Sieve (optional; see Step 1; e.g., 40-µm cell strainers; Falcon 352340)
Standard laboratory pipettes (200-µL and 1000-µL; e.g., Gilson Pipetman P200 and P1000) and plastic tips
Vacuum chamber (optional; see Step 1)

METHOD

Rear and maintain mosquitoes under standard laboratory conditions.

1. Collect G1 larvae directly after hatching by pipetting or filtering.

   G1 larvae are the progeny of the G0 transformed individuals crossed to wild-type mosquitoes. Anopheles *larvae tend to hatch synchronously 2 d after egg laying. For* Aedes *mosquitoes, hatching can lag in time, especially after longer periods of storing dry eggs. In this case, applying a vacuum (25% of atmospheric pressure in a vacuum chamber for 30 min to a few hours) can promote more immediate and synchronous hatching. For promoters driving fluorescent expression from early developmental stages, screening as soon as G1 larvae hatch, and before they are fed for the first time, is the most efficient procedure (a higher number of larvae can be screened at a time, without interference from food contamination). To allow more time for screening, larvae can be maintained in the neonate stage for some time (preferably not exceeding 36 h) by depriving them of food.*

   The most convenient method is to collect larvae directly from the dish where eggs were laid by female mosquitoes, leaving the eggshells stuck to the wet paper. For this method, an egg dish made of a glass bowl filled with water, on top of which a conically folded piece of paper is fitted, with a small hole at the cone's summit dipped into the water allowing escape of the larvae into the water, is convenient. To maximize the speed and efficiency of screening, larvae can be concentrated on a fine sieve (Fig. 1A) either by carefully pouring water containing larvae in the sieve or by pipetting them onto it using a P1000 pipette fitted with a cut tip, with care taken not to squish larvae with the pipette tip (after initiating the flow of water through the filter mesh, larvae-containing water can be dropped onto the filter without pressing against it).

2. With a P200 pipette, move droplets containing 10–100 larvae to the transparent lid of a Petri dish (Fig. 1B).

   Droplet size should be kept small to fit in the field of the binocular microscope at medium magnification (higher magnification may be required if expression from the fluorescence marker is weak). Also, prepare a few empty water droplets on the same lid, to transfer the discovered transgenics.

   In case the marker gene expression is very strong and easy to detect, the totality of larvae (up to several thousands) concentrated in a small sieve or in a larger drop of water can also be screened collectively.

3. Moving the Petri dish lid from one droplet to the next in the optical field, examine the larvae under the correct fluorescence light, at a magnification at which the water droplets occupy the entire field of vision.

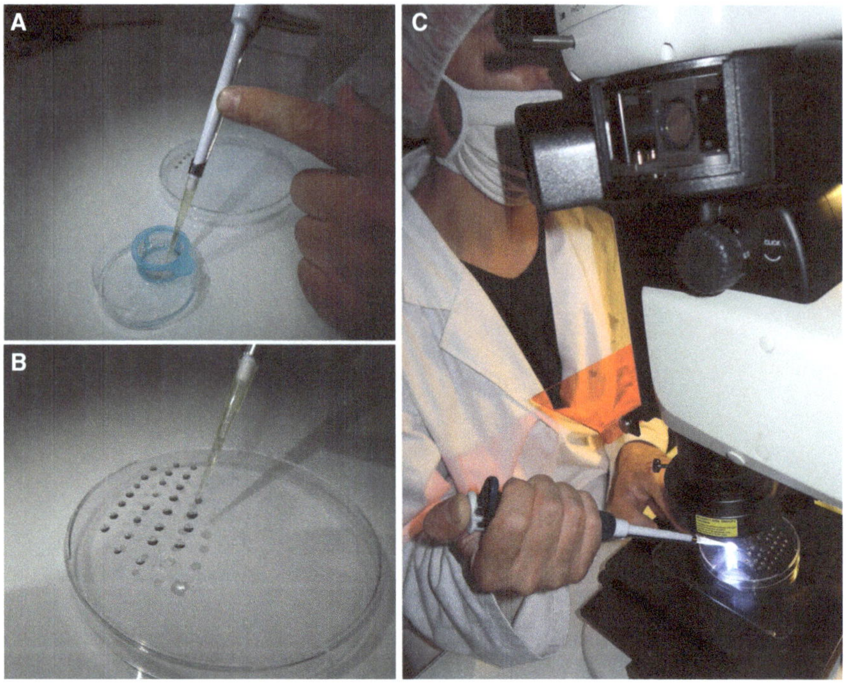

FIGURE 1. Screening for fluorescent mosquito larvae under the binocular microscope. (*A*,*B*) Neonate larvae are filtered in a fine sieve and transferred by groups of 20–100 in small drops of water. (*C*) Each drop of water containing larvae is examined under fluorescent light, and positive larvae are collected with a pipette and transferred to an empty drop of water.

If necessary, increase the magnification to ensure that weakly expressing transgenics are not missed, especially if using the 3xP3 promoter in Aedes *larvae. Learn to distinguish natural background larva autofluorescence from the fluorescence of a true fluorophore, which is usually much brighter and well-defined within tissue boundaries. A fluorescent transgenic strain obtained from elsewhere may be useful in initial experiments, as a comparison reference.*

4. When a fluorescent larva is detected, catch it in a fresh pipette tip fitted on a P200 pipette (Fig. 1C) by following the steps below. If necessary, cut the plastic pipette tip with scissors to slightly enlarge its opening.

   i. To stabilize the pipette and avoid trembling, apply light pressure with your index finger just above the pipette tip while pipetting with your other hand.

      This operation requires sufficient working distance between the objective and the sample, which is a crucial parameter when selecting a binocular microscope (we find that a working distance of 7 cm is convenient).

   ii. Transfer the positive larva into an empty drop of water prepared on the same or on a second, neighboring Petri dish lid. If negative larvae were caught together with the positive larva, repeat this operation or remove negatives until the fluorescent larva is isolated.

      This operation can also be performed on several positive larvae simultaneously.

      Correctly performing this gesture may require some training, which can be quickly acquired by practicing on control or negative larvae.

      To better visualize negative larvae at the same time as positives, allow sufficient white light to shine through the sample at the same time as the fluorescence excitation light, by appropriate shutter opening and adjustment of white light intensity as permitted by the optical system.

      See Troubleshooting.

5. Transfer the selected larvae to a breeding container and provide them with larval food. Rear larvae normally and use for future transgenic line amplification by mating the obtained adult mosquitoes to wild-type partners.

 Cite this protocol as *Cold Spring Harb Protoc*; doi:10.1101/pdb.prot108306

If larvae are few (<50), they can be kept in a small Petri dish with small amounts of food for the first 24–48 h, before being transferred to a larger breeding container. Avoid overfeeding, as bacterial blooms may asphyxiate the larvae, which is a risk that is elevated when larvae are fewer.

Some mosquito strains grow better when raised at high larval density. If this is the case, it may be preferable to grow positive larvae with a sufficient number of negatives and repeat the screening process at the pupal stage after rinsing pupae under running water in a sieve to eliminate autofluorescent food particles and debris. In this case, plan to hatch and grow wild-type larvae at the same time as the selected G1 transgenics to serve as crossing partners to establish and amplify transgenic lines.

TROUBLESHOOTING

Problem (Step 4): Autofluorescent tissue in larvae confuses the screening.

Solution: Tissue autofluorescence can have diverse origins, as follows: (1) Larvae that are only a few minutes old often display midgut fluorescence. Unless the expected pattern of marker activity is intestinal, ignore it and focus on the tissue of expected expression or resume screening after 1 h. (2) Some but not all *Anopheles* strains display epidermal CFP-like fluorescence on larvae, which is visible as small dots and patches of blue fluorescence, especially on the thorax. Ignore blue fluorescence that is limited to the surface of larvae and focus your search on the internal tissue of expected expression. (3) Likewise, strong blue autofluorescence in the larval Malpighian tubules in the CFP channel is natural and can be ignored. (4) Some young *Aedes* larvae (often 1%–5% of the total) can display GFP-like autofluorescence in the epidermis of abdominal segments. This phenomenon is observed very specifically in the GFP channel and can be mistaken for true marker gene expression. Ignore weak fluorescence that is limited to the epidermis. Experience helps to distinguish these natural patterns from true transgenic expression, which will usually be stronger and better defined. Also ignore fluorescence that can be seen nonspecifically in all channels.

DISCUSSION

For fluorescent markers that become visible only later in development (such as those under control of the *OpIE2* promoter in some lines), these methods can be applied to larger larvae and pupae. In this case, it is important to rinse the animals under running tap water in a strainer before screening to eliminate autofluorescent food particles and debris. Screening pupae rather than late-stage larvae is preferred because the autofluorescent midgut content is expelled at the pupal molt.

ACKNOWLEDGMENTS

In our laboratory, research on mosquitoes is supported by funding from Inserm, CNRS, the University of Strasbourg, the city of Strasbourg, through contrat triennal "Strasbourg capitale européenne" 2018–2020 and Agence Nationale de la Recherche (ANR) through grants ANR-11-EQPX-0022, ANR-19-CE35-0007, and ANR-11-JSV3-0001.

REFERENCES

Benedict MQ. 2015. Methods in *Anopheles* Research—MR4; https://www.beiresources.org/portals/2/MR4/MR4_Publications/Methods%20in%20Anopheles%20Research%202014/2014MethodsinAnophelesResearchManualFullVersionv2tso.pdf [Accessed August 4, 2023]

Carballar-Lejarazú R, Tushar T, Pham TB, James AA. 2021. Microinjection method for *Anopheles gambiae* embryos. *J Vis Exp* doi:10.3791/62591

Fuchs S, Nolan T, Crisanti A. 2013. Mosquito transgenic technologies to reduce *Plasmodium* transmission. *Methods Mol Biol (Clifton NJ)* **923:** 601–622. doi:10.1007/978-1-62703-026-7_41

Pondeville E, Puchot N, Meredith JM, Lynd A, Vernick KD, Lycett GJ, Eggleston P, Bourgouin C. 2014. Efficient ΦC31 integrase-mediated site-specific germline transformation of *Anopheles gambiae*. *Nat Protoc* **9:** 1698–1712. doi:10.1038/nprot.2014.117

Volohonsky G, Terenzi O, Soichot J, Naujoks DA, Nolan T, Windbichler N, Kapps D, Smidler AL, Vittu A, Costa G, et al. 2015. Tools for *Anopheles gambiae* transgenesis. *G3 (Bethesda Md)* **5:** 1151–1163. doi:10.1534/g3.115.016808

Cite this protocol as *Cold Spring Harb Protoc*; doi:10.1101/pdb.prot108306

Protocol 2

Sorting Mosquito Larvae with a COPAS Machine

Eric Marois[1]

CNRS UPR9022, INSERM U1257, Institut de Biologie Moléculaire et Cellulaire, Université de Strasbourg, 67000 Strasbourg, France

Transgenic mosquitoes are widely used in mosquito research. To distinguish transgenic individuals from wild types, genes for fluorescent proteins are the most commonly used genetic markers in transgenic constructs, offering all the advantages of visual selection. Although manual selection under a fluorescence binocular microscope is perfect for the selection of first-generation transgenics, managing established fluorescent lines can be facilitated by complex object parametric analyzer and sorter (COPAS) sorting, which we describe in this protocol. COPAS sorting allows researchers to purify large mosquito larval populations containing only homozygous transgenic individuals, only heterozygotes, or a mix of homozygous, wild types, and heterozygotes in desired proportions. Sorting large populations of a single sex is also possible. Finally, especially when several transgenes of different fluorescence colors are inserted in the same docking site (a recombination site previously inserted in the mosquito genome, which can be used to insert new transgenes into the same locus), they can be maintained together in a single mosquito population to save insectarium space and labor. COPAS sorting can then be used to extract the desired genotype when needed and to readjust transgene frequencies every few generations in case drift is observed.

MATERIALS

It is essential that you consult the appropriate Material Safety Data Sheets and your institution's Environmental Health and Safety Office for proper handling of equipment and hazardous material used in this protocol.

Reagents

Demineralized water

Transgenic and wild-type mosquito larvae in water (freshly hatched, without food; preferably within 24 h of their hatching)

For detailed protocols for mosquito rearing, see Benedict (2015).

For the detailed process of microinjection of Anopheles *eggs and transgenesis, refer to the following references: Fuchs et al. (2013); Pondeville et al. (2014); Volohonsky et al. (2015); Carballar-Lejarazú et al. (2021).*

Equipment

Collection flask (any cup made of glass, plastic, or paper)

Complex object parametric analyzer and sorter (COPAS) Select or COPAS 500 instrument (Union Biometrica)

The COPAS Select (or FP-500) instrument is best-suited to screen and sort neonate mosquito larvae. Later stages of larval growth are unsuitable for larva sorting because their high autofluorescence and the presence of food

[1]Correspondence: e.marois@unistra.fr

Cite this protocol as *Cold Spring Harb Protoc*; doi:10.1101/pdb.prot108307

particles confuse fluorescence analysis. In addition, sorting throughput is orders of magnitude higher when using neonate larvae. Their small size allows sorting rates of 10–20 larvae/sec with high accuracy or up to 50 larvae/sec with some loss of efficiency (good purity but decreased recovery rate). Therefore, 20,000 neonate larvae of a desired genotype can easily be purified in 30 min. Larger larvae cannot be as highly concentrated in the sample water and therefore would be sorted at much lower speeds.

Fluorescence binocular microscope with illumination source
Small Petri dishes (5.5-cm-diameter) or 24- or 96-well plates (see Step 14)
Tissue (i.e., ordinary laboratory wipes)
Wastewater container

METHOD

For more detailed information about how a COPAS instrument works (e.g., how samples pass through the machine and where and how detection takes place), see information provided by manufacturer (Union Biometrica) for your instrument of choice via the user manual or manufacturer website.

1. Ensure all water reservoirs of the COPAS instrument are filled with demineralized water.

   Using demineralized water helps to avoid salty deposits.

   For COPAS instrument use with mosquitoes, "sheath" medium (used for Caenorhabditis elegans*), which is mentioned in the manual, is replaced with water.*

2. Ensure that waste containers are not already full and empty them as needed.
3. Switch on all controls, including COPAS instrument, laser, compressor or compressed air, and computer.
4. Start the COPAS analysis program and click "start."
5. Verify that pressures have reached the optimal values determined during instrument calibration and then check the "pressure OK" button to initiate "prime flow cell."
6. In the meantime, open the sample cup lid and visually verify the absence of larvae in the reservoir that may have been left by a previous user. If the reservoir is not clean, empty the reservoir water into a wastewater container, wipe the cup with a tissue, and add clean water. Also, clean the tip of the waste tube where eggs tend to stick.
7. Press the "clean" button once or twice and reverify the absence of larvae in the sample cup.
8. Activate the laser within the Tools menu. Click "Run external laser," click "Run," and, once the power reaches 10 mW, click "Done."
9. In the meantime, load the mosquito larvae into the sample cup by pouring the larvae-containing water in the cup.

   Minimize the presence of contaminating debris (mosquito body parts, eggs, and eggshells) that may clog the tubing or confuse data analysis.

10. Choose the correct program and the correct dichroic filter (488 nm for GFP and dsRed or 514 nm for YFP and dsRed), as a function of the fluorophores in the larvae. Ensure the "pure" mode is selected in the Configuration > Configure Coincidence menu (if the presence of larvae of other genotypes is acceptable, "enrich" mode can be used).

    New programs optimized for the specific fluorescence of new strains must be defined empirically by analyzing larvae while adjusting photomultiplier (PMT) values, compensation values (in the Configuration menu), and the scale of the graphs (using the scroll menus in the "full scale" window of the control panel). It is helpful to obtain training from an experienced user to learn how to perform these adjustments. See Figure 1.

11. To see the cloud diagrams representing larval populations, double-click the default graphs.

    A new program optimal for the sorted materials can be defined empirically by varying the PMT outputs in the desired fluorescence channels, as well as the compensation parameters (in the configuration menu) and the

 Cite this protocol as *Cold Spring Harb Protoc*; doi:10.1101/pdb.prot108307

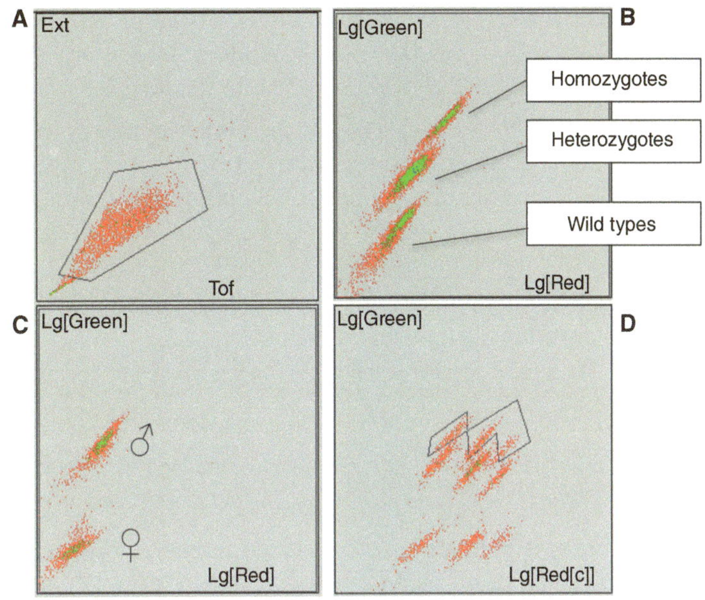

FIGURE 1. Examples of COPAS diagrams from the sorting of mosquito larval populations of various genotypes. (*A,C*) Time of flight (Tof) versus extinction diagram (*A*), reflecting the size and opacity of the analyzed objects, with a gate around ~2000 first-instar *Aedes albopictus* larvae of a sexing strain expressing GFP in males. Only the gated objects appear again in the red autofluorescence versus green fluorescence diagram (*C*), which shows clear separation between GFP-expressing males and nonfluorescent female larvae. (*B*) Fluorescence diagram from an analysis of ~7000 *Anopheles* larvae of a population containing negative, heterozygous, and homozygous individuals of an autosomal transgene, forming three distinct clouds as indicated. (*D*) Fluorescence diagram from the analysis of a compound strain carrying two transgenes, expressing either GFP or DsRed fluorescence. Nine genotypes are resolved, according to each transgene's copy number (0, 1, or 2). The shown gating served to sort homozygous GFP larvae regardless of the DsRed transgene. Green patches in diagrams show the superposition of many objects with the same fluorescence intensity.

scale of the axes (adjust log values to bring the relevant clouds of detected larvae toward the center of the graphs).

12. Click "acquire" to analyze a subset of larvae and to obtain the profile of your population on the diagrams. Collect the flowthrough larvae in a clean flask to be able to run them again (some COPAS machines possess a flowthrough collection vessel). Adjust the number of events per second to be 15–20. Stop this pilot test when enough data have been collected for the population profile to become evident enough (i.e., well-defined clouds of larvae on the diagram).
13. Click "regions" on the menu bar, "define gate region," and "define sort region" to gate the larvae you desire to collect in the time of flight/extinction diagram as well as in the fluorescence diagram.
14. Ensure the number of larvae per well in the upper right field of the dialog window is set to the desired number of larvae to collect (e.g., 4000).

    Dispensing larvae in a 24- or 96-well plate is also possible. In the Plate Formats menu (calibration may be necessary), indicate the desired number of larvae per well as above.

15. Click "load plate."

    This step is essential so that the program takes the new gating and sorting parameters into account.

16. Orient the waste tube to the waste bottle or to a collection flask if you want to reuse the flowthrough larvae.
17. Place a small Petri dish below the flow cell and click "fill plate." While the plate is filling, make sure the number of events per second is not too high (e.g., 20/sec); if necessary, dilute your sample or manually reduce sample pressure. If it is too low, concentrate your larvae or increase sample pressure but not much higher than the sheath pressure (larvae could backflow and contaminate the sheath bottle).

18. During sorting, verify visually that one larva is present in each falling drop. If not, stop and clean the sorter. In particular, wipe the white collection tray and surrounding structure with a tissue, and insert a piece of twisted tissue into the flow cell outlet hole.
19. (Optional) In case of clogging or aberrant dripping, stop the program, use the metal stylet to clean the flow cell carefully, press "clean" once or twice, wipe flow cell elements, and turn on sheath again by clicking the corresponding button. If no more dripping occurs, resume sorting.

    Dripping may result in the contamination of the collected larvae with ones of undesired genotypes. If such an accident occurs, resort the already collected larvae.
20. When the last milliliters of the sample have been reached, tilt the reservoir to gather the last water without allowing bubbles into the tube.

    Do not damage the tubing by pulling too hard on the sample reservoir.
21. Click "stop" before bubbles have a chance to alter the quality of the diagram.
22. To save the data with a possibility to analyze the data in FACS software, click "store."

    "Save as" only saves program parameters, and "save screen image" provides a screenshot.
23. Before switching off the COPAS instrument, empty the reservoir and wipe it well with a tissue. Add clean water. Tick the "sheath" and then "sample" buttons to pass clean water into the tubing. Click the "clean" button twice.
24. Untick the "sample" button, open the reservoir, and visually check for the absence of larvae. If some larvae are present, empty the liquid, wipe the reservoir, add new water, and repeat. Verify again the absence of larvae. If the laser is still on (it may be switched off earlier to start cooldown), click "erase" and "acquire" to see if any object appears on the diagram. Repeat this cleaning until no object is detected.
25. Wipe the white collection tray and surrounding parts with a tissue and insert twisted tissue into the flow cell outlet. Run water again (both sheath and sample) for ~1 min.
26. Unless already done, stop "external laser" in the "tools" menu. Allow it to cool for 5 min before switching off the laser block itself. Exit the program and switch off all components.
27. Visually verify sorted larvae under a binocular microscope.

    This step confirms that no accidental dispensing of wild-type larvae among the transgenics occurred during the sorting process. It may happen if clogging or pressure variations caused a disturbance in the waste stream within the flow cell.

ACKNOWLEDGMENTS

In our laboratory, research on mosquitoes is supported by funding from Inserm, CNRS, the University of Strasbourg, the city of Strasbourg, through contrat triennal "Strasbourg capitale européenne" 2018–2020 and Agence Nationale de la Recherche (ANR) through grants ANR-11-EQPX-0022, ANR-19-CE35-0007, and ANR-11-JSV3-0001.

REFERENCES

Benedict MQ. 2015. Methods in *Anopheles* Research—MR4; https://www.beiresources.org/portals/2/MR4/MR4_Publications/Methods%20in%20Anopheles%20Research%202014/2014MethodsinAnophelesResearchManualFullVersionv2tso.pdf [Accessed August 4, 2023]

Carballar-Lejarazú R, Tushar T, Pham TB, James AA. 2021. Microinjection method for *Anopheles gambiae* embryos. *J Vis Exp* 173. doi:10.3791/62591

Fuchs S, Nolan T, Crisanti A. 2013. Mosquito transgenic technologies to reduce *Plasmodium* transmission. *Methods Mol Biol* **923:** 601–622. doi:10.1007/978-1-62703-026-7_41

Pondeville E, Puchot N, Meredith JM, Lynd A, Vernick KD, Lycett GJ, Eggleston P, Bourgouin C. 2014. Efficient ΦC31 integrase-mediated site-specific germline transformation of *Anopheles gambiae*. *Nat Protoc* **9:** 1698–1712. doi:10.1038/nprot.2014.117

Volohonsky G, Terenzi O, Soichot J, Naujoks DA, Nolan T, Windbichler N, Kapps D, Smidler AL, Vittu A, Costa G, et al. 2015. Tools for *Anopheles gambiae* transgenesis. *G3 (Bethesda Md)* **5:** 1151–1163. doi:10.1534/g3.115.016808

Protocol 3

Selecting Transgenic Mosquito Larvae with Puromycin

Eric Marois[1]

CNRS UPR9022, INSERM U1257, Institut de Biologie Moléculaire et Cellulaire, Université de Strasbourg, 67000 Strasbourg, France

The selection of mosquito transgenic larvae using a nonfluorescent approach can be advantageous to reserve fluorophores for downstream applications, such as immunostaining or for the study of promoter activity by cloning a fluorescence reporter gene under the control of that promoter. We previously reported that puromycin selection is efficient in transgenic *Anopheles gambiae* or *Anopheles coluzzii* larvae expressing an *OpIE2-pac* selection marker. A concentration of puromycin of >10 µg/mL is lethal for *Anopheles* larvae, unless they carry the resistance gene, conferring them resistance to puromycin concentrations of 25–80 µg/mL. A drawback of this fully dominant selection marker is that, unlike with fluorescence markers, homozygous transgenics cannot be distinguished from heterozygotes. Here, we outline the procedure for selecting puromycin-resistant transgenic *Anopheles* larvae.

MATERIALS

It is essential that you consult the appropriate Material Safety Data Sheets and your institution's Environmental Health and Safety Office for proper handling of equipment and hazardous material used in this protocol.

Reagents

Mosquito larval food (e.g., pulverized TetraMin goldfish flakes)

Puromycin stock solution (Sigma-Aldrich P7255)

Make a 25 mg/mL solution of puromycin dihydrochloride in water, aliquot in tubes of 200 µL, and store at −20°C. A loss of larvicide activity can be observed after several freeze–thaw cycles.

Transgenic and wild-type *Anopheles* larvae (freshly hatched, unfed neonates)

For detailed protocols for mosquito rearing, see Benedict (2015).

For the detailed process of microinjection of Anopheles *eggs and transgenesis, refer to the following references: Fuchs et al. (2013); Pondeville et al. (2014); Volohonsky et al. (2015); Carballar-Lejarazú et al. (2021).*

Equipment

Sieves (40-µm cell strainers; e.g., Falcon 352340)

Small Petri dishes (5.5-cm-diameter)

Standard laboratory 20-µL, 200-µL, and 1-mL pipettes (e.g., Gilson P20, P200, and P1000)

[1]Correspondence: e.marois@unistra.fr

Cite this protocol as *Cold Spring Harb Protoc*; doi:10.1101/pdb.prot108308

METHOD

Rear and maintain mosquitoes under standard laboratory conditions.

1. Collect up to 1000 freshly hatched, neonate mosquito larvae (using a sieve if necessary) into 8 mL of water in a Petri dish.

   Collect larvae by direct pipetting or after straining them from the water of the oviposition cup, measuring the volume with a P1000 pipette.

   It is important to collect larvae early after hatching and before they receive any food. The development of gut microbiota will otherwise increase background resistance to puromycin.

2. Supplement the water with 8–12 µL of puromycin stock solution and mix well, using gentle agitation with the tip of the pipette.
3. Feed the larvae every day with a very small amount of mosquito larval food, if necessary (i.e., if older food has been consumed; overfeeding should be avoided as many larvae are arresting development because of the puromycin).
4. Observe larvae daily for mortality.

   On day 3, mortality of nontransgenic larvae is expected, but larvae should still be maintained in the same solution, as relieving drug pressure at this time may result in the survival of nontransgenic escaper individuals.

5. On day 4, collect puromycin-resistant larvae using a P200 pipette and transfer them to regular water, with ordinary culture conditions.

   At this point, resistant transgenic larvae that have progressed in their growth can be clearly distinguished from dead or dying wild types, in which growth was arrested (Fig. 1). They will proceed with normal development. Resulting adults should be crossed with wild-type partners for line amplification. Puromycin treatment in the next generations should be repeated to enrich for transgenics. Note that it is not possible to purify homozygotes, except by polymerase chain reaction (PCR) genotyping or if the transgene can be genetically balanced by crossing to another transgenic line carrying a fluorescent transgene at the same locus. In the latter case, F1 larvae from this cross should be selected to carry both the fluorescence and puromycin resistance, and then intercrossed; fluorescence can then be counterselected in the F2 generation. This procedure yields a homozygous puromycin-resistant line. If using a commonly used docking site for transgenesis, fluorescent lines that can serve as genetic balancers will be readily available.

   See Troubleshooting.

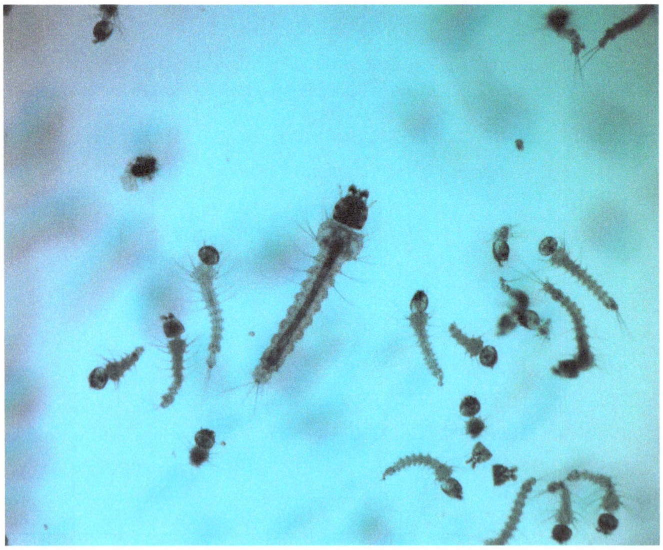

FIGURE 1. A surviving puromycin-resistant transgenic *Anopheles* larva on treatment day 4 is in the center of the image and surrounded by dying or dead nontransgenic siblings.

Cite this protocol as *Cold Spring Harb Protoc*; doi:10.1101/pdb.prot108308

TROUBLESHOOTING

Problem (Step 5): Too many larvae survived puromycin treatment compared to the expected frequency of transgenesis success.

Solution: The puromycin solution has probably lost activity because of too many freeze–thaw cycles. Avoid defrosting the solution more than three times. Repeat the screening on new larvae using a fresh 25 mg/mL stock solution. If some larvae from the failed treatment are much larger than the others, they may be bona fide transgenics; therefore, outcross them to wild types and rescreen their progeny (half will be puromycin-resistant if the parent carried the transgene).

RELATED INFORMATION

For more details about the advantages and limitations of puromycin selection and reasons for its use, see Introduction: Techniques for Identifying and Sorting Transgenic Mosquito Larvae (Marois 2023). For information about puromycin selection in transgenic *Anopheles gambiae* or *Anopheles coluzzii* larvae expressing an *OpIE2-pac* selection marker, see Volohonsky et al. (2015).

ACKNOWLEDGMENTS

In our laboratory, research on mosquitoes is supported by funding from Inserm, CNRS, the University of Strasbourg, the city of Strasbourg, through contrat triennal "Strasbourg capitale européenne" 2019–2021 and Agence Nationale de la Recherche (ANR) through grants ANR-11-EQPX-0022, ANR-19-CE35-0007, and ANR-11-JSV3-0001.

REFERENCES

Benedict MQ. 2015. Methods in *Anopheles* Research—MR4; https://www.beiresources.org/portals/2/MR4/MR4_Publications/Methods%20in%20Anopheles%20Research%202014/2014MethodsinAnophelesResearchManualFullVersionv2tso.pdf [Accessed August 4, 2023]

Carballar-Lejarazú R, Tushar T, Pham TB, James AA. 2021. Microinjection method for *Anopheles gambiae* embryos. *J Vis Exp* 173. doi:10.3791/62591

Fuchs S, Nolan T, Crisanti A. 2013. Mosquito transgenic technologies to reduce *Plasmodium* transmission. *Methods Mol Biol* **923:** 601–622. doi:10.1007/978-1-62703-026-7_41

Marois E. 2023. Techniques for identifying and sorting transgenic mosquito larvae. *Cold Spring Harb Protoc* doi:10.1101/pdb.top107694

Pondeville E, Puchot N, Meredith JM, Lynd A, Vernick KD, Lycett GJ, Eggleston P, Bourgouin C. 2014. Efficient ΦC31 integrase-mediated site-specific germline transformation of *Anopheles gambiae. Nat Protoc* **9:** 1698–1712. doi:10.1038/nprot.2014.117

Volohonsky G, Terenzi O, Soichot J, Naujoks DA, Nolan T, Windbichler N, Kapps D, Smidler AL, Vittu A, Costa G, et al. 2015. Tools for *Anopheles gambiae* Transgenesis. *G3 (Bethesda Md)* **5:** 1151–1163. doi:10.1534/g3.115.016808

CHAPTER 39

The Golgi Method

Rickard Ignell[1] and Sharon Rose Hill

Disease Vector Group, Unit of Chemical Ecology, Department of Plant Protection Biology, Swedish University of Agricultural Sciences, 23422 Alnarp, Sweden

Silver staining by the Golgi method is a classical procedure to identify the three-dimensional morphology of individual neurons. Although the method was developed in the 1870s, it is still used to study the morphological characteristics of neuron types in the central nervous system, either alone or in combination with other neuroanatomical techniques. The neuropil of insects is fully accessible to the original refractive staining procedure, and modifications of the Golgi technique have paved the path for modern structural research on the insect central nervous system. Here, we provide an introduction to this easy and low-cost method.

BACKGROUND

In 1873, Camillo Golgi developed a staining method, originally called "*la reazione nero*" or the black reaction, which uses silver nitrate impregnation of fixed brain tissue to study the morphological characteristics of neuron types in the central nervous system. Kenyon (1896) was the first to use (a modification of) Golgi's method to stain neurons of the brain of an insect—a bee. Together with contemporary contributions by others (e.g., Cajal and Sánchez 1915), these accounts constitute the foundation of the modern structural research on the insect central nervous system.

THE GOLGI METHOD

Silver staining of neurons using the Golgi method is attained by impregnating formaldehyde-, glutaraldehyde- and/or osmium tetroxide–fixed nervous tissue with potassium dichromate and silver nitrate. This results in microcrystallization of silver chromate in the neurons, including the soma, axon, dendrites, and dendritic spines. The seemingly stochastic selectivity of the method has received critique but has also stimulated the development of modified procedures from Golgi's original method, which should be considered when some neuron types are seemingly refractory to staining (Strausfeld 1980). There are three interdependent stages in the procedure that can be modified: (1) fixation, (2) chromation, and (3) metal impregnation.

The role of tissue fixation is to maintain the natural, living morphology of the structures under investigation. Fixation with either a monoaldehyde or a dialdehyde, often formaldehyde and glutaraldehyde, respectively, ensures preservation of fine structures. Monoaldehydes alone fail to promote metal impregnation; however, this fixative can be used to control how many neurons are labeled. It is

[1]Correspondence: rickard.ignell@slu.se

Cite this introduction as *Cold Spring Harb Protoc*; doi:10.1101/pdb.top107695

common practice to use a combination of formaldehyde and glutaraldehyde and/or osmium tetroxide prior to chromation, with osmium tetroxide often used in conjunction with chromation in a second cycle.

Chromation prepares the tissues to receive and react to metal staining. Dichromate salts in solution, either alone or in combination with glutaraldehyde or osmium tetroxide, are used for chromation. In this step, pH is an important factor to consider. For example, solutions of dichromate with osmium tetroxide at a low pH level yields fewer neurons stained compared to those at a more alkaline pH. The inclusion of osmium tetroxide is required for scanning electron microscopy visualization. To achieve optimal impregnation with silver nitrate, buffered chromating solutions are recommended. Chromation is undoubtedly the stage in the process in which selective staining of neurons can be best controlled, by using different chromation solutions at different pH levels.

Metal impregnation (i.e., the treatment of chromated tissue with silver nitrate) initiates the reaction to produce visualizable microcrystals and thus stain the neural tissue. The manipulation of pH, time, and concentration of silver nitrate can be customized to achieve a variety of results. When working with small tissues (e.g., insect brains), impregnation occurs within 1–2 h.

MODERN USES OF GOLGI SILVER STAINING IN INSECTS

The Golgi method can be used to stain neurons within the insect central nervous system. Our associated protocol describes this in detail (see Protocol 1: Golgi Silver Staining of Mosquito Central Neuropils [Ignell and Hill 2022]). Although neurons are seemingly stained stochastically, a more nonrandom selection of neuronal staining can be achieved by guiding the local influx of silver ions to select populations of neurons. This can be achieved, for example, by opening up small windows in the cuticle close to the neuropil of interest, such as the primary olfactory center, the antennal lobe, the calyces of the mushroom bodies, and the optic lobes. By doing this, the chance of staining neurons within these neuropils is greatly enhanced.

Modern neuroanatomical techniques are now more commonly used for research on insect nervous tissue (e.g., see Daul et al. 2010; Gonzalez-Bellido and Wardill 2012). However, Golgi silver staining is still used in many cases in which, for example, antibodies against proteins are not available or in cases in which these proteins are degraded. Moreover, the Golgi method may be used to complement immunostaining techniques, guide the analysis, and provide complementary information about the neuronal structure in both situational and fine structural detail.

REFERENCES

Cajal S, Sánchez D. 1915. Contribución al conocimiento de los centros nerviosos de los insectos. *Trabajos de Laboratorio de Investigaciones Biologicas* **13:** 1–68.

Daul AL, Komori H, Lee CY. 2010. Immunofluorescent staining of *Drosophila* larval brain tissue. *Cold Spring Harb Protoc* doi:10.1101/pdb.prot5460

Gonzalez-Bellido PT, Wardill TJ. 2012. Labeling and confocal imaging of neurons in thick invertebrate tissue samples. *Cold Spring Harb Protoc* doi:10.1101/pdb.prot069625

Ignell R, Hill SR. 2022. Golgi silver staining of mosquito central neuropils. *Cold Spring Harb Protoc* doi:10.1101/pdb.prot107840

Kenyon FC. 1896. The brain of the bee. A preliminary contribution to the morphology of the nervous system of the arthropoda. *J Comp Neurol* **6:** 133–210. doi:10.1002/cne.910060302

Strausfeld NJ. 1980. The Golgi method: its application to the insect nervous system and the phenomenon of stochastic impregnation. In *Neuroanatomical techniques: insect nervous system* (ed. Strausfeld NJ, Miller TA), pp. 131–203. Springer-Verlag, New York.

Cite this introduction as *Cold Spring Harb Protoc*; doi:10.1101/pdb.top107695

Protocol 1

Golgi Silver Staining of Mosquito Central Neuropils

Rickard Ignell[1] and Sharon Rose Hill

Disease Vector Group, Unit of Chemical Ecology, Department of Plant Protection Biology, Swedish University of Agricultural Sciences, 23422 Alnarp, Sweden

The Golgi silver staining procedure relies on three interdependent stages: fixation, chromation, and metal impregnation. Each of these stages can be modified. This protocol describes a method demonstrated to stain neurons within the mosquito central nervous system. The resulting preparations are stable at room temperature.

MATERIALS

It is essential that you consult the appropriate Material Safety Data Sheets and your institution's Environmental Health and Safety Office for proper handling of equipment and hazardous materials used in this protocol.

RECIPES: Please see the end of this protocol for recipes indicated by <R>. Additional recipes can be found online at http://cshprotocols.cshlp.org/site/recipes.

Reagents

CO_2 (optional; see Step 1)

Distilled H_2O

Durcupan ACM epoxy resin <R>

Ethanol (50%, 75%, 95%, and 100%)

Mosquitoes

For rearing common laboratory mosquitoes, see Gerberg (1970), and for rearing malaria mosquitoes, see Moll et al. (2013).

The age (days post-emergence), sex, and physiological state (e.g., starved, sugar-fed, blood-fed, time following a blood meal, and parity) of the mosquito, as well as the time of day, can potentially affect the neural circuitry. Choose these parameters carefully to match the hypothesis tested and be consistent in the implementation.

Potassium dichromate (2.5%)

Use the highest possible grade (>99%).

Potassium dichromate with glutaraldehyde and sucrose <R>

Potassium dichromate with osmium tetroxide <R>

Potassium dichromate with sucrose <R>

Propylene oxide

Silver nitrate (0.75%)

Use the highest possible grade (>99%).

[1]Correspondence: rickard.ignell@slu.se

Cite this protocol as *Cold Spring Harb Protoc*; doi:10.1101/pdb.prot107840

Equipment

BEEM capsules (e.g., BEEM Embedding Capsules for Electron Microscopy [VWR 102096])
Container for anesthetizing mosquitoes (e.g., *Drosophila* rearing vials)
Disposable plastic transfer pipettes (e.g., 4.6-mL thin-stem transfer pipettes with fine tips [VWR 16001-178])
Dissection dish (e.g., a glass or plastic Petri dish; 5-cm-diameter)
Fine forceps, straight with sharp edges (Dumont #5 preferred)
Hot plate
Incubator or similar dry heat source that can maintain a temperature of 60°C
Microcentrifuge tubes (1.5-mL)
Micro-scissors or razor blade (optional; see Step 3)
Microscope coverslips
Microscope slides
Rotary mixer with suitable carousel to fit the chosen container (e.g., Clifton RM2 rotary mixer with carousel [Fisher Scientific 117-965-87 and 123-17-498])
Sledge microtome (a modern sliding microtome, e.g., Leica SM2010R, or comparable older versions of sledge microtomes)
Standard aspirator (John W. Hock PN 419)
Syringe
Toothpicks
Upright microscope equipped with objectives of variable magnification

These microscopes can be equipped with cameras for direct documentation or camera lucida systems, to permit manual documentation.

Wide-mouthed container, small (similar to the BEEM capsules)

METHOD

The following method is based on the routine Golgi Colonnier method (Strausfeld 1980) and has been optimized to stain neurons within the neuropil of mosquitoes (Ignell et al. 2005). Other Golgi silver staining procedures were evaluated by Ignell et al. (2005), but yielded no reliable staining of neurons.

When you dissect under potassium dichromate solution, place the dissection microscope in a fume hood, maintaining a safe distance for viewing from outside the fume hood. Avoid creating a backwash of air from the fume hood by cluttering the opening. Minimize the time spent dissecting, as much as possible. Never dissect under osmium.

Dissection

1. Transfer mosquitoes by aspiration into suitable containers (e.g., *Drosophila* vials). Anesthetize mosquitoes on ice (30 sec–2 min) or by using CO_2 (1–2 min). To remove cuticular hydrocarbons, use fine forceps to briefly dip them in a dissection dish filled with 70% ethanol at room temperature.

Chromation—Part 1

2. Using fine forceps, immerse the mosquitoes in cold (4°C) potassium dichromate with sucrose in a dissection dish.
3. Remove the head appendages, including antennae and mouthparts, using a razor blade or micro-scissors, or create openings in the head capsule with fine forceps, to allow local influx of Ag^+ in chromated tissue.

   By creating these openings near the neuropils of interest, the likelihood of obtaining penetration of the Ag^+ into select neuropil and individual neurons will increase.

Cite this protocol as *Cold Spring Harb Protoc*; doi:10.1101/pdb.prot107840

Fixation

4. Immerse the opened heads in potassium dichromate with glutaraldehyde and sucrose in a microcentrifuge tube for 4–5 d at room temperature or for 7 d at 4°C (until the solution turns chestnut brown; this will take longer at a lower temperature). Keep the opened heads in these tubes until Step 12.

   Do not put more than 10 heads in a single tube. Use a volume of 1 mL of solution per tube (at this step and in subsequent steps).

   Specimens will be brittle following this step.

Chromation—Part 2

5. Rinse the specimens in 2.5% potassium dichromate several times (e.g., for eight washes of 15 min each) at room temperature. Remove and replace the solution using disposable plastic transfer pipettes, which will allow easy removal of the solution without removing the heads.
6. Remove the final solution from Step 5 using a disposable plastic transfer pipette. Incubate specimens in potassium dichromate with osmium tetroxide for 4 d in the dark at 4°C.
7. Briefly (<10 sec) rinse specimens in distilled H_2O at room temperature.

Metal Impregnation

8. Rinse the specimens in a series of 0.75% silver nitrate solutions at room temperature until no red precipitation appears.

   The red precipitate will form rapidly, and the silver nitrate solution may be exchanged as quickly as every min, depending on the number of specimens processed at one time.

9. Incubate the specimens in 0.75% silver nitrate solution in the dark for 2–3 d at 4°C.

Mounting

Use the fume hood when mixing embedding material.

10. Wash the specimens in distilled H_2O for 10 min at room temperature. Dehydrate the specimens serially in ethanol (once in 50% ethanol, once in 70% ethanol, once in 95% ethanol, and twice in 100% ethanol) for 10 min per wash at room temperature. Transfer the specimens to propylene oxide for two washes of 10 min each at room temperature.
11. Exchange the solution from Step 10 with 1:1 propylene oxide:Durcupan and leave in open microcentrifuge tubes on a cantilevered rotator overnight in a fume hood at room temperature. To assure that the liquid contents remain in the microcentrifuge tubes during rotation, set the rotator to a suitable angle.

    This will allow the propylene oxide to evaporate, slowly increasing the concentration of Durcupan.

12. Using toothpicks, transfer the specimens to fresh Durcupan into a small wide-mouthed container, which will improve access to the tissue in future steps.

    Alternatively, exchange the solution from Step 11 with fresh Durcupan directly in the original microcentrifuge tubes.

13. Incubate specimens in fresh Durcupan for 6 h at room temperature.
14. Remove the cap from the cast of a BEEM capsule. Place the cap on a clean surface, fill it with Durcupan, and transfer either a single specimen, or several, into the cap of the BEEM capsule. Orient the specimen(s) to the center of the cap using toothpicks and ensure that they are mounted to provide consistent coronal, transverse, or sagittal sections.
15. Fill the BEEM capsule with Durcupan, and place the inverted capsule over the cap.
16. Place the filled BEEM capsules in a plastic weighing boat or other appropriate dish, and cure the embedded specimens overnight in an incubator at 60°C.

17. After polymerization, allow the block to cool slowly to room temperature.
18. Strip off the BEEM capsule from the cured block and mount the block in a suitable holder. Trim as much plastic as possible from around the specimen, so that at the top it has a trapezoid shape.
19. Clamp the holder to the platform of a slide microtome. Orient the holder to allow consistent coronal, transverse, or sagittal sectioning of the samples.
20. Using a C knife, cut sections of between 15 and 40 µm.
21. With a hypodermic syringe, apply drops of water on a clean microscope slide. Place the drops close together in rows. Place a specimen section on each drop.
22. Warm the slide to dry the sections to the glass.

    The sections can be oriented early during this procedure.
23. Leave the slide on the hot plate for 5–10 min at 40°C.
24. Apply fresh Durcupan at the center line of the microscope slide. Gently apply a coverslip edge first and then gently lower it down onto the slide, for a permanent preparation, and cure overnight at 60°C.

    These preparations are permanent and will last for an extended period of time. They should be stored in the dark.

RECIPES

Durcupan ACM Epoxy Resin

Component:	
A (M epoxy resin; CY 212)	21.6 g
B (Hardener; NY 964)	17.8 g
C (Accelerator 960; DY 060)	1.0 g
D (DBP)	1.9 g

Combine the components in the order given using a disposable plastic container with a tightly fitting lid. By shaking the container, mix vigorously until the mixture is foamy. Next, fill 20–50-mL plastic disposable syringes and seal the syringes using Parafilm. Store the syringes at −20°C. When thawing for use, do not open the syringe until it has reached room temperature to avoid condensation on the cold plastic, which may result in fractures during sectioning.

Potassium Dichromate with Glutaraldehyde and Sucrose

1. Prepare a 2.5% solution of potassium dichromate using the highest possible grade (>99%).
2. Mix the 2.5% potassium dichromate solution with 25% glutaraldehyde (Grade I, 25% in H_2O, specially purified for use as an electron microscopy fixative) at a ratio of 5:1 (2.5% potassium dichromate:25% glutaraldehyde).
3. Add 1.3 g of sucrose/100 mL of the potassium dichromate-glutaraldehyde solution.

Store for up to 6 mo at 4°C.

Cite this protocol as *Cold Spring Harb Protoc*; doi:10.1101/pdb.prot107840

Potassium Dichromate with Osmium Tetroxide

1. Prepare a 2.5% solution of potassium dichromate using the highest possible grade (>99%).
2. Mix the 2.5% potassium dichromate solution with 1% osmium tetroxide (suitable for electron microscopy) at a ratio of 99:1 (2.5% potassium dichromate:1% osmium tetroxide).

   Never open osmium tetroxide (or an osmium admixer) outside of the fume hood.

Store for up to 12 mo at 4°C.

Potassium Dichromate with Sucrose

1. Prepare 100 mL of a 2.5% solution of potassium dichromate using the highest possible grade (>99%).
2. Add 3 g of sucrose/100 mL of the 2.5% potassium dichromate solution.

Store for up to 6 mo at 4°C.

REFERENCES

Gerberg EJ. 1970. Manual for mosquito rearing and experimental techniques. *Am Mosq Control Assoc Bull* **5:** 72–73.

Ignell R, Dekker T, Ghaninia M, Hansson BS. 2005. Neuronal architecture of the mosquito deutocerebrum. *J Comp Neurol* **493:** 207–240. doi:10.1002/cne.20800

Moll K, Ljungström I, Perlmann H, Scherf A, Wahlgren M, eds. 2013. *Methods in malaria research*, 6th ed. EVIMalaR, Glasgow. https://www.beiresources.org/portals/2/MR4/Methods_In_Malaria_Research-6th_edition.pdf (accessed on 2021-12-06)

Strausfeld NJ. 1980. The Golgi method: its application to the insect nervous system and the phenomenon of stochastic impregnation. In *Neuroanatomical techniques: insect nervous system* (ed. Strausfeld NJ, Miller TA), pp. 131–203. Springer-Verlag, New York.

CHAPTER 40

Introduction to Techniques Used to Study Mosquito Neuroanatomy and Neural Circuitry

Florence V. Guerina,[1,2] Ameya P. Patkar,[1] and Meg A. Younger[1,3,4,5,6]

[1]Department of Biology, [2]Program in Molecular Biology, Cell Biology, and Biochemistry, [3]Center for Systems Neuroscience, [4]Department of Biomedical Engineering, [5]Center for Neurophotonics, Boston University, Boston, Massachusetts 02215, USA

Mosquitoes transmit deadly pathogens from person to person as they obtain the blood meal that is essential for their life cycle. Female mosquitoes of many species are unable to reproduce without consuming protein that they obtain from blood. This developmental stage makes them highly efficient disease vectors of deadly pathogens. They can transmit pathogens between members of the same species and different species that can provide a route for evolving zoonotic viruses to jump from animals to humans. One possible way to develop novel strategies to combat pathogen transmission by mosquitoes is to study the sensory systems that drive mosquito reproductive behaviors, in particular the neural architecture and circuits of mosquito sensory afferent neurons, the central circuits that process sensory information, and the downstream circuits that drive reproductive behaviors. The study of mosquito neuroanatomy and circuitry also benefits basic neuroscience, allowing for comparative neuroanatomy in insect species, which has great value in the current model species-heavy landscape of neuroscience. Here, we introduce two important techniques that are used to study neuroanatomy and neural circuitry—namely, immunofluorescent labeling and neural tracing. We describe how to apply these approaches to study mosquito neuroanatomy and describe considerations for researchers using the techniques.

BACKGROUND AND SIGNIFICANCE

Mosquito-borne diseases are among the leading global threats to human health, infecting hundreds of millions and killing hundreds of thousands of people worldwide every year (CDC 2020). Multiple species of mosquitoes belonging to the *Aedes*, *Culex*, and *Anopheles* genera serve as vectors for the viruses and parasites that are responsible for malaria, Zika virus infection, dengue fever, yellow fever, West Nile fever, chikungunya fever, and many other diseases (CDC 2020). Pathogens are transmitted from mosquitoes to humans during blood feeding. This behavior is performed only by females of many species, which require blood to obtain the protein needed for egg development (Duvall et al. 2019). Once eggs mature, a female mosquito will search out an appropriate oviposition site to lay her eggs, often in proximity to water (Day 2016), after which point, she will again search for a blood meal to mature another clutch of eggs.

These essential mosquito reproductive behaviors, and others such as mating, rely on sensory cues, including chemosensory, mechanosensory, thermal, and visual cues (McBride 2016; Raji and

[6]Correspondence: myounger@bu.edu

Cite this introduction as *Cold Spring Harb Protoc*; doi:10.1101/pdb.top108305

DeGennaro 2017). Sensory signals are detected by receptor-bearing sensory neurons. They include the olfactory sensory neurons that are housed in the sensilla (sensory hairs) that cover external organs such as the maxillary palps and antennae (Mclver 1982). Axonal projections of these sensory neurons innervate dedicated brain regions associated with each stimulus, such as the olfactory glomeruli located in the antennal lobe, from which the information is sent to higher brain regions for processing (Ignell et al. 2005; Herre et al. 2022). Untangling the complex pathways and neural networks involved in transmitting and processing sensory information is a pivotal part of understanding how mosquitoes receive and interpret cues collectively to evoke specific behaviors, including biting behavior. This information could be used to develop evidence-based strategies to interfere with mosquito attraction to humans, such as novel repellant compounds. Understanding these processes necessitates techniques to examine the neuroanatomy of the mosquito central nervous system (CNS) and the sensory inputs from which it receives information.

THE MOSQUITO CENTRAL NERVOUS SYSTEM

The nervous system of the mosquito comprises the brain and the ventral nerve cord (VNC), which are located in the head capsule and thorax, respectively (Fig. 1). The insect brain contains regions with localized function (Ito et al. 2014; Court et al. 2020), the size and shape of which vary between species in a manner that is believed to reflect the sensory and behavioral repertoires of each insect (Carle 2021). These regions contain areas of synaptic neuropil, which are dense with synaptic connections. Techniques to visualize the neuropil, in particular the synaptic sites, have played a major role in defining regions of the insect CNS. Unlike much of the mammalian brain, the cell bodies of insect neurons do not reside in the vicinity of the neuropil, but rather the cell bodies line the outside of the brain to form a region called the cell body rind. The VNC can be thought of as loosely analogous to the vertebrate spinal cord, in that it is primarily responsible for innervation of the thoracic segments. The VNC consists of six neuromeres, which innervate and receive sensory input from each leg, and contains ascending and descending tracts that project to the brain and abdominal segments, respectively (Court et al. 2020).

Whole-brain and whole-VNC immunofluorescent labeling are essential techniques for examining the neuroanatomy of the CNS. Immunofluorescent labeling allows proteins with little or no optical contrast to be visualized using fluorescence microscopy. This technique relies on the use of a primary antibody to recognize a specific protein or region of a protein (epitope) and a secondary antibody, which is conjugated to a fluorophore, to recognize the primary antibody. This two-component system allows for amplification of the fluorescent signal because many secondary antibodies will bind to a single primary antibody. Indirect labeling with a secondary antibody also creates flexibility in the detection approach the researcher can use. A single primary antibody can be used with secondary antibodies that can be conjugated to a variety of fluorophores with different spectral properties or to

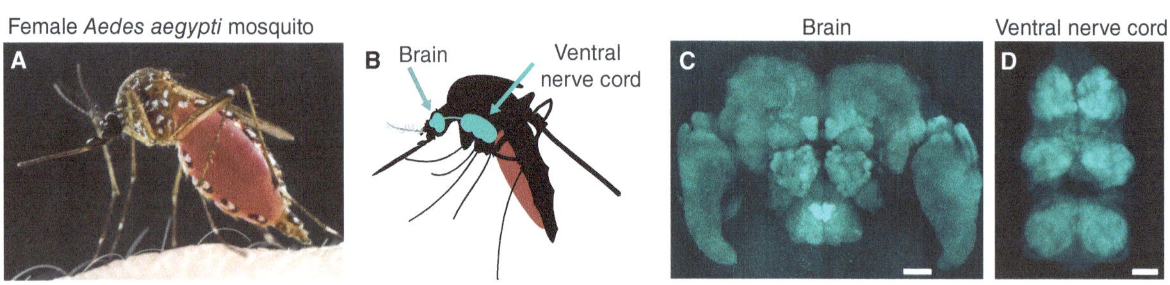

FIGURE 1. (*A*) A female *Aedes aegypti* mosquito that has consumed a blood meal. (*B*) Cartoon representation of a mosquito with the brain and ventral nerve cord (VNC) in cyan. (*C,D*) A whole mosquito brain (*C*) and VNC (*D*) labeled with the NC82 antibody to define the structure of the brain. Scale bars, 50 µm.

Cite this introduction as *Cold Spring Harb Protoc*; doi:10.1101/pdb.top108305

other types of labels, including enzyme labels (such as horseradish peroxidase). The selection of a secondary antibody should take into account the microscope that will be used to image the tissue.

A second useful technique for examining neuroanatomy and neural projections is neural tracing. This method involves the transport of a substance throughout the neuron in either the anterograde or retrograde direction or bidirectionally (Lanciego and Wouterlood 2020). Neurons can be loaded with fluorescent dyes that are transported to fill in the entire neural cytosol, rendering visible all of the processes of a neuron. This so-called dye-filling approach can be used to visualize the morphology of single neurons and groups of neurons, including the projection patterns from peripheral sense organs into the brain and VNC. Fluorescent dye filling can be coupled with immunofluorescent labeling to simultaneously visualize the structure of a neuron or neurons and the location of proteins of interest.

Our associated protocols (see below) describe in detail how to conduct both neural immunofluorescent labeling and neural tracing in mosquitoes.

IMMUNOSTAINING METHOD

One antibody that is commonly used to label the synaptic neuropil, to visualize regions of the brain, and to examine insect neuroanatomy is NC82. This antibody from the Developmental Studies Hybridoma Bank (DSHB) was generated against the *Drosophila melanogaster* Bruchpilot (Brp) protein. Brp is a component of the presynaptic active zone (Wagh et al. 2006). The NC82 antibody can be used to immunolabel *Aedes aegypti* (Matthews et al. 2019), *Anopheles gambiae* (Riabinina et al. 2016), and *Culex quinquefasciatus* CNS tissues (Ye et al. 2021) and would likely be useful for other mosquito species as well. NC82 labels the mosquito neuropil in each of these three mosquito species in a pattern similar to what is seen in *D. melanogaster* (Mysore et al. 2011). Such labeling allows the researcher to delineate clear boundaries between areas of the brain and VNC. Distinguishing brain regions by morphology has allowed researchers to examine the localization of different functions in the insect brain, leading to the identification of regions with different roles, and to compare the anatomy and function of brain regions between species to gain insight into the conservation or divergence of function. The NC82 antibody likely cross-reacts with mosquito Brp orthologs, although this has not been shown definitively as no loss-of-function mutations in *brp* exist in mosquitoes. Our first accompanying protocol (see Protocol 1: Whole-Mount Immunofluorescent Labeling of the Mosquito Central Nervous System [Younger 2023a]) describes a method for whole-mount brain or VNC immunostaining in mosquitoes. This protocol has been optimized for *Ae. aegypti* but can be used as a starting point for immunostaining CNS tissues collected from other mosquito species. This protocol works well with the NC82 antibody and other antibodies, although the number of available antibodies targeted to mosquito proteins is relatively limited compared to what is available for *Drosophila*. Nonetheless, new antibodies are being generated often, allowing for new insights into the mosquito nervous system.

DYE-FILLING METHOD

A method for dye-filling neurons that have processes in a peripheral tissue, such as the chemosensory organs, the antennae, and maxillary palps, is described in our second accompanying protocol (see Protocol 2: Dextran Amine–Conjugated Neural Tracing in Mosquitoes [Younger 2023b]). It allows the researcher to visualize both afferent sensory neurons that arise from the tissue as well as efferent fibers that innervate the tissue. These efferent and afferent projections for the periphery can serve as a starting point from which to examine the central circuits involved in sensory perception and behavior.

Specifically, this procedure details dextran-conjugated dye filling of neurons. Conjugated dextran amines are commonly used bidirectional tracers that readily fill the intracellular matrix, and unlike anterograde- or retrograde-only tracers, they spread without regard to direction from the loading site

(Saleeba et al. 2020). Dextran amines are easily conjugated to a variety of fluorophores, of which many are commercially available. They are well-suited for visualizing neurons in live tissue because they are nontoxic to cells and they cannot pass though the plasma membrane (Mobbs et al. 1994). Last, they are aldehyde fixable and will withstand many protocols for immunolabeling tissue following dye filling (Lanciego and Wouterlood 2020). Taken together, these traits have led to the wide use of these compounds in a variety of species. We note that although we only describe one method for neural tracing, it is possible that others may also work well in the mosquito.

Neural tracing can be used in combination with immunofluorescent labeling to identify the regions of the CNS that are innervated by filled neurons. One approach would be to follow up dye filling with immunofluorescent labeling with the NC82 primary antibody to label synaptic neuropil and define which brain regions contain the neurites of the dye-filled neurons. This method is described in Protocol 1: Whole-Mount Immunofluorescent Labeling of the Mosquito Central Nervous System (Younger 2023a) to define regions of the brain and VNC that receive innervation from a particular sense organ.

ACKNOWLEDGMENTS

We thank Olena Riabinina for immunostaining advice. The mosquito photo in Figure 1 was taken by Alex Wild. M.A.Y. is supported by funding from the Searle Scholars Program, the Richard and Susan Smith Family Foundation, the Esther A. & Joseph Klingenstein Fund, and the Simons Foundation.

REFERENCES

Carle T. 2021. A review of effects of environment on brain size in insects. *Insects* **12:** 461. doi:10.3390/insects12050461

Centers for Disease Control and Prevention. 2020. *Mosquitoes in the United States*. CDC, Atlanta.

Court R, Namiki S, Armstrong JD, Borner J, Card G, Costa M, Dickinson M, Duch C, Korff W, Mann R, et al. 2020. A systematic nomenclature for the *Drosophila* ventral nerve cord. *Neuron* **107:** 1071–1079.e1072. doi:10.1016/j.neuron.2020.08.005

Day JF. 2016. Mosquito oviposition behavior and vector control. *Insects* **7:** 65. doi:10.3390/insects7040065

Duvall LB, Ramos-Espiritu L, Barsoum KE, Glickman JF, Vosshall LB. 2019. Small-molecule agonists of *Ae. aegypti* neuropeptide Y receptor block mosquito biting. *Cell* **176:** 687–701.e685. doi:10.1016/j.cell.2018.12.004

Herre M, Goldman OV, Lu TC, Caballero-Vidal G, Qi Y, Gilbert ZN, Gong Z, Morita T, Rahiel S, Ghaninia M, et al. 2022. Non-canonical odor coding in the mosquito. *Cell* **185:** 3104–3123.e3128. doi:10.1016/j.cell.2022.07.024

Ignell R, Dekker T, Ghaninia M, Hansson BS. 2005. Neuronal architecture of the mosquito deutocerebrum. *J Comp Neurol* **493:** 207–240. doi:10.1002/cne.20800

Ito K, Shinomiya K, Ito M, Armstrong JD, Boyan G, Hartenstein V, Harzsch S, Heisenberg M, Homberg U, Jenett A, et al. 2014. A systematic nomenclature for the insect brain. *Neuron* **81:** 755–765. doi:10.1016/j.neuron.2013.12.017

Lanciego JL, Wouterlood FG. 2020. Neuroanatomical tract-tracing techniques that did go viral. *Brain Struct Funct* **225:** 1193–1224. doi:10.1007/s00429-020-02041-6

Matthews BJ, Younger MA, Vosshall LB. 2019. The ion channel *ppk301* controls freshwater egg-laying in the mosquito *Aedes aegypti*. *eLife* **8:** e43963. doi:10.7554/eLife.43963

McBride CS. 2016. Genes and odors underlying the recent evolution of mosquito preference for humans. *Curr Biol* **26:** R41–R46. doi:10.1016/j.cub.2015.11.032

McIver SB. 1982. Sensilla of mosquitoes (Diptera: Culicidae)1, 2. *J Med Entomol* **19:** 489–535. doi:10.1093/jmedent/19.5.489

Mobbs P, Becker D, Williamson R, Bate M, Warner A. 1994. Techniques for dye injection and cell labelling. In *Microelectrode techniques: the Plymouth Workshop handbook* (ed. Ogden D), pp. 361–387. The Company of Biologists Ltd, Cambridge.

Mysore K, Flister S, Muller P, Rodrigues V, Reichert H. 2011. Brain development in the yellow fever mosquito *Aedes aegypti*: a comparative immunocytochemical analysis using cross-reacting antibodies from *Drosophila melanogaster*. *Dev Genes Evol* **221:** 281–296. doi:10.1007/s00427-011-0376-2

Raji JI, DeGennaro M. 2017. Genetic analysis of mosquito detection of humans. *Curr Opin Insect Sci* **20:** 34–38. doi:10.1016/j.cois.2017.03.003

Riabinina O, Task D, Marr E, Lin CC, Alford R, O'Brochta DA, Potter CJ. 2016. Organization of olfactory centres in the malaria mosquito *Anopheles gambiae*. *Nat Commun* **7:** 13010. doi:10.1038/ncomms13010

Saleeba C, Dempsey B, Le S, Goodchild A, McMullan S. 2020. A student's guide to neural circuit tracing (vol. 13, 897, 2019). *Front Neurosci* **14:** 177. doi:10.3389/fnins.2020.00177

Smarandache-Wellmann CR. 2016. Arthropod neurons and nervous system. *Curr Biol* **26:** R960–R965. doi:10.1016/j.cub.2016.07.063

Wagh DA, Rasse TM, Asan E, Hofbauer A, Schwenkert I, Durrbeck H, Buchner S, Dabauvalle MC, Schmidt M, Qin G, et al. 2006. Bruchpilot, a protein with homology to ELKS/CAST, is required for structural integrity and function of synaptic active zones in *Drosophila*. *Neuron* **49:** 833–844. doi:10.1016/j.neuron.2006.02.008

Ye Z, Liu F, Liu N. 2021. Three-dimensional structure of the antennal lobe in the Southern house mosquito *Culex quinquefasciatus*. *Insect Sci* **28:** 93–102. doi:10.1111/1744-7917.12767

Younger MA. 2023a. Whole-mount immunofluorescent labeling of the mosquito central nervous system. *Cold Spring Harb Protoc* doi:10.1101/pdb.prot108336

Younger MA. 2023b. Dextran amine–conjugated neural tracing in mosquitoes. *Cold Spring Harb Protoc* doi:10.1101/pdb.prot108337

Cite this introduction as *Cold Spring Harb Protoc*; doi:10.1101/pdb.top108305

Protocol 1

Whole-Mount Immunofluorescent Labeling of the Mosquito Central Nervous System

Meg A. Younger[1,2,3,4,5]

[1]*Department of Biology,* [2]*Center for Systems Neuroscience,* [3]*Department of Biomedical Engineering,* [4]*Center for Neurophotonics, Boston University, Boston, Massachusetts 02215, USA*

Mosquito-borne disease is a major global public health issue. One path toward the development of evidence-based strategies to limit mosquito biting is the study of the mosquito nervous system—in particular, the sensory systems that drive biting behavior. The central nervous system of insects consists of the brain and the ventral nerve cord. Here, we describe a protocol for dissecting, immunofluorescent labeling, and imaging both of these structures in the mosquito. This protocol was optimized for *Aedes aegypti* and works well on *Anopheles gambiae* tissue. It has not been tested in other mosquito species, but we anticipate that it would work on a range of mosquitoes, and, if not, our protocol will provide a starting point from which to optimize. Notably, a limited number of antibodies cross-react with *Ae. aegypti* proteins. This protocol is intended for use with validated antibodies and can also be used to test new antibodies as they are generated. It has been successfully used to visualize protein tags, such as green fluorescent protein, that have been introduced into the mosquito to amplify or detect their presence.

MATERIALS

It is essential that you consult the appropriate Material Safety Data Sheets and your institution's Environmental Health and Safety Office for proper handling of equipment and hazardous material used in this protocol.

RECIPES: Please see the end of this protocol for recipes indicated by <R>. Additional recipes can be found online at http://cshprotocols.cshlp.org/site/recipes.

Reagents

Aluminum foil

Antibody buffer <R>

Fixative solution for mosquito tissues <R>

Glycerol

Ice

Mosquitoes (male or female adults in a small container [e.g., soup cup covered with insect mesh])

This protocol is designed for use in Aedes aegypti*, although we anticipate the procedure will work across genera.*

Mounting medium such as VectaShield (Vector Labs H-1000-10) or SlowFade Diamond (Thermo Fisher S36972)

[5]Correspondence: myounger@bu.edu

Cite this protocol as *Cold Spring Harb Protoc;* doi:10.1101/pdb.prot108336

Nail polish
Permeabilization solution for mosquito tissues <R>
Phosphate-buffered saline (10× without calcium or magnesium [Lonza 17-517Q])

Dilute 1/10 in double-distilled H_2O for 1× PBS.

Primary antibody

A common antibody that is used to stain the neuropil is anti-NC82 (Developmental Studies Hybridoma Bank [DSHB]) (Wagh et al. 2006) used at a 1/50 dilution. Notably, some Drosophila melanogaster *primary antibodies can be used because they have cross-reactivity with mosquito proteins, or an antibody custom-made for mosquitoes can be used.*

Secondary antibody

One commonly used secondary antibody is anti-mouse-488 (Life Technologies A11029) at a 1/500 dilution. Standard secondary antibodies work well, including Life Technologies Alexa Fluor antibodies.

Wash solution for mosquito tissues <R>

Equipment

Cell strainers

Prepare these items by cutting the rim off of a cell strainer (Corning 352235) so that it resembles a small basket that can fit inside a well of the 24-well plate. Prepare a basket for each experimental condition.

Confocal microscope and imaging software

Other microscopes can be used to image the brains as long as they are equipped for fluorescence and are capable of optical sectioning. The imaging parameters will depend on the desired image quality and resolution.

Culture dish (35-mm)
Dissecting stereoscope with light source
Flat-bottom plates (24-well; Corning 353047)
Forceps (very sharp)
Glass Pasteur pipette (long-tip; VWR 14673-043)
Glass slides and coverslips
Ice bucket
Microcentrifuge tube (1.5-mL)
Nutator (set at 4°C)
Orbital shaker (one set at room temperature and another set at 4°C)
Slide holder
Standard freezer box or aluminum foil (see Step 19)
Sylgard 184 silicone elastomer (WPI SYLG184)

METHOD

This protocol was modified from Riabinina et al. (2016) and Matthews et al. (2019).

Perform all dissection steps under a dissecting stereoscope with a light source.

Preparation of Sylgard Plates

1. A few days before dissection, pour a thin Sylgard slab in a 35-mm culture dish lid according to the manufacturer's instructions.

   This lid is used to dissect because of the shorter lip. The Sylgard slab is used to cushion the head or ventral nerve cord (VNC) and minimize damage to the tissues and the forceps. The Sylgard should be mixed according to the manufacturer's instructions and poured as thin as possible.

Tissue Removal and Fixation

2. Anesthetize mosquitoes by placing a small container filled with mosquitoes on ice in an ice bucket. It will take ~5 min until they are anesthetized.
3. Once the mosquitoes are anesthetized, perform the following steps under a dissecting stereoscope to begin isolating tissues.
   i. Place the mosquito on the 35-mm culture dish lid containing a Sylgard slab.

For Brain Dissection

   ii. Carefully remove the heads of adult mosquitoes from the body by pinching at the neck with sharp forceps.
   iii. Quickly place heads in a 1.5-mL tube containing 1.25 mL of fixative solution on ice. Between one and 30 heads can be placed in a tube, depending on the experimental needs.

For Ventral Nerve Cord Dissection

   iv. Carefully remove the body of the adult mosquito from the head by pinching at the neck with sharp forceps.
   v. Quickly place the entire body (without the head) in a 1.5-mL tube containing 1.25 mL of fixative solution on ice. Between one and five bodies can be placed in a tube, depending on the experimental needs.
4. Place the samples on a nutator for 3 h at 4°C.

Dissection and Blocking

5. Prepare a 24-well plate by adding a single cell strainer and 0.5 mL of wash solution to each well in the first column of the 24-well plate.
   Multiple samples can be processed together in one cell strainer, but different genotypes or conditions should be kept in separate strainers and labeled for identification.
6. Perform dissections as follows under a dissecting stereoscope.

For Brain Dissection

   i. Fill the 35-mm culture dish lid containing a Sylgard slab with ice-cold PBS and transfer a single head to the dish with forceps.
   ii. Dissect the brain out of the head capsule in ice-cold PBS with very sharp forceps.
   Take care not to puncture the brain and to delicately remove the esophagus.
   iii. Transfer the dissected brain to a cell strainer containing wash solution in the 24-well plate by using a long glass Pasteur pipette.
   Be careful to only suction the brain into the first inch of the pipette, otherwise it may get stuck to the side and lost.

For Ventral Nerve Cord Dissection

   iv. Fill the 35-mm culture dish lid containing a Sylgard slab with ice-cold PBS and transfer a single body to the dish with forceps.
   v. Dissect the VNC out of the thorax in ice-cold PBS with very sharp forceps.
   Take care not to puncture the tissue as you delicately remove the surrounding musculature.
   vi. Transfer the dissected VNC (which will resemble a translucent gummy bear) to a cell strainer containing wash solution in the 24-well plate by using a long glass Pasteur pipette.

Be careful to suction the VNC only into the first inch of the pipette, otherwise it may get stuck to the side and lost.

7. Wash the tissue in wash solution for 15 min at room temperature by placing the plate on the orbital shaker set to a low speed.
8. For each strainer in use, fill the next available well in the plate with 0.5 mL of wash solution and transfer the cell strainer containing the tissue to the well containing the fresh wash solution. Wash the sample for 15 min at room temperature by placing the plate on the orbital shaker.
9. Repeat Step 8 for six total washes.
10. Transfer the cell strainer containing tissues to a well containing 0.5 mL of permeabilization solution and place the plate on an orbital shaker for 48 h at 4°C.

Primary Antibody Incubation

11. Add 0.5 mL of wash solution to a clean well and wash the tissue for 15 min at room temperature by placing the plate on the orbital shaker.
12. Transfer the cell strainer containing the tissue to the next well in the plate that contains fresh wash solution. Carry out six total 15-min washes in fresh wash solution at room temperature on the orbital shaker.
13. Dilute primary antibodies in antibody buffer in a 1.5-mL microcentrifuge tube on ice and mix by pipetting or inverting tube. The dilution of the antibody will vary for each primary antibody used.
14. Add 0.5 mL of the primary antibody solution to a clean well. Transfer the cell strainer containing the tissue into the primary antibody solution and place the plate on an orbital shaker for 48 h at 4°C.

Secondary Antibody Incubation

15. Transfer the cell strainer containing the tissue to the next well in the plate that contains 0.5 mL of fresh wash solution. Place the plate on the orbital shaker for 15 min at room temperature.
16. Repeat Step 15 for six total 15-min washes.
17. Dilute secondary antibodies in antibody buffer in a 1.5-mL microcentrifuge tube on ice and mix by pipetting or inverting the tube. The dilution of the antibody will vary for each secondary antibody used.
18. Add 0.5 mL of the secondary antibody solution to a clean well. Transfer the cell strainer containing the tissue into the secondary antibody solution, and place the plate on an orbital shaker for 48 h at 4°C.
19. From this point onward, keep the plate covered to minimize exposure of the fluorophores to light.

    This can be achieved by covering the plate with the lid of a standard freezer box or by using aluminum foil.

Tissue Mounting and Visualization

20. Transfer the cell strainer containing the tissue to the next well in the plate that contains 0.5 mL of fresh wash solution. Place the plate on the orbital shaker to wash the tissue for 15 min at room temperature.
21. Repeat Step 20 for six total 15-min washes.
22. Add 0.5 mL of mounting medium to a clean well. Transfer the cell strainer containing the tissue into the well containing mounting medium and let it sit briefly.

    This step is to remove excess liquid that would dilute the mounting medium.

 Cite this protocol as *Cold Spring Harb Protoc*; doi:10.1101/pdb.prot108336

23. Add 0.5 mL of mounting medium to another clean well. Transfer the cell strainer containing the tissue into the well containing mounting medium and let it sit for between 12 and 20 h at 4°C.
24. Mount the tissue on slides in a small amount of fresh mounting medium. Elevate the coverslip by ~150 µm to avoid distorting the tissue, which can be accomplished by creating a small pedestal with two additional coverslips. Transfer the tissue to the slide using a long Pasteur pipette and then position the tissue gently with forceps between the two pedestals. Gently lower the coverslip onto the pedestals.

    Be careful to suction the brain or VNC only into the first inch of the pipette, otherwise it may get stuck to the side and lost.
25. Seal the slide with nail polish by sealing all of the exposed edges of the pedestal and the coverslip and allow it to dry. Be careful to avoid mixing the nail polish with mounting media or else it will not harden. Once the nail polish has completely hardened at room temperature, the slides can be placed in a slide holder and either stored at 4°C or you can proceed to Step 26. If using fluorescent secondary antibodies, it is recommended to image the tissue in the week after mounting, because the fluorescence will decay over time.
26. View the whole-mount brain or VNC on a confocal microscope.
    - For imaging the entire brain, we frequently use a 25×/0.8 numerical aperture immersion corrected objective, used with glycerol as the immersion medium to match the refractive index of the mounting medium most closely. We conduct a tiled scan with 10% overlap to capture the brain in four tiles that can be stitched together.
    - We recommend imaging the brain using a microscope that is equipped with laser compensation that will allow the user to increase the laser intensity and photomultiplier tube gain as the imaging depth increases. Insect tissue in general, and mosquito tissue in particular, scatters light to a larger degree than mammalian tissue, so laser and gain compensation are particularly important when imaging whole-mount tissue.
    - The boundaries of the brain and between regions can be used to generate three-dimensional reconstructions using both open-source and commercially available software. *Ae. aegypti* neuroanatomy resources that include brain reconstructions are available at mosquitobrains.org and insectbraindb.org (Matthews et al. 2019; Heinze et al. 2021).

RELATED INFORMATION

This protocol was optimized for *Ae. aegypti* (Matthews et al. 2019) and has been adapted for *Anopheles gambiae* tissue (Riabinina et al. 2016). For a discussion of antibodies that cross-react with *Ae. aegypti* proteins, see Mysore et al. (2011). For a discussion of the central nervous system in insects, see Ito et al. (2014); Smarandache-Wellmann (2016); and Court et al. (2020).

RECIPES

Antibody Buffer

Reagent	Amount	Final concentration
10× Phosphate-buffered saline without calcium or magnesium (Lonza 17-517Q)	1 mL	1×
10% Triton X-100 (Sigma-Aldrich 93443 diluted 1/10 in H_2O)	0.25 mL	0.25%
Normal goat serum (Jackson ImmunoResearch 005-000-121 rehydrated in 10 mL double-distilled H_2O)	0.2 mL	2%
Double-distilled H_2O	8.55 mL	

Makes 10 mL. Split into 1-mL aliquots in 1.5-mL tubes on ice, and store for up to 6 mo at −20°C.

Fixative Solution for Mosquito Tissues

Reagent	Amount	Final concentration
20× Paraformaldehyde aqueous solution (PFA) (EMS 15713-S)	10 mL	4%
0.2 M Millonig's phosphate buffer (EMS 11582-05)	25 mL	0.1 M
10% Triton X-100 (Sigma-Aldrich 93443 diluted 1/10 in H_2O)	1.25 mL	0.25%
Double-distilled H_2O	13.75 mL	

Makes 50 mL. Store for up to 2 wk at 4°C.

Permeabilization Solution for Mosquito Tissues

Reagent	Amount	Final concentration
10× Phosphate-buffered saline without calcium or magnesium (Lonza 17-517Q)	1 mL	1×
10% Triton X-100 (Sigma-Aldrich 93443 diluted 1/10 in H_2O)	4 mL	4%
Normal goat serum (Jackson ImmunoResearch 005-000-121; rehydrated in 10 mL double-distilled H_2O)	0.2 mL	2%
Double-distilled H_2O	4.8 mL	

Makes 10 mL. Split into 1-mL aliquots in 1.5-mL tubes on ice, and store for up to 6 mo at −20°C.

Wash Solution for Mosquito Tissues

Reagent	Amount	Final concentration
10× Phosphate-buffered saline without calcium or magnesium (Lonza 17-517Q)	5 mL	1×
10% Triton X-100 (Sigma-Aldrich 93443 diluted 1/10 in H_2O)	1.25 mL	0.25%
Double-distilled H_2O	43.75 mL	

Makes 50 mL. Store for up to 6 mo at room temperature.

Cite this protocol as *Cold Spring Harb Protoc*; doi:10.1101/pdb.prot108336

ACKNOWLEDGMENTS

We thank Olena Riabinina, Stanley Heinze, Nicolas Renier, and members of the Rockefeller University Bio-Imaging Resource Center and its staff members (RRID:SCR_017791) for help and advice. M.A.Y. is supported by funding from the Searle Scholars Program, the Richard and Susan Smith Family Foundation, the Esther A. & Joseph Klingenstein Fund, and the Simons Foundation.

REFERENCES

Court R, Namiki S, Armstrong JD, Borner J, Card G, Costa M, Dickinson M, Duch C, Korff W, Mann R, et al. 2020. A systematic nomenclature for the *Drosophila* ventral nerve cord. *Neuron* **107:** 1071–1079.e1072. doi:10.1016/j.neuron.2020.08.005

Heinze S, El Jundi B, Berg BG, Homberg U, Menzel R, Pfeiffer K, Hensgen R, Zittrell F, Dacke M, Warrant E, et al. 2021. A unified platform to manage, share, and archive morphological and functional data in insect neuroscience. *eLife* **10:** e65376. doi:10.7554/eLife.65376

Ito K, Shinomiya K, Ito M, Armstrong JD, Boyan G, Hartenstein V, Harzsch S, Heisenberg M, Homberg U, Jenett A, et al. 2014. A systematic nomenclature for the insect brain. *Neuron* **81:** 755–765. doi:10.1016/j.neuron.2013.12.017

Matthews BJ, Younger MA, Vosshall LB. 2019. The ion channel *ppk301* controls freshwater egg-laying in the mosquito *Aedes aegypti*. *eLife* **8:** e43963. doi:10.7554/eLife.43963

Mysore K, Flister S, Muller P, Rodrigues V, Reichert H. 2011. Brain development in the yellow fever mosquito *Aedes aegypti*: a comparative immunocytochemical analysis using cross-reacting antibodies from *Drosophila melanogaster*. *Dev Genes Evol* **221:** 281–296. doi:10.1007/s00427-011-0376-2

Riabinina O, Task D, Marr E, Lin CC, Alford R, O'Brochta DA, Potter CJ. 2016. Organization of olfactory centres in the malaria mosquito *Anopheles gambiae*. *Nat Commun* **7:** 13010. doi:10.1038/ncomms13010

Smarandache-Wellmann CR. 2016. Arthropod neurons and nervous system. *Curr Biol* **26:** R960–R965. doi:10.1016/j.cub.2016.07.063

Wagh DA, Rasse TM, Asan E, Hofbauer A, Schwenkert I, Durrbeck H, Buchner S, Dabauvalle MC, Schmidt M, Qin G, et al. 2006. Bruchpilot, a protein with homology to ELKS/CAST, is required for structural integrity and function of synaptic active zones in *Drosophila*. *Neuron* **49:** 833–844. doi:10.1016/j.neuron.2006.02.008

Protocol 2

Dextran Amine–Conjugated Neural Tracing in Mosquitoes

Meg A. Younger[1,2,3,4,5]

[1]*Department of Biology,* [2]*Center for Systems Neuroscience,* [3]*Department of Biomedical Engineering,* [4]*Center for Neurophotonics, Boston University, Boston, Massachusetts 02215, USA*

To understand the circuitry of the brain, it is often advantageous to visualize the processes of a single neuron or population of neurons. Identifying sites where a neuron, or neurons, originates and where it projects can allow a researcher to begin to map the circuitry underlying various processes, including sensory-guided behaviors. Furthermore, neural tracing allows one to map locations where processes terminate onto regions of the brain that may have known functions and sometimes to identify candidate upstream or downstream connections, based on proximity. Many methods of neural tracing are available; here, we focus on loading fluorescent dyes into a neuron (fluorescent dye filling). Different options for dyes exist to label neurites. Among the most versatile and easy to use are dextran amine–conjugated dyes. They fill neurons bidirectionally, not discriminating between anterograde or retrograde loading direction. Dye filling must be done in unfixed tissue, as the dye needs to move through the neurons; however, dextran amine conjugates are aldehyde-fixable and once cells have been fully loaded with dye the tissue can be fixed and subjected to immunostaining. Coupling neural tracing with immunofluorescence is a useful way to determine specific brain or ventral nerve cord (VNC) regions where a neuron projects. This protocol describes methods for loading dextran amine conjugated dyes into a sensory tissue in the mosquito to visualize sites of sensory neuron innervation in the central nervous system, as well as efferent projections to these structures. This protocol is described for *Aedes aegypti*, for which it was optimized, but it also works across a variety of insects.

MATERIALS

It is essential that you consult the appropriate Material Safety Data Sheets and your institution's Environmental Health and Safety Office for proper handling of equipment and hazardous material used in this protocol.

RECIPES: Please see the end of this protocol for recipes indicated by <R>. Additional recipes can be found online at http://cshprotocols.cshlp.org/site/recipes.

Reagents

Dextran dye solution

Dilute 10 mg of dye (one commonly used example is Texas-Red Dextran, Molecular Probes D3328) in 100 µL of External saline <R>, mix well, aliquot, and freeze at −20°C if not using the day of the experiment.

Ice

Mosquitoes (any genera, male or female adults in a rearing container [e.g., soup cup covered with insect mesh])

Sucrose (10%, w/v, solution)

[5]Correspondence: myounger@bu.edu

Cite this protocol as *Cold Spring Harb Protoc*; doi:10.1101/pdb.prot108337

Equipment

Cotton wick
Dissecting stereoscope
Forceps
Ice bucket
Large Petri dish (150-mm × 15-mm; Fisherbrand FB0875714)
Microscalpel (Fine Science Tools 10315-12)
Rearing container (e.g., soup cup covered in mesh)

METHOD

Dye Loading

This process can be applied to many sensory tissues, including the antennae, maxillary palps, stylet, and proboscis. It is described here for an antenna but could be adapted to other tissues as needed.

1. Anesthetize mosquitoes in a rearing container on ice in an ice bucket for 10 min.
2. Fill a Petri dish with ice to create a cold surface and close it with a lid. Place this under the dissecting stereoscope.
3. With forceps, gently transfer an individual mosquito to the top of the Petri dish lid.

   The ice that is inside of the closed dish will keep the mosquito cool and immobile during dye loading.
4. Place a small drop (0.5–1 µm) of dextran dye solution onto the surface of the dish near the immobilized mosquito.
5. With forceps, position a single antenna into the drop of dye so that the area that will be cut is submerged. Make sure that only the area where you want to load the dye is in contact with the dye droplet and that the rest of the animal is kept clean.
6. Sever the antenna where it is submerged in dye with a sharp microscalpel. There are 13 segments in the antenna of *Aedes aegypti*. If you wish to fill the neurons that innervate the entire antenna, sever the proximal antenna near the antennal pedicle. If you wish to fill neurons that innervate only certain segments of the antenna, sever the antenna at the segment junction proximal to that segment.

   This will allow the neurons to take the dextran amine–conjugated dye into the cytosol. It is important to sever the tissue while it is submerged in the dye to avoid the possibility that the axon will reseal before taking up dye.
7. Leave the antenna submerged in the dye for between 2 and 5 min on the ice-cold Petri dish lid.
8. With forceps, gently transfer the mosquito to a rearing container. Supply the mosquito with H_2O or 10% sucrose to drink with a cotton wick. Either is suitable, and the choice should be made based on your scientific question.
9. Allow dye to traffic while the mosquito recovers in the normal rearing environment for an hour up to a few hours or overnight.

   Smaller neurons may need less time and larger neurons may need more time to take up the dye, depending on the diameter and shape. The exact amount of time will need to be determined empirically for each tissue.
10. Proceed with the next steps depending on experimental needs.

    After the dye has had time to diffuse through the neuron, the next steps will depend on your applications. Some options include (1) remove mosquito heads and fix tissue as in Protocol 1: Whole-Mount Immunofluorescent Labeling of the Mosquito Central Nervous System (Younger 2023) or (2) dissect live tissue for immediate imaging.

    See Troubleshooting.

TROUBLESHOOTING

Problem (Step 10): Neural degeneration or membrane blebbing is seen then the neurons are imaged.
Solution: Decrease the time allotted for the diffusion of the dye (Step 9). The dye will optimally have enough time to travel throughout the neuron but not so much time that the neuron will become unhealthy from the damage created by severing the nerve. The time required for the dye to disperse and the time that the neuron remains healthy will depend on the temperature of the animal and the size of the cell.

Problem (Step 10): Not enough dye was loaded into the neurons to visualize them during imaging.
Solution: This issue could be due to the conditions under which the neuron was severed. Neurites will often reseal after damage. For this reason, the nerve must be exposed to the dye while it is damaged. In other words, the sensory appendage must be severed within the dye, so that dye can be taken up into the neurons before they reseal.

RELATED INFORMATION

This protocol was optimized in *Ae. aegypti* (Herre et al. 2022). Couple this protocol with Protocol 1: Whole-Mount Immunofluorescent Labeling of the Mosquito Central Nervous System (Younger 2023) to determine specific brain or ventral nerve cord regions where a neuron projects (Ito et al. 2014; Smarandache-Wellmann 2016; Court et al. 2020).

For a discussion of the neuron-filling properties of dextran amine–conjugated dyes, see Lanciego and Wouterlood (2020).

RECIPE

External Saline

Reagent	Final concentration
NaCl	103 mM
KCl	3 mM
$CaCl_2$	1.5 mM
$MgCl_2$	4 mM
NaH_2PO_4	1 mM
$NaHCO_3$	26 mM
TES (*N*-tris[hydroxymethyl]methyl-2-aminoethane sulfonic acid)	5 mM
Trehalose	10 mM
Glucose	10 mM
Sucrose	9 mM

1. Adjust the osmolarity of the solution to 275–280 mOsm by changing the concentration of sucrose, which is not metabolized by the brain and acts as an inert solute here.
2. Oxygenate by bubbling 95% O_2/5% CO_2 continuously through the solution. The pH should equilibrate to 7.3.

 Cite this protocol as *Cold Spring Harb Protoc*; doi:10.1101/pdb.prot108337

ACKNOWLEDGMENTS

We thank Vanessa Ruta and Olena Riabinina for helpful advice on dye filling in insects. M.A.Y. is supported by funding from the Searle Scholars Program, the Richard and Susan Smith Family Foundation, the Esther A. & Joseph Klingenstein Fund, and the Simons Foundation.

REFERENCES

Court R, Namiki S, Armstrong JD, Borner J, Card G, Costa M, Dickinson M, Duch C, Korff W, Mann R, et al. 2020. A systematic nomenclature for the *Drosophila* ventral nerve cord. *Neuron* **107:** 1071–1079.e1072. doi:10.1016/j.neuron.2020.08.005

Herre M, Goldman OV, Lu TC, Caballero-Vidal G, Qi Y, Gilbert ZN, Gong Z, Morita T, Rahiel S, Ghaninia M, et al. 2022. Non-canonical odor coding in the mosquito. *Cell* **185:** 3104–3123.e3128. doi:10.1016/j.cell.2022.07.024

Ito K, Shinomiya K, Ito M, Armstrong JD, Boyan G, Hartenstein V, Harzsch S, Heisenberg M, Homberg U, Jenett A, et al. 2014. A systematic nomenclature for the insect brain. *Neuron* **81:** 755–765. doi:10.1016/j.neuron.2013.12.017

Lanciego JL, Wouterlood FG. 2020. Neuroanatomical tract-tracing techniques that did go viral. *Brain Struct Funct* **225:** 1193–1224. doi:10.1007/s00429-020-02041-6

Smarandache-Wellmann CR. 2016. Arthropod neurons and nervous system. *Curr Biol* **26:** R960–R965. doi:10.1016/j.cub.2016.07.063

Younger MA. 2023. Whole-mount immunofluorescent labeling of the mosquito central nervous system. *Cold Spring Harb Protoc* doi:10.1101/pdb.prot108336

CHAPTER 41

Mapping Transcript Cell-Specific Localization and Protein Subcellular Localization in the Adult Mosquito *Aedes aegypti*

Farwa Sajadi,[1] Salwa Afifi,[1] Britney Picinic,[1] and Jean-Paul V. Paluzzi[2]

Department of Biology, York University, Toronto, Ontario M3J 1P3, Canada

This introduction reviews techniques used to examine the distribution and expression of gene transcripts and proteins in a variety of tissues/organs in the medically important global disease vector mosquito, *Aedes aegypti*. Specifically, these methods allow the detection of cell-specific transcript expression by fluorescent in situ hybridization; facilitate immunohistochemical mapping of a protein of interest in whole-mount small tissue/organ samples; examine the subcellular localization of proteins, such as membrane transporters, through sectioning of paraffin-embedded tissue/organ samples; and finally, enable the efficient separation of cytosolic and membrane proteins for western blot analysis without the need for specialized equipment (e.g., ultracentrifuge) in the mosquito *Ae. aegypti*. Such techniques are useful to help answer fundamental questions in mosquito scientific research including (but not limited to) the identification of specific cells in an organ responsible for expressing a receptor of particular interest and necessary for eliciting a response to exogenous signals, including hormones. Moreover, changes in the subcellular localization of specific targets of interest can be assessed both qualitatively and quantitatively, providing insight into transient or long-term physiologically relevant regulation necessary for activity under experimental treatments or varied internal (e.g., development) or external (e.g., environmental stress) factors that might be normally experienced by the organism.

VISUALIZING TRANSCRIPTS AND PROTEIN DISTRIBUTION HELPS ADVANCE STUDIES IN MOSQUITO PHYSIOLOGY

Mosquitoes are the deadliest animals on the planet, as they transmit several pathogens that lead to human morbidity and mortality. The expansion of vector distributions is projected to further increase over the coming decades (Kraemer et al. 2019), resulting in a large majority of the human population being at risk of at least one mosquito-borne disease (WHO 2021). The mosquito *Aedes aegypti* is a resilient species found in tropical, subtropical, and even some temperate regions and is a notorious vector of several pathogens that threaten human health, including yellow fever, dengue, chikungunya, and Zika viruses (Bhatt et al. 2013; Kraemer et al. 2015; Kotsakiozi et al. 2017). Studies to discern the physiology of this infamous human disease vector are tasked with broadening our knowledge to better understand the natural world; however, fundamental advancements may also provide opportunities to interfere with the normal biology of these organisms and impede their vector competence or lessen their burden as human disease vectors.

[1]These authors contributed equally to this work.

[2]Correspondence: paluzzi@yorku.ca

Cite this introduction as *Cold Spring Harb Protoc*; doi:10.1101/pdb.top107698

The physiology of *Ae. aegypti* mosquitoes has been studied for decades with many areas of interest. One branch of research has focused on renal excretory processes (see Piermarini 2016) including epithelial transporters critical for the rapid diuresis following blood gorging. The endocrine regulation of this rapid diuresis has also been studied (see Sajadi and Paluzzi 2021) as has the endocrine control of female reproductive biology (Raikhel et al. 2002, 2005). Additional areas of study include blood-feeding behavior and the consequences for transmission of mosquito-borne pathogens (Harrington et al. 2001; League et al. 2021), mosquito development (Zhu and Noriega 2016), and vector competence (see Souza-Neto et al. 2019). The techniques summarized below have helped to advance insight into mosquito physiology and behavior, providing pathways toward a more comprehensive understanding of their biology and raising the prospect for new strategies better suited for controlling these disease vectors and reducing their burden on humans.

Fluorescent In Situ Hybridization

Many studies in mosquitoes have examined the cell-specific localization of transcripts for genes of interest by using techniques related to or derived from in situ hybridization. Although in situ hybridization is not typically capable of providing the type of quantitative data that is achieved using quantitative reverse transcription polymerase chain reaction (RT-qPCR) or RNA sequencing techniques, in situ hybridization of mRNA transcripts contextualizes the expression of specific genes in heterogeneous cell populations within complex tissues and organs. Notably, more recent technologies such as RNAscope have enabled significant improvements to transcript detection and can be used to quantify transcript levels (Wang et al. 2012); however, there is significant cost associated with the patented probe design. At the same time, some target proteins of interest may lack available antibodies or, moreover, antibodies used may not differentiate between isoforms or structurally related proteins. Thus, in situ hybridization of mRNA transcripts helps to fill the knowledge gap for targets whose protein products are unable to be reliably immunolocalized or enhance characterization by complementing protein detection and mapping. Specifically, fluorescent in situ hybridization has been routinely used in *Ae. aegypti* to map cell-specific neuropeptide expression in the nervous system and to localize neuropeptide receptors, which are often members of the G-protein-coupled receptor superfamily, to cells in peripheral organ target sites (Rocco et al. 2019; Rocco and Paluzzi 2020; Sajadi et al. 2020). Our first accompanying protocol (see Protocol 1: Examining Cell-Specific Localization in *Aedes aegypti* Tissues with Fluorescence In Situ Hybridization [Sajadi and Paluzzi 2023]), describes this technique.

Immunohistochemistry of Whole-Mount and Paraffin-Embedded Tissue Sections

Immunohistochemistry (IHC) uses antibodies to detect target proteins (or specific antigens) and provides insight about the distribution of the protein expression, including cellular and subcellular localization. IHC can be performed on (1) whole-mount small tissues/organs as well as (2) paraffin-embedded preparations to immunolocalize proteins and other antigens in sectioned tissues/organs. Whole-mount IHC provides an avenue with which to visualize complex immunoreactive patterns while providing a greater analysis of three-dimensional tissue/organ structure. For example, axonal projections originating from a neuron in a distinct region of the nervous system or innervation outside the nervous system onto peripheral organs can be visualized via IHC. This technique was recently implemented to localize CAPA neuropeptide immunoreactivity within a pair of neurosecretory cells in each of the abdominal ganglia of adult *Ae. aegypti* that project axonal processes to the perivisceral organs for the release of these neurohormones into the insect hemolymph (Sajadi et al. 2020). Similarly, subunits of the heterodimeric glycoprotein hormone were also immunolocalized to laterally positioned neurons within the abdominal ganglia (Rocco and Paluzzi 2020) with its cognate receptor immunolocalized to peripheral organs including males' testes (Rocco et al. 2017), where it appears to regulate spermatogenesis (Rocco et al. 2019).

Two of our accompanying protocols describe various IHC techniques. First, we describe a technique for whole-mount IHC in adult *Ae. aegypti* (see Protocol 2: Examining Protein Localization in

 Cite this introduction as *Cold Spring Harb Protoc*; doi:10.1101/pdb.top107698

Aedes aegypti Cells, Tissues, and Organs: Whole-Mount Immunohistochemistry [Afifi et al. 2023]), which provides three-dimensional spatial information in small biological samples including the nervous system, intestine, and reproductive organs in the mosquito. We also present a protocol for performing IHC in cross sections of tissue from adult *Ae. aegypti* (see Protocol 3: Examining Apical or Basolateral Protein Localization in *Aedes aegypti* Tissues: Cross-Section Immunohistochemistry [Sajadi et al. 2023]). This protocol involves tissue/organ sections that allow an analysis of large complex preparations that may otherwise not be accessible to reagents essential for successful immunohistochemical detection. Furthermore, sectioning allows cell-specific and subcellular distribution of target antigens to be immunolocalized, providing critical insight related to protein localization and possible function in the context of an intact tissue/organ preparation. Although fluorescence in situ hybridization and IHC often require some level of target-specific optimization, they can be combined (Sajadi et al. 2020) to colocalize gene transcripts and proteins, which helps to better elucidate the dynamics of gene product expression.

MEMBRANE PROTEIN ISOLATION AND ENRICHMENT

The isolation of membrane proteins using traditional methods is often complicated, is time-consuming, requires specialized equipment, and is inefficient as the yield of membrane proteins is low. Membrane proteins comprise an important and large component of eukaryotic proteomes representing 20%–30% of all genes (Wallin and von Heijne 1998). As in other organisms, membrane proteins in the mosquito *Ae. aegypti* are fundamentally important in cell adhesion and paracellular transport (Jonusaite et al. 2017); cell signaling and transduction including that mediated by G-protein-coupled receptors (Duvall et al. 2019; Rocco and Paluzzi 2020; Sajadi et al. 2020) and vast chemosensory receptors (Bohbot et al. 2007; DeGennaro et al. 2013; Matthews et al. 2016; Matthews et al. 2018); vector competence (Mercado-Curiel et al. 2008); and water, solute, and ion transport (Durant et al. 2017, 2021; Misyura et al. 2017, 2020; Durant and Donini 2019). Therefore, having a reliable technique that is efficient and easy to reproduce, but does not rely on expensive, complicated equipment and long laborious protocols, will greatly accelerate research aimed at isolating and characterizing membrane proteins in the mosquito *Ae. aegypti.* Our accompanying protocol (see Protocol 4: Membrane Protein Fractionation and Analysis through Western Blot in *Aedes aegypti* Malpighian Tubules [Picinic and Paluzzi 2023]) describes an efficient technique to partition hydrophobic membrane proteins from hydrophilic cytosolic proteins using a commercial extraction kit (Qoronfleh et al. 2003) with some modifications following optimization for mosquito protein isolation. Thereafter, membrane proteins are quantified and ultimately the target membrane proteins can be analyzed using sodium dodecyl-sulfate polyacrylamide gel electrophoresis (SDS-PAGE) and western blot analysis.

CONCLUSION

Taken together, the techniques introduced here help to accelerate studies in insect physiology, including those of disease vectors such as *Ae. aegypti.* Once the detailed expression patterns of target genes are known, either at the mRNA or protein level (or both), researchers of subsequent studies can use this knowledge to design new or to implement established techniques to test and validate the physiological or overall biological role of the gene or protein of interest. Although many of the early advancements in insect physiology were centered on (neuro)endocrine control of metamorphosis, development and molting (Wigglesworth 1934, 1936, 1952; Scharrer 1946; Williams 1951, 1965), diuresis (Maddrell 1963, 1964, 1966), diapause and embryonic development (Hasegawa 1957; Riddiford and Williams 1967), and reproductive biology (Hagedorn 1985; Simonet et al. 2004; Raikhel et al. 2005), many of the signaling components regulating these fundamental processes were not comprehensively identified and remained elusive. Thus, the protocols described here have greatly advanced knowledge both in the pregenomic and postgenomic era, allowing insect physiologists to map the distribution and association of effectors within tissues and/or organs to discern the critical roles each gene product holds in relation to the biology of the insect.

REFERENCES

Afifi S, Sajadi F, Paluzzi J-PV. 2023. Examining protein localization in *Aedes aegypti* cells, tissues, and organs: whole-mount immunohistochemistry. *Cold Spring Harb Protoc* doi:10.1101/pdb.prot108281

Bhatt S, Gething PW, Brady OJ, Messina JP, Farlow AW, Moyes CL, Drake JM, Brownstein JS, Hoen AG, Sankoh O, et al. 2013. The global distribution and burden of dengue. *Nature* **496:** 504–507. doi:10.1038/nature12060

Bohbot J, Pitts RJ, Kwon HW, Rützler M, Robertson HM, Zwiebel LJ. 2007. Molecular characterization of the *Aedes aegypti* odorant receptor gene family. *Insect Mol Biol* **16:** 525–537. doi:10.1111/j.1365-2583.2007.00748.x

DeGennaro M, McBride CS, Seeholzer L, Nakagawa T, Dennis EJ, Goldman C, Jasinskiene N, James AA, Vosshall LB. 2013. Orco mutant mosquitoes lose strong preference for humans and are not repelled by volatile DEET. *Nature* **498:** 487–491. doi:10.1038/nature12206

Durant AC, Donini A. 2019. Development of *Aedes aegypti* (Diptera: Culicidae) mosquito larvae in high ammonia sewage in septic tanks causes alterations in ammonia excretion, ammonia transporter expression, and osmoregulation. *Sci Rep* **9:** 19028. doi:10.1038/s41598-019-54413-6

Durant AC, Chasiotis H, Misyura L, Donini A. 2017. *Aedes aegypti* Rhesus glycoproteins contribute to ammonia excretion by larval anal papillae. *J Exp Biol* **220:** 588–596. doi:10.1242/jeb.151084

Durant AC, Grieco Guardian E, Kolosov D, Donini A. 2021. The transcriptome of anal papillae of *Aedes aegypti* reveals their importance in xenobiotic detoxification and adds significant knowledge on ion, water and ammonia transport mechanisms. *J Insect Physiol* **132:** 104269. doi:10.1016/j.jinsphys.2021.104269

Duvall LB, Ramos-Espiritu L, Barsoum KE, Glickman JF, Vosshall LB. 2019. Small-molecule agonists of *Ae. aegypti* neuropeptide Y receptor block mosquito biting. *Cell* **176:** 687–701.e5. doi:10.1016/j.cell.2018.12.004

Hagedorn HH. 1985. The role of ecdysteroids in reproduction. In *Endocrinology II*, pp. 205–262. Elsevier, New York.

Harrington LC, Edman JD, Scott TW. 2001. Why do female *Aedes aegypti* (Diptera: Culicidae) feed preferentially and frequently on human blood? *J Med Entomol* **38:** 411–422. doi:10.1603/0022-2585-38.3.411

Hasegawa K. 1957. The diapause hormone of the silkworm, *Bombyx mori*. *Nature* **179:** 1300–1301. doi:10.1038/1791300b0

Jonusaite S, Donini A, Kelly SP. 2017. Salinity alters snakeskin and mesh transcript abundance and permeability in midgut and Malpighian tubules of larval mosquito, *Aedes aegypti*. *Comp Biochem Physiol A Mol Integr Physiol* **205:** 58–67. doi:10.1016/j.cbpa.2016.12.015

Kotsakiozi P, Gloria-Soria A, Caccone A, Evans B, Schama R, Martins AJ, Powell JR. 2017. Tracking the return of *Aedes aegypti* to Brazil, the major vector of the dengue, chikungunya and Zika viruses. *PLoS Negl Trop Dis* **11:** e0005653. doi:10.1371/journal.pntd.0005653

Kraemer MUG, Sinka ME, Duda KA, Mylne A, Shearer FM, Barker CM, Moore CG, Carvalho RG, Coelho GE, Van Bortel W, et al. 2015. The global distribution of the arbovirus vectors *Aedes aegypti* and *Ae. albopictus*. *eLife* **4:** e08347. doi:10.7554/eLife.08347

Kraemer MUG, Reiner RC, Brady OJ, Messina JP, Gilbert M, Pigott DM, Yi D, Johnson K, Earl L, Marczak LB, et al. 2019. Past and future spread of the arbovirus vectors *Aedes aegypti* and *Aedes albopictus*. *Nat Microbiol* **4:** 854–863. doi:10.1038/s41564-019-0376-y

League GP, Degner EC, Pitcher SA, Hafezi Y, Tennant E, Cruz PC, Krishnan RS, Garcia Castillo SS, Alfonso-Parra C, Avila FW, et al. 2021. The impact of mating and sugar feeding on blood-feeding physiology and behavior in the arbovirus vector mosquito *Aedes aegypti*. *PLoS Negl Trop Dis* **15:** e0009815. doi:10.1371/journal.pntd.0009815

Maddrell SH. 1963. Excretion in the blood-sucking bug, *Rhodnius prolixus* Stal. I. The control of diuresis. *J Exp Biol* **40:** 247–256. doi:10.1242/jeb.40.2.247

Maddrell SH. 1964. Excretion in the blood-sucking bug, *Rhodnius prolixus* Stal. III. The control of the release of the diuretic hormone. *J Exp Biol* **41:** 459–472. doi:10.1242/jeb.41.3.459

Maddrell SH. 1966. Nervous control of the mechanical properties of the abdominal wall at feeding in *Rhodnius*. *J Exp Biol* **44:** 59–68. doi:10.1242/jeb.44.1.59

Matthews BJ, McBride CS, DeGennaro M, Despo O, Vosshall LB. 2016. The neurotranscriptome of the *Aedes aegypti* mosquito. *BMC Genomics* **17:** 32. doi:10.1186/s12864-015-2239-0

Matthews BJ, Dudchenko O, Kingan SB, Koren S, Antoshechkin I, Crawford JE, Glassford WJ, Herre M, Redmond SN, Rose NH, et al. 2018. Improved reference genome of *Aedes aegypti* informs arbovirus vector control. *Nature* **563:** 501–507. doi:10.1038/s41586-018-0692-z

Mercado-Curiel RF, Black IV WC, Mũoz MDL. 2008. A dengue receptor as possible genetic marker of vector competence in *Aedes aegypti*. *BMC Microbiol* **8:** 118. doi:10.1186/1471-2180-8-118

Misyura L, Yerushalmi GY, Donini A. 2017. A mosquito entomoglyceroporin, *Aedes aegypti* AQP5, participates in water transport across the Malpighian tubules of larvae. *J Exp Biol* **220:** 3536–3544. doi:10.1242/jeb.158352

Misyura L, Guardian EG, Durant AC, Donini A. 2020. A comparison of aquaporin expression in mosquito larvae (*Aedes aegypti*) that develop in hypo-osmotic freshwater and iso-osmotic brackish water. *PLoS ONE* **15:** e0234892. doi:10.1371/journal.pone.0234892

Picinic B, Paluzzi J-PV. 2023. Membrane protein fractionation and analysis through western blot in *Aedes aegypti* Malpighian tubules. *Cold Spring Harb Protoc* doi:10.1101/pdb.prot108283

Piermarini PM. 2016. Renal excretory processes in mosquitoes. *Adv Insect Physiol* **51:** 393–433. doi:10.1016/bs.aiip.2016.04.003

Qoronfleh MW, Benton B, Ignacio R, Kaboord B. 2003. Selective enrichment of membrane proteins by partition phase separation for proteomic studies. *J Biomed Biotechnol* **2003:** 249–255. doi:10.1155/S1110724303209244

Raikhel AS, Kokoza VA, Zhu J, Martin D, Wang SF, Li C, Sun G, Ahmed A, Dittmer N, Attardo G. 2002. Molecular biology of mosquito vitellogenesis: from basic studies to genetic engineering of antipathogen immunity. *Insect Biochem Mol Biol* **32:** 1275–1286. doi:10.1016/S0965-1748(02)00090-5

Raikhel ASS, Brown MRR, Belles X. 2005. Hormonal control of reproductive processes. In *Comprehensive molecular insect science* (ed. Gilbert LI), pp. 433–491. Elsevier, New York.

Riddiford LM, Williams CM. 1967. The effects of juvenile hormone analogues on the embryonic development of silkworms. *Proc Natl Acad Sci* **57:** 595–601. doi:10.1073/pnas.57.3.595

Rocco DA, Paluzzi J-PV. 2020. Expression profiling, downstream signaling, and inter-subunit interactions of GPA2/GPB5 in the adult mosquito *Aedes aegypti*. *Front Endocrinol (Lausanne)* **11:** 158. doi:10.3389/fendo.2020.00158

Rocco DA, Kim DH, Paluzzi J-PV. 2017. Immunohistochemical mapping and transcript expression of the GPA2/GPB5 receptor in tissues of the adult mosquito, *Aedes aegypti*. *Cell Tissue Res* **369:** 313–330. doi:10.1007/s00441-017-2610-3

Rocco DA, Garcia ASG, Scudeler EL, dos Santos DC, Nóbrega RH, Paluzzi J-P. 2019. Glycoprotein hormone receptor knockdown leads to reduced reproductive success in male *Aedes aegypti*. *Front Physiol* **10:** 266. doi:10.3389/fphys.2019.00266

Sajadi F, Paluzzi J-PV. 2021. Hormonal regulation and functional role of the "renal" tubules in the disease vector, *Aedes aegypti*. *Vitam Horm* **117:** 189–225. doi:10.1016/bs.vh.2021.06.007

Sajadi F, Paluzzi J-PV. 2023. Examining cell-specific localization in *Aedes aegypti* tissues with fluorescence in situ hybridization. *Cold Spring Harb Protoc* doi:10.1101/pdb.prot108280

 Cite this introduction as *Cold Spring Harb Protoc*; doi:10.1101/pdb.top107698

Sajadi F, Uyuklu A, Paputsis C, Lajevardi A, Wahedi A, Ber LT, Matei A, Paluzzi J-PV. 2020. CAPA neuropeptides and their receptor form an anti-diuretic hormone signaling system in the human disease vector, *Aedes aegypti*. *Sci Rep* **10:** 1755. doi:10.1038/s41598-020-58731-y
Sajadi F, Afifi S, Paluzzi J-PV. 2023. Examining apical or basolateral protein localization in *Aedes aegypti* tissues: cross-section immunohistochemistry. *Cold Spring Harb Protoc* doi:10.1101/pdb.prot108282
Scharrer B. 1946. The role of the corpora allata in the development of *Leucophaea* (Orthoptera). *Endocrinology* **38:** 35–45. doi:10.1210/endo-38-1-35
Simonet G, Poels J, Claeys I, Van Loy T, Franssens V, De Loof A, Vanden Broeck J. 2004. Neuroendocrinological and molecular aspects of insect reproduction. *J Neuroendocrinol* **16:** 649–659. doi:10.1111/j.1365-2826.2004.01222.x
Souza-Neto JA, Powell JR, Bonizzoni M. 2019. *Aedes aegypti* vector competence studies: a review. *Infect Genet Evol* **67:** 191–209. doi:10.1016/j.meegid.2018.11.009
Wallin E, Von Heijne G. 1998. Genome-wide analysis of integral membrane proteins from eubacterial, archaean, and eukaryotic organisms. *Protein Sci* **7:** 1029–1038. doi:10.1002/pro.5560070420
Wang F, Flanagan J, Su N, Wang LC, Bui S, Nielson A, Wu X, Vo HT, Ma XJ, Luo Y. 2012. RNAscope: a novel in situ RNA analysis platform for formalin-fixed, paraffin-embedded tissues. *J Mol Diagn* **14:** 22–29. doi:10.1016/j.jmoldx.2011.08.002
WHO. 2021. WHO issues new guidance for research on genetically modified mosquitoes to fight malaria and other vector-borne diseases. https://www.who.int/news/item/19-05-2021-who-issues-new-guidance-for-research-on-genetically-modified-mosquitoes-to-fight-malaria-and-other-vector-borne-diseases (accessed 12.1.21).
Wigglesworth VB. 1934. The physiology of ecdysis in *Rhodnius prolixus* (Hemiptera). II. Factors controlling moulting and metamorphosis. *J Cell Sci* **s2-77:** 191–222. doi:10.1242/jcs.s2-77.306.191
Wigglesworth VB. 1936. The function of the corpus allatum in the growth and reproduction of *Rhodnius prolixus* (Hemiptera). *J Cell Sci* **s2-79:** 91–121. doi:10.1242/jcs.s2-79.313.91
Wigglesworth VB. 1952. The thoracic gland in *Rhodnius prolixus* (Hemiptera) and its role in moulting. *J Exp Biol* **29:** 561–570. doi:10.1242/jeb.29.4.561
Williams CM. 1951. Biochemical mechanisms in insect growth and metamorphosis. *Fed Proc* **10:** 546–552.
Williams CM. 1965. DNA synthesis and hormonal control of insect metamorphosis. *Science* **148:** 670. doi:10.1126/science.148.3670.670-a
Zhu J, Noriega FG. 2016. The role of juvenile hormone in mosquito development and reproduction. *Adv Insect Phys* **51:** 93–113. doi:10.1016/bs.aiip.2016.04.005

Protocol 1

Examining Cell-Specific Localization in *Aedes aegypti* Tissues with Fluorescence In Situ Hybridization

Farwa Sajadi and Jean-Paul V. Paluzzi[1]

Department of Biology, York University, Toronto, Ontario M3J 1P3, Canada

Fluorescence in situ hybridization (FISH) is a macromolecular recognition tool that uses RNA or DNA fragments combined with fluorophore- or digoxigenin-coupled nucleotides as probes to examine transcript localization through the presence or absence of complementary sequences in fixed tissues or samples under a fluorescent microscope. FISH technology has been highly effective for mapping genes and constructing a visual map of animal genomes. Here, we describe the application of FISH technology in the *Aedes aegypti* mosquito, where it is specifically used to localize receptor transcripts in gut tissues/organs. The methods presented highlight the synthesis of RNA probes and describe the 2-d process of incubating the tissues/organs with the RNA probes. We also describe tyramide signal amplification for improved signal detection.

MATERIALS

It is essential that you consult the appropriate Material Safety Data Sheets and your institution's Environmental Health and Safety Office for proper handling of equipment and hazardous material used in this protocol.

RECIPES: Please see the end of this protocol for recipes indicated by <R>. Additional recipes can be found online at http://cshprotocols.cshlp.org/site/recipes.

Reagents

Aedes aegypti pupae (raised as described in Rocco et al. 2017)

Agarose

Alexa Fluor 555 Tyramide SuperBoost Kit, streptavidin (Invitrogen)

Any labeled probe can be used but this example uses this particular kit.

cDNA synthesized from total or messenger RNA purified (as described in Sajadi et al. 2020) from tissue/organ where the gene of interest is enriched

Digoxigenin (DIG) RNA labeling mix (10×; Roche Applied Science)

DNase I (RNase-free; New England BioLabs)

Ethanol (100%; MilliporeSigma)

Ethidium bromide

H_2O_2 (1%, v/v, diluted in Dulbecco's phosphate-buffered saline [DPBS]), freshly made from a 3% (v/v) solution (Invitrogen)

[1]Correspondence: paluzzi@yorku.ca

Cite this protocol as *Cold Spring Harb Protoc*; doi:10.1101/pdb.prot108280

HiScribe T7 high-yield RNA synthesis kit (New England BioLabs)
Ice
Monarch PCR Cleanup kit (New England Biolabs)
Mounting medium

Use the ProLong Diamond Antifade Mountant with 4′,4-diamidino-2-phenylindole dihydrochloride (DAPI; Life Technologies) or a similar mounting medium comprised of DPBS with 50% (v/v) glycerol containing 4 µg/mL of DAPI.

Mouse anti-DIG biotin-conjugated antibody (Jackson ImmunoResearch Laboratories Inc.)
Nail polish
Nuclease-free Dulbecco's phosphate-buffered saline (DPBS; 1×; Wisent)

DPBS acts as an isotonic buffer solution during dissections.

Paraformaldehyde (4%, v/v, in DPBS), freshly prepared
PBT buffer for FISH <R>
PBT buffer with block (PBTB) <R> (optional; see Steps 22–24)
Primers described in Steps 1 and 2 from the preferred custom oligonucleotide supplier (e.g., Sigma-Aldrich, Integrated DNA Technologies)
RNA hybridization solution <R>
RNA standards for gel electrophoresis
Sucrose solution (10%, w/v)
Taq or Q5 high-fidelity DNA polymerase (New England BioLabs) and corresponding PCR reagents
Tris-acetate-EDTA (TAE) buffer (50×; Wisent Bioproducts)
Triton X-100 (4%, v/v, in PBT; Sigma-Aldrich), freshly prepared

Equipment

CO_2 anesthetizing pad (Genesee Scientific)
Confocal or standard fluorescence microscope
Cotton ball
Forceps, straight with sharp edges (Dumont #5 preferred)
Gel electrophoresis chamber
Glass beaker (25-mL; VWR)
Microcentrifuge tubes (200-µL)

These small microcentrifuge tubes are recommended for this protocol to better visualize small tissues.

Microscope slides with coverslips
Minutien pin
Pasteur (transfer) pipette
Rocker
Silicone-lined 35-mm Petri dish (prepared as described by Schellinger and Rodan 2015)
Slide storage boxes (optional; see Step 31)
Small insect rearing container (Bugdorm, MegaView Science Co.)
Stereomicroscope
Synergy Multimode Microplate Reader for UV Spectroscopy with Take3 plate for nucleic acid quantification (BioTek)
Thermocycler
Vial for mosquito feeding (Sarstedt)

METHOD

This method has been successfully conducted in Ae. aegypti *tissues previously; following are some papers for more images and examples of fluorescence in situ hybridization (FISH) (*Rocco et al. 2017; Sajadi et al. 2020*). See sample results in Figure 1.*

RNA Probe Template Preparation

1. To obtain a template for synthesizing DIG-labeled RNA probes for use in FISH, design and order gene-specific primers (Primer3, https://primer3.ut.ee/) for the target gene(s) and perform PCR in a thermocycler to amplify the target by using either a Taq or Q5 high-fidelity DNA polymerase by following manufacturer-recommended conditions. Use high-fidelity polymerase if the target is difficult to amplify with standard Taq polymerase. Empirically determine the specificity of the PCR based on the gene of interest and reagents used. Use cDNA synthesized from total or messenger RNA purified (as described in Sajadi et al. 2020) from the tissue/organ where your gene of interest is enriched as the template for this PCR. In general, because of being relatively inexpensive and easily synthesized by in vitro transcription (see below) in a basic molecular biology laboratory, single-stranded RNA probes are used that range between 500 and 800 bases. Verify the DNA template for synthesizing RNA probes for use in FISH by Sanger sequencing.
2. Purify the above PCR products using a commercial kit (Monarch PCR Cleanup kit) and use 1–10 ng of the purified product as a template to generate the RNA probe template with the addition of a T7 promoter sequence (5′-TAATACGACTCACTATAGGG-3′) to either (i) the 5′ end of the sense (forward) gene-specific primer sequence for synthesis of a sense probe (control) or (ii) the 5′ end of the antisense (reverse) gene-specific primer sequence for synthesis of antisense probe templates. Order oligonucleotides that include the T7 promoter sequence on the 5′ end of gene-specific primers. Determine specific PCR conditions empirically, perform PCR as in Step 1, and then purify the new PCR products, which now contain a single T7 promoter sequence (either sense or antisense), using a commercial kit (Monarch PCR Cleanup kit).

   Quantify DNA template for RNA probe synthesis using the Synergy Multimode Microplate Reader to determine concentration.
3. Generate the DIG-labeled antisense and sense RNA probes using 500 ng of the appropriate DNA template prepared in Step 2 by in vitro transcription. Include the T7 RNA polymerase mix and 10× reaction buffer from the HiScribe T7 high-yield RNA synthesis kit following the manufac-

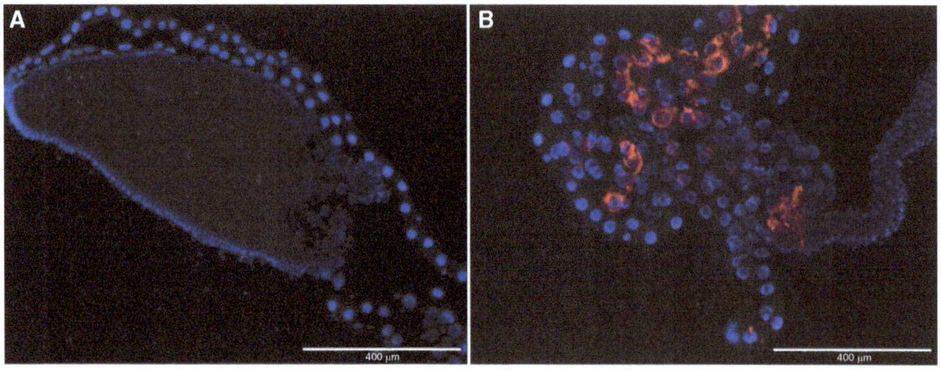

FIGURE 1. Localization of an antidiuretic hormone receptor transcript in the adult male *Aedes aegypti* Malpighian tubules (MTs) as reported previously (Sajadi et al. 2020). (*A*) Sense and (*B*) antisense RNA probes were used to determine cell-specific expression of the receptor mRNA in the mosquito MTs. Scale bars, 400 μm.

turer's protocol but substitute the unlabeled nucleotides provided in the kit with the DIG RNA labeling mix.

4. Once DIG-labeled RNA synthesis is complete, remove template DNA with DNase I (RNase-free) following the manufacturer's protocol for removal of template DNA in transcription reactions.

   Quantify RNA probes using the Synergy Multimode Microplate Reader to determine RNA concentration.

5. Run an aliquot of ~0.5 µg RNA on a nondenaturing 1% (w/v) agarose gel prepared with 1× TAE in a gel electrophoresis chamber to confirm RNA probe integrity. RNA probe size and quality can be assessed by comparing the relative fluorescence intensity of the RNA probe band to RNA standards following staining with ethidium bromide. To avoid RNA degradation, ensure all reagents and equipment used are DNase/RNase-free. Store DIG-labeled RNA probes for up to 1 wk at −20°C or for up to several months at −80°C.

Mosquito Collection and Dissections

6. Collect pupae using a transfer pipette and place in a beaker within a small insect rearing container. Provide sucrose solution in a small vial fitted with a cotton ball for feeding adult mosquitoes after emergence.

   If working with a specific age or sex, isolate hatched immature mosquitoes and place them in separate jars as needed, noting both the age and sex of mosquitoes for further processing.

7. Once adult mosquitoes emerge and are at the desired age, briefly anesthetize mosquitoes within the small rearing container using a CO_2 pad and wait until they are unresponsive. If CO_2 pads are not available, cold anesthesia can be used as an alternative. Clean fine forceps with ethanol for sterility then use them to carefully lift a mosquito from its wing or leg and place it on a silicone-lined dish (lateral side up). Using a minutien pin, impale into the thorax through to the silicone to secure the mosquito in place.

   Pin placement can be modified based on the target tissue to be dissected.

8. Using a Pasteur pipette, add a drop of room temperature DPBS to immerse the mosquito in the solution, and place the silicone-lined dish under a stereoscopic microscope. Using the fine forceps, dissect target tissue(s) and or organ(s) as desired.

   This protocol has been used successfully with nervous tissue, intestine, and reproductive organs (Rocco et al. 2019, 2020; Rocco and Paluzzi 2020; Sajadi et al. 2020).

   A small piece of cuticle can be left with the dissected tissue(s) or organ(s) to help visualize tissues during washes. This cuticle can be removed on the final day before mounting specimens on slides.

Tissue Fixation and Washes

For washes and incubations, if a 200-µL microcentrifuge tube is used, add enough solution to ensure all tissues or organs are immersed in solution, ideally 100–150 µL. Washes should be performed carefully and under a stereomicroscope to ensure no tissues or organs are damaged or lost during pipetting.

9. Immediately after dissection, place tissues or organs in a microcentrifuge tube containing 4% paraformaldehyde in DPBS and fix them for 1–2 h at room temperature.

   Fixation may need to be extended overnight (and at 4°C) if working with larger organs, and fixation timing should be determined empirically.

10. Following fixation, remove the fixative and wash the tissue/organ samples with PBT five times for 1–2 min each on a rocker at room temperature.

11. After the washes, treat samples with 1% H_2O_2 for 10–20 min at room temperature on a rocker to quench any endogenous peroxidase activity.

12. Remove the 1% H_2O_2 and incubate samples in 4% Triton X-100 in PBT for 1 h at room temperature on a rocker to allow for the permeabilization of tissues/organs.

13. After incubation, wash the samples with PBT five times for 1–2 min each on a rocker at room temperature.
14. Fix the samples again using 4% paraformaldehyde for 20 min on a rocker at room temperature and then wash once for 1–2 min using PBT to remove all traces of fixative.

Adding the Labeled Probe

15. Place the RNA hybridization solution in a thermocycler for 5 min at 100°C to allow for denaturation and then immediately place it on ice for 5 min to let it cool. Leave it at room temperature after this process.

    Prepare enough for the number of microcentrifuge tubes and all treatments below involving stock or diluted RNA hybridization solution. For all steps below, use this denatured hybridization solution.

16. Rinse samples with a 1:1 mixture of PBT:RNA hybridization solution and replace the solution with room temperature RNA hybridization solution. Incubate samples for 1–2 h at 56°C.

    This step is the prehybridization treatment.

17. During the last 15 min of the incubation in Step 16, prepare the DIG-labeled RNA probe solutions as follows: Add the labeled RNA probe (antisense for experimental or sense for control) to an aliquot of the RNA hybridization solution to a final concentration of ~2–4 ng/µL. Heat the mixture for 3 min at 80°C to allow for denaturation of the single-stranded RNA probes and then subsequently cool on ice for 5 min.

    The probe amount for each target gene needs to be determined empirically; however, add ~200 ng (for Malpighian tubules, hindgut tissues, or central nervous system) or ~400 ng (for midgut tissues) of probe per 100 µL of hybridization solution.

18. Once the prehybridization is complete (Step 16), incubate samples overnight (~16 h) in the hybridization solution containing the DIG-labeled RNA probe at 56°C.
19. Prepare 3:1, 1:1, and 1:3 (vol/vol) mixtures of RNA hybridization solution:PBT and PBT alone, and incubate the solutions overnight at 56°C.

    These mixtures will be used the following day after hybridization with gene-specific RNA probes. Prepare enough for each tube/treatment (i.e., if you have three tubes, prepare 700 µL of each mixture to pipette 200 µL in each tube. Always prepare an additional 100 µL to account for pipetting error).

Washes and Blocking Solution

20. After the overnight hybridization with gene-specific DIG-labeled RNA probes, wash samples at 56°C (15 min each wash) with prewarmed fresh hybridization solution and subsequently with the prewarmed 3:1, 1:1, and 1:3 hybridization solution:PBT mixtures prepared in Step 19.
21. Next, wash samples with prewarmed PBT from Step 19 (for 15 min) and then allow to cool to room temperature.
22. To reduce nonspecific staining, block samples with 1× Blocking buffer (from the Tyramide SuperBoost kit) or PBTB for 1 h and place on a rocker.

Antibody Washes

23. Incubate the samples with a mouse anti-DIG biotin-conjugated antibody for 1.5 h on a rocker at room temperature, protected from light.

    We recommend starting with a 1:400 dilution in Blocking buffer or PBTB.

24. Following the anti-DIG primary antibody incubation, wash samples in 1× Blocking buffer or PBTB over the course of 1 h on a rocker at room temperature.

    Up to five washes can be performed.

 Cite this protocol as *Cold Spring Harb Protoc*; doi:10.1101/pdb.prot108280

25. Incubate samples in a horseradish peroxidase (HRP)-conjugated streptavidin (100–150 µL if using HRP-conjugated streptavidin from Tyramide SuperBoost kit, or 1:100 dilution if using from another compatible kit) for 1 h on rocker at room temperature.
26. Repeat Step 24.
27. Briefly (~5 min) wash samples twice with PBT and once with DPBS at room temperature.
28. Following the Tyramide SuperBoost kit guidelines, prepare an Alexa Fluor tyramide working solution in 1× Reaction buffer with 0.01× H_2O_2 and 0.01× Tyramide stock solution (e.g., for every 100 µL of 1× Reaction buffer, add 1 µL of 3% H_2O_2 and 1 µL of Tyramide stock solution [are provided in the Tyramide SuperBoost kit]). Once the last DPBS wash is completely removed from tissues/organs, add the freshly prepared Tyramide working solution and incubate for 2–10 min at room temperature, protected from light.

    We recommend optimizing the incubation period using positive and negative control samples at various incubation time points when conducting the experiment for the first time.

29. Add 100 µL of Reaction stop reagent to tissues/organs and wash samples with DPBS 10 times over the course of 1 h at room temperature.

    Longer extended washes may be necessary in some cases to improve the signal-to-noise ratio.

30. Store samples in DPBS overnight at 4°C (protected from light).
31. The following day, mount tissues/organs on microscope slides in a mounting medium containing DAPI onto coverslips and seal with nail polish. View slides under a confocal or standard fluorescence microscope. Slides can also be stored at 4°C in storage boxes and viewed under fluorescent microscope at a later date. In some cases in which strong target signals are detected, slides can be stored for up to several months and reimaged as necessary.

RECIPES

PBT Buffer for FISH

Dilute 0.1% (v/v) Tween-20 in Dulbecco's phosphate-buffered saline (DPBS; 1×; Wisent).

Freshly prepare.

PBT Buffer with Block (PBTB)

Dulbecco's phosphate-buffered saline (DPBS; 1×; Wisent)
0.1% (v/v) Tween-20
1% (w/v) bovine serum albumin (Invitrogen)

Freshly prepare; can be stored for up to 1 wk at room temperature.

RNA Hybridization Solution

Reagent	Final concentration
Formaldehyde	50% (v/v)
Saline sodium citrate (SSC), 20× (Invitrogen 15557044)	5×
Heparin	0.1 mg/mL
Sonicated salmon sperm DNA	0.1 mg/mL
Tween-20	0.1% (v/v)

Store for up to 1 yr at −20°C.

REFERENCES

Rocco DA, Paluzzi J-PV. 2020. Expression profiling, downstream signaling, and inter-subunit interactions of GPA2/GPB5 in the adult mosquito, *Aedes aegypti*. *Front Endocrinol (Lausanne)* **11:** 158. doi:10.3389/fendo.2020.00158

Rocco DA, Kim DH, Paluzzi J-PV. 2017. Immunohistochemical mapping and transcript expression of the GPA2/GPB5 receptor in tissues of the adult mosquito, *Aedes aegypti*. *Cell Tissue Res* **369:** 313–330. doi:10.1007/s00441-017-2610-3

Rocco DA, Garcia ASG, Scudeler EL, dos Santos DC, Nóbrega RH, Paluzzi J-PV. 2019. Glycoprotein hormone receptor knockdown leads to reduced reproductive success in male *Aedes aegypti*. *Front Physiol* **10:** 266. doi:10.3389/fphys.2019.00266

Sajadi F, Uyuklu A, Paputsis C, Lajevardi A, Wahedi A, Ber LT, Matei A, Paluzzi J-PV. 2020. CAPA neuropeptides and their receptor form an anti-diuretic hormone signaling system in the human disease vector, *Aedes aegypti*. *Sci Rep* **10:** 1755. doi:10.1038/s41598-020-58731-y

Schellinger JN, Rodan AR. 2015. Use of the Ramsay assay to measure fluid secretion and ion flux rates in the *Drosophila melanogaster* Malpighian tubule. *J Vis Exp* 53144. doi:10.3791/53144

 Cite this protocol as *Cold Spring Harb Protoc*; doi:10.1101/pdb.prot108280

Protocol 2

Examining Protein Localization in *Aedes aegypti* Cells, Tissues, and Organs: Whole-Mount Immunohistochemistry

Salwa Afifi, Farwa Sajadi, and Jean-Paul V. Paluzzi[1]

Department of Biology, York University, Toronto, Ontario M3J 1P3, Canada

Immunohistochemistry (IHC) is a powerful technique used for visualizing cellular components and determining the presence and/or location of proteins or other macromolecules in tissue samples. The classical IHC process involves the detection of epitopes using a highly specific primary antibody. This is followed by a secondary antibody that is coupled to a reporter molecule or fluorophore and capable of binding to the primary antibody and allowing for protein immunodetection. Although IHC does not routinely provide quantitative results compared to an enzyme-linked immunoassay or western blotting, it permits the localization of the proteins in intact tissues. This protocol describes an IHC assay for whole-body *Aedes aegypti* mosquito tissues that is used to detect small proteins, specifically neuropeptide hormones. This method is useful for protein detection in whole-mount preparations; however, cross-section IHC is recommended to determine if a protein is localized in the apical versus basolateral membrane of tissues/organs or to visualize immunological distribution in larger, more complex preparations.

MATERIALS

It is essential that you consult the appropriate Material Safety Data Sheets and your institution's Environmental Health and Safety Office for proper handling of equipment and hazardous material used in this protocol.

RECIPES: Please see the end of this protocol for recipes indicated by <R>. Additional recipes can be found online at http://cshprotocols.cshlp.org/site/recipes.

Reagents

Aedes aegypti pupae (raised as described in Rocco et al. 2017)

Antibody dilution buffer for whole-mount immunohistochemistry <R>

Mounting medium

Use the ProLong Diamond Antifade Mountant with 4′,4-diamidino-2-phenylindole dihydrochloride (DAPI; Life Technologies) or a similar mounting medium comprised of Dulbecco's phosphate-buffered saline (DPBS) with 50% glycerol containing 4 µg/mL of DAPI.

Nail polish (optional; see Step 15)

Normal sheep serum (Jackson ImmunoResearch Inc.)

Nuclease-free DPBS (Wisent)

DPBS acts as an isotonic buffer solution during dissections.

[1]Correspondence: paluzzi@yorku.ca

Cite this protocol as *Cold Spring Harb Protoc*; doi:10.1101/pdb.prot108281

Paraformaldehyde (4%, v/v, in DPBS), freshly prepared
Permeabilization solution for immunohistochemistry <R>
Primary antibody (based on the target protein of interest)
Secondary antibody (Alexa Fluor)

Any labeled secondary antibody can be used, with varying working concentrations (e.g., 1:200 to 1:1000) depending on the abundance of the target protein.

Sucrose solution (10%, w/v)

Equipment

Aluminum foil (optional; see Step 13)
CO_2 anesthetizing pad (Genesee Scientific)
Confocal or fluorescence microscope
Cotton ball
Forceps, straight with sharp edges (Dumont #5 preferred)
Glass beaker (25-mL; VWR)
Microcentrifuge tubes (1.5-mL)
Microscope slides and coverslips
Minutien pin
Pasteur (transfer) pipette
Plates with lids (12-well or 24-well; optional, see Step 5)
Rocker
Silicone-lined 35-mm Petri dish (prepared as described by Schellinger and Rodan 2015)
Small insect rearing container (Bugdorm, MegaView Science Co.)
Stereomicroscope
Vial for mosquito feeding (Sarstedt)

METHOD

This method was adapted from previously published protocols (Rocco et al. 2017; Wahedi and Paluzzi 2018; Lajevardi and Paluzzi 2020; Sajadi et al. 2020). Sample images are provided in Figure 1.

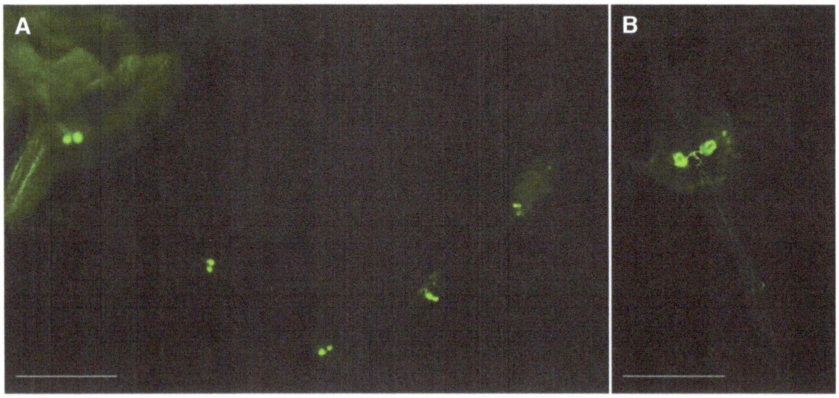

FIGURE 1. Immunohistochemical distribution of an antidiuretic neuropeptide in the abdominal ganglia of the adult *Aedes aegypti* as shown previously (Sajadi et al. 2020). (*A*) A pair of neurosecretory cells observed in each abdominal ganglia with a higher magnification observed in (*B*) demonstrating peptide immunoreactivity within ventrally located neurosecretory cells with axonal projections from the cells. Scale bars, 400 µm (*A*); 50 µm (*B*).

 Cite this protocol as *Cold Spring Harb Protoc*; doi:10.1101/pdb.prot108281

Mosquito Dissections

1. Collect pupae using a transfer pipette and place them in a beaker within a small insect rearing container. Provide sucrose solution in a small vial fitted with a cotton ball for feeding.

   If working on a specific age or sex, isolate hatched mosquitoes and place them in separate jars, noting the age and sex of mosquitoes.

2. Once adult mosquitoes emerge and are at the desired age, briefly anesthetize mosquitoes within the small rearing container using a CO_2 pad and wait until they are unresponsive.

   Alternatively, mosquitoes can be anesthetized by briefly placing mosquitoes at 4°C to minimize activity.

3. Using forceps, carefully lift a mosquito with its wing or leg and place it on a silicone-lined dish (lateral side up). Using a Minutien pin, impale the thorax through to the silicone to secure the mosquito in place.

   Pin placement can be modified based on the target tissue to be dissected.

4. Using a transfer pipette, add a drop of room temperature DPBS to immerse the mosquito in solution, and place the silicone-lined dish under a stereomicroscope. Using the forceps, dissect target tissue(s)/organ(s) such as nervous tissue, intestine, or reproductive organs as previously described (Rocco et al. 2017; Lajevardi and Paluzzi 2020; Sajadi et al. 2020).

Fixation and Washes

For washes and incubations, the volume depends on whether a 1.5-mL centrifuge tube or a 12- or 24-well plate is used. Ensure that tissues/organs are fully immersed in solution; ideally volumes should be 500 µL–1 mL per tube or well.

5. Place tissues of interest in a 1.5-mL microcentrifuge tube or a well of a 12- or 24-well plate containing 4% paraformaldehyde in DPBS and incubate for 1 h on a rocker at room temperature to fix samples.

   Tissues may be incubated in fixation solution overnight at 4°C and the procedure continued the following day (optional). Note that fixation duration might need to be optimized for the particular target antigen or antibody used.

   If dissecting the central nervous system, incubate whole mosquitoes with legs and wings removed in 4% paraformaldehyde in DPBS overnight at 4°C before dissection because brain and thoracic ganglia may be easier to dissect if fixed overnight.

6. Discard the paraformaldehyde and wash tissues/organs once with DPBS on a rocker for 30 min at room temperature.

   If tissues/organs were fixed overnight before dissection (as suggested above for the central nervous system), dissect and isolate desired tissues/organs before proceeding to the next step.

7. Incubate samples with permeabilization solution on a rocker for 1 h at room temperature.

8. Immediately after placing the samples in permeabilization solution in Step 7, prepare a dilution of the primary antibody of interest in antibody dilution buffer, and allow it to incubate for at least 1 h at 4°C. Prepare solutions for control treatments as described below, which may include omission of primary antibody or antigen pre-adsorption.

   Primary antibody dilutions depend on the antibody used; a recommended dilution may be provided by the antibody manufacturer. If there is no recommended dilution, start with 1 µg purified antibody per mL of antibody dilution buffer (recommended range is 0.2–5.0 µg/mL). To improve the signal-to-noise ratio and to reduce nonspecific binding, prepare the primary antibody dilution the day before and allow it to incubate at 4°C.

   For controls, incubate the samples in antibody dilution buffer lacking the primary antibody or with antibody that has been pre-adsorbed with a >10-fold molar excess of antigen, if available.

9. Remove the permeabilization solution and wash samples three times with DPBS (10 min for each wash) on a rocker at room temperature.

10. Discard the last wash of DPBS and replace with the preincubated primary antibody/antiserum. As an example of volume needed, we usually use between 0.5 and 1 mL of primary antibody solution in a 1.5-mL microcentrifuge tube containing dozens of replicate biological samples/organs. Incubate samples in primary antibody solution for 48 h at 4°C on rocker.

    In some cases, to improve signal-to-noise ratio and detection of low abundance protein targets, we increase the duration of the primary antibody incubation to 4–5 d.

Secondary Antibody Incubation

11. Following the primary antibody incubation, remove the primary antibody solution, and wash samples four times for 15 min each with DPBS, on a rocker at room temperature.

    It may be possible to save and reuse the primary antibody solution (label tubes accordingly and store at −20°C).

    Samples can also be incubated in DPBS overnight at 4°C (optional).

12. During the last DPBS wash, prepare the secondary antibody solution in DPBS containing 10% (v/v) normal sheep serum.

    The secondary antibody must be directed against the species in which the primary antibody was raised (i.e., if using a primary antibody raised in mouse, an anti-mouse secondary antibody raised in a species other than mouse must be used).

    Serum is added together with secondary antibody as a blocking agent because it contains antibodies that bind to nonspecific sites. If possible, it is recommended to use a serum that matches the host species of the secondary antibody.

    Ensure the secondary antibody mixture is always protected from light! The concentration of the secondary antibody will vary depending on which Alexa Fluor (or other conjugate) reagent is used. If the secondary antibody used is highly cross-adsorbed, preincubation is not required; however, troubleshooting may be needed to determine the optimal dilution. We recommend a dilution ranging from 1:200 to 1:1000 for secondary antibody solution.

13. Allow samples to incubate in secondary antibody solution overnight at 4°C on a rocker.

    Ensure samples are covered to protect from light (e.g., wrap in aluminum foil).

Mounting Samples

14. Discard the secondary antibody solution and wash samples four times for 15 min each with DPBS on a rocker at room temperature.

    Samples can be incubated in DPBS overnight at 4°C (optional).

15. After the last DPBS wash, mount tissues/organs on slides in mounting medium containing DAPI and cover with a coverslip. If the slides will be kept for an extended period (more than a couple of days), seal the cover slips using nail polish.
16. View slides under a confocal or a fluorescence microscope.

RECIPES

Antibody Dilution Buffer for Whole-Mount Immunohistochemistry

1. In a 15-mL Falcon tube, combine 200 mg of bovine serum albumin (final concentration 2%, w/v), 9.56 mL of Dulbecco's phosphate-buffered saline (DPBS; Wisent), 200 µL of 10% (v/v) normal sheep serum (Jackson ImmunoResearch Inc.), and 40 µL of Triton X-100.
2. Vortex this mixture well. Makes 10 mL.

    For optimal results, freshly prepare; however, aliquots can be made and stored long term (up to 1 yr) at −20°C.

 Cite this protocol as *Cold Spring Harb Protoc*; doi:10.1101/pdb.prot108281

Permeabilization Solution for Immunohistochemistry

1. In a 15-mL Falcon tube, combine 200 mg of bovine serum albumin (final concentration 2%, w/v), 8.4 mL of Dulbecco's phosphate-buffered saline (DPBS; Wisent), 1 mL of 10% (v/v) normal sheep serum (Jackson ImmunoResearch Inc.), and 400 µL of Triton X-100.
2. Vortex this mixture well. Makes 10 mL.

   For optimal results, freshly prepare; however, aliquots can be made and stored long term (up to 1 yr) at −20°C.

REFERENCES

Lajevardi A, Paluzzi J-PV. 2020. Receptor characterization and functional activity of pyrokinins on the hindgut in the adult mosquito, *Aedes aegypti*. *Front Physiol* **11:** 490. doi:10.3389/fphys.2020.00490

Rocco DA, Kim DH, Paluzzi J-PV. 2017. Immunohistochemical mapping and transcript expression of the GPA2/GPB5 receptor in tissues of the adult mosquito, *Aedes aegypti*. *Cell Tissue Res* **369:** 313–330. doi:10.1007/s00441-017-2610-3

Sajadi F, Uyuklu A, Paputsis C, Lajevardi A, Wahedi A, Ber LT, Matei A, Paluzzi J-PV. 2020. CAPA neuropeptides and their receptor form an anti-diuretic hormone signaling system in the human disease vector, *Aedes aegypti*. *Sci Rep* **10:** 1755. doi:10.1038/s41598-020-58731-y

Schellinger JN, Rodan AR. 2015. Use of the Ramsay assay to measure fluid secretion and ion flux rates in the *Drosophila melanogaster* Malpighian tubule. *J Vis Exp* 53144. doi:10.3791/53144

Wahedi A, Paluzzi J-PV. 2018. Molecular identification, transcript expression, and functional deorphanization of the adipokinetic hormone/corazonin-related peptide receptor in the disease vector, *Aedes aegypti*. *Sci Rep* **8:** 2146.

Protocol 3

Examining Apical or Basolateral Protein Localization in *Aedes aegypti* Tissues: Cross-Section Immunohistochemistry

Farwa Sajadi, Britney Picinic, Salwa Afifi, and Jean-Paul V. Paluzzi[1]

Department of Biology, York University, Toronto, Ontario M3J 1P3, Canada

Immunohistochemistry (IHC) is an important technique that permits visualization of cellular components and for determining the presence and/or distribution of proteins or other macromolecules in tissue samples. Normally, IHC involves the detection of epitopes using an antigen-specific primary antibody and a secondary antibody coupled with a reporter molecule or fluorophore that can bind to the primary antibody, allowing for the spatial distribution of a protein of interest to be detected. Although normally IHC does not provide quantitative results compared to techniques such as enzyme-linked immunoassay or western blotting, it permits the localization, expression mapping, and distribution of target proteins in intact tissues. Here, we describe an IHC protocol for examining apical versus basolateral protein staining through sectioning tissue samples from fixed, embedded tissues (e.g., IHC-paraffin) and adding primary antibodies against a target protein. This IHC protocol provides a guide for tissue fixation, sectioning, and staining of tissue samples.

MATERIALS

It is essential that you consult the appropriate Material Safety Data Sheets and your institution's Environmental Health and Safety Office for proper handling of equipment and hazardous material used in this protocol.

RECIPES: Please see the end of this protocol for recipes indicated by <R>. Additional recipes can be found online at http://cshprotocols.cshlp.org/site/recipes.

Reagents

Aedes aegypti pupae (raised as described in Rocco et al. 2017)

Antibody dilution buffer (ADB) for cross-section immunohistochemistry <R>

Bouin's fixative for in situ hybridization <R>

Distilled H_2O

Ethanol (30%, v/v; 70%, v/v; 95%, v/v; and 100%)

Hormonal or other treatment (optional; see Step 5)

Kodak Photo-Flo 200 in PBS (PF/PBS; 0.4%, v/v)

Mounting medium

Use ProLong Diamond Antifade mountant with 4′,4-diamidino-2-phenylindole dihydrochloride (DAPI; LifeTechnologies) or similar mounting medium comprised of Dulbecco's phosphate-buffered saline (DPBS) with 50% glycerol and containing 4 µg/mL of DAPI.

[1]Correspondence: paluzzi@yorku.ca

Cite this protocol as *Cold Spring Harb Protoc*; doi:10.1101/pdb.prot108282

Nail polish (optional; see Step 45)
Nuclease-free DPBS (1×; Wisent)
Paraplast Plus tissue embedding medium (Oxford Worldwide, LLC)
Primary antibody mixture, freshly prepared

Prepare the mixture with a protein of interest primary antibody in ADB. Dilutions can range depending on the antibody and abundance of target protein; generally, for cross-section immunohistochemistry, use a 10× greater concentration than that used for a western blot. Approximately 100–150 µL of diluted antibody is required for each slide; thus, prepare enough to cover each sample on your slide.

Secondary antibody mixture, freshly prepared

Prepare the mixture with an Alexa Fluor secondary antibody in ADB. Generally, a 1:500 dilution is used for a secondary antibody. Approximately 100–150 µL of diluted antibody is required for each slide; thus, prepare enough to cover each sample on your slide.

Sodium citrate buffer <R>
Sucrose solution (10%, w/v)
Triton X-100 in PBS (TX/PBS; 0.05%, v/v)
Xylene (absolute, 100%)

Equipment

CO_2 anesthetizing pad (Genesee Scientific)
Compound microscope (optional; see Step 18)
Cotton ball
EasyDip slide staining rack
Embedding rings and molds

Size can differ based on target tissue/organ.

Fluorescence microscope
Forceps, straight with sharp edges (Dumont #5 preferred)
Fume hood
Glass beaker (2-mL; VWR)
Glass scintillation vials
Hot plate (set at 60°C)
Hydrophobic (PAP) pen for immunostaining
Incubator set at 60°C
Kimwipes
Manual rotary microtome
Microcentrifuge tube or multiwell plate (optional; see Step 5)
Microscope slides and coverslips
Microwave oven
Minutien pin (optional; see Step 3)
Paper towels
Pasteur (transfer) pipette

Its narrow tip is ideal for washes to ensure no tissue/organs are damaged during pipetting.

Plastic container with lid (1-L; Tupperware Brand)
Razor
Scissors
Silicone-lined 35-mm Petri dish (prepare as described by Schellinger and Rodan 2015) (optional; see Step 3)
Slide warmer
Small insect rearing container (Bugdorm, MegaView Science Co.)
Stereomicroscope (optional; see Step 4)
Thermal gloves

Vial for mosquito feeding (Sarstedt)
Water bath (set to ~37°C)
Wheaton glass staining dish with removable rack

METHOD

This method was adapted from previously published protocols (Chasiotis et al. 2016; Misyura et al. 2017). An example is provided in Figure 1. All steps are performed at room temperature unless indicated otherwise.

Mosquito Dissections

1. Collect *Ae. aegypti* pupae in a beaker within a small insect rearing container. Provide sucrose solution in a small vial fitted with a cotton ball for feeding.

   If working on a specific age or sex, isolate hatched mosquitoes and place them in a separate beaker noting the age and sex of mosquitoes.

2. Once the mosquitoes are at the desired age, briefly anesthetize mosquitoes on a CO_2 pad and wait until they are unresponsive.

   Alternatively, mosquitoes can be anesthetized by briefly placing mosquitoes at 4°C to minimize activity.

   If sectioning the whole mosquito, proceed to Step 6. Otherwise, continue to Step 3.

3. Using forceps, carefully lift a mosquito from its wing or leg and place it on a silicone-lined dish (lateral side up). Using a Minutien pin, impale into the thorax through to the silicone to secure the mosquito in place.

   Pin placement can be modified based on the target tissue to be dissected.

4. Using a Pasteur pipette, add a drop of DPBS to immerse the mosquito in solution, and place a silicone-lined dish under a stereomicroscope. Using the forceps, dissect target tissue(s)/organ(s). Using the forceps, dissect target tissue(s)/organ(s) such as nervous tissue, intestine, or

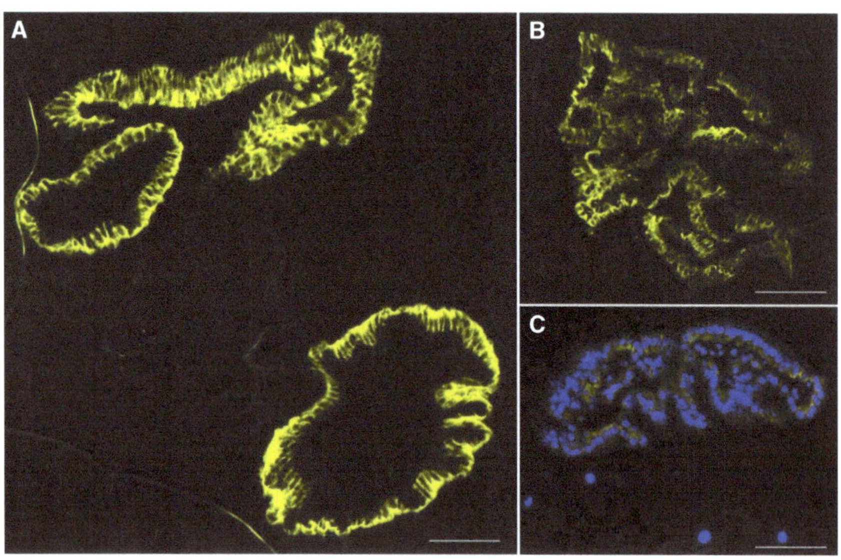

FIGURE 1. Membrane localization of Na^+/K^+-ATPase (NKA) in the midgut of *Aedes aegypti* mosquitoes. Similar findings were reported previously by Patrick et al. (2006). Transverse and cross sections of the adult female midgut reveal NKA (yellow) distribution and localization along the entire length of the membrane (*A*,*B*). (*C*) Display of the NKA immunolocalization (yellow) with nuclei counterstained with DAPI (blue). Scale bars, 100 µm.

Cite this protocol as *Cold Spring Harb Protoc*; doi:10.1101/pdb.prot108282

reproductive organs as previously described (Rocco et al. 2017; Lajevardi and Paluzzi 2020; Sajadi et al. 2020).

5. (Optional) If testing hormonal effects or other treatments that might influence subcellular localization of your target protein, incubate your tissues/organs in a microcentrifuge tube (or in a multiwell plate) containing this treatment immediately following dissection for 30 min to 1 h (optimal dose and incubation time must be determined empirically) before proceeding with Step 6. See Sajadi et al. (2022) for an example.

Mosquito Fixation

For washes and incubations, ensure that tissues/organs are fully immersed in solution; ideally, volumes should be 3–5 mL per glass scintillation vial.

6. Following mosquito dissection (or hormonal incubation), place your tissues/organs in Bouin's fixative in glass scintillation vials and fix for 1.5–2 h at room temperature.

   Be careful; Bouin's fixative is toxic! This step can be done under fume hood (optional).

7. Following fixation, rinse tissues/organs three times with 70% ethanol at room temperature using a transfer pipette to exchange solutions.

   Tissues/organs can be stored for a few days to weeks in 70% ethanol at 4°C until further processing.

Sample Washes

8. Dehydrate the fixed tissues through a series of ethanol washes using a transfer pipette, as follows: 70% ethanol for 30 min, 95% ethanol for 30 min, and 100% ethanol three times for 30 min each at room temperature.

   These incubation times are ideal for isolated/dissected tissues/organs; if you are sectioning whole bodies, increase wash times to 45 min.

   If tissues have been stored long term in 70% ethanol, the first 70% ethanol wash can be skipped.

9. Clear tissues with the following washes: 100% ethanol:xylene (1:1) for 30 min and then 100% xylene three times for 30 min each at room temperature.

   Be careful; xylene is toxic! Perform this step under a fume hood.

10. After the last xylene wash, incubate samples in preheated xylene:paraffin wax (Paraplast Plus tissue embedding medium) (1:1) for 60 min in an incubator set at 60°C.
11. Next, rinse samples in pure paraffin wax twice for 1 h each at 60°C.

    Samples may be left overnight in a 60°C incubator or stored at room temperature in hardened wax and reheated later for final embedding.

Embedding Samples

12. To embed the samples, prepare the embedding rings and mold in place and carefully (with a pipette cut with scissors into a blunt end) place the sample in the middle of the ring with some melted paraffin wax. Fill up the entire ring with paraffin wax and allow it to solidify at room temperature.

    This process should be done quickly and consistently. Do not allow the paraffin wax to solidify before the embedding ring is filled. To ensure optimal results, place glass scintillation vials and embedding rings over a hot plate (60°C) before the embedding process. Once the embedding ring is filled with the sample and paraffin wax, do not move the sample until the wax has solidified (the wax will start to become concave in the middle).

13. Incubate samples overnight at 4°C.

Sectioning Samples

14. To section across a mosquito's whole body or tissue, use a razor to cut around the sample embedded in the paraffin wax to allow for easier sections.
15. Use a manual rotary microtome and make 3- to 5-µm sections of the samples.

    If the tissue/sample is embedded in the middle of the embedding ring, multiple sections will have to be made until the sample is reached. Cut multiple sections of sample for greater technical replicates.
16. Add a drop of 30% ethanol on a microscope slide and transfer sections onto the slide using forceps.
17. Next, transfer the slide into a lukewarm (~37°C) bath to allow the wax sections to expand. Note that the sections will float as the slide is transferred into the bath.
18. After a few seconds, lower the microscope slide under the floating sections and slowly raise the slide, allowing the tissue slices to adhere to the slide.

    To confirm the sections contain the desired sample, inspect the slides under a compound microscope.
19. Incubate slides overnight at 37°C on a slide warmer.

Washes and Primary Antibody Incubation

20. Place slides in an EasyDip slide staining rack and rinse the slides in 100% xylene twice for 5 min each to deparaffinize sections.
21. Next, rinse the slides via a descending series of ethanol washes, as follows: 100% ethanol twice for 2 min each, 95% ethanol for 2 min, 70% ethanol for 2 min, and 50% ethanol for 2 min.
22. Finally, incubate slides in distilled H_2O for 20 min.
23. To promote epitope availability and enhance immunogenicity, subject sections to heat-induced epitope retrieval (HIER) by immersing slides in a Wheaton glass staining dish filled with room temperature 10 mM sodium citrate buffer and heat both slides and solution in a microwave oven for 4 min.
24. Allow the solution and slides to cool for 20 min by leaving both in the microwave with the door open.
25. Reheat the solution and slides again in the microwave for 2 min.
26. Allow the solution and slides to cool for 15 min at room temperature by removing them from the microwave.

    Make sure to wear proper thermal gloves, as the Wheaton glass dish will get hot!
27. Wash slides in a Wheaton glass staining dish filled with 0.4% PF/PBS for 10 min.
28. Wash slides in a Wheaton glass staining dish filled with 0.05% TX/PBS for 10 min.
29. Wash slides in a Wheaton glass staining dish filled with 10% (v/v) ADB in PBS (ADB/PBS) for 10 min.
30. Place slides on paper towels and allow them to dry for 15–20 min at room temperature.
31. Wipe down and dry slides carefully with Kimwipes.
32. Draw a circle around the sections with a hydrophobic (PAP) pen.

    Ensure there are no breaks in the circle to avoid the antibody solution spilling out.
33. Pipette the primary antibody mixture into the PAP-drawn circles on slides.

    Carefully, with the pipette tip, make sure the antibody is covering the entire circle.
34. Place slides in a 1-L size plastic container with two to three distilled H_2O-soaked paper towels on the bottom, close the lid, and place in a dark drawer overnight at room temperature, protecting from light.

 Cite this protocol as *Cold Spring Harb Protoc*; doi:10.1101/pdb.prot108282

Washing and Probing with Secondary Antibody

35. Following the overnight incubation in Step 34, wash slides briefly by dipping them in a beaker of distilled H_2O.
36. Repeat Steps 27–29.
37. Repeat Steps 31 and 32.

    Do not remove the circle drawn previously with the PAP pen. Redraw circles if necessary.
38. Pipette the secondary antibody mixture into the PAP-drawn circles on the slides.
39. Repeat Step 34 but incubate for 60 min rather than overnight.

Washing and Mounting Slides

40. Wash slides briefly by dipping them in a beaker of distilled H_2O.
41. Repeat Steps 27–29.
42. Wash slides with 0.4% PF/PBS three times for 1 min each.
43. Wash slides in distilled H_2O for 1 min.
44. Repeat Steps 31 and 32 and allow slides to air-dry for 30 min.
45. Pipette a few drops of mounting medium containing DAPI (ensure the entire circle is covered) and place a coverslip on top. If the slides will be kept for an extended period (more than a couple of days), seal the coverslips using nail polish.
46. View slides under a fluorescence microscope.

RECIPES

Antibody Dilution Buffer (ADB) for Cross-Section Immunohistochemistry

1. In a 1-L beaker, combine 50 mL of normal goat serum (10%, v/v, final concentration, EMD Millipore Corp.), 15 g of bovine serum albumin (3%, w/v, final concentration), 250 µL of Triton X-100 (0.05%, v/v, final concentration), and 450 mL of Dulbecco's phosphate-buffered saline (DPBS; 1×; Wisent).
2. Aliquot ADB into 50-mL aliquots in 50-mL centrifuge tubes.
3. Store for up to 1 yr at −20°C.

Bouin's Fixative for In Situ Hybridization

Reagent	Volume (for 100 mL)	Final concentration
Picric acid (saturated)	70 mL	70%
Formaldehyde (37%)	25 mL	9.25%
Acetic acid (glacial)	5 mL	5%

Store in a glass container in a corrosive cabinet at room temperature.

Sodium Citrate Buffer

Reagent	Quantity (for 1 L)	Final concentration
Trisodium citrate (dehydrate)	2.94 g	10 mM
H_2O	1 L	

Dissolve trisodium citrate in H_2O and adjust the pH to 6.0 with 1 N HCl.
Store at room temperature for 3 mo or at 4°C for 6 mo.

REFERENCES

Chasiotis H, Ionescu A, Misyura L, Bui P, Fazio K, Wang J, Patrick M, Weihrauch D, Donini A. 2016. An animal homolog of plant Mep/Amt transporters promotes ammonia excretion by the anal papillae of the disease vector mosquito *Aedes aegypti*. *J Exp Biol* **219:** 1346–1355. doi:10.1242/jeb.134494

Lajevardi A, Paluzzi J-PV. 2020. Receptor characterization and functional activity of pyrokinins on the hindgut in the adult mosquito, *Aedes aegypti*. *Front Physiol* **11:** 490. doi:10.3389/fphys.2020.00490

Misyura L, Yerushalmi GY, Donini A. 2017. A mosquito entomoglyceroporin, *Aedes aegypti* AQP5, participates in water transport across the Malpighian tubules of larvae. *J Exp Biol* **220:** 3536–3544. doi:10.1242/jeb.158352

Patrick ML, Aimanova K, Sanders HR, Gill SS. 2006. P-type Na^+/K^+-ATPase and V-type H^+-ATPase expression patterns in the osmoregulatory organs of larval and the adult mosquito, *Aedes aegypti*. *J Exp Biol* **209:** 4638–4651. doi:10.1242/jeb.02551

Rocco DA, Kim DH, Paluzzi J-PV. 2017. Immunohistochemical mapping and transcript expression of the GPA2/GPB5 receptor in tissues of the adult mosquito, *Aedes aegypti*. *Cell Tissue Res* **369:** 313–330. doi:10.1007/s00441-017-2610-3

Sajadi F, Uyuklu A, Paputsis C, Lajevardi A, Wahedi A, Ber LT, Matei A, Paluzzi J-PV. 2020. CAPA neuropeptides and their receptor form an anti-diuretic hormone signaling system in the human disease vector, *Aedes aegypti*. *Sci Rep* **10:** 1755. doi:10.1038/s41598-020-58731-y

Sajadi F, Vergara-Martínez MF, Paluzzi J-PV. 2022. The V-type H^+-ATPase is targeted in anti-diuretic hormone control of the Malpighian 'renal' tubules. bioRxiv doi:10.1101/2022.02.13.480270

Schellinger JN, Rodan AR. 2015. Use of the Ramsay assay to measure fluid secretion and ion flux rates in the *Drosophila melanogaster* Malpighian tubule. *J Vis Exp* 53144. doi:10.3791/53144

Cite this protocol as *Cold Spring Harb Protoc*; doi:10.1101/pdb.prot108282

Protocol 4

Membrane Protein Fractionation and Analysis through Western Blot in *Aedes aegypti* Malpighian Tubules

Britney Picinic and Jean-Paul V. Paluzzi[1]

Department of Biology, York University, Toronto, Ontario M3J 1P3, Canada

Western blot analysis is a well-known and dependable technique used to quantify protein abundance in a wide variety of samples. A major consideration for running a successful western blot is ensuring that the protein to be analyzed is purified appropriately. For work with membrane-bound proteins, traditional methods of protein processing such as the use of high-frequency sonication and ultracentrifugation to separate proteins from the membrane are being replaced with less time-consuming approaches. The use of a membrane fractionation kit, which involves the separation of membrane proteins from soluble (cytosolic) proteins, is effective in allowing for the quantification and analysis of membrane-bound proteins. In this protocol, we describe use of the membrane fractionation kit to isolate membrane-bound proteins, followed by western blot analysis, to observe protein abundance. The protocol involves methods that require organ (or tissue) collection, followed by protein processing, and a 2-d western blot procedure.

MATERIALS

It is essential that you consult the appropriate Material Safety Data Sheets and your institution's Environmental Health and Safety Office for proper handling of equipment and hazardous material used in this protocol.

RECIPES: Please see the end of this protocol for recipes indicated by <R>. Additional recipes can be found online at http://cshprotocols.cshlp.org/site/recipes.

Reagents

10× Running buffer <R>

Dilute to 1× and store at room temperature.

Aedes adult saline <R>

Aedes aegypti pupae (raised as described in Rocco et al. 2017)

Ammonium persulfate (APS; 10%, w/v)

Antibody dilutions (primary and secondary), freshly prepared

Prepare antibody dilutions in milk solution in a 15-mL disposable tube. Approximately 6 mL of diluted antibody is required per membrane; calculate antibody dilutions accordingly. Vortex solutions well.

If antibody is limited, 4–5 mL total of diluted antibody in milk solution is possible.

Bovine serum albumin (BSA; 5 mg/mL in H_2O)

[1]Correspondence: paluzzi@yorku.ca

Cite this protocol as *Cold Spring Harb Protoc*; doi:10.1101/pdb.prot108283

Bradford reagent (Bio-Rad 5000006)
Coomassie Blue/methanol solution <R>
De-stain solution (50%, v/v, methanol)
Double-distilled H_2O
Enhanced chemiluminescence (ECL) kit
Ethanol (70%, v/v, and 75%, v/v)
Ice
Loading buffer (6×) <R>
Mem-PER Plus Membrane Protein Extraction Kit (Thermo Scientific 89842)
Methanol (100%; Sigma-Aldrich)
Milk solution (5%, w/v, in TBS-T)

To block nonspecific staining, membranes should be blocked using a 5% (w/v) milk solution (5 g of skim milk powder in 100 mL of TBS-T).

Protease inhibitor (Sigma-Aldrich 029M4044V)
Protein ladder (e.g., Bio-Rad Precision Plus Protein Dual Color Standards)
Separation/resolving gel mixture (12%) <R>
Stacking gel mixture (4%) <R>
Sucrose solution (10%, w/v)
TBS-T <R>
Tetramethylethylenediamine (TEMED; Bio-Rad Laboratories 161-0801)
Transfer buffer (1×) <R>

Equipment

Bidirectional rocker

Ideally, this item should be moveable, as it needs to be placed at both room temperature and 4°C (or have one at each temperature)

CO_2 anesthetizing pad (Genesee Scientific)
Cotton ball
Glass beaker (25-mL; VWR)
Disposable tubes (15- and 50-mL)
Forceps (two pairs, fine-tip and blunt)
Gel casting plates
Gel electrophoresis chamber
Glass Petri dish (150-mm)
Heat block (set to 100°C)
Heat sealer
Ice pack (placed in a −20°C freezer; optional; see Step 37)
ImageJ computer program
Imager system (e.g., ChemiDoc gel imager; Bio-Rad)
Large, clear plastic sandwich bag
Large glass dish (40-cm × 20-cm, or similar)
Magnetic stirrer and stir bar
Mesh sponge
Microcentrifuge
Microcentrifuge tubes (1.5-mL)
Stereo-dissecting microscope (Nikon)
Overhead transparency sheet
Paper towels
Plastic bag

Cite this protocol as *Cold Spring Harb Protoc*; doi:10.1101/pdb.prot108283

Plastic pipette
PowerPac high-current power supply
Polyvinylidene fluoride (PVDF) membrane
Rotator
Scissors
Silicone-lined 35-mm Petri dish (prepared as described by Schellinger and Rodan 2015)
Small container (such as reusable food storage containers; see Step 38)
Small insect rearing container (Bugdorm, MegaView Science Co.)
Sodium dodecyl-sulfate polyacrylamide gel electrophoresis (SDS-PAGE) gel casting apparatus
Spectrophotometer
Thick blotting filter paper
Vial for mosquito feeding (Sarstedt)
Vortex mixer
Wet membrane transfer cassettes

METHOD

This method was adapted from previously published protocol (Qoronfleh et al. 2003). See Figure 1 for an example of a western blot. Perform all steps at room temperature unless otherwise indicated.

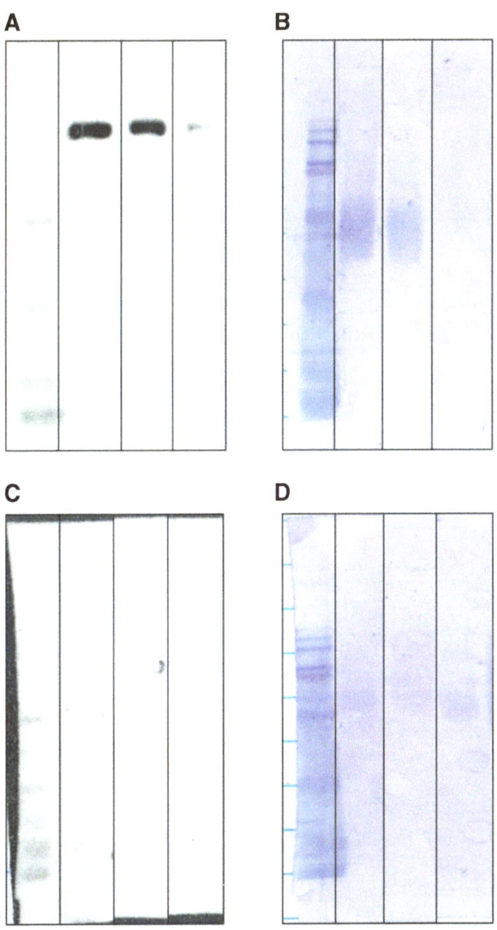

FIGURE 1. AaAQP1 immunoreactive bands visualized in adult female *Aedes aegypti* Malpighian tubule membrane protein fraction. Western blot analysis was used to visualize the protein band of AaAQP1 in the (*A*) membrane fraction of tubule protein and (*C*) cytosolic fraction of the tubule protein. Coomassie staining was used to visualize total protein for the (*B*) membrane fraction and (*D*) cytosolic fraction of tubule protein. Malpighian tubules were isolated in saline, $n = 3$. AaAQP1 immunoreactive bands only appeared in the membrane protein fraction and not the cytosolic fraction, as expected because AaAQP1 is a membrane protein (Misyura et al. 2020).

Tissue Collection

1. If using adult *Ae. aegypti* mosquitoes for the assay, collect pupae with a plastic transfer pipette, and place them in a glass beaker (~25-mL). Place the beaker in a small insect container (or cage) to easily isolate the adult mosquitoes once they have hatched. Provide sucrose solution in a small vial fitted with a cotton ball for feeding.
2. At the desired age, use a CO_2 pad to gently anesthetize the mosquitoes by placing it on a mesh opening of the cage until all mosquitoes have stopped moving. Remove the mosquitoes from the cage using forceps and place them in a Petri dish. Immediately place the dish on ice to prevent mosquitoes from waking up.

   Alternatively, if a CO_2 pad is not available, mosquitoes can be anesthetized by briefly placing mosquitoes at 4°C to minimize activity.
3. Using fine-tip forceps, grasp the leg of the mosquito, and briefly dip it in a Petri dish containing 70% ethanol. Then, use the forceps to place the mosquito directly in a silicone-lined Petri dish with a sufficient volume of *Aedes* adult saline to completely submerge the mosquito.
4. Using fine forceps, dissect the tissues/organs of interest from the animal under a stereo (dissecting) microscope and place them in a 1.5-mL microcentrifuge tube containing 1 mL of *Aedes* adult saline, placed on ice.

   The number of mosquitoes needed depends on the size of tissues/organs being collected. For example, when dissecting Malpighian tubules from adult female Ae. aegypti *(see Lajevardi et al. 2021 for description of dissection of different mosquito organs), each biological replicate is normally composed of 75–100 sets of tubules (a greater number would be required if working with males, which are smaller than females).*
5. When the desired number of animals have been dissected, centrifuge the tube containing the tissue of interest at 10,000*g* for 1 min. Carefully remove and discard the supernatant without disrupting the pellet. If not proceeding directly to the next step, store the tube containing the organ/tissue pellet for up to 1 yr at −80°C.

Protein Processing: Membrane Fractionation

The Mem-PER Plus Membrane Protein Extraction Kit is used in Steps 6 to 14 to complete the membrane protein extraction from the tissue of interest.

6. Add approximately 1 mL of permeabilization buffer per 75 mg of tissue to the tube containing your samples.

   The volume of permeabilization buffer can be decreased to achieve a more concentrated final product. In the example provided in Step 4 above, if dissecting 100 sets of adult female Ae. aegypti *Malpighian tubules, this is equivalent to ~6–7.5 mg of tissue; therefore, add ~100 µL of permeabilization buffer directly to the tissue pellets that were stored at −80°C in a microcentrifuge tube.*
7. To the same tube, add protease inhibitor at a 1:200 dilution.
8. Incubate tubes for 10 min at 4°C with constant mixing on a bidirectional rocker.
9. Centrifuge the tube(s) at 16,000*g* for 15 min at 4°C. Carefully remove the supernatant and place it in a new, labeled tube.

   This is the cytosolic (soluble) protein fraction of the tissue/organ sample.
10. Resuspend the remaining pellet in solubilization buffer using the same volume as used for the permeabilization buffer in Step 6.
11. Add protease inhibitor at a 1:200 dilution to the resuspended pellet and vortex to ensure the pellet has loosened.
12. Incubate the tube for 30 min at 4°C with constant mixing on a bidirectional rocker.

 Cite this protocol as *Cold Spring Harb Protoc*; doi:10.1101/pdb.prot108283

13. Centrifuge the tube(s) at 16,000*g* for 15 min at 4°C. Remove the supernatant and transfer it to a new, labeled tube. Discard the tube containing only the pellet.

    This is the membrane protein fraction of the tissue/organ sample.

14. Store cytosolic and membrane protein fractions in microcentrifuge tubes at −80°C for later use.

    Protein concentrations of both fractions must be determined before proceeding to the next step. Use your method of choice for protein quantification, such as a Bradford assay, which is described in Step 15. This assay is effective for relatively low protein levels if using a microtiter or microassay protocol.

Bradford Assay for Protein Quantification

15. If proceeding with a Bradford assay for protein quantification, prepare standards with BSA.
    i. If working with presumed low protein concentrations, prepare standards at 100, 80, 60, 40, 20, 10, 5, 2.5, and 0 µg/mL.
    ii. If working with presumed moderate to high protein concentrations, prepare standards at 2.0, 1.5, 1.0, 0.5, 0.25, and 0 mg/mL.
    iii. Add 8 µL of each standard to a microcentrifuge tube and add 2 µL of Bradford reagent to each tube. Pipette to mix.
    iv. Incubate the mixtures for at least 5 min at room temperature.
    v. Measure the wavelength reading with a spectrophotometer using 2 µL of the mixture.

16. For protein samples from Step 14, add 1 µL of protein to 9 µL of double-distilled H_2O and 2.5 µL of Bradford reagent, incubate at room temperature as was done for standards, and measure the wavelength reading with a spectrophotometer using 2 µL of the mixture.

    Be sure to account for the dilution factor when calculating protein concentrations.

    A Nanodrop instrument can be used for measuring wavelength readings of standards and processed protein. To normalize against the reagent used for protein processing, add a sample with 1 µL of solubilization buffer, 9 µL of double-distilled H_2O, and 2.5 µL of Bradford reagent. The reading from the Nanodrop of the buffer mix alone should be subtracted from all protein sample readings. Quantify protein concentrations for experimental samples according to the instructions of the Bradford reagent.

Western Blotting: Gel Preparation

17. Set up an SDS-PAGE gel casting apparatus and fill the plates with double-distilled H_2O. Leave the setup for ~5 min to see if there are any leaks. If a leak has occurred, reposition plates and repeat the leak test.

18. Prepare a 12% separation/resolving gel mixture in a 50-mL disposable tube, leaving out the TEMED. Check the gel apparatus for leaks and dispose of the H_2O between the well plates by tilting the apparatus over a sink.

    The gel percentage can be adjusted based on the target protein molecular weight.

19. Add 10 µL TEMED to the gel mixture and using a plastic pipette, pipette up and down quickly. Add the resolving gel mixture between the glass plates, leaving room for the combs (~3 cm). To help remove any bubbles at the top of the resolving gel, top off the gel with 75% ethanol and leave it to polymerize for ~40 min.

20. Prepare the 4% stacking gel mixture in a 15-mL disposable microcentrifuge tube, leaving out the APS and TEMED until the resolving gel has completely polymerized and the 75% ethanol has been washed out. Add 25 µL 10% APS and 5 µL TEMED to the gel mixture. Add the stacking gel mixture between the glass plates and place a comb with the desired number of wells. Leave it to polymerize for 30 min.

    Gels can be made up to ~24 h in advance. Wrap each set of glass plates in a damp paper towel, place in a plastic bag, and store in a fridge at 4°C.

Western Blotting: Day 1

Sample Preparation

21. Remove protein samples from the freezer and place on ice to thaw.
22. Prepare an appropriate number of microcentrifuge tubes for each sample. Add an equal amount of protein to each tube with an appropriate amount of sample buffer with loading dye for denaturing and loading of protein samples (e.g., 6× loading buffer). Use the same final volume of sample in each lane.

    If the protein yield is high, 10–20 µg of protein can be added. However, if protein yield is low, 5 µg can be used but must be empirically optimized for the protein target.
23. Vortex all samples and then place them for 5 min in a heat block set to 100°C to fully denature the protein.

    It is very important that the time is not exceeded.
24. Once the protein is denatured, place protein samples on ice and briefly centrifuge (10,000*g* for 1 min at room temperature) to ensure the sample collects at the base of the tube.

SDS-PAGE

25. Remove gel plates from the casting apparatus or refrigerator, and set them up in a gel electrophoresis chamber. Add 1× running buffer to the electrophoresis chamber before loading protein samples and ladder.

    If running a single gel, a buffer dam must be inserted into the gel electrophoresis chamber for the gel to run properly.
26. Load ~5–7 µL of an appropriate protein ladder in the first well on the left-hand side of the gel.
27. Begin with loading the entirety of each sample into individual wells, from left to right. Repeat for a second gel if necessary.
28. Connect the chamber to appropriate power supply and begin running the gel at a low voltage (~70 V). Monitor the gel until the bands have migrated into a straight, thin line.
29. Increase the voltage to ~110 V and run until protein bands have reached the required separation for the protein target of interest.

    The time it takes to complete running a gel depends upon the weight of the protein of interest. Therefore, it is important to monitor the gel periodically to prevent over running it.
30. While the gel is running, place 1× transfer buffer (at least 1 L) in a −20°C freezer.

Wet Transfer

31. Set up a large glass dish by placing the wet membrane transfer cassettes down and open, with the clear side above the black side. Pour in the entirety of the 1× transfer buffer from the freezer into the glass dish to cover the cassettes. Start by placing a mesh sponge on each side of the cassette, followed by a prewet thick blotting filter paper on top of each sponge.

    If only thin blotting papers are available, use two for each side of the cassette. This setup should be repeated for each cassette.
32. With scissors, cut a piece of PVDF membrane (~2″ × 1.5″) for each cassette (adjust as needed for gel size). Make a small cut to the bottom right corner of each membrane to provide an orientation landmark to the membrane during the transfer process.

    If transferring more than one membrane, make the bottom corner cuts different in size to distinguish each blot.
33. Activate each membrane before placing it in transfer cassettes by briefly dipping each one in methanol, grasping the bottom right corner of the membrane with blunt forceps. Place the PVDF

Cite this protocol as *Cold Spring Harb Protoc*; doi:10.1101/pdb.prot108283

membrane onto the filter paper over top the clear (or red) side of the cassette that will face the anode (+), with the snipped corner on the bottom right.

34. Remove gels from between glass plates once they have finished running and place them directly onto the filter paper over top the black side of the cassette that will face the cathode (−) the same way it was removed from the plate. Then, using blunt forceps, grasp the snipped corner of the membrane from the clear side of the cassette and flip it over on top of the gel on the black side of the cassette.

    The snipped corner should now be backward, on the bottom left-hand corner of the gel.

    Orientation here is very important, as the transfer process uses electric current to direct the negatively charged protein (due to SDS) from the gel and onto the membrane.

35. Close each transfer cassette and insert into the transfer apparatus, facing the black side of each cassette to the black side of the transfer apparatus (cathode side).
36. With distilled H_2O, rinse off excess SDS in the chamber that was used for the SDS-PAGE. Fill the chamber with the chilled transfer buffer by pouring it from the glass dish. Place the transfer apparatus into the same chamber.
37. Place the lid on the electrophoresis chamber and connect the electrodes to the power supply in the proper orientation, matching colors from the chamber to the power supply. Set the voltage to 100 V and run the gel for 1 h, with constant mixing using a stir bar on a magnetic stirrer and keeping the transfer cool (~4°C) by either running in a cold room or adding an ice pack.

Blocking and Antibody Incubation

38. Remove the membrane(s) from the transfer cassette with blunt forceps and place it into a small container with a 5% milk solution. Incubate the membrane for 1 h with constant mixing on a rotator at room temperature.

    The milk solution should cover the membrane entirely.

39. Using a large clear plastic sandwich bag, cut out a rectangle that is slightly larger than the membrane. Using a heat sealer, seal one of the sides of the rectangle, leaving the rest open to place the membrane in.

    An alternative method to using a plastic bag is to use a small container that fits the membrane and seals with a lid.

40. After blocking is complete, remove membrane(s) from the milk solution and place it in between the plastic sheets. Seal two more of the sides, leaving one side open to add 6 mL of the diluted primary antibody. Seal the final side and secure the plastic pouch containing the membrane; move the pouch to a rotator and leave overnight with constant mixing at 4°C. Repeat this step with additional membranes.

Western Blotting: Day 2

41. Remove membrane(s) from the plastic pouch and place them in a container filled with copious TBS-T. Incubate for 15 min three times with constant mixing on the rotator to remove nonspecific antibody binding.
42. Prepare plastic pouches as described in Step 39.
43. Place the membrane(s) in the plastic pouch, seal two more of the sides, and add 6 mL of the diluted secondary antibody solution. Seal the final side of the pouch and place it on a rotator for 1 h at room temperature with constant mixing. Repeat this step with additional membranes.
44. Remove membrane(s) following secondary antibody incubation and repeat Step 41.
45. Using an ECL kit, mix the chemiluminescent reagents at a ratio of 1:1 to a total of 4 mL. Place the membrane(s) on an overhead transparency sheet and add the reagent mixture to the

membrane(s), ensuring that the reagents cover the entirety of the blot. Incubate for 5 min at room temperature. Remove membrane(s) from the film and place them between two clear sheets of overhead transparency film.

If your laboratory or facility has an imager capable of fluorescence detection (for example, if required for multiplexing capabilities), this detection method could be used as an alternative to chemiluminescence.

46. Perform imaging using an imager system.

Imaging can also be done using a traditional photo developing process.

Be sure to select the image just before any individual band becomes over saturated.

Total Protein Quantification

47. Immediately after imaging, place membranes in a container with TBS-T on a rocker for 5 min and then incubate them in Coomassie Blue/methanol solution for 5 min, also on a rocker.
48. Using blunt forceps, rinse the membrane in de-stain solution for 30 sec to 1 min, until the dark-blue color of the membrane has lightened.
49. Quickly dip the membrane in double distilled H_2O before placing it on an overhead transparency sheet to dry for 1 h at room temperature.
50. Scan the membrane as done above in Step 46 to acquire a total protein loading image for quantification using the ImageJ computer program.

RECIPES

10× Running Buffer

Dissolve 30.0 g of Tris base, 144.0 g of glycine, and 10.0 g of SDS in 1000 mL of H_2O. The pH of the buffer should be 8.3 and no pH adjustment is required. Store the running buffer at room temperature and dilute to 1× before use.

Aedes *Adult Saline*

150 mM NaCl
3.4 mM KCl
1.7 mM $CaCl_2$
0.6 mM $MgCl_2$
1.8 mM $NaHCO_3$
5.0 mM glucose
25 mM HEPES, pH 7.1

Aliquot and store for up to a week at 4°C. Bring aliquots to room temperature before use.

Coomassie Blue/Methanol Solution

0.1% (w/v) Coomassie Blue R250
50% (v/v) methanol

Loading Buffer (6×)

Add 1.8 mL of Tris-HCl (1.0 M, pH 6.8) <R> to 1.5 mL of glycerol, 0.6 g SDS, 0.464 g DTT, and 0.15 mL of 1% (w/v) bromophenol blue. Adjust the volume to 5 mL with double distilled H_2O and store for up to 1 yr at −20°C.

 Cite this protocol as *Cold Spring Harb Protoc*; doi:10.1101/pdb.prot108283

Separation/Resolving Gel Mixture (12%)

Tris-HCl (1.5 M, pH 8.8) <R>	2.5 mL
Double distilled H_2O	4.29 mL
Acrylamide/Bis solution 29:1 (30%; Bio-Rad)	3 mL
SDS, 10% (w/v)	0.1 mL
Ammonium persulfate, 10% (w/v)	0.1 mL
Tetramethylethylenediamine (TEMED; Bio-Rad Laboratories 161-0801)	0.01 mL

Freshly prepare. Add TEMED last, immediately before pouring the gel solution into the gel casting apparatus.

Stacking Gel Mixture (4%)

Tris-HCl (1.0 M, pH 6.8) <R>	0.313 mL
Double-distilled H_2O	1.913 mL
Acrylamide/Bis solution 29:1 (30%; Bio-Rad)	0.25 mL
SDS, 10% (w/v)	0.025 mL
Ammonium persulfate, 10% (w/v)	0.025 mL
Tetramethylethylenediamine (TEMED; Bio-Rad Laboratories 161-0801)	0.005 mL

Freshly prepare. Add TEMED last, immediately before pouring the gel solution into the gel casting apparatus.

TBS-T

Prepare a 10× stock as follows:

Add 12 g Tris and 90 g NaCl to a 1-L plastic bottle with a lid.

Add 10 mL of Tween-20 (Bio-Rad) and 10 mL of NP-40 (Thermo Fisher Scientific) to the plastic bottle.

Add double-distilled H_2O to 1 L and titrate to pH 7.4 with HCl. Store for up to 6 mo at 4°C.

Make a 1× working stock (100 mL 10× TBST-T in 900 mL double-distilled H_2O) and store for up to 1 mo at room temperature.

Transfer Buffer (1×)

Prepare 10× transfer buffer as follows:

22.5 g Tris

105 g Glycine

Add double-distilled H_2O to 1 L. Store at room temperature.

Freshly prepare the 1× working buffer by adding:

100 mL 10× stock transfer buffer

200 mL methanol (100%; Sigma-Aldrich)

700 mL double-distilled H_2O

Chill to 4°C before use.

Tris-HCl (1.0 M, pH 6.8)

Add 6.06 g Tris to 40 mL double-distilled H_2O, titrate to pH 6.8 with HCl, and add double-distilled H_2O to 50 mL.

Store for up to 1 yr at room temperature.

Tris-HCl (1.5 M, pH 8.8)

Add 9.08 g Tris to 40 mL double-distilled H_2O, titrate to pH 8.8 with HCl and add double-distilled H_2O to 50 mL.

Store at room temperature for up to a year.

REFERENCES

Lajevardi A, Sajadi F, Donini A, Paluzzi J-PV. 2021. Studying the activity of neuropeptides and other regulators of the excretory system in the adult mosquito. *J Vis Exp*. doi:10.3791/61849

Misyura L, Grieco Guardian E, Durant AC, Donini A. 2020. A comparison of aquaporin expression in mosquito larvae (*Aedes aegypti*) that develop in hypo-osmotic freshwater and iso-osmotic brackish water. *PLoS ONE* **15:** e0234892. doi:10.1371/journal.pone.0234892

Qoronfleh MW, Benton B, Ignacio R, Kaboord B. 2003. Selective enrichment of membrane proteins by partition phase separation for proteomic studies. *J Biomed Biotechnol* **2003:** 249–255. doi:10.1155/S1110724303209244

Rocco DA, Kim DH, Paluzzi J-PV. 2017. Immunohistochemical mapping and transcript expression of the GPA2/GPB5 receptor in tissues of the adult mosquito, *Aedes aegypti*. *Cell Tissue Res* **369:** 313–330. doi:10.1007/s00441-017-2610-3

Schellinger JN, Rodan AR. 2015. Use of the Ramsay assay to measure fluid secretion and ion flux rates in the *Drosophila melanogaster* Malpighian tubule. *J Vis Exp* 53144. doi:10.3791/53144

 Cite this protocol as *Cold Spring Harb Protoc*; doi:10.1101/pdb.prot108283

CHAPTER 42

Performing Immunohistochemistry in Mosquito Salivary Glands

Ines Martin-Martin,[1] Thiago Luiz Alves E. Silva,[1] Adeline E. Williams,[1,2] Joel Vega-Rodriguez,[1] and Eric Calvo[1,3]

[1]*Laboratory of Malaria and Vector Research, National Institute of Allergy and Infectious Diseases, National Institutes of Health, Rockville, Maryland 20852, USA;* [2]*Department of Microbiology, Immunology, and Pathology, Colorado State University, Fort Collins, Colorado 80523, USA*

Studying protein localization in mosquito salivary glands provides novel insights on the function and physiological relevance of salivary proteins and also provides an avenue to study interactions between mosquitoes and pathogens. Salivary proteins display compartmentalization. For example, proteins involved in blood feeding are stored in the medial and distal lateral lobes, whereas proteins related to sugar metabolism localize to the proximal portion of the lateral lobes. Immunohistochemistry assays use antibodies raised against recombinant salivary proteins to reveal the protein localization and interactions within the tissue. In this assay, permeabilization of the salivary glands allows the antibodies to enter the cells and bind their target proteins. The primary antibody–antigen complexes are later marked with fluorescently labeled secondary antibodies. Antibodies that recognize pathogen-specific proteins can also be incorporated in these assays, providing information about pathogen localization within the salivary glands or pathogen interactions with mosquito salivary proteins. Here, we introduce immunohistochemistry assays for use in mosquito salivary glands.

MOSQUITO SALIVARY GLANDS

Salivary glands are a paired organ located in the anterior portion of the thoracic hemocele (Jobling 1978). Each gland contains three cylindrical lobes: one medial lobe and two lateral lobes divided into proximal and distal regions. Salivary proteins are secreted from gland cells to the cell secretory cavity and into the salivary duct that connects both glands and carries the saliva through the length of the proboscis. Saliva is injected into the host skin during mosquito probing, and salivary proteins facilitate blood feeding by preventing hemostasis (Ribeiro and Arca 2009).

COMPARTMENTALIZATION OF SALIVARY PROTEINS

Mosquito salivary proteins are produced by the salivary gland epithelial cells (Sanchez-Vargas et al. 2021). Compartmentalization of salivary proteins within the salivary glands has been observed in many studies (Ribeiro 1992; Juhn et al. 2011; Chagas et al. 2014; Wells and Andrew 2015; Martin-Martin et al. 2020; Kern et al. 2021). Salivary proteins that facilitate blood feeding are produced and stored in the medial and distal lateral lobes, whereas salivary proteins involved in sugar metabolism localize to the proximal portion of the lateral lobes. Neuron innervation varies among the salivary

[3]Correspondence: ecalvo@niaid.nih.gov

Cite this introduction as *Cold Spring Harb Protoc*; doi:10.1101/pdb.top107699

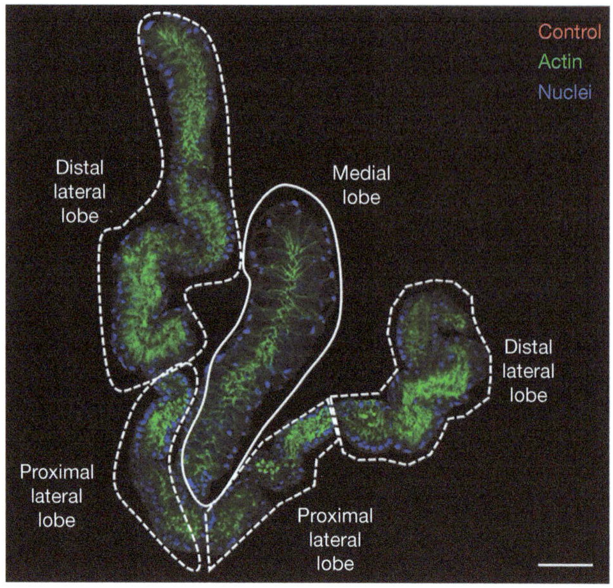

FIGURE 1. Anatomy of salivary glands. Salivary glands of *Aedes aegypti* stained with mouse antibodies raised against adjuvant only. Distal and proximal lateral lobes are indicated with dashed white lines. The medial lobe is surrounded by a solid white line. Actin, stained with Phalloidin 488, and nuclei, stained with DAPI, are shown in green and blue, respectively. Scale bar, 50 µm.

gland lobes, leading to specific responses to different stimuli. For example, mosquito feeding on blood or sugar triggers different responses that release neurotransmitters to activate specific sets of neurons (Soohoo-Hui et al. 2021). These mechanisms are poorly understood, and salivary protein compartmentalization remains a mystery.

The localization of salivary transcripts and proteins within the glands has been investigated using RNA in situ hybridization (James et al. 1991; Juhn et al. 2011) and immunohistochemistry, respectively (Ribeiro 1992; Chagas et al. 2014; Wells and Andrew 2015; Islam et al. 2019; O'Brochta et al. 2019; Martin-Martin et al. 2020). Juhn et al. (2011) reported the in situ hybridization patterns of 30 transcripts expressed in the salivary glands of adult *Aedes aegypti* females, providing the first catalog for salivary transcript localization in *Aedes* mosquitoes. Immunohistochemistry has been used to study the localization of specific proteins in the salivary glands of *Ae. aegypti* (James et al. 1991; Ribeiro 1992; Chagas et al. 2014; Wells and Andrew 2015; Islam et al. 2019; O'Brochta et al. 2019; Martin-Martin et al. 2020; Soohoo-Hui et al. 2021), *Anopheles* spp. (Wells and Andrew 2015; Islam et al. 2019; O'Brochta et al. 2019; Nozaki et al. 2020), and the less well-studied *Culex quinquefasciatus* (Martin-Martin et al. 2020; Kern et al. 2021). Comparing transcript and protein localization among different species could provide insights on translational regulation or protein transportation. Figure 1 illustrates the morphology of *Ae. aegypti* salivary glands.

MORPHOLOGY AND CELLULAR ORGANIZATION

Confocal microscopy following immunohistochemistry using antibodies against well-conserved structural and organellar markers can be used to define the morphology and cellular organization of the salivary glands. Immunohistochemistry has validated published electron microscopy data showing morphological details of mosquito salivary glands, including the specific localization of proteins and organelles. Each salivary gland lobe consists of a basal lamina surrounding a single layer of acinar cells that are distributed around a central salivary duct. Each cell is basally compressed and contains a cup-like apical cavity for saliva storage (Sánchez-Vargas et al. 2021). Wells and colleagues described the morphology and development of *An. stephensi* and *An. gambiae* female

 Cite this introduction as *Cold Spring Harb Protoc*; doi:10.1101/pdb.top107699

and male salivary glands showing a high degree of sex-specific morphological variation. They extensively described the architecture of the salivary duct, which is the passageway for salivary proteins and mosquito-borne pathogens from the salivary glands to the host (Wells and Andrew 2015; Wells et al. 2017).

MOSQUITO-BORNE PATHOGENS

Salivary glands are critical organs for mosquito-borne pathogen transmission. There are several tissue barriers that, together with mosquito genetic factors, affect the ability of a mosquito to acquire, maintain, and transmit a pathogen, therefore reducing vector competence (Sánchez-Vargas et al. 2021). Immunofluorescence studies in infected mosquito vectors have facilitated the characterization of salivary gland tissue barriers. Barriers in the salivary glands include factors that prevent arbovirus invasion or replication in the salivary gland epithelial cells, and factors that prevent arbovirus release from the salivary glands into the saliva (Franz et al. 2015). Salivary gland barriers against other medically relevant viruses including dengue, Chikungunya, and Venezuelan equine encephalomyelitis viruses in *Ae. aegypti* have been characterized (Gaidamovich et al. 1973; Salazar et al. 2007; Le Coupanec et al. 2017; Raquin and Lambrechts 2017). The insect-specific Sindbis virus has also been imaged in the salivary glands of *Ae. aegypti* and *Ae. albopictus* (Gaidamovich et al. 1973; Bowers et al. 1995; Bowers et al. 2003). Different viruses preferentially replicate in specific lobes or regions of the salivary glands (Salazar et al. 2007). For example, dengue and Chikungunya viruses infect the medial and lateral lobes of *Ae. aegypti* (Tchankouo-Nguetcheu et al. 2012), whereas Sindbis virus does not infect the medial lobes of *Aedes* spp. (Bowers et al. 2003; Ciano et al. 2014).

Immunofluorescence confocal microscopy has provided new insights into the biology of *Plasmodium* sporozoite invasion of *Anopheles* salivary glands. Immunohistochemistry showed that *Plasmodium berghei* sporozoites form clusters within the secretory cavities of *An. stephensi* salivary gland cells. Sporozoites were most often detected in the distal lateral lobes, although they occasionally invaded the medial and proximal lateral lobes (Wells and Andrew 2019).

Immunohistochemistry, coupled with technologies like fluorescence lifetime imaging microscopy (FLIM), can also be used to characterize protein–protein interactions. FLIM is an imaging technique based on the differences in the exponential decay rate of the photon emission of a fluorophore. As the fluorescence lifetime does not depend on concentration, absorption by the sample, sample thickness, photobleaching, and/or excitation intensity, it is more robust than intensity-based methods (Datta et al. 2020). The production of specific antibodies against salivary recombinant proteins coupled with the wide availability of commercial antibodies against mosquito-borne pathogens, such as *Plasmodium* parasites and dengue and Zika viruses, make the study of host–pathogen interactions possible.

IMMUNOHISTOCHEMISTRY TECHNIQUE

Immunohistochemistry of mosquito salivary glands is a technique that requires meticulous attention to detail. The most critical step is the dissection and isolation of intact salivary glands that are free of other mosquito tissues, like the fat body. Salivary glands are localized within the thorax and are surrounded by fat bodies. Dissection of intact and fat body–free salivary glands requires expertise and practice. For salivary gland dissection protocols, see Coleman et al. (2007) and Schmid et al. (2017). Salivary glands are small and delicate tissues that must be manipulated with care under a stereoscope to avoid lost or broken glands. Thus, removing fat bodies surrounding salivary glands, antibody washing, and slide-mounting steps require extra care. Our accompanying protocol (see Protocol 1: A Simple Method for Immunohistochemistry and Imaging of Mosquito Salivary Glands [Martin-Martin et al. 2022]) describes in detail the methodology for successful immunohistochemistry of mosquito salivary glands.

In conclusion, immunohistochemistry provides valuable information on the overall structure of mosquito salivary glands as well as the cellular and subcellular localization of mosquito- and pathogen-specific proteins within the salivary glands. These studies improve our understanding of salivary gland protein function and show how saliva impacts vector competence and pathogen transmission.

ACKNOWLEDGMENTS

This research was supported by the Intramural Research Program of the National Institutes of Health (NIH)/National Institute of Allergy and Infectious Diseases (NIAID) (AI001246).

REFERENCES

Bowers DF, Abell BA, Brown DT. 1995. Replication and tissue tropism of the alphavirus Sindbis in the mosquito *Aedes albopictus*. *Virology* **212**: 1–12. doi:10.1006/viro.1995.1447

Bowers DF, Coleman CG, Brown DT. 2003. Sindbis virus-associated pathology in *Aedes albopictus* (Diptera: Culicidae). *J Med Entomol* **40**: 698–705. doi:10.1603/0022-2585-40.5.698

Chagas AC, Ramirez JL, Jasinskiene N, James AA, Ribeiro JM, Marinotti O, Calvo E. 2014. Collagen-binding protein, Aegyptin, regulates probing time and blood feeding success in the dengue vector mosquito, *Aedes aegypti*. *Proc Natl Acad Sci* **111**: 6946–6951. doi:10.1073/pnas.1404179111

Ciano KA, Saredy JJ, Bowers DF. 2014. Heparan sulfate proteoglycan: an arbovirus attachment factor integral to mosquito salivary gland ducts. *Viruses* **6**: 5182–5197. doi:10.3390/v6125182

Coleman J, Juhn J, James AA. 2007. Dissection of midgut and salivary glands from *Ae. aegypti* mosquitoes. *J Vis Exp* **2007**: 228. doi:10.3791/228

Datta R, Heaster TM, Sharick JT, Gillette AA, Skala MC. 2020. Fluorescence lifetime imaging microscopy: fundamentals and advances in instrumentation, analysis, and applications. *J Biomed Opt* **25**: 071203. doi:10.1117/1.JBO.25.7.071203

Franz AWE, Kantor AM, Passarelli AL, Clem RJ. 2015. Tissue barriers to arbovirus infection in mosquitoes. *Viruses* **7**: 3741–3767. doi:10.3390/v7072795

Gaidamovich SY, Khutoretskaya NV, Lvova AI, Sveshnikova NA. 1973. Immunofluorescent staining study of the salivary glands of mosquitoes infected with group A arboviruses. *Intervirology* **1**: 193–200. doi:10.1159/000148846

Islam A, Emran TB, Yamamoto DS, Iyori M, Amelia F, Yusuf Y, Yamaguchi R, Alam MS, Silveira H, Yoshida S. 2019. Anopheline antiplatelet protein from mosquito saliva regulates blood feeding behavior. *Sci Rep* **9**: 3129. doi:10.1038/s41598-019-39960-2

James AA, Blackmer K, Marinotti O, Ghosn CR, Racioppi JV. 1991. Isolation and characterization of the gene expressing the major salivary gland protein of the female mosquito, *Aedes aegypti*. *Mol Biochem Parasitol* **44**: 245–253. doi:10.1016/0166-6851(91)90010-4

Jobling B. 1978. *Anatomical drawings of biting flies*. British Museum (Natural History) and Wellcome Trust, London.

Juhn J, Naeem-Ullah U, Maciel Guedes BA, Majid A, Coleman J, Paolucci Pimenta PF, Akram W, James AA, Marinotti O. 2011. Spatial mapping of gene expression in the salivary glands of the dengue vector mosquito, *Aedes aegypti*. *Parasit Vectors* **4**: 1. doi:10.1186/1756-3305-4-1

Kern O, Valenzuela Leon PC, Gittis AG, Bonilla B, Cruz P, Chagas AC, Ganesan S, Ribeiro JMC, Garboczi DN, Martin-Martin I, et al. 2021. The structures of two salivary proteins from the West Nile vector *Culex quinquefasciatus* reveal a β-trefoil fold with putative sugar binding properties. *Curr Res Struc Biol* **3**: 95–105. doi:10.1016/j.crstbi.2021.03.001

Le Coupanec A, Tchankouo-Nguetcheu S, Roux P, Khun H, Huerre M, Morales-Vargas R, Enguehard M, Lavillette D, Missé D, Choumet V. 2017. Co-infection of mosquitoes with chikungunya and dengue viruses reveals modulation of the replication of both viruses in midguts and salivary glands of *Aedes aegypti* mosquitoes. *Int J Mol Sci* **18**: 1708. doi:10.3390/ijms18081708

Martin-Martin I, Paige A, Valenzuela Leon PC, Gittis AG, Kern O, Bonilla B, Chagas AC, Ganesan S, Smith LB, Garboczi DN, et al. 2020. ADP binding by the *Culex quinquefasciatus* mosquito D7 salivary protein enhances blood feeding on mammals. *Nat Commun* **11**: 2911. doi:10.1038/s41467-020-16665-z

Martin-Martin I, Alves E Silva TL, Williams AE, Vega-Rodriguez J, Calvo E. 2022. A simple method for immunohistochemistry and imaging of mosquito salivary glands. *Cold Spring Harb Protoc* doi:10.1101/pdb.prot107990

Nozaki M, Baba M, Tachibana M, Tokunaga N, Torii M, Ishino T. 2020. Detection of the rhoptry neck protein complex in *Plasmodium* sporozoites and its contribution to sporozoite invasion of salivary glands. *mSphere* **5**: e00325–20. doi:10.1128/mSphere.00325-20

O'Brochta DA, Alford R, Harrell R, Aluvihare C, Eappen AG, Li T, Chakravarty S, Sim BKL, Hoffman SL, Billingsley PF. 2019. Is Saglin a mosquito salivary gland receptor for *Plasmodium falciparum*? *Malar J* **18**: 2. doi:10.1186/s12936-018-2634-5

Raquin V, Lambrechts L. 2017. Dengue virus replicates and accumulates in *Aedes aegypti* salivary glands. *Virology* **507**: 75–81. doi:10.1016/j.virol.2017.04.009

Ribeiro JM. 1992. Characterization of a vasodilator from the salivary glands of the yellow fever mosquito *Aedes aegypti*. *J Exp Bio* **165**: 61–71. doi:10.1242/jeb.165.1.61

Ribeiro JMC, Arca B. 2009. From sialomes to the sialoverse: an insight into salivary potion of blood-feeding insects. *Adv In Insect Phys* **37**: 59–118. doi:10.1016/S0065-2806(09)37002-2

Salazar MI, Richardson JH, Sánchez-Vargas I, Olson KE, Beaty BJ. 2007. Dengue virus type 2: replication and tropisms in orally infected *Aedes aegypti* mosquitoes. *BMC Microbiol* **7**: 9. doi:10.1186/1471-2180-7-9

Sánchez-Vargas I, Olson KE, Black WC. 2021. The genetic basis for salivary gland barriers to arboviral transmission. *Insects* **12**: 73. doi:10.3390/insects12010073

Schmid MA, Kauffman E, Payne A, Harris E, Kramer LD. 2017. Preparation of mosquito salivary gland extract and intradermal inoculation of mice. *Bio Protoc* **7**: e2407. doi:10.21769/BioProtoc.2407

Soohoo-Hui A, Li Z, Maldonado-Ruiz LP, Zhang G, Swale DR. 2021. Neurochemical regulation of *Aedes aegypti* salivary gland function. *J Insect Physiol* **129**: 104193. doi:10.1016/j.jinsphys.2021.104193

Tchankouo-Nguetcheu S, Bourguet E, Lenormand P, Rousselle J-C, Namane A, Choumet V. 2012. Infection by chikungunya virus modulates the expression of several proteins in *Aedes aegypti* salivary glands. *Parasit Vectors* **5**: 264. doi:10.1186/1756-3305-5-264

Wells MB, Andrew DJ. 2015. Salivary gland cellular architecture in the Asian malaria vector mosquito *Anopheles stephensi*. *Parasit Vectors* **8**: 617. doi:10.1186/s13071-015-1229-z

Wells MB, Andrew DJ. 2019. Anopheles salivary gland architecture shapes *Plasmodium* sporozoite availability for transmission. *mBio* **10**: e01238-19. doi:10.1128/mBio.01238-19

Wells MB, Villamor J, Andrew DJ. 2017. Salivary gland maturation and duct formation in the African malaria mosquito *Anopheles gambiae*. *Sci Rep* **7**: 601. doi:10.1038/s41598-017-00672-0

Protocol 1

A Simple Method for Immunohistochemistry and Imaging of Mosquito Salivary Glands

Ines Martin-Martin,[1] Thiago Luiz Alves E. Silva,[1] Adeline E. Williams,[1,2] Joel Vega-Rodriguez,[1] and Eric Calvo[1,3]

[1]*Laboratory of Malaria and Vector Research, National Institute of Allergy and Infectious Diseases, National Institutes of Health, Rockville, Maryland 20852, USA;* [2]*Department of Microbiology, Immunology, and Pathology, Colorado State University, Fort Collins, Colorado 80523, USA*

Immunohistochemistry is a valuable technique that provides information on protein localization and interactions in tissues. Mosquito salivary gland immunohistochemistry requires the meticulous dissection of a delicate tissue. The integrity of the salivary glands must be closely monitored throughout the entire process to prevent structural damage and loss of saliva. This protocol describes a series of simple steps to perform salivary gland immunohistochemistry including tissue dissection, permeabilization, immunostaining, mounting, and imaging by confocal microscopy.

MATERIALS

It is essential that you consult the appropriate Material Safety Data Sheets and your institution's Environmental Health and Safety Office for proper handling of equipment and hazardous materials used in this protocol.

RECIPES: Please see the end of this protocol for recipes indicated by <R>. Additional recipes can be found online at http://cshprotocols.cshlp.org/site/recipes.

Reagents

Alexa Fluor 488 phalloidin (2-µg/mL; Thermo Fisher Scientific A12379)

CO_2 for anesthetizing mosquitoes (see Step 1)

Custom primary antibodies raised in mouse against salivary proteins of interest such as maltases, D7 long form proteins, or CqDVP proteins

These are available from the authors upon request.

DAPI (1-mg/mL; Sigma-Aldrich D9542)

Ethanol (70%)

Immunohistochemistry blocking buffer <R>

Immunohistochemistry washing buffer <R>

Mosquitoes (5- to 7-d-old females, raised as described in Ote and Kanuka 2018)

The protocol can be used for both female and male mosquitoes. However, female mosquitoes are the only ones that transmit diseases, so only salivary glands from females have been described here.

Paraformaldehyde (Electron Microscopy Science 15710), diluted to 4% (v/v) in distilled H_2O

[3]Correspondence: ecalvo@niaid.nih.gov

Cite this protocol as *Cold Spring Harb Protoc*; doi:10.1101/pdb.prot107990

Phosphate buffered saline (PBS), pH 7.4 (Gibco 10010-023)
ProLong Gold antifade reagent with DAPI mounting medium (Invitrogen P36935)
Texas Red-X goat anti-mouse IgG (H + L) fluorescent secondary antibody (Thermo Fisher T6390)

Equipment

Aluminum sealing foil
Computer with Imaris software version 9.2.1
Corning PYREX 9 Depression Glass Spot Plates, 85 × 100-mm, with well dimensions of 22-mm (outer diameter) × 7-mm (deep) (Thomas Scientific 7812G17)
Entomological tweezers (Dumont No. 5)
Glass concave well slides
Homemade dissection needles

To prepare these needles, insert and glue entomological BioQuip minutien stainless steel short pins, number 20, with a 0.20-mm diameter (EntoSPHINX manufactured for BioQuip 1208SB) to a wooden applicator (Puritan 807) using regular superglue (e.g., Loctite).

Leica SP8 Microscope with a 20× oil immersion objective and a photomultiplier tube/hybrid detector
Microscope coverslips
Microscope glass slides
Mosquito aspirator, electric (PN 419, John W. Hock Company)
Petri dishes
Stereo microscope (Stemi 508, Zeiss)

METHOD

Salivary Gland Dissection

See Coleman et al. (2007) and Fig. 1 for details on dissection.

1. Pick up approximately 20 mosquitoes with an electronic aspirator. Anesthetize the mosquitoes by applying CO_2 for 20–30 sec to the aspirator. Immediately transfer the anesthetized mosquitoes to a Petri dish set on ice.
2. Using tweezers, transfer one mosquito to a concave well slide containing three to four drops of cold PBS (Fig 1C).
3. Under the stereomicroscope, separate the head from the thorax and body using fine homemade needles (Fig 1D).
4. Gently press down on the thorax with one needle while scooping the salivary glands out of the thorax with the other needle.

   For salivary gland recognition, see Coleman et al. (2007) and Figure 1I–L.
5. Using a needle, transfer the glands to a new drop of cold PBS within the same concave well to separate and remove the fat bodies, which are visualized as diffuse white tissue surrounding the salivary glands (Fig 1I).

   See Troubleshooting.
6. Transfer the glands to a nine-depression glass spot plate containing cold PBS. Place the plate on ice (Fig 1E–H).

   The optimal number of glands is five to 10 glands/well.

Salivary Gland Immunostaining

Perform all buffer exchanges under the microscope to monitor that no glands are lost while pipetting (Fig. 1M). Use incubation volumes of 500 µL unless stated otherwise.

 Cite this protocol as *Cold Spring Harb Protoc*; doi:10.1101/pdb.prot107990

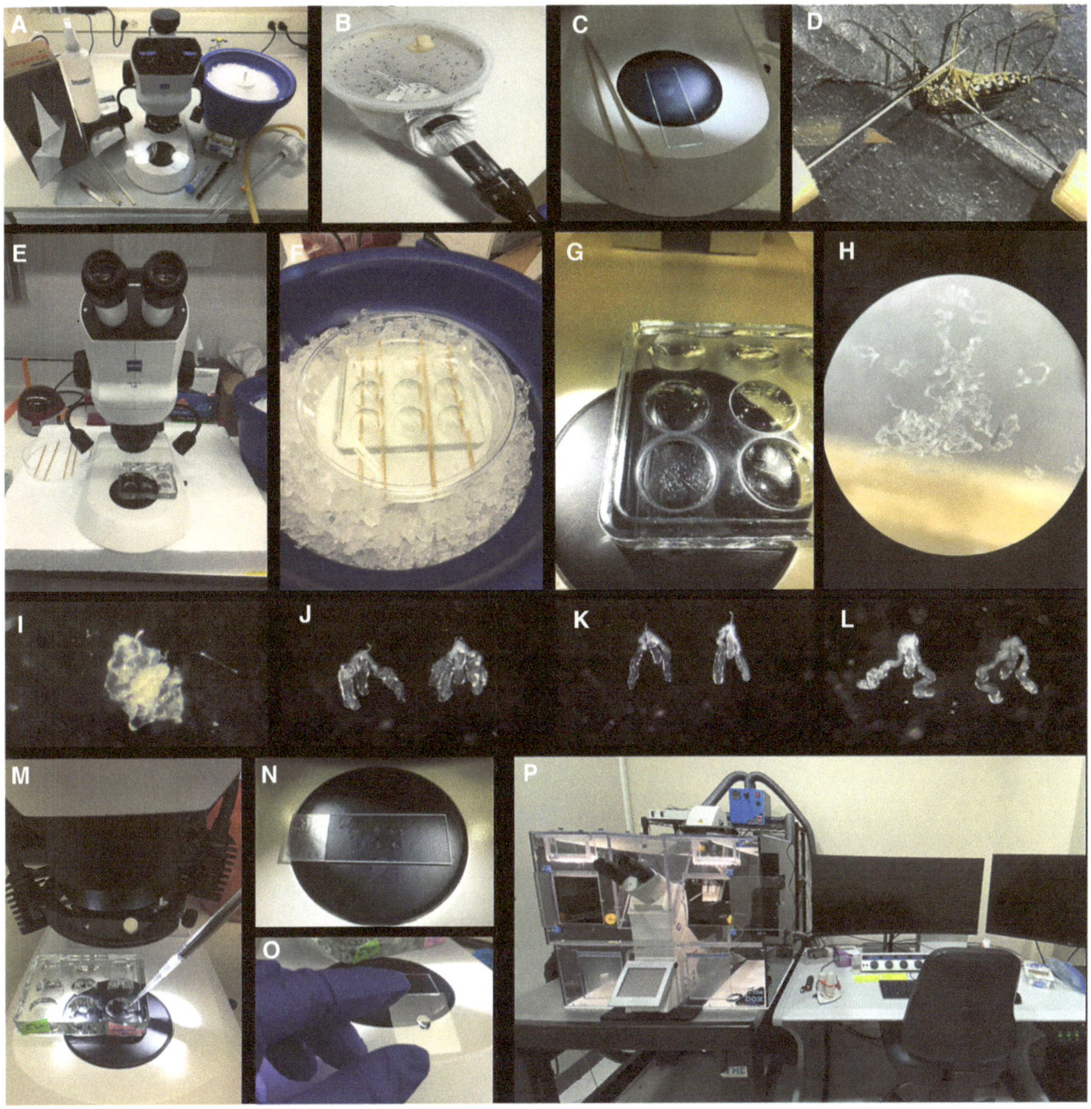

FIGURE 1. Immunohistochemistry of mosquito salivary glands. (*A*) Dissection station. (*B*) Mosquito aspiration. (*C*) Fine needles and concave well slide for dissections (Step 2/3). (*D*) *Aedes aegypti* female mosquito dissection (Step 3). (*E*) Stereomicroscope setup. (*F*) While dissecting, salivary glands are deposited into a glass concave well plate and maintained on ice (Step 6). (*G,H*) Concave well plate with dissected salivary glands. (*I*) *Ae. aegypti* salivary glands surrounded by fat body tissue (Step 5). (*J–L*) Clean *Ae. aegypti*, *Anopheles gambiae*, and *Culex quinquefasciatus* salivary glands, respectively. (*M*) Washing step monitored under stereoscope. (*N*) Salivary glands are transferred to a drop of PBS onto a slide (Step 17). (*O*) Prolong Gold mounting media is added to a coverslip (Step 19). (*P*) Leica SP8 confocal microscope (Step 22).

7. Remove PBS carefully and fix the tissues with 4% paraformaldehyde for 30 min at room temperature.

   No agitation is required.

   See Troubleshooting.

8. Remove the paraformaldehyde solution and wash the glands three times with PBS for 10 min each wash at room temperature.

   The fixed tissues can be stored in PBS at 4°C until further processing (up to 1 yr).

   See Troubleshooting.

9. Remove PBS and incubate the tissues with blocking buffer for 30 min at room temperature.

   The Triton X-100 in the blocking buffer permeabilizes the cell membrane, thus facilitating antibody access into the tissue.

   See Troubleshooting.

10. Remove blocking buffer by washing three times for 10 min each with washing buffer at room temperature to remove traces of Triton X-100 that might interfere with the antibody binding.

    See Troubleshooting.

11. Using a needle, separate glands into wells labeled with the appropriate primary antibody name, replace the washing buffer with the primary antibody diluted in washing buffer, and incubate overnight at 4°C.

    Most primary antibodies usually work at a concentration of 1–10 µL/mL, although the optimal antibody concentration for each primary antibody should be determined empirically.

    Five salivary gland pairs per well is good number with which to start.

    See Troubleshooting.

12. On the following day, remove the primary antibody solution and wash with PBS three times for 10 min each at room temperature. Incubate for 1 h at room temperature with the appropriate fluorescent secondary antibody diluted in washing buffer.

    From this step on, keep concave well dishes protected from the light by covering them with aluminum foil.

    See Troubleshooting.

13. Remove the secondary antibody solution and wash tissues three times for 10 min each with washing buffer at room temperature.
14. Stain DNA and actin, respectively, with 1 µg/mL DAPI and 0.04 µg/mL phalloidin Alexa 488 diluted in PBS for 20 min at room temperature.
15. Remove the solution with DAPI and phalloidin and wash tissues three times for 10 min each with washing buffer at room temperature.

    Stained salivary glands can be stored in PBS for at least 6 mo at 4°C.

Salivary Gland Mounting

16. Clean microscope slides with 70% ethanol, and label them with preparation date and sample information.
17. Place three to four individual drops of ~5 µL of PBS onto the slide within a range that would fit under the coverslip (Fig. 1N).
18. Carefully, using fine needles, transfer one salivary gland per drop of PBS. Make sure the lobes are well-separated and the medial lobe, which is shorter in length, is located between the two lateral lobes (see Fig. 1 in Introduction: Performing Immunohistochemistry in Mosquito Salivary Glands [Martin-Martin et al. 2022]).
19. Once there are three to four glands on the slide, pipette 20–25 µL of ProLong Gold antifade mounting medium onto a coverslip (Fig. 1O). Remove any resulting air bubbles in the mounting medium using the pipette tip.
20. To dry the glands and make sure the glands remain well-positioned, remove the remaining PBS by pipetting. Avoid overdrying by immediately covering the glands entirely with the coverslip.
21. Leave the slides to dry at room temperature overnight and then preserve at 4°C until imaging.

Imaging

See Figures 2 and 3 for examples of acquired images of salivary glands.

 Cite this protocol as *Cold Spring Harb Protoc*; doi:10.1101/pdb.prot107990

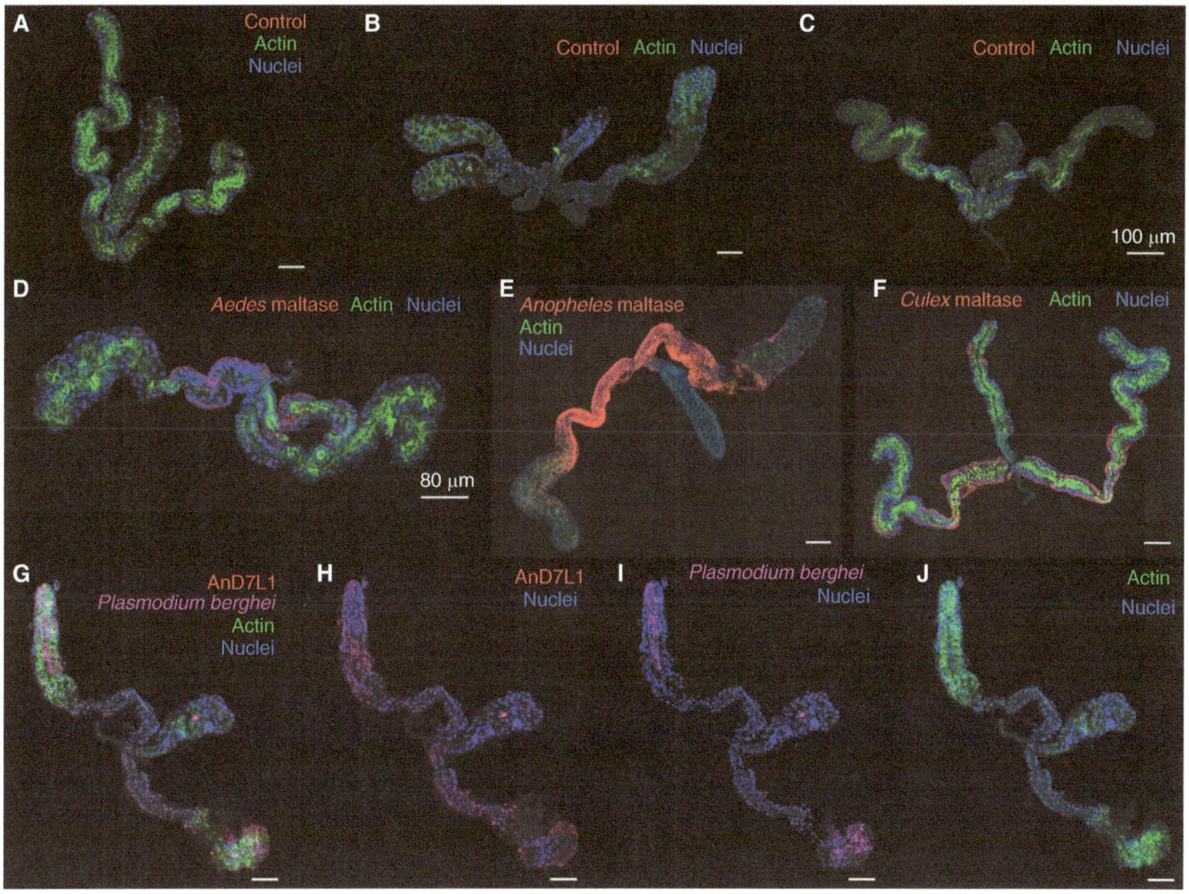

FIGURE 2. Confocal microscopy images of mosquito salivary glands. Salivary glands of *Aedes aegypti* (*A*,*D*), *Anopheles gambiae* (*B*,*E*), and *Culex quinquefasciatus* (*C*,*F*) stained with mouse antibodies raised against adjuvant only (*A–C*) or *Aedes* maltase (AAEL009524) (*D*), *Anopheles* maltase (AGAP002102) (*E*), and *Culex* maltase (CPIJ016362) (*F*). Maltases are localized in the proximal lateral lobes. (*G–J*) *An. gambiae* salivary gland infected with *Plasmodium berghei*-green fluorescent protein (GFP). Salivary glands were dissected 24 d after parasite infection and stained with Alexa Fluor 647 mouse anti-GFP fluorescent antibody. (*G*) *An. gambiae* salivary D7 long form protein 1 (AGAP008278) showed an ubiquitous localization, whereas *P. berghei* invaded the distal lateral lobes, as previously described (Wells and Andrew 2019) (*H*). (*I*) Merged image showing 498 E and 405 channels. Texas Red-X anti-mouse IgG; actin, stained with Phalloidin 488; and nuclei, stained with DAPI, are shown in red, green, and blue, respectively. Scale bars, 50 µm, unless otherwise indicated in the images.

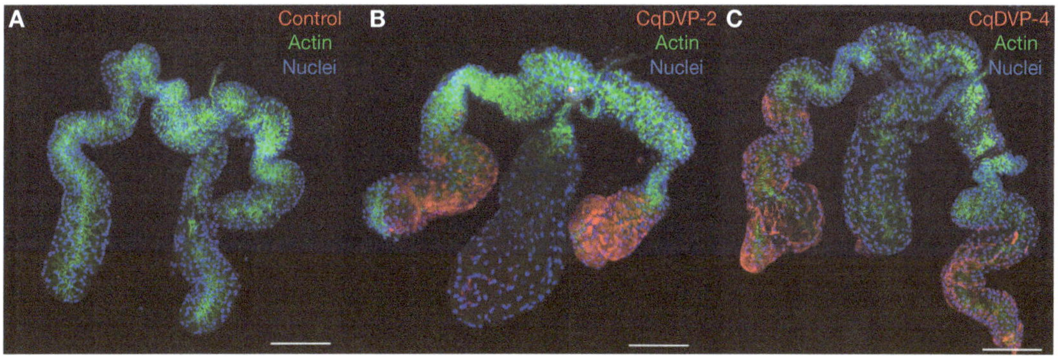

FIGURE 3. Salivary CqDVP-2 and CqDVP-4 localize to the distal lateral lobes of *Culex quinquefasciatus* salivary glands. Glands incubated with sera from mice immunized against only adjuvant, CqDVP-2, and CqDVP-4 are shown in *A–C*, respectively. Texas Red-X anti-mouse IgG; actin, stained with Phalloidin 488; and nuclei, stained with DAPI, are shown in red, green, and blue, respectively. Scale bars, 100 µm. (Modified from Kern et al. 2021.)

22. Capture fluorescent images with a Leica SP8 Microscope using a 20× oil immersion objective (using zoom factor 2 or 3; numerical aperture 1.25) equipped with a photomultiplier tube/hybrid detector (Fig. 1P).

    See Troubleshooting.

23. Visualize salivary glands using a 405 laser (for DAPI) and white light laser. Set peaks of excitation for each fluorophore at 498, 595, and 665 nm when Alexa 488, Texas red, and Alexa 647, respectively, are used as the fluorophores conjugated to the secondary antibodies (Figs. 2 and 3).

24. Acquire images using sequential mode and variable *z*-steps.

TROUBLESHOOTING

Problem (Step 5): The salivary glands have too much fat tissue attached.
Solution: Certain mosquito species have a higher content of thoracic fat body tissues surrounding the salivary glands that must be removed as follows before initiating the immunohistochemistry protocol.

- Using the highest magnification of the stereomicroscope, remove fat body tissues with fine needles. It might take practice, as salivary gland lobes might break during this process.
- Alternatively, fat body tissues can be washed away. While the salivary glands are in PBS in the well, gently pipette the salivary gland in and out several times with a 10-µL pipette tip.

Problem (Steps 7–12): Salivary glands are lost during the washing steps.
Solution: Always perform the washing steps under the stereomicroscope to monitor that no glands are accidentally removed during washings. Remove air bubbles due to traces of Triton X-100 or BSA that may entrap salivary glands, thus making it difficult to visualize them. Working with a small number of glands per well will minimize this issue; the optimal number is five to 10 glands/well.

Problem (Step 22): Staining is impaired as a result of the attachment of the salivary glands to the well surface during immunostaining.
Solution: During Steps 7–12, use fine needles to separate the glands from the walls under the stereoscope at high magnification (40×).

Problem (Step 22): There is high background during imaging.
Solution: Make sure that glands are well-washed and ensure the following.

- Remove all buffer contents in the well before adding the washing buffer.
- Change tips between different sample washings.
- Optimize the antibody concentration.

RECIPES

Immunohistochemistry Blocking Buffer

Reagents	Concentration
Bovine serum albumin (BSA) (Sigma-Aldrich A7409)	2%
Triton X-100 (Sigma-Aldrich T8787)	0.5%

Prepare in phosphate buffered saline (PBS) with a pH of 7.4 (Gibco 10010-023). Store for up to 6 mo at 4°C.

Immunohistochemistry Washing Buffer

Reagent	Concentration
Bovine serum albumin (BSA) (Sigma-Aldrich A7409)	2%

Prepare in phosphate buffered saline (PBS) with a pH of 7.4 (Gibco 10010-023). Store for up to 6 mo at 4°C.

ACKNOWLEDGMENTS

The authors thank Brian Bonilla for excellent salivary gland dissections and photos and Olivia Kern and Sundar Ganesan for their input in the protocol optimization. We also thank Leticia B. Smith for providing antibodies against *Anopheles gambiae* D7L1.

REFERENCES

Coleman J, Juhn J, James AA. 2007. Dissection of midgut and salivary glands from *Ae. aegypti* mosquitoes. *J Vis Exp* **5:** 228. doi:10.3791/228

Kern O, Valenzuela Leon PC, Gittis AG, Bonilla B, Cruz P, Chagas AC, Ganesan S, Ribeiro JMC, Garboczi DN, Martin-Martin I, et al. 2021. The structures of two salivary proteins from the West Nile vector *Culex quinquefasciatus* reveal a β-trefoil fold with putative sugar binding properties. *Curr Res Struct Biol* **3:** 95–105. doi:10.1016/j.crstbi.2021.03.001

Martin-Martin I, Alves E Silva TL, Williams AE, Vega-Rodriguez J, Calvo E. 2022. Performing immunohistochemistry in mosquito salivary glands. *Cold Spring Harb Protoc.* doi:10.1101/pdb.top107699

Ote M, Kanuka H. 2018. A highly secure method for rearing *Aedes aegypti* mosquitoes. *Trop Med Health* **46:** 16. doi:10.1186/s41182-018-0098-5

Wells MB, Andrew DJ. 2019. *Anopheles* salivary gland architecture shapes *Plasmodium* sporozoite availability for transmission. *mBio* **10:** e01238-19. doi:10.1128/mBio.01238-19

CHAPTER 43

RNA In Situ Hybridization and Immunohistochemistry to Visualize Gene Expression in Peripheral Chemosensory Tissues of Mosquitoes

Margaret Herre[1,2,4] and Chloe Greppi[3,4]

[1]*The Laboratory of Neurogenetics and Behavior, The Rockefeller University, New York, New York 10065, USA;* [2]*Kavli Neural Systems Institute, New York, New York 10065, USA;* [3]*Department of Biological Sciences, Columbia University, New York, New York 10027, USA*

Mosquito-borne diseases such as malaria, Zika virus, and dengue virus are a menace to the human population. Although many mosquito species are not attracted to humans and do not feed on blood, human-biting female mosquitoes are strongly attracted to people and use chemosensory cues to identify a suitable host for a blood meal. Mosquitoes need blood components to reproduce, rendering them excellent vectors for blood-borne diseases. The three genera (*Culex*, *Anopheles*, and *Aedes*) responsible for most of these diseases find hosts by using their peripheral sensory organs. These organs include the antennae, maxillary palps, and proboscis. All three contain diverse populations of highly sensitive neurons that express sensory receptors that can detect odorants, temperature, chemicals, and tastants. Although these organs are essential to the host-seeking behavior that results in biting, their small size and thick outer cuticle can hinder typical histochemical analyses. Here, we briefly review the role the peripheral sensory organs play in mosquito behavior. Then, we introduce how to investigate their gene expression profiles using immunohistochemical and RNA in situ approaches for both whole-mount and frozen-section preparations.

BACKGROUND

The antennae and maxillary palps of the female mosquito sense host-associated cues such as odors, carbon dioxide, heat, and moisture (Brown 1966; DeGennaro et al. 2013; McMeniman et al. 2014; Corfas and Vosshall 2015; Raji and DeGennaro 2017; Greppi et al. 2020). These cues allow a mosquito to locate hosts on which to blood feed, a step that provides proteins necessary for egg production. These cues are detected by different classes of sensory receptors that are expressed in olfactory sensory neurons throughout these tissues. Three large gene families—namely, odorant receptors (ORs), ionotropic receptors (IRs), and gustatory receptors (GRs)—encode ionotropic ligand-gated ion channels, and mosquitoes express hundreds of these genes in their olfactory tissues. ORs detect odors, IRs sense heat and some volatile chemicals, and GRs detect the important host cue carbon dioxide. In model systems, odor identity is encoded by the combined activation of subpopulations of olfactory neurons in specific combinatorial patterns (Rahm 1957; Hallem et al. 2004; Andersson et al. 2015). This property of combinatorial coding likely enables mosquitoes to detect and distinguish many olfactory stimuli, including components of human odor. In addition to host-associated cues, these appendages

[4]Correspondence: margo.herre@gmail.com; cg3251@columbia.edu

Cite this introduction as *Cold Spring Harb Protoc*; doi:10.1101/pdb.top107700

are important for the detection and evaluation of other stimuli such as nectar or sugar sources, oviposition sites, and auditory signals from potential mates. Thus, these appendages are of broad interest given their varied roles in mosquito reproduction and ecology. The accompanying protocols, which are introduced below, can be adapted to identify a range of transcripts and protein types.

IMPROVED METHODS TO CHARACTERIZE PROTEIN AND GENE EXPRESSION

Robust methods of visualizing transcript expression and protein localization are important tools in the characterization of sensory receptor expression in mosquito peripheral tissues and for the development of vector control strategies that target this deadly olfactory behavior. In the accompanying protocols, we present procedures for whole-mount and frozen-section immunohistochemistry for reliable protein labeling and for whole-mount RNA in situ hybridization. The peripheral tissues have a thick layer of cuticle that produces background fluorescence and hinders reagent penetration. These methods circumvent this problem by using sections or digestion steps to reduce cuticle-related complications.

The first protocol (see Protocol 1: Cryosectioning and Immunohistochemistry of Peripheral Tissues of Mosquitoes [Greppi 2022]) describes a method to cryosection mosquito peripheral tissues followed by immunofluorescence. This strategy enables the detection of protein expression in both superficial and deep antennal structures and does not rely on digestion steps to achieve antibody penetration into tissue. In addition, sections can provide detailed information on the distribution of different cell types and are useful for cell-counting experiments.

The second (see Protocol 2: Immunostaining Whole-Mount Olfactory Tissues of Mosquitoes [Herre 2022a]) and third procedures (see Protocol 3: RNA In Situ Hybridization of Whole-Mount Mosquito Olfactory Tissues [Herre 2022b]) are whole-mount protocols that use a chitinase-chymotrypsin digestion step before fixation to enable RNA in situ probes and antibodies to penetrate the cuticle (Fig. 1). A whole-mount immunostaining preparation may be warranted when it is desirable to keep the entire antenna intact for cell counting purposes or to examine the spatial distribution of proteins across many segments of the antenna. Finally, RNA in situ by hybridization chain reaction (Choi et al. 2018) enables the localization of gene expression when antibodies are not available. This method has been particularly useful for use in emerging model organisms, such as mosquitoes, that have sequenced genomes but relatively few genetic tools and available antibodies to enable fluorescent visualization of proteins of interest.

These methods have been optimized to enable the detection of chemosensory genes; however, they are generalizable to other targets in these tissues, including but not limited to transcription factors, cytosolic proteins, and transmembrane proteins. Furthermore, RNA in situ hybridization can be used to detect noncoding RNAs and alternative splicing events. Although the selection of antibodies and probes is out of the scope of the associated protocols, users are advised to optimize staining procedures by using antibodies and/or probes against targets that are relatively highly expressed in bulk sequencing data (e.g., >100 transcripts per million), as these chitinous tissues can show high background and noise can be easily mistaken for true signal. It should be noted that an extensive study of any transcript detection threshold has not been conducted using this protocol.

Additionally, the procedures described in the accompanying protocols have been optimized for maxillary palps and antennae; however, these techniques can likely be applied to other peripheral appendages such as the legs or proboscis.

SUMMARY

Blood-feeding mosquitoes transmit deadly pathogens to millions of people annually. The localization and expression levels of the genes and proteins required for chemosensory neuron function and

Cite this introduction as *Cold Spring Harb Protoc*; doi:10.1101/pdb.top107700

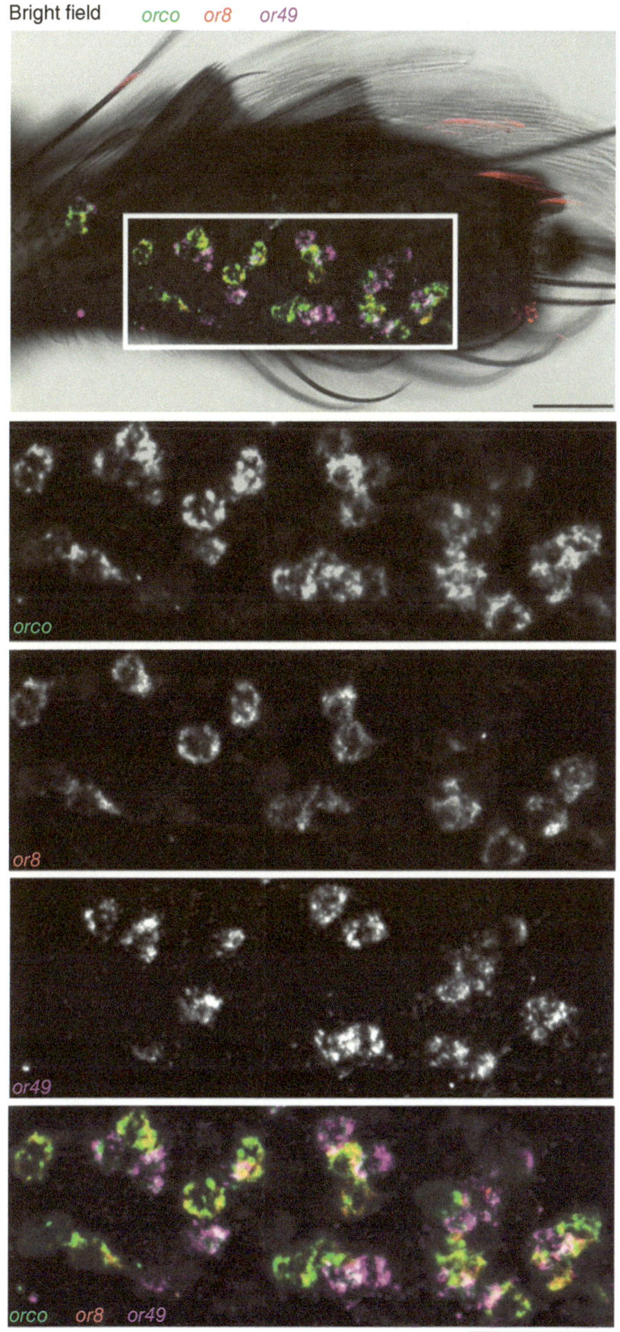

FIGURE 1. Confocal maximum projection of a female *Aedes aegypti* maxillary palp using hybridization chain reaction RNA in situ hybridization (Protocol 3: RNA In Situ Hybridization of Whole-Mount Mosquito Olfactory Tissues [Herre 2022b]) with probes against *orco*, *or8*, and *or49*. The white inset in the *top* panel is enlarged to show individual and merged channels in the panels below. Scale bars, 25 µm.

downstream behavior are crucial to our understanding of how the mosquito olfactory system detects human odor. The accompanying protocols provide an entry point toward answering these questions by identifying the expression profiles and molecular repertoire of chemosensory neurons. Understanding this complex system central to host-seeking and therefore disease transmission is crucial in the fight to reduce the burden of mosquito-borne diseases.

REFERENCES

Andersson MN, Löfstedt C, Newcomb RD. 2015. Insect olfaction and the evolution of receptor tuning. *Front Ecol Evol* **3:** 1–13.

Brown AWA. 1966. The attraction of mosquitoes to hosts. *J Am Med Assoc* **196:** 249–252. doi:10.1001/jama.1966.03100160099028

Choi HMT, Schwarzkopf M, Fornace ME, Acharya A, Artavanis G, Stegmaier J, Cunha A, Pierce NA. 2018. Third-generation in situ hybridization chain reaction: multiplexed, quantitative, sensitive, versatile, robust. *Development* **145:** dev165753. doi:10.1242/dev.165753

Corfas RA, Vosshall LB. 2015. The cation channel TRPA1 tunes mosquito thermotaxis to host temperatures. *eLife* **4:** e11750. doi:10.7554/eLife.11750

DeGennaro M, McBride CS, Seeholzer L, Nakagawa T, Dennis EJ, Goldman C, Jasinskiene N, James AA, Vosshall LB. 2013. *orco* mutant mosquitoes lose strong preference for humans and are not repelled by volatile DEET. *Nature* **498:** 487–491. doi:10.1038/nature12206

Greppi C. 2022. Cryosectioning and immunohistochemistry of peripheral tissues of mosquitoes. *Cold Spring Harb Protoc* doi:10.1101/pdb.prot107914.

Greppi C, Laursen WJ, Budelli G, Chang EC, Daniels AM, van Giesen L, Smidler AL, Catteruccia F, Garrity PA. 2020. Mosquito heat seeking is driven by an ancient cooling receptor. *Science* **367:** 681–684. doi:10.1126/science.aay9847

Hallem EA, Ho MG, Carlson JR. 2004. The molecular basis of odor coding in the *Drosophila* antenna. *Cell* **117:** 965–979. doi:10.1016/j.cell.2004.05.012

Herre M. 2022a. Immunostaining whole-mount olfactory tissues of mosquitoes. *Cold Spring Harb Protoc* doi:10.1101/pdb.prot107915.

Herre M. 2022b. RNA in situ hybridization of whole-mount mosquito olfactory tissues. *Cold Spring Harb Protoc* doi:10.1101/pdb.prot107916.

McMeniman CJ, Corfas RA, Matthews BJ, Ritchie SA, Vosshall LB. 2014. Multimodal integration of carbon dioxide and other sensory cues drives mosquito attraction to humans. *Cell* **156:** 1060–1071. doi:10.1016/j.cell.2013.12.044

Rahm U. 1957. On the significance of odor and sweat in the attraction of *Aedes aegypti* to man. *Acta Trop* **14:** 208–217.

Raji JI, DeGennaro M. 2017. Genetic analysis of mosquito detection of humans. *Curr Opin Insect Sci* **20:** 34–38. doi:10.1016/j.cois.2017.03.003

Cite this introduction as *Cold Spring Harb Protoc*; doi:10.1101/pdb.top107700

Protocol 1

Cryosectioning and Immunohistochemistry of Peripheral Tissues of Mosquitoes

Chloe Greppi[1,2]

[1]*Department of Biological Sciences, Columbia University, New York, New York 10027, USA*

Immunohistochemistry analysis of mosquitoes is complicated by the outer cuticle that prevents reagents from penetrating peripheral tissues. This protocol incorporates a cryosectioning method that provides a higher resolution of the internal architecture of mosquito peripheral sensory tissues and enables the visualization of protein expression. This eliminates the need for enzymatic steps to digest the outer cuticle that encases these tissues. This protocol can also be adapted for other tissues, such as the brain and the legs, as chitin exoskeleton thickness does not affect antibody penetration once the sample is sectioned.

MATERIALS

It is essential that you consult the appropriate Material Safety Data Sheets and your institution's Environmental Health and Safety Office for proper handling of equipment and hazardous materials used in this protocol.

RECIPES: Please see the end of this protocol for recipes indicated by <R>. Additional recipes can be found online at http://cshprotocols.cshlp.org/site/recipes.

Reagents

Blocking solution for immunostaining <R>
Cryoprotection solution <R>
Dissection surface (e.g., a slide or Sylgard dish)
Ethanol (70%)
Mosquitoes (any genera, male or female)
Mounting medium (e.g., VectaShield [Vector Labs H-1000-10] or SlowFade Diamond [Thermo Fisher S36972], with or without 4′,6-diamidino-2-phenylindole [DAPI])
Optimal cutting medium (e.g., Tissue-Tek O.C.T. Compound [Sakura 4583])
Paraformaldehyde (PFA; 4%) <R>, prechilled on ice
Phosphate-buffered saline (PBS; 10×) (Thermo Scientific J62036-K2)
Phosphate-buffered saline with Triton X-100 (PBST) <R>
Primary antibody

Many Drosophila melanogaster *primary antibodies can be used because they have cross-reactivity, or an antibody can be custom-made for the mosquito.*

[2]Correspondence: cg3251@columbia.edu

Cite this protocol as *Cold Spring Harb Protoc*; doi:10.1101/pdb.prot107914

Secondary antibody

Any standard secondary antibody can be used, such as Life Technologies Alexa Fluor secondaries.

Equipment

Closed container, such as an empty pipette box (optional; see Step 8)

Ideally it should be airtight to prevent the sample from drying out.

Confocal microscope
Coverslips

Select the appropriate coverslip thickness for the microscope being used.

Dissecting forceps
Dissecting microscope
Dissecting well (optional; see Step 6)
Dry ice
Frozen microtome or cryostat (see Step 9)
Glass Pasteur pipette (VWR 14673)
Glass slides, charged (e.g., Superfrost Plus)
Humidified chamber

Humidify a slide chamber by adding wet Kimwipes.

Ice
Kimwipes
Microcentrifuge tubes (1.5-mL)
Microtome blades
Nail polish
Nutator or rocker
Parafilm (optional; see Step 13)
Petri dish and lid
Tissue embedding molds (ideally plastic; VWR 25608-922)

METHOD

This protocol was adapted from DeGennaro et al. (2013).

Tissue Fixation and Cryoprotection

1. Using forceps, place approximately 25 mosquitoes in a covered Petri dish. Place the dish on ice for 10 min to anesthetize the mosquitoes.

   Mosquitoes are fully anesthetized when legs and wings are no longer moving.

2. Working quickly, place a mosquito under a dissecting microscope and use forceps to remove the head from the body. Place the head in 1 mL of ice-cold 4% PFA in a 1.5-mL microcentrifuge tube.
3. Fix heads for 30 min on ice.
4. Remove the 4% PFA solution and wash heads with a glass Pasteur pipette two times with 1 mL of PBS.
5. Place heads in a new 1.5-mL tube containing 1 mL cryoprotection solution and incubate on a nutator or rocker overnight for 12–16 h at 4°C.

Cite this protocol as *Cold Spring Harb Protoc*; doi:10.1101/pdb.prot107914

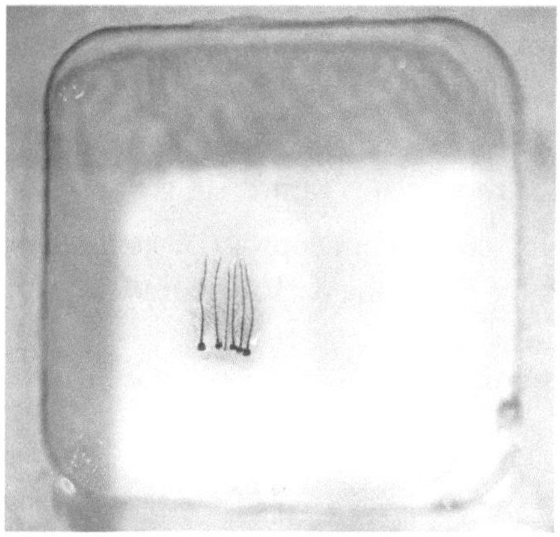

FIGURE 1. Female mosquito antennae lined up in a tissue embedding mold. Note that the antennae are close together but not overlapping.

Tissue Dissection and Sectioning

6. Dissect the peripheral tissue of choice on a slide or dissecting well in ice-cold 70% ethanol, and carefully transfer the tissue using forceps to the tissue embedding mold. Arrange the antennae in a single row of approximately 15 to 20 antennae per mold, without letting them overlap each other (Fig. 1).

   The cryoprotection solution evaporates quickly and renders the tissue sticky and prone to breaking. Take care to keep the tissue in enough 70% ethanol until you are satisfied with its position in the mold. It is very common to lose tissue during sectioning and staining. Up to 30 antennae can fit comfortably in a mold.

7. Carefully use a Kimwipe to soak up any excess ethanol on the tissue embedding mold, and air-dry the tissue on the benchtop for 5 min.
8. Cover the tissue with optimal cutting medium (enough to completely cover and surround tissue, at least ~250 µL) and immediately freeze on dry ice.

   At this point, samples can be stored in a closed, ideally airtight container for several months at −80°C.

9. Use a microtome blade to section the samples sagittally (along the entire length of the antenna) at 10–15 µm on a frozen microtome or cryostat. Let the sections dry completely (at least 30 min) at room temperature before proceeding.

   Typically, two to three slides are needed per block of sectioned tissue. At this point, slides with sections can be stored for several weeks at −20°C.

Immunostaining

All solutions are applied directly to the slides unless otherwise noted.

10. If slides were previously frozen, warm them up for ~30 min to room temperature. Then fix samples by adding 250 µL of 4% PFA per slide for 15 min at room temperature.

    Be sure that all of the tissue is covered by the solution.

11. Discard the 4% PFA by blotting the slides, by tilting them gently by ~45° on a Kimwipe. Wash slides by adding 200 µL of PBST to each slide. Wash three times for 5 min each at room temperature, slowly rocking on a benchtop rocker.

    Take care that the slides do not fall or slide around.

12. Blot the slides on a Kimwipe, and block tissue in 250 µL of blocking solution for immunostaining for 1 h at room temperature, slowly rocking.
13. Incubate slides for 12–16 h in 200 µL of primary antibody diluted (typically 1/50–1/150) in blocking solution for immunostaining at 4°C in a humidified chamber.

To minimize evaporation, coverslips or thin rectangles of Parafilm can be carefully placed on the slides. Alternatively, the humidified chamber can be sealed with Parafilm. If primary antibody incubation needs to be >2 d, it may be necessary to refresh the solution to avoid evaporation.

14. Remove coverslips or Parafilm carefully, if using. Blot excess primary solution and wash six times for 20 min each with 200 µL of PBST at room temperature, slowly rocking.
15. Incubate slides overnight for 12–16 h in 200 µL of an appropriate secondary antibody diluted in blocking solution for immunostaining for 3 h in the dark at room temperature in a humidified chamber.

    Most commercially available secondary antibodies are effective at a dilution of 1/200–1/500.
16. Wash slides three times for 10 min each with 200 µL of PBST at room temperature, slowly rocking.
17. Blot excess solution with a Kimwipe, add 20–40 µL of mounting medium on the sample, and add a coverslip. Let dry for ~20 min in the dark at room temperature, and then apply nail polish to the outer edge of the coverslip to seal the sample.

    At this point, slides can be kept for short-term storage (~2–3 mo) at 4°C until imaging.
18. Image samples on a confocal microscope.

    See Troubleshooting.

TROUBLESHOOTING

Problem (Step 18): The staining signal is weak.
Solution: The primary antibody incubation in Step 13 can be extended from overnight to several days. Alternatively, the antibody concentration can be increased. If incubating for several days, take care to make sure the chamber remains humidified to ensure that the solution does not evaporate (sealing with Parafilm can help, as can refreshing the antibody solution after 2 d).

DISCUSSION

One major advantage of sectioning tissue for immunostaining is that it allows researchers to examine detailed neuron architecture and projections on a micron-level scale. Although this would not be necessarily advantageous for larger tissues like the brain, this technique is useful given the small width of the olfactory tissues. Sectioning also eliminates the need to image through the cuticle, which may be advantageous if confocal microscopy is not readily available.

RECIPES

Blocking Solution for Immunostaining

Reagent	Amount
Normal goat serum (NGS; 100%) (e.g., Jackson ImmunoResearch 005-000-001)	200 µL
Phosphate-buffered saline with Triton X-100 <R>	1.8 mL

Store in 1-mL aliquots for up to 1 yr at −20°C.

Cite this protocol as *Cold Spring Harb Protoc*; doi:10.1101/pdb.prot107914

Cryoprotection Solution

Reagent	Amount
Sucrose (e.g., Sigma-Aldrich S0389)	500 mg
Phosphate-buffered saline with Triton X-100 <R>	1.5 mL

Store for up to 6 mo at room temperature.

Paraformaldehyde (PFA; 4%)

Reagent	Quantity (for 100 mL)	Final concentration
Paraformaldehyde (PFA)	4 g	4%
Phosphate-buffered saline for immunohistochemistry (10×), diluted to 1× <R>	100 mL	

Slowly dissolve PFA in 1× phosphate-buffered saline over low heat until solution clears. Cool. Use immediately or aliquot and store for up to 1 yr at −20°C.

Phosphate-Buffered Saline with Triton X-100 (PBST)

Reagent	Amount
Phosphate-buffered saline (PBS; 10×) (e.g., Thermo Fisher AM9625)	1 mL
Triton X-100 (10%) (e.g., Sigma-Aldrich 93443)	100 µL
Deionized H_2O	8.9 mL

Store for up to 6 mo at room temperature. Discard and prepare fresh if any contaminants or floating debris is seen.

REFERENCES

DeGennaro M, McBride CS, Seeholzer L, Nakagawa T, Dennis EJ, Goldman C, Jasinskiene N, James AA, Vosshall LB. 2013. *orco* mutant mosquitoes lose strong preference for humans and are not repelled by volatile DEET. *Nature* **498:** 487–491. doi:10.1038/nature12206

Protocol 2

Immunostaining Whole-Mount Olfactory Tissues of Mosquitoes

Margaret Herre[1,2]

[1]*The Laboratory of Neurogenetics and Behavior, The Rockefeller University, New York, New York 10065, USA;*
[2]*Kavli Neural Systems Institute, New York, New York 10065, USA*

Mosquito olfactory tissues are covered in fine hair-like sensory structures called sensilla that house olfactory sensory neurons. This protocol for immunostaining whole-mounted mosquito antennae and maxillary palps enables the visualization of these neurons.

MATERIALS

It is essential that you consult the appropriate Material Safety Data Sheets and your institution's Environmental Health and Safety Office for proper handling of equipment and hazardous materials used in this protocol.

RECIPES: Please see the end of this protocol for recipes indicated by <R>. Additional recipes can be found online at http://cshprotocols.cshlp.org/site/recipes.

Reagents

Antibody incubation buffer <R>
Blocking buffer for immunostaining <R>
Chitinase-chymotrypsin dimethyl sulfoxide (DMSO) buffer (CCD) <R>
DMSO (e.g., Sigma-Aldrich 472301-100ML)
Fixative solution for immunostaining <R>
Methanol (80%; molecular biology grade) (e.g., Fisher A412-500)
Mosquitoes (any genus, male or female)
PBTx for immunostaining <R>
Phosphate-buffered saline (PBS; 10×) (e.g., Thermo Fisher AM9625), divided 1:10 in diH_2O
Primary antibody solution

Many Drosophila melanogaster *primary antibodies can be used because they have cross-reactivity, or an antibody can be custom-made for the mosquito. When troubleshooting a new antibody, test concentrations from 1/50 to 1/500. We use many antibodies at 1/50 because of low penetration through the cuticle. Very potent commercial antibodies can be used at 1/200.*

Secondary antibody solution

We use many commercially available secondary antibodies (e.g., Life Technologies Alexa Fluor secondaries) at 1/200.

SlowFade Diamond (Thermo Fisher S36972)

This reagent is preferred and tested. Users may optimize the protocol for their preferred mounting medium.

[2]Correspondence: mherre@gmail.com

Cite this protocol as *Cold Spring Harb Protoc*; doi:10.1101/pdb.prot107915

Equipment

Confocal microscope
Culture dish (35-mm)
Dissecting microscope
Forceps
Glass slides and coverslips
Greiner HLA Terasaki multiwell plates (VWR 82050-706)
Heat block
Hybridization oven
Ice
Microcentrifuge tubes (1.5-mL and 0.5-mL)
Nail polish
Petri dish with cover
Rotator

METHOD

The CCD digestion step was adapted from Manning and Doe (2017) and originally described in Edgar and McGhee (1988). The whole-mount immunofluorescence protocol is derived from Riabinina et al. (2016).

Perform all incubation steps with samples on a gentle rotator (~40 rpm).

Antennae will sink to the bottom of the tube, facilitating buffer exchange, with the exception of steps involving hybridization buffer and amplification buffer. Antennae should be pelleted on a desktop mini centrifuge for these steps. Maxillary palps can be pelleted between any step to facilitate buffer exchange.

Chitinase Digestion and Fixation

1. Using forceps, place approximately 100 mosquitoes in a covered Petri dish. Place the dish on ice for 10 min to anesthetize the mosquitoes.

   Mosquitoes are fully anesthetized when legs and wings are no longer moving.

2. Working quickly, place a mosquito under a dissecting microscope and use forceps to remove whole heads. Place heads into 1 mL of CCD buffer in a 1.5-mL microcentrifuge tube on ice.
3. Heat heads and CCD in the 1.5-mL tube for 5 min at 37°C on a heat block and then incubate them at 37°C rotating in a hybridization oven.

   The CCD incubation time will vary depending on the tissue type. See Discussion.

4. Remove CCD buffer and save it by storing at −20°C in 1-mL aliquots in 1.5-mL tubes.

   CCD can be reused three times.

5. (Optional) To remove all remaining CCD buffer, wash the heads with 0.5 mL of fixative solution for immunostaining for 5 min at 4°C on a rotator.
6. Fix heads by adding 1 mL of first fixative solution for immunostaining and then rotating in a hybridization oven for 24 h at 4°C.

Dissection, Methanol Dehydration, Blocking, and Antibody Incubation

7. Wash the sample three times for 30 min each in 400 µL of PBTx, and rotate it at room temperature.
8. Transfer heads to a 35-mm culture dish in PBTx by dumping heads onto a clean dish. Remove antennae and/or palps with forceps, and place them in 400 µL of PBTx in a 0.5-mL tube.

9. Add 400 µL of 80% methanol and 20% DMSO to the tissue and dehydrate it for 1 h, rotating at room temperature.
10. Wash the sample three times for 5 min each in 400 µL of PBTx in a 0.5-mL tube.
11. Block the sample overnight for 12–16 h in 400 µL of blocking buffer for immunostaining rotating at 4°C.
12. Wash the sample twice for 5 min each with 400 µL of antibody incubation buffer (without antibodies).
13. Remove antibody incubation buffer (without antibodies) and replace with 400 µL of primary antibody solution.
14. Incubate the sample in primary antibody solution while rotating for ~96 h at 4°C.
15. Wash the sample five times for 30 min each in 400 µL of PBTx rotating at 4°C.
16. Incubate in 400 µL of secondary antibody solution while rotating for ~96 h at 4°C.
17. Wash the sample five times for 30 min each in 400 µL of PBTx, rotating at 4°C.
18. Wash the sample once for 10 min with PBS, rotating at room temperature.

Tissue Mounting and Visualization

Before imaging, it is necessary to perform several washes with mounting medium such as SlowFade Diamond to ensure tissues are completely enveloped in mounting medium because peripheral chemosensory tissues are covered in many sensillar hairs.

19. Mount sample types as follows.

Palps

i. Remove as much PBS as possible with a P200 pipette.
ii. Wash the sample three times for 10 min each with 100 µL of SlowFade Diamond.
iii. Remove as much SlowFade Diamond as possible with a P200 pipette, and resuspend the palps in 20 µL of SlowFade Diamond.
iv. Mount the sample onto two slides, dividing the sample equally between them.
   The approximate volume of sample (SlowFade Diamond with palps) to pipette onto each slide will be 12 µL.
v. Cover with coverslip, and seal with nail polish.

Antennae

vi. Remove as much PBS as possible with a P200 pipette.
vii. Wash the sample twice for 10 min each with 100 µL of SlowFade Diamond.
viii. Add 100 µL of SlowFade Diamond to each of about five wells of a Greiner HLA Terasaki multiwell plate.
ix. Use a pipette to transfer antennae in SlowFade to a different well of the plate.
x. Use forceps to move antennae from the well containing old SlowFade Diamond to one of the wells containing fresh SlowFade Diamond, and incubate for ~1 min.
xi. Repeat Step 19.x at least three times, for a total of about four 1-min washes in wells containing fresh SlowFade Diamond.
xii. Add 10 µL SlowFade to the slide. Transfer washed antennae into the SlowFade bubble, and spread the antennae on the slide in an area no larger than that of the coverslip.
xiii. Cover with coverslip, and seal with nail polish.

 Cite this protocol as *Cold Spring Harb Protoc*; doi:10.1101/pdb.prot107915

20. Visualize samples on a confocal microscope.

Palps and one to two antennal segments with chemosensory neurons can be visualized using a 25× objective, although 40× and 63× objectives may be used as well.

Antennal segments are cylindrical structures with chemosensory neurons lining the periphery. Neurons anterior to the objective are most easily visualized, but users should be able to capture neurons posterior to the objective as well. Chemosensory neurons are located on the dorsal region of the fourth maxillary palp segment.

See Troubleshooting.

TROUBLESHOOTING

Problem (Step 20): Superficial staining is observed or staining is observed only in thin segments of peripheral tissues and not in thicker segments.
Solution: Increase CCD digestion time (see Step 3 and Discussion). Alternatively, modifying the methanol dehydration step (Step 9) for the whole-mount immunofluorescence protocol to a graded methanol/PBT series with overnight methanol treatment (e.g., see Steps 9 and 10 in Protocol 3: RNA In Situ Hybridization of Whole-Mount Mosquito Olfactory Tissues [Herre 2022b]) may also increase permeabilization.

Problem (Step 20): Air bubbles are observed on tissues after mounting.
Solution: Make sure to keep antennae and palps submerged in mounting medium throughout the mounting process in Step 19.

DISCUSSION

To immunostain whole-mounted tissues, it is important to optimize the duration of CCD digestion. For example, *Aedes aegypti* female antennae require ~1 h of CCD digestion time and male antennae require 30 min, whereas *Anopheles gambiae* female antennae require ~1 h 20 min of CCD digestion. We have heated tissues in CCD buffer for as long as 2 h without seeing degradation of the antibody signal. Additionally, this protocol can be adapted to visualize proteins in non-chitin-encased tissues, such as the brain or midgut. For these applications, removal of the CCD digestion step and reduction of the proteinase K digestion concentration and duration are helpful to keep these delicate tissues intact.

RECIPES

Antibody Incubation Buffer

Reagent	Amount	Final concentration
Deionized H_2O	16.8 mL	
Phosphate-buffered saline (10×) (e.g., Thermo Fisher AM9625)	2 mL	1×
Dimethyl sulfoxide (e.g., Sigma-Aldrich 472301)	200 µL	1%
Normal goat serum (e.g., Jackson ImmunoResearch 005-000-001)	1 mL	5%
Triton X-100 (10%) (e.g., Sigma-Aldrich 93443)	60 µL	0.03%

Prepare fresh.

Blocking Buffer for Immunostaining

Reagent	Amount
Triton X-100 (10%) (e.g., Sigma-Aldrich 93443)	4 mL
Deionized H_2O	4.5 mL
Phosphate-buffered saline (10×) (e.g., Thermo Fisher AM9625)	1 mL
Dimethyl sulfoxide (e.g., Sigma-Aldrich 472301)	100 µL
Normal goat serum (e.g., Jackson ImmunoResearch 005-000-001)	0.5 mL

Prepare fresh.

Chitinase-Chymotrypsin Dimethyl Sulfoxide Buffer (CCD)

Reagent	Amount	Final concentration
HEPES larval buffer (HLB) <R>	10 mL	
Chitinase (Sigma-Aldrich C6137)	50 U	5 U/mL
Chymotrypsin (Sigma-Aldrich CHY5S-10VL)	1000 U	100 U/mL
Dimethyl sulfoxide (DMSO) (e.g., Sigma-Aldrich 472301)	100 µL	1%

Add HLB to chitinase. Resuspend chymotrypsin (five vials of 200 U/vial) in HLB/chitinase solution and DMSO. Aliquot and store at −20°C.

Fixative Solution for Immunostaining

Reagent	Amount
Phosphate-buffered saline (10×) (e.g., Thermo Fisher AM9625)	1 mL
Deionized H_2O	7 mL
Paraformaldehyde (20%)	2 mL
Triton X-100 (10%) (e.g., Sigma-Aldrich 93443)	30 µL

Prepare fresh.

HEPES Larval Buffer (HLB)

Reagent	Amount	Final concentration
1 M HEPES (e.g., Sigma–Aldrich H0887)	12.5 mL	25 mM
5 M NaCl (e.g., Invitrogen AM9760G)	11.8 mL	119 mM
2 M KCl (e.g., Thermo Fisher AM9640G)	12 mL	48 mM
1 M MgCl2 (e.g., Thermo Fisher AM9530G)	1 mL	2 mM
1 M CaCl2 (e.g., Sigma-Aldrich 21115)	1 mL	2 mM
Deionized H2O	461.7 mL	

Filter-sterilize and store for up to 6 mo at 4°C.

Cite this protocol as *Cold Spring Harb Protoc*; doi:10.1101/pdb.prot107915

PBTx for Immunostaining

Reagent	Amount
Deionized H_2O	44.35 mL
Phosphate-buffered saline (10×) (e.g., Thermo Fisher AM9625)	5 mL
Dimethyl sulfoxide (e.g., Sigma-Aldrich 472301)	500 µL
Triton X-100 (10%) (e.g., Sigma-Aldrich 93443)	300 µL

Store for up to 1 mo at room temperature.

REFERENCES

Edgar LG, McGhee JD. 1988. DNA synthesis and the control of embryonic gene expression in *C. elegans*. *Cell* **53:** 589–599. doi:10.1016/0092-8674(88)90575-2

Herre M. 2022b. RNA *in situ* hybridization of whole-mount mosquito olfactory tissues. *Cold Spring Harb Protoc* doi:10.1101/pdb.prot107916

Manning L, Doe CQ. 2017. Immunofluorescent antibody staining of intact *Drosophila* larvae. *Nat Protoc* **12:** 1–14. doi:10.1038/nprot.2016.162

Riabinina O, Task D, Marr E, Lin CC, Alford R, O'Brochta DA, Potter CJ. 2016. Organization of olfactory centres in the malaria mosquito *Anopheles gambiae*. *Nat Commun* **7:** 13010. doi:10.1038/ncomms13010

Protocol 3

RNA In Situ Hybridization of Whole-Mount Mosquito Olfactory Tissues

Margaret Herre[1,2,3]

[1]*The Laboratory of Neurogenetics and Behavior, The Rockefeller University, New York, New York 10065, USA;*
[2]*Kavli Neural Systems Institute, New York, New York 10065, USA*

This protocol presents a method for RNA in situ hybridization of whole-mounted preparations of mosquito antennae and maxillary palps. It can be adapted for other peripheral appendages such as the legs or proboscis.

MATERIALS

It is essential that you consult the appropriate Material Safety Data Sheets and your institution's Environmental Health and Safety Office for proper handling of equipment and hazardous materials used in this protocol.

RECIPES: Please see the end of this protocol for recipes indicated by <R>. Additional recipes can be found online at http://cshprotocols.cshlp.org/site/recipes.

Reagents

Chitinase-chymotrypsin dimethyl sulfoxide buffer (CCD) <R>
First fixative solution <R>
Hybridization chain reaction (HCR) buffers and reagents

- HCR amplification buffer (Molecular Instruments)
- HCR amplifier hairpins (Molecular Instruments)
- HCR probe hybridization buffer (Molecular Instruments)
- HCR probe wash buffer (Molecular Instruments)

Although we use reagents purchased from Molecular Instruments, users may be able to reduce costs by making buffers and reagents in house as described in Choi et al. (2018).

HCR probes (Molecular Instruments)

For users designing probes with Molecular Instruments (recommended), users submit desired sequences to www.molecularinstruments.com. We recommend ordering the maximum number of probes for each desired sequence (up to 40) to enable maximum visualization. Users may submit full-length RNA sequences or sequences corresponding to a specific region.

Methanol (molecular biology grade; e.g., Fisher A412-500)
Mosquitoes (any genus, male or female)

[3]Correspondence: mherre@gmail.com

Cite this protocol as *Cold Spring Harb Protoc*; doi:10.1101/pdb.prot107916

Phosphate-buffered saline with Tween 20 (PBT) for mosquito RNA-FISH <R>
Proteinase K (20 mg/mL; Thermo Fisher AM2548)

Use it as a 20 μg/mL solution. Other tissues may require different dilutions. Try 0–50 μg/mL when optimizing a new protocol.

Second fixative solution <R>
SlowFade Diamond (Thermo Fisher S36972)
Sodium chloride–sodium citrate with Tween 20 (SSCT) <R>

Equipment

Confocal microscope
Culture dish (35-mm)
Dissecting microscope
Forceps (two pairs, including one dull; see Step 8)
Glass slides and coverslips
Greiner HLA Terasaki multiwell plates (VWR 82050-706)
Heat block
Hybridization oven
Ice
Microcentrifuge tubes (1.5-mL and 0.5-mL; RNase/DNase-free preferred)
Minicentrifuge with inserts for 0.5-mL tubes
Nail polish
PCR thermocycler
Petri dish
Rotator

METHOD

This protocol uses the hybridization chain reaction RNA fluorescent in situ hybridization (HCR RNA-fluorescence in situ hybridization [FISH]) method described in Choi et al. (2018).

Perform all steps with samples on a gentle rotator (~40 rpm).

Chitinase Digestion and Fixation

1. Using forceps, place approximately 100 mosquitoes in a covered Petri dish. Place the dish on ice for 10 min to anesthetize the mosquitoes.

   Mosquitoes are fully anesthetized when legs and wings are no longer moving.

2. Working quickly, place a mosquito under a dissecting microscope and use forceps to remove whole heads. Place heads into 1 mL of CCD buffer in a 1.5-mL microcentrifuge tube on ice.
3. Heat the heads and CCD in the 1.5-mL tube on a heat block for 5 min at 37°C, and then incubate the sample at 37°C, rotating in a hybridization oven.

   The CCD incubation time will vary among tissues. For Aedes aegypti*, male antennae require 30 min, female antennae require 60 min, and female palps require 1–1.5 h.*

4. Remove CCD buffer and save at −20°C in 1-mL aliquots in 1.5-mL tubes.

   CCD can be reused three times.

5. (Optional) To remove all remaining CCD buffer, wash the heads with 0.5 mL of fixative solution for 5 min at 4°C on a rotator. Heads will sink to the bottom of the tube.
6. Fix heads by adding 1 mL of first fixative solution and then rotating for 24 h at 4°C.

Dissection and Methanol Dehydration

7. Wash heads four times for 5 min each in 400 µL of PBT, rotating on ice or at 4°C.

   You may observe some antennae detaching from heads during the wash steps; this is normal.

8. Transfer heads to a new 35-mm culture dish in 400 µL of PBT by inverting the tube above the culture dish. Remove all antennae and/or palps in PBT on ice with dull forceps (newly sharpened forceps will bend easily on the culture dishes), and place them in 0.5-mL tubes.
9. Dehydrate tissues stepwise, as follows, with a graded series of 400 µL methanol/PBT washes for 10 min each rotating on ice or at 4°C:
   i. 25% methanol and 75% PBT,
   ii. 50% methanol and 50% PBT,
   iii. 75% methanol and 25% PBT,
   iv. 100% methanol, and
   v. 100% methanol.

   Antennae will sink to the bottom of the tube, facilitating buffer exchange. Maxillary palps can be pelleted or allowed to sink to the bottom of the tube.

10. Incubate in 400 µL of 100% methanol for 12–16 h at −20°C.

Rehydration, Proteinase K Digestion, Second Fixation, and Probe Hybridization

11. Thaw HCR probe hybridization buffer and HCR probes on ice. Prepare a total of three 400 µL aliquots of the HCR probe hybridization buffer per sample (i.e., tube of antennae and/or palps) in 1.5-mL tubes. Equilibrate one of the aliquots for each sample at room temperature, and preheat the other two aliquots at 37°C on a heat block.

    Make sure all solutions are mixed well before aliquoting.

12. Rehydrate tissues stepwise with a graded series of 400 µL of methanol and PBT for 10 min while rotating on ice or at 4°C, as follows:
    i. 75% methanol and 25% PBT,
    ii. 50% methanol and 50% PBT,
    iii. 25% methanol and 75% PBT, and
    iv. 100% PBT.
13. Wash tissues with 400 µL of PBT for 10 min, rotating at room temperature.
14. Incubate tissues in 400 µL of 20 µg/mL proteinase K solution for 30 min, rotating at room temperature.
15. Wash the sample twice with 400 µL of PBT for 10 min each time, rotating at room temperature.
16. Fix tissues in 400 µL of the second fixative solution for 20 min, rotating at room temperature.
17. Wash tissues three times for 15 min with 400 µL of PBT, rotating at room temperature.
18. Incubate tissues in 400 µL of HCR probe hybridization buffer for 5 min rotating at room temperature.

    Probe hybridization buffer is very viscous. The tissue can be pelleted on a desktop mini centrifuge to facilitate buffer exchange.

19. Remove buffer and prehybridize with 400 µL of preheated HCR probe hybridization buffer for 30 min, rotating in a 37°C hybridization oven.
20. Prepare a heated probe solution by adding 4 pmol of each probe (4 µL of 1 µM stock) to 250 µL of preheated HCR probe hybridization buffer, and then keep it at 37°C on a heat block or in a 37°C hybridization oven until use in Step 21.

 Cite this protocol as *Cold Spring Harb Protoc*; doi:10.1101/pdb.prot107916

21. Remove the HCR probe hybridization buffer, and replace it with 250 µL of the heated probe solution from Step 20. Incubate tissue for ~48 h at 37°C, rotating in the hybridization oven.

Amplification

22. Thaw HCR probe wash buffer on ice. Aliquot ~500 µL of HCR probe wash buffer into each 1.5-mL tube (for a total of five tubes per sample), and warm to 37°C in the hybridization oven or on a heat block.

    Probe wash buffer is very viscous. Make sure it is mixed well before aliquoting.

23. Equilibrate two aliquots of HCR amplification buffer to room temperature, and thaw HCR amplifier hairpins on ice in the dark.
24. Prepare the HCR amplifier hairpin solution.
    i. Transfer 18 pmol of hairpin h1 and 18 pmol of hairpin h2 (6 µL of 3 µM stock per sample) for each fluorophore into separate PCR tubes.
    ii. Heat hairpins for 90 sec at 95°C in a polymerase chain reaction (PCR) thermocycler, and cool for 30 min to room temperature in a dark drawer.
    iii. Add all hairpin solutions to 100 µL of room-temperature HCR amplification buffer, and keep in the dark until Step 28.
25. Wash the tissue five times for 10 min each with 400 µL of prewarmed HCR probe wash buffer, rotating in the 37°C hybridization oven.
26. Wash samples twice for 5 min each with 400 µL of SSCT, rotating at room temperature.
27. Preamplify tissue with 400 µL of room-temperature HCR amplification buffer for 10 min, rotating at room temperature.
28. Remove HCR amplification buffer, transfer samples into the HCR amplifier hairpin solution, and incubate 12–16 h in the dark rotating at room temperature.

    Amplification buffer is very viscous. Tissue can be pelleted on a desktop minicentrifuge to facilitate buffer exchange.

29. Add 300 µL of SSCT to the tissue in the HCR amplifier hairpin solution and immediately pellet the tissue in a desktop minicentrifuge.

    The addition of SSCT to the HCR amplifier hairpin solution makes the solution much less viscous and facilitates easy pelleting in a desktop minicentrifuge.

30. Wash five times for 10 min each with 400 µL SSCT, rotating at room temperature in the dark.

    Keep at 4°C in the dark until the sample is ready to be mounted.

31. Mount the samples as follows.

    Maxillary palps

    i. Pellet the palps on a microcentrifuge, and remove as much SSCT as possible with a P200 pipette.
    ii. Wash three times for 10 min each with 100 µL of SlowFade Diamond.
    iii. Remove as much SlowFade Diamond as possible, and resuspend palps in 20 µL of SlowFade Diamond.
    iv. Mount the sample onto two slides, dividing the sample equally between them.

        The approximate volume of sample (SlowFade Diamond with palps) to pipette onto each slide will be 12 µL.

    v. Cover the slide with a coverslip, and seal with nail polish.

Antennae

vi. Remove as much SSCT as possible with a P200 pipette.

vii. Wash twice with 100 µL of SlowFade Diamond for ~10 min each.

viii. Add 100 µL of fresh SlowFade Diamond to each of about five wells of a Greiner HLA Terasaki multiwell plate.

ix. Use a pipette to transfer antennae in SlowFade Diamond to a different well of the plate.

x. Use forceps to move antennae from the well containing old SlowFade Diamond to one of the wells containing fresh SlowFade Diamond, and incubate for ~1 min.

xi. Repeat Step 31.x at least three times, for a total of about four 1-min washes in wells containing fresh SlowFade Diamond.

xii. Add 10 µL of SlowFade Diamond to the slide. Transfer washed antennae into the SlowFade bubble, and spread the antennae on the slide in an area no larger than that of the coverslip.

xiii. Cover with coverslip, and seal with nail polish.

32. Visualize samples on a confocal microscope.

Palps and one to two antennal segments with chemosensory neurons can be visualized using a 25× objective, although 40×, and 63× objectives may be used as well.

Antennal segments are cylindrical structures with chemosensory neurons lining the periphery. Neurons anterior to the objective are most easily visualized, but users should be able to capture neurons posterior to the objective as well. Chemosensory neurons are located on the dorsal region of the fourth maxillary palp segment.

See Troubleshooting.

TROUBLESHOOTING

Problem (Step 32): Superficial staining is observed or staining is observed only in thin segments of peripheral tissues and not thicker segments.
Solution: Increase CCD digestion time (Step 3).

Problem (Step 32): Air bubbles are observed on tissues after mounting.
Solution: Make sure to keep antennae and palps submerged in mounting medium throughout the mounting process in Step 31.

RECIPES

Chitinase-Chymotrypsin Dimethyl Sulfoxide Buffer (CCD)

Reagent	Amount	Final concentration
HEPES larval buffer (HLB) <R>	10 mL	
Chitinase (Sigma-Aldrich C6137)	50 U	5 U/mL
Chymotrypsin (Sigma-Aldrich CHY5S-10VL)	1000 U	100 U/mL
Dimethyl sulfoxide (DMSO) (e.g., Sigma-Aldrich 472301)	100 µL	1%

Add HLB to chitinase. Resuspend chymotrypsin (five vials of 200 U/vial) in HLB/chitinase solution and DMSO. Aliquot and store at −20°C.

Cite this protocol as *Cold Spring Harb Protoc*; doi:10.1101/pdb.prot107916

First Fixative Solution

Reagent	Amount	Final concentration
Phosphate-buffered saline (10×) (e.g., Thermo Fisher AM9625)	1 mL	1×
Nuclease-free H_2O	7 mL	
Paraformaldehyde (20%)	2 mL	4%
Triton X-100 (10%) (e.g., Sigma-Aldrich 93443)	30 µL	0.03%

Prepare fresh.

HEPES Larval Buffer (HLB)

Reagent	Amount	Final concentration
1 M HEPES (e.g., Sigma–Aldrich H0887)	12.5 mL	25 mM
5 M NaCl (e.g., Invitrogen AM9760G)	11.8 mL	119 mM
2 M KCl (e.g., Thermo Fisher AM9640G)	12 mL	48 mM
1 M $MgCl_2$ (e.g., Thermo Fisher AM9530G)	1 mL	2 mM
1 M $CaCl_2$ (e.g., Sigma-Aldrich 21115)	1 mL	2 mM
Deionized H_2O	461.7 mL	

Filter-sterilize and store for up to 6 mo at 4°C.

Phosphate-Buffered Saline with Tween 20 for Mosquito RNA-FISH

Reagent	Amount
Phosphate-buffered saline (10×) (e.g., Thermo Fisher AM9625)	5 mL
Nuclease-free H_2O	44.5 mL
Tween 20 (10%)	500 µL

Prepare fresh.

Second Fixative Solution

Reagent	Amount
Phosphate-buffered saline with Tween 20 for mosquito RNA-FISH <R>	4 mL
Paraformaldehyde (20%)	1 mL

Prepare fresh.

Sodium Chloride–Sodium Citrate with Tween 20 (SSCT)

Reagent	Amount	Final concentration
SSC (20×)	10 mL	5×
Nuclease-free H_2O	30 mL	
Tween 20 (10%)	400 µL	0.1%

Prepare fresh.

ACKNOWLEDGMENTS

Thank you to Nipun Basrur, Saher Rahiel, Harry Choi, and the Molecular Instruments team for help and advice on optimizing the HCR RNA-FISH protocol. Funding to M.H. was provided by a predoctoral fellowship from the Kavli Neural Systems Institute and a National Institutes of Health (NIH) National Institute on Deafness and Other Communication Disorders (NIDCD) grant F30DC017658. M.H. is also supported by a Medical Scientist Training Program grant from the National Institute of General Medical Sciences of the NIH under award number T32GM007739 to the Weill Cornell/Rockefeller/Sloan Kettering Tri-Institutional MD-PhD Program.

REFERENCES

Choi HMT, Schwarzkopf M, Fornace ME, Acharya A, Artavanis G, Stegmaier J, Cunha A, Pierce NA. 2018. Third-generation *in situ* hybridization chain reaction: multiplexed, quantitative, sensitive, versatile, robust. *Development* **145:** dev165753. doi:10.1242/dev.165753

CHAPTER 44

Osmoregulatory Challenges Faced by Mosquitoes and Techniques for Studying the Transporters Involved in Their Adaptations

Andrea C. Durant[1] and Andrew Donini[2,3]

[1]*Department of Marine Biology and Ecology, University of Miami-Rosenstiel School of Marine, Atmospheric and Earth Science, Miami, Florida 33149-1031, USA;* [2]*Department of Biology, York University, Toronto, Ontario M3J 1P3, Canada*

Mosquitoes are the most important disease vector in the world, and gaining knowledge of their physiology to develop novel population control strategies has been a focus of research for some time. Both aquatic larvae and terrestrial adults face harsh environmental factors that severely challenge their salt and water balance, which are regulated by the function of epithelia of various organs. The regulated passage of water and solutes across epithelia occurs, in part, through transporters expressed in epithelial cell membranes. Identifying these transporters and their localization is necessary to understand how mosquitoes regulate salt and water balance. Here, we review environmental challenges faced by mosquitoes and how they cope with them, in addition to introducing techniques used to identify organ epithelial transporters.

ENVIRONMENTAL CHALLENGES FACED BY MOSQUITOES

Mosquitoes place a significant burden on human populations because they transmit pathogens that cause disease. The pathogens that are transmitted by *Aedes aegypti* are considered a major global threat (Leta et al. 2018). Normally, immature stages of *Ae. aegypti* complete their development to adulthood in freshwater; however, ample evidence demonstrates that they also inhabit brackish water and sewage water (Irving-Bell et al. 1987; Burke et al. 2010; Ramasamy et al. 2011; Ramasamy and Surendran 2016; Durant and Donini 2019). These habitats present mosquito larvae with different challenges to survival because sewage water contains high levels of ammonia, which is toxic; freshwater tends to dilute mosquito larvae body fluids; and brackish water tends to salt load mosquito larvae. Larval mosquitoes are well adapted to survive under these difficult conditions (Patrick and Bradley 2000; Ramasamy et al. 2011; Surendran et al. 2012, 2018; Chasiotis et al. 2016; Durant et al. 2017; Durant and Donini 2019). Freshwater is the native ancestral habitat for mosquito larvae, and their osmoregulatory physiology is adapted for living in dilute water where the conservation of salt and excretion of excess water are prioritized (Bradley 1987; Albers and Bradley 2011). However, these mosquitoes that normally live in freshwater are increasingly exploiting salinated habitats where their physiology now must prioritize the excretion of salt (Ramasamy et al. 2011, 2021; Surendran et al. 2012; Ramasamy and Surendran 2016). Furthermore, larvae of the genus *Aedes*, which are responsible for transmitting the viral diseases dengue and chikungunya, have been found inhabiting septic tanks, sewage drains, and sewage

[3]Correspondence: adonini@yorku.ca

Cite this introduction as *Cold Spring Harb Protoc*; doi:10.1101/pdb.top107701

water (Banerjee et al. 2015; Chitolina et al. 2016; Durant and Donini 2019; Martini et al. 2019). In sewage water, larvae are successful in completing development at neutral to basic pH up to pH ~9.3, at which point they can no longer survive to adulthood (Martini et al. 2019). Adults can face similar challenges, especially females when blood feeding, with the digestion of proteins that liberates ammonia and the high NaCl and water content of the ingested blood plasma. Upon blood feeding, female mosquitoes rely on a rapidly induced and hormonally coordinated physiological response in which the majority of NaCl and water imbibed from blood plasma is excreted (Beyenbach and Petzel 1987). Once proteins from the blood meal are digested, the liberation of high amounts of ammonia requires mechanisms to detoxify ammonia and/or excrete it (Scaraffia et al. 2005, 2010).

SURVIVAL SYSTEMS

Mosquitoes survive these challenges because they have physiological mechanisms that detoxify and excrete ammonia and regulate the uptake and excretion of water and salts. The epithelia of various organs are major components of these mechanisms because they are sites of ammonia, salt, and water flux that occur through membrane-associated transporter proteins. A part of understanding these critical physiological processes is to identify these transporters and localize them to cellular membranes where they carry out their functions. For example, studies on ammonia transporters in the anal papillae of mosquito larvae have implicated these organs as important sites for ammonia excretion (Chasiotis et al. 2016; Durant et al. 2017; Durant and Donini 2018). Furthermore, the ammonia transporters are critical for this important process to occur. Aquaporins, which are water channels expressed in the Malpighian tubules (the functional analog of the vertebrate kidney), are instrumental in mosquito diuresis and thus survival (Drake et al. 2015; Misyura et al. 2017, 2020). At least some of these aquaporins are salinity-responsive in the anal papillae and gastric caeca of mosquito larvae, suggesting that they play an important role in larvae acclimating to changes in osmolarity (Akhter et al. 2017; Misyura et al. 2020). Because these membrane-associated transporters participate in these vital physiological functions, studying them can ultimately lead to the discovery of novel methods for controlling mosquito populations.

LABORATORY TECHNIQUES

Immunohistochemistry (IHC) was first reported in 1941 and continues to be an excellent technique to identify transporters (Coons et al. 1941; Patrick et al. 2006; Linser et al. 2012; Chasiotis et al. 2016). IHC relies on the ability of immunoglobulins to bind antigens and other antibodies, and the bound antibodies are then visualized with a chemical reaction that can be viewed with either light or fluorescence microscopy (Ramos-Vara and Miller 2014). Organs of interest can be processed and viewed whole in preparations that are referred to as whole mounts, and, in this case, a confocal microscope is required to pinpoint the localization of membrane transporters. Another approach that is arguably superior at localizing membrane transporters, particularly in epithelial cells that exhibit basal polarity, is embedding samples in paraffin wax and slicing samples into thin sections for processing and viewing on a conventional fluorescence microscope. This latter technique combines histology with IHC and has been refined and improved on over the years (Ramos-Vara and Miller 2014). For example, one ingenuity that has improved this technique is antigen retrieval that can be achieved by heating samples, countering the masking of antigens by formalin-based fixatives (Shi et al. 1997). Another example is methods developed for increasing sensitivity that use avidin–biotin binding. One of these methods uses a secondary antibody that is biotinylated with a third solution containing a labeled avidin. These methods amplify the signal because they create a complex that has many more labeled molecules than in the standard procedure of using a primary antibody with a labeled secondary antibody (Ramos-Vara and Miller 2014).

 Cite this introduction as *Cold Spring Harb Protoc*; doi:10.1101/pdb.top107701

No matter which procedure for IHC is used, the most important factor that determines success is the specificity and affinity of the antibody to its intended antigen. The specificity of the antibody is determined by the antibody's ability to bind an epitope, which is a part of the intended antigen, in the presence of other epitopes from other native proteins in the tissue or sample (Lipman et al. 2005). Antibodies with higher specificity will result in less cross-reactivity with unintended targets and therefore will be better at localizing the intended antigen in cells and tissues. The affinity of an antibody is a measure of how strongly the antibody binds to the epitope of the intended antigen. Therefore, an antibody with a higher affinity will bind a greater amount of antigen in a more stable manner over a shorter period of time than an antibody with lower affinity (Lipman et al. 2005). Commercial, off-the-shelf antibodies do not generally exist for mosquito solute transporters. However, the highly conserved sequences of the primary ionomotive enzymes Na^+/K^+-ATPase and V-type H^+-ATPase mean that these two transporters can be localized with commercial antibodies. Otherwise, it is usually necessary to contract a company to produce a custom antibody against the transporter of interest. In this case, an antigenic site in the protein is chosen, and it is best if this antigen is from an area of the protein that is not embedded in the membrane such that the native antigen is readily available to the antibody.

In an accompanying protocol (see Protocol 1: Localizing Protein Expression of Transporters in Epithelial Cell Membranes of Mosquitoes [Durant and Donini 2023]), we detail tissue processing, immunohistochemistry, and visualizing target proteins in sections of mosquitoes and their organs.

REFERENCES

Akhter H, Misyura L, Bui P, Donini A. 2017. Salinity responsive aquaporins in the anal papillae of the larval mosquito, *Aedes aegypti*. *Comp Biochem Physiol A Mol Integr Physiol* **203:** 144–151. doi:10.1016/j.cbpa.2016.09.008

Albers MA, Bradley TJ. 2011. On the evolution of saline tolerance in the larvae of mosquitoes in the genus *Ochlerotatus*. *Physiol Biochem Zool* **84:** 258–267. doi:10.1086/659769

Banerjee S, Mohan S, Saha N, Mohanty SP, Saha GK, Aditya G. 2015. Pupal productivity and nutrient reserves of *Aedes* mosquitoes breeding in sewage drains and other habitats of Kolkata, India: implications for habitat expansion and vector management. *Indian J Med Res* **142:** 87–94. doi:10.4103/0971-5916.176632

Beyenbach K, Petzel D. 1987. Diuresis in mosquitoes: role of a natriuretic factor. *Physiology* **2:** 171–175. doi:10.1152/physiologyonline.1987.2.5.171

Bradley TJ. 1987. Physiology of osmoregulation in mosquitoes. *Annu Rev Entomol* **32:** 439–462. doi:10.1146/annurev.en.32.010187.002255

Burke RL, Barrera R, Lewis M, Kluchinsky T, Claborn D. 2010. Septic tanks as larval habitats for the mosquitoes *Aedes aegypti* and *Culex quinquefasciatus* in Playa-Playita, Puerto Rico. *Med Vet Entomol* **24:** 117–123. doi:10.1111/j.1365-2915.2010.00864.x

Chasiotis H, Ionescu A, Misyura L, Bui P, Fazio K, Wang J, Patrick M, Weihrauch D, Donini A. 2016. An animal homolog of plant Mep/Amt transporters promotes ammonia excretion by the anal papillae of the disease vector mosquito *Aedes aegypti*. *J Exp Biol* **219:** 1346–1355. doi:10.1242/jeb.134494

Chitolina RF, Anjos FA, Lima TS, Castro EA, Costa-Ribeiro MCV. 2016. Raw sewage as breeding site to *Aedes* (Stegomyia) *aegypti* (Diptera, culicidae). *Acta Trop* **164:** 290–296. doi:10.1016/j.actatropica.2016.07.013

Coons A, Creech H, Jones N. 1941. Immunological properties of an antibody containing a fluorescent group. *Exp Biol Med* **47:** 200–202. doi:10.3181/00379727-47-13084P

Drake LL, Rodriguez SD, Hansen IA. 2015. Functional characterization of aquaporins and aquaglyceroporins of the yellow fever mosquito, *Aedes aegypti*. *Sci Rep* **5:** 7795. doi:10.1038/srep07795

Durant AC, Donini A. 2018. Evidence that Rh proteins in the anal papillae of the freshwater mosquito *Aedes aegypti* are involved in the regulation of acid–base balance in elevated salt and ammonia environments. *J Exp Biol* **221:** jeb186866. doi:10.1242/jeb.186866

Durant AC, Donini A. 2019. Development of *Aedes aegypti* (Diptera: Culicidae) mosquito larvae in high ammonia sewage in septic tanks causes alterations in ammonia excretion, ammonia transporter expression, and osmoregulation. *Sci Rep* **9:** 19028. doi:10.1038/s41598-019-54413-6

Durant AC, Donini A. 2023. Localizing protein expression of transporters in epithelial cell membranes of mosquitoes. *Cold Spring Harb Protoc* doi:10.1101/pdb.prot108326

Durant AC, Chasiotis H, Misyura L, Donini A. 2017. *Aedes aegypti* Rhesus glycoproteins contribute to ammonia excretion by larval anal papillae. *J Exp Biol* **220:** 588–596. doi:10.1242/jeb.151084

Irving-Bell RJ, Okoli EI, Diyelong DY, Lyimo EO, Onyia OC. 1987. Septic tank mosquitoes: competition between species in central Nigeria. *Med Vet Entomol* **1:** 243–250. doi:10.1111/j.1365-2915.1987.tb00350.x

Leta S, Beyene TJ, De Clercq EM, Amenu K, Revie CW, Kraemer MUG. 2018. Global risk mapping for major diseases transmitted by *Aedes aegypti* and *Aedes albopictus*. *Int J Infect Dis* **67:** 25–35. doi:10.1016/j.ijid.2017.11.026

Linser PJ, Neira Oviedo M, Hirata T, Seron TJ, Smith KE, Piermarini PM, Romero MF. 2012. Slc4-like anion transporters of the larval mosquito alimentary canal. *J Insect Physiol* **58:** 551–562. doi:10.1016/j.jinsphys.2012.01.002

Lipman NS, Jackson LR, Trudel LJ, Weis-Garcia F. 2005. Monoclonal versus polyclonal antibodies: distinguishing characteristics, applications, and information resources. *ILAR J* **46:** 258–267. doi:10.1093/ilar.46.3.258

Martini M, Triasputri Y, Hestiningsih R, Yuliawati S, Purwantisasi S. 2019. Longevity and development of *Aedes aegypti* larvae to imago in domestic sewage water. *J Med Sci (Berkala Ilmu Kedokteran)* **51:** 325–332. doi:10.19106/JMedSci005104201906

Misyura L, Yerushalmi GY, Donini A. 2017. A mosquito entomoglyceroporin, *Aedes aegypti* AQP5, participates in water transport across the Malpighian tubules of larvae. *J Exp Biol* **220:** 3536–3544. doi:10.1242/jeb.158352

Misyura L, Durant AC, Grieco Guardian E, Donini A. 2020. A comparison of aquaporin expression in mosquito larvae (*Aedes aegypti*) that develop in hypo-osmotic freshwater and iso-osmotic brackish water. *PLoS ONE* **15:** e0234892. doi:10.1371/journal.pone.0234892

Patrick ML, Bradley TJ. 2000. The physiology of salinity tolerance in larvae of two species of culex mosquitoes: the role of compatible solutes. *J Exp Biol* **203:** 821–830. doi:10.1242/jeb.203.4.821

Patrick ML, Aimanova K, Sanders HR, Gill SS. 2006. P-type Na^+/K^+-ATPase and V-type H^+-ATPase expression patterns in the osmoregulatory organs of larval and adult mosquito *Aedes aegypti*. *J Exp Biol* **209:** 4638–4651. doi:10.1242/jeb.02551

Ramasamy R, Surendran SN. 2016. Mosquito vectors developing in atypical anthropogenic habitats: global overview of recent observations, mechanisms and impact on disease transmission. *J Vector Borne Dis* **53:** 91–98.

Ramasamy R, Surendran SN, Jude PJ, Dharshini S, Vinobaba M. 2011. Larval development of *Aedes aegypti* and *Aedes albopictus* in peri-urban brackish water and its implications for transmission of arboviral diseases. *PLoS Negl Trop Dis* **5:** e1369. doi:10.1371/journal.pntd.0001369

Ramasamy R, Thiruchenthooran V, Jayadas TTP, Eswaramohan T, Santhirasegaram S, Sivabalakrishnan K, Naguleswaran A, Uzest M, Cayrol B, Voisin SN, et al. 2021. Transcriptomic, proteomic and ultrastructural studies on salinity-tolerant *Aedes aegypti* in the context of rising sea levels and arboviral disease epidemiology. *BMC Genomics* **22:** 1–16. doi:10.1186/s12864-021-07564-8

Ramos-Vara JA, Miller MA. 2014. When tissue antigens and antibodies get along: revisiting the technical aspects of immunohistochemistry—the red, brown, and blue technique. *Vet Pathol* **51:** 42–87. doi:10.1177/0300985813505879

Scaraffia PY, Isoe J, Murillo A, Wells MA. 2005. Ammonia metabolism in *Aedes aegypti*. *Insect Biochem Mol Biol* **35:** 491–503. doi:10.1016/j.ibmb.2005.01.012

Scaraffia PY, Zhang Q, Thorson K, Wysocki VH, Miesfeld RL. 2010. Differential ammonia metabolism in *Aedes aegypti* fat body and midgut tissues. *J Insect Physiol* **56:** 1040–1049. doi:10.1016/j.jinsphys.2010.02.016

Shi SR, Cote RJ, Young LL, Taylor CR. 1997. Antigen retrieval immunohistochemistry: practice and development. *J Histotechnol* **20:** 145–154. doi:10.1179/his.1997.20.2.145

Surendran SN, Jude PJ, Thabothiny V, Raveendran S, Ramasamy R. 2012. Pre-imaginal development of *Aedes aegypti* in brackish and fresh water urban domestic wells in Sri Lanka. *J Vector Ecol* **37:** 471–473. doi:10.1111/j.1948-7134.2012.00254.x

Surendran SN, Sivabalakrishnan K, Jayadas TTP, Santhirasegaram S, Laheetharan A, Senthilnanthanan M, Ramasamy R. 2018. Adaptation of *Aedes aegypti* to salinity: characterized by larger anal papillae in larvae. *J Vector Borne Dis* **55:** 235–238. doi:10.4103/0972-9062.249482

Protocol 1

Localizing Protein Expression of Transporters in Epithelial Cell Membranes of Mosquitoes

Andrea C. Durant[1] and Andrew Donini[2,3]

[1]*Department of Marine Biology and Ecology, University of Miami-Rosenstiel School of Marine, Atmospheric and Earth Science, Miami, Florida 33149-1031, USA;* [2]*Department of Biology, York University, Toronto, Ontario M3J 1P3, Canada*

Identifying and localizing ion transport proteins in epithelia is important for understanding how these tissues can regulate salt and water balance in animals. Mosquitoes face distinct challenges regarding salt and water balance as larvae live in water of varying ionic composition while adult female mosquitoes must deal with periodic large blood meals. This protocol will explain how to localize ion transporters in epithelia of mosquitoes. Antibodies raised against specific proteins can be used to carry out immunohistochemistry to tag and visualize native proteins in tissues and cells. This technique may be used for whole tissues and organs or can also be used on histological thin sections of fixed tissue. This protocol will detail the use of immunohistochemistry to localize membrane proteins in mosquito organs.

MATERIALS

It is essential that you consult the appropriate Material Safety Data Sheets and your institution's Environmental Health and Safety Office for proper handling of equipment and hazardous material used in this protocol.

RECIPES: Please see the end of this protocol for recipes indicated by <R>. Additional recipes can be found online at http://cshprotocols.cshlp.org/site/recipes.

Reagents

Antibodies (primary and secondary)

Antibodies for highly conserved membrane transporters are commercially available (e.g., V-type H^+-ATPase and Na^+/K^+-ATPase); however, most transporters in mosquitoes will require custom antibodies. Many companies provide this service. We have used products from GenScript and New England Peptide with success. For the localization of protein using immunohistochemistry, it is recommended that the antigen sequence be a part of the protein that is not embedded in the membrane. A variety of secondary antibodies are commercially available from many different companies. We have purchased them from Jackson ImmunoResearch and used them with success. See Introduction: Osmoregulatory Challenges Faced by Mosquitoes and Techniques for Studying the Transporters Involved in Their Adaptations (Durant and Donini 2023) for more details.

Bouin's fixative for IHC <R>

Bouin's fixative solution is also commercially available (e.g., Sigma-Aldrich HT101128).

Distilled H_2O

Reverse-osmosis H_2O is sufficient, but ultrapure H_2O such as from a MilliQ H_2O purification system is recommended.

[3]Correspondence: adonini@yorku.ca

Cite this protocol as *Cold Spring Harb Protoc*; doi:10.1101/pdb.prot108326

Ethanol

We recommend using 100% ethyl alcohol, from which the necessary dilutions (30%, 50%, 70%, and 95%, v/v) can be made.

Mosquito antibody dilution buffer (MADB) <R>

Mosquitoes (larvae or adults depending on experimental need)

We have successfully used this protocol with three mosquito species from the genera Aedes, Culex, *and* Ochlerotatus. *Also, we expect that it can be used with other insects as well.*

Mounting medium

We use ProLong Gold Antifade Mountant with 4′,6-diamidino-2-phenylindole (DAPI) (Thermo Fisher Scientific P36941).

Paraffin wax chips (Thermo Fisher Scientific 22-900-700 3)

Phosphate-buffered saline (PBS; 1×) <R>

Sodium citrate buffer

Dissolve sodium citrate ($Na_3C_6H_5O_7$) in distilled H_2O to a concentration of 10 mmol L^{-1} and adjust to a pH of 6.0. Store in refrigerator (for up to 6 mo at 2°C–8°C, checking the pH periodically).

Triton X-100 (Bioshop Life Science Products TRX506)

Wetting agent (e.g., Kodak Photo-Flo 200 Solution; Thermo Fisher Scientific 50-268-05)

Xylene (Sigma-Aldrich 214736)

Equipment

Beaker (1-L)

Computer with software

We use a desktop computer running Windows 10 and CellSens software that controls the Olympus IX70 inverted microscope and digital camera functions. The computer and software you use depend on the chosen microscope.

Fine forceps (e.g., Dumont #5)

Fine paintbrush

Fluorescence microscope

We use an Olympus IX70 inverted microscope with camera.

Glass scintillation vials (20-mL)

Heated water bath

Heat the water bath to 5°C–10°C below the melting point of your paraffin wax.

Hot plate

Heat the hot plate slightly above the melting point of the paraffin wax (see Step 10).

Hydrophobic barrier pen

ImageJ program

Laboratory oven set to 60°C

Magnetic stirrer plate and stir bar

Microwave

Nail polish

Refrigerator (4°C)

Rotary microtome (e.g., Leica RM2125RT)

Single-edge razor blades

Slide staining jars with removable staining racks (e.g., Thermo Fisher Scientific 22-309-247)

These jars are microwave safe so we can also perform target antigen retrieval steps in the microwave with them (see Step 18).

Use of at least eight individual staining jars is recommended.

Slide warmer set at 45°C

 Cite this protocol as *Cold Spring Harb Protoc*; doi:10.1101/pdb.prot108326

Slides with coverslips

Adhesion slides are strongly recommended to avoid sections from detaching during processing. We have used Fisherbrand Superfrost slides with success.

Tissue-embedding (base) molds
Tissue-embedding rings
Tissue paper
Transfer pipette (Thermo Fisher Scientific 13-711-7M)
Tupperware container with lid (medium-sized)

METHOD

Conduct all steps at room temperature unless otherwise indicated.

1. Using forceps, immerse whole mosquito larvae or adults in Bouin's fixative (the volume should be at least 20 times the amount of animal tissues) in a 20-mL glass scintillation vial and fix for 2 h.

   If desired, this protocol can be used for specific organs extracted through dissection. In this case, we recommend extracting a relatively large number of the organs of interest. For example, 20 Malpighian tubules (from four animals) can be extracted and fixed for embedding together into a single block. See Discussion for more information.

2. Rinse mosquitoes twice for 5 min each in 70% ethanol. Use fresh 70% ethanol for each rinse. Carry out this step in the same scintillation vial.

   It is best to use a fine-tip transfer pipette to carefully remove used 70% ethanol at the end of the rinse, taking care not to inadvertently transfer the mosquitoes. After the rinse steps, store the mosquitoes in fresh 70% ethanol at 4°C in the scintillation vial until you are ready to proceed with Step 3.

3. Melt the paraffin wax chips by placing the chips in a scintillation vial and heating until molten in a laboratory oven at 60°C.

   To avoid the issue of brittle ribboning while sectioning, do not heat paraffin wax above the specified melting point. Keep the molten paraffin in the laboratory oven at 60°C so that it is ready when needed. Furthermore, if you have many tissues to process, you can melt 100 mL of wax in a beaker and then transfer it to individual scintillation vials.

4. Dehydrate fixed mosquitoes through incubation in an ascending series of ethanol beginning with 70% ethanol, then 95% ethanol, followed by three separate incubations in 100% ethanol. Carry out this step in the scintillation vial.

   If mosquitoes were previously fixed and stored in 70% ethanol at 4°C, the initial incubation in 70% ethanol can be skipped.

   i. For each step in the series, immerse the mosquitoes or organs in solution that is 20 times the volume of animals or isolated organs (e.g., 1 g of animals in 20 mL of ethanol) for 45 min.

   ii. Use transfer pipettes to carefully remove the solution after incubation and to add the next solution in the series.

5. Clear mosquitoes with one 45-min incubation in a 1:1 mixture of 100% ethanol and xylene.

   It is recommended that the 1:1 mixture of 100% ethanol and xylene is made in advance for a uniform mixture. This mixture can be made in bulk for long-term storage in a chemical fume hood.

6. Incubate mosquitoes in three sequential 45-min incubations in pure xylene in the glass scintillation vial.

   Animals or organs should become translucent during these incubations.

   Use glass for holding xylene because xylene will dissolve plastic.

7. Submerge the mosquito in a 1:1 ratio of xylene and molten paraffin solution in the scintillation vial. Place the vial in the laboratory oven for 1 h at 60°C.

   The 1:1 mixture of xylene and molten paraffin can be made in bulk before this step and kept at 60°C to allow for a uniform blending.

8. Working quickly, remove the scintillation vial with your mosquito tissues from the laboratory oven and use one plastic transfer pipette to remove as much of the xylene and molten paraffin solution as possible, without disturbing the tissues.
9. Next, use another clean transfer pipette to add fresh molten paraffin wax to the scintillation vial (at least 25 times the volume of the tissue in a scintillation vial), and return the vial back to the oven for 1 h at 60°C. Perform this step twice.

    This step should be performed quickly, otherwise the molten paraffin will solidify before it is removed.

    We find it useful to cut the tip (~1 cm) off the plastic transfer pipette to create a wider opening to take up the wax, as some of the wax will inevitably solidify within the transfer pipette.

    It is critical that all the xylene be cleared from the scintillation vial, as xylene contamination in the wax can greatly interfere with tissue sectioning. An additional step with pure molten paraffin wax (1 h at 60°C) can be performed if not all of the xylene:paraffin wax mixture was removed during the two pure molten paraffin wax infiltration steps.
10. Embed mosquitoes in the molten paraffin by pouring molten paraffin into embedding rings and placing mosquitoes or organs in the center using fine forceps.

    This step is best performed on a hot plate that is heated slightly above the melting point of the paraffin wax to allow for proper positioning of the animals or organs in the embedding mold before the paraffin wax solidifies.
11. Allow the paraffin in the embedding rings to solidify by resting until the paraffin wax is cloudy and then at 4°C overnight (at least 12 h) or for long-term storage.
12. Remove the base mold to expose the embedded mosquito. Use a single-edge razor blade to remove excess wax around the tissue.

    This step results in a smaller area of each individual tissue section, such that more sections can fit onto a single adhesion microscope slide, if desired.
13. Mount embedding rings onto the manual rotary microtome and cut sections of 4- to 5-µm thickness.

    Follow the manufacturer's instructions for using the microtome.
14. Use a fine paintbrush and fine forceps to place the ribbon of tissue sections that are cut with the microtome on a coated microscope slide that has been covered with 30% ethanol. Then float the ribbons on distilled H_2O that has been heated to 5°C–10°C below the melting point of your paraffin wax in a water bath.

    The 30% ethanol prevents the ribbons from sticking to the slides until after the paraffin wax sections have been heated and smoothed out in the water bath.
15. Place the sections onto adhesion slides by using a clean slide to carefully scoop up the floated ribbon in the water bath (see Fig. 1).

    Sections can be arranged onto slides as desired using the paintbrush while they are floating on distilled H_2O.
16. Incubate slides with sections on a slide warmer for 24 h at 45°C.

    After this incubation, slides can be stored indefinitely.
17. Submerge slides in slide staining jars with xylene and a descending ethanol series, as follows: Submerging slides with sections in xylenes will deparaffinize them and the ethanol series will rehydrate the tissue sections. Perform washes at room temperature.
    i. Xylene twice for 5 min.
    ii. 100% ethanol twice for 5 min.
    iii. 95% ethanol once for 2 min.
    iv. 70% ethanol once for 2 min.
    v. 50% ethanol once for 2 min.
    vi. Distilled H_2O once for 30 min.

Cite this protocol as *Cold Spring Harb Protoc*; doi:10.1101/pdb.prot108326

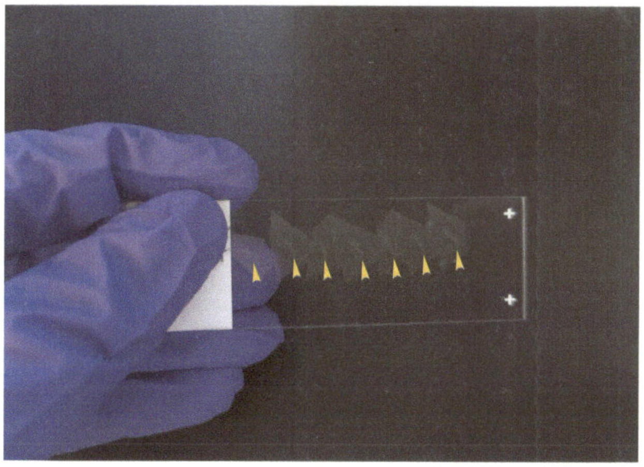

FIGURE 1. A slide with paraffin-embedded tissue sections. Arrows point to tissue sections.

18. Subject sections on slides to heat-induced epitope retrieval by immersing slides in 10 mM sodium citrate buffer in glass, microwave-safe staining jars and heating in a microwave.

    We use medium power for 4 min. The power level of the microwave and the time the slides are heated in the microwave will vary for individual microwaves. The sodium citrate buffer should heat up without boiling. You will need to test your microwave settings to establish the appropriate power and time settings.

19. After the first heating cycle, cool immersed slides with sections for 20 min in the microwave with the microwave door ajar.
20. Reheat the immersed slides in the microwave for half the time of the first heating cycle.

    We use medium power for 2 min.

21. Remove the immersed slides with sections from the microwave, and allow them to cool for 15 min inside the staining jar.
22. Wash slides with sections with PBS for 3 min in a staining jar placed on a magnetic stirrer plate with a magnetic stir bar below the slide rack.

    Slides with sections should be completely submerged in PBS; the volume of PBS will vary depending on the size of the staining jar. The jars that we used required 175 mL of solution to completely submerge the slides.

23. In staining jars, carry out a series of incubations by immersing sections on slides in different solutions as follows:

    i. 0.4% (v/v) wetting agent in PBS for 10 min.

    ii. 0.05% (v/v) Triton X-100 in PBS for 10 min.

    iii. MADB for 10 min.

24. Incubate sections on slides with a desired primary antibody.

    To save on antibody volumes, use a hydrophobic barrier pen to draw a hydrophobic barrier around the sections on the slide that will be incubated with antibody.

    i. Dilute the primary antibody to the desired dilution (concentration) in MADB.

    At this stage, if you would like to tag two different targets, you can do so if your primary antibodies are produced in different host animals. For example, one antibody may be produced in rabbit and the other may be produced in mouse. In this case, dilute both antibodies to their desired dilutions into the same MADB.

    ii. Pipette a sufficient volume of diluted antibody in MADB solution onto all sections so that they are covered within the hydrophobic barrier.

    Normally we use 100–200 µL of diluted antibody solution for one slide.

    Do not allow the antibody solution to flow over the hydrophobic barrier.

iii. Place slides in a Tupperware container that has been lined with moistened tissue paper using distilled H_2O, and cover with a lid. Place the Tupperware container in a refrigerator for 24 h at 4°C.

This step prevents the antibody solution from evaporating during incubation.

25. In staining jars, remove the primary antibody solution by immersing slides in a beaker of distilled H_2O and incubating in the following solutions:
    i. 0.4% (v/v) wetting agent in PBS for 10 min.
    ii. 0.05% (v/v) Triton X-100 in PBS for 10 min.
    iii. MADB for 10 min.
26. Incubate sections on slides with a desired secondary antibody.
    i. Dilute the secondary antibody to the desired dilution (concentration) in MADB.

    We use fluorescent tagged antibodies that recognize the animal host in which the primary antibody was produced. For example, if the primary antibody was produced in rabbit, use an anti-rabbit secondary antibody. If you have used two primary antibodies to tag two different targets, you can dilute two different secondary antibodies into the same MADB to apply them at the same time.

    ii. Pipette a sufficient volume of diluted secondary antibody in MADB to cover the sections within the hydrophobic barrier on the slides and incubate the sections for 1 h.

    You may need to carefully dry each slide and redraw the hydrophobic barrier before applying the secondary antibody.

27. Remove the secondary antibody solution from slides by dipping each slide into a beaker of distilled H_2O, and then immersing slides in the following solutions in staining jars:
    i. 0.4% (v/v) wetting agent in PBS for 10 min.
    ii. 0.05% (v/v) Triton X-100 in PBS for 10 min.
    iii. 0.4% (v/v) wetting agent in PBS for 10 min.
    iv. 0.4% (v/v) wetting agent in distilled H_2O three times for 1 min each.
    v. Distilled H_2O for 1 min.
28. Air-dry slides with sections in a dark place for 1 h.
29. Apply 5 µL of mounting medium to each tissue section and then cover sections with coverslips.

    Take care not to touch the tissue section during this step.

    Some commercially available mounting media are provided in dropper bottles; in this case, dispense a single droplet on top of each tissue section.

30. Secure the coverslips to the microscope slide using clear nail polish.
31. View sections with a fluorescence microscope using the appropriate filter for the tag on the secondary antibody.
32. Process images using ImageJ.

    For ethical reasons and many journal requirements, we do not recommend adjusting the exposure or brightness/contrast beyond what was optimized when imaging the sections using a fluorescence microscope and the microscope-accompanying software. Using ImageJ, the fluorescence of monochrome images can be colored as desired. Additionally, identical images of the same sections stained with different antibodies (if desired) can be merged to visualize immunolocalization of different proteins within the same tissue. Furthermore, various tools via ImageJ allow for quantitative assessments of staining intensity and cell/tissue sizes.

Cite this protocol as *Cold Spring Harb Protoc*; doi:10.1101/pdb.prot108326

DISCUSSION

This protocol works well for whole fixed mosquitoes of any life stage and can work, although with additional challenges, for individual mosquito organs. Whole mosquitoes are large enough to handle during the many processing stages without unintentionally removing them when changing solutions. Embedding whole mosquitoes is also relatively easy compared with isolated organs because they can be more readily positioned in the desired orientation for slicing during the embedding procedure. For example, Figure 2 shows sections from slices of whole mosquitoes that were positioned during the embedding procedure for slicing in different planes (Fig. 2A,B). Individual organs can also be processed using this protocol, but several challenges exist. Because mosquito organs are small, it is easy to lose them during one of the many solution changes during the protocol. Embedding organs is also a challenge because they cannot be as readily manipulated as whole mosquitoes and, therefore, positioning the organs in a controlled manner is not feasible. Nevertheless, sometimes it is desirable to have a specific organ to view on a section. To mitigate the challenges, we advise one to extract several organs of interest. For example, if one is interested in Malpighian tubules, then extract at least 20 individual tubules and as many as you can together. This will result in sections that will contain multiple tubules that will likely have different planes such that a single section could provide the desired visual information.

Importantly, include controls every time this protocol is used. Control tissue sections are produced and mounted on slides following the same protocol. The slides are then treated in the identical manner as the experimental slides except for the primary antibody incubation step. At this stage, incubate the control sections with MADB that does not contain antibody and continue the remaining protocol as described. Ideally, it is better to preincubate the diluted antibody in MADB with the antigenic/target protein. Add antigenic peptide to the diluted antibody solution at a molarity that is at least three times that of the antibody and then incubate before applying to the control sections. Usually, the ideal molarity and incubation time will be different for each different antibody, and this will need to be experimentally determined. Control sections are required to compare staining patterns and identifying any background, nonspecific staining that may appear.

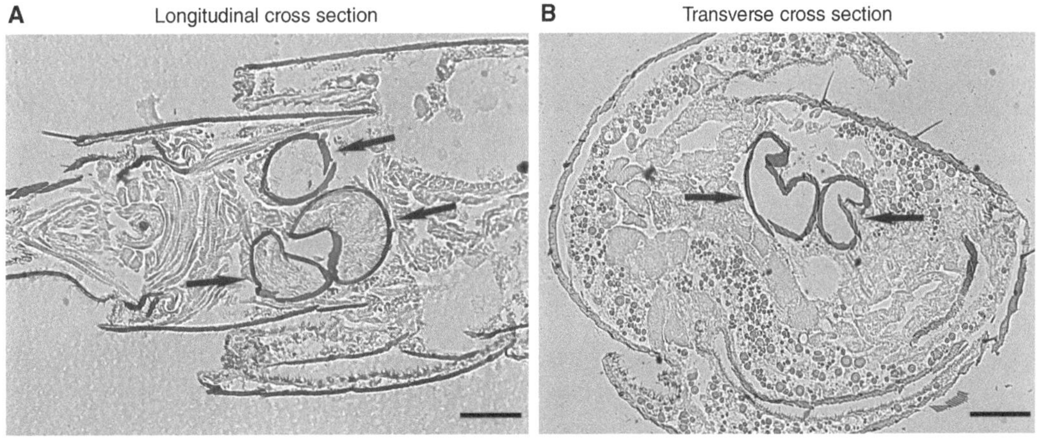

FIGURE 2. Bright-field images of paraffin-embedded cross sections of posterior abdominal segments of adult female *Aedes aegypti*. (*A*) Longitudinal and (*B*) transverse cross sections through the terminal abdominal segment housing the spermathecae (black arrows). Scale bars, 50 µm.

RECIPES

Bouin's Fixative for IHC

Picric acid (saturated, aqueous)	75 mL
Formalin (40%, w/v, formaldehyde)	25 mL
Glacial acetic acid	5 mL

First add the formalin to the saturated picric acid in a glass bottle. Next, add the glacial acetic acid to the mixture and tightly seal bottle with a nonmetal lid or cap. The solution can be mixed by gently and very carefully inverting the bottle two to three times.

Bouin's fixative solution should be stored in a dark, well-ventilated area at room temperature. It can be stored indefinitely if hydrated.

Bouin's fixative solution should always be made in a fume hood and handled with appropriate personal protective equipment (gloves, safety goggles, and lab coat). Aqueous picric acid must always be saturated with H_2O in a sealed nonmetal container and stored in a dark, flame-proof area. Picric acid and solutions containing picric acid should never come in long-term contact with metal as metal picrates pose a greater explosive risk than picric acid crystals. The hydration status of aqueous picric acid should be checked regularly and replenished to at least 30% H_2O by volume.

Mosquito Antibody Dilution Buffer (MADB)

Reagent	Final concentration	Amount
Goat serum	10% (v/v)	50 mL
Bovine serum albumin	3% (w/v)	15 g
Triton-X	0.05% (v/v)	250 µL
Phosphate-buffered saline <R>	90% (v/v)	450 mL

This recipe makes a 500-mL solution that can be stored as 50-mL aliquots for up to 6 mo at −20°C.

Phosphate-Buffered Saline (PBS)

Reagent	Amount to add (for 1× solution)	Final concentration (1×)	Amount to add (for 10× stock)	Final concentration (10×)
NaCl	8 g	137 mM	80 g	1.37 M
KCl	0.2 g	2.7 mM	2 g	27 mM
Na_2HPO_4	1.44 g	10 mM	14.4 g	100 mM
KH_2PO_4	0.24 g	1.8 mM	2.4 g	18 mM
If necessary, PBS may be supplemented with the following:				
$CaCl_2 \cdot 2H_2O$	0.133 g	1 mM	1.33 g	10 mM
$MgCl_2 \cdot 6H_2O$	0.10 g	0.5 mM	1.0 g	5 mM

PBS can be made as a 1× solution or as a 10× stock. To prepare 1 L of either 1× or 10× PBS, dissolve the reagents listed above in 800 mL of H_2O. Adjust the pH to 7.4 (or 7.2, if required) with HCl, and then add H_2O to 1 L. Dispense the solution into aliquots and sterilize them by autoclaving for 20 min at 15 psi (1.05 kg/cm^2) on liquid cycle or by filter sterilization. Store PBS at room temperature.

REFERENCES

Durant AC, Donini A. 2023. Osmoregulatory challenges faced by mosquitoes and techniques for studying the transporters involved in their adaptations. *Cold Spring Harb Protoc* doi:10.1101/pdb.top107701

CHAPTER 45

Visualization of the Linear and Spatial Organization of Chromosomes in Mosquitoes

Jiangtao Liang,[1] Simon M. Bondarenko,[1,2] Igor V. Sharakhov,[1,2] and Maria V. Sharakhova[1,3,4]

[1]*Department of Entomology, Virginia Polytechnic Institute and State University, Blacksburg, Virginia 24060, USA;* [2]*Department of Genetics and Cell Biology, Tomsk State University, Tomsk 634050, Russia;* [3]*Laboratory of Evolutionary Genomics of Insects, Institute of Cytology and Genetics, Siberian Branch of the Russian Academy of Sciences, Novosibirsk 630090, Russia*

Mosquitoes are vectors of dangerous human diseases such as malaria, dengue, Zika, West Nile fever, and lymphatic filariasis. Visualization of the linear and spatial organization of mosquito chromosomes is important for understanding genome structure and function. Utilization of chromosomal inversions as markers for population genetics studies yields insights into mosquito adaptation and evolution. Cytogenetic approaches assist with the development of chromosome-scale genome assemblies that are useful tools for studying mosquito biology and for designing novel vector control strategies. Fluorescence in situ hybridization is a powerful technique for localizing specific DNA sequences within the linear chromosome structure and within the spatial organization of the cell nucleus. Here, we introduce protocols used in our laboratories for chromosome visualization and their application in mosquitoes.

LINEAR ORGANIZATION OF MOSQUITO CHROMOSOMES

Genetic material in all eukaryotic organisms is organized into chromosomes, which represent the units of inheritance at the cellular level. Chromosomes play essential parts in vital organismal functions such as cell division and genetic recombination during meiosis. Cytogenetic studies have been important in understanding genome function and evolution (Zhimulev 1996). Mosquito cytogenetics is a relatively well-developed discipline for several reasons: (1) Mosquitoes have received an increased attention from researchers because of their epidemiological importance; (2) mosquitoes conveniently have a low number of chromosomes (2n = 6) (Rai 1963); and (3) malaria mosquitoes (subfamily Anophelinae) can develop giant polytene chromosomes in several tissues (salivary glands, ovarian nurse cells, Malpighian tubules) (Sharakhov and Sharakhova 2008). Sex chromosomes are represented as typical X and Y chromosomes in species from the subfamily Anophelinae. In the subfamily Culicinae, sex-determining chromosomes are homomorphic, and male sex is determined by a locus found on chromosome 1. Polytene chromosomes are usually paired in mosquito species and the chromosomal compliment is represented by five chromosomal arms: X, 2R, 2L, 3R, and 3L (Fig. 1A). DNA in one of the arms of the X chromosome and in the entire Y chromosome is underreplicated, and this DNA is not visible in the polytene chromosome complement. In classical cytogenetics studies, visualization of polytene chromosome banding patterns was used to develop cytogenetic maps for about 50 species of

[4]Correspondence: msharakh@vt.edu

Cite this introduction as *Cold Spring Harb Protoc*; doi:10.1101/pdb.top107732

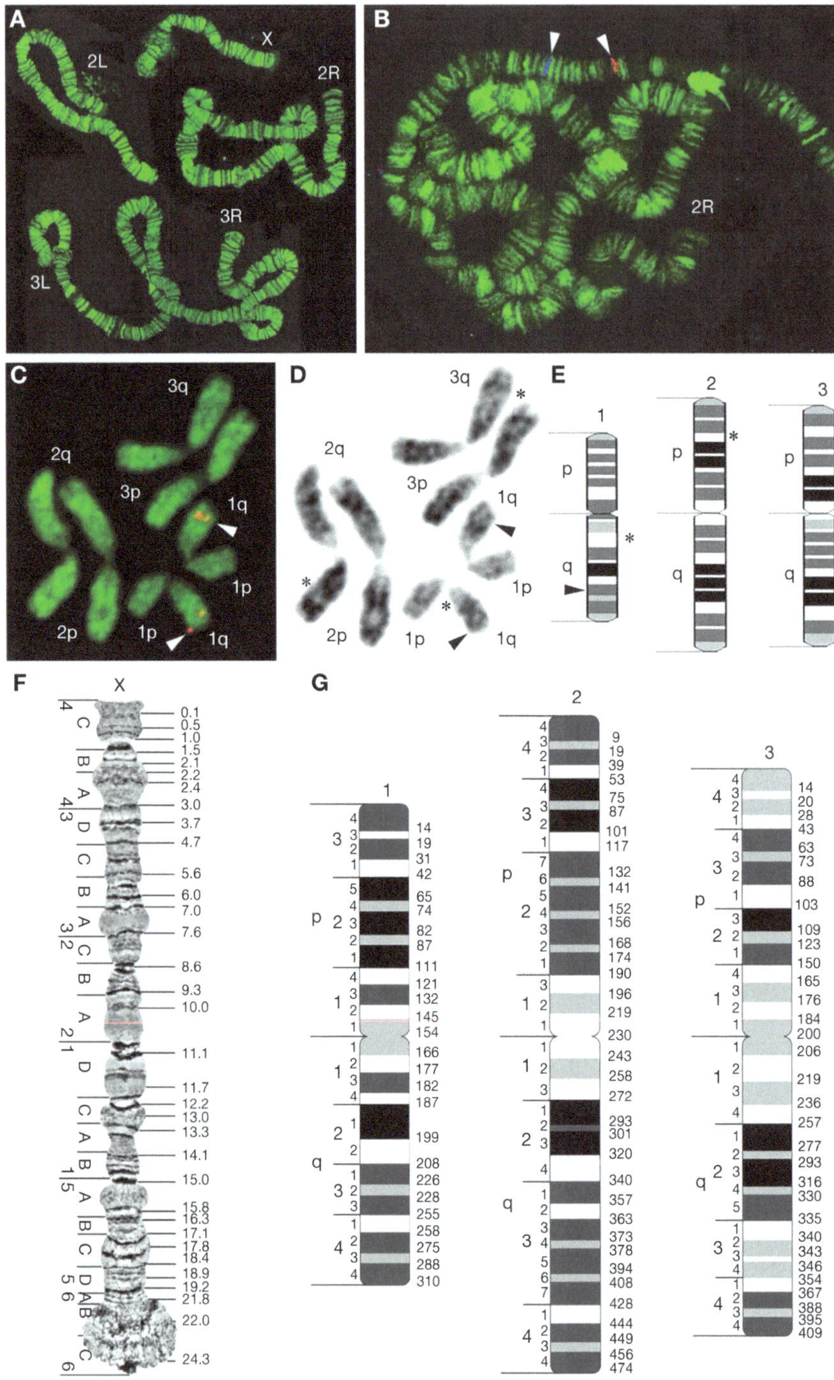

FIGURE 1. Visualizing polytene (*A,B,F*) and mitotic (*C–E,G*) chromosomes in an application for the physical mapping of mosquito genomes. The polytene chromosome complement for *Anopheles stephensi* is shown in *A*. X, 2R, 2L, 3R, and 3L stand for the chromosome arms. An example of fluorescence in situ hybridization (FISH) in polytene chromosome arm 2R in *Anopheles albimanus* is showed in *B* (Artemov et al. 2017). The positions of the signals are indicated by arrows. (*C,D*) Different steps in physical mapping of the mitotic chromosomes of *Culex quinquefasciatus* (Naumenko et al. 2015). First, the position of the fluorescence signal is recognized on chromosomes that are distinguished from each other based on their sizes, with chromosomes 1, 2, and 3 being the smallest, biggest, and intermediate, respectively (*C*). Second, chromosome arms are recognized in black and white images using cytogenetic landmarks indicated by asterisks (*D*). Finally, the position of the probe is placed to a particular band on the chromosome idiogram (*E*). Chromosome arms are shown as 1p, 1q, 2p, 2q, 3p, and 3q. The position of a signal is indicated by an arrow. (*F,G*) The X-chromosome physical genome map of *Anopheles gambiae* (George et al. 2010) and the entire genome map for *Aedes aegypti* (Matthews et al. 2018), respectively. Chromosome regions and genomic coordinates are shown on the right and left sides of the chromosomes, respectively.

 Cite this introduction as *Cold Spring Harb Protoc*; doi:10.1101/pdb.top107732

the genus *Anopheles* (Sharakhov and Sharakhova 2008). Such maps were primarily used in population genetics studies using the analysis of inversion polymorphism in mosquito populations (Stegniy 1991; Coluzzi et al. 2002). In addition, chromosomal maps have been helpful for mosquito taxonomy and systematics as tools for species discovery and validation (Kitzmiller 1977; Stegniy et al. 1978; Green and Miles 1980). Finally, such maps have been used to study chromosomal evolution and phylogeny in several groups of malaria mosquitoes (Stegniy 1991; Green et al. 1992; Coluzzi et al. 2002).

Polytene chromosomes in anopheline mosquitoes allow researchers to map genes with high-resolution using fluorescence in situ hybridization (FISH) (Sharakhova et al. 2015, 2019). An example of a two-color FISH of *Anopheles albimanus* chromosomes (Artemov et al. 2017) is shown in Figure 1B. A combination of highly developed polytene chromosomes and the in situ hybridization technique helped to create detailed physical genome maps for malaria mosquitoes including *Anopheles gambiae* (Holt et al. 2002; Sharakhova et al. 2007; George et al. 2010), *Anopheles stephensi* (Jiang et al. 2014; Waterhouse et al. 2020), *An. albimanus* (Artemov et al. 2017), *Anopheles atroparvus* (Artemov et al. 2018), and *An. funestus* (Sharakhov et al. 2002; Waterhouse et al. 2020). Among these species, the most populated physical map was developed for the African malaria vector *An. gambiae.* This map included positions for ~approximately 2000 bacterial artificial chromosome (BAC) clones and had a resolution of 0.5–1 Mb. A fragment of the *An. gambiae* X chromosome physical map is shown in Figure 1F.

Cytogenetic mapping in Culicine mosquitoes is difficult because of the absence of good-quality, easily spreadable polytene chromosomes (Sharma et al. 1978; Campos et al. 2003a,b). Thus, most of the cytogenetic studies for these mosquitoes were performed on meiotic chromosomes from male testes or mitotic chromosomes from brain ganglia (Rai 1963; Newton et al. 1974; Motara and Rai 1978). These studies showed differences in the C-band staining pattern among different species of *Aedes,* suggesting the presence of chromosomal rearrangements (Motara and Rai 1977, 1978; Wallace and Newton 1987). A chromosomal analysis was used to study genome organization and evolution in different mosquito lineages including both Anophelinae and Culicinae subfamilies (Rai and Black 1999). It showed that Anophelinae mosquitoes with small chromosome lengths and genome sizes have long-period interspersion of repetitive elements in their genomes with heterochromatin accumulated around the centromeres. A dramatic increase in chromosome lengths and genome sizes in the Culicinae subfamily is related to the short-period interspersion of repeats in their genomes. Chromosomes from the brain ganglia of *Aedes aegypti* have been used for FISH for positioning of ribosomal genes (Kumar and Rai 1990) and cloned DNA probes on chromosomes in *Ae. aegypti* (Brown et al. 1995). In addition, mapping of cDNA-based genetic markers was used to integrate the genetic and physical maps for this species (Brown et al. 2001). The discovery of a new source of mitotic chromosomes in imaginal discs of *Ae. aegypti* (Sharakhova et al. 2011) enabled the creation of the first physical genome map for this mosquito species (Fig. 1G) (Timoshevskiy et al. 2014) and its integration with the genetic map (Timoshevskiy et al. 2013). Later, a similar study was performed on chromosomes from *Culex quiquefasciatus* (Naumenko et al. 2015). Figure 1C–E show the process of physical mapping in mitotic chromosomes in this species.

Further progress in mosquito genomics has resulted in significantly increased contiguity of genomic scaffolds up to chromosome-level assemblies for mosquitoes. Such advancements have been achieved through the combination of cytogenetic FISH-based physical mapping with newly developed molecular approaches, such as long-read single molecule sequencing (Berlin et al. 2015; Goodwin et al. 2015; Miller et al. 2018), optical mapping using high-resolution restriction maps from single, stained molecules (Deschamps et al. 2018), and the Hi-C scaffolding method based on chromosome conformation capture to create proximity-guided assemblies (Belton et al. 2012; Dudchenko et al. 2017). Chromosome-level genome assemblies have been created for *An. albimanus* (Artemov et al. 2017; Compton et al. 2020), *An. atroparvus* (Artemov et al. 2018; Waterhouse et al. 2020; Lukyanchikova et al. 2022), *An. stephensi* (Waterhouse et al. 2020; Chakraborty et al. 2021; Lukyanchikova et al. 2022), *An. funestus* (Ghurye et al. 2019; Waterhouse et al. 2020), *Anopheles merus* (Lukyanchikova et al. 2022), *Anopheles arabiensis* (Zamyatin et al. 2021), *Anopheles coluzzii* (Zamyatin et al. 2021), and *Ae. aegypti* (Fig. 1G; Matthews et al. 2018). Recently, 57% of the genome was mapped to the chromosomes of the Asian tiger mosquito *Aedes albopictus* (Palatini et al. 2020). Although optical mapping and Hi-C scaffolding can significantly increase contiguity of genomic scaffolds, only FISH-based cytogenetic mapping provides physical evidence of a scaffold location in a specific chromosomal region (Sharakhova et al. 2019). Moreover, this technique directly relates DNA probes of known sequence to a particular band or structure on a chromosome. Chromosome-scale genome assemblies for mosquitoes offer new opportunities to better understand genome organization and function (Matthews et al. 2018; Palatini et al. 2020) and population structure and adaptation (Rose et al. 2020), and, finally, enable development of new genome-based strategies for mosquito control (Ruzzante et al. 2019; Aryan et al. 2020).

SPATIAL ORGANIZATION OF MOSQUITO CHROMOSOMES

The nonrandom hierarchical folding of the DNA molecule in the nuclear space plays a critical role in the regulation and coordination of gene expression (Sharakhov et al. 2018; Lezcano et al. 2020). The three-dimensional (3D) organization of chromosomes in the interphase nucleus is governed by many mechanisms. For example, the specific radial distribution of chromatin is maintained by the attachment of chromosomal regions to the peripheral protein lamin (Bondarenko and Sharakhov 2020). A microscopy study showed that the percentage of the chromosome–nuclear envelope contacts and the number of interchromosomal interactions are cell-type-specific and conserved among the species of the *An. gambiae* complex. Moreover, at the genome-wide level, there is an inverse relationship between chromosome–nuclear envelope attachments and chromosome–chromosome interactions (George et al. 2020). Recent work used Hi-C and microscopy to study 3D chromatin architecture in five *Anopheles* species (Lukyanchikova et al. 2022). It revealed that, like all eukaryotic genomes, mosquito genomes are hierarchically folded into chromatin loops, topologically associating domains (TADs), compartments, and chromosome territories. Chromosome territories in *Anopheles* assume a Rabl-like configuration (telomeres and centromeres are at the opposite poles of the nucleus) that deteriorates in some adult tissues. Separation into active (A) and inactive (B) chromatin compartments is stronger in adult mosquitoes than in embryos. In addition, TADs within synteny blocks and specific long-range looping interactions have been conserved for tens of millions of years of evolution (Lukyanchikova et al. 2022).

3D-FISH is an inexpensive method for direct microscopic visualization of specific chromatin targets in the nuclear space with single-cell resolution. These targets include entire chromosome territories, individual chromosome arms, TADs, specific chromosomal regions, and individual genes. Mosquitoes are convenient model organisms for studying the 3D organization of chromatin. Many species have large nuclei with polytene or polyploid chromosomes in multiple cell types. This feature makes it easier to perform 3D-microscopic studies because the normal nuclear size poses limitations on the quality and resolution of microscopic analyses. In addition, the highly reproducible patterns of bands in polytene chromosomes of anophelines provide a great opportunity for recognizing chromosomal segments for producing region-specific fluorescent probes by microdissection (George et al. 2014).

CONCLUSION

The continued development of cytogenetic methods is important for characterizing genome organization. The two associated protocols are used in our laboratories. The first protocol describes how to obtain polytene, meiotic, and mitotic mosquito chromosomes for visualization (see Protocol 1: Obtaining Polytene, Meiotic, and Mitotic Chromosomes from Mosquitoes for Cytogenetic Analysis [Liang et al. 2022]). The second protocol describes how to hybridize and visualize fluorescent probes in the cell nuclei of whole-mount mosquito tissues (see Protocol 2: Visualization of Polytene Chromatin in Mosquito Cell Nuclei Using Three-Dimensional Fluorescence In Situ Hybridization [Bondarenko et al. 2022]).

 Cite this introduction as *Cold Spring Harb Protoc*; doi:10.1101/pdb.top107732

ACKNOWLEDGMENTS

Our work on linear chromosome visualization was supported by grant 19-14-00130 from the Russian Science Foundation (to M.V.S.) and our work on spatial chromosome visualization was performed using grant 21-14-00182 from the Russian Science Foundation (to I.V.S.). We thank Janet Webster for the text editing.

REFERENCES

Artemov GN, Peery AN, Jiang X, Tu Z, Stegniy VN, Sharakhova MV, Sharakhova IV. 2017. The physical genome mapping of *Anopheles albimanus* corrected scaffold misassemblies and identified interarm rearrangements in genus *Anopheles*. *G3 (Bethesda)* **7:** 155–164. doi:10.1534/g3.116.034959

Artemov GN, Bondarenko SM, Naumenko AN, Stegniy VN, Sharakhova MV, Sharakhova IV. 2018. Partial-arm translocations in evolution of malaria mosquitoes revealed by high-coverage physical mapping of the *Anopheles atroparvus* genome. *BMC Genomics* **19:** 278. doi:10.1186/s12864-018-4663-4

Aryan A, Anderson MAE, Biedler JK, Qi Y, Overcash JM, Naumenko AN, Sharakhova MV, Mao C, Adelman ZN, Tu Z. 2020. Nix alone is sufficient to convert female *Aedes aegypti* into fertile males and myo-sex is needed for male flight. *Proc Natl Acad Sci* **117:** 17702–17709. doi:10.1073/pnas.2001132117

Belton JM, McCord RP, Gibcus JH, Naumova N, Zhan Y, Dekker J. 2012. Hi-C: a comprehensive technique to capture the conformation of genomes. *Methods* **58:** 268–276. doi:10.1016/j.ymeth.2012.05.001

Berlin K, Koren S, Chin CS, Drake JP, Landolin JM, Phillippy AM. 2015. Assembling large genomes with single-molecule sequencing and locality-sensitive hashing. *Nat Biotechnol* **33:** 623–630. doi:10.1038/nbt.3238

Bondarenko SM, Sharakhov IV. 2020. Reorganization of the nuclear architecture in the *Drosophila melanogaster* Lamin B mutant lacking the CaaX box. *Nucleus* **11:** 283–298. doi:10.1080/19491034.2020.1819704

Bondarenko SM, Liang J, Sharakhova MV, Sharakhov IV. 2022. Visualization of polytene chromatin in mosquito cell nuclei using three-dimensional fluorescence in situ hybridization. *Cold Spring Harb Protoc* doi:10.1101/pdb.prot107873

Brown SE, Menninger J, Difillipantonio M, Beaty BJ, Ward DC, Knudson DL. 1995. Toward a physical map of *Aedes aegypti*. *Insect Mol Biol* **4:** 161–167. doi:10.1111/j.1365-2583.1995.tb00021.x

Brown SE, Severson DW, Smith LA, Knudson DL. 2001. Integration of the *Aedes aegypti* mosquito genetic linkage and physical maps. *Genetics* **157:** 1299–1305. doi:10.1093/genetics/157.3.1299

Campos J, Andrade CF, Recco-Pimentel SM. 2003a. Malpighian tubule polytene chromosomes of *Culex quinquefasciatus* (Diptera, Culicinae). *Mem Inst Oswaldo Cruz* **98:** 383–386. doi:10.1590/S0074-02762003000300016

Campos J, Andrade CF, Recco-Pimentel SM. 2003b. A technique for preparing polytene chromosomes from *Aedes aegypti* (Diptera, Culicinae). *Mem Inst Oswaldo Cruz* **98:** 387–390. doi:10.1590/S0074-02762003000300017

Chakraborty M, Ramaiah A, Adolfi A, Halas P, Kaduskar B, Ngo LT, Jayaprasad S, Paul K, Whadgar S, Srinivasan S, et al. 2021. Hidden genomic features of an invasive malaria vector, *Anopheles stephensi*, revealed by a chromosome-level genome assembly. *BMC Biol* **19:** 28. doi:10.1186/s12915-021-00963-z

Coluzzi M, Sabatini A, della Torre A, Di Deco MA, Petrarca V. 2002. A polytene chromosome analysis of the *Anopheles gambiae* species complex. *Science* **298:** 1415–1418. doi:10.1126/science.1077769

Compton A, Liang J, Chen C, Lukyanchikova V, Qi Y, Chen HZ, Chen Y, Compton A, Elkhatib R, Geiger L, et al. 2020. The beginning of the end: a chromosomal assembly of the new world malaria mosquito ends with a novel telomere. *G3 (Bethesda)* **10:** 3811–3819. doi:10.1534/g3.120.401654

Deschamps S, Zhang Y, Llaca V, Ye L, Sanyal A, King M, May G, Lin H. 2018. A chromosome-scale assembly of the sorghum genome using nanopore sequencing and optical mapping. *Nat Commun* **9:** 4844. doi:10.1038/s41467-018-07271-1

Dudchenko O, Batra SS, Omer AD, Nyquist SK, Hoeger M, Durand NC, Shamim MS, Machol I, Lander ES, Aiden AP, et al. 2017. *De novo* assembly of the *Aedes aegypti* genome using Hi-C yields chromosome-length scaffolds. *Science* **356:** 92–95. doi:10.1126/science.aal3327

George P, Sharakhova MV, Sharakhov IV. 2010. High-resolution cytogenetic map for the African malaria vector *Anopheles gambiae*. *Insect Mol Biol* **19:** 675–682.

George P, Sharma A, Sharakhov IV. 2014. 2D and 3D chromosome painting in malaria mosquitoes. *J Vis Exp* e51173. doi:10.3791/51173

George P, Kinney NA, Liang J, Onufriev AV, Sharakhov IV. 2020. Three-dimensional organization of polytene chromosomes in somatic and germline tissues of malaria mosquitoes. *Cells* **9:** 339. doi:10.3390/cells9020339

Ghurye J, Koren S, Small ST, Redmond S, Howell P, Phillippy AM, Besansky NJ. 2019. A chromosome-scale assembly of the major African malaria vector *Anopheles funestus*. *Gigascience* **8:** giz063. doi:10.1093/gigascience/giz063

Goodwin S, Gurtowski J, Ethe-Sayers S, Deshpande P, Schatz MC, McCombie WR. 2015. Oxford Nanopore sequencing, hybrid error correction, and de novo assembly of a eukaryotic genome. *Genome Res* **25:** 1750–1756. doi:10.1101/gr.191395.115

Green CA, Miles SJ. 1980. Chromosomal evidence for sibling species of the malaria vector *Anopheles* (Cellia) *culicifacies* Giles. *J Trop Med Hyg* **83:** 75–78.

Green CA, Rattanarithikul R, Charoensub A. 1992. Population genetic confirmation of species status of the malaria vectors *Anopheles willmori* and *An. pseudowillmori* in Thailand and chromosome phylogeny of the Maculatus group of mosquitoes. *Med Vet Entomol* **6:** 335–341. doi:10.1111/j.1365-2915.1992.tb00629.x

Holt RA, Subramanian GM, Halpern A, Sutton GG, Charlab R, Nusskern DR, Wincker P, Clark AG, Ribeiro JM, Wides R, et al. 2002. The genome sequence of the malaria mosquito *Anopheles gambiae*. *Science* **298:** 129–149. doi:10.1126/science.1076181

Jiang X, Peery A, Hall A, Sharma A, Chen XG, Waterhouse RM, Komissarov A, Riehle MM, Shouche Y, Sharakhova MV, et al. 2014. Genome analysis of a major urban malaria vector mosquito, *Anopheles stephensi*. *Genome Biol* **15:** 459. doi:10.1186/s13059-014-0459-2

Kitzmiller JB. 1977. Chromosomal differences among species of anopheles mosquitoes. *Mosq Syst* **9:** 112–122.

Kumar A, Rai KS. 1990. Chromosomal localization and copy number of 18S + 28S ribosomal RNA genes in evolutionarily diverse mosquitoes (Diptera, Culicidae). *Hereditas* **113:** 277–289. doi:10.1111/j.1601-5223.1990.tb00094.x

Lezcano OM, Sanchez-Polo M, Ruiz JL, Gomez-Diaz E. 2020. Chromatin structure and function in mosquitoes. *Front Genet* **11:** 602949. doi:10.3389/fgene.2020.602949

Liang J, Bondarenko SM, Sharakhov IV, Sharakhova MV. 2022. Obtaining polytene, meiotic, and mitotic chromosomes from mosquitoes for cytogenetic analysis. *Cold Spring Harb Protoc* doi: 10.1101/pdb.prot107872

Lukyanchikova V, Nuriddinov M, Belokopytova P, Taskina A, Liang J, Reijnders MJMF, Ruzzante L, Feron R, Waterhouse RM, Wu Y, et al. 2022. *Anopheles* mosquitoes revealed new principles of 3D genome organization in insects. *Nat Commun* **13:** 1960. doi:10.1038/s41467-022-29599-5

Matthews BJ, Dudchenko O, Kingan SB, Koren S, Antoshechkin I, Crawford JE, Glassford WJ, Herre M, Redmond SN, Rose NH, et al. 2018.

Improved reference genome of *Aedes aegypti* informs arbovirus vector control. *Nature* **563:** 501–507. doi:10.1038/s41586-018-0692-z

Miller DE, Staber C, Zeitlinger J, Hawley RS. 2018. Genome Report: highly contiguous genome assemblies of 15 *Drosophila* species generated using nanopore sequencing. *G3 (Bethesda)* **8:** 3131–3141. doi:10.1534/g3.118.200160

Motara MAR, Rai KS. 1977. Chromosomal differentiation in two species of *Aedes* and their hybrids revealed by Giemsa C-banding. *Chromosoma* **64:** 125–132. doi:10.1007/BF00327052

Motara MAR, Rai KS. 1978. Giemsa C-banding patterns in *Aedes* (Stegomia) mosquitoes. *Chromosoma* **70:** 51–58. doi:10.1007/BF00292215

Naumenko AN, Timoshevskiy VA, Kinney NA, Kokhanenko AA, de Bruyn BS, Lovin DD, Stegniy VN, Severson DW, Sharakhov IV, Sharakhova MV. 2015. Mitotic-chromosome-based physical mapping of the *Culex quinquefasciatus* genome. *PLoS One* **10:** e0115737. doi:10.1371/journal.pone.0115737

Newton ME, Southern DI, Wood RJ. 1974. X and Y chromosomes of *Aedes aegypti* (L.) distinguished by Giemsa C-banding. *Chromosoma* **49:** 41–49. doi:10.1007/BF00284986

Palatini U, Masri RA, Cosme LV, Koren S, Thibaud-Nissen F, Biedler JK, Krsticevic F, Johnston JS, Halbach R, Crawford JE, et al. 2020. Improved reference genome of the arboviral vector *Aedes albopictus*. *Genome Biol* **21:** 215. doi:10.1186/s13059-020-02141-w

Rai KS. 1963. A comparative study of mosquito karyotypes. *Ann Ent Soc Am* **56:** 160–170. doi:10.1093/aesa/56.2.160

Rai KS, Black WC. 1999. Mosquito genomes: structure, organization, and evolution. *Adv Genet* **41:** 1–33. doi:10.1016/S0065-2660(08)60149-2

Rose NH, Sylla M, Badolo A, Lutomiah J, Ayala D, Aribodor OB, Ibe N, Akorli J, Otoo S, Mutebi JP, et al. 2020. Climate and urbanization drive mosquito preference for humans. *Curr Biol* **30:** 3570–3579.e3576. doi:10.1016/j.cub.2020.06.092

Ruzzante L, Reijnders M, Waterhouse RM. 2019. Of genes and genomes: mosquito evolution and diversity. *Trends Parasitol* **35:** 32–51. doi:10.1016/j.pt.2018.10.003

Sharakhov IV, Sharakhova MV. 2008. Cytogenetic physical mapping of mosquito genomes. In *Chromosome mapping research developments* (ed. Verrity JF, Abbington LE), pp. 35–76. Nova Science Publishers, New York.

Sharakhov IV, Serazin AC, Grushko OG, Dana A, Lobo N, Hillenmeyer ME, Westerman R, Romero-Severson J, Costantini C, Sagnon N, et al. 2002. Inversions and gene order shuffling in *Anopheles gambiae* and *A. funestus*. *Science* **298:** 182–185. doi:10.1126/science.1076803

Sharakhova MV, Hammond MP, Lobo NF, Krzywinski J, Unger MF, Hillenmeyer ME, Bruggner RV, Birney E, Collins FH. 2007. Update of the *Anopheles gambiae* PEST genome assembly. *Genome Biol* **8:** R5. doi:10.1186/gb-2007-8-1-r5

Sharakhova MV, Timoshevskiy VA, Yang F, Demin S, Severson DW, Sharakhov IV. 2011. Imaginal discs: a new source of chromosomes for genome mapping of the yellow fever mosquito *Aedes aegypti*. *PLoS Negl Trop Dis* **5:** e1335. doi:10.1371/journal.pntd.0001335

Sharakhova MV, George P, Timoshevskiy VA, Sharma A, Peery A, Sharakhov, IV. 2015. Mosquitoes (Diptera). In *Protocols for cytogenetic mapping of arthropod genomes* (ed. Sharakhov IV), pp. 93–170. CRC Press, Boca Raton, Florida.

Sharakhov IV, Bondarenko SM, Artemov GN, Onufriev AV. 2018. The role of chromosome-nuclear envelope attachments in 3D genome organization. *Biochemistry (Mosc)* **83:** 350–358. doi:10.1134/S0006297918040065

Sharakhova MV, Artemov GN, Timoshevskiy VA, Sharakhov IV. 2019. Physical genome mapping using fluorescence *in situ* hybridization with mosquito chromosomes. *Methods Mol Biol* **1858:** 177–194. doi:10.1007/978-1-4939-8775-7_13

Sharma GP, Mittal OP, Chaudhry S, Pal V. 1978. A preliminary map of the salivary gland chromosomes of *Aedes* (Stegomyia) *aegypti* (Culicadae, Diptera). *Cytobios* **22:** 169–178.

Stegniy VN. 1991. *Population genetics and evolution of malaria mosquitoes.* Tomsk State University Publisher, Tomsk, Russia.

Stegniy VN, Novikov Iu M, Kabanova VM. 1978. Cytogenetic analysis and distribution of *Anopheles beklemishevi*. *Zool J* **57:** 873–876.

Timoshevskiy VA, Severson DW, Debruyn BS, Black WC, Sharakhov IV, Sharakhov MV. 2013. An integrated linkage, chromosome, and genome map for the yellow fever mosquito *Aedes aegypti*. *PLoS Negl Trop Dis* **7:** e2052. doi:10.1371/journal.pntd.0002052

Timoshevskiy VA, Kinney NA, de Bruyn BS, Mao C, Tu Z, Severson DW, Sharakhov IV, Sharakhova MV. 2014. Genomic composition and evolution of *Aedes aegypti* chromosomes revealed by the analysis of physically mapped supercontigs. *BMC Biol* **12:** 27. doi:10.1186/1741-7007-12-27

Wallace AJ, Newton ME. 1987. Heterochromatin diversity and cyclic responses to selective silver staining in *Aedes aegypti* (L.). *Chromosoma* **95:** 89–93. doi:10.1007/BF00293847

Waterhouse RM, Aganezov S, Anselmetti Y, Lee J, Ruzzante L, Reijnders MJMF, Feron R, Bérard S, George P, Hahn MW, et al. 2020. Evolutionary superscaffolding and chromosome anchoring to improve *Anopheles* genome assemblies. *BMC Biol* **18:** 1. doi:10.1186/s12915-019-0728-3

Zamyatin A, Avdeyev P, Liang J, Sharma A, Chen C, Lukyanchikova V, Alexeev N, Tu Z, Alekseyev MA, Sharakhov IV. 2021. Chromosome-level genome assemblies of the malaria vectors *Anopheles coluzzii* and *Anopheles arabiensis*. *Gigascience* **10:** giab017. doi:10.1093/gigascience/giab017

Zhimulev IF. 1996. Morphology and structure of polytene chromosomes. *Adv Genet* **34:** 1–497. doi:10.1016/S0065-2660(08)60533-7

Protocol 1

Obtaining Polytene, Meiotic, and Mitotic Chromosomes from Mosquitoes for Cytogenetic Analysis

Jiangtao Liang,[1] Simon M. Bondarenko,[1,2] Igor V. Sharakhov,[1,2] and Maria V. Sharakhova[1,3,4]

[1]*Department of Entomology, Virginia Polytechnic Institute and State University, Blacksburg, Virginia 24060, USA;* [2]*Department of Genetics and Cell Biology, Tomsk State University, Tomsk 634050, Russia;* [3]*Laboratory of Evolutionary Genomics of Insects, Institute of Cytology and Genetics, Siberian Branch of the Russian Academy of Sciences, Novosibirsk 630090, Russia*

Chromosome visualization is a key step for developing cytogenetic maps and idiograms, analyzing inversion polymorphisms, and identifying mosquito species. Three types of chromosomes—polytene, mitotic, and meiotic—are used in cytogenetic studies of mosquitoes. Here, we describe a detailed method for obtaining high-quality polytene chromosome preparations from the salivary glands of larvae and the ovaries of females for *Anopheles* mosquitoes. We also describe how to obtain mitotic chromosomes from imaginal discs of fourth-instar larvae and meiotic chromosomes from the testes of male pupae for all mosquitoes. These chromosomes can be used for fluorescence in situ hybridization (FISH), a fundamental technique in cytogenetic research that is used for physical genome mapping, detecting chromosomal rearrangements, and studying chromosome organization.

MATERIALS

It is essential that you consult the appropriate Material Safety Data Sheets and your institution's Environmental Health and Safety Office for proper handling of equipment and hazardous materials used in this protocol.

RECIPES: Please see the end of this protocol for recipes indicated by <R>. Additional recipes can be found online at http://cshprotocols.cshlp.org/site/recipes.

Reagents

Carnoy's solution <R>

This solution is used for the preparation of all tissues except testes.

Ethanol (50%, 70%, 90%, and 100%)

Hypotonic solution (0.075 M potassium chloride or 0.5% sodium citrate) (for the preparation of imaginal discs and testes only; see Steps 15 and 22)

Liquid nitrogen

Mosquitoes from the Anophelinae or Culicinae subfamilies

The chromosome type (polytene, mitotic, or meiotic) needed for downstream applications will determine which mosquito species, stage of development, and sex must be selected for this protocol. See Steps 4, 8, 13, and 20, as well as the Discussion section, for considerations.

Propionic acid (50%)

[4]Correspondence: msharakh@vt.edu

Cite this protocol as *Cold Spring Harb Protoc*; doi:10.1101/pdb.prot107872

Equipment

Dissecting scissors with straight tips (2-mm cutting edge; Fine Science Tools 15000-03)
Filter paper
Forceps with straight and super fine tips (tip dimensions: 0.025-mm × 0.005-mm; Fine Science Tools 11252-00)
Glass or plastic slide jar (e.g., Wheaton Coplin staining jar; MilliporeSigma S5891-6EA)
Incubator at 16°C or 22°C (optional; see Step 2)
Incubator or warm room at a temperature of 27°C ± 1°C and 70% ± 5% relative humidity
Microcentrifuge tubes (1.5-mL)
Microscope coverslip (dust-free, 22-mm × 22-mm; Thermo Fisher Scientific)
Microscope slides (76-mm × 26-mm; Thermo Fisher Scientific)
Pencil with eraser
Petri dish
Phase-contrast microscope (Olympus Corporation)
Razor blade
Stereomicroscope (Olympus Corporation)

We suggest utilizing a stereomicroscope to perform tissue dissection.

Tungsten dissecting needles (0.125-mm rod diameter; Fine Science Tools 10130-05)

METHOD

Our protocol is modified from previously described chromosome preparation procedures, including those for polytene chromosomes (Xia et al. 2013; Sharakhova et al. 2015, 2019) and mitotic chromosomes (Timoshevskiy et al. 2012; Sharakhova et al. 2019). These protocols can be referenced for additional information.

Perform all steps at room temperature unless otherwise noted.

Mosquito Rearing

1. Rear mosquito colonies in an incubator or warm room at a temperature of 27°C ± 1°C with a 12-h photoperiod. Keep larvae in H_2O and adult mosquitoes at 70% ± 5% relative humidity (Benedict and Dotson 2015).
2. To obtain a large number of mitotic and meiotic metaphase/prometaphase chromosomes, transfer second- or third-instar larvae to a lower temperature, usually 22°C for Anophelinae and 16°C for Culicinae mosquitoes, to slow down cell division. The larvae need to be left at the lower temperature until they get to the right stage before dissection.

   This step is not needed for polytene chromosome preparation.

Tissue Dissection and Treatment

Here, we describe four options for tissue preparation. Ovaries of female adults (Steps 3–7) and salivary glands of fourth-instar larvae (Steps 8–12) are used to obtain high-quality polytene chromosomes. Imaginal discs of fourth-instar larvae (Steps 13–19) are used to prepare mitotic chromosomes and testes of male pupae (Steps 20–25) are used to prepare meiotic chromosomes.

Preparation of Ovaries from Female Adults

3. Place mosquitoes that are in a container (such as a cage or smaller container) into a −20°C freezer for a few minutes to immobilize them. Using forceps, transfer them to a Petri dish, and keep the dish on ice.

Cite this protocol as *Cold Spring Harb Protoc*; doi:10.1101/pdb.prot107872

4. Under a dissecting microscope using dissecting scissors and forceps, dissect the ovaries from half-gravid females.

    See Sharakhova et al. (2015) for specific directions about how to identify, localize, and dissect ovaries.

    It typically takes >12 h for Anopheles *females to reach the half-gravid stage (Fig. 1A,B), and the development period varies among species. To get high-quality polytene chromosomes, after the first oviposition,*

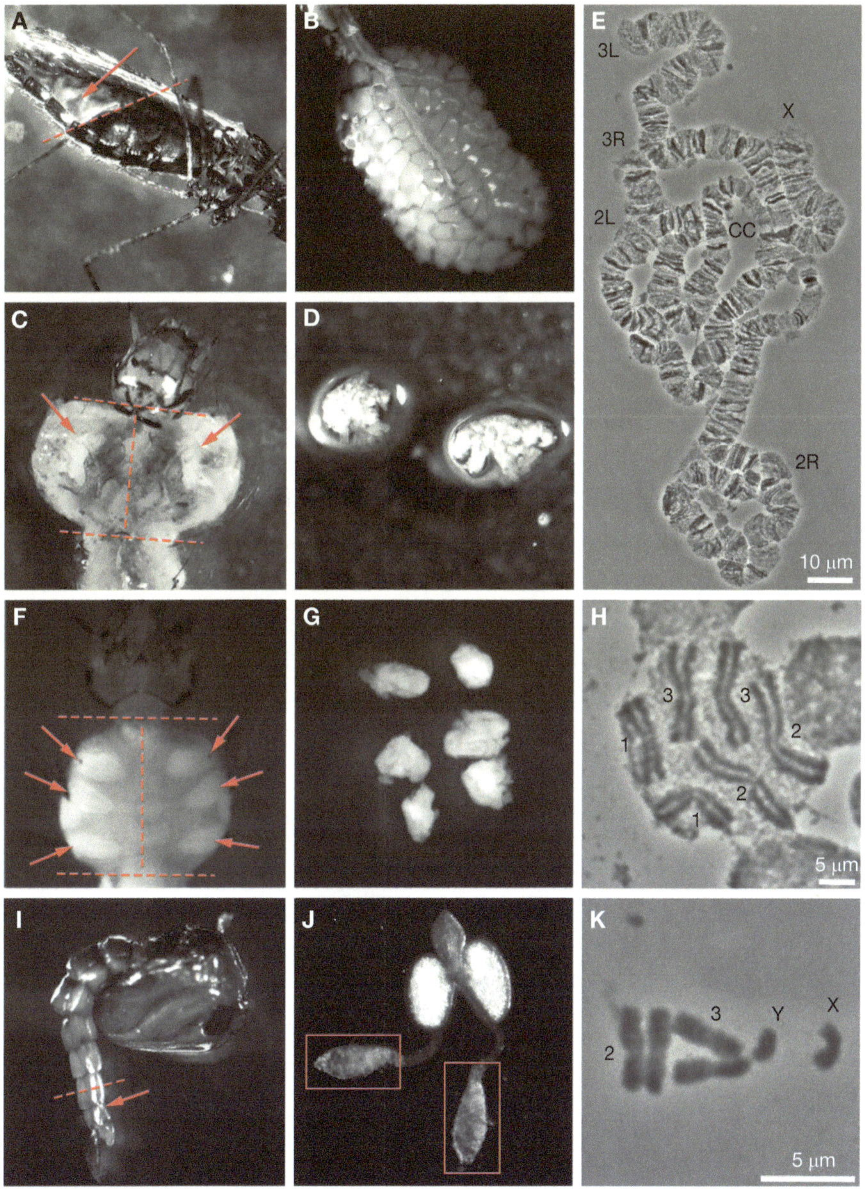

FIGURE 1. Linear visualization of polytene chromosomes, mitotic chromosomes, and meiotic chromosomes from mosquitoes. Dissections of polytene chromosomes from an adult female with ovaries at the Christophers' stage III (*A,B*), polytene chromosomes from salivary glands of the fourth-instar larva (*C,D*), and mitotic chromosomes from imaginal discs of the fourth-instar larva (*F,G*) are shown. Dashed lines in *A*, *C*, *F*, and *I* indicate the cutting positions of dissections. Arrows in panels *A*, *C*, *F*, and *I* point to the position of ovaries (*B*), salivary glands (*D*), imaginal discs (*G*), and the male reproductive system with testes in the rectangles (*J*), respectively. Examples of a high-quality polytene chromosome from the salivary glands of *Anopheles albimanus* (*E*), the mitotic chromosome from *Aedes aegypti* at metaphase stage (*H*), and the meiotic chromosome at the metaphase I stage from *Anopheles arabiensis* (*K*) are demonstrated. Chromosomes are labeled as 1, 2, 3, X, and Y. Letters L and R stands for the left and right chromosomal arms, respectively. Sources: (*A,B,C,D*) Sharakhova et al. 2019; (*E*) Artemov et al. 2017; (*F,G,H*) Timoshevskiy et al. 2012; (*J*) Liang et al. 2021.

blood-feed the females again and use these ovaries for the experiment. In addition, the ovaries must be developed to Christophers' stage III, when follicles have an oval shape and the transparent area of nurse cells within follicles has a round shape (Sharakhova et al. 2015).

5. Transfer four to five ovaries into a 1.5-mL tube with 1 mL of Carnoy's solution.

   Before dissection, the ovaries are suggested to be kept at room temperature for 24 h to get polytene chromosomes with clear bands. Alternatively, they can be stored long term at−20°C.

6. Using forceps or dissecting needles, take one ovary out of the tube and place it on a glass slide. Immediately add one drop of fresh Carnoy's solution, and remove all other tissue except for the follicles.

   To prevent the tissue from drying out, add additional drops of Carnoy's solution if needed. Follicles from one ovary can be divided into two to four pieces to make more chromosome preparations.

7. Separate individual follicles using the dissecting needles to make sure they are equally distributed on the slide.

   Proceed to Step 26.

Preparation of Salivary Glands from Fourth-Instar Larvae

8. Using a pipette, place early fourth-instar larvae onto filter paper and let them dry.

   We recommend using early fourth-instar larvae with the biggest head capsule as the quality of polytene chromosomes is the best at this stage (Benedict and Dotson 2015).

9. Using forceps, transfer larvae to a tube or small container containing fresh Carnoy's solution. Fix larvae in the Carnoy's solution for at least 2 d before dissection at −20°C.

10. Using forceps, place one larva with its back facing upward on a dust-free microscope slide and immediately add a drop of fresh Carnoy's solution.

    To prevent the tissue from drying out, add more drops of Carnoy's solution if needed during dissection.

11. After removing the larvae's head and abdomen with dissecting needles, open the thorax cuticle along the dorsal line with dissecting scissors.

    The cutting positions for dissections are shown in Figure 1C. See Xia et al. (2013) for detailed guidance.

12. Open the thorax cuticle, separate the salivary glands from the other tissue, and discard the carcass.

    The position and shape of the salivary glands is shown by arrows in Figure 1C,D. Usually, after treatment with Carnoy's solution, the salivary glands are whiter in color than the other tissue.

    Proceed to Step 26.

Preparation of Imaginal Discs from Fourth-Instar Larvae

13. Using a pipette, place live fourth-instar mosquito larvae in a small container with water, and place the container on ice for a few minutes to immobilize the larvae.

    Only early fourth-instar larvae with oval imaginal discs have a high number of cells undergoing mitosis (Fig. 1F) and can be used for further dissection (Sharakhova et al. 2011).

14. Place the larvae on filter paper to remove any H_2O.

15. Put one larva on a slide and add one drop of cold hypotonic solution (0.075 M potassium chloride or 0.5% sodium citrate) that has been kept on ice.

16. Remove the head from the larva using dissecting scissors, and cut the cuticle of the larval thorax from the ventral side. Then, using needles, open the cuticle of the thorax on both sides, and remove the fat body and the gut from the open thorax.

    The cutting positions of dissections are shown by the dashed lines in Figure 1F. See Timoshevskiy et al. (2012) for detailed guidance for this step.

17. Place a new drop of hypotonic solution on the thorax, and for 10 min keep it at room temperature.

Cite this protocol as *Cold Spring Harb Protoc*; doi:10.1101/pdb.prot107872

18. Remove the hypotonic solution using filter paper, and add a drop of fresh Carnoy's solution.

    The imaginal discs will turn white in color (Fig. 1G) in fixative and can be easily distinguished from other tissues under the microscope.

19. Remove other thorax tissues, except for imaginal discs, as much as possible.

    As an alternative to Steps 18 and 19, the imaginal discs can be dissected in the hypotonic solution and then treated with Carnoy's solution.

    Proceed to Step 26.

Preparation of Testes from Male Pupae

20. Using a pipette, place one pupa (Fig. 1I) in a small container with water, and place the container on ice for a few minutes to immobilize the pupa.

    Pupae can be recognized by their specific "comma" shape (Benedict and Dotson 2015). For some species (e.g., An. arabiensis*), early-emerged males must be used to obtain good-quality meiotic chromosomes. The use of early-emerged males is not described here but is similar to what we discuss below. We suggest checking both pupae and early emerged adults in the beginning of experiments as the speed of spermatogenesis may vary between species and based on rearing conditions.*

21. Transfer the pupa to filter paper to remove H_2O from the surface of the pupa.
22. Place the pupa on a slide, and add one drop of cold hypotonic solution (0.075 M potassium chloride or 0.5% sodium citrate).
23. Cut off the bottom of the pupa at the second or third somite (Fig. 1I) of the abdomen, and remove the carcass.
24. Hold the top part of the pupa's abdomen using one dissecting needle, and squeeze out the whole reproductive organ (Fig. 1J) from the abdomen with another dissecting needle.

    A whole reproductive organ includes two testes, two male accessary glands, one ejaculatory duct, and two vas deferens connecting the testes and the ejaculatory duct. The chromosomes of interest are located in the testes.

25. Use dissecting needles to separate the two testes by breaking the vas deferens, and discard the rest of the reproductive organ.

    For the best-quality preparation, remove most of the hypotonic solution during this step and add a new drop of hypotonic solution. We suggest keeping the two testes on the same slide.

    Proceed to Step 26.

Chromosome Preparation

26. If Carnoy's solution was used to fix the tissues, remove the Carnoy's solution from the slide.
27. Add one drop of fresh 50% propionic acid onto the target tissue. Cover the tissue with a dust-free coverslip. Incubate for 5 min (for salivary glands and ovaries) or for 10 min (for imaginal discs and testes) at room temperature.

    The tissue becomes transparent after this treatment.

28. Tap the coverslip gently with a pencil eraser to squash the tissue.

    This will release the chromosomes from the cells and flatten them. After squashing, there should be no space between the coverslip and the slide.

29. Examine the quality of the chromosome preparation using a phase-contrast microscope.

    Good polytene chromosome preparations should have well-spread chromosomes with clear banding patterns (Fig. 1E). Good preparations of mitotic and meiotic chromosomes should have high numbers (~50) of chromosome spreads (Fig. 1H,K, respectively).

30. Immerse the slide in liquid nitrogen until it stops bubbling, and then remove the coverslip from the slide with a razor blade.
31. Transfer the slide into a slide jar with 50% ethanol chilled to −20°C, and store overnight at −20°C.

32. Dehydrate the slide in a series of 70%, 90%, and 100% ethanol for 5 min each at room temperature.

    Slide jars with different concentrations of ethanol can be stored at room temperature and used up to 10 times within 1 mo.

33. Air-dry the slides, and then store them at room temperature for further use or at −20°C for longer storage.

    Such slides can be used for FISH. (See Xia et al. 2013 and Sharakhova et al. 2015, 2019 for polytene chromosomes and Timoshevskiy et al. 2012 and Sharakhova et al. 2019 for mitotic chromosomes.) Slides that have been dehydrated for at least 2 weeks at room temperature have higher-quality chromosome images after performing FISH. For best results, we recommend using dehydrated slides that have been stored at room temperature rather than slides that have been stored at −20°C. The FISH method for meiotic chromosomes is the same as for mitotic chromosomes. The first description of the FISH method for meiotic chromosomes is in Liang and Sharakhov (2019).

DISCUSSION

In this protocol, we describe techniques for the preparation of polytene, mitotic, and meiotic chromosomes from mosquito species in the Anophelinae and Culicinae subfamilies. We have included modifications of the chromosome preparation procedure from previously described protocols, including those for polytene chromosomes (Xia et al. 2013; Sharakhova et al. 2019) and mitotic chromosomes (Timoshevskiy et al. 2012; Sharakhova et al. 2019). We also provide a detailed description of meiotic chromosome preparations. We standardized the whole process based on our experience, combining overlapping steps that can be universally used for different types of chromosomes.

A detailed comparison of the different types of chromosomes used for cytogenetic studies in mosquitoes is presented in Table 1. Each type of chromosome has its own advantages and limitations. Polytene chromosomes have superior quality because they contain a thousand DNA strands combined together that form specific and easily recognizable banding patterns (Zhimulev 1996). However, good-quality polytene chromosomes are only found in mosquitoes from the subfamily Anophelinae. Moreover, for some species in this subfamily, such as *Anopheles gambiae*, high-quality chromosomal preparations can only be obtained using chromosomes from ovarian nurse cells (Coluzzi 1968;

TABLE 1. A comparison of three types of chromosomes for cytogenetic applications

Chromosome type	Source	Applications	Probe types for FISH	Advantages	Limitations
Polytene chromosome	Salivary glands from fourth-instar larva; ovaries from female adult	Genome mapping; direct chromosome rearrangement observation; species identification	BAC clones and other cloned DNA; PCR product of gene exons; repetitive DNA sequences	Chromosomes are giant and show clear banding patterns; easy to get good chromosome preparations; easy to perform FISH; suitable for high resolution of physical mapping (100 kb distance between the probes)	Good chromosome quality with clear banding patterns only present in Anophelinae mosquitoes; in some species, high-quality chromosomes are present only in female ovaries
Mitotic chromosome	Imaginal discs from fourth-instar larva	Genome mapping; chromosome rearrangement observation using FISH	BAC clones and other cloned DNA; PCR products of gene exons or cDNA; repetitive DNA sequences	Easy to get high-quality chromosome preparations; present in both Anophelinae and Culicinae mosquitoes	Requires large size probe for FISH; low resolution of physical mapping (1–5 Mb distance between the probes)
Meiotic chromosome	Testes from male pupa and adult	Meiotic chromosome behavior observation	Repetitive DNA	Used for studying chromosome behavior in meiosis and X and Y heterochromatin in *Anopheles*	Chromosomes are restricted to germline cells

(BAC) bacterial artificial chromosome, (PCR) polymerase chain reaction, (FISH) fluorescence in situ hybridization.

Coluzzi et al. 2002). Several species, such as *Anopheles funestus* (Sharakhov et al. 2001, 2002), *Anopheles moucheti* (Sharakhova et al. 2014), and Culicine mosquitoes (Unger et al. 2015; Campos et al. 2003b) have low polyteny, making the preparation of polytene chromosomes from these organisms difficult or nearly impossible. Thus, mitotic chromosomes from the brain ganglia (Brown et al. 1995, 2001) and from imaginal discs of fourth-instar larvae (Sharakhova et al. 2011; Timoshevskiy et al. 2013, 2014) are used in most of the cytogenetic studies in these mosquitoes. The major limitation for the use of mitotic chromosomes is their relatively small size, which provides only lower mapping resolution and requires large probes for FISH. For example, the distance between two probes that are distinguishable from each other is ~100 kb on polytene chromosomes of *An. gambiae* (Sharakhova et al. 2007) but ~1–5 Mb on metaphase or up to 0.5 Mb on prophase mitotic chromosomes of *Aedes aegypti* (Timoshevskiy et al. 2013). Mitotic chromosomes from the imaginal discs can be used for different applications, including FISH, in both mosquito subfamilies Anophelinae and Culicinae (Timoshevskiy et al. 2012; Sharakhova et al. 2019). The use of meiotic chromosomes in males is mostly limited to studies that are focused on observing specific chromosomal behaviors during spermatogenesis (Liang and Sharakhov 2019) or studying heterochromatin of the X and Y chromosome in Anophelinae mosquitoes (Hall et al. 2016; Sharma et al. 2020).

To conclude, all cytogenetic techniques described in this protocol require extensive training and labor, and thus further improvements (Masri et al. 2021), and the development of automation for these methods are important (George et al. 2012).

RECIPE

Carnoy's Solution

Reagent	Quantity
100% ethanol	3 parts
Glacial acetic acid	1 part

Store at room temperature for up to a week.

ACKNOWLEDGMENTS

Our work on linear chromosome visualization was supported by grant 19-14-00130 from the Russian Science Foundation to M.V.S. We thank Janet Webster for the text editing.

REFERENCES

Artemov GN, Peery AN, Jiang X, Tu Z, Stegniy VN, Sharakhova MV, Sharakhova IV. 2017. The physical genome mapping of *Anopheles albimanus* corrected scaffold misassemblies and identified interarm rearrangements in genus anopheles. *G3 (Bethesda)* **7:** 155–164. doi:10.1534/g3.116.034959

Benedict M, Dotson EM. 2015. Methods in *Anopheles* research, p. 408. https://www.beiresources.org/Portals/2/VectorResources/2016%20Methods%20in%20Anopheles%20Research%20full%20manual.pdf. [Accessed Feb. 18, 2022].

Brown SE, Menninger J, Difillipantonio M, Beaty BJ, Ward DC, Knudson DL. 1995. Toward a physical map of *Aedes aegypti. Insect Mol Biol* **4:** 161–167. doi:10.1111/j.1365-2583.1995.tb00021.x

Brown SE, Severson DW, Smith LA, Knudson DL. 2001. Integration of the *Aedes aegypti* mosquito genetic linkage and physical maps. *Genetics* **157:** 1299–1305. doi:10.1093/genetics/157.3.1299

Campos J, Andrade CF, Recco-Pimentel SM. 2003b. A technique for preparing polytene chromosomes from *Aedes aegypti* (Diptera, Culicinae). *Mem Inst Oswaldo Cruz* **98:** 387–390. doi:10.1590/S0074-02762003000300017

Coluzzi M. 1968. Chromosomi politenici cellule nutrici ovariche nel complesso *gambiae* del genere *Anopheles. Parassitologia* **10:** 179–183.

Coluzzi M, Sabatini A, della Torre A, Di Deco MA, Petrarca V. 2002. A polytene chromosome analysis of the *Anopheles gambiae* species complex. *Science* **298:** 1415–1418. doi:10.1126/science.1077769

Hall AB, Papathanos PA, Sharma A, Cheng C, Akbari OS, Assour L, Bergman NH, Cagnetti A, Crisanti A, Dottorini T, et al. 2016. Radical remodeling of the Y chromosome in a recent radiation of malaria mosquitoes. *Proc Natl Acad Sci* **113:** E2114–E2123. doi:10.1073/pnas.1525164113

Liang J, Sharakhov IV. 2019. Premeiotic and meiotic failures lead to hybrid male sterility in the *Anopheles gambiae* complex. *Proc R Soc B Biol Sci* **286:** 20191080. doi:10.1098/rspb.2019.1080

Liang J, Hodge M, Sharakhov IV. 2021. Asymmetric phenotypes of sterile hybrid males from reciprocal crosses between species of the

Anopheles gambiae complex. *Front Ecol Evol* **9:** 375. doi:10.3389/fevo.2021.660207

Masri RA, Karagodin DA, Sharma A, Sharakhova MV. 2021. A gene-based method for cytogenetic mapping of repeat-rich mosquito genomes. *Insects* **12:** 138. doi:10.3390/insects12020138

Naumenko AN, Timoshevskiy VA, Kinney NA, Kokhanenko AA, deBruyn BS, Lovin DD, Stegniy VN, Severson DW, Sharakhov IV, Sharakhova MV. 2015. Mitotic-chromosome-based physical mapping of the *Culex quinquefasciatus* genome. *PLoS One* **10:** e0115737. doi:10.1371/journal.pone.0115737

Sharakhov IV, Sharakhova MV, Mbogo CM, Koekemoer LL, Yan G. 2001. Linear and spatial organization of polytene chromosomes of the African malaria mosquito *Anopheles funestus*. *Genetics* **159:** 211–218. doi:10.1093/genetics/159.1.211

Sharakhov IV, Serazin AC, Grushko OG, Dana A, Lobo N, Hillenmeyer ME, Westerman R, Romero-Severson J, Costantini C, Sagnon N, et al. 2002. Inversions and gene order shuffling in *Anopheles gambiae* and *A. funestus*. *Science* **298:** 182–185. doi:10.1126/science.1076803

Sharakhova MV, Hammond MP, Lobo NF, Krzywinski J, Unger MF, Hillenmeyer ME, Bruggner RV, Birney E, Collins FH. 2007. Update of the *Anopheles gambiae* PEST genome assembly. *Genome Biol* **8:** R5. doi:10.1186/gb-2007-8-1-r5

Sharakhova MV, Timoshevskiy VA, Yang F, Demin S, Severson DW, Sharakhov IV. 2011. Imaginal discs: a new source of chromosomes for genome mapping of the yellow fever mosquito *Aedes aegypti*. *PLoS Negl Trop Dis* **5:** e1335. doi:10.1371/journal.pntd.0001335

Sharakhova MV, Antonio-Nkondjio C, Xia A, Ndo C, Awono-Ambene P, Simard F, Sharakhov IV. 2014. Polymorphic chromosomal inversions in *Anopheles moucheti*, a major malaria vector in Central Africa. *Med Vet Entomol* **28:** 337–340. doi:10.1111/mve.12037

Sharakhova MV, George P, Timoshevskiy VA, Sharma A, Peery A, Sharakhov, IV. 2015. Mosquitoes (Diptera). In *Protocols for cytogenetic mapping of arthropod genomes* (ed. Sharakhov IV), pp. 93–170. CRC Press, Boca Raton, Florida.

Sharakhova MV, Artemov GN, Timoshevskiy VA, Sharakhov IV. 2019. Physical genome mapping using fluorescence *in situ* hybridization with mosquito chromosomes. *Methods Mol Biol* **1858:** 177–194. doi:10.1007/978-1-4939-8775-7_13

Sharma A, Kinney NA, Timoshevskiy VA, Sharakhova MV, Sharakhov IV. 2020. Structural variation of the X chromosome heterochromatin in the *Anopheles gambiae* complex. *Genes (Basel)* **11:** 327. doi:10.3390/genes11030327

Timoshevskiy VA, Sharma A, Sharakhov IV, Sharakhova MV. 2012. Fluorescent *in situ* hybridization on mitotic chromosomes of mosquitoes. *J Vis Exp* e4215. doi:10.3791/4215.

Timoshevskiy VA, Severson DW, Debruyn BS, Black WC, Sharakhov IV, Sharakhov MV. 2013. An integrated linkage, chromosome, and genome map for the yellow fever mosquito *Aedes aegypti*. *PLoS Negl Trop Dis* **7:** e2052. doi:10.1371/journal.pntd.0002052

Timoshevskiy VA, Kinney NA, deBruyn BS, Mao C, Tu Z, Severson DW, Sharakhov IV, Sharakhova MV. 2014. Genomic composition and evolution of *Aedes aegypti* chromosomes revealed by the analysis of physically mapped supercontigs. *BMC Biol* **12:** 27. doi:10.1186/1741-7007-12-27

Unger MF, Sharakhova MV, Harshbarger AJ, Glass P, Collins FH. 2015. A standard cytogenetic map of Culex quinquefasciatus polytene chromosomes in application for fine-scale physical mapping. *Parasit Vectors* **8:** 307. doi:10.1186/s13071-015-0912-4

Xia A, Peery A, Kamali M, Liang J, Sharakhova MV, Sharakhova IV. 2013. Fluorescence *in situ* hybridization to the polytene chromosomes of anopheles mosquitoes. *Bio-protocol* **3:** e860. doi:10.21769/BioProtoc.860

Zhimulev IF. 1996. Morphology and structure of polytene chromosomes. *Adv Genet* **34:** 1–497. doi:10.1016/S0065-2660(08)60533-7

 Cite this protocol as *Cold Spring Harb Protoc*; doi:10.1101/pdb.prot107872

Protocol 2

Visualization of Polytene Chromatin in Mosquito Cell Nuclei Using Three-Dimensional Fluorescence In Situ Hybridization

Simon M. Bondarenko,[1,2] Jiangtao Liang,[1] Maria V. Sharakhova,[1,3] and Igor V. Sharakhov[1,2,4]

[1]*Department of Entomology, Virginia Polytechnic Institute and State University, Blacksburg, Virginia 24060, USA;* [2]*Department of Genetics and Cell Biology, Tomsk State University, Tomsk 634050, Russia;* [3]*Laboratory of Evolutionary Genomics of Insects, Institute of Cytology and Genetics, Siberian Branch of the Russian Academy of Sciences, Novosibirsk 630090, Russia*

Chromosomes are intricately folded within the cell nucleus and interact with peripheral nuclear proteins. The chromatin architecture has a profound effect on how the genome is organized. 3D-FISH is a powerful technique that can reveal the structural and functional organization of chromosomes in the nuclear space. Here, we present a protocol for visualizing specific genomic regions in whole-mount paraformaldehyde-fixed cell nuclei of *Anopheles* mosquitoes. This protocol was tested in our laboratories and has been showed to be effective and reliable for visualizing genomic regions of various lengths—from 1-kb gene-scale fragments to chromosome-scale segments of DNA.

MATERIALS

It is essential that you consult the appropriate Material Safety Data Sheets and your institution's Environmental Health and Safety Office for proper handling of equipment and hazardous materials used in this protocol.

RECIPES: Please see the end of this protocol for recipes indicated by <R>. Additional recipes can be found online at http://cshprotocols.cshlp.org/site/recipes.

Reagents

2× SSCT <R>
2× SSCT +50% formamide <R>
Anopheles gambiae mosquitoes, adult
Defibrinated sheep blood (Colorado Serum Co.)
Deionized H_2O
Hybridization buffer for fluorescence in situ hybridization (FISH) <R>
PBT for FISH <R>
PBT with RNase A <R>
PBTr <R>
Probe (see Box 1)
Prolong Gold Antifade Reagent with DAPI (Invitrogen 2155583)
Rubber cement
Tissue fixation solution <R>

[4]Correspondence: igor@vt.edu

Cite this protocol as *Cold Spring Harb Protoc*; doi:10.1101/pdb.prot107873

BOX 1. PREPARATION OF FLUORESCENT PROBES FOR 3D-FISH

Conceptually, 3D-FISH is a variation of DNA:DNA hybridization methods, in which a relatively short fluorescently labeled DNA sequence (probe) is hybridized with an immobilized DNA (target) on the basis of complementarity. The essential feature of 3D-FISH is the utilization of the cell components as an immobilization phase (the solid phase that serves for holding the target DNA) for the reaction. The process of probe preparation does not significantly differ from that of other hybridization methods such as northern and Southern analyses. Here, we list the most common probe preparation techniques and provide a brief description and general recommendations for choosing the appropriate method.

There are two possible sources of DNA for probe preparation: DNA can be extracted from a living organism or it can be artificially synthesized. In the first method, the preparation of probes begins with the isolation of source DNA complementary to the region of interest. DNA can be isolated by a laser or by mechanical microdissection of either the entire chromosome or part of it with further whole genome amplification (WGA) of the dissected sample (George et al. 2014; Artemov et al. 2015; Bondarenko et al. 2017). Alternatively, a single or several short (<15-kb) sequences of DNA can be amplified by PCR using primers flanking specific region(s) using the total DNA as a template. This artificial method of probe preparation requires the design and synthesis of a homogeneous or heterogeneous oligoprobe library, which is complementary to any particular genomic region(s) of interest (George et al. 2020).

To identify the probe using fluorescent microscopy, it must carry a fluorophore. The selection of a particular probe-labeling system, which provides conjugation of a fluorophore with a DNA probe, depends on many factors. The most common labeling techniques are listed below.

- ***Nick translation*** is a traditional technique of probe labeling. One of the key components of the reaction is DNase I, which at extremely low concentrations nicks the backbone of a double-stranded DNA, producing a free 3′-OH group at a random nucleotide position. The nick is then recognized by DNA polymerase I, which replaces the native nucleotides with fluorescently labeled ones in the $5' \rightarrow 3'$ direction (Farrell 2017). Because the reaction does not involve DNA amplification, a sufficient amount of source DNA is required (~1 µg of DNA per reaction). Nick translation can be used for all sources of DNA and provides good stability of the labeled probes. All major biotech companies offer reagents for nick translation in a kit form. These include the following: Nick Translation System (Thermo Fisher Scientific 18160-010), Nick Translation DNA Labeling System 2.0 (Enzo ENZ-GEN111-0050), AF488 NT Labeling Kit (Jena Bioscience PP-305S-AF488), and Amersham Nick Translation Kit (GE Healthcare N5500).
- The ***PCR-based labeling system*** combines amplification of probes with labeling. Continuously labeled probes can be produced by PCR in the presence of the appropriate labeled nucleotides, or end-labeled probes can be produced in the presence of the 5′-labeled primers (because the primers become a part of the PCR product) (Farrell 2017). The common PCR-related strategy, which is useful for amplification of repetitive DNA, is degenerate oligonucleotide–primed polymerase chain reaction (DOP-PCR) (Bondarenko et al. 2017). The commercially available GenomePlex Single Cell Whole Genome Amplification Kit (Sigma-Aldrich WGA4-10RXN) has also been tested for amplification of laser-microdissected DNA sources (George et al. 2014).
- ***Random priming*** is based on the ability of the Klenow fragment of DNA polymerase I to incorporate labeled nucleotides when it extends the DNA. The ideal length of the probe is between 200 and 2000 bp, which makes random priming the best option for labeling probes in a single gene. The commercially available kits for this reaction include the following: Random Primers DNA Labeling System (Thermo Fisher Scientific 18187013), Random Primed DNA Labeling Kit (Sigma-Aldrich 11004760001), Random Primer DNA Labeling Kit Ver. 2 (Takara 6045), and Prime-a-Gene Labeling System (Promega U1100).
- *The **artificial generation of probes*** is used for a large variety of applications and provides the greatest flexibility. Probe libraries can be designed computationally, which allows a hybridization cocktail (a mixture of heterogeneous probes) to specifically label predefined genomic regions with great precision (George et al. 2020). Probes can now be designed and manufactured by biotech companies. Oligonucleotide probes can be designed using the program OligoArray 2.0 (Rouillard et al. 2003). The program uses a thermodynamic approach to compute oligonucleotides free of stable secondary structure, which would otherwise interfere with probe hybridization. MyTags oligonucleotide libraries can be synthesized by Arbor Biosciences.

This protocol requires the availability of labeled probe dissolved in a hybridization buffer. The hybridization procedure does not depend on the particular way the probe was prepared.

Cite this protocol as *Cold Spring Harb Protoc*; doi:10.1101/pdb.prot107873

Equipment

Cover glass (18-mm × 18-mm)
Dissecting microscope (e.g., Leica M60)
Entomological needles and forceps for tissue dissection
Eppendorf ThermoMixer C
Fluorescent or confocal microscope
Incubator or warm room at a temperature of 27°C ± 1°C and 70% ± 5% relative humidity
Microcentrifuge
Microcentrifuge tubes (1.5-mL)
Microscope slides (25-mm × 40-mm)
Petri dish

METHOD

This protocol is designed for visualizing polytene chromosomes from ovarian nurse cell nuclei in An. gambiae. *Specific modifications may be required for other mosquito species or tissues (see Discussion).*

Tissue Dissection

1. Rear adult *An. gambiae* mosquitoes at 27°C ± 1°C, with 12-h light–dark cycles and 70% ± 5% relative humidity. To induce the development of ovaries, feed female mosquitoes with defibrinated sheep blood as described in Benedict and Dotson (2015).
2. Approximately 28 h after blood feeding, immobilize the mosquitoes by placing their cage into a −20°C freezer for a few minutes until they stop moving. Using forceps, move immobilized females to a Petri dish and keep the dish on ice.
3. Use a pair of entomological needles and/or forceps with extremely sharp tips to remove Christophers' stage III (Clements 1992) ovaries from 10 to 15 females. Work under a dissecting microscope, and use any microscope slide as a surface for the dissection. Discard all undesirable body parts, especially the chitin exoskeleton.

   See Sharakhova et al. (2015) for specific directions about how to identify, localize, and dissect ovaries.
4. Collect the tissues in a 1.5-mL microcentrifuge tube containing 1 mL of PBT. Maintain the samples on ice until fixation.

Tissue Fixation

From this point forward, all procedures will be performed in the tube used for tissue collection. Before each solution change, centrifuge the tube for 1 min at 2000 rpm to allow the tissues to settle to the bottom of the microcentrifuge tube, and then carefully discard all but 50–100 µL of the solution using a 1000-µL pipette. This will minimize the loss of the sample. See Troubleshooting.

5. Fix tissues by incubating in 1 mL of freshly prepared tissue fixation solution for 20 min at room temperature while mixing at 450 rpm on the ThermoMixer.
6. Wash tissues in 1 mL of PBT three times for 5 min each at room temperature.

   After completing this step, the tissues can be stored in PBT overnight at 4°C.
7. Incubate tissues in 1 mL of PBT with RNase A for 2 h at 37°C with mixing at 450 rpm.
8. Wash tissues in 1 mL of PBT three times for 5 min each at room temperature.
9. Incubate tissues in 1 mL of PBTr for 15 min at room temperature.
10. Wash tissues in 1 mL of PBT three times for 5 min each at room temperature.
11. Incubate tissues in 1 mL of 2× SSCT +50% formamide for 30 min at 55°C with mixing at 450 rpm.

Probe Hybridization

12. Preheat 250–1000 ng of probe in 40 µL of hybridization buffer to 37°C.

    The working concentration of the probe varies and must be determined empirically. It depends on the brightness of the fluorophore and the affinity and specificity of the probe for the target DNA.

    The intensity of the fluorescent probes may be affected by direct and continuous exposure to light. We recommend performing all of the remaining steps in the dark.

13. Incubate the tissues with the probe-hybridization buffer mixture for 5 min at 92°C with mixing at 450 rpm and then overnight or for 2–3 d at 37°C with mixing at 450 rpm.

Washing and Counterstaining

14. Preheat freshly prepared 2× SSCT +50% formamide and 2× SSCT solutions to 37°C.
15. Wash the tissues with 1 mL of 2× SSCT +50% formamide solution three times for 15 min each at 37°C.
16. Wash the tissues with 1 mL of 2× SSCT solution three times for 15 min each at 37°C.
17. Remove as much supernatant as possible. Add 20 µL of Prolong Gold Antifade Reagent with DAPI. Incubate overnight at 4°C.

Mounting and Visualization

18. The next day, draw the outline of an 18-mm × 18-mm rectangle on a standard 25-mm × 40-mm slide using a thin layer of rubber cement. Pipette 20 µL of the sample mixed with Prolong Gold Antifade Reagent with DAPI and put it in the center of the drawn rectangle. Cover the sample with an 18-mm × 18-mm cover glass of the appropriate thickness for your microscope.
19. Use a fluorescent or confocal microscope to visualize the probe in the nuclei.

    The slides can be stored in the dark for months at 4°C.

TROUBLESHOOTING

Problem: Because the tissues are freely floating in the tube and not attached to the surface, the tissue can be accidentally sucked into the pipette tip during buffer changing.

Solution: The precipitation of tissues by a short centrifugation at <4000 rpm partially solves this problem; however, the process of changing the buffer requires special attention and practice, especially when the tissue is incubated in the PBTr or formaldehyde buffers, in which the tissues become transparent. The presence of 0.3% Tween 20 in some buffers significantly improves the precipitation rate and decreases the chances of losing the material. The phosphate buffer 1× PBS in this protocol can be substituted with buffer A (60 mM KCl, 15 mM NaCl, 0.5 mM spermidine, 0.15 mM spermine, 2 mM EDTA, 0.5 mM EGTA, and 15 mM PIPES).

DISCUSSION

The described protocol can be applied for staining of any polytene, polyploid, or diploid insect nuclei. However, because the physical properties of the insect tissues vary, some steps may require minor adjustments. For example, the incubation time in the tissue-fixation solution may vary from 10 to 20 min. The concentration of Triton X-100 in the PBTr solution can be changed from

0.7% to 2.0%. The best parameters for these steps should be determined experimentally for different tissues.

The described procedure can be used for studying the dynamics of chromosome territories during nuclear growth (Fig. 1A,B) (Bondarenko et al. 2017), for measuring the tissue-specific distances between chromosomal regions (Artemov et al. 2015; George et al. 2020), and for examining the contact frequencies between the chromosome territories and the nuclear envelope (Fig. 1C,D; George et al. 2020).

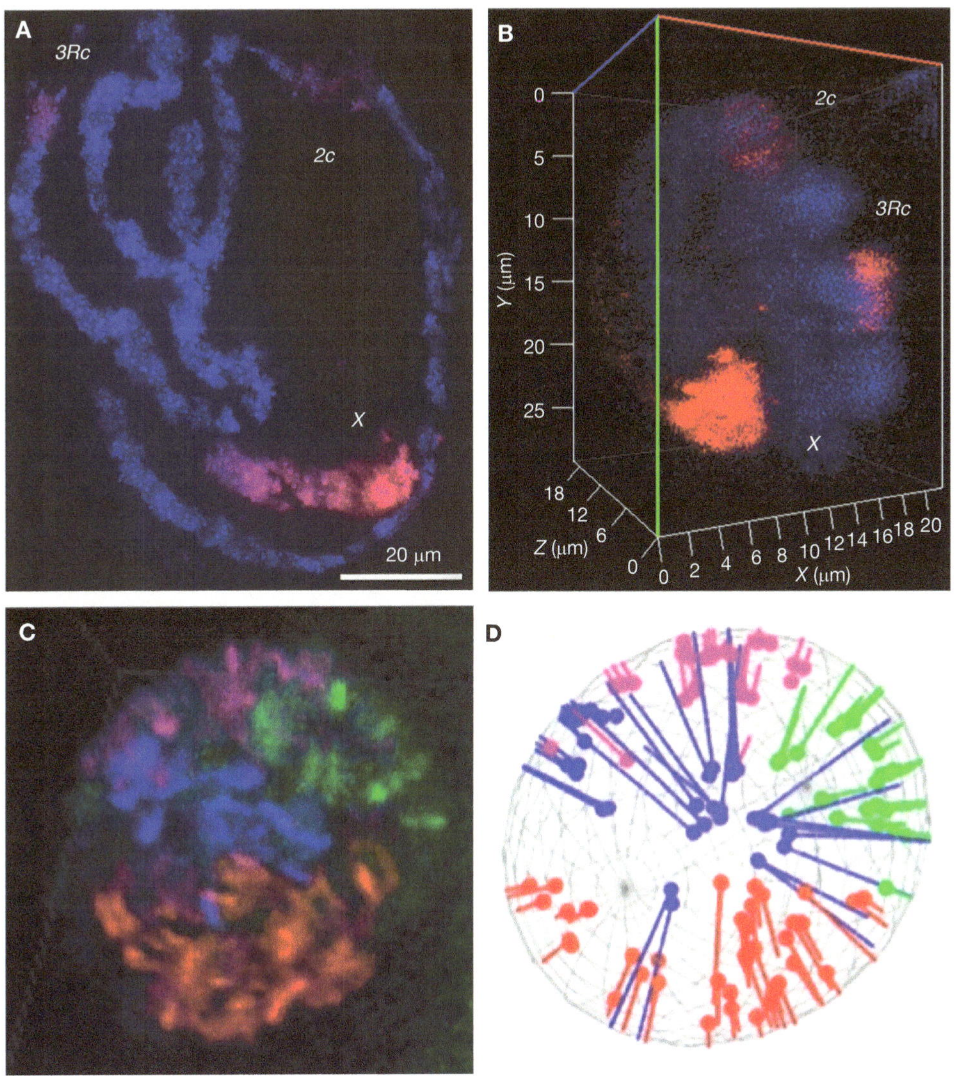

FIGURE 1. Visualization of polytene chromatin in mosquito cell nuclei. Fluorescence in situ hybridization (FISH) of the X-chromosome-specific probe with the squashed preparation of the *Anopheles atroparvus* nurse cell nucleus (*A*) and 3D-FISH of the X-chromosome-specific probe with the whole-mount preparation of the *An. atroparvus* nurse cell nucleus (*B*) are shown. The probe was produced from mechanically microdissected DNA labeled with degenerate oligonucleotid–primed polymerase chain reaction (DOP-PCR). The presence of highly repetitive heterochromatic DNA in the probe caused nonspecific labeling of the pericentromeric heterochromatic regions *3Rc* and *2c* enriched with complementary repeats (Bondarenko et al. 2017). Four-color 3D-FISH of arm-specific oligopainting probes visualizing the banding pattern of polytene chromosomes in ovarian nurse cells of *Anopheles gambiae* is showed in *C*. Arms X, 2L, 3R, and 3L are stained in blue, red, green, and purple, respectively. Distances between the oligopainted bands and the projected nuclear envelope in ovarian nurse cells of *An. gambiae* were determined by analysis in MATLAB (*D*). The oligopainting probes were produced by the design and synthesis of the sequences complementary to regions of interest (George et al. 2020).

RECIPES

2× SSCT

Reagent	Quantity	Final concentration
Tween 20 (Fisher 181617)	75 µL	0.3%
20× SSC (Invitrogen 00765425)	2.5 mL	2×

Combine the reagents and then bring the volume to 25 mL with double-distilled H_2O. Prepare fresh before use.

2× SSCT +50% Formamide

Reagent	Quantity	Final concentration
Tween 20 (Fisher 181617)	75 µL	0.3%
Deionized formamide (Thermo Fisher AM 9344)	12.5 mL	50%
20× SSC (Invitrogen 00765425)	2.5 mL	2×

Combine the reagents and then bring the volume to 25 mL with double-distilled H_2O. Prepare fresh before use.

Hybridization Buffer for FISH

Reagent	Quantity	Final concentration
Deionized formamide (ThermoFisher AM9344)	600 µL	60%
Dextran sulfate sodium salt (MilliporeSigma 67578)	0.1 g	10%
20× SSC (Invitrogen 00765425)	100 µL	2×

FISH probe of interest, dissolved in H_2O

Determine the working concentration of the probe empirically, taking into account the brightness of the fluorophore and the affinity and specificity of the probe for the target DNA. Combine the reagents and add double-distilled H_2O to 1 mL. Store for up to several months at −20°C.

PBT for FISH

Reagent	Quantity	Final concentration
Tween 20 (Fisher 181617)	150 µL	0.3%
10× PBS (Gibco 2000522)	5 mL	1×

Combine the reagents and then bring the volume to 50 mL with double-distilled H_2O. Prepare fresh before use.

Cite this protocol as *Cold Spring Harb Protoc*; doi:10.1101/pdb.prot107873

PBT with RNase A

Reagent	Quantity
20 µg/mL RNAse A (Invitrogen 2031592)	5 µL
10× PBS (Gibco 2000522)	100 µL
Tween 20 (Fisher 181617)	3 µL
Double-distilled H_2O	892 µL

Prepare fresh before use.

PBTr

Reagent	Quantity	Final concentration
Triton X-100 (Sigma-Aldrich SLBT2255)	70–200 µL	0.7%–2%
10× PBS (Gibco 2000522)	1 mL	1×

Determine the appropriate amount of Triton X-100 empirically. Add double-distilled H_2O to 10 mL. Prepare fresh before use.

Tissue Fixation Solution

Reagent	Quantity	Final concentration
Paraformaldehyde (Sigma-Aldrich 158127)	1 g	4%
10× PBS (Gibco 2000522)	2.5 mL	1×
Tween 20 (Fisher 181617)	75 µL	0.3%

Combine the first two ingredients and then add the third. Bring the volume to 25 mL with double-distilled H_2O. Dissolve by heating at 60°C, and adjust pH to 7.4. Store for up to 2 mo at −20°C.

ACKNOWLEDGMENTS

This work was performed using grant 21-14-00182 from the Russian Science Foundation (to I.V.S.). We thank Janet Webster for the text editing.

REFERENCES

Artemov G, Bondarenko S, Sapunov G, Stegniy V. 2015. Tissue-specific differences in the spatial interposition of X-chromosome and 3R chromosome regions in the malaria mosquito *Anopheles messeae* fall. *PLoS ONE* **10:** e0115281. doi:10.1371/journal.pone.0115281

Benedict M, Dotson EM. 2015. Methods in *Anopheles* research, pp. 408. https://www.beiresources.org/Portals/2/VectorResources/2016%20Methods%20in%20Anopheles%20Research%20full%20manual.pdf. [Accessed Feb. 18, 2022].

Bondarenko SM, Artemov GN, Sharakhov IV, Stegniy VN. 2017. Tissue-specific features of the X chromosome and nucleolus spatial dynamics in a malaria mosquito, *Anopheles atroparvus*. *PLoS ONE* **12:** e0171290. doi:10.1371/journal.pone.0171290

Clements AN. 1992. *The biology of mosquitoes: development, nutrition and reproduction*. Chapman & Hall, London.

Farrell RE. 2017. *Nucleic acid probe technology. In RNA methodologies: laboratory guide for isolation and characterization*, 5th ed., pp. 469–526. Elsevier, New York.

George P, Sharma A, Sharakhov IV. 2014. 2D and 3D chromosome painting in malaria mosquitoes. *J Vis Exp* e51173. doi:10.3791/51173

George P, Kinney NA, Liang J, Onufriev AV, Sharakhov IV. 2020. Three-dimensional organization of polytene chromosomes in somatic and germline tissues of malaria mosquitoes. *Cells* **9:** 339. doi:10.3390/cells9020339

Rouillard JM, Zuker M, Gulari E. 2003. OligoArray 2.0: design of oligonucleotide probes for DNA microarrays using a thermodynamic approach. *Nucl Acids Res* **31:** 3057–3062. doi:10.1093/nar/gkg426

Sharakhova MV, George P, Timoshevskiy VA, Sharma A, Peery A., Sharakhov IV. 2015. Mosquitoes (Diptera). In *Protocols for cytogenetic mapping of arthropod genomes* (ed. Sharakhov IV), pp. 93–170. CRC Press, Boca Raton, Florida.

CHAPTER 46

Assays to Measure Insecticide Toxicity and Insecticide Resistance in Mosquitoes

Nannan Liu,[1,3] Yifan Wang,[1] and Ting Li[1,2]

[1]*Department of Entomology and Plant Pathology, Auburn University, Auburn, Alabama 36849, USA;*
[2]*Department of Biological Sciences, Alabama State University, Montgomery, Alabama 36104, USA*

Mosquitoes' resistance to commonly used insecticides is now widespread, hampering control efforts and leading to substantial increases in human illness and mortality rates in many areas of the world. Insecticide bioassays are quantitative methodologies used to determine the dose–response relationship of insects to insecticides and to evaluate the susceptibility or resistance of mosquitoes to specific insecticides. They are frequently used to monitor the development of insecticide resistance in mosquitoes for both field resistance diagnoses (surveillance assays), in which the ability of mosquitoes to survive exposure to a standard dose or concentration of an insecticide is measured, and laboratory bioassays, in which responses to insecticides are tested in parallel populations of resistant (field) populations and laboratory susceptible strains using serial doses or concentrations. Metabolic detoxification, in which insecticides are metabolized by enzymes, including cytochrome P450s, hydrolases, and glutathione-*S*-transferases (GSTs), to become more polar and less toxic, is one resistance mechanism. Piperonyl butoxide (PBO), *S,S,S*-tributyl phosphorotrithioate (DEF), and diethyl maleate (DEM) are the inhibitors of P450s, hydrolases, and GSTs, respectively, and act as synergists for rapidly testing the involvement of these enzymes in insecticide resistance. Such synergistic assays are used to identify the detoxification enzyme that leads to resistance to a specific insecticide. This introduction and its associated protocols present a detailed discussion of appropriate methodologies and procedures for laboratory larval, adult, and synergistic bioassays and introduces the field surveillance tests used to monitor insecticide resistance as recommended by the latest World Health Organization (WHO) and U.S. Centers for Disease Control (CDC) guidelines.

MOSQUITO VECTORS AND INSECTICIDE RESISTANCE

Vector control of mosquitoes has long been a critical part of the current global strategy to control mosquito-associated diseases, with insecticides such as pyrethroids, carbamates, and organophosphates (Liu 2015; Choi et al. 2019; Sahu et al. 2020; World Health Organization 2020) being the most important component in these efforts. In the past, massive spraying of insecticides significantly limited outbreaks of mosquito-borne diseases and even eradicated malaria in a few areas (Hemingway et al. 2002). However, the widespread development of resistance to the most commonly used insecticides in mosquitoes is now hampering mosquito control efforts worldwide (Hemingway et al. 2002; Zaim and Guillet 2002; World Health Organization 2020), leading to substantial increases in human illness and mortality rates in many areas of the world (World Health Organization 2020). Research on insecticide resistance in mosquitoes has proliferated since the 1950s, when the first report of resistance

[3]Correspondence: liunann@auburn.edu

Cite this introduction as *Cold Spring Harb Protoc*; doi:10.1101/pdb.top107705

to chlorinated-hydrocarbon insecticides in mosquitoes was published (Giullin and Peters 1952). Investigators worldwide are attempting to elucidate the toxicity of insecticides and the sensitivity of mosquitoes to the insecticides currently used to understand the development and status of insecticide resistance in field populations. Doing so is a vital first step toward the creation of new, more effective strategies that prevent resistance development, control resistant mosquitoes, and, ultimately, reduce the prevalence of mosquito-borne diseases.

THE USE OF INSECTICIDE ASSAYS TO MEASURE MOSQUITOES' RESPONSE TO INSECTICIDES AND MONITOR THE DEVELOPMENT OF INSECTICIDE RESISTANCE

Insecticide assays are quantitative methodologies used for determining the relationship between the amount (i.e., dose or concentration) of an insecticide administered in insects and their subsequent response. This relationship, which is referred to as the dose–response relationship, is a fundamental feature of insecticide bioassays (Fig. 1) and the information revealed is critical for evaluating the susceptibility or resistance of mosquitoes to insecticides. The insecticide bioassays most frequently used in mosquito research include both insecticide surveillance/diagnosis assays, which examine the ability of mosquitoes to survive exposure to a standard dose or concentration of an insecticide, and

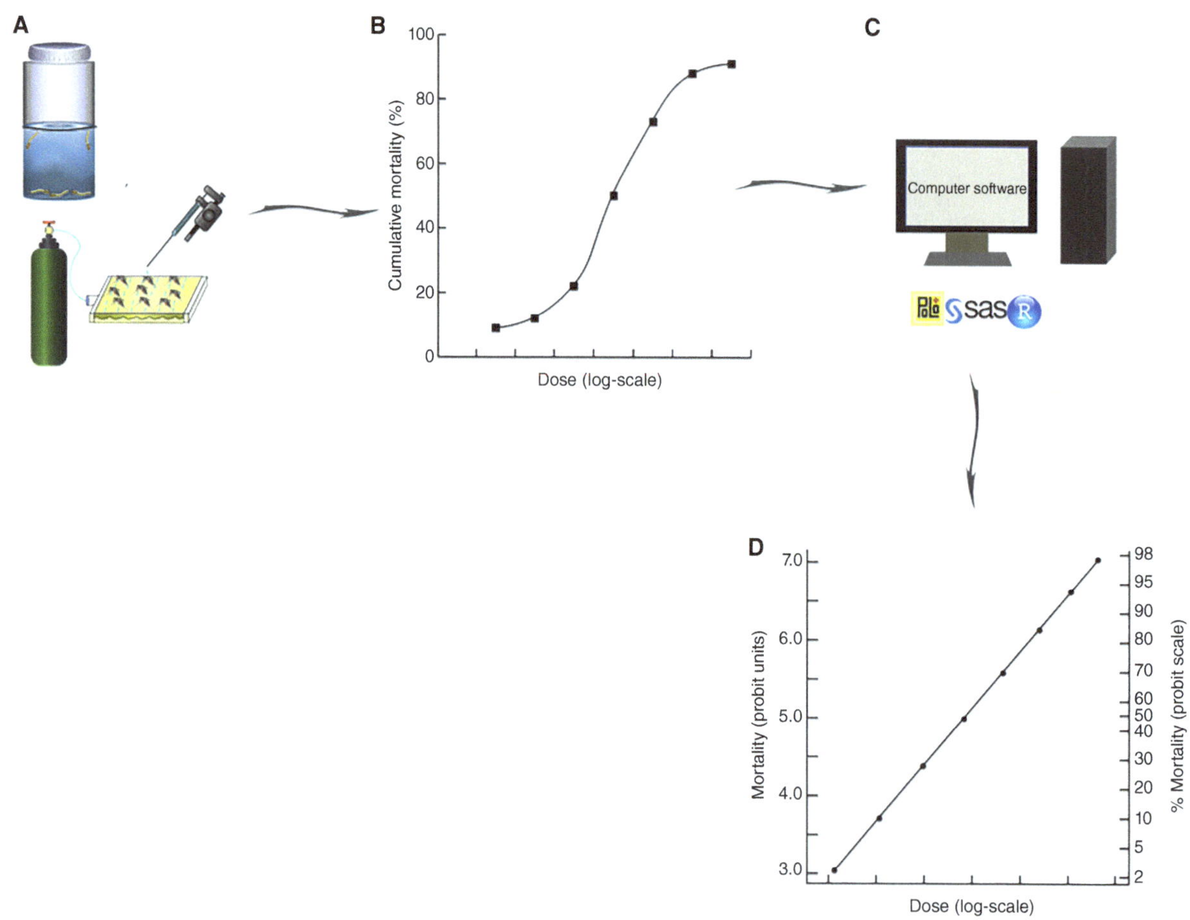

FIGURE 1. Schematic diagram of quantal dose–response relationship of insects' response to chemicals. (*A*) Insecticide bioassays for mosquito larvae and adults. (*B*) Dose–response curve with a cumulative mortality plotted along the *y*-axis and the logarithm of the dose or concentration being plotted along the *x*-axis. (*C*) Mathematical transformation of probit (probability unit) analysis using appropriate computer software. (*D*) Dose–response curve of mortality in probits with the logarithm of the dose or concentration being plotted along the *x*-axis and the percent mortalities, converted to probits, being plotted along the *y*-axis.

Cite this introduction as *Cold Spring Harb Protoc*; doi:10.1101/pdb.top107705

laboratory bioassays, in which the responses to insecticides are tested in parallel resistant (field) populations and laboratory susceptible strains with a series of doses or concentrations across a range from 0% to 100% mortality (Liu et al. 2004).

INSECTICIDE RESISTANCE SURVEILLANCE

Monitoring insecticide resistance in the field is critical for understanding the resistance status of field populations. To achieve this, a field population is exposed to a concentration of insecticide that reliably kills susceptible insects, so any survivors are presumed to be resistant. The World Health Organization's (WHO's) standard tube-test method and the U.S. Centers for Disease Control's (CDC's) bottle bioassays are commonly used for insecticide resistance surveillance tests. Both tests are designed to distinguish between baseline susceptibility and resistance to insecticides in adult mosquitoes by exposing the mosquitoes to known concentrations of an insecticide for a fixed period of time. The CDC bottle assay uses a bottle coated with a known amount of insecticide at the diagnostic concentration, which is the concentration of the insecticide known to kill 100% of susceptible mosquitoes within a given time (Centers for Disease Control and Prevention 2019). Resistance is assumed to be present if a significant portion of the test population survives the diagnostic concentration within the diagnostic time, with the level of resistance to that insecticide being determined based on the percentage of mosquitoes that die (mortality rate) before the cutoff time. In the WHO susceptibility assays, mosquitoes are loaded into a set of holding tubes containing paper coated with a "discriminating concentration" of an insecticide, which is the concentration of an insecticide known to kill 100% of susceptible mosquitoes in a given amount of time. Intensity assays test susceptibility of mosquitoes at 5× and 10× concentration of the discriminating concentrations of insecticides (World Health Organization 2016). The detailed diagnostic concentrations, threshold time, and standard determination of susceptibility/resistance for the majority of commonly used insecticides are listed in the guidelines issued by the World Health Organization (2016) and the Centers for Disease Control and Prevention (2019) for their respective resistance surveillance protocols.

INSECTICIDE LABORATORY BIOASSAYS

Insecticide bioassays of insects are often used to estimate the median lethal dose (LD_{50}) or concentration (LC_{50}) and/or the 90% lethal dose (LD_{90}) or concentration (LC_{90}) of an insecticide along with the associated 95% confidence intervals (95% CI), in dose–response models. The LD_{50} or LC_{50} is the dose or concentration of an insecticide required to kill 50% of a given population or strain under the specified conditions in a given time period. The 95% confidence interval signifies a range of values where the true mean of the population is 95% certain (Burgess et al. 2020). The LD_{50} (or LC_{50}) and the 95% CI obtained for the dose–response relationship is generated via a probit analysis (Finney 1971; Raymond 1985), a method commonly used to analyze the relationship between a stimulus (dose) and the quantal response (Fig. 1). A probit analysis allows researchers to mathematically transform the sigmoidal normal distribution of dose–response data (in this case, the mortality of the insects) into a cumulative normal distribution for the response that can then be analyzed by performing a linear regression, with the logarithm of the dose or concentration being plotted along the x-axis and the percent mortalities, converted to probits (probability units), being plotted along the y-axis (Aldrich and Nelson 1984; Agresti 2015). Levels of insecticide resistance in insects are characterized by comparing the LD_{50} or LC_{50} measured for the resistant and susceptible strains (populations that have not been exposed to insecticides) (Liu et al. 2004). It has been suggested that an insect population should be referred to as resistant when the resistance ratio is as low as >4 (Scott and Wen 1997); others have argued that insects should not be considered resistant until a resistance ratio of 10 is reached (Valles et al. 1997). Here, we consider an insect population with a resistance ratio for a specific insecticide between 1 and 10 to be tolerant rather than resistant, and the resistance is defined when the resistance ratio is >10. (Liu et al. 2004).

Insecticide bioassays provide a reliable and informative methodology with which to measure the median lethal dose, lethal concentration, and toxicity of a specific insecticide for a specific insect population and are thus essential for the evaluation of levels of resistance of an insect population to an insecticide. Laboratory insecticide bioassays are frequently used by researchers to evaluate the susceptibility and resistance of mosquitoes to insecticides and are increasingly being used to generate comparative toxicity data for different chemicals and different populations, often within a short time period. This introduction and the associated protocols therefore discuss both adult and larval bioassays (see Protocol 1: Mosquito Larval Bioassays [Wang et al. 2023] and Protocol 2: Mosquito Adult Bioassays [Li et al. 2023a]).

SYNERGISM STUDY

Insecticide resistance in mosquitoes develops as a result of complex interactions among multiple resistance mechanisms or genes. Increased metabolic detoxification is one such mechanism in mosquitoes. The detoxification of insecticides in mosquitoes involves three major metabolic detoxification gene families: cytochrome P450s (P450s), hydrolases, and glutathione-*S*-transferases (GSTs), which can metabolize/detoxify the insecticides, making them more polar and less toxic. Piperonyl butoxide (PBO), S,S,S,-tributyl phosphorotrithioate (DEF), and diethyl maleate (DEM) are the inhibitors of P450s, hydrolases, and GSTs, respectively, and act as synergists for rapidly testing the involvement of these enzymes in insecticide resistance of mosquitoes. Using synergists in insecticide bioassays to examine the corresponding detoxification enzymes in the development of insecticide resistance is called a synergism study (see Protocol 3: Synergism Study for Investigating Possible Mechanisms of Insecticide Resistance in Mosquitoes [Li et al. 2023b]). In these studies, a synergist(s) is applied to mosquitoes together with an insecticide. For example, synergism studies of pyrethroid insecticide resistance in mosquitoes have enabled researchers to draw diagnostic conclusions regarding whether and which detoxification mechanisms are involved in the development of pyrethroid resistance (Kasai et al. 1998; Brogdon et al. 1999; Enayati et al. 2003; Liu et al. 2004; Xu et al. 2005). However, conclusions drawn from synergistic studies alone must be treated with caution; although using synergists may often correctly indicate the role of detoxification proteins in insecticide resistance, in some cases the synergists may be only imperfect inhibitors of some of the detoxification enzymes responsible for resistance (Feyereisen 2005). Further work is thus needed to support the findings from synergistic studies. The detailed methodology required for such studies is described in our associated protocol (see Protocol 3: Synergism Study for Investigating Possible Mechanisms of Insecticide Resistance in Mosquitoes [Li et al. 2023b]).

CONCLUSION

In this introduction and the associated protocols, we provide detailed methodologies for the most commonly used types of laboratory bioassays, including the larval bioassays and adult bioassays (topical and bottle residual) used to characterize the toxicity of insecticides and evaluate levels of insecticide resistance. We also introduce synergism assays used to test possible mechanisms of metabolic detoxification-mediated resistance conferred by important metabolic enzymes, especially P450s, GSTs and hydrolases. Synergism studies provide a useful way to measure the target proteins involved in the development of insecticide resistance and other proteins that may be involved in the regulation of insecticide resistance once the inhibitors of the target proteins are known. Each protocol provides a detailed methodology, describing how the assays should be conducted and ways of dealing with problems that may arise that could influence the accuracy of the data analyses and results. Finally,

 Cite this introduction as *Cold Spring Harb Protoc*; doi:10.1101/pdb.top107705

we introduce the field surveillance tests used to monitor insecticide resistance based on the protocols presented in the World Health Organization (2016) and Centers for Disease Control and Prevention (2019) recommendations.

REFERENCES

Agresti A. 2015. *Foundations of linear and generalized linear models.* Wiley, New York.

Aldrich JH, Nelson FD. 1984. *Linear probability, logit, and probit models, Vol. 45.* Sage Publications, Los Angeles.

Brogdon WG, McAllister JC, Corwin AM, Cordon-Rosales C. 1999. Independent selection of multiple mechanisms for pyrethroid resistance in Guatemalan *Anopheles albimanus* (Diptera: Culicidae). *J Econ Entomol* **92:** 298–302. doi:10.1093/jee/92.2.298

Burgess ER IV, King BH, Geden CJ. 2020. Oral and topical insecticide response bioassays and associated statistical analyses used commonly in veterinary and medical entomology. *J Insect Sci* **20:** 6. doi:10.1093/jisesa/ieaa041

Centers for Disease Control and Prevention (U.S.). 2019. *CONUS manual for evaluating insecticide resistance in mosquitoes using the CDC bottle bioassay kit* (updated 2019). https://www.cdc.gov/mosquitoes/pdfs/CONUS-508.pdf [accessed June 27, 2021].

Choi L, Pryce J, Garner P. 2019. Indoor residual spraying for preventing malaria in communities using insecticide-treated nets. *Cochrane Database Syst Rev* **5:** 1–65. doi:10.1002/14651858.CD012688.pub2

Enayati AA, Vatandoost H, Ladonni H, Townson H, Hemingway J. 2003. Molecular evidence for a kdr-like pyrethroid resistance mechanism in the malaria vector mosquito *Anopheles stephensi. Med Vet Entomol* **17:** 138–144. doi:10.1046/j.1365-2915.2003.00418.x

Feyereisen R. 2005. Insect cytochrome P450. *Comprehensive Mol Insect Sci* **4:** 1–77. doi:10.1016/B0-44-451924-6/00049-1

Finney DJ. 1971. *Probit analysis.* Cambridge University Press, Cambridge, UK.

Giullin C, Peters RF. 1952. Recent studies of mosquito resistance to insecticides in California. *Mosq News* **12:** 1–7.

Hemingway J, Field L, Vontas J. 2002. An overview of insecticide resistance. *Science* **298:** 96–97. doi:10.1126/science.1078052

Kasai S, Weerashinghe IS, Shono T. 1998. P450 monooxygenases are an important mechanism of permethrin resistance in *Culex quinquefasciatus* Say larvae. *Arch Insect Biochem Physiol* **37:** 47–56. doi:10.1002/(SICI)1520-6327(1998)37:1<47::AID-ARCH6>3.0.CO;2-S

Li T, Wang Y, Liu N. 2023a. Mosquito adult bioassays. *Cold Spring Harb Protoc* doi:10.1101/pdb.prot108041

Li T, Wang Y, Liu N. 2023b. Synergism study for investigating possible mechanisms of insecticide resistance in mosquitoes. *Cold Spring Harb Protoc* doi:10.1101/pdb.prot108042

Liu N. 2015. Insecticide resistance in mosquitoes: impact, mechanisms, and research directions. *Annu Rev Entomol* **60:** 537–559. doi:10.1146/annurev-ento-010814-020828

Liu H, Cupp EW, Micher KM, Guo A, Liu N. 2004. Insecticide resistance and cross-resistance in Alabama and Florida strains of *Culex quinquefaciatus. J Med Entomol* **41:** 408–413. doi:10.1603/0022-2585-41.3.408

Raymond M. 1985. Presentation d'un programme basic d'analyse log-probit pour micro-ordinateur. *Cah ORSTOM Ser Entomol Med Parasitol* **23:** 117–121.

Sahu SS, Thankachy S, Dash S, Nallan K, Swaminathan S, Kasinathan G, Purushothaman J. 2020. Evaluation of long-lasting indoor residual spraying of deltamethrin 62.5 SC-PE against malaria vectors in India. *Malar J* **19:** 19. doi:10.1186/s12936-020-3112-4

Scott JG, Wen Z. 1997. Toxicity of fipronil to susceptible and resistant strains of german cockroaches (Dictyoptera: Blattellidae) and house flies (Diptera: Muscidae). *J Econ Entomol* **90:** 1152–1156. doi:10.1093/jee/90.5.1152

Valles SM, Koehler PG, Brenner RJ. 1997. Antagonism of fipronil toxicity by piperonyl butoxide and *S, S, S*-tributyl phosphorotrithioate in the German cockroach (Dictyoptera: Blattellidae). *J Econ Entomol* **90:** 1254–1258. doi:10.1093/jee/90.5.1254

Wang Y, Li T, Liu N. 2023. Mosquito larval bioassays. *Cold Spring Harb Protoc* doi:10.1101/pdb.prot108040

World Health Organization. 2016. *Test procedures for insecticide resistance monitoring in malaria vector mosquitoes.* 2nd ed. (updated 2018). https://apps.who.int/iris/bitstream/handle/10665/250677/9789241511575-eng.pdf?sequence=1 [accessed June 27, 2021].

World Health Organization. 2020. *Insecticide resistance.* Last update: February 19, 2020. https://www.who.int/teams/global-malaria-programme/prevention/insecticide-resistance [accessed June 27, 2021].

Xu Q, Liu H, Zhang L, Liu N. 2005. Resistance in the mosquito, *Culex quinquefasciatus*, and possible mechanisms for resistance. *Pest Manag Sci* **61:** 1096–1102. doi:10.1002/ps.1090

Zaim M, Guillet P. 2002. Alternative insecticides: an urgent need. *Trends Parasitol* **18:** 161–163. doi:10.1016/S1471-4922(01)02220-6

Protocol 1

Mosquito Larval Bioassays

Yifan Wang,[1] Ting Li,[1,2] and Nannan Liu[1,3]

[1]*Department of Entomology and Plant Pathology, Auburn University, Auburn, Alabama 36849, USA;*
[2]*Department of Biological Sciences, Alabama State University, Montgomery, Alabama 36104, USA*

Insecticide bioassays are frequently used to measure levels of insecticide resistance in mosquito populations, examining the ability of mosquitoes to survive exposure to insecticides. Laboratory bioassays measure insects' responses to insecticides in resistant (field) populations and laboratory susceptible strains using serial doses or concentrations across the range of >0 and <100% mortality. This protocol measures the toxicity of insecticides to mosquito larvae and determines the level of insecticide resistance. Generally, laboratory-reared mosquito larvae of known age or instar are exposed to H_2O containing various concentrations of an insecticide, and the response (mortality) is recorded 24 h after the test. Larval bioassay tests can (1) identify the lethal concentrations of larvicide that cause 50% and 90% mortality (LC_{50} and LC_{90}, respectively); (2) determine the diagnostic concentration needed to monitor susceptibility in mosquito larvae in the field; and (3) investigate the resistance status and the mechanisms governing insecticide resistance to a specific insecticide.

MATERIALS

It is essential that you consult the appropriate Material Safety Data Sheets and your institution's Environmental Health and Safety Office for proper handling of equipment and hazardous materials used in this protocol.

RECIPES: Please see the end of this protocol for recipes indicated by <R>. Additional recipes can be found online at http://cshprotocols.cshlp.org/site/recipes.

Reagents

Acetone, DMSO, or other organic relatively nontoxic solvents (technical grade quality)

Acetone is generally used as the solvent to dissolve the insecticide. If an insecticide is not soluble in acetone, other solvents, such as ethanol, methanol, ethyl acetate, and DMSO, can be used. Since solvents like acetone are volatile liquids, prepare the insecticide stock solution and the desired concentrations quickly.

Insecticides to be tested (technical grade quality)

Many insecticides are organic compounds and are thus insoluble in H_2O. An organic solvent must be used to dissolve organic insecticides to prepare dilution series concentrations of insecticide for laboratory tests.

Mosquito larvae of the species of choice, third- to fourth-instar

Rear mosquito larvae in a six-quart storage container with 3 L of tap H_2O, covered with a lid, maintained in an insectary room at 25°C ± 2°C, 70% ± 5% relative humidity, and with a light:dark cycle of 12 h (Kauffman et al. 2017). Feed 2 mL of larval food every other day. Determine the mosquito larval instar by measuring the width of the thorax and width of the head capsule (Timmermann and Briegel 1999).

[3]Correspondence: liunann@auburn.edu

Cite this protocol as *Cold Spring Harb Protoc*; doi:10.1101/pdb.prot108040

This bioassay is conducted on both lab-susceptible and field-collected strains simultaneously to examine the levels of resistance in the field-collected mosquitoes according to the resistance ratio(s). Lab-susceptible mosquito strains are those that have never been exposed to insecticides.

Mosquito larval food <R>
Tap H_2O

Equipment

Aluminum foil
Analytical balance
Disposable plastic transfer pipettes, 15-mL
Glass scintillation vials (20-mL; VWR 66022-065)

Disposable vials are used to avoid contamination. Use vials from a single vendor to ensure consistency.

Insectary room or environmentally controlled incubator set at 25°C ± 2°C, 70% ± 5% relative humidity, with an L:D cycle of 12 h
Log-probit software (e.g., https://leora-software.com) or paper (e.g., R program, SAS, POLO-PC)
Notebook for recording the number of larval deaths
Screw-top amber glass vials with lids (4-mL)

METHOD

Accuracy is important in this experiment, so ensure your pipettes are accurate. This larval bioassay follows the protocol developed by the World Health Organization (WHO) for mosquitoes (WHO 2005, 2016). For investigations of the underlying mechanisms of insecticide resistance, carry out larval bioassays of both lab-susceptible and field-collected (resistant) strains. The resistance ratio between the two is used to measure the resistance level of a resistant strain (Liu and Yue 2000).

Dose Range Determination

1. Determine the appropriate concentration of insecticide for an initial bioassay to determine the dose that covers 0%–100% mortality.
   - For the first time performing the bioassay with a specific insecticide, plan to use a wide range of insecticide concentrations for the serial dilution to ensure it covers the entire range of insect mortality from 0% to 100%. A typical initial series would be 0.0003, 0.003, 0.03, 0.3, 3, and 30 ppm, for example.
   - After ascertaining the mortality using this initial dose range, focus on a narrower range that includes a concentration that causes just over 0% mortality and one that causes just under 100% mortality, with three or four intermediate concentrations, to obtain precise values for LC_{50} and LC_{90} (World Health Organization 2005, 2016; Brito-Sierra et al. 2019).

The Preparation of Insecticide Working Solutions

The first step in preparing a serial set of the working solutions is to make a stock solution and then dilute it as appropriate.

2. Using an analytical balance, weigh the required amount of insecticide and dissolve it in an appropriate solvent, generally acetone, DMSO, or another suitable organic solvent according to the insecticide's solubility characteristics. Prepare stock solutions and dilutions in amber glass screw-cap vials.
   - Prepare stock solutions at a concentration of 20,000 ppm (20 mg/mL) in a total volume of 4 mL solvent (World Health Organization 2005, 2016; Brito-Sierra et al. 2019). This is a general rule and may differ for a given insecticide.

- Account for the purity of the insecticide (Brito-Sierra et al. 2019) by using the following equation:

$$\text{Weight of insecticide} = \frac{\text{Desired concentration of stock solution} \times \text{Volume required}}{\%\ \text{purity of insecticide}}.$$

For example: To make a 4 mL 20,000 ppm (20 mg/mL) stock solution of an insecticide with 98% purity: (4 mL × 20 mg/mL)/98% = 81.63 mg. Therefore, add 81.63 mg insecticide into 4 mL solution. Shake or stir the vial gently to dissolve or disperse the insecticide in the solvent; no special equipment is required.

3. Cover tightly closed amber glass screw-cap vials containing stock solutions with aluminum foil to avoid light. Store solutions for up to 2 months at −20°C.
4. Prepare working solutions the day of the bioassay by preparing 10-fold serial dilutions of the stock solution in acetone or another appropriate solvent (Fig. 1A). Cover tightly closed screw-cap vials containing working solutions with aluminum foil to avoid light, and keep them at room temperature.

Larval Bioassay Procedures

5. Mark the 10-mL level on the wall of each 20-mL disposable scintillation vial and add 5 mL tap H_2O to the vials. Prepare two vials for each experimental concentration of pesticide, plus two control vials.

   Add each concentration of solution to two parallel bottles for a total of 20 mosquitos per treatment; each vial will contain 10 mosquito larvae.

6. Using a 15-mL disposable plastic transfer pipette, transfer 10 late-third-/early-fourth-instar larvae into each vial that has a premarked 10-mL scale. Fill each vial with tap H_2O to a total volume of 10 mL (Fig. 1B).

   Ensure the sizes of the mosquito larvae are similar. Larvae of different sizes respond differently to the same concentration of an insecticide; hence a tiny mistake may produce a completely different result.

7. Add ~1% of the total volume (not >100 µL) of insecticide solution to each 20-mL disposable vial to achieve the experimental concentration. For the control group, add 1% of the total volume of acetone (or the appropriate organic solvent) to each of two vials. Swirl the vials gently to ensure that the chemicals are mixed well.
8. Maintain the vials at a room temperature between 25°C and 28°C, and subject them to a light regime of 12 h dark followed by 12 h light (12L:12D).
9. Record larval mortality after 24 h of exposure.

   Dying larvae are those that do not rise to the surface or show an active response when the vial is shaken. If the mortality rate in the control group is >20%, the entire experiment needs to be redone.

   Repeat each bioassay at least three times on different days, thus testing a total of 60 mosquito larvae with each concentration of insecticide.

Data Analysis

10. Calculate the corrected mortality of treated groups according to Abbott's formula:

$$\text{Corrected mortality}\ (\%) = \frac{\text{mortality rate in test}\ (\%) - \text{mortality rate in control}\ (\%)}{100 - \text{mortality rate in control}\ (\%)} \times 100.$$

For example, if the mortality rate in the test group is 90% and the mortality rate in the control group is 10%, then the corrected mortality is (90%–10%)/(100%–10%) × 100 = 88.89%.

See Troubleshooting.

 Cite this protocol as *Cold Spring Harb Protoc*; doi:10.1101/pdb.prot108040

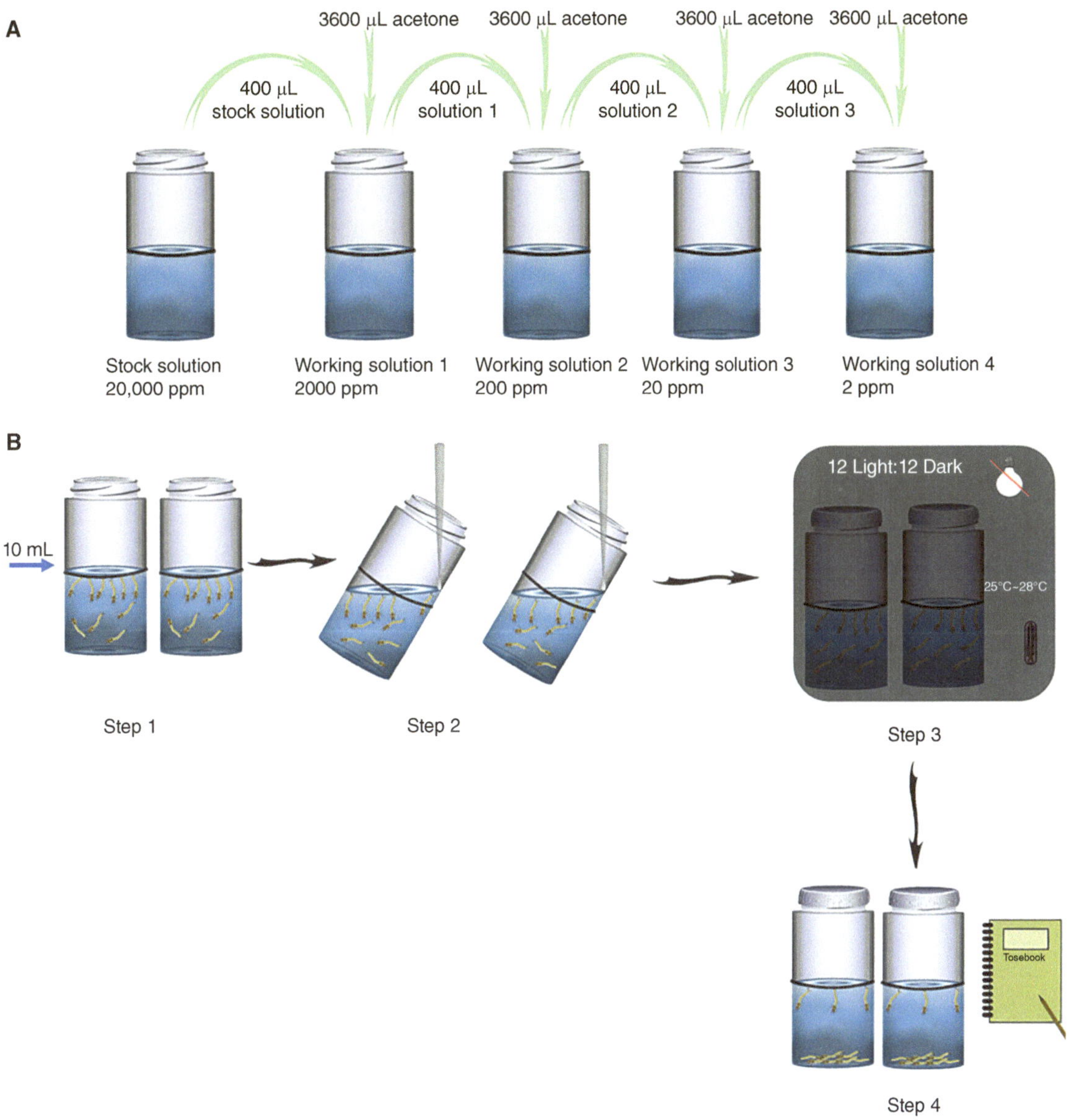

FIGURE 1. Schematic diagram of mosquito larval bioassay. (*A*) Procedure for preparing 10-fold serial dilutions from insecticide stock solution. (*B*) Procedure for mosquito larval bioassay: (Step 1) Add 10 larvae into the 20-mL bottle and fill with tap H_2O up to 10 mL. (Step 2) Add appropriate volume (not >100-µL) of the insecticide solution to get the desired test concentration. (Step 3) Place the test bottles in an insectary room (12 h L:12 h D) or in an environmentally controlled incubator. (Step 4) Record the mortality 24 h after treatment.

11. Calculate the LC_{50}, LC_{90}, and 95% CI (confidence interval) values via probit analysis using appropriate computer software (https://leora-software.com).
12. Calculate the resistance ratio as

$$\text{Resistance ratio} = \frac{LC_{50}\text{ in resistant strain}}{LC_{50}\text{ in susceptible strain}}.$$

TROUBLESHOOTING

Problem (Step 10): Significant differences are observed between replicates (i.e., variation in the mortality rate is too high). Variation is considered to be high if the mortality rate of the control is >20% and if it is not possible to obtain an acceptable value for 95% CI.

Solution: High variation could be due to the parameters of (1) the pipetting technique used, (2) the pipette calibration, (3) solution preparation, (4) mosquito size, sex, and age, and (5) the number of individuals tested per dose. Conduct additional replications to obtain an accurate result while correcting the parameters causing the variation.

DISCUSSION

This protocol describes the procedures used to measure the toxicity of insecticides to mosquito larvae. Laboratory bioassays measure the insects' responses to insecticides in resistant (field) populations and laboratory susceptible strains using serial doses or concentrations across the range of >0 and <100% mortality (Russell et al. 1977; Liu and Yue 2000; Liu et al. 2004). Larval bioassay tests can (1) identify the lethal concentrations of larvicide that cause 50% and 90% mortality (LC_{50} and LC_{90}, respectively); (2) determine the diagnostic concentration needed to monitor susceptibility in mosquito larvae in the field; and (3) investigate the resistance status and the mechanisms governing insecticide resistance to a specific insecticide (World Health Organization 2005, 2016).

In contrast to the single dose bioassay (Brito-Sierra et al. 2019), which defines the mortality of mosquito larvae at a particular concentration, a multiple dose bioassay determines the LC_{50}, LC_{90}, and 95% CI, thus providing more accurate values to guide field applications of insecticide and more accurate comparisons of the responses of field populations to different insecticides. Compared to the method developed by the World Health Organization (2005, 2016), in which 20–25 larvae are assayed in 100–200 mL H_2O, and the method developed by Brito-Sierra et al. (2019), which uses five larvae in 1 mL H_2O, we prefer to choose an intermediate density for the mosquito larvae, reasoning that 10 larvae in 10 mL H_2O provides sufficient space for the mosquito larvae to survive while still reducing the chance of the sample becoming contaminated. If the level of insecticide resistance is the main focus of the investigation, bioassays of the field and susceptible strains must be conducted for comparison.

RECIPE

Mosquito Larval Food

4 g flake fish food (Tetra TetraMin Tropical Flakes)
20 g yeast (Fleischmann's RapidRise Instant Yeast Fast Acting)

Mix in 300 mL tap H_2O and store for up to 1 mo at 4°C.

REFERENCES

Brito-Sierra CA, Kaur J, Hill CA. 2019. Protocols for testing the toxicity of novel insecticidal chemistries to mosquitoes. *J Vis Exp* **2019:** e57768. doi:10.3791/57768

Kauffman E, Payne A, Franke MA, Schmid MA, Harris E, Kramer LD. 2017. Rearing of *Culex* spp. and *Aedes* spp. mosquitoes. *Bio-protocol* **7:** 17. doi:10.21769/BioProtoc.2542

Liu N, Yue X. 2000. Insecticide resistance and cross-resistance in the house fly (Diptera: Muscidae). *J Econ Entomol* **93:** 1269–1275. doi:10.1603/0022-0493-93.4.1269

Liu H, Cupp EW, Micher KM, Guo A, Liu N. 2004. Insecticide resistance and cross-resistance in Alabama and Florida strains of *Culex quinquefaciatus*. *J Med Entomol* **41:** 408–413. doi:10.1603/0022-2585-41.3.408

Russell RM, Robertson JL, Savin NE. 1977. POLO: a new computer program for probit analysis. *Ann Entomol Soc Am* **23:** 209–213. doi:10.1093/besa/23.3.209

Timmermann SE, Briegel H. 1999. Larval growth and biosynthesis of reserves in mosquitoes. *J Insect Physiol* **45:** 461–470. doi:10.1016/S0022-1910(98)00147-4

World Health Organization. 2005. *Guidelines for laboratory and field testing of mosquito larvicides.* https://apps.who.int/iris/handle/10665/69101 [accessed June 27, 2021].

World Health Organization. 2016. *Monitoring and managing insecticide resistance in* Aedes *mosquito populations: interim guidance for entomologists.* https://apps.who.int/iris/handle/10665/204588 [accessed June 27, 2021].

Protocol 2

Mosquito Adult Bioassays

Ting Li,[1,2] Yifan Wang,[1] and Nannan Liu[1,3]

[1]*Department of Entomology and Plant Pathology, Auburn University, Auburn, Alabama 36849, USA;*
[2]*Department of Biological Sciences, Alabama State University, Montgomery, Alabama 36104, USA*

Topical application and bottle bioassays measure the responses of adult mosquitoes to specific doses of an insecticide (dose–response). Topical application bioassays are generally used to measure the dose–response of adult mosquitoes to insecticides in which the amount (dose) of insecticides the mosquitoes receive is known in the laboratory. Here, a 0.5-µL drop of the insecticide dissolved in a relatively nontoxic solvent, such as acetone, is applied to the thorax of insects, and the insects' susceptibility to the insecticide is determined in terms of either the median lethal dose (LD_{50}) or 90% of the mortality lethal dose (LD_{90}). Bottle bioassays measure the dose–responses in which the exact amount of insecticide in a bottle is known while the exact amount of insecticide that mosquitoes (field-collected or laboratory-susceptible) receive is not known. Bottle bioassays can be either single dose tests or multiple dose applications. The bottle bioassay described in this protocol is a modified form of the World Health Organization (WHO) and U.S. Centers for Disease Control (CDC) bottle bioassays. For the single bottle assay, a detailed protocol with the amount (dose/bottle) of each insecticide and threshold times is provided by the CDC; here we provide protocols for topical and bottle bioassay applications with multiple doses.

MATERIALS

It is essential that you consult the appropriate Material Safety Data Sheets and your institution's Environmental Health and Safety Office for proper handling of equipment and hazardous materials used in this protocol.

Reagents

Acetone, DMSO (dimethyl sulfoxide), or other organic relatively nontoxic solvents (technical grade quality)

Acetone is generally used as the solvent to dissolve the insecticide. If an insecticide is not soluble in acetone, other solvents, such as ethanol, methanol, ethyl acetate, and DMSO, can be used. Because solvents like acetone are volatile liquids, prepare the insecticide stock solution and the desired doses quickly.

CO_2 gas tank

Insecticides to be tested (technical grade quality)

Many insecticides are organic compounds and are thus insoluble in H_2O. An organic solvent must be used to dissolve organic insecticides to prepare dilution series doses of insecticide for laboratory tests.

Mosquitoes of the species of choice (2- to 3-d-old; female or male)

[3]Correspondence: liunann@auburn.edu

Cite this protocol as *Cold Spring Harb Protoc*; doi:10.1101/pdb.prot108041

Maintain mosquito colonies in an insectary room at 25°C ±2°C, 70% ±5% relative humidity, and with a light: dark cycle of 12 h (Kauffman et al. 2017) both before and after the bioassays.

This bioassay is conducted on both lab-susceptible and field-collected strains simultaneously to examine the levels of resistance in the field-collected mosquitoes according to the resistance ratio(s). Lab-susceptible mosquito strains are those that have never been exposed to insecticides.

Sugar H_2O (10% sucrose)

Equipment

Cheesecloth
Cotton balls
D-cell powered aspirator and 2″H collecting tubes (Gemplers)
Drosophila CO_2 Fly Pad (Tritech Research, Inc.)
Featherweight forceps, wide-tip
Glass scintillation bottles (20-mL; VWR 66022-065)
Hamilton gastight syringe (25-µL; Thermo Fisher Scientific)
Ice cream cups (6-oz, Sweetheart ice cream cup only)
Insectary room or environmentally controlled incubator at 25°C ± 2°C, 70% ± 5% humidity, with an L:D cycle of 12 h
Mosquito cages (BioQuip Products, Inc.)
Screw-top amber glass vials with lids (4-mL)

METHOD

Dose Range Determination

1. Determine the appropriate doses of insecticide for an initial bioassay over a wide dose range to determine the doses that covers 0% ~ 100% mortality.
   - If performing a topical application assay with a particular insecticide on a mosquito strain with an unknown level of sensitivity, plan an initial dose range with doses of 0.00001, 0.0001, 0.001, 0.01, 0.1, 1, and 10 µg/mosquito.
   - If performing a bottle bioassay with a particular insecticide on a mosquito strain with an unknown level of sensitivity, plan an initial dose range with doses of 0.001, 0.01, 0.1, 1, 10, and 100 µg/bottle.
   - Consider doses that cause just more than 0% mortality and less than 100% mortality as the lowest and the highest doses, respectively, for the main bioassay experiment. Add several doses (typically three to five) that fall in between these highest and lowest doses.

     For example, if the lowest dose is 0.001 µg/mosquito, causing ~5% mortality, and the highest is 1 µg/mosquito, causing ~95% mortality, the final dose range could be 0.001, 0.003, 0.01, 0.03, 0.1, 0.3, and 1 µg/mosquito.

Preparing the Stock Solution

2. Calculate the composition of the stock solution using the following equation:

$$\text{Amount of weighed insecticide} = \frac{\text{Dose of stock solution} \times \text{Volume to be made}}{\text{\% purity of AI (active ingredient)}}.$$

Dissolve the calculated amount of insecticide in an appropriate solvent, generally acetone, DMSO, or some other suitable organic solvent according to the insecticide's solubility characteristics.

For example, to make a 10 mg/mL stock solution of an insecticide with 94.34% purity, the amount of the insecticide in a 1 mL stock solution will be (10 mg/mL × 1 mL)/94.34% = 10.60 mg.

Cover tightly closed amber glass screw-cap vials containing stock solutions with aluminum foil to avoid light. Store solutions for up to 2 mo at −20°C.

Making Serial Insecticide Doses

3. Prepare a broad range of working solutions as described in Step 1, aiming to cover a mortality range of 0% to 100% for mosquitoes (Li and Liu 2010).

   Prepare these working solutions the day of the bioassay at room temperature. The working solutions can be stored for up to 1 wk at −20°C.

Mosquito Adult Preparation

As male and female mosquitoes may show different insecticide susceptibilities, conduct adult bioassays on males and females separately (most researchers use only female mosquitoes to conduct their bioassays [Liu and Yue 2000; Li and Liu 2010]).

4. Collect test mosquito and lab-susceptible strain pupae on the same day as follows.
   i. Collect newly emerging male and female mosquitoes at <8 h postemergence using a D-cell powered aspirator and 2″H collecting tube (Li and Liu 2010).
   ii. Maintain the separated male and female adults in cages separately in the insectary and rear with cotton balls containing 10% sucrose H_2O for 2 d before bioassay.

      Test 2-d-old adult mosquitoes in the bioassays below. For mosquito insecticide resistance studies, conduct bioassays on both resistant and susceptible strains at the same time. Proceed to Step 5 for the topical application bioassay or Step 13 for the bottle bioassay.

Topical Application Bioassays

This assay is mainly used in laboratory experiments that focus on insecticide-resistant and -susceptible strains or to evaluate the toxicity of a new insecticide. As described in Step 1, the doses of each insecticide are designed to cause >0% and <100% mortality.

Run all tests at 25°C with at least three replications on different days.

5. Using a D-cell powered aspirator and 2″H collecting tube, transfer 2-d-old adult mosquitoes of the same size to a *Drosophila* CO_2 Fly Pad for anesthesia and the bioassay.

   Once mosquitoes are anesthetized, treat them immediately with insecticide (Fig. 1) to avoid damage caused by long-term anesthesia.

   Transfer 30 mosquitoes for each experimental condition including controls.

6. Deliver a 0.5-µL drop of acetone (or the corresponding solvent) to 30 adult mosquitoes to serve as solvent control mosquitoes.
7. Beginning with the lowest dose, use a 25-µL Hamilton gastight syringe to deliver a 0.5-µL drop of insecticide solution to the dorsal portion of the thoracic segment (thoracic notum) of each adult mosquito (Fig. 1A). Apply each dose of insecticide to 30 adult mosquitoes as one treatment. Repeat with each dose, proceeding from lowest dose to highest dose.
8. Immediately after the insecticide application, use featherweight forceps to transfer anesthetized mosquitoes to Sweetheart ice cream cups so that they contain 15 mosquitoes each. Cover each cup with cheesecloth and label with the corresponding insecticide dose.
9. Rinse the Hamilton syringe 10 times with full occupation of solvent after the final dose of insecticide is delivered.
10. Place a cotton ball with 10% sugar H_2O on top of each cup for nutrition. Do not exceed 100% saturation of the cotton ball to avoid sugar H_2O dripping into the cup.
11. Transfer the insecticide-treated and control groups to an insectary room or to an environmentally controlled incubator.

 Cite this protocol as *Cold Spring Harb Protoc*; doi:10.1101/pdb.prot108041

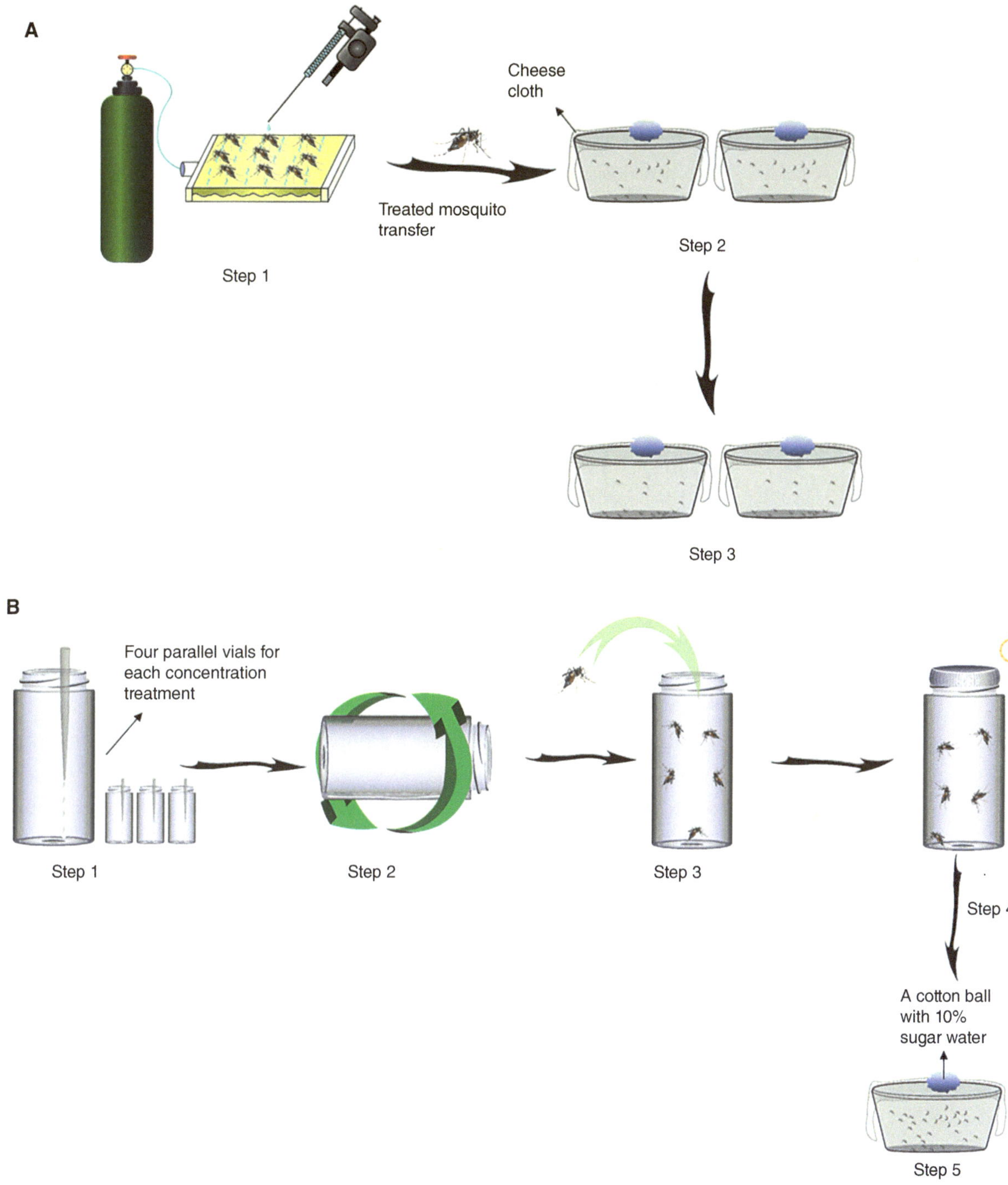

FIGURE 1. Insecticide bioassays for adult mosquitoes. (*A*) Topical application. (Step 1) A 0.5-µL drop of solution is delivered to the thoracic notum of each mosquito on a *Drosophila* CO_2 Fly Pad using a 25-µL Hamilton gastight syringe. (Step 2) The treated mosquitoes are transferred to a Sweetheart ice cream cup after insecticide or solvent is applied. Each cup is covered by cheesecloth with a cotton ball saturated with 10% sugar H_2O on the top. (Step 3) Treated mosquitoes are placed in an insectary room or in an environmentally controlled incubator and mortality is recorded 24 h after treatment. (*B*) Bottle assay. (Step 1) 100 µL of insecticide at a certain dose or a solvent control is placed into 20-mL glass bottles. (Step 2) Bottles are rolled for one min and dried completely in the fume hood. (Step 3) Five adult mosquitoes are transferred to each insecticide- or solvent-coated glass bottle. (Step 4) Mosquitoes are incubated in the bottle. (Step 5) Mosquitoes are transferred to ice cream cups and kept in an insectary room or in an environmentally controlled incubator. Mortality is recorded 24 h after treatment.

12. Assess mortality 24 h after treatment by lightly shaking each cup for ~5 sec. The absence of movement is the criteria for death.

    Proceed to Step 19.

Laboratory Bottle Bioassay with Serial Doses of Insecticide

The following protocol is modified from World Health Organization (2016). Perform this assay at 25°C.

13. Place 100 µL of each dose of insecticide solution in a 20-mL glass scintillation bottle and 100 µL of acetone or the appropriate solvent in the control bottle. Prepare four bottles for each dose of insecticide and four bottles for the control. Roll bottles for 1 min to ensure the insecticide is evenly coated on the bottle walls (Fig. 1B).

    In the CDC protocol, 1 mL of insecticide is placed into a 250-mL bottle.

14. Dry the bottles completely in a fume hood before transferring the adult mosquitoes into the bottles.
15. Keeping the sexes separated, transfer five 2-d-old female (or male) adult mosquitoes to each bottle containing an insecticide or solvent using a D-cell powered aspirator and 2″H collecting tube. Cover the lid of each bottle loosely and keep the bottles for 1 h at room temperature.

    Each dose of insecticide was added in four parallel bottles in Step 13, such that a total of 20 adult mosquitos will be treated per dose.

16. After a 1-h exposure to the insecticide in the glass bottles, anesthetize mosquitoes by CO_2 in the bottle (usually takes a few seconds) and gently pour the mosquitoes into Sweetheart ice cream cups by inverting the bottle, keeping five mosquitoes per cup. Label each cup and cover with cheesecloth. Place a cotton ball with 10% sugar H_2O on top of each cup. Do not exceed 100% saturation of the cotton ball to avoid sugar H_2O dripping into the cup.
17. Transfer the insecticide-treated and control groups to an insectary room or an environmentally controlled incubator.
18. Assess mortality 24 h after treatment. The criterion for death is the mosquitoes' inability to move when their cup is shaken.

    Repeat this bioassay at least three times on different days.

Data Analysis of Insecticide Resistance Ratio

19. If the mortality in the control group is between 5% and 20%, calculate the corrected mortality of treated groups according to Abbott's formula as in Protocol 1: Mosquito Larval Bioassays (Wang et al. 2023):

$$\text{Corrected mortality (\%)} = \frac{\text{mortality rate in test (\%)} - \text{mortality rate in control (\%)}}{100 - \text{mortality rate in control (\%)}} \times 100.$$

20. Calculate LD_{50} and LD_{90} (for adult topical application), LC_{50} and LC_{90} (for adult bottle bioassay), and 95% CI (confidence interval) values via probit analysis using appropriate computer software (https://leora-software.com) as in Protocol 1: Mosquito Larval Bioassays (Wang et al. 2023).
21. Calculate the resistance ratio as in Protocol 1: Mosquito Larval Bioassays (Wang et al. 2023):

$$\text{Resistance ratio} = \frac{LD_{50}\ \text{in resistant strain}}{LD_{50}\ \text{in susceptible strain}}.$$

    See Troubleshooting.

22. Compare the ICs (confidence intervals) between the target population(s) and the susceptible strain. If the 95% ICs between the target population(s) and the susceptible strain do not overlap, it

Cite this protocol as *Cold Spring Harb Protoc*; doi:10.1101/pdb.prot108041

can be concluded that there is a significant difference in insecticide resistance between the two populations (Robertson and Preisler 1992). If the mortality rate in the control group is >20%, repeat the experiment.

TROUBLESHOOTING

Problem (Step 21): Significant differences are observed between replicates of an experiment.
Solution: Various possible sources of error must be considered, including the dilution processes used to create the desired doses, syringe calibration, contamination of insecticide doses, or different size of mosquitoes in each replication. Additional replications are needed.

DISCUSSION

The bottle bioassay described in this protocol is a modification of the World Health Organization (2005, 2016) and Centers for Disease Control and Prevention (2016, 2019) bottle bioassays. The CDC bottle bioassay with a single dose is generally used for initial tests of insecticide effectiveness in mosquitoes. Although the original CDC bottle assay uses a 250-mL bottle, the 20-mL bottle used in our protocol speeds up the preparation process in the laboratory and is safer to handle because of less insecticide exposure. The CDC "Bottle bioassay threshold times and amounts" should still be followed when observing insecticide susceptibility in field populations (Centers for Disease Control and Prevention 2016, 2019). The original WHO protocol uses filter paper impregnated with insecticide, whereas our modified bottle bioassay uses insecticide covered bottles to speed up the preparation process and reduce contamination in the laboratory.

REFERENCES

Centers for Disease Control and Prevention. 2016. *CDC bottle bioassay*. https://www.cdc.gov/mosquitoes/mosquito-control/professionals/cdc-bottle-bioassay.html [accessed January 14, 2021].

Centers for Disease Control and Prevention. 2019. *CONUS manual for evaluating insecticide resistance in mosquitoes using the CDC bottle bioassay kit* (updated 2019). https://www.cdc.gov/mosquitoes/pdfs/CONUS-508.pdf [accessed June 27, 2021].

Kauffman E, Payne A, Franke MA, Schmid MA, Harris E, Kramer LD. 2017. Rearing of *Culex* spp. and *Aedes* spp. mosquitoes. *Bio-protocol* **7**: e2542. doi:10.21769/BioProtoc.2542

Li T, Liu N. 2010. Inheritance of permethrin resistance in *Culex quinquefasciatus*. *J Med Entomol* **47**: 1127–1134. doi:10.1603/ME10142

Liu N, Yue X. 2000. Insecticide resistance and cross-resistance in the house fly (Diptera: Muscidae). *J Econ Entomol* **93**: 1269–1275. doi:10.1603/0022-0493-93.4.1269

Robertson JL, Preisler HK. 1992. *Pesticide bioassays with arthropods*. CRC, Boca Raton, FL.

Wang Y, Li T, Liu N. 2023. Mosquito larval bioassays. *Cold Spring Harb Protoc* doi:10.1101/pdb.prot108040

World Health Organization. 2005. *Guidelines for laboratory and field testing of mosquito larvicides*. https://apps.who.int/iris/handle/10665/69101 [accessed June 27, 2021].

World Health Organization. 2016. *Test procedures for insecticide resistance monitoring in malaria vector mosquitoes*. 2nd ed. (updated 2018) https://apps.who.int/iris/bitstream/handle/10665/250677/9789241511575-eng.pdf?sequence=1 [accessed June 27, 2021].

Protocol 3

Synergism Study for Investigating Possible Mechanisms of Insecticide Resistance in Mosquitoes

Ting Li,[1,2] Yifan Wang,[1] and Nannan Liu[1,3]

[1]*Department of Entomology and Plant Pathology, Auburn University, Auburn, Alabama 36849, USA;*
[2]*Department of Biological Sciences, Alabama State University, Montgomery, Alabama 36104, USA*

Metabolic detoxification, in which insecticides are metabolized by enzymes, including cytochrome P450s, hydrolases, and glutathione-*S*-transferases (GSTs), to become more polar and less toxic, is one of the major mechanisms involved in the development of insecticide resistance. Piperonyl butoxide (PBO), *S,S,S*,-tributylphosphorotrithioate (DEF), and diethyl maleate (DEM) are inhibitors of P450s, hydrolases, and GSTs, respectively, and are frequently used as insecticide synergists in assessing the metabolic mechanisms that may be involved in the detoxification of insecticides and in the development of insecticide resistance. Synergistic assays can be used to identify the detoxification enzyme that leads to resistance to a specific insecticide. Here, we describe the procedures used in synergist studies of insecticides in both mosquito larvae and adults. The synergist is applied at a maximum sublethal concentration, which is the highest concentration that produces no apparent mortality in the experimental population, above which mortality appears. Insecticide synergistic experiments measure (1) the synergism ratio (SR), which is the difference in the levels of toxicity of a specific insecticide to a strain with and without the presence of the synergists; and (2) the synergism resistance ratio (SRR), which compares SR in the resistant strain with SR in a susceptible strain. In effect, SR indicates the levels of specific enzymes involved in the detoxification of insecticide and SRR identifies the detoxification enzymes/mechanisms that may be involved in the insecticide resistance of insects.

MATERIALS

It is essential that you consult the appropriate Material Safety Data Sheets and your institution's Environmental Health and Safety Office for proper handling of equipment and hazardous materials used in this protocol.

RECIPES: Please see the end of this protocol for recipes indicated by <R>. Additional recipes can be found online at http://cshprotocols.cshlp.org/site/recipes.

Reagents

Acetone, DMSO, or other organic relatively nontoxic solvents (technical grade quality)

Acetone is generally used solvent to dissolve the insecticide. If an insecticide is not soluble in acetone, other solvents, such as ethanol, methanol, ethyl acetate, and DMSO, can be used. Because solvents like acetone are volatile liquids, prepare the insecticide stock solution and the desired concentrations quickly.

CO_2 gas tank

Insecticides to be tested (technical grade quality)

[3]Correspondence: liunann@auburn.edu

Cite this protocol as *Cold Spring Harb Protoc*; doi:10.1101/pdb.prot108042

Many insecticides are organic compounds and are thus insoluble in H_2O. An organic solvent must be used to dissolve organic insecticides to prepare dilution series concentrations of insecticide working solutions for laboratory tests.

Mosquito adults of the species of choice

Maintain mosquito colonies in an insectary room at 25°C ± 2°C, 70% ± 5% relative humidity, and with a light: dark cycle of 12 h (Kauffman et al. 2017).

This bioassay is conducted on both lab-susceptible and field-collected strains simultaneously to examine the levels of resistance in the field-collected mosquitoes according to the resistance ratio(s). Lab-susceptible mosquito strains are those that have never been exposed to insecticides.

Mosquito larvae of the species of choice, third- to fourth-instar

Rear mosquito larvae in a six-quart storage container with 3 L of tap H_2O, covered with a lid, maintained in an insectary room at 25°C ± 2°C, 70% ± 5% relative humidity, and with a light:dark cycle of 12 h (Kauffman et al. 2017). Feed 2 mL of larval food every other day. Determine the mosquito larval instar by measuring the width of the thorax and width of the head capsule (Timmermann and Briegel 1999).

Mosquito larval food <R>

Sugar H_2O (10% sucrose)

Synergists

Examples include diethyl maleate (DEM), piperonyl butoxide (PBO), and S,S,S,-tributylphosphorotrithioate (DEF) (all from MilliporeSigma; all at 1 mg/mL in acetone).

Equipment

Cheesecloth

Cotton balls

D-cell powered aspirator and 2″H collecting tubes (Gemplers)

Disposable plastic pipettes (15-mL)

Drosophila CO_2 Fly Pad (Tritech Research, Inc.)

Featherweight forceps, wide-tip

Glass scintillation bottles (20-mL; VWR 66022-065)

Hamilton gastight syringe (25-µL; Thermo Fisher Scientific)

Insectary room or an environmentally controlled incubator at 25°C ± 2°C, 70%±5% humidity, with a L:D cycle of 12 h

Maintain mosquitoes in an insectary room before and after bioassays.

Screw-top amber glass vials with lids (4-mL)

Sweetheart ice cream cups (6-oz)

METHOD

This assay compares the LD_{50} of synergist + insecticide to the LD_{50} of insecticide alone.

Begin at Step 1 for the larval synergism bioassay, Step 9 for the adult topical synergism bioassay, or Step 19 for the adult bottle synergism bioassay.

Synergism Bioassay in Mosquito Larvae

Identifying Synergist and Insecticide Concentrations

1. To identify the maximum sublethal concentration of a synergist, prepare serial concentrations of the desired synergist in acetone. Dilute the synergist as described for the dilution of insecticide in Protocol 1: Mosquito Larval Bioassays (Wang et al. 2023). Prepare at least five concentrations with sequential serial dilutions for this initial selection.

- For example, initial synergist concentrations of 0.01, 0.03, 0.1, 0.3, 1, and 3 ppm could be prepared.

  The maximum sublethal concentrations of PBO, DEF, or DEM for Culex quinquefasciatus *larvae are 5, 2, and 3 ppm, respectively (Xu et al. 2005).*

2. Treat mosquito larvae with synergist dilutions according to Steps 5–9 of Protocol 1: Mosquito Larval Bioassays (Wang et al. 2023). Using the concentrations above, if the 100% of the larvae survive at concentrations up to 0.3 ppm and mortality starts to be observed in the 1-ppm solution, then choose the 0.3-ppm solution as the maximum sublethal concentration (Liu and Scott 1996; Liu and Yue 2000).
3. Determine the appropriate range of concentrations of insecticide alone to use, ensuring it covers the entire range of insect mortality from 0% to 100% as described in Protocol 1: Mosquito Larval Bioassays (Wang et al. 2023).
4. Conduct the initial synergist-insecticide bioassays to determine the concentration range of insecticide + the maximum sublethal concentration of synergist that covers 0 to 100% mortality. Treat mosquito larvae with synergist-insecticide dilutions according to Steps 5–9 of Protocol 1: Mosquito Larval Bioassays (Wang et al. 2023).
   - We suggest using the concentration range insect mortality from 0% to 100% identified for insecticide alone in Step 3 and adding two to three lower concentrations. For example, if the concentration range for insecticide alone is 0.003, 0.01, 0.03, 0.1, and 0.3 ppm, here also include 0.0003 and 0.001 ppm.
   - Add an appropriate volume (not >50 µL) of each concentration of the insecticide to the H_2O and larvae present in the 20-mL glass scintillation vials to achieve the desired concentrations, and then add the appropriate volume of the synergist (not >50 µL) to the same vials to achieve the maximum sublethal concentration of the synergist identified in Step 2. The volume of insecticide + synergist solution should not be >100 µL (1% of the total volume) in each 20-mL disposable vial.
5. After ascertaining the mortality using this initial dose range, choose a narrower range that includes an insecticide concentration that causes just >0% mortality and one that causes just <100% mortality, with three or four intermediate concentrations in between, to obtain precise values for LC_{50} and LC_{90} (World Health Organization 2005, 2016; Brito-Sierra et al. 2019) in the next section.

Larval Synergism Assay

6. Label an appropriate number of 20-mL vials to test the following conditions: insecticide concentration #1, insecticide concentration #1 plus synergist, insecticide concentration #2, insecticide concentration #2 plus synergist, and so on to the last concentration of insecticide, synergist control, and solvent control.

   Perform each condition in duplicate (two bottles) which will result in 20 tested mosquitoes for each concentration.

7. Conduct larval assays according to Steps 5–9 of Protocol 1: Mosquito Larval Bioassays (Wang et al. 2023) with the following modifications (Fig. 1A; Xu et al. 2005).
   - Add an appropriate volume of each concentration of the insecticide and the appropriate volume of the synergist (not >50 µL each) to the 20-mL disposable vials to achieve the desired concentrations as described in Step 4.
   - For the insecticide alone control vial, add 1% of the total volume of acetone (or the appropriate organic solvent) to each of vials.

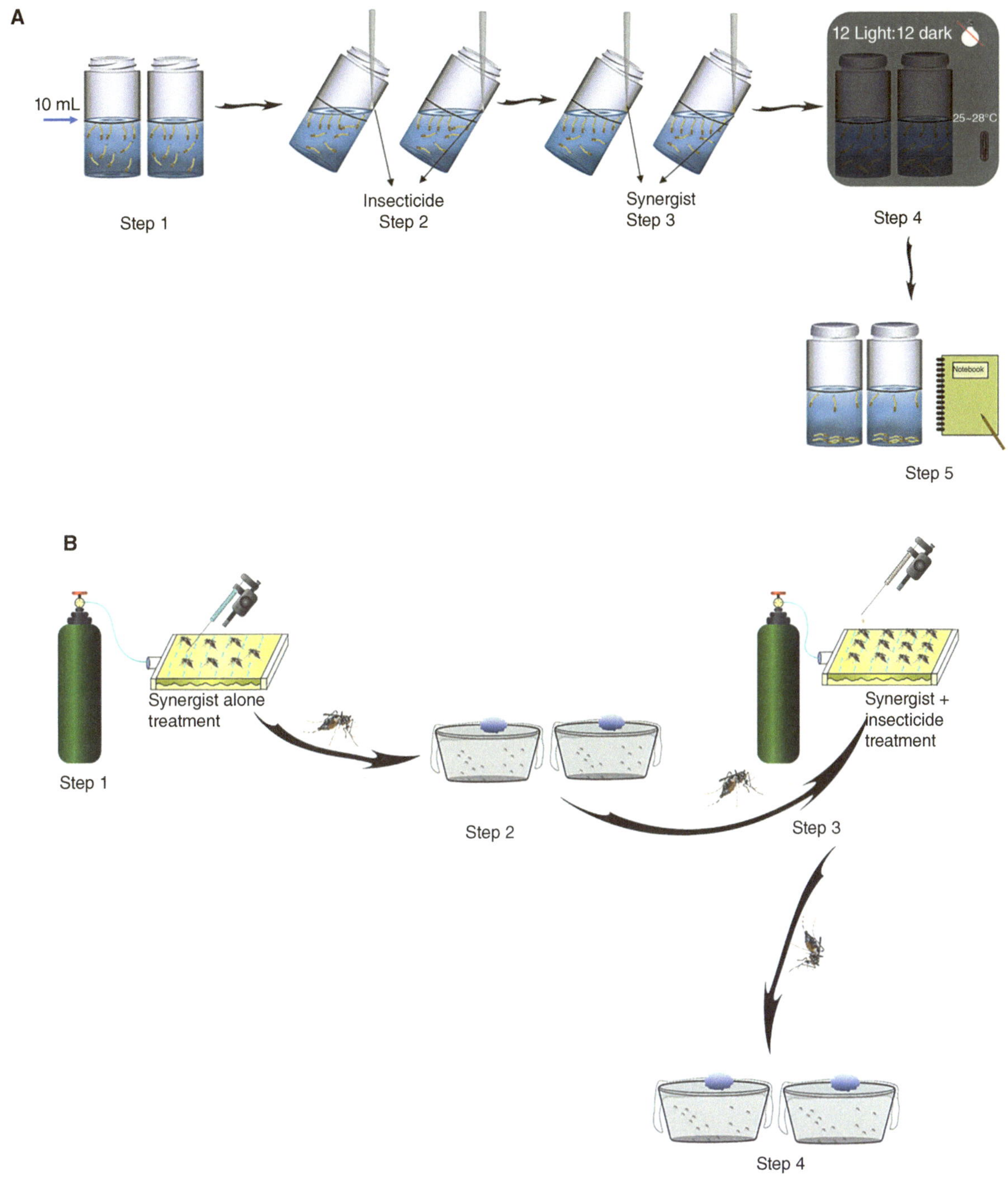

FIGURE 1. Schematic diagram of mosquito synergism bioassays. (*A*) Mosquito larvae synergism assay. (Step 1) Add 10 larvae into a 20-mL bottle and fill with tap H_2O up to 10 mL. (Step 2) Add an appropriate volume (not more than 50 µL) of insecticide solution to get the desired test concentration. (Step 3) Add an appropriate volume (not >50 µL) of synergist solution to get the desired test concentration. (Step 4) Place the test vials in an insectary room (12L:12D h) or in an environmentally controlled incubator. (Step 5) Record the mortality 24 h after treatment. (*B*) Synergism bioassay in adult mosquito—topical application. (Step 1) A 0.5-µL drop of synergist solution (with maximal sublethal dose) or solvent only is delivered to the thoracic notum of each mosquito on a *Drosophila* CO_2 Fly Pad using a 25-µL Hamilton gastight syringe. (Step 2) Mosquitoes are transferred to a Sweetheart ice cream cup and incubated for an hour. (Step 3) Mosquitoes are then treated with a serial dose of insecticide (the procedure is same as Step 1) and labeled as "Synergist + Insecticide." (Step 4) Treated mosquitoes are immediately transferred to Sweetheart ice cream cups, covered with cheesecloth. Mortality is recorded 24 h after treatment.

- For the synergist control vial, add the same volume of synergist solution as that in the treatment plus the same volume of solvent as that added in the insecticide treatment groups. Ensure that the total volume of synergist solution plus solvent solution is not >100 µL (1% of the total volume).

8. Record the larval mortality after 24 h treatment.

   Repeat each bioassay at least three times on different days.

   Proceed to Steps 20–24.

   See Troubleshooting.

Synergism Bioassay in Adult Mosquitoes—Topical Application

Perform all tests at 25°C with at least three replications.

Identifying Synergist and Insecticide Concentrations

9. To identify the maximum sublethal concentration of a synergist, prepare serial concentrations of the desired synergist as described in Step 1.
10. Treat adult mosquitoes with the serial dilutions of synergist as described in Steps 4–12 of Protocol 2: Mosquito Adult Bioassays (Li et al. 2023). Designate the maximum sublethal concentration as the highest concentration that produces no apparent mortality in the experimental population, above which mortality appears.
11. Determine the appropriate range of concentrations of insecticide alone to use, ensuring it covers the entire range of insect mortality from 0% to 100% as described in the topical application section of Protocol 2: Mosquito Adult Bioassays (Li et al. 2023).
12. Conduct the initial synergist-insecticide bioassays to determine the concentration range of insecticide + the maximum sublethal concentration of synergist that covers 0% to 100% mortality. Treat adult mosquitos with synergist-insecticide dilutions according to Steps 4–12 of Protocol 2: Mosquito Adult Bioassays (Li et al. 2023).
    - We suggest using the dose range insect mortality from 0% to 100% identified for insecticide alone in Step 11 and adding two to three lower doses. For example, if the dose range for insecticide alone is 0.01, 0.03, 0.1, 0.3, and 1 µg/mosquito, here also include 0.001 and 0.003 µg/mosquito.
13. After ascertaining the mortality using this initial dose range, choose a narrower range that includes an insecticide concentration that causes just >0% mortality and one that causes just <100% mortality, with three or four intermediate concentrations in between, to obtain precise values for LC_{50} and LC_{90} (World Health Organization 2005, 2016; Brito-Sierra et al. 2019) in the next section.

Topical Synergism Assay in Adult Mosquitoes

Two mosquito groups will be set up; one group will be treated with synergist and insecticides (synergist + insecticide group, Steps 14–16), the other group will be treated with insecticides and no synergist (insecticide alone group, Step 17)

14. Apply a 0.5-µL drop of synergist solution with the maximum sublethal dose using a 25-µL Hamilton gastight syringe to thoracic notum of each mosquito adult according to Steps 4–7 of Protocol 2: Mosquito Adult Bioassays (Li et al. 2023) (and see Fig. 1B).

    Treat a total of 150 to 180 adult mosquitoes with the maximum sublethal dose of synergist at a time.

15. Transfer synergist-treated mosquitoes to ice cream cups as in Steps 8 and 10 of Protocol 2: Mosquito Adult Bioassays (Li et al. 2023). Use 5–6 cups for a total of 150 to 180 synergist-

 Cite this protocol as *Cold Spring Harb Protoc*; doi:10.1101/pdb.prot108042

treated mosquitoes. Incubate for 1 h in the insectary room or in an environmentally controlled incubator before insecticide treatment (Liu and Yue 2000).

16. Following the 1 h incubation, treat mosquitoes with insecticide according to Steps 4–12 of Protocol 2: Mosquito Adult Bioassays (Li et al. 2023) with the following modifications.
    - Treat groups of 30 mosquitoes with each concentration of insecticide identified in Step 13, proceeding from low to high concentration.
    - Place a 0.5-µL drop of acetone on each of 30 mosquitoes that have been pretreated with synergist. These mosquitoes are the synergist only control.
17. For the group of insecticide-only mosquitoes, deliver a 0.5-µL drop of acetone (or the corresponding solvent) to 30 adult mosquitoes that are not treated with synergist to serve as solvent control mosquitoes. Then treat each group of 30 mosquitoes that are not treated with synergist with each concentration of insecticide identified in Step 13, proceeding from low concentration to high concentration.
18. Record the adult mosquito mortality after 24 h treatment.

    Repeat each bioassay at least three times on different days.

    See Troubleshooting.

    Proceed to Steps 20–24.

Synergist Bioassay in Mosquito Adults—Bottle Bioassay

A synergist bottle bioassay protocol was developed by the World Health Organization (WHO) (World Health Organization 2016) with detailed methodology and procedures.

19. Follow the WHO's synergist bottle bioassay protocol, which can be carried out with either a series of concentrations or a single concentration of insecticide.

Data Analysis for the Synergist Bioassay of Insecticide Resistance

Synergism is expected to increase the susceptibility of mosquito strains to insecticide.

20. Calculate the LC_{50} for the larval bioassay or the LD_{50} for adult bioassay analysis of insecticide only and synergist + insecticide as in Protocol 1: Mosquito Larval Bioassays (Wang et al. 2023) and Protocol 2: Mosquito Adult Bioassays (Li et al. 2023), respectively.
21. If the 95% CIs (confidence intervals) between the target population(s) and the susceptible strain do not overlap, it can be concluded that there is a significant difference in insecticide resistance between the two populations (Robertson and Preisler 1992). If the mortality rate in the control group is >20%, the entire experiments need to be redone.
22. If the mortality in the control group is between 5% and 20%, adjust the mortality of treated groups according to Abbott's formula

$$\text{Corrected mortality (\%)} = \frac{\text{mortality rate in test (\%)} - \text{mortality rate in control (\%)}}{100 - \text{mortality rate in control (\%)}} \times 100.$$

23. Calculate the synergist ratios for insecticides using the formula

$$\text{Synergist ratio (SR)} = \frac{LC_{50} \text{ or } LD_{50} \text{ of insecticide alone}}{LC_{50} \text{ or } LD_{50} \text{ of synergist + insecticide}}.$$

    For example, if the LD_{50} of the insecticide alone is 5 µg/mosquito, and the LD_{50} of synergist + insecticide is 2.5 µg/mosquito, the SR = 5 µg/mosquito/2.5 µg/mosquito = 2.

24. Calculate the synergism resistance ratio using the formula

$$\text{Synergism resistance ratio (SRR)} = \frac{\text{SR of resistant strain}}{\text{SR of susceptible strain}}.$$

For example, if the SR of the resistant strain is 2, and the SR of susceptible strain is 1, the SRR = 2/1 = 2.

TROUBLESHOOTING

Problem (Steps 8 and 18): Significant differences are observed between replicates.

Solution: Consider possible sources of error, including the dilution processes used to obtain the desired concentrations, syringe calibration, and contamination of the synergist or insecticide concentrations. These errors could also be caused by excessive natural variation, in which case increasing the number of replications and controls should be considered.

DISCUSSION

For additional information on synergism studies, see Introduction: Assays to Measure Insecticide Toxicity and Insecticide Resistance in Mosquitoes (Liu et al. 2023).

Synergistic studies have been used by researchers to initially test the involvement of metabolic detoxification in the development of insecticide resistance in mosquitoes (Liu 2015). Using chemical synergists of PBO, DEF, and DEM, the inhibitors of cytochrome P450 monooxygenases, hydrolases, and GSTs, respectively, synergistic studies have revealed several possible mechanisms involving metabolic detoxification in insecticide resistance. Comparing resistance levels with or without the synergists can help researchers make initial diagnostic conclusions on whether and/or which detoxification mechanisms or enzymes are involved in the development of resistance. Synergism studies on pyrethroid resistance in different insect species strongly support the importance of metabolic detoxification in insecticide resistance (Brogdon et al. 1999; Scott 1999; Liu and Yue 2000; Liu et al. 2004; Feyereisen 2005; Cuamba et al. 2010; Jones et al. 2012; Snoeck et al. 2017). However, conclusions drawn from synergistic studies must be treated with caution: Although in many cases using synergists may correctly indicate the role of detoxification proteins in insecticide resistance, in some cases synergists may be only imperfect inhibitors for some of the detoxification enzymes responsible for resistance (Feyereisen 2005). Further work is needed to support the findings from synergistic studies.

This protocol was developed to identify the metabolic enzymes that are engaged in the development of insecticide resistance, as well as to investigate the effectiveness of using novel synergists to boost insecticide potency in mosquito larvae and adults. Two methodologies are used in synergist-insecticide bioassays. In the first assay a synergist combined with a single dose of insecticide is applied to the field-collected mosquitoes. This assay is referred to by WHO as a synergist-insecticide bioassay (World Health Organization 2016). The second assay compares the LD_{50} of synergist + insecticide to the LD_{50} of insecticide alone and is described here. The synergist bottle bioassay has been developed from the WHO (World Health Organization 2016) synergist assay. The WHO synergist assay uses filter papers impregnated with synergist or insecticide and calculates the mortality after synergist and insecticide exposure, whereas in this protocol bottles coated with synergist or serial concentrations of insecticide are used instead of the impregnated paper. Our new approach makes it easier to prepare samples and reduces the risk of contamination. SR and SRR are used to identify synergism effectiveness accurately and determine the impact of synergism on insecticide resistance.

Cite this protocol as *Cold Spring Harb Protoc*; doi:10.1101/pdb.prot108042

RECIPE

Mosquito Larval Food

4 g flake fish food (Tetra TetraMin Tropical Flakes)
20 g yeast (Fleischmann's RapidRise Instant Yeast Fast Acting)

Mix in 300 mL tap H_2O and store for up to 1 mo at 4°C.

REFERENCES

Brito-Sierra CA, Kaur J, Hill CA. 2019. Protocols for testing the toxicity of novel insecticidal chemistries to mosquitoes. *J Vis Exp* **2019:** e57768. doi:10.3791/57768

Brogdon WG, McAllister JC, Corwin AM, Cordon-Rosales C. 1999. Independent selection of multiple mechanisms for pyrethroid resistance in Guatemalan *Anopheles albimanus* (Diptera: Culicidae). *J Econ Entomol* **92:** 298–302. doi:10.1093/jee/92.2.298

Cuamba N, Morgan JC, Irving H, Steven A, Wondji CS. 2010. High level of pyrethroid resistance in an *Anopheles funestus* population of the Chokwe District in Mozambique. *PLoS ONE* **5:** e11010. doi:10.1371/journal.pone.0011010

Desneux N, Decourtye A, Delpuech J-M. 2007. The sublethal effects of pesticides on beneficial arthropods. *Annu Rev Entomol* **52:** 81–106. doi:10.1146/annurev.ento.52.110405.091440

Feyereisen R. 2005. Insect cytochrome P450. In *Comprehensive molecular insect science* (ed. Gilbert L, et al.), Vol. **4**, pp. 1–77. Elsevier, Oxford.

Jones CM, Liyanapathirana M, Agossa FR, Weetman D, Ranson H, Donnelly MJ, Wilding CS. 2012. Footprints of positive selection associated with a mutation (*N1575Y*) in the voltage-gated sodium channel of *Anopheles gambiae*. *Proc Natl Acad Sci* **109:** 6614–6619. doi:10.1073/pnas.1201475109

Kauffman E, Payne A, Franke MA, Schmid MA, Harris E, Kramer LD. 2017. Rearing of *Culex* spp. and *Aedes* spp. mosquitoes. *BioProtoc* **7:** 17. doi:10.21769/BioProtoc.2542

Li T, Wang Y, Liu N. 2023. Mosquito adult bioassays. *Cold Spring Harb Protoc* doi:10.1101/pdb.prot108041

Liu N. 2015. Insecticide resistance in mosquitoes: impact, mechanisms, and research directions. *Annu Rev Entomol* **60:** 537–559. doi:10.1146/annurev-ento-010814-020828

Liu N, Scott JG. 1996. Genetic analysis of factors controlling high-level expression of cytochrome P450, CYP6D1, cytochrome b(5), P450 reductase, and monooxygenase activities in LPR house flies, *Musca domestica*. *Biochem Genet* **34:** 133–148. doi:10.1007/BF02396246

Liu N, Yue X. 2000. Insecticide resistance and cross-resistance in the house fly (Diptera: Muscidae). *J Econ Entomol* **93:** 1269–1275. doi:10.1603/0022-0493-93.4.1269

Liu H, Cupp EW, Micher KM, Guo A, Liu N. 2004. Insecticide resistance and cross-resistance in Alabama and Florida strains of *Culex quinquefasciatus*. *J Med Entomol* **41:** 408–413. doi:10.1603/0022-2585-41.3.408

Liu N, Wang Y, Li T. 2023. Assays to measure insecticide toxicity and insecticide resistance in mosquitoes. *Cold Spring Harb Protoc* doi:10.1101/pdb.top107705

Robertson JL, Preisler HK. 1992. *Pesticide bioassays with arthropods*. CRC, Boca Raton, FL.

Scott JG. 1999. Cytochromes P450 and insecticide resistance. *Insect Biochem Mol Biol* **29:** 757–777. doi:10.1016/S0965-1748(99)00038-7

Snoeck S, Greenhalgh R, Tirry L, Clark RM, Van Leeuwen T, Dermauw W. 2017. The effect of insecticide synergist treatment on genome-wide gene expression in a polyphagous pest. *Sci Rep* **7:** 1–12. doi:10.1038/s41598-016-0028-x

Timmermann SE, Briegel H. 1999. Larval growth and biosynthesis of reserves in mosquitoes. *J Insect Physiol* **45:** 461–470. doi:10.1016/S0022-1910(98)00147-4

Wang Y, Li T, Liu N. 2023. Mosquito larval bioassays. *Cold Spring Harb Protoc* doi:10.1101/pdb.prot108040

World Health Organization. 2005. *Guidelines for laboratory and field testing of mosquito larvicides*. https://apps.who.int/iris/handle/10665/69101 [Accessed June 27, 2021].

World Health Organization (WHO). 2016. *Test procedures for insecticide resistance monitoring in malaria vector mosquitoes*. 2nd edition. (Updated 2018) https://apps.who.int/iris/bitstream/handle/10665/250677/9789241511575-eng.pdf?sequence=1 [Accessed June 27, 2021].

Xu Q, Liu H, Zhang L, Liu N. 2005. Resistance in the mosquito, *Culex quinquefasciatus*, and possible mechanisms for resistance. *Pest Manag Sci* **61:** 1096–1102. doi:10.1002/ps.1090

Appendix

General Safety and Hazardous Material Information

Cold Spring Harbor Laboratory Manuals should be used by laboratory personnel with experience in laboratory and chemical safety or students under the supervision of trained personnel. The procedures, chemicals, and equipment referenced in these manuals may be hazardous and could cause serious injury unless performed, handled, and used with care and in a manner consistent with safe laboratory practices. Students and researchers using the procedures in these manuals do so at their own risk. It is essential for your safety that you consult the appropriate Material Safety Data Sheets, the manufacturers' manuals provided with the relevant equipment, and your institution's Environmental Health and Safety Office (hereafter referred to as safety office), as well as the General Safety and Disposal Cautions in this appendix for proper handling of hazardous materials. Cold Spring Harbor Laboratory makes no representations or warranties with respect to the material set forth in its manuals and has no liability in connection with the use of these materials.

All registered trademarks, trade names, and brand names mentioned in these manuals are the property of the respective owners. Readers should consult individual manufacturers and other resources for current and specific product information.

Appropriate sources for obtaining safety information and general guidelines for laboratory safety are provided in the General Safety and Hazardous Material Information Appendix below.

Users should always consult individual manufacturers, the manufacturers' safety guidelines, and other resources, including local safety offices, for current and specific product information and for guidance regarding the use and disposal of hazardous materials.

THE PRIMARY SAFETY INFORMATION RESOURCES FOR LABORATORY PERSONNEL ARE THE FOLLOWING

Institutional Safety Office. The best source of toxicity, hazard, storage, and disposal information is your institutional safety office, which maintains and makes available the most current information. Always consult this office for proper use and disposal procedures.

Post the phone numbers for your local safety office, security office, poison control center, and laboratory emergency personnel in an obvious place in your laboratory.

Material Safety Data Sheets (MSDSs). The Occupational Safety and Health Administration (OSHA) requires that MSDSs accompany all hazardous products that are shipped. These data sheets contain detailed safety information. MSDSs should be filed in the laboratory in a central location as a reference guide.

GENERAL SAFETY AND DISPOSAL CAUTIONS

The guidance offered here is intended to be generally applicable. However, proper waste disposal procedures vary among institutions; therefore, always consult your local safety office for specific instructions. All chemical waste must be disposed of in a suitable container clearly labeled with the type of material it contains and the date the waste was initiated.

It is essential for laboratory workers to be familiar with the potential hazards of materials used in laboratory experiments, and to follow recommended procedures for their use, handling, storage, and disposal.

The following general cautions should always be observed.

- **Before beginning the procedure**, become completely familiar with the properties of substances to be used.
- **The absence of a warning** does not necessarily mean that the material is safe, because information may not always be complete or available.
- **If exposed** to toxic substances, contact your local safety office immediately for instructions.
- **Use proper disposal procedures** for all chemical, biological, and radioactive waste.
- **For specific guidelines on appropriate gloves to use**, consult your local safety office.
- **Handle concentrated acids and bases** with great care. Wear goggles and appropriate gloves. A face shield should be worn when handling large quantities.

  Do not mix strong acids with organic solvents because they may react. Sulfuric acid and nitric acid especially may react highly exothermically and cause fires and explosions.

  Do not mix strong bases with halogenated solvents because they may form reactive carbenes that can lead to explosions.

- **Handle and store pressurized gas containers** with caution because they may contain flammable, toxic, or corrosive gases; asphyxiants; or oxidizers. For proper procedures, consult the Material Safety Data Sheet that is required to be provided by your vendor.
- **Never pipette** solutions using mouth suction. This method is not sterile and can be dangerous. Always use a pipette aid or bulb.
- **Keep halogenated and nonhalogenated** solvents separately (e.g., mixing chloroform and acetone can cause unexpected reactions in the presence of bases). Halogenated solvents are organic solvents such as chloroform, dichloromethane, trichlorotrifluoroethane, and dichloroethane. Nonhalogenated solvents include pentane, heptane, ethanol, methanol, benzene, toluene, *N,N*-dimethylformamide (DMF), dimethylsulfoxide (DMSO), and acetonitrile.
- **Laser radiation**, visible or invisible, can cause severe damage to the eyes and skin. Take proper precautions to prevent exposure to direct and reflected beams. Always follow the manufacturer's safety guidelines and consult your local safety office. See caution below for more detailed information.
- **Flash lamps**, because of their light intensity, can be harmful to the eyes. They also may explode on occasion. Wear appropriate eye protection and follow the manufacturer's guidelines.
- **Photographic fixatives, developers, and photoresists** contain chemicals that can be harmful. Handle them with care and follow the manufacturer's directions.
- **Power supplies and electrophoresis equipment** pose serious fire hazard and electrical shock hazards if not used properly.
- **Microwave ovens and autoclaves** in the laboratory require certain precautions. Accidents have occurred involving their use (e.g., when melting agar or Bacto Agar stored in bottles or when sterilizing). If the screw top is not completely removed and there is inadequate space for the steam

to vent, the bottles can explode and cause severe injury when the containers are removed from the microwave or autoclave. Always completely remove bottle caps before microwaving or autoclaving. An alternative method for routine agarose gels that do not require sterile agar is to weigh out the agar and place the solution in a flask.

- **Ultrasonicators** use high-frequency sound waves (16–100 kHz) for cell disruption and other purposes. This "ultrasound," conducted through air, does not pose a direct hazard to humans, but the associated high volumes of audible sound can cause a variety of effects, including headache, nausea, and tinnitus. Direct contact of the body with high-intensity ultrasound (not medical imaging equipment) should be avoided. Use appropriate ear protection and display signs on the door(s) of laboratories where the units are used.
- **Use extreme caution when handling cutting devices**, such as microtome blades, scalpels, razor blades, or needles. Microtome blades are extremely sharp. Use care when sectioning. If unfamiliar with their use, have an experienced user demonstrate proper procedures. For proper disposal, use the "sharps" disposal container in your laboratory. Discard used needles *unshielded*, with the syringe still attached. This prevents injuries and possible infections when manipulating used needles because many accidents occur while trying to replace the needle shield. Injuries may also be caused by broken Pasteur pipettes, coverslips, or slides.
- **Procedures for the humane treatment of animals** must be observed at all times. Consult your local animal facility for guidelines. Animals such as rats are known to induce allergies that can increase in intensity with repeated exposure. Always wear a lab coat and gloves when handling these animals. If allergies to dander or saliva are known, wear a mask.

DISPOSAL OF LABORATORY WASTE

There are specific regulatory requirements for the disposal of all medical waste and biological samples. In the United States, these are mandated by the U.S. Environmental Protection Agency (see http://www.epa.gov/hw) and regulated by the individual states and territories (see http://www.epa.gov/hw/state-universal-waste-programs-united-states). Medical and biological samples that require special handling and disposal are generally termed Medical Pathological Waste (MPW), and medical, veterinary, and biological facilities will have programs for the collection of MPW and its disposal. Restrictions on how radioactive waste can be disposed of as regulated in the United States by the U.S. Nuclear Regulatory Commission can be found in 10 CFR 20.2001, General requirements for waste disposal (see http://www.nrc.gov/reading-rm/doc-collections/cfr/part020/part020-2001.html) or the individual **Agreement States.** The preferred method for the disposal of radioactively contaminated MPW is decay-in-storage (see http://www.nrc.gov/reading-rm/doc-collections/cfr/part035/part035-0092.html). Contact your institution's safety office and local regulatory agencies for instructions and requirements.

Waste and any materials contaminated with biohazardous materials must be decontaminated and disposed of as regulated medical waste. No harmful substances should be released into the environment in an uncontrolled manner. This includes all tissue samples, needles, syringes, scalpels, etc. Be sure to contact your institution's safety office concerning the proper practices associated with the handling and disposal of biohazardous waste.

Some basic rules are outlined below. For treatment of radioactive and biological waste, see sections on Radioactive Safety Procedures and Biological Safety Procedures.

- In practice, only **neutral aqueous solutions** without heavy metal ions and without organic solvents can be poured down the drain (e.g., most buffers). Acid and basic aqueous solutions need to be neutralized cautiously before their disposal by this method.
- For proper disposal of **strong acids and bases**, dilute them by placing the acid or base onto ice and neutralize them. Do not pour water into them. If the solution does not contain any other toxic compound, the salts can be flushed down the drain.

- For disposal of **other liquid waste**, similar chemicals can be collected and disposed of together, whereas chemically different wastes should be collected separately. This avoids chemical reactions between components of the mixture (see above). Collect at least inorganic aqueous waste, non-halogenated solvents, and halogenated solvents separately.
- Waste **from photo processing and automatic developers** should be collected separately to recycle the silver traces found in it.

RADIOACTIVE SAFETY PROCEDURES

In the United States and other countries, the access to radioactive substances is strictly controlled. You may be required to become a registered user (e.g., by attending a mandatory seminar and receiving a personal dosimeter). A convenient calculator to perform routine radioactivity calculations can also be found at http://www.graphpad.com/quickcalcs/ChemMenu.cfm.

If you have never worked with radioactivity before, follow the steps below.

- **Avoid it when possible.** Many experiments that are traditionally performed with the help of radioactivity can now be done using alternatives based on fluorescence or chemiluminescence and colorimetric assays, including, for example, DNA sequencing, Southern and northern blots, and protein kinase assays. However, in other cases (e.g., metabolic labeling of cells), use of radioactivity cannot be avoided.
- **Be informed.** While planning an experiment that involves the use of radioactivity, consider the physicochemical properties of the isotope (half-life, emission type, and energy), the chemical form of the radioactivity, its radioactive concentration (specific activity), total amount, and its chemical concentration. Order and use only as much as is really needed.
- **Familiarize yourself** with the designated working area. Perform a mental and practical dry run (replacing radioactivity with a colored solution) to make sure that all equipment needed is available and to get used to working behind a shield. Handle your samples as if sterility would be required to avoid contamination.
- **Always wear appropriate gloves**, lab coat, and safety goggles when handling radioactive material.
- **Check the work area** for contamination before, during, and after your experiment (including your lab coat, hands, and shoes).
- **Localize your radioactivity.** Avoid formation of aerosols or contamination of large volumes of buffers.
- **Liquid scintillation cocktails** are often used to quantitate radioactivity. They contain organic solvents and small amounts of organic compounds. Try to avoid contact with the skin. After use, they should be regarded as radioactive waste; the filled vials are usually collected in designated containers, separate from other (aqueous) liquid radioactive waste.
- **Dispose of radioactive waste** only into designated, shielded containers (separated by isotope, physical form [dry/liquid], and chemical form [aqueous/organic solvent phase]). Always consult your safety office for further guidance in the appropriate disposal of radioactive materials.
- Among the experiments requiring **special precautions** are those that use [^{35}S]methionine and ^{125}I, because of the dangers of airborne radioactivity. [^{35}S]methionine decomposes during storage into sulfoxide gases, which are released when the vial is opened. The isotope ^{125}I accumulates in the thyroid and is a potential health hazard. ^{125}I is used for the preparation of Bolton–Hunter reagent to radioiodinate proteins. Consult your local safety office for further guidance in the appropriate use and disposal of these radioactive materials before initiating any experiments. Wear appropriate gloves when handling potentially volatile radioactive substances, and work only in a radioiodine fume hood.

BIOLOGICAL SAFETY PROCEDURES

Biological safety fulfills three purposes: to avoid contamination of your biological sample with other species; to avoid exposure of the researcher to the sample; and to avoid release of living material into the environment. Biological safety begins with the receipt of the living sample; continues with its storage, handling, and propagation; and ends only with the proper disposal of all contaminated materials. A catalog of operations known as "sterile handling" is usually employed in manipulating living matter. However, the correct procedures largely depend on the samples, which can be quite diverse: *Escherichia coli* and other bacterial strains, yeasts, tissues of animal or plant origin, cultures of mammalian cells, or even derivatives from human blood are routinely handled in a biological laboratory. Two of these, bacteria and human blood products, are discussed in more detail below.

The U.S. Department of Health, Education, and Welfare (HEW) has classified various bacteria into different categories with regard to shipping requirements (see Sanderson and Zeigler, *Methods Enzymol* **204:** 248–264 [1991]). Nonpathogenic strains of *E. coli* (such as K12) and *Bacillus subtilis* are in Class 1 and are considered to present no or minimal hazard under normal shipping conditions. However, *Salmonella*, *Haemophilus*, and certain strains of *Streptomyces* and *Pseudomonas* are in Class 2. Class 2 bacteria are "Agents of ordinary potential hazard: agents which produce disease of varying degrees of severity... but which are contained by ordinary laboratory techniques." Contact your institution's safety office concerning shipping biological material.

Human blood, blood products, and tissues may contain occult infectious materials such as hepatitis B virus and human immunodeficiency virus (HIV) that may result in laboratory-acquired infections. Investigators working with lymphoblast cell lines transformed by Epstein–Barr virus (EBV) are also at risk of EBV infection. Any human blood, blood products, or tissues should be considered a biohazard and should be handled accordingly until proved otherwise. Wear appropriate disposable gloves, use mechanical pipetting devices, work in a biological safety cabinet, protect against the possibility of aerosol generation, and disinfect all waste materials before disposal. Autoclave contaminated plasticware before disposal; autoclave contaminated liquids or treat with bleach (10% [v/v] final concentration) for at least 30 minutes before disposal (this is valid also for used bacterial media).

Always consult your local institutional safety officer about specific handling and disposal procedures for samples. Further information can be found in the Frequently Asked Questions of the ATCC homepage (http://www.atcc.org) and is also available from the U.S. National Institute of Environmental Health and Human Services, Biological Safety (http://www.niehs.nih.gov/about/stewardship).

GENERAL PROPERTIES OF COMMON HAZARDOUS CHEMICALS

The hazardous materials list can be summarized in the following categories:

- Inorganic acids, such as hydrochloric, sulfuric, nitric, or phosphoric, are colorless liquids with stinging vapors. Avoid spills on skin or clothing. Spills should be diluted with large amounts of water. The concentrated forms of these acids can destroy paper, textiles, and skin and cause serious injury to the eyes.
- Inorganic bases, such as sodium hydroxide, are white solids that dissolve in water and under heat development. Concentrated solutions will slowly dissolve skin and even fingernails.
- Salts of heavy metals are usually colored, powdered solids that dissolve in water. Many of them are potent enzyme inhibitors and, therefore, toxic to humans and the environment (e.g., fish and algae).
- Most organic solvents are flammable volatile liquids. Avoid breathing the vapors, which can cause nausea or dizziness. Also avoid skin contact.
- Other organic compounds, including organosulfur compounds such as mercaptoethanol or organic amines, can have very unpleasant odors. Others are highly reactive and should be handled with appropriate care.

- If improperly handled, dyes and their solutions can stain not only your sample but also your skin and clothing. Some are also mutagenic (e.g., ethidium bromide), carcinogenic, and toxic.
- Nearly all names ending with "ase" (e.g., catalase, β-glucuronidase, or zymolyase) refer to enzymes. There are also other enzymes with nonsystematic names such as pepsin. Many of them are provided by manufacturers in preparations containing buffering substances, etc. Be aware of the individual properties of materials contained in these substances.
- Toxic compounds are often used to manipulate cells. They can be dangerous and should be handled appropriately.
- Be aware that several of the compounds listed have not been thoroughly studied with respect to their toxicological properties. Handle each chemical with appropriate care. Although the toxic effects of a compound can be quantified (e.g., LD_{50} values), this is not possible for carcinogens or mutagens, where one single exposure can have an effect. Also realize that dangers related to a given compound may also depend on its physical state (fine powder vs. large crystals/diethyl ether vs. glycerol/dry ice vs. carbon dioxide under pressure in a gas bomb). Anticipate under which circumstances during an experiment exposure is most likely to occur and how best to protect yourself and your environment.

Cold Spring Harbor Laboratory Press (CSHLP) has used its best efforts in collecting and preparing the material contained herein but does not assume, and hereby disclaims, any liability for any loss or damage caused by errors and omissions in the publication, whether such errors and omissions result from negligence, accident, or any other cause. CSHLP does not assume responsibility for the user's failure to consult more complete information regarding the hazardous substances listed in this publication.

REFERENCE

Sanderson KE, Zeigler DR. 1991. Storing, shipping, and maintaining records on bacterial strains. *Methods Enzymol* **204:** 248–264. doi:10.1016/0076-6879(91)04012-D

WWW RESOURCES

ATCC Home page. http://www.atcc.org
ATCC, for Sample Handling (in Frequently Asked Questions). http://www.atcc.org/CulturesandProducts/TechnicalSupport/FrequentlyAskedQuestions/tabid/469/Default.aspx
GraphPad Software, Radioactivity Calculations. http://www.graphpad.com/quickcalcs/ChemMenu.cfm
National Institute of Environmental Health and Human Services, Biological Safety (NIEHS). http://www.niehs.nih.gov/about/stewardship
U.S. Environmental Protection Agency (EPA), Federal waste disposal regulations, Laboratory. http://www.epa.gov/epawaste/hazard/tsd/index.htm
U.S. Environmental Protection Agency (EPA), Individual States and Territories. http://www.epa.gov/epawaste/wyl/stateprograms.htm
U.S. Nuclear Regulatory Commission (NRC), Medical Pathological Radioactively Contaminated Waste (Decay-in-Storage). http://www.nrc.gov/reading-rm/doc-collections/cfr/part035/part035-0092.html
U.S. Nuclear Regulatory Commission (NRC), Radioactive Waste Disposal Regulations: General Requirements. http://www.nrc.gov/reading-rm/doc-collections/cfr/part020/part020-2001.html

Index

Note: Page numbers followed by f, t, or b denote a figure, table, or box, respectively, on the corresponding page.

B

D

I

Q

R

S

T

U